Reaction Mechanism in Organic Chemistry

REACTION MECHANISM IN ORGANIC CHEMISTRY

(REVISED EDITION)

S.M. MUKHERJI & S.P. SINGH

Revised by : S.P. Singh & Om Prakshan

TRIUNITY PRESS

(An Imprint of Laxmi Publications Pvt. Ltd.)

An ISO 9001:2015 Company

BENGALURU • CHENNAI • GUWAHATI • HYDERABAD

KOCHI • KOLKATA • LUCKNOW • RANCHI

NEW DELHI

REACTION MECHANISM IN ORGANIC CHEMISTRY

Printed and bound in India
Typeset at : Shubham Composer, Delhi
Revised Edition : 2015, *Reprint* : 2016, 2017, 2020, 2023; *Edition* : 2026
ISBN : 978-93-5138-178-5

Branches

✆	Bengaluru	080-26 75 69 30
✆	Chennai	044-24 34 47 26
✆	Guwahati	0361-254 36 69
✆	Hyderabad	040-27 55 53 83
✆	Kochi	0484-405 13 03
✆	Kolkata	033-40 04 77 79
✆	Lucknow	0522-430 36 13
✆	Ranchi	0651-224 24 64

PUBLISHED IN INDIA BY

TRIUNITY PRESS
(An Imprint of Laxmi Publications Pvt. Ltd.)

An ISO 9001:2015 Company
113, GOLDEN HOUSE, GURUDWARA ROAD, DARYAGANJ,
NEW DELHI - 110002, INDIA

Corporate Office : B-33, Sector-5, Noida,
Gautambuddha Nagar (U.P.) - 201 301, INDIA
Telephone : +91-120-44 46 900
www.laxmipublications.com
info@laxmipublications.com

C—00000/025/012
Printed at: Ajit Printing Press, Delhi.

This book is dedicated
to
Our grand children

Poorva, Uday, Adi, and Veer
—S.P. Singh

Khushi, Avi, Aadya, and Aarna
—Om Prakash

Contents

Preface to the Revised Edition *xi*
Preface to the Previous Edition *xiii*

Chapter 1: Nature of Bonding in Organic Molecules **1**
1.1 Localized Chemical Bonding 2
1.2 Delocalized Chemical Bonding 12
1.3 Bonding Weaker Than Covalent 42
Problems and Review Questions 54
Solutions 58

Chapter 2: Stereochemistry-I **62**
2.1 Stereoisomerism 63
2.2 Molecular Representation of Stereoisomers 65
2.3 Chirality 70
2.4 Geometrical Isomerism 100
2.5 Conformational Isomerism 104
Problems and Review Questions 110
Solutions 115

Chapter 3: Stereochemistry-II **118**
3.1 Prostereoisomerism 119
3.2 Stereoselective and Stereospecific Reactions 127
3.3 Asymmetric Synthesis 129
3.4 Optical Activity in the Absence of a Chiral Carbon 142
3.5 Chirality Due to Helical Shape 151
3.6 Stereochemistry of the Compounds Containing Nitrogen, Phosphorus and Sulphur 153
3.7 Conformations of Cyclohexanes and Decalins 159
3.8 Effect of Conformation on Reactivity 177
3.9 Conformational Analysis of Monosaccharides 185
Problems and Review Questions 189
Solutions 194

Chapter 4: Organic Reactions, Reagents and Reaction Intermediate **199**
4.1 Types of Organic Reactions 201
4.2 Types of Reagents 204
4.3 Organic Acids and Bases 207
4.4 Reaction Intermediates 207
Problems and Review Questions 242
Solutions 245

Chapter 5: General Principles and Methods of Determination of Reaction Mechanism **247**
5.1 Energy Considerations 248
5.2 Methods of Determination of Reaction Mechanisms 265
Problems and Review Questions 277
Solutions 279

Chapter 6: Aliphatic Nucleophilic Substitution **280**
6.1 Direct Displacement Process: SN2 Reactions 281
6.3 Stereochemistry of Substitution Reactions 285
6.4 Neighbouring Group Participation 289
6.5 S_Ni and SET Mechanisms 308
6.6 Factors Affecting Reactivity in Substitution Reactions 311
6.7 Ambident Nucleophiles 317
Problems and Review Questions 319
Solutions 324

Chapter 7: Addition to Carbon-Carbon Multiple Bond **328**
7.1 Electrophilic Additions 329
7.2 Nucleophilic Additions 355
Problems and Review Questions 357
Solutions 360

Chapter 8: Elimination Reactions **365**
8.1 β-Elimination 366
8.2 α-Elimination: Carbenes 384
Problems and Review Questions 392
Solutions 395

Chapter 9: Reactions of the Carbonyl Group **397**
9.1 Reactions of Aldehydes and Ketones 398
9.2 Reactions of Carboxylic Acid Derivatives 432
Problems and Review Questions 447
Solutions 452

Chapter 10: Aromatic Electrophilic Substitution **456**
10.1 Electrophilic Substitution Reactions of Benzene 462
10.2 Reactivity and Orientation 472
10.3 Quantitative Treatment 484
10.4 Some Synthetically Useful Aromatic Electrophilic Substitution Reactions 486
Problems and Review Questions 491
Solutions 495

Chapter 11: Aromatic Nucleophilic Substitution **498**
11.1 Mechanisms 499
11.2 Aromatic Nucleophilic Rearrangements 507
Problems and Review Questions 512
Solutions 515

Chapter 12: Molecular Rearrangements **518**
12.1 Molecular Rearrangements Involving Electron-Deficient Species 519
12.2 Further Molecular Rearrangements 541
Problems and Review Questions 556
Solutions 560

Chapter 13: Free Radicals **564**
13.1 Short-Lived Radicals 565
13.2 Some Synthetically Useful Reactions 593
13.3 Stable Free Radicals 596
13.4 Biradicals 600
Problems and Review Questions 601
Solutions 604

Chapter 14: Organic Photochemistry **607**
14.1 Principles of Photochemistry 608
14.2 Photochemistry of Carbonyl Compounds 617
14.3 Photochemistry of Olefins 628
14.4 Photochemistry of Aromatic Compounds 634
14.5 Photorearrangement of Cyclohexadienones 637
Problems and Review Questions 641
Solutions 644

Chapter 15: Pericyclic Reactions **646**
15.1 Conservation of Molecular Orbital Symmetry 647
15.2 Electrocyclic Reactions 652
15.3 Cycloaddition Reactions 661
15.4 Sigmatropic Rearrangements 674
15.5 Applications of PMO Method to Pericyclic Reactions 680
Problems and Review Questions 685
Solutions 688

Index **695**

Preface to the Revised Edition

Understanding the mechanisms of reactions constitutes the core of the study of organic chemistry. We made an effort to meet this objective by writing a book in 1976 which was the first book on this topic published from India. It is gratifying to know that the book was used by hundreds of thousands students and researchers in the last four decades hopefully to their benefit.

However, it was felt that the book written in 1970s on such a vibrant and dynamic topic needs to be thoroughly reworked, updated and enlarged to reflect new facts and advances in different areas. This became particularly relevant when the UGC Committee on Curriculum Development in Chemistry-with one of the authors (SPS) being the Convener- recommended a thorough revision of this topic in particular. Hence we have undertaken the exercise of the complete rewriting of the book with lot of addition and deletion taking into account the new developments. Some of the key features of the revised edition are:

We have incorporated two chapters on stereochemistry. While the first chapter deals with the basic concepts of chirality, the second chapter describes stereoheterotopic ligands and faces, stereoselective and stereospecific reactions, asymmetric synthesis, effect of conformation on reactivity and finally conformational analysis of monosaccharides. The objective is to understand reaction mechanisms with particular focus on spatial relationship among the reacting atoms which is one of the important factors in determining how fast a reaction goes and what product it yields.

As the concept of the carbocation has served to relate reactions of many different kinds, chemistry of this important reactive intermediate has received special attention. Neighbouring group participation of a lone pair, a double bond and a phenyl group along with a discussion of sigma bond participation and nonclassical cations have been introduced. Topics such as photochemistry of aromatic compounds, cheletropic reactions, hydrogenation of unsaturated systems, ambident nucleophiles, Hammette and Taft equations are incorporated in appropriate chapters. A large number of energy profile diagrams have been sketched in many chapters to illustrate different processes, and many synthetically important reactions and rearrangements have been added in chapters dealing with aromatic electrophilic and nucleophilic substitutions. Units for expressing energies are given in IUPAC mandated joules as well as in calories which are still being used by many organic chemists.

We feel confident that this reorganization and revision along with a large number of problems will assist the students to develop a knowledge and appreciation of structure and mechanisms of reactions in organic chemistry. Besides helping a large number of post-graduate students (Chemistry and pharmacy), this revised edition will cater to the needs of those students who are preparing for NET and other competitive examinations.

We will appreciate all those who read and use this revised edition to contact us directly with their comments, and point out errors for rectification in the next reprint. Our email addresses are: shivpsingh@rediffmail.com; dromprakash50@rediffmail.com

We thank Ms. Loveena Arora, who had worked as Technical Assistant in the project, for her assistance in the preparation of the manuscript. Without her support, this work would not have been completed in time. We thank our former students, Prof. Devinder Kumar (GJU, Hisar) and Dr. Vinod Kumar (MMU, Mullana) for their help in revising the chapter on Pericyclic Reactions. We take this opportunity to acknowledge the contributions of all our former students and colleagues who had helped us in the development of this text. We especially thank Prof. Ranjana Aggarwal, Prof. Pawan K. Sharma, Prof. Neera Raghav and Dr. Rashmi Pundeer of our Department and Dr. Ravi Kumar of D. S. College, Karnal for their help at one stage or the other.

Finally, the authors gratefully acknowledge that the book was revised under a project catalyzed and supported by the Department of Science and Technology, Govt. of India under its Utilization of Scientific Expertise of Retired Scientists (USERS) Scheme.

Kurukshetra,
June, 2014

S.P. SINGH
OM PRAKASH

Preface to the Previous Edition

In the course of teaching post-graduate classes, we have constantly been feeling the need of a concise volume that gives outlines of most of the important organic reactions, together with a critical examination of the evidence leading to the accepted mechanisms. The students have often expressed their difficulty caused by the absence of such a book. Our present effort is to meet this long-felt need and is designed primarily for the students who had a basic course in organic chemistry at the undergraduate level. It is an attempt to bridge the gap between an elementary treatment and the more masterly treatises on the subject.

The book opens with a chapter on the Chemical Bond which apart from describing the different types of chemical bonds, highlights the importance of the concepts of the electronegativity and effect of the state of hybridization of the bonding orbitals on the physical properties of the molecules. In addition to Hückel's treatment of aromaticity, a new outlook for determining aromaticity and resonance energy has been introduced based on the Perturbational Molecular Orbital (PMO) method developed by M. J. S. Dewar. This chapter coupled with the second and third chapters- which deal with the reaction intermediates, reagents, energies, stereochemical terms and methods for determining reaction mechanisms- form the background materials for the subsequent chapters.

In dealing with the conventional topics such as substitution reactions, addition and elimination reactions, reactions of carbonyl group, free radicals, etc., we have tried to project current developments, wherever possible. The purpose of giving specific examples in these chapters is to illustrate principles rather than giving a review-type coverage. We had to be selective rather than comprehensive because of our anxiety to limit the book to a reasonable size and price.

The last two chapters, namely, Organic Photochemistry and Pericyclic Reactions, have been included due to our conviction that these topics should form an integral part of the course on reaction mechanism in recognition of their growing importance. These topics should also be helpful to those research students whose training in these fascinating areas of organic chemistry was not up to the current standard. In addition to Woodward Hoffmann's orbital correlation method based on the principles of molecular orbital symmetry, and the frontier orbital method, we have discussed in reasonable detail the PMO method for the analysis of pericyclic reactions. This latter method should be particularly welcome to the readers as it offers simple mnemonics of selection rules for prediction of the stereochemical courses of this class of reactions.

An effort to guide the readers to relevant review articles and books has also been made for further consultation. It is hoped that this book would be useful and effective in helping the students of chemistry develop an insight into the mechanistic aspects of organic chemistry as a whole. We shall welcome constructive criticisms and suggestions with a view to improve upon our present effort.

We are grateful to Drs. R. P. Gandhi, R. P. Kapoor and Krishan Lal for going through some of the chapters and to our many friends and colleagues who have contributed valuable comments and criticisms.

We also appreciate the good work done by Mr. R. G. Gupta in typing the manuscript in his spare time.

Kurukshetra,
June, 1976

S.M. MUKHERJI
S. P. SINGH

Chapter 1

Nature of Bonding in Organic Molecules

Chapter Outline

1.1 LOCALIZED CHEMICAL BONDING
- A. Hybridization
- B. Hybridization and Bond Properties
- C. Cyclopropane: Banana Bond

1.2 DELOCALIZED CHEMICAL BONDING
- A. Conjugation
- B. Resonance (Mesomerism)
- C. Hyperconjugation
- D. Aromaticity
- E. Alternant and Nonalternant Hydrocarbons
- F. Bonding in Fullerenes
- G. Tautomerism

1.3 BONDING WEAKER THAN COVALENT
- A. Inclusion Compounds and Clatherates
- B. Crown Ethers
- C. Cryptands
- D. Cyclodextrins
- E. The Hydrogen Bond
- F. Charge-Transfer Complexes
- G. van der Waals Interactions

A molecule is formed by the combination of atoms, and in the process they attain the electronic configuration of an inert gas. A *chemical bond* is formed as a result of attraction between the combining atoms. To explain the formation of bonds in diatomic molecules such as H_2, O_2, N_2 and Cl_2, it was first suggested by GN Lewis that atoms of such molecules attain an inert gas electronic configuration by sharing one or more pairs of electrons, each atom contributing one or more electrons. The atoms are then held together by the shared pair(s) of electrons forming what is called a *covalent bond*. The structures in which the shared electrons are shown by *dots* were first used by Lewis, and are known as Lewis structures. The system of using *dashes* to represent the shared electrons was suggested by Couper.

The structure of a compound represents how different atoms in a molecule are joined to each other and are arranged in a three-dimensional space. The properties of different compounds depend on their structures. Structural concepts of an organic compound can be understood by the nature and strength of bonds present in the molecule.

Now the question arises as to why two or more atoms combine together to form molecules? This is because, as a result of bond formation, the atoms reach a state of lower energy. In other words, a certain amount of energy is released during the bond formation. The same amount of energy, called *bond dissociation energy*, is needed to break the molecule into its component atoms. This Chapter discusses the nature of bonding and its types, the parameters of molecular structure, and the various types of bonds and their characteristics.

1.1 LOCALIZED CHEMICAL BONDING

Atomic or hybrid orbitals overlap to form covalent bonds where the electrons are either localized or delocalized. When the electrons forming a bond are distributed in the space between the two bonded atoms, they are called localized electrons, and such a bond is called a localized bond, *e.g.* all σ-bonds are localized bonds.

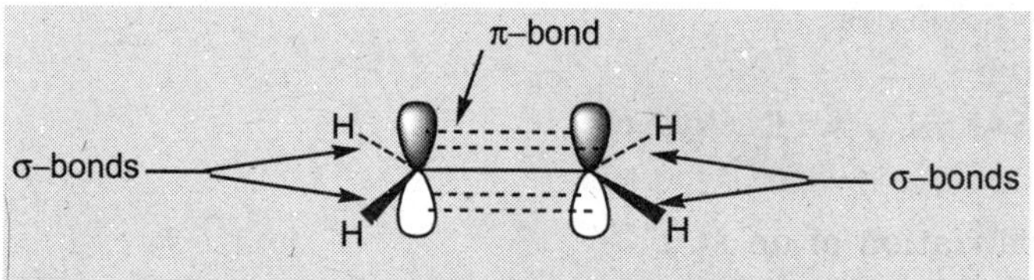

Fig. 1.1: Localized bonds in ethylene

Electrons forming π-bonds may be localized or delocalized. For example, π-bond of ethylene is localized because the electrons forming this bond are confined to the space between two carbon atoms in such a way that these are distributed equally in the space above and below the plane of C–C σ-bond (**Fig. 1.1**).

Sigma (σ) bond

The molecular orbital which is formed by the end-on overlap of two atomic orbitals, and is cylindrically symmetrical about (*i.e.* having the highest electron density along) the molecular axis, is called a sigma (σ) bond. A sigma bond may be formed by the overlap of *s-s* orbitals as in hydrogen molecule, by *s-p* overlap as in HF molecule, by *p-p* overlap as in fluorine molecule, by the overlap of sp^3, sp^2 and sp hybrid orbitals of two or more atoms, and also by the overlap of any of these hybrid orbitals with an s or a p orbital.

Pi (π) bond

A pi (π) bond is the result of the sideways (lateral) overlap of two p orbitals approaching each other in parallel or almost parallel fashion. In contrast to the sigma orbital, the pi molecular orbital has zero electron density along the molecular axis, but has the maximum electron density above and below the internuclear line. All the sigma bonds around the pi bond are coplanar, the bond angles being 120°. The plane of the molecule is the *nodal plane* of the pi bond.

A. Hybridization

Hybridization may be defined as the mixing of two or more than two atomic orbitals of an atom, having comparable energy, to give an equal number of identical orbitals having same energy and shape. All hybrid orbitals are oriented symmetrically to have maximum distance from each other.

Thus, the molecule of methane may be represented by the overlap of a hydrogen 1*s* orbital with each of the four sp^3 orbitals of carbon. Hybrid orbitals of other molecules may similarly be represented. Linear combination of a 2*s* and two of the 2*p* orbitals, for example in ethylene, gives rise to three trigonal sp^2 orbitals directed towards the corners of an equilateral triangle. The plane defined by the two original 2*p* orbitals leaves the remaining 2*p* orbital, perpendicular to the plane of the triangle.

The planar structure of methyl radical can be represented by the overlap of each of the three sp^2 orbitals of carbon with an *s* orbital of hydrogen, forming three C–H bonds, leaving the odd electron on the third unhybridized 2*p* orbital free (**Fig. 1.2**).

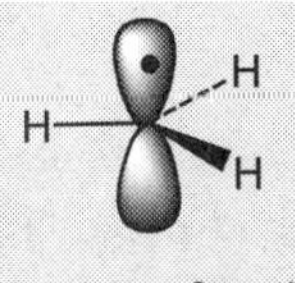

Fig. 1.2: Structure of methyl radical

Likewise, hybridization of an *s* and *p* orbital gives two diagonal (*sp*) orbitals, directed towards the opposite ends of the line defined by the *p* orbital.

Methane, ethylene, and acetylene are the classic examples of sp^3, sp^2 and *sp* hybridized carbon atom, respectively. The pictorial representation of hybridized orbitals of methane is given in **Fig. 1.3.** Ethylene is represented by two carbon atoms combining through two sp^2 orbitals, and overlapping of the remaining two sp^2 orbitals on each carbon atom with 1*s* orbitals of two hydrogen atoms. The unhybridized parallel 2*p* orbitals, one on each of the trigonal carbon atoms overlap each other sideways to form a π bond. The electrons involved in such a bonding are called π-electrons. The π-electron cloud (pi bond) is distributed above and below the plane of the molecule, which is the nodal plane of the pi cloud. The bond energy of a carbon-carbon pi bond is about 60 kcal or 250.8 kJ, and, is therefore, weaker than a C–C sigma bond which is 83 kcal or 346.9 kJ of energy. As the carbon atoms are held more tightly, the carbon-carbon bond distance in ethylene is shorter (1.34 Å) than the C–C sigma bond length in ethane (1.54 Å). The angle between the bonds is 120°, and the molecule is planar.

Likewise, in acetylene, as represented in **Fig. 1.3**, each carbon atom is bonded diagonally to two other atoms, a carbon and hydrogen, through the overlap of two *sp*-hybridized orbitals of the carbon atoms, and of the remaining two *sp* orbitals of carbon atoms with two 1*s* orbitals of hydrogen. This leaves two *p* orbitals on each carbon atom, perpendicular to each other, as also to the *sp* hybrid orbital. The sideways overlap of the two parallel pairs of *p* orbitals leads to the formation of two π bonds, which merge into something like a cylindrical π electron cloud.

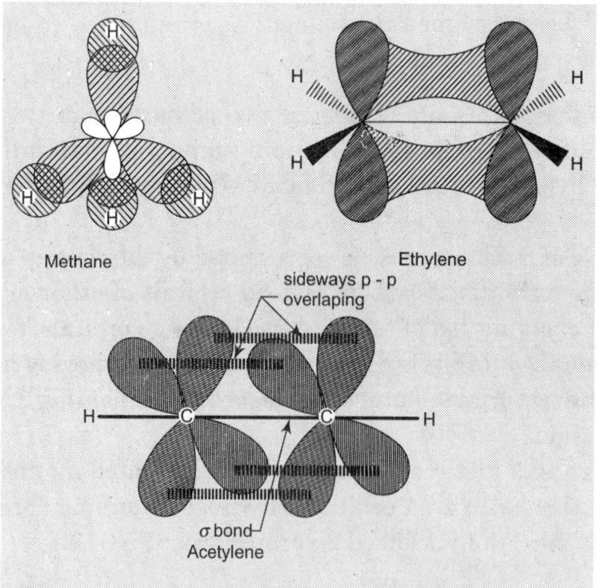

Fig.1.3: Orbital picture of methane, ethylene and acetylene

Electronegativity (Partial Ionic Character of Covalent Bond)

We know that an ionic bond involves complete transfer of one or more electrons from one atom to another. On the other hand, a covalent bond between two atoms of the same element involves equal sharing of the pair of bonding electrons between them. However, it has been observed that in many covalent compounds, the electron pair binding the two atoms is not equally shared between these atoms. This can be explained on the basis of the fact that atoms of different elements have different tendencies to attract electrons. *Electronegativity of an element is a measure of this tendency*. It is maximum for atoms in upper right corner of the periodic table, and minimum for atoms in the lower left corner. In a family of elements, the smaller atoms have higher electronegativities than the larger ones. For instance, in the halogen family, fluorine has the highest electronegativity. When two atoms with different electronegativities form a covalent bond, there is a greater probability of finding the bonding electrons near the more electronegative atom. But, this difference in electronegativity of the atoms is not enough to form ions, as would have been the situation in the case of an ionic bond between these atoms. Such a bond is called a *polar covalent bond*, and can be considered to be intermediate between an ionic bond and a pure covalent bond. Unequal sharing of bonding electrons is sometimes referred to in terms of *per cent ionic character* of a covalent bond.

Several attempts have been made to develop a quantitative scale of electronegativity of different elements. Mulliken suggested that the electronegativity of an atom would be an average of ionization energy. But, average of ionization energy is very difficult to determine experimentally. Pauling has suggested an alternative scale of electronegativity based on the difference in bond energies of diatomic molecules.

Electronegativities of different elements have been assigned values between 0 and 4 with hydrogen, which has been assigned the value 2.1, as reference standard. **Table 1.1** lists the electronegativities of some elements.

Table 1.1 Electronegativities of some elements (Pauling)

H 2.1						
Li 0.97	Be 1.47	B 2.0	C 2.5	N 3.0	O 3.5	F 4.0
Na 1.0	Mg 1.23	Al 1.5	Si 1.8	P 2.09	S 2.5	Cl 3.0
K 0.91	Ca 1.04			As 2.0	Se 2.4	Br 2.8
Rb 0.89	Sr 0.99			Sb 1.9	Te 2.1	I 2.5
Cs 0.86	Ba 0.97					

Dipole Moment

The unequal sharing of bonding electrons of a covalent bond gives rise to a *dipole moment* in molecules formed by the atoms of different electronegativity. The dipole moment of a bond X–Y, in Debye units (D), is roughly equal to the difference in the electronegativities of X and Y. The dipole moment of an individual bond cannot be measured, but it is possible to measure the total moment of the molecule, which is the vectorial sum of the individual bond moments. For example, $CHCl_3$ has a dipole moment of 1.87 D, whereas CCl_4 has no dipole moment as the resultant of four C–Cl bond moments is zero. The difference in electronegativities of carbon and hydrogen is very small, hence the hydrocarbons such as propane and ethane have no dipole moment. On the other hand, HCl serves as an example of difference in electronegativities with a dipole moment of 1.03 D. Hydrogen atom constitutes the positive end and chlorine atom the negative end of HCl dipole. The dipole in a molecule is usually represented by the symbol $\leftarrow\!\!\!+\!\!\!\longrightarrow$, the arrowhead pointing towards the negative end, for example, H$+\!\!\!\longrightarrow$Cl. The dipole moment of methyl chloride is 1.87 D, whereas that of ethyl chloride and *n*-propyl chloride is 2.05 and 2.10 D, respectively. This shows that the substitution of a hydrogen atom in the methyl group of methyl chloride by alkyl groups produces little change in the dipole moment.

B. Hybridization and Bond Properties

Bond properties, such as bond length and bond energy, are greatly influenced by the state of hybridization in which the atom exists. An *s* orbital is at a lower level than a *p* atomic orbital (AO), which is at a lower level than a *d* AO. Therefore, the greater the *s* contribution in the hybrid AOs of the valence state, the greater is the electronegativity of the atom. Further, since hybridization may be of different types, the electronegativity of the atom relative to a second atom is determined by the electronegativity of the hybridized AO with which they enter into bonding. For instance, toluene has a dipole moment of 0.37 D, which is explained by the fact that the aromatic ring carbon (sp^2) to which the methyl carbon is attached is more electronegative than the latter, although both are of the same element.

The dissociation energy of a bond increases with the difference in electronegativities of the bonded atoms. Therefore, it depends on the state of hybridization of the bonded atoms. The electronegativity of carbon is greatest in *sp* hybridized state, and least in sp^3 state. As a result, the C–H bond formed with a carbon orbital of high *s*-character will be shorter than the one formed with carbon orbital of high *p*-character. *The change in hybridization of AOs in carbon, thus, produces a change in the size of covalent atomic radius, decreasing from the tetrahedral (sp^3) to the diagonal type (sp).* In fact, the state of hybridization in which the bonded atoms exist is the most important factor in determining bond length. **Table 1.2** shows the average bond lengths depending on the state of hybridization of the bonded atoms.

Table 1.2 Bond lengths and hybridization

Bond type	**Bond length Å**	**Bond type**	**Bond length Å**	**Bond type**	**Bond length Å**
C–C		C=C		C–H	
sp^3–sp^3	1.54	sp^2–sp^2	1.34	sp^3–*H*	1.11
sp^3–sp^2	1.50	sp^2–*sp*	1.31	sp^2–*H*	1.10
sp^3–*sp*	1.46	*sp*–*sp*	1.28	*sp*–*H*	1.08
sp^2–sp^2	1.48				
sp^2–*sp*	1.43	C≡C			
sp–*sp*	1.38	*sp*–*sp*	1.205		

It was pointed out that *bond dissociation energy* (D) is related to the state of hybridization. It is defined as the energy required to break a bond between two atoms. The bond dissociation energy of diatomic molecules such as H_2 and N_2, can be easily determined. But, for a polyatomic molecule containing more than one covalent bond, as in methane, the energy for each of the steps of C–H bond dissociation will be different. This is because bond dissociation energy not only includes the energy needed for rupturing the bond, besides the steric interactions in higher molecules, but also the changes accompanying the change in hybridization at the carbon atom, which are different for each step. Therefore, it is difficult to determine bond dissociation energy. But to have an idea of the order of magnitude of the energy of a particular type of bond, the average value (*E*) from the heat of atomization of a molecule is taken as the *bond energy*.

$CH_4 \rightarrow CH_3 + H$ D(CH_3–H) = 102 kcal/mol or 426 kJ/mol
$CH_3 \rightarrow CH_2 + H$ D(CH_2–H) = 105 kcal/mol or 439 kJ/mol
$CH_2 \rightarrow CH + H$ D(CH–H) = 108 kcal/mol or 451 kJ/mol
$CH \rightarrow C + H$ D(C–H) = 83 kcal/mol or 347 kJ/mol

The bond energy (*E*) as per C–H bond in methane = 398/4 = 99.5 kcal/mol or 416 kJ/mol.

Table 1.3 lists the bond energies (E) of some of the common covalent bonds averaged over a large number of compounds. **Table 1.4** lists the theoretical values of bond dissociation energies showing their variation with the state of hybridization.

Table 1.3 Selected bond energies

Bond	**Bond energy (E) (kcal)**	**Bond energy (E) (kJ)**
O–H	110-11	460-465
C–H	96-99	401-414
C–C	83-85	347-355
C=C	146-151	610-631
C≡C	199-200	832-836

Table 1.4 Effect of hybridization on bond dissociation energy

Hybridized bond	**Bond dissociation energy (D) (theoretical) (kcal)**	**Bond dissociation energy (D) (kJ)**
sp^3–sp^3 C–C	85.00	355.3
sp^3–sp^2 C–C	89.9	375.7
sp^2–sp^2 C–C	100.31	419.2
sp^2–sp^2 C=C	127.7	533.7
sp^3 C–H	97.2	406.2
sp^2 C–H	102.3	427.6
sp C–H	104.0	434.7

Inductive Effect: We have seen that the difference in electronegativities of two unlike atoms joined by a covalent bond induces a permanent dipole moment in the molecule. This polarization is a permanent state of the molecule, and reflected as the *inductive effect*. We know that many chemical reactions are initiated by the cleavage of a covalent bond. Carbon of methyl chloride, for example, becomes partially positive due to inductive effect of chlorine. Thus, it is susceptible to attack by a negatively charged hydroxide ion (nucleophile).

$$HO^{\ominus} + \overset{\delta^+}{CH_3} \rightarrow\!\!- \overset{\delta^-}{Cl} \longrightarrow HO{-}CH_3 + Cl^{\ominus}$$

It may be noted that the partial ionic character of a covalent bond due to inductive effect is shown by the symbol $\rightarrow\!\!-$, the arrowhead pointing towards the more electronegative atom. The partial positive and negative charges, developed due to the difference in the electronegativities and inductive effect, are shown by δ^+ and δ^-, respectively.

Inductive effect, being the result of difference in electronegativities, may also be due to the difference in the state of hybridization of atoms linked by a covalent bond. For instance, the acid strength of acrylic acid (CH_2=CH–COOH) is considerably higher than that of propionic acid (CH_3CH_2COOH) due to the electron withdrawing inductive effect of α,β-double bond, even though the conjugative effect of the α,β-unsaturated carbonyl system would tend to decrease the acid strength.

Coordinate Covalent Bond: It is a special type of covalent bond with a certain amount of charge separation. In this type of covalent bond, the shared pair of electrons is donated by one of the two atoms. In other words, in the formation of such a bond, one atom *donates* a pair of electrons and the other atom, which is deficient in electrons, *accepts* it. The former atom is a *donor* and the latter an *acceptor*, and the bond formed between them is called a *coordinate bond* or a *dative* bond or a *semipolar* bond. The pair of electrons shared in this type of bond is the lone pair donor. A coordinate covalent bond is represented by two alternative symbols, X→Y or X^+—Y^-. The donor develops a positive charge and the acceptor, a negative charge. It may be noted that the coordinate covalent bond is somewhere between a covalent bond and an ionic bond. Although, coordinate compounds do not ionize in water but most of them have physical properties intermediate between those of ionic and covalent compounds. Generally, they are as stable as covalent compounds, but those which are formed from donor and acceptor compounds of independent existence are less stable, *e.g.* $H_3N \rightarrow BF_3$. A coordinate bond is rigid and directional like a covalent bond. Coordination compounds, therefore, exhibit *stereoisomerism*.

Molecular Orbital Method-LCAO Treatment

The molecular orbital approach for understanding chemical bonding envisages the set of electronic orbitals belonging to the molecule as a whole. Like an isolated atom, a molecule is assumed to have orbitals associated with it. The only difference is that after the bond formation, the valence electrons of the constituent atoms occupy a new set of orbitals known as *molecular orbitals*. The Pauli exclusion principle applies to the MOs in the same way as it applies to the AOs. In other words, a MO cannot accommodate more than two electrons, and these paired electrons must have opposite spins.

Molecular orbitals have been represented mathematically by an approximation known as *Linear Combination of Atomic Orbitals* (LCAO). In LCAO, the combination of any number of AOs gives rise to an equal number of MOs of different energies. Half of these MOs, which are associated with lower energies than the AOs, are filled with electron pairs, while the remaining MOs of higher energies (than the AOs) are unoccupied in the ground state of the molecule. The former are known as *bonding orbitals,* and the latter as *antibonding orbitals*. In case of radicals, carbocations, and carbanions, one (or more) of the molecular orbitals may possess the same energy as that of the atomic orbitals. Such an orbital is said to be a *nonbonding orbital*.

Let us, for instance, consider the linear combination of AO wave functions of two atoms X and Y forming a molecule XY. This can be done either by *addition* or *subtraction* of two wave functions under consideration.

$$\psi = \varphi X + \lambda\varphi Y \quad ; \quad \psi' = \varphi X - \lambda\varphi Y$$

Here ψ and ψ′ are MO wave functions, and are obtained by combination (addition or subtraction) of AO wave functions φX and φY, λ representing the ionic character of the bond between X and Y. The value of λ is unity when X is equal to Y.

Since ψ^2 is a measure of the electron charge density in a given volume element, the charge densities between the two atoms in the MOs will be given by:

$$\psi^2 = (\varphi X + \lambda\varphi Y)^2 = \varphi X^2 + \lambda^2\varphi Y^2 + 2\lambda\phi X\phi Y$$
$$\psi'^2 = (\varphi X - \lambda\varphi Y)^2 = \varphi X^2 + \lambda^2\varphi Y^2 - 2\lambda\phi X\phi Y$$

The term $2\lambda\phi X\phi Y$ will, therefore, determine the energy difference between the two MOs. It follows that the charge density between the bonded atoms in the MO, ψ, will be more than that between that two nonbonded atoms by the amount $2\lambda\phi X\phi Y$. Similarly, the charge density between the bonded atoms in the MO, ψ', will be less than that between the two nonbonded atoms by the amount $2\lambda\phi X\phi Y$. The MO, ψ, is called a *bonding orbital* since in this orbital the probability of finding the two electrons along the molecular axis is maximum. On the other hand, the probability of finding the two electrons along the molecular axis in the MO, ψ', is minimum. It is, therefore, called *antibonding orbital*. The bonding orbital is associated with lower energy as compared to any of the AOs, and is designated by σ. The antibonding orbital is associated with higher energy as compared to any of the AOs, and is designated by σ^*. It remains unfilled in the ground state of the molecule.

Figure 1.4 depicts the linear combination of two hydrogen 1*s* AOs resulting in one bonding and one antibonding MO.

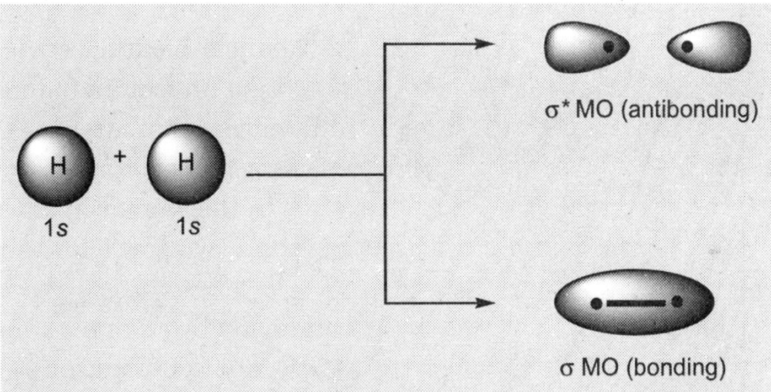

Fig. 1.4: LCAO of two 1*s* atomic orbitals

Figure 1.5 shows the relative energies of the AOs and MOs of hydrogen. It must be remembered that the sigma orbital is cylindrically symmetrical about the internuclear axis of the two atoms and has no nodal plane, *i.e.* the electron density is the highest along the internuclear axis.

Thus, there are two electrons of paired spins in the ground state of hydrogen molecule. The energy of this system is at a minimum when the separation between the two nuclei is 0.74 Å, a distance that represents the *bond length*. The situation may be illustrated by plotting the energy of two hydrogen atoms, as a function of distance (r) between the nuclei, when the electrons are in bonding and *antibonding orbitals* (**Fig. 1.6**). Since a less energetic system is more stable, the depth of the potential well represents the strength of the bond. *Antibonding orbitals* tend to push the nuclei apart.

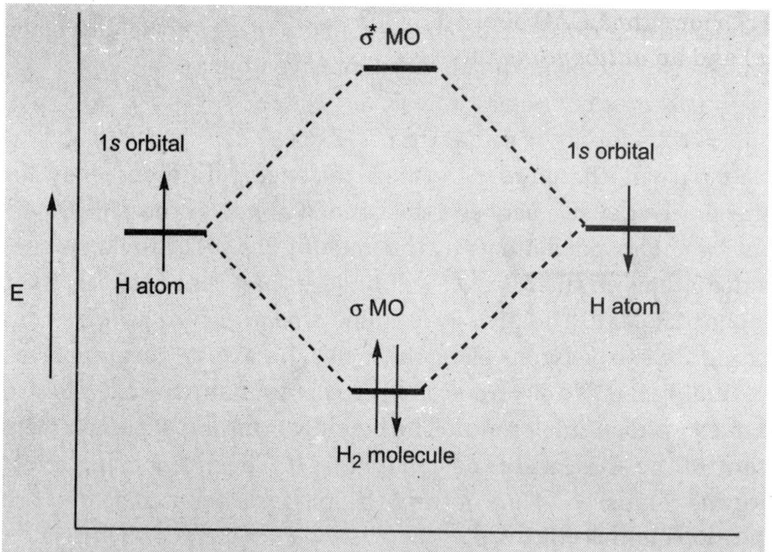

Fig. 1.5: Relative energies of atomic and molecular orbitals for hydrogen

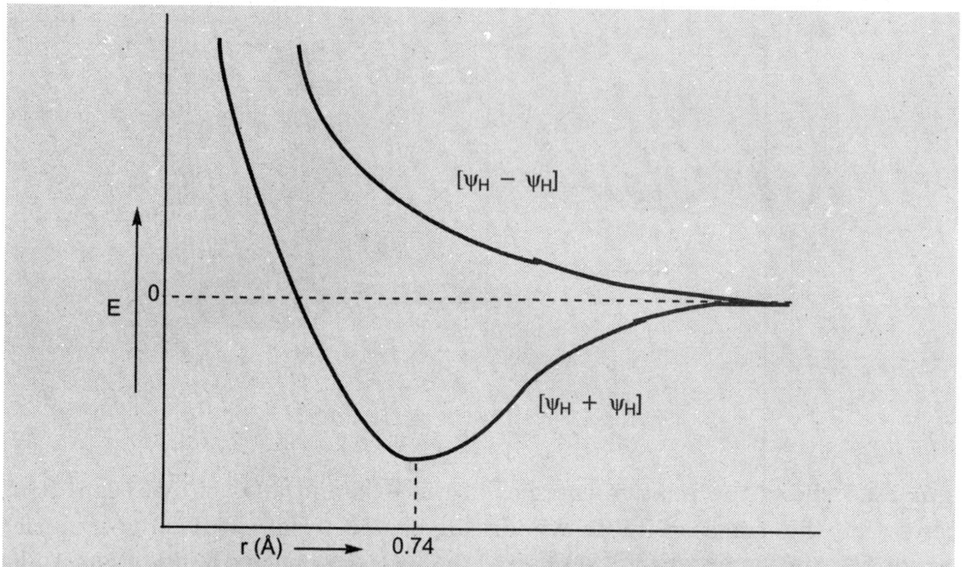

Fig. 1.6: Potential energy curves for the formation of σ and σ* bonds between two hydrogen atoms

Similarly, the LCAO of s and *p* AOs, and of two *p* AOs would also result in the formation of σ and σ* MOs each. If, however, two *p* AOs overlap sideways, the LCAO approximation of the wave functions would lead to the formation of two MOs, one of which, the bonding orbital called pi (π) orbital, is of lower energy than any of the isolated *p* orbitals, and is occupied by two pi electrons in the ground state. Beside this, an *antibonding orbital*, called pi-starred (π*) orbital, is of higher energy than either of the *p* orbitals, and hence remains unfilled in the ground state of the molecule.

Figure 1.7 shows the LCAO of two *p* AOs (by sideways overlap) resulting in the formation of a *bonding* (π) and an *antibonding* (π*) MO.

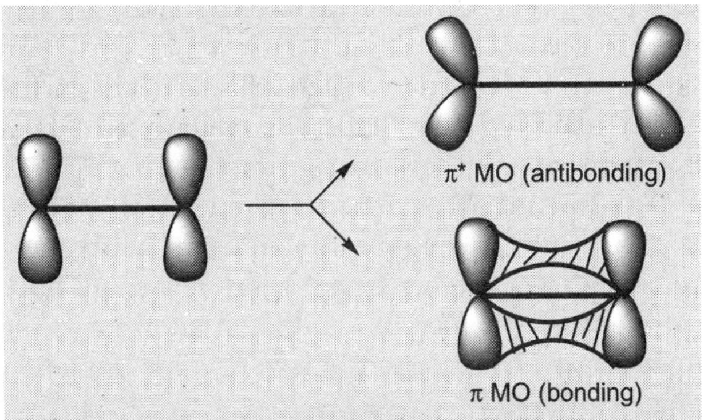

Fig. 1.7: LCAO of two *p* atomic orbitals

Figure 1.8 depicts the relative energies of these AOs and MOs. It should be noted that the bonding π orbital has a nodal plane (plane of the molecule) and has, therefore, more energy than a bonding σ orbital.

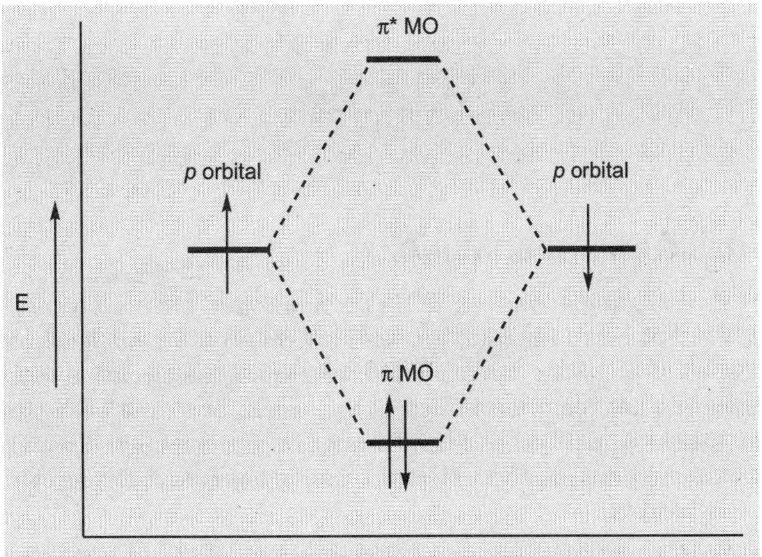

Fig. 1.8: Relative energies of *p* and π orbitals

C. Cyclopropane: Banana Bond

Cyclopropane (**1**) offers an interesting case of bent C–C covalent bond. It is planar, and internal bond angles of 60°(θ) mean a considerable deviation from the tetrahedral angle of 109.5°. But, carbon cannot have hybridized orbitals oriented at less than 90° to each other, and since there are four bonds on each carbon, the angles have to be more than even 90°. This would mean that in **1**, the ring σ bonds cannot be formed by the bonding atomic orbitals pointing directly at one another. It has been shown that the orbitals are hybridized in such a way that the ring-forming orbitals have a slightly greater *p* character than those forming the normal sigma bonds between carbon atoms in saturated molecules.

This explains why the angle between two such orbitals of one carbon atom in **1** is 104°(ϕ), somewhat less than the tetrahedral angle. Thus, the major axes of symmetry of the orbitals forming the C–C bonds in **1** do not point towards one another, but outwards from the molecule. As a consequence, the C–C bonding MOs are not symmetrical about the internuclear axes, but are bent giving the appearance of bananas. A significant point arising out of the higher *p* character of AOs of the bonded C–C atoms is that there is a small but appreciable degree of pi delocalization around the ring which occurs, unlike in an aromatic system, in the plane of the ring, and not above and below the carbon framework. This in-plane pi delocalization adds some stability to the ring.

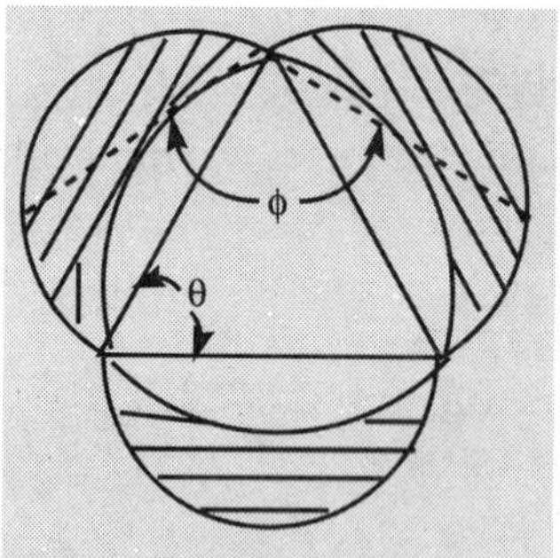

1

1.2 DELOCALIZED CHEMICAL BONDING

When the electrons are moving in and out of the space between the two bonded atoms, they are called delocalized electrons, and the bond formed by them is called a delocalized bond.

The two π-bonds of acetylene are delocalized, because the electrons forming the π-bond in the plane of paper do not remain confined to the space above and below the plane of C–C σ-bond. Similarly, the electrons of other π-bond formed in perpendicular direction do not remain confined to that space. Actually, all these electrons merge together to form a cylindrical electron cloud around C–C σ-bond (**2**).

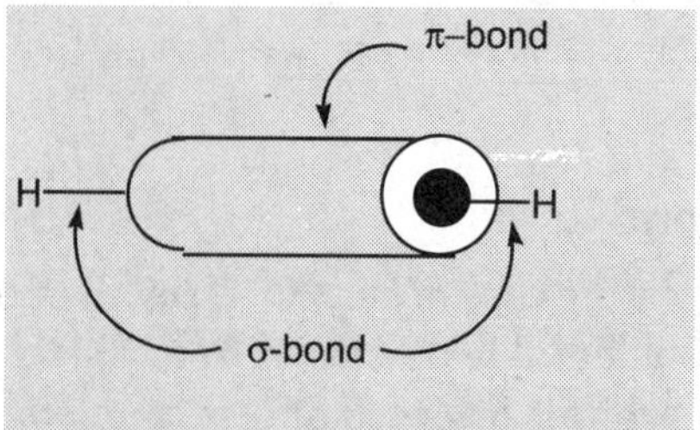

2

Other well-known examples of delocalized bonds are 1,3-butadiene and benzene, where two or more than two π-bonds are in conjugation. The electrons of one π-bond are delocalized into the space of other π-bond and *vice-versa*. This delocalization occurs through the overlap of *p* orbitals present on each sp^2 carbon.

(a) 1,3-Butadiene: All the four carbon atoms of 1,3-butadiene lie in the same plane, due to which all the *p* orbitals at four carbon atoms overlap with each other, and the π-electrons can move to a limited extent over all the four carbon atoms, *i.e.* π-electrons of 1,3-butadiene are delocalized. Hence, the π-bonds of butadiene are delocalized bonds. In contrast, the π-*electrons (and hence the* π-*bonds) are localized in isolated dienes, such as 1,4-pentadiene,* where each pair of π- electrons is confined to the space between two carbon atoms.

$CH_2{=}CH{-}CH{=}CH_2$ $CH_2{=}CH{-}CH_2{-}CH{=}CH_2$

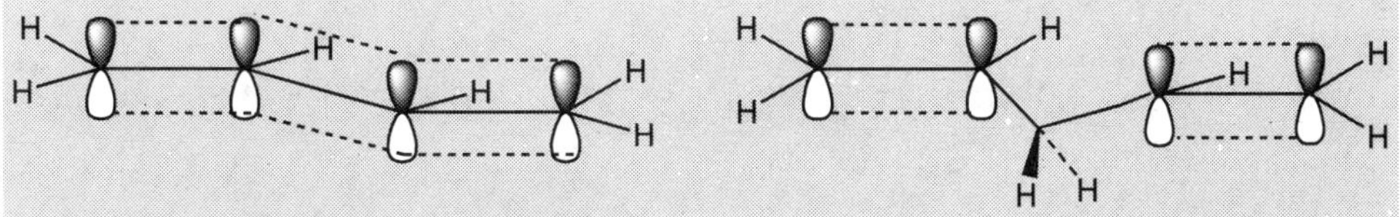

1,3-Butadiene (delocalized π-bonds) 1,4-Pentadiene (localized π-bonds)

(b) Benzene: In case of benzene, there are six sp^2 hybridized carbon atoms, and each sp^2 carbon has one unhybridized *p*-orbital containing one electron. These *p* orbitals are so close to each other that they can overlap sideways to form a π-bond. There are two modes, **3a** and **3b** for overlap of adjacent *p* orbitals. Actually, each *p* orbital overlaps equally with *p* orbitals on two adjacent carbon atoms on both sides, as shown in structure **4** which has three delocalized π-bonds (**Fig. 1.9**).

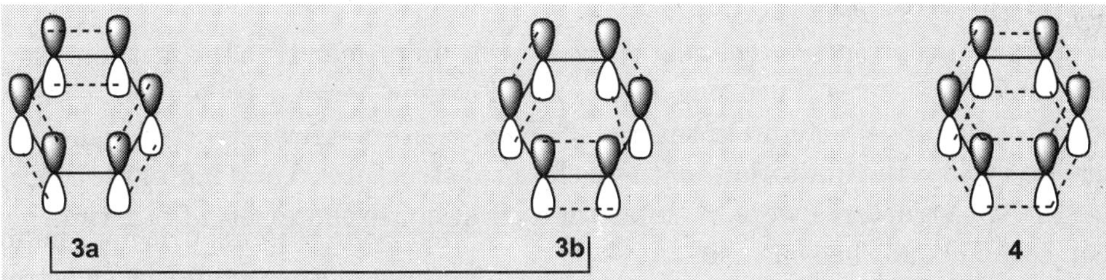

localized three π-bonds delocalized three π-bonds

Fig. 1.9: Localized and delocalized three π-bonds in benzene

A. Conjugation

Conjugation is the overlap of one *p* orbital with another across an intervening sigma bond. It is possible by means of not only alternating single and double bonds, but also in other ways. This allows delocalization of π-electrons across all the adjacent aligned *p* orbitals. The π-electrons do not belong to a single bond or atom, but rather to a group of atoms. Some well-known compounds containing conjugated systems are: 1,3-dienes, benzene, and α,β-unsaturated carbonyl compounds such as cinnamaldehyde.

$CH_2{=}CH{-}CH{=}CH{-}CH_3$

1,3-Pentadiene,
a conjugated diene

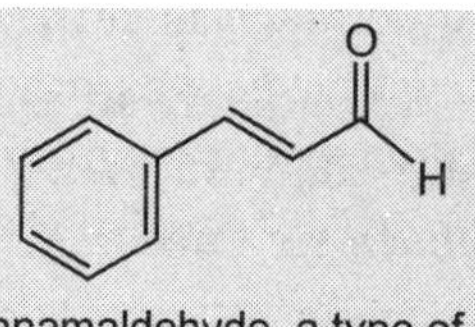

Cinnamaldehyde, a type of
α,β-unsaturated carbonyl compound

Largest conjugated systems are found in graphite, conductive polymers, and carbon nanotubes.

Cross Conjugation: In cross conjugated compounds, three groups are present, two of which are not conjugated with each other, although each is conjugated with the third. Examples are compounds **5-7**.

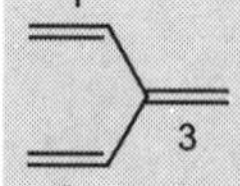

5

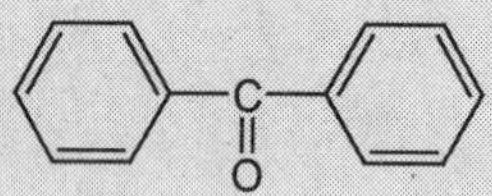

6

$CH_2{=}CH{-}\ddot{N}H{-}CH{=}CH_2$

7

In example **5**, 1 and 2 are not conjugated, but 3 is conjugated with 1 as well as 2. Cross conjugated trienes are known as *dendralenes*. In example **6**, carbonyl group is in conjugation with both the rings, but both rings are not in conjugation to each other. In example **7**, lone pair present on nitrogen is in conjugation with both the double bonds, but these are not in conjugation with each other.

B. Resonance (Mesomerism)

The concept of resonance in organic chemistry owes its origin primarily to the fact that isomers of *ortho*-substituted derivatives of benzene do not exist. Further, benzene is found to be resistant unlike ethylenic compounds to addition reactions, although its classical structures show three double bonds. To explain the limitations of these classical structures **8** and **9**, Kekule conceived the idea that the two structures were in a state of oscillation, so that over a finite time period all the carbon-carbon bonds become equal.

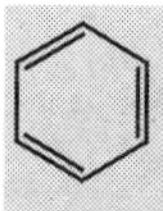

8

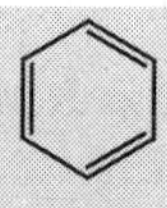

9

Therefore, the real structure of benzene cannot be represented by either. This would mean that two adjacent carbon atoms of the benzene molecule are joined neither by a single nor a double bond. In terms of theory of resonance propounded by Pauling, the two valence bond structures 'must be superimposed, fused together, to represent the molecule of benzene, with consideration given also to the stabilizing effect of resonance'. According to this theory, several possible Lewis structures for benzene, such as **8**, **9** and others called *canonical forms*, are drawn, and energy of each is calculated from the wave equation for each structure. The concept of the real structure as weighted average of these canonical forms is called *resonance*, and the difference in energy between the actual molecule and the canonical form of the lowest energy is called *resonance energy*. An idea of the *resonance energy* is conveyed in **Fig. 1.10**.

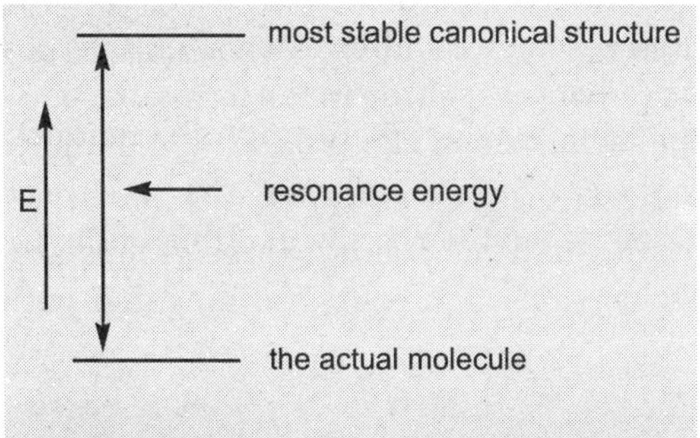

Fig.1.10: Measurement of resonance energy

In benzene, the resonance energy has been calculated from the heats of combustion, and from the heats of hydrogenation. The values obtained by either approach are close to each other, and work out to about 36 kcal/mol or 150 kJ/mol. To illustrate, the heat of hydrogenation of cyclohexene is –28.6 kcal/mol or –119.5 kJ/mol, and that of 1,3-cyclohexadiene is –55.6 kcal/mol or –232.4 kJ/mol. On this basis, the heat of hydrogenation of benzene should be $-28.6 \times 3 = -85.8$ kcal/mol or –358.6 kJ/mol. But the experimental value for benzene is –49.8 kcal/mol or 208 kJ/mol. Thus, the molecule of benzene is said to be stabilized by resonance to the extent of 36 kcal/mole or 150kJ/mol.

Besides benzene, there are a number of compounds and reaction intermediates, whose chemical and physical properties suggest that localized valence bond structures do not adequately describe the actual electron distribution in them. For instance, in carbonate anion CO_3^{2-}, all the bond lengths have been found to be equal (1.3Å). According to resonance theory, this is explained by writing three different Lewis structures in which the position of carbon and oxygen atoms remain unchanged; only the electrons have changed their positions.

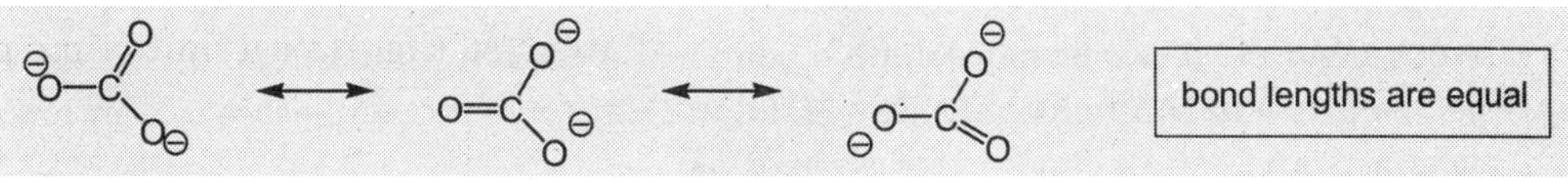

As shown above, the resonance is depicted using a double-headed arrow (↔) which is placed between each pair of the contributing forms. It is important to note that the double-headed arrow does not signify the oscillation of molecule from one Lewis structure to another. In fact, the word resonance does not mean the mixing of formal structures. Therefore, a less confusing term, *π-electron delocalization,* has been coined for this phenomenon.

The true structure of such a species is considered to be a hybrid of the various Lewis structures that can be written for it. The structures which differ only in the arrangement of electrons are called *contributing* or *resonance* structures, and we have only an idea of the true structure by considering the properties of two or more Lewis structures, which contain the same π electrons localized in various ways. This way of studying the properties of a molecule is called *resonance method*.

The concept of resonance stabilization is of particular importance for a qualitative understanding of reaction mechanism. As we shall see in subsequent chapters, most of the organic reactions involve formation of reactive intermediates such as carbocations, carbanions, and free radicals. Any variation in their resonance stabilization often changes the course of the reaction. Addition of bromine to 1,3-butadiene, resulting in the formation of two dibromo products, offers an illustrative example. The formation of the product of 1,4-addition can be rationalized by assuming the intermediacy of a carbocation, in which the positive charge is divided between 2 and 4 carbons. Addition of bromide ion then occurs at either of the two partially charged carbon atoms.

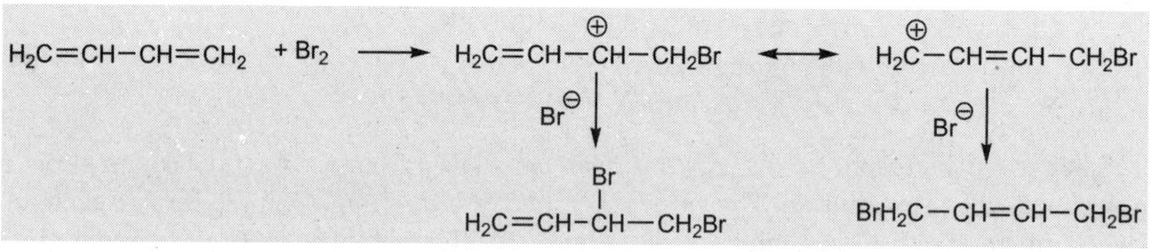

It is significant that initial attack of Br^+ takes place at the position 1 in preference to the 2-position of the diene, as in the latter case the intermediate carbocation (CH_2=CH–CHBr–CH_2^+) is unable to gain resonance stabilization.

I. Rules for the Resonance Method

The following rules for the use of resonance method enable us to predict when and to what extent resonance is important in determining the properties of various molecules:

(i) Structures contributing towards the hybrid must conform to real Lewis structures, *i.e.* there can be no structure with pentacovalent carbon and bicovalent hydrogen.

(ii) Resonance involves movement of only π or *n*-electrons, and a resonance structure can be derived from another by a series of one or more short electron shifts without changing the relative position of atoms.

In other words, if two or more double bonds, or a double bond and an unshared pair of electrons are in conjugation, there is delocalization of π or *n*-electrons. For example, cyclopropenyl cation can be represented as:

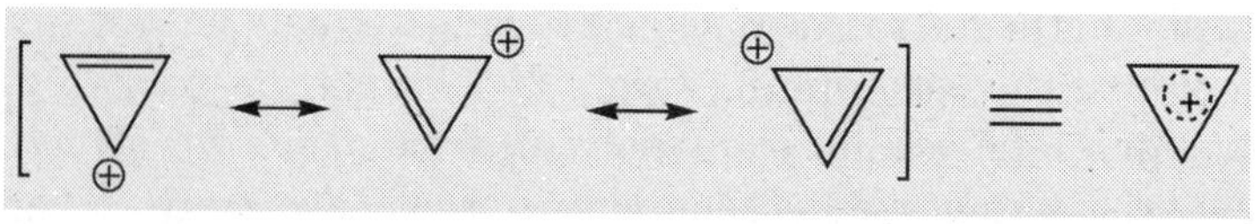

It may be mentioned that in these canonical forms the relative positions of atomic nuclei remain unchanged, otherwise resonance would imply tautomerism (see Section 1.2 G of this Chapter).

(iii) All resonance structures must have same number of paired electrons. Thus, $\dot{C}H_2$–$\dot{C}H_2$ is not a resonance form of CH_2=CH_2.

(iv) All resonance structures do not contribute equally to the hybrid. The relative energies of various contributing structures are estimated by considering bond energies. Since each bond adds about 50-100 kcal/mol or 209-418 kJ/mol to the stability of the system, it is to be expected that structures with more covalent bonds are more stable than those with fewer. The nonpolar form of butadiene (**10**), for instance, is the most stable of all the resonance structures. The other forms, **11** and **12** make only minor contribution, if at all, to the resonance hybrid.

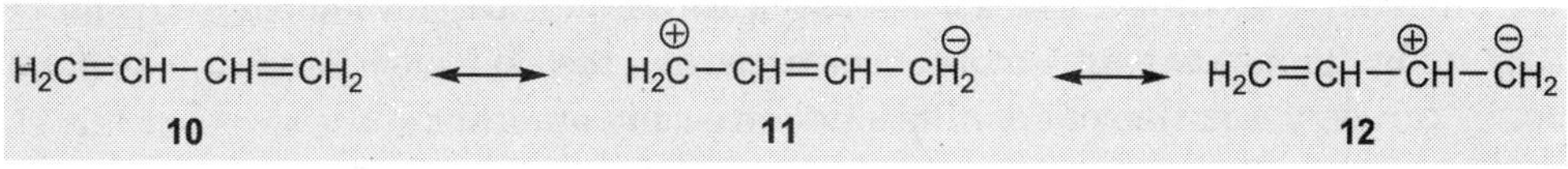

(v) When atoms of different electronegativities are involved, the structure which places a negative charge on the most electronegative atom would be more important. Thus, for a ketone, **13** is more important than **14**.

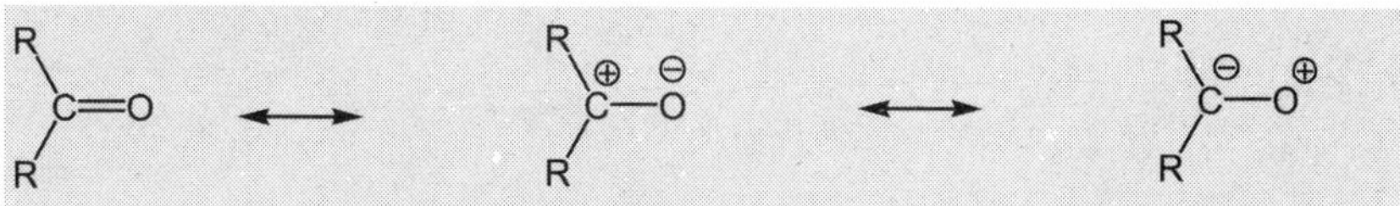

13 (major contributor) **14** (minor contributor)

(vi) Resonance contribution is the greatest when there are two or more equivalent contributing forms of lower energy. The species involved may be neutral or ionic. For example, resonance in allyl cation, cyclopentadienyl anion, and benzene involves contributing forms of comparable energies and renders them quite stable.

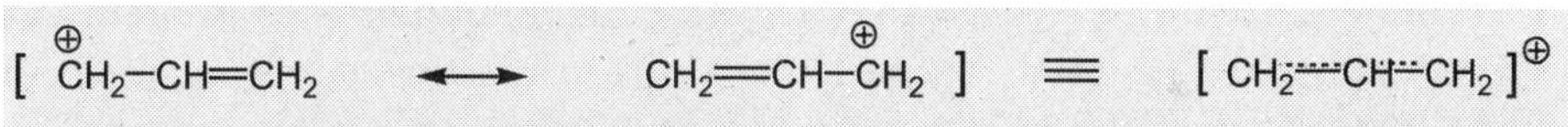

Allyl cation (a stable cation)

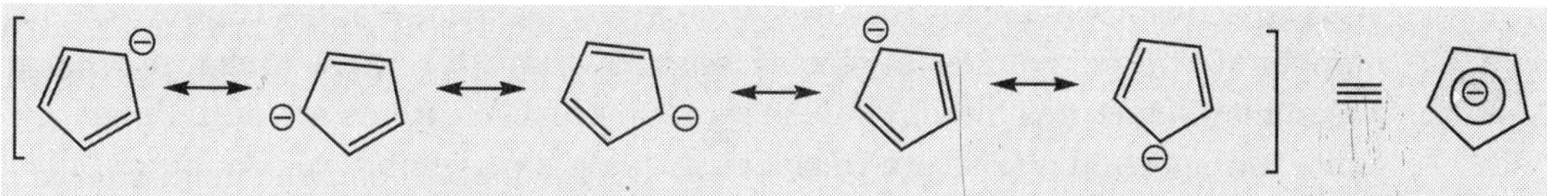

Cyclopentadienyl anion (a stable anion)

It may be noted that ionic species are especially stabilized if resonance is possible in them. This is due to the movement of electrons, there is a dispersal of charge over the whole system rather than being localized at a particular atom.

(vii) If there is only a single contributing structure of lowest energy, the resonance hybrid possesses properties expected for that structure. Ethylene and butadiene are, thus, well represented by their conventional structures, but benzene and carbonate anion are not.

(viii) In order to achieve maximum overlap of *p* orbitals, which is necessary for delocalization, it is essential that the skeleton should remain planar in conjugated system.

In other words, any structural feature that destroys the coplanarity of the conjugated system inhibits resonance. This inhibition is referred to as *steric inhibition of resonance*. Picryl iodide (**15**) provides a good example. A comparison of C–N bond lengths of the nitro group in the *para* position (135 pm) with that of the nitro groups in the *ortho* positions (145 pm) reveals that the former has some double bond character. This is explained on the basis of involvement of the *para* nitro group in resonance because of coplanarity of benzene ring and nitro group. On the other hand, the *ortho* nitro groups do not participate in resonance as they are pushed out of the plane of the ring by the bulky iodine atom. Consequently, the C–N bond length of *ortho* nitro groups is very close to that of aliphatic nitro compounds.

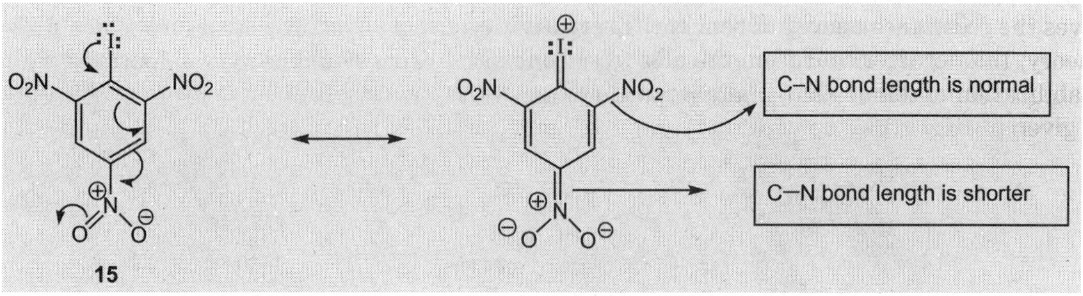

(ix) Resonance structures involving charge separation are less stable than nonpolar structures.

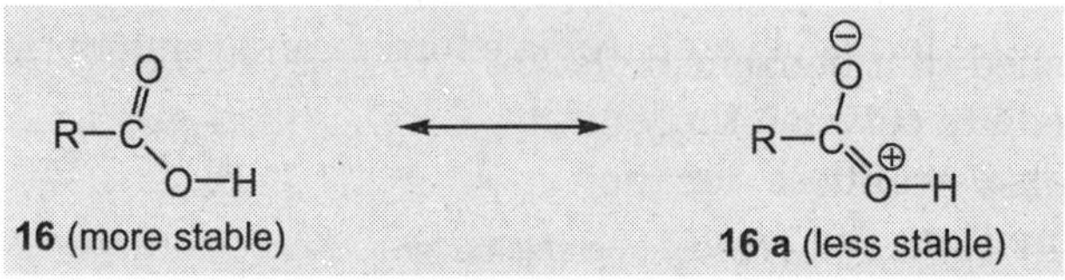

Thus, the two resonance structures of a carboxylic acid, **16** and **16a** have the same number of covalent bonds, but **16a** is less stable than **16** because of charge separation.

(x) Resonance structures having like charges on adjacent atoms are of high energy and hence unimportant. Structure **18** obviously plays a very minor role in the stabilization of **17**.

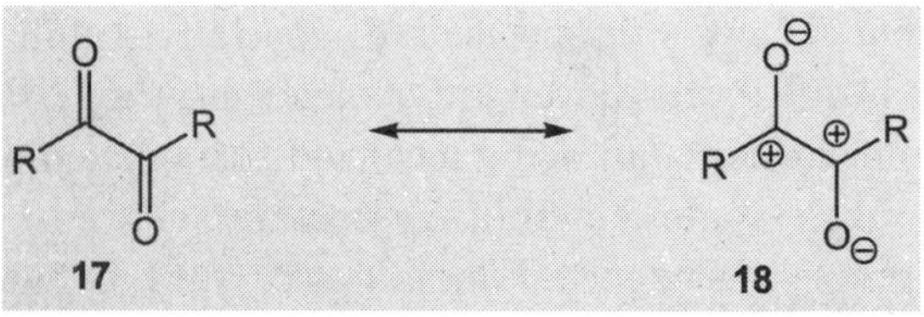

Stabilization through Resonance

(i) Benzene: The Hückel molecular orbital (HMO) treatment of benzene considers it as composed of six sp^2 carbon atoms, each being connected to three other atoms through sp^2 orbitals with the result that all the atoms are in the same plane, making angles of 120° to one another. Each of the remaining six 2*p* orbitals, containing an electron each, overlaps with two adjacent 2*p* orbitals, producing a total of six pi molecular orbitals, three *bonding* and three *antibonding*. The energy of these six electrons has been calculated in terms of α and β, where α, called the *Coulomb* integral, represents the energy of the atomic orbital before overlap, and β, called the *resonance* integral, represents the energy of stabilization due to pi orbital overlap. β vanishes between two atoms not linked directly, but has a finite value which remains uniform between similar atoms in a given pi system. The energy of the six pi orbitals is in the increasing order: $\alpha+2\beta$; $\alpha+\beta$; $\alpha+\beta$; $\alpha-\beta$; $\alpha-\beta$; $\alpha-2\beta$.

There are two electrons in each of the occupied orbitals (first three), and the total energy is given by $6\alpha+8\beta$, whereas, the total energy of six electrons in three independent double bonds is $6\alpha+6\beta$, each double bond with two pi electrons having the energy $2\alpha+2\beta$. The difference, 2β, gives the resonance energy of benzene. There is no way of calculating β from molecular orbital theory; this is determined empirically from the value of resonance energy. β is negative for stabilization of the system. The graphical representation of three bonding orbitals of benzene is given in **Fig. 1.11** (top views).

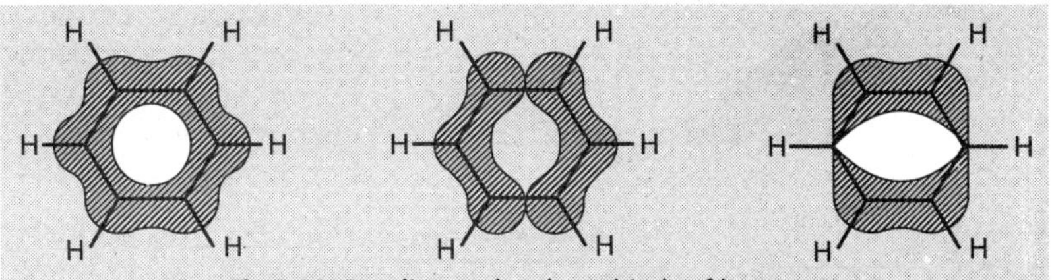

Fig.1.11: Bonding molecular orbitals of benzene

It is established that benzene molecule forms a flat, regular hexagon with C–C distance of 1.40 Å, intermediate between the length of a single bond, 1.54 Å., and a double bond, 1.34 Å. The molecule is also strainless, for the three valencies of each trigonal carbon atom (sp^2) are planar and inclined at 120° to one another, the natural angle of a regular hexagon. The increased stability of benzene is attributed to the interaction or delocalization of the orbitals of the conjugated double bonds in a cyclic structure.

It was earlier recognized by Robinson that benzenoid aromatic compounds are characterized by the presence of six π electrons in a cyclic structure, called 'the aromatic sextet'. The structure, **19** and a simpler symbolic representation **20**, very much in use these days, were suggested by Robinson, depicting delocalization of the pi electron sextet.

19 **20**

Further developments emphasized the importance of aromatic sextet not only in six-membered rings, but also in five- and seven-membered carbocyclic and heterocyclic systems. There have been several attempts at defining an aromatic compound, but the most concise definition is due to MJS Dewar: 'an aromatic compound is a cyclic compound with a large resonance energy where all the annular atoms take part in a single conjugated system'.

Aromaticity may also be defined as the ability to sustain an induced 'ring current' in nuclear magnetic resonance. The π electrons in aromatic compounds, due to the induced ring current under an applied magnetic field, produce characteristic *deshielding effect* on the outer protons (**Fig. 1.12**). All the protons of benzene fall in deshielding region (outer) and, therefore, its NMR spectrum shows a single peak at about δ 7.2.

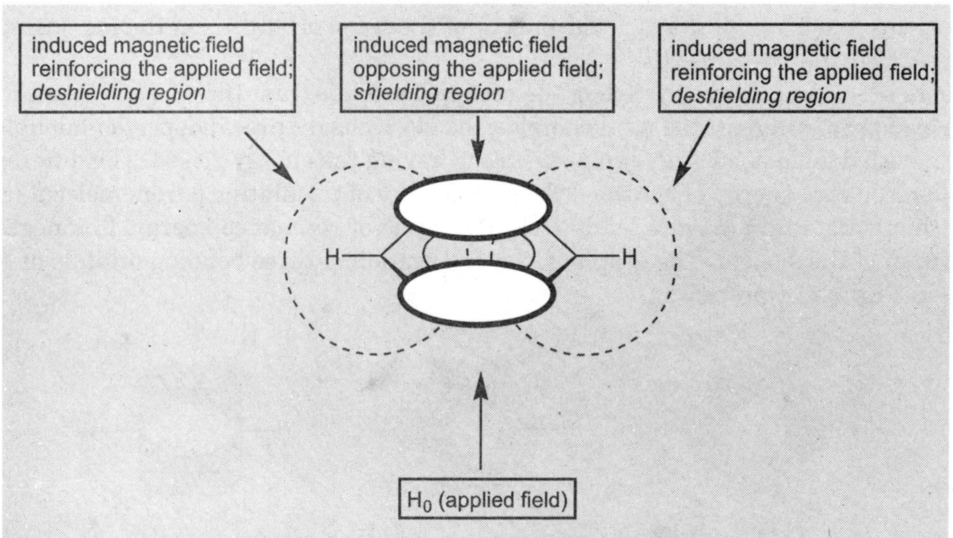

Fig. 1.12: Induced π-electron ring current in benzene

(ii) Acyclic Polyenes: Butadiene shows stabilization to the extent of about 4 kcal/mol (or 17 kJ/mol), over and above the sum of the π energies of two independent ethylenic bonds. This is attributed to the delocalization (π-π overlap) of the two double bonds in conjugation resulting in shortening of the middle (σ) bond and lowering of the π energy of the whole system.

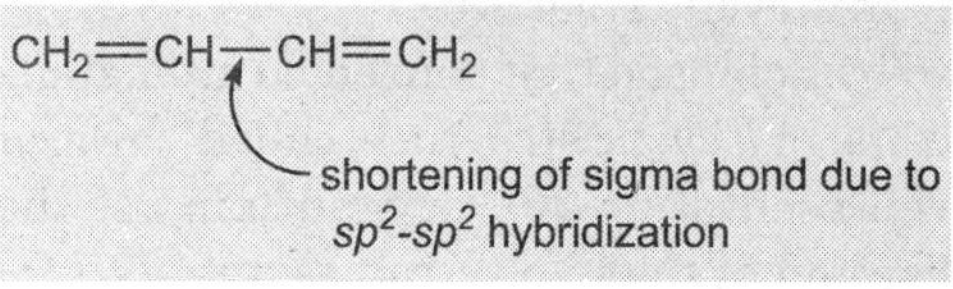

The same argument holds for polyenes and benzenoid compounds. Recently it has been pointed out, however that delocalization (resonance) of π electrons in butadiene and other acyclic conjugated hydrocarbons (polyenes) is negligible, and whatever stability they show is largely due to hybridization of the σ bond from sp^3-sp^3 *to* sp^2-sp^2, and second-order π-π interaction across the sp^2-sp^2 C–C single bonds.

(iii) Vinylic and Allylic Systems: Systems in which the double bond is conjugated with the *p* orbital of an adjacent atom, as in vinyl chloride (CH_2=CH–Cl), also show resonance stabilization. In vinyl chloride, *p* electrons on the chlorine atom interact with π electrons of the double bond

(p-π overlap), thereby shortening the C–Cl distance. This explains relative inactivity of chlorine atom in substitution reactions (Chapter 6).

Similar effect is seen in structures having *allyl anion* moiety, in which two electrons in $2p$ orbital of the adjacent atom can interact with π electrons of the double bond thus stabilizing the system through delocalization.

$$\overset{\ominus}{C}H_2{-}CH{=}CH_2 \longleftrightarrow CH_2{=}CH{-}\overset{\ominus}{C}H_2$$

Again, when the $2p$ orbital of the adjacent atom is partially filled as in the *allyl radical*, or vacant as in the *allyl cation*, delocalization results due to p-π overlap (**Fig. 1.13**).

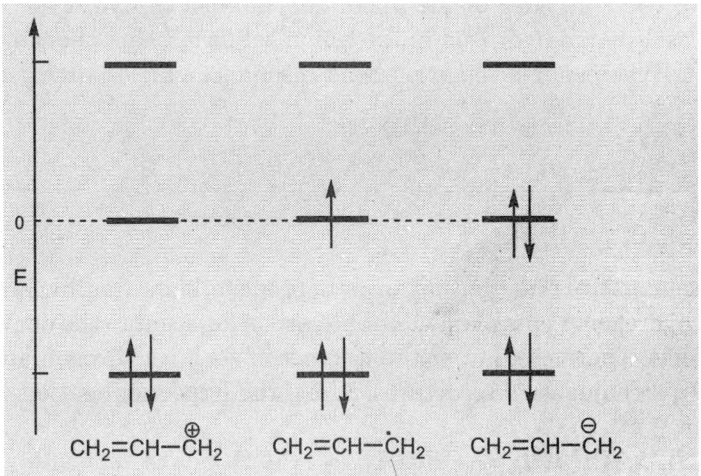

Fig. 1.13: HMO π-energy level diagrams for the allyl system

C. Hyperconjugation

There is another kind of delocalization, called *hyperconjugation*, which is a special type of resonance. It is attributed to σ-π orbital overlap (**Fig. 1.14**) against the π-π overlap in butadiene or benzene.

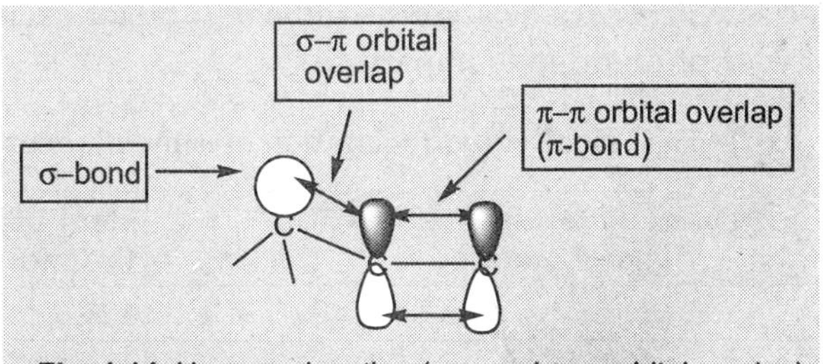

Fig. 1.14: Hyperconjugation (σ–π and π–π orbital overlap)

Baker and Nathan observed that the rate of reaction of p-alkybenzyl bromides with pyridine was contrary to what would be expected from the order of the inductive effect of the substituent alkyl groups. The order of inductive effect is: R = *tert*-butyl > *iso*propyl > ethyl >

methyl, whereas the rate of the above reaction follows the order: methyl > ethyl > *iso*propyl > *tert*-butyl, *i.e.* when these alkyl groups are attached to an unsaturated system, the order of +I effect is reversed.

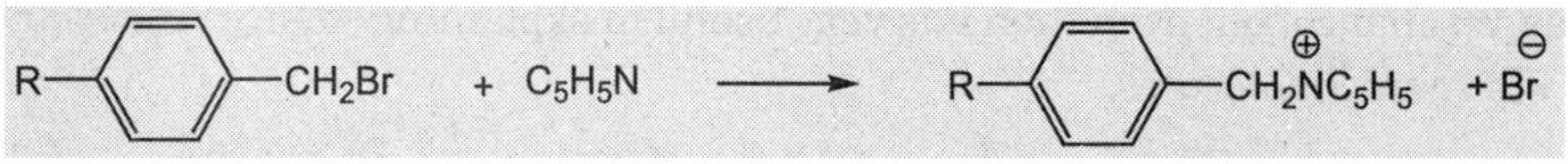

This effect is known as the *Baker-Nathan effect* or *no-bond resonance* or *hyperconjugation* effect. It was explained on the basis of hyperconjugative forms, contributing to the overall electron-donating effect for the substituent having hydrogen attached to it, methyl group showing maximum effect because of the maximum number of hydrogens. Since there is no bond between the carbon and hydrogen in the canonical forms, this phenomenon is also called *no-bond resonance*. Hyperconjugation or no-bond resonance may be shown as:

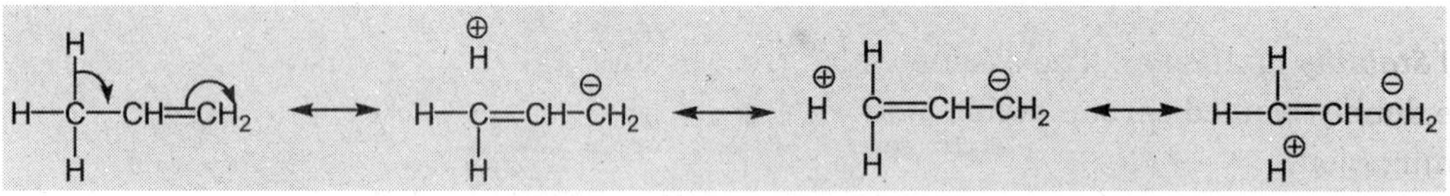

These hyperconjugative contributing forms of propene, show that hyperconjugation occurs through σ bonded hydrogen atoms present on the carbon adjacent to the doubly bonded carbon, *i.e.* α-hydrogen atoms. Therefore, larger the number of such α-hydrogen atoms, more are the number of such hyperconjugative structures. Thus, the hyperconjugation effect decreases in the following order:

$$-CH_3 > -CH_2CH_3 > -CH(CH_3)_2 > -C(CH_3)_3$$

The fact that the heat of hydrogenation of alkyl-substituted ethylenes decreases with the degree of substitution (**Table 1.6**) is explained on the basis of hyperconjugation or no-bond resonance. Although such effects, where hyperconjugative effect is greater than the inductive effect, are observed in some processes, there is a great deal of controversy about the above explanation based on no-bond structures just as the idea of resonance is rejected in butadiene.

The energy change in ground state may be the result of σ bond stabilization. In other words, sp^2-sp^2 C–C bond may be sufficiently more stable than a sp^3-sp^3 C–C bond to account for this difference. However, for carbocations and free radicals, and for excited states of molecules, there is evidence that hyperconjugation is important.

Table 1.6 Heat of hydrogenation of some alkenes

Olefin	Heat of hydrogenation (kcal/mol)	Heat of hydrogenation (kJ/mol)
$CH_2=CH_2$	31.0	129.5
$R-CH=CH_2$	28.5	119.1
R–CH=CH–R (*cis*)	27.0	112.8
R–CH=CH–R (*trans*)	25.7	107.4
$R_2C=CH_2$	26.1	109.0

I. Importance of Hyperconjugation Effect

Both resonance and hyperconjugation are the results of delocalization of electrons involving π–π conjugation and $-\pi$ conjugation, respectively. The conjugation effect due to hyperconjugation is weaker than resonance, but even then it is very useful in explaining some physical and chemical properties of organic molecules:

(i) Stability of Carbocations and Free Radicals: The order of stability of carbocations and free radicals can be explained on the basis of hyperconjugation. The order of stability of carbocations and free radicals is: 3° > 2° > 1°

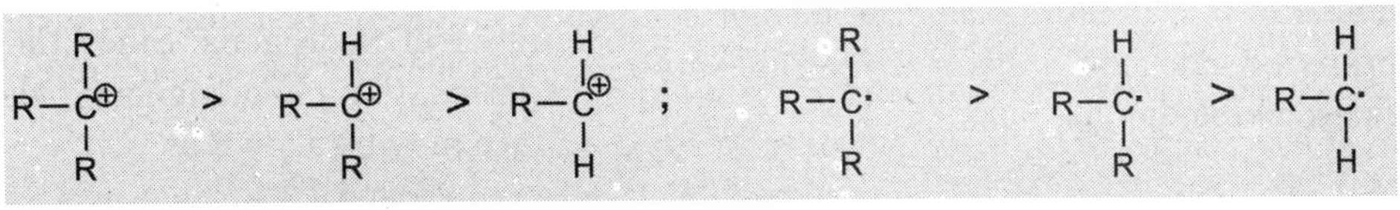

(ii) Stability of Alkenes: The alkene which has maximum number of alkyl groups attached to the doubly bonded carbon atoms is most stable. This can be explained on the basis of hyperconjugation.

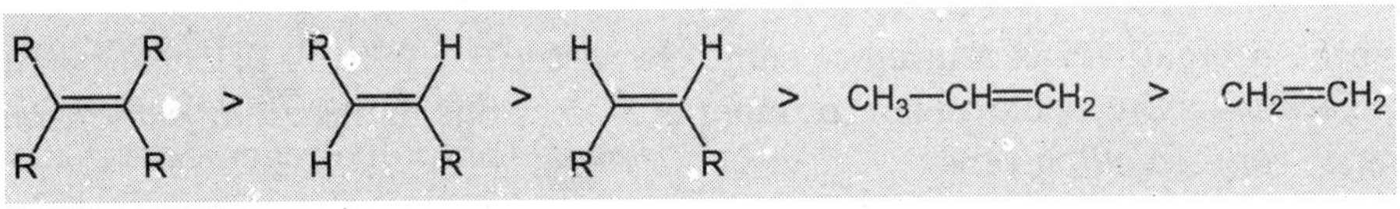

More the number of alkyl groups, more is the number of hyperconjugation structures that can be written for each one of the alkene. Therefore, greater the number of such structures, more stable is the alkene. For 2-butene, *trans* form is more stable than the *cis* form due to electronic repulsion present in *cis* arrangement where both methyl groups are on the same side, *i.e.* close to each other.

trans-2-butene (more stable) *cis*-2-butene (less stable)

(iii) Directive Influence of Alkyl Groups: The *o*, *p*-directing influence of alkyl groups can be explained on the basis of hyperconjugation. For example, toluene can be depicted as a hybrid of the following contributing forms:

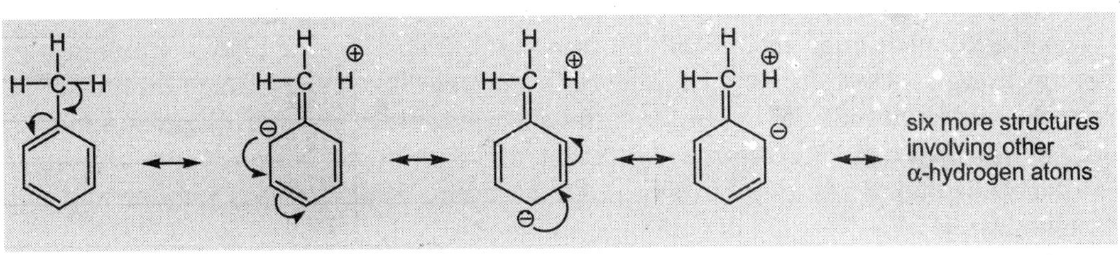

D. Aromaticity

The most significant feature of benzene and other compounds containing this ring system is that they are *more stable* as compared to the corresponding acyclic polyenes. The stability of the ring compounds is due to the fact that they possess large resonance energies and are called *aromatics* or *aromatic compounds*. The extra stability of such compounds is due to a special property known as *aromatic character* or *aromaticity*.

Since benzene is a cyclic conjugated polyene, earlier chemists thought that the lower and higher analogs of benzene, for instance, cyclobutadiene, and cyclooctatetraene can be prepared, and they should also exhibit aromaticity. Attempts were made to synthesize these types of compounds, and it was found that the former compound is highly unstable while the latter, though prepared, does not possess comparable stability, and behaves like an ordinary polyene with isolated double bonds.

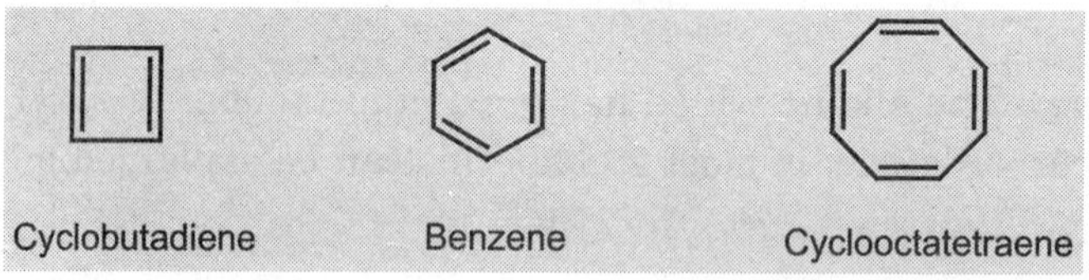

Evidently, aromaticity is related to some characteristic feature of benzene and related systems, generally having some 'magic' number of π-electrons in a regular *planar, monocyclic, and completely conjugated polyene.*

I. Hückel's Rule for Aromaticity

Molecular orbital method developed by Hückel led him to suggest that the aromatic sextet is only one of many π electron systems, which can confer aromatic stability and character upon suitable molecules. The cyclic conjugated polymethines should also be aromatic if, they contain $(4q + 2)$ π electrons, where q is zero or a whole number. This is what is known as Huckel's rule for aromaticity.

The annulenes are ring compounds, having an even number of CH units, represented as neutral monocyclic polyenes, $(\text{–CH=CH–})_n$. They have one bonding orbital of lowest energy, and other bonding orbitals are in pairs having the same energy (degenerate). According to Hund's rule, when n is *even*, as in square cyclobutadiene, two π electrons will be in the lowest orbital and the other two will be unpaired, one in each orbital of the degenerate pair.

When n is *odd*, as in benzene, there will be no orbital with unpaired electrons. This explains the stability of $(4q + 2)$ *vis-à-vis* $4q\pi$ electron systems, which are *antiaromatic* as they are of higher energy than the hypothetical structures having independent double bonds without mutual overlap, that is, without delocalization.

The Hückel molecular orbital (HMO) treatment of the regular planar annulenes gives the orbital energies which follow a characteristic pattern. There are pairs of orbitals of equal energies flanked on either sides by single orbital of lowest energy and one of highest energy. The orbital energy diagrams for cyclopropenyl cation, cyclobutadiene, benzene, cyclooctatetraene, and cyclodecapentaene are given in **Fig. 1.15**, assuming that they are regular and planar (cyclooctatetraene is, however, a nonplanar compound).

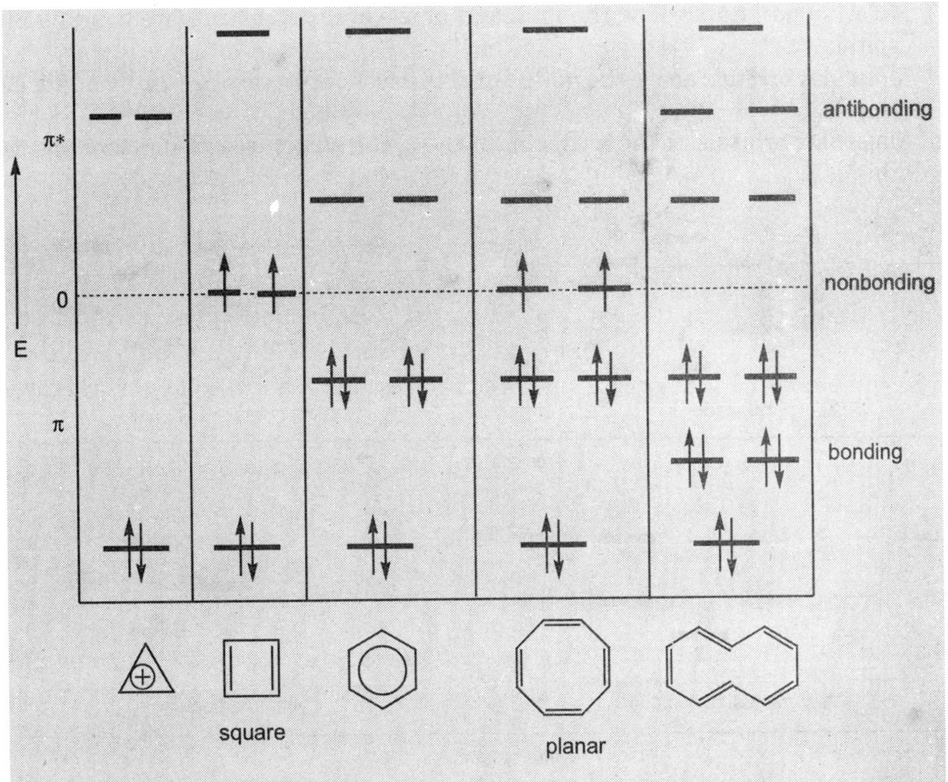

Fig.1.15: HMO π-energy level diagram of some annulenes

It has been found that conjugated rings with 2, 6, 10, and 14 π electrons are aromatic, while those with 4, 8, 12, 16, and 20 surely are not. It has also been shown that bond alteration starts when $q = 6$, that is from the 26-annulene.

It is clear from preceding discussion on Hückel's rule for aromaticity that it is not necessary for aromatic compounds to have a benzene ring. All the cyclic planar and completely conjugated polyene systems having $(4q + 2)$ π-electrons are aromatic, and those having $4q$ π-electrons are either antiaromatic or nonaromatic.

The classification of cyclic conjugated systems is based on the energy criterion. A compound is aromatic if it is more stable than its open chain conjugate analogue, and is antiaromatic if it is less stable than its open chain conjugated analogue. If a compound neither shows stabilization nor destabilization, it is known as nonaromatic.

An alternative approach has been evolved to determine the relative energies of π-molecular orbitals of planar molecules with uninterrupted cyclic π electron clouds without any calculations. In this approach, the structure of a cyclic compound is drawn with one of its vertices pointed down as shown in the **Fig. 1.16**. The relative energies of π-molecular orbitals correspond to the relative levels of the vertices. The following points are noticeable:

(i) The number of π orbitals is equal to the number of atoms in the ring, since each ring atom contributes p orbital.

(ii) Molecular orbitals below the midpoint of the cyclic structure are bonding molecular orbitals.

(iii) Molecular orbitals above the midpoint of the cyclic structure are antibonding molecular orbitals.

(iv) Molecular orbitals at the midpoint of the cyclic structure are nonbonding molecular orbitals.

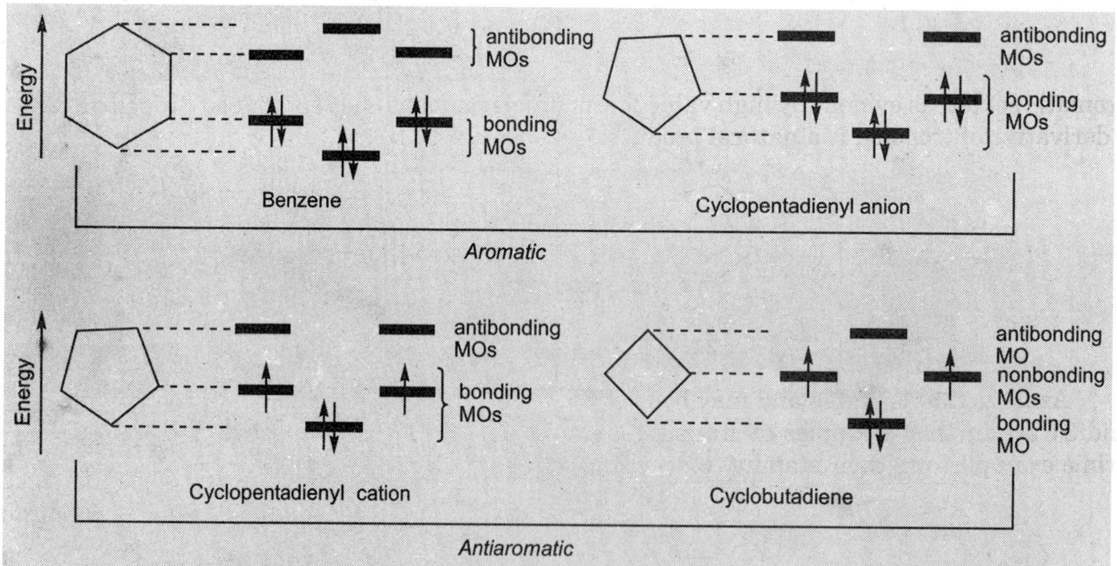

Fig. 1.16 : Aromaticity and antiaromaticity on the basis of distribution in the π molecular orbitals of benzene, cyclopentadienyl anion, cyclopentadienyl cation and cyclobutadiene

Aromatic Compounds

Six π electrons of benzene and cyclopentadienyl anion occupy three bonding π molecular orbitals. It is to be mentioned that there is always an odd number of bonding orbitals in such cases, because one corresponds to the lowest vertex, and others come in degenerate pairs. This means that aromatic compounds (such as benzene and the cyclopentadienyl anion) with an odd number of pairs ($4q + 2$) of π electrons have all completely filled bonding orbitals. This is what gives aromatic compounds their stability.

Antiaromatic Compounds

In case of antiaromatic compounds having even number of pairs, $4q$ π electrons, either they are unable to fill their bonding orbitals (cyclopentadienyl cation) or they have a pair of π electrons left over after the bonding orbitals are filled (cyclobutadiene).

(i) Aromatic Compounds (systems with 4q + 2 π electrons)

Aromatic Duet ($q = 0$; 2 π-electron systems): Examples of aromaticity in 2π-electron systems are found in **21** (as perchlorate), **22** (as perchlorate), and **23** in spite of the bond angle of 60°. The dication (**24**) (as fluoroborate) is an important addition to this class of stable systems.

Aromatic Sextet ($q = 1$; 6 π-electron systems): Benzene is the best known example of aromatic sextet. Cyclopentadienyl anion (**25**) and cycloheptatrienyl cation (**26**) are also well known.

Further illustrations are provided by tropone (**27**), which is stable since the electronegative oxygen atom attracts the carbonyl π-electrons leaving the seven-membered ring with the

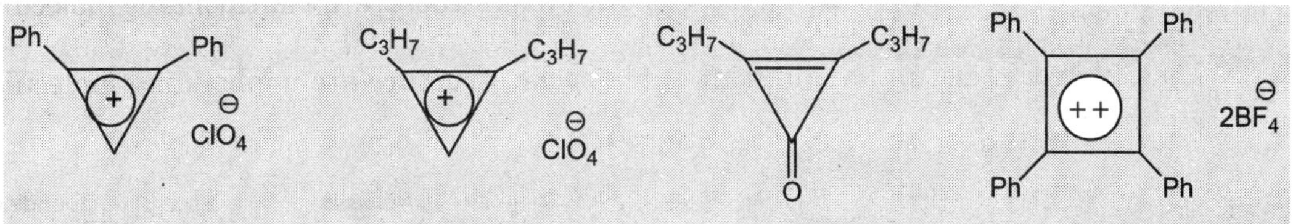

21 **22** **23** **24**

aromatic sextet, as evident by high value for its dipole moment (4.17D). Indeed, tropolone (**28**), a derivative of tropone, is a natural product.

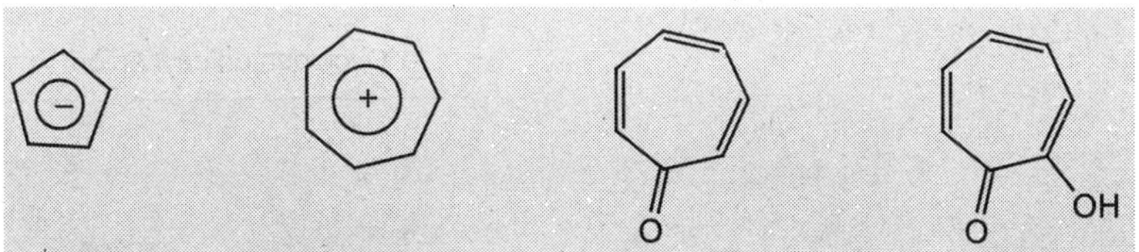

25 **26** **27** **28**

Azulene (**29**) is stable and may be regarded as a combination of **25** and **26**. The ylids **30** and **31** are further examples of aromatic sextet. Metallocenes, of which ferrocene (**32**) is the prime example, owe their stability to the aromatic sextet.

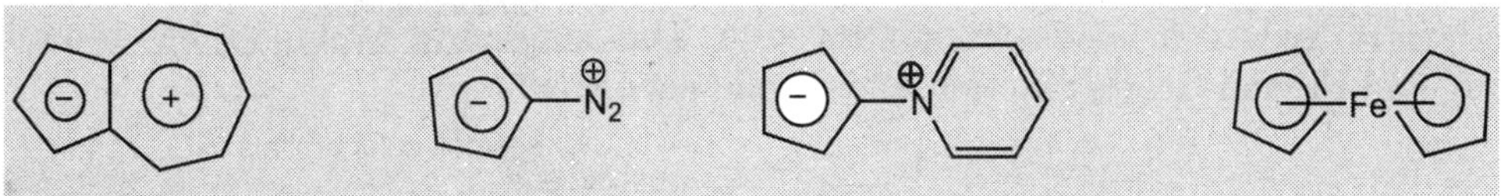

29 **30** **31** **32**

Hückel's rule also applies to heterocyclic systems such as furan, thiophene, and pyrrole, each of which contains a group of six annular electrons including the lone pair from oxygen, sulphur, and nitrogen, respectively. Pyridine and pyrylium cation offer two other examples of aromatic sextet.

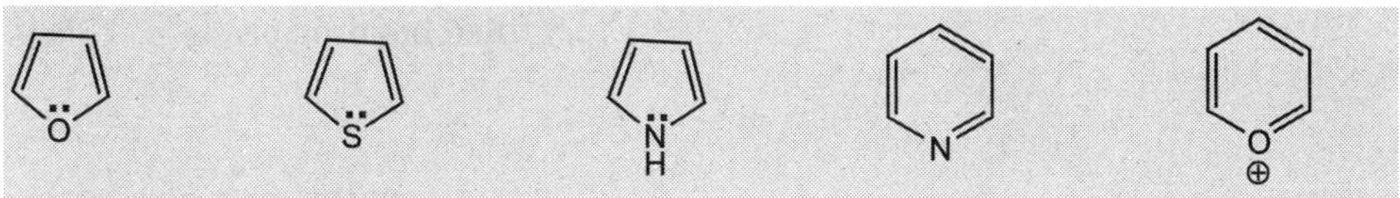

Furan Thiophene Pyrrole Pyridine Pyrylium cation

However, the stable inorganic compounds, **33** and **34** as well as sydnones (**35**), though apparently having π electron sextet, are not considered aromatic.

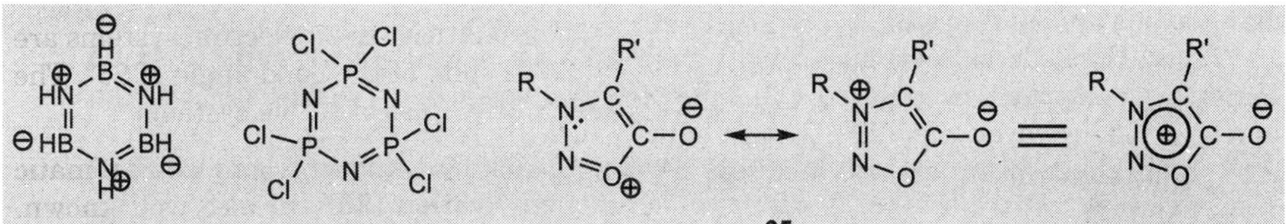

33 **34** **35**

Ten π-electron Systems (q = 2) and other Annulenes: [10]Annulene, **36** is very unstable apparently, because of the non-bonded interaction of the hydrogens inside the planar ring, but stable derivatives (**37**; X = CH_2, O, NH) of 10-annulene are known in which the hydrogen interference is absent. These substances undergo aromatic substitution and sustain a ring current in their NMR spectra.

36 **37**

Cyclooctatrienyl dianion (**38**) and cyclononatetraenyl anion (**39**) have been prepared as potassium and lithium salts, respectively.

38 **39**

Fourteen π-electron Systems (q = 3): One of the spectacular examples of Hückel's rule is the discovery that *trans*-dimethyldihydropyrene derivative, (**40**), displays aromatic stability of a much higher order than 14-annulene, (**41**). This is explained by the absence of non-bonded interactions of hydrogen inside the planar ring. The compound **40** undergoes bromination, nitration, and Friedel-Crafts reaction under mild conditions.

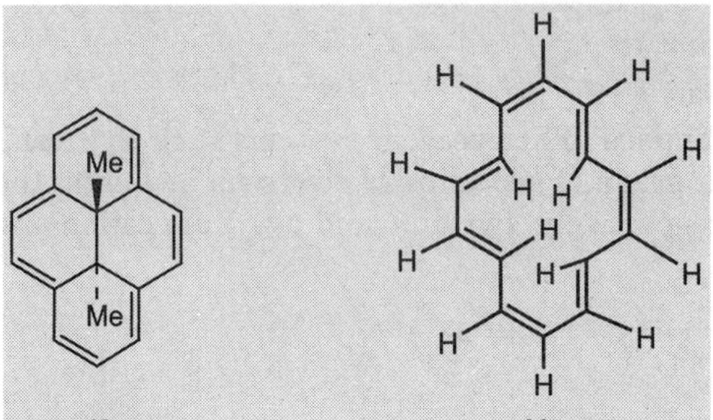

40 **41**

Eighteen and Twenty two π-electron Systems: 18- and 22-annulenes, **42** and **43**, are aromatic if we go by their NMR spectra. The π electrons in aromatic compounds, produce characterisitic shielding effect on the inner protons, and deshielding effect on the outer protons. These annulenes show similar NMR characteristics (in structure **42**; the inner protons appear at δ 1.9, and the outer protons appear at δ 8.8).*

Thus, Hückel's rule seems to hold good upto an extent, but it fails to explain how aromaticity disappears even in (4*q* + 2) π-electron systems above a certain critical ring size. It must be noted, however, that (4*q* + 2) annulenes, which are not aromatic, still remain relatively more stable than *4q* annulenes. Medium size annulenes are aromatic only on the basis of

* Annulines are monocyclic compounds of the general formula $[-CH{=}CH-]_n$

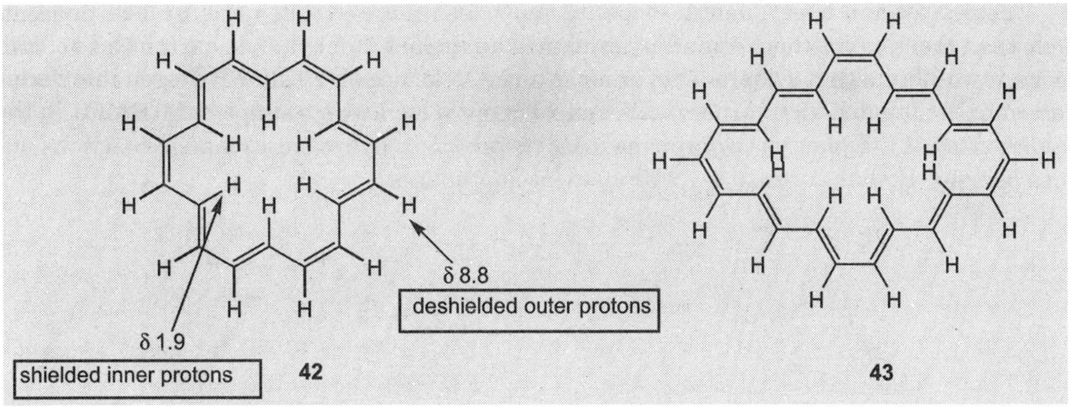

42 43

NMR characteristics. Chemically, they show no aromatic stability; rather they are very reactive, and polymerize readily like polyenes. The decrease in aromaticity may be explained, firstly, by increasing tendency of σ-bonds with increasing ring size to favour alternating structures with localized bonds, and secondly, by the non-bonded interaction among the hydrogens inside the planar rings of the medium size annulenes.

Hückel's rule has been extended to polycyclic ring system as well. Aromatic stability, however, decreases in this order: benzene, naphthalene, anthracene, tetracene, pentacene, hexacene, and heptacenes with 6, 10, 14, 18, 22, 26, and 30 carbons atoms. The higher members of this series (acenes) show extreme reactivity, and are very unstable. Further, the rule does not account for the difference in aromatic stability of the above hydrocarbons, and the angularly fused isomeric hydrocarbons which are more stable. Again, it is applied only to half of the most stable aromatic hydrocarbons having $6q$ π-electrons, such as biphenyl, triphenyl and triphenylene dibenzpyrene, which may be regarded as derivatives of polyphenyls. *Hückel's rule is, therefore, not suitable for predicting aromatic stability in polycyclic hydrocarbons.*

***(ii) Antiaromatic and Nonaromatic Systems with 4 π-electrons*:** According to Hückel's rule, planar cyclic systems such as butadiene with four electrons should be *antiaromatic*. In fact, cyclobutadiene is so unstable that all attempts to synthesize it have failed. However, when generated at –77°C, it undergoes dimerization to give at least two products.

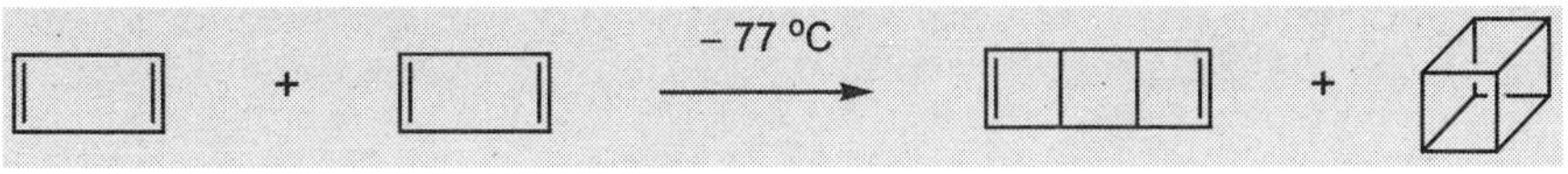

Other important examples of such system include cyclopropenyl anion and cyclopentadienyl cation.

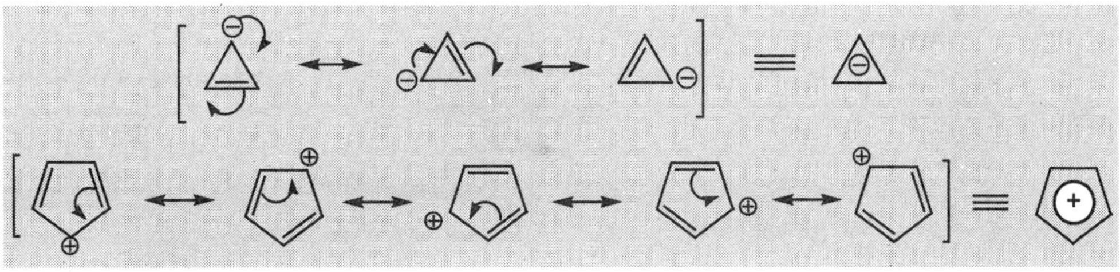

Theoretical and experimental evidence show that these systems are, in fact, much less stable than their acyclic (nonaromatic) conjugated analogues. For example, on the basis of HMO theory, it was found that cylopropenyl anion is about19 kcal/mol (79 kJ/mol) less stable than an open-chain conjugated allyl carbanion. An experimental evidence to support the HMO theory is that 3-benzoyl-1,2-diphenylcyclopropene loses its proton in hydrogen-exchange reactions 6000 times less slowly than 3-benzoyl-1,2-diphenylcyclopropane.

3-Benzoyl-1,2-diphenylcyclopropene (leads to the formation of an unstable carbanion)

3-Benzoyl-1,2-diphenylcyclopropane (leads to the formation of a normal carbanion)

Systems with 8 π-electrons-Cyclooctatetraene (q = 2)

According to Hückel's rule, planar cyclic system having 8π electrons should be antiaromatic. Cyclooctatetraene, however,escapes its antiaromaticity by abandoning planar geometry. In fact, it has a tub-shaped structure in which each double bond is almost perpendicular to its neighbouring double bond, thereby reducing the π-overlap to a great extent. This explains its relative stability compared to cyclobutadiene, which enjoys no such option. Cyclobutadiene, however, can acquire some stability by assuming a rectangular form, though even in this form it may be unstable due to ring strain (**Fig. 1.17**).

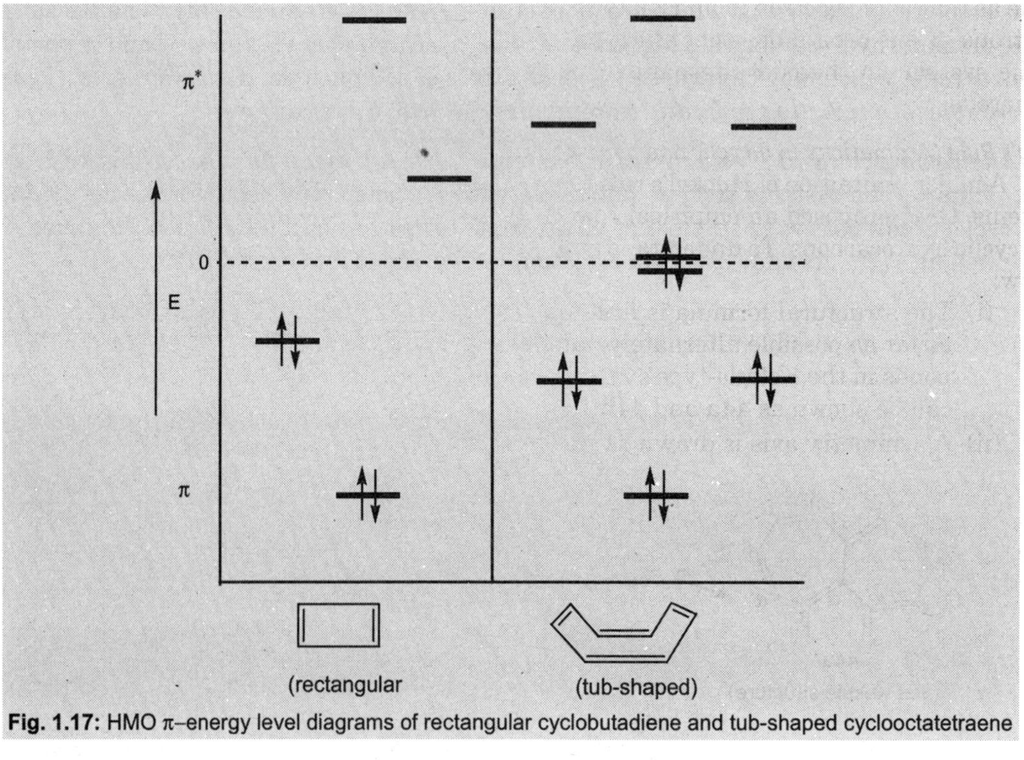

Fig. 1.17: HMO π–energy level diagrams of rectangular cyclobutadiene and tub-shaped cyclooctatetraene

E. Alternant and Nonalternant Hydrocarbons

Conjugated hydrocarbons may be divided into two types, *alternant* and *nonalternant*. In the alternant type, the atoms can be divided into two sets, one starred and the other unstarred, in such a way that no two atoms of the same set are directly linked. Benzene and naphthalene are alternants, whereas azulene is a nonalternant hydrocarbon.

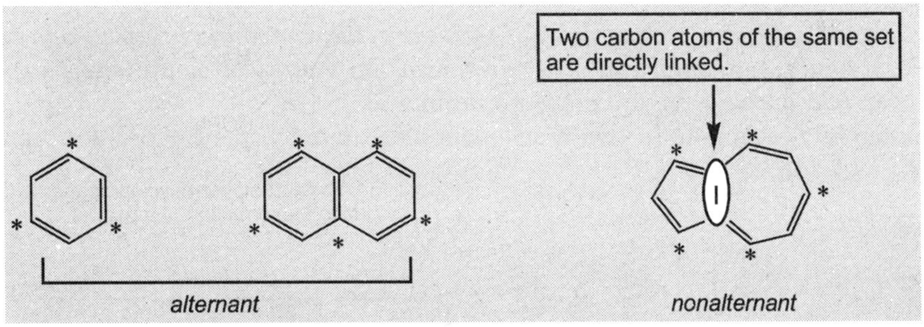

The even alternant hydrocarbons, namely, those with equal number of starred and unstarred atoms, have equal number of bonding and antibonding π molecular orbitals. The bonding orbitals are filled in the ground state. The odd alternant hydrocarbons, namely, those having odd number of conjugated atoms must occur as radicals, carbocations, or carbanions. They have a nonbonding π molecular orbital (NBMO) of zero bonding energy in addition to the usual bonding and antibonding orbitals. For example, the NBMO of allyl cation, radical, and anion have 0, 1, and 2 electrons, respectively. The energy of an electron in the NBMO is the same as that of an electron in an isolated *p* orbital, and consequently the presence or absence of electrons in this orbital does not affect the total bonding energy, and thus the overall stability of the system. In the odd alternant hydrocarbon, the starred atoms are taken as the more numerous.

Crag's Rule (Aromaticity in bicyclic and polycylic polyenes)

A major limitaition of Hückel's rule for aromaticity is that it is applicable only to monocyclic systems. Crag proposed an empirical rule to determine the aromaticity in certain bicyclic and polycyclic hydrocarbons. To understand Crag's rule, it is necessary to follow the steps mentioned below:

(i) The structural formula is first labelled with equal number of spin symbols α and β, as far as possible alternately, different symbols being given to the ends of all double bonds in the Kekule-type structure. For example, two Kekule structures of heptalene can be shown as **44a** and **44b**.

(ii) A symmetry axis is drawn as shown below:

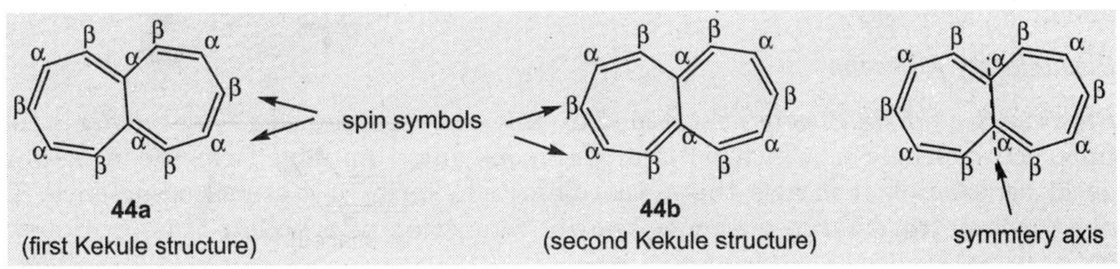

44a (first Kekule structure) **44b** (second Kekule structure)

(iii) The sum of the number of symmetrically related π-centres, *which do not lie on* the symmetry axis, is denoted as '*f*'. In structures **44a** and **44b** *f* is 5 (two pairs of α and three pairs of β).

(iv) The structure is rotated about symmetry (*e.g.* conversion of **44a** into **44b** and *vice versa*). This may result in some interconversions of spin symbols. The number of such interconversions is denoted as '*g*'. In the conversion **44a** → **44b**, there is no change, so *g* is equal to zero.

According to Crag's rule, *"if the sum of f + g is even, the compound is expected to be aromatic, and if this sum is odd, the compound is nonaromatic"*. In the case of heptalene, $f + g = 5 + 0 = 5$. Since this is an odd number, heptalene is nonaromatic.

Let us apply Crag's rule in two more examples, namely, pentalene (**45a** and **45b**) and azulene (**46a** and **46b**).

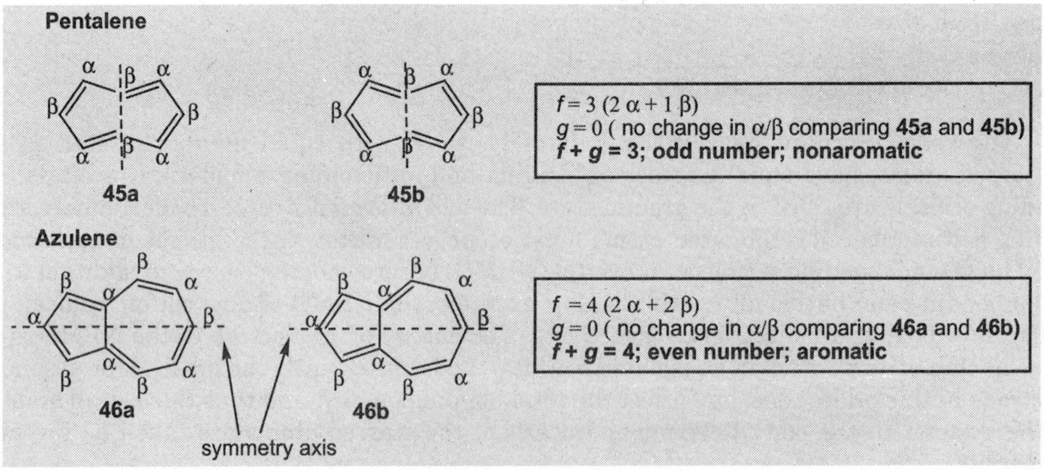

It is to be mentioned that Crag's rule also suffers from the limitation that it is applicable to those bicyclic and polycyclic systems, in which at least two centres lie on a symmetry axis that converts one Kekule type structure to another.

F. Bonding in Fullerenes

The first fullerene to be discovered is *buckminster-fullerene* (C_{60}) (**Fig. 1.18**), which was prepared by Kratschmer, Huffman and co-workers. The discovery of fullerenes expanded the number of known carbon allotropes, which were limited to graphite, diamond, and amorphous carbon such as soot and charcoal. Buckyballs and buckytubes have been the subject of intense research for their unique chemistry, and technological applications, especially in material science, electronics and nanotechnology.

Buckminster-fullerene

Buckminster fullerene (often shortened to fullerene; nicknamed as *buckyball*) is the smallest fullerene molecule in which no two pentagons share an edge (which can be destabilizing as in pentalene). It is also the most common in terms of natural occurrence. This is the most symmetrical large molecule known, consisting of a hollow cluster of sixty carbons (C_{60}). Its structure resembles a ball made of twenty hexagons and twelve pentagons, with a

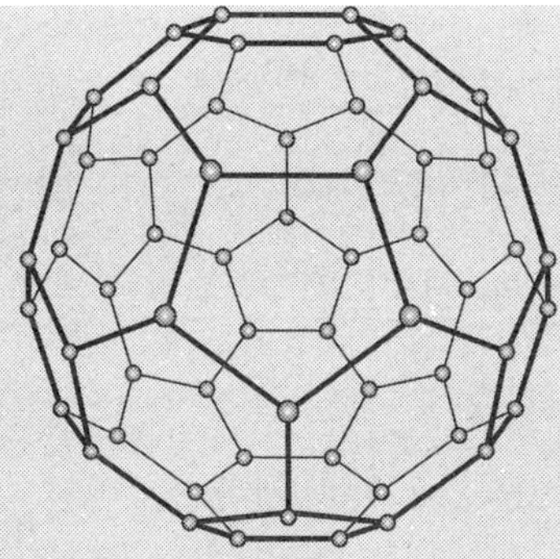

Fig.1.18: Structure of *Buckminster-fullerene* C_{60} (*buckyball*)

carbon atom at the vertices of each polygon, and a bond along each polygon edge. The van der Waals diameter of a C_{60} molecule is about 1.1 nanometers (nm). The nucleus to nucleus diameter of a C_{60} molecule is about 0.71 nm.

At first glance, buckyball would appear to be aromatic because of the presence of benzene-like rings. However, it does not undergo electrophilic substitution reactions. Instead, it undergoes electrophilic addition reactions at double bonds, like an alkene, which reduces angle strain by changing sp^2-hybridized carbons into sp^3-hybridized ones. The change in hybridized orbitals causes the bond angles to decrease from about 120° in the sp^2 orbitals to about 109.5° in the sp^3 orbitals. This decrease in bond angles allows the bonds to bend less when closing the sphere or tube, and thus the molecule becomes more stable.

Fullerenes are stable compounds, and have extraordinary chemical and physical properties. Because of hollow cages, they can be manipulated to make materials never known before. An interesting, and very useful example is the development of an organic superconductor by doping the buckyball by inserting potassium or cesium into its cavity.

G. Tautomerism

The reversible formation of enols from enolizable ketones brings into focus a phenomenon termed as *tautomerism*. In a broad sense, two structural isomers, which differ in the relative positions of their atoms, and are in rapid equilibrium, are called tautomers. The interconvertibility of tautomers is a chemical reaction, which involves making and breaking of a σ-bond. This must be clearly distinguished from *resonance,* which involves the delocalization of π electrons while the atomic nuclei remain in identical relative positions in all resonance structures. The familiar symbol for *resonance* is a double-headed arrow (↔), whereas tautomerism, being an equilibrium process, is symbolized by double arrows ($\leftrightarrows$ or $\rightleftharpoons$).

Tautomerism is classified into cationotropy and anionotropy depending upon whether atoms or groups of atoms shift as cations or anions. A great majority of cases of cationotropy, however, is prototropy in which protons are involved.

Prototropy

The classical example of prototropy is the keto-enol tautomerism exhibited by ethyl acetoacetate. The compound forms a cyanohydrin with HCN, an oxime with hydroxylamine, and phenylhydrazone with phenylhydrazine. Therefore, it appears to be a saturated keto ester.

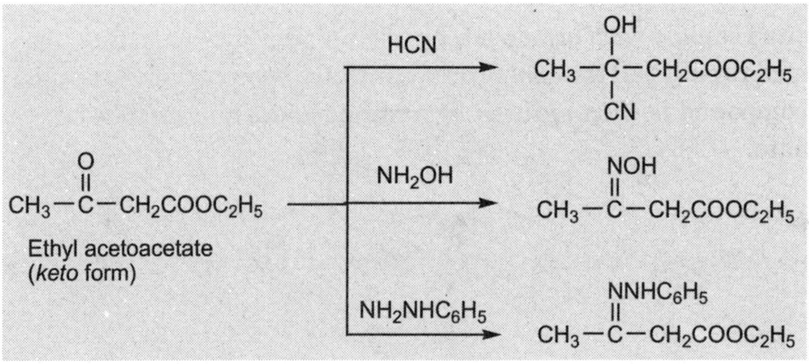

On the other hand, it appears to be an unsaturated hydroxyl ester (**47**) since it gives an intense red colour with ferric chloride, rapidly absorbs bromine, and reacts with diazomethane.

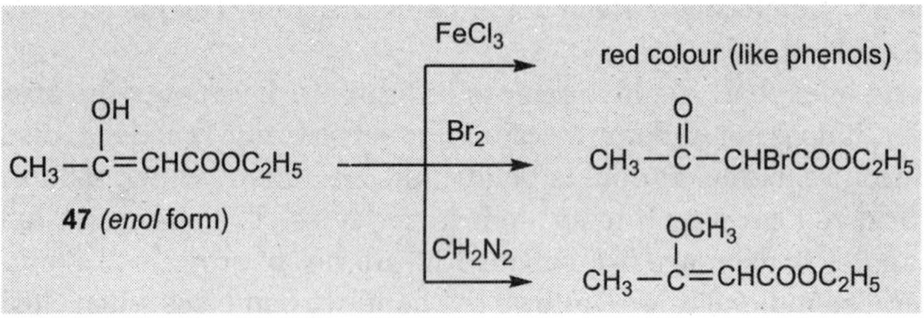

The IR spectral data also shows the presence of *enol* form, as C=O stretching band appears at much shorter frequency (~ 1650 cm^{-1}) than the expected frequency (~ 1715 cm^{-1}). Moreover, it displays an O-H stretching band at around 2700 cm^{-1}, which is much shorter than the expected value (~3600 cm^{-1}).

To explain this anomalous behaviour, it was proposed that ethyl acetoacetate is an equilibrium mixture of two distinct substances, the *keto* form and the *enol* form.

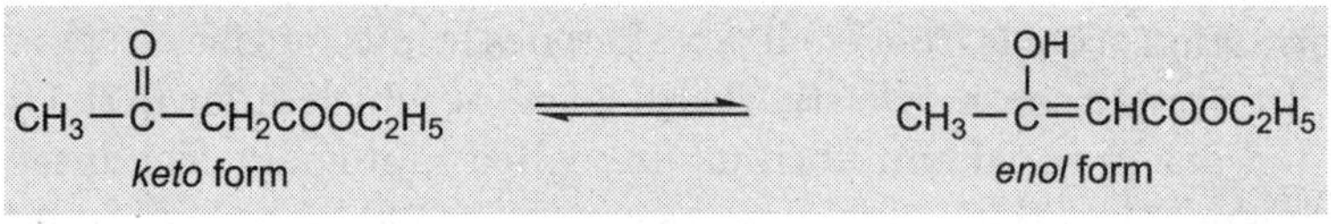

These components were separated by Knorr in 1911, who found that the pure *keto* form (m.p. –39°C) can be crystallized from light petroleum at –78°C, and the *enol* form (liquid) can be obtained by decomposing the sodium salt of ethyl acetoacetate suspended in light petroleum at –78°C by the addition of slightly less than the calculated quantity of hydrogen chloride. Either form is stable at low temperatures, and can be preserved for several days in the absence of acids and bases. A more convenient method of preparing pure samples of *keto* and *enol* forms was devised by Meyer, who placed the equilibrium mixture in a quartz (since glass has a catalytic effect on interconversion) distilling flask and separated the *enol* form, which is more volatile than the *keto* form, by distillation.

The estimation of the *enol* content of *keto-enol* mixture is based on the fact that the enol forms absorb bromine rapidly, whereas the keto form does not. Thus, if an equilibrium *keto-enol* mixture is titrated rapidly with bromine then at the end point (when the colour of bromine is no longer discharged rapidly) all the *enol* form would have been converted to the bromo compound. The α-bromo compound is then reduced by adding iodide, and the liberated iodine is titrated with thiosulphate.

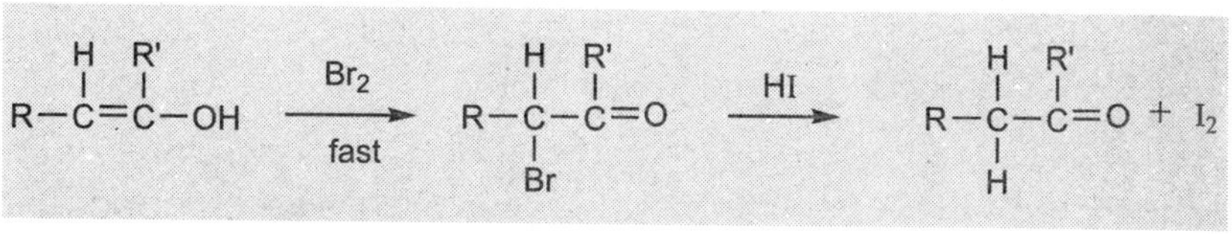

In principle, any molecule containing a *keto* grouping and α-hydrogen(s) is capable of undergoing tautomeric change to an isomeric *enol* molecule.

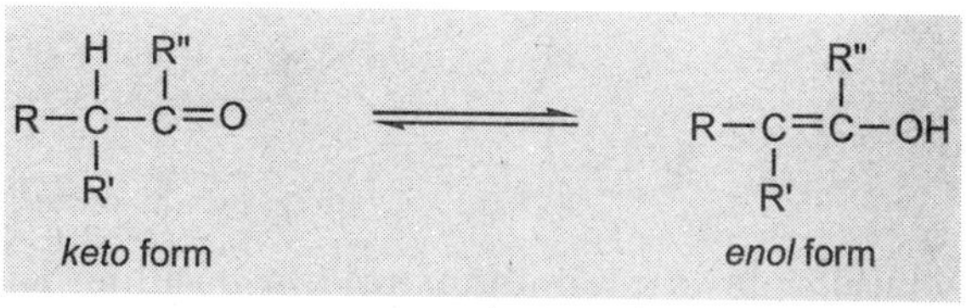

The *enol* forms of simple aldehydes and ketones have, however, not been isolated because of their easy conversion into the more stable carbonyl forms. However, in the presence of more than one carbonyl function on the same carbon in the molecule, the enol content becomes more significant, and the *enol* forms become isolable. The *enol* contents of some ketones, diketones and keto esters are listed in **Table 1.7**.

Table 1.7 Enol contents of some ketones, diketones, and keto esters

Compound	% Enol
CH_3COCH_3	0.00025
$CH_3COCH_2COCH_3$	80.4
$CH_3COCH_2COOC_2H_5$	7.5
$CH_3COCOCH_3$	0.0056
1,2-Cyclopentanedione	99.0
Cyclohexanone	0.020

Acetone exists in equilibrium almost wholly in its *keto* form, whereas ethyl acetoacetate and acetylacetone have significant *enol* contents. This is to be expected, since the electron-withdrawing effect of the cabethoxy or the acetyl substituent increases the acidity of the α-hydrogen. Since the carbonyl group of a ketone is an effective electron-withdrawing group, much more than the carbonyl group of an ester, acetylacetone has more *enol* content than ethyl acetoacetate. These substituents increase the stability of the *enol* form by conjugative effect. In addition, intramolecular hydrogen bonding also occurs in the *enol* forms stabilizing them with respect to the *keto* forms by 6 kcal/ mol or 25 kJ/mol.

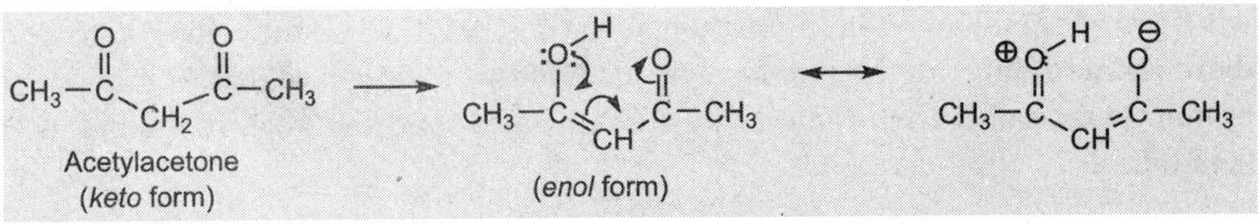

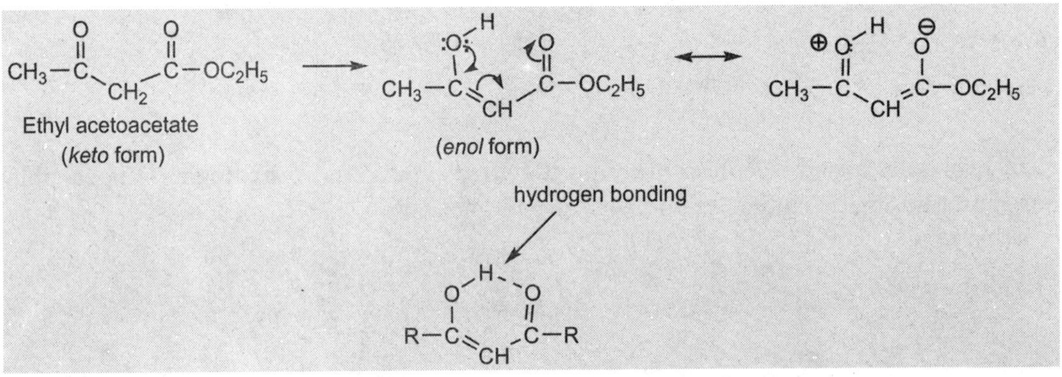

Sometimes, conformation plays a dominant role in deciding the extent of enolization in certain α-diketones. Biacetyl, for instance, has very little enol content in contrast to 1,2-cyclopentanedione (**48**) which exists mostly in the enolic form. The carbonyl groups of biacetyl are oriented in opposite directions so as to reduce dipole-dipole repulsion, whereas being a cyclic compound **48** has no such choice, and has to resort to enolization in order to avoid this repulsion.

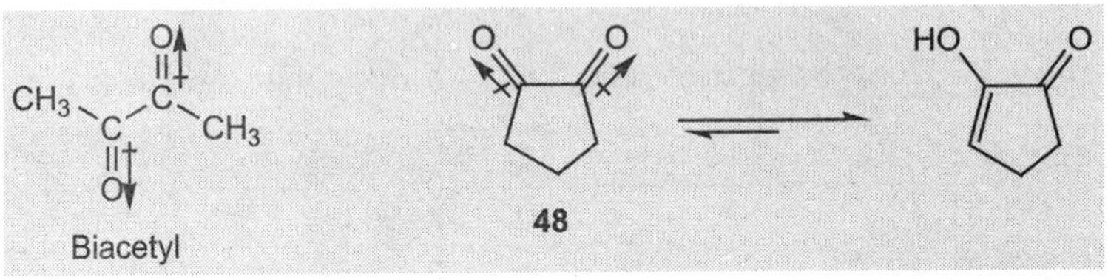

The percentage of *enol* present in a *keto-enol* equilibrium varies markedly with the nature of the solvent. This is to be expected, since the *keto* form of a diketone or a keto ester is more polar than the *enol* form, and hence the percentage of the *enol* form should increase as the polarity of the solvent decreases. The positions of the equilibria in ethyl acetoacetate in various solvents are shown in **Table 1.8.**

Table 1.8 Influence of solvents on the keto-enol equilibria of ethyl acetoacetate

Solvent	% Enol	Solvent	% Enol
Water	0.4	Chloroform	8.2
Acetic acid	5.7	Ethanol	10.5
Methanol	6.9	Benzene	16.2
Acetone	7.3	Ethyl Ether	27.1
Pure Ester	7.7	Hexane	46.4

Various other classes of organic compounds, which show prototropy are outlined in **Fig.1.19**.

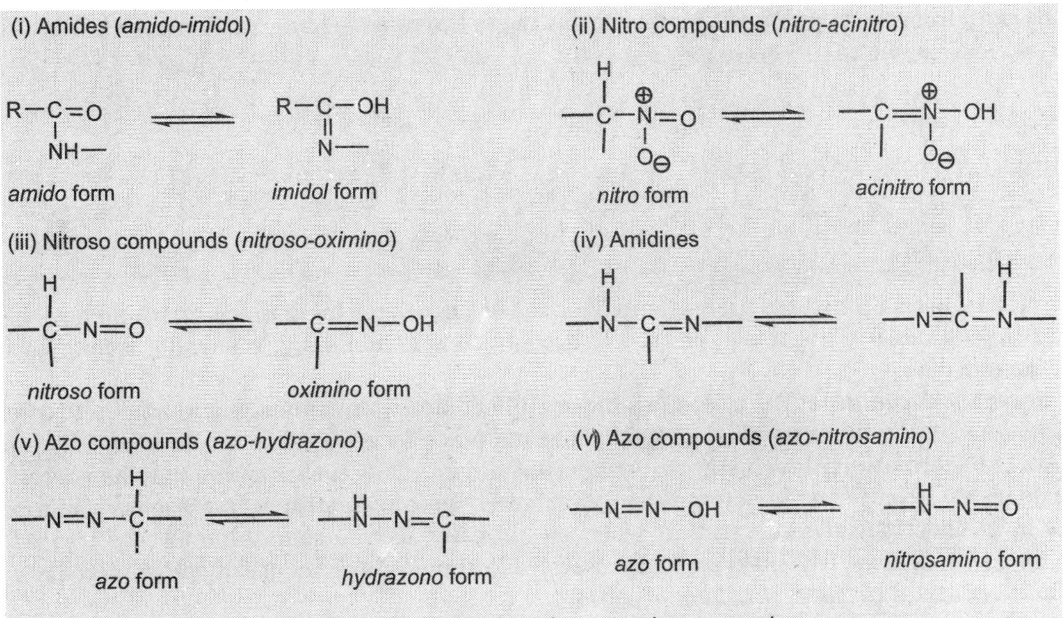

Fig. 1.19: Prototropy in various organic compounds

I. Mechanism of Prototropy

Prototropic interconversions may involve a *stepwise mechanism* in which the abstraction and addition of protons are separate steps, or it may be a *concerted process* in which abstraction and addition of protons are simultaneous. In practice, both these mechanisms operate, and it is often possible to distinguish between them by the study of changes of optical activity accompanying prototropic change of an optically active compound, and also by the study of the accompanying hydrogen-deuterium exchange.

(i) Stepwise Mechanism: In base-catalyzed halogenations (Chapter 9), it has been shown that formation of an enolate anion is the rate-determining step. In the absence of halogen, however, the enolate anion is protonated much more rapidly on oxygen than on carbon, thus producing an *enol* in a stepwise mechanism.

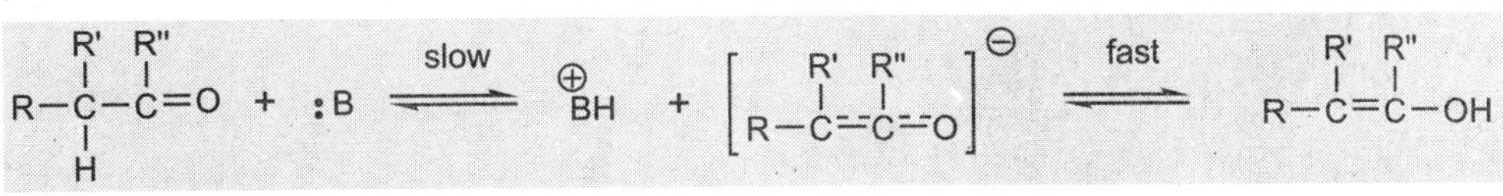

A study of base-catalyzed isomerization of β,γ-unsaturated nitrile (**49**) to α,β-unsaturated tautomer (**50**) also leads to the same conclusion. When the reaction was carried out in C_2H_5OD in the presence of $C_2H_5O^-$, it was observed that the rate of uptake of deuterium by **49** was much higher than its rate of isomerization to **50**. These results are explained by a mechanism in which the ethoxide ion reversibly transforms **49** into a resonance-stabilized anion **51** that may combine with a hydrogen ion (or a deuterium) to yield **49** or **50**. It also becomes clear that a considerable proportion of the anion reverts back to the starting tautomer by accepting deuterium from the medium without contributing to the rate of isomerization and thus making isotope exchange a faster process.

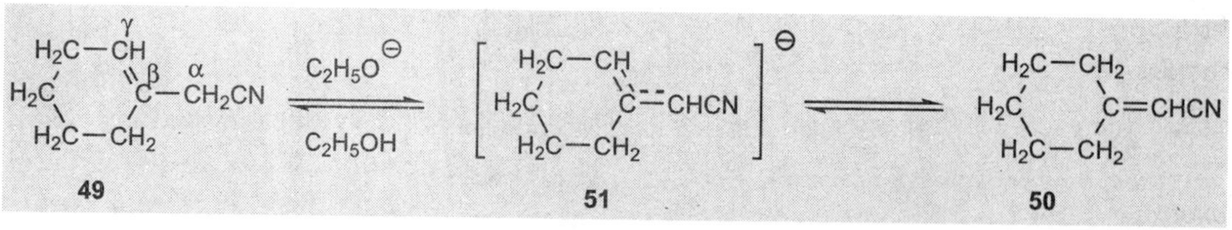

Had it been a concerted process, introduction of γ-hydrogen (or deuterium) would have been simultaneous with the removal of an α-hydrogen, thus establishing a parallel isomerization and deuteration.

B. Concerted Mechanism: Although tautomerization of most compounds including β-keto esters, β-diketones, and aliphatic nitro compounds takes place by a stepwise mechanism, there are a few transformations involving concerted mechanism. One such system that has received considerable attention is the tautomerization of methylene azomethines in presence of ethoxide ions in alcohol. It was observed that when an optically active azomethine **52** was treated in the above conditions, it yielded optically inactive tautomer **53**. The loss of optical activity was found to be equal to the rate of isomerization.

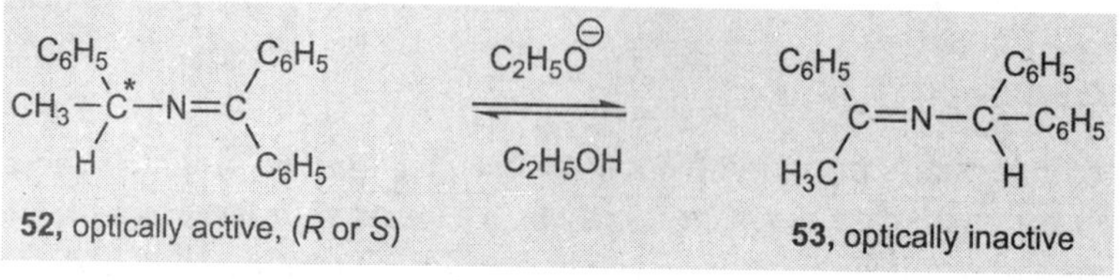

This observation rules out a stepwise mechanism in which *reversible* formation of an intermediate anion, and its subsequent conversion either to **52** or to **53** would involve *more racemization* than isomerization.

Isomerization of methylene azomethines in presence of C_2H_5OD provided additional evidence against intermediacy of an anion. The rate of isomerization was found to be *equal* to that of deuterium uptake, whereas in a stepwise mechanism, isotope exchange should be *faster* than isomerization. All these observations are consistent with a concerted mechanism in which each removal of hydrogen is accompanied by an addition of a hydrogen or deuterium, as the case may be.

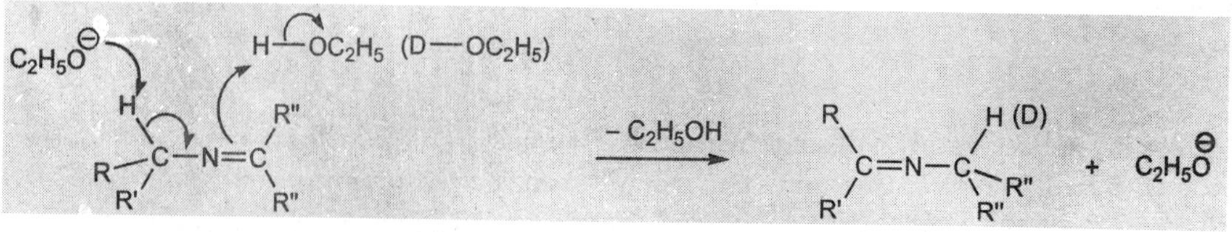

The validity of this mechanism has been questioned. It was found that the isomerization of **54** to **55** in the presence of *tert*-butyl alcohol-O–D and potassium *tert*-butoxide is *much slower*

than isotope exchange. This result indicates intermediacy of carbanion in the conversion of **54** to **55**.

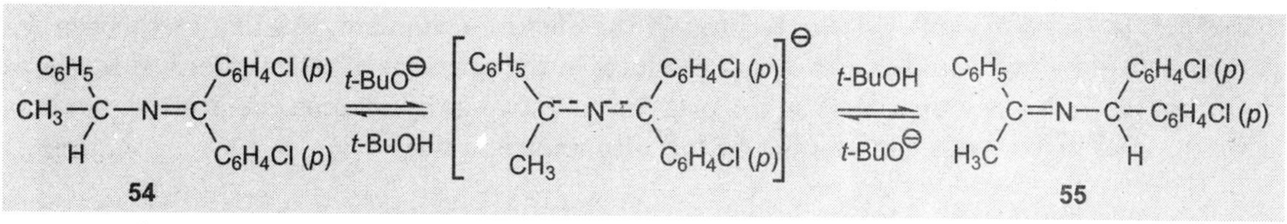

Anionotropy

There are a number of examples of tautomerism involving movement of atoms or groups of atoms as anions. These processes are collectively known as *anionotropy*. A simple example is the transformation of crotyl chloride into methylvinylcarbinyl chloride.

$$CH_3CH{=}CHCH_2Cl \rightleftharpoons CH_3CHClCH{=}CH_2$$

Either of these substances gives the reactions of both, and either can be converted into an equilibrium mixture. Other well-known examples include the conversion of 1-phenylallyl-*p*-nitrobenzoate (**56**) to **57** and that of geraniol (**58**) to linalool (**59**).

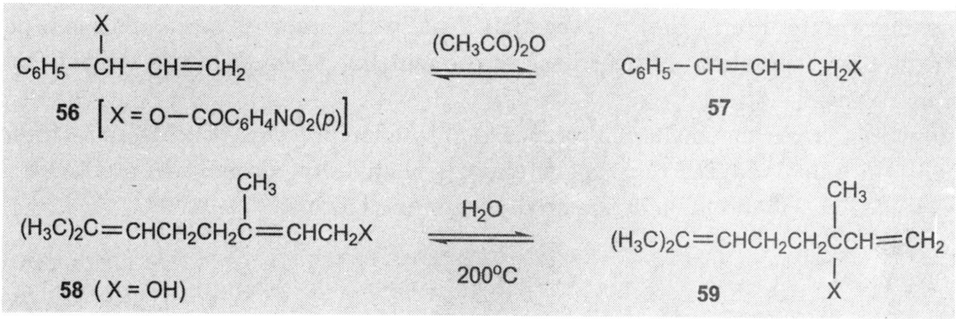

These transformations are also known as *allylic rearrangements*.

II. Mechanism of Anionotropy

(i) Stepwise Mechanism: Most of the anionotropic changes involve a stepwise or unimolecular mechanism in which complete elimination of X as anion occurs with the formation of resonance-stabilized cation (**60**). The attachment of X^- at a different carbon atom of the cation then yields the product.

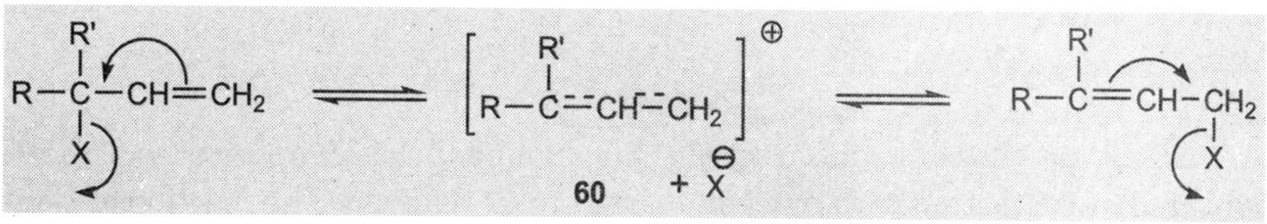

This mechanism resembles the S_N1 mechanism very closely, since both involve a carbocation intermediate. In the S_N1 mechanism, the anion gets attached to that carbon of the carbocation, which has parted from the leaving group, whereas in the above mechanism the anion is attached to the other end of the resonating carbocation. The latter mechanism has been termed S_N1' mechanism.

There is a good deal of evidence to support the above mechanism. Firstly, if a carbocation is an intermediate, we would expect mixed products in the presence of other anions. It has been found that the rearrangement of **56** in the presence of tetramethylammonium acetate, not only yields the expected rearranged product **57** but also an acetate **61**.

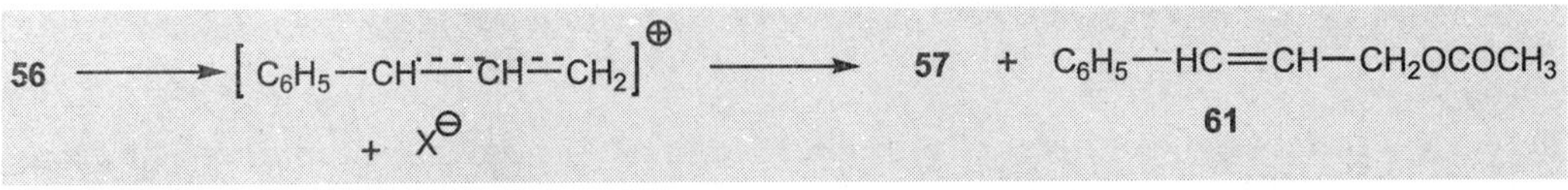

Secondly, the rate of rearrangement increases with the increase in the anionoid character of X. For example, the interconversion of type **56** $\rightleftharpoons$ **57** is quite difficult when X = OH, but becomes easy with the acetates and bromides. Thirdly, when X is kept constant, the interconversion is facilitated with the increase in stability of the intermediate carbocation formed after the separation of the anion. Thus, in the system **56** $\rightleftharpoons$ **57** (X = $OCOCH_3$), mobility is increased by a *p*-chloro and *p*-methyl substituent in the phenyl ring, and decreased by replacing phenyl by a methyl substituent. These observations are in agreement with the relative stability of the intermediate carbocation. Finally, the stepwise mechanism gets powerful support from the observation that increase in the ionizing strength of the solvent facilitates anionotropic transformations.

Anionotropic rearrangement of alcohol is facilitated in presence of acids. In an interesting experiment, the acid-catalyzed rearrangement of 1-phenylallyl alcohol was studied in $H_2{}^{18}O$. As expected, ^{18}O was found to be in the product cinnamyl alcohol.

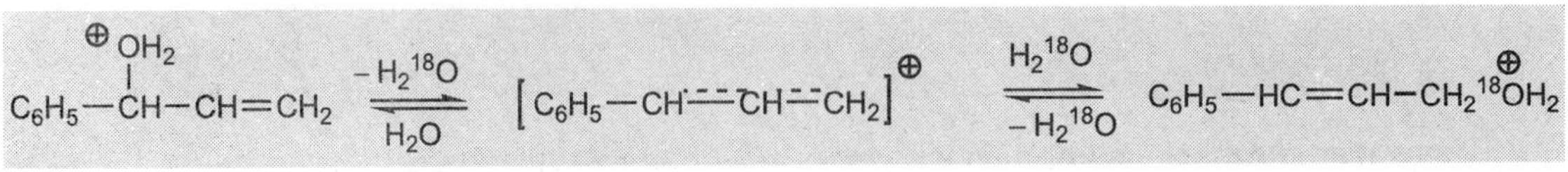

(ii) Concerted Mechanism: A number of anionotropic transformations involve a bimolecular mechanism, which is termed S_N2' in analogy with S_N2 reactions. The reaction may be depicted as follows:

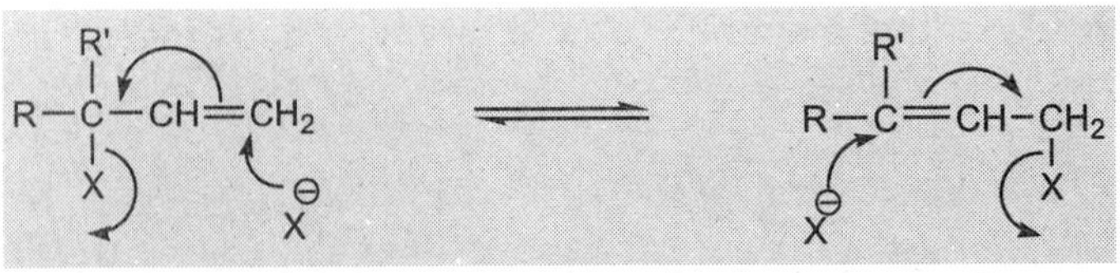

The incoming and outgoing anions may or may not be the same species. S_N2' reactions are often complicated by competing with much more rapid S_N2 reaction. The reaction of 1-chloroallyl chloride with ethoxide ions in ethanol offers an interesting example, in which products of S_N2 and S_N2' are formed in equal amounts.

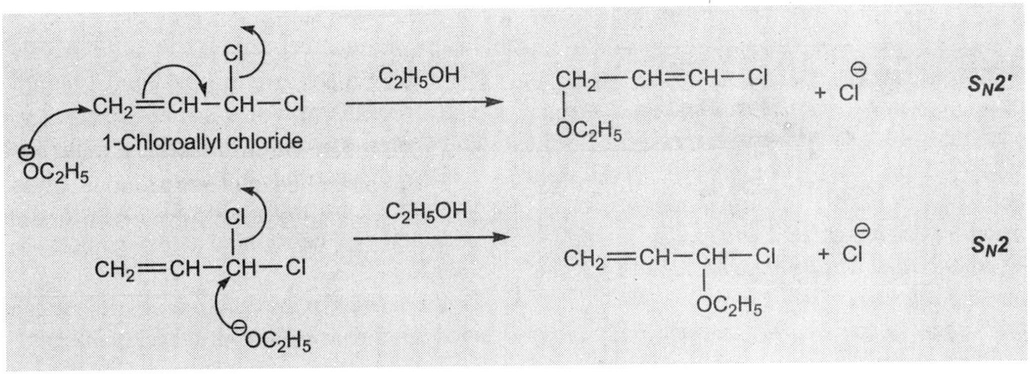

However, a similarly constituted halide, 1-*tert*-butylallyl chloride yields the product of S_N2' reaction only.

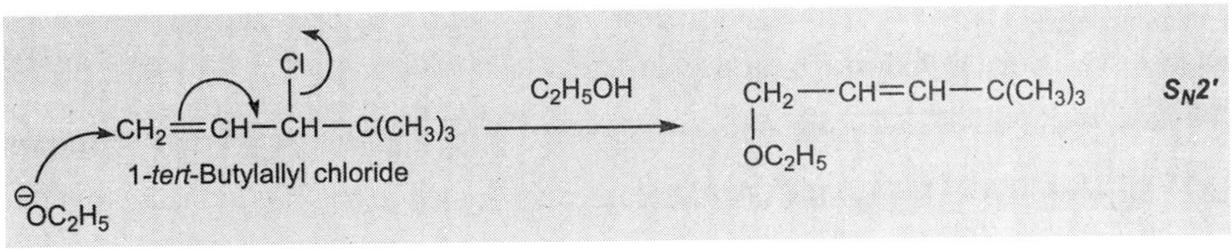

(a) C. $S_N i$ Mechanism: Nucleophilic substitution on a saturated carbon atom, sometimes, takes place with retention of configuration, and this phenomenon is explained by *an internal mechanism known as $S_N i$ reaction* (Chapter 6). In an analogous reaction, a substituted allyl alcohol is converted into two anionotropic halides by treatment with thionyl chloride. Thus, a secondary alcohol yields a mixture of primary and secondary chlorides.

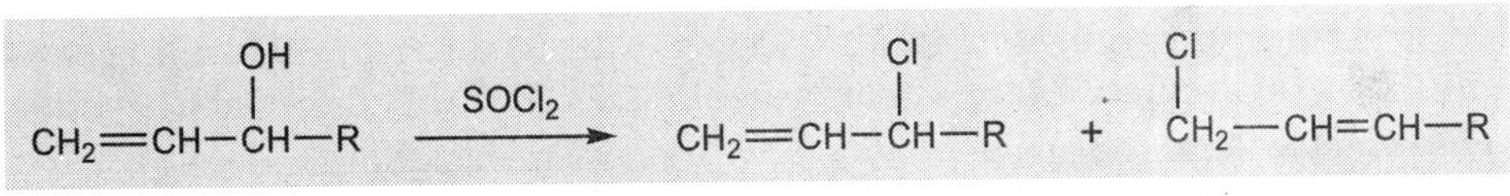

The formation of a secondary chloride with retention of configuration from a secondary alcohol is explained on the basis of $S_N i$ mechanism, but a modified mechanism is needed to explain the formation of primary chloride from secondary alcohol. The most plausible mechanism may be depicted as below:

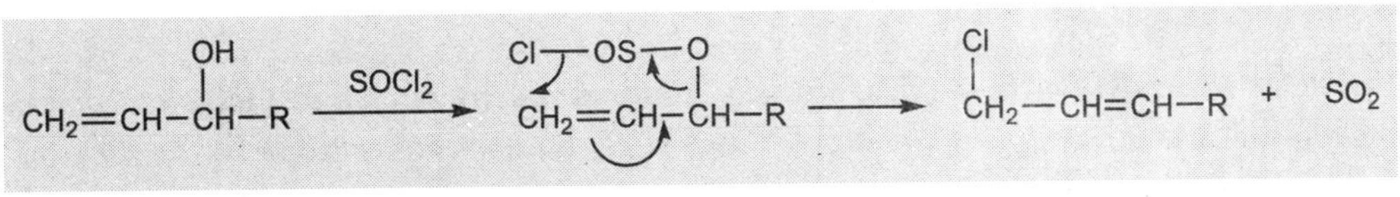

Tautomerism vs. Resonance

For the sake of revision and clear understanding, it is worthwhile to summarize the main differences between resonance and tautomerism (**Table 1.9**). Benzene and ethyl acetoacetate can be considered as an example for 'resonance' and 'tautomerism', respectively.

Table 1.9 Difference between resonance and tautomerism

Resonance	Tautomerism
1. Resonance involves delocalization of π electrons. and hence resonance structures (canonical forms or contributing forms) differ only in the position of electrons. The position of atoms remains fixed. 2. Various resonance structures are only imaginary, and do not exist. The real molecule is a hybrid of these structures. 3. The resonance structures are not in dynamic equilibrium as they are not real structures. 4. Resonance leads to stabilization of the molecule. The actual molecule (hybrid, which cannot be shown by real structures) is more stable than any canonical structure. 5. Resonance structures are shown by a double-headed arrow (↔).	1. Tautomerism involves movement of both electrons and atoms, and hence tautomers differ both in position of electrons and atoms. 2. Tautomers are real molecules, which actually exist and can be isolated under suitable conditions. 3. Tautomers exist in dynamic equilibrium with each other as they are actual compounds. 4. Tautomerism does not lead to stabilization of the molecule, although tautomeric forms have different stability. 5. Tautomeric structures are shown using equilibrium arrows ($\leftrightarrows$ or $\rightleftharpoons$).

1.3 BONDING WEAKER THAN COVALENT

A. Inclusion Compounds and Clatherates

Certain organic solids such as urea, thiourea and hydroquinone have crystalline shapes. However, in some cases, the shapes of these compounds undergo a change in presence of certain other compounds. *The former compounds, whose crystalline structures change, are called* hosts, *while the latter ones, in whose presence the crystalline structures of the former change, are called* guests. The crystal lattice of host compounds, forming long channels or cage-like structures, has enough space to trap the guest molecules. In fact, there is no chemical bonding between the host and the guest molecules. There are only weak van der Waals forces of attraction, which hold the guest molecule in the space provided by the host molecule.

I. Types of Host-Guest Addition Compounds

Depending upon the type of the space available within the crystal lattice of the host molecule, the host-guest addition compounds are divided into two categories:

(a) Inclusion Compounds

(b) Clatherates.

1. Inclusion Compounds: *The host-guest addition compounds are known as inclusion compounds when the space available within the crystal lattice of the host molecule is in the form of long channels.* Urea is one of the most common host molecules used for the formation of inclusion compounds. Ordinarily, urea crystallizes in tetragonal shape. However, in presence of a guest, urea crystallizes in the hexagonal lattice trapping the guest molecules (such as *n*-alkanes) in the space (**Fig.1.20**). Since hexagonal type of lattice is formed only in the presence of a specific guest molecule, it may be reasonable to believe that the van der Waals forces of attraction between the host and the guest molecule are essential for the stability of inclusion compounds.

The diameter of urea channel is about 500 pm, and the type of guest molecules will depend upon their shape and size.

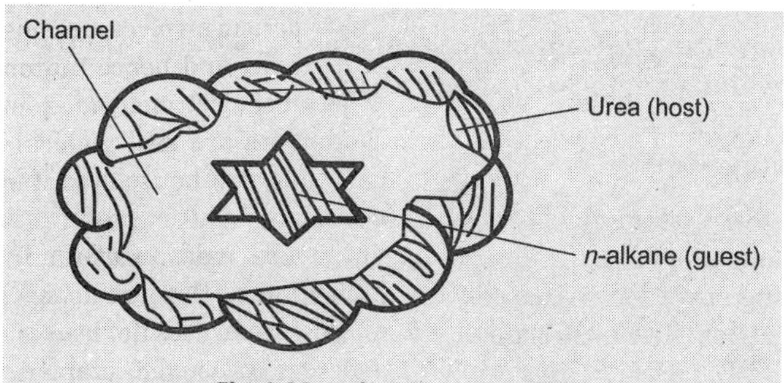

Fig. 1.20: Inclusion compounds

Thiourea also forms inclusion compounds with channels of larger diameter. Consequently, *n*-alkanes cannot act as guests for thiourea. However, 2-bromooctane, cyclohexane, and chloroform have the right size and shape to be trapped in the channels of thiourea.

Another example of the formation of an inclusion compound is the appearance of blue colour on the treatment of starch with iodine. Starch consists of a mixture of two components, *i.e.* amylose, which is water-soluble, and amylopectin, which is water-insoluble. Amylose, having helical structure, has the inside of the helix of the appropriate size and polarity to accommodate an iodine molecule. Thus, the blue colour is due to the formation of an inclusion complex between amylose and iodine (**Fig.1.21**).

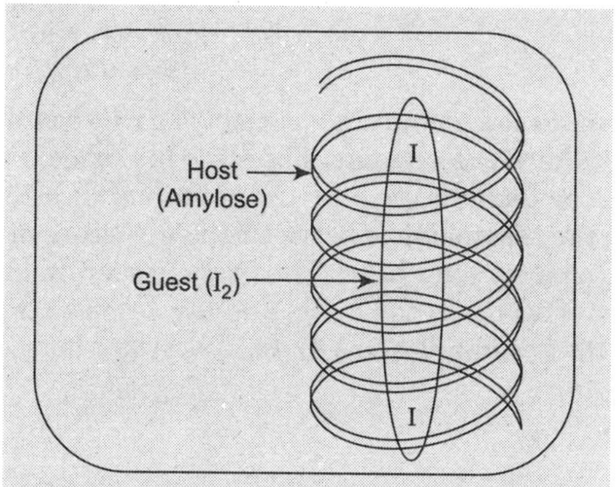

Fig. 1.21: Inclusion complex of amylose and iodine

A few common applications of the formation of inclusion compounds are as follows:

(i) These complexes are quite useful in separating certain isomers that would be otherwise difficult to separate. For example, *n*-octane can be separated from its branched-chain isomers, because only *n*-octane can form inclusion compound with urea. (ii) Urea inclusion compounds have also been used for resolving racemic mixtures. For example, (±) racemic mixture of 2-chlorooctane forms two different inclusion compounds (diastereomers), which can be separated by fractional crystallization.

(iv) Clatherates: *The adducts, in which the host molecule has cage-like cavities, are called clatherates or cage compounds.* Unlike inclusion compounds, the crystal lattice of clatherates is not fully occupied by the guest. Quinol or hydroquinone, water, sodium chloride, and some alkali metal halides are the most important hosts for clatherates. On the basis of the nature of host, clatherates are divided into two categories namely,

- Quinol Clatherates
- Water Clatheratest

(a) Quinol Clatherates: Quinol clatherates have molecule of hydroquinone as host. In these complexes, three molecules of hydroquinone, which are held together by hydrogen bonding form a cage containing several cavities of 400 pm diameter. The guest molecules of appropriate size fit into these cavities. The best typical guests for quinol clatherates are methanol (but not ethanol), SO_2, CO_2, H_2S, CH_3CN, and noble gases such as Ar, Kr, and Xe.

(b)Water Clatherates: Water is another important host. Generally, six molecules of water form the cage, in which guest molecules of appropriate shape and size, *i.e.* Cl_2, propane, CH_3I and noble gases such as Ar, Kr, and Xe, can get trapped. When water is frozen under a high pressure of any one of the above noble gases, solid crystalline addition compounds of the formula H_2O: gas (5.75:1) are obtained. These clatherates are called *noble gas hydrates*. These are normally preserved at low temperature, because at room temperature they decompose, whereby trapped gases get released. He and Ne, however, do not form clatherates, since their atoms are too small to be get entrapped.

Two important uses of clatherates of noble gases are: (i) He and Ne can be separated from other noble gases present in a mixture, because they do not form clatherates, and (ii) radioactive isotopes of Kr and Xe, which are produced in nuclear reactors can be stored as clatherates.

B. Crown Ethers

Crown ethers are cyclic chemical compounds that consist of a ring containing several *ether* groups. The most common crown ethers are oligomers of ethylene oxide, $(-CH_2CH_2O-)_n$. Important members of this series are the tetramer (n = 4), the pentamer (n = 5), and the hexamer (n = 6). The term 'crown' refers to the resemblance between the structure of a crown ether bound to a *cation*, and a *crown* sitting on a person's head. The first number (in brackets) in a crown ether's name refers to the number of atoms in the cycle, and the second number refers to the number of oxygen atoms. Some of these examples are **62-65**.

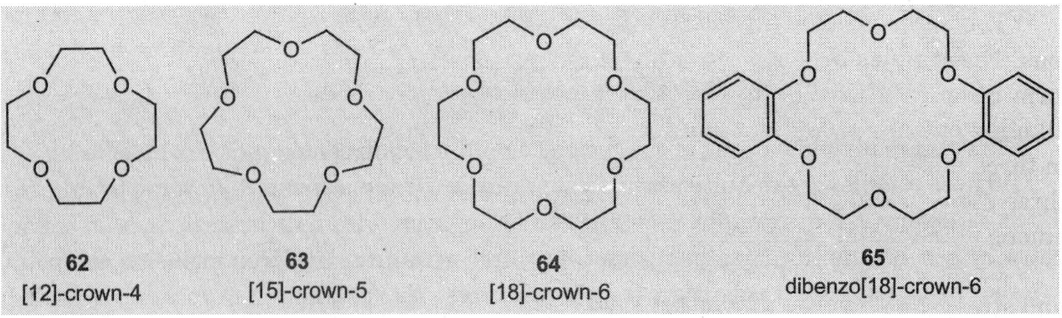

62 [12]-crown-4 **63** [15]-crown-5 **64** [18]-crown-6 **65** dibenzo[18]-crown-6

The most important property of crown ethers is that *they strongly bind certain cations (metallic or ammonium) to form complexes*. The oxygen atoms are well situated to coordinate with a cation located at the interior of the ring, whereas the exterior of the ring is hydrophobic. The resulting cations often form salts that are soluble in nonpolar solvents, and for this reason crown ethers are useful as *phase transfer catalysts*. The *denticity* of the polyether influences the affinity of the crown ether for various cations. For example, **62** binds with Li^+ but not with K^+, while 15-crown-5 (**63**) and 18-crown-6 (**64**) bind with Na^+ and K^+, respectively. The high affinity of 18-crown-6 for potassium ions contributes towards its toxicity.

The term crown ethers, now, includes macrocyclic compounds containing nitrogen and sulphur atoms, as well as macrocyclics with more than one kind of heteroatoms. Two such examples are **66** and **67**.

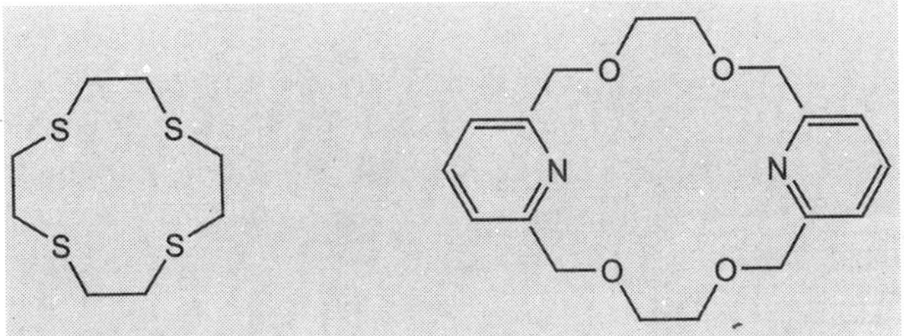

66 **67**

C. Cryptands

Cryptands are a family of synthetic bi- and polycyclic multidentate ligands for a variety of cations. These molecules are three-dimensional analogues of crown ethers, and complex the guest ions more strongly and selectively.

The most common and important cryptand is 1,10-diaza-4,7,13,16,21,24 hexaoxabicyclo [8.8.8] hexacosane (**68**). The compound is termed [2.2.2] cryptand, where the numbers indicate the number of ether oxygen atoms in each of the three bridges between the amine nitrogen 'caps'. Many cryptands exhibit particularly high affinity for alkali metal cations, thus allowing the isolation of salts of K^+.

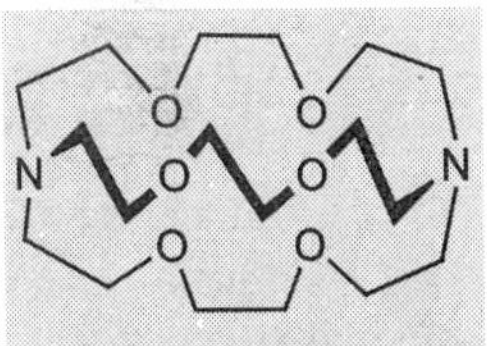

68

The 3-dimensional interior cavity of cryptand provides a binding site- or nook- for 'guest' ions. The complex between the cationic guest and the cryptand is called a *cryptate*. Cryptands form complexes with many 'hard cations' including $^+NH_4$, lanthanoids, alkali metals, and alkaline earth metals. For example, bicyclic crown ether (**68**) can surround the enclosed ion in three dimensions, binding it more strongly than the monocyclic crown ethers. Further, the three-dimensional encapsulation mode confers some size-selectivity, enabling discrimination among alkali metal cations (*e.g.* Na^+ vs. K^+).

Cryptands are more expensive and difficult to prepare, but offer much better selectivity and strength of binding than other complexants for alkali metals, such as crown ethers. They are able to bind otherwise insoluble salts into organic solvents. They can also be used as *phase transfer catalysts*.

D. Cyclodextrins

Cyclodextrins are a family of macrocyclic compounds made up of sugar molecules bound together in a ring (cyclic oligosaccharides). Cyclodextrins are produced from starch by means of enzymatic conversion and, therefore, these are sometimes also called as cycloamylose. They are used in food, pharmaceuticals, and chemical industries, as well as agriculture and environmental engineering.

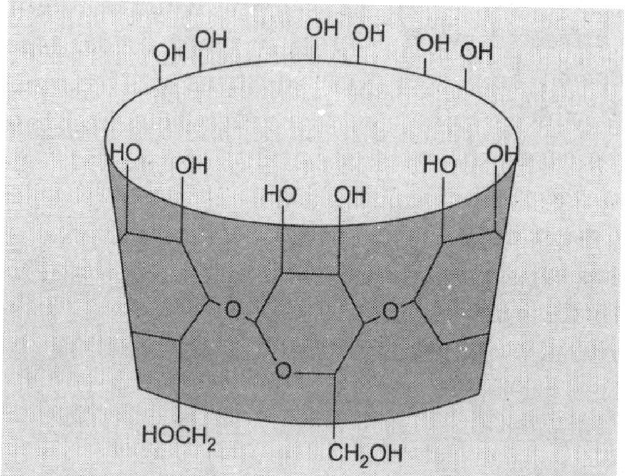

Fig.1.22: A schematic representation of α-cyclodextrin

Cyclodextrins are composed of five or more α-D-glucopyranoside units linked through 1,4-alpha linkages in such a way as to form a ring bracelet, each link of which is pyranose hexagon. Typical cyclodextrins contain a number of glucose monomers ranging from six to eight units in a ring, named as α-cyclodextrin (containing six glucose units), β-cyclodextrin (containing seven glucose units), and γ-cyclodextrin (containing eight glucose units).

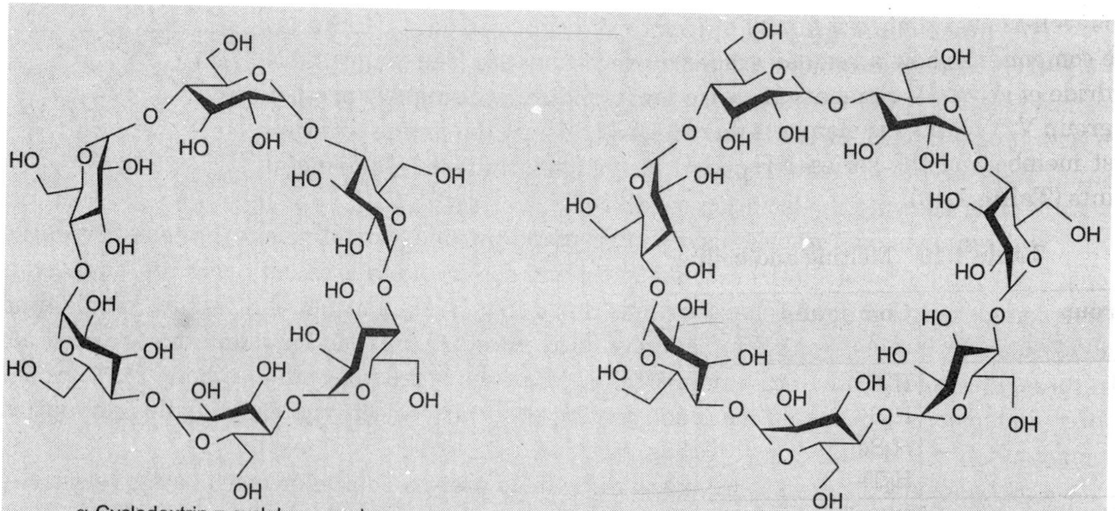

α-Cyclodextrin = cyclohexaamylose   β-Cyclodextrin = cycloheptaamylose

They can be topologically represented as toroids with the larger and the smaller openings of the toroid exposing to the solvent secondary and primary hydroxyl groups, respectively.

Because of this arrangement, the interior of the toroids is not *hydrophobic*, but considerably less hydrophilic than the aqueous environment, and thus able to host other hydrophobic molecules. In contrast, the exterior is sufficiently hydrophilic to impart cyclodextrins (or their complexes) water solubility.

The formation of the *inclusion compounds* greatly modifies the physical and chemical properties of the guest molecule, mostly in terms of water solubility. This is the reason why cyclodextrins have attracted much interest in many fields, especially pharmaceutical applications, because inclusion compounds of cyclodextrins with hydrophobic molecules are able to penetrate body tissues, which can be used to release biologically active compounds under specific conditions. In most cases, the mechanism of controlled degradation of such complexes is based on pH change of water solutions leading to the cleavage of hydrogen or ionic bonds between the host and the guest molecules.

Cyclodextrins are used in host-guest complexes with hydrophobic molecules because of the unique nature imparted by their structures. As a result, these molecules have found a number of applications in a wide range of fields. Other than the above-mentioned pharmaceutical applications for drug release, cyclodextrins can be employed in environmental protection; these molecules can effectively immobilise toxic compounds, like trichloroethane or heavy metals, or can form complexes with stable substances, like trichlorfon (an organophosphorus insecticide) or sewage sludge inside their rings, thus enhancing their decomposition.

In food industry, cyclodextrins are employed for the preparation of cholesterol-free products; the bulky and hydrophobic cholesterol molecule is easily lodged inside cyclodextrin rings that are then removed.

E. The Hydrogen Bond

It is well-known that in a family of elements in the periodic table, the physical properties of the compounds show a regular gradation. For example, the melting and boiling points of the hydride of group IV elements increase progressively with molecular mass. However, hydrides of group V, VI, and VII elements do not follow the generalization as in each of the groups, the first members of the series–NH_3, H_2O, and HF–show abnormally high melting and boiling points (**Table 1.10**).

Table 1.10 Melting and boiling points of hydrides of group VI and VII elements

Group	Compound	Molecular weight	Melting point (°C)	Boiling point (°C)
VI	H_2O	18	0	100
	H_2S	34	–83	-59.6
	H_2Se	81	–64	-42.1
	H_2Te	130	–54	-1.8
VII	HF	20	–92	19.4
	HCl	36	–112	-85.1
	HBr	81	–88	-67.2
	HI	128	–51	-35.4

The above deviations have been traced to the special ability of hydrogen atoms in the molecules of ammonia, water and hydrogen fluoride to form a link between two electronegative atoms, one of which is joined to it by a polar covalent bond and the other by electrostatic attraction. The electrostatic attraction comes into play due to the uneven sharing of the pair of bonding electrons between hydrogen and a highly electronegative atom (F, O, and N). Thus, a partial positive charge develops on hydrogen, which attracts the negative end of a group of the same or other molecules forming a relatively weak bond. This type of bond, called the *hydrogen bond*, is the most powerful kind of dipole-dipole attraction. Being electrostatic in nature, it is much weaker than a covalent bond; the hydrogen bond energy is only 3-10 kcal/mol (12.5-42 kJ/mol) as compared to the normal covalent bond energy of 50-100 kcal/mol (210-420 kJ/mol). The hydrogen bonds are usually represented by *dotted lines,* in contrast to a normal covalent bond, which is shown by a dash. For instance, the structure of HF_2^- ion in KHF_2 can be written as:

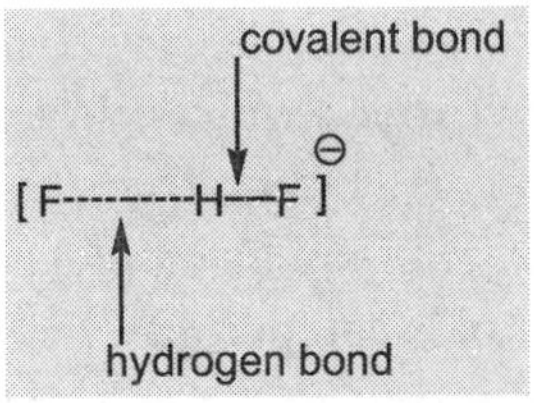

There are two types of hydrogen bonds–*intermolecular* and *intramolecular*. Intermolecular hydrogen bond exists between atoms of two or more molecules resulting in their association. For instance, water and alcohols are associated as polymeric aggregates in the solid and liquid state, whereas carboxylic acids and amides exist as dimers due to intermolecular hydrogen bonding.

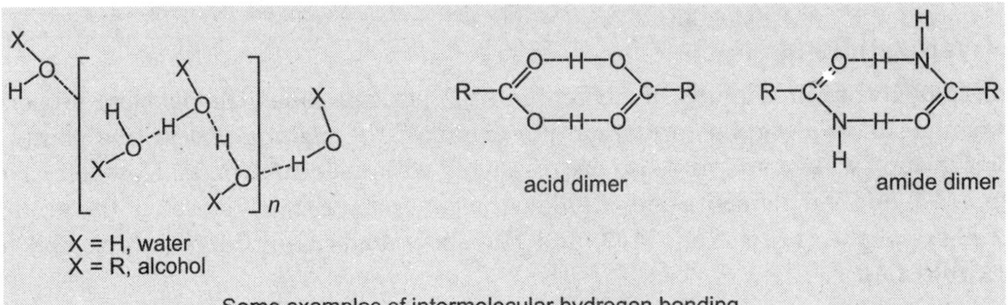

Some examples of intermolecular hydrogen bonding

Intramolecular hydrogen bond is formed between two atoms of the same molecule. As a consequence, a five or six-membered ring, called a chelate ring, is formed. Examples of compounds exhibiting this type of hydrogen bond are *o*-nitrophenol, salicylaldehyde, and *o*-chlorophenol.

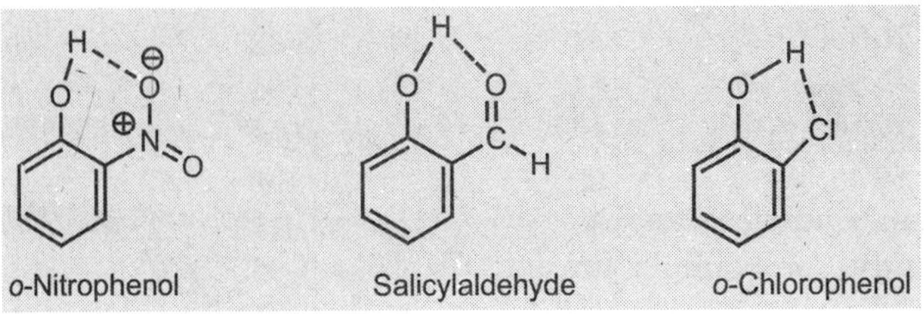

For forming intramolecular hydrogen bonds, the molecule must satisfy the following conditions:

(i) The molecule must have two groups in such a way that one group contains H-atom linked to a highly electronegative atom, and the other group also contains an electronegative atom. (ii) The molecule must be planar. (iii) The hydrogen bond must lead to the formation of a five or six-membered ring.

Effect of Hydrogen Bond on Physical Properties

1. As some extra energy is needed to break the hydrogen bond, intermolecular hydrogen bonding raises the melting and boiling points of organic substances. **Table 1.11** lists the boiling points and molecular weights of a few substances to illustrate this point.

Table 1.11 Boiling points of alkanes, ethers and alcohols of comparable molecular weights

Compound	Molecular Weight	Boiling Point (°C)
$CH_3CH_2CH_3$	44	44
CH_3OCH_3	46	25
CH_3CH_2OH	46	78
$CH_3CH_2CH_2CH_2CH_3$	72	36
$CH_3CH_2OCH_2CH_3$	74	35
$CH_3CH_2CH_2CH_2OH$	74	118

The effect of intramolecular hydrogen bonding on transition temperatures is the reverse of that of intermolecular hydrogen bonding. *o*-Nitrophenol, for instance, melts at 44°C while the m.p. of *p*-nitrophenol is 114°C. Similarly, salicylic acid melts at 158°C as compared to *p*-hydroxybenzoic acid, which melts at 214°C.

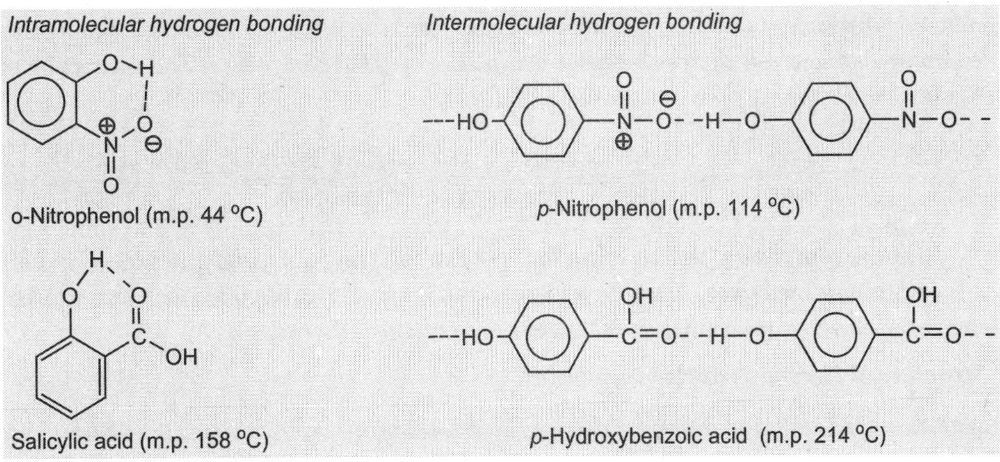

2. Hydrogen bonding shifts the position of bands in infrared and NMR spectra of organic compounds. For example, infrared spectrum of ethyl alcohol in vapour phase shows a sharp absorption band at 3700 cm^{-1} owing to a free hydroxyl group. However, in

solution this band is completely replaced by a broad band around 3350 cm^{-1}, which is characteristic of hydrogen bonded hydroxyl groups. This type of shift in absorption frequency is to be expected as hydrogen bonding weakens the O–H bond. The NMR spectrum of neat ethyl alcohol displays hydroxyl absorption signal at δ 5.35, but when the degree of hydrogen bonding is reduced by dilution with carbon tetrachloride, the –OH signal is observed around δ 2.5. This large shift in the position of hydroxyl proton has been explained by the existence of hydrogen bonding in neat alcohol, which decreases the electron density around proton, and thus pushes the absorption to lower field. The downfield shift is also observed when spectrum is run in dimethyl sulphoxide.

3. Hydrogen bonding also plays an important role in biological systems. In plants and animals, most of the water is present as attached to protein molecules by hydrogen bonding. Furthermore, the secondary and tertiary structures of proteins are stabilized by hydrogen bonding. For instance, the unique stability of α-helical structure of certain proteins such as keratin in wool, silk, nails and horns is due to intermolecular hydrogen bonding between the C=O of one amino acid and the NH of the other (fourth) amino acid residue in the chain. The specific double helical structure of DNA is also due to intermolecular hydrogen bonding, which involves base moieties.

F. Charge-Transfer Complexes

When an electron-rich compound (called the donor molecule and designated as D) is mixed with an electron-deficient compound (called the acceptor molecule and designated as A) in 1:1 ratio, it forms a complex called *charge-transfer complex*. In the formation of charge-transfer complexes, the donor molecule may donate either a non-bonding or an unshared pair of electrons (n), or a pair of electrons of π-system (an olefinic double bond or the aromatic ring system) to the acceptor molecule. The former donor is called an n-donor, and the latter as a π-donor. Such complexes are also known as *electron-donor–acceptor complexes or EDA complexes*. The formation of a charge transfer complex may be represented as follows:

$$\underset{\text{donor}}{D} + \underset{\text{acceptor}}{A} \longrightarrow \underset{\text{charge transfer complex}}{[D\text{------}A]} \longleftrightarrow D^{+}—A^{-}$$

In these complexes, the bonding forces between the donor and the acceptor molecules are the weak coloumbic forces. A bonding of such type, which results from attractive forces between donor and acceptor molecules is called *charge-transfer interaction*.

I. Structure of Charge-Transfer Complexes

Formation of so called 'picrates' by the action of 2,4,6-trinitrophenol (picric acid), and an aromatic hydrocarbon such as naphthalene, anthracene and mesitylene, represents the most common example of charge-transfer complexes. The picrates are yellow, orange, or red-coloured solids having definite melting points, and hence are often used as derivatives for characterization of aromatic hydrocarbons.

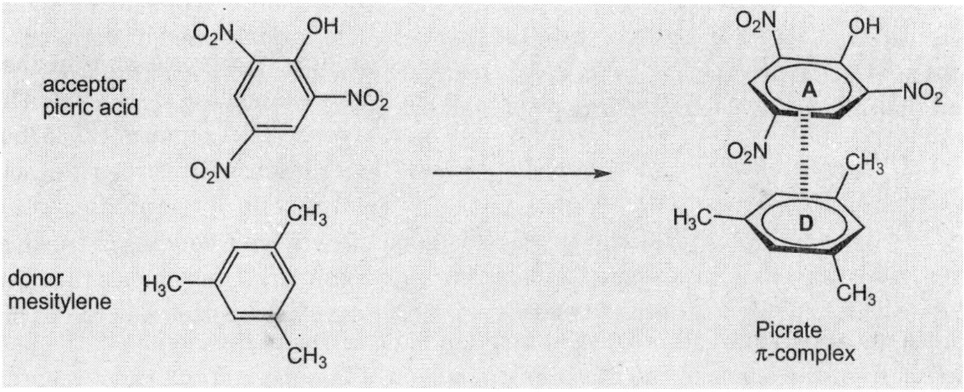

The nitro groups, rather than a hydroxyl group, are essential for the success of the reaction, as not only picric acid, but other polynitro aromatic compounds such 1,3,5-trinitrobenzene form similar addition compounds. The formation of charge-transfer complexes is facilitated by the presence of electron-donating substituents in D, and electron-withdrawing substituents in A. Donor has a low ionization potential, and acceptor has a high electron-affinity with a low energy vacant orbital. Due to parallel arrangement of the two rings, overlap occurs between molecular orbitals of D and A causing electron-transfer (D $\rightarrow$ A).

The distance between the donor and the acceptor compounds is 300-350 pm, which is much larger than the typical covalent bond length and shows that though there is a slight overlap of π-orbitals of both the rings, it is not so strong as to form a pure covalent bond. Determination of crystal structure has shown that in such a complex the aromatic rings are held in *parallel planes*. This is further supported by the observation that compounds such as **69** in which the two rings cannot attain *coplanarity* because of steric interactions (due to the presence of nitro groups at *ortho*-positions), fail to form charge-transfer complexes.

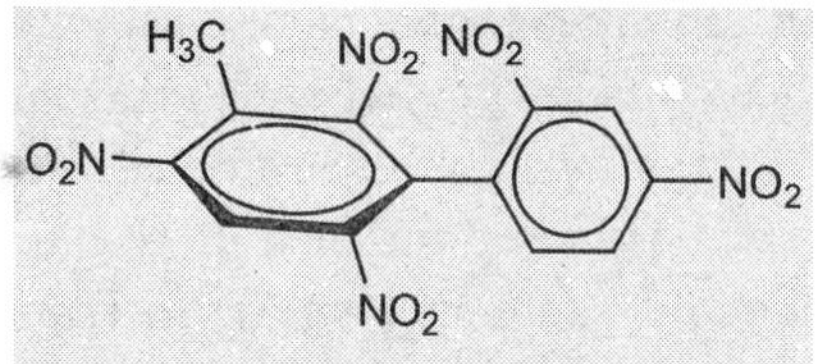

69

The extent of charge separation in a charge-transfer complex is often reflected in ground state properties such as dipole moment. The dipole moment of pyridine (2.28 D), for instance, is raised significantly (4.90 D) in the pyridine-iodine complex. Similarly, 1,3,5-trinitrobenzene does not display dipole moment in carbon tetrachloride, but shows a significant dipole moment in benzene due to formation of charge-transfer complex.

II. Charge-Transfer Spectra

The spectroscopic properties of charge-transfer complexes are changed due to weak ground-state interactions, and results in the appearance of a new type of electronic transition called a *charge-transfer transition*. The absorption spectra of these complexes are, thus, quite characteristic as there is a new absorption band in addition to the original bands of the *donor* and *acceptor* components. In the case of 1-alkylpyridinium iodides, where the ground state is an ion-pair, the light absorption process has been shown to involve the loss of an electron from the iodide ion, and the gain of the electron by the pyridinium ring.

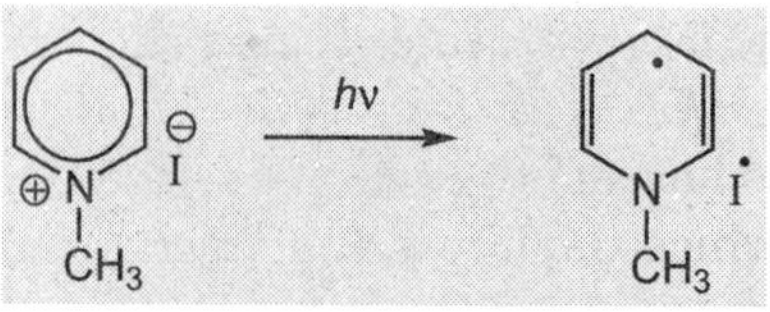

Since the appearance of this new absorption band is due to the change in the position of an electron from the donor to the acceptor component, the nature of the substituents in both the components should affect its position. Indeed, its position has been found to vary with the ionization potential of the donor and with the electron affinity of the acceptor.

G. van der Waals Interactions

van der Waals, a Dutch chemist, proposed the existence of some sort of attractive forces between the molecules of non-polar compounds. These attractive forces are called van der Waals forces. These are short range forces.

Normally, in a non-polar neutral molecule, the centres of positive and negative charge densities coincide with each other, and consequently these molecules should not have any dipole moment. But, because of the constant random movement of electron clouds around the nucleus, it is expected that at some point these clouds may get slightly distorted in one or some of the molecules due to mutual collisions or some other factors. In such a situation, the centres of positive and negative charge densities no longer coincide at that point, and result in the creation of a small instantaneous local dipole. This instantaneous dipole moment induces oppositely oriented *dipole moment*, known as induced dipole moment, in the neighbouring molecules, and when they are close enough these dipoles attract each other and cause the molecules to cling together. Because of this, such attractions are also referred to as *instantaneous dipole-induced dipole attractions*. These instantaneous and induced dipoles are constantly changing, but the overall result is the attraction between such molecules. Such attractions are also called *London forces or dispersive forces.*

These forces are very weak, and are at maximum in the solid and minimum in the gases because of difference in the intermolecular distances in the two cases.

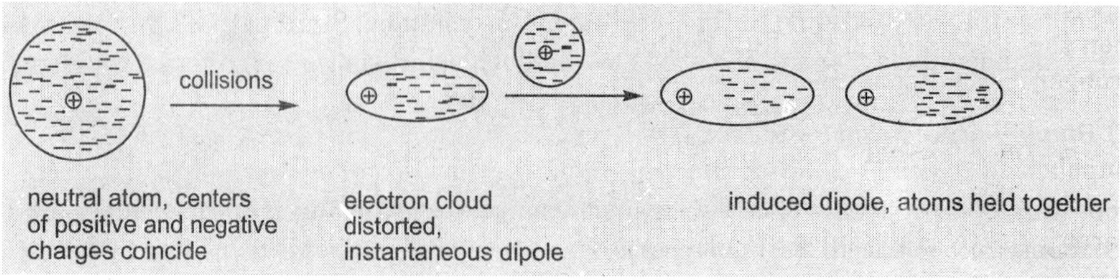

I. van der Waals Repulsion

As mentioned above, the van der Waals forces of attraction increase in magnitude with the decrease [illegible] distance between non-bonded atoms or group of atoms (molecules). But, this attra[illegible] up to a certain point when such moieties are at a minimum distance so that they càn touch each other, and *if they are forced closer than this minimum distance, the attractive forces turn into great repulsive forces, called van der Waals* repulsions. Thus, it can be said that every atom or group of atoms has an effective size referred to as *van der Waals radius,* and the minimum distance beyond which the forces of attraction turn into forces of repulsion is equal to the sum of the van der Waals radii of two atoms or group of atoms. In **Table 1.12** are listed van der Waals radii of some atoms and groups of atoms. It may be mentioned that the van der Waals radii of various atoms are greater than their covalent radii by about 80 pm.

Table 1.12 van der Waals radii of some atoms and groups of atoms.

Atoms and Groups	van der Waals radii (pm)	Atoms and Groups	van der Waals radii (pm)
H	120	Cl	180
N	150	Br	195
O	140	I	220
S	190	CH_3	200
F	135	half the thickness of benzene nucleus	170

van der Waals repulsive forces greatly influence the shapes and geometries of the molecules. Because of the directional nature of covalent bonds, the molecules have certain spatial requirements, and if the atoms or groups are situated at such positions that the distance between them is smaller than the sum of their van der Waals radii, they are said to be *crowding* together. In order to achieve maximum stability, a situation where such repulsive forces could be avoided, the molecule orients itself in such a way that the distance between these atoms increases.

II. van der Waals Attraction in Polar Molecules

Apart from non-polar molecules, the van der Waals forces of attraction also occur in polar molecules. These may originate from *dipole-dipole* and *dipole-induced dipole* interactions.

(i) Dipole-dipole interaction: Polar molecules having permanent dipole moment are held together by dipole-dipole interactions. For instance, in gases such as NH_3, HCl, HF and SO_2, there are significant dipole-dipole attractions between the molecules of these gases due to the presence of permanent dipole moment in the molecules. The extent of this type of interaction depends upon the magnitude of the dipole-moment of the molecule. Thus, greater the dipole moment, stronger is the dipole-dipole interaction.

(ii) Dipole-induced dipole interaction: These are attractive interactions between polar and nonpolar molecules. A polar molecule may induce polarization in a nonpolar molecule present in its vicinity. This induced dipole then interacts with the dipole moment of the first molecule, and thereby, the two molecules are attracted to each other. The extent of such interaction depends upon the magnitude of the dipole moment of the polar molecule, and the polarizability of the non-polar molecule. For instance, the increasing order of water solubility of noble gases, from He to Rn, is attributed to increase in the magnitude of dipole-induced dipole interaction, due to increase in their polarizability in the same order.

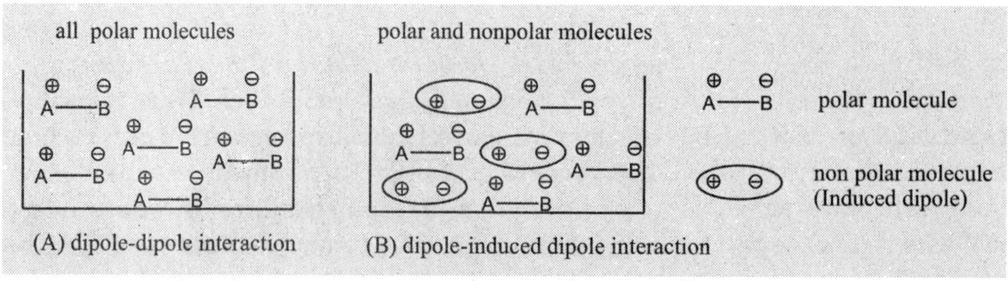

PROBLEMS AND REVIEW QUESTIONS

1. (i) An aqueous solution of tropylium bromide (C_7H_7Br) on treatment with silver nitrate gives a precipitate of silver bromide. How do you account for this observation?
   (ii) Cycloheptatrienyl cation has a low π-electron energy than its open-chain counterpart. Explain.
2. Cyclopentadienyl cation is an antiaromatic compound, whereas cyclopropenyl cation is an aromatic compound. What do these observations mean in terms of π-electron energies of these systems?
3. Write the resonance structures for cyclopentadienyl cation. Can the equivalence of five hydrogen atoms in this compound be explained by resonance?
4. Explain why cyclopropenyl anion is antiaromatic though a number of resonance structures can be written for it.
5. 1,3,5-Cycloheptatriene (**A**), though a cyclic planar triene containing 6π electrons (4q+2), does not exhibit usual stability of aromatic compounds, whereas tropolone (**B**), a derivative of **A**, behaves like a typical phenol. Explain.

A **B**

6. *p*-Aminobenzaldehyde behaves like an amide, as evident by the fact that it is not easily attacked by the nucleophile at the carbonyl carbon. Explain.
7. Which one of the following species is aromatic?
   (i) [10] Annulene (ii) [20] Annulene
   (iii) [8] Annulene (iv) C_9H_{10}
   (v) $C_9H_9^+$ (vi) $C_9H_9^-$
8. Offer suitable explanations for each of the following observations:
   (i) Biacetyl has very little enol content, whereas 1,2-cyclopentanedione exists mostly in the enolic form.

(ii) Acetone exists almost completely in its keto form, whereas acetylacetone has significant enol content.

(iii) Ethyl acetoacetate has about 10% enol content in ethanoic solution, which increases to 46% when dissolved in hexane.

(iv) Compound **I** easily dissolves in aqueous sodium hydroxide, and gives a red colour with $FeCl_3$ solution while compound **II** does neither.

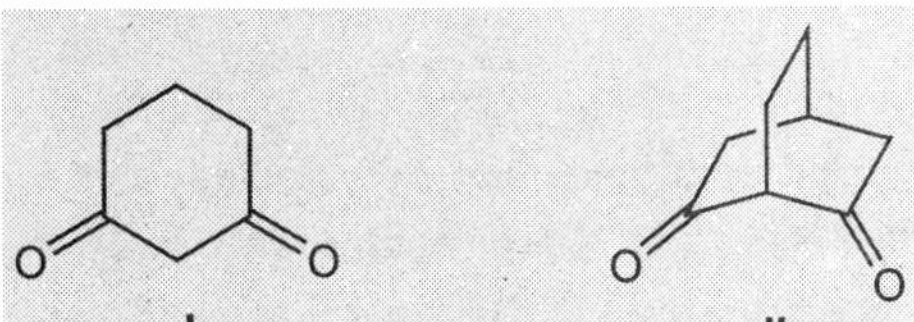

I   II

9. Offer explanation for the following observations related to amides **I** and **II**.

I   II

(i) Compound **I** is more basic than compound **II**.

(ii) Compound **I** behaves as a ketone and compound **II** as an amide.

10. Which of the following compounds (**I** or **II**) shows greater basicity towards Lewis acid, trimethylborane?

I   II

11. The NMR spectral data of compounds **A** and **B** are given along with their structures. What conclusion do you draw from these data?

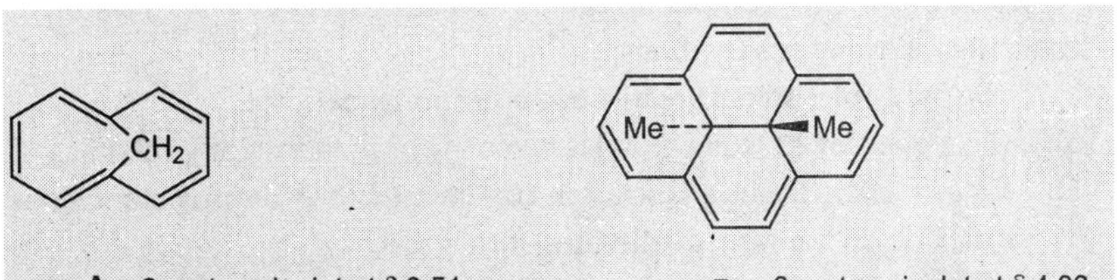

A 2 proton singlet at δ 0.51
8 proton multiplet at δ 7.08

B 6 proton singlet at δ 4.03
10 proton singlet at δ 8.47

12. (i) Illustrate the role of NMR spectroscopy in establishing the aromaticity of a compound.

(ii) Cyclooctatetraene reacts with two moles of potassium to yield a stable compound whose NMR spectrum indicates a single line. Comment on the formation of this compound.

13. When 3-chlorocyclopropene, C_3H_3Cl (**A**) is treated with antimony pentachloride, it gives a stable salt (**B**), $C_3H_3SbCl_6$, which is highly polar.

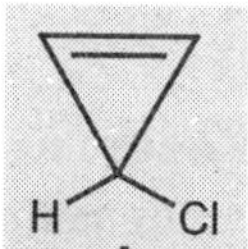

A

The NMR spectrum of **B** exhibits only one sharp singlet. Further, when **A** is treated with $AgBF_4$, it gives a white precipitate of AgCl, and a crystalline salt **C** whose NMR spectrum was exactly the same as that of **B**. Both **B** and **C** are much more stable than corresponding allylic substances. Identify **B** and **C** giving reasons for their unusual stability.

14. Compound **A** reacts with water much more slowly than the resonance stabilized allyl cation or *tert*-butyl cation. Similarly, compound **B**, though having a strained ring system, is extremely stable and does not participate in many nucleophilic additions.
    (i) Explain why **A** is highly stable?
    (ii) Draw resonance structures of **B** which account for these observations.

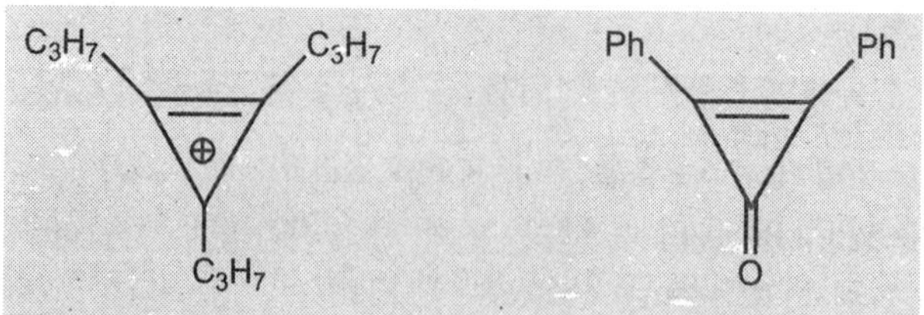

15. Account for the following facts:
    (i) The boiling point of methanol (CH_3OH) is 66°C and that of methyl mercaptan (CH_3SH) is 6°C, whereas the boiling points of diethyl ether and diethyl sulphide are 35°C and 92°C, respectively.
    (ii) 8-Hydroxyquinoline can be separated from 4-hydroxyquinoline by steam distillation.
    (iii) Ethanol (M.W. 45) boils at 78°C, whereas ethylamine (M.W. 46) has b.p. of 17°C.
    (iv) Oxygen-containing organic compounds dissolve in cold conc. H_2SO_4 to yield a solution from which the original compound can be liberated by dilution with water.

16. Explain the following observations:
    (a) *o*-Nitrophenol is steam volatile but *p*-nitrophenol is not.
    (b) Boiling point of ethylene glycol is much lower than that of glycerol.
    (c) 2-Chlorooctane forms an inclusion compound but 2- bromooctane does not.
    (d) He and Ne do not form clatherates with water.
    (e) Acetamide is a weaker base than ethylamine.
    (f) Methyl chloride has higher dipole moment than chloroform.
    (g) Allyl and benzyl halides are more reactive than alkyl halides towards nucleophilic substitution reactions.
    (h) All the carbon-carbon bond lengths in benzene are equal.

17. The order of basicity of primary, secondary and tertiary amines in aqueous medium is $Me_2NH > MeNH_2 > Me_3N$, whereas in gas phase the order is $Me_3N > Me_2NH > MeNH_2$. Justify the observation.
18. Cyclopentadienone is less stable than cyclopropenone and cycloheptatrienone. Explain.
19. How will you explain the following facts?
    (i) Cyclooctatetraene possesses $4q$ π electrons, but it is not antiaromatic.
    (ii) Tropone possesses high dipole moment.
20. Explain the following observations:
    (i) The dipole moment of allyl bromide is higher than that of vinyl bromide.
    (ii) The dipole moment of *p*-nitroaniline is much higher than 2,3,5,6-tetramethyl-4-nitroaniline.
    (iii) Cyclopropenone has higher dipole moment than cyclopropanone.
    (iv) Although fluorine is more electronegative than chlorine, dipole moment of chlorobenzene is higher than fluorobenzene.
    (v) *Cis*-1,2-dichloroethylene possesses considerable value of dipole moment ($\mu = 1.89$ D), whereas *trans*-1,2-dichloroethylene shows no dipole moment ($\mu = 0$ D).
21. Explain the following observations:
    (i) Although benzene contains three double bonds, it resists addition reactions.
    (ii) Cyclopentadiene is acidic while cyclopropene is not.
    (iii) Solvolysis of cyclopentadienyl iodide is 1000 times slower than that of cyclopropenyl iodide.
22. The NMR spectrum of cyclooctatetraene exhibits a sharp singlet at δ 5.7. With the help of this observation, explain its aromatic nature.
23. Offer explanation for each of the following facts:
    (i) Amides are less basic than alkylamines.
    (ii) Aniline is a weaker base than cyclohexylamine.
    (iii) Urea acts as monoacidic base.
    (iv) Guanidine behaves as a strong base.
24. Cyclopropenyl cation is much more stable than allyl cation while cyclopropenyl anion is less stable than allyl anion. Explain.
25. How do you explain the unstability of vinyl alcohol and stability of phenol, although both of these possess an OH group attached to a double bond?
26. Define and illustrate what you understand by localized and delocalized bonds.
27. What are the inclusion compounds and clatherates? Mention some of their important uses.
28. What are charge-transfer complexes? Discuss the nature of bonding and structure of these complexes.
29. Define and illustrate with the help of suitable examples the following terms:
    (i) bond energy (ii) bond length
    (iii) bond angle.
30. Explain why chloroacetic acid is stronger acid than acetic acid.
31. Define aromaticity. Mention some characteristics of aromatic compounds.
32. What is hyperconjugation? How does it explain the relative order of the stability of alkenes? Why is it called no-bond resonance?

33. Write notes on:
    (i) resonance energy
    (ii) banana bond
    (iii) alternant and non-alternant hydrocarbons
    (iv) picrates
34. Draw the structures of a hydrocarbon having eight carbon atoms with (i) three isolated double bonds, (ii) two conjugated and one isolated double bond and (iii) three conjugated double bonds.
35. Considering the example of benzene and ethyl acetoacetete, explain the difference between resonance and tautomerism, respectively.
36. (i) Draw a labelled diagram of molecular orbital picture of benzene.
    (ii) Justify the statement "cyclohexatriene is different from benzene"
37. What are annulenes? Discuss their aromatic, nonaromatic, and antiaromatic character.
38. (a) What are nonbenzenoid aromatic compounds? Give three examples along with structures.
    (b) What is Hückel's rule for aromaticity? What are its limitations?
    (c) What is Crag's rule? How does it overcome the limitations of Hückel's rule? Explain with examples.
39. What are the implications of Linear Combination of Atomic Orbitals (LACO method) in the formation of covalent bonds?
40. Discuss various mechanisms of anionotropy. How do these mechanisms resemble those of substitution reactions?
41. Discuss the phenomenon of tautomerism with particular reference to the mechanisms involved in prototropy.
42. What are crown ethers? Discuss their applications particularly as phase transfer catalysts.
43. Write short note on the followings.
    (i) dextrins (ii) fullerenes
    (iii) clatharates (iv) dipole-dipole interactions

SOLUTIONS

1. (*i*) It is largely an ionic compound consisting of a cation and an anion.
   (*ii*) The cation is aromatic in character having $(4q + 2)\pi$ electrons delocalized over a seven carbon planar ring system.
2. The former has $4q\pi$ electrons whereas the latter has $(4q + 2)\pi$ electrons- an important criterion for aromaticity.
3. 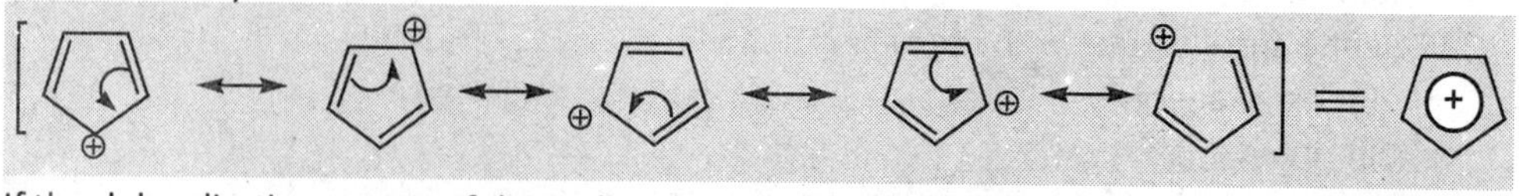
4. If the delocalization energy of the cyclic, planar, conjugated polyene is less than that of its acyclic conjugated analogue, it is said to be antiaromatic. Molecular orbital calculations have revealed

that cyclopropenyl anion has higher energy than any one of the contributing forms. Hence, it is antiaromatic ($4q\pi$ electrons).

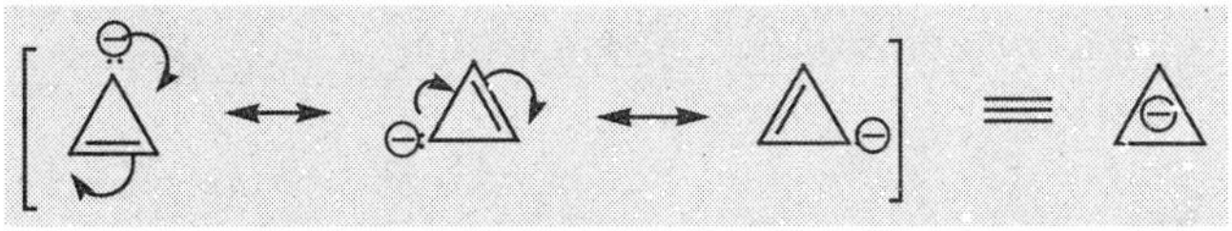

5. Cycloheptatriene does not possess the usual stability of aromatic compounds as it does not have a continuous closed planar ring. The methylene carbon does not offer a parallel *p*-orbital for overlap, thereby preventing the π-electrons from being delocalized over the whole ring system. On the other hand, tropolone can be represented as a hybrid of polar structures possessing a seven-membered ring similar to cycloheptatrienyl cation having a continuous system of seven carbons delocalizing 6π electrons.

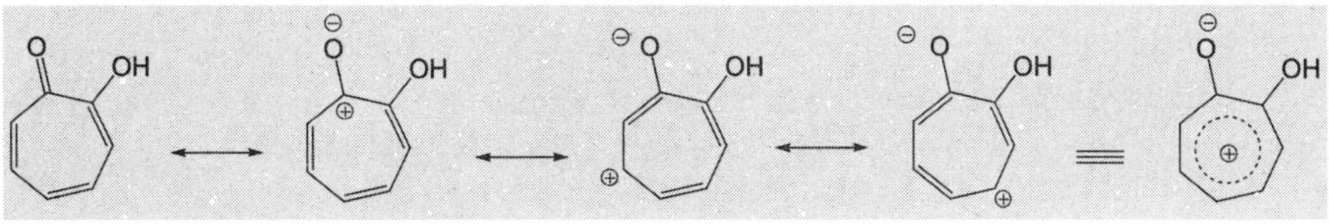

6. The partial positive charge on the carbonyl carbon, which is responsible for nucleophilic addition, is considerably neutralized by resonance feedback of nonbonding electrons of the amino group. This resonance is comparable to that of amide.

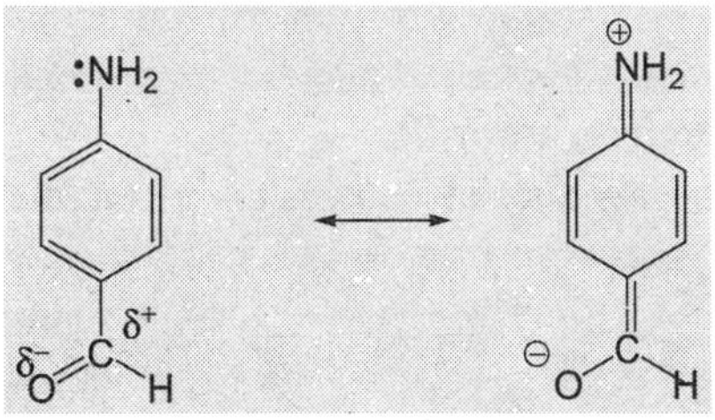

7. (*i*) [10] Annulene; This compound having $(4q + 2)\pi$ electrons would have been aromatic but it is not. This is because of the fact that the two 'inner' hydrogens are crowded. The molecule, therefore, buckles in order to supply space to the hydrogen atoms and because of this disturbance the coplanarity of the molecule is lost making the compound nonaromatic.
   (*ii*) It is not aromatic as it does not statisfy the $(4q + 2)\pi$ requirement.
   (*iii*) It is cyclooctatetraene having $4q\pi$ electrons. So it is not aromatic.
   (*iv*) and (*v*) are not aromatic as they have $4q\pi$ electrons (9-membered cyclic compounds with 4 double bonds).
   (*vi*) It is an aromatic compound as it has $(4q + 2)\pi$ electrons.
8. (iv) Compound **I** can exist in enol form **I A** which gives red colour with $FeCl_3$ solution. On the other hand, such a structure **II A** in case of compound **II** seems to be unlikely because of the presence of double bond at the bridgehead carbon.

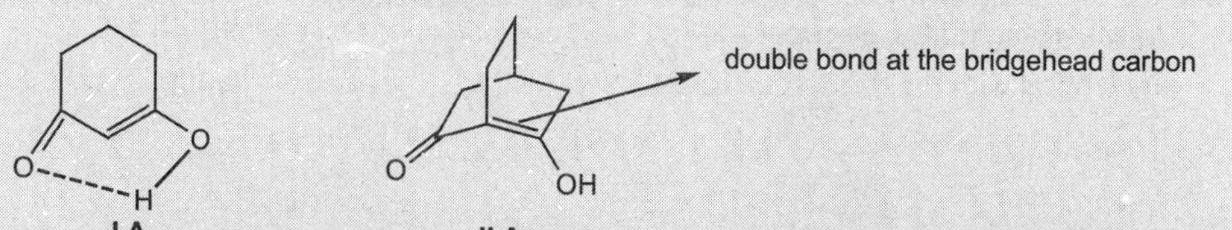

I A II A

9. (*i*) In case of compound **I**, lone pair of electrons on nitrogen is freely available for protonation.

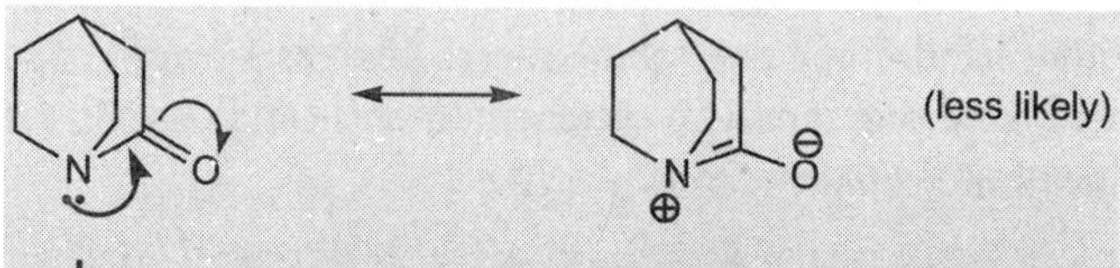

I Ia

(*ii*) Compound **I** cannot show properties of amide because resonance structure **Ia** has double bond at bridge nitrogen.

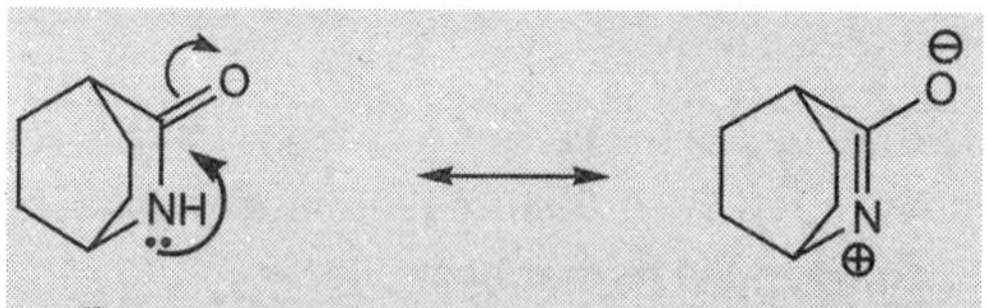

II IIa

11. Central region below and above the plane of an aromatic ring is highly shielded. Both of these compounds are, therefore, aromatic.
12. (*ii*) Cyclooctatetraene is reduced by potassium metal to a dianion:

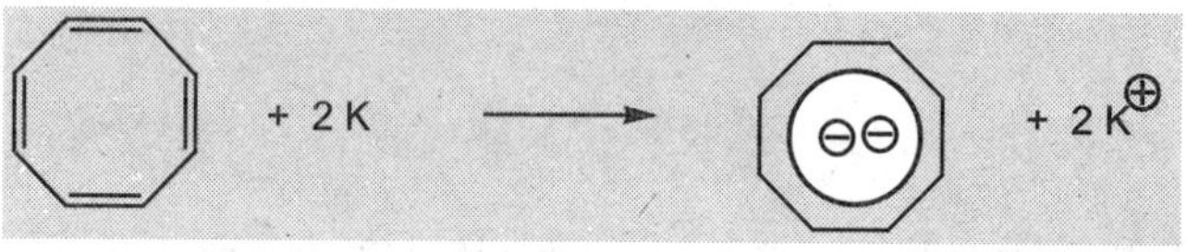

The doubly negatively charged anion now has 10 electrons (8π and $2p$) fitting in the formula of $(4q + 2)\pi$ electrons spread over the carbon atoms of a planar ring system. Due to the delocalization of the negative charge over the whole system, all the hydrogen atoms become equivalent. Hence, one singlet in the NMR spectrum is observed.

13. In both the cases, salts of symmetrical cyclopropenyl cation are formed. This being a planar cyclic system having $(4q + 2)\pi$ electrons ($q = 0$) is aromatic in character and hence shows the stability. The cation being symmetrical, exhibits only one singlet in the NMR spectrum.

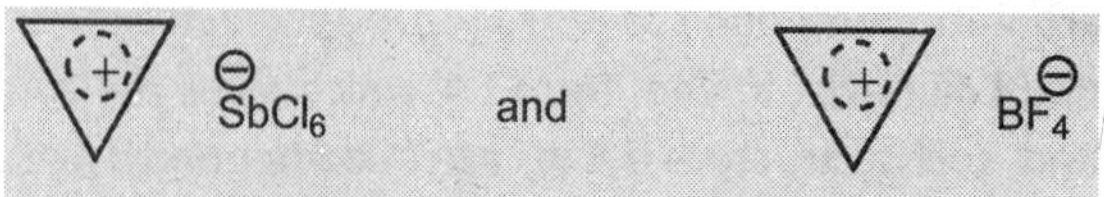

B C

14. For the stability of **A**, see problem 13. The ketone **B**, though has a strained ring system, is stable. This is due to the fact that the carbonyl resonance introduces positive charge on the ring making it identical with the cyclopropenyl cation having $(4q + 2)\pi$ electrons ($q = 0$).

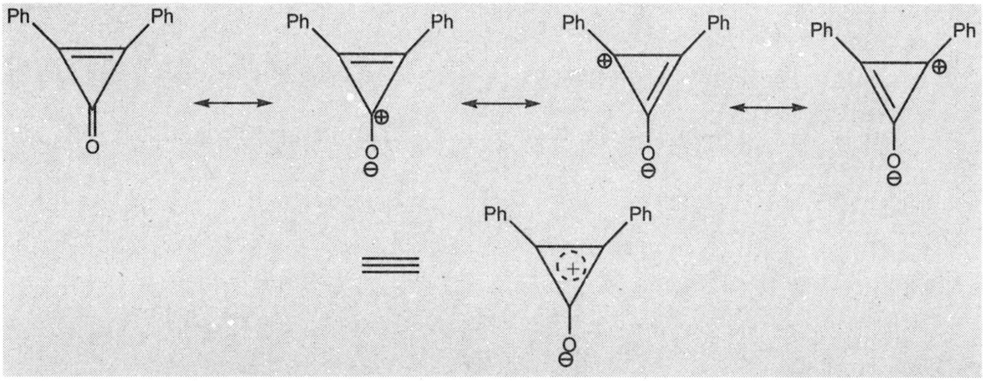

15. (*i*) Sulphur is much less electronegative than oxygen and thiols consequently forms only very weak hydrogen bonds and have lower boiling points than alcohols. In the absence of hydrogen bonding, large molecular weight sulphur compounds have higher boiling points than oxygen compounds.

(*ii*) Due to intramolecular hydrogen bonding in 8-hydroxyquinoline.

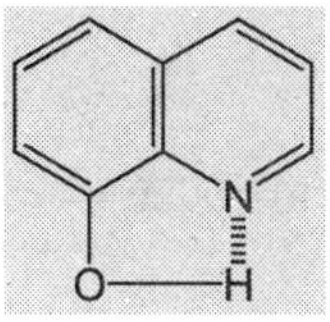

(*iv*) There is reversible protonation of non bonding electron pair located on oxygen atom.

Chapter 2

Stereochemistry-I

Chapter Outline

2.1 STEREOISOMERISM
- A. Configurational Isomerism
- B. Conformational Isomerism

2.2 MOLECULAR REPRESENTATION OF STEREOISOMERS
- A. Fischer Projection
- B. Newman Projection
- C. Sawhorse Projection
- D. Flying Wedge Projection
- E. Interconversions of Projection Formulae

2.3 CHIRALITY
- A. Chirality-Optical Activity
- B. Elements of Symmetry
- C. Enantiomerism
- D. Molecules (Enantiomers and Diastereomers) with Several Chirality Centres (Stereogenic Centres)
- E. Absolute and Relative Configurations
- F. Retention and Inversion of Configuration
- G. Racemic Modification or Racemic Mixture
- H. Resolution

2.4 GEOMETRICAL ISOMERISM
- A. Definition and General Aspects
- B. *E* and *Z* Nomenclature of Geometrical Isomers

2.5 CONFORMATIONAL ISOMERISM
- A. Conformational Isomers of Ethane
- B. Conformational Isomers of Propane
- C. Conformational Isomers of *n*-Butane
- D. Conformational Isomers of 1,2-Dibromoethane
- E. Conformational Isomers of 1,2-Glycols: Ethylene Glycol

The literal meaning of stereochemistry is *chemistry in space*. More specifically, the branch of chemistry, which deals with three-dimensional structures of molecules and their effect on physical and chemical properties, is known as stereochemistry.

2.1 STEREOISOMERISM

Isomers having same molecular and same structural formula, but differing in the manner in which their atoms or groups are arranged in space are called *stereoisomers,* and the phenomenon is called *stereoisomerism.* Stereoisomerism is classified as shown in **Fig. 2.1.**

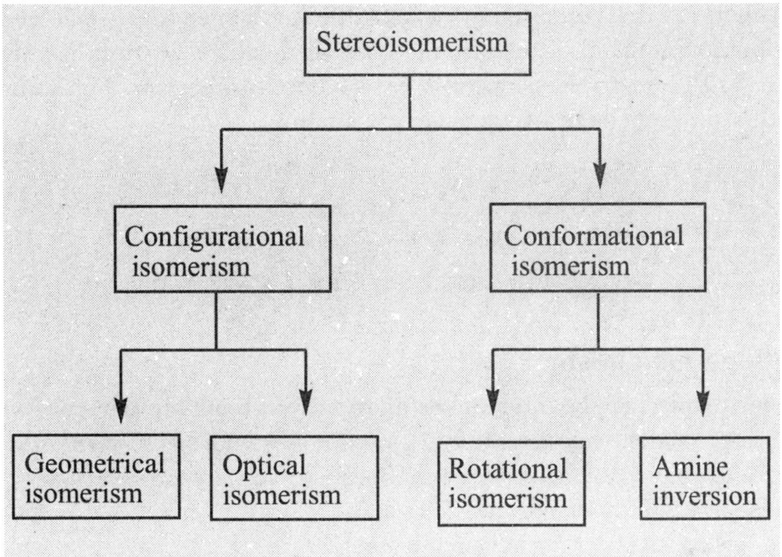

Fig. 2.1: Classification of stereoisomerism

A. Configurational Isomerism

The stereoisomers, which cannot be interconverted unless a covalent bond is broken, are called configurational isomers. This type of isomerism is called *configurational isomerism*. These isomers can be separated under normal conditions. Configurational isomerism can be further classified as

- Optical Isomerism
- Geometrical Isomerism

I. Optical Isomerism

The stereoisomers, which are related to each other as an object and its non-superimposable mirror image are called *optical isomers* or *enantiomers* (Greek: *enantion* means *opposite*). This type of isomerism is called *optical isomerism or enantiomerism*.

Optical isomers rotate the plane of plane-polarized light to an equal degree but in opposite directions. The property of rotating the plane of plane-polarized light is known as *optical activity*. Optical isomers have similar physical and chemical properties. For example, molecular formula $C_3H_6O_3$ represents two enantiomeric lactic acids as shown below:

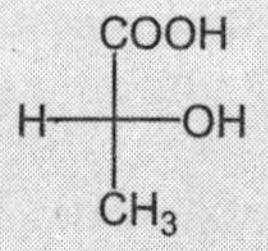

(–)-Lactic acid

(rotates the plane of plane-polarized light towards left hand side i.e., anticlockwise)

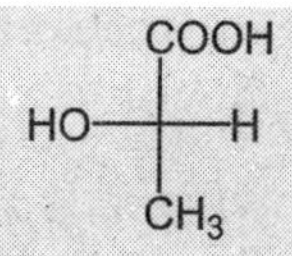

(+)-Lactic acid

(rotates the plane of plane-polarized light towards right hand side i.e., clockwise)

II. Geometrical Isomerism

Geometrical isomers are the stereoisomers which differ in their spatial geometry due to restricted rotation across a double bond. These isomers are also called *cis-trans* isomers. This type of isomerism is called *geometrical isomerism* or *cis-trans isomerism*. For example, molecular formula $C_2H_2Br_2$ corresponds to two geometric isomers.

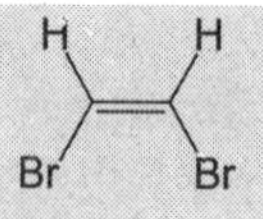

cis-1,2-Dibromoethene

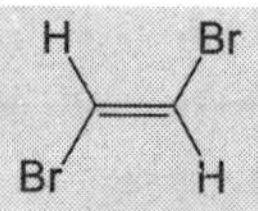

trans-1,2-Dibromoethene

B. Conformational Isomerism

The stereoisomers, which can be interconverted rapidly at room temperature without breaking a covalent bond, are called *conformational isomers* or *conformers* or *rotamers*. This type of isomerism is called *conformational isomerism*. Because such isomers are readily interconverted, they cannot be separated under normal conditions.

There are two types of conformational isomers:

(i) Resulting from Rotation about a Single Bond: Because single bond in a molecule rotates continuously, the compounds containing single bonds have many interconvertible conformational isomers. For example, *'boat'* and *'chair'* forms of cyclohexane.

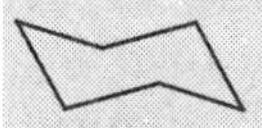

Chair

Boat

(ii) Arising from Amine Inversion: This type of conformational isomers exists because nitrogen atom of amines has a pair of non-bonding electrons, which allow the molecule to turn *inside out* rapidly at room temperature. This is called *amine inversion*.

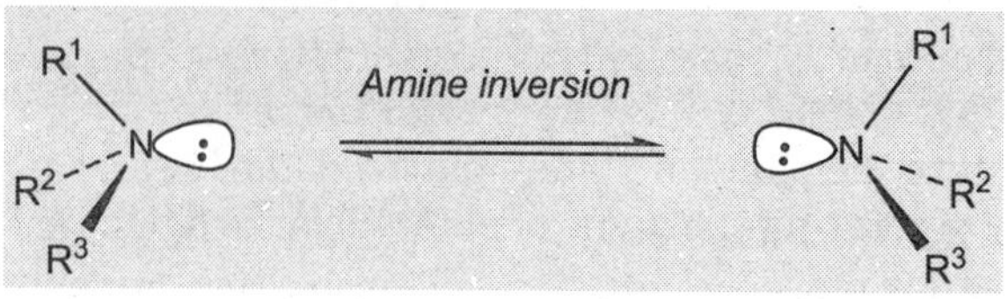

2.2 MOLECULAR REPRESENTATION OF STEREOISOMERS

Configurational and conformational isomers have three-dimensional structures, and is not easy to depict on a paper having only two dimensions. To overcome this problem, the following four two- dimensional structures known as projections have been evolved.

- Fischer Projection
- Newman Projection
- Sawhorse Projection
- Flying Wedge Projection

A. Fischer Projection

Fischer projections are always generated by orienting the molecule in such a manner that vertical bonds connected to the chirality centre are directed *away* from the viewer, and the horizontal bonds point *towards* the viewer. On paper, the projection of bonds is in shape of a cross with the chiral centre at the centre of the cross.

Fischer projection of one stereoisomer of glyceric acid (HOOC–CHOH–CH_2OH) is shown as structure **1**.

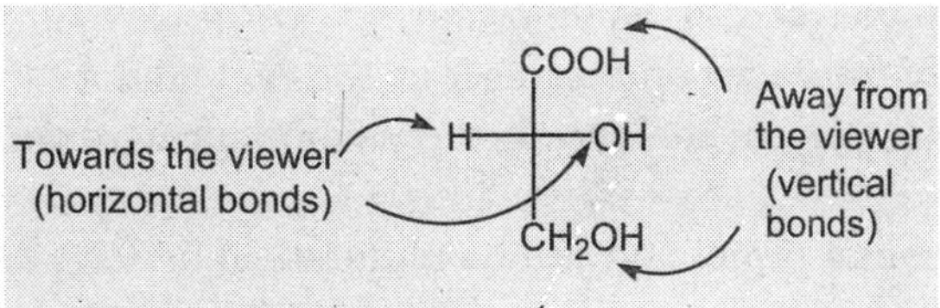

1

I. Characteristic Features of Fischer Projection

(i) Fischer projection of a chiral molecule can be changed into another identical Fischer projection by interchanging the position of any two pairs of substituents (two interchanges). Thus, structure **1** after interchange of two pairs of substituents generates identical structure **3**.

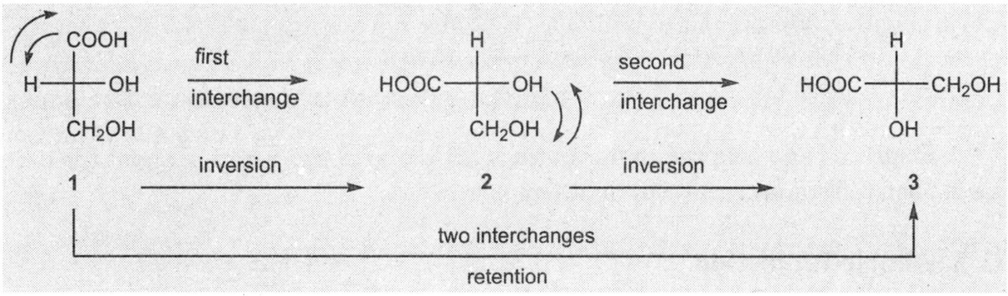

Only one such interchange produces an enantiomer (inversion of configuration). As we will see later, structures **1** and **3** are identical, and **1** and **2** are a pair of enantiomers.

(ii) Rotation of a Fischer projection by an angle of 180° about the axis which is perpendicular to the plane of the paper gives identical structure. However, similar rotation by an angle of 90° produces non-identical structure.

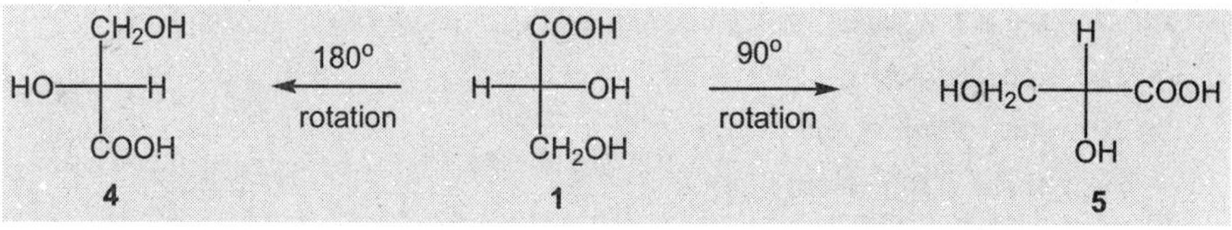

Thus, **1** and **4** are identical, whereas **1** and **5** are non-identical (enantiomers). This is to be remembered that during rotation of a Fischer projection formula, it should not be lifted out of plane.

B. Newman Projection

In Newman projection, we look at the molecule down the length of a particular carbon-carbon bond. The carbon atom away from the viewer is called *rear* carbon, and is represented by a circle. The carbon atom facing the viewer is called *front* carbon, and is represented as the centre of the above circle, which is shown by a dot. The remaining bonds on each carbon are shown by small straight lines at angles of 120° as follows:

(i) Bonds joined to *front* carbon intersect at the central dot.

(ii) Bonds joined to *rear* carbon are shown as emanating from the circumference of the circle.

The concept of Newman projection can be understood by **Fig. 2.2**.

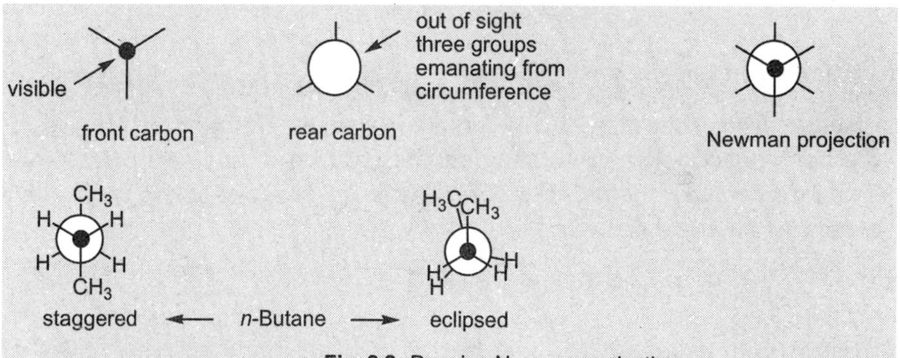

Fig. 2.2: Drawing Newman projections

Staggered and eclipsed conformations arise due to free rotation about the carbon–carbon single bond (front and rear carbon atoms).

C. Sawhorse Projection

The Sawhorse projection is very useful for understanding the mechanism and stereoelectronic aspects of reactions. In this projection, the bond between two carbon atoms is shown by a longer diagonal line, because we are looking at this bond from an oblique angle. The bonds linking other substituents to these carbons are shown projecting above or below this line. Thus, Sawhorse projections of staggered and eclipsed conformations of n-butane can be drawn as shown in **Fig. 2.3**.

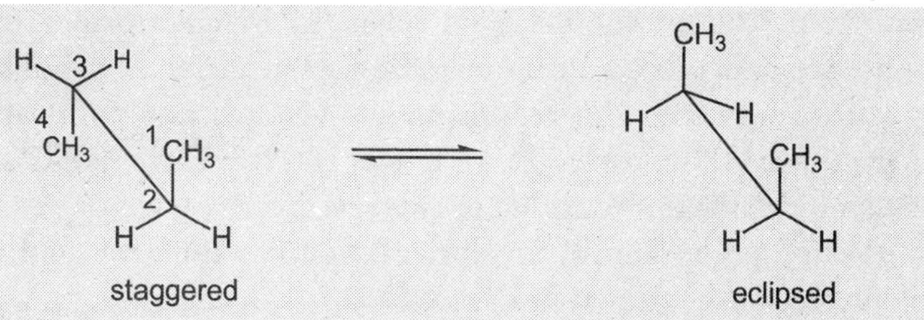

Fig. 2.3: Sawhorse projections of staggered and eclipsed forms of n-butane

Due to free rotation along the central bond, two extreme conformations are possible, staggered and the eclipsed (for details see Section 2.5C of this Chapter)

D. Flying Wedge Projection

It is a three-dimensional representation, and is particularly useful for chiral molecules. The procedure for drawing this type of projection formula is as follows:

Out of four bonds connecting substituents to a chiral carbon, two bonds are drawn in the plane of the paper, which are shown by normal lines (——), third bond lies above this plane, which is shown by a solid cone (▬), and the fourth bond lies below the plane of the paper which is shown by broken lines (----). The flying wedge formulae of two enantiomeric lactic acids are shown in **Fig. 2.4**.

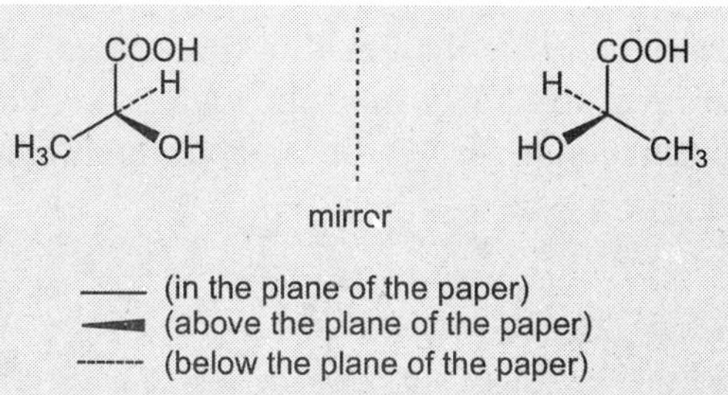

Fig. 2.4: Flying wedge projections of enantiomeric pair of lactic acid

Both these structures are mirror images of each other. In such representation, the main functional group is, generally, held on the upper side in the vertical plane.

E. Interconversions of Projection Formulae

It is important to have a clear idea of the inter-relationship between various projections of a given organic molecule to understand the stereochemical implications of the reactions. For this, we should be well conversant with the methods of translating one projection into another. The methods used for converting one projection into another without changing the configuration are given in the following description.

I. Conversion of Fischer Projection into Sawhorse Projection and vice-versa

(i) Fischer Projection to Sawhorse Projection: Fischer projection of a compound can be converted into Sawhorse projection; first in the eclipsed form (in Fischer projection the groups on neighbouring carbons are considered to be eclipsing each other), by holding the model in horizontal plane in such a way that the groups on the vertical line point above, and the last numbered chiral carbon faces the viewer. Then, one of the two carbons is rotated by an angle of 180° to get the staggered form (*more stable* or *relaxed form*).

For example, Fischer projection of an optically active tartaric acid is converted into staggered Sawhorse projection as shown in **Fig. 2.5**.

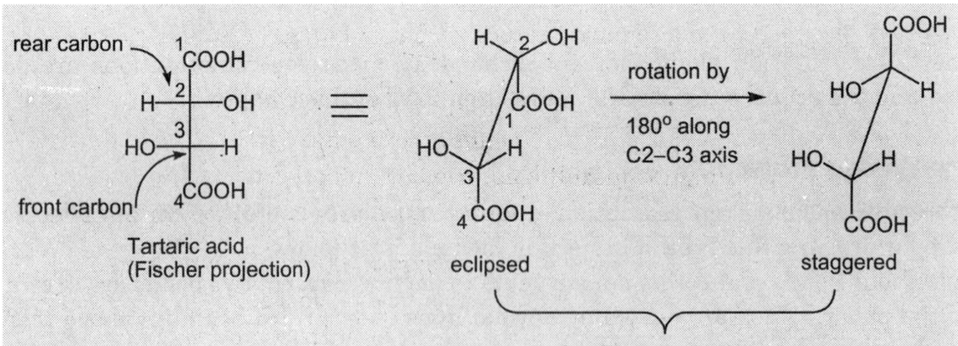

Fig. 2.5: Conversion of Fischer projection into Sawhorse projection

(ii) Sawhorse Projection to Fischer Projection: First, the staggered Sawhorse projection is converted to an eclipsed projection. It is then held in the vertical plane in such a manner that the two groups pointing upwards are away from the viewer, *i.e.* both these groups are shown on the vertical line. Such a conversion for 2,3-dibromobutane is shown in **Fig. 2.6**.

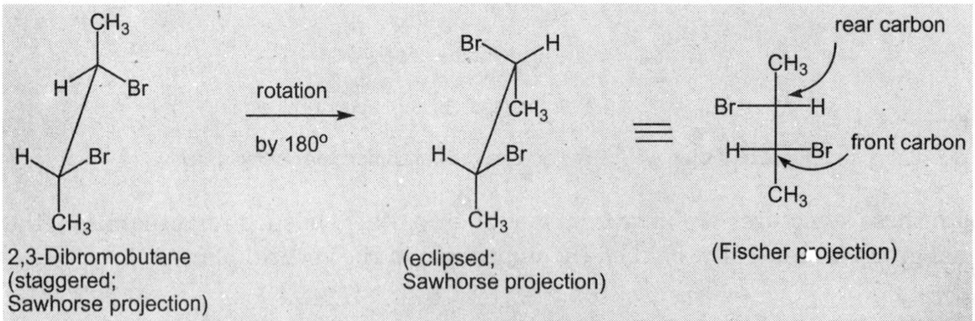

Fig. 2.6: Conversion of Sawhorse projection into Fischer projection

II. Conversion of Sawhorse Projection to Fischer Projection via Newman Projection and vice-versa

(i) Sawhorse Projection to Newman Projection and then Fischer Projection: Conversion of Sawhorse projection to Newman projection is quite easy. The molecule is viewed from front carbon (the central C—C bond being invisible) to get the staggered Newman projection. The rear carbon is rotated by 180° to get eclipsed Newman projection. Then, the molecule is held in vertical plane, *i.e.* central bond is visible in vertical plane in such a manner that front carbon is the lowest carbon (**Fig. 2.7**).

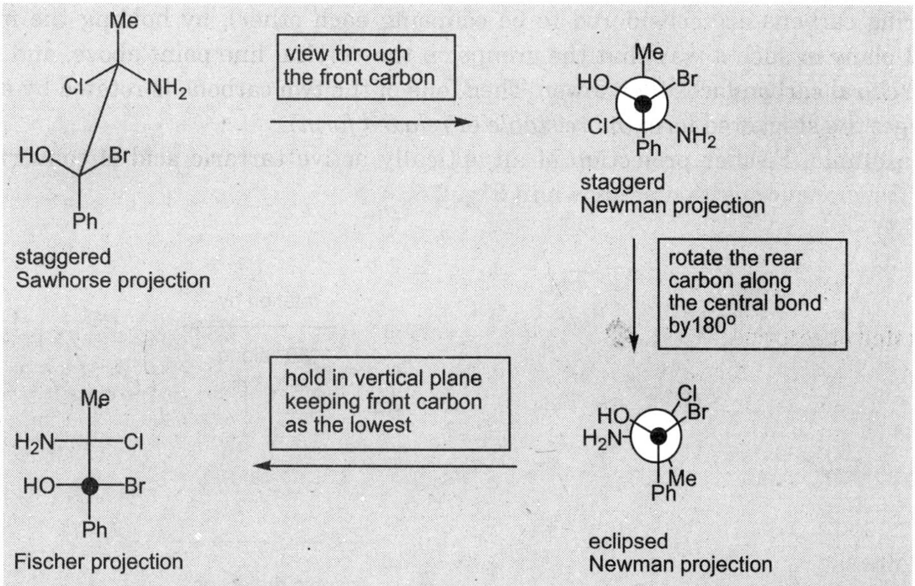

Fig. 2.7: Conversion of Sawhorse projection into Fischer projection *via* Newman projection

(ii) Fischer Projection to Newman Projection and then Sawhorse Projection: The molecule is viewed through the lowest chiral carbon, which becomes the front carbon, and thus eclipsed Newman projection is drawn. It is then converted into staggered conformation. Finally, the molecule is viewed through the bond connecting the front carbon with rear carbon. Such a conversion of *D*-erythrose is illustrated in the following scheme (**Fig. 2.8**).

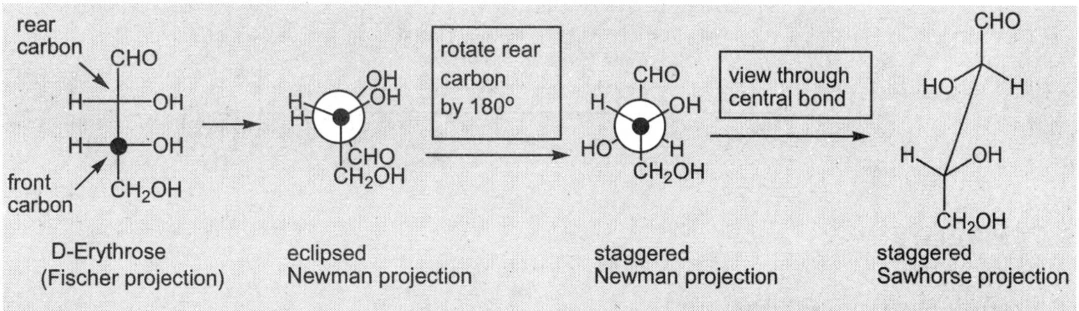

Fig. 2.8: Conversion of Fischer projection into Sawhorse projection *via* Newman projection

III. Conversion of Fischer Projection into Flying Wedge Projection and vice versa

(i) Fischer Projection to Flying Wedge Projection: The vertical bonds in the Fischer projection are drawn in the plane of the paper using simple lines (—). Consequently, horizontal bonds will project above and below the plane ('*a*' and '*b*' in **Fig. 2.9**). Conversion of Fischer projection of one of the enantiomers of α-bromopropanoic acid into five flying wedge formulae (without changing the configuration) is illustrated in **Fig. 2.9**.

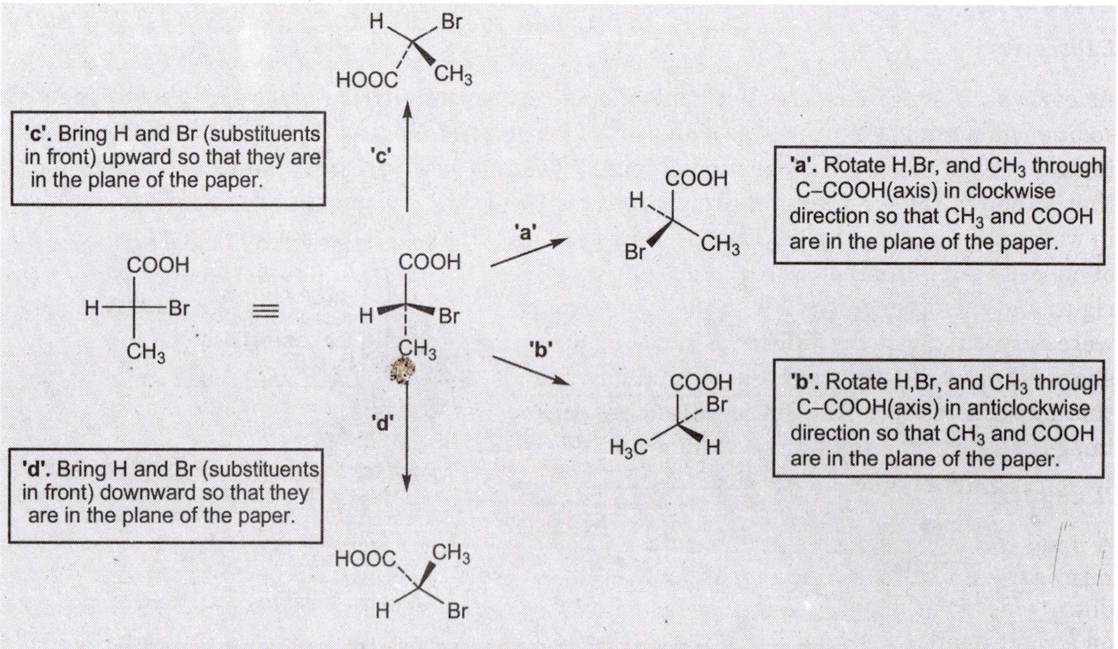

Fig. 2.9: Conversion of Fischer projection into flying wedge projections

(ii) Flying Wedge Projection to Fischer Projection: The molecule is rotated (in the vertical plane) in such a way that the bonds shown in the plane of the paper go away from the viewer, and are vertical.

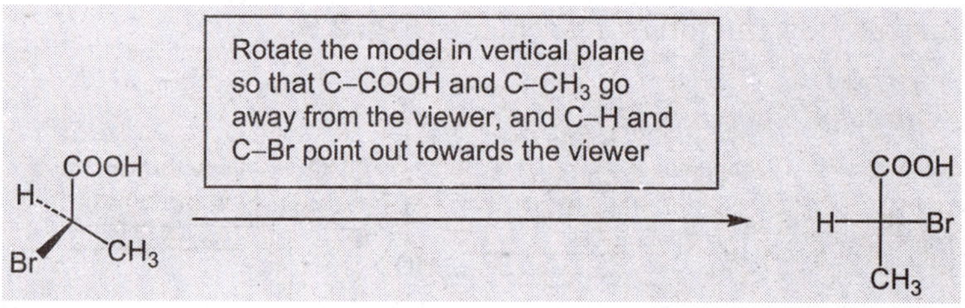

2.3 CHIRALITY

A. Chirality: Optical Activity

Enantiomers are known to possess same physical and chemical properties, but they differ from each other in the way they interact with the *plane of plane-polarized light*.

Substances which can rotate the plane of plane-polarized light are said to be *optically active*. When such a substance rotates the plane of plane-polarized light in a clockwise direction or to the right, it is called *dextrorotatory* (Latin: *dextre* means *right*) and is indicated by (+) sign, whereas the one which rotates it in an anticlockwise (or counter-clockwise) direction or to the left is called *laevorotatory* (Latin: *laevo* means *left*) and is indicated by (–) sign. Those substances which do not rotate the plane of plane-polarized light are said to be *optically inactive*.

I. Discovery

As early as 1848, French chemist Louis Pasteur discovered that crystalline, optically inactive sodium ammonium tartarate was a mixture of two different kinds of crystals, which were mirror images of each other. On separation of these two kinds of crystals mechanically, Pasteur found that although the original mixture was *optically inactive*, each set of crystals when dissolved in water was *optically active.* The specific rotations of the two solutions were exactly equal, but of opposite sign. In other words, one solution rotated the plane of plane-polarized light to the right, and the other to the left to the same extent. In all other properties, the two substances were identical. Since the difference in rotation in the two samples was found in solution in which shapes of crystals disappear, Pasteur laid the foundation of stereochemistry when he proposed that like the two sets of crystals, the molecules making up the crystals were themselves mirror images of each other, and the difference in rotation was due to *molecular dissymmetry.*

II. Chirality

A hand cannot be superimposed on its mirror image. If you hold your left hand in front of a mirror, the image looks like a right hand. When you try to superimpose your hands (both palms down) you simply cannot do it. Have you ever thought why you cannot wear a left-hand glove on a right hand?

Such an object, which cannot be superimposed on its mirror-image is said to be chiral (kyral) [Greek: *cheir 'handedness'*], and the property of non-superimposability is called *chirality*. Thus, our hands are chiral. A molecule (or an object) which is non-superimposable on its mirror image is called *chiral* (dissymmetric or asymmetric).

On the other hand, a molecule (or an object) which is superimposable on its mirror image is called *achiral* (non-dissymmetric or symmetric). We can also say that objects having a plane of symmetry are achiral, whereas objects not having a plane of symmetry are, usually, chiral.

(i) Molecular Chirality and Asymmetric Carbon: There is a close relationship between *chirality* and *optical activity*, and quite often the descriptions *optically active* and *chiral* are used interchangeably. It should, however, be remembered that optical activity referes to one property of chiral molecules, namely, the ability to rotate the plane of plane-polarized light.

Chirality in molecules is usually due to the presence of an sp^3 carbon atom with four different groups attached to it. Such a carbon atom is called a *chiral carbon* or a *chirality centre*. The presence of a chirality centre usually leads to *molecular chirality*. Such a molecule has no plane of symmetry, and exists as a pair of enantiomers. Such a carbon atom is, sometimes, also referred to as an *asymmetric carbon* atom.

It is known through electron diffraction, X-ray diffraction, and spectroscopic studies supported by quantum mechanical considerations that the four valencies of sp^3-hybridized carbon are directed towards the four corners of a regular tetrahedron with carbon atom at the centre. Let us construct a ball and stick model of a compound C_{WXYZ}, a derivative of methane, where W, X, Y, and Z are all different atoms or groups. We also construct a model of its mirror image. These mirror images can be drawn as follows:

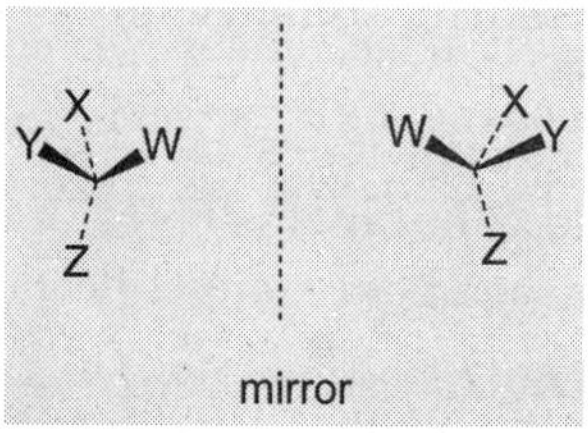

We may twist and turn the above two representations in any way we like so long as we do not break any bond, yet we find that the two are not superimposable. Therefore, they must represent two isomers, *i.e.* two enantiomers.

Although we have distinguished the two enantiomers of C_{WXYZ} with the help of the above picture of a three-dimensional model, to show the tetrahedral arrangement of the groups attached with the chiral carbon, it is not convenient to draw such picture in all the cases. It, therefore, became necessary to have a simpler convention for distinguishing between enantiomers. The following projection formulae (Fischer projection formulae, Section 2.2A) represent the two enantiomers in two dimensions with the assumption that the two horizontal bonds (C–Y and C–W) project towards viewer out of the plane of the paper, and the two vertical bonds (C–X and C–Z) project away from viewer behind the paper. The superimposability of two such flat, two-dimensional structures is tested by rotating end-to-end without raising them out of the plane of the paper. The asymmetric carbon atom is, generally, not inserted in such projection formulae, in which case it is assumed that it is located at the junction of the crossed lines.

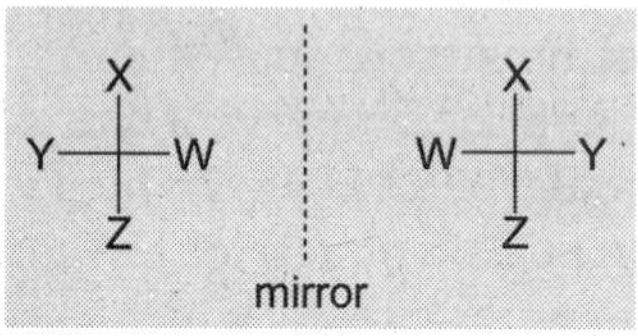

The dextrorotatory (+) and laevorotatory (–)-glyceraldehydes and lactic acids are simple examples to illustrate the non-superimposability of enantiomers.

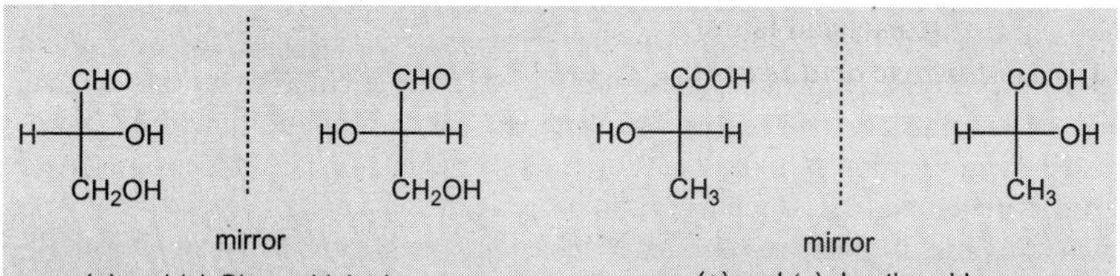

(+) and (–)-Glyceraldehydes (+) and (–) -Lactic acids

Therefore, chirality or molecular dissymmetry is the necessary and sufficient condition for a molecule to be optically active.

(ii) Stereogenic Centre or Chirality Centre: As we have seen, molecules having tetrahedral carbon atoms attached to four different atoms or groups are called asymmetric or chiral carbon atoms. In 1996, the IUPAC recommended that a tetrahedral carbon atom bearing four different atoms or groups may be called a *chirality centre*. However, several earlier terms including asymmetric centre, asymmetric carbon, chiral centre, stereogenic centre, and stereocentre are still widely used.

Here, it may be mentioned that if a molecule has one chirality centre, it is very easy to know that it is chiral. For example, 2-butanol is chiral having C2 as the chirality centre. But 1-butanol is achiral in which no carbon bears four different atoms.

B. Elements of Symmetry

We have seen that enantiomerism will depend on whether a molecule is superimposable on its mirror image or not. If it is superimposable, the molecule is optically inactive. If it is not superimposable, the molecule is optically active. The most convenient method of inspecting superimposability consists in determining whether the molecule has any one of the four elements of symmetry. These elements are: plane of symmetry (σ), centre of symmetry (i), simple or proper axis of symmetry (C_n), and alternating or improper or rotation reflection axis of symmetry (S_n)

This method can be applied to the Fischer projection and tetrahedral formula alike. If there is at least one symmetry element present, the molecule will not show enantiomerism. We will now discuss these elements in the following subsections.

I. Plane of Symmetry (σ)

A plane of symmetry is defined as an imaginary plane, which divides a molecule in such a way that one half of it is a mirror image of the other half. A molecule with at least one plane of symmetry can be superimposed on its mirror image and is achiral. A molecule that does not have a plane of symmetry is usually *chiral*; it cannot be superimposed upon its mirror image. A plane of symmetry may pass through atoms, between atoms, or both. For example, the lower half of the molecule of *meso*-tartaric acid is mirror image of the upper half and both are superimposable. Therefore, *meso-tartaric acid is achiral, and hence optically inactive*.

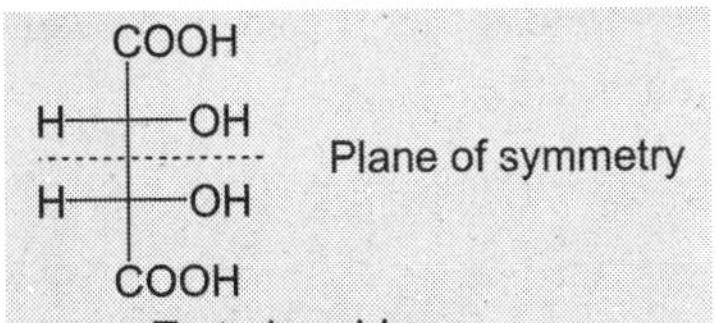

meso-Tartaric acid

II. Centre of Symmetry (i)

A centre of symmetry, which is usually present in an even-membered ring, is also called *centre of inversion*, and is defined as a point within the molecule *where an atom, joined to it by a straight line, meets an equivalent atom if the line is extrapolated in equal distance beyond it in opposite direction*. In other words, *it is a point at which all the straight lines joining identical points in the molecule cross each other*.

For instance, the molecule of 2,4-dimethylcyclobutane-1,3-dicarboxylic acid has a centre of symmetry (centre of ring) and is, therefore, optically inactive.

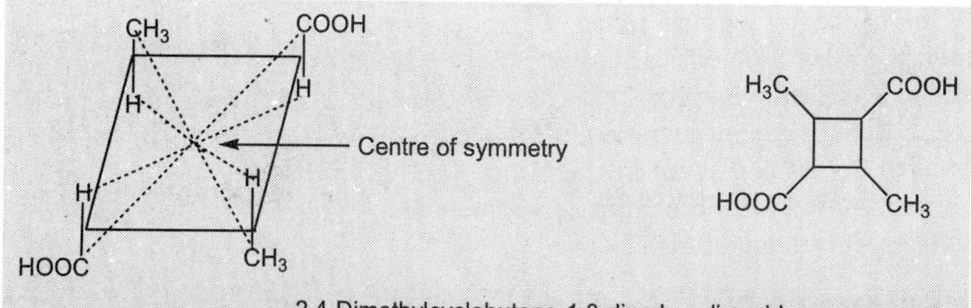

2,4-Dimethylcyclobutane-1,3-dicarboxylic acid

Another interesting example is dimethyldiketopiperazine, which exists as two geometric isomers, *cis* and *trans* depending on whether the methyl groups are on the same side or opposite side. The *cis*-isomer has no centre of symmetry, and can exist in optically active forms. The *trans*-isomer, on the other hand, is optically inactive, because it has a centre of symmetry.

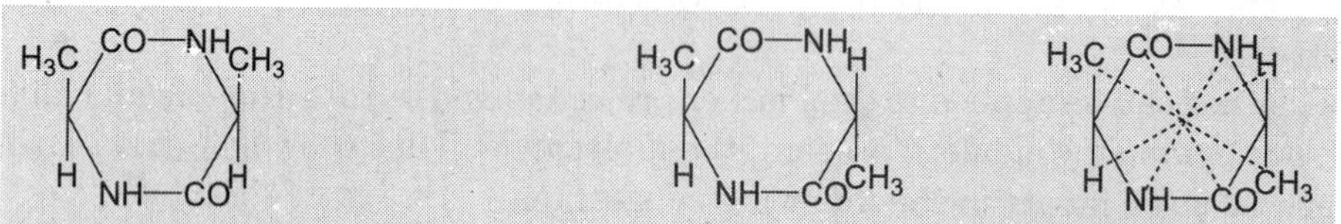

cis-isomer
(chiral)

trans-isomer
(achiral due to centre of symmetry)

Dimethyldiketopiperizine

III. Simple or Proper Axis of Symmetry (C_n)

An axis of symmetry is defined as an imaginary line passing through the molecule in such a way that when the molecule is rotated about it by an angle of 360°/*n*, an arrangement indistinguishable from the original is obtained. Such an axis is called *n*-fold axis of symmetry. For example, *cis*-1,3-dimethylcyclobutane has a two-fold axis of symmetry (C_2), *i.e.* rotation by 180° gives indistinguishable appearance.

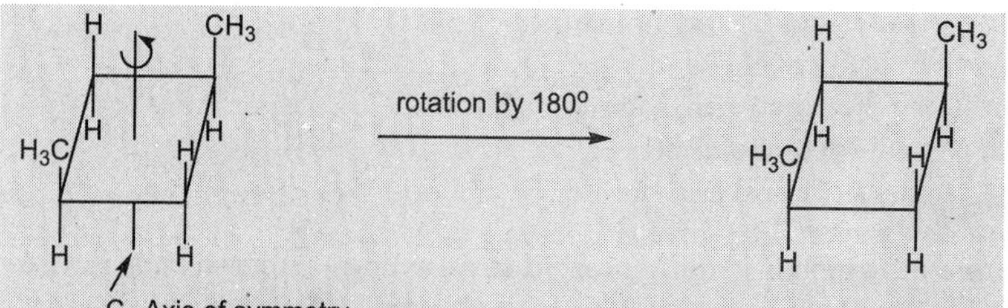

C_2-Axis of symmetry
cis-1,3-Dimethylcyclobutane

(achiral)

IV. Alternating or Improper or Rotation Reflection Axis of Symmetry (S_n)

Alternating axis of symmetry is rather rare. It is also called improper or rotation-reflection axis of symmetry. To determine whether an n-fold improper axis of symmetry is present, the molecule is rotated through an angle of $360°/n$ about this axis, followed by reflection in a plane perpendicular to the axis. If we obtain the original molecule, it is said to have an n-fold improper axis of symmetry, and will be optically inactive. For instance, 1,2,3,4-tetramethylcyclobutane (a) has a four-fold axis of symmetry, because its rotation by 90° about the axis AB (passing through the centre of the ring perpendicular to its plane) gives (b) (**Fig. 2.10**). The form (a) is the mirror image of (b) by its reflection in the plane parallel to the ring, *i.e.* perpendicular to the axis AB. We should note that if it so happens, the molecule in this case has two vertical planes of symmetry through the diagonals of the ring.

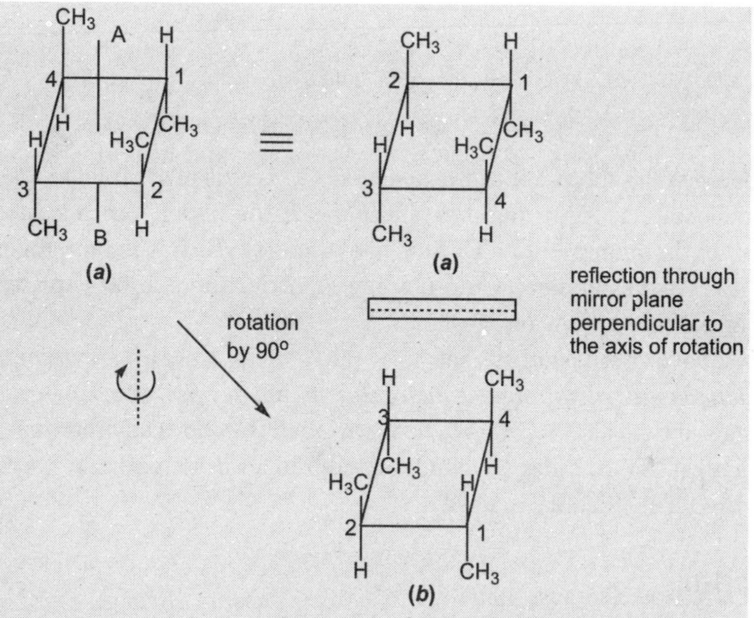

Fig. 2.10: Alternating or improper or rotation-reflection axis of symmetry (S*n*)

It can be concluded that the necessary and sufficient condition for a molecule to show optical activity is that it is not superimposable on its mirror image. In other words, *a molecule can exist in enantiomeric forms only when it has molecular dissymmetry*. The concept of molecular dissymmetry is important, since the phenomenon of optical rotation is not related to chiral carbon atoms only. Even when a chiral carbon atom is present, a compound need not necessarily have molecular dissymmetry. For instance, some compounds, such as *meso*-compounds, which have two or more chiral carbons do not exhibit optical activity. On the other hand, there are optically active compounds having no chiral carbon. For instance, allenes having no chiral carbons show optical activity due to dissymmetry of the molecule (see Chapter 3).

Asymmetry vs Dissymmetry

We know that only those substances, which are related to each other as non-superimposable mirror images are optically active. Such substances may be dissymmetric or asymmetric. So far, these two terms have been used interchangeably (without distinction) to describe optical activity. But this is not so, and there exists a subtle difference between them. In general, the term *asymmetry is used for those optically active compounds, which have none of the four elements of symmetry*. In contrast, the term *dissymmetry is used for all stereoisomeric compounds, which are capable of existing as pairs of non-superimposable mirror images despite the presence of some elements of symmetry*. Thus, the term dissymmetry is applicable to all stereoisomers, which are related to each other as non-superimposable mirror images of each other. The following example is illustrative:

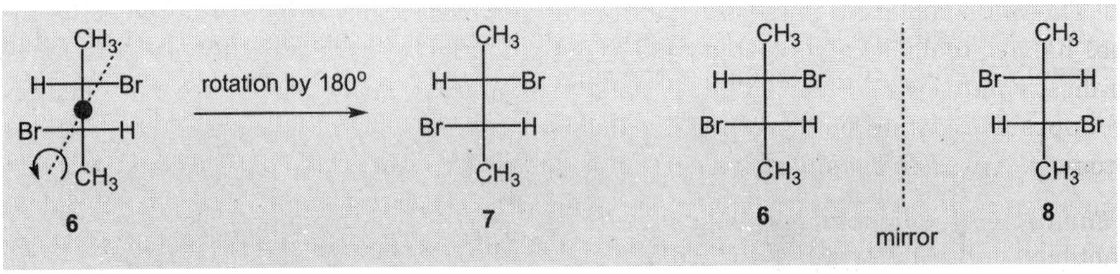

The molecule of 2,3-dibromobutane possesses a C_2 axis of symmetry, passing through the centre of the molecule, at right angle to the plane of the paper. Since structures **6** and **7** are indistinguishable, the molecule has C_2 axis of symmetry. But, it is non-superimposable on its mirror image, so it is dissymmetric and not asymmetric, and exhibits optical activity. **6** and **8** are non-superimposable mirror images.

From the above discussion, it becomes quite clear that *all asymmetric molecules are dissymmetric, whereas all dissymmetric molecules are not asymmetric*. However, both these types of molecules show optical activity. It is to be mentioned that both asymmetric and dissymmetric molecules are chiral. So, in order to avoid any confusion in using these terms– asymmetry or dissymmetry, the term chirality is used.

C. Enantiomerism

The term 'enantiomers' and 'enantiomerism' have already been introduced in Section 2.1A. Here, we shall discuss the necessary and sufficient condition for enantiomerism and properties of enantiomers.

I. Necessary and Sufficient Condition for Enantiomerism

The necessary and sufficient condition for a compound to exhibit enantiomerism is that it should be chiral or dissymmetric, *i.e.* the molecule and its mirror image should be non-superimposable, even if it may not have an asymmetric carbon or stereocentre. In general, it has been observed that compounds having one or more chirality centre show enantiomerism, and are therefore optically active. However, this statement does not hold good for all such molecules like,

(i) Compounds having Chirality Centre(s) but not Enantiomeric: *Meso*-2,3-dibromobutane contains two chirality centres (marked with *), but it does not exhibit enantiomerism, and hence is optically inactive.

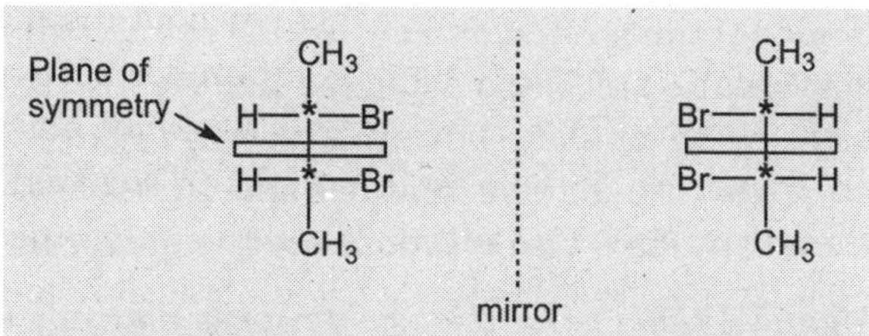

Meso-2,3-dibromobutane (optically inactive)

The *meso* molecule possesses a plane of symmetry that divides the molecule into two equal halves, which are superimposable mirror images of each other. Thus, it is an achiral, non-dissymmetric molecule. Actually, the optical activity due to one half is cancelled by equal and opposite rotation of the other half. This is called *internal compensation*. So *meso*-2,3-dibromobutane is an example of an optically inactive compound containing two chirality centers.

(ii) Enantiomeric compounds having no chirality centre: Such compounds are also known which do not have a chirality centre, but still they are optically active. These molecules show chirality or dissymmetry, and hence enantiomerism. Examples of such compounds are *o*-substituted biphenyls and allenes having even number of double bonds. (For a detailed discussion see Chapter 3).

It thus emerges that the necessary and sufficient condition for compounds to exhibit enantiomerism is that they should possess chirality or dissymmetry rather than asymmetry.

II. Properties of Enantiomers

Enantiomers have same physical and chemical properties. They differ in their sign of rotation of plane of plane-polarized light, and *rates of reactions with chiral compounds.* Some important properties of enantiomers are:

(i) They have identical physical properties, but differ in the direction of the rotation of plane polarized light: Although the two enantiomers rotate the plane of plane-polarized light in two opposite directions, the magnitude of rotation is the same. The direction of rotation is usually denoted by the prefix (+) for dextrorotatory and (–) for laevorotatory enantiomers. **Table 2.1** shows the physical properties of (+) and (–) forms of tartaric acid, and 2-methyl-1-butanol.

Table 2.1 Physical properties of (+) and (–) forms of tartaric acid and 2-methyl-1-butanol

Tartaric acids Enantiomer	**Specific rotation**	**M.P.**	**Density**
(+)	+11.98°	170°C	1.76
(–)	–11.98°	170°C	1.76
2-Methyl-1-butanols Enantiomer	**Specific rotation**	**B.P.**	**Refractive index**
(+)	+5.75°	129°C	1.41
(–)	–5.75°	129°C	1.41

It is clear that the two enantiomers have same melting points, boiling points, and refractive indices. The magnitude of rotation of plane-polarized light is also the same, but in opposite direction. The molecules being so similar, they are expected to have same degree of rotation. But, being mirror images of each other, their rotations should also be the mirror images of each other. Indeed, clockwise and anticlockwise rotations are the mirror images of each other.

(ii) Enantiomers have identical chemical properties towards optically inactive reagents: Since the structural environment in the two enantiomers is same, the optically inactive reagents, such as H_2SO_4, HBr, and CH_3COOH encounter the *same environment* while approaching either enantiomer. Enantiomers, such as those of 2-methyl-1-butanol are converted to 2-methylbutene when treated with strong H_2SO_4, to 2-methylbutyl bromide on treatment with HBr, and give the same ester on esterification with CH_3COOH. The rates of above transformations are also same.

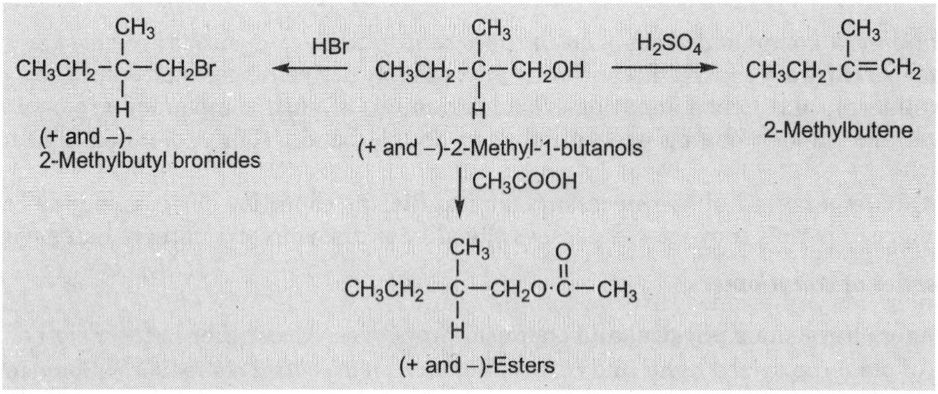

(iii) Enantiomers have different chemical properties towards optically active reagents: If we use an optically active reagent, the reaction rates will be different with different enantiomers. If we esterify the two enantiomers of 2-methyl-1-butanol with (–)-lactic acid, the influence exerted by the reagent will not be identical due to *different spatial disposition* of the –OH group in the two enantiomers, in relation to the groups attached to the chirality centre of (–)-lactic acid. It, therefore, follows that the rate of esterification of (+)-2-methyl-1-butanol will be different from that of (–)-2-methyl-1-butanol.

(iv) Enantiomers have different biological properties: In biological systems, this difference is a rule to the extent that more often reaction with one of the enantiomers does not take place at all. This is quite understandable as the biological catalysts, enzymes, and most of the compounds with which they react, are *optically active*. The following examples are illustrative:

(a) (+)-Glucose plays an important role in animal metabolism and fermentation, but (–) glucose is not metabolized by animals, and furthermore cannot be fermented by yeasts.
(b) *Penicillium glaucum*, consumes only the (+)-enantiomer when fed with a mixture of equal quantities of (+)-and (–)-tartaric acids.
(c) Hormonal activity of (–)-adrenaline is many times more than that of its enantiomer.

We can also understand the differential activity of optically active reagents towards enantiomers by *transition-state approach*. The transition states in the reaction of two enantiomers (+)-A and (–)-A with an optically inactive reagent B will be mirror images of each other, and hence will be of the same energy. On the other hand, the transition states in the reaction of the enantiomers with an optically active reagent (*e.g.* (–)-B) will not be mirror images, and therefore will not have the same energy. Thus, the energy of activation will be different, and so will be the rates of reaction (For detailed discussion, see Chapter 3).

D. Molecules (Enantiomers and Diastereomers) with Several Chirality Centres (Stereogenic Centres)

In general, the optical rotation of asymmetric compounds depends upon the presence of a chirality centre. Therefore, the number of optically active forms increases with the number of chiral carbon atoms. If there are n chiral carbons, the compound will exist in 2^n optically active forms, provided chiral atoms are *not identically substituted*. Thus, 2-bromo-3-hydroxybutanedioic acid, HOOC–CH(OH)–CH(Br)–COOH, in which the two chiral carbon atoms are dissimilar, exists in $2^2 = 4$ optically active forms (For structures, see **9-12**).

The two chiral carbon atoms of tartaric acid, HOOC-CHOH-CHOH-COOH, on the other hand, are *identically substituted* (similar), and hence the total number of optically active isomers cannot be predicted by using 2^n formula. We will now discuss few examples of such molecules.

I. Compounds with Two Dissimilar Chirality Centres

Compounds such as 2-bromo-3-hydroxybutanedioic acid have two dissimilar chiral carbon atoms, and hence exist in four stereoisomeric forms, *i.e.* as two pairs of enantiomers.

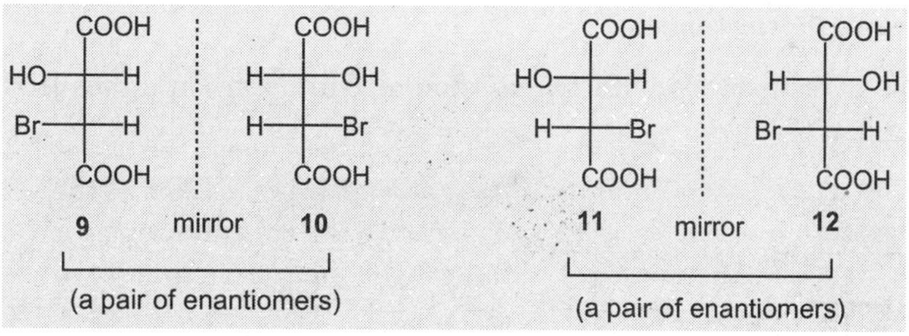

The two enantiomers of any pair, (**9,10** or **11,12**), have equal and opposite rotations, and if mixed in equal amounts, **9** and **10** will give rise to one racemic modification, while **11** and **12** will give another. Now, the question is the relationship between **9** and **11** or **9** and **12** or between **9** and **11** or **10** and **12**. They are optically active, but are not mirror images. Such stereoisomers are referred to as *diastereomers*. For example, stereoisomer **9** or **10** is a diastereomer of **11** or **12**.

Diastereomers are stereoisomers, which have the same configuration at one chirality centre, but different configuration at the other. In other words, diastereomers are stereoisomers, which are not mirror images of each other (enantiomers).

Properties of Diastereomers

As diastereomers are not related to each other as mirror images, they have different properties. Some of the properties are:

(i) Physical properties: The melting points, boiling points, water solubility, and relative densities of diastereomers are different. The physical properties of tartaric acids as given in **Table 2.2** are illustrative.

Table 2.2 Physical properties of tartaric acids

	(+)	(–)	(±)	**meso**
$[\alpha]^{20°}{}_D$	+12°	–12°	0°	0°
Melting point (°C)	170	170	205	140
Water solubility (g/100 mL)	147	147	25	120
Relative density	1.760	1.760	1.687	1.666

As diastereomers have different solubilities, boiling points, and Rf values, they can easily be separated using techniques such as fractional crystallization, fractional distillation, and chromatography.

(ii) Behaviour towards plane-polarized light: If the diastereomers happen to be optically active, they have different values of $[\alpha]^{20°}$D.

(iii) Chemical properties: Diastereomers have similar, but non-identical chemical properties. In particular, they react with chiral or achiral reagents at different rates. This may be due to the fact that neither the reagents nor the transition states are mirror images of each other. Thus, diastereomers differ in their energies of activation for a particular reaction, and hence react at different rates.

II. Threo and Erythro Diastereomers

Another example of compounds with two dissimilar chiral carbon atoms is the four carbon sugars, *erythrose* and *threose*.

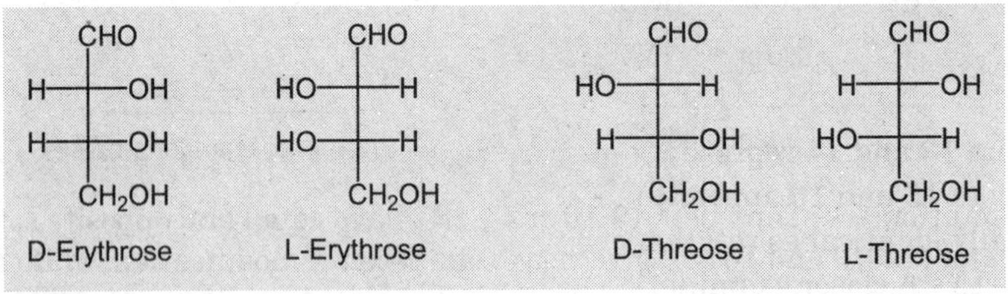

By analogy, diastereomers of compounds possessing two dissimilarly constituted chiral carbon atoms are commonly distinguished from each other by using the perfix *threo* or *erythro*. These prefixes are used to designate those compounds, which contain two stereocentres having two identical and a third non-identical group attached. *The erythro diastereomer is the one in which two like groups on the chiral carbon atoms are on the same side in the Fischer projection formula as in erythrose.* In contrast, the *threo* diastereomer resembles *threose* as the two like groups are on opposite sides. This nomenclature is illustrated by taking the example of 3-bromo-2-butanol.

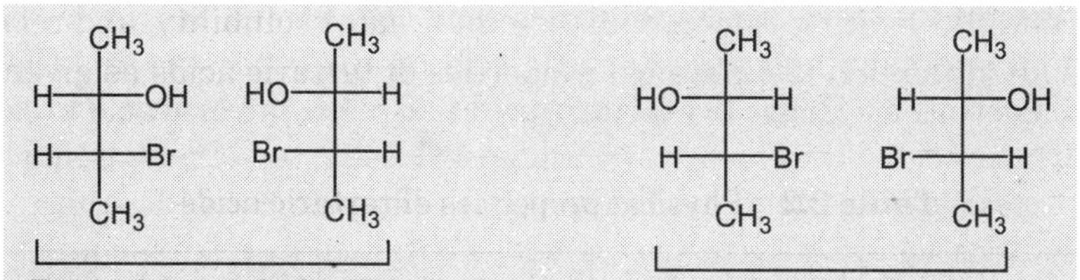

Erytrho-3-bromo-2-butanol (pair of enantiomers) *Threo*-3-bromo-2-butanol (pair of enantiomers)

Although the Fischer projection formulas are conveniently constructed, and show at a glance whether the given compound is *erythro* or *threo*, they, however, give the impression that the molecule exists in the *eclipsed* form. Actually, it exists in the *staggered* form in which the bulky substituents are as far apart as possible.

Thus, it can be said that an *erythro* isomer corresponds to that diastereomer, which when viewed along the bond connecting the chiral carbons has a rotational isomer in which all similar groups are eclipsed. The *threo* diastereomers, on the other hand, do not have an isomer in which all similar groups are eclipsed.

III. Compounds with Two Similar Chirality Centres: Meso compounds

When a compound contains chiral carbon atoms that are similar, the number of optical isomers cannot be predicted by the application of 2^n formula. The best known example of this class of compounds is tartaric acid, HOOC–*CH(OH)–*CH(OH)–COOH, with two chiral carbon atoms being denoted by asterisks. We can write four projection formulae for tartaric acid.

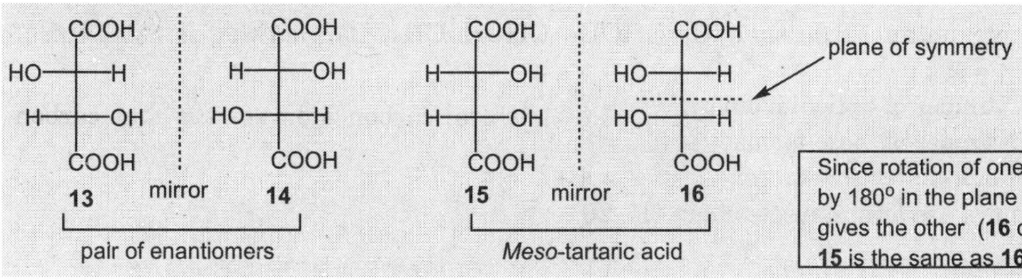

13 mirror **14** — pair of enantiomers; **15** mirror **16** — *Meso*-tartaric acid

Since rotation of one (**15** or **16**) by 180° in the plane of the paper gives the other (**16** or **15**), 15 is the same as 16.

The isomers **13** and **14** represent a pair of enantiomers, and each is diastereomeric to **15** or **16**. The isomers **15** and **16** appear to be another pair of enantiomers by a casual inspection, but they are really identical as rotation of one (**15** or **16**) by 180° in the plane of the paper gives the other (**16** or **15**). A closer examination of **15** (or **16**) reveals that it has a plane of symmetry that bisects the molecule between the two chiral carbon atoms. Such compounds are called *meso* compounds. Thus, the combined activity of each chiral centre in **15** (or **16**) is zero by *internal compensation*. This is in contrast with *external compensation,* where the combined optical activity of equimolar proportions of an optically active molecule and its enantiomer is zero. This is commonly referred to as *racemization*.

The isomers having two similar chirality centres, such as **15**, are optically inactive due to the presence of a plane of symmetry, and are termed *meso* compounds. Hence, *meso* compounds are optically inactive as a molecule is superimposable on its mirror image.

A look at the three stereoisomers of tartaric acid (**13, 14** and **15**), *i.e.* (+)-tartaric acid or (–)-tartaric acid on one hand and *meso*-tartaric acid on the other, reveals that they are not mirror images of each other. Hence (+)-tartaric acid and *meso*-tartaric acid are diastereomers. Similarly, (–)-tartaric acid and *meso*-tartaric acid are also diastereomers. Being diastereomers, the properties of *meso*-tartaric acid are different than those of (+)- and (–)-tartaric acids.

Racemic modification (±) of tartaric acid can be obtained by mixing equimolar amounts of (+) and (–)-tartaric acids (**13** and **14**). (±)-Tartaric acid is a diastereomer of *meso*-tartaric acid, and hence will have different physical properties.

Racemic modification, being a mixture of two enantiomeric compounds, can be resolved. In contrast, a *meso*-compound is only one stereoisomer.

Prediction of the number of stereoisomers (optical isomers and meso forms)

The number of optical isomers and *meso* forms in compounds containing chirality centres depend upon the following:

(*i*) Number of chirality centres, and

(*ii*) Whether the chirality centres are similar or dissimilar.

Following formulae are useful for making predictions about the number of optical isomers and *meso* forms:

1. *For molecules having dissimilar chirality centres*

Suppose the number of chirality centres = n

Now, (*a*) Number of optical isomers $= 2^n$

(*b*) Number of *meso*-forms $= 0$

Total number of stereoisomers = Number of optical isomers + Number of *meso* forms.

$= 2^n + 0 = 2^n$

For example, butan-1,2,3-triol ($CH_3{*}CHOH–{*}CHOH–CH_2OH$) has two dissimilar chiral carbon atoms. $n = 2$

Now, (*a*) Number of optical isomers $= 2^2 = 4$

(*b*) Number of *meso* forms $= 0$

Total number of stereoisomers $= 4 + 0 = 4$

Indeed, there are four stereoisomers (**17-20**).

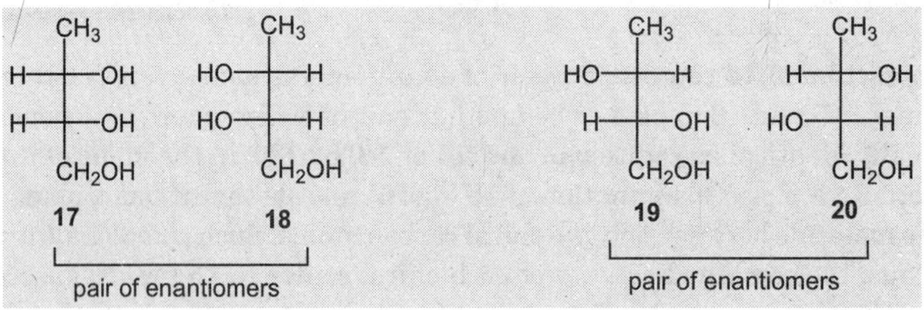

The compounds **17** and **18** as well as **19** and **20** are enantiomers, whereas compounds **17** and **19** and **18** and **19** (or **20**) are diastereomers.

2. *For molecules having similar chirality centres:* These molecules are of two types:

(*i*) Molecules having even number of chiral carbons.

(*ii*) Molecules having odd number of chiral carbons.

(*i*) *For molecules having even number of chiral carbons*, the number of stereoisomers is calculated as follows:

(*a*) Number of optical isomers $= 2^{(n-1)}$

(*b*) Number of *meso* forms $= 2^{(n/2-1)}$

Total number of stereoisomers $= 2^{n-1} + 2^{n/2-1}$

For example, tartaric acid (HOOC–*CHOH–*CHOH–COOH) has two similar chiral carbon atoms, $n = 2$

(*a*) Number of optical isomers $= 2^{n-1} = 2^{2-1} = 2^1 = 2$

(*b*) Number of *meso* forms $= 2^{n/2-1} = 2^{1-1} = 2^0 = 1$

Total number of stereoisomers $= 2 + 1 = 3$.

Indeed, there are three stereoisomers (**13-15**).

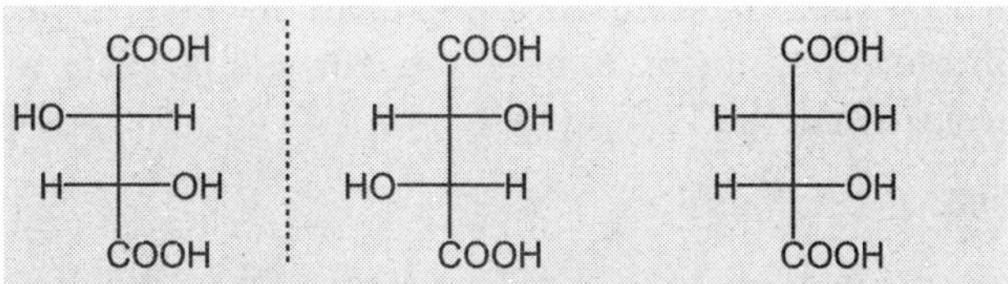

13 mirror **14** **15**

pair of enantiomers *meso*-Tartaric acid

(*ii*) *For molecules having odd number of chiral carbon atoms*, the number of stereoisomers is calculated as follows:

(*a*) Number of optical isomers $= 2^{n-1} - 2^{(n-1)/2}$

(b) Number of *meso* forms $= 2^{(n-1)/2}$

Total number of stereoisomers $= [2^{n-1} - 2^{(n-1)/2}] + [2^{(n-1)/2}] = 2^{n-1}$

For example, trihydroxyglutaric acid (HOOC–*CHOH–*CHOH–*CHOH–COOH) has three chiral carbon atoms, *i.e.* $n = 3$.

Now, (*a*) Number of optical isomers $= 2^{3-1} - 2^{(3-1)/2}$

$= 2^2 - 2^1 = 4 - 2 = 2$

(*b*) Number of *meso*-forms $= 2^{(3-1)/2} = 2^1 = 2$

Total number of stereoisomers $= 2 + 2 = 4$ (or $2^{3-1} = 2^2 = 4$)

There are actually four stereoisomers (**21-24**).

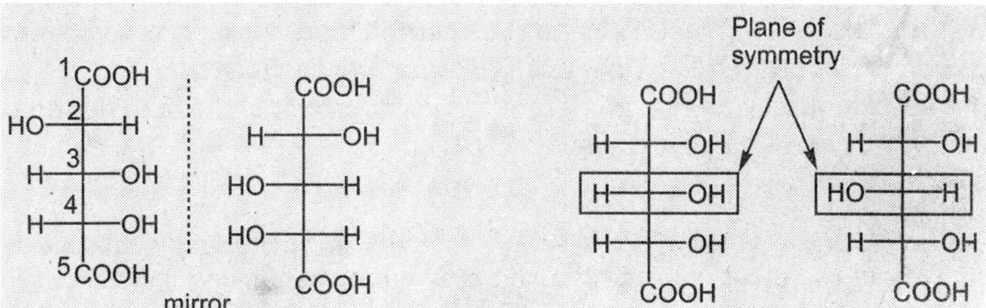

mirror

21 **22** **23** **24**

pair of enantiomers *meso*-forms

Here, **21** and **22** are optically active and are enantiomers, whereas **23** and **24** are *meso* compounds having a plane of symmetry passing through central carbon C3. In **21** and **22**, the carbon C3 is called a *pseudo chiral centre* or *pseudo asymmetric centre*, where C3 *is a stereogenic centre, but not a chirality centre.*

E. Absolute and Relative Configurations

For describing the configurations of different stereoisomers, the following two terms are used:

I. Absolute configuration
II. Relative configuration

I. Absolute Configuration

It denotes the actual arrangement of atoms or group of atoms, of a particular stereoisomer of a compound, in space. Absolute configuration of a molecule can be ascertained by X-ray studies of the crystals of pure compound.

II. Relative Configuration

It denotes the arrangement of atoms (or group of atoms), of a particular stereoisomer relative to the atoms or groups of atoms of another compound chosen as arbitrary standard, in space. For example, the configuration **A** was arbitrarily assigned to designate the configuration of (+)-glyceraldehyde. Taking this as standard, the relative configuration of (–)-lactic acid was assigned.

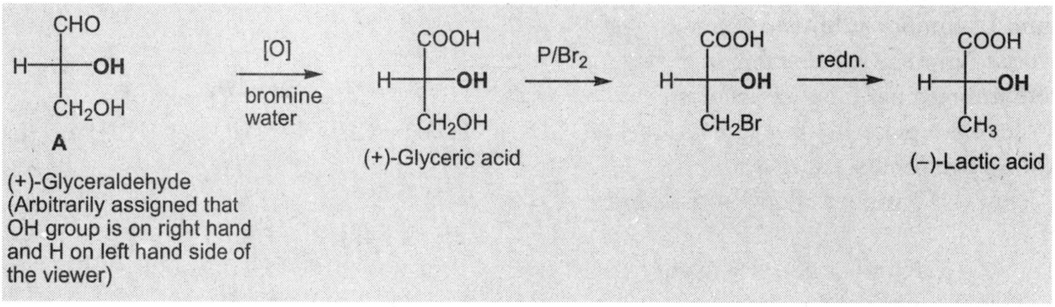

In the above reactions, the bond between chiral carbon and the substituents has not broken in any step. Hence, the configuration of (–)-lactic acid is the same as that of (+)-glyceraldehyde. The relative configuration of (–)-lactic acid is thus established. Here, it is to be mentioned that the *sign of rotation is not related with the configuration of the molecule.*

Now, the question arises as how to define the configuration of the given enantiomer of a chiral substance. Two commonly used systems are:

(*i*) D/L System
(*ii*) R/S System

(i) D/L System: This is the most commonly used system for assigning configuration to a given enantiomer. It is based upon the comparison of the projection formula of one enantiomer to which the name is to be assigned, with that of a standard substance arbitrarily chosen for comparison.

The following two conventions are used for this purpose:

- Hydroxy Acid or Amino Acid Convention
- Sugar Convention

(a) Hydroxy Acid or Amino Acid Convention: According to this convention, the prefix D- and L- refer to the configuration of the *lowest numbered chirality centre* of α-hydroxy or α-amino acids in the Fischer projection formula. If the α-OH or α-NH_2 group is on the right hand side (of the viewer), the prefix D- is used, whereas if these groups are on the left hand side, the prefix L- is used.

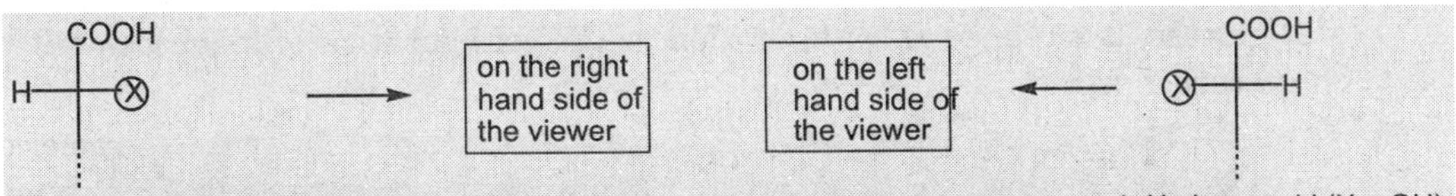

D-Hydroxy acid (X = OH)
D-Amino acid (X = NH_2)

L-Hydroxy acid (X = OH)
L-Amino acid (X = NH_2)

D-(–)-Lactic acid L-(+)-Lactic acid D-(–)-Alanine L-(+)-Alanine

(b) Sugar Convention: This convention was evolved by Emil Fischer who arbitrarily assigned D- and L- configurations to (+) and (–)-glyceraldehydes, respectively, as they were dextro and laevo rotatory. He further assigned D-configuration (OH on the right) to (+)-glyceraldehyde and L-configuration (OH on the left) to (–)-glyceraldehyde, without any evidence.

D-(+)-Glyceraldehyde L-(–)-Glyceraldehyde

The relative configurations of a large number of compounds were determined by correlating them with D-(+)- or L-(–)-glyceraldehyde, *e.g.* relative configuration of (–)-lactic acid was designated as D-(–)-lactic acid as it had the same configuration as D⁻ (+)-glyceraldehyde.

For compounds containing several chiral carbon atoms, the configuration at one centre is related to glyceraldehyde, and the configurations at other carbon atoms are determined relative to that carbon. In case of glucose, this carbon atom is C5, which is next to the CH_2OH group, *i.e. the highest numbered chiral carbon*. Since naturally occurring glucose was assumed to have the OH group at C5 projecting at right hand side, it belongs to the D series of compounds, and hence designated D-glucose. In case of the compounds having the OH group on the highest numbered chiral carbon on the left side, notation L is used.

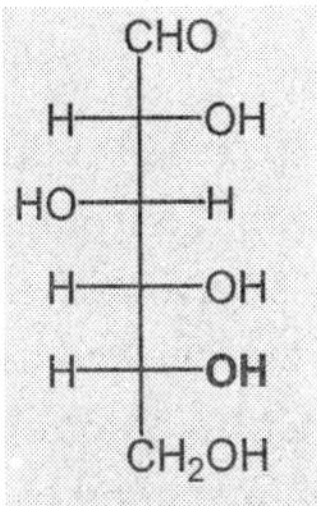

D-(+)-Glucose

In 1951, however, Bijvoet determined the absolute configuration of (+)-sodium rubidium tartarate using X-ray crystallography. It was found to be the same as assigned arbitrarily by Fischer. Before Bijvoet's work, configurations of thousands of compounds were assigned relative to (+)-tartaric acid. As a result of this finding, all the relative configurations of a large number of compounds became their absolute configurations.

Limitations of Sugar Convention

(*i*) Sugar molecules have more than one chirality centre. The configuration of only the highest numbered chirality centre is assigned, and that of the other centres are not shown (hidden in their names).

(*ii*) Another complication is encountered when compounds of one series are converted into the other series. For example, saccharic acid, which is obtained by the oxidation of D-glucose, is assigned D-configuration as the configuration of highest numbered chiral carbon is D.

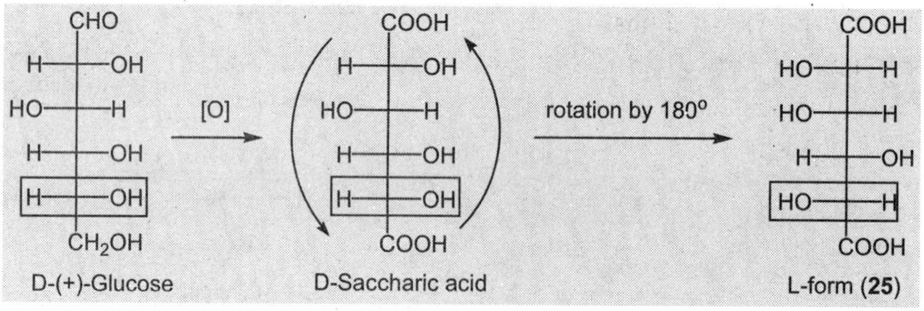

Now, if we rotate the molecule by 180° in the plane of the paper we get formula **25**, which should be designated as L. This is a very serious drawback in the system, *i.e.* the same molecule can have both D- and L - configurations.

(*iii*) (+)-Tartaric acid and (–)-threonine can be designated as D as well as L depending upon whether the reference compound is glyceraldehyde (highest numbered chiral carbon) or hydroxy or amino acid (lowest numbered chiral carbon).

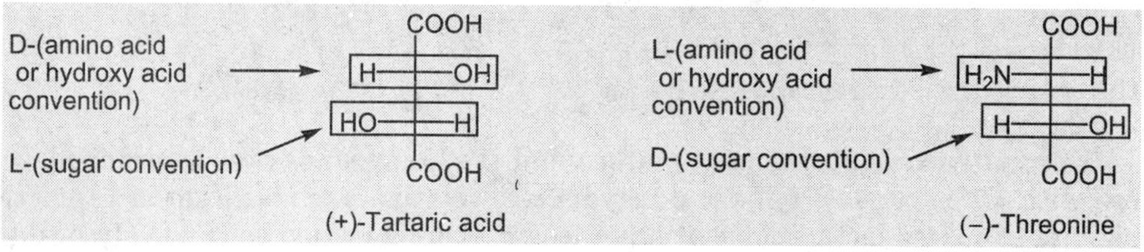

In conclusion, it may be added that this system suffers from several drawbacks, and is of limited use as it is confined only to sugars, hydroxy acids, and amino acids.

(ii) R/S System: It is, thus, clear that D and L system for specification of configuration around a chiral carbon was not found to be suitable for molecules which are not related with a particular glyceraldehyde. To overcome this problem, Cahn, Ingold, and Prelog evolved an unambiguous system for assigning absolute configuration to chiral molecules. This system is named as CIP (**C**ahn, **I**ngold, **P**relog) system after their names. It is called *R/S* system, as the prefixes *R*- and *S*- are used to designate the configuration at a particular chirality centre. A racemic mixture is named as (*RS*). This system is based on certain rules called as *sequence rules* or *CIP rules*.

The nomenclature of a chirality centre is done in the following steps:

Step I: Identify the atoms or groups directly attached to the chirality centre, and assign them a sequence of priority by using Greek numerals 1, 2, 3, and 4, where number 1 is assigned to an atom or a group of highest priority, and 4 is assigned to the group of lowest priority.

The priorities of the groups are assigned as per CIP rules, *e.g.*

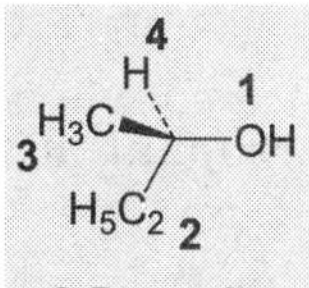

2-Butanol

The priorities of the substituents as determined by CIP rules are: –OH is **1**, CH_3CH_2– is **2,** CH_3– is **3** and H is **4**. Thus –OH has the highest priority and H has the lowest priority.

Step II: View the molecule in such a way that the lowest ranked group (priority **4**) points away from you.

In the above formulation, the molecule has already been appropriately oriented. H atom, having lowest priority (**4**), is away from viewer.

Step III: Draw the remaining three groups, as they appear to you, in the following manner when the molecule is oriented as specified in ***step II***.

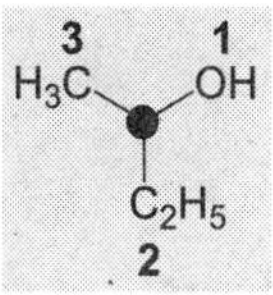

Step IV: Move your eye from the group of priority number **1** to the group of priority number **3** *via* the group of priority number **2** (1→2→3). If during this movement your eye travels in the clockwise direction, the molecule is designated as *R* (Latin: *rectus* meaning right), and if it moves in the anticlockwise direction, it is designated as *S* (Latin: *sinister* meaning left). The letters *R* and *S* are written in parenthesis.

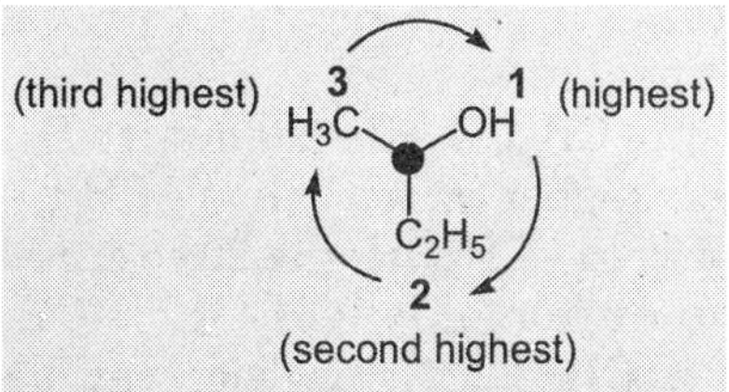

Clockwise; ***R*** configuration

In the above arrangement, our eye moves in clockwise direction, so the absolute configuration of (–)-2-butanol is ***R***. Here, it may be mentioned that the complete name of the enantiomer specifies both configuration and direction of rotation. Thus, the full name of the above molecule is (*R*)-(–)-2-butanol.

A comprehensive list of some common groups in increasing order of priority is given:

Non-bonding electrons on N (At.N. = 0), –H, –D, $–CH_3$, $–CH_2CH_3$, $–CH_2(CH_2)_nCH_3$, $–CH_2–CH{=}CH_2$, $–CH_2–C{\equiv}CH$, $–CH_2C_6H_5$, $–CH(CH_3)_2$, $–CH{=}CH_2$, $–C(CH_3)_3$, $–C{\equiv}CH$, $–C_6H_5$, $–CH_2OH$, –CHO, –COR, $–CONH_2$, –COOH, –COOR, $–NH_2$, $–NHCH_3$, $–N(CH_3)_2$, –NO, $–NO_2$, –OH, $–OCH_3$, $–OC_6H_5$, –OCOR, –F, –SH, –SR, –SOR, –Cl, –Br, –I.

Sequence Rules

For fixing priorities of various atoms/groups attached to a chirality centre, the following sequence rules have been evolved:

- Sequence Rule I: If four atoms/groups attached to the chirality centre are all different, the atom with the highest atomic number is given the highest priority. For example, in bromochlorofluoroiodomethane, the priorities are:

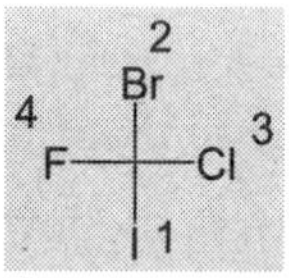

I = 1; Br = 2; Cl = 3; F = 4, *i.e.* the sequence of priorities is I > Br > Cl > F.
However, if two isotopes of the same element are attached to the chirality centre, the atom with higher mass number is given higher priority, for example, in α-deuterioethyl bromide.

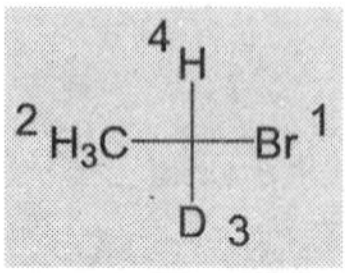

In this case, two isotopes of hydrogen, H and D are attached to the chiral carbon; D having higher mass number is given higher priority than H. Hence, the sequence of priority of the four groups will be: Br > CH_3 > D > H

- Sequence Rule 2: If on basis of the *sequence rule 1* the priorities of two groups cannot be decided, it can be determined by a similar comparison of the next atoms in both groups. If by doing so the priority cannot be decided, one goes to 'next' atom, and continues moving outwards commencing with the chiral atom till one reaches the first point of difference.

Some examples are as follows:

(*i*) **$–CH_2CH_3$ and $–CH_3$**: In $–CH_2CH_3$, the atoms are H, H, C, and in $–CH_3$, they are H, H, H. So, $CH_3CH_2–$ gets higher priority than $–CH_3$.

(*ii*) **$–CH_2OH$ and $–CH_2NH_2$**: In $–CH_2OH$, the atoms are H, H, O, and in $–CH_2NH_2$, they are H, H, N. So, the priority order is $–CH_2OH > –CH_2NH_2$.

(*iii*) **$–CH_2CHFBr$ and $–CH_2CHFCl$**: No decision can be taken as the first groups are CH_2 in both cases. Now going to the next carbon, the groups are H, F, Br in the first case, and in the second case, they are H, F, Cl. So, the priority order is $–CH_2CHFBr > –CH_2CHFCl$.

(*iv*) **$–CH_2CH_3$ and $–CH(CH_3)_2$:** In $–CH_2CH_3$, the groups are H, H, C, and in $–CH(CH_3)_2$ they are H, C, C. So, the priority order is $–CH(CH_3)_2 > –CH_2CH_3$.

(*v*) **$-CH(CH_3)_2$ and $-CH_2Cl$:** In $-CH(CH_3)_2$, atoms are H,C,C, and in $-CH_2Cl$, they are H, H, Cl. So, the priority order is $-CH_2Cl > -CH(CH_3)_2$.

It is worth mentioning that the decision about priorty should be made at the very first point of difference, and should not be effected from the consideration of substituents further along the chain.

- *Sequence Rule 3:* In case the group attached to the chiral carbon contains a double bond or a triple bond, both atoms joined by multiple bonds are considered to be duplicated (in case of a double bond) and triplicated (in case of a triple bond). For example,

1. Group containing a double bond say $>C=X$ is considered to be equal to 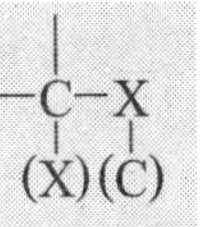
2. Group containing a triple bond, say $-C\equiv X$ is considered to be equal to 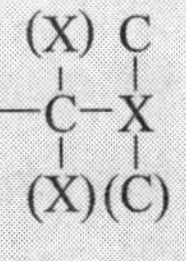
3. The phenyl substituent 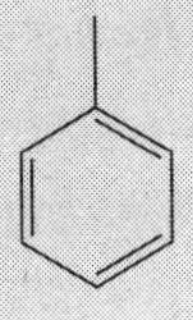 is considered equal to 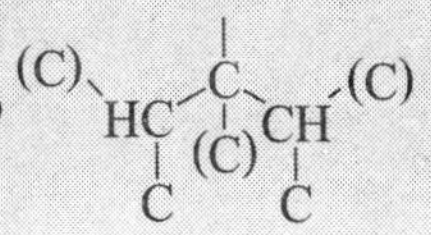

This rule is illustrated by the following examples:

(i) Glyceraldehyde

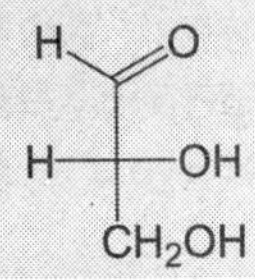

The problem here is to decide the priorities of $-CH_2OH$ and $-CHO$.

Now, $-CHO$ is regarded as equal to 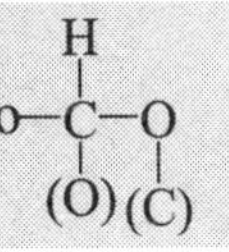, and $-CH_2OH$ is 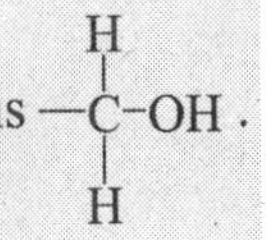.

Here, carbon atom of $-CHO$ having H, O, O takes priority over that of $-CH_2OH$ having H, H, O.

(ii) 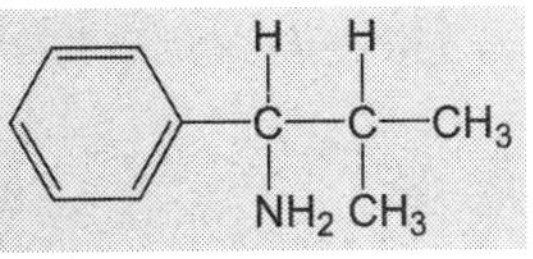

In this case, priorities of phenyl ring and *iso*propyl group are to be decided.

Now, phenyl substituent 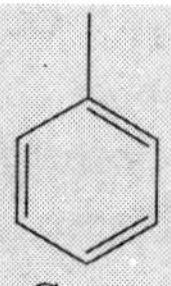 is considered equal to 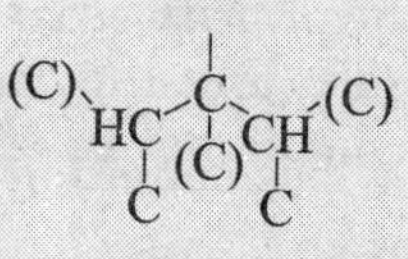

and *iso*propyl group is 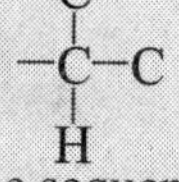 , so phenyl group having C, C, C takes priority over *iso*propyl group having C, C, H. Thus, the sequence of priority of the four groups is $-NH_2$ > –Ph > *iso*propyl > H.

Now, let us apply the above-mentioned Steps to assign the configuration of some compounds. The structures are shown as flying wedge and Fischer projections. While working with these structures, there are two possibilities: *(i)* group of lowest priority is away from us (observer) (shown by broken bonds in flying wedge projection, and by vertical bond in Fischer projections), and *(ii)* the group of lowest priority is not away from us. Therefore, the examples have been discussed in two groups:

(i) Group of lowest priority is away, and shown by broken bonds in flying wedge projection

(1)

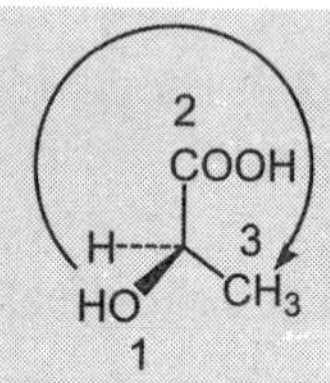

Lactic acid (clockwise, ***R***-configuration)

I. The priorities are: OH > COOH > CH_3 > H.
II. The group of *lowest* priority (H) is *away* from us. This is *correct* situation.
III. Movement of eye from 1 → 2 → 3 takes place in clockwise direction. So, the configuration is *R*.

Absolute configuration of the other enantiomer of lactic acid can be similarly assigned.

(2)

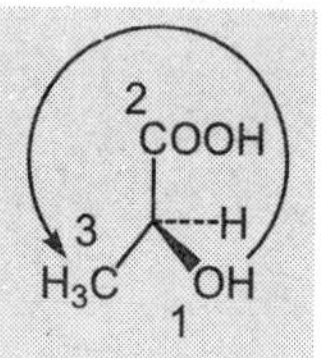

anticlockwise
S-configuration

(ii) Group of lowest priority is not away and is shown by solid thickened bonds in flying wedge projection

(3)

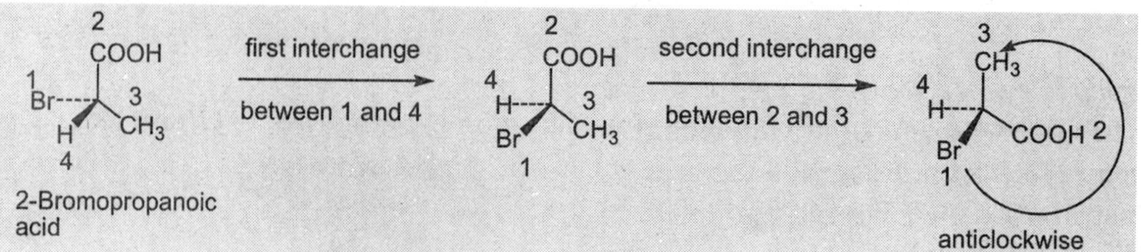

2-Bromopropanoic acid

anticlockwise
S-configuration

I. The priorities are: Br (1) > COOH (2) > CH_3 (3) > H (4).

II. The group of *lowest* priority (H) is not *away* from us. This is *incorrect* situation.

III. Make two interchanges-one between Br (1) and H (4), and another between COOH (2) and CH_3 (3). This operation gives original configuration with lowest priority 'H' in *correct* situation.

IV. Movement of eye from 1 → 2 → 3 takes place in anticlockwise direction. So, the configuration is *S*.

Absolute configuration of the other enantiomer of 2-bromopropanoic acid (*R*) can be similarly assigned.

(4)

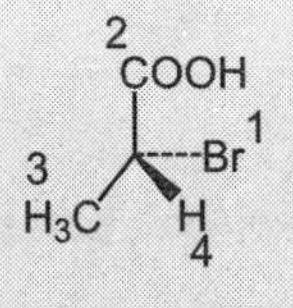

R-2-Bromopropanoic acid

Similar steps are followed for the structures drawn as Fischer projections.

After assigning priorities to four atoms attached to the chiral carbon, the assignment of configuration will depend upon the fact whether the group of lowest priority falls on the vertical line *(i)* or on horizontal line *(ii)*.

(i) Group of lowest priority is located on vertical line

It does not matter whether the group is locateds upward or downwards as in Fischer projection, both the groups on the vertical line are always away from the viewer.

The eye is moved in the 1 → 2→ 3 direction. If it is clockwise, the configuration is *R,* and if it is anticlockwise, the configuration is *S*.

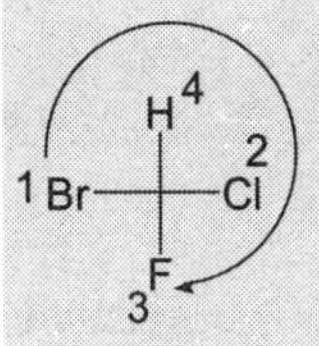

1 ⟶ 2 ⟶ 3 movement is clockwise; configuration is ***R***.

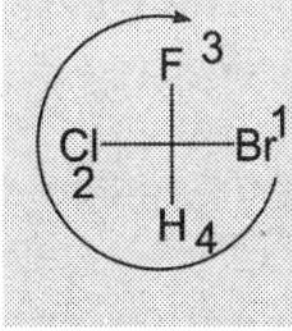

1 ⟶ 2 ⟶ 3 movement is clockwise; configuration is ***R***.

(ii) Group of lowest priority is located on the horizontal line

In such a case, make two interchanges of the atoms or groups in such a way that the atom or group of lowest priority falls on vertical line either upwards or downwards. This procedure is illustrated by the following examples:

(1) D-(+)-Glyceraldehyde

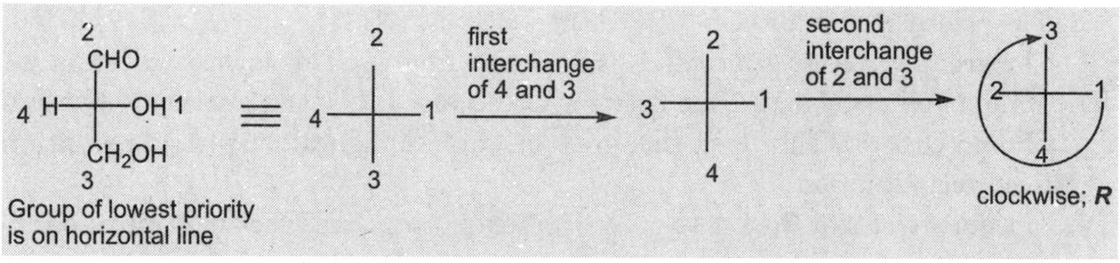

(2) D-(-)-Lactic acid

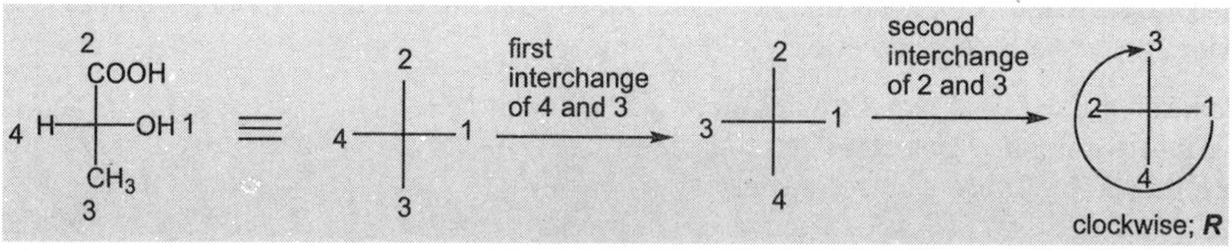

(3) 2-Butanol

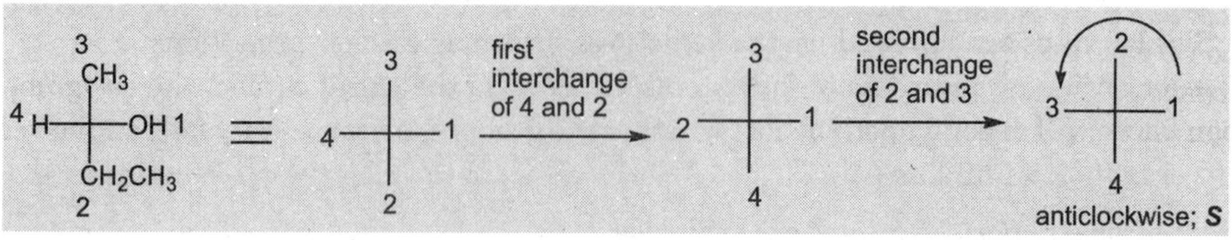

(4) L-Alanine

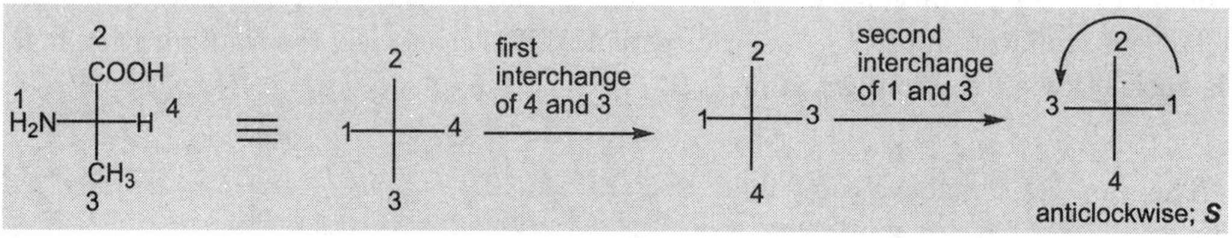

Alternate Procedure for Assigning Configuration: Very Good Mnemonic

Out of a number of alternative procedures for assigning *R*- or *S*- configuration on the basis of Fischer projection, the simplest and most widely accepted one was suggested by Eliel, who coined the term *very good* for the expression *vertical-good!* for his procedure which consists of two operations.

Fix up priorities of the groups, and move your eye from $1 \rightarrow 2 \rightarrow 3$ ignoring 4. Now proceed further as follows:

(*i*) If the group of lowest priority (4) is on the vertical line (whether on top or bottom), the sequence $1 \rightarrow 2 \rightarrow 3$ is in clockwise direction and the configuration is *R*, and if it is in anticlockwise direction, the configuration is *S*.

(*ii*) When the group of lowest priority (4) is on the horizontal line, assign the configuration which is opposite to what you see. If the movement of the eye from $1 \rightarrow 2 \rightarrow 3$ is in clockwise direction, assign *S*-configuration, and if it moves in anticlockwise direction assign *R*-configuration. The example of *R*/*S* lactic acid is given.

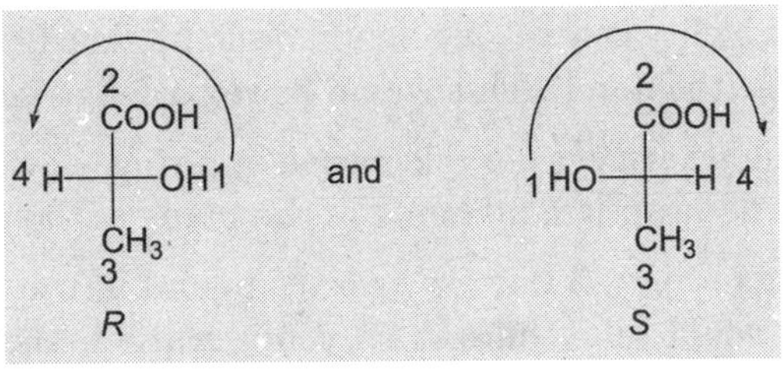

In the first situation, movement of eye from 1 → 2 → 3 is in anticlockwise direction, so it has *R*-configuration. In the second situation, movement of eye from 1 → 2 → 3 is in clockwise direction, so it has *S*-configuration.

R/S Nomenclature of Compounds Having More than One Chiral Carbon

When a compound possesses two different chiral carbons, each of these chirality centres may be either *R* or *S*. Four different stereoisomers are possible having the following configurations; (1*R*, 2*R*); (1*R*, 2*S*); (1*S*, 2*R*) and (1*S*, 2*S*). Each chiral carbon is considered separately for assigning (*R*) or (*S*) configuration. Let us consider the configuration around chiral carbon atoms number 2 and 3 in the 2,3,4-trihydroxybutanal, commonly referred to as D-erythrose.

For carbon 2, movement of eye from ***1 → 2 → 3*** is anticlockwise, so as per ***very good*** device carbon 2 has *R*-configuration as the lowest priority (4) group is on horizontal line.

For carbon 3, movement of eye from ***1→ 2 → 3*** is anticlockwise, so as per ***very good*** device carbon 3 also has *R*-configuration.

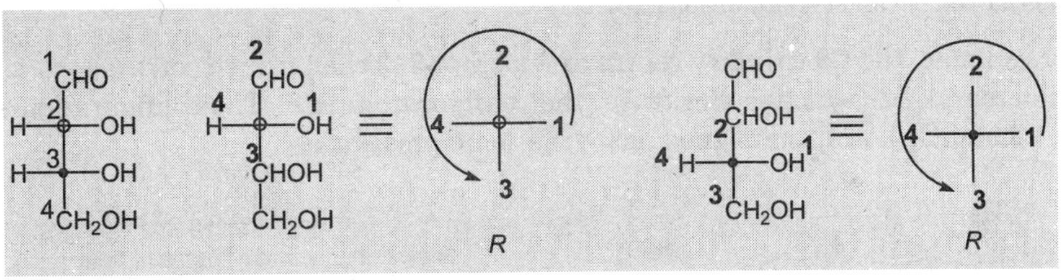

The name of the above compound will, therefore, be (2*R*,3*R*)-2,3-4-trihydroxybutanal or 2*R*,3*R*-erythrose. We can, similarly, assign configurations to the other compounds.

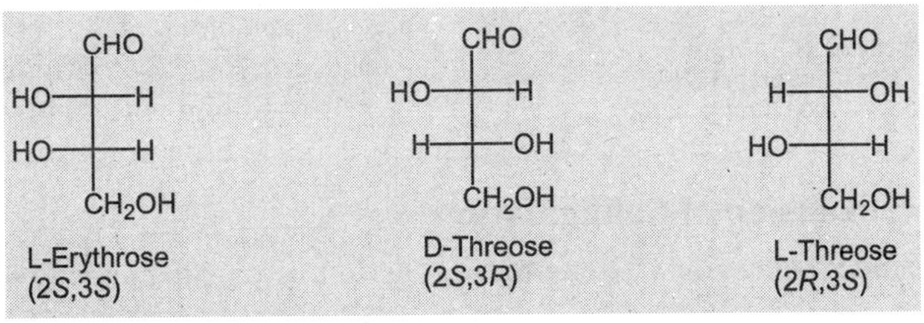

(iii) Additional Rule for Cyclic Molecules: The above-mentioned rules cannot be applied to the cyclic molecules such as steroids and terpenes, which are usually projected on the plane of the paper. Hydrogens (or substituents) located below and above the plane are assigned α and β descriptions, respectively (α represented by dotted and β by thick lines). An additional method was suggested by Eliel to determine the configuration (*R*/*S*) of such molecules. The steps in this method are as follows:

Step-I: At any particular chirality centre, one ligand must be clearly in the front (F, thick bond) or clearly in the back (B, dotted bond). This would be regarded as the reference ligand.

Step-II: The sequence of the priority order (clockwise or anticlockwise) of the remaining three ligands can be very easily determined, all being in the plane of the paper.

Step-III: If the reference ligand is 4/B (lowest priority ligand in the back), the sequence of the remaining three ligands would give the correct configuration. Using 4/B(+), [(+) stands for correct] as mnemonic, it is possible to determine the configuration with other combinations as shown in **Table 2.3**.

Step-IV: Finally, configuration is assigned using CIP rules. If it is a correct situation (+), and sequence of the remaining ligands arranged in increasing number of order of their priority is clockwise, the configuration of the chirality centre will be *R*. Accordingly, when this sequence is anticlockwise, the configuration will be *S*. In the incorrect situation (–), the resulting configuration will be reversed to get the correct answer.

Table 2.3 Additional rule for absolute configuration of cyclic molecules

Projection of ligand	Priority of ligand (sign)*
Back (B, -----)	4(+) 3(–) 2(+) 1(–)
Front (F, ▬▬)	4(–) 3(+) 2(–) 1(+)

*(+) stands for correct and (–) stands for incorrect.

For example, the C9 chirality centre of 9-methyl-3-decalol, which corresponds to 4/F (–), gives the order 1,2,3, which is clockwise, the configuration is *S*. The configuration at other centres, C3 and C10, is also assigned using the same procedure.

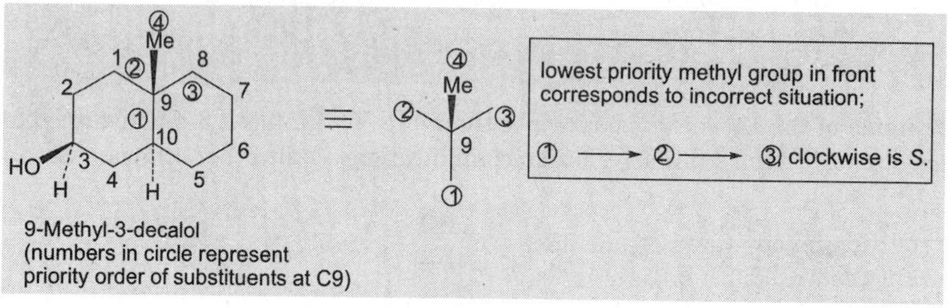

9-Methyl-3-decalol
(numbers in circle represent
priority order of substituents at C9)

F. Retention and Inversion of Configuration

Stereochemical consequences of the chemical reaction of chiral molecules focusses on the following.

- The configuration of the product may be the same as that of the reactant. In such a case, the reaction is said to proceed with *retention of configuration*.
- The configuration of the product may be opposite to that of the reactant. In such a case, the reaction is said to proceed with *inversion of configuration*.

Retention or inversion of configuration in the reaction of a chiral molecule depends upon the following two factors:

(*i*) The side of the molecule from which the reagent attacks the reactant.

(*ii*) Manner of bond cleavage in the reaction, *i.e.* whether the bond between the substituent and chirality centre is broken or not.

We shall now discuss the stereochemical consequences of some of the important reactions of chiral molecules, particularly in the light of the manner of bond cleavage.

1. If the bond between the chirality centre and the substituent remains intact throughout the course of the reaction, the reagent will attack at some other site, and the reaction will proceed with retention of configuration. For example, conversion of optically active 2-butanol to 2-butyl tosylate takes place with retention of configuration as the C–O bond remains intact.

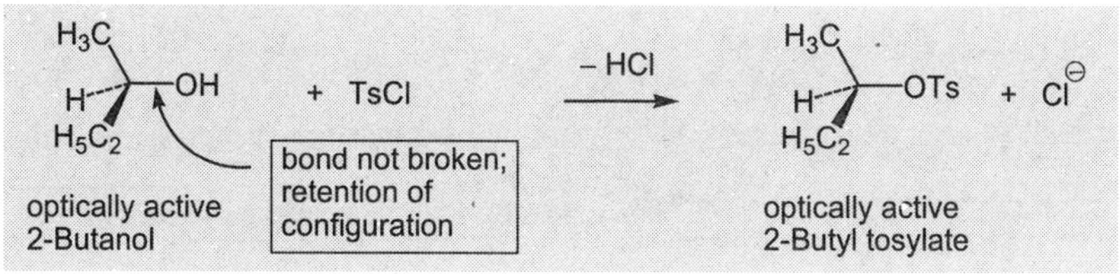

2. If the bond between the chirality centre and the substituent is cleaved, the consequence may be either *inversion* or *retention* of configuration. For example,

I. Inversion of Configuration

All S_N2 reactions proceed with inversion of configuration at the chirality centre. For instance, conversion of 2-butyl chloride into 2-butanol by treating with aq. NaOH is accompanied by inversion of configuration.

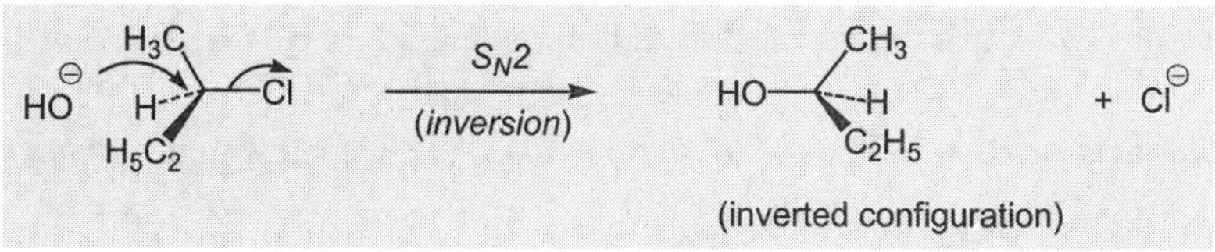

The reaction involves cleavage of C–Cl bond with simultaneous backside attack of $^-$OH on the chirality centre. Since the process of this inversion was discovered by Walden, it is also called as Walden inversion.

II. Retention of Configuration

Retention of configuration may be the result of the following two mechanisms:

(*i*) The S_N2 reaction takes place twice on the same chirality centre. This will result in the inversion followed by another inversion, which means no inversion, *i.e.* retention of configuration. For example, treatment of 2-bromopropionate with dilute aq. NaOH produces lactate ion with retention of configuration. The steps involved in the process are as follows:

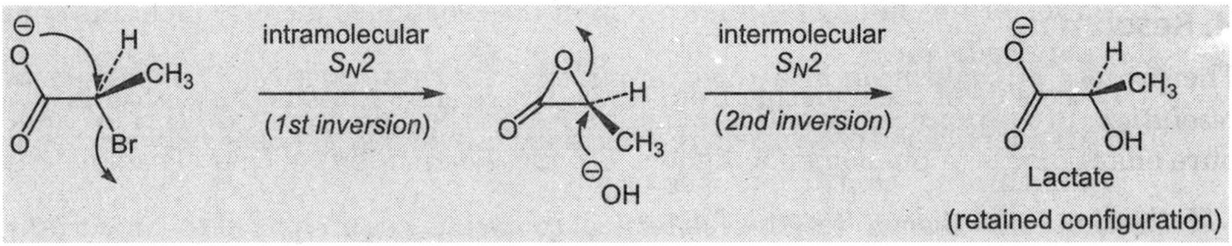

(*ii*) Bond breaking and bond making may occur at the front side of the molecule. For example, reaction of a chiral alcohol with thionyl chloride to give a chiral alkyl halide with retention of configuration. The reaction involves a chlorosulphite as intermediate, which decomposes to supply Cl^- to produce alkyl chloride.

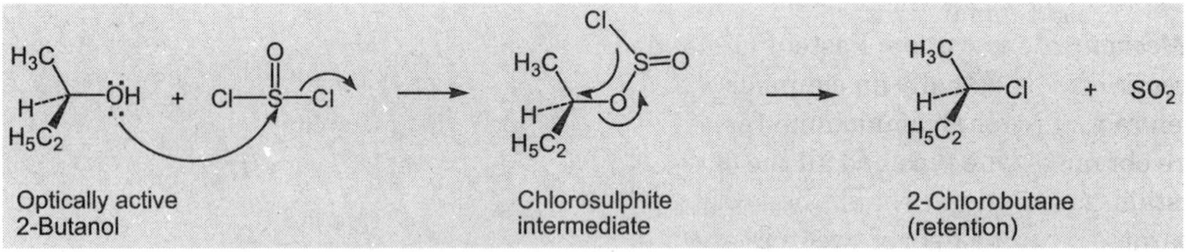

G. Racemic Modification or Racemic Mixture

An equimolar mixture of two enantiomers does not possess optical activity, and is called racemic mixture or racemic modification or conglomerate. The loss of optical activity is due to the fact that the rotation due to molecules of one enantiomer is cancelled by an equal and opposite rotation due to the same number of molecules of the other enantiomer.

Prefixes such as (*dl*) or (±) or (*RS*) are used before the name of the compound to specify that it is racemic. For example, if equal amounts of (+) and (–) tartaric acids are mixed together, it gives rise to a racemic mixture designated as (±) tartaric acid, or *dl*-tartaric acid, or *RS*-tartaric acid.

[50% (+)-Tartaric acid] + [50% (–)-Tartaric acid] = (±)-Tartaric acid (a *racemic mixture*).

$[\alpha]^{293}{}_D = +12°$ $\quad$ $[\alpha]^{293}{}_D = -12°$ $\quad$ $[\alpha]^{293}{}_D = 0°$

The optical rotation as well as other physical properties of the racemic mixture such as melting point, boiling point, and solubility in a given solvent is also different from those of enantiomers.

Racemization

Conversion of an enantiomer into a racemic mixture is known as racemization. In this process, the optical activity becomes zero. It has been found that attempts to synthesize chiral molecules using achiral reactants and achiral reagents result in the formation of a racemic mixture. In addition, many reactions of chiral molecules also produce racemic mixture. Thus, racemization is an undesirable process, and if the product molecule is chiral, it becomes necessary to avoid the use of reaction conditions and reagents which bring about racemization.

H. Resolution

The process of separation of a racemic mixture into its constituent enantiomers is called *resolution*. In the process, the optical activity is said to be enhanced, and becomes equal to that of pure enantiomers (a phenomenon known as *optical activation*) in case of complete resolution.

I. Methods of Resolution of Racemic Mixture

Enantiomers do not differ in their physical properties such as boiling points, solubility, and adsorptivity. Therefore, usual methods of separation such as fractional distillation, fractional crystallization, or adsorption techniques cannot be used for the separation of enantiomers. Therefore, some special procedures have been evolved for the resolution of racemic mixtures. Some of the more important methods are discussed.

(i) Mechanical Separation: Pasteur (1948) used this method to prove that the compound called *'racemic acid'* is actually an equimolecular mixture of (+) and (–)-tartaric acids. He found that when racemic sodium ammonium tartarate was crystallized below 300K, two types of crystals were obtained. One type had all the (+)-molecules, and other type had all (–)-molecules. These crystals had distinguishable hemihedral faces and were non-superimposable. They were separated with the help of tweezers and a magnifying glass.

This method has the following limitations:

(*i*) This method is painstaking and time consuming.

(*ii*) It is of limited use being applicable only to those compounds which can crystallize to give two well-defined types of crystals.

(ii) Preferential Crystallization: Preferential crystallization is closely related to mechanical separation of crystals. In this technique, a supersaturated solution of the racemic mixture is inoculated with a crystal of one of the enantiomers or an isomorphous crystal of another chiral compound. For example, when the saturated solution of (±)-sodium ammonium tartarate is seeded with the crystal of one enantiomer or a crystal of (–)-asparagine, (–)-sodium ammonium tartarate crystallizes out first.

This method is called as *entrainment,* and the seed crystal is called *entrainer*.

(iii) Biochemical Method: Microorganisms or enzymes are known to be highly stereoselective in their reactions. Pasteur first found that when a solution of ammonium salt of (±)-tartaric acid is fermented in presence of yeast or a mold, *e.g. Pencillium glaucum*, the (+)-tartaric acid is completely consumed leaving behind (–)-tartaric acid. This method has also found wide applicability in the resolution of (±)-amino acids. Thus, when acetyl derivative of a (±)-amino acid is treated with *hog-kidney acylase*, only acetyl derivative of L-amino acid is hydrolyzed, leaving behind acetyl derivative of D-amino acid.

However, this method suffers from the following limitations:

(*i*) As these reactions are to be carried out in dilute solutions, isolation of the products becomes difficult.

(ii) There is loss of one enantiomer, which is consumed by the microorganisms. Hence, only half of the compound is isolated. In other words, this procedure is *partially destructive*.

(iv) Resolution Through the Formation of Diastereomers- The Chemical Method: This method was also discovered by Pasteur. It is based upon the principle that treatment of a racemic mixture (±)-**A** with an optically pure reagent (+) or (–)-**B** results in the formation of a mixture of two products, which are diastereomers. The reagent (+) or (–)-**B** is called the resolving agent.

(±)-**A** + (+)-**B** → (+)**A**(+)**B** + (–)**A**(+)**B**

We know that the physical properties such as solubility, boiling point, and adsorption coefficient of the diastereomers are different because they are not mirror images of each other. Thus, the mixture of diastereomers obtained can be separated by using the methods of fractional distillation or fractional crystallization. The pure diastereomers are then decomposed into the corresponding enantiomers and the original optically active reagent, which are then separated.

In practice, the racemic mixture is treated with an optically active pure resolving agent. The product so obtained is separated into diastereomeric derivatives by fractional crystallization if it is a solid, and by fractional distillation if it is a liquid.

To illustrate, let us take the case of (±)-tartaric acid. When treated with (–)-cinchonidine, (±)-tartaric acid gives a mixture of two diastereomeric salts: (+)-tartarate of (–) chinchonidine and (–)-tartarate of (–)-cinchonidine, which are diastereomers and have different solubility. The latter separates out first when the mixture is subjected to fractional crystallization. After the separation and purification of the two diastereomeric salts, each is treated with dilute sulphuric acid, which decomposes the salt into the (+)-or (–)-tartaric acid as the case may be, and the base cinchonidine remains in solution as its soluble sulphate (**Fig. 2.11**).

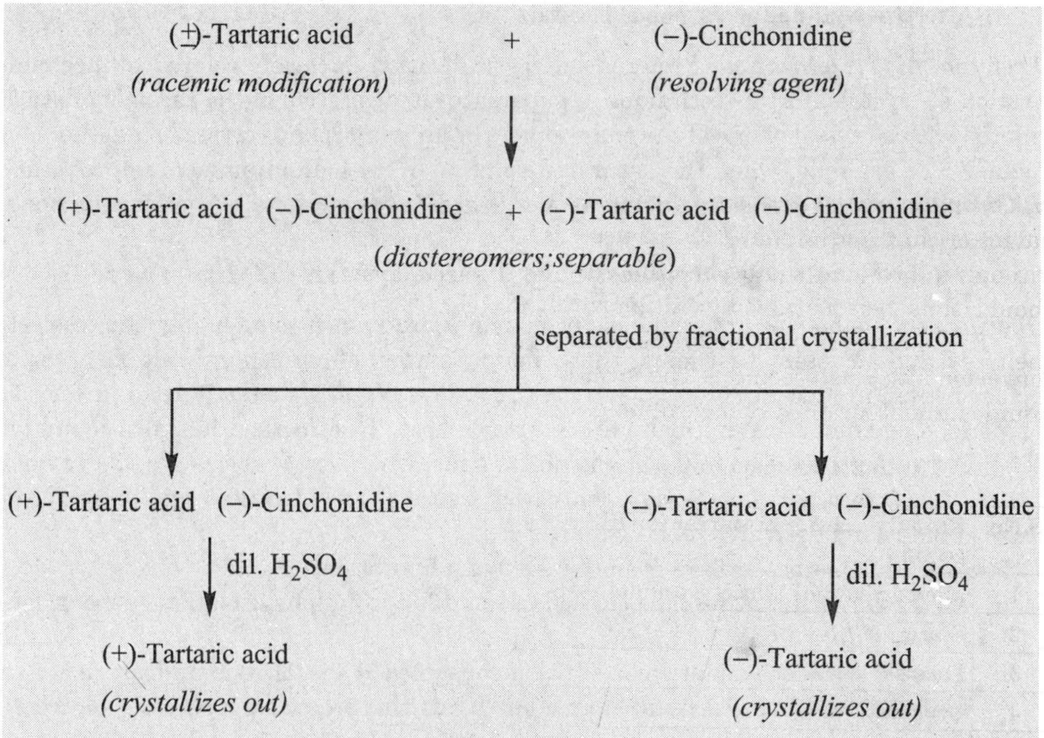

Fig. 2.11: Resolution of racemic modification of tartaric acid through the formation of diastereomers

The same procedure may be adopted for the resolution of a (±)-base with an optically active acid.

The choice of the reagent will depend upon the reactive group in the racemic compound. For example, a racemic alcohol can be resolved as outlined in **Fig. 2.12**.

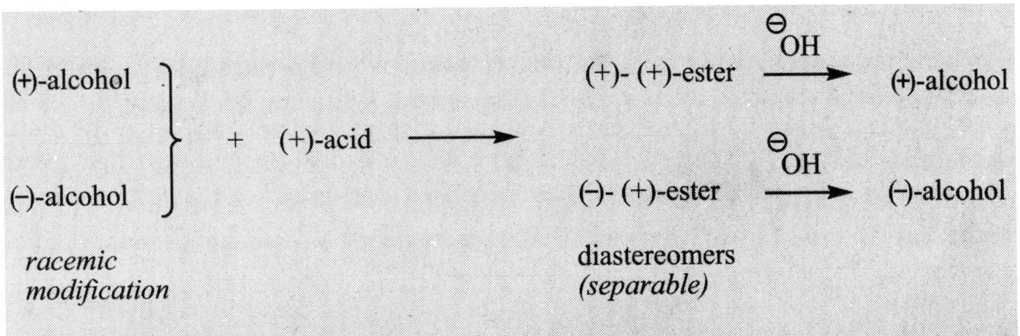

Fig. 2.12: Resolution of racemic modification of an alcohol

The chemical method of resolution is widely used and has the advantage that both the enantiomers are obtained. However, success of the reaction depends upon the following factors.

(*i*) The resolving agent should be optically pure.

(*ii*) The substrate (racemic mixture) and the resolving agent should have suitable functional groups for reaction with each other.

(*iii*) The resolving agent should be cheap, and should be capable of regeneration and recycling.

(*iv*) The resolving agent should be such which produces easily crystallizable diastereomeric products.

(*v*) The resolving agent should be easily separable from pure enantiomers.

(v) Chromatographic Method: When a racemic mixture is placed on a chromatographic column having a chiral adsorbent then in principle, the rates of movement of the two enantiomers through the column should be different (due to difference in the extent of adsorption). They should, thus, be separable by elution with suitable solvent.

This method has an advantage over chemical separation as the enantiomers need not be converted into diastereomers. The techniques used include paper, column, thin layer, gas, and liquid chromatography. Some examples are given in **Table 2.4**.

Table 2.4 Some examples of chromatographic separation of racemic mixtures

S.No.	Chiral packing material for filling the column	Compounds whose racemic mixtures are separated
1.	Microcrystalline triacetyl cellulose (MCTC)	Alcohols, olefins, ketones, acids, esters, biphenyls
2.	Starch	Mandelic acid
3.	Lactose	Chloromycetin, Troger's base
4.	Brucine	2-Octyl hydrogen phthalate

Optical Purity and Enantiomeric Excess

We can easily determine the amount of enantiomers present in a mixture. It has been shown that for an enantiomerically pure sample (only one enantiomer), the value of specific rotation [α] is the highest. Any contamination by the other enantiomer lowers the value of specific rotation proportionately.

For example, the observed specific rotation of enantiomerically pure (*S*)-(+)-2-bromobutane has been found to be +23.1°. Now, if a given sample of the compound has [α] = +23.1°, it can be concluded that the sample is 100% (*S*)-(+)-enantiomer. If the specific rotation of the sample is zero, we will say that it is a racemic mixture. But, if the value of [α] lies between 0 and +23.1°, it means that the sample is a mixture of unequal amounts of the two enantiomers. The positive sign of the observed specific rotation means that the mixture has excess of (*S*)-enantiomer over (*R*)-enantiomer. This excess is known as *enantiomeric excess (ee).*

The amount of each enantiomer present in the mixture can be calculated in two steps from the observed specific rotation.

Step I: The *optical purity* of the sample is determined using the following formula:

$$\text{Optical purity (OP)} = \frac{\text{observed specific rotation, } [\alpha]_{obs}}{\text{specific rotation of the pure enantiomer, } [\alpha]_{max}}$$

Step-II: Now suppose a sample of 2-bromobutane has observed specific rotation of +9.2°. We know that for the pure sample [α]max is + 23.1°.

$$\text{Optical purity} = \frac{+9.2^{o}}{+23.1^{o}} = 0.4 \text{ or } 40\%$$

The positive sign of observed specific rotation indicates that the mixture has excess of *S*-enantiomer. The amount is calculated from optical purity as follows:

As calculated above, the optical purity of the sample is 40%. It means that 40% of the mixture is excess of *S*-isomer, and 100 – 40 = 60% is racemic mixture. Thus, total amount of enantiomer-*S* in the mixture will be 40 + 60/2 = 40 + 30 = 70%, and, therefore the amount of enantiomer-*R* in the mixture = 100 – 70 = 30%.

2.4 GEOMETRICAL ISOMERISM

A. Definition and General Aspects

Isomers having the same structural formula, but different relative spatial arrangement of atoms or groups around a double bond are called *geometrical isomers, cis-trans isomers* or *E-Z isomers,* and this phenomenon is known as *geometrical isomerism, cis-trans or E-Z isomerism*. This type of isomerism arises if we assume that there is no free rotation about the double bond.

As there is different arrangement of atoms or groups in the space, geometrical isomerism is designated as stereoisomerism. These geometrical isomers belong to the category of configurational isomers because they cannot be interconverted without breaking bond. Further, geometric isomers are examples of diastereomers because they are not mirror images of each other.

Due to the rigidity of double bond towards free rotation, these isomers have two spatial arrangements. In one arrangement, called *cis* (Latin: *cis* meaning on same side), two similar or identical atoms or groups are on the same side of the double bond. In the other arrangement called *trans* (Latin: *trans* meaning across), two similar or identical atoms or groups are on the opposite side of the double bond. For example, the two geometrical isomers of butenedioic acid are:

cis-Butenedioic acid (*Maleic acid*) *trans*-Butenedioic acid (*Fumaric acid*)

In general, geometrical isomerism is exhibited by the compounds of the type A having non-identical groups attached to doubly-bonded carbon atoms. It is not exhibited by compounds of the type B because in these types of molecules, at least one of the doubly-bonded carbon atoms bears two identical groups or atoms.

A B

Geometrical isomerism is not confined only to the compounds having carbon–carbon double bonds. In fact, the following compounds also exhibit this type of isomerism:

(*i*) Compounds having a double bond, *i.e.* imines (C=N) and azo compounds (N=N)
(*ii*) Cyclic compounds
(*iii*) Compounds exhibiting geometrical isomerism due to restricted rotation about carbon-carbon single bond

B. *E* and *Z* Nomenclature of Geometrical Isomers

The *cis* and *trans* designations have been used for many years to indicate the spatial relationship of atoms or groups of atoms around a double bond. For compounds of the type abC=Cab or abC=Cac, the prefixes *cis* and *trans* are quite unambiguous. However, when there are three or four different atoms or groups attached to the carbon atoms of a double bond, *cis-trans* nomenclature leads to ambiguity, and sometimes creates a lot of confusion. Let us consider a molecule in which the two carbon atoms of a double bond are attached with four different halogens.

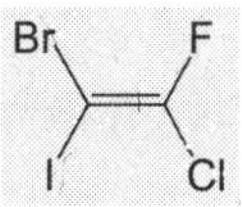

When we say that Br and CI are *trans* to each other, we can also say with equal degree of confidence that I and CI are *cis* to each other. It is, thus, difficult to name such a substance as the *cis* or the *trans* isomer.

To eliminate this confusion, a more general and easy system for designating configuration about a double bond has been adopted. This method, which is called the *E* and *Z system*, is based on *a priority system* originally developed by Cahn, Ingold, and Prelog for use with optically active substances. In olefins, we assign a first (1) and a second (2) priority to the atoms or groups attached to each carbon of the double bond. Priority is then compared at one carbon relative to the other. If the higher-priority atoms or groups are on opposite side of the double bond, the configuration is designated as *E* (German: *entgegen* meaning across). If the higher-priority atoms or groups are on the same side of the double bond, the configuration is designated as *Z* (German: *zusammen* meaning together).

Rules for fixing the priorities of atoms or groups attached to doubly bonded carbon atoms are the same as discussed earlier. Some examples are as follows:

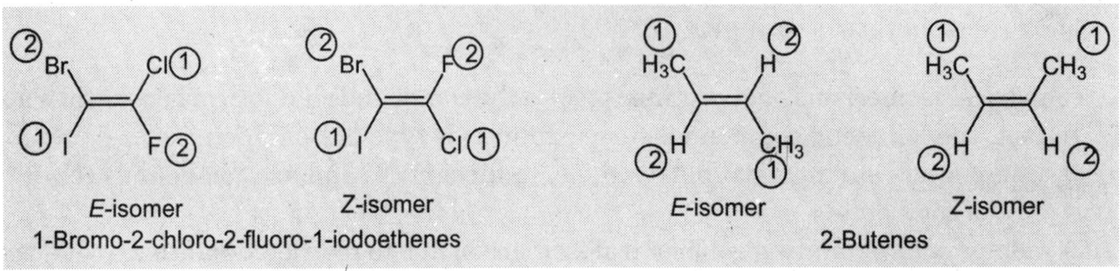

If there are two or more double bonds, prefixes *E* and Z are used for designating configuration around each double bond. For example,

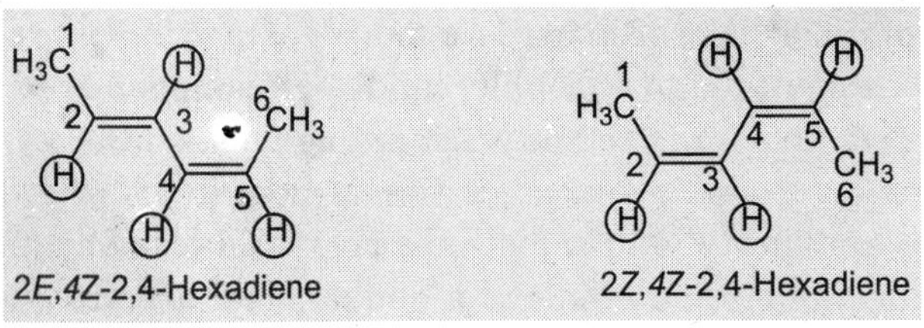

Number of geometrical isomers of compounds containing two or more double bonds with non-equivalent terminii:

Dienes in which the two terminii are different (XHC=CH–CH=CHY) exist as four geometrical isomers.

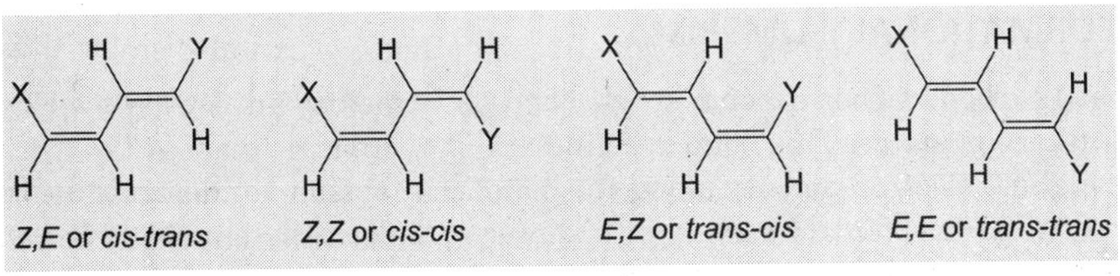

Thus, it can be concluded that for a given molecule the number of geometrical isomers is 2^n, where n is the number of double bonds. It may be noted here that this formula applies only to those molecules in which the two terminii are different. However, in identical terminii (XHC=CH-CH=CHX) the number of geometrical isomers is given by the following expressions:

(*i*) **If *n* is even**, the number of geometric isomers = $2^{(n-1)} + 2^{(n/2-1)}$

(*ii*) **If *n* is odd**, the number of geometric isomers = $2^{(n-1)} + 2^{(n-1/2)}$

Geometrical (cis-trans) Isomerism in Alicyclic Compounds

In addition to the compounds containing a double bond polysubstituted alicyclic compounds such as disubstituted derivatives of cyclopropane, cyclopentane, cyclobutane, and cyclohexane can also show *cis-trans* isomerism. In such cases, *cis-trans* isomerism arises due to restricted rotation about a single bond.

To study geometric isomerism in cyclic compounds, it is necessary to understand the relationship between the substituents in terms of *cis* and *trans* orientation. Although rings and substituents can be represented by using planar bonds, flying wedge projections are often used for a better understanding.

Some representative examples are:

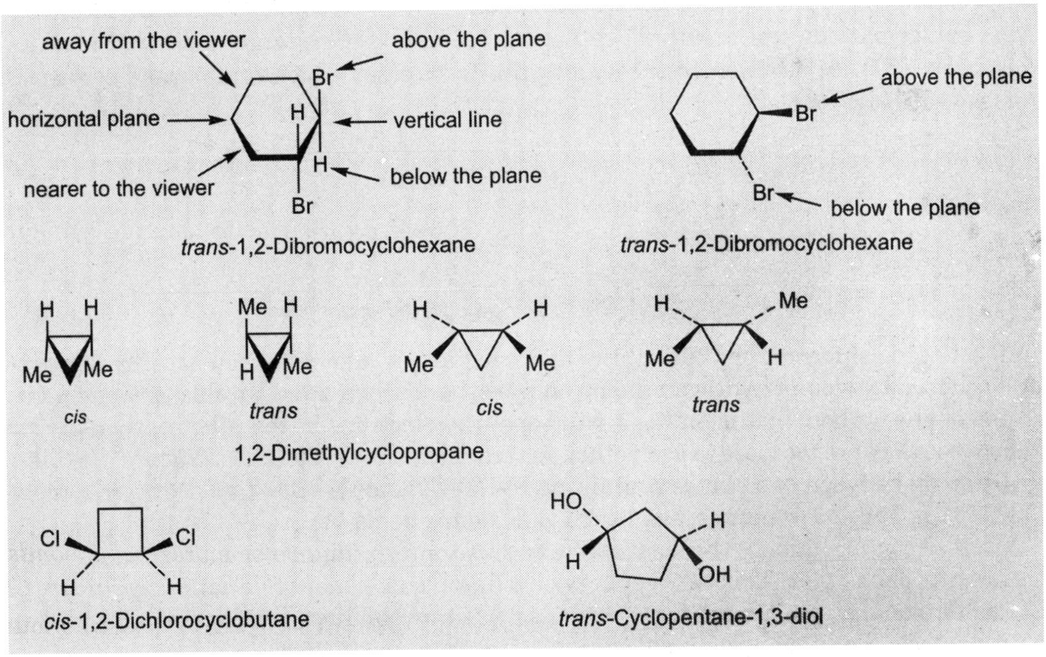

2.5 CONFORMATIONAL ISOMERISM

We know that a carbon–carbon σ-bond (single bond) is formed by an 'end-on' overlapping of sp^3-orbitals of two carbon atoms. This bond is cylindrically symmetrical about the axis **(Fig. 2.13)**, and has the highest electron density along the bond axis (this is an imaginary line connecting the centres of two atoms joined by a σ-bond). Due to cylindrical symmetry, free rotation about a carbon–carbon single bond is possible without affecting the amount of orbital overlap. Almost an infinite number of spatial arrangements of atoms about the carbon-carbon single bond exist. All those arrangements, which result from free rotation about a single bond are called *conformations* or *conformers* or *rotational isomers* or *simply rotamers.*

Unlike configurational isomers, the isolation of conformers in non-rigid systems is not feasible at ordinary temperature. This is due to very low potential energy barrier 2.87-3.34 kcal (10-12 kJ)/mol between two conformers.

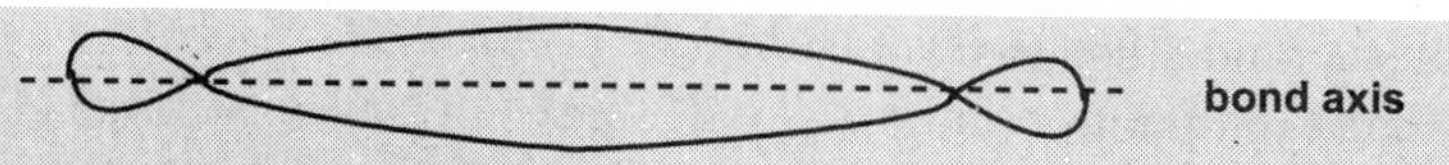

Fig. 2.13: A cylindrically symmetric MO bond obtained by sp^3-sp^3 overlap of two carbon atoms

A. Conformational Isomers of Ethane

In acyclic hydrocarbons such as ethane, it was originally assumed that there is complete freedom of rotation about the carbon-carbon single bond. But, Pitzer postulated that there exists a potential energy barrier, which causes restriction in free rotation.

Let us consider the rotation of one of the methyl groups about the central C—C bond in ethane keeping the other methyl group stationary. As a consequence of such rotations, two extreme conformations can result. They are an *eclipsed conformation* **(26)** and a *staggered conformation* **(27)**. In addition to these, an infinite number of conformations between these extremes are also possible.

26 eclipsed conformation

27 staggered conformation

In eclipsed conformation (**26**), all the hydrogen atoms on one carbon are directly opposed to, and eclipsed by vicinal hydrogen atoms on the other carbon atom. In this arrangement, the electrons of one carbon–hydrogen bond will repel the electrons of the other carbon–hydrogen bond because they are too close to each other (internuclear H—H distance 229 pm); the dihedral angle (θ, angle between two planes containing X–C1–C2 and C1–C2–Y as shown in a molecule X–C1–C2–Y θ, **Fig. 2.14**) between two such C–H bonds being 0°.

Eclipsed conformation is the least stable because only in this conformation the C—H bonds are nearest to each other. This extra energy of eclipsed conformation is called *torsional strain*. The term torsional strain is used for the repulsion felt by bonding electrons on one substituent when it passes close to the bonding electrons of another substituent.

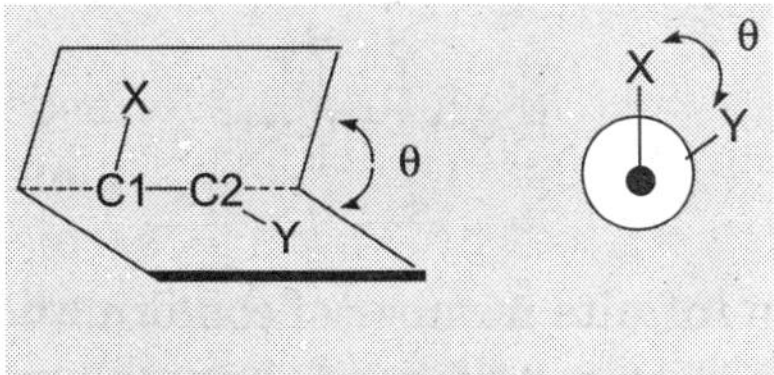

Fig. 2.14: Dihedral angle (θ)

In contrast to the eclipsed conformation, the staggered conformation is the most stable because the distance between C—H bonds on neighbouring carbons is maximum (internuclear H—H distance is 310 pm); dihedral angle between two C—H bonds being 60°. Experimentally also, it has been found to be true. The staggered conformation of ethane is more stable by about 2.87 kcal(12 kJ)/mol than eclipsed conformation. In fact, this energy difference 2.87 kcal (12 kJ)/mol between eclipsed and staggered conformations is the energy barrier to free rotation about the C—C single bond in ethane.

The variation of energy in conformations of ethane with rotation about C—C single bond is shown in **Fig. 2.15**. A look at the figure reveals that at the initial stage, *i.e.* the eclipsed conformation with dihedral angle 0°, the molecule has maximum energy. A rotation of front carbon by 60° generates a staggered conformation having minimum energy. Another rotation by 60° (rotation by 120° from initial conformation) generates another eclipsed form, and so on. Thus, every rotation of 60° alternatively generates eclipsed and staggered conformations.

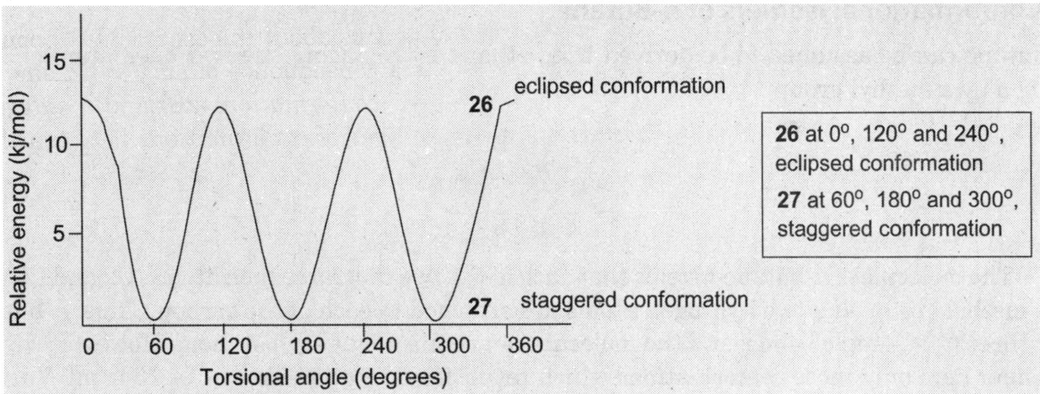

Fig. 2.15: Rotational or torsional energy in ethane

It may be noted here that all the staggered conformations are of equal energy. Similarly, all the eclipsed conformations are of equal energy.

The investigation of the various conformations of a compound, and their relative stabilities is called *conformational analysis*.

B. Conformational Isomers of Propane

For its conformational analysis, propane can be considered as a derivative of ethane in which one hydrogen atom of ethane is replaced by a methyl group as shown below:

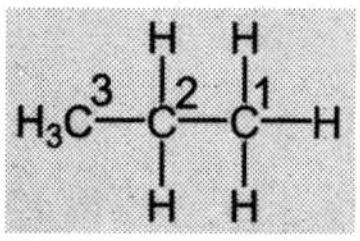

In case of propane also, an infinite number of conformations are possible due to rotation of one methyl group around the C—C single bond through a complete rotation of 360°. Out of these only two extreme conformations, *i.e.* eclipsed and staggered are important which are shown as follows:

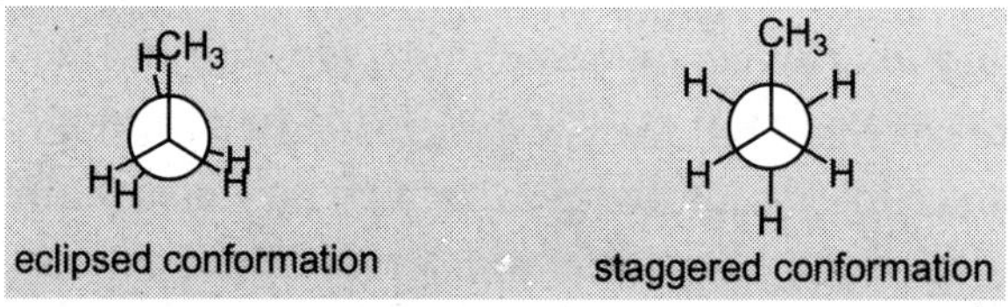

Similar to ethane, the staggered conformation of propane is more stable than the eclipsed conformation. The energy difference 3.39 kcal (14.2 kJ)/mol between these two conformations is slightly more than that in ethane 2.87 kcal (12 kJ)/mol. Even then, this energy barrier to free rotation about C—C single bond is so small that it cannot prevent rotation completely. Like ethane, at room temperature the two conformations are not readily isolable, and are inter-converted by rotation through an angle of 60°.

C. Conformational Isomers of *n*-Butane

n-Butane can be assumed to be derived from ethane by replacing one hydrogen atom of each carbon by a methyl group.

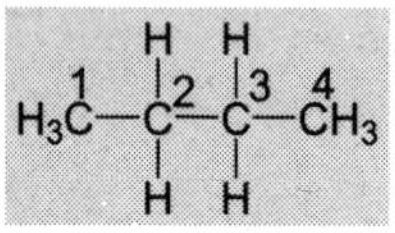

The molecule of *n*-butane reveals that each of the two central carbon atoms, C2 and C3, has one methyl group, and two hydrogen atoms are attached to each of the carbon. Thus, *n*-butane has three C—C single bonds, and the molecule can rotate about each of them. However, we will consider here only those conformations which result from rotation about C2–C3 bond. Various conformations derived due to rotation about this bond are shown in **Fig. 2.16**.

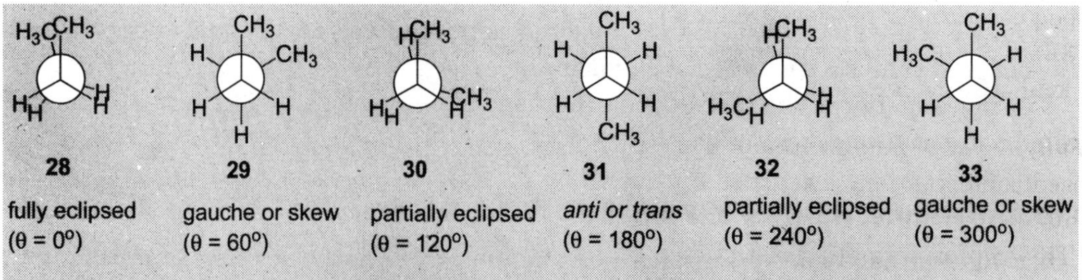

Fig. 2.16: Conformations of *n*-butane

Thus, unlike ethane in which the staggered and eclipsed conformations are equivalent, two different staggered and two different eclipsed conformations occur in *n*-butane. These are discussed below:

I. Staggered Conformations

Staggered conformations can be classified into two different ways.

(a) *Anti* Conformation

(b) *Gauche* Conformation

(i) Anti (or trans) Conformation (Greek: anti meaning opposite): In this conformation, the two methyl groups are farthest apart (**31**, θ = 180°), and is the most stable **(Fig. 2.17)**.

(ii) Gauche (or skew) conformation (French: Gauche meaning left): Both these conformations (**29** and **33**) are staggered, so they are free from torsional strain. However, the hydrogens of methyl groups in *gauche* conformation lie within 210 pm of each other. This distance is less than the sum of their van der Waals radii of 240 pm. Due to this congestion in space, a repulsive force acts between the methyl groups. This type of destabilization is called *van der Waals strain* or *steric hindrance*. In case of butane, due to van der Waals strains the *gauche* conformation is less stable than *anti*-conformation by ~ 1 kcal (4 kJ)/mol. This is also called as *n-butane gauche (or skew) interaction*. Further, if these two forms could be isolated, they would show different physical properties. However, their isolation is not possible because the energy barrier of ~1 kcal (4 kJ)/mol is too small to permit isolation of the two forms at normal temperature. It may, however, be possible to separate them at very low temperature (–230°C).

Finally, it may be mentioned that conformations **29** and **33** are non-superimposable mirror images of each other. Such pairs of conformations are called as *conformational enantiomers*. However, due to low energy barrier these enantiomers cannot be resolved.

II. Eclipsed Conformations

Eclipsed conformations can be classified in two different ways:

(a) Partially Eclipsed Conformation

(b) Fully Eclipsed Conformation

(i) Partially Eclipsed Conformation: The first type of eclipsed conformation, also called partially eclipsed conformation, is reached during rotation from *anti* to *gauche* conformation. In the eclisped conformation, there are two CH_3:H and one H:H eclipsed interactions (structures **30** and **32**). This eclipsed conformation in less stable than *anti* conformation by approximately 3.82 kcal (16 kJ)/ mol.

(ii) Fully Eclipsed Conformation: The second type of eclipsed conformation, also called fully eclipsed conformation, is reached during rotation from one *gauche* conformation to the other. In this conformation, there is one CH_3:CH_3 and two H:H eclipsed interactions (structure **28**). This conformation is least stable, and has higher energy than *anti*-conformation (**31**) by approximately 4.7 kcal (20 kJ)/mol. Thus, the molecule in this conformation possesses highest potential energy.

The potential energy relationship of various conformations of *n*-butane is illustrated in **Fig. 2.17**. The staggered conformations representing energy minima are more stable than eclipsed conformations representing energy maxima.

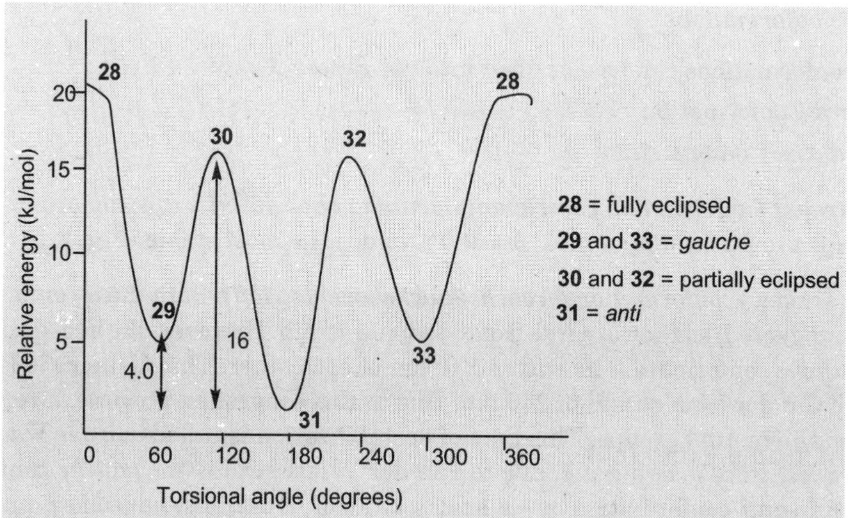

Fig. 2.17: Rotational or torsional energy in *n*-butane

At room temperature, almost all molecules exist in staggered conformation, and amongst staggered conformations 78% exist in *anti* and 22% in *gauche* conformations. The point of maximum potential energy (fully eclipsed) lies about 4.7 kcal (20 kJ)/mol above the *anti*-conformation. The total strain of this form is divided between torsional strains and *van der Waals* strain between two methyl groups.

Conformers are also described in terms of dihedral angle between specified groups. For instance, dihedral angles for the skew and staggered (*anti*) forms of *n*-butane are 60° and 180°, respectively. When the exact dihedral angle is not known, the following nomenclature in terms of approximate dihedral angles has been proposed. The terms *syn* and *anti* are used to indicate, respectively, a dihedral angle smaller or greater than 90°. The terms 'periplanar' and 'clinal', respectively, are used for approximately planar (0° ± 30° and 180° ± 30°) and inclined positions (all other angles). These designations are summarized in **Fig. 2.18.**

Dihedral angle	**Designation**	**Symbol**
–30° to +30°	± *syn*-periplanar	±sp
+30° to +90°	+ *syn*-clinal	+sc
+90° to +150°	+ *anti*-clinal	+ac
+150° to –150°	± *anti*-periplanar	±ap
–30° to –90°	– *syn*-clinal	–sc
–90° to –150°	– *anti*-clinal	–ac

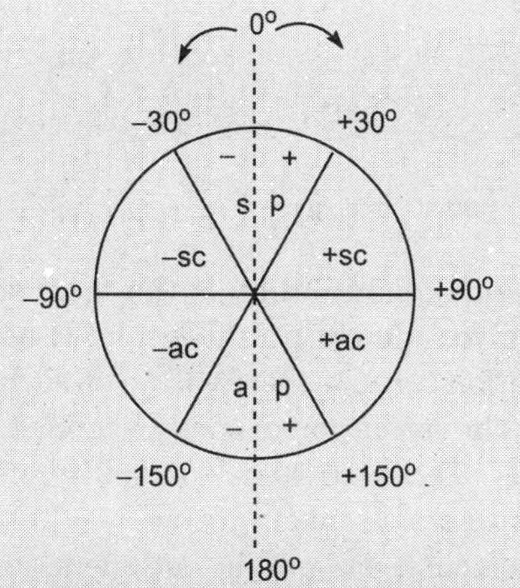

Fig. 2.18: Designations of conformations based on dihedral angle

D. Conformational Isomers of 1,2-Dibromoethane

Like *n*-butane, 1,2-dibromoethane can exist in four different conformations.

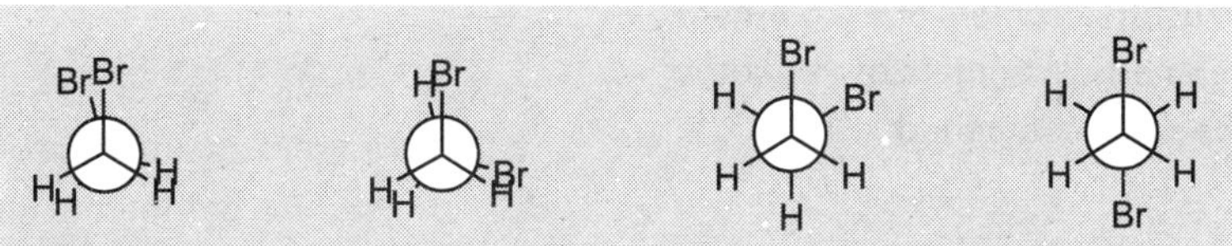

On the basis of *torsional strain* and *van der Waals steric hindrance*, staggered (*anti*) conformation of 1,2-dibromoethane is the most stable, which is followed by *gauche*.

A look at these conformations shows that the dipole moment of *anti*-conformation is zero, because the two C–Br dipoles are equal but opposite. However, the *gauche* conformation has some finite dipole moment, since the two C–Br dipoles are at an angle of 60° to each other.

Experimentally, it has been observed that the dipole moment of 1,2-dibromoethane is 1.0 D, therefore, the molecule cannot exist entirely in the *anti* form. Further, the molecule also cannot exist entirely in the *gauche* form as the dipole moment varies with the temperature. Therefore, to justify these observations, it is assumed that there is equilibrium between the staggered and the gauche conformations, and with an increase in temperature the equilibrium shifts in favour of the gauche conformation, and hence the dipole moment increases with temperature.

E. Conformational Isomers of 1,2-Glycols: *Ethylene Glycol*

Similar to *n*-butane and 1,2-dibromoethane, ethylene glycol can exist in four different conformations.

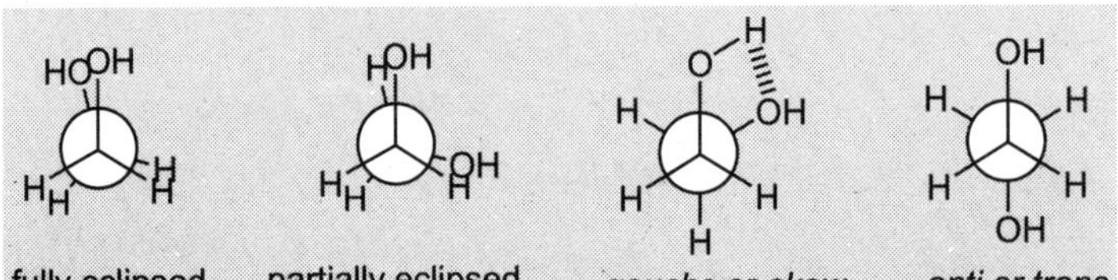

fully eclipsed partially eclipsed *gauche* or *skew* *anti* or *trans*

We know that normally *anti*-conformation is the most stable conformation followed by *gauche*. But in case of ethylene glycol, due to intramolecular H-bonding, the *gauche* form becomes more stable than *anti*-conformation because there will be no such H-bonding possible in *anti* conformation. The formation of such H-bond stabilizes the molecule by approximately 4.7-7.1kcal (20-30 kJ)/ mol. Therefore, *gauche form of ethylene glycol is the most stable conformation.*

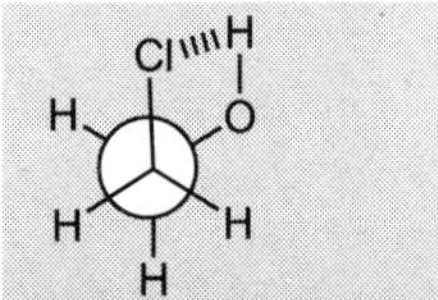

gauche conformation (*more stable*)

Similarly due to intramolecular H-bonding, ethylene chlorohydrin ($CH_2Cl–CH_2OH$) exists in *gauche* conformation, which is more stable than its *anti*-conformation.

PROBLEMS AND REVIEW QUESTIONS

1. Label each of the following pair of structures as:
   (*i*) structural isomers of each other
   (*ii*) geometrical isomers of each other
   (*iii*) the same compound:

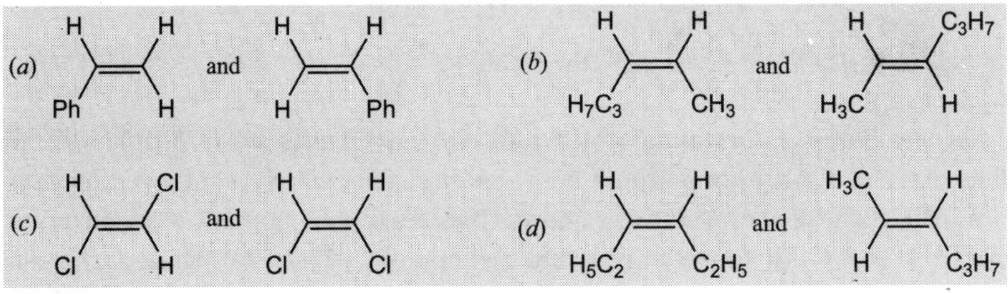

2. (*i*) A carboxylic acid having the molecular formula $C_3H_5O_2Br$ is optically active. Write its structure.
   (*ii*) An ester of tartaric acid is optically active. However, it yields an optically inactive acid on hydrolysis. Give its structure.
   (*iii*) What is the structure of an optically active alcohol having the molecular formula $C_4H_{10}O$?
   (iv) Write structures for all the possible olefins of molecular formula C_5H_{10} that exhibit geometrical isomerism.
   (v) An acid having molecular formula $C_5H_{10}O_2$ is optically active. What is its structure?
3. What is the lowest molecular weight alkane which is chiral? Is there another alkane of the same molecular formula that is also chiral? If so, then give its structure also.
4. Listed below are several chiral and achiral molecules.
   Select which of them are chiral and which are achiral modecules.

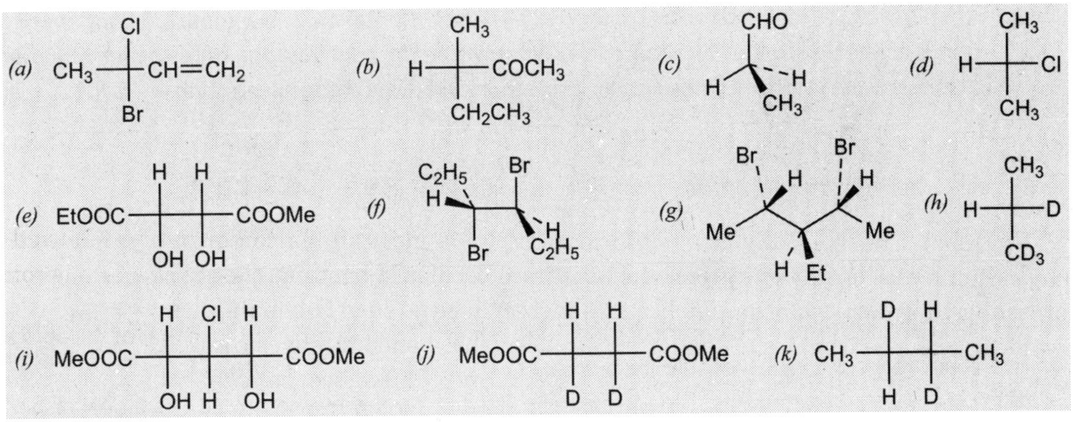

5. Consider the following synthesis of citric acid from oxaloacetic acid and acetyl CoA. Answer the following.

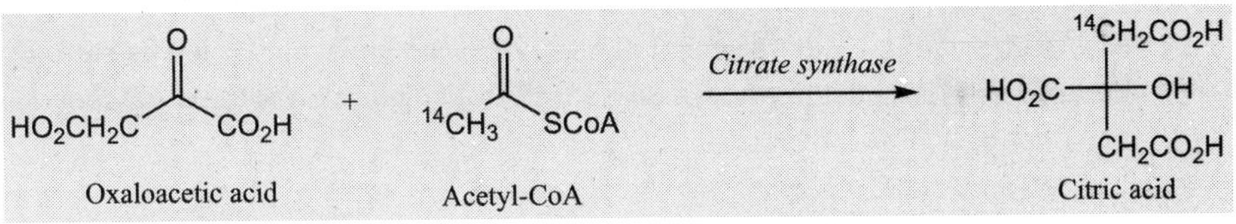

(*i*) What is the configuration of citric acid?
(*ii*) Why is the other stereoisomer not obtained?
(*iii*) If the acetyl-CoA used in the synthesis does not contain ^{14}C, will the product of the reaction be chiral or achiral?

6. Assign *R* or *S* designation to the chirality centres of each of the following molecules:

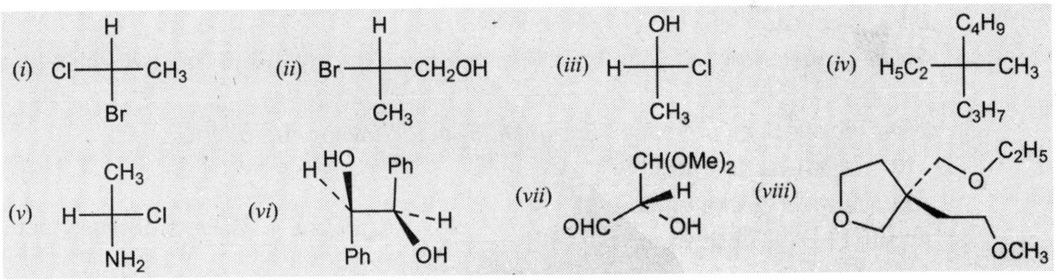

7. Draw Fischer projections for the two enantiomers of each of the non-cyclic isomers of (*i*) $C_5H_{11}Cl$, and (*ii*) C_6H_{12} which possess a chiral carbon. Designate the configuration of each enantiomer by *R* and *S* convention.

8. Draw flying wedge and Fisher projections of all the possible stereoisomers of 2-aminopropanoic acid and 3-iodopentan-2-ol. Comment upon the stereochemical relationships between their respective stereoisomers.

9. Draw suitable stereostructures for the following molecules:
(*i*) *S*-2-(2-Cyclopentenyl)acetic acid,
(*ii*) *R*-2-Phenylpropanoic acid
(*iii*) Most preferred conformation of ethane-1,2-diol.

10. *R*-(+)-2-Methyl-1-butanol can be oxidized to (–)-2-methylbutanoic acid without affecting any of the bonds connected to the chirality centre. What will be the configuration of (+)-2-methylbutanoic acid? Show its structure using Fischer projection and flying wedge projection.

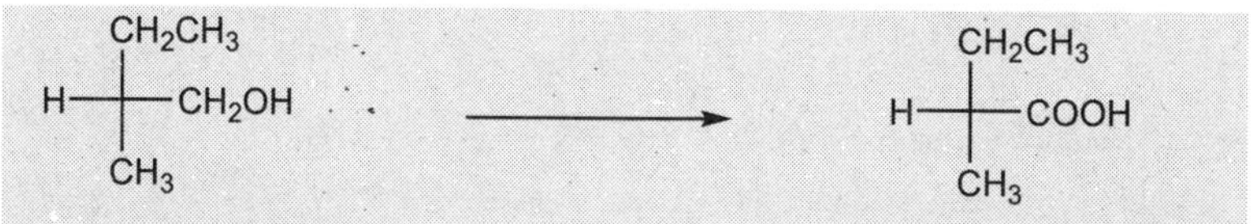

R-(+)-2-Methyl-1-butanol (–)-2-Methylbutanoic acid

11. Explain why the dipole moment of *meso* form is much lower (μ = 1.27 D) than that of the optically active form (μ = 2.75D) of stilbene dichloride.
12. How many asymmetric carbon atoms are created during the complete reduction of benzil (PhCOCOPh) with $LiAlH_4$? Write the number of possible stereoisomers of the product.
13. Butane on monobromination gives a mixture of 1-bromobutane and 2-bromobutane. Draw suitable structures of their stereoisomers, and comment on their chirality.
14. How is *R*/*S* configuration at chiral centres of cyclic molecules, such as steroids and terpenes, determined? Using this method, assign absolute configuration at the chiral centers of the following compounds:

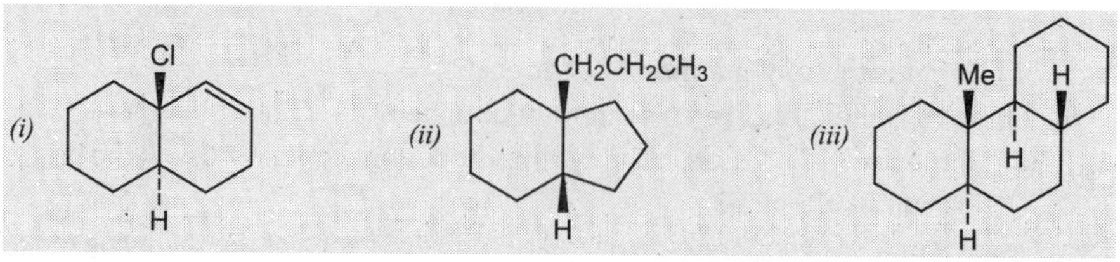

15. Discuss optical isomerism in 3,4-dichlorohexane.
16. Giving proper structures of all possible stereoisomers, discuss chirality in Ph-CH(Cl)-CH=CH-(Cl) CH-Ph.
17. Draw the Newman projection formulae for the following compounds:
    (*i*) (2*R*,3*S*)-Tartaric acid
    (*ii*) (2*S*,3*R*)-3-Bromobutan-2-ol
18. Account for the observations:
    (*i*) (*R*)-2-Butyl acetate is hydrolyzed in basic conditions to give (*R*)-2-butanol.
    (*ii*) (*R*) (or *S*)-2-Butanol on reaction with hydrochloric acid gives a mixture of (*R*)- and (*S*)-2-butyl chloride.
    (*iii*) (*R*) (or *S*)-2-Butanol on reaction with thionyl chloride gives optically active 2-chlorobutane with retention of configuration.
19. Define and explain enantiomeric excess (*ee*). The observed specific rotation of pure *R* enantiomer of compound **A** is + 24°. If the given sample of the same compound shows specific rotation +20°, calculate its *ee*.

20. Indicate whether each of the following compounds is *E* or *Z*:

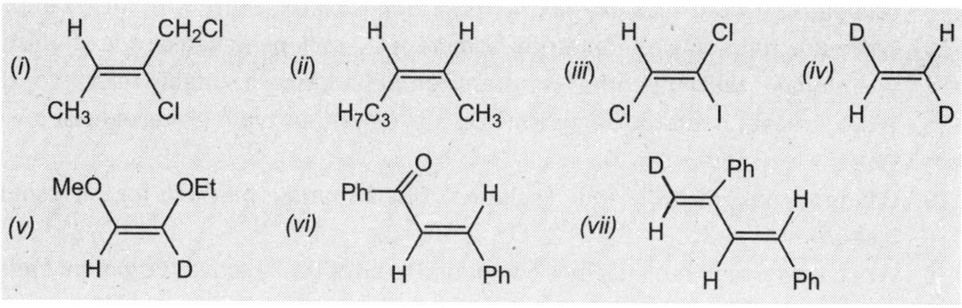

21. Which of the following compounds can exhibit geometrical isomerism?
    (i) 2,3-Dimethyl-2-butene
    (ii) 2-Hexene
    (iii) 2,3-Dimethyl-1-butene
    (iv) 4-Methyl-2-pentene
22. Draw Sawhorse, Newman, Fischer, and Flying Wedge projections of 3-bromo-2-butanol.
23. Write Fischer projection formulae for all the optical isomers of 3-bromo-2,5-hexanediol (**A**).

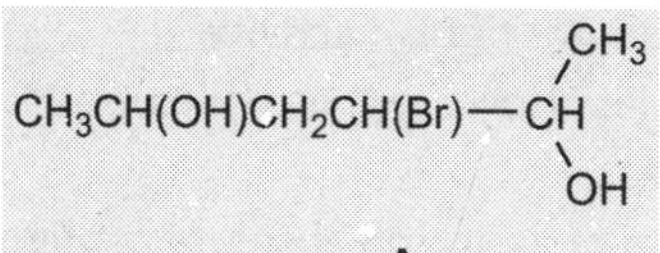

A

24. Examine each of the following pairs and find out whether they are identical. If they are different, indicate the relationship that they have to each other (enantiomeric or diastereomeric).

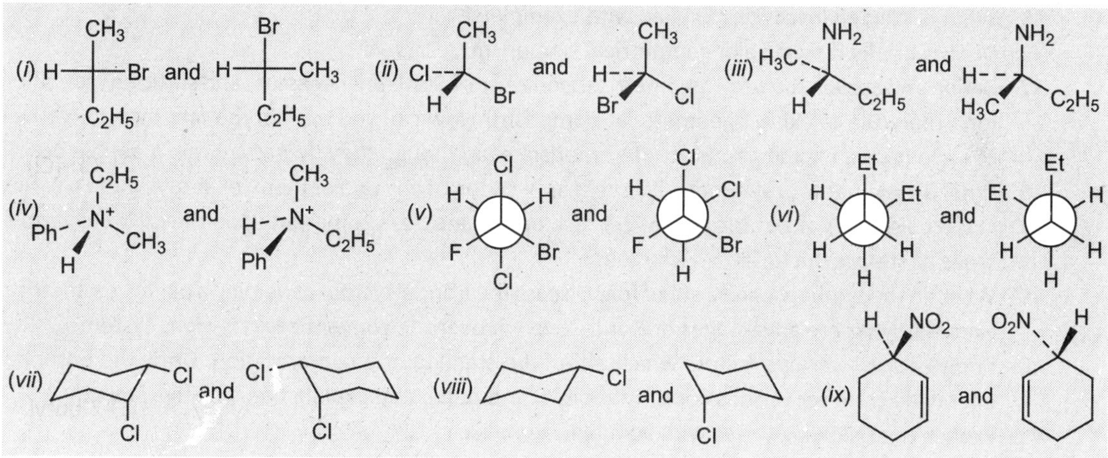

25. Draw Newman projections of the following Fischer projections. Also indicate which of these is/are chiral or achiral.

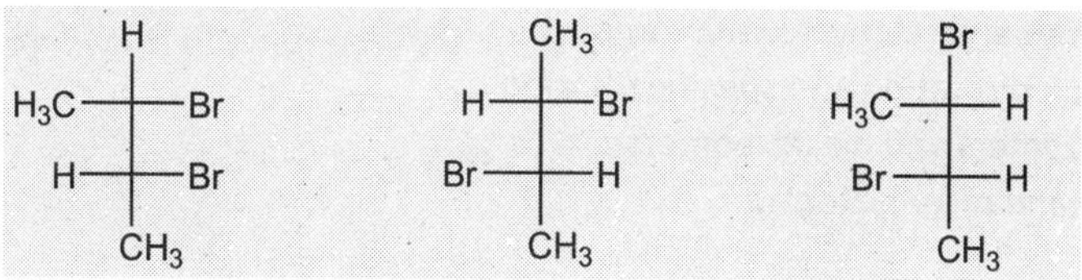

26. Draw all configurational isomers of trihydroxyglutaric acid [HOOC(CHOH)$_3$COOH]. Indicate which of them are chiral? Also explain the term *pseudo chiral centre* with this example.
27. Write structures of two geometrical isomers of 1,2-cyclopentanediol, one of which exists as a pair of enantiomers, while the other is a *meso* compound. Give a suitable reason for their existence.
28. What is meant by stereoisomerism? Discuss its various types giving at least one example in each case.
29. Define 'racemic mixture' and 'resolution'. Discuss various methods for the resolution of a racemic mixture.
30. What are the requirements for a compound to be optically active? Comment upon the observation that the presence of a chiral carbon is not always essential for a compound to exhibit optical activity.
31. (*i*) What is meant by the terms: (a) Chirality, (b) Diastereomers, (c) Configurational isomers, (d) Conformational isomers, (e) *R* and *S* convention, (f) Plane of plane-polarized light (g) Specific rotation?
    (*ii*) Differentiate between asymmetry, dissymmetry, and chirality. Explain why chirality is the essential condition for optical activity.
32. What are sequence rules? How are these rules used for assigning *R*- and *S*- configurations? Discuss with suitable examples.
33. Explain D and L system of assigning absolute configuration to a pair of stereoisomers. What are its limitations, and how are these overcome in *R* and *S* system?
34. (*i*) What is racemization? Discuss various mechanisms by which racemization can occur.
    (*ii*) Discuss the characteristics of enantiomers towards chiral and achiral reagents.
35. What is Walden inversion? Explain with example(s).
36. Discuss briefly the cause of geometrical isomerism.
37. Define and explain the term 'geometrical isomerism'? State the necessary and sufficient condition for a molecule to exhibit geometrical isomerism? How can you explain the fact that derivatives of cyclohexane exhibit geometrical isomerism even though they do not contain a double bond?
38. What are elements of symmetry? Discuss application of any two of them for studying the chirality in stereoisomers of tartaric acid or 2,3-dichlorobutane. Also comment upon the utility of various types of stereostructures in such cases.
39. What do the symbols *E* and *Z* stand for? Discuss the *E* and *Z* system of naming a pair of geometrical isomers. What are advantages of *E* and *Z* system over the conventional *cis-trans* system?
40. Briefly discuss various factors which affect the stability of a conformation. Draw the potential energy diagram for various conformations of *n*-butane, and explain their relative stability.
41. Explain the following terms with suitable examples:
    (*i*) Angle strain (*ii*) Torsional strain (*iii*) Steric strain
42. Distinguish between the following terms:
    (*i*) Configuration and conformation
    (*ii*) Absolute and relative configuration
    (*iii*) Observed rotation and specific rotation
    (*iv*) Enantiomers and diastereomers
    (*v*) *Meso* and racemic forms

SOLUTIONS

1. (a) same (b) geometrical isomers (c) geometrical isomers (d) structural isomers

2. 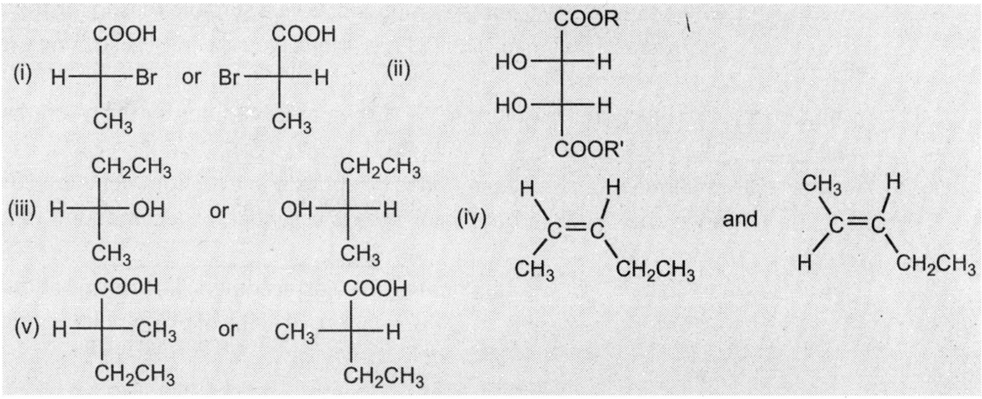

3. Lowest molecular weight is 100. Two possible structures are:

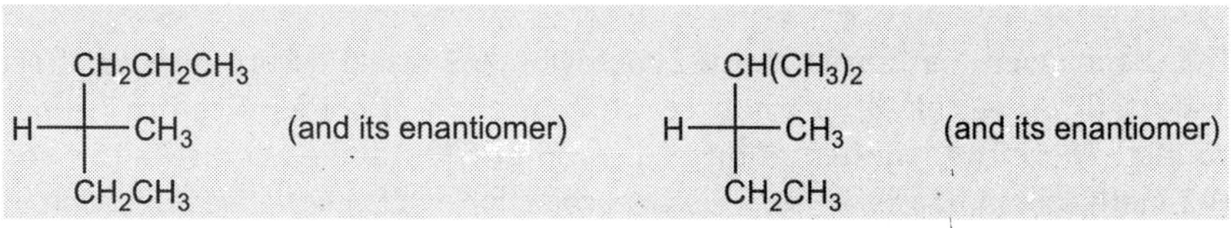

4. *(a)* chiral *(b)* chiral *(c)* achiral *(d)* achiral
   (e) chiral *(f)* achiral *(g)* achiral *(h)* chiral
   (i) achiral *(j)* achiral *(k)* chiral

5. (i) (*R*) (ii) reactions with enzymes are stereoselective (for discussion, see Chapter 3) (iii) achiral.

6. (i) *R* (ii) *R* (iii) *R* (iv) *S*
   (v) *S* (vi) *S,S* (vii) *R* (viii) *R*

7. 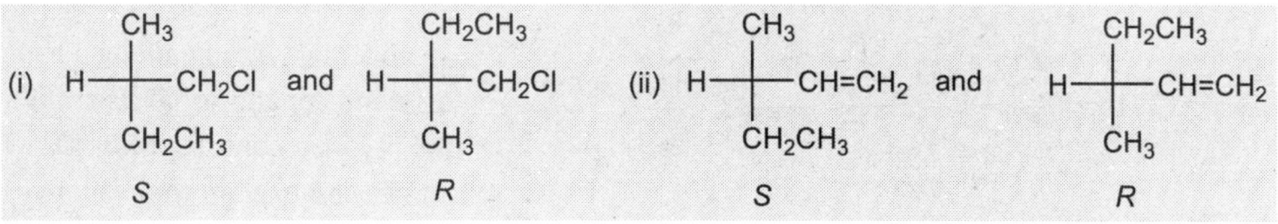

8. Only Fischer projections are shown:

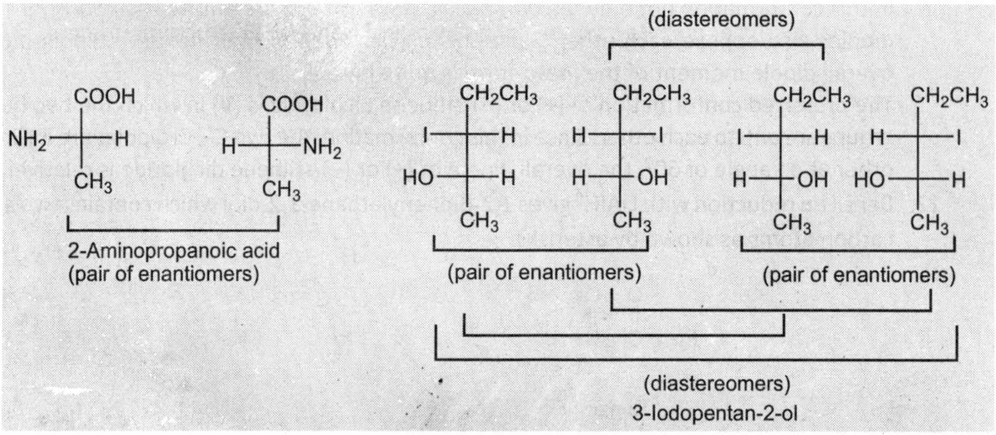

9. (i) 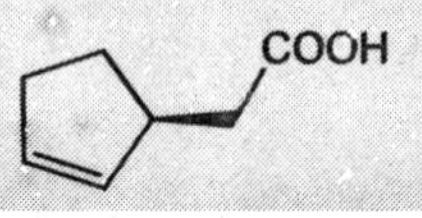 (ii) 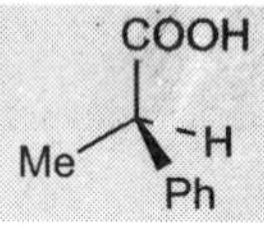 (iii) Skew conformation is the most preferred. (for structure see text)

10. It has been given that formation of 2-methylbutanoic acid occurs without affecting any bond connected to the chirality carbon *i.e.* configuration of chirality centre is unaffected in the product. Therefore, configuration of (–)-2-methylbutanoic acid is *R*. The configuration of (+)-2-methylbutanoic acid will be *S*. Fischer projection and flying wedge projection of *S*-enantiomer are **A** and **B**, respectively.

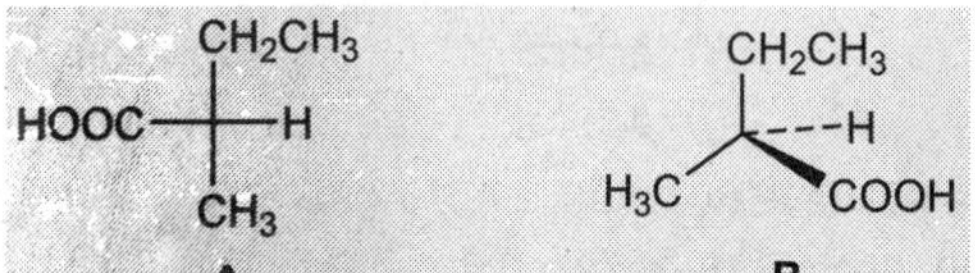

A **B**

11. The three **important conformations** of *meso*-form of stilbene dichloride (μ = 1.27D) are:

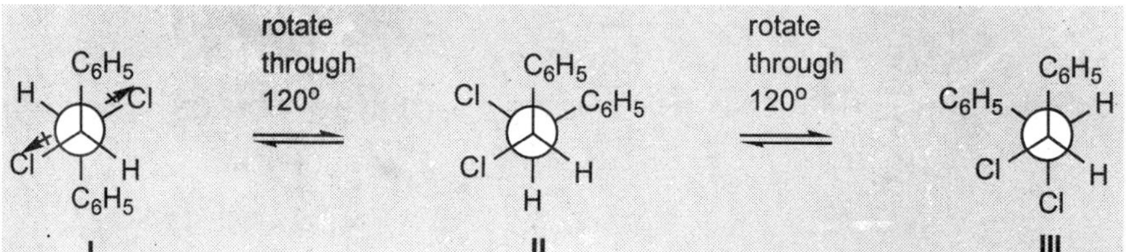

The conformations of optically active (+ or –) form of stilbene dichloride (μ = 2.75D) are:

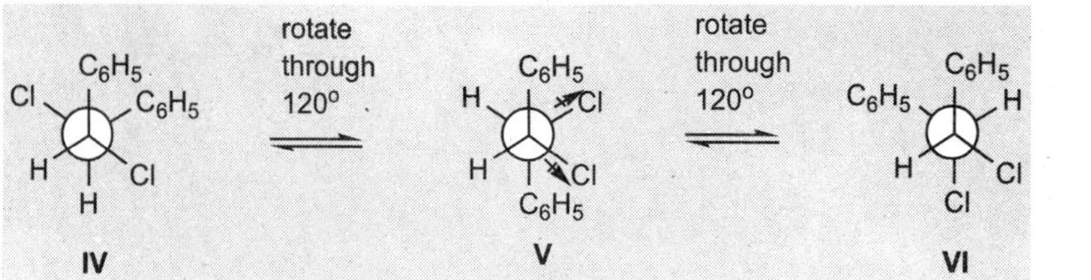

The most favoured conformation of *meso*-stilbene dichloride is (**I**). This is due to the reason that in this conformation not only the two bulky phenyl groups are *anti* to each other, the two C—Cl dipoles also oppose each other. Since this particular conformer has zero dipole moment, the overall dipole moment of the *meso*-form is quite low.

The preferred conformation of (+) or (–)-stilbene dichloride is (**V**) in which the two bulky phenyl groups are *anti* to each other. Since in this conformation, the two C—Cl dipoles are inclined to each other at an angle of 60°, the overall dipole of (+) or (–)-stilbene dichloride is relatively large.

12. Benzil on reduction with $LiAlH^4$ gives 1,2-diphenylethane-1,2-diol which contains two asymmetric carbon atoms as shown by asterisks.

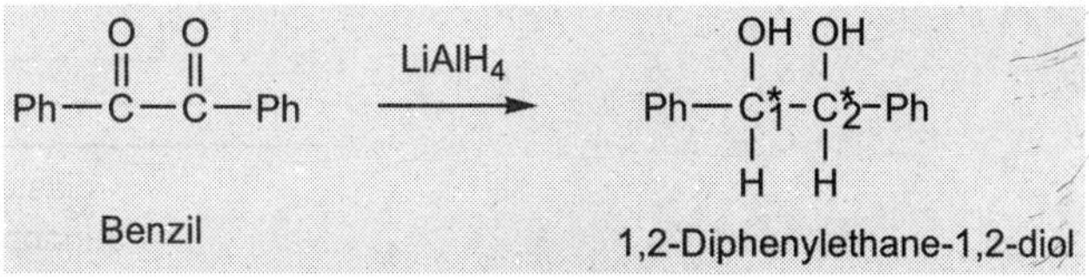

Since 1,2-diphenylethane -1,2-diol contains two similar asymmetric carbons (*cf*. tartaric acid) it exists in the following three stereoisomeric forms (**I, II** and **III**):

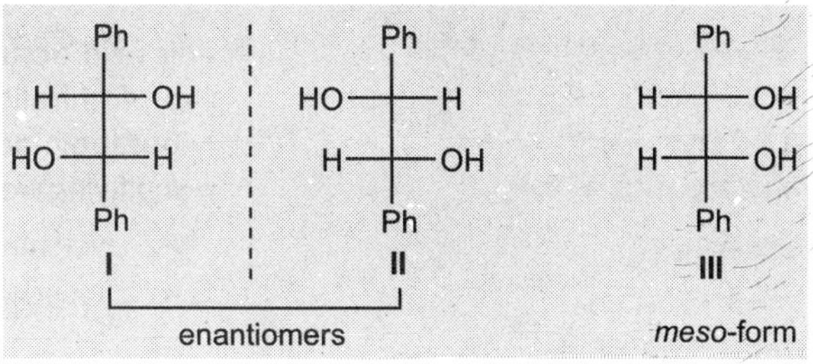

13. 2-Bromobutane will have two optical isomers (enantiomers) and 1-bromobutane will be optically inactive (achiral). Both of these bromobutanes will show conformational isomerism.

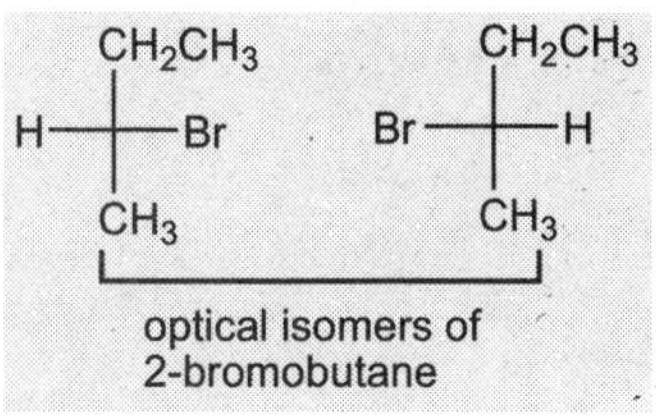

Chapter 3

Stereochemistry-II

Chapter Outline

3.1 **PROSTEREOISOMERISM**

- **A.** Definition and Classification
- **B.** Nomenclature of Stereoheterotopic Ligands and Faces

3.2 **STEREOSELECTIVE AND STEREOSPECIFIC REACTIONS**

- **A.** Stereoselectivity
- **B.** Stereospecificity

3.3 **ASYMMETRIC SYNTHESIS**

- **A.** Definition and Classification
- **B.** Principle of Stereoselectivity
- **C.** Partial Asymmetric Synthesis

3.4 **OPTICAL ACTIVITY IN THE ABSENCE OF A CHIRAL CARBON**

- **A.** Stereochemistry of Biphenyls (Atropisomerism)
- **B.** Stereochemistry of Allenes
- **C.** Stereochemistry of Spiranes
- **D.** Configurational Nomenclature of Biphenyls, Allenes, and Spiranes

3.5 **CHIRALITY DUE TO HELICAL SHAPE**

3.6 **STEREOCHEMISTRY OF THE COMPOUNDS CONTAINING NITROGEN, PHOSPHORUS AND SULPHUR**

- **A.** Stereochemistry of Nitrogen Compounds
- **B.** Stereochemistry of Phosphorus Compounds
- **C.** Stereochemistry of Sulphur Compounds

3.7 **CONFORMATIONS OF CYCLOHEXANES AND DECALINS**

- **A.** Cyclohexane and its Derivatives
- **B.** Decalins

3.8 **EFFECT OF CONFORMATION ON REACTIVITY**

- **A.** Nucleophilic Substitutions
- **B.** Elimination Reactions
- **C.** Esterification and Ester Hydrolysis

3.9 **CONFORMATIONAL ANALYSIS OF MONOSACCHARIDES**

- **A.** Fischer Projections of Open Chain Form of Monosaccharides
- **B.** Cyclic Structures of Monosaccharides and Their Conformational Analysis
- **C.** Relative Stability Based on Instability Factors/Units

3.1 PROSTEREOISOMERISM

A. Definition and Classification

The molecules which contain a *nonstereogenic centre* or plane or axis, and can be converted into the molecules with *stereogenic centre* or plane or axis, are known to be associated with *prostereoisomerism*. To understand this concept, it is necessary to learn the meaning of certain other terms used in such cases.

A classification diagram is given in **Fig. 3.1.**

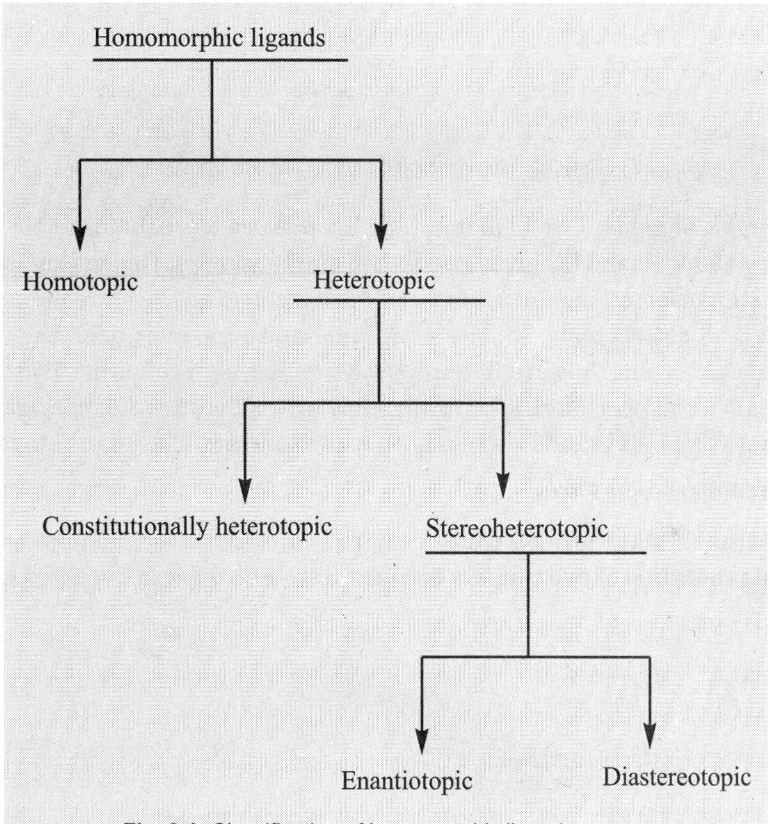

Fig. 3.1: Classification of homomorphic ligands

I. Homomorphic Ligands

Ligands (atoms or groups in a molecule) that are structurally identical when detached are called homomorphic (Greek *homos* means same, and *morphe* means form) ligands. For example, two hydrogen atoms of methylene group of ethanol or propanoic acid or n-pentane are homomorphic atoms. Relationship between identical (homomorphic) ligands within intact molecules is called the *topic relationship* (topicity or topism).

(i) Homotopic Ligands: Homomorphic ligands in constitutionally and configurationally equivalent positions are known as homotopic ligands. Separate replacement of one of such ligands by a new ligand gives identical products. For example, the two hydrogen atoms H_A and H_B in molecules like propane are homotopic as replacement of either hydrogen by any new ligand, *e.g.* Z gives the same compound (structures **1** and **2** in **Fig. 3.2**).

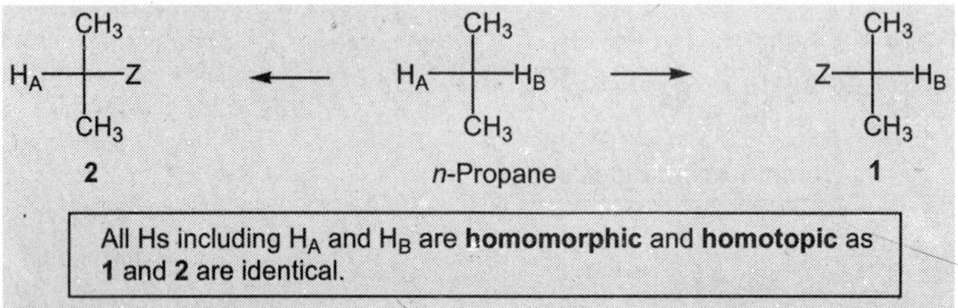

Fig. 3.2: Homomorphic and homotopic ligands

(ii) Stereoheterotopic Ligands: Two homomorphic ligands are stereoheterotopic if replacement of each in turn by a new ligand (Z) gives rise to two stereoisomers. Depending upon the nature of the resulting stereoisomers (enantiomers and diastereomers), the stereoheterotopic ligands are enantiotopic or diastereotopic. In case of ethanol and propanoic acid, the two methylene hydrogens are homomorphic, but are heterotopic as replacement of either hydrogen by a new ligand Z produces *enantiomers*. More specifically, these are called *enantiotopic ligands*. Likewise, the geminal Hs at C2 (H_A/H_B) and C4 (H_E/H_F) of *n*-pentane are also enantiotopic (**Fig. 3.3**).

II. Homotopic and Heterotopic Faces

When two stereoisomers are formed from a compound containing a double bond, there is a possibility of the generation of two isomers because of the existence of two faces (front and rear)

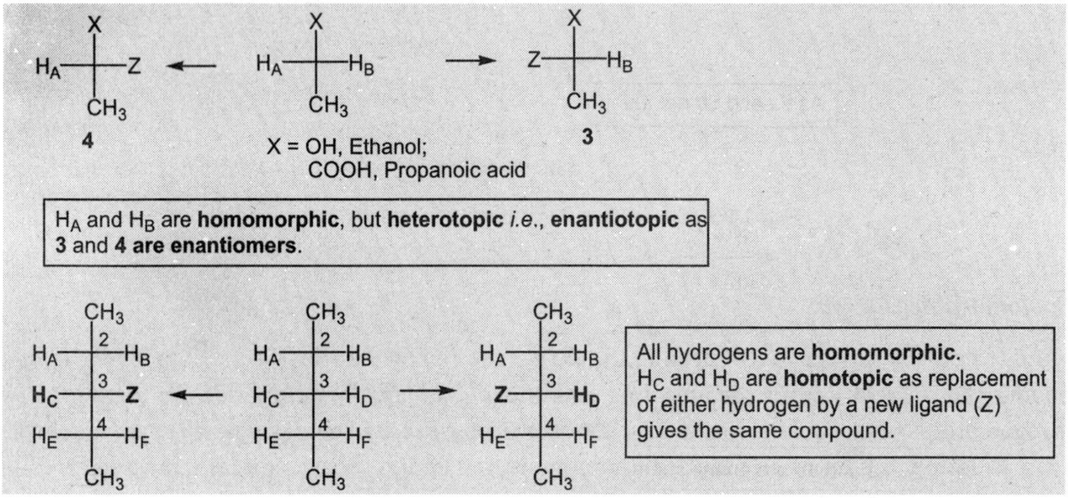

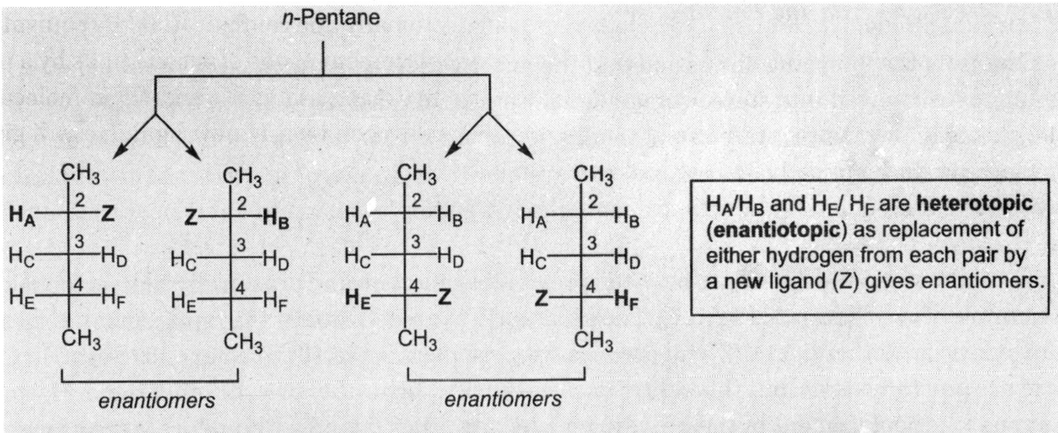

Fig. 3.3: Homomorphic, homotopic and heterotopic ligands

in the molecule. These faces are known as *homotopic* and *heterotopic* faces, which are used in similar sense as that of ligands. For example, addition of a nucleophile on the double bond of a carbonyl compound can occur from two faces of the plane of the molecule. If such an attack generates identical compounds, the two faces of the carbonyl are homotopic. However, when there is possibility of generating two stereoisomers (enantiomers or diastereomers), the two faces of the molecules are heterotopic. This situation has been explained with the help of **Fig. 3.4**. Obviously, symmetrical structures such as formaldehyde and acetone (**5**) have homotopic faces, whereas unsymmetrical structures such as acetaldehyde and acetophenone (**7**) have heterotopic faces.

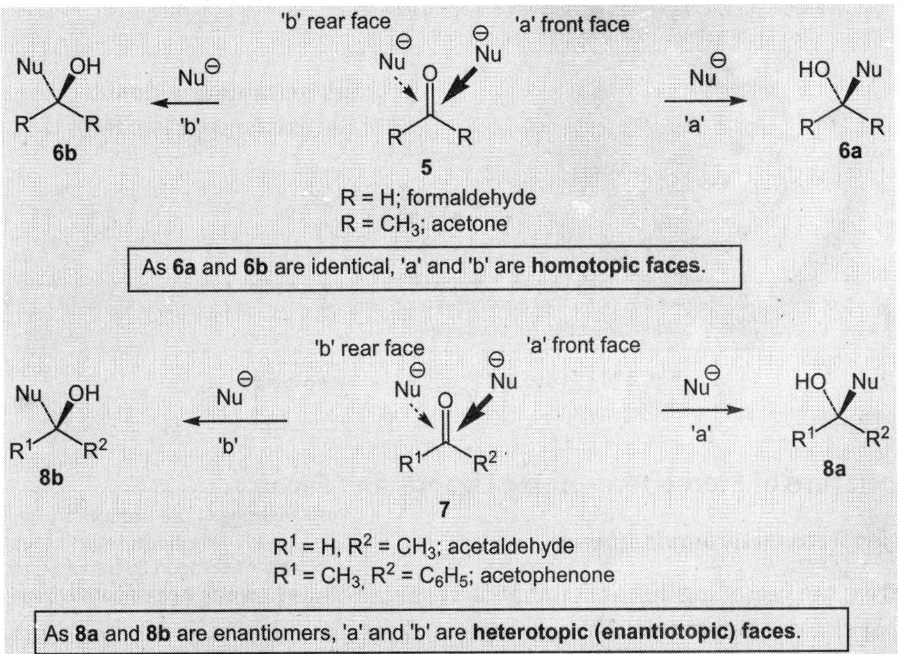

Fig. 3.4: Homotopic and heterotopic faces

III. Prostereoisomerism and Prochirality

It is clear from the foregoing discussion that the molecules having stereoheterotopic ligands and faces can be converted into stereoisomers (enantiomers or diastereomers). This property of the molecules is known as prostereoisomerism, and covers *prochirality* as well. Those molecules, which can produce optically active products, are known as *prochiral* molecules. The remaining cases, where stereomeric products are not chiral, are simply known to have prostereoisomerism.

To understand the difference between prostereoisomerism and prochirality, let us consider the example of cyclobutanol (**Fig. 3.5**). The H_A and H_B hydrogen atoms (the replacement of these hydrogens by any other ligand 'Z' will produce stereoisomers **9** and **10** which are not chiral) at C3 centre are prostereogenic, but this centre does not show prochirality. On the other hand, H_C and H_D atoms (the replacement of these hydrogens by any other ligand will produce stereoisomers **11** and **12**, which are chiral) at C2 exhibits prostereoisomerism, which is prochirality as well. Thus, prostereoisomerism is a more general term.

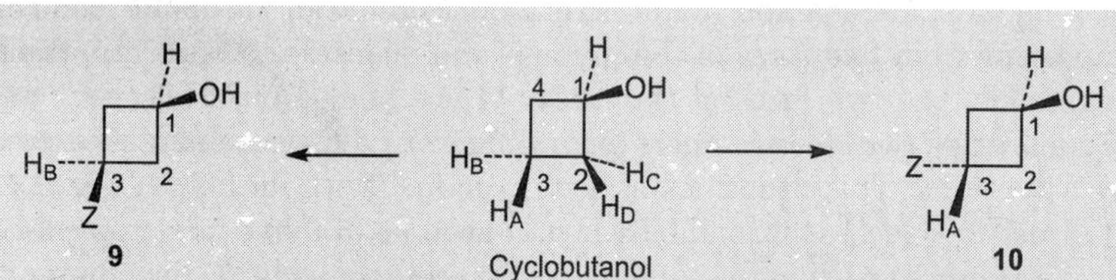

In case of C3, H_A and H_B exhibit **prostereoisomerism**, but **not prochirality** as **9** and **10** are stereoisomers (diastereomers, but not chiral).

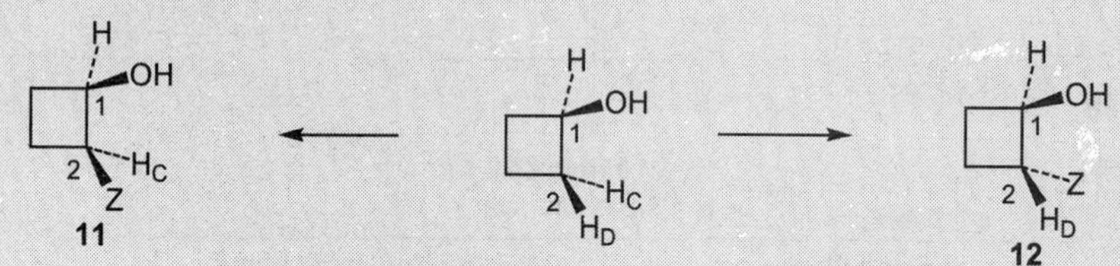

In case of C2, H_C and H_D exhibit **prostereoisomerism**, which is **prochirality** as well as **11** and **12** are chiral molecules (diastereomers).

Fig. 3.5: Prostereoisomerism and prochirality

B. Nomenclature of Stereoheterotopic Ligands and Faces

I. Symbols for Stereoheterotopic Ligands

It is clear from the preceding discussion that stereoheterotopic ligands are of two types, namely, enantiotopic and diastereotopic. The symbols used for these ligands are *pro-R* and *pro-S*. A general method for determining the name of the heterotopic ligands at a prochiral centre of the given molecule with general formula CabXX is as follows.

Step I The molecule is depicted by Fischer projection/flying wedge projection.

Step II The priority of the two ligands (a,b) is determined by CIP rules.

Step III The priority of one of the ligands to be named (say X_A) is arbitrarily assigned higher than the other one (X_B).

Step-IV As the resultant molecule having four different ligands, it will be a chiral compound. The configuration of this molecule can be determined by applying CIP rules. If the configuration is R, the ligand will be known as *pro-R,* and *pro-S* in case the configuration of the molecule is S (**Fig. 3.6**).

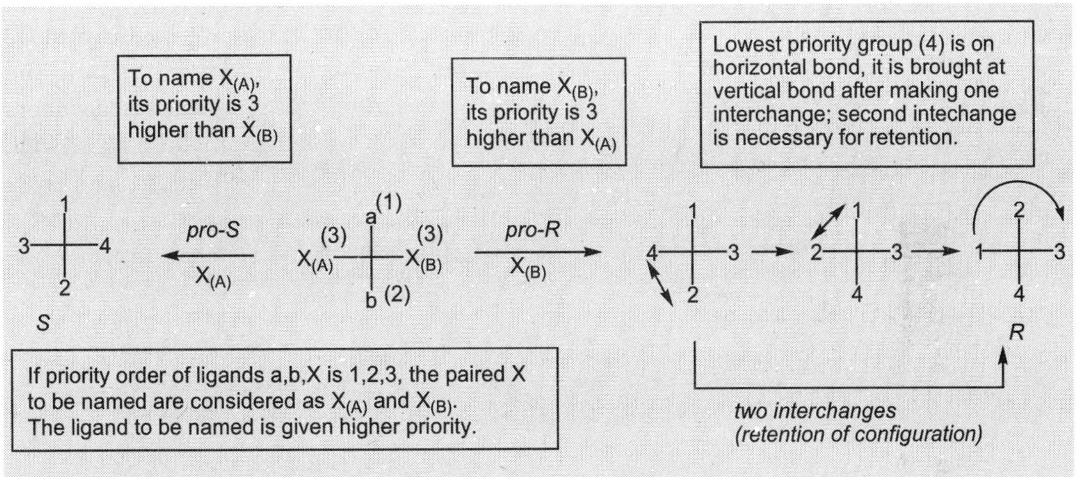

Fig. 3.6: Nomenclature of enantiotopic ligands

The method is illustrated by the examples of ethanol and 2,2-dichloropropanoic acid, as shown in the **Fig. 3.7** and **Fig. 3.8**, respectively.

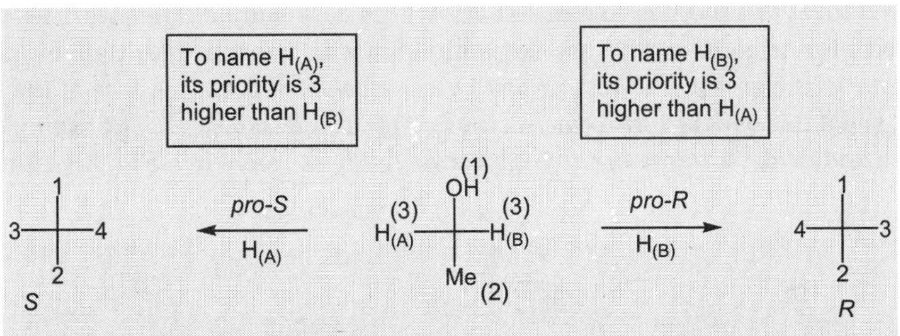

Fig. 3.7: Nomenclature of enantiotopic ligands of ethanol

The same method can be applied to name *diastereotopic ligands.* To explain diastereotopic ligands, we can consider the example of Ha and Hb diastereotopic hydrogens situated at C1 of 2-*R*-chloropropanol. Applying the general procedure, Ha and Hb can be named as *pro-S* and *pro-R*, respectively (**Fig. 3.9**).

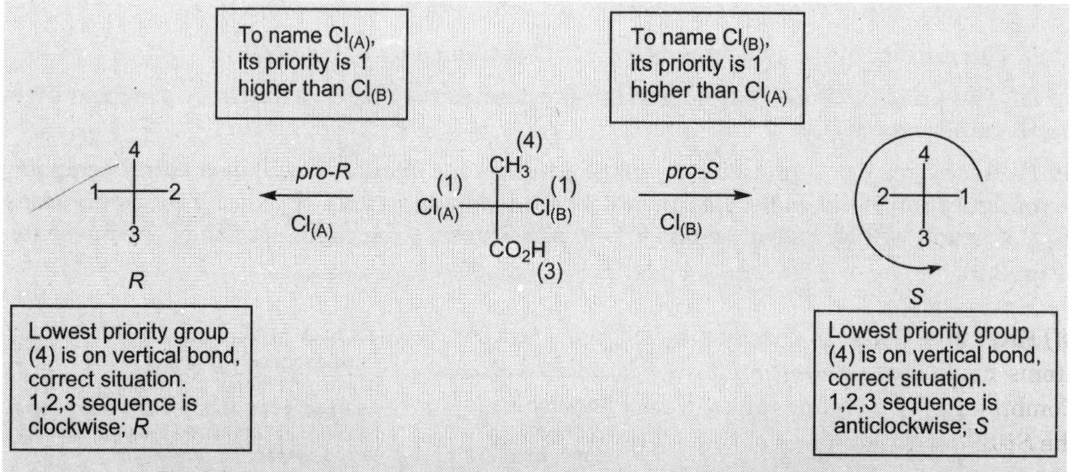

Fig. 3.8: Nomenclature of enantiotopic ligands of 2,2-dichloropropanoic acid

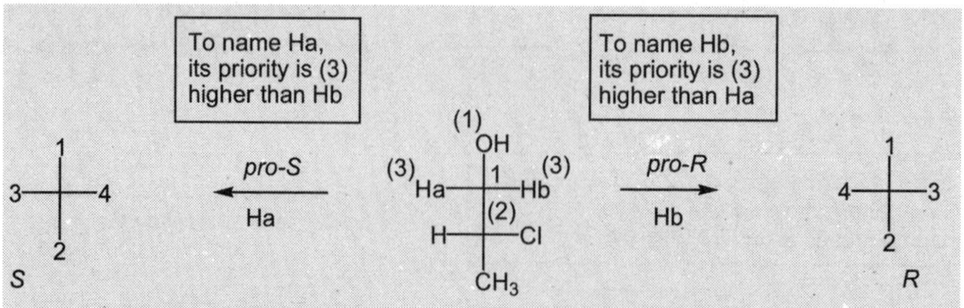

Fig. 3.9: Nomenclature of diastereotopic ligands Ha and Hb at C1 of 2-*R*-chloropropanol

II. Symbols for Stereoheterotopic Faces

(i) Faces of Carbonyl Compounds: The two faces of carbonyl compounds are differentiated by using *Re*-*Si* nomenclature. The method to name such faces is very simple. The priorities of the three groups connected to trigonal centre are determined by CIP rules. If moving from the group of highest priority to the group of lowest priority gives *clockwise* path, the face of the molecule is known as *Re*; and if this path is *anticlockwise*, the face of the molecule is *Si*. The examples of *Re* and *Si* faces of benzaldehyde, butanone, and *S*-glyceraldehyde are shown in **Fig. 3.10** and **Fig. 3.11**.

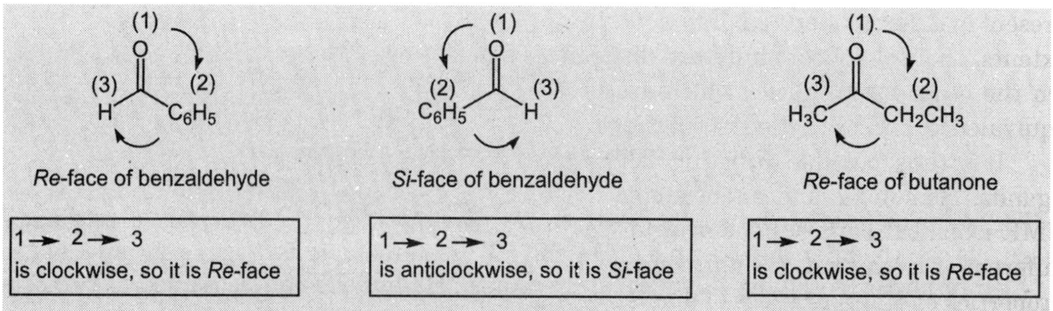

Fig. 3.10: Enantiotopic faces of benzaldehye and butanone

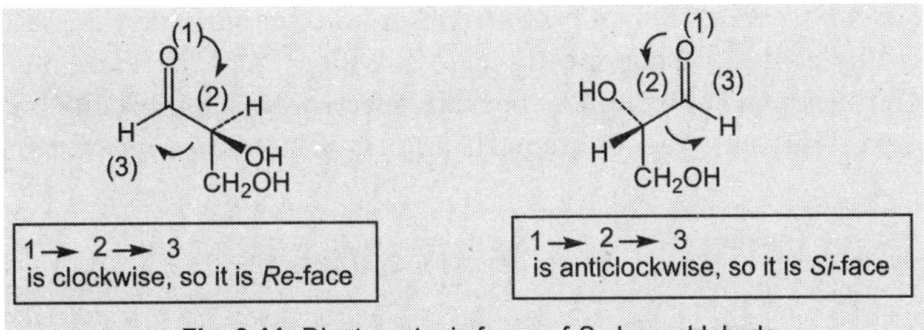

Fig. 3.11: Diastereotopic faces of *S*-glyceraldehyde

(ii) Faces of Alkenes: In case of alkenes, there are two sp^2carbon atoms. Each of the sp^2 carbon atoms has its face and, therefore, it is necessary to name both the carbon atoms separately. Combining the two names gives overall name of the face of the molecule. **Figure 3.12** illustrates the *Si-Si* and *Re-Re* faces of *trans*-2-butene.

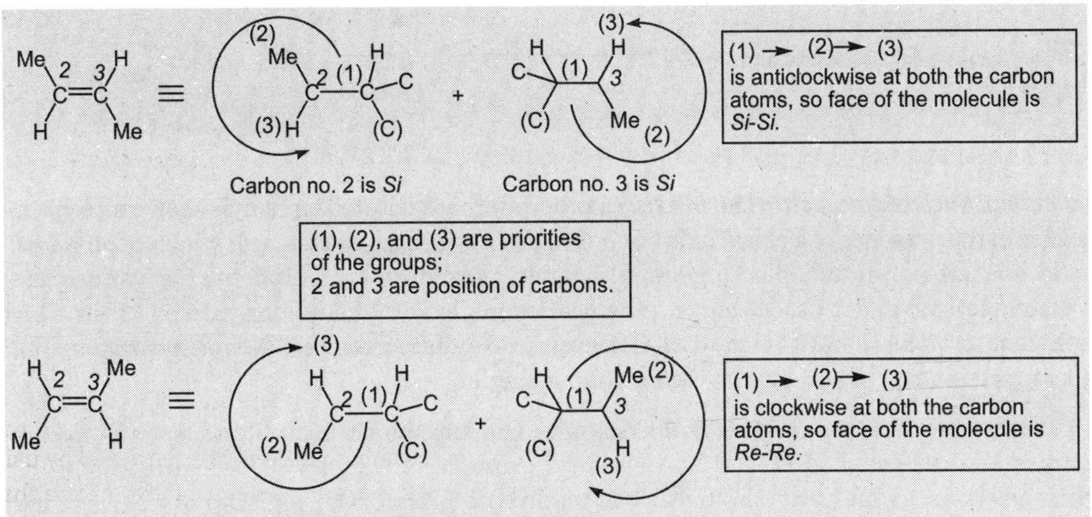

Fig. 3.12: *Si-Si* and *Re-Re* faces of *trans*-2-butene

III. NMR Spectral characteristics of Stereoheterotopic Ligands

NMR spectroscopy has been successfully employed to distinguish between the nuclei, which are present in different steric environments. Since these nuclei are shielded or deshielded to different extents, their chemical shifts are different, and are called non-equivalent or *anisochronous*. On the other hand, nuclei which are present in identical environment show chemical shift equivalence, and are called *isochronous*.

It is, thus, possible to distinguish between diastereotopic and enantiotopic (and homotopic) ligands. Homotopic and enantiotopic nuclei are *isochronous* and, therefore, show identical NMR chemical shifts under achiral conditions. In contrast, diastereotopic nuclei, being in different environment, are *anisochronous,* and show different NMR characteristics. The NMR properties of different types of diastereotopic nuclei (particularly protons) are as follows.

(i) Geminal Ligands Adjacent to a Chirality Centre: Geminal ligands, such as methylene protons, adjacent to a chirality centre are chemically non-equivalent, and therefore do not appear as a singlet in the NMR spectrum. For example, the NMR spectrum of 2,3-dibromo-2-methylpropanoic acid (**Fig. 3.12a**) exhibits methylene protons H_A and H_B, which are *anisochronous* as a quartet (J_{AB} = 10 Hz).

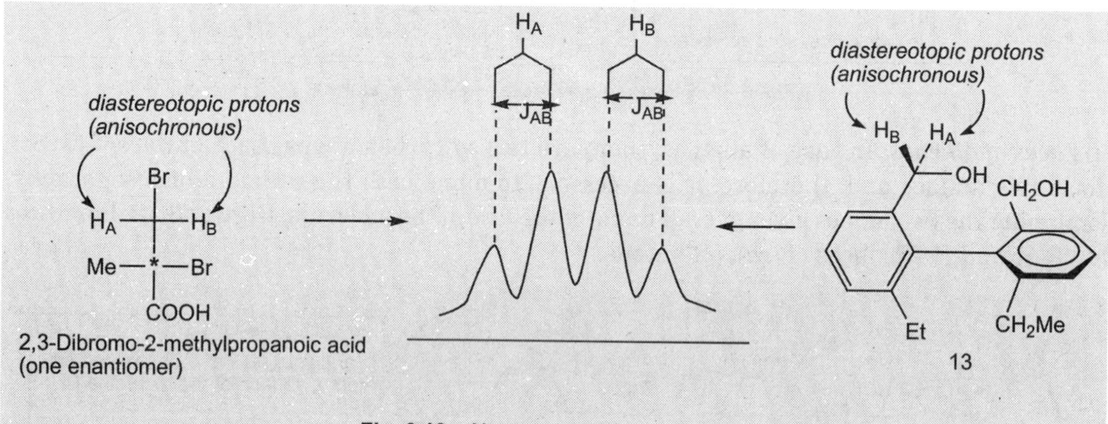

Fig. 3.12a: Non-equivalence of geminal protons

(ii) Geminal Nuclei Adjacent to a Chiral Axis: Appropriately substituted biphenyls show anisochrony, because of the presence of a chiral axis (*for a detailed discussion on stereochemistry of biphenyls, see Section 3.4A*). Suitably placed geminal ligands, present in such biphenyl derivatives, may be diastereotopic and *anisochronous*. A typical example is biphenyl derivative (**13**) in which methylene protons in each of the CH_2OH groups are diastereotopic, *i.e. anisochronous*, and appear as a quartet (J_{AB} = 12 Hz) rather than a singlet.

(iii) Diastereotopic Ligands in cis-trans Isomers: Diastereotopic ligands in *cis-trans* isomers can also be distinguished by NMR. A classical example is *N*-benzyl-2,6-dimethylpiperidine, which exists as *cis* and *trans* isomers. The methylene protons (H_A and H_B) of the *cis* isomer are enantiotopic, *i.e. isochronous*, and therefore, appear as a sharp singlet at δ 3.70. On the other hand, methylene protons (H_A and H_B) of the *trans* isomer, which are diastereotopic (*anisochronous*) appear as an AB-quartet (J_{AB} = 14 Hz) at δ 3.63.

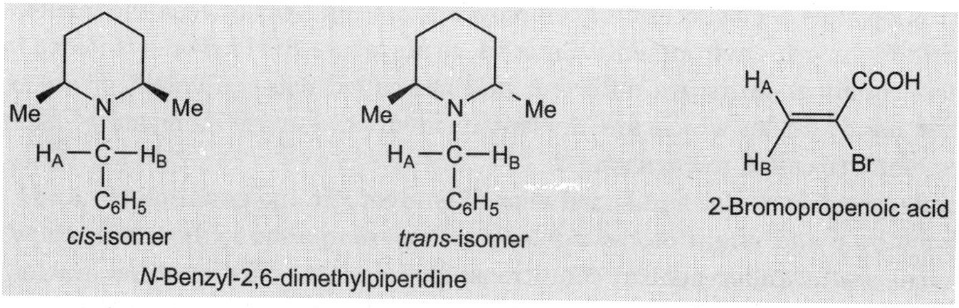

A similar type of geminal anisochrony is shown by olefinic compounds. For example, NMR of 2-bromopropenoic acid displays stereotopic protons H_A and H_B as an AB-quartet.

3.2 STEREOSELECTIVE AND STEREOSPECIFIC REACTIONS

A. Stereoselectivity

The preferential formation of one stereoisomer over another in a chemical reaction is known as stereoselectivity. Depending on the degree of preference for a particular isomer, a reaction can be described as being *moderately stereoselective, highly stereoselective,* or *completely stereoselective*.

A *stereoselective* reaction generates more of one stereoisomer than of another

A ⟶ **B** + **C**

(B + C: stereoisomers)

If **B** is more than **C** or *vice-versa*, the reaction will be *stereoselective.*

In case the products are enantiomers, one speaks of *enantioselectivity*; if they are diastereomers, one speaks of *diastereoselectivity*.

The term *enantioselective* may be applied to the ultimate outcome of a sequence of reactions. For example, reduction of pyruvic acid could, in principle, produce a pair of two enantiomers, *i.e. R*- and *S*-lactic acids. When the reaction is carried out with coenzyme NADH, the process is completely stereoselective (enantioselective), and results in the formation of *S*-lactic acid exclusively. On the other hand, such reductions with achiral hydride donors ($NaBH_4$/$LiAlH_4$) are not stereoselective as they lead to the formation of equal amount of enantiomers (racemic mixtures) (**Fig. 3.13**).

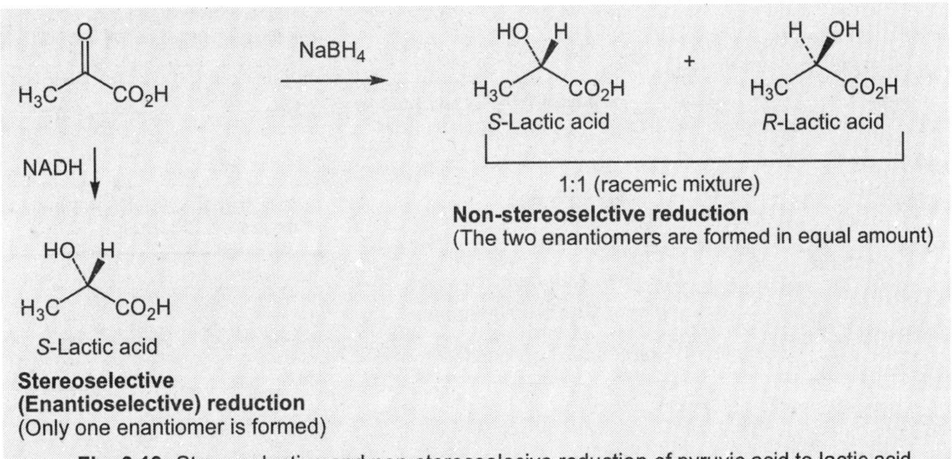

Fig. 3.13: Stereoselective and non-stereoselecive reduction of pyruvic acid to lactic acid

B. Stereospecificity

While dealing with *stereospecific* reactions, it is necessary to have stereoisomerism in the substrate as well as the product. If in a given reaction, a stereoisomer exclusively generates one stereoisomer (or a set of stereoisomers), and the other stereoisomer gives different stereoisomer (or a different set of stereoisomers), then the reaction is known to have *stereospecificity*. All the stereospecific reactions are stereoselective, but the converse is not true.

In a *stereospecific* reaction, each stereomeric reactant forms a different stereoisomeric product OR a different set of stereoisomeric products

A ⟶ **B**

C ⟶ **D**

Reactants **A** and **C** are stereoisomers
Products **B** and **D** are stereoisomers

Addition of bromine to alkenes is a *stereospecific* process. It can be illustrated by considering the bromination of *E*/*Z*-2-butenes (**Fig. 3.14**). All S_N2 reactions are stereospecific, and proceed with the inversion of configuration. For example, the reaction of two enantiomers of 2-bromopropanoic acid with azide ions leads to the formation of 2-azidopropanoic acids in a *stereospecific* manner (**Fig. 3.15**)

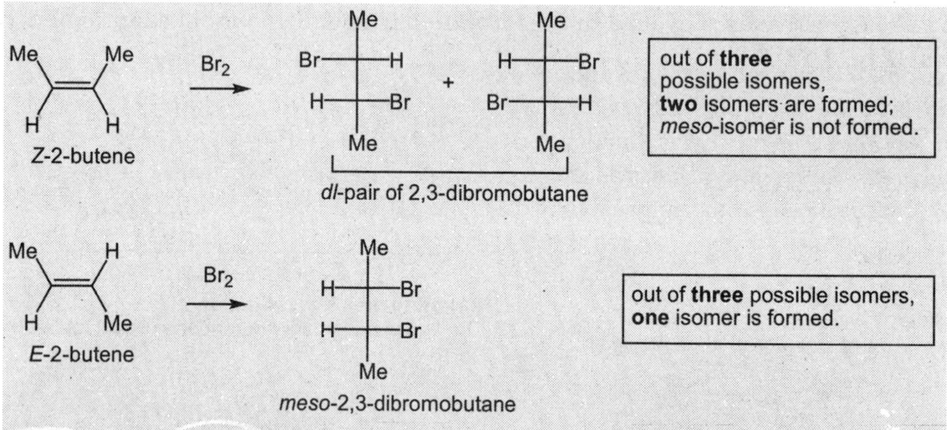

Fig. 3.14: Stereospecific bromination of *E/Z* 2-butenes

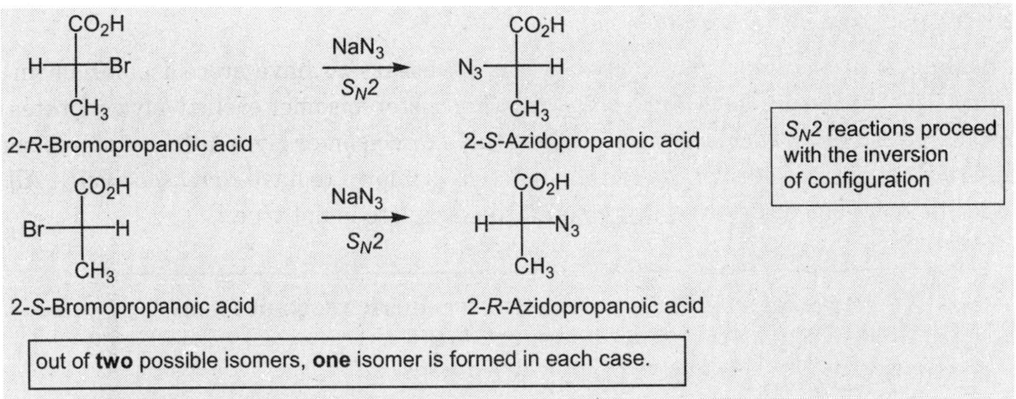

Fig. 3.15: Stereospecific azidation of enantiomers of 2-bromopropanoic acid

3.3 ASYMMETRIC SYNTHESIS

A. Definition and Classification

Although 'asymmetric synthesis' simply means synthesis of asymmetric (chiral) molecules, there is some confusion in defining this term. Marcwald defined an asymmetric synthesis as *a reaction, which produces optically active substances from symmetrically constituted compounds with the intermediate use of optically active materials*. However, a broader definition by Morrison and Mosher defines an asymmetric synthesis as *a reaction, which converts a prochiral unit (having enantiotopic and diastereotopic faces / groups) into a chiral unit in such a manner that unequal amounts of stereoisomers are produced.*

Thus, asymmetric syntheses comprise both enantioselective and diastereoselective reactions. While all enantioselective syntheses are well covered under the given definitions, all diastereoselective syntheses cannot be covered. There is no problem as far as a reaction produces optical diastereomers. However, confusion arises when reactions give two achiral diastereomers, *e.g.* reduction of 4-*tert*-butylcyclohexanone to *cis*- and *trans*-4-*t*-butylcyclohexanols (**Fig. 3.16**; for nomenclature of cyclohexanes, see section 3.7A). According to the above definitions, this is not an asymmetric synthesis, but the analogous reduction of 2-methylcyclohexanone is considered an asymmetric synthesis as an additional chiral centre is being created in this process (**Fig. 3.17**) (Also see **Fig. 3.5**, and relevant description).

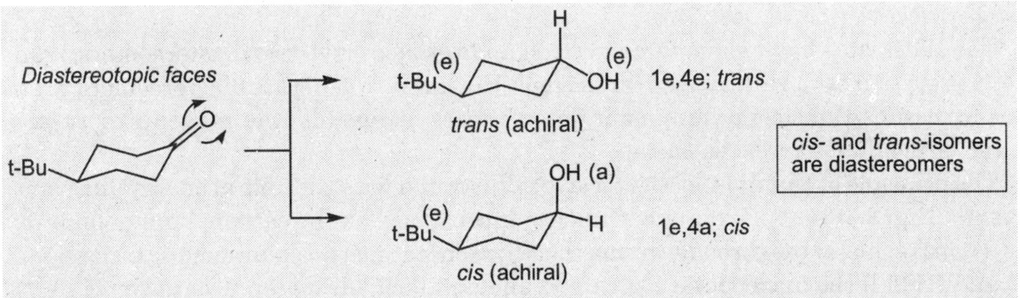

Fig. 3.16: Diastereoselective, but not asymmetric synthesis

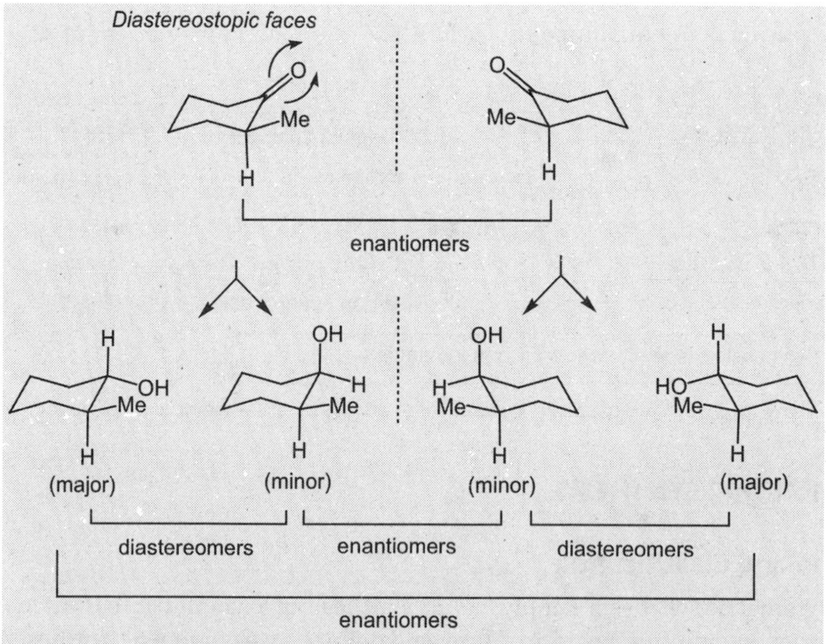

Fig. 3.17: Diastereoselective asymmetric synthesis

Asymmetric synthesis can be of two types – 'partial asymmetric synthesis' and 'absolute asymmetric synthesis'. 'Partial asymmetric synthesis' can be distinguished from 'absolute asymmetric synthesis' in the way that former involves optically active compounds instead of a physical reagent (circularly polarized light) used in the latter case.

B. Principle of Stereoselectivity

For an asymmetric synthesis, the necessary condition is that the *reactions involved must be stereoselective, and the substrate must have prostereogenic elements.*

The principle of stereoselectivity, which includes enantioselectivity and diastereoselectivity, in a kinetically controlled reaction is the same. The two stereomeric products whether enantiomers or diastereomers must be formed through two diastereomeric transition states. Because of the diastereomeric nature, the two transition states would differ in their free energies, thereby, giving the products in different amounts. The *greater is the difference in energy, higher will be the stereoselectivity of the reaction.*

The substrates having diastereomeric ligands or faces give two diastereomeric transition states on reaction even with an achiral reagent. However, to generate diastereomeric transition states from substrates containing enantiotopic faces or ligands, the reagents or catalysts or solvents or medium have to be chiral.

The principle of enantioselectivity can be illustrated with the help of energy diagrams (**Fig. 3.18a** and **Fig. 3.18b**). For example, a nucleophile can attack the carbonyl compound from two faces (enantiotopic) of the carbonyl group, thereby, generating two enantiomeric transition states, TS-1 and TS-2. If the nucleophile, *e.g.*, an organometallic hydride represented by H-M (L) (M = metal, L = an achiral organic ligand) is achiral, TS-1 and TS-2 being equienergetic produce two enantiomeric alcohols in equal amounts, and thus, the reaction is not enantioselective (case I; **Fig. 3.18a**).

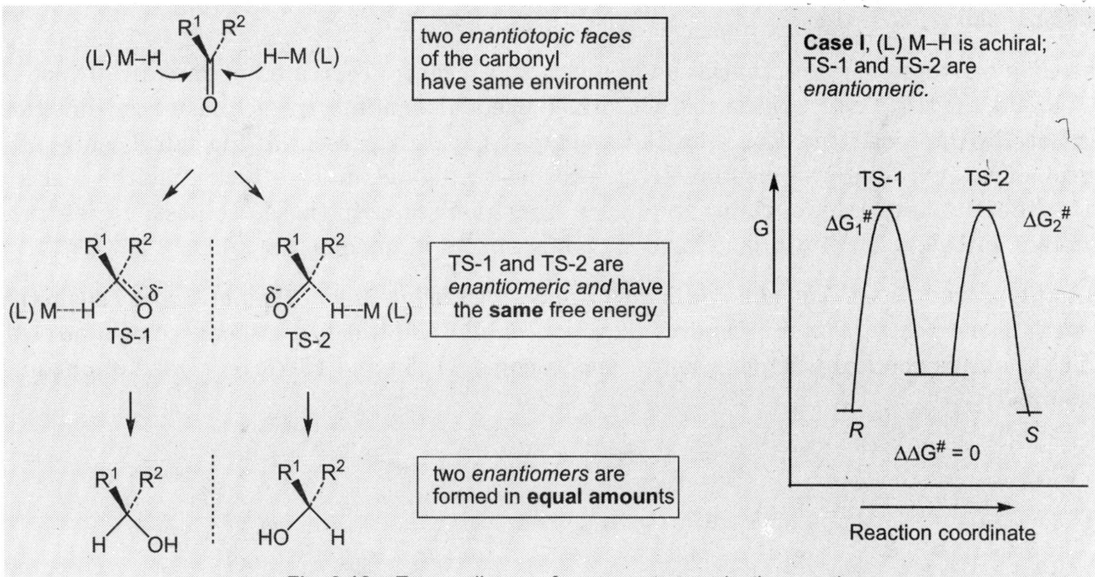

Fig. 3.18a: Energy diagram for a non-stereoselective reaction

But, if the L is replaced with a chiral ligand (L*), the two transition states will be diastereomeric rather than being enantiomeric. Obviously, the free energies of the two diastereomeric transition states will be different and, therefore, two enantiomeric products are formed in unequal amounts (case II; **Fig. 3.18b**). Such a reaction is thus enantioselective, and the stereoselectivity is achieved through the formation of diastereomeric transition states (diastereotopic differentiation).

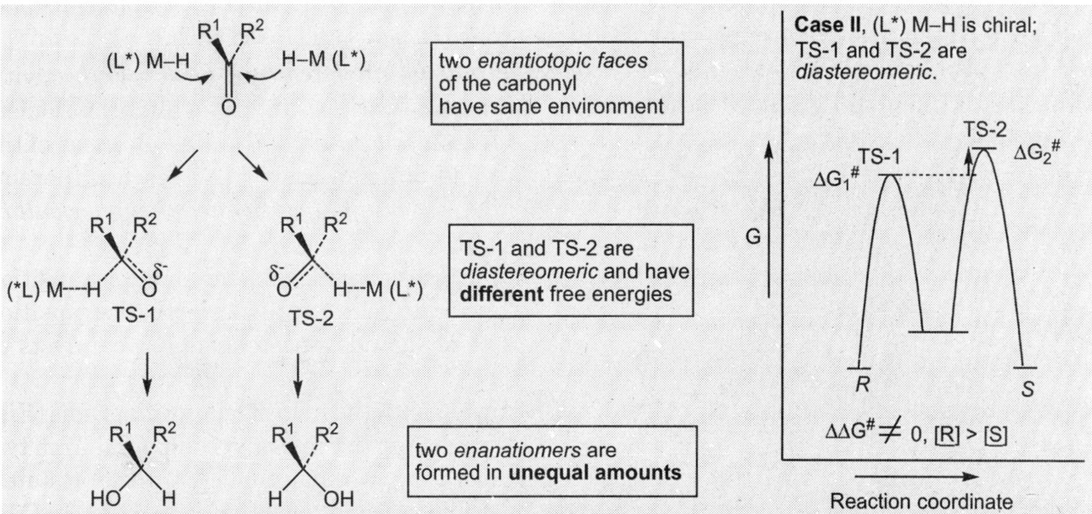

Fig. 3.18b: Principle of enantioselectivity- enantioselective synthesis through diastereotopic differentiation

C. Partial Asymmetric Synthesis

Partial asymmetric synthesis can be discussed under five headings:

I. Using a Chiral Substrate

Based on diastereotopic differentiation, if a new chiral centre is created in an optically active compound, the two diastereomers are not formed in equal amounts. In such cases, the direction of attack by the reagent is determined by the groups already present at the chiral centre of a chiral substrate. Thus, two diastereotopic faces or two ligands to be attacked by the reagent have different steric environment, and hence transition states in these two possibilities have different energies, thereby, producing two diastereomers in unequal amounts.

(i) Addition to Carbon–Oxygen Double Bond (Cram's Rule, Chelate Model, Dipolar Model, Felkin-Anh Model): If a new chiral center is generated by nucleophilic addition to a carbonyl compound (**14**), which is already optically active, the two diastereomers (**15a** and **15b**) are formed in unequal amounts.

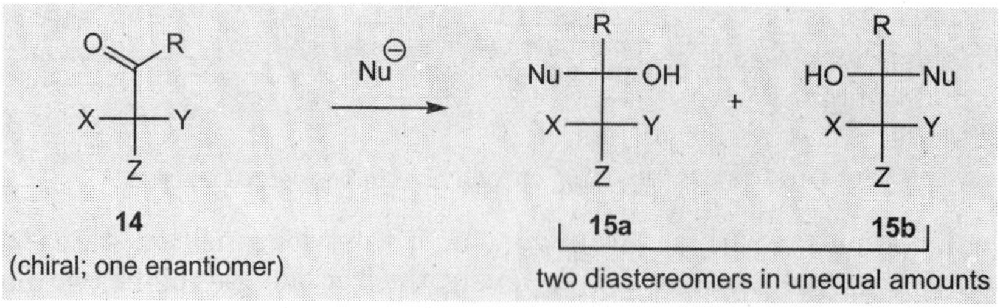

(a) Cram's Rule: Cram's rule, based on some arbitrary models, predicts which diastereomer will predominate in such transition states. To understand this rule, three groups (other than carbonyl) attached to the chiral centre are indicated as S, M, and L where they stand for small, medium, and large groups, respectively. In the open chain model, the carbonyl oxygen orients itself between the S and M groups, with the large group (L) nearly eclipsed with R as shown in the **Fig. 3.19**. Cram's rule *states that the incoming group will preferentially attack on the least hindered side (the small group S) of the plane of the double bond.*

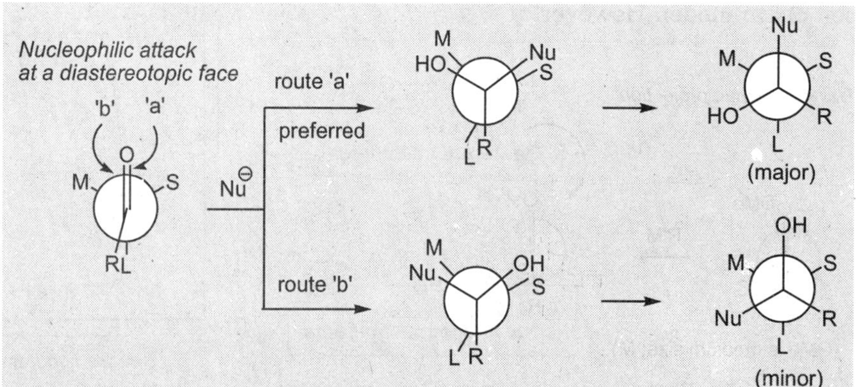

Fig. 3.19: Cram's rule

Let us understand Cram's rule by studying the reaction between α-phenylpropionaldehyde (S = H, M = Me, L = Ph) and methylmagnesium bromide, generating the *erythro*- and the *threo*-products. The major product in this case is *erythro*- compound as shown in **Fig. 3.20**.

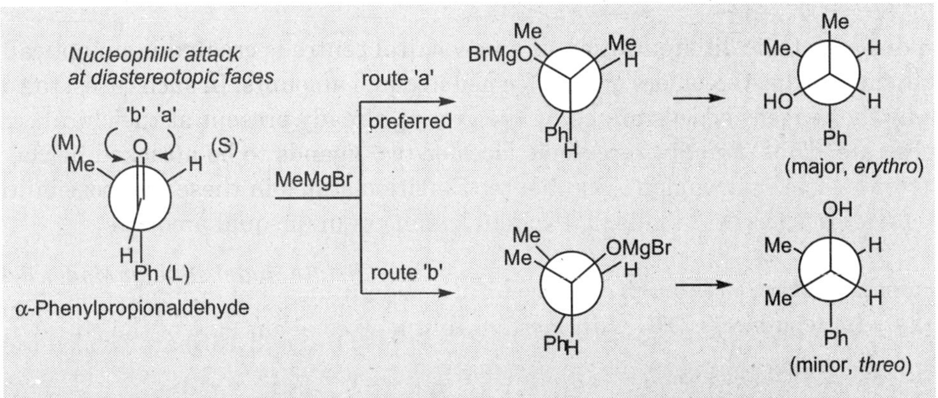

Fig. 3.20: Diastereoselective formation of alcohols following Cram's rule

The following points need to be considered while applying Cram's rule:

(i) The major/minor ratio is greatly affected by the difference in size of groups. Some exceptions to the predictions based on Cram's rule have been observed, particularly when the group L is very large.

(ii) Cram's rule does not apply in catalytic reductions.

(iii) It does not apply to the compounds in which asymmetric carbon atom is joined to a group capable of forming complex with the reagent, *e.g.* –OH, $-NH_2$, and –OMe.

In addition to Cram's rule, several other models have been proposed to predict the stereochemistry of the products formed in such reactions.

(b) Chelate Model: For example, in the α-substituted ketones, which contain an –OH, $-NH_2$, and –OMe groups at the α position (point iii), the stereochemistry of the product is predicted by modified Cram's rule based on a rigid (chelate) *cyclic model* (**Fig. 3.21**). In this model, the metallic part of the reagent is doubly coordinated to form a five-membered ring. The nucleophile preferentially approaches the carbonyl carbon from the side of small group S. This is to be noted that if the chelating group is M (medium), the cyclic model predicts the same stereochemistry as in the open chain model. However, if it is S or L, products of the opposite stereochemistry are formed.

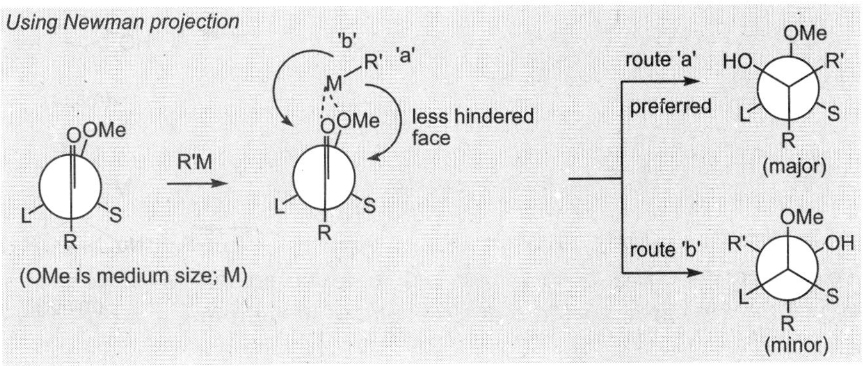

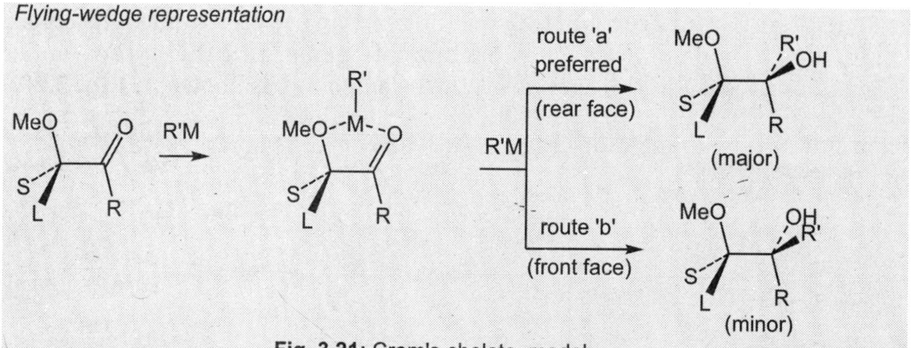

Fig. 3.21: Cram's chelate model

Asymmetric induction through chelate model is usually high. For example, reaction of S-1,2-diphenyl-2-hydroxybutanone (**16**) with methyllithium gives about 90% of the chiral diol (**17**).

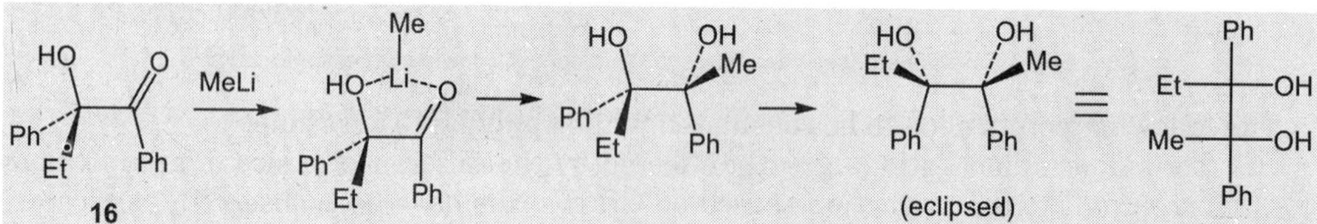

16 **17**

(c) *Dipolar Model:* Another model, known as *dipolar model,* has been suggested for the prediction of stereochemistry of the products when a strong electronegative group, *e.g.* a halogen atom is present at α-position of carbonyl group of the reactant. In such cases, the dipoles of the carbonyl bond and C–X bond oppose each other. To minimize the dipole repulsion, the two groups are placed *anti* to each other. The nucleophile adds from less hindered side (S group) giving the major product (**Fig. 3.22**).

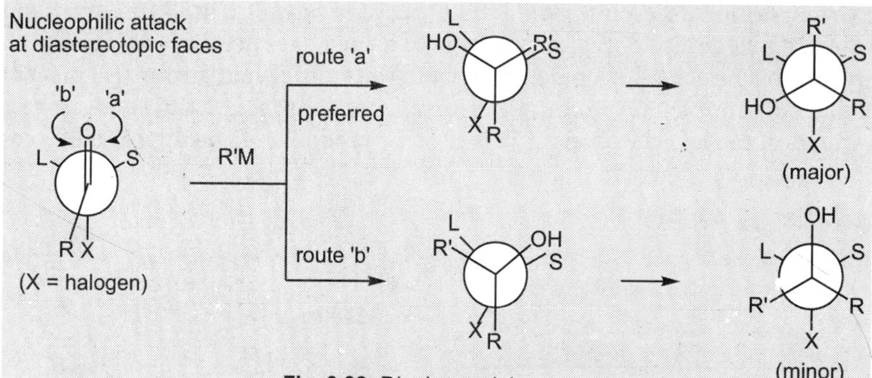

Fig. 3.22: Dipolar model

(d) *Felkin-Anh Model:* One of the widely accepted models is Felkin-Anh model. In this model, two reactive conformations have been considered in which either the large group L, or the most electron-withdrawing group present at Cα is placed at right angle to the carbonyl double bond. Between the two possible conformations, one with S (small group) gauche to R is usually preferred. The non-bonded interactions involving R and S are minimized (as compared to R and M in the other situation). The major product is formed by the nucleophilic attack on the carbonyl from side *anti* to large group (**Fig. 3.23**). This model predicts the same stereochemistry as predicted by Cram's rule.

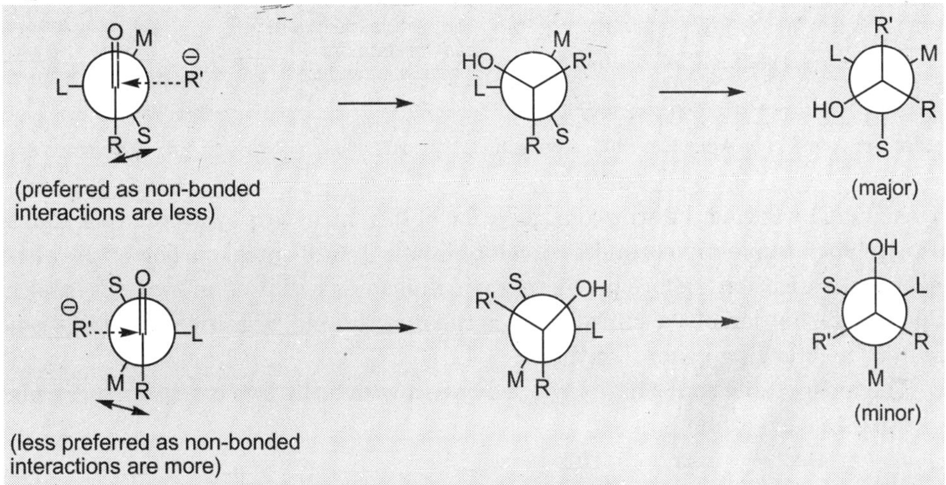

Fig. 3.23: Felkin-Anh model

II. Using Chiral Auxillary to Modify Achiral Substrate

An asymmetric synthesis is often carried out by the initial addition of a chiral group to the achiral substrate. The added group, which is removed in the end of the synthesis, is called a *chiral auxillary*. Of the various examples of such syntheses, one based on Prelog's rule involving 1,4-asymmetric induction is discussed herein.

Prelog's Rule (1,4-asymmetric induction)

Nucleophilic reactions of α-keto esters with metal hydrides and Grignard reagents can generate a chiral centre with the formation of hydroxy esters, and which on hydrolysis give the corresponding hydroxy acids. But, the products are racemic mixtures because the two enantiotopic faces of the carbonyl ketone have same steric environment (**Fig. 3.24a**).

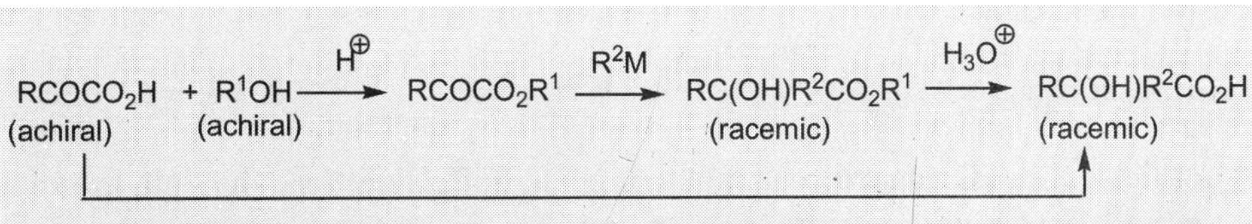

Fig. 3.24a: Conversion of α-keto acids into racemic α-hydroxy acids using an achiral alcohol

To carry out the asymmetric synthesis of hydroxyl esters and acids, achiral α-keto acids are converted into chiral keto esters by using chiral alcohols for the esterification of acids. Then, resulting chiral keto esters are subjected to nucleophilic addition. Since the products in such situations are diastereomeric, they are formed in unequal amounts. Thus, it is possible to synthesize chiral α-hydroxy esters, which on hydrolysis provide chiral α-hydroxy acids. The overall reaction sequence can be considered an asymmetric synthesis (enantioselective) of α-hydroxy acids starting from corresponding achiral α-keto acids (**Fig. 3.24b**).

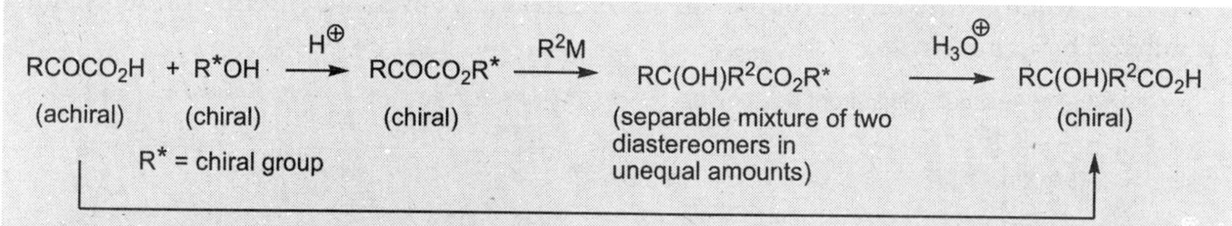

Fig. 3.24b: Conversion of α-keto acids into chiral α-hydroxy acids using a chiral alcohol (asymmetric synthesis)

It was found by Prelog that the configuration of the chiral carbon atom in the diastereomer dominating in such conversions can be correlated with that of chiral carbon of the alcohol used in the initial esterification. Before predicting which isomer will predominate, it is necessary to know which conformation of the chiral ester is the most stable. Significant studies reported by Prelog led to the following generalizations.

(*i*) The most stable conformation is the one in which the two oxo groups are planar and *trans* to each other.

(*ii*) The atoms O–C also lie in this plane.

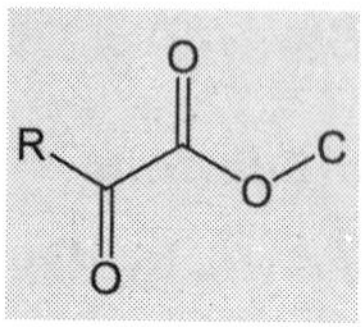

(*iii*) The rotation of the three groups (indicated as S, M, and L) attached to the chiral carbon about the single C–O bond can give several conformations; out of which three are energetically favourable (preferred conformations). These are shown as structures **18a, 18b,** and **18c** in which, groups S, M, and L lie in the plane, respectively.

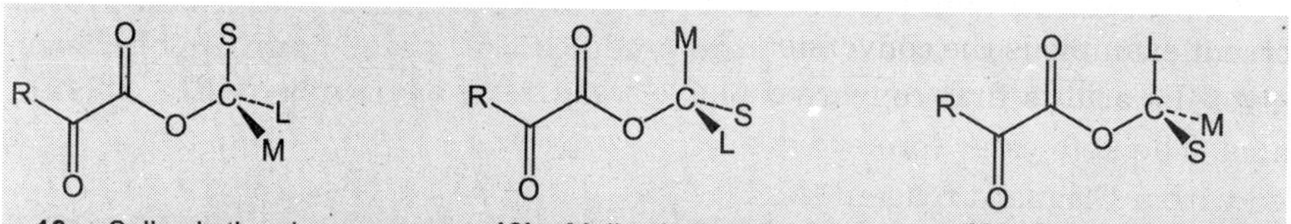

18a, S lies in the plane **18b**, M lies in the plane **18c**, L lies in the plane

(iv) A nucleophile (Grignard reagent) will preferentially attack the keto group (of RCO) from the less hindered side, thereby giving major product.

Following the above generalizations, it is possible to correlate the configuration of the product hydroxyl acid with that of the alcohol. However, predictions based on three favourable conformations do not give the same results, as shown in **Fig. 3.25**. Conformations ***a*** and ***c*** lead to the same results, whereas conformation ***b*** when M lies in the plane gives opposite result. Therefore, predictions based on conformations ***a*** and ***c*** are more reliable.

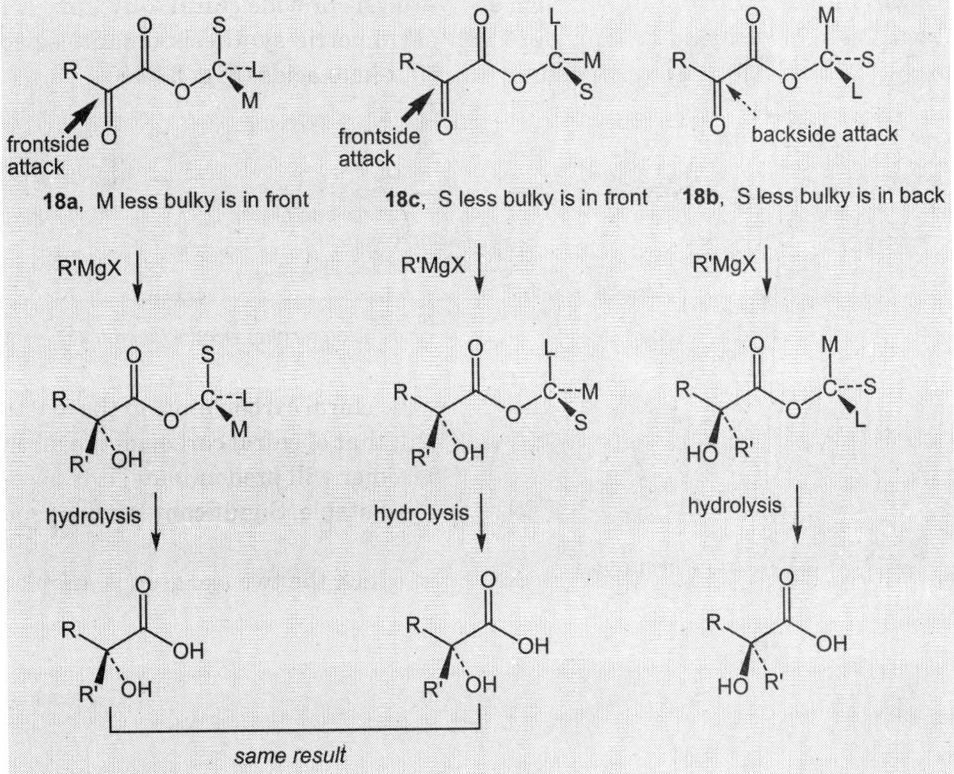

Fig. 3.25: Asymmetric synthesis of α-hydroxy acids based on Prelog's generalizations

Thus, to predict the configuration of α-hydroxy acid on the basis of Prelog's generalizations, it is advisable *to start with preferred conformation of the chiral keto ester with either small group or large group in the plane. Then, nucleophile preferentially attacks the keto group (of RCO) from the less hindered side, thereby giving major product*

Among the various examples of asymmetric syntheses based on Prelog's generalizations, an important example is the conversion of phenylglyoxylic acid into atrolactic acid (**Fig. 3.26**). Phenylglyoxylic acid is first converted to the chiral ester by using a chiral alcohol, such as *S*-2-octanol. The keto ester (preferred conformation with small group in the plane is shown) is treated with a Grignard reagent (MeMgI) to give a mixture of two diastereomers formed in unequal amounts. The one which is formed by the attack on less hindered side (front side; as methyl group present in front side is less bulky than hexyl group) is obtained in major amount, and the other one in minor amount. Hydrolysis of esters gives the corresponding acids.

Analysis of these results makes it clear that predominant *S*-enantiomer of atrolactic acid produced is related to the configuration of the chiral alcohol. In order to synthesize *R*-atrolactic acid, the other enantiomer of alcohol, *R*-2-octanol, will be required for the esterification of the keto acid.

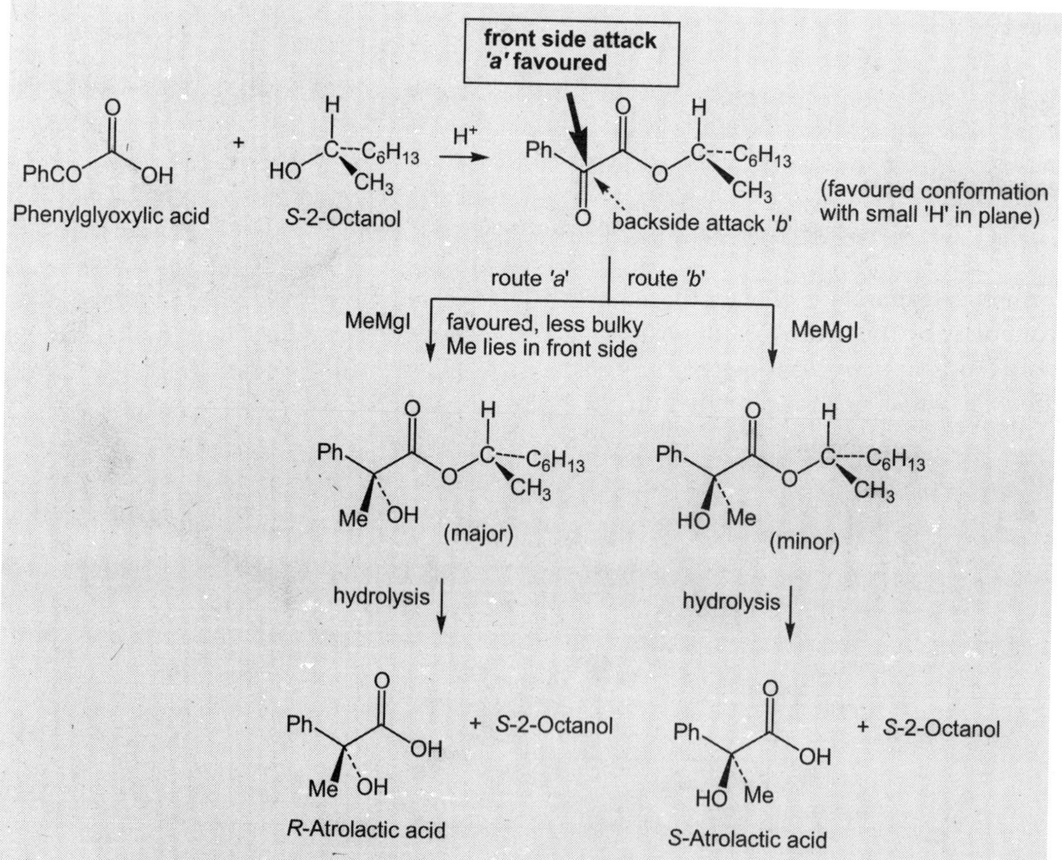

Fig. 3.26: Asymmetric synthesis of atrolactic acid based on Prelog's generalizations

Asymmetric induction in these reactions is, usually, low to moderate. However, high asymmetric induction (98%) has, so far, been observed when (–)-8-phenylmenthyl glyoxylate is used as the substrate. Other chiral alcohols, such as borneol and amyl alcohol, can also be used in these reactions.

The optical yield in these reactions is not only affected by the nature of alcohol, but also by the structure of keto acid and the nucleophile.

III. Using a Chiral Reagent (Enantiotopic Differentiation)

Third approach for asymmetric synthesis involves an achiral substrate and a chiral reagent. In such a situation, one enantiomer is formed in preference to the other, and these asymmetric syntheses are known as enantioselective syntheses (based on *enantiotopic differentiation*). The examples of this category include synthesis of compounds, such as chiral alcohols and hydroxy acids, using chiral reducing agents.

An illustrative example is that of the reduction of phenylglyoxylic acid using optically active *N*-benzyl-3-(hydroxymethyl)-4-methyl-1,4-dihydropyridine (**19**), which produces 97% of *S*-(+)-mandelic acid. It is to be noted that the other product, **20** is not chiral.

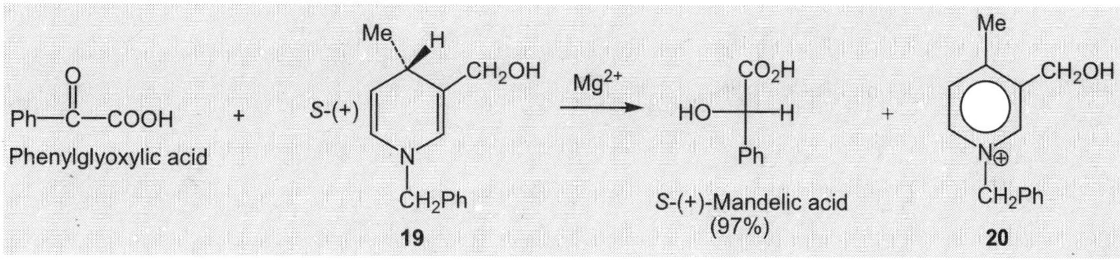

Another example is the synthesis of chiral 2-butanol which has been obtained in 91% optical purity by the hydroboration-oxidation of *cis*-2-butene using diisopinocamphenylborane [$(Ipc)_2BH$, prepared by treating optically active α-pinene with BH_3] (**Fig. 3.27**). These types of reactions, in which one reagent gives up its chirality to another are called *self-immolative*.

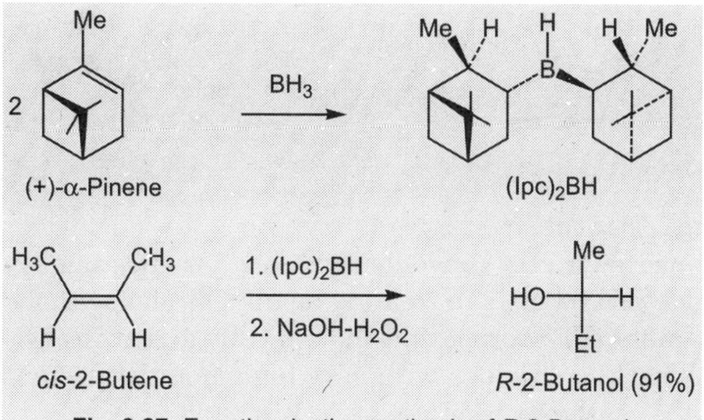

Fig. 3.27: Enantioselective synthesis of *R*-2-Butanol

IV. Using a Chiral Catalyst or Solvent

Many examples of asymmetric synthesis using a chiral catalyst or solvent are known in literature. Both homogenous and heterogenous catalytic hydrogenations have been employed for the enantioselective reduction of a double bond. These include: (*i*) reduction of ketones and substituted alkenes to optically active secondary alcohols and substituted alkanes, respectively, by treatment with hydrogen and chiral homogenous hydrogenation catalyst, (*ii*) the nucleophilic addition of organometallic compounds to aldehydes or ketones in the presence of chiral catalyst, (*iii*) conversion of alkenes to optically active epoxides by treatment with a hydroperoxide and a chiral catalyst and (*iv*) the reaction between secondary alkyl Grignard reagents and vinyl halides in the presence of chiral transition-metal complexes.

Two important examples of asymmetric reduction with very high enantiomeric excess (*ee*) values are:

1. Homogenous catalytic hydrogenation with 2,2'-bis(diphenylphosphino)-1,1'-binaphthyl-ruthenium acetate [BINAP-Ru$(OAc)_2$; **21**], which reduces β-keto esters giving > 98% *ee*.
2. Reduction with BH_3-THF or catechol-borane using an oxazoborolidine (**22**; R = H, Me, or *n*-Bu; Ar = Ph or β-naphthyl) as a catalyst, which gives high *ee* values with various types of ketones, especially α,β-unsaturated ketones.

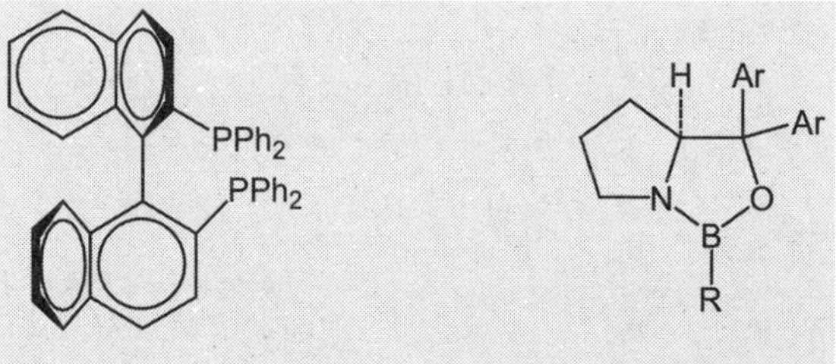

21 22

An example of the use of heterogeneous catalyst, such as tartaric acid (2*R*, 3*R*)-modified Raney nickel (MRN), is the reduction of methyl acetoacetate to methyl *R*-3-hydroxybutyrate in 88% enantiomeric excess (*ee*).

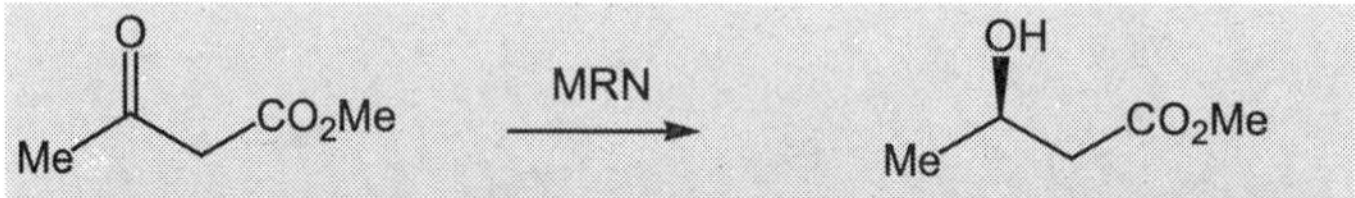

Sharpless Enantioselective Epoxidation

Sharpless (1980) reported a very convenient method for effecting asymmetric epoxidation of allyl alcohols. The method consists in treating an allyl alcohol with *tert*-butylhydroperoxide in the presence of titanium(IV) *iso*propoxide and optically active diethyl tartarate (DET). The reaction gives optically active epoxides with very high enantioselectivity (91-95%), and it is possible to prepare either of the two enantiomers (**Fig. 3.28**).

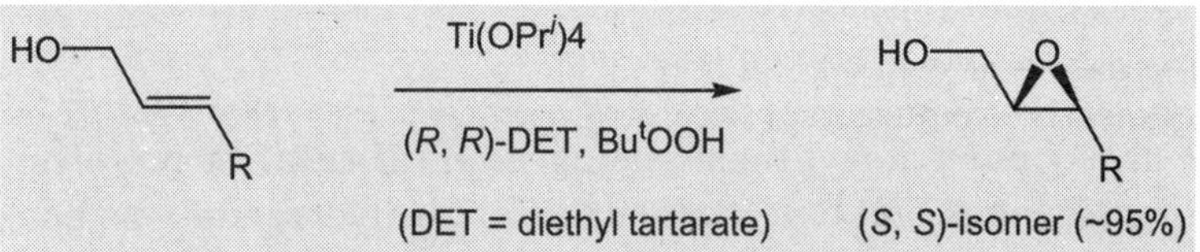

Fig. 3.28: Sharpless enantioselective epoxidation

A plausible explanation for the high stereoselectivity in Sharpless reaction is given in the following steps:

(*i*) Two molecules of *isopropyl* alcohol are released from titanium(IV) *iso*propoxide forming ten-membered intermediate (**23**). The structure of this intermediate is fluxional, because of a rapid exchange of coordination of titanium with two (out of four) carbonyl groups at a time. The structure possessing C_2 symmetry is ideal for enantioselectivity.

(*ii*) The allyl alcohol and *tert*-butylhydroperoxide displace the two remaining *iso*proppxide groups to form a chiral matrix in which the substrate is rigidly fixed.

(*iii*) The absolute configuration of the epoxide is predictable from the model (**24** in **Fig. 3.29**). The CH_2OH group of allyl alcohol is shown on lower right in the rectangular plane.

(*iv*) With (–) DET, oxygen approaches the double bond from the frontside giving the epoxide **25**, while with (+) DET, oxygen approaches from the backside, thus producing the other enantiomer **26**.

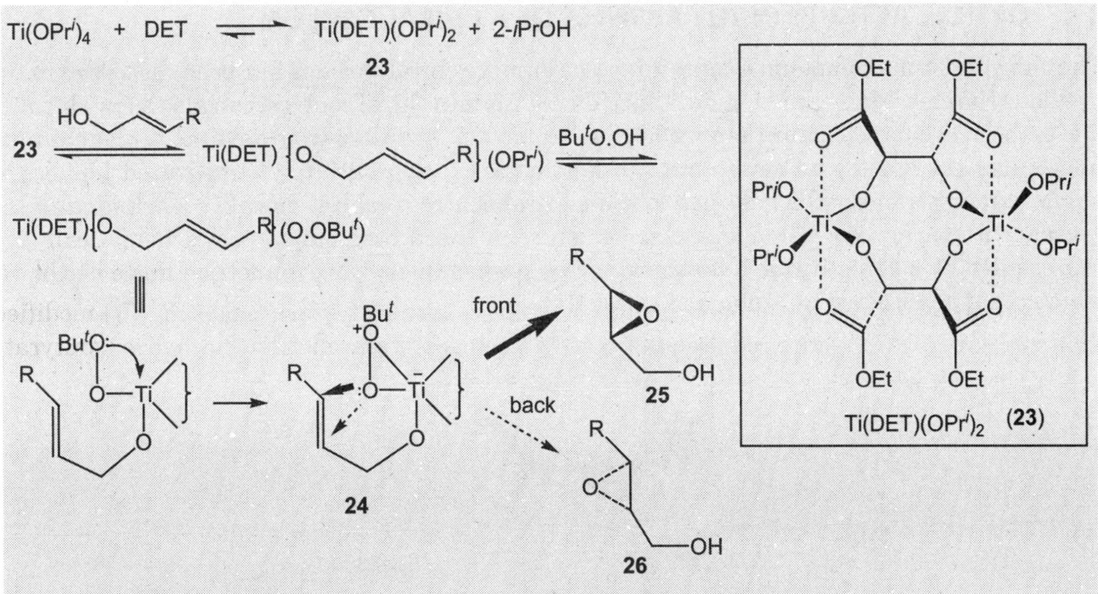

Fig. 3.29: A plausible mechanism for Sharpless enantioselective epoxidation

V. Using Enzymes as Catalyst

Enzymes, which are large protein molecules and have several chiral centres, can act as catalysts in various asymmetric syntheses occurring in biochemical systems or in the laboratory. These asymmetric syntheses are completely stereoselective. Of the various examples of enzymatic asymmetric syntheses, enantioselective reduction of ketones to alcohols using Baker's yeast is well-known. The first example is the conversion of 2-butanone to *S*- and *R*- 2-butanols, while the second one is the selective reduction of the carbonyl group of a β-keto ester (ethyl acetoacetate).

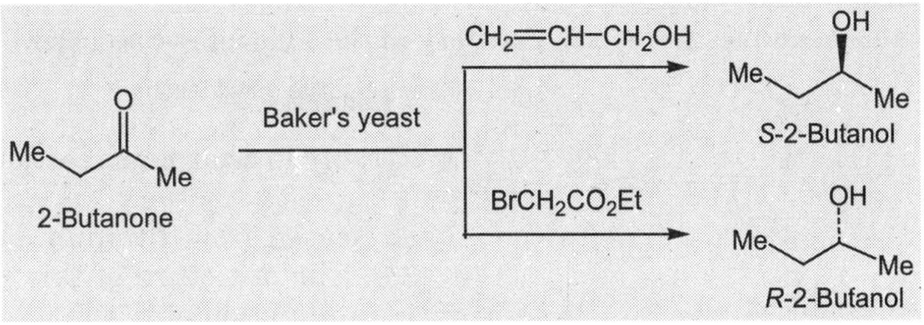

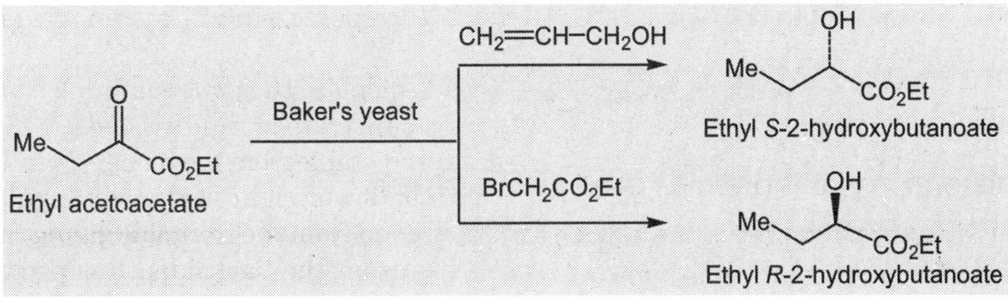

3.4 OPTICAL ACTIVITY IN THE ABSENCE OF A CHIRAL CARBON

The chirality in the compounds containing one or more chiral carbons has been discussed in the previous chapter. Many compounds, which do not contain chiral centres, can also show chirality due to the molecular structure as a whole. Such compounds are known to have molecular chirality (molecular asymmetry or dissymmetry). For instance, appropriately substituted biphenyls, allenes, and spiranes, which do not contain chiral centres, exhibit chirality attributed to the presence of chiral axis. The molecular structure of these compounds can, schematically, be represented by a general **Fig. 3.30** showing two perpendicular dissymmetric planes neither of which could be bisected by a plane of symmetry.

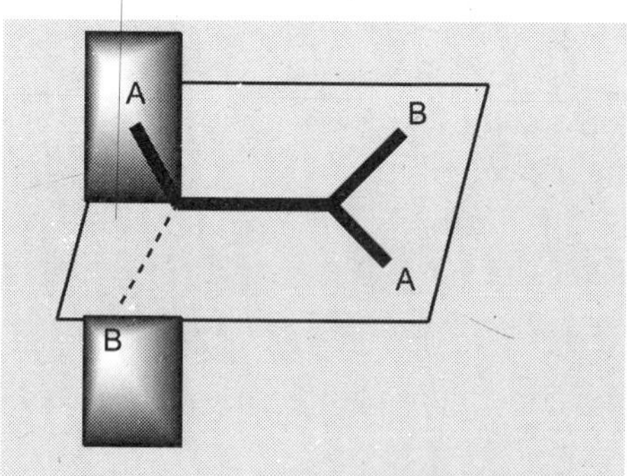

Fig. 3.30: Perpendicular dissymmetric planes

Elongated Tetrahedron Approach

Such compounds show chirality due to the presence of a *chiral axis* (*axis of asymmetry*). This chiral axis may be derived from chiral centre *Z* in a usual tetrahedral compound by extending the point into line AB, which now passes through elongated tetrahedron (**Fig. 3.31**). For *Z* to be a chiral centre, *a, a', b,* and *b'* must all be different, but for AB to be chiral axis, it is sufficient that *a* and *b* be different, and *a'* and *b'* be different. However, it is not necessary that *a* should be different from *a',* and *b* should be different from *b'*. These concepts will be explained by considering suitable examples from each category in the following discussion.

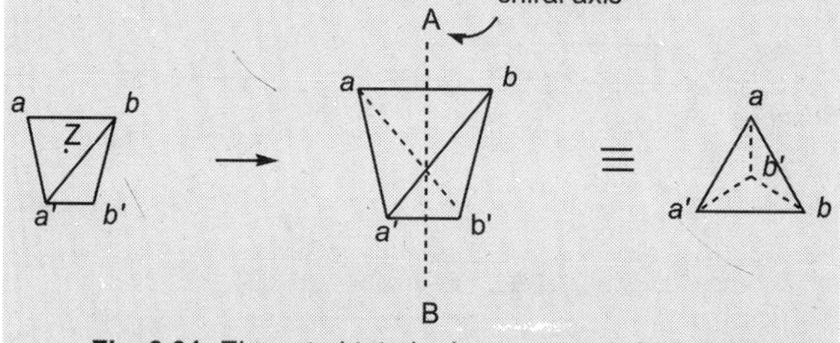

Fig. 3.31: Elongated tetrahedron approach (chiral axis)

A. Stereochemistry of Biphenyls (Atropisomerism)

The planar structure of biphenyl is shown as **27**. In this structure, two phenyl rings are joined by a single bond (sp^2-sp^2), which is known as the *pivotal* bond. Since the distance between *ortho* hydrogens in adjacent rings in this conformation is greater than twice the van der Waals radius of hydrogen, free rotation about the pivotal bond is not hindered.

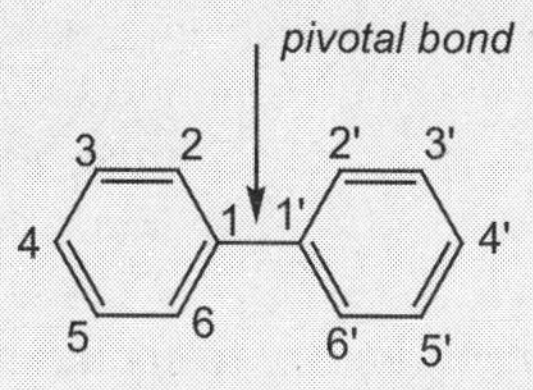

27

planar conformation of biphenyl

Free rotation about pivotal bond can be prevented by introducing bulky groups at the *ortho* positions. Apart from the size of groups, number of groups also affects the free rotation. Biphenyls containing at least three large groups in the *ortho* positions cannot freely rotate about the pivotal bond because of steric hindrance. Due to the presence of dissymmetry (chiral axis), such biphenyl derivatives are resolvable. This type of isomerism, which occurs due to restricted rotation around a single bond, is known as *atropisomerism,* and the isomers are called *atropisomers.*

There are two necessary conditions for biphenyls to exhibit optical activity:

(*i*) Neither ring must be symmetrically substituted.

(*ii*) The *ortho* positions must be occupied by large substituents so that the molecule cannot attain planar conformation due to steric repulsion.

When one or both the rings are symmetrically substituted, the biphenyl compounds will have a plane of symmetry whether the rings are coplanar (**28** and **29**) or lie in different planes (perpendicular, **30**). It is, thus, not possible to resolve such biphenyl derivatives. In a similar manner, when one ring is symmetrically substituted and the other is unsymmetrically substituted, the compounds are achiral (**Fig. 3.32**).

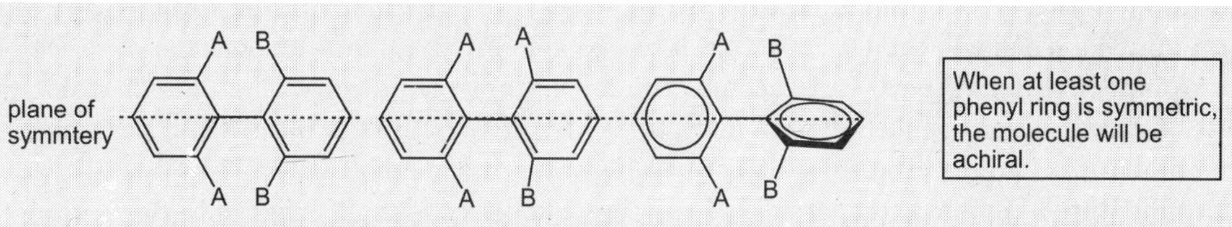

28 **29** **30**

planar conformations

rings are perpendicular to each other

Fig. 3.32: Plane of symmetry in biphenyls when at least one phenyl ring is symmetric

When both the rings are unsymmetrically substituted, even then there is possibility of the presence of a plane of symmetry, as shown in **31** and **32**.

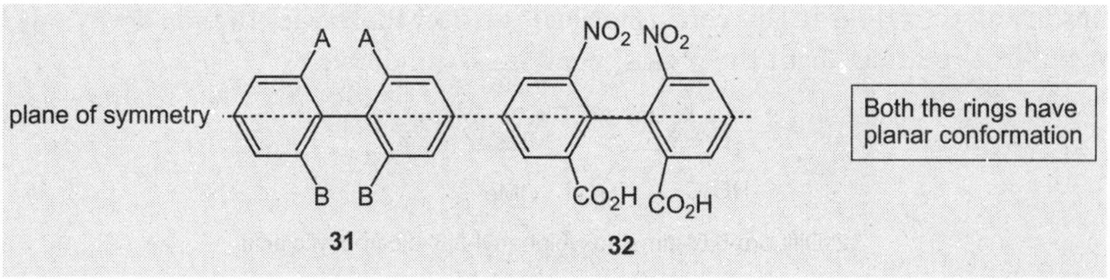

31 **32**

However, if the two rings are perpendicular to each other due to restricted rotation around pivotal bond, these compounds could show optical activity. Many such compounds having sufficiently bulky groups, for example, 6-nitrodiphenic acid, 6,6'-dinitrodiphenic acid, and 6,6'-dichlorodiphenic acid have been resolved. As can be seen in **Fig. 3.33**, enantiomeric pairs of chiral compounds like, 6,6'-dinitrodiphenic acid (**32**), can be represented by two different ways as structures **32a/ 32b** and **32a'/32b'**.

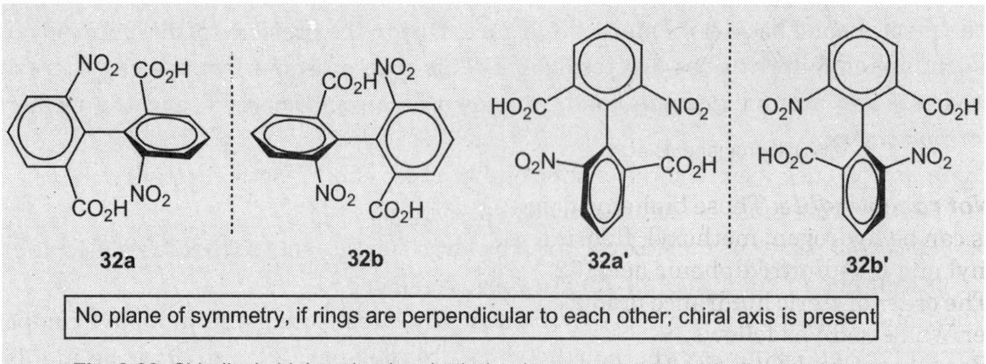

Fig. 3.33: Chirality in biphenyls when both the rings are unsymmetrically substituted

Optical Stability of Chiral Biphenyl Derivatives and Racemization

It is clear from the preceding discussion that chiralty in *atropisomers* arises from restricted rotation around a single bond. In practice, it has been found that many atropisomers (optically active biphenyl derivatives) can racemize under suitable conditions. The optical stability (rate of racemization) of such enantiomers is greatly affected by the nature and positions of the substituent groups. The majority of atropisomers in biphenyl derivatives are 2,2',6,6'-tetrasubstituted biphenyl compounds. These compounds may be classified into three groups according to the nature of substituents:

(i) Non-resolvable: Biphenyl compounds containing hydrogen, methoxyl or fluorine are not resolvable. For example, 2,2'-difluoro-6,6'dimethoxybiphenyl-3,3'-dicarboxylic acid is not resolvable. The effective volumes of the *ortho* groups are too small to prevent rotation about the single bond. Carboxyl groups present at the *meta* positions do not affect rotation about the single bond.

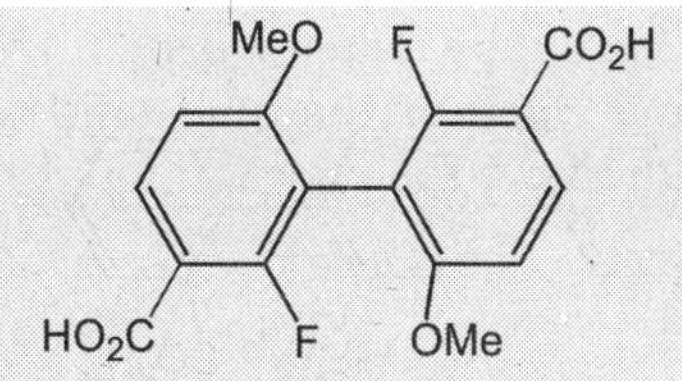

2,2'-Difluoro-6,6'-dimethoxybiphenyl-3,3'-dicarboxylic acid
(non-resolvable)

(ii) Resolvable, but easily racemized: These compounds must contain at least two amino groups, or two carboxyl groups, or one amino and one carboxyl group at *ortho* positions; the remaining groups may be methoxyl or fluorine. For example, 6,6'-difluorodiphenic acid and 6,6'-dimethoxybiphenyl-2,2'-diamine are resolvable, but are readily racemized.

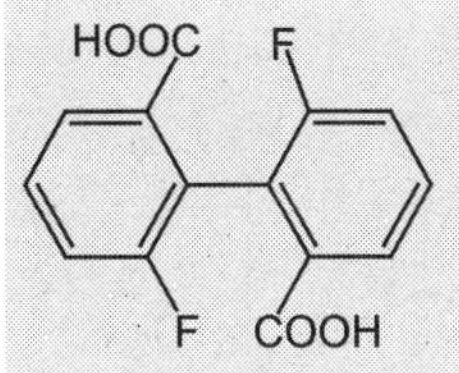

6,6-Difluorodiphenic acid

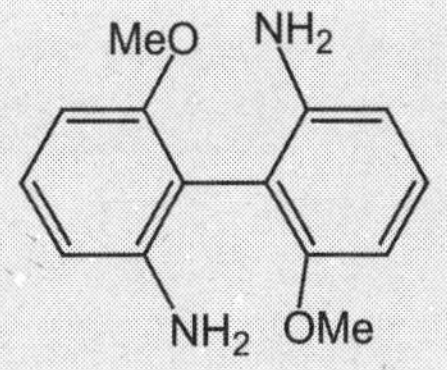

6,6'-Dimethoxybiphenyl-2,2'-diamine

(iii) Not racemizable: These biphenyl derivatives contain at least two nitro groups; the other groups can be hydrogen, methoxyl, fluorine, amino or carboxyl. Thus, 2,2'-difluoro-6,6'-dinitrobiphenyl and 6,6'-dinitrodiphenic acid (**32**) are resolvable, and cannot be racemized.

The order of steric hindrance produced by various groups follows roughly the order of their van der Waals radii as follows:

Br > Me > Cl > NO_2 > CO_2H > OMe > F

(a) Buttressing Effect: In addition to the size of the groups in *ortho*-position, the nature and position of other substituents also determine the rate of racemization. For example, it has been found that the rate of racemization of compound **33** (planar structure) is much slower than that of compound **34**. This clearly indicates that the nitro group in position 3' has a much greater stabilizing influence than in position 5'.

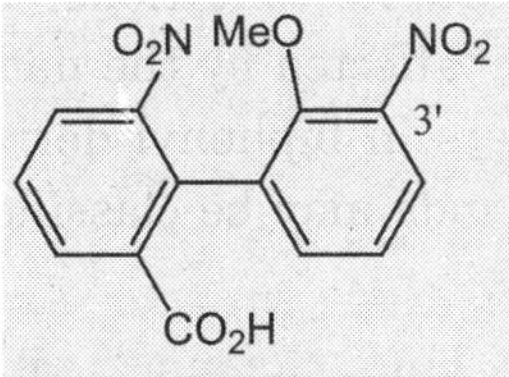

33
more stable

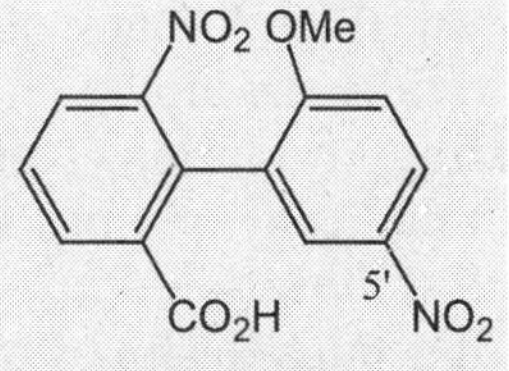

34
less stable

Although reason for this observation in not clear, one possible explanation is as follows. In compound **34**, the methyl of the methoxyl group takes up the position as shown in **34a** because of the absence of any substituent at position 3'. On the other hand, compound **33**, which contains the nitro group at the position 3', cannot take up the configuration **33a** or **33b** similar to **34a**.

Instead, the methyl of the methoxyl group is forced by the nitro group to take up the position directed towards the other phenyl ring. Such a situation would not allow planar conformation of the biphenyl derivative, thereby stabilizing the non-planar (chiral) structure **33c** (**Fig. 3.34**).

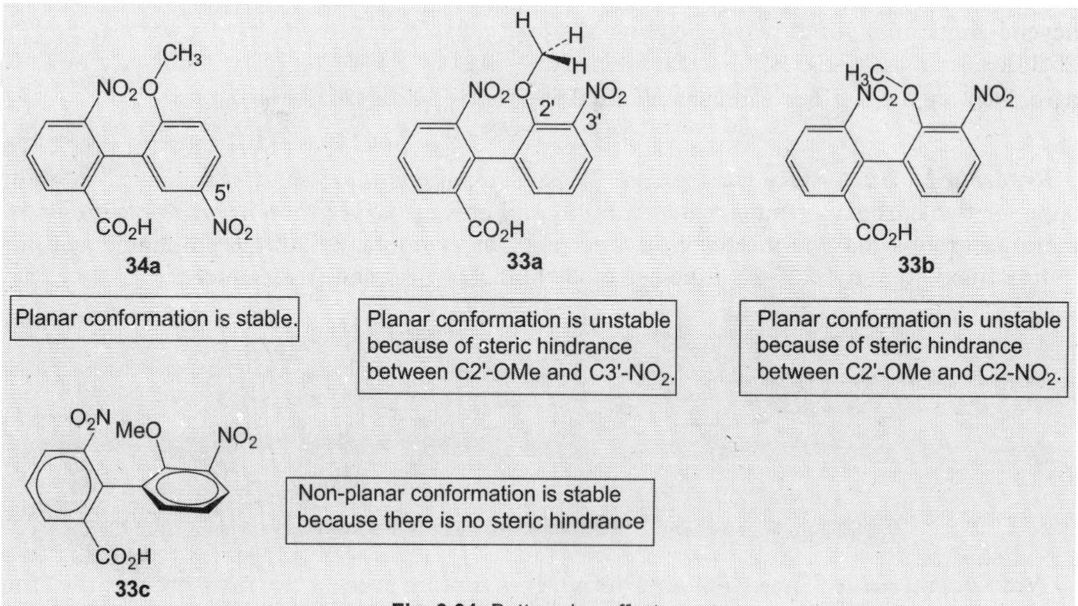

Fig. 3.34: Buttressing effect

Thus, it can be understood why the substituent at position 5' would not exert much influence on the optical stability. This type of stabilizing influence of various groups is known as *buttressing effect*.

The buttressing effect of some groups is in the following order:

$NO_2 > Br > Cl > Me$

It is interesting to note that the above order of buttressing effect is very different from that of the steric hindrance produced by these groups in the *ortho* position.

(b) Racemization of Biphenyl Compounds: In the preceding discussion, it has been shown that optical stability of biphenyl compounds is greatly affected by the nature and position of the substituent groups. In addition, physical conditions, such as temperature, are also responsible for the optical stability of biphenyl derivatives. Many such compounds can undergo racemization under suitable conditions, *e.g.* on heating in solution. The most suitable explanation for racemization is that heating increases the amplitude of vibrations of the substituent groups at 2,2',6,6' positions, and also the amplitude of vibration of the two benzene rings with respect to each other. This permits the substituent groups to slip by one another. Thus, the nuclei probably pass through a common plane, thereby, producing (+)- and (–)- forms in equimolecular amount.

The racemization of biphenyl compounds may easily be understood with the help of an energy profile diagram. An approximate energy profile diagram is shown in **Fig. 3.35** for a 360° (θ) rotation around the pivotal bond. Since inter-ring interactions are minimum at 90°/270°, the rings are perpendicular to each other in the preferred conformations of enantiomers. Thus, the enantiomers A and A' have the same energy content. When they approach the planar configuration (B or C), the energy content increases maximum due to steric repulsion. Thus, size of the groups will greatly affect the energy content. Larger the groups *a* and *b*, higher will be the energy barrier, that means slower will be the rate of racemization. As a matter of fact, when phenyl group rotate with respect to each other about pivotal bond, the energy barrier is very low, and the various positions would correspond to different conformers analogous to acyclic compounds. In the case of resolvable biphenyls, the energy barrier is large enough (17.9-25.11kcal/mol or 75-105 kJ/ mol) to prevent the rings to pass through a planar structure. Thus, atropisomers are neither conformational isomers nor configurational isomers.

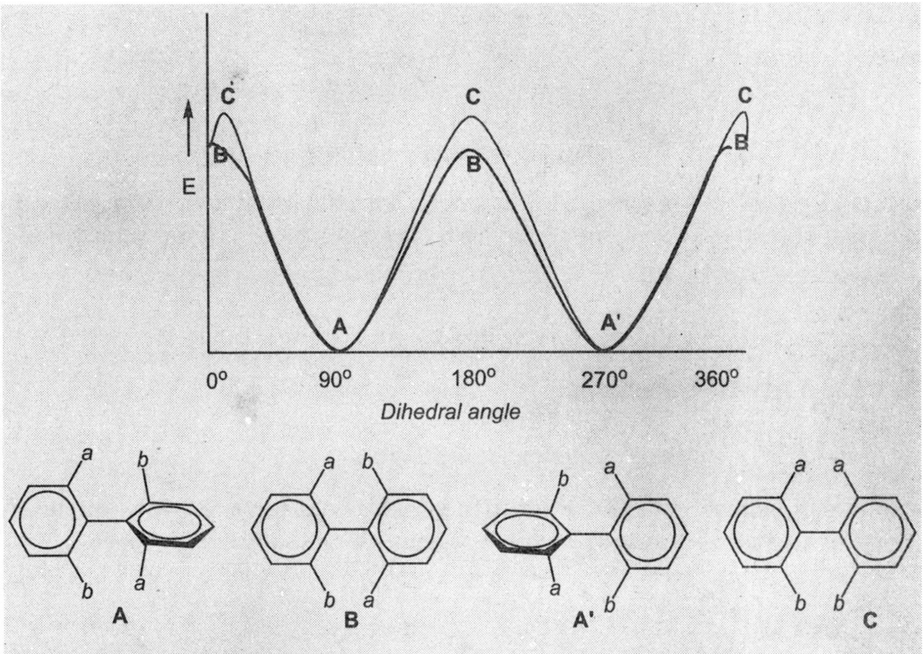

Fig. 3.35: Racemization of biphenyls

Atropisomerism in compounds other than biphenyls

Many other cases are also known, where optical activity arises due to the hindered rotation of other types of aromatic ring, such as binaphthyls, bipyrryls, and bipyridyls. Important examples include 2,2'-binaphthol (constituent of chiral reagent BINAL-H) , bipyrryl (**35**), and bipyridyl (**36**).

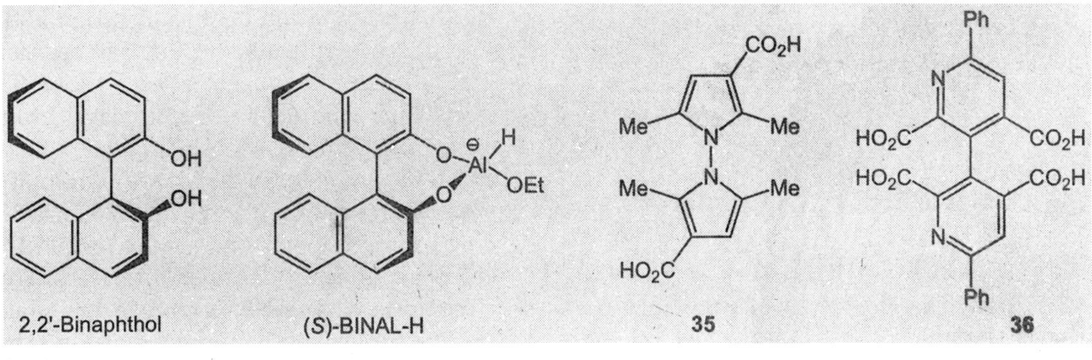

2,2'-Binaphthol (*S*)-BINAL-H **35** **36**

B. Stereochemistry of Allenes

In allenes, the central carbon atom is *sp*-bonded, and the remaining two *p* orbitals are perpendicular to each other. Each *p* orbital overlaps with a *p* orbital of the adjacent carbon atom (sp^2), forcing the two remaining bonds of each carbon into perpendicular planes. Thus, structure of allenes may be represented as shown in **Fig. 3.36.**

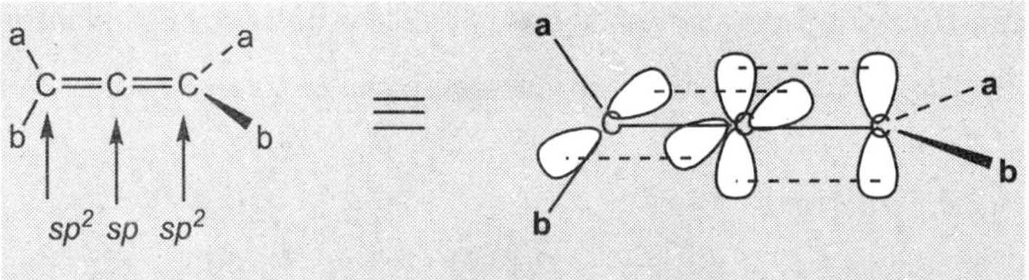

Fig. 3.36: Structure of allenes

Like biphenyls, allenes display chirality only if both the terminal carbon atoms are unsymmetrically substituted (two substituents on each carbon *must not* be the same) (**Fig. 3.37**).

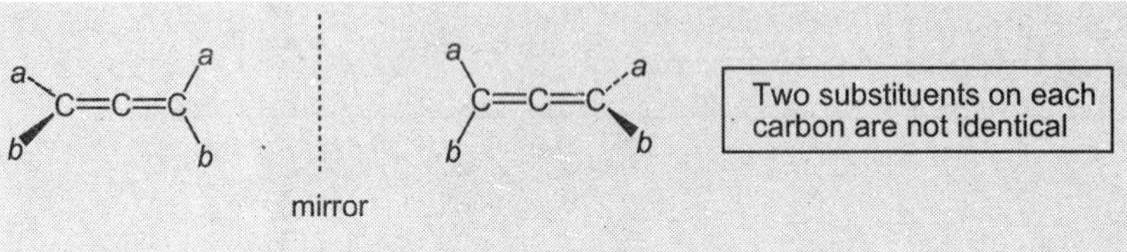

Fig.3.37: Enantiomerism in allenes

But unlike biphenyls, size of the substituents does not have any role for the chirality of allenes. Some examples of chiral (**37**)/achiral allenes (**38** and **39**) are given.

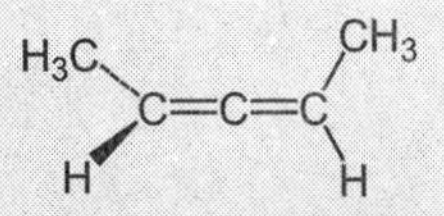

37 (chiral)
(both the planes are dissymmetric)

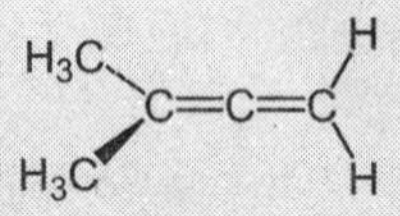

38 (achiral)
(both the planes are symmetric)

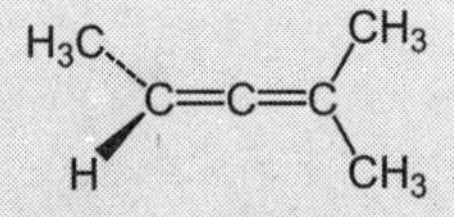

39 (achiral)
(one plane is symmetric)

It must be noted that these cases are entirely different from the geometrical isomerism shown by compounds having a double bond. All the four groups are in *one plane* in geometrical isomers, whereas in allenes the groups are in *perpendicular planes*.

The situation becomes different when odd numbers like 3, 5 and 7 of cumulative double bonds are present in a cumulene. In these cases, orbital overlap causes four groups to occupy one plane, and as a result geometrical isomerism is observed (**Fig. 3.38**).

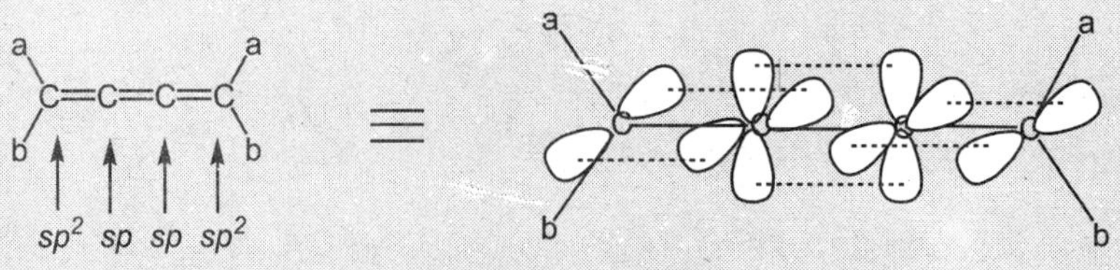

Fig.3.38: Geometrical isomerism in cumulenes containing three (odd) cumulative double bonds

When even number like 4 and 6 of cumulative double bonds are present, the situation is analogous to that in the allenes (two cumulative double bonds), and hence it is possible to resolve such compounds (**Fig. 3.39**).

Fig. 3.39: Enantiomerism in cumulenes containing four cumulative double bonds

C. Stereochemistry of Spiranes

Spiranes are named by using the alkane name to designate the total number of carbons and prefix "spiro", which is followed by brackets indicating the number of carbon atoms in each ring. These are listed in the order of decreasing ring size. To determine the position of a substituent, the carbons of the rings are numbered starting at a shared carbon, and going around the longer ring first as shown in the examples in **Figs. 3.40-3.42**.

Because of the sharing of a carbon atom, the two rings of spiranes are in perpendicular plane. In case both the rings are dissymmetric, such compounds will display chirality. A few examples of spiranes, having symmetric or dissymmetric rings are illustrated in **Figs. 3.40-3.42**.

Spiro[2.2]pentane   Spiro[5.4]decane

Fig. 3.40: Achiral spiranes (both the rings are symmetric)

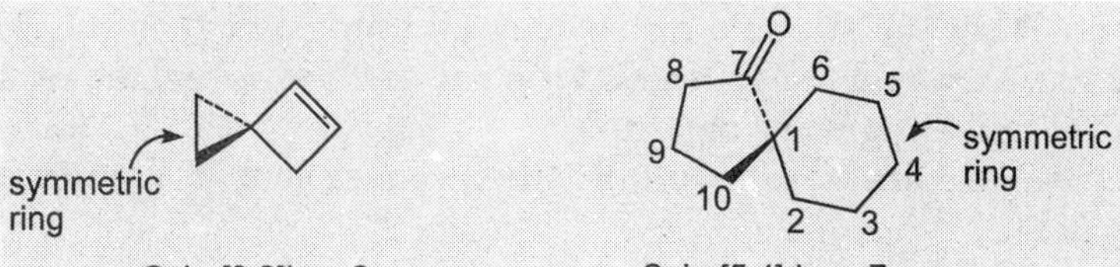

Spiro[3.2]hex-2-ene   Spiro[5.4]deca-7-one

Fig. 3.41: Achiral spiranes (one ring is symmetric)

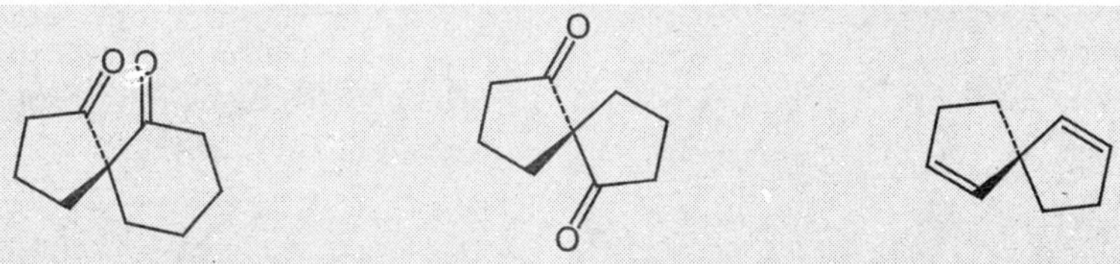

Spiro[5.4]deca-2,10-dione   Spiro[4.4]nona-2,6-dione   Spiro[4.4]nona-2,6-diene

Fig. 3.42: Chiral spiranes (both the rings are dissymmetric)

D. Configurational Nomenclature of Biphenyls, Allenes, and Spiranes

The assignment of absolute configuration of these classes of compounds can be carried out by a common procedure. Biphenyl, allene, and spirane are converted into an elongated tetrahedron as shown in **Fig. 3.43**, which can be used for assignment of the configuration. To apply sequence rule in such cases, an additional sub-rule has to be designed. The near groups are given priority over the far groups. If the tetrahedron is viewed from point A, then the pair *a*-*b* being nearer A, precedes the pair *a'*-*b'*. If the order of priority is $a > b$ and $a' > b'$, then priority of *a*-*b* will be 1 and 2, and *a'*-*b'* will be 3 and 4 as shown in **40**, which is better shown as **41** and **42**. After carrying out the conversion of the given structures, we can see that situation is similar to that of Fischer projection of a compound containing one chiral carbon with lowest priority group (4) away from the observer. Now applying sequence rule, it can be seen that 1, 2, 3 is clockwise, and therefore configuration is *R*.

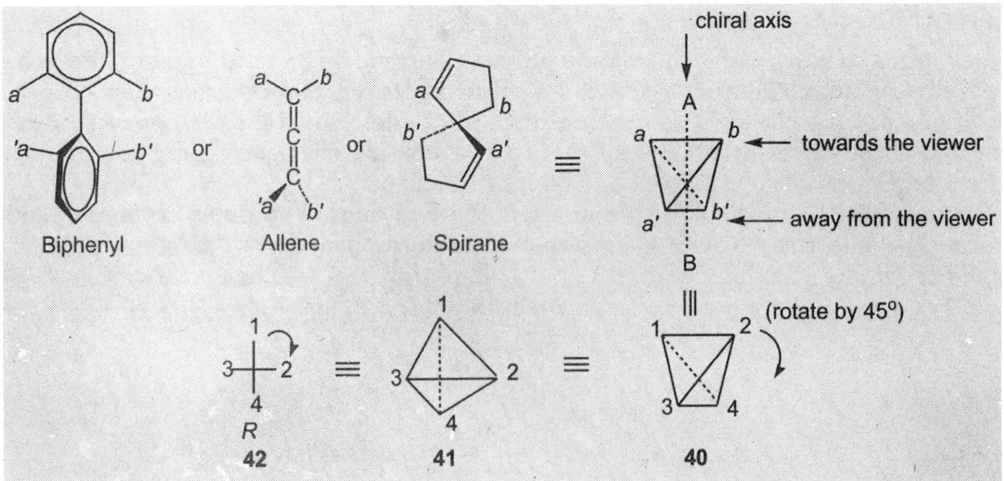

Fig. 3.43: Configurational nomenclature of biphenyls, allenes and spiranes

It is to be noted that the **40** could also be viewed from point B on the chiral axis. In such a situation, the pair *a'*-*b'* would have preceded the pair *a*-*b*. However, the result after following the whole procedure will be same as before.

If the structures are drawn horizontally, then conversions of the formulas are followed according to **Fig. 3.44**.

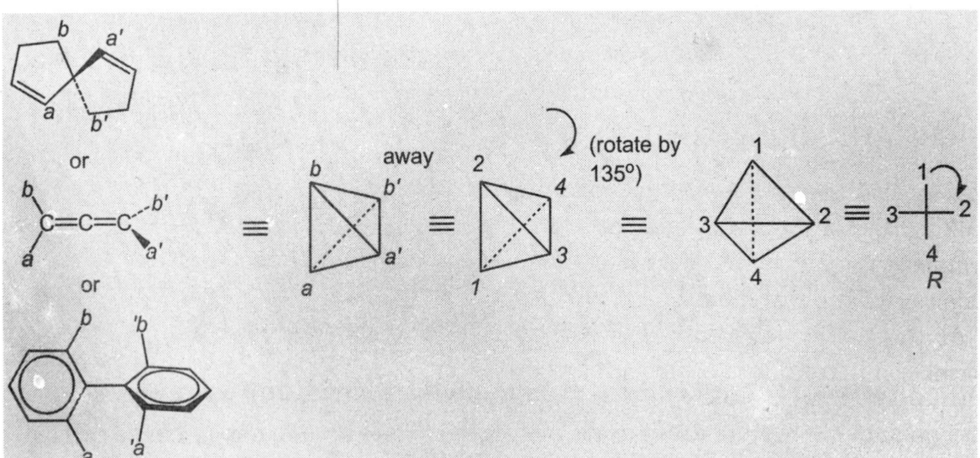

Fig. 3.44: Configurational nomenclature of biphenyls, allenes and spiranes

There is another method through which it is possible to determine the configuration of the compounds containing a chiral axis without having above conversions. A simplified procedure is to draw flying wedge formula of the given structure, and then directly apply sequence rule. For drawing flying wedge formula, pivotal bond (and other bonds) of biphenyl, carbon-carbon double bonds of allene, and ring carbons of spirane are replaced by a carbon centre as shown in **Fig. 3.45**. Then, the groups that are present in the plane of the paper are given higher priority over the groups of perpendicular plane. In these cases, either 3 or 4 priority will be away from the observer. The method can easily be understood with the help of **Fig. 3.45**.

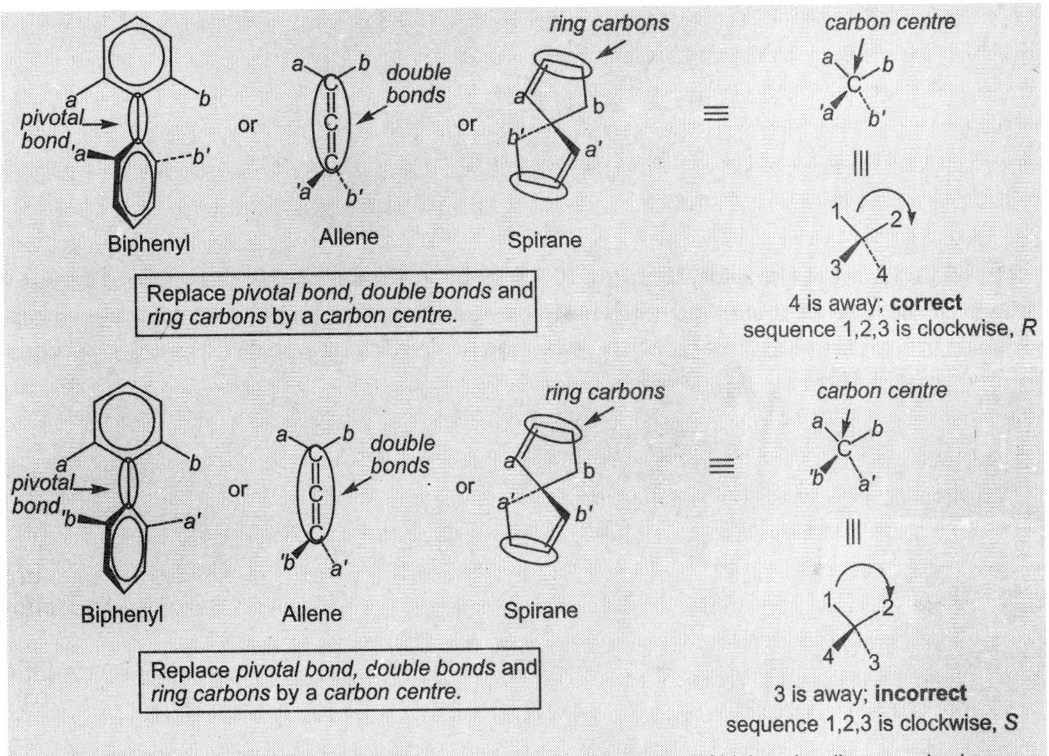

Fig. 3.45: Alternative approach for determination of configuration of biphenyls, allenes and spiranes

3.5 CHIRALITY DUE TO HELICAL SHAPE

Helical molecules are those in which the arrangement of atoms or groups is an imaginary helix. A helix is inherently chiral, and therefore the helical molecules are known to possess *helical dissymmetry or helicity*. A helix may also be considered as manifesting axial chirality, when its axis serves the chiral axis.

At the molecular level, helicity often arises from the secondary structure, and is conformational in origin. The helical structure has intensively been studied in protein molecules. The polypeptide chain structure derived from naturally occurring L-amino acids coils to form α-helix. Naturally occurring DNA also exists in three different forms A-helix, B-helix, and Z helix. The skeleton of a double helix present in the structure of nucleic acids is shown in **Fig. 3.46(a)**.

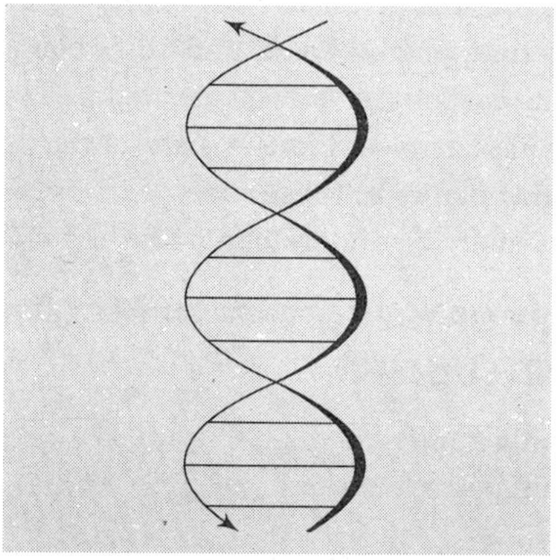

Fig. 3.46a: Double helical structure of DNA

The assignment of chiral designation to the helix is simple, and is done in a different way as follows. If moving from one end to the other along the axis the helix describes a clockwise direction, it is designated *P* (plus), and if it describes an anticlockwise direction, it is designated *M* (minus) (**Fig. 3.46b**).

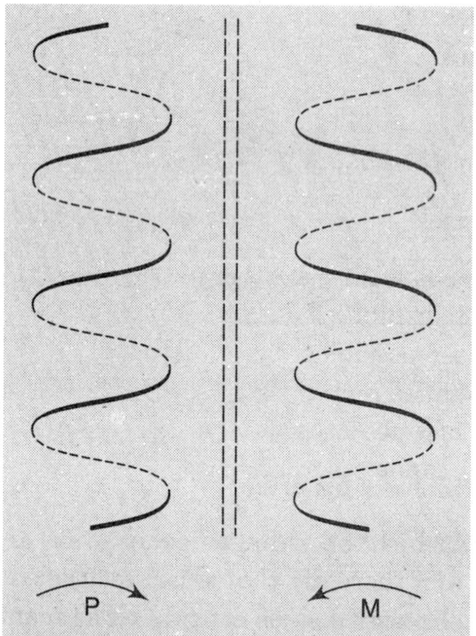

Fig. 3.46b: *P*/*M* designation to the helix

Helicity due to Molecular Overcrowding: Helicity may also result due to molecular overcrowding. Some examples include helicenes (*e.g.* hexahelicene, **43**), benzphenanthrenes (**44**), and phenanthrenes (**45**). These molecules are normally expected to be planar, but the ring structures assume a helical shape due to molecular overcrowding.

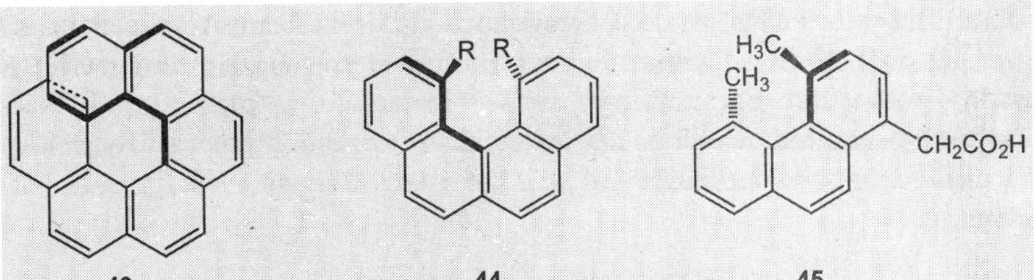

43 44 45

3.6 STEREOCHEMISTRY OF THE COMPOUNDS CONTAINING NITROGEN, PHOSPHORUS AND SULPHUR

The carbon atom is not the only atom that creates chiral centres in the molecules. Many nitrogen, sulphur, and phosphorus compounds are also known to exhibit stereoisomerism.

A. Stereochemistry of Nitrogen Compounds

The tetrahedral concept of carbon can also be applied to nitrogen. The only difference is that one of the sp^3 orbitals of nitrogen usually contains a lone pair of electrons which is not involved in bonding. Thus, nitrogen may be considered either tetrahedral or pyramidal. Many nitrogen compounds have been found to exhibit stereoisomerism. These compounds may be classified into the following four categories: (1) tertiary amines, (2) quaternary ammonium salts, (3) tertiary amine oxides and (4) oximes.

I. Tertiary Amines

Assuming the pyramidal or tetrahedral shape of the nitrogen atom, amines containing three different ligands are expected to give rise to chirality. Since the unshared pair of electrons of nitrogen atom may be treated as the fourth group, tertiary amines of the type $R^1R^2R^3N$ (or secondary amine of the type R^1R^2NH) may be treated as having a chiral centre, and therefore, such amines are expected to exhibit enantiomerism.

However, it has not been possible to resolve open-chain amines. A plausible reason for this is the *umbrella effect* (also called *pyramidal inversion*), in which the unshared pair of electrons rapidly oscillates (millions of times a second) from one side of the $R^1R^2R^3$ plane to the other, thus converting one enantiomer to the other. These stereoisomers are also called **invertomers**.

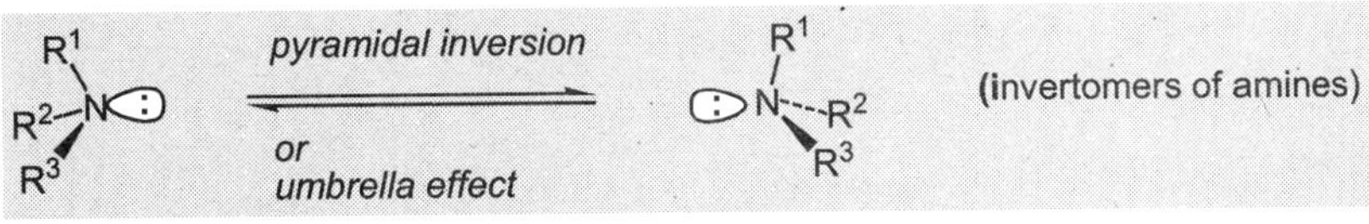

There are two types of nitrogen atom which invert slowly- i) when nitrogen atom is a part of the ring, particularly a three-membered ring, and ii) when a nitrogen atom is connected to another atom bearing an unshared pair.

However, umbrella effect in such compounds is still too rapid to permit isolation of enantiomers. The goal of isolating enantiomers could be achieved only when both the features are present in a compound. For example, the two isomers of 1-chloro-2-methylaziridine (**46** and **47**) have been separated, which do not interconvert at room temperature. In none of these cases, the optical activity of the compound was not solely because of the presence of tervalent nitrogen atom.

46 **47**

The examples of compounds which are chiral only because of asymmetric tervalent nitrogen atom include oxaziridines and aziridines of the type **48** and **49** (**Fig. 3.47**). While enantiomers of oxaziridine **48** (**48a** and **48b**) are stable at room temperature, enantiomers of aziridine **49** (**49a** and **49b**) undergo complete racemization upon standing..

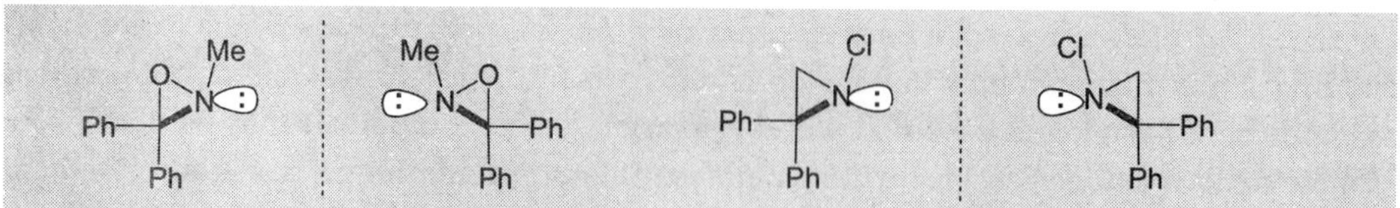

48a **48b** **49a** **49b**

Fig. 3.47: Enantiomerism in oxaziridines and aziridines due to the presence of asymmetric tervalent nitrogen

When the nitrogen is at a bridgehead of the molecules, pyramidal inversion is prevented. Such molecules, if chiral, can be resolved even without the presence of the above mentioned two structural features. For example, Troger's base (**50**) has been resolved by chromatographic adsorption on D-lactose. The frequency of oscillation has been brought to zero in such compounds by having the three valencies of nitrogen as part of the ring system.

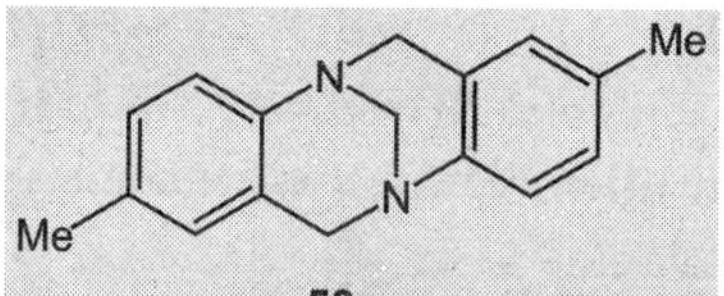

50

II. Quaternary Ammonium Salts

When four different groups are attached to the nitrogen of quaternary ammonium salts ($R^1R^2R^3R^4N^+$ X^-), the compounds of this type show enantiomerism (**51a** and **51b**; enantiomers).

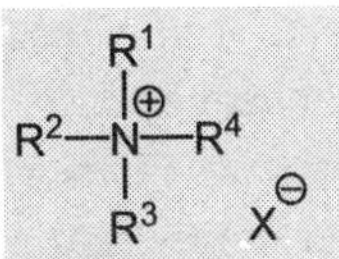

51a

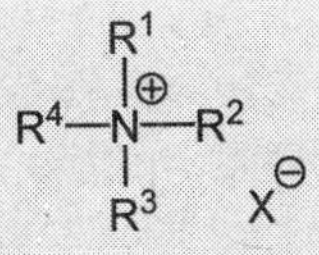

51b

Assuming that charge has no effect on the configuration of the cation, it may be considered as five-point system similar to that of carbon in compounds of the type *Cabde*. If the cation is not planar, it must have either pyramidal or tetrahedral shape, and therefore, quaternary ammonium salts containing four different groups are resolvable. The first example of such chiral compounds was reported by Pope and Peachey who resolved allylbenzylmethylphenylammonium iodide (**52**) by using (+)-bromocamphorsulphonic acid.

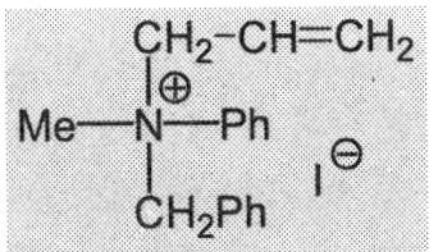

52

Racemization of Optically Active Quaternary Ammonium Salts

Optically active quaternary ammonium salts undergo racemization far more readily than carbon compounds containing a chiral carbon. The racemization of such chiral ammonium salts can be explained by their dissociation into the amines, which are not resolvable, and subsequently react with alkyl halide to give racemic mixture (**Fig. 3.48**).

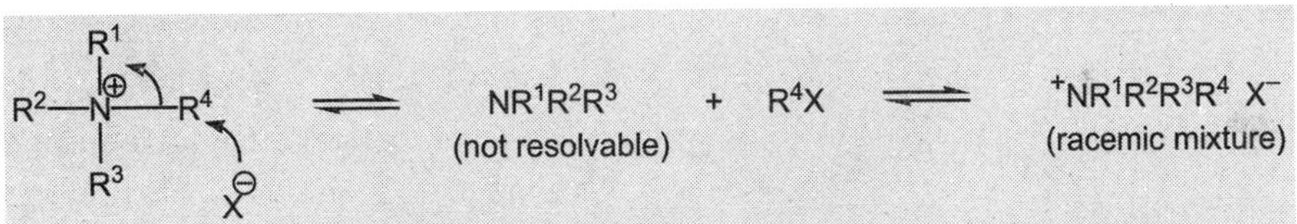

Fig. 3.48: Racemization of quaternary ammonium salts

III. Tertiary Amine Oxides

If nitrogen atom of tertiary amine oxides of the type $R^1R^2R^3N^+O^-$ is joined to three different groups and its configuration is tetrahedral, such compounds can show enantiomerism (**53a** and **53b** are enantiomers), and are resolvable.

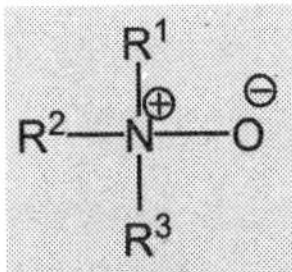

53a

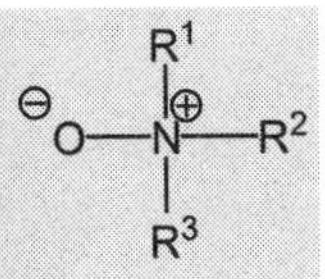

53b

Some of these compounds are ethylmethylphenylamine oxide (**54**), ethylmethyl-1-naphthylamine oxide (**55**), and kairoline oxide (**56**).

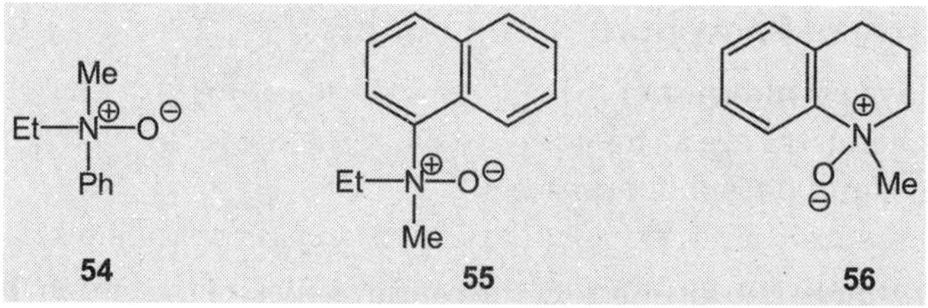

IV. Oximes

The oximes which contain >C=NOH functionality exhibit geometrical isomerism. Both carbon and nitrogen atoms in the oximes are sp^2 hybridized. This is analogous to the hybridization of the carbon atoms in olefinic compounds. Thus, the oximes are coplanar about the double bond, and there is no free rotation. The configurations of two geometrical isomers are shown as **57** and **58**.

Nomenclature of Geometrical Isomers of Oximes

Two types of prefix have been used to indicate the configuration of oximes. These are: *syn*/*anti* and *E*/*Z*.

In case of aldoximes, the *syn* isomer is the one in which both, the hydroxyl group and the hydrogen atom, are on the same side; when these groups are on opposite sides, the configuration of the oxime is named as *anti*.

While considering the case of ketoximes, the prefix (*syn*/*anti*) indicates the spatial relationship between the first group named and the hydroxyl group. Thus, the oximes **59a** and **59b** may be named as *anti*-*p*-anisyl phenyl ketoxime (or *syn*-phenyl *p*-anisyl ketoxime) and *syn*-*p*-chlorophenyl *p*-anisyl ketoxime (or *anti*-*p*-anisyl *p*-chlorophenyl ketoxime), respectively.

The *E*-*Z* system of nomenclature is also applicable to oximes and is found to be more convenient and clear. If the group with the greater priority and the hydroxyl group are on the same side of the double bond, the configuration is *Z*; if these are on the opposite sides, it is called *E*. For example, the oximes **59a** and **59b** are the *E* and *Z* isomers, respectively.

B. Stereochemistry of Phosphorus Compounds

Phosphorus displays covalencies of 3, 4, 5, and 6. Important types of phosphorus compounds displaying stereochemistry are as follows: phosphines, phosphine oxides, phosphonic esters and related compounds, and phosphonium salts.

The valency disposition in tercovalent compounds is tetrahedral (*sp*^{3}), one orbital being occupied by a lone-pair. In quinquevalent compounds, the valency disposition is trigonal bipyramidal (*sp*^{3}*d*). In case of quadricovalent univalent compounds, one electron is transferred from the phosphorus atom to the anion, and the valency disposition is tetrahedral (*sp*^{3}).

I. Phosphines

Assuming the trigonal pyramidal configuration of tercovalent compounds, R_3P, it is to be expected to resolve phophines of the type $R^1R^2R^3P$ in which phosphorus is linked to three different groups. The phosphorus atom is in a state of oscillation. However, frequency of oscillation in phosphine is *much slower* than that of nitrogen. If frequency of the oscillation could be brought to zero (when *umbrella effect* is absent), then tertiary phosphines would be resolvable (**Fig. 3.49**). One important factor which can slow down the oscillation in phosphorus compounds is the weight of the groups.

Fig.3.49: Enantiomerism in phosphine (*umbrella effect* is absent)

Heavier the weight of the groups, slower will be the frequency of oscillation. For example, replacement of the three hydrogen atoms by deuterium atoms slows down the frequency of oscillation.

Phosphines and diphosphines such as EtMePhP, MePhPrP and $MePhPCH_2CH_2PPhMe$ have been resolved. The absolute configuration of methylphenylpropylphosphine was determined through the formation of benzylmethylphenylpropylphosphonium bromide. Optically active phosphines are somewhat optically stable, and racemize on heating (probably through oscillation).

II. Phosphine Oxides

Phosphine oxides of the type $R^1R^2R^3P{=}O$ are among the earliest phosphorus compounds to be resolved (**Fig. 3.50**).

Fig. 3.50: Enantiomerism in phosphine oxides

The examples of this category include ethylmethylphenylphosphine oxide (**60a/60b**) and benzylmethylphenylphosphine oxide (**61a/61b**).

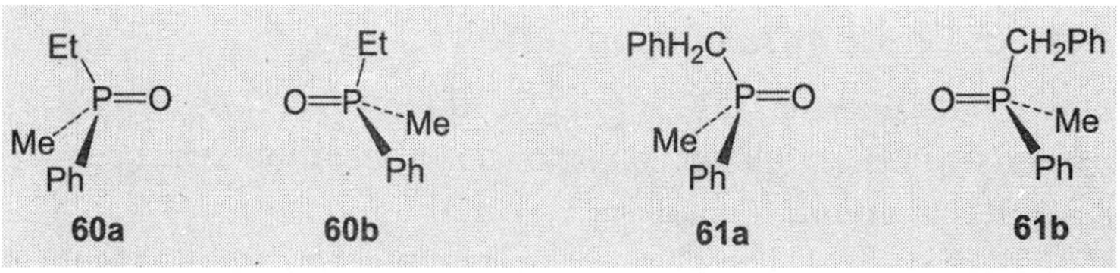

Phosphines (**62**) undergo oxidation to phosphine oxides (**63**) by using various oxidizing agents, such as hydrogen peroxide, with retention of configuration. It is, thus, possible to prepare an enantiomer of a phosphine oxide with a known configuration.

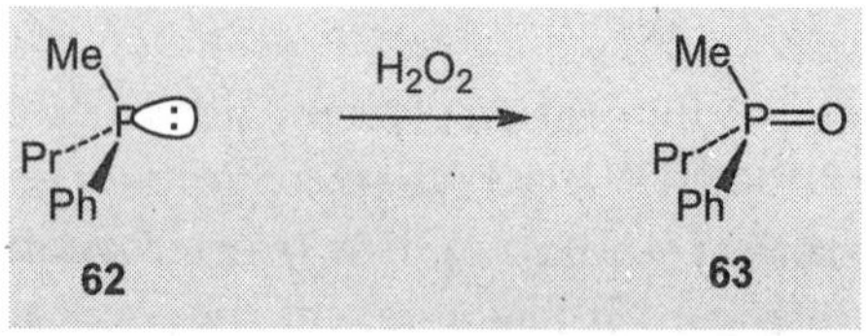

Racemization of Optically Active Phosphine Oxides

Optically active phosphine oxides undergo racemization in the presence of acids. A possible mechanistic explanation involves the formation of symmetrical hydrate which has a trigonal bipyramid configuration (**Fig. 3.51**).

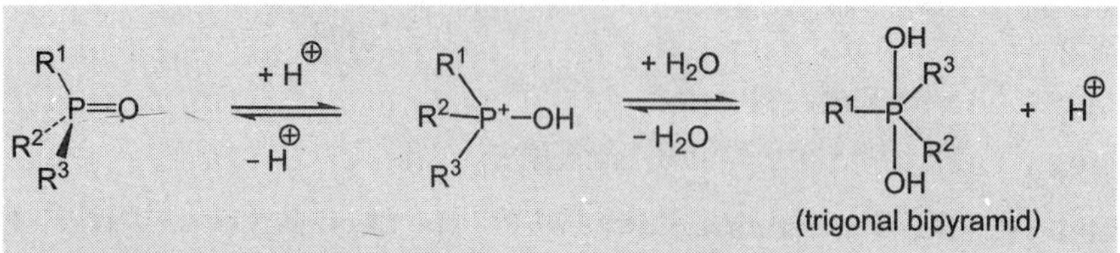

Fig. 3.51: Racemization of phosphine oxides

III. Phosphonic Esters and Related Compounds

Based on the chirality of phosphine oxides, the chirality of phosphonic esters and related compounds which contain quinquevalent phosphorus can be explained. Some of these compounds are **64-66**.

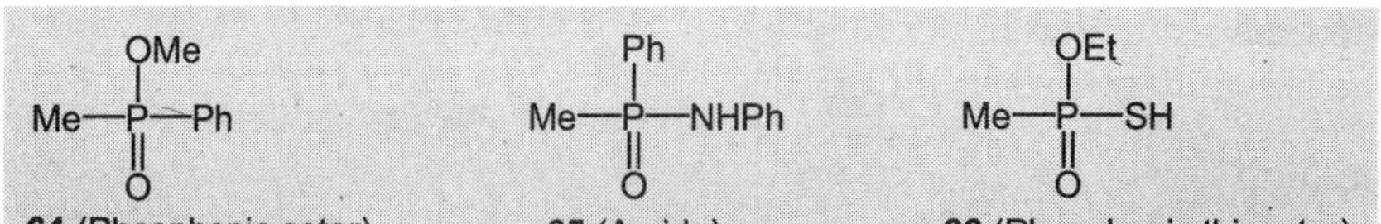

VI. Quaternary Phosphonium Salts

The compounds which can be represented by the general formula $R^1R^2R^3R^4P^+X^-$ are also expected to exhibit chirality (**Fig. 3.52**). Apart from open-chain compounds, cyclic phosphonium salts have also been resolved. For example, 2-*p*-hydroxyphenyl-2-phenyl-1,2,3,4-tetrahydroiso-phosphinolinium bromide (**67**) has been resolved.

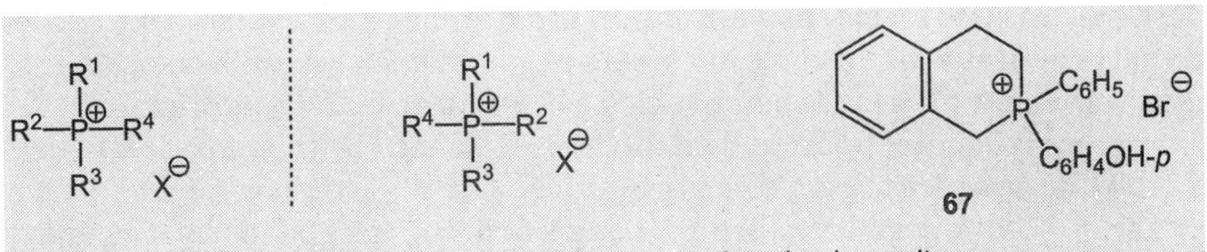

67

Fig. 3.52: Enantiomerism in quateranary phosphonium salts

C. Stereochemistry of Sulphur Compounds

Different types of sulphur compounds, which can be obtained in optically active forms include sulphonium salts ($R^1R^2R^3S^+X^-$), sulphoxides (R^1SOR^2), sulphinic esters (RSO_2R'), and sulfites (RSO_3R'). The structures of representative chiral compounds are **68-70b**.

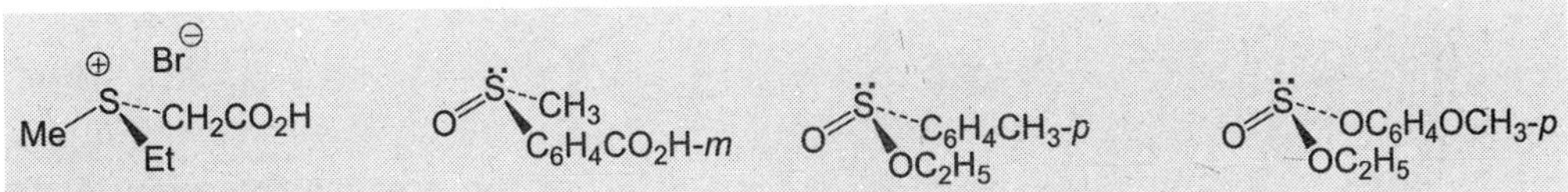

68 (Sulphonium salt) **69** (Sulphoxide) **70a** (Sulphinic ester) **70b** (Sulphite)

3.7 CONFORMATIONS OF CYCLOHEXANES AND DECALINS

A. Cyclohexane and its Derivatives

I. Cyclohexane

Determination of the heats of combustion of alicyclic hydrocarbons shows absence of substantial Baeyer strain in rings of six or more carbon atoms. Such strain, when present, should have a direct effect on the thermodynamic stability of a molecule, and will be reflected in the heat of combustion. For instance, heat of combustion per $-CH_2-$ group in acyclic alkanes is 157.9 kcal/mol (660 kJ /mol), whereas in various cycloparaffins is as follows: cyclopropane, 166.7 kcal/mol (697.5 kJ/mol), cyclobutane, 164.2kcal/mol (686.3kJ/mol), cyclopentane, 159.1 kcal/mol (665 kJ/mol), cyclohexane, 157.4 kcal/mol (658 kJ/mol), cycloheptane, 158.6 kcal/mol (663 kJ/mol), cyclooctane, 158.9 kcal/mol (664 kJ/mol), cyclononane, 159.1kcal/mol (665 kJ/mol).

The noticeable points are:

(*i*) The greatest strain is in the three-membered ring, cyclopropane.

(*ii*) The strain decreases rapidly with ring size, but reaches a minimum for cyclohexane (not cyclopentane) as predicted from the bond angle calculations.

(*iii*) Cyclohexane has heat of combustion per $-CH_2-$ group of around 157.4 kcal/mol (658 kJ/mol), nearly the same value as that of a $-CH_2-$ in a straight-chain alkane. That means, *cyclohexane is essentially strain-free*.

Sachse proposed two non-planar conformations of cyclohexane which are free of Baeyer strain. One of these was a fairly rigid form *shaped like a chair,* and the other was a flexible form shaped like a boat.

Chair Conformation

It has been found that the Raman spectrum of cyclohexane can be best explained by the chair structure. A similar conclusion was reached by a study of infrared spectral studies. It was in

1947 that Hassel, through an extensive study of cyclohexane by the electron diffraction method, confirmed that practically all the molecules were in symmetrical chair forms, and that all C—C bond lengths were normal, and all bond angles were approximately tetrahedral.

The above experimental findings were further supported by the consideration of n-butane skew and eclipsing interactions as discussed in Chapter 2. Cyclohexane may be regarded as composed of six n-butane segments. A view along two of the C–C bonds by Newman projection is shown in **Fig. 3.53**. Thus, the chair form has six n-butane skew interactions, and the boat form has four n-butane skew and two eclipsing interactions (2,3 and 5,6). The energy difference between the boat and chair forms will therefore be about 7 kcal/mol (29 kJ/mol). In other words, cyclohexane chair form is stabler than the boat form by 7 kcal/mol (29 kJ/mol) (see **Fig. 3.59** for energy diagram). Thus, cyclohexane will spend most of its time in chair conformation.

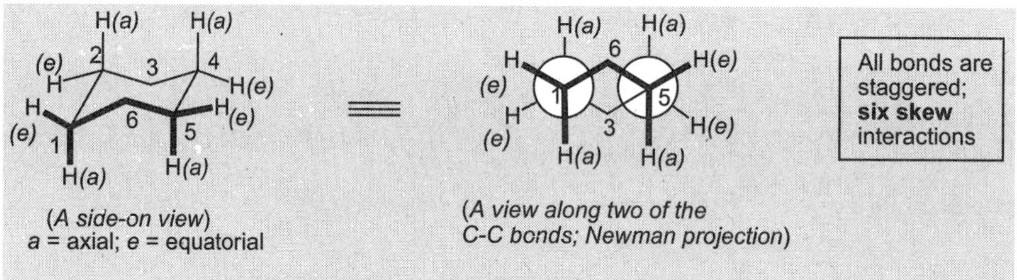

Fig. 3.53: Chair conformation of cyclohexane

Boat and Twist-Boat Conformations

If we flip the "left" end (C1) of the chair cyclohexane up, we can make boat conformation. The hydrogen atoms are disposed as shown in **Fig. 3.54**.

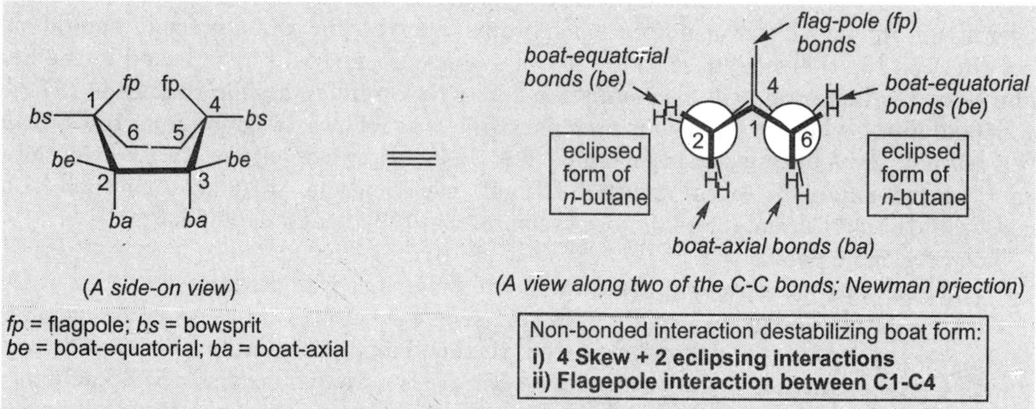

Fig. 3.54: Boat conformation of cyclohexane

There are considerable non-bonded interactions between the flagpole (*fp*) hydrogens, and also between the other eight eclipsed hydrogens. As a consequence, this form of cyclohexane can flex into what is known as twist-boat conformation (flexible form), which is stabler by about 1.4-1.7 kcal/mol (6-7 kJ/mol) than the regular boat, because none of the n-butane units in the twist form is either exactly eclipsed or exactly staggered (skew). The process of conversion of boat to twist-boat can be understood with the help of a model. For this, we hold a model of the boat conformation by keeping C2 and C3 in the right hand and C5 and C6 in the left hand. Then we *twist* the molecule so that C2 and C5 go *down,* and C3 and C6 come *up*. By doing so, two flagpole hydrogens, H*a* and H*b,* move diagonally apart. When the distance between H*a* and H*b* is equal to the distance between H*c* and H*d*, we have brought to minimum the flagpole–flagpole interaction as well as the eclipsing interaction (torsional strain) at the C2—C3 and C5—C6 bonds, which makes the *twist-boat more stable than the regular boat form* (**Fig. 3.55**).

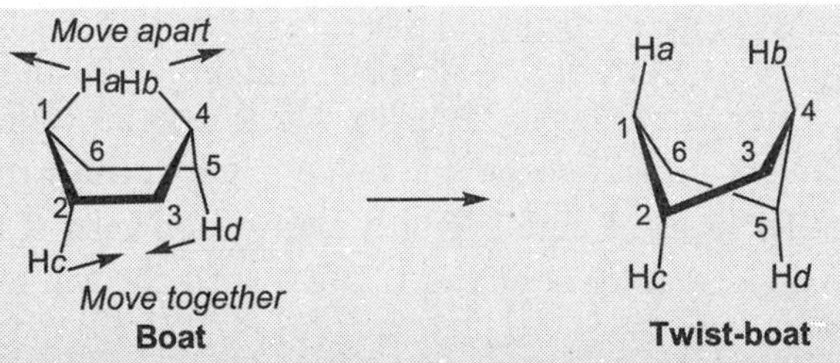

The eclipsing interactions of the boat conformation are relieved slightly; **Twist-boat** conformation is **lower in energy** (by 1.4-1.7 kcal/mol or 6-7 kJ/mol) than the **boat** conformation.

Fig. 3.55: Conversion of a boat conformation of cyclohexane to a twist-boat

(i) Drawing Chair Conformation of Cyclohexane: The chair conformation of cyclohexane could be drawn in several ways. Since trying to draw chair conformation in one continuous line (at least in the beginning) can lead to some dreadful diagrams, one should try to follow some steps for this purpose. Two methods are given in the following description.

Method-1: The following are the steps/guidelines:

1. Draw two parallel lines of the same length, slanting upward. Both lines should start at the same height, and distance between these should be more (about 1.5 times) than the length of original lines (number of carbon atoms may be given for better understanding).

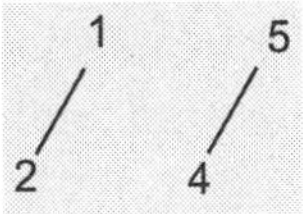

2. Draw a line (C1-C6) from the top point (C1) of left hand side slanted downward. Draw a parallel line (C4-C3) from the bottom (C4) of right hand side slanted upward, of equal length.

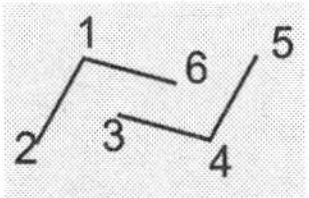

3. Connect the remaining points (C2-C3 and C5-C6) to complete the chair.

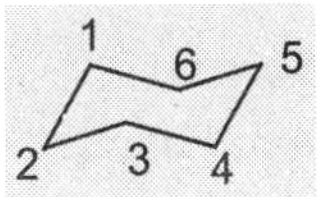

The same method is followed to draw another chair (obtained after flipping).

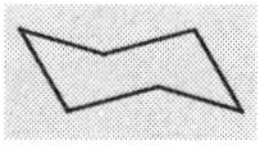

Method-2: The following are the steps/guidelines:

1. Start with one end, *i.e.* C1 flanked by C2 and C6 as shown below.

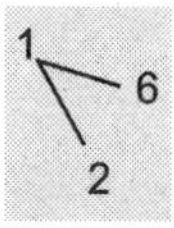

2. Draw two parallel lines of equal length slanted upward (C6-C5 and C2-C3). The top of the new lines should be at the LEVEL with the top of original pair (C1 and C5).

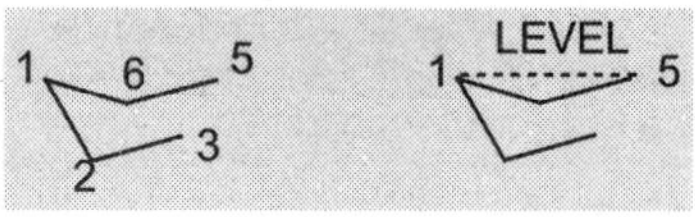

3. Draw the last two lines (C3-C4 and C4-C5). The lowest points (C2 and C4) should also be at the LEVEL.

line C3-C4 should be parallel to line C1-C6
line C4-C5 should be parallel to line C1-C2

Axial and Equatorial Bonds

The twelve carbon-hydrogen bonds in the chair form of cyclohexane are of two types. The six C–H bonds parallel to the axis in the structures shown in **Fig. 3.56** are called axial (*a*) bonds, and the remaining six C–H bonds which extend radially outward at an angle of 109° 28' to the axis are called equatorial (*e*).

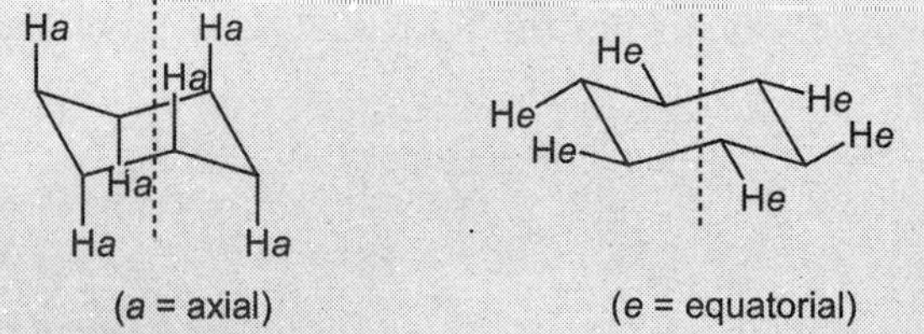

Fig. 3.56: Chair conformation of cyclohexane with axial and equatorial bonds

(ii) Drawing Axial and Equatorial Bonds

Axial bonds: It is relatively easy to draw axial bonds as they are vertically aligned, and alternate up and down all around the ring.

Equatorial bonds: A little more care should be taken to draw equatorial bonds. An important thing to remember is that each equatorial bond must be parallel to two C–C bonds as shown in **Fig. 3.57.**

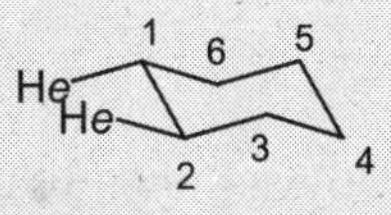

He = equatorial hydrogen

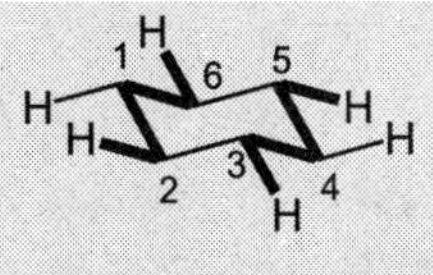

C1-He is parallel to C2-C3 and C5-C6;
C2-He is parallel to C1-C6 and C3-C4;
C1-He (downward and C2-He (upward) are in opposite direction.

Equatorial hydrogen atoms at C2 and C6 look like 'W' shape
Equatorial hydrogen atoms at C3 and C5 look like 'M' shape

Fig. 3.57: Drawing equatorial bonds

(iii) Ring Flipping (inversion) of Cyclohexane: The chair form (**C1**) of cyclohexane can easily flip (by rotation about its single bonds) into the mirror-image chair conformation (**C2**). Since the energy barrier for flipping process is only 10.3 kcal/mol (43 kJ/mol), it occurs rapidly at room temperature.

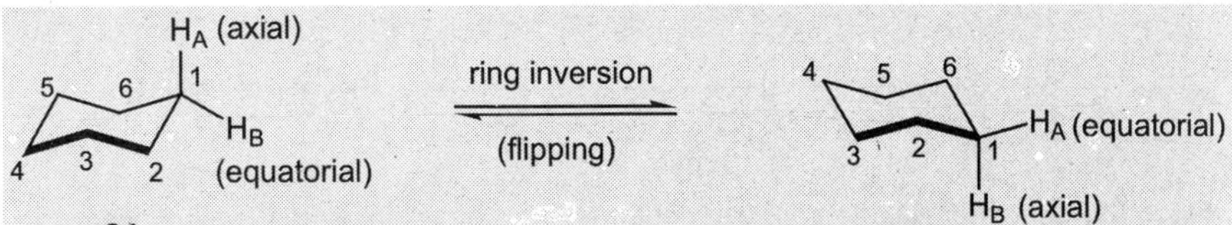

This interconversion of chair conformations occurs *via* various conformations, and notable of them are half-chair (**HC1** and **HC2**; twist-chair), boat (**B**), and twist-boat (**TB1** and **TB2**; skew boat) as shown in **Fig. 3.58.** *It is important to mention that all the equatorial bonds become axial, and all the axial bonds become equatorial after the flipping process.*

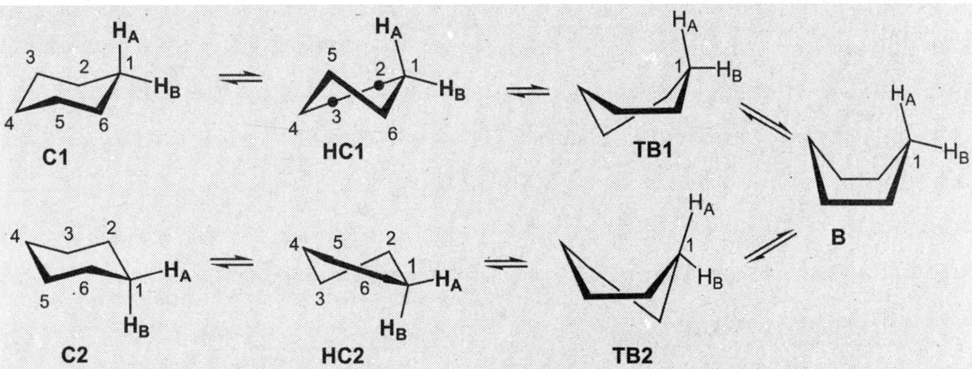

Fig. 3.58: Ring inversion of cyclohexane chair to another chair *via* various conformations

The energy profile (**Fig. 3.59**) for the ring flipping shows that the half-chair has *more energy* than the chair, whereas twist-boat is of *lower energy* than the boat.

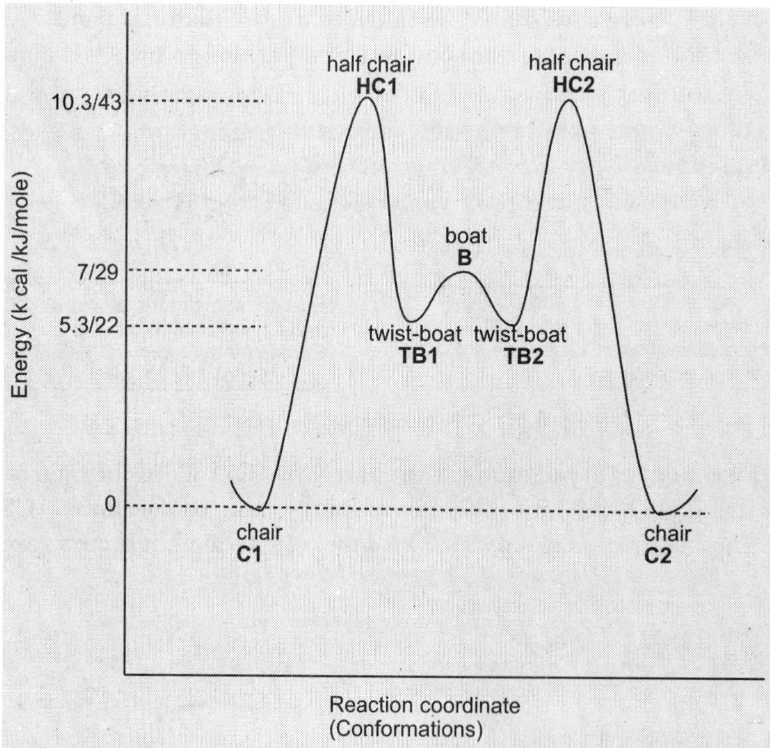

Fig. 3.59: Energy profile diagram and conformations of cyclohexane (flipping/inversion)

The distances between the various hydrogen atoms and carbon atoms in both chair and boat conformations have been calculated. The data are helpful in studying the effectiveness of interactions among various groups. In order to avoid crowding, chair and boat conformations are drawn somewhat bigger in **Fig. 3.60.**

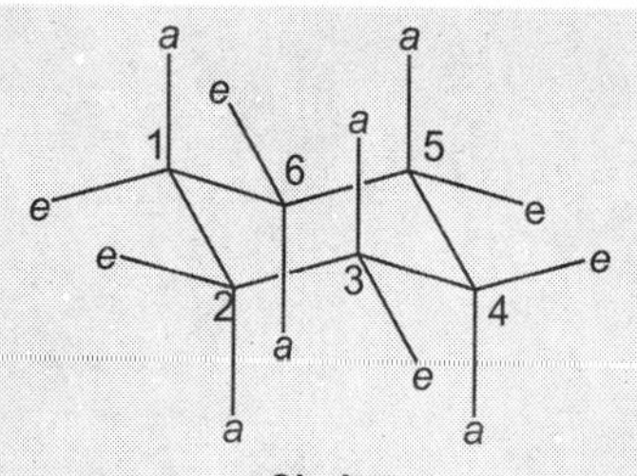

Chair

***a* = axial; *e* = equatorial**

Position	H-H (Å)
1*e*,2*e*	2.49
1*e*,2*a*	2.49
1*a*,2*a*	3.06
1*a*,3*a*	2.51

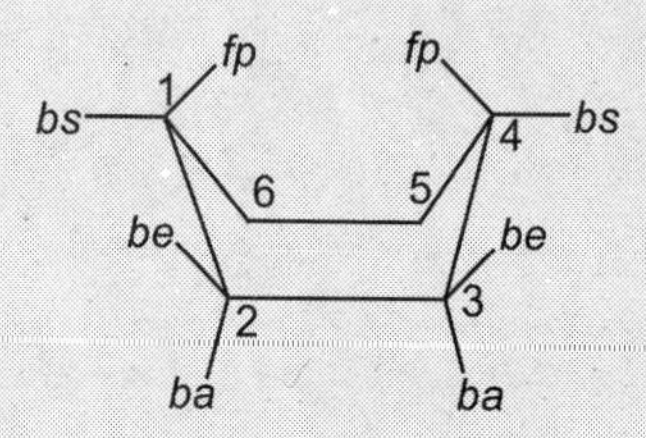

Boat

***fp* = flagpole; *bs* = bowsprit**
***be* = boat-equatorial; *ba* = boat-axial**

Position	H-H (Å)
2*ba*,3*ba*	2.27
2*be*,3*be*	2.27
1*fp*,4*fp*	1.83

Fig. 3.60: The distances between the various hydrogen atoms of chair and boat conformations

II. Monosubstituted Cyclohexanes

Monosubstituted cyclohexanes exist in two non-equivalent chair conformations (**71** and **72**). These two diasteromeric conformations are obtained by substituting equatorial and axial hydrogen atoms by other atoms or groups. The two conformations are interconvertible by flipping process. The rate of flipping and composition of the two conformers depend upon the nature of the group present in cyclohexane.

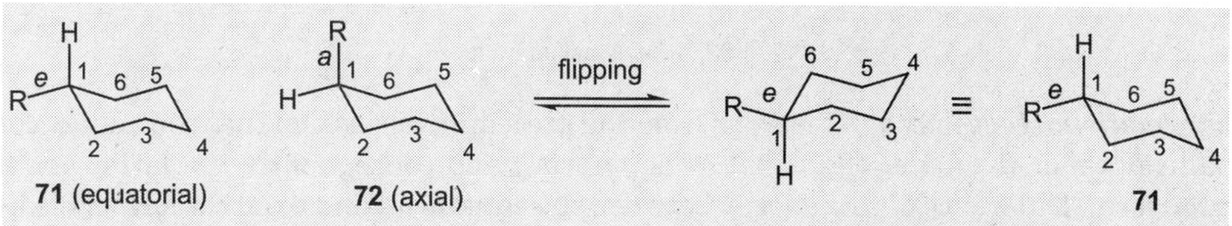

71 (equatorial) **72** (axial) **71**

An equatorial conformer is *more stable* than an axial conformer. Two explanations have been offered:

1. When the substituent is located in an equatorial position, it is *anti* to C3 and C5 carbons. In contrast, when it is in an axial position, it is skew (gauche) to C3 and C5 carbons. Thus, the axial conformer has two more skew interactions (1.80 kcal/mol or 7.54 kJ/mol when R = Me) than the equatorial conformer (**Fig. 3.61**).

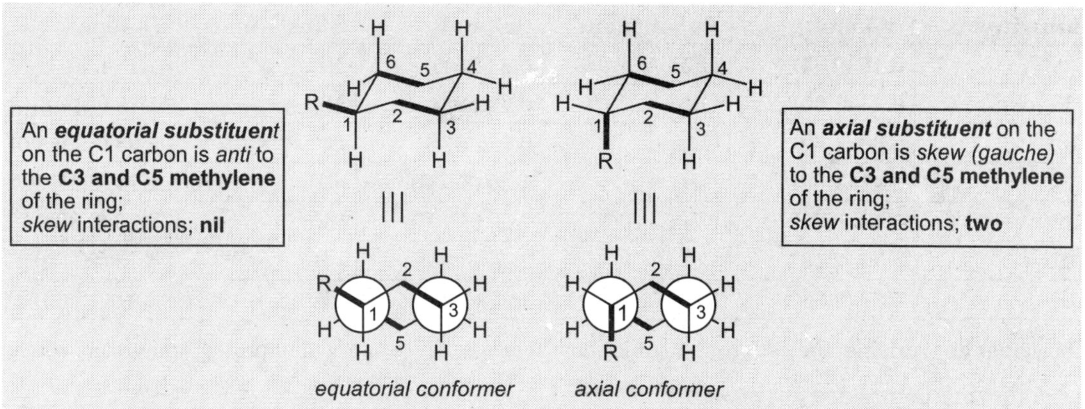

Fig. 3.61: Comparison of skew interactions between equatorial and axial conformers of monosubstituted cyclohexane

2. As the three axial bonds are on the same side and parallel to each other, any axial substituent will be too close to the axial substituents (here hydrogens) on the other two carbons (C3 and C5). These interactions, which are unfavourable in monosubstituted cyclohexanes, are called 1,3-diaxial interactions. Thus, there are two 1,3-diaxial interactions between the axial substituent and axial hydrogens at C3 and C5 in monosubstituted cyclohexane, *e.g.* methylcyclohexane. These interactions are absent in the equatorial conformer (**Fig. 3.62**).

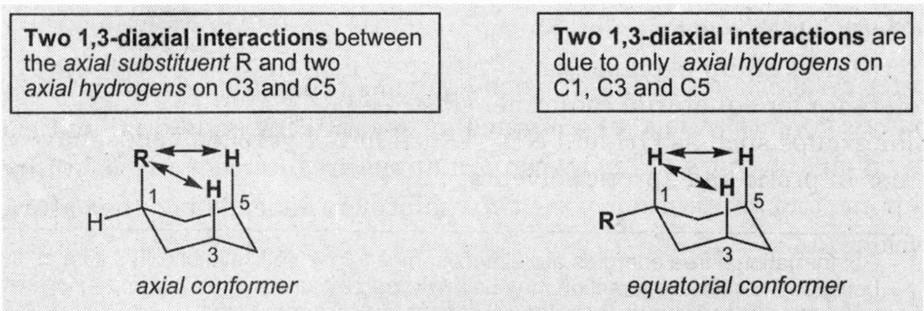

Fig. 3.62: 1,3-Diaxial interactions

(i) Conformational Free Energy: It is clear from the preceding discussion that the energy content of axial conformer is *greater* than that of equatorial, and consequently the latter one is the preferred form. Difference of free energy between the equatorial and axial conformers is known as conformational free energy of the substituent, and is denoted as $-\Delta G°$. It determines the preference of equatorial substituent over the axial, and is based on steric consideration. The conformational free energies of a number of common substituents are listed in **Table 3.1.**

Table 3.1 Conformational free energies ($-\Delta G°$) in kcal/mol or (kJ/mol) for some common substituents#

Substituent	$-\Delta G°$ kcal/mol (kJ/mol)	Substituent	$-\Delta G°$ kcal/mol (kJ/mol)	Substituent	$-\Delta G°$ kcal/mol (kJ/mol)
F	0.15 (0.63)	OMe	0.6 (2.51)	NO_2	1.1 (4.60)
Cl	0.43 (1.80)	OEt	0.7 (2.96)	Me	1.79 (7.50)
Br	0.38 (1.59)	OAc	0.6 (2.51)	Et	1.93 (8.10)
I	0.43 (1.80)	NH_2	1.20 (5.03[a])	*n*-Pr	1.94 (8.15)
CN	0.16 (0.71)	NH_2	1.60 (6.70[b])	*i*-Pr	2.15 (9.00)
OH	0.52 (2.18[a])	CO_2H	1.35 (5.65)	*t*-Bu	4.78 (20.00)
OH	0.87 (3.65[b])	CO_2Me	1.27 (5.32)	Ph	3.01 (12.60)

#The values are variable; the data in the Table show average of free energy in aprotic[a] and protic[b] solvent.

The following points are noteworthy:

(*i*) In thc halogen series, F has the least effective bulk. The conformational free energies of Cl, Br, and I are almost equal (**Table 3.1**). One possible explanation for this is that with the increase of size of the halogens, the bond length also increases, and with it the distance from axial hydrogens.

(*ii*) Number and nature of substituents at the first atom connected to the ring are more important to affect the effective bulk of the substituent, and hence conformational free energies of groups OCH_3, OC_2H_5, and OCOR are almost the same.

(*iii*) The $\Delta G°$ values for Me, Et, *n*-Pr (and to some extent *i*-Pr) do not differ much (as was expected on the basis of their size) (**Fig. 3.63**), whereas the value of $\Delta G°$ increases sharply for the *tert*-butyl, which has the highest effective bulk in the series. In case of ethyl and *iso*propyl, the 1,3-diaxial interaction is minimized by turning the α-hydrogen inwards, which is not possible for the *tert*-butyl group. The high conformational energy of *tert*-butyl group explains why this group has an almost total (more than 99.9%) preference for equatorial conformer (Fig. 3.64).

(*iv*) Polar groups such as OH and NH_2, which form hydrogen bonds, have different $\Delta G°$ values in protic and aprotic solvents.

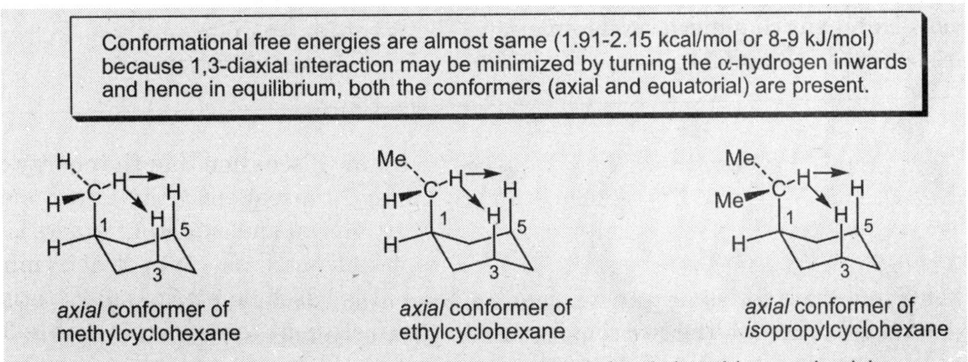

Fig. 3.63: 1,3-Diiaxial interactions in axial conformers of methyl-, ethyl- and *iso*propylcyclohexanes

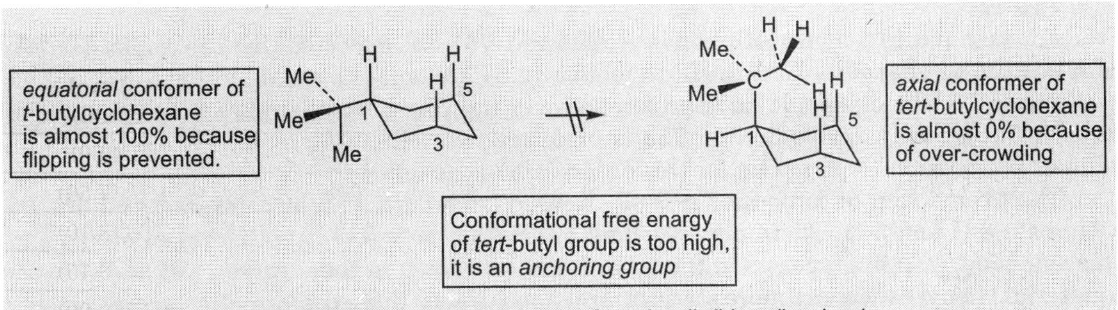

Fig. 3.64: *t*-Butylcyclohexane, conformaionally 'biased' molecule

(ii) Anchoring Groups and Conformationally Biased Molecules: The *tert*-butyl group, because of its large size, is far more stable in the equatorial form than in the axial form, and this situation is like that of conformationally locked (or frozen) molecule as flipping is prevented. *tert*-Butyl group is referred to as an anchor or *'anchoring group'*. The compounds containing *'anchoring group'* are said to be conformationally 'biased'.

III. Disubstituted Cyclohexanes

The conformational studies of disubstituted cyclohexanes can be discussed by putting them into four categories, depending upon the position of the substituents – (a) 1,1-disubstituted cyclohexanes, (b) 1,2-disubstituted cyclohexanes, (c) 1,3-disubstituted cyclohexanes, and (d) 1,4-disubstituted cyclohexanes.

Generally speaking, the principle of conformational analysis of monosubstituted cyclohexanes may be extended to disubstituted cyclohexanes as well. However, due to the presence of two substituents, some additional factors (besides the relative size of the two substituents and their nature) need to be considered. These factors are: position isomerism (mentioned above as four categories i-iv), stereoisomerism (geometrical and optical), and interactions between the substituents. *If the other forces such as dipole-dipole interaction and hydrogen bonding are absent, the chair conformation with the maximum number of equatorial substituents will be the preferred conformation.*

We will now draw preferred comformations, and name them as *cis-* and *trans-* isomers in cases where needed. Afterwards, their relative stability based on various interactions, and then problems related to chirality will be discussed.

(i) 1,1-Disubstituted Cyclohexanes: It is relatively easy to study 1,1-disubstituted cyclohexanes as they do not exhibit any configurational isomerism. They exist in two interconvertible conformers, which are shown as **73** and **74**.

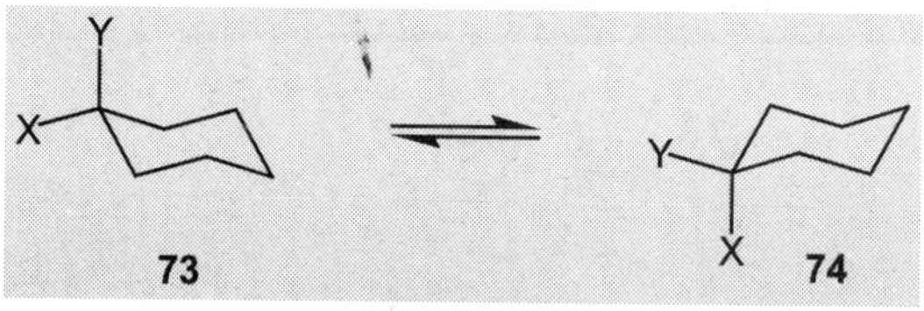

When X and Y are the same, the two conformers are identical, and are known as *topomers*. When X and Y are different, the two conformers are diastereomers, and are present in unequal amounts. In principle, the ratio of the two conformers should correspond to the differences in the conformational free energies of the two substituents. But in practice, it may not be true in all the cases.

For example, in 1-methyl-1-phenylcyclohexane (**75**), it is expected that the conformer **75b** with equatorial phenyl ($-\Delta G°$ = 3.01kcal/mol or 12.60 kJ/mol) and axial methyl ($-\Delta G°$ = 1.79 kcal/mol or 7.5 kJ/mol) should have preference over the **75a** with axial phenyl and equatorial methyl. But actually, the conformer **75a** is preferred over other **75b** by 0.31 kcal/mol or 1.33 kJ/mol. The possible explanation for this observation is as follows:

The phenyl ring in conformer **A** is so oriented (structure **A**) that it interacts minimally with axial 3-H and 5-H. There may be some interaction between *ortho* hydrogen atoms and adjacent equatorial hydrogens. On the other hand, as shown in the structures **B** and **C** (with equatorial phenyl), there are more steric interactions. Hence these conformation are less stable (**Fig. 3.65**).

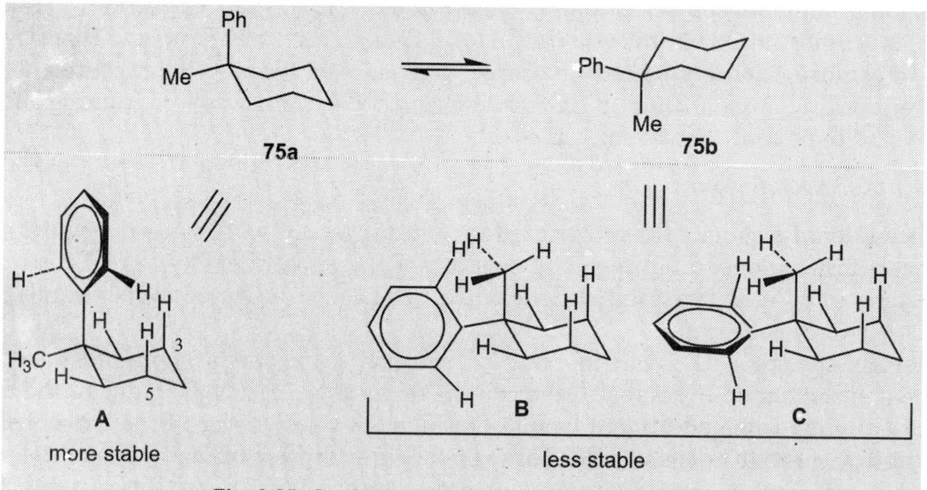

Fig. 3.65: Conformers of 1-methyl-1-phenylcyclohexane

(ii) 1,2-Disubstituted Cyclohexanes:

Structure and Name of Conformers

Disregarding chirality, four conformations are possible for a 1,2-disubstituted cyclohexane (pairs of *cis* and *trans* isomers); one pair is *trans* (**76** and **77**), and the other one *cis* (**78** and **79**). But, when X = Y, the number is reduced to three. For drawing their structures, it is convenient to draw, first 1,2-diaxially substituted cyclohexane derivative, which has two substituents in *antiperiplanar* form, and therefore, is a straight forward case of a *trans* isomer (**76**). Treating this as a reference, other conformers can be drawn and named accordingly. The planar hexagon structures of *cis* and *trans* isomers may also be drawn. However, they are not helpful in studying conformational details. Furthermore, planar structures may lead to wrong conclusions related to chirality of various conformations.

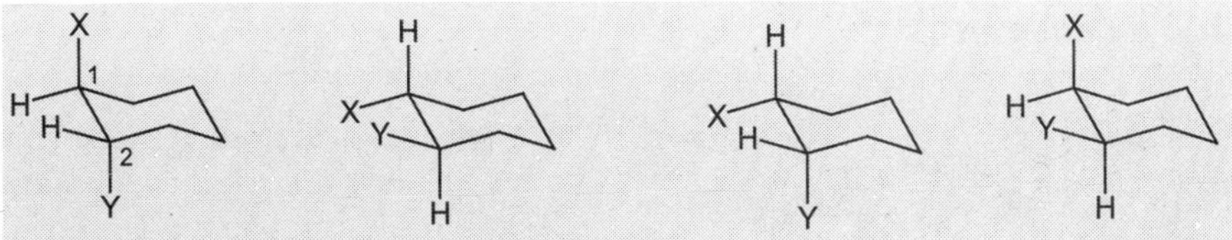

76 (1*a*,2*a*) **77** (1*e*,2*e*) — *trans*; **78** (1*e*,2*a*) **79** (1*a*,2*e*) — *cis*

Stability

Out of the two *cis* and two *trans* isomers, one in each pair is more stable than the other. Applying the general principle of the stability of monosubstituted cyclohexanes, it can easily be predicted that conformer with two equatorial substituents (*trans*) should be the *most stable*, and the other having two axial substituents (*trans*) should be the *least stable*. The stability of the remaining two isomers, both *cis* would lie in between the stability of the two *trans* isomers. Relative stability of the *cis* isomers depends upon the nature of the substituents. In practice, these predictions are generally true. However, some other factors are also to be taken into account while studying population of conformers in different cases.

Flipping of chair is likely to occur in *cis* and *trans* isomers, and population of the two conformers is determined by their relative stability. If X = Y, the two *cis* conformers will exist in 1:1 ratio. However, the same is not true for *trans* conformations as shown in **Fig. 3.66**.

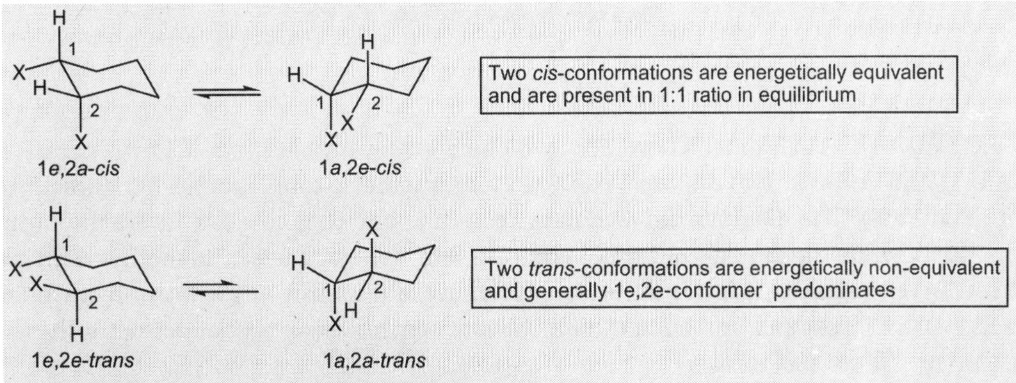

Fig. 3.66: Relative proportion/stability of *cis* and *trans* isomers of 1,2-disubstituted cyclohexane with identical subtituents

Chirality

Since none of the conformers have any symmetry element, all of them are chiral, and can exist as resolvable racemic mixtures. Thus, total number of stereoisomers is *eight* (*four* enantiomeric pairs). When X and Y are the same, situation becomes somewhat different.

All the above points can be elaborated by considering typical examples. Let us first take the examples when X = Y. 1,2-Dimethylcyclohexane offers a typical example, which exists in *cis* and *trans* forms.

cis-1,2-Dimethylcyclohexane: It exists in two energetically equivalent conformations (1*e*,2*a*; 1*a*,2*e*; X = Me; **80** and **81**, **Fig. 3.67**). They are non-superimposable mirror images of each other and constitute a pair of enanatiomers, which are readily interconvertible by flipping the cyclohexane ring. Thus, it is not possible to resolve such racemic mixtures, and such cases may be called *conformational enantiomers*.

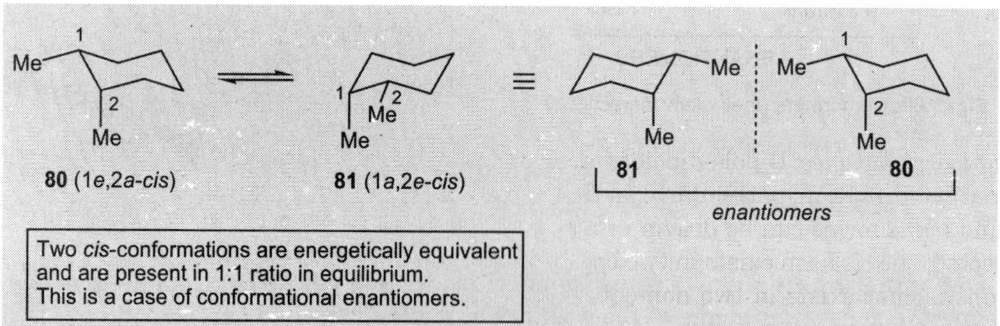

Fig. 3.67: Conformational enantiomers of *cis*-1,2-dimethylcyclohexane

On the other hand, the planar structure for *cis*-1,2-dimethylcyclohexane shows the presence of a plane of symmetry, suggesting it to be a *meso* isomer (**Fig. 3.68**). Therefore, it is advisable to work with conformational structures to avoid any confusion.

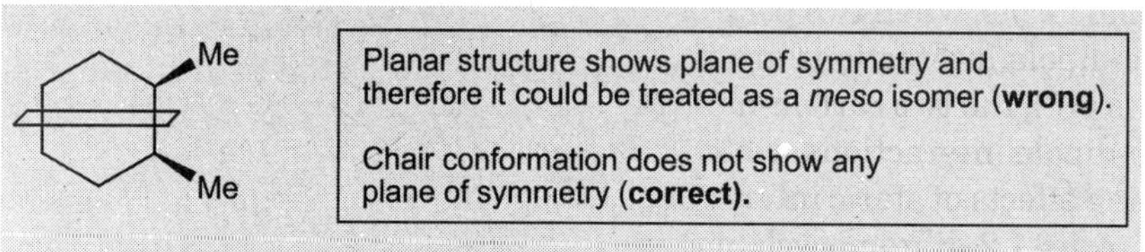

Fig. 3.68: Planar structure for *cis*-1,2-dimethylcyclohexane

trans-1,2-Dimethylcyclohexane: There are two chair conformations (*e,e*; *a,a*) which are non-equivalent, and occur in two distinct resolvable enantiomers (total number of stereoisomers is four; **Fig. 3.69**). The diequatorial conformer contains one butane skew interaction (due to *e,e*,-dimethyl), while the diaxial conformer contains four (two for each axial methyl as in mono-substituted cyclohexanes), which makes the preference of former over the latter by three skew interactions (2.67 kcal/mol or 11.2 kJ/ mol). Consequently, *trans* isomer contains these two conformers in the ratio of 99 to 1.

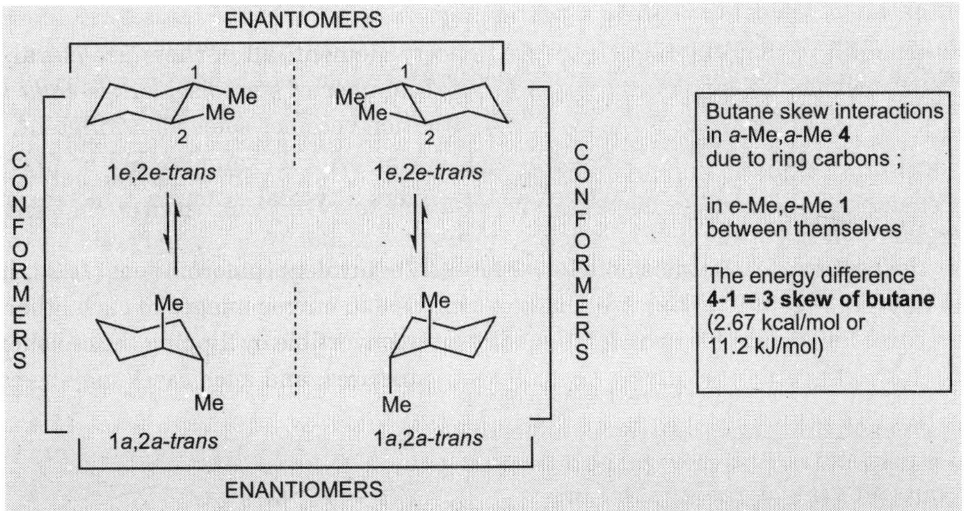

Fig. 3.69: Conformers (their *skew* ineractions) and enantiomers of *trans*-1,2-dimethylcyclohexane

1,2-Dihalocyclohexanes: Dipole-dipole interactions play major role in determining the ratio of the preferred conformation, particularly, in the *trans* form of 1,2-dihalocyclohexanes. The structure of *cis* and *trans* forms can be drawn in a similar way as shown for 1,2-dimethylcyclohexanes. As expected, the *cis* form exists in two energetically equivalent interconvertible conformations. The *trans* isomer exists in two non-equivalent chair conformations, which in turn exist as a pair of enantiomers.

Unlike 1,2-dimethylcyclohexanes, in which the diaxial conformer is almost non-existent, this conformer is present in significant amount (sometimes more than 1,2-diequatorial), in 1,2-dihalocyclohexanes. The ratio depends upon the nature of the halogen and the solvent. The ratio of the diaxial form increases across the series: Cl < Br < I. The primary reason for this unexpected behavior is as follows-

In the diaxial forms, when two polar halogens are oppositely placed (with a dihedral angle 180°), the dipole-dipole interactions are absent. On the other hand, in diequatorial conformer (dihedral angle is 60°); the two carbon-halogen bonds (C–X) are closer enough to produce strong repulsive dipole-dipole interactions, which destabilize this isomer.

The opposing effects of steric interaction and dipole-dipole repulsion are illustrated in the equilibration of *trans*-1,2-dibromocyclohexane (**Fig. 3.70**) in benzene. The equilibrium consists of almost equal amounts of the two conformers (*e,e*- and *a,a*-) in benzene.

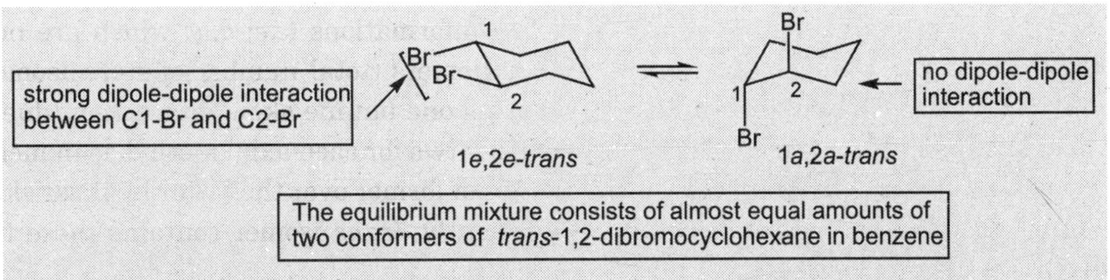

Fig. 3.70: Population of conformers of *trans*-1,2-dibromocyclohexane

The nature of solvents and state of aggregation can also affect the ratio of *a,a*- and *e,e*-conformers. The stability of diequatorial form increases with the increase in polarity of the solvent.

For the compounds having 1,2-substituents on cyclohexane, there is no problem with the *trans* isomer, because the favoured conformation will be the one that places these two substituents in equatorial position. But, when we deal with the *cis* isomer where one substituent is axial and the other one is equatorial, the preferred conformation will be the one that has larger group in equatorial position. In such cases, one has to make predictions based on conformational free energies of the substituents as discussed in the case of monosubstituted cyclohexanes. For example, the preferred conformation of *cis*-1-bromo-2-methylcyclohexane is the one (**82**) that contains equatorial methyl ($-\Delta G^\circ$ = 1.79 kcal/mol or 7.50 kJ/mol) and axial bromo ($-\Delta G^\circ$ = 0.38 kcal/mol or 1.59 kJ/mol).

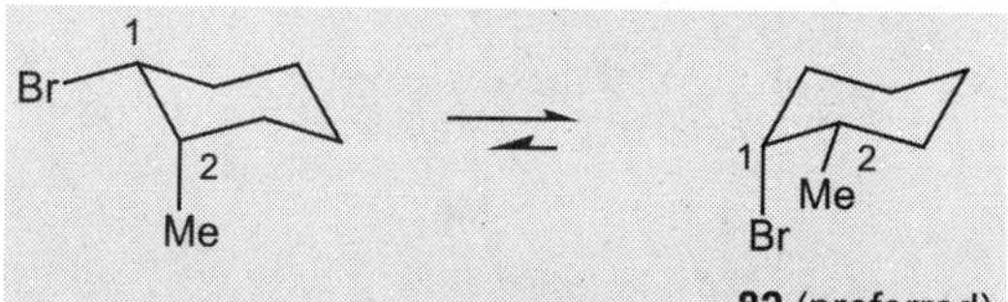

82 (preferred)

As expected, if the sizes of the two substituents (conformational free energies) are comparable, the two conformers of the *cis* isomer will be present in almost equal amounts in the equilibrium.

III. 1,3-Disubstituted Cyclohexanes

Like 1,2-disubstituted cyclohexanes, *cis* and *trans* conformers are resolvable chiral compounds when two groups at C1 and C3 are different, and total number of stereoisomers is *eight* (*four* enantiomeric pairs) (**Fig. 3.71**).

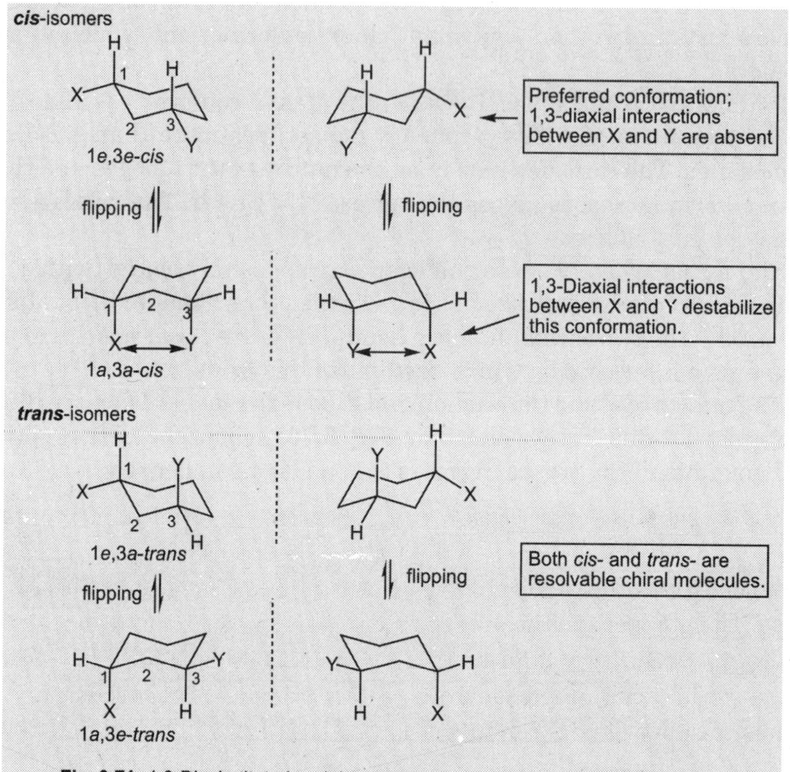

Fig. 3.71: 1,3-Disubstituted cyclohexanes when two groups are different

On the other hand when X = Y, the total number of stereoisomers is reduced to *four* (one enatiomeric pair of *trans* isomers, and two *meso* conformers of *cis* form). This situation can be explained by considering *trans* and *cis* isomers separately.

Enantiomeric pairs of the two *trans* conformations are identical as the ring inversion in *trans* isomer with identical groups regenerates the same enantiomer (**Fig. 3.72**). *It is advisable to take the help of models for clear understanding.*

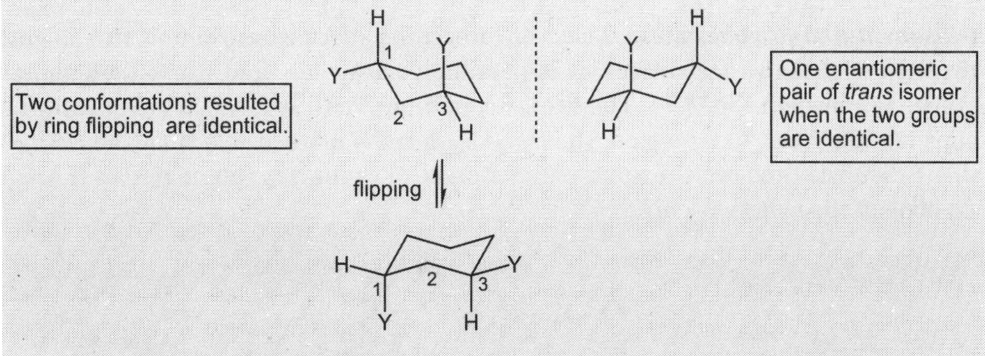

Fig. 3.72: *Trans*-1,3-disubstituted cyclohexanes when two groups are same

There are only two *cis* conformations which are obtained by ring flipping. These are achiral (*meso*) because of the presence of a plane of symmetry (**Fig. 3.73** and structure **83**).

The *cis* conformer with two equatorial groups is more stable than the *cis-a,a*-conformer and the *trans* isomer. For example, the most stable conformation of 1,3-dimethylcyclohexane has been shown to be *cis*-1,3-*e,e*,-form. This is due to the presence of steric repulsions between the two diaxial methyl groups (**Fig. 3.73**).

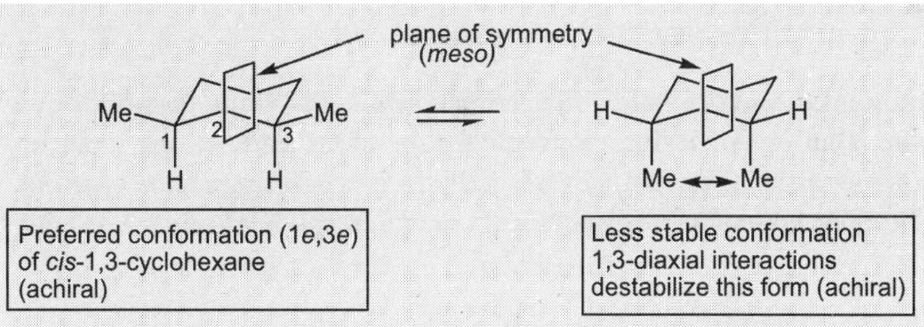

Fig. 3.73: *Cis* confomers of 1,3-dimetylcyclohexane (two *meso* forms)

A preferred 1,3-*cis-e,e*-form can adopt a diaxial conformation by ring inversion, if the situation demands so. This happens in the case of *cis*-cyclohexane-1,3-diol (**83**) where intramolecular hydrogen bonding between two axial hydroxyl groups stabilizes the diaxial form. Similarly, during the formation of cyclic anhydride from *cis*-1,3-cyclohexanedicarboxylic acid, diequatorial conformation has to adopt diaxial conformation (**Fig. 3.81**).

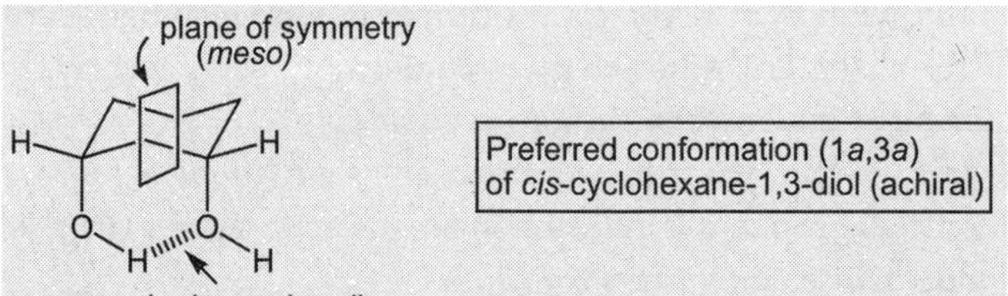

83 *hydrogen bonding*

(iv) 1,4-Disubstituted Cyclohexanes: Based on preceding discussion related to 1,2- and 1,3-disubstituted cyclohexane derivatives, it is possible to draw *cis*- and *trans*-1,4-disubstituted cyclohexanes as shown in **Fig. 3.74**. The two *cis*-conformations (*e,a*-form) are identical when the two groups are identical. As expected, the *trans-e,e*-form is more stable than the *cis-a,e*-form. Since all 1,4-disubstituted cyclohexanes possess plane of symmetry (plane passes through 1,4-carbons), none of the isomer is chiral.

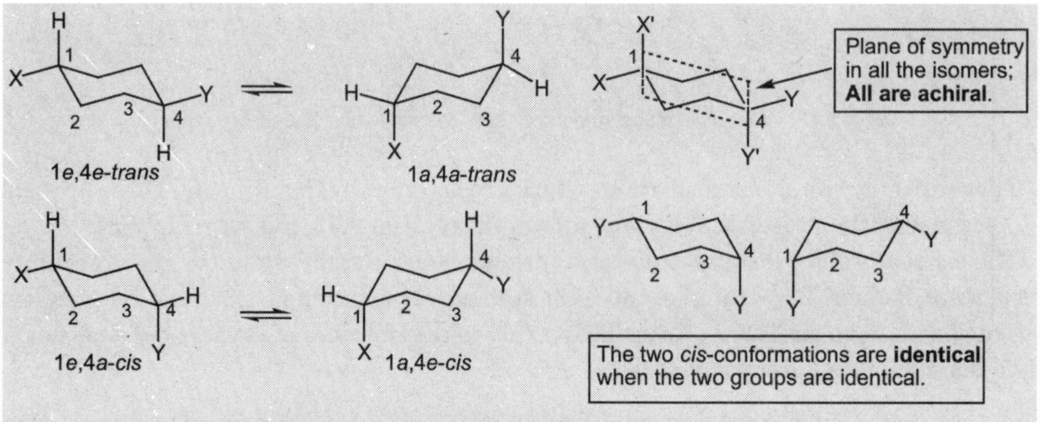

Fig. 3.74: 1,4-Disubstituted cyclohexanes

In the examples discussed so far, the conformation of the ring is chair, but when intramolecular hydrogen bonding is possible between groups at C1 and C4, the molecule may assume a boat or twist-boat rather than chair in which hydrogen bonding is not possible. For example, *cis*-2,5-di-*tert*-butylcyclohexane-1,4-diol mainly exists in the boat and twist-boat forms.

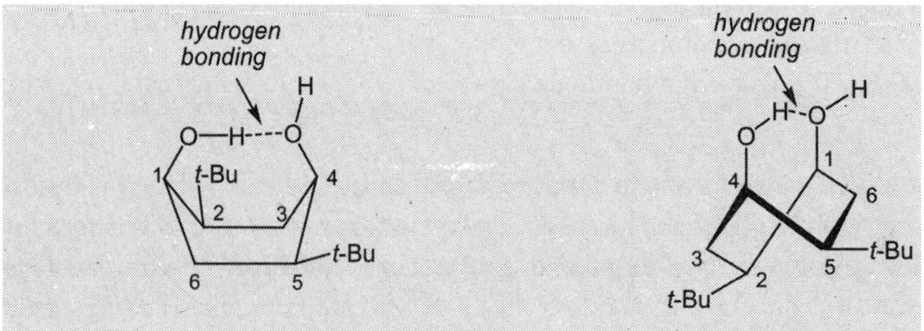

cis-2,5-Di-*tert*-butylcyclohexane-1,4-diol **(boat)** **(twist-boat)**

For the sake of convenience, a summary of the nomenclature and chirality of disubstituted cyclohexanes is given in **Fig. 3.75**.

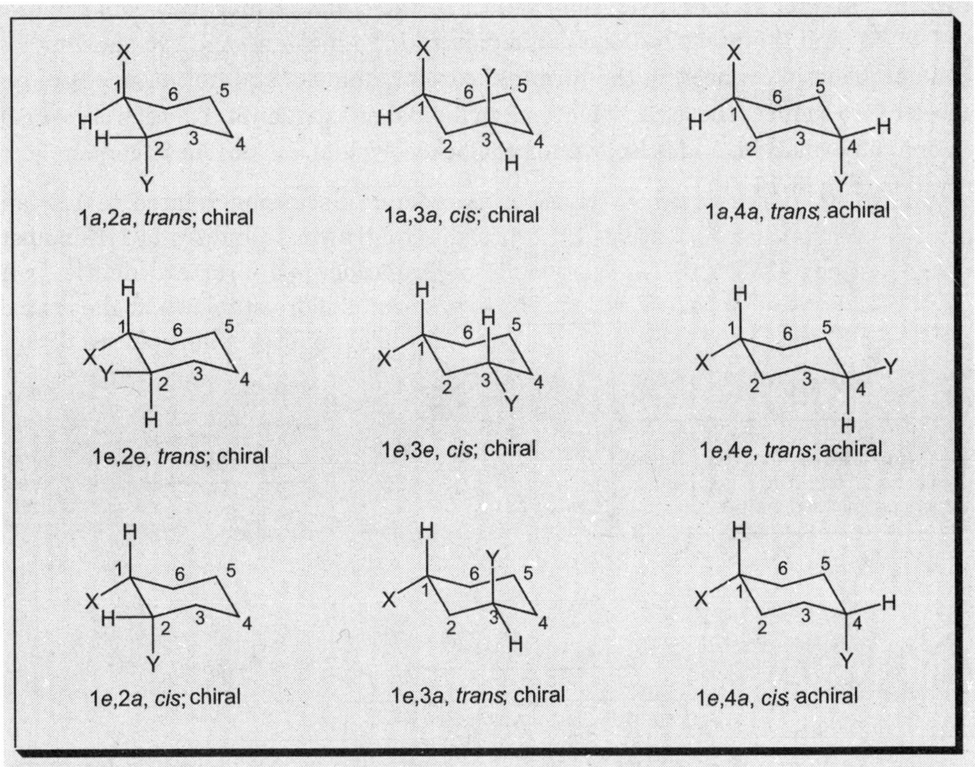

Fig. 3.75: A summary of *cis* and *trans* isomers of disubstituted cyclohexanes and their chirality

B. Decalins

Fusion of two cyclohexane rings at a common C–C bond generates a decalin. The study of conformational isomerism of decalins is very much related with the conformational study of 1,2-cyclohexanes. Two diastereomers (*cis* and *trans*) are possible for the decalin, depending on whether the hydrogen atoms at ring junction are *cis* or *trans*.

As drawing stereochemical structures of decalins is relatively a difficult task, classical hexagonal structures of cyclohexanes are often used. *Cis* and *trans* fusions at C–C bonds are indicated by using flying-wedge bonds as shown in structures **84** and **85**, respectively.

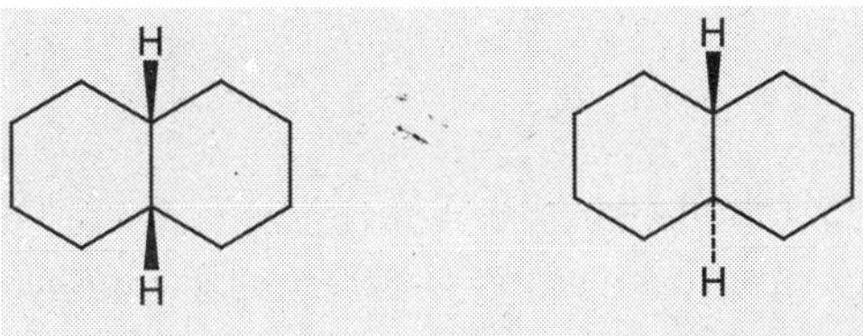

84 **85**

For an understanding of the conformational analysis of such systems, it is necessary to draw preferred conformations (generally chair) of cyclohexane. We can compare the situation with the conformation of 1,2-ethylcyclohexanes. In the *trans* isomer, two conformations are 1*e*,2*e* and 1*a*,2*a*, and the preferred one is diequatorial. To make second cyclohexane ring (**B**), there is no problem in connecting the terminal carbon atoms of equatorial ethyl groups. But, it is impossible to connect these carbon atoms in diaxial substituents. Therefore, *trans*-decalin exists in one conformation, and is known to be a locked system, *i.e.* no ring inversion is possible in this system (**Fig. 3.76**).

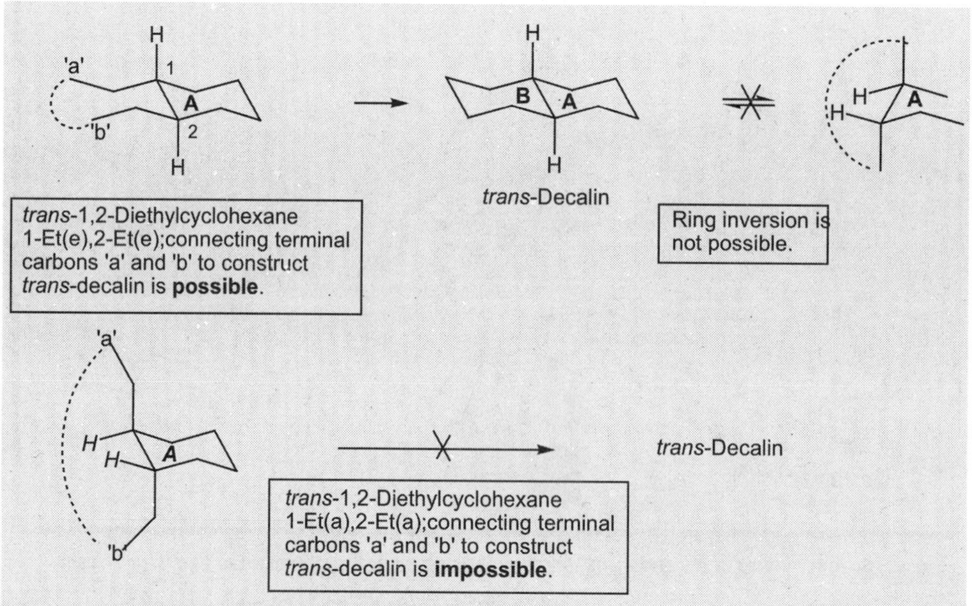

Fig. 3.76: *trans*-Decalin

For *cis*-decalin, the second ring (**B**) has to be connected to the first (**A**), so that it is axial at one carbon (fusion point), and equatorial at the other. Thus, we have two alternate possibilities obtained by the ring flipping of 1,2-diethylcyclohexane as shown in **Fig. 3.77**. In contrast to *trans*-decalin, there is no problem of ring inversion in *cis* decalin, and therefore, this isomer is flexible. The two *cis*-isomers are energetically equivalent, and are non-superimposable mirror images of each other. It is also evident that due to rapid interconversion of one enantiomer to the other by ring flipping, the resultant racemic mixture is not resolvable (conformational enantiomers).

I. Relative Stability of Decalins

Relative stability of *cis* and *trans* decalins can be determined by counting the skew (gauche) butane units. The *cis* isomer has three extra skew interactions involving two axial bonds, as shown by solid thickened bonds in structure **86**. On the basis of the fact that the skew interaction of the hydrogens in *n*-butane is 0.80 kcal (3.35 kJ), the total energy difference between *cis* and *trans* forms is $3 \times 0.80 = 2.4$ kcal ($3 \times 3.35 = 10.05$ kJ). Obviously, the *trans* isomer is more stable than the *cis* by 2.4 kcal/mol (10.05 kJ/mol).

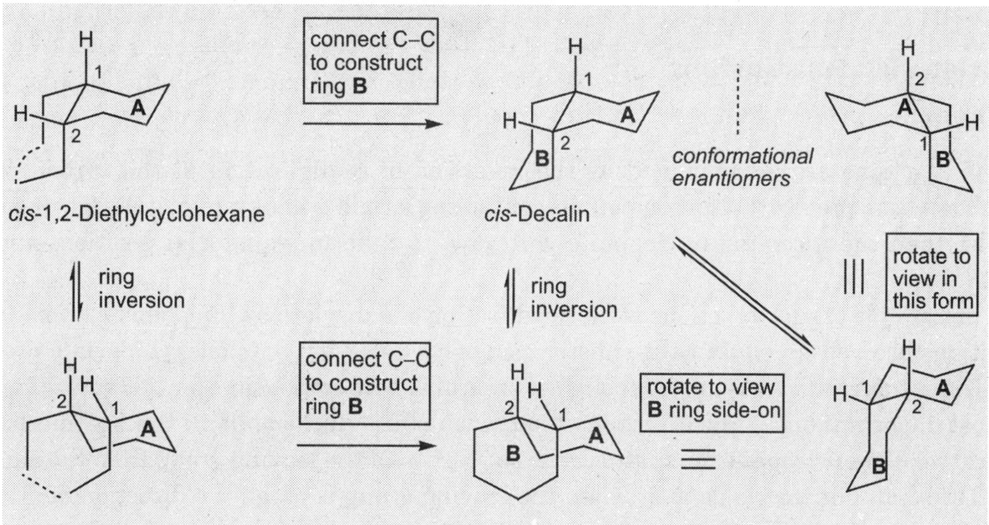

Fig. 3.77: Conformations of *cis*-1,2-diethylcyclohexane and *cis*-decalin. Ring inversion and conformational enantiomers

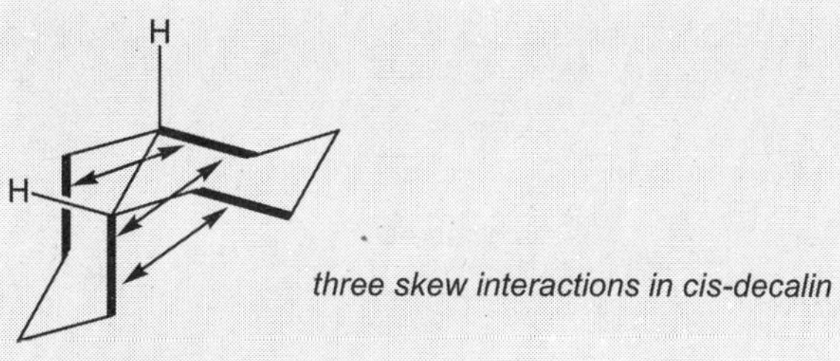

86

II. Distinction Between cis- and trans-Decalins by NMR

In the ^{1}H NMR spectrum of the rigid *trans*-decalin, the equatorial and axial protons are distinguishable, and appear as two broad bands due to spin-spin coupling. On the other hand, in the case of *cis*-decalin, the equatorial and axial protons appear as one broad signal (averaged) due to flipping of the rings at room temperature. It is possible to split up this signal into two broad signals by scanning the NMR at low temperature (–18°C).

3.8 EFFECT OF CONFORMATION ON REACTIVITY

Since steric environment of axial and equatorial substituents is not the same, substituted rings having these substituents react at different rates. Here, we shall consider a few situations in which the outcome of a reaction will depend upon whether a functional group is located at an equatorial or axial position. In many cases, the functional group is held deliberately at the desired position by 'locking' the ring using an *anchoring* group (*tert*-butyl), or a fused ring system such as *trans*-decalin. The examples discussed here are based on the type of reactions.

A. Nucleophilic Substitutions

I. S_N2 Reactions

All S_N2 reactions are accompanied by the inversion of configuration at the carbon centre. This allows us to predict that in compounds containing a rigid cyclic system, if the leaving group is axial, then the incoming nucleophile will have to take up equatorial position, and *vice versa*.

An example of such reactions is the substitution of a tosylate by thiophenolate ion. In this case, it has been observed that the substitution of *cis*-4-*tert*-butylcyclohexyl tosylate proceeds much faster than its *trans* isomer. Among the several factors responsible for this rate difference, the most important one is the direction of approach of the nucleophile. If the S_N2 mechanism is operative, the nucleophile must attack the sp^3 carbon of the leaving group from the backside of C–OTs bond. For the *cis* isomer, when the leaving group is axial, the direction of attack is parallel with the axial hydrogens, and *antiperiplanar* to the leaving group, and the reaction is faster. In the case of *trans* isomer, when the leaving tosylate group is equatorial, this line of attack is hindered by the axial hydrogens (**Fig. 3.78**).

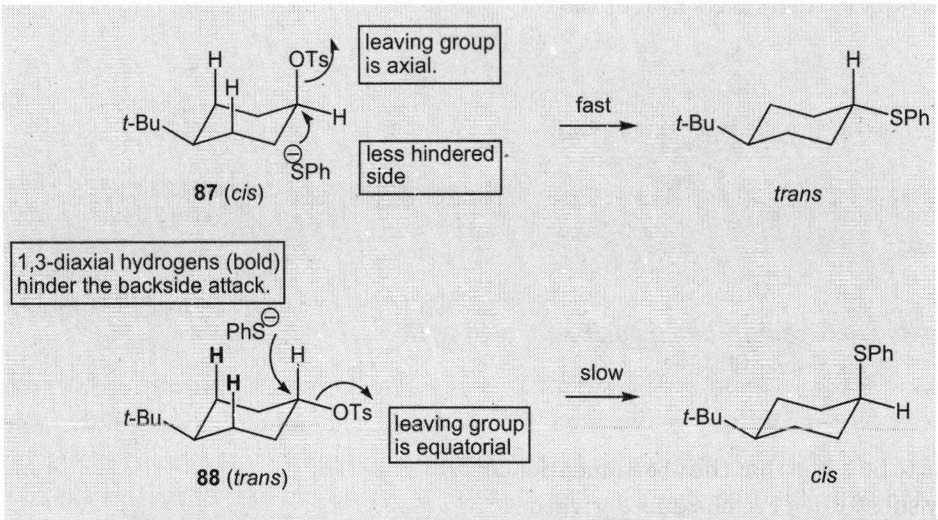

Fig. 3.78: S_N2 reaction of *cis*- and *trans*- isomers of 4-*tert*-butylcyclohexyl tosylate

S_N2 Type Reactions Involving Internal Nucleophiles: The S_N2-type reactions involving internal nucleophiles (neighbouring group participation) occur at much faster rates than those involving external nucleophile. Geometry of the two groups (nucleophile and leaving group) plays an important role in such reactions. The essential requirement for intramolecular participation of a nucleophile in S_N2-type reactions is that nucleophile attacks from the backside. In other words, such an attack in cyclohexane derivatives is possible only if both the groups are axially positioned. This geometric requirement is well illustrated by the observation that the rate of acetolysis of *trans*-2-iodocyclohexyl brosylate (**89**) is many times faster that the acetolysis of the *cis* isomer (**Fig. 3.79**). Since this process involves two S_N2-type reactions, there is retention of configuration in the product **90** (*trans*).

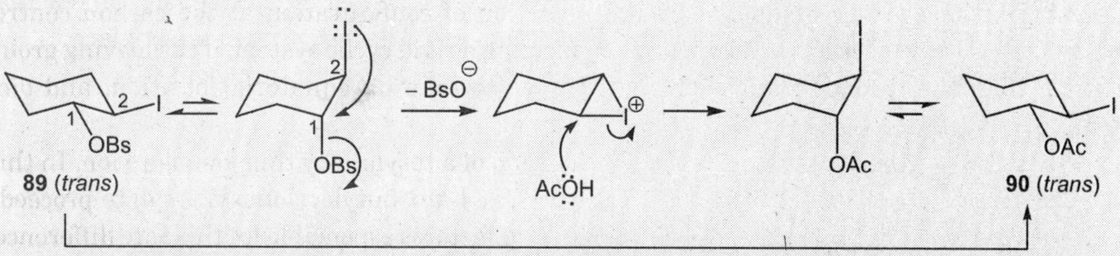

Acetolysis of **89** *(trans) involving two S_N2- type reactions lead to retention of configuration (trans product)*

Fig. 3.79: Acetolysis of *trans*-2-iodocyclohexyl brosylate

Another example is the formation of cyclohexene epoxide (**92**) from 2-chlorocyclohexanol (**91**) by treatment with a base. The reaction involves an intramolecular participation by hydroxyl nucleophile in the S_N2-type reaction. As expected, *trans*-2-chlorocyclohexanol (**91**) forms epoxide far more rapidly than its *cis* isomer (**93**).

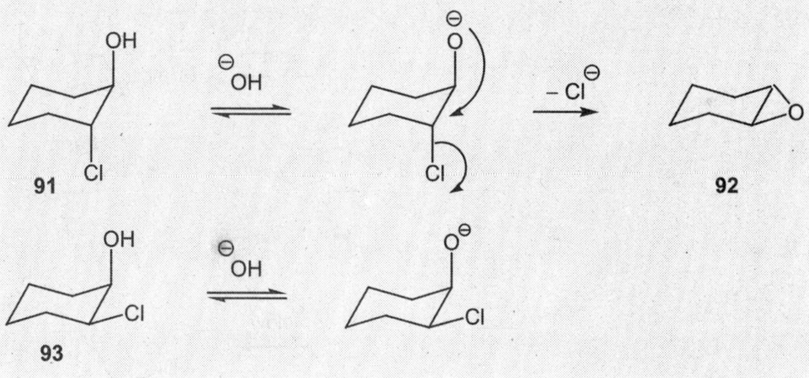

It is to be noted that the above-mentioned examples involve energetically unfavourable 1,2-diaxially substituted cyclohexane derivatives. In such cases, as the diaxial conformer undergoes reaction, rapid ring inversion of the major diequatorial isomer ensures that it is replaced, and equilibrium is maintained.

The compounds containing bicyclic or polycylic rigid systems, such as *trans*-decalins, also undergo intramolecular nucleophilic substitutions, readily, if the two internal groups (nucleophile and leaving) are *antiperiplanar* to each other. For example, *trans*-1-β-bromo-2-α-hydroxydecalin (**94**, both, Br and OH are axial) is smoothly converted into *trans*-1α,2α-epoxydecalin (**95**), whereas the 1α,2β-isomer (**96**; both Br and OH are equatorial) reacts extremely slowly.

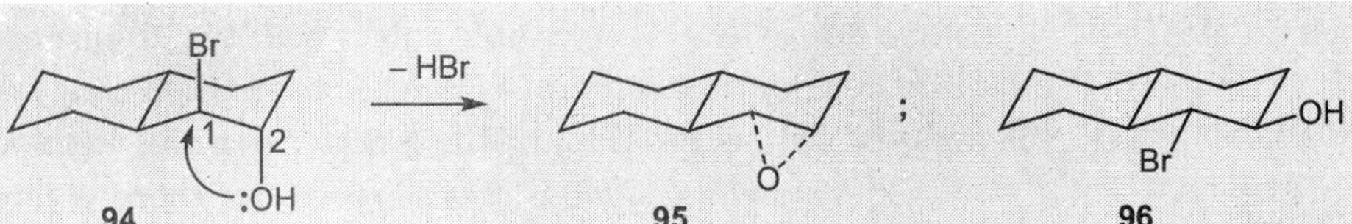

94 **95** **96**

The effect of conformation on reactivity can also be illustrated by considering the reaction of 1,3- and 1,4-disubstituted cyclohexanes. Formation of cyclic anhydrides takes place in *cis*-cyclohexane 1,3-dicarboxylic acids only when the two carboxyl groups are axial (**Fig. 3.80**).

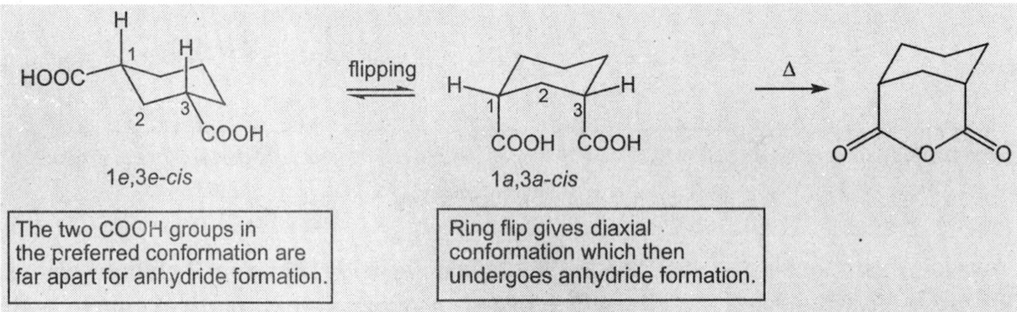

Fig. 3.80: Formation of anhydride from *cis*-1,3-cyclohexanedicarboxylic acid

Similarly, formation of the lactone is possible from *cis*-3-hydroxycyclohexanecarboxylic acid when alcoholic and acid groups are axially oriented. (**Fig. 3.81**).

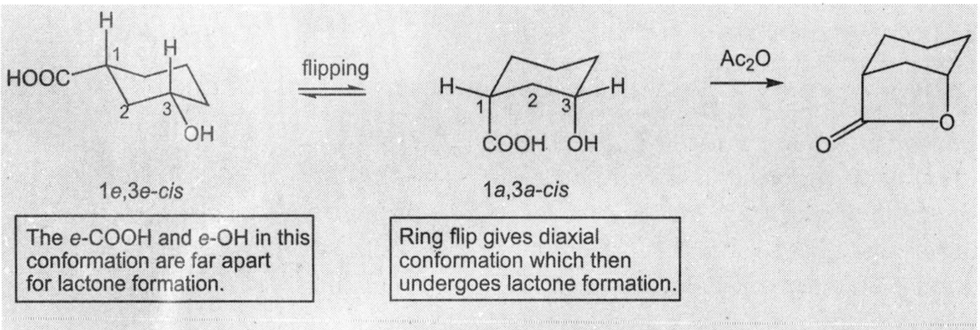

Fig.3.81: Formation of lactone from *cis*-3-hydroxy-cyclohexanecarboxylic acid

Such transformations play key role in establishing the stereochemistry of natural products such as steroids and alkaloids.

II. S_N1 Reactions

S_N1 reactions involve a two-steps mechanism, and the rate-determining step of these reactions is the formation of the intermediate carbocation. Since the formation of the cation would relieve the steric strain of the axial conformer due to 1,3-diaxial interactions, it is expected that an axially substituted conformer would undergo *faster S_N1* reaction than the equatorial conformer.

An example of the S_N1 reaction, which illustrates that these reactions are assisted by an axial substituent, is the acetolysis of *cis*-4-*tert*-butylcyclohexyl tosylate (axial) (**87**) which occurs about 3.4 times faster than that of the *trans* isomer (**88**) (**Fig. 3.82**).

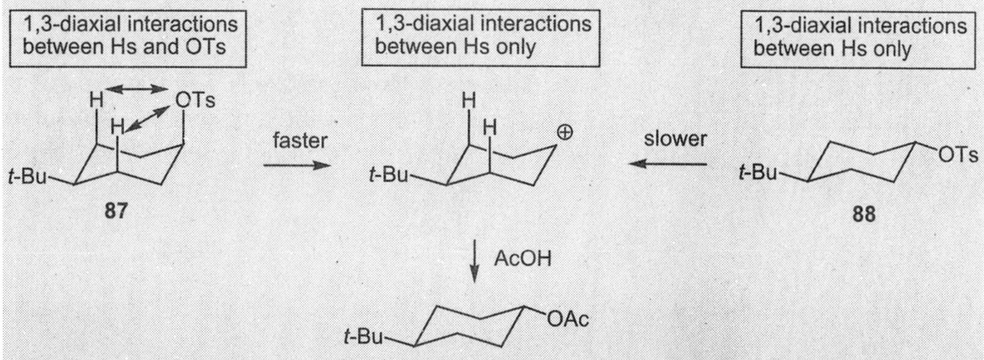

Fig. 3.82: Acetolysis of 4-*tert*-butylcyclohexyl tosylates involving S_N1 mechanism

B. Elimination Reactions

Three types of elimination reactions can be discussed to illustrate the conformation-reactivity relationship. These are *E2*, *syn* (pyrolytic) and 1,4-eliminations.

I. E2 Reactions

Most *E2* reactions are stereospecific, and proceed with greater ease when the departing groups are *trans* rather than when they are *cis*. Elimination reactions of cyclic compounds have been studied in considerable detail to shed more light on stereochemistry of *E2* reactions. The stereoelectronic requirement for *E2* elimination reactions in cyclohexane derivatives is that the two leaving groups must be *antiperiplanar* (1,2-diaxial).

The classical example of *E2* reactions differentiating the reactivities of epimeric pairs is dehydrochlorination of menthyl and *neo*menthyl chlorides with ethoxide ions. The two leaving groups (H and Cl) in the former having stable conformation are equatorial (not diaxial), and cannot be antiperiplanar. To fulfill the antiperiplanarity requirement, the cyclohexane ring has to be flipped to its unstable conformation, which on elimination of one molecule of HCl gives 2-menthene, exclusively. On the other hand, in the case of *neo*menthyl chloride the leaving groups on adjacent carbon atoms are antiperiplanar in the stable conformer. It can, thus, be predicted that the latter will undergo *E2* reaction more rapidly than the former (**Fig. 3.83**).

Another interesting observation in the *E2* elimination reactions is that *neo*menthyl chloride provides a mixture of 2-menthene and 3-menthene because of the availability of axial hydrogens at both the adjacent carbon with respect to axial chlorine.

Similarly, it is possible to explain why rate of *E2* elimination of *cis*-4-*tert*-butylcyclohexyl tosylate (**87**) to 4-*tert*-butylcyclohexene (**97**) is many times faster than its *trans* isomer (**88**).

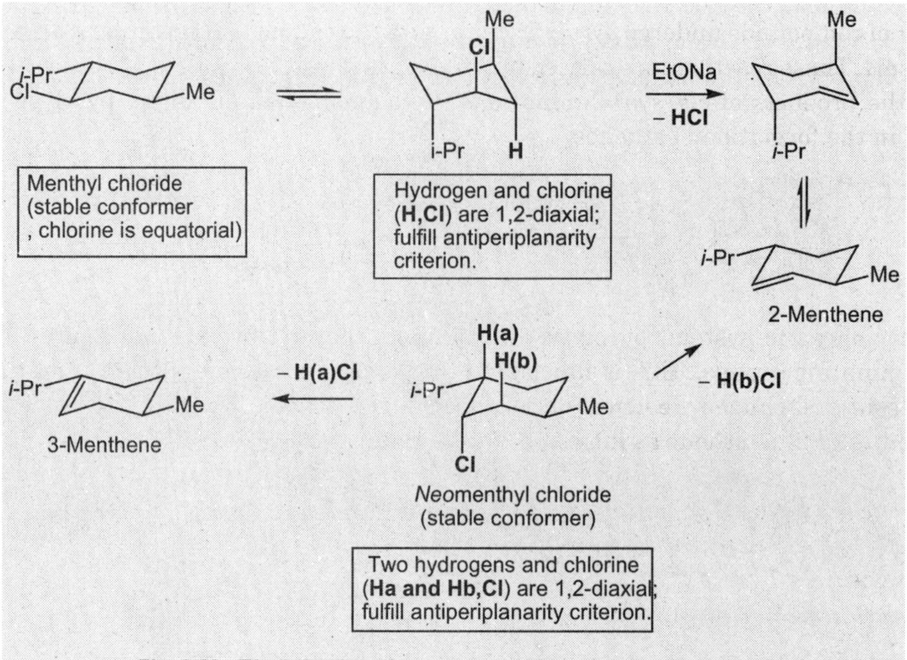

Fig. 3.83: *E2* elimination of menthyl chloride and *neo*menthyl chloride

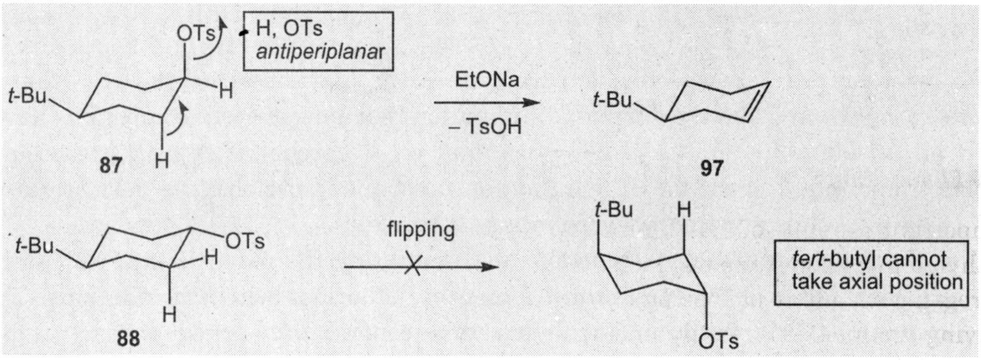

Although majority of *E2* reactions of cyclic compounds involve *trans* elimination, certain cyclic compounds undergo *E2* elimination when two groups to be eliminated are *cis* and *coplanar*. An important example of *cis* elimination, where dihedral angle of 0° serves the same purpose as that of 180°, is found in 2,3-dichloronorbornane series. For example, *cis* elimination from **98** is faster than *trans* elimination from **99** (for further discussion see Chapter 8).

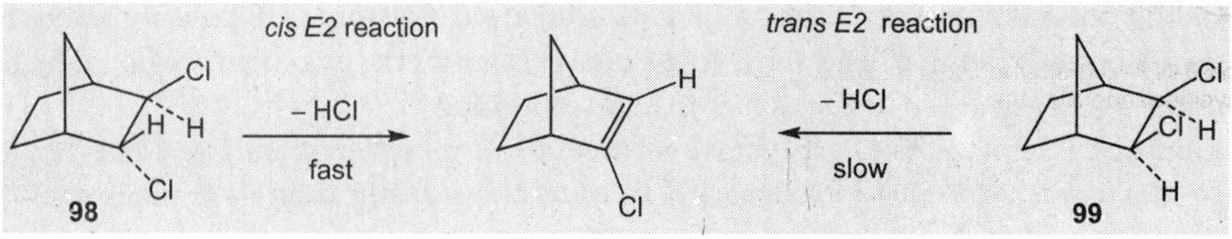

II. Pyrolytic Eliminations (cis/syn)

A number of compounds undergo *pyrolytic eliminations,* reacting unimolecularly through cyclic mechanisms. These reactions are carried out in the gas phase, and proceed in concerted fashion yielding the products of *cis*/*syn* eliminations. A common example is the pyrolysis of esters resulting in the formation of alkenes.

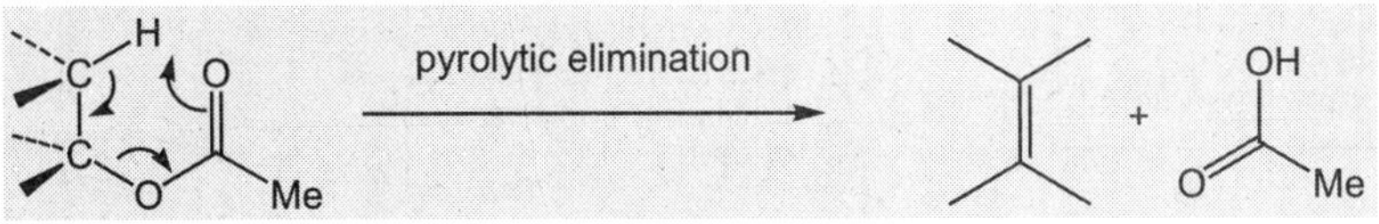

In cyclohexane system, pyrolytic eliminations require an axial-equatorial alignment of the eliminating groups. This is illustrated by the product decomposition from the methyl xanthate esters (Tschugaev reaction) derived from menthol and *neo*menthol. These isomers give 3-menthene and 2-menthene as major products, respectively (**Fig. 3.84**).

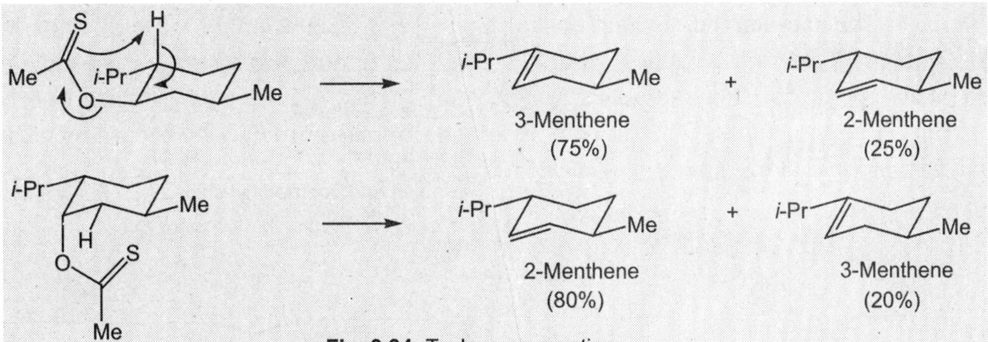

Fig. 3.84: Tschugaev reaction

III. 1,4-Eliminations

An important example of 1,4-elimination reactions is the comparison of two epimeric 3-*N,N*-dimethylaminocyclohexane tosylates. It has been found that the *cis*- isomer in which *p* orbital of nitrogen with a lone pair being *antiperiplanar* to 3-2 bond, which in turn is antiperiplanar to leaving group (OTs), readily undergoes 1,4-elimination leading to the carbon-carbon ring cleavage. On the other hand, *trans* isomer lacks this geometric requirement, and hence does not undergo 1,4-elimination under similar conditions.

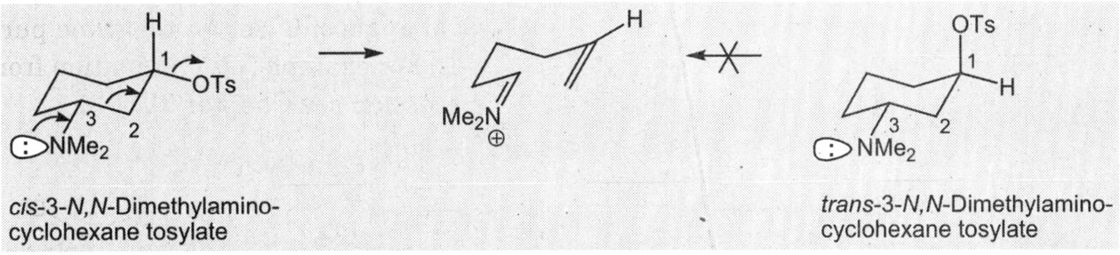

C. Esterification and Ester Hydrolysis

In the preceding sections, we have discussed the reactions in which a bond linked to the cyclohexane ring carbon is broken. Now, we shall consider reactions in which this does not happen, *e.g.* esterification and ester hydrolysis involving $A_{AC}2$ mechanism as shown in the following general scheme:

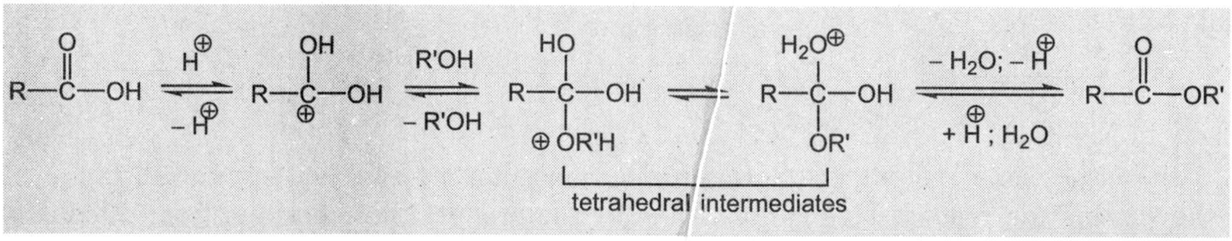

The equatorial and axial conformational isomers react at different rates in esterification and hydrolysis. In esterification, replacement of carboxylic (or hydroxyl) hydrogen by the large group increases the non-bonded interactions far more for an axial than for an equatorial –COOH (or OH). Obviously, the former (axial) would be expected to undergo esterification more slowly than the latter (equatorial). For example, esterification of *trans*-4-*tert*-butylcyclohexanecarboxylic acid (having an equatorial –COOH in conformationally biased molecule) occurs more readily than the *cis*-isomer (having an axial –COOH) (**Fig. 3.85**).

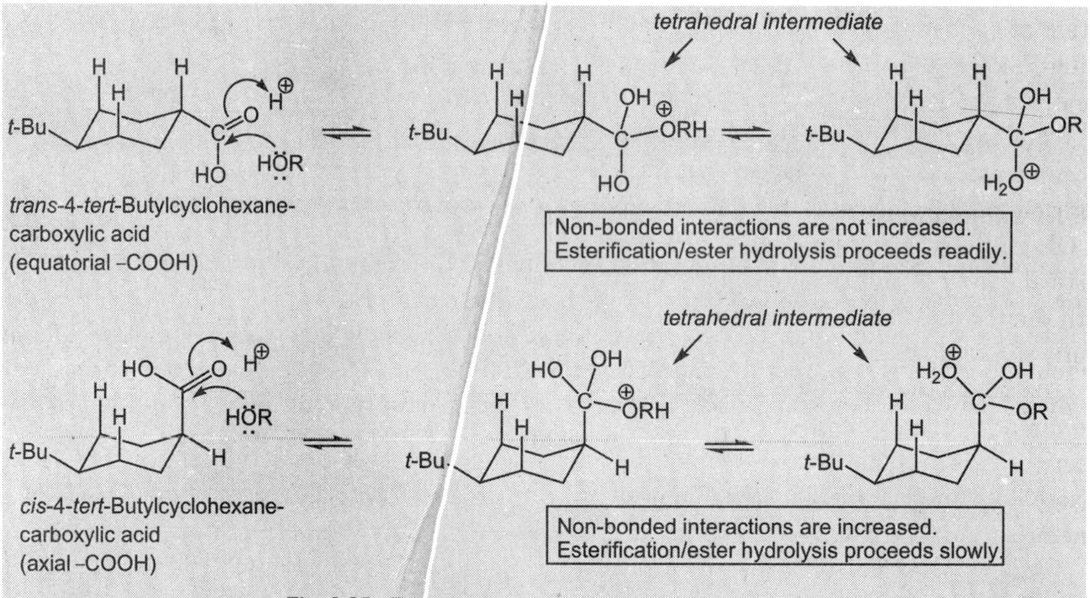

Fig. 3.85: Esterification and ester hydrolysis

Since both esterification and ester hydrolysis proceed through the same tetrahedral intermediate, the axial conformer of the ester such as *cis*-4-*tert*-butylcyclohexanecarboxylate, would be expected to undergo hydrolysis more slowly than the *trans* isomer (having an equatorial COOR).

Using similar arguments, the difference in the reactivity of *cis* and *trans* isomers in ester hydrolysis involving $B_{AC}2$ mechanism can be explained. An important example from steroids is cholestanyl acetate (having acetate group in the equatorial position), which undergoes hydrolysis with alkali more readily than its isomer epicholestanyl acetate (having acetate group in the axial position).

Cholestanyl acetate (equatorial acetate)

Epicholestanyl acetate (axial acetate)

3.9 CONFORMATIONAL ANALYSIS OF MONOSACCHARIDES

A. Fischer Projections of Open Chain Form of Monosaccharides

To study the conformational analysis of monosaccharides, it is necessary to know the configuration at each chiral carbon of a monosaccharide. Normally, open-chain formulae are drawn using Fischer projections. Thus, it is important to develop methodology for drawing Fischer projections of monosaccharides.

The scheme given in **Fig. 3.86** shows D-enantiomers and starts from D-glyceraldehyde, which is drawn according to the accepted convention. Stepping up by one carbon, two diastereomeric tetroses, namely, D-erythrose and D-threose are obtained; their structures are drawn by putting an OH group at the new carbon on the right side, and then the second one on the left side of the vertical bonds. Similarly, structures of the pentoses, obtained by stepping up by one carbon in both the tetroses independently, can be drawn. They are ribose, arabinose, xylose, and lyxose. Using the same technique, the structures of aldohexoses are drawn (**Fig. 3.86**).

Although the above scheme works nicely to assign the configuration of the monoschharides, the main problem is to memorize the names in the desired order, particularly in the case of hexoses. This problem has been simplified by using an alternative scheme, which is meant for hexoses and does not include lower monosaccharides. For this, one has to memorize the following mnemonic, and then derive the names of the aldohexoses in the desired sequence.

All	**Altr**uists	**Gl**adly	**Ma**ke	**Gu**m	**I**n	**Gal**lon	**Ta**nks
Allose	**Altr**ose	**Gl**ucose	**Ma**nose	**Gu**lose	**I**dose	**Gal**actose	**Ta**lose

The structure of these aldohexoses can also be drawn by using an independent method. The method is as follows-

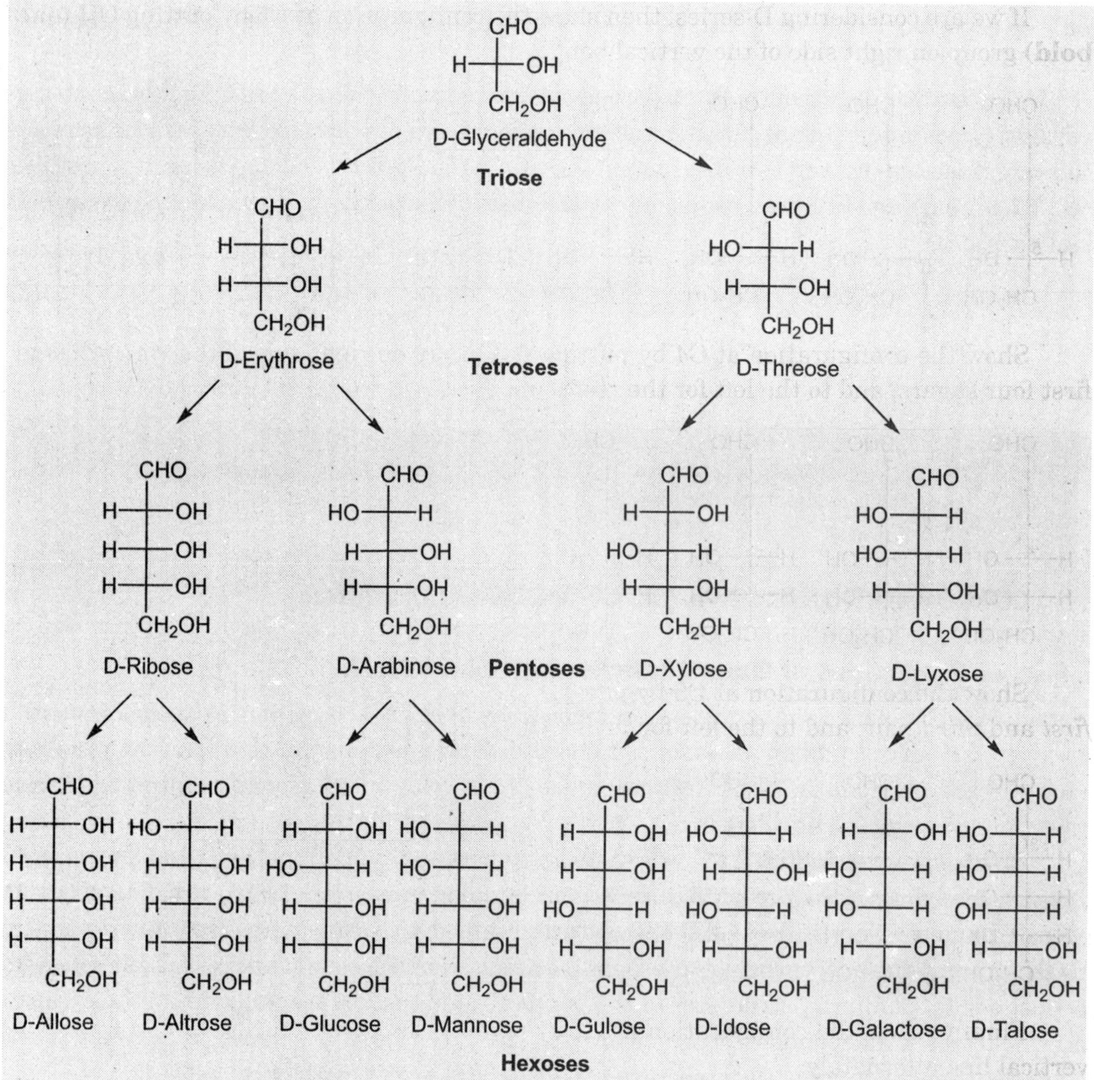

Fig. 3.86: Fischer projections of some monoschharides (Aldoses)

First, draw 8 vertical lines indicating CHO at the top, and CH_2OH at the bottom.

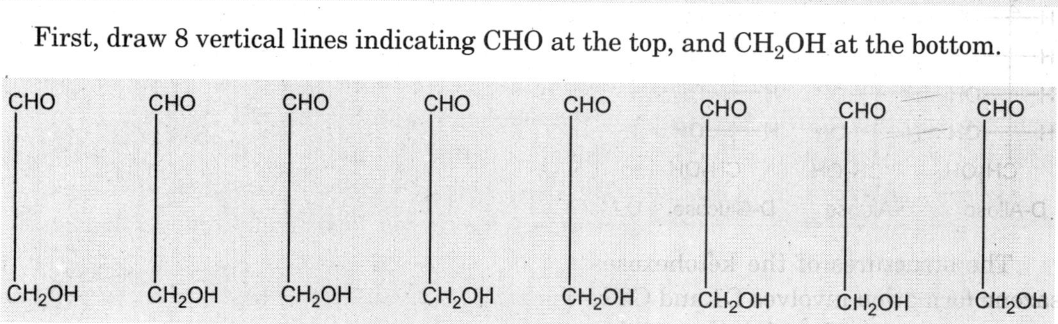

If we are considering D-series, then show the configuration at C5 by putting OH **(shown bold)** group on right side of the vertical bond.

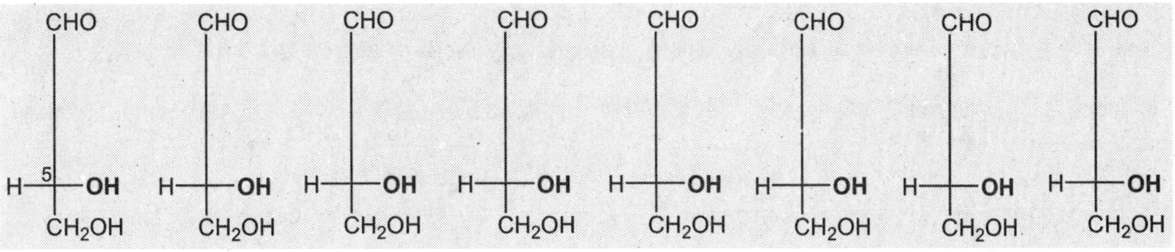

Show the configuration at C4 by putting OH group on right side of the vertical bond for first four sugars, and to the left for the rest four.

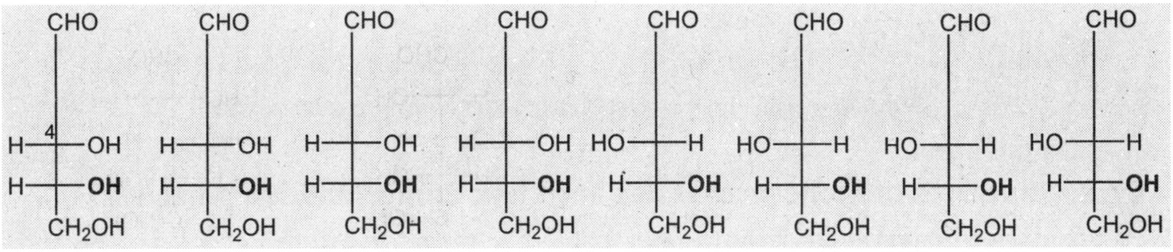

Show the configuration at C3 by putting OH group on right side of the vertical bond for *first* and *third* pair, and to the left for the *second* and *fourth* pair.

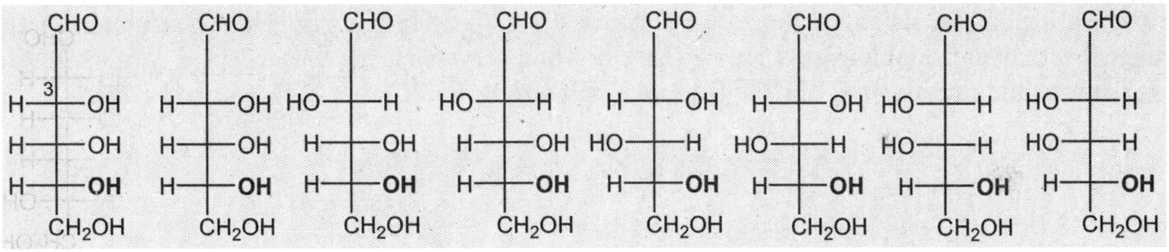

Finally, show the configuration at C2 by putting OH group on right and left side of the vertical line alternately.

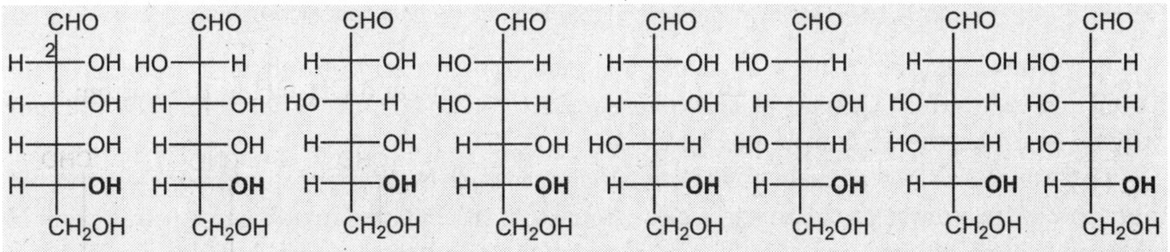

D-Allose D-Altrose D-Glucose D-Mannose D-Gulose D-Idose D-Galactose D-Talose

The structures of the ketohexoses are based on aldohexoses. In view of the fact that osazone formation involves C1 and C2 carbons, and the configuration of the remaining carbon atoms remains unaffected, it is possible to determine the configuration of ketohexoses from the configuration of the pair of the aldohexoses that give identical osazone. Thus, first pair (allose and alltrose) gives an osazone, which is identical to the one that is obtained from allulose. Similarly, glucose and mannose are related to fructose, whereas gulose/idose and galactose/talose are related to sorbose and tagatose, respectively as shown in **Fig. 3.87**.

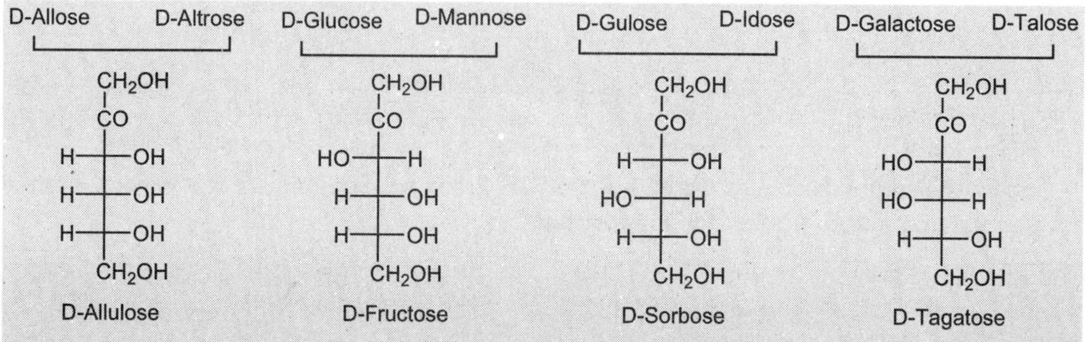

Fig. 3.87: Fischer projections of ketohexoses relating to aldohexoses

B. Cyclic Structures of Monosaccharides and Their Conformational Analysis

Cyclic structures (hemiacetals) of the sugars exist as five-membered (furanose) or six-membered-ring (pyranose). Contribution of pyranose structures is generally more than that of the furanose form. It has been established that the pyranose sugars assume a chair form in preference to the boat form, wherever both are structurally possible. Substitution of an oxygen atom for a carbon atom in cyclohexane causes only minor distortions in the ring, and consequently the general conformational features are retained in the pyranose form. Two regular conformations are drawn, and named as C1 or $^4C_1$ (the *normal chair*) and 1C or $^1C_4$ (the *reverse chair*).

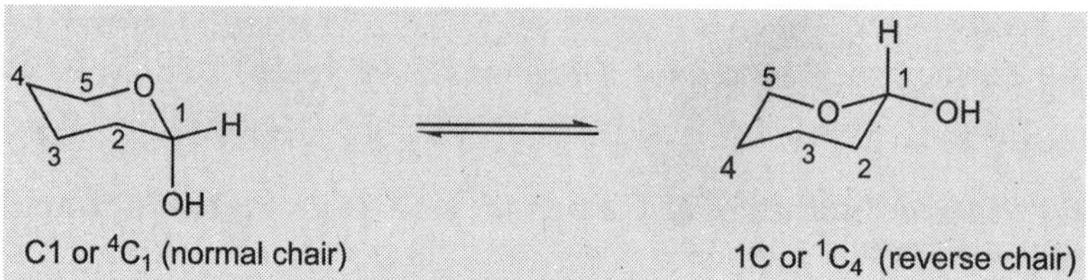

Pyranose structures

For D-aldohexoses in normal chair (C1), the α-form has the anomeric hydroxyl group (C1–OH) in the axial position; in the reverse chair form (1C), the α-form has the anomeric hydroxyl group in the equatorial position.

To draw the conformations of various aldohexoses, β-D-glucopyranose, which is the most stable one, is considered as the reference compound. In this, all the substituents (other than Hs) are present in equatorial position. The conformations of other monosaccharides are drawn by comparing the configuration of the sugar at each carbon atom shown in the open-chain formula (Fischer). The examples of anomers of glucopyranose and xylopyranose, which are *homomorphous* (have the same configurations at different chiral centres) are drawn. In both the cases, β-anomer is present in major amount as all the groups are located in equatorial position (**Fig. 3.88**).

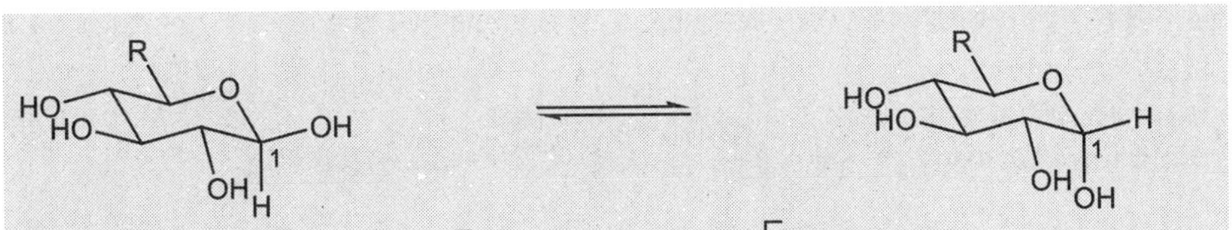

R = CH_2OH β-D-Glucopyranose (64%); R = H β-D-Xylopyranose (71%) More stable as all the substituents are in the equatorial position	*Homomorphous*	R = CH_2OH α-D-Glucopyranose (36%); R = H α-D-Xylopyranose (29%) Less stable as anomeric hydroxyl is in the axial position

Fig. 3.88: Anomers of D-glucopyranose/D-xylopyranose

C. Relative Stability Based on Instability Factors/Units

Relative stability of conformation of monosaccharides is investigated by using various methods. One method involves **instability factors**, which are based on the following rules:

(*i*) Axial hydroxyl groups (or any substituent other than hydrogen) increase the instability of the molecule. Each axial hydroxyl group results in one *instability unit*.

(*ii*) 1,3-Interactions involving axial hydroxyl result in 0.5 instability.

(*iii*) An axial CH_2OH group (at C5) results in two instability units, if only axial hydrogens are on C1 and C3. If an axial substituent (other than hydrogen) is on C1 or C3, the instability factor is 2.5 units. Because of this large value, it is unusual to have CH_2OH in the axial position.

(*iv*) If the hydroxyl (or an oxygen substituent) on C2 is axial and the hydroxyl (or oxygen substituent) is equatorial, this results in 2.5 instability units. This situation is referred to as the *delta-2 (Δ2) condition* or the *delta-2 instability factor*.

These rules can be used to solve several problems in sugar chemistry. For example, in contrast to glucose and xylose, it has been seen that the α-anomers of mannose and lyxose are more stable than the corresponding β-anomers. This is due the presence of delta-2 instability factor in the β-anomers of mannose and lyxose.

PROBLEMS AND REVIEW QUESTIONS

1. Assuming that addition of bromine occurs in an *anti* manner, what configurational stereoisomers would you obtain from each one of the following reactions?

a. *cis*-3-heptene + Br_2 ⟶	**c.** 1,2-dimethylcyclopentene + Br_2 ⟶
b. *trans*-3-hexene + Br_2 ⟶	**d.** *trans*-3-methyl-3-hexene + Br_2 ⟶

2. Consider the following *E2* reactions and answer the questions.

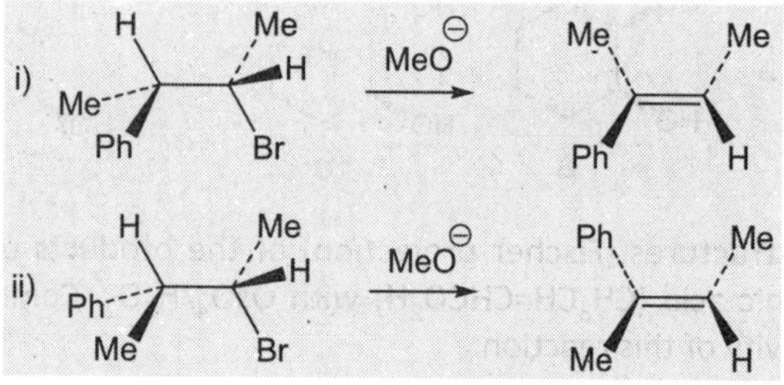

(*a*) Assign the absolute configuration (*R/S*) at the chiral centres of the starting bromides.

(*b*) Give the configuration (*E/Z*) of the resulting alkenes.

(*c*) Are these reactions stereospecific/stereoselective/none?

(*d*) Explain why other regioisomers and stereoisomers are not obtained.

3. Draw the conformational isomers of sodium 3-bromocyclohexylcarboxylate (BrC_6H_{10} COONa). It has been observed that the *trans* isomer spontaneously eliminates bromide ion to give the lactone, whereas the *cis* isomer is inert. Explain.

4. Give the products when the following bromides are treated with methoxide ions in S_N2 reactions. If the product can exist as stereoisomers, show which isomers are formed.

(*a*) (3*R*,4*R*)3-bromo-4-methylhexane

(*b*) (3*S*,4*S*)3-bromo-4-methylhexane

(*c*) (3*R*,4*S*)3-bromo-4-methylhexane

(*d*) (3*S*,4*R*)3-bromo-4-methylhexane.

5. When the following stereoisomer of 2-chloro-1,3-dimehtylcyclohexane reacts with methoxide ion in a solvent that favours $S_N2/E2$ reactions, only one product is formed. When the same compound reacts with methoxide ion in a solvent that favors $S_N1/E1$ reactions, a mixture of several products is formed. Identify the products that are formed under the two set of conditions.

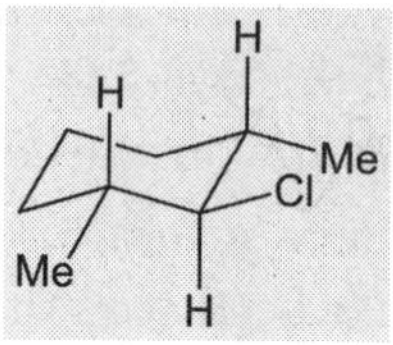

6. Draw the preferred conformation of the following compounds, and rank them in the order of decreasing rate in an *E2* reaction. Provide explanation to your answer.

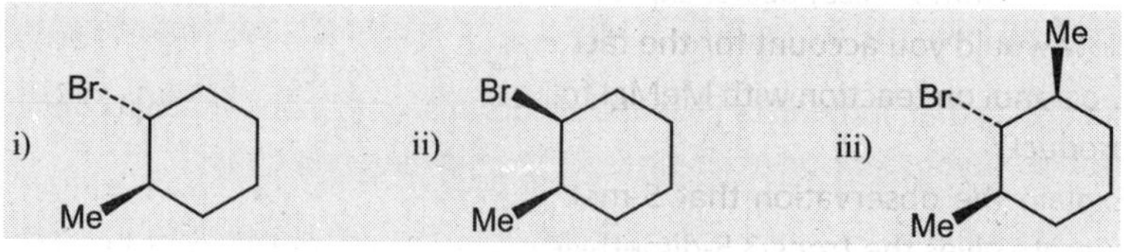

7. Draw the preferred conformation of the compounds **A-E**. Out of these, only **C** cannot be cleaved by periodic acid involving a cyclic intermediate. Explain.

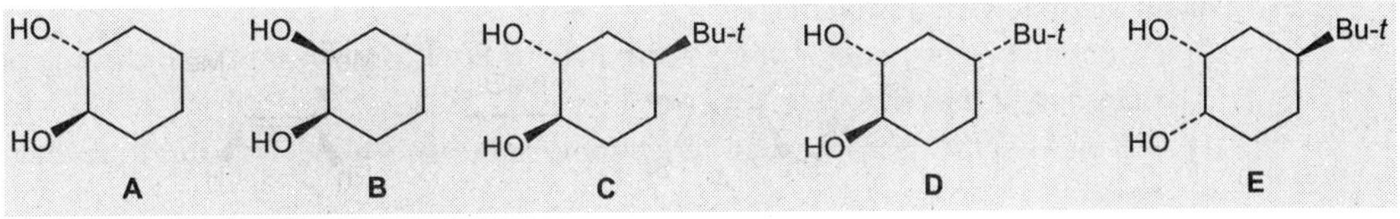

8. Write the structures (Fischer projection) of the products obtained by the hydroxylation of *trans*-crotonic acid ($CH_3CH{=}CHCO_2H$) with OsO_4/H_2O_2. Comment upon the stereospecificity/steroselectivity of this reaction.
9. Draw the structures of *cis*- and *trans*-cyclodecenes. What will be the outcome of *anti*-addition of bromine to these isomers? Explain.
10. Designate the hydrogen atoms in the following compounds as homotopic/enantiotopic/diastereotopic.
    Propanoic acid, 2-chlorobutane, *cis*-dichlorocyclopropane, and *meso*-2,3-dibromobutane.
11. On saponification optically active biphenyl derivative **A**, gives an optically inactive compound **B**. Give reason. What is the configuration of **A** (*R* or *S*)?

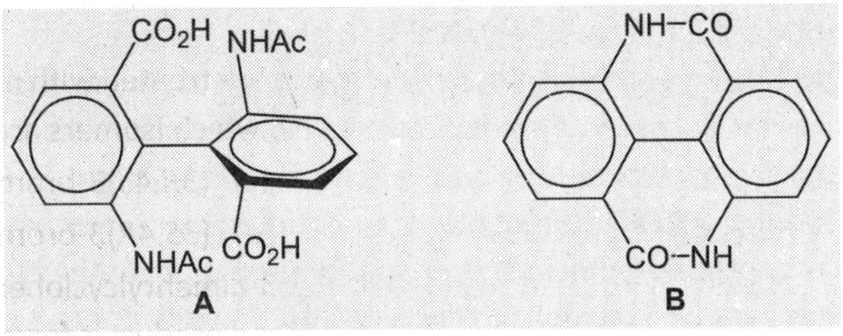

12. What are *syn* axial interactions? Which conformation (**A** or **B**) is preferred? Offer explanation to your choice.

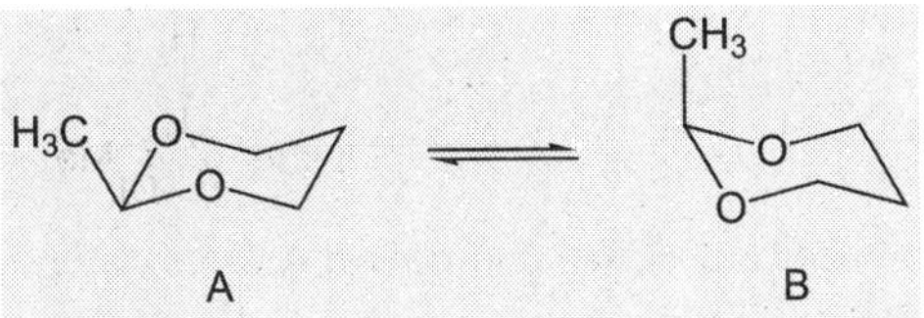

13. The dilactone of benzophenone-2,2′,4,4′-tetracarboxylic acid is resolvable. When an optically active isomer of this dilactone (any enantiomer, *R* or *S*) is treated with sodium hydroxide, its optical rotation disappears. Explain.
14. How would you account for the fact that the ester obtained from benzoylformic acid and R-(–)-2-octanol on reaction with MeMgI followed by hydrolysis gives R-(–)-atrolactic acid as the major product?
15. Explain the observation that 5-methylcyclohex-2-en-one on treatment with lithium dimethyl cuprate gives the *trans*-3,5-dimethylcyclohexanone to the extent of 98%.
16. 1,2-Diphenyl-1-propanol may be prepared by either of the two ways:
    (*i*) Reduction of PhCH(Me)COPh with LAH.
    (*ii*) Reaction of PhCH(Me)CHO with PhMgBr.
    Which method would you choose to prepare the *erythro*-isomer? Give reason to your choice.

17. Explain the following observations:
    (*i*) Oxidation of *trans* isomer of 3,3-dimethyl-5-methylcyclohexanol with chromic acid is faster than that of the *cis* isomer.
    (*ii*) Solvolysis of the *cis* isomer of 4-*tert*-butylcyclohexyl tosylate is greater than that of the *trans* isomer.
    (*iii*) *Cis*-1,2-dimethylcyclohexane is optically inactive at room temperature even though the molecule lacks any element of symmetry.
    (*iv*) Stilbene dichloride (racemate) gives *Z*-chlorostilbene on heating at 200°C with pyridine, whereas *meso*-isomer does not react under these conditions.
    (*v*) Addition of bromine to maleic acid gives 2,3-dibromosuccinic acid (racemic mixture), whereas addition of bromine to fumaric acid gives *meso*-2,3-dibromosuccinic acid.
18. Draw preferred conformations of *cis*- and *trans*- isomers of 2-acetoxycyclohexyl *p*-toluenesulphonate, and account for the following observations-
    (*i*) The acetolysis of *trans* isomer is faster than the *cis* isomer, even though the concentration of the reactive conformation (*trans*) is low.
    (*ii*) The acetolysis of *cis* isomer is accompanied with inversion of configuration, while acetolysis of *trans* isomer occurs with retention of configuration.
    (*iii*) On acetolysis, optically active *trans* isomer gives the racemic mixture of *trans*-diacetate.
19. Draw Fischer projections of configurational isomers of 2,3-dibromosuccinic acid. Indicate their relationship as enantiomeric or diastereomeric. Also designate whether bromine atoms of 2,3-dibromosuccinic acid are homotopic or enantiotopic or diastereotopic.
20. How many conformers (including enanatiomers) of 1-methyl-2-phenylcyclohexane can exist? Discuss their relative stability and chiralty with the help of chair conformation of cyclohexane.
21. Explain Cram's rule. What are its modifications? Predict the major diastereomer formed in the following reaction using lithium aluminium hydride (LAH).

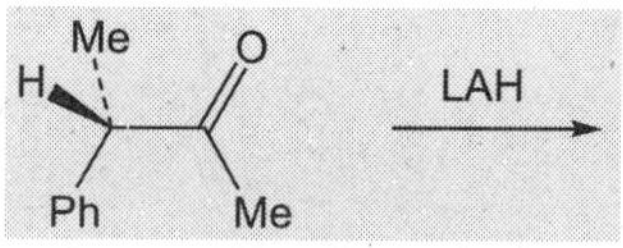

22. The acetolysis of optically active compound **A** gives racemic product, while that of compound **B** proceeds with the retention of configuration. Identify the product, and explain its formation.

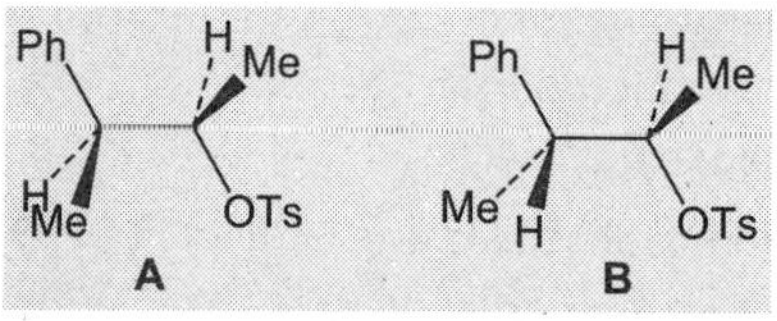

23. What type of compounds can show chirality due to axially dissymmetric planes? Discuss the stereochemistry of allenes.
24. Draw preferred conformations of α/β-anomers of D-glucopyranose and D-mannopyranose. What is Δ-2 (delta two) condition? Discuss its role in comparing the stability of monosaccharides, *e.g.* glucose and mannose.
25. What are instability factors? Using these, discuss relative stability of monosaccharides with suitable examples.

26. In aqueous solution, the amount of β-anomer of carbohydrates is: glucose (64%); mannose (36%); xylose (71%), and lyxose (31%). Why (draw preferred conformations for your explanation)?
27. Addition of bromine to *cis*-2-pentene gives two *threo* enantiomers in equal amounts. Explain.
28. Draw the preferred conformation of *trans*-1,2-dibromocyclohexane. What will be the effect of change of polarity of solvent on the preferred conformation? Also discuss what will happen if bromine is replaced with the other halogen atom.
29. Bromine is a larger atom than chlorine, but the equilibrium constants (2.2 and 2.4) indicate that a chloro substituent has greater preference for the equatorial position. Offer an explanation.
30. Explain why it is not possible to resolve the amines even though three different groups are attached to the nitrogen.
31. Explain the cause of optical activity of biphenyls. Giving examples, discuss how absolute configuration of biphenyls is determined.
32. Define asymmetric synthesis. What are various methods available for carrying out an asymmetric synthesis? Explain chiral auxiliary control (second generation) method of asymmetric synthesis.
33. What is atropisomerism? Discuss racemization of chiral biphenyl compounds.
34. Explain *buttressing effect* with suitable examples.
35. Discuss briefly geometrical isomerism in oximes. How *E/Z* nomenclature is used in these isomers?
36. Explain chirality in quaternary ammonium salts. Why do they undergo racemization on heating?
37. Tertiary amines with three different groups are not resolvable, whereas many phosphines can be resolved. Explain with examples.
38. Discuss stereochemistry of quadrivalent and quinquevalent phosphorus compounds. Explain why optically active phosphine oxides undergo racemization with acids.
39. Define and explain the following terms-
    Homotopic ligands/faces, enantiotopic ligands/faces, diastereotopic ligands/faces, prochirality, prochiral centre.
40. What is a prochiral centre? Designate the hydrogen atoms of fluorochloromethane as *pro-R*/*pro-S*.
41. How are enantiotopic and diastereotopic faces named? Explain with suitable examples.
42. Draw the potential energy diagram for various conformations of cyclohexane and explain their relative stability? Why is the chair conformation of cyclohexane more stable than the boat conformation?
43. What are axial and equatorial bonds in cyclohexane? Discuss briefly the conformations of monosubstituted cyclohexanes, and their relative stability.
44. Draw Newman projections of axial and equatorial methylcyclohexanes, and comment upon their relative stability.
45. Bromination of cyclohexene with bromine in carbon tetrachloride occurs in *anti* manner, thereby giving predominantly diaxial conformer of 1,2-dibromocyclohexane. Explain by writing suitable structures. What will be the effect of polarity of the solvent?
46. Define and explain *self-immolative* reactions with suitable examples.
47. Discuss the role of NMR spectroscopy to differentiate between enantiotopic and diastereotopic ligands.
48. (*a*) What are conformationally biased molecules?
    (*b*) Draw preferred conformations of the following compounds:

(*i*) 4-*cis*-*tert*-Butylcyclohexyl bromide
(*ii*) *cis*-Cyclohexane-1,3-diol
(*iii*) *cis*-2,5-Di-*tert*-Butylcyclohexane-1,4-diol.

49. Discuss relative stability of *cis*- and *trans*-decalins. Which one is rigid and why? How will you differentiate them using NMR spectroscopy?

SOLUTIONS

1. (a) The reactant is *cis* and addition is *anti*, the product containing two chirality centres will be a mixture of the two *threo* enantiomers.

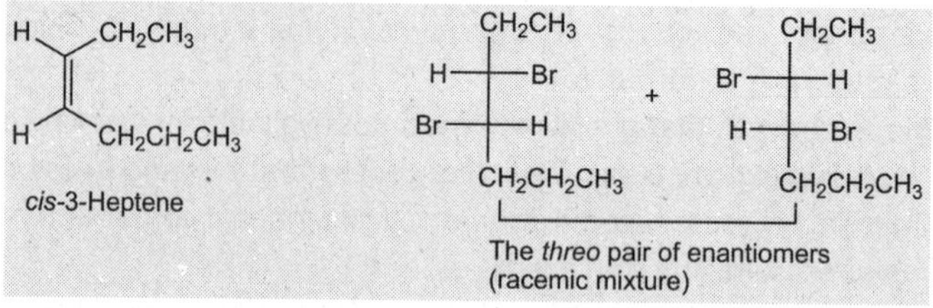

(b) Since two chirality centres created after addition of bromine on the *trans* reactant are similar, the *erythro* product is *meso* isomer.

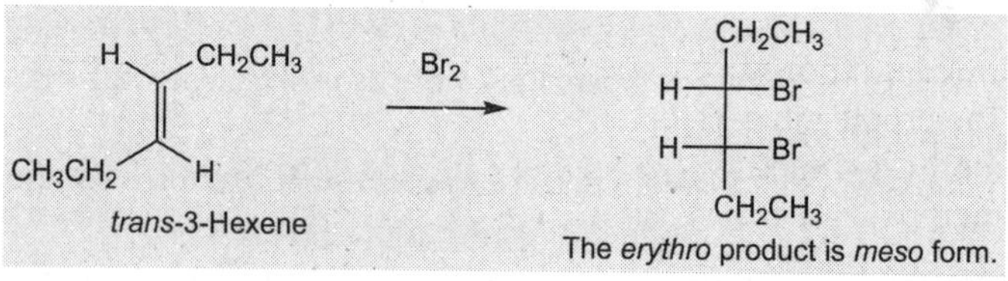

(c)

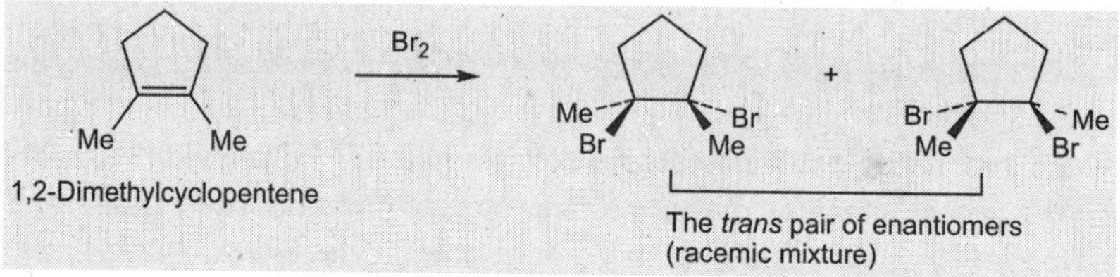

(d) Since two chirality centres created after addition of bromine on the *trans* reactant are dissimilar, racemic mixture of the *erythro* enantiomers is produced.

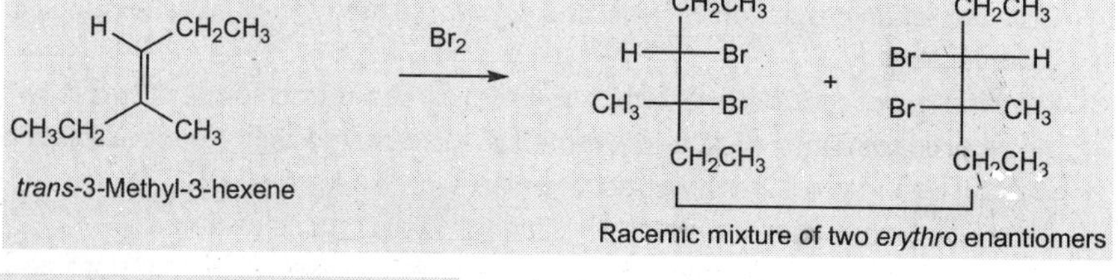

2. (a)

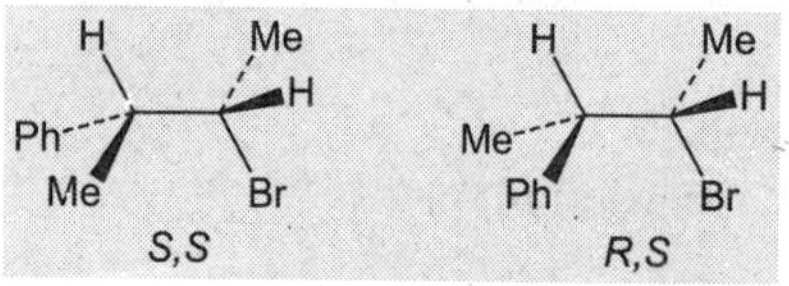

(b) 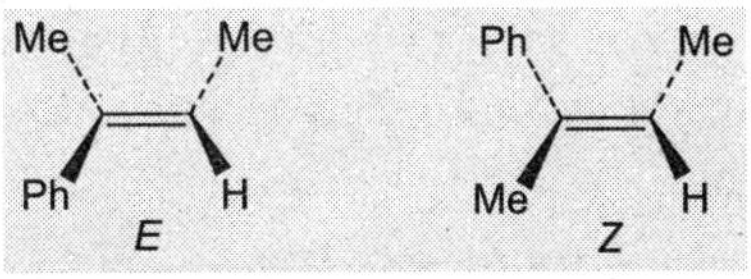

(c) These reactions are stereospecific (stereoselective as all stereospecific reactions are stereoselective).

(d) *E2* reactions are both regioselective and stereoselective.

3. See the text for drawing all conformations of $BrC_6H_{10}COONa$. Two conformers **A** and **B** of the *trans* isomer equilibrate. In the conformer **B**, carboxylate anion can attack from the backside of C–Br (equatorial) in S_N2-type manner and , therefore, this isomer spontaneously eliminates bromide ion to give the lactone. In the case of *cis* isomer, none of the conformer fulfils this condition and hence it is inert.

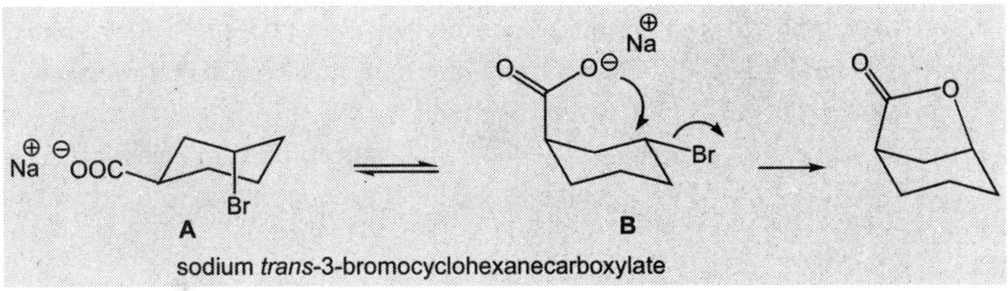

4. Substitution of bromine atom by methoxyl group will be accompanied by inversion of configuration at C3 position. The product in each case will be-
   (a) (3*S*,4*R*)-3-methoxy-4-methylhexane
   (b) (3*R*,4*S*)-3-methoxy-4-methylhexane
   (c) (3*S*,4*S*)-3-methoxy-4-methylhexane
   (d) (3*R*,4*R*)-3-methoxy-4-methylhexane.

5. Under the $S_N2/E2$ conditions, there is only one possibility *i.e.* substitution with inversion of configuration. *E2* elimination is not possible as leaving chlorine atom is equatorial and ring flipping will be difficult in this case.

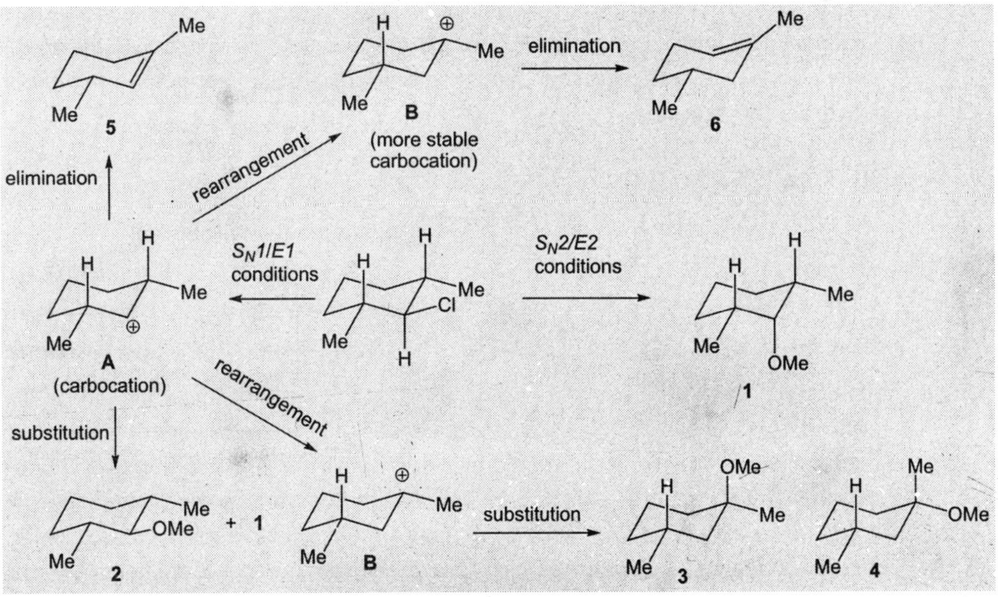

Under $S_N1/E1$ conditions, reaction involves formation of carbocation (**A**) and then rearranged *tertiary* carbocation (**B**, more stable). These carbocations undergo substitution and elimination reactions to give a mixture of several products.

6. Preferred conformations of i)-iii) are:

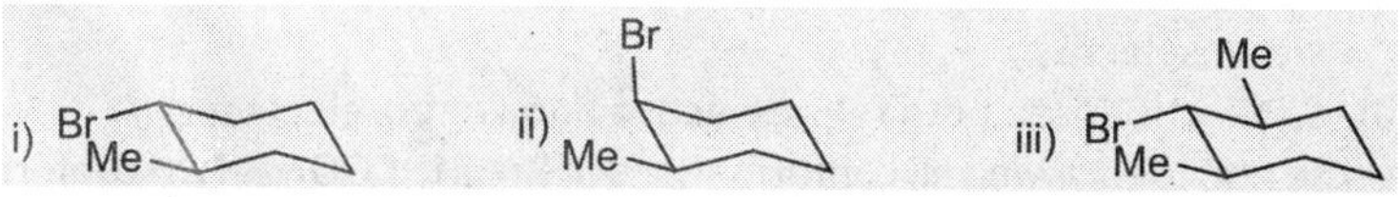

Order of reactivity in *E2* reaction will be:

(ii) > i) > iii).

For *E2* elimination, the two groups (H, Br) to be eliminated must have *antiperiplanarity* (H*a* and Br*a*). Since, example ii) already fulfils this requirement, it is the most reactive one. In case of (i) and (ii), to fulfil this requirement, the ring flipping is necessary. Ring flipping in (i) will result in two axial substituents (Br and Me), whereas in iii) it will result in three axial substituents (Br, Me, Me). Obviously, i) will react faster than (iii).

7. To solve such problems, it is essentially important to translate classical cyclohexanes into chair conformations with correct relationship of the equatorial and axial substituents at different positions.

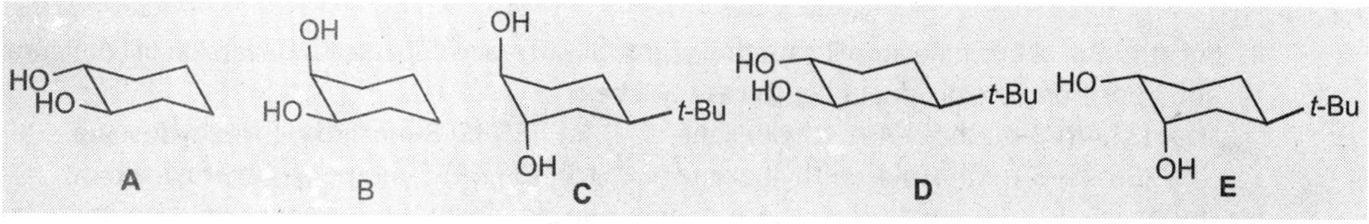

The sterochemical requiremens of the oxidative cleavage of 1,2-diols with periodic acid is that the hydroxyl groups must be positioned so as to form the cyclic intermediate. In the structures **A-E**, all except **C** (having two OH groups axially positioned) meet this requirement.

8. Reaction of olefinic compounds with OsO_4/H_2O_2 gives *syn* hydroxylation of the double bond. This reaction is having *stereospecifity*. *Trans*-crotonic acid gives enantiomeric pair of only *threo*-2,3-dihydroxybutanoic acid.

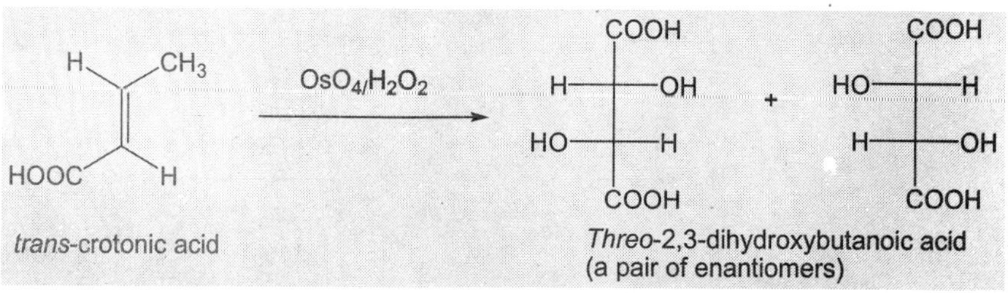

9. Cycloalkenes with larger rings such as cyclodecene can exist as *cis* and *trans* isomers (geometrical isomers) and their structures can be drawn as shown in fig. *Anti* addition of bromine to *cis* alkene gives mixture of two *trans* isomers which are enantiomers and formed in the same ratio. Likewise, *trans* isomer on *anti* addition produces *cis* dibromo product which is a *meso* isomer in this case. For clear understanding of formation of the *meso* isomer, models are helpful.

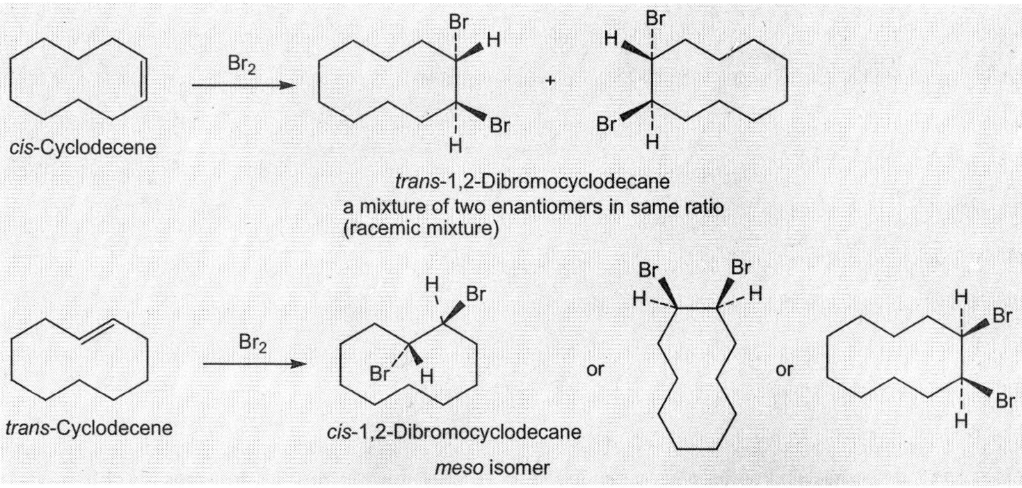

10. Problems concerning with propanoic acid and 2-chlorobutane are relatively simple and are predictable with the following schemes.

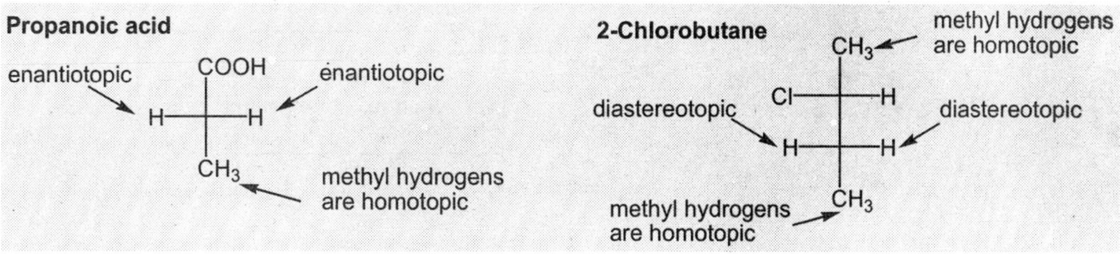

In the other two cases, it will be easier to visualize replacement of hydrogens by different group 'Z' and then compare the relationship between the resulting structures. These situations have been explained with the help of following schemes for *cis*-dichlorocyclopropane and *meso*-2,3-dibromobutane individually.

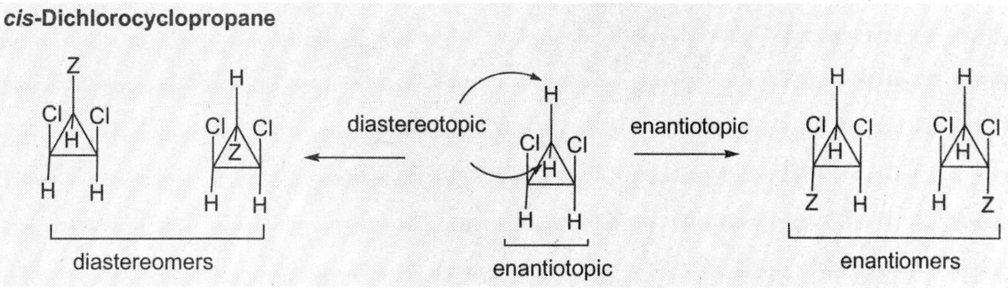

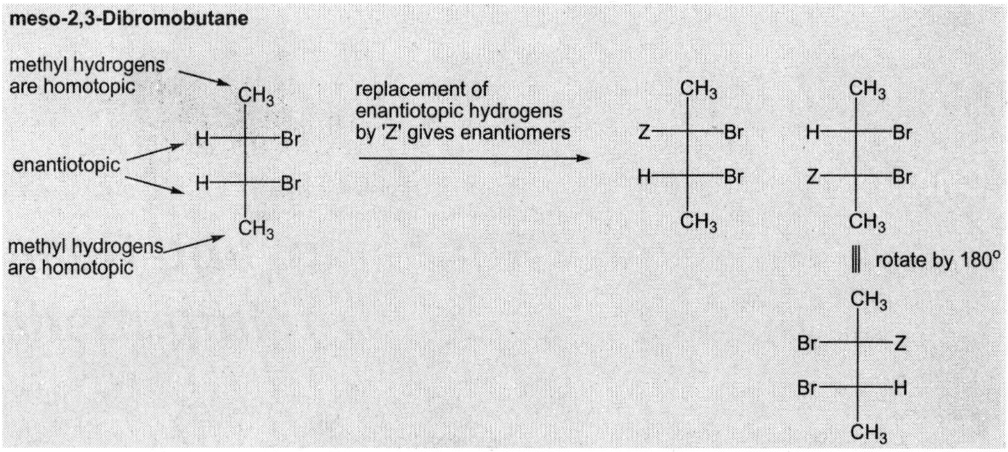

11. Biphenyl derivative **A** is optically active (single enantiomer) as it possesses chiral axis. However, the product **B** is optically inactive because it is a cyclic amide which has planar structure (non-resolvable).

    The configuration of **A** is *S*.

12. 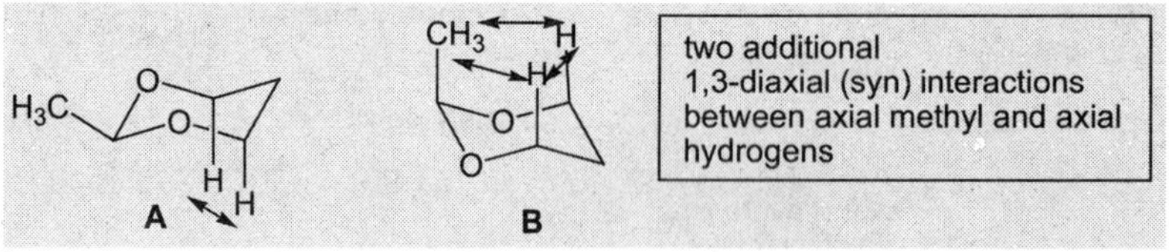

Chapter 4

Organic Reactions, Reagents, and Reaction Intermediates

Chapter Outline

4.1 TYPES OF ORGANIC REACTIONS

- **A.** Substitution Reactions
- **B.** Addition Reactions
- **C.** Elimination Reactions
- **D.** Molecular Rearrangements
- **E.** Molecular Reactions (or Pericyclic Reactions)

4.2 TYPES OF REAGENTS

- **A.** Nucleophilic Reagents (Nucleophiles)
- **B.** Electrophilic Reagents (Electrophiles)

4.3 ORGANIC ACIDS AND BASES

- **A.** Strength of Acids and Bases
- **B.** Influence of Structure on pK_a Values
- **C.** General Acid and Base Catalysis
- **D.** HSAB Principle

4.4 REACTION INTERMEDIATES

- **A.** Carbocations
- **B.** Carbanions
- **C.** Free Radicals
- **D.** Carbenes
- **E.** Nitrenes
- **F.** Arynes

Organic reactions usually involve fission and formation of covalent bonds. The covalent bond is often represented by a dash (—), and the movement of a pair of electrons is shown by a *curved arrow* (), whereas the movement of a single electron is shown by a *fishhook arrow* ().

There are two ways in which a covalent bond may be broken:

- Homolytic Fission
- Heterolytic Fission

Homolytic Fission

Suppose the atoms X and Y are bonded through a covalent bond. This bond may break in such a way that each fragment carries one unpaired electron. The species formed by this fission are known as *free radical intermediates,* and this manner of cleavage is known as *homolytic fission*. Homolytic fission is also called as homolysis. Generally, non-polar bonds undergo homolytic fission, which is favoured by high temperature, UV-irradiation, or by the presence of radical initiators such as peroxides.

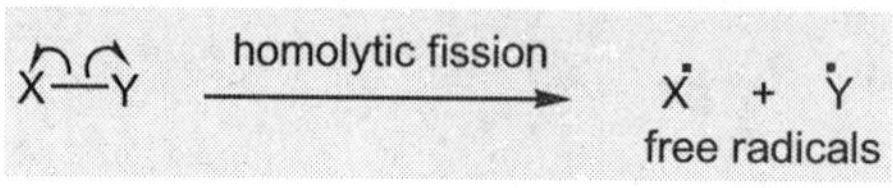

Heterolytic Fission

When the covalent bond joining the atoms X and Y breaks in such a way that the pair of electrons stays on any one of the atoms, such fission is called as *heterolytic fission*. This type of fission results in the formation of *ionic intermediates*. Heterolytic fission is also called *heterolysis*, which occurs in polar bonds, and is favoured by polar solvents.

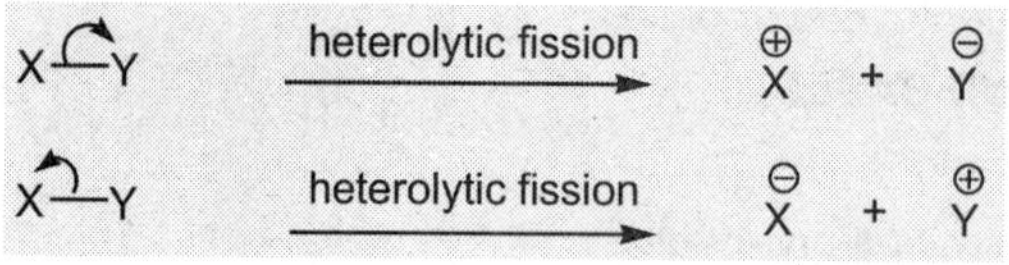

If the initially bonded atoms are of different electronegativities, the electron pair becomes attached to the more electronegative atom. Further, when as a result of heterolytic fission a carbon atom develops negative charge, it is known as *carbanion*. On the other hand, a positively charged carbon atom is called *carbocation*.

It is important to note that depending upon the reaction conditions, the same covalent bond can undergo homolytic or heterolytic fission. Generally, the bond dissociation energy for homolytic fission is lower than that for heterolytic fission, *e.g.*

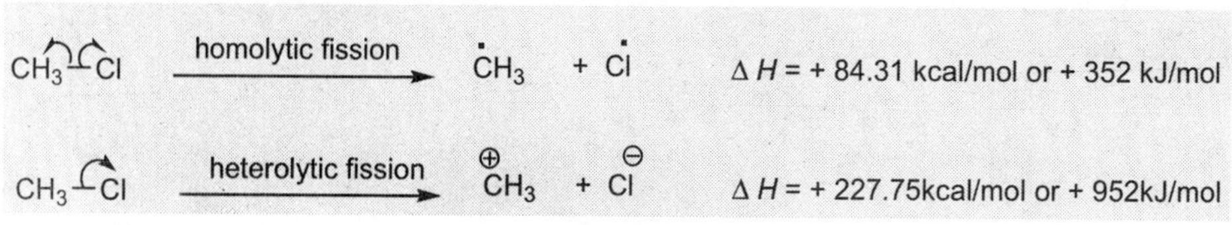

4.1 TYPES OF ORGANIC REACTIONS

Broadly speaking, organic compounds may undergo five kinds of reactions, namely:

(A) Substitution Reactions
(B) Addition Reactions
(C) Elimination Reactions
(D) Molecular Rearrangements
(E) Molecular Reactions

A. Substitution Reactions

A *substitution reaction* is the one in which an atom or a group in a molecule is replaced by another. The incoming group becomes attached to the same carbon to which the outgoing group was attached. The substituting species may be a nucleophile, an electrophile, or a free radical.

I. Nucleophilic Substitution Reactions

In nucleophilic substitution (Chapter 6; *substituting species is a nucleophile*), a nucleophile brings an electron pair to the substrate, and the leaving group departs *with* an electron pair.

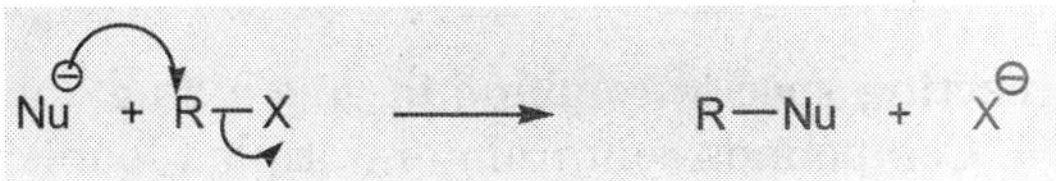

II. Electrophilic Substitution Reactions

In electrophilic substitution (Chapter 10; *the attacking species is an electrophile)*, the leaving group departs *without* the pair of bonding electrons. This type of reaction occurs mostly in aromatic systems where high electron density attracts electrophiles.

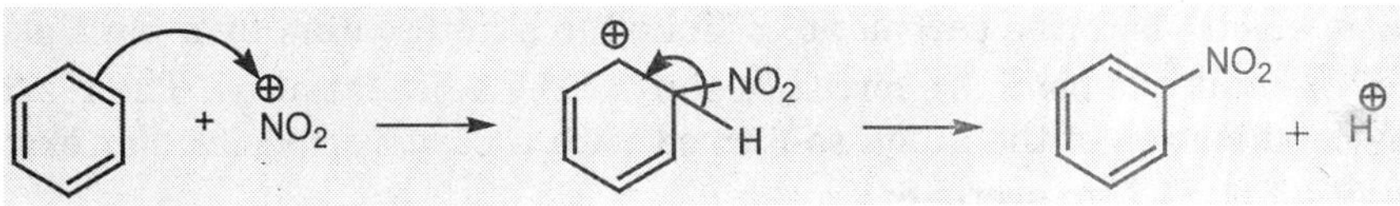

III. Free Radical Substitution Reactions

The reaction in which substituting species is a free radical is called a free radical substitution reaction. Such reactions are common in alkanes and other saturated compounds. For example, chlorination of methane in presence of heat or light gives methyl chloride and hydrogen chloride.

$$CH_3{-}H + Cl_2 \xrightarrow[\text{or light}]{247\text{–}397\ ^\circ C} CH_3{-}Cl + HCl$$

It is a chain reaction, which takes place through a free radical mechanism, and involves the following steps.

Step 1: Chain initiation

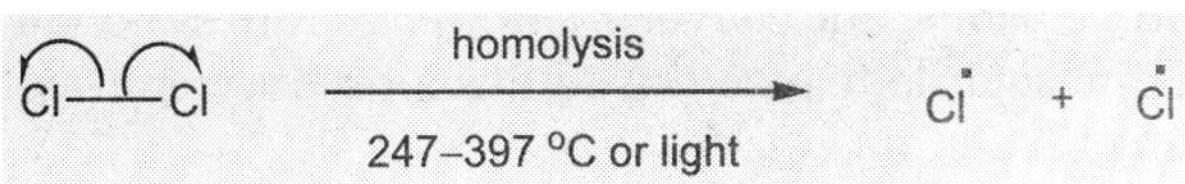

Step 2: Chain propagation

$$CH_3{-}H + \dot{C}l \longrightarrow \dot{C}H_3 + HCl$$

$$\dot{C}H_3 + Cl_2 \longrightarrow CH_3{-}Cl + \dot{C}l$$

Step 3: Chain termination

$$\dot{C}l + \dot{C}l \longrightarrow Cl_2$$

$$\dot{C}H_3 + \dot{C}H_3 \longrightarrow CH_3{-}CH_3$$

It may be noted that the steps involved in chain propagation occur over and over again until the reaction is complete.

B. Addition Reactions

A reaction in which two reacting species combine to give a single molecule of the product is called *an addition reaction*. Compounds containing multiple (double or triple) bonds undergo this type of reaction. As in substitution reactions, in these reactions also the attacking species may be a nucleophile, an electrophile, or a free radical.

I. Nucleophilic Addition Reactions

Addition reactions in which the attacking species is a nucleophile are called *nucleophilic additions*. In such reactions, which are exhibited by aldehydes and ketones, a nucleophile brings a pair of electrons to the carbon atom linked to a heteroatom by a multiple bond. The π-electrons of the multiple bond are forced to shift to the adjacent atom. The reaction is then completed by combination of the anion so formed with a positive species. For example, base-catalyzed addition of HCN to acetone:

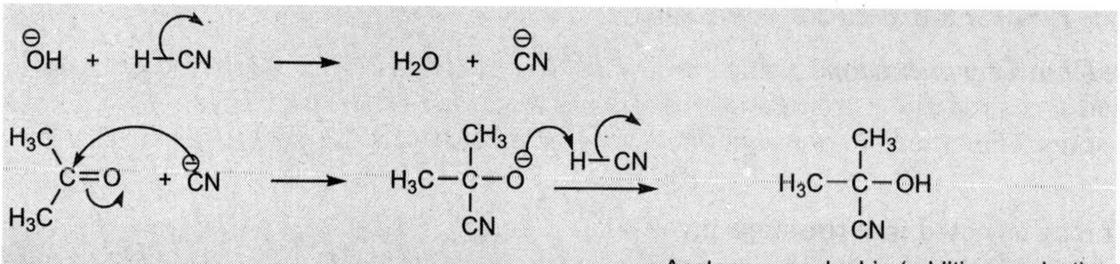

Acetone cyanohydrin (addition product)

II. Electrophilic Addition Reactions

Addition reactions in which the attacking species is an electrophile are called electrophilic additions. In these reactions, a positive or electron-deficient species (electrophile) adds to one of the atoms joined by multiple bonds, and converts the π-bond into a σ-bond forming a carbocation, which combines with a nucleophile to give the addition product. For example, addition of HBr to propylene:

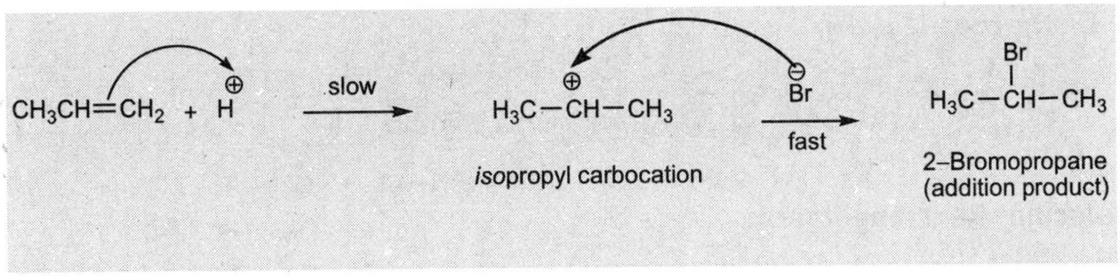

III. Free Radical Addition Reactions

Addition reactions in which the attacking species is a free radical are called free radical addition reactions (Chapter 13). For example, addition of HBr to alkenes in presence of peroxides:

$$CH_3CH{=}CH_2 \;+\; HBr \xrightarrow{\text{peroxides}} CH_3{-}CH_2{-}CH_2Br$$

It is a chain reaction which occurs in the following three stages:

Step 1: Chain initiation

$$\underset{\text{peroxide}}{RO{-}OR} \xrightarrow[\text{homolytic fission}]{\text{heat or light}} \underset{\text{alkoxy free radical}}{2\,R\dot{O}}$$

$$R\dot{O} \;+\; HBr \longrightarrow ROH \;+\; \dot{Br}$$

Step 2: Chain propagation

$$CH_3{-}CH{=}CH_2 \;+\; \dot{Br} \xrightarrow{\text{slow}} \underset{\text{free radical}}{CH_3\dot{C}HCH_2Br}$$

$$CH_3{-}\dot{C}H{-}CH_2Br \;+\; HBr \xrightarrow{\text{fast}} CH_3{-}CH_2{-}CH_2Br \;+\; \dot{Br}$$

Step 3: Chain termination

$$\dot{Br} \;+\; \dot{Br} \longrightarrow Br_2$$

It may be noted that the steps involved in the chain propagation stage are repeated again and again until the reaction is complete.

C. Elimination Reactions

An elimination reaction involves removal of two atoms or groups from the same molecule. In many such reactions, two atoms or groups are lost from adjacent carbon atoms, so that a new double (or triple) bond is formed. Acid-catalyzed dehydration of alcohols and base-catalyzed dehydrohalogenation of alkyl halides are typical reactions.

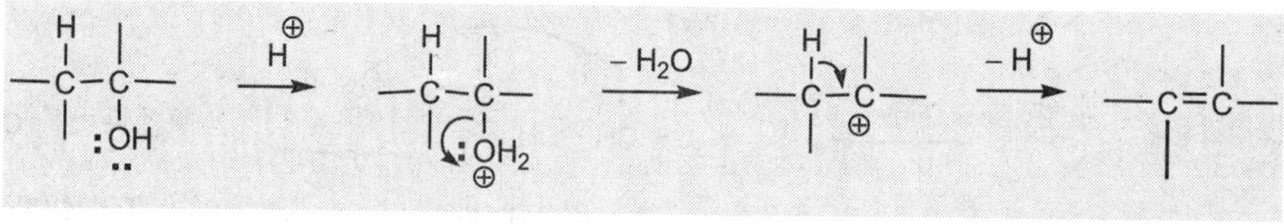

D. Molecular Rearrangements

A rearrangement reaction is the one in which an atom or a group moves from one atom to another within the molecule. Rearrangements involving shift of a group from a carbon atom to an adjacent carbon are quite frequent, and are referred to as 1,2-shifts. Acid-catalyzed rearrangement of 1,2-glycols yields carbonyl compounds, the classical example being the acid-catalyzed rearrangement of pinacol to pinacolone.

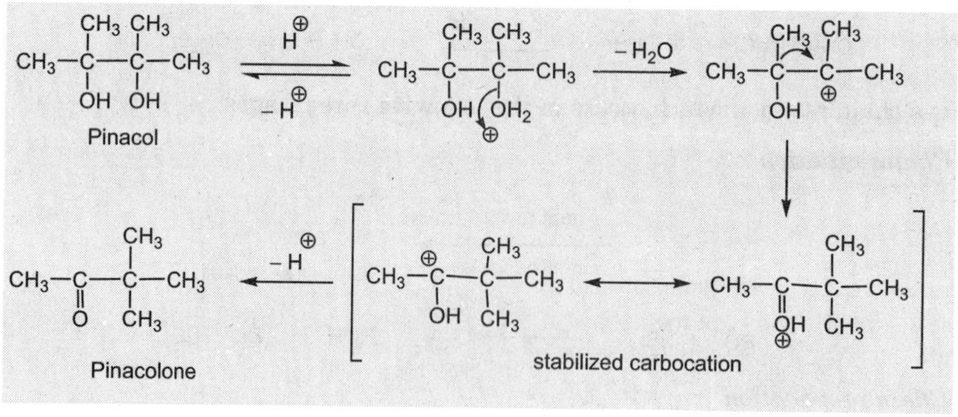

E. Molecular Reactions (or Pericyclic Reactions)

These are the reactions in which the covalent bonds undergo reorganization, either intramolecularly or intermolecularly through *cyclic transition states,* and without the intervention of ionic or radical intermediates; acids and bases having no effect. These reactions take place under the influence of heat or light, and are characterized by a high degree of stereospecificity.

Electrocylic and *cycloaddition* reactions are typical examples.

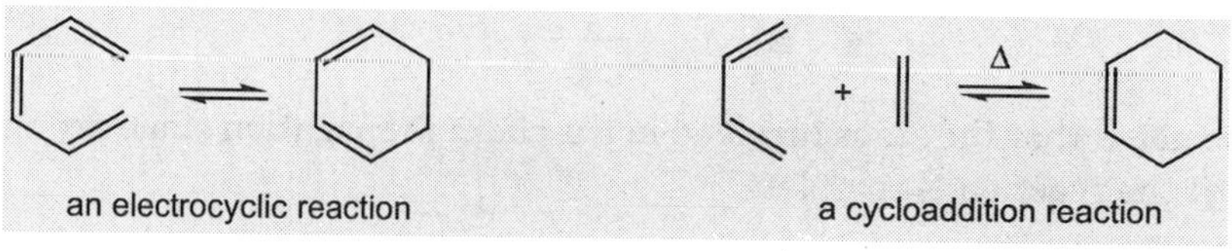

4.2 TYPES OF REAGENTS

For an organic reaction to take place, some starting materials are required. These materials are called *reactants* or *reacting molecules*. The reactants are classified into two categories, the *substrates* and the *reagents*. In fact, no clear cut distinction between these two types of reactants is possible. However, by convention the more reactive reactant is called the reagent, and the other one is called the substrate.

A substrate can also be defined as a molecule in which old bonds are broken and new bonds are formed, or in other words a molecule in which one functional group changes into another. For example, in the following reaction, ethyl alcohol is designated as the substrate (*i.e.* the compound in which the old C—O bond is broken and a new C—Br bond is formed, or the functional group OH is replaced by Br).

$$C_2H_5OH + HBr \longrightarrow C_2H_5Br + H_2O$$

With this background let us discuss the various types of reagents. They are broadly classified as nucleophilic and electrophilic reagents (nucleophiles and electrophiles, respectively).

A. Nucleophilic Reagents (Nucleophiles)

A nucleophilic reagent or nucleophile (Latin: *nucleus* meaning kernel; break: *philos* meaning loving, hence 'nucleus loving') is an atom or a group of atoms that has an available unshared pair of electrons.

Thus, nucleophiles must be electron-rich species. These are of two types:

I. Charged Nucleophiles

These negatively charged species are also called as **anionoid reagents**. Examples of charged nucleophiles or anionoid reagents are chloride ion (Cl^-), bromide ion (Br^-), iodide ion (I^-), hydroxide ion (HO^-), alkoxide ion (RO^-), amide ion (NH_2^-), cyanide ion (CN^-), carbanions (R^-), hydride ion (H^-), carboxylate ion (COO^-), acetylide ion ($R—C{\equiv}C^-$), etc.

II. Neutral Nucleophiles

Many neutral compounds containing oxygen, nitrogen, and sulphur are also nucleophilic because of the presence of unshared pair(s) of electron(s) on these atoms. Examples of neutral nucleophiles are: R_3N (tertiary amines), NH_3, H_2O, R–OH (alcohols), R–S–R (alkyl sulphides), and R–SH (mercaptans).

As the nucleophiles (neutral and charged both) contain at least one unshared pair of electrons, they can easily donate this pair to electron-deficient atoms or substrates, and hence behave as Lewis bases.

Nucleophilicity

The ability of a nucleophile to donate its unshared pair of electrons is called nucleophilicity. Fully charged ions, such as hydroxide (HO^-) and alkoxide (RO^-) ions, have more nucleophilicity than neutral nucleophiles such as water (H_2O) or alcohols (ROH). As expected, nucleophilicity increases in the presence of electron-donating groups, such as alkyl groups, which intensify negative charge on the atom having unshared pair of electrons.

Thus, methoxide ion, (CH_3O^-) has greater nucleophilicity than hydroxide ion (HO^-) due to +I-effect of CH_3 group in CH_3O^-. Some common nucleophiles are arranged in the increasing order of nucleophilicity:

$H_2O < ROH < Cl^- < Br^- < OH^- < OR^- < I^- < CN^-$

In general, *nucleophilicity* of a reagent is related to its *basic property*. However, this is not always true. For example, methyl mercaptide (CH_3S^-) is a stronger nucleophile than methoxide (CH_3O^-), although the former is much weaker base than the latter.

B. Electrophilic Reagents (Electrophiles)

An electrophilic reagent or electrophile (*electron-loving*) is a reagent which is deficient in electrons (usually a pair). The electrophiles are electron-poor species. They are of two types:

I. Charged Electrophiles

These reagents are deficient in a pair of electrons, and therefore, positively charged. Due to positive charge, they are also called as *cationoid reagents*. Examples of charged electrophiles are: proton (H^+), hydronium ion (H_3O^+), nitronium ion (NO_2^+), bromonium ion (Br^+), and carbocations (R^+).

II. Neutral Electrophiles

These are reagents in which the electron-deficient atom does not carry any charge. Some examples of neutral electrophiles are: boron trifluoride (BF_3), zinc chloride ($ZnCl_2$), aluminium chloride ($AlCl_3$), sulphur trioxide (SO_3), and dichlorocarbene ($:CCl_2$). These reagents behave as Lewis acids.

Electrophiles react with substrates at a negative or partially negative site. The charged electrophiles are more reactive than neutral ones. The presence of electron-withdrawing substituents intesifies the positive charge on the reactive site of the electrophile.

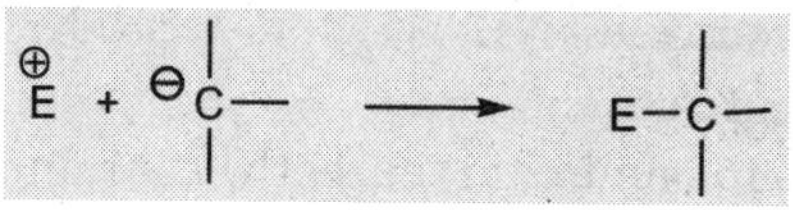

Nucleophilic and Electrophilic Reactions

A reaction is said to be nucleophilic if the reagent is a nucleophile, and electrophilic if the reagent is an electrophile.

Electron density on the carbon atoms of organic compounds may increase or decrease depending upon the electronegativities of the bonded atoms, and also on resonance effects. If the electron density on a given carbon is reduced, a nucleophilic reagent (*nucleophile*) may bring a pair of electrons to the carbon. Such a reaction is called a *nucleophilic reaction*. The carbon atom of methyl chloride, for instance, becomes partially positive due to the inductive effect of chlorine, and a nucleophile such as hydroxide ion attacks this carbon. The whole sequence may be written as:

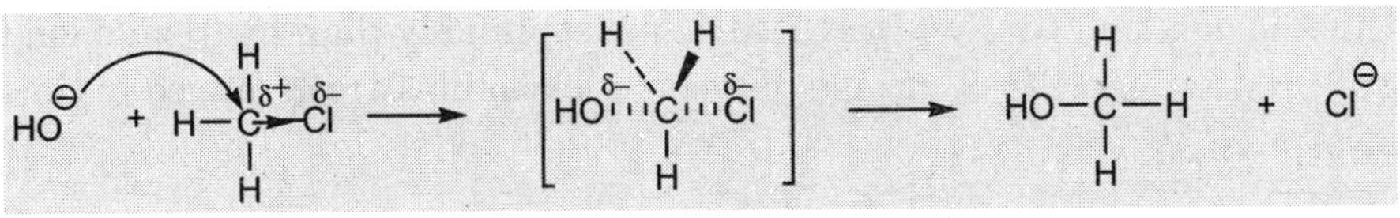

In an *electrophilic reaction*, the reagent takes a pair of electrons from the substrate. A common example is the addition of hydrogen bromide to the double bond of an alkene through the initial electrophilic attack of a proton to form a carbocation, which subsequently combines with a nucleophile, the bromide ion.

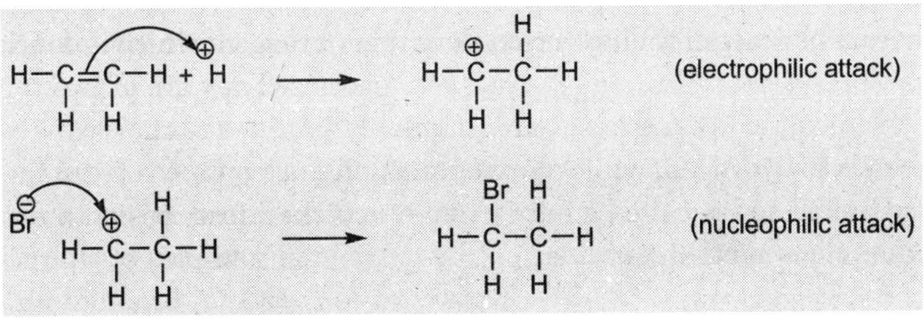

4.3 ORGANIC ACIDS AND BASES

According to the Lowry-Bronsted theory, an acid is a substance which can donate a proton, whereas a base is a substance which can accept a proton. An acid-base reaction, thus, involves transfer of a proton from an acid to a base. For example,

$$\underset{\text{acid}}{HCl} + \underset{\text{base}}{H_2O} \rightleftharpoons \underset{\text{conjugate base}}{Cl^{\ominus}} + \underset{\text{conjugate acid}}{H_3O^{\oplus}}$$

Here, the chloride ion produced in the reaction still possesses the electron pair to which the proton was formerly attached, and is called the conjugate base of HCl, while water acts as a base by accepting a proton and gets converted to the hydronium ion, which is the conjugate acid of the water molecule. Each acid has its conjugate base, and each base has its conjugate acid. Members of a conjugate pair are related to each other by the gain or loss of one proton.

As the Lowry-Bronsted theory was considered to be somewhat restrictive, Lewis defined acids as substances capable of accepting electron pairs, and bases as substances having available unshared electron pairs. It should be noted that a base in Lewis theory is the same as in the Lowry-Bronsted theory; both are substances with an available pair of electrons. However, Lewis definition of acids includes many substances, such as boron trifluoride, zinc chloride, and aluminium chloride, which have only six electrons in the outer shell, whereas they can accommodate eight electrons, and consequently react with bases.

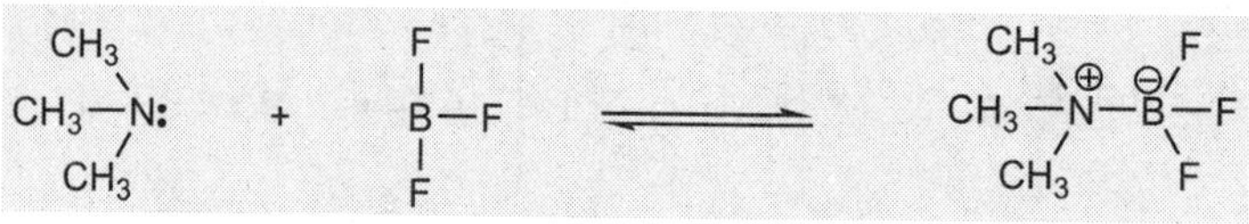

These electron-deficient substances are called Lewis acids to distinguish them from proton donating acids, and are especially useful as catalysts in many chemical reactions.

Nucleophiles and electrophiles may also be looked upon as bases and acids, respectively. By Lewis definition, an acid is a substance capable of accepting an electron pair, and a base is an electron pair donor. Since electrophiles are electron seeking reagents, they may be called Lewis acids, whereas because of their electron-donating nature, nucleophiles may be considered as Lewis bases.

However, the terms acids and bases are commonly confined to their capacity for breaking and making of bonds to hydrogen, while electrophile and nucleophile refer to their reactivity at carbon. Thus, hydroxide ion is called a base when it attacks a proton, and a nucleophile when it attacks carbon atom of methyl chloride.

A. Strength of Acids and Bases

All protonic acids ionize in aqueous solution by transferring protons to the solvent molecules; stronger acids give up most of their ionizable protons, while weaker acids transfer only smaller proportion. The equilibrium constant (K_a) of the process is given by the following expression:

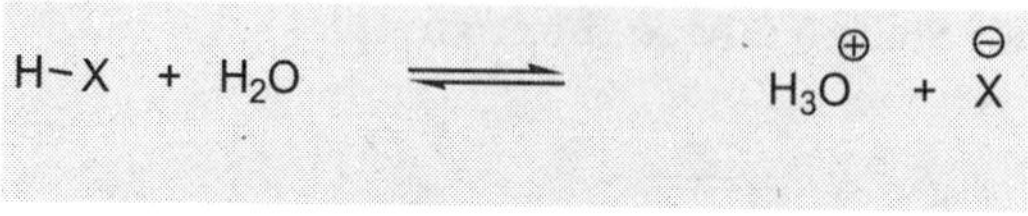

$$K_a = \frac{[H_3\overset{\oplus}{O}]\,[\overset{\ominus}{X}]}{[H-X]}$$

Water is present in excess, and its concentration has been taken as a constant. A large value of K_a signifies high degree of ionization, and therefore, a stronger acid. Since the position of equilibrium depends on the relative stabilities of reactants and products, the acidity constant (K_a) of an acid is related to standard free energy change (ΔG^o) by the following expression:

$$\Delta G^o = -\ 2.303\ RT \log K_a$$

Often, it is more convenient to express the strength of an acid by pK_a which is the negative logarithm of K_a. Thus, for acetic acid, whose K_a is 1.8×10^{-5}, the pK_a can be conveniently calculated.

$$pK_a = -\log 1.8 \times 10^{-5} = -\ 0.3 + 5 = 4.7$$

A stronger acid has larger K_a, but smaller pK_a value.

The strength of a base is expressed by an equilibrium constant, K_b. The relationship between pK_a and pK_b in water is given by the expression, $pK_a + pK_b = 14$. It has been found more convenient to express the base strength by K_a of the conjugate acid, B^+:H. The weaker is base B:, the stronger its conjugate acid B^+:H.

$$\overset{\oplus}{\text{B:H}} + H_2O \rightleftharpoons \text{B:} + H_3\overset{\oplus}{O}$$

$$K_a = \frac{[H_3\overset{\oplus}{O}]\;[\text{B:}]}{[\overset{\oplus}{\text{B:}}\text{H}]}$$

Thus, the strength of acids and bases may be compared by pK_a values, which express the relative acidity of the conjugate acid for any conjugate pair. Larger the value of pK_a, stronger the base; smaller the value of pK_a, stronger the acid. Since water can act both as an acid and a base, the pK_a values are not related to the degree of ionization beyond 0 to 14 range. Values for such acids and bases are related to water by indirect methods, and are often less accurate than those between 0 and 14.

Table 4.1 gives pK_a values of several organic compounds in increasing order of acid strength, and hence necessarily in decreasing order of base strength.

B. Influence of Structure on pK_a Values

All hydrogen-containing compounds are potentially proton donors, and may be considered as acids in the Lowry-Bronsted theory. However, significant acidity is displayed by only those compounds in which hydrogens are linked with the atoms of elements of fifth, sixth, and seventh groups of the periodic table. Compounds containing hydrogen attached to carbon can act as acids when functional groups such as carbonyl, nitrile, or nitro groups are also attached to the same carbon atom.

Generally, factors that stabilize X^- more than XH make a base B: stronger. It should be noted that stabilization of X^- more than XH leads to lower pK_a values, whereas stabilization of B^+:H more than B: leads to higher pK_a values. Thus, factors affecting acid strength are related to those affecting base strength. The major structural effects that operate in any acid-base conjugate pair are I-V:

I. Resonance Effects

The lower pK_a values of acetone and nitromethane (20 and 10.2, respectively) are explained on the basis of resonance-stabilization of their anions:

Table 4.1 pK_a values of some organic compounds.

Conjugate acid	pK_a	Conjugate base
$CH_3–CH_3$	42	$CH_3–CH_2^-$
C_6H_6	37	$C_6H_5^-$
$CH_2=CH_2$	36	$CH_2=CH^-$
$CH≡CH$	25	$CH≡C^-$
CH_3COCH_3	20	$CH_3COCH_2^-$
$C_6H_5C≡CH$	19	$C_6H_5C≡C^-$
CH_3OH	16	CH_3O^-
H_2O*	15.7	HO^-
$CH_2(COOEt)_2$	13	$^-CH(COOEt)_2$
$CH_2(CN)_2$	12	$^-CH(CN)_2$
CH_3COCH_2COOEt	11	CH_3COCH^-COOEt
R^+NH_3	10–11	RNH_2
$R_2N^+H_2$	10–11	R_2NH
R_3N^+H	10–11	R_3N
CH_3NO_2	10.2	$CH_2^-NO_2$
m-$O_2NC_6H_4OH$	9.3	*m*-$O_2NC_6H_4O^-$
N^+H_4	9.3	NH_3*
HCN	9.1	CN^-
$CH_2(COCH_3)_2$	9.0	$^-CH(COCH_3)_2$
C_6H_5SH	8.0	$C_6H_5S^-$
p-$O_2NC_6H_4OH$	7.2	*p*-$O_2NC_6H_4O^-$
Pyridinium ion	5.2	C_5H_5N
RCOOH	4.5	$RCOO^-$
$C_6H_5N^+H_3$	4.6	$C_6H_5NH_2$
$CH_2ClCH_2CH_2COOH$	4.5	$CH_2ClCH_2CH_2COO^-$
$CH_3CHClCH_2COOH$	4.0	$CH_3CHClCH_2COO^-$
2,4-Dinitrophenol	4.0	$(O_2N)_2C_6H_3O^-$
HCOOH	3.7	$HCOO^-$
$CH_2ClCOOH$	2.8	CH_2ClCOO^-
$CH_3CH_2CHClCOOH$	2.8	$CH_3CH_2CHClCOO^-$
$CHCl_2COOH$	1.3	$CHCl_2COO^-$
CCl_3COOH	0.9	CCl_3COO^-
CF_3COOH	0	CF_3COO^-
$C_6H_5CON^+H_3$	–2	$C_6H_5CONH_2$
$CH_3O^+H_2$	–2	CH_3OH
$(CH_3)_2{}^+OH$	–4	$(CH_3)_2O$

* water and ammonia are taken as reference compounds.

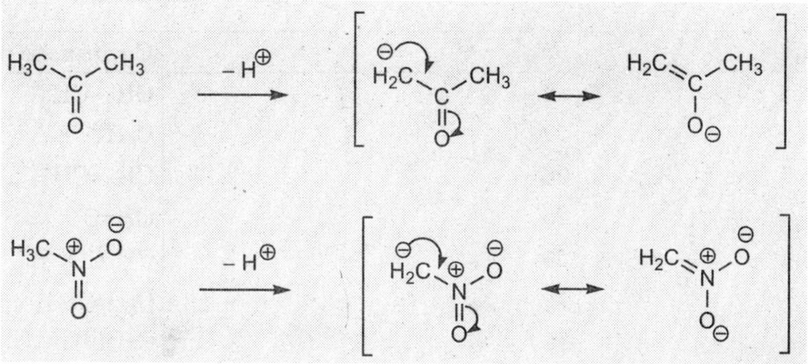

The pK_a values of acetylacetone, ethyl malonate, and ethyl acetoacetate illustrate the resonance stabilization of the carboxylate anion in which the two contributing forms are equivalent. We can also write two resonance structures for the acid, but since the second structure involves separation of charge, its contribution is minor, and hence resonance in the acid is not as stabilizing as that in the anion.

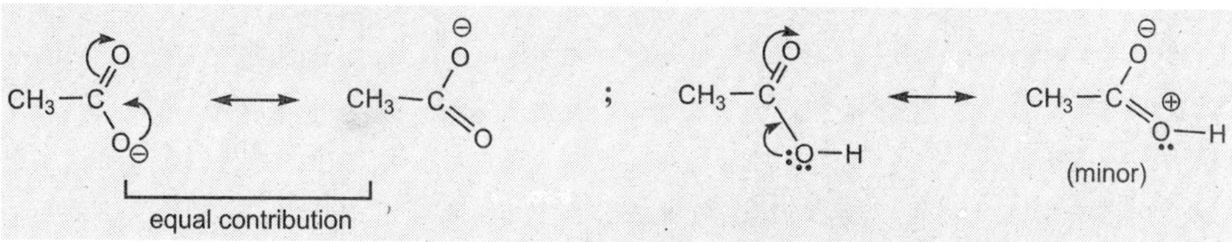

The concept of resonance stabilization of the anion has received strong support from the measurement of bond lengths. There is only one bond-length for both C–O bonds (1.28 Å) in the anion in contrast to two C–O bond lengths (1.24 Å for C=O and 1.37 Å for C–OH) found in the acid. As expected, alcohols (pK_a 16) do not exhibit significant acidity, because negative charge in the alkoxide ion is localized on a single oxygen atom.

Phenols (pK_a 10) are stronger acids than alcohols, because the electron pair of the anion is delocalized by resonance interaction with the π-electron cloud of the benzene ring (**2**). There is resonance stabilization in phenol (**1**) as well, but it involves charge separation, and so is less favoured. Ionization of phenol, therefore, is an energetically gainful process.

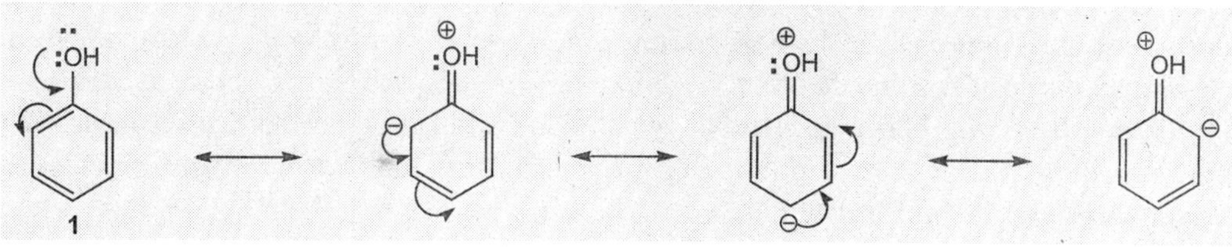
1

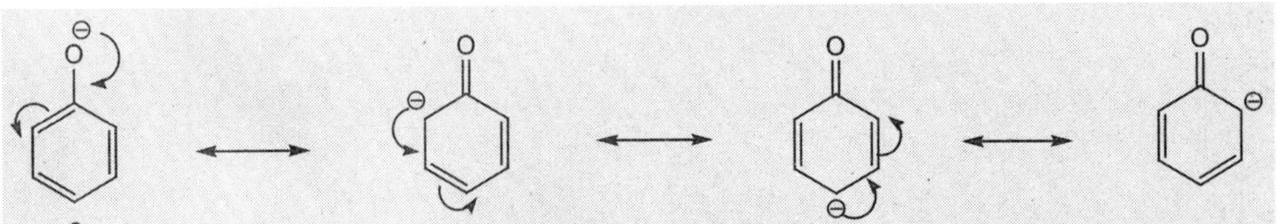

2

The relative acidity of nitrophenols further highlights the importance of resonance in determining the pK_a values. *o*- and *p*-Nitrophenols (pK_a 7.2) are relatively more acidic, as stabilization of the resultant anion is promoted by a resonance form (**3**) in which the negative charge is placed on the nitro group. The anion of *m*-nitrophenol has no such conjugative relationship with the electrons on the nitro group, and hence is a weaker acid (pK_a 9.3).

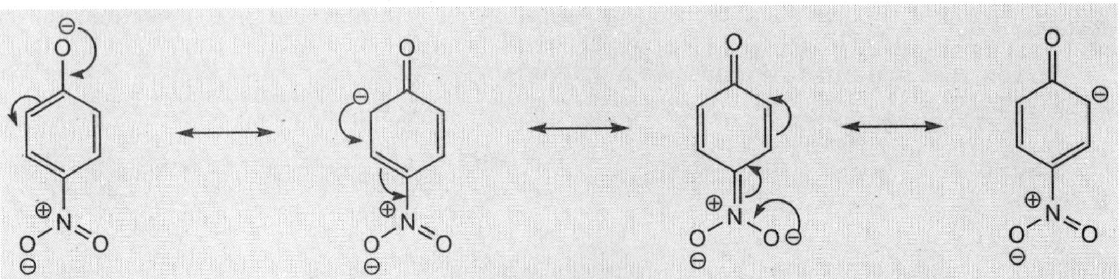

3

Aniline (pK_a 4.6) (**4**) is a far weaker base than aliphatic amines (pK_a 10–11), primarily due to its stabilization, which can be ascribed to the delocalization of unshared electron pair of nitrogen over the aromatic ring. Anilinium cation (**5**) loses this stabilization as the electron pair becomes localized when the nitrogen-proton bond is formed.

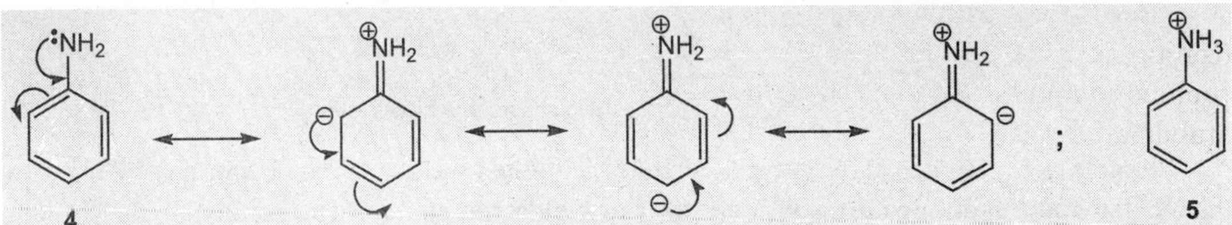

4 5

The base strength of the amine nitrogen is further reduced in nitroanilines. This can again be explained by the fact that the substituted anilines are still more stabilized than substituted anilinium cations. Comparison of base strengths of *p*-and *m*-nitroanilines provides another interesting example of resonance effect. *m*-Nitroaniline (pK_a 2.5) is a somewhat stronger base than *p*-nitroaniline (pK_a 1) because the latter has one additional resonance structure (**6**). The conjugation energy lost in the formation of *p*-nitroanilinium ion is greater than that lost in the formation of the *m*-nitroanilinium ion.

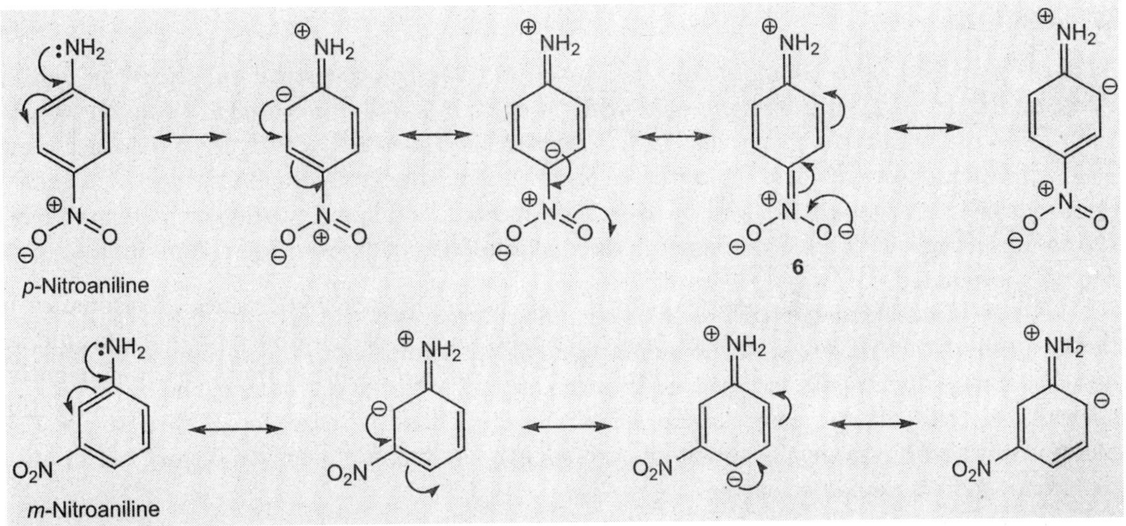

p-Nitroaniline

m-Nitroaniline

In contrast to amines, amides are much less basic because of the resonance stabilization of the molecules.

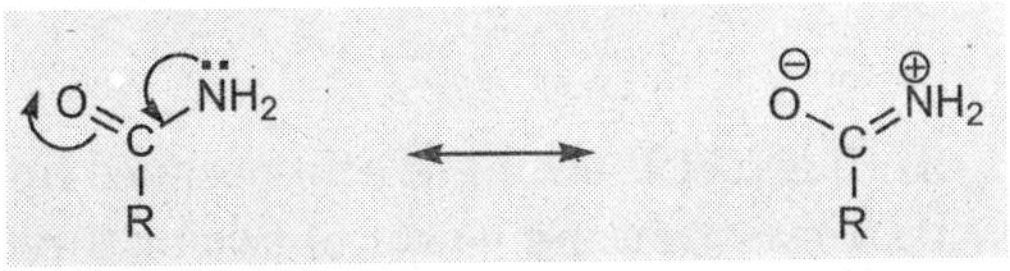

However, amides can be protonated in the presence of strong acids, but the protonation occurs on oxygen rather than on nitrogen. For example, the NMR spectrum of acetamide displays three signals with areas in the ratio 3:1:2, corresponding to the system $[CH_3C(OH)NH_2]^+$rather than $[CH_3CONH_3]^+$. Thus, the basic properties of nitrogen are transferred to oxygen because of resonance in amide grouping.

Base strength may sometimes increase on account of resonance stabilization in cation (conjugate acid). Guanidine (**7**), which is stabilized by resonance, is a very strong base because protonation on the imino nitrogen leads to guanidinium ion (**8**), which has more resonance stabilization than **7** as all its contributing structures are equivalent.

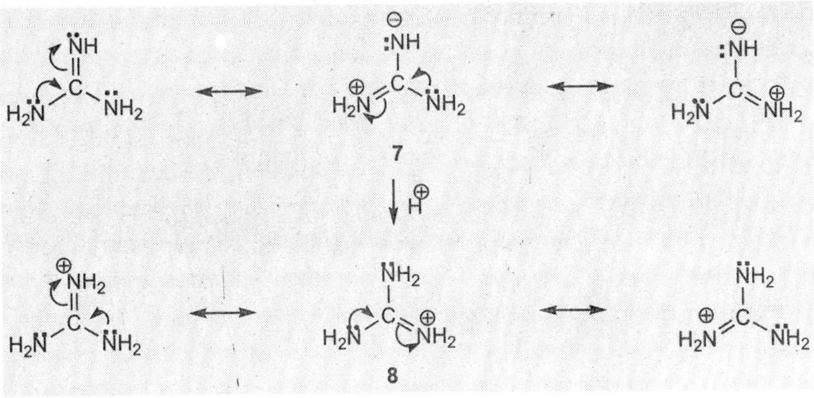

II. Inductive Effects

Generally, acidity of a compound is decreased by electron-donating substituents, while electron-attracting substituents increase acid strength. It is well illustrated in Table 2.1, where substitution of hydrogen of formic acid (pK_a 3.7) by an alkyl group (RCOOH, pK_a 4.5), which is an electron donor, lowers the acid strength. An electron-donating group increases electron density on the carboxylic grouping (**9**), and the hydrogen shows less tendency to depart. Alternatively, the anions of such acids have increased electron availability on oxygen, and recombination with proton is promoted.

The increased acid strength of acids such as chloroacetic acid (**10**, pK_a 2.8) results from the electron-attracting power of the substituent halogen relative to the carbon to which it is attached. Such substituents decrease electron density on the carboxyl grouping, and the hydrogen shows more tendency to depart as the polarity of O–H bond is increased. Alternatively, the anion is more stable as the oxygen atoms are slightly electron-deficient, and thus, better able to accommodate a negative charge.

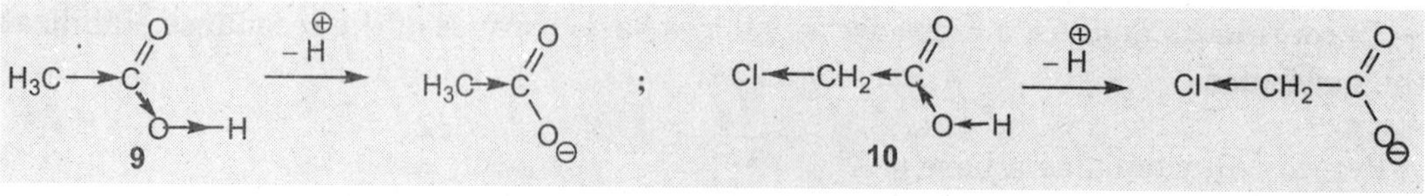

Fluoroacetic acid (pK_a 2.66) is a still stronger acid because fluorine is the most electron-attracting halogen. As the electron-attracting power of two or three halogen atoms is greater than one halogen, acids such as bi- and trichloroacetic acid (pK_a 1.3 and 0.9, respectively) are progressively stronger than chloroacetic acid.

The inductive effect progressively falls off with increasing distance of the electron-attracting substituent from the carboxyl group. This is well exemplified by the significant difference in pK_a values of 2-, 3-, and 4-chlorobutanoic acids (pK_a 2.8, 4.0 and 4.5, respectively).

Since, base strength of the usual organic compounds depends upon the availability of the unshared pair of electrons on the nitrogen atom, the electron-donating inductive effect of alkyl groups causes alkyl and dialkylamines to be somewhat more basic than ammonia. Although tertiary amines are more basic than ammonia, they are less basic than primary and secondary amines. This anomaly has been explained on the basis of solvation, and the consequent stabilization of conjugate acids of amines by water molecules through the formation of hydrogen bonds. The conjugate acid of a tertiary amine has less capacity to enter into hydrogen bonding as it has only one hydrogen as compared to two and three hydrogens in the conjugate acids of secondary and primary amines, respectively. The presence of the third alkyl group in a tertiary amine further reduces the extent of solvation of its conjugate acid.

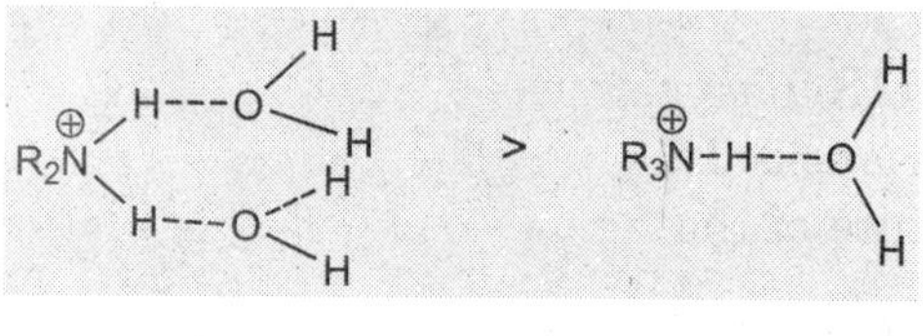

The low basic strength of tertiary amines has also been explained on the basis of the strain involved in contracting the bond angle in tertiary amines (about 112°) to the tetrahedral value (109.5°) in the corresponding conjugate acids.

III. Hybridization

We have already seen that *s* orbitals have higher electronegativity than *p* orbitals, and therefore, electronegativity of a hybrid orbital increases with increasing *s* character. Thus, hydrogens attached to a carbon through *sp* and sp^2 orbitals should be removed more easily as protons than hydrogens bonded to a carbon through sp^3 orbitals. This explains the order of acidity of hydrocarbons: ethane (pK_a 42), ethylene (pK_a 36), and acetylene (pK_a 25).

Similar arguments have been advanced to explain the higher acidity of acrylic acid relative to propionic acid. Since the unsaturated α-carbon atom of acrylic acid is sp^2 hybridized, it is less electron-donating than the sp^3 hybridized α-carbon atom of propionic acid, and thus, the former is a stronger acid. As expected, propiolic acid is the strongest of the three acids.

CH_2=CHCOOH (pKa 4.25)	CH_3CH_2COOH (pKa 4.88)	CH≡CCOOH (pKa 1.85)
Acrylic acid	Propionic acid	Propiolic acid

Pyridine functions as a base (pK_a 5.2), but is much weaker than aliphatic amines (pK_a 10-11).This difference can be accounted for by an inspection of the state of hybridization of nitrogen atom in both cases. The unshared pair of electrons in aliphatic amines is in an sp^3 orbital, while in pyridine this pair occupies an sp^2 orbital. Obviously, the nitrogen atom in the sp^2 state retains a greater hold on the unshared pair of electrons than a nitrogen atom in the sp^3 state, and is, therefore, less readily available for bonding to protons.

IV. Hydrogen Bonding

Intramolecular hydrogen bonding sometimes plays an important role in determining the strengths of acids and bases. Salicylic acid (**11**, pK_a 3), for example, is a much stronger acid than *p*-hydroxybenzoic acid (**12**, pK_a 4.6). This has been attributed to internal hydrogen bonding in the salicylate ion, which stabilizes it substantially with respect to the acid.

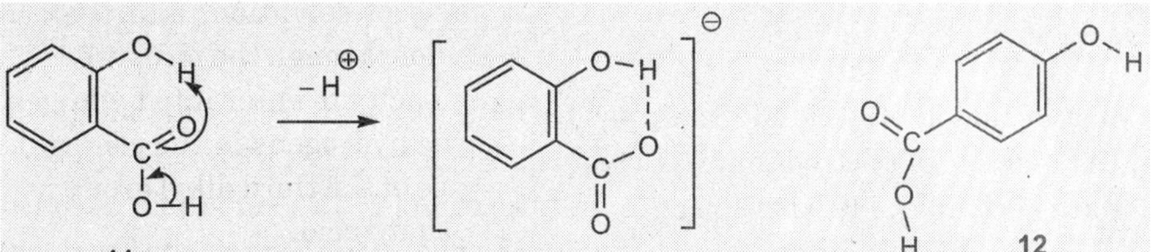

11 12

Similarly, maleic acid (**13**) has a lower pK_a value (1.92) for the first ionization constant than the *trans* isomer, fumaric acid (**14**, pK_a 3.0), because the electrostatic effect of one carbonyl dipole facilitates the departure of the first proton from the other carboxyl group which is in its proximity; this is coupled with the fact that the singly charged *cis* anion is stabilized by intramolecular hydrogen bonding.

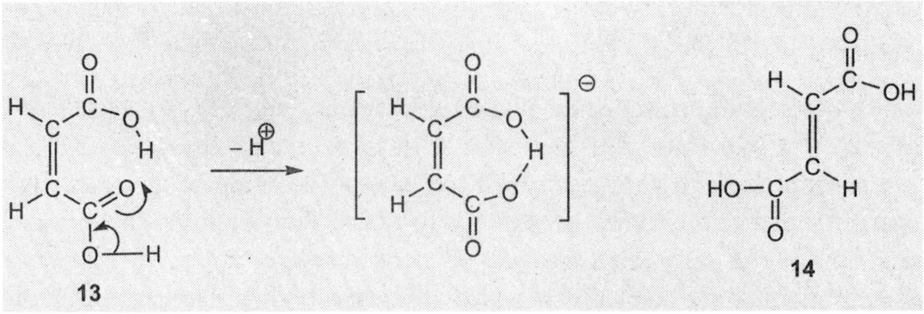

V. Steric Effects

Since a proton occupies a very small volume, steric effects are not of much significance in proton transfer reactions. However, steric effects may indirectly influence acid or base strength by interfering in resonance stabilization. Since all the atoms of a delocalized system must be in a plane, any steric factor that prevents atoms from assuming coplanarity reduces resonance, and therefore, destabilizes the whole system. For example, steric hindrance increases the base strength of *N,N*-dimethyl-*o*-toluidine (**16**, *pKa* 5.9) relative to *N, N*-dimethylaniline (**15**, pK_a 5.1). The *ortho* methyl group in **16** compresses one of the methyl groups attached to the nitrogen in the conformation needed for maximum delocalization of the unshared pair of electrons on nitrogen with the aromatic ring, and thus, coplanarity of the ring and the dimethylamino substituent is prevented. No such steric factor operates in **15** as there is enough space to accommodate methyl groups attached to the nitrogen and the *ortho* hydrogen atoms.

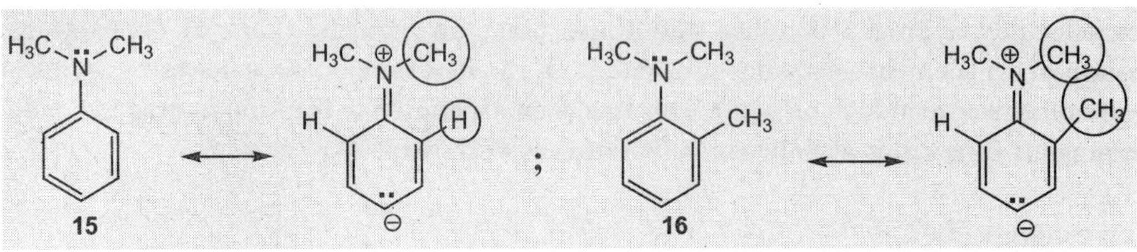

A dramatic example of steric effects is provided by a study of the relative stability of complexes formed by the reaction of triethylamine (**17**) and quinuclidine (**18**) with trimethylborane. Both the amines are of comparable base strength, but the complex **19** is unstable in contrast to the very stable complex **20**. The difference in behavior can be rationalized on the consideration that the substituents in **18** are held back by the ring system, while freely rotating substituents of **17** block the approach to the nitrogen.

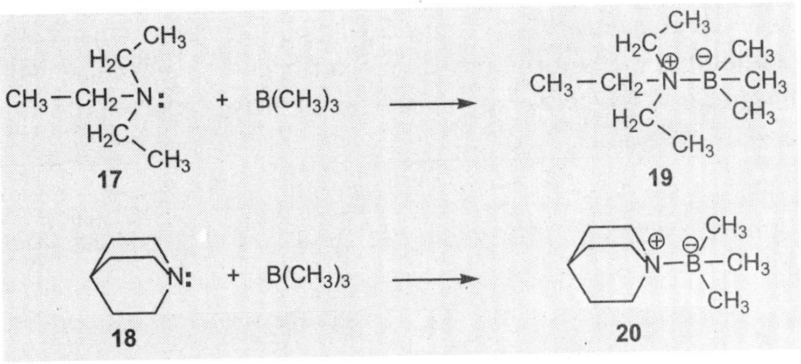

17 18 19 20

C. General Acid and Base Catalysis

As we shall see in subsequent chapters, acids and bases are the most important catalysts in organic chemistry. Hydration of alkenes, dehydration of alcohols, formation and hydrolysis of esters are some common reactions catalyzed by acids, whereas important base-catalyzed reactions are aldol condensations and alkylation of ketones, Mechanism of acid- and base-catalyzed reactions involve the presence of an acid or base at the initial and final stage, as the catalyst is not used up in the reaction. In an acid-catalyzed reaction, for instance, a proton is donated to the reactant, and subsequently a proton must also be removed.

In many cases, such as acid-catalyzed hydrolysis of acetals, the rate of reaction in aqueous medium depends only on the *pH* of a solution.

$$CH_3CH(OC_2H_5)_2 + H_2O \xrightleftharpoons{+H^{\oplus},\ -H^{\oplus}} CH_3CH(OH)OC_2H_5 + C_2H_5OH$$

All such reactions, where the proton-donating acid and proton-abstracting base are assumed to be H^+ (or H_3O^+) and OH^-, respectively, are called *specific acid- and specific base-catalyzed reactions,* and their rate depends only on the *pH* of the solution. In contrast to this, there are a large number of reactions catalyzed not only by H^+ and OH^-, but by all acids and bases which are present in solution. In the solution of acetic acid in water, for instance, CH_3COOH and CH_3COO^- are present besides H^+ and OH^-, and if each of the acids can catalyze a reaction, it is an example of *general acid catalysis*. Similarly, in *general base catalysis* each of the bases catalyzes the reaction.

Two types of catalysis can be distinguished by measuring the rate of reaction in solutions, all buffered to the same *pH*, but containing different concentrations of an undissociated acid. For example, the *pH* of solutions containing different concentrations of undissociated acetic acid can be kept constant by the use of acetic acid-acetate buffers. An increase in the rate of reaction with increase in the concentration of undissociated acid, while keeping the *pH* constant, signifies general acid catalysis. On the other hand, if the reaction rate remains the same in different concentrations of undissociated acids at the same *pH*, it is a case of *specific acid catalysis*.

An example of *general acid catalysis* is provided by the bromination of ketones, which takes place through the enol form. When the reaction is performed in solutions containing different concentrations of acetic acid, the rate of reaction is found to be proportional to the concentrations of hydrogen ion and acetic acid.

$$\text{Rate} = k_1\,[\text{ketone}]\,[H_3O^+] + k_2\,[\text{ketone}]\,[CH_3COOH]$$

Thus, in the above reaction, proton donor may be any acid (H–A), while any base (B:) may remove the proton.

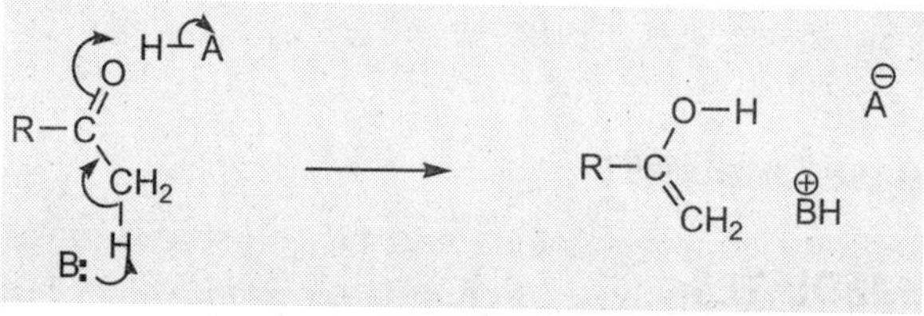

D. HSAB Principle

Hard and soft acid-base concept (HSAB), also known as Pearson concept, is a qualitative scheme in which each Lewis acid and base is categorized on the basis of two parameters, strength and softness. This concept is widely used in chemistry for explaining stability of compounds, reaction mechanism, and pathways.

This principle implies that for a simple acid-base reaction, **A** + **B** $\longrightarrow$ **A**——**B**, acid (**A**) with strength S_A and softness σ_A, and base (**B**) with strength S_B and softness σ_B, the equilibrium constant K is given as:

$$\log K = S_A\,S_B + \sigma_A\,\sigma_B.$$

The gist of this concept is that soft acids react faster and form stronger bonds with soft bases, whereas hard acids react faster and form stronger bonds with hard bases, all other factors being equal.

Hard acids and hard bases have the following characteristics: (i) small atomic/ionic radius (ii) high oxidation state (iii) low polarizability (iv) high electronegativity (v) bases possess highest-occupied molecular orbitals (HOMO) of low energy and acids with lowest-unoccupied molecular orbitals (LUMO) of high energy (see Chapter 15).

Soft acids and soft bases have the following characteristics: (i) large atomic /ionic radius.(ii) low oxidation state. (iii) high polarizability (v) low electronegativity (vi) bases possess HOMO of high energy and acids LUMO of lower energy.

Some examples of hard and soft acids and bases are:

Hard acids = H^+, Li^+, Ti^+, Cr^{3+}, Cr^{6+}

Hard bases = HO^-, F^-, Cl^-, NH_3, CH_3COO^-

Soft acids = Pt^{2+}, Ag^+, Au^+, Hg^{2+}, $Hg_2{}^{2+}$, Cd^{2+}, BH_3

Soft bases = H^-, R_3P, SCN^-, I^-

In nucleophilic addition reaction of α,β-unsaturated ketones, the possibility of formation of either 1,2-addition product or /and 1,4-addition product arises. The carbonyl group has a high partial positive charge on the carbonyl carbon, and will tend to behave like a harder site, *i.e.* react with hard nucleophile such as Grignard's reagent and organolithium compounds. Conversely, the β-carbon of α,β-unsaturated carbonyl compounds does not have high partial positive charge and behaves like a softer site, and it attacks the soft nucleophiles, like amines and thiols.

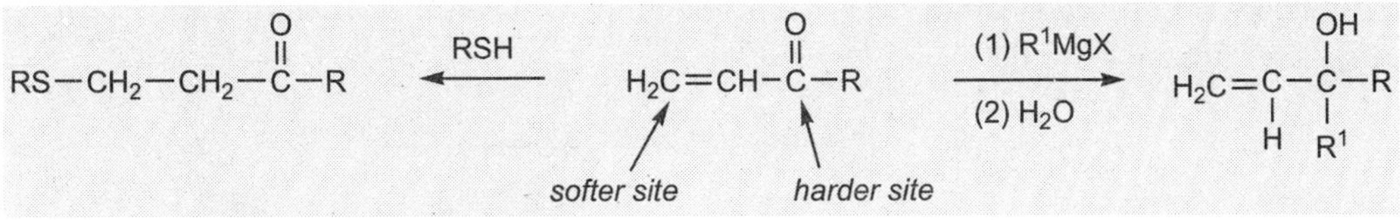

4.4 REACTION INTERMEDIATES

A great majority of organic reactions are multistep processes involving intermediates whose lifetimes are usually very short. These short-lived, highly reactive chemical species are called *reactive intermediates* or *reaction intermediates.* An understanding of the structures and properties of these intermediates, which are normal covalent compounds, ions, or radicals is of paramount importance for understanding organic reaction mechanisms.

The intermediates possess sufficiently low energy to be *formed* under the reaction conditions, but in most of the cases are *not stable* enough to permit isolation. This is especially true of those intermediates in which carbon has fewer than four covalent bonds. The reactive intermediates are: carbocations, carbanions, free radicals, carbenes, nitrenes and arynes.

A. Carbocations

Carbocations are cations, which are normally obtained by the removal of one of the four groups attached to the carbon atom along with its bonding electrons. Thus, they may be defined as organic species, which have a positive charge on a carbon atom bearing only six electrons. They are short-lived and highly reactive species. Until the early 1970s, the ions with positively charged carbon atoms were known as carbonium ions. As this name does not follow the usual pattern of nomenclature, the term *carbocation* was introduced to describe such ions. The term carbocation is consistent with the well-accepted name for related anions, *carbanions*. However, both these terms-carbocation and carbonium ion are used to describe an ion with a carbon atom having no unshared pair of electrons, only three covalent bonds, and a positive charge.

In most cases, carbocations cannot be isolated, but this instability does not minimize their importance as reactive intermediates in a great number of chemical reactions.

I. Classification and Structure

Alkyl carbocations are classified into three categories *i.e.* primary (1°), secondary (2°), and tertiary (3°) depending upon whether the cationic carbon is primary, secondary, or tertiary. Methyl carbocation with one carbon is a special case.

$\overset{\oplus}{C}H_3$	$CH_3\overset{\oplus}{C}H_2$	$(CH_3)_2\overset{\oplus}{C}H$	$(CH_3)_3\overset{\oplus}{C}$
Methyl carbocation	Ethyl carbocation (1°)	*iso*-Propyl carbocation (2°)	*tert*-Butyl carbocation (3°)

The carbon atom of the carbocation is *sp^2 hybridized,* and uses the three hybrid orbitals for single bonding to three substituents; the remaining unhybridized *p*-orbital is empty. The carobcation, thus, has a planar structure having all the three covalent bonds in one plane with bond angle of 120° between them (**Fig. 4.1**). The empty *p*-orbital is at right angle to the plane of the sp^2-hybridized orbitals.

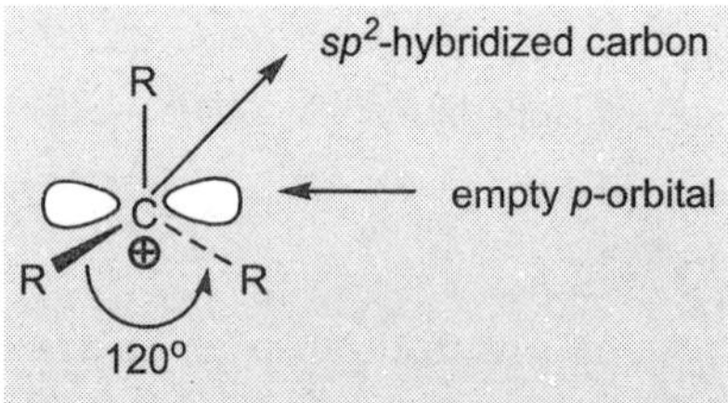

Fig.4.1: Orbital picture of a carbocation

Carbocations have a strong tendency to complete the octet of carbon, and consequently they are highly reactive species and participate in many chemical reactions *i.e.* in unimolecular substitution and elimination reactions, in the decomposition of diazonium salts, and in various rearrangements.

II. Stability

Any factor, which tends to disperse the positive charge, must increase the stability of carbocations, while any factor, which tends to intensify the positive charge, must decrease the stability of the carbocation.

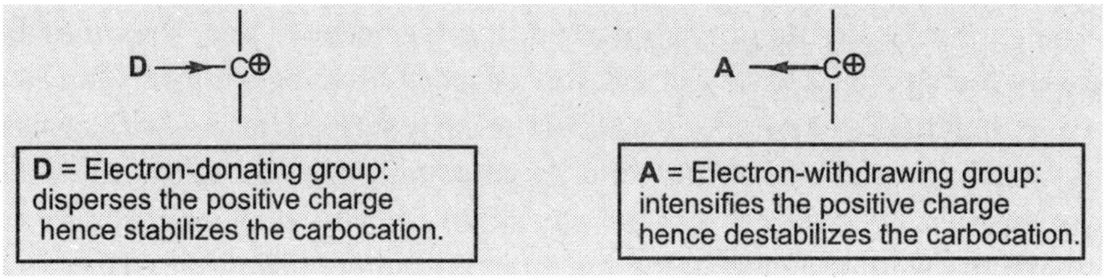

(i) Stability of Saturated Carbocations: The stability of alkyl carbocations decreases in the following order:

$$\text{Tertiary} > \text{Secondary} > \text{Primary}$$

$$(R_3\overset{\oplus}{C}) > (R_2\overset{\oplus}{C}H) > (R\overset{\oplus}{C}H_2)$$

This order can be explained on the basis of the following factors:

(a) Inductive Effect: There is strong evidence, both physical and chemical, that alkyl groups are more electron-donating than hydrogen. Thus, when these groups are attached to the electron-

deficient carbon of the carbocation, they release electrons towards this carbon. The positive charge, thus, gets dispersed over the alkyl groups, and this dispersal of charge increases the stability of the carbocation. Obviously, the greater the number of such alkyl groups attached to the cationic carbon, the more is the dispersal of positive charge, and hence stabler will be the carbocation. Thus, a *tertiary* carbocation having three alkyl groups, should be more stable than a *secondary* having two alkyl groups, which in turn should be more stable than a *primary* having only one alkyl group.

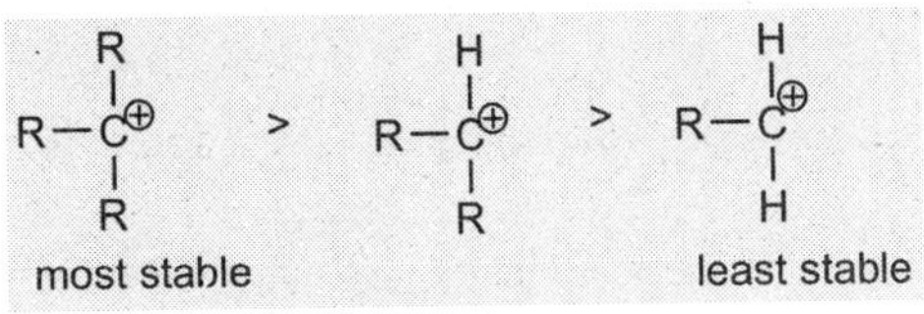

(b) Hyperconjugation Effect: The greater stability of alkyl substituted carbocations can also be explained on the basis of the phenomenon of *hyperconjugation*. According to this, the σ-electrons of alpha C–H bonds are delocalized into vacant *p*-orbital of the positively charged carbon.

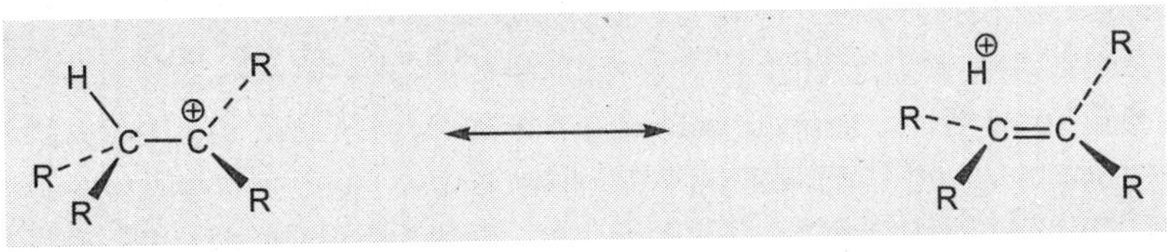

In agreement with this, the *tert*-butyl carbocation having nine such σ C–H bonds is most stable, followed by the *i*sopropyl carbocation having six such σ C–H bonds. The ethyl carbocation, having only three such σ C–H bonds, is the least stable in this series. These cations are regarded as hybrids of various hyperconjugative resonance contributing forms, and the number of these contributing forms depends upon the number of alpha σ C–H bonds. Thus, greater the number of alpha σ C–H bonds, greater is the number of hyperconjugative resonance contributing forms, and hence stabler is the carbocation.

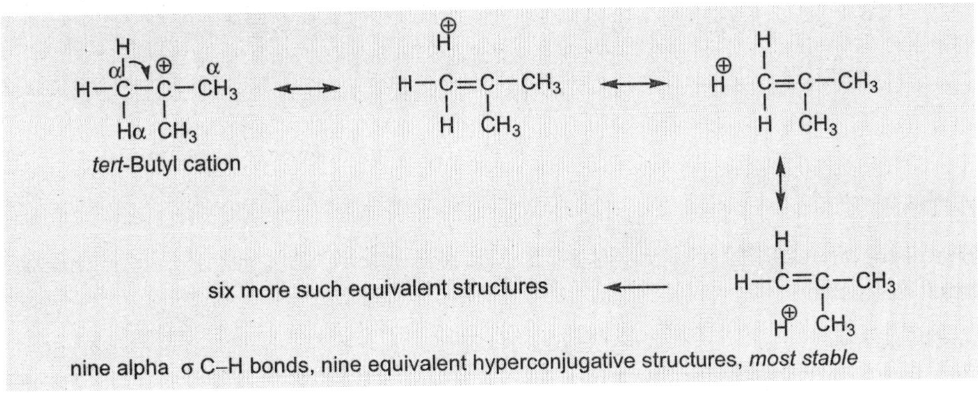

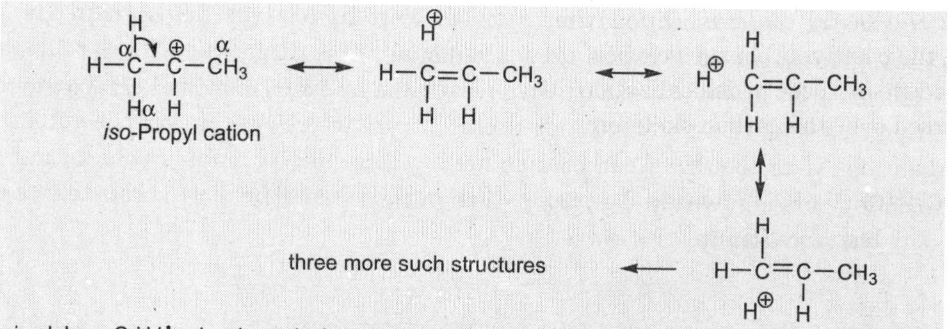

iso-Propyl cation

six alpha σ C-H bonds, six equivalent hyperconjugative structures, *less stable* than *tert*-butylcation

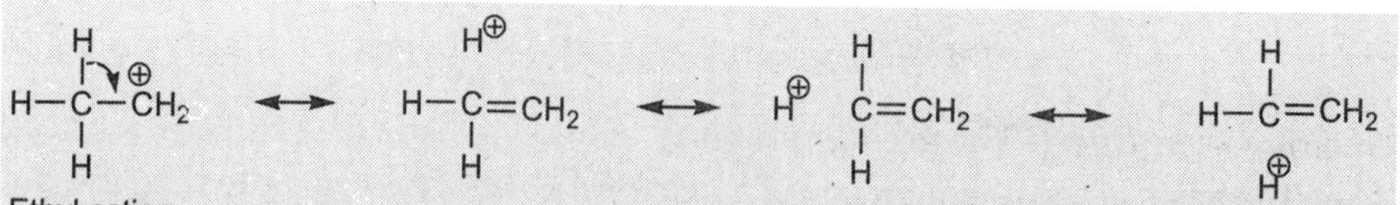
Ethyl cation

three alpha σ C-H bonds, three equivalent hyperconjugative structures, *least stable* of the three

(c) Steric Effects: Steric effects play a key role in the ease of formation as well stability of carbocations. We know that carbocations are planar with bond angles of 120° (*i.e.* having maximum distance between the substituents attached to the positively charged carbon), but the substrates from which they are formed have sp^3 hybridized carbon and have bond angles of 109.5° (*i.e.* lesser distance between substituents, thus causing congestion or *steric crowding*). Now, if the four substitutents attached to an sp^3 hybridized carbon are bulky, the molecule will be highly strained due to congestion. However, when such molecule ionizes to produce a carbocation, the bond angles expand from 109.5° (sp^3) to 120° (sp^2), thereby resulting in relief (removal of congestion) due to increase in space between the bulky substituents. A typical case of steric relief is the ionization of tri-*iso*propylmethyl chloride.

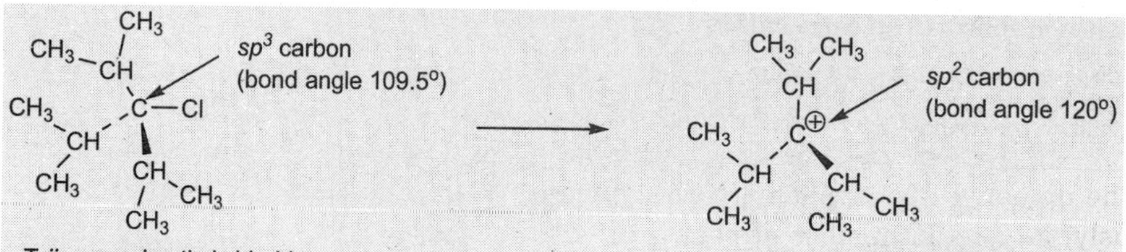

Tri*iso*propylmethyl chloride
steric crowding

Tri*iso*propylmethyl cation
steric relief

The order of stability of simpler 3°, 2°, 1°, and methyl cations is also in tune with the concept of steric relief.

(ii) Stability of Unsaturated and Other Types of Carbocations: The carbocations having a double bond or a non-bonding pair of electrons next to the positively charged carbon are stabilized by *resonance,* which involves the delocalization of π or *n*-electrons into vacant *p*-orbital of this carbon.

(a) Allyl and Benzyl Cations, Diphenylmethyl, and Triphenylmethyl Carbocations: As in these cations, the positively charged carbon atom is in conjugation with a double bond. This arrangement leads to increase in delocalization due to resonance, and because of this, the positive charge is dispersed over the carbon skeleton.

$$H_2C{=}CH{-}\overset{\oplus}{C}H_2 \longleftrightarrow \overset{\oplus}{C}H_2{-}CH{=}CH_2$$

resonance stabilization of allyl cation

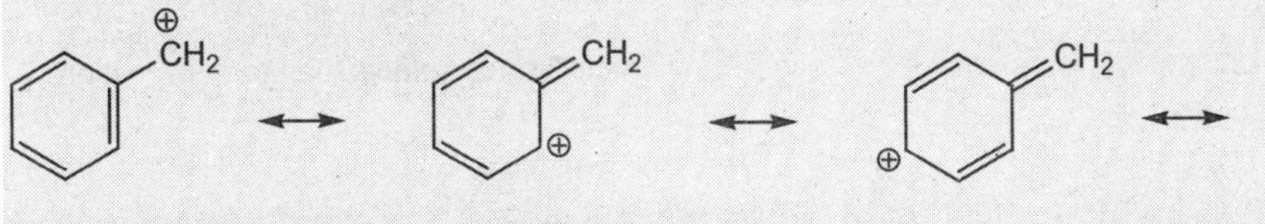

resonance stabilization of benzyl cation

The presence of various substituents at *o*-and *p*-positions of benzene ring in benzyl carbocation greatly influences the stability of substituted benzyl carbocations. *Electron-donating groups* such as alkyl groups, —OH, —OR, —NH_2, —NHR, and —NR_2 *stabilize* these cations. This may be attributed to the dispersal of positive charge by resonance (+R effect) in case of groups having unshared pair of electrons, and hyperconjugative effect of the alkyl groups (CH_3 group having highest effect). In contrast, *electron-withdrawing groups* such as –CN, –NO_2, –CHO, –COR, –COOH, and –COOR *destabilize* these cations. This may be due to the fact that electron-withdrawl by these substituents would intensify positive charge on the cationic carbon. Thus, for example, *p*-methoxybenzyl cation is stabler than benzyl cation, whereas *p*-nitrobenzyl cation is less stable.

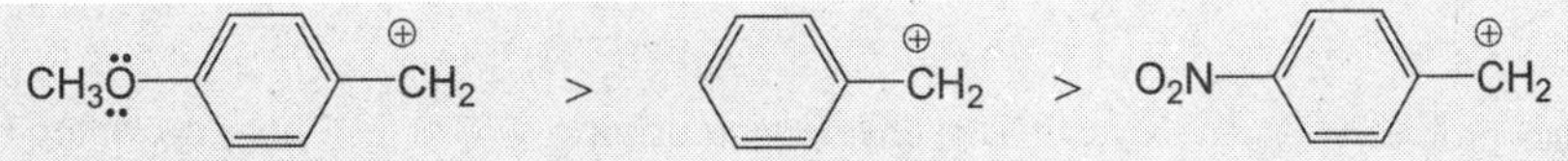

p-Methoxybenzyl cation　　Benzyl cation　　*p*-Nitrobenzyl cation

dispersal of charge by methoxyl group due to resonance, more stable than benzyl cation

intensification of charge due to electron-withdrawal by nitro group, less stable than benzyl cation

The diphenyl and triphenyl carbocations are much more stable than benzyl cation. Obviously, larger the number of phenyl groups attached to the cationic carbon, greater is the dispersal of positive charge. Hence, more stable are such carbocations. The decreasing order of stability of benzyl, diphenylmethyl, and triphenylmethyl carbocations is as follows:

$$(C_6H_5)_3\overset{\oplus}{C} > (C_6H_5)_2\overset{\oplus}{C}H > C_6H_5\overset{\oplus}{C}H_2$$

Greater stability of diphenymethyl carbocation as compared to benzyl carbocation is due to greater number of resonating structures.

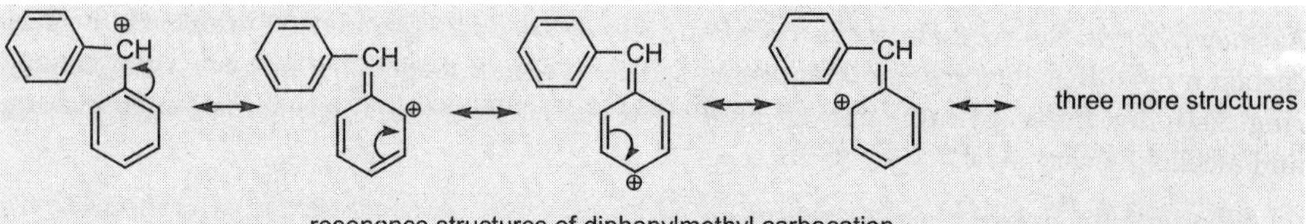

resonance structures of diphenylmethyl carbocation

Similarly, triphenylmethyl carbocation is more stable than diphenylmethyl carbocation due to larger number of resonating structures, which can be written as shown below:

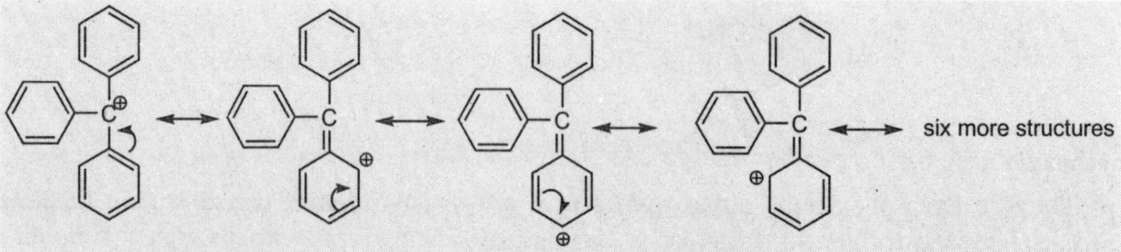

resonance structures of triphenylmethyl carbocation

High stability of triphenylmethyl carbocation is reflected in the fact that salts like perchlorates and fluoroborates, have been isolated and stored for a very long time. Similarly, triphenylcyclopropenyl cation is more stable than diphenylcyclopropenyl cation, which in turn is more stable than phenylcyclopropenyl cation.

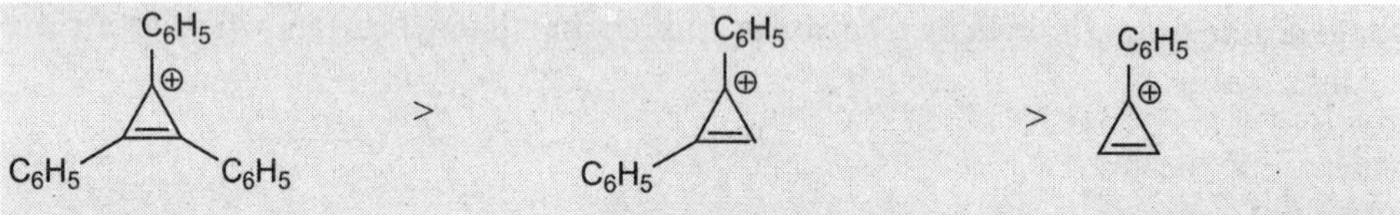

(b) Tropylium Ion or Cycloheptatrienyl Cation: Tropylium ion is very stable. The reason for its unusual stability is that it is an aromatic cation containing six π-electrons.

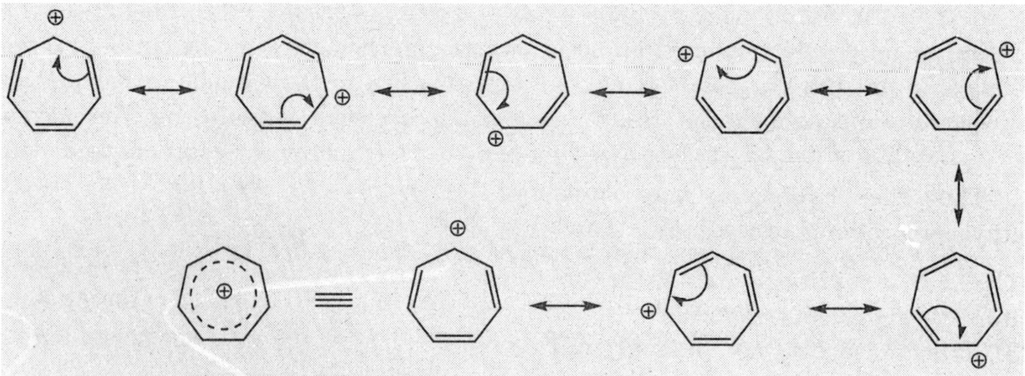

resonance structures of tropylium ion

As shown below, the six π-electrons are completely delocalized over all the seven sp^2 hybridized carbon atoms, just in the same manner as the six π-electrons are delocalized in the benzene ring. Actually, the tropylium ion is so stable that its reaction with nucleophiles, such as water and alcohol, is very slow.

(c) Alkoxyalkyl Carbocations: Carbocations, in which the positively charged carbon atom is attached to a heteroatom (carrying non-bonding pair (s) of electrons) like oxygen, nitrogen and halogen are also stabilized by resonance.

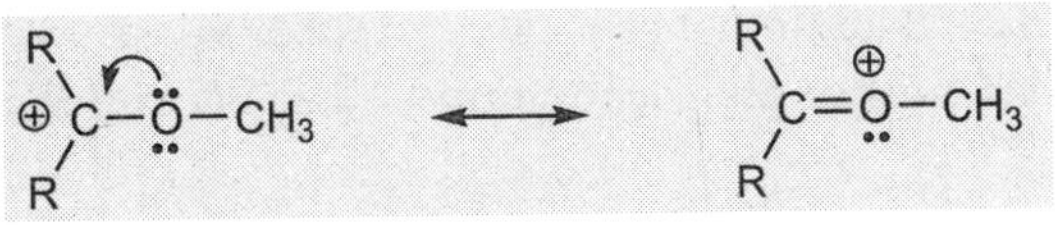

The stability conferred by resonance on such carbocations is reflected in the fact that methoxymethyl cation can be isolated as a stable salt, $[CH_3OC^+H_2SbF^-_6]$

(d) Acyl Cations: Like alkoxyalkyl cations, acyl cations can also be regarded as hybrid of the following contributing forms, and hence are stabilized.

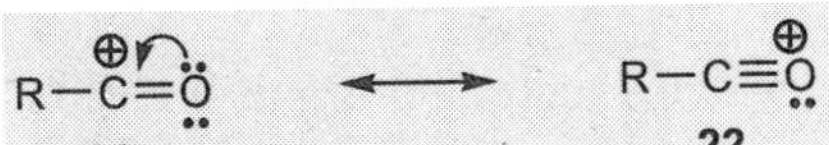

21 **22**

Although having positive charge on the more electronegative oxygen atom, the contribution of canonical structure **22** is less than structure **21** to the resonance hybrid, acetyl cation CH_3CO^+ is almost as stable as *tert*-butyl carbocation.

(e) Vinyl and Phenyl Carbocations: Carbocations having positive charge on the doubly bonded carbon atom, as in the case of vinyl and phenyl cations, are not stabilized by resonance, because the positively charged sp^2-orbital of the carbon atom is perpendicular to the *p*-orbitals of the π-bonds.

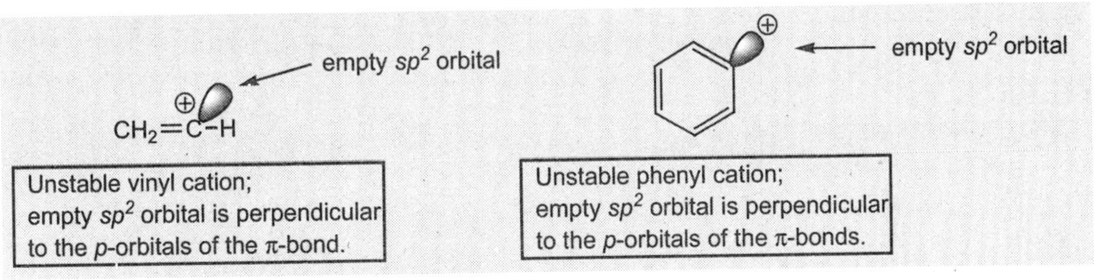

Thus, these cations are very short-lived. However, vinyl cations are more stable than methyl carbocation, but less stable than other primary carbocations (Table 4.2).

(f) Carbocations at Bridgehead: As the carbocations are sp^2 hybridized and planar, any structural feature which prevents rehybridization and planarity will resist its formation. 1-Chlorocamphane (**23**), for instance, does not undergo substitution reactions; S_N1 reaction is prevented because of the difficulty in the formation of a planar carbocation, and S_N2 reaction is prevented because

of the bridgehead location of C–Cl bond.

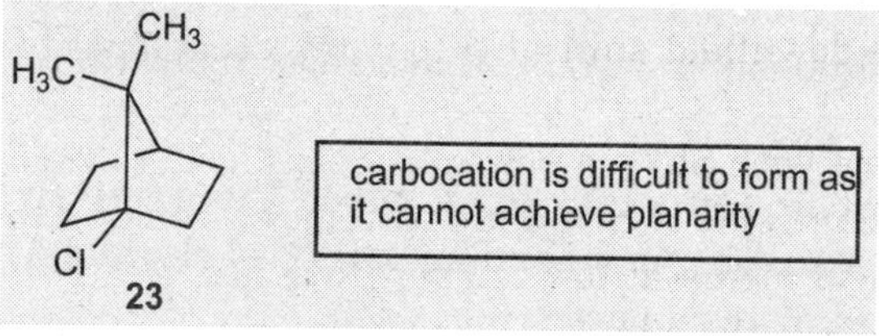

However, bridgehead carbocations are formed in cases where there is more flexibility. The relative rates for solvolysis in ethanol for compounds, **24**, **25** and **26** are illustrative.

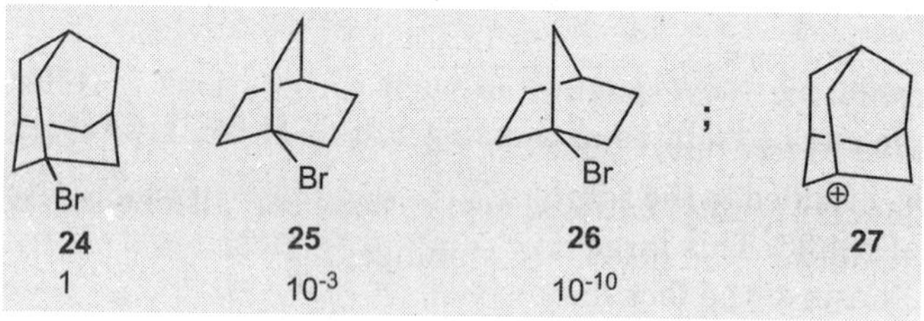

Larger bridgehead carbocations, such as admantyl cation (**27**), have been prepared and characterized by NMR spectroscopy.

Relative Stability of Carbocations

Relative stability of carbocations can be easily assessed by determining bond dissociation energy for heterolytic fission. Lower the value of this energy, greater is the stability of the carbocation. The bond dissociation energies of some alkyl/aryl halides are given in **Table 4.2.**

Table 4.2 Bond dissociation energies for some alkyl / aryl chlorides for heterolytic fission

Alkyl/ Aryl chloride	**Bond Dissociation Energy (kcal/mol)**	**Bond Dissociation Energy (kJ/mol)**
CH_3–Cl	227.75	952
CH_3CH_2–Cl	191.14	799
$CH_3CH_2CH_2$–Cl	177.99	744
$(CH_3)_2CH$–Cl	170.09	711
$(CH_3)_3C$–Cl	157.17	657
CH_2=CH–Cl	207.17	866
CH_2=$CHCH_2$–Cl	173.20	724
C_6H_5–Cl	219.13	916
$C_6H_5CH_2$–Cl	166.26	695

A look at the data listed in Table 4.2 reveals that the order of stability of carbocations is: *tert*-butyl > benzyl > *iso*propyl > allyl > *n*-propyl > ethyl > vinyl > phenyl > methyl.

III. Detection

Simple carbocations can be detected by making use of NMR spectroscopy. Their proton NMR spectra show extensive downfield shift of the hydrogen atoms (deshielding) attached to the cationic carbon.

Alkyl fluorides react with SbF_5, a powerful Lewis acid, to yield the carbocations. SbF_5 in liq. SO_2 is more acidic than 100% H_2SO_4, and is commonly referred to as *super acid*. Generating carbocations in super acid media at low temperature has greatly helped in the investigation of rearrangement of these cations by NMR spectroscopy. Another combination that has been used for the NMR studies of carbocations is FSO_3H-SbF_5-SO_2, which is referred to as *magic acid* because of its powerful protonating ability.

$$R{-}F + SbF_5 \xrightarrow{FSO_3H\text{-}SbF_5\text{-liq. }SO_2} \overset{\oplus}{R}\ \overset{\ominus}{SbF_6}$$

The nine protons in *tert*-butyl fluoride (**28**) in sulphur dioxide exhibit a singlet at δ 1.3. However, when SbF_5 is added to the solution, the protons show a large downfield shift appearing as a sharp singlet at δ 4.35. This large downfield shift of the signal due to methyl protons can be explained on the basis of the fact that positively charged carbon exerts large inductive and hyperconjugative effect on these protons.

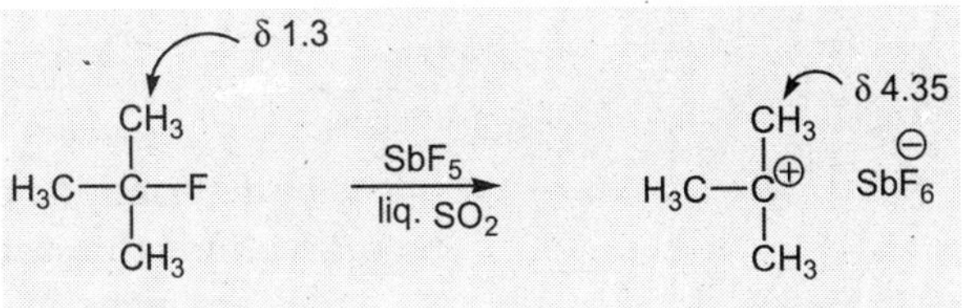

28

Similarly, the NMR spectrum of *iso*propyl fluoride (**29**) shows methyl protons at δ 1.23, while in SbF_5-SO_2 medium they show up at δ 5.06. The secondary proton in *iso*propyl fluoride gives a signal at δ 4.64, whereas in the carbocation, it is shifted to δ 13.50.

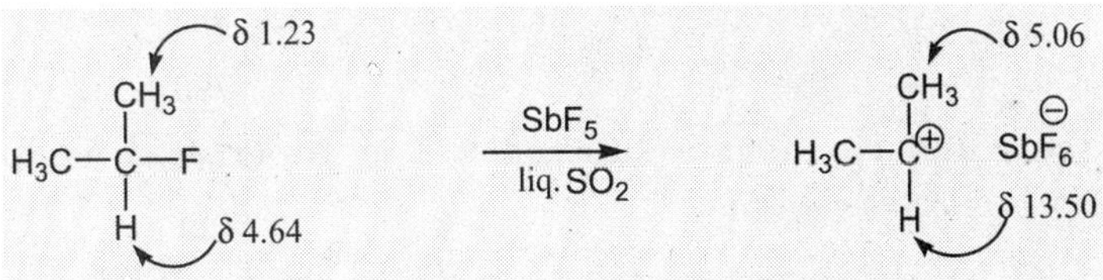

29

The NMR spectroscopy has also provided evidence for the rearrangement of carbocations to yield more stable ions. Thus, *n*-butyl (**30**), *iso*-butyl (**31**), and *sec*-butyl fluorides (**32**) all display the same NMR spectrum in antimony pentafluoride and liquid SO_2, as **28**. Since the NMR spectrum for *tert*-butyl carbocation is unequivocal (one signal at δ 4.35), the other three primary and secondary carbocations (generated from isomeric alkyl fluorides) must have rearranged to a more stable tertiary carbocation.

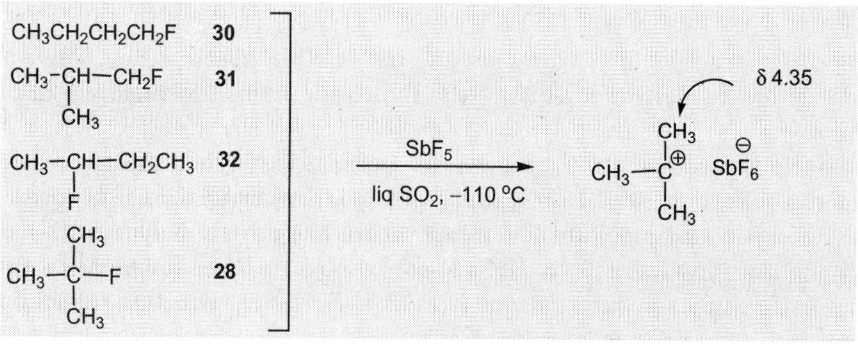

Similarly, the NMR spectra of *n*-propyl and *iso*propyl fluorides in SbF_5/liq SO_2 are similar; presumably, the less stable *n*-propyl carbocation rearranges to the more stable *iso* propyl carbocation.

Many cyclic systems generate the same carbocation, which was characterized using NMR spectroscopy.

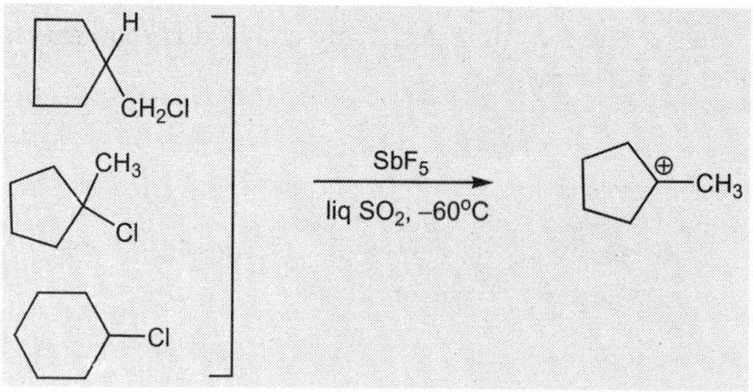

Cyclohexanol also gives the same carbocation, apparently by ring contraction.

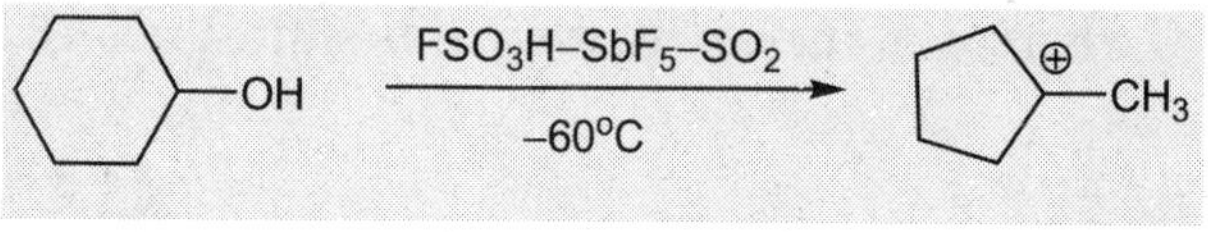

Bicyclic systems undergoing rearrangements in super acid media have also been characterized by NMR spectroscopy.

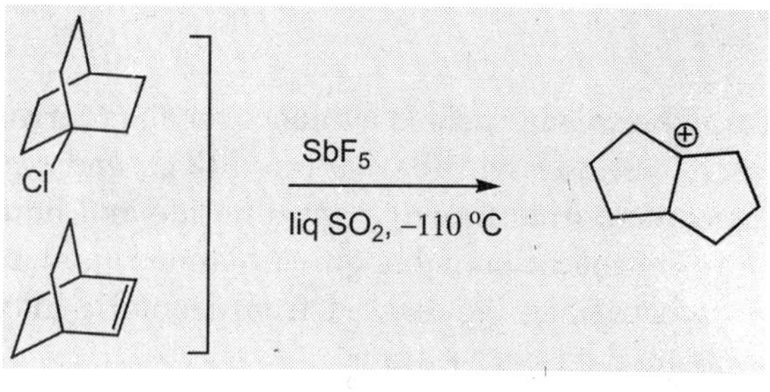

IV. Formation

Carbocations can be generated in a variety of ways.

(*i*) Solvolysis of C—X bond (X = halogens and OBs) *i.e.* ionization of these compounds in polar solvents.

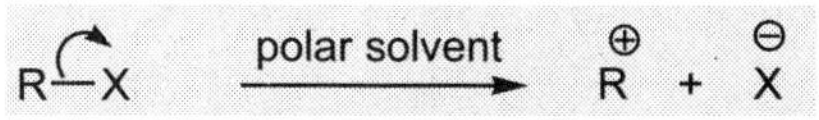

(*ii*) Deamination of aliphatic amines by nitrous acid.

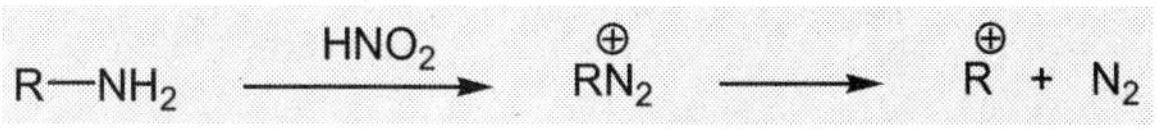

(*iii*) Reaction of halides with super acids (like SbF_5 and FSO_3H)

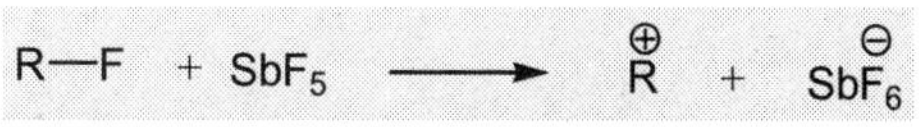

(*iv*) Protonation of alcohols followed by dehydration.

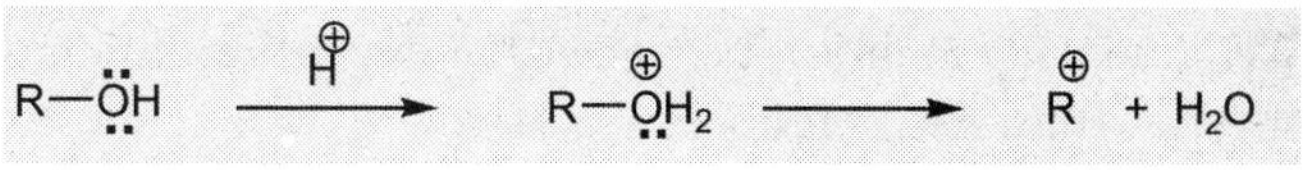

(*v*) Protonation of alkenes.

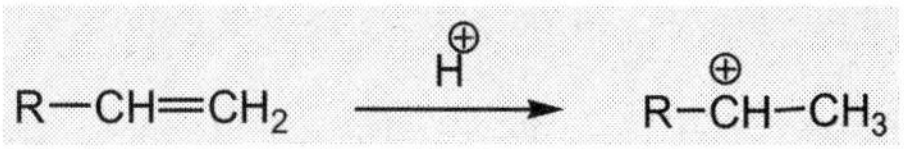

(*vi*) Protonation of carbonyl compounds

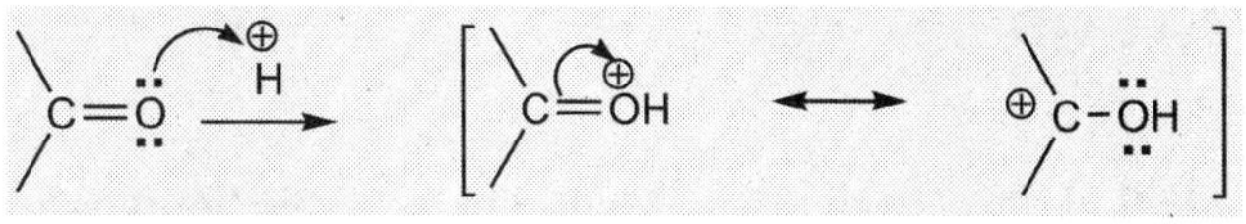

(*vii*) Protonation of nitriles

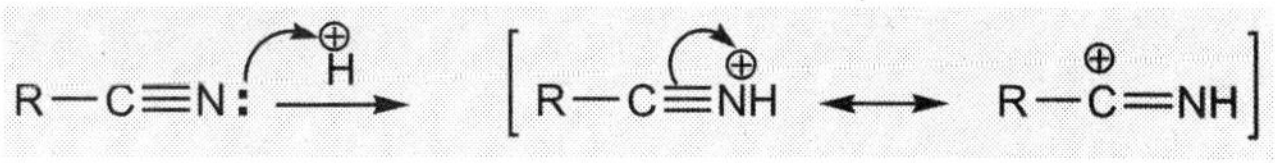

V. Reactions

Broadly, the carbocations undergo four basic types of reactions:

(*a*) Combination with a nucleophile
(*b*) Elimination of a neighbouring proton
(*c*) Addition to a multiple bond
(*d*) Rearrangements

***(a) Combination with a Nucleophile*:** A carbocation may combine with a species having unshared pair of electrons.

$$\overset{\ominus}{Nu} + \overset{\oplus}{R} \longrightarrow Nu{-}R$$

For example, S_N1 reaction of *tert*-butyl chloride to produce *tert*-butyl alcohol. The reaction takes place in the following two steps:

(*i*) Ionization of *tert*-butyl chloride to give *tert*-butyl carbocation.

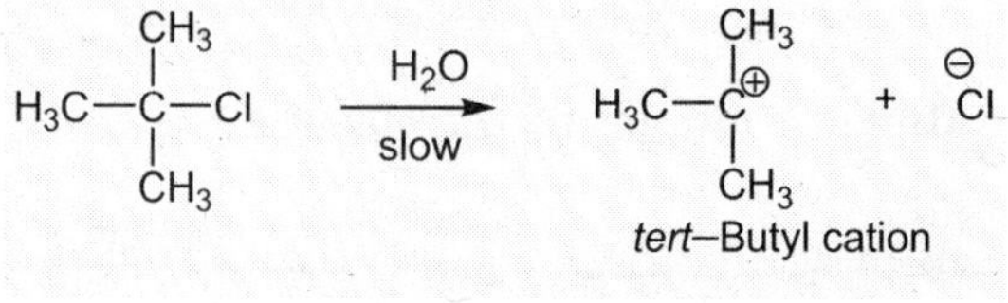

(*ii*) Combination with H_2O (a weak nucleophile to produce *tert*-butyl alcohol).

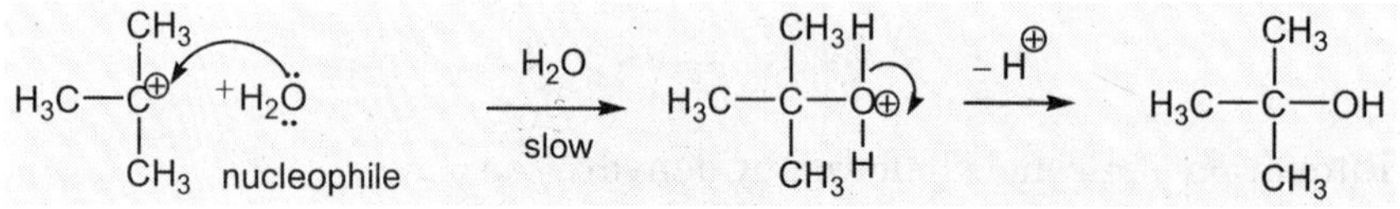

***(b) Elimination of a Neighbouring Proton*:** A carbocation may eliminate a proton (or some other positively charged species) from the neighbouring atom.

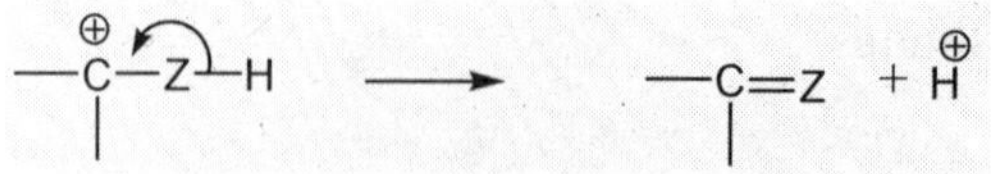

For example, elimination of a proton from the *tert*-butyl cation gives isobutylene.

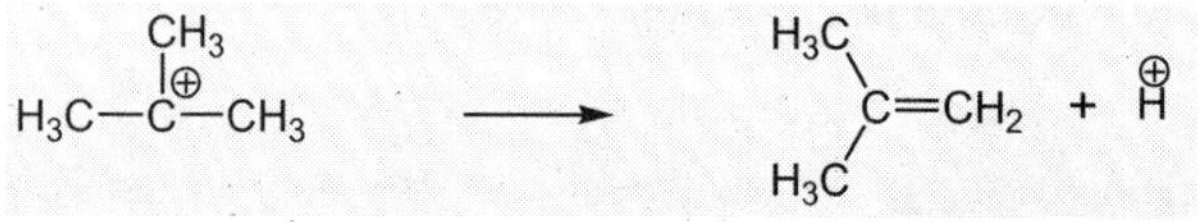

***(c) Addition to a Multiple Bond*:** A carbocation can add to a double bond generating positive charge at a new position (a new carbocation).

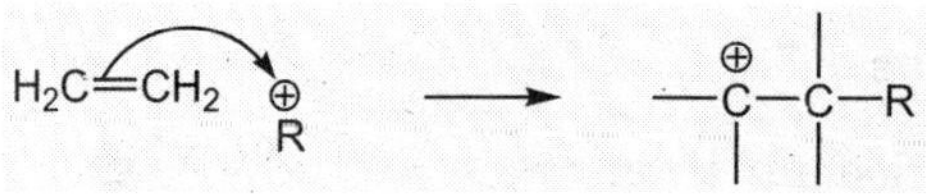

***(d) Rearrangements*:** A carbocation may undergo 1,2-shift (migration to neighbouring carbon) of hydrogen/alkyl groups/aryl groups to generate a more stable carbocation. Some examples are as follows.

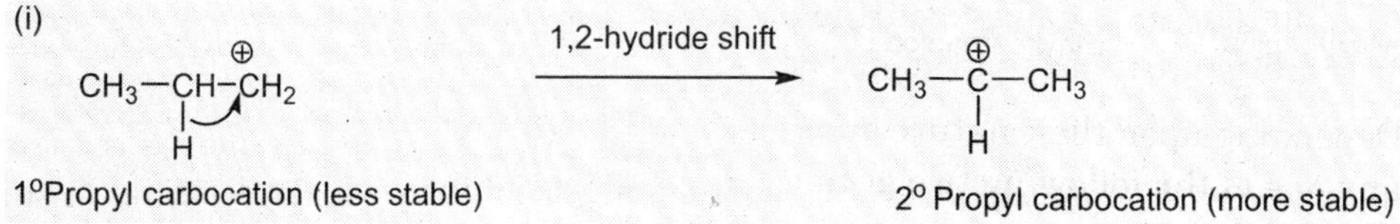

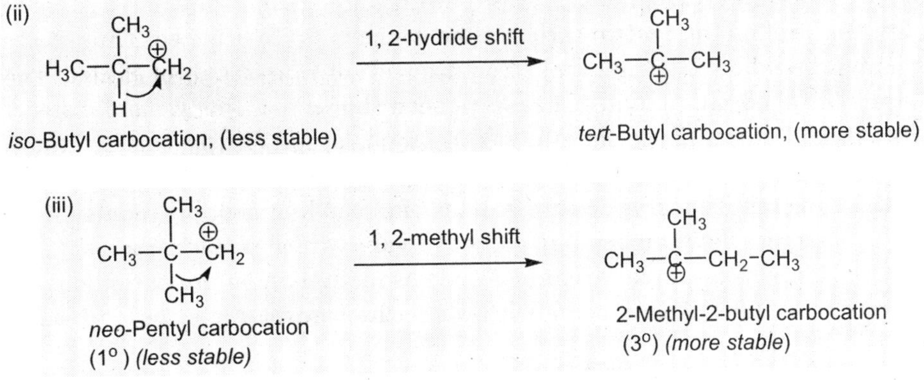

The new carbocation produced can subsequently add a nucleophile or eliminate a proton.

B. Carbanions

Carbanions are anions of carbon which are obtained by the removal of one of the groups attached to the carbon leaving behind the bonding electrons on it. Thus, an organic species carrying negative charge on one of the carbon atoms is called a carbanion.

Nitromethane, for example, reacts with an alkali to yield an anion, which is stabilized by resonance.

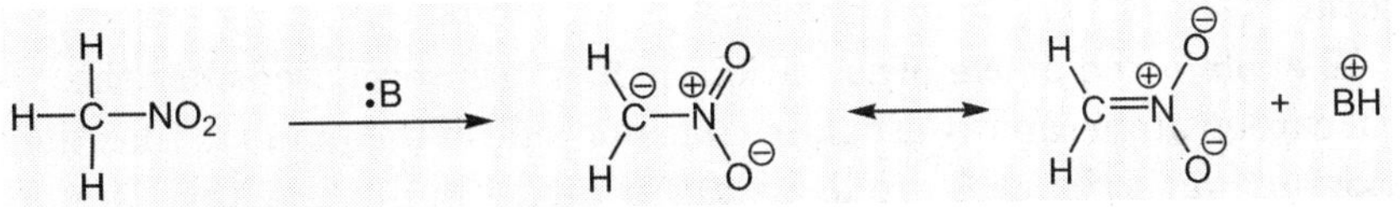

In contrast to carbocations in which the positively charged carbon atom has six electrons, in these anions the negatively charged carbon atom has eight electrons (6 bonding and 2 non-bonding). These ions are also short-lived, highly reactive species. Their reactivity is believed to be due to the presence of formal negative charge on carbon atom.

I. Classification and Structure

Alkyl carbanions are classified into three categories, *i.e.* primary (1°), secondary (2°), and tertiary (3°) depending upon whether the anionic carbon is primary, secondary or tertiary. Methyl carbanion with one carbon atom is a special case, *e.g.*

Depending upon their nature and source from which they are formed, the carbanions can have any one of the following two geometries: ***(a)*** pyramidal or ***(b)*** planar

(a) Pyramidal: Simple alkyl carbanions possess one unshared pair of electrons and three pairs of bonding electrons around the central carbon atom which is sp^3 hybridized. Thus, it is expected that a carbanion would have tetrahedral structure with unshared pair of electrons occupying one apex of the tetrahedron. Actually, they have been shown to have pyramidal structure.

The pyramidal structure of carbanion can be explained on the basis of the fact that it possesses one pair of non-bonding electrons and three pairs of bonding electrons. Due to nonbonding-bonding repulsions, which are greater than bonding-bonding repulsions, there is a reduction of bond angles between bonding pairs to a value which is lesser than the tetrahedral value of 109.5°. Hence, the simple carbanions such as methyl anion are pyramidal (with bond angles between 97° – 100°) like ammonia with which they are isoelectronic (**Fig. 4.2**).

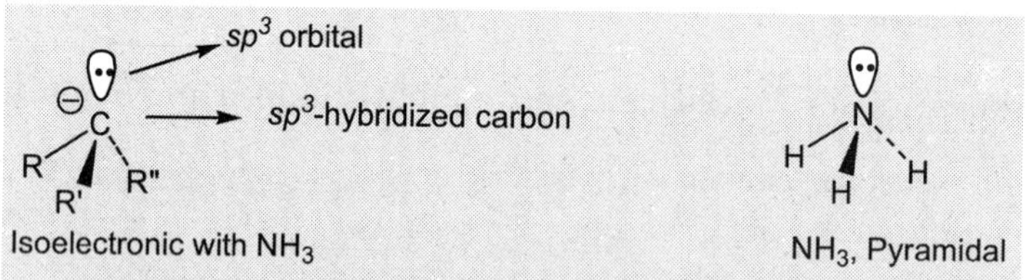

Fig.4.2: Pyramidal structure of a carbanion and ammonia

If this is the picture of a carbanion, then a carbanion in which three groups are different (R, R', and R" as in **Fig. 4.2**) should be chiral, and any reaction in which such a carbanion is an intermediate should proceed with retention of configuration. However, it has been shown that the reactions involving such anions result in loss of optical activity, *i.e.* they produce a racemic mixture.

In order to explain the above observation, it has been proposed that the unshared pair and the central carbon atom rapidly oscillate from one side of the plane to the other, *i.e.* the two enantiomeric pyramidal structures are in a state of rapid equilibrium (*pyramidal inversion* or *umbrella effect*) due to which there is a loss of optical activity.

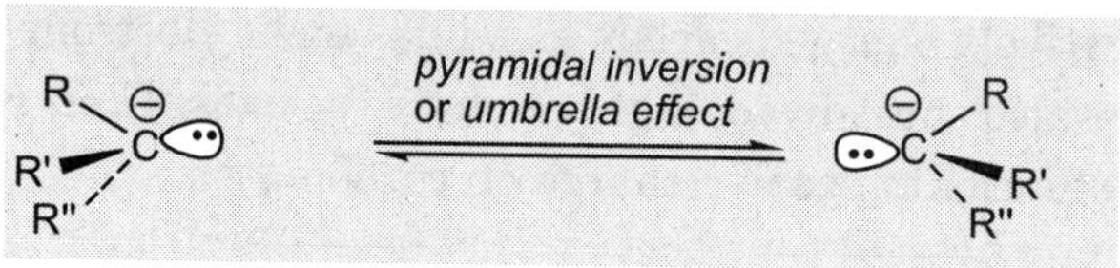

(b) Planar: Resonance-stabilized carbanions, in which the orbital containing unshared pair of electrons overlaps with the π-orbital of the neighbouring multiple bond, are essentially planar. This is due to the fact that planarity is essential for resonance to occur. Thus, in this type of carbanions the anionic carbon is sp^2 hybridized.

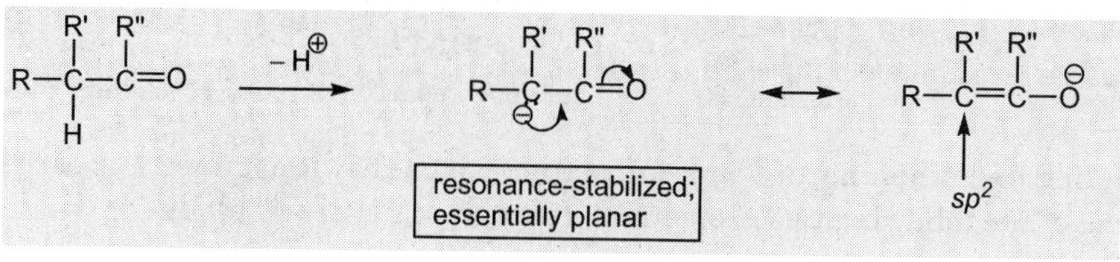

Similarly, allyl anion is also resonance-stabilized.

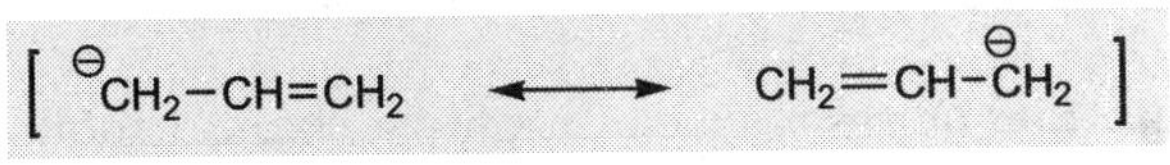

The following molecular orbital depiction of the above type of overlapping makes this concept more clear. All carbon atoms are sp^2 hybridized. Hence, the anion is planar.

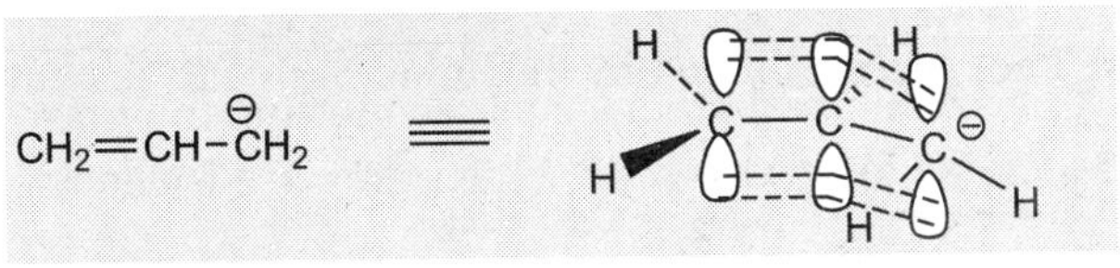

It has been shown that if this type of anion is generated on a chiral carbon atom, there is loss of optical activity. The reason for this is that the chirality of anionic carbon is destroyed by the formation of a carbon–carbon double bond. This can be illustrated by base catalyzed racemization of (+)- phenyl *sec*-butyl ketone (generation of carbanion **34**).

That the coplanarity is an essential requirement for the formation of carbanions stabilized by a neighbouring group is well illustrated by the negligible acidity of bicyclo [2.2.2] octanedione-2,6 (**33**), although it has a bridgehead hydrogen flanked by two carbonyl groups. It is to be expected, since the bridgehead prevents the coplanarity of the atoms that would be involved in the charge delocalization of the potential carbanion.

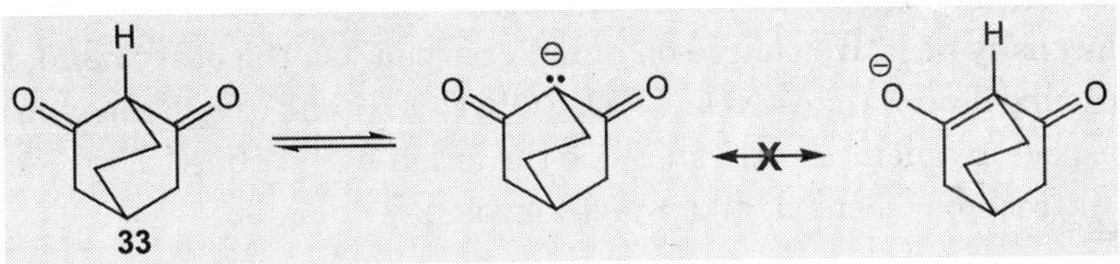

Thus, it is quite reasonable to believe that the delocalization of electrons in a conjugated carbanion requires sp^2 hybridization, and the unshared pair of electrons on the α-carbon then overlaps with the *p*-orbital of the carbonyl carbon, and as a result four groups or atoms attached to it lie in the same plane making that portion of the molecule flat.

In conclusion, it may be said that the carbon atom of an unconjugated carbanion is sp^3 hybridized, and has a pyramidal structure. On the other hand, in a resonance-stabilized carbanion, the carbon is sp^2 hybridized, and the anion is essentially planar. In both cases, there is a loss of optical activity in the reactions involving optically active reactants.

II. Stability

As we have already seen, carbanions are stabilized by electron-withdrawing substituents. An illustrative example is the ready conversion of triphenylmethane to triphenylmethyl carbanion.

$$(C_6H_5)_3CH \xrightarrow{Na} (C_6H_5)_3\overset{\ominus}{C}\overset{\oplus}{Na}$$

Obviously, delccalization or resonance of negative charge stabilizes the carbanion to a great extent. If a multiple bond (like C=C, C=O, NO_2, and C≡N) is located alpha to the anionic carbon, it stabilizes the anion by resonance involving dispersal of charge. Some examples of resonance-stabilized carbanions are:

Allyl and Benzyl Anions: They are stabilized by dispersal of negative charge due to resonance.

$$CH_2{=}CH{-}\overset{\ominus}{C}H_2 \longleftrightarrow \overset{\ominus}{C}H_2{-}CH{=}CH_2$$

resonance stabilization of allyl anion

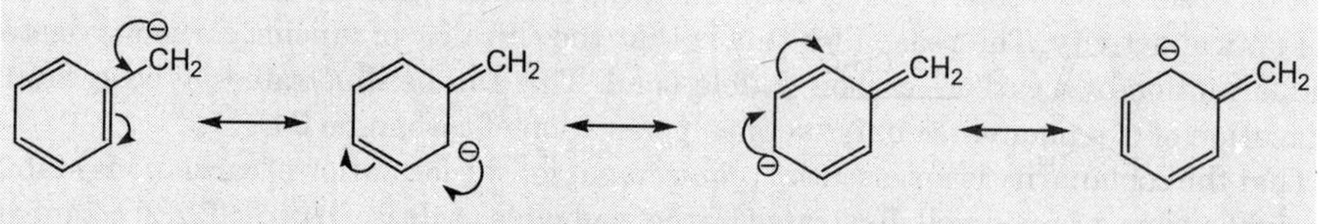

resonance stabilization of benzyl anion

The presence of electron-donating groups (such as alkyl, OH, NH_2 having +R effect) at *o*-and *p*-positions of the ring with respect to anionic centre results in destabilization of the anion, as they intensify negative charge on anionic carbon. On the other hand, the presence of electron-withdrawing groups (like $-NO_2$, CHO, COR, COOH and COOR) on *o*-and *p*-positions of the ring with respect to anionic centre stabilize the anion by –R effect. Thus, decreasing order of stability of *p*-nitrobenzyl, benzyl, and *p*-methoxybenzyl anion is:

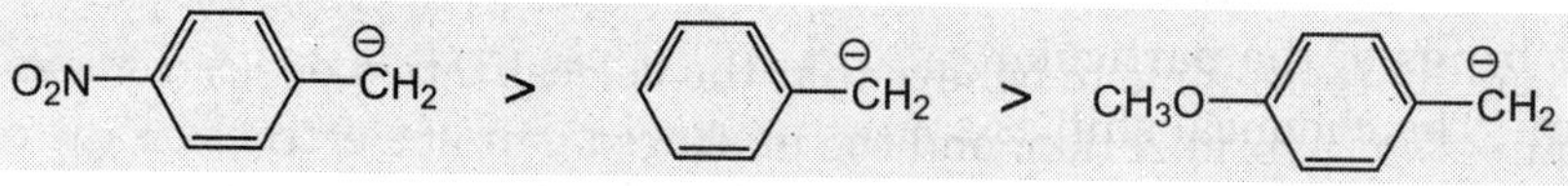

The resonance stabilization of benzyl anion suggests that greater is number of benzene rings attached to the anionic carbon, stabler is the anion. Thus, diphenylmethyl anion is more stable than benzyl anion and triphenylmethyl anion is more stable than diphenylmethyl anion. This is due to increasing number of resonating structures.

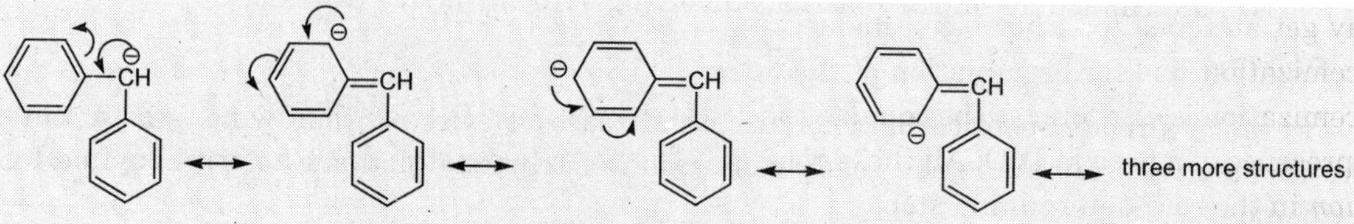

resonance stabilization of diphenylmethyl anion

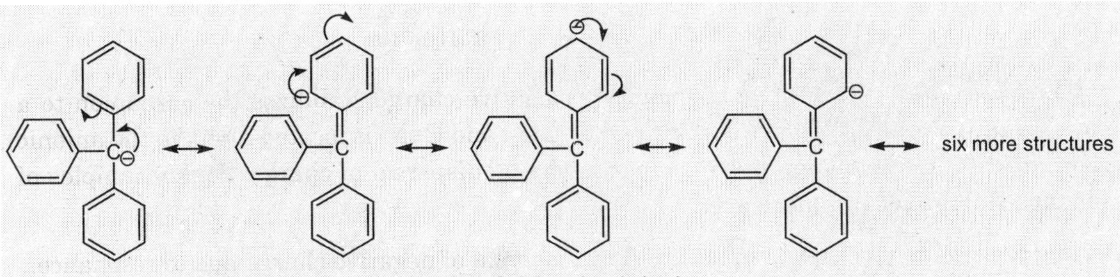

resonance stabilization of triphenylmethyl anion

An interesting example is cyclopentadiene, which readily loses a proton to form the cyclopentadienyl anion.

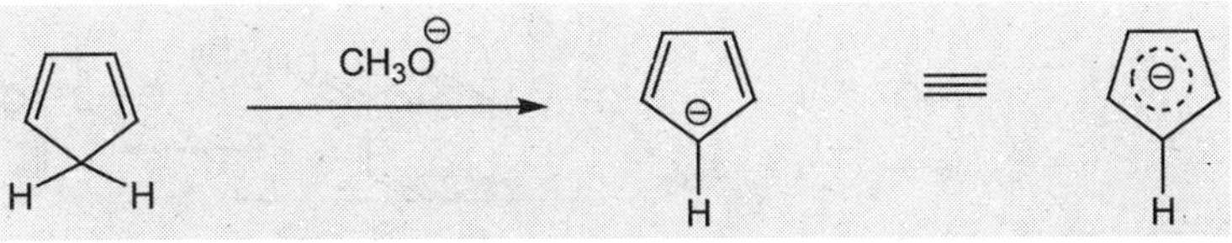

The unusual stability of cyclopentadienyl anion is explained by the fact that although it has only four π electrons, two more electrons become available to it from the heterolysis of C–H bond, thus forming a system of six electrons. These electrons are spread over all the five carbon atoms like the delocalized aromatic system of benzene ring, thus conferring on it the stability so characteristic of the aromatic compounds.

III. Detection

(a) Spectroscopic Methods: for the detection of carbanions have not been found to be as successful as in the case of carbocations. Quite often, an NMR spectrum of a carbanion is complicated by the covalent bonds of the particular compound, which, in turn, are influenced by solvent and temperature. The chemical shift is also influenced by these factors. However, the extent of negative charge localized at carbanionic carbon can be related to the degree of *upfield* shift experienced by the neighbouring protons.

(b) Chemical Evidence: for the existence of carbanions is provided by the base-catalyzed racemization of (+)-phenyl *sec*-butyl ketone. Here, the abstraction of a proton by the base yields the carbanion **34** and since it has a flat structure and the reaction is reversible, the proton may get attached to either face, producing the racemic mixture. Also, the rate constants for racemization and for bromination of the above ketone are found to be the same. The rate of racemization is also found to be equal to the rate of deuteration when the reaction is carried out in presence of a base in D_2O. All these observations strongly support the formation of a *planar anion* in the rate-determining step.

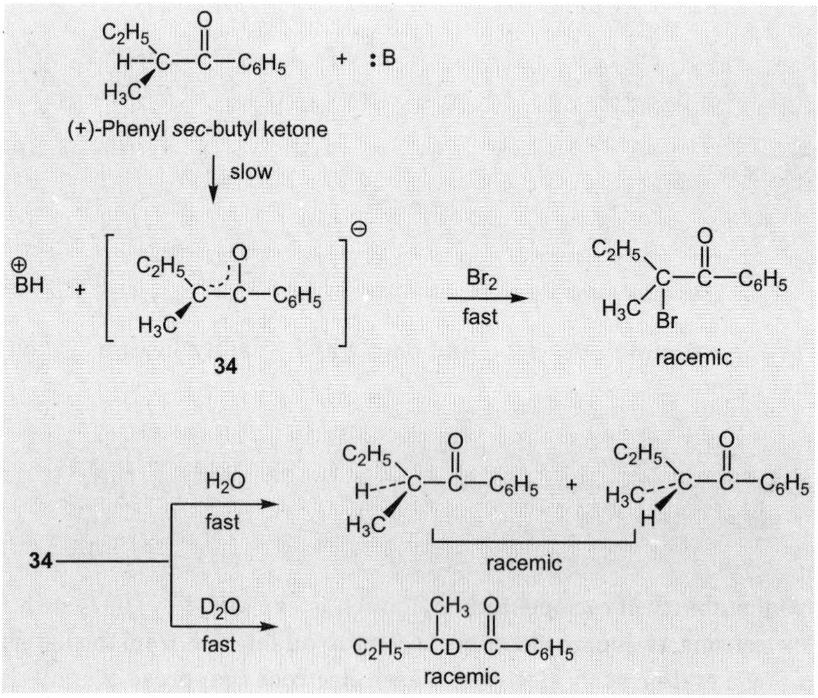

IV. Formation

As there is little difference between the electronegativity of carbon and hydrogen (2.5 and 2.1, respectively), the polarity of the C–H bond is very small, therefore, the heterolytic fission of this bond to form an anion and a proton should be a very difficult process. In other words, a hydrogen atom bound to an sp^3 carbon atom shows negligible acidity.

However, the presence of electron-attracting substituents such as nitro, cyano, or carbonyl groups on the same carbon renders the hydrogen relatively acidic. The increase in acidic character is not only due to the electron-withdrawing ability of these substituents, but also due to their ability to delocalize the negative charge of the anion. Thus, hydrogens on the carbon atom alpha to nitro, cyano or carbonyl groups have acidic character, and can be removed as protons, leaving behind resonance-stabilized anions.

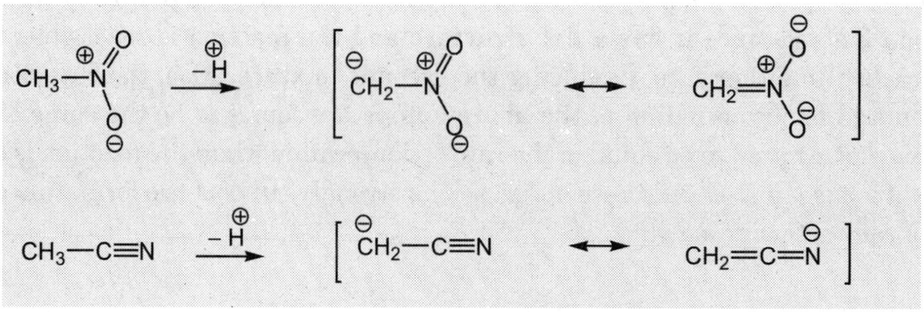

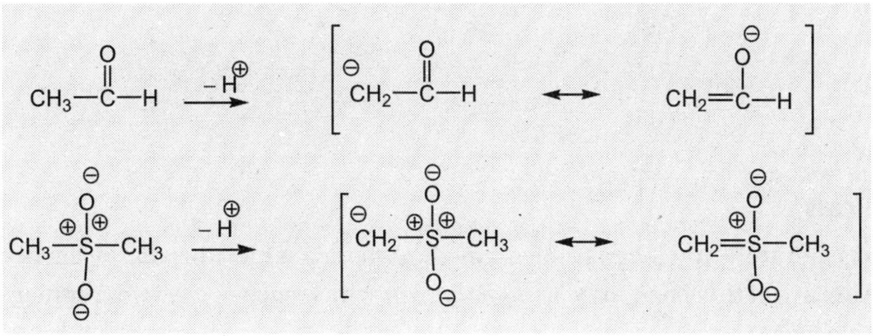

V. Reactions

Carbanions undergo three basic types of reactions.

(*a*) Combination with an electrophile
(*b*) Nucleophilic substitution and addition
(*c*) Rearrangements

(a) Combination with an Electrophile: The most common reaction of carbanions is combination with a positive species (electrophile), usually a proton.

(b) Nucleophilic Substitution and Addition: (i) Carbanions may act as nucleophiles, and push out one of the four groups on a saturated carbon atom (S_N2 reactions).

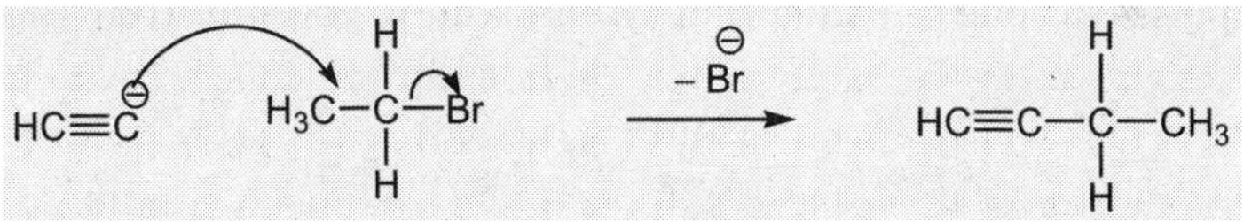

(ii) Carbanions frequently add to the carbonyl double bond (aldol condensation)

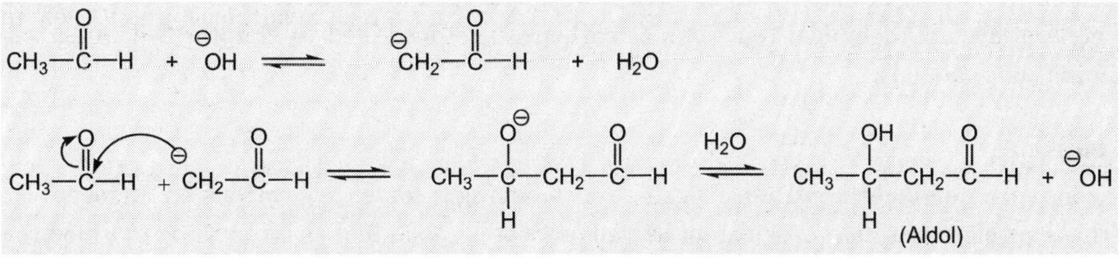

(c) Rearrangements: Carbanions also undergo rearrangement reactions, though they are rare.

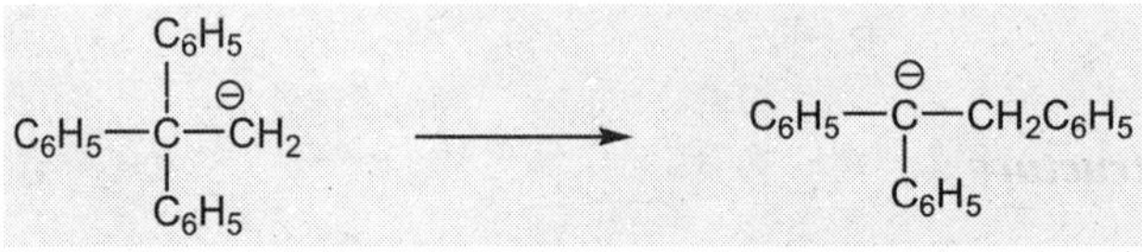

C. Free Radicals

In addition to heterolytic cleavage of a covalent bond, which produces carbocations and carbanions, rupture of this bond may be *homolytic* in which each of the two fragments possesses an unpaired electron.

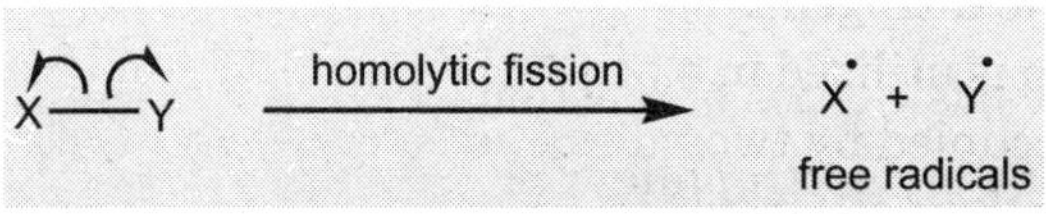

These fragments, carrying an odd or unpaired electron, have no charge, and are called free radicals. Free radicals are electrically neutral, but due to the presence of an odd electron, they are weakly attracted by the magnetic field, and hence are paramagnetic in nature. They are extremely reactive, and yield a variety of products depending upon the experimental conditions.

Chapter 13 is devoted for a complete discussion of the chemistry of carbon free radicals.

D. Carbenes

Many organic reactions involve neutral, electron-deficient carbon intermediates known as carbenes. The simplest carbene is called methylene, and is generated by the decomposition of diazomethane.

$$CH_2N_2 \longrightarrow N_2 + :CH_2$$

In a formal sense, methylene may be regarded as derived by having methyl cation lose a proton (H^+), or by having methyl radical lose a hydrogen atom ($H^\cdot$). The structure, methods of generation, and chemical reactions of carbenes are described in chapter 8 dealing with elimination reactions.

E. Nitrenes

Nitrenes are univalent neutral nitrogen species having a sextet of electrons on nitrogen. Out of these six electrons, four are present as two pairs of non-bonding electrons, and two electrons are used for bonding with a monovalent atom or group. They can be called as nitrogen analogs of carbenes. Like carbenes, they are also short-lived, highly reactive chemical species. Thus, like carbenes, they also behave as electrophiles (Lewis acids), and are intermediates in many organic reactions. They are represented by the following structure:

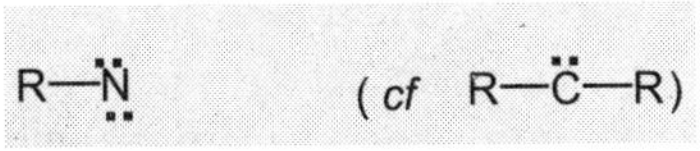

I. Classification and Structure

Like carbenes, these intermediates also exist in two states, *i.e.* singlet and triplet. These states are depicted below:

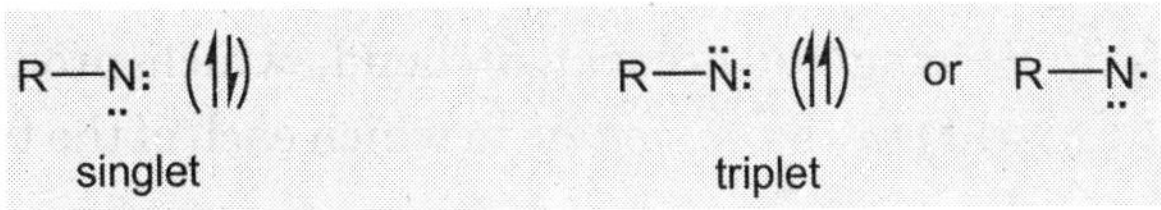

In *singlet nitrene*, the nitrogen is sp^2 hybridized. Out of the three sp^2-hybridized orbitals, one having single electron is utilized in σ bond formation with a monovalent atom or group, and the remaining two are occupied by two pairs of electrons. An unhybridized *p*-orbital is empty, and this is responsible for their electrophilic character. On the other hand, in triplet nitrene, the nitrogen is *sp* hybridized. Out of two *sp*-hybrid orbitals, one having single electron is used in σ bond formation, and the other is occupied by a pair of electrons. The two unhybridized *p*-orbitals are occupied by one electron each. The orbital picture of both the species is as follows:

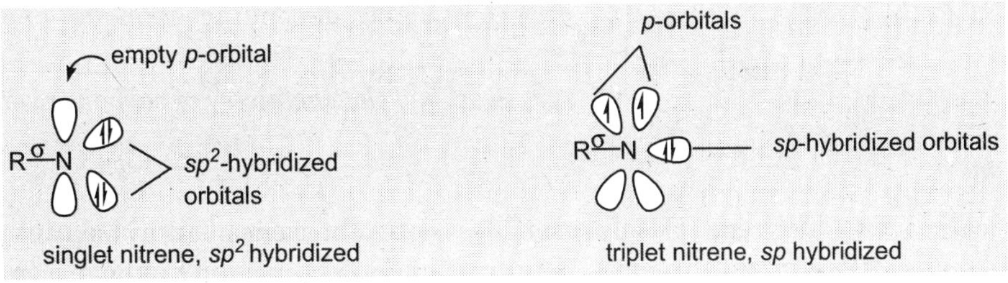

II. Detection

Nitrenes have been detected as reactive intermediates by applying the techniques of spectroscopy and trapping.

(a) Spectroscopic Techniques: The triplet nitrenes can be detected by electron spin resonance (ESR) spectroscopy.

(b) Trapping: The involvement of nitrenes as intermediates has been established by trapping experiments. Two examples are given below:

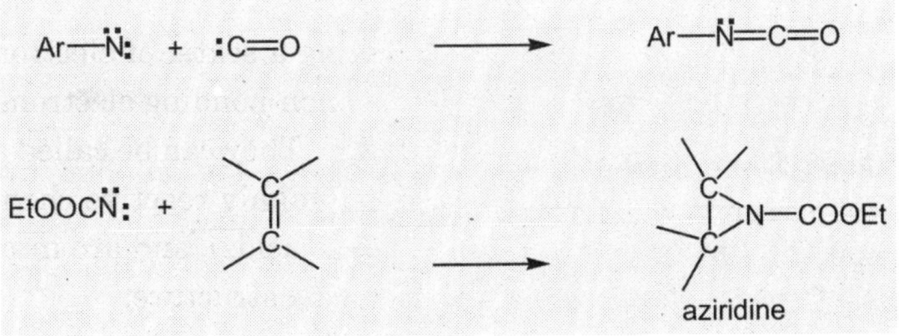

III. Formation

Methods of formation of nitrenes are analogous to those used for the formation of carbenes.

(a) α- Elimination: Treatment of hydroxylamine–O–sulphonates with alkalis results in the formation of nitrenes by the elimination of sulphonic acid group from nitrogen.

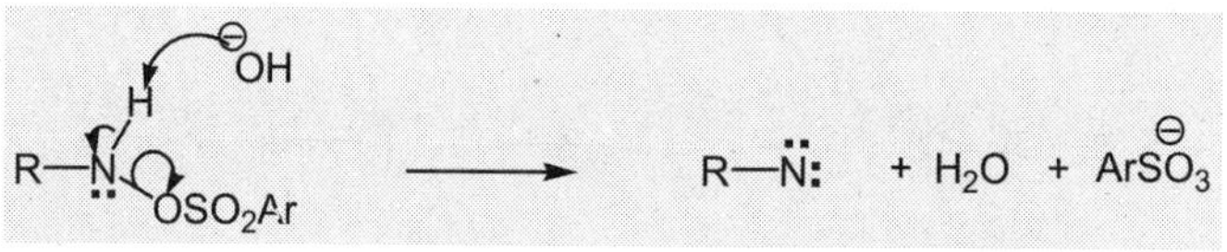

(b) Decomposition of Azides: Azides eliminate nitrogen, either on heating or irradiating with light, to produce nitrenes.

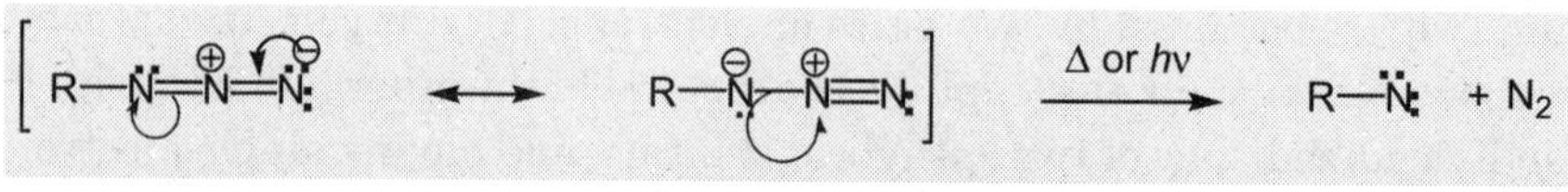

(c) Decomposition of isocyanates: Isocyanate, on heating or irradiation by sunlight, gets decomposed into nitrene and carbon monoxide.

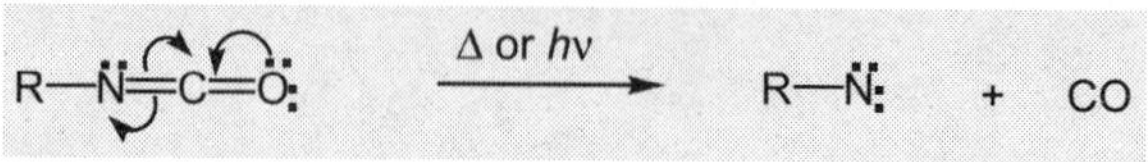

IV. Reactions

The chemical reaction of nitrenes closely resembles those of carbenes. It may be pointed out that many reactions where a nitrene intermediate can be proposed may not involve a free nitrene. However, some of the reactions in which nitrenes may be intermediates are:

(a) Addition reactions
(b) Insertion reactions
(c) Rearrangements
(d) Abstraction of hydrogen
(e) Dimerization

(a) Addition Reactions: Both singlet and triplet nitrenes add on to C=C bonds forming aziridines.

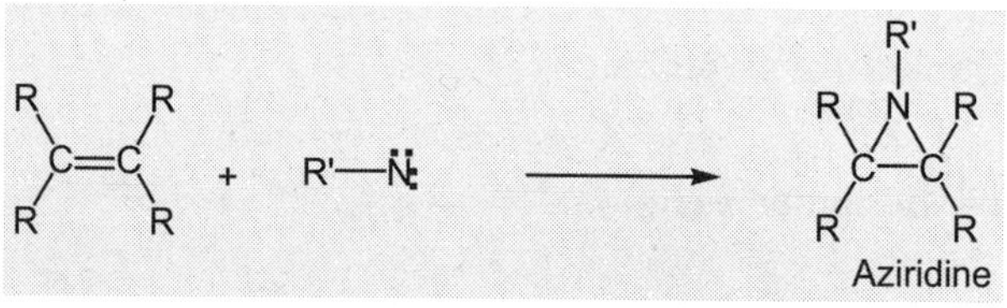

(b) Insertion Reactions: Some nitrenes, *e.g.* acyl nitrenes and sulphonyl nitrenes can insert into C–H or other sigma bonds, *e.g.* formation of amides when an acyl nitrene is generated in a hydrocarbon solvent.

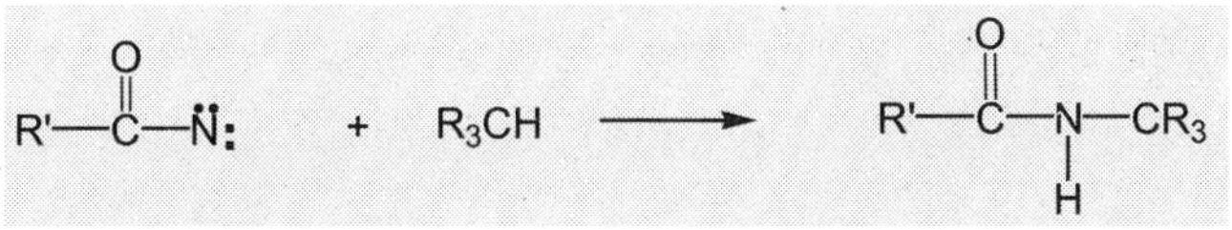

(c) Rearrangements: Alkyl nitrenes usually prefer to undergo rearrangement to give imines instead of undergoing the above two reactions.

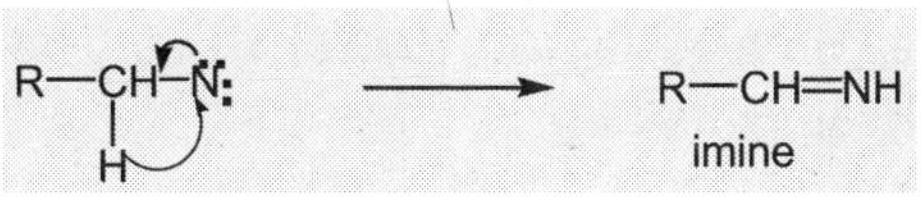

A typical example of a rearrangement where a nitrene seems to be involved as an intermediate is Hofmann rearrangement for which the following mechanism has been proposed:

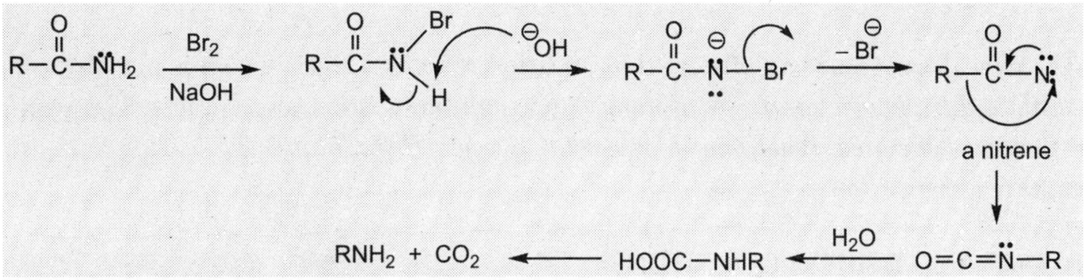

However, it may be pointed out here that as nitrene has not yet been detected in this reaction, the following concerted mechanism has also been proposed.

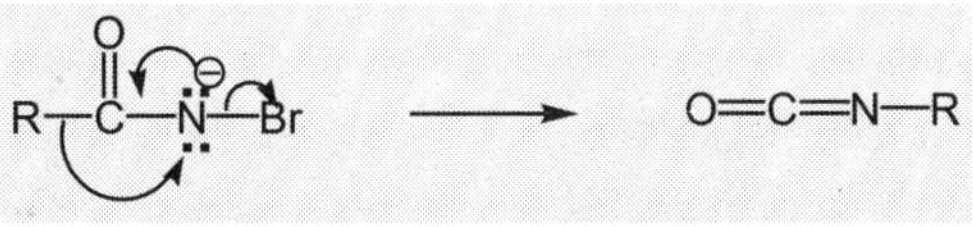

(d) Abstraction of hydrogen: The triplet nitrenes (like triplet carbenes) are capable of abstracting a hydrogen atom from an alpha C—H bond or other such group to form a pair of radicals, which may couple.

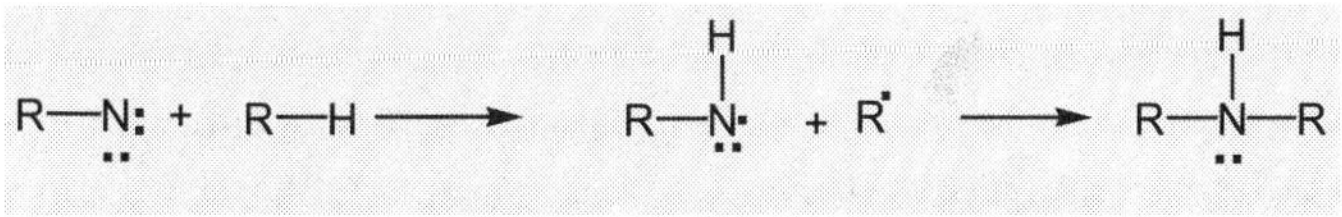

(e) Dimerization: One of the major reactions of the parent nitrene is its dimerization to give diimide.

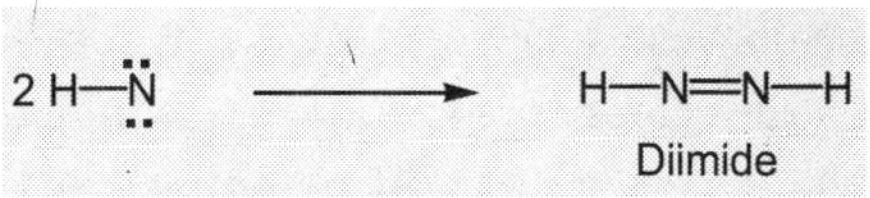

Similarly, arylnitrenes produce azoarene in the reactions where these nitrenes are involved.

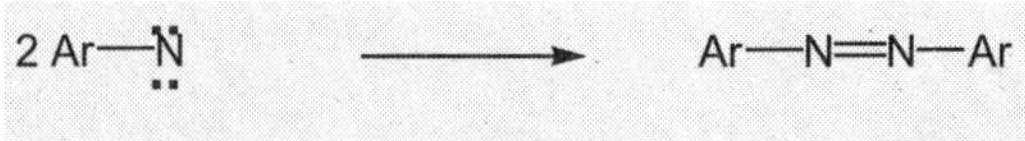

F. Arynes

When an aryl halide, such as chlorobenzene, is treated with a strong base such as potassium amide in liquid ammonia, it is converted into aniline. It is not a displacement in a conventional sense as it involves two stages, elimination and then addition. The intermediate is the molecule called aryne (benzyne or dehydrobenzene).

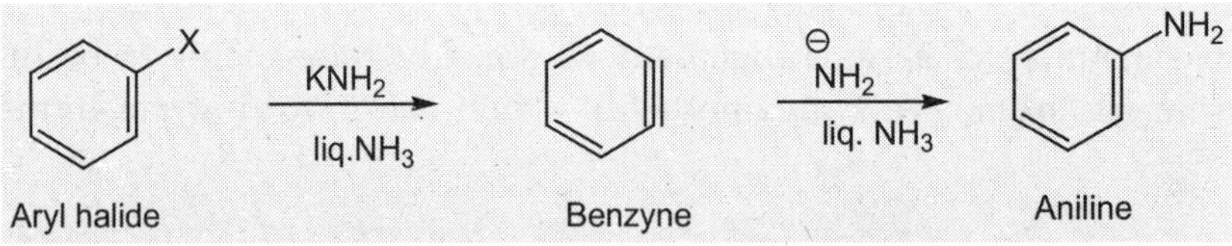

The overall process is an elimination-addition reaction, and hence it is also known as the *elimination-addition mechanism* of aromatic nucleophilic substitution. *The structure, methods of generation, and chemical reactions of arynes are described in Chapter 11 dealing with aromatic nucleophilic substitution.*

PROBLEMS AND REVIEW QUESTIONS

1. Classify the following reactions as nucleophilic or electrophilic. In each case, identify the reagent as electrophile or nucleophile.
   (i) $C_6H_6 + HNO_3 \rightarrow C_6H_5NO_2 + H_2O$
   (ii) $(CH_3)_2C=CH_2 + HCl \rightarrow (CH_3)_3\,CCl$
   (iii) $(CH_3)_3CBr + H_2O \rightarrow (CH_3)_3COH$
   (iv) $CH_3CHO + HCN \rightarrow CH_3CH(OH)CN$
2. (i) Explain with suitable examples the role of the NMR spectroscopy in the detection of carbocations.
   (ii) Arrange the following carbocations in order of increasing stability.
   $C_6H_5\overset{+}{C}H_2$, $(C_6H_5)_2\overset{+}{C}H$, $(CH_3)_2\overset{+}{C}H$, $(CH_3)_3\overset{+}{C}$, $CH_3\overset{+}{C}H_2$, $\overset{+}{C}H_3$
3. When propyl fluoride and *iso*propyl fluoride are dissolved in SbF_5/SO_2, only *iso*propyl cation is formed. What is the relative stability of propyl and *iso*propyl cations?
4. (i) How are hyperconjugative effects used to explain stability of carbocations.
   (ii) Benzyl cation is said to be stabilized by resonance involving dispersal of charge on the benzene ring. Can you write a contributing form having positive charge on the ring at *meta* position?
5. How do you explain the unusual stability of benzyl and tropylium cations?
6. Arrange the following acids in order of increasing acid strength (weakest first):
   Benzoic acid (**A**), *p*-*tert*-butylbenzoic acid (**B**), *p*-nitrobenzoic acid (**C**).
7. Which of the compounds (**A** or **B**) would you expect to be the stronger acid? Give reasons.

A B

8. Offer a suitable explanation for the following:
   (i) 2,6-Di-*tert*-butylphenol is much weaker acid than phenol.
   (ii) Compound **A** is stronger acid than **B**:

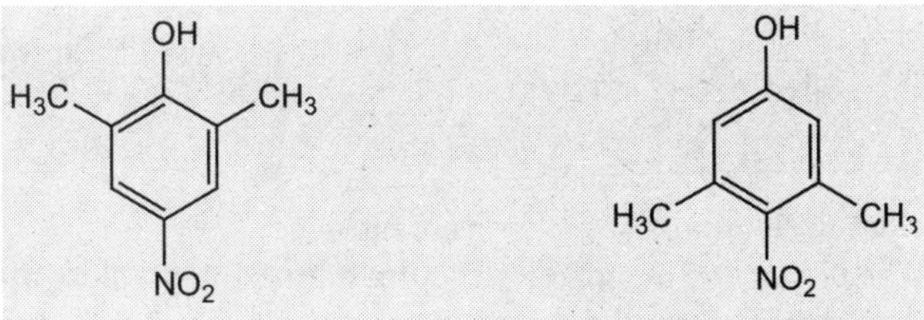

A B

9. Arrange the following compounds in order of increasing acidity (weakest acid first): (a) Phenol, (b) cyclohexanol, (c) *p*-bromophenol, (d) *p*-methoxyphenol.
10. (i) Why is acetic acid weaker than formic acid, and *p*-methoxybenzoic acid weaker than benzoic acid?
    (ii) Offer an explanation for the fact that *p*-nitrophenol is much stronger than *m*-nitrophenol, whereas the acidity of *p*-nitrobenzoic acid is only slightly more than that of *m*-nitrobenzoic acid.
11. What is the structure of a carbocation? Discuss the factors influencing the stability of carbocations.
12. (i) Why are the cyanoanilinium ions stronger acids than the anilinium ions?
    (ii) Rank the three cyanoanilinium ions (*o*-, *m*- and *p*-) in the order of decreasing acidity. Give reasons.
13. Arrange the following in decreasing order of acidity (strongest acid first):
    Benzene, propyne, ethane.
    Suggest reasons for this order.
14. (i) The pK_a of benzoic acid is 4.2, whereas that of *p*-hydroxybenzoic acid is 4.5. Explain.
    (ii) The pK_a of *p*-fluorobenzoic acid is 4.14, whereas that of *p*-chlorobenzoic acid is 3.99.
    (iii) Arrange the following compounds in order of increasing acidity (weakest acid first). Give reason for your answer.
    (a) *p*-Nitrobenzoic acid (b) 2,4-dinitrobenzoic acid
    (c) 2,4,6-trimethylbenzoic acid (d) 3,4-dinitrobenzoic acid.
15. Triphenylmethane is a stronger acid than diphenylmethane, which in turn is a stronger acid than toluene. Explain.
16. Arrange the following sets of compounds in the order of increasing acidity (weakest first):
    (i) 2-fluorobutanoic acid, 2-iodobutanoic acid, 2-bromobutanoic acid, butanoic acid.
    (ii) 2-chlorobutanoic acid, 4-chlorobutanoic acid, 3-chlorobutanoic acid, butanoic acid

(iii) propanoic acid, propiolic acid, acrylic acid.

17. Explain the following observations:
    (i) Guanidine is a very strong base.
    (ii) Amides are less basic than amines.
    (iii) Pyridine is a much weaker base than aliphatic amines.
    (iv) *N,N*-dimethyl-*o*-toluidine is a stronger base than *N,N*-dimethylaniline.
18. Listed below are pairs of carbocations which are interconvertible by hydride shift. Which cation in each pair will have tendency to undergo rearrangement?

	A		B
(i)	$(CH_3)_2CH\overset{\oplus}{C}H_2$	,	$(CH_3)_3\overset{\oplus}{C}$
(ii)	$CH_3CH_2\overset{\oplus}{C}H_2$	,	$CH_3\overset{\oplus}{C}HCH_3$
(iii)	$CH_3CH_2CH_2\overset{\oplus}{C}HCH_3$	,	$CH_3CH_2CH_2CH_2\overset{\oplus}{C}H_2$

19. Discuss structure, formation and reactions of nitrenes.
20. (i) Discuss the planar and pyramidal structure of carbanions. Illustrate with suitable examples how these intermediates are detected.
    (ii) Arrange the following carbanions in the order of decreasing stability (most stable first).

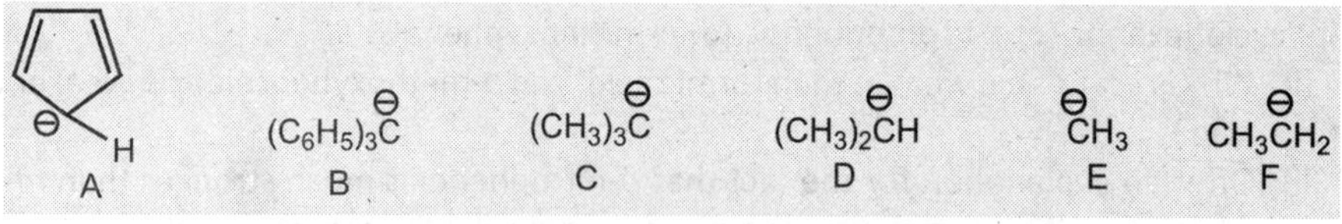

21. (i) What are carbanions? How are they formed? Comment upon the statement 'Carbanions can be either planar or pyramidal'.
    (ii) Carbanions in which the unshared pair of electrons is conjugated with a carbon-carbon, carbon-oxygen, and carbon-nitrogen double bond are much more stable than simple carbanions. Justify the statement by giving one example in each case.
22. Why aniline is a weaker base than aliphatic amines? Compare the base strengths of *p*- and *m*-nitroaniline with aniline.
23. What are free radicals? Discuss important methods for their generation. Comment upon the statement, 'primary, secondary and tertiary carbocations, and free radicals follow the same order of stability, but the former are prone to rearrangements and the latter are not.'
24. Comment upon the structure of free radicals. How are they detected? Pick up the most stable and least stable radicals from the following: $\dot{C}H_3$, $(CH_3)_3\dot{C}$, $C_6H_5\dot{C}H_2$, $CH_3\dot{C}H_2$, $(C_6H_5)_3\dot{C}$.
25. Comment upon the unusual stability of the following:
    (i) Cyclopentadienyl anion
    (ii) Triphenylmethyl anion
26. Explain why carbocations are always planar, while carbanions and free radicals may be either planar or pyramidal?
27. How do hyperconjugation and resonance effects explain the stability of free radicals? Arrange the following free radicals in the increasing order of their stability (least stable first):

$C_6H_5-\dot{C}H-CH{=}CH_2$, $C_6H_5-\dot{C}H-CH_3$, $C_6H_5CH_2\dot{C}H_2$, $C_6H_5\dot{C}H_2$

28. Discuss the structure of benzyne? Why is the triple bond in benzyne much more reactive than that of acetylene?
29. Give an example of a reaction which is believed to occur through a nitrene intermediate.
30. Discuss briefly the various factors which affect the stability of (i) carbocations, (ii) carbanions, and (iii) free radicals.
31. Discuss the structure of carbenes, nitrenes, and arynes. Illustrate the technique of 'trapping' for the detection of these intermediates.
32. Define and illustrate the terms (i) electrophile, (ii) nucleophile (iii) intermediate, and (iv) transition state.
33. In how many ways can fission of a covalent bond take place? Discuss briefly the conditions under which each type of fission occurs.
34. Give a brief account of the common type of organic reactions with suitable examples.

SOLUTIONS

1. (i) electrophilic, the electrophile is NO_2^+
   (ii) electrophilic, the electrophile is H^+
   (iii) nucleophilic, the nucleophile is H_2O
   (iv) nucleophilic, the nucleophile is CN^-
2. (i) See the text.
   (ii) $^+CH_3 < {}^+CH_2CH_3 < {}^+CH(CH_3)_2 < (CH_3)_3C^+ < C_6H_5\overset{+}{C}H_2 < {}^+CH(C_6H_5)_2$
3. *Iso*propyl cation is more stable than *n*-propyl cation.
5. Benzyl cation gets rearranged to tropylium ion which is very stable. The reason for its unusual stability is that it is an aromatic cation containing six π-electrons. As shown, the six π- electrons are completely delocalized over all the seven sp^2- hybridized carbon atoms just in the same manner as the six π-electrons are delocalized in the benzene ring. Various resonance contributing forms of tropylium ion are as under:

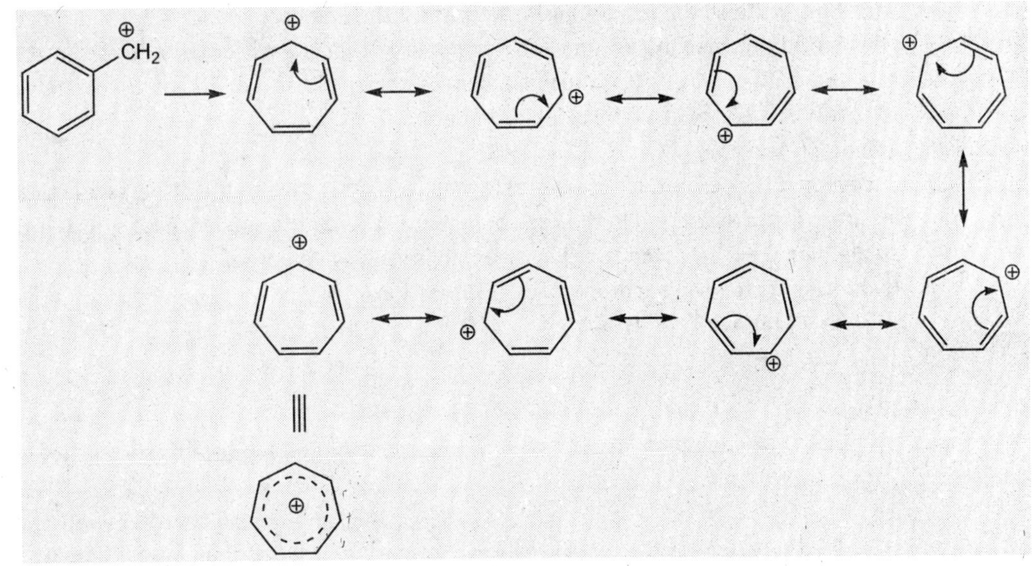

Actually the tropylium ion is so stable that its reaction with nucleophiles such as water and alcohols is very slow.

6. **B** < **A** < **C**
7. **A**, the carbanion derived from **A** is better stabilized by the aromatic rings which have to be coplanar than is the carbanion from **B** in which the aromatic rings prefer to be in different planes for steric reasons. Coplanarity allows optimum *p* orbital overlap.
8. (i) Bulky *o-tert*-butyl groups prevent solvation of the corresponding phenoxide anion thus making it less stable than phenoxide ion.
   (ii) Steric inhibition of resonance in the anion derived from **B** makes it less stable than the anion derived from **A**.
9. Cyclohexanol < *p*-methoxyphenol < phenol < *p*-bromophenol.
10. (ii) Resonance-stabilization of *p*-nitrophenoxide ion. No such resonance is possible in either of the benzoic acid ions. The difference is small because of the operation of inductive effect alone.
12. (i) The cyano group is electron-withdrawing because of the resonance and inductive effect. This explains the stronger acidity of cyanoanilinium ions.
   (ii) The *o*-cyano derivative is the strongest acid because of the resonance effect and the strongest inductive effect followed by *p*-cyano compound which has only resonance effect (equal to that of *ortho* isomer). The *meta* cyano compound is the weakest acid because the resonance effect is weakest.
13. $CH_3C{\equiv}CH > C_6H_6 > C_2H_6$

   The anions $CH_3C{\equiv}C^-$, $C_6H_5^-$ and $CH_3CH_2^-$, derived from propyne, benzene and ethane, involve nonbonding electrons with *sp*, sp^2 and sp^3 hybrid orbitals, respectively. The electrons in *sp* hybrid orbital are held more tightly than those in sp^2 which are held more tightly than those in sp^3 hybrid orbital. Therefore, the anion of first is more stable than that of second which in turn is more stable than the third. The acidic strength depends upon the relative stability of the corresponding anion.
14. (i) Hint, write all the resonating structures for the anions.
   (ii) Hint, in former the +R and –I effects nearly cancel each other whereas in the latter the intensity of +R effect is much smaller. Consider the size of *2p* and *3p* orbitals of F and Cl, respectively, which they offer for + R effect.
   (iii) (c), (a), (d), (b).
15. The anion $(C_6H_5)_3C^-$ is much more stable than $(C_6H_5)_2CH^-$ due to the dispersal of negative charge in three rings against two in the case of $(C_6H_5)_2CH^-$. For the similar reason, $(C_6H_5)_2CH^-$ is more stable than $C_6H_5CH_2^-$. Hence, acidic strength is in the given order.
18. Hydride shifts to produce more stable carbocations.
   (i) **A→B**; (ii) **A→B**; (iii) **B→ A**;
20. (ii) **A, B, E, F, D, C.**

Chapter 5

General Methods of Determination of Reaction Mechanism

Chapter Outline

5.1 ENERGY CONSIDERATIONS

- **A.** Kinetics
- **B.** Reactivity and Rates of Reactions
- **C.** Catalysis
- **D.** Competitive Reactions: Thermodynamic and Kinetic Control
- **E.** Hammett Equation
- **F.** Curtin-Hammett Principle
- **G.** Taft Equation

5.2 METHODS OF DETERMINATION OF REACTION MECHANISMS

- **A.** Product Analysis
- **B.** Determination of the presence of Intermediates
- **C.** Crossover Experiments
- **D.** Isotopic Labelling
- **E.** Stereochemical Evidence
- **F.** Kinetic Evidence

A molecule may be converted into another molecule by a chemical reaction. The new molecule (product) has an arrangement of atoms different from the starting molecule (reactant). The change is accompanied by a redistribution of electrons. In most of the organic reactions, transformation of reactant to product takes place through some definite steps. An organic reaction mechanism is a detailed description of these steps. It not only tells us the number of steps involved in the reaction, but also gives us information regarding the sequence of breaking and making of bonds.

5.1 ENERGY CONSIDERATIONS

Chemical reactions may be regarded as *equilibrium processes*, the point of equilibrium being dictated by the relative energies of the reactants and the products. Reactions, which do not proceed at all, are also equilibrium processes for which the equilibrium constant has a value near zero. A chemical reaction proceeds when it has a favourable equilibrium or, in other words, when the products are more stable than the reactants.

The equilibrium constant of a reaction depends on the enthalpy change (ΔH), which is the quantity of heat released or absorbed when a mole of reactant is converted to a mole of product, and on the entropy change (ΔS), which is a measure of the energy consumed (or released) in the reaction to create more order (or disorder) in products relative to reactants. When entropy change is taken with absolute temperature (T), it gives free energy change (ΔG) according to the relationship $\Delta G = \Delta H - T\Delta S$.

It should be noted that a reaction proceeds in the forward direction only when ΔG is negative (*exergonic reaction*). A positive ΔG (*endergonic reaction*), on the other hand, indicates that reaction has a tendency to go backwards. Since entropy is a measure of the degree of freedom (or disorder) of a system, a system is most stable when it has maximum entropy. Reactions in which there is loss of freedom should, therefore, not be favourable processes, and it has indeed been found that reactions involving a *decrease* in the number of molecular species on going from reactants to products proceeds only if ΔH is negative (*exothermic reaction*), and large enough to give overall negative value for ΔG. On the other hand, when the number of molecular species *increases* on going from reactants to products, there is increase in entropy, and the $-T\Delta S$ term then becomes large to make the reaction move in the forward direction even when it is not an exothermic reaction. When all the substances are in their standard states (at 25°C and 1 atmosphere pressure), the equation may be written as

$$\Delta G^o = \Delta H^o - T\,\Delta S^o$$

in which the superscript zero is used to denote standard states, and ΔG^o is the difference in standard free energies of products and reactants.

$$\Delta G^o = G^o\,(\text{products}) - G^o\,(\text{reactants}).$$

The relationship between the equilibrium constant (K) and ΔG^o is given by the following expression:

$$\Delta G^o = -RT\,ln\,K$$

where, $ln\,K$ = natural logarithm of K

R = gas constant

T = absolute temperature.

A. Kinetics

Although a negative value of ΔG tells us that the thermodynamics is favourable for the reaction to proceed in the forward direction, it does not necessarily tell us that the reaction will proceed at a measurable rate. This is quite understandable as thermodynamics only predicts the position of equilibrium between the reactants and products without giving any information regarding the rate of the reaction.

Kinetic studies deal with *the rates of reactions* and their dependence on the concentration of the various reacting species. In contrast to thermodynamics, time is the most important variable of kinetics. Kinetics is often used to gain an insight into reaction mechanisms. A typical organic reaction is a multistep process in which one of the steps, which is the slowest, is the rate-determining step; kinetic studies give us the information that helps us in identifying the species that participate in this step. Generally speaking, the rate of a reaction is proportional to the product of concentrations of the species that are involved in the rate-determining step.

Order of Reaction: The rates of many organic reactions are found to vary with the concentrations of one or more of the reactants. This dependence is usually determined by measuring the rates of a reaction by performing several experiments, in each of which, all reactants except one are present in such a large excess that their concentrations may be regarded as constant. Obviously, the rate of reaction will vary only with the concentration of the species that are involved in the rate determining step. Let us illustrate it by considering a reaction between A and B to give C:

$$A + B \rightleftharpoons C$$

When the concentration of either A or B is doubled, the rate is doubled; or concentration of either of them is halved, the rate is halved. Thus, the rate of this reaction depends upon the concentration of both A and B. The reaction rate may be measured by observing either (a) the disappearance of reactants A or B with time, or (b) the appearance of product C with time. A rate equation can thus be written as follows:

$$\text{Rate} = \frac{-d[A]}{dt} = \frac{-d[B]}{dt} = \frac{+d[C]}{dt} = k[A][B]$$

where, k is the rate constant, dt is the change in time, and the terms in square brackets denote concentrations. Minus sign is used to signify that concentration of the reactants decreases with time.

The above rate equation expresses the kinetic order of a reaction, and the reaction is said to be a second order reaction as the right hand side of the equation is the product of two concentrations [A] and [B]. In addition to the overall order of a reaction, it is often useful to indicate the order of a reaction with respect to a specific reactant that reflects the power to which the concentration of that particular reagent must be raised to have a direct proportionality between concentration and reaction rates. Thus, the above reaction is first order with respect to A, and first order with respect to B.

Order of Reaction and Molecularity: *Order of a reaction, n,* is the sum of all the exponents of the concentrations of various reactants, A, B, C..... in the rate equation.

$$\text{Rate} = k\,[A]^{n_1}\,[B]^{n_2}\,[C]^{n_3}\,.....$$

$$n = n_1 + n_2 + n_3 + ----$$

The reaction is said to be of total order n, and each individual exponent is the order with respect to that component, *e.g.* n_1^{th} order with respect to A, n_2^{th} order with respect to B, and n_3^{th} order with respect to C. The powers, n_1, n_2, and n_3 may not necessarily be whole numbers. The decomposition of acetaldehyde into methane and carbon monoxide, for instance, is of three halves order, conforming to the relation.

$$\frac{-d[CH_3CHO]}{dt} = k[CH_3CHO]^{3/2}$$

The *molecularity* of a reaction, on the other hand, may be defined as the number of species that are undergoing covalency changes in the rate-determining step. The order of a reaction is determined experimentally, whereas molecularity is a theoretical concept. Further, the order of a reaction can be a fraction, whereas molecularity shall always be a whole number. The order of reaction is the same as the molecularity for reactions taking place in one step, while for a multistep reaction, the order for each step is the same as the molecularity for that step.

B. Reactivity and Rates of Reactions

From the above discussion, it is clear that a reaction will proceed only when (a) the products are more stable than, or of the same order of stability as, the reactants, and (b) there is a suitable pathway between reactants and products, *i.e.* the rate constant (k) is favourable. There is a characteristic rate constant associated with every reaction; the larger the value of k, the faster the reaction. Now, the question arises as to what factors make one rate constant larger than another. Two theories have been proposed to account for this variation.

- Collision Theory
- Transition State Theory

I. Collision Theory

According to collision theory, molecules react only when they collide. Thus, information regarding the rate of a reaction can be gained by knowing the number of collisions taking place per unit of time. Obviously, the frequency of collisions will increase by increasing the concentration of reactants. As the frequency of collisions also depends upon molecular motions, which in turn depend upon temperature, it is expected that the rate of reaction will increase with rise in temperature.

The rate of reaction, however, does not depend only on frequency of collisions, as it has been found that only one in every 10^{15} collisions leads to products. This has been explained by assuming that collision between molecules will be successful only when they contain at least a minimum amount of energy, called *the activation* energy *(Ea)*. Activation energy is required even for a thermodynamically favoured reaction, as energy is needed for breaking the bonds in the reactants before new bonds are formed in the products. Thus, there is a geometrical arrangement of atoms possessing higher energy than that of either reactants or products. Reactants have to pass through this highest-energy state, called an *activated complex*, before their conversion to products. However, there is a further requirement, *i.e.* the colliding activated molecules must be effectively oriented with respect to each other for a successful combination. In other words, the activated molecules must be suitably oriented to react when they collide. This limitation is referred to as the probability factor, and is symbolized by P. The rate is expressed in the following form:

$$\text{Rate} = P\,Z\,exp\,(-Ea\,/RT)$$

where, Z is the number of collisions per unit time. Other symbols have their usual meanings.

II. Transition State Theory

The transition state theory of reactions has been found to be more useful in understanding the mechanisms of organic reactions. It takes into account the energy changes of the reactant molecules as they are brought together, and are ready to collapse into products. Like collision theory, this theory also assumes the existence of an energy barrier (transition state) through which reactants must pass before their conversion to products. The rate of reaction depends on the rate at which the reactants reach the transition state, instead of *frequency of collisions* as envisaged in the collision theory. The transition state species is in equilibrium with reactants, and hence its concentration is determined by an equilibrium constant denoted as $K^{\ddagger}$, which is related to the enthalpy and entropy of activation in the usual way.

$$\Delta G^{\ddagger} = \Delta H^{\ddagger} - T\,\Delta S^{\ddagger} = -\,2.303\,RT\log K^{\ddagger}$$

where, transition state is denoted by ‡ and $\Delta G^{\ddagger}$ is the free energy of activation, $\Delta H^{\ddagger}$ is the heat of activation, $\Delta S^{\ddagger}$ is the entropy of activation.

Let us exemplify transition state theory by considering a reaction without involving an intermediate.

$$\text{A} + \text{B–C} \longrightarrow \text{A–B} + \text{C}$$

It has been shown, by making certain approximations, that the minimum activation energy is required when A approaches B–C along the bonding line of B–C, and on the side remote from C. The bond A–B begins to form, whereas bond B–C begins to stretch. During the course of the reaction, each bond may be regarded as a partial one with both A–B and B–C bond distances greater than those in the product and the reactant. However, the amount of energy released during the partial formation of bond A–B is less than that absorbed by the corresponding partial separation of the bond B–C, and as a result, the energy of the system begins to rise with the process of bond transfer. The rise in energy finally reaches a maximum, where the system is least stable, followed by a fall. At the end of the reaction, the bond B–C has been completely replaced by the bond A–B. The transition state is the state of highest potential energy acquired by the reactants during the transformation. The system at this point is also called an activated complex. It possesses a definite geometry and charge distribution. However, due to the associated high energy, it is highly unstable, and cannot be isolated. We study it indirectly with the help of models.

The activation energy, besides other factors, depends upon the strengths of B–C and A–B bonds. Stronger the B–C bond, greater will be $\Delta G^{\ddagger}$, and conversely, greater the strength of A–B bond, lower is $\Delta G^{\ddagger}$. The energy changes that accompany a chemical reaction are conveniently represented in the form of an *energy diagram* with *free energy* (*G*) plotted against the pathway of the reaction coordinate (**Fig. 5.1**).

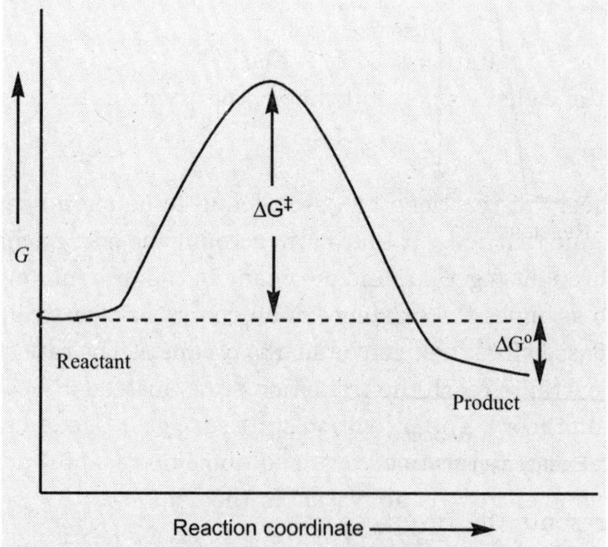

Fig. 5.1: Energy diagram for a single-step reaction

Since $\Delta G^{\ddagger}$ represents the *energy barrier*, therefore, greater the height, slower will be the reaction. ΔG^{o}, which reflects the difference in standard free energy of the reactants and the products, is a thermodynamic quantity, and does not affect the rate of reaction, which is governed solely by $\Delta G^{\ddagger}$.

Many organic reactions proceed through intermediates which may or may not possess enough stability to permit isolation. There are two transition states in a two-step reaction, and it will have one intermediate between reactants and products represented by a minimum in the energy profile diagram. Deeper the depression, more stable will be the intermediate, and conversely, a shallow depression in the energy curve indicates a less stable intermediate. Although there are two transition states each with energy higher than that of the intermediate, the overall rate of the reaction is reflected in the energy difference between the reactants and the highest energy transition state ($\Delta G^{\ddagger}$ in **Fig. 5.2**). The intermediate is very unstable, and is transformed to products more rapidly than the reactants generate the intermediate.

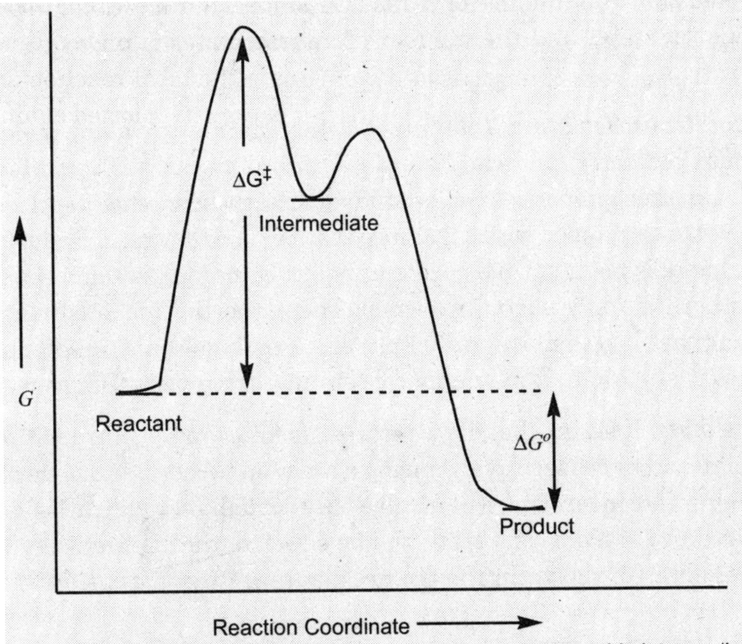

Fig. 5.2: Energy diagram for a two-step reaction with an unstable intermediate

Figure 5.3 represents the reverse situation in a two-step reaction. In this case, the intermediate is very stable, and the second step is thus rate-determining.

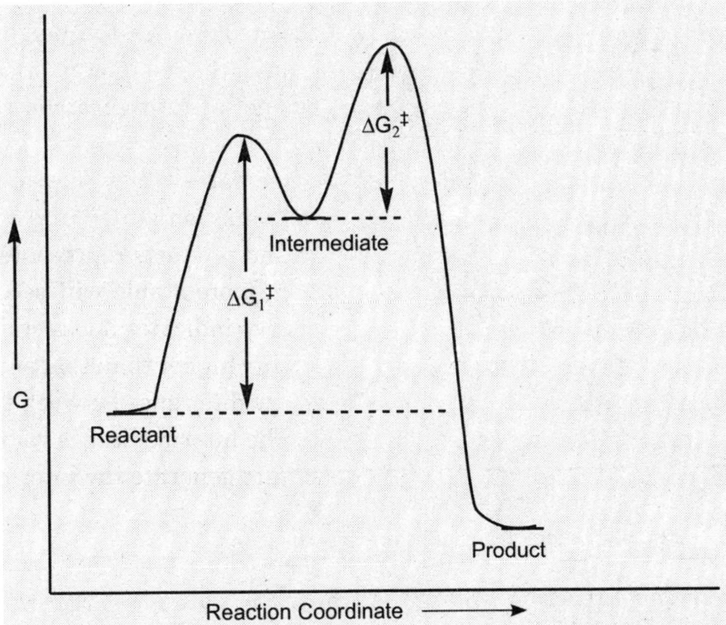

Fig. 5.3: Energy diagram for a two-step reaction with a stable intermediate

It is to be noted that the number of transition states in a multistep reaction is always equal to the number of steps, and the number of intermediates is always one less than the number of steps.

(i) Intermediate and transition state: *Intermediates* are normal covalent molecules (or ions) possessing a definite geometry, and can be isolated in many cases. Even when they are too unstable to be isolated, their properties can be studied directly by methods such as spectroscopy, or alternatively they can be trapped with substances with which they react rapidly. *The transition state*, on the other hand, is not a normal covalent molecule (or ion) as some bonds are partially broken, whereas others are only partially formed. The properties of transition state are often studied by assuming that it resembles, rather closely, a real molecule (reactant, intermediate, or product) that lies very close to it in energy content in a free energy diagram.

(ii) Hammond's Postulate: It states that 'if two states, as for example, a transition state and an unstable intermediate, occur consecutively during a reaction process, and have nearly the same energy content, their interconversions will involve only a small reorganization of the molecular structure'. (The structure of transition state will be closer to the energetically closer species). There are three possibilities shown in the energy diagram (**Fig. 5.4**) as Case-I, Case-II, and Case-III.

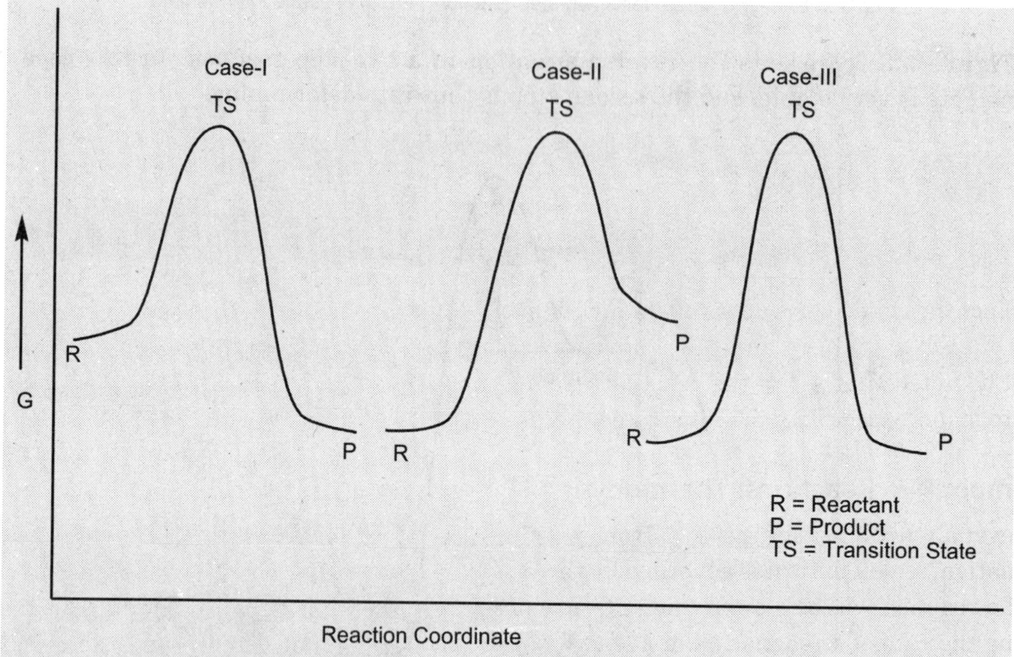

Fig. 5.4: Different structures of transition state

Case I is an *exergonic* reaction, where the transition state is energetically closer to the reactant than the product. Therefore, the structure of the transition state will resemble more closely to the structure of the reactant than the structure of the product.

In Case II, which is an *endergonic* reaction, the structure of transition state will resemble to the structure of the product.

The third possibility (Case III) shows a situation when the transition state is nearly halfway between the reactant and the product. Therefore, the structure of the transition state will be halfway between the reactant and the product.

C. Catalysis

A catalyst may be defined as a substance that increases the rate of a reaction without itself being consumed in the process. It combines with the reactants in the initial step, and is regenerated in a later stage of the reaction. We have already seen that the rate of the reaction is determined by $\Delta G^{\ddagger}$; a catalyst increases the rate of reaction by providing an alternate path from reactants to products with a lower energy transition state. In other words, the catalyst involves itself with reactants in such a way that the energy of the transition state of a catalyzed reaction is always lower than that of the transition state for the corresponding uncatalyzed reaction.

The catalytic effect of iodide ions on the hydrolysis of methyl chloride provides an illustrative example. Methyl chloride reacts with hydroxide ion in a one-step process, and as expected the rate depends on $\Delta G^{\ddagger}$.

$$CH_3Cl + \overset{\ominus}{O}H \longrightarrow CH_3OH + \overset{\ominus}{Cl}$$

Addition of iodide ion to the reaction mixture provides an alternative path consisting of a two-step mechanism in which methyl iodide is an intermediate.

$$CH_3Cl + \overset{\ominus}{I} \longrightarrow CH_3I + \overset{\ominus}{Cl}$$

$$CH_3I + \overset{\ominus}{O}H \longrightarrow CH_3OH + \overset{\ominus}{I}$$

Since the iodide ion is an excellent nucleophile and an excellent leaving group, the transition states of each step ($\Delta G_1^{\ddagger}$ and $\Delta G_2^{\ddagger}$) have lower energy than that for the one-step process. It should be noted that ΔH of the reaction remains unchanged by addition of the catalyst; only the rate is increased. **Fig. 5.5** illustrates these points.

D. Competitive Reactions: Thermodynamic and Kinetic Control

Whenever a reaction yields more than one product, their relative amounts depend on their ease of formation, which in turn is governed by the relative values of $\Delta G^{\ddagger}$ for each process. Thus, the rate of a reaction will be slower when $\Delta G^{\ddagger}$ is relatively higher. Since $\Delta G^{\ddagger}$ is more important in deciding the rate of a reaction than ΔG^{o} (which is a thermodynamic description of a reaction), these reactions are called *kinetically controlled reactions*.

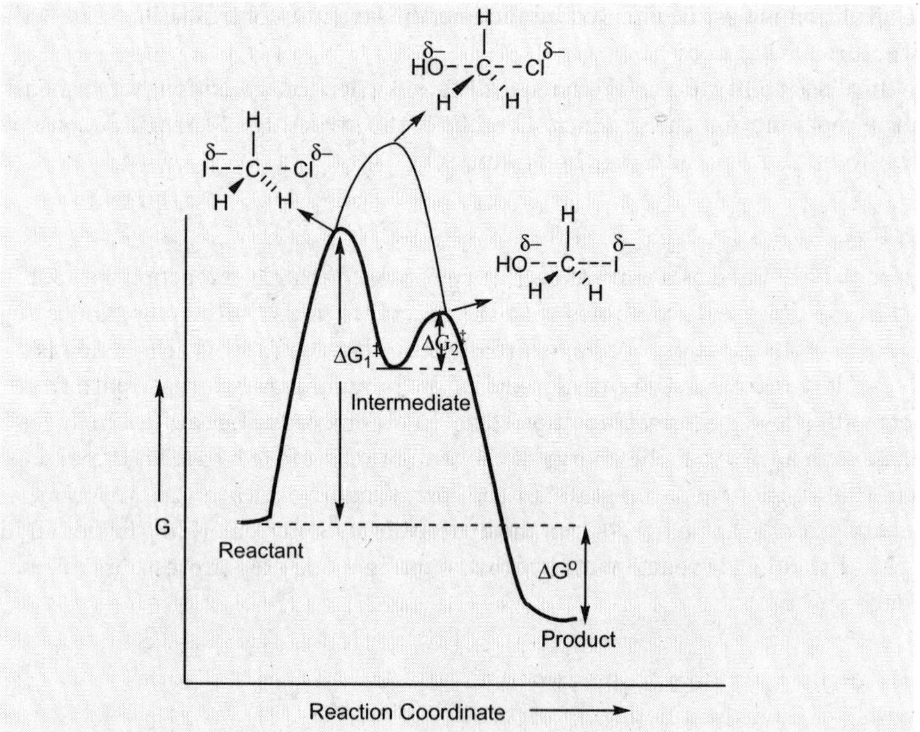

Fig.5.5: Energy diagram for the uncatalyzed (thinner line) and catalyzed (thicker line) reactions

Many reactions produce a mixture of products formed from a common intermediate. Very often, the initially formed product (kinetic control) may possess higher energy than the other product, and under the conditions of the reaction may attain equilibrium with it, which ultimately leads to the predominance of the most stable product. Such reactions are said to be *thermodynamically controlled*. Addition of HBr to butadiene (**1**) elegantly exemplifies the importance of kinetic and thermodynamic control in product composition.

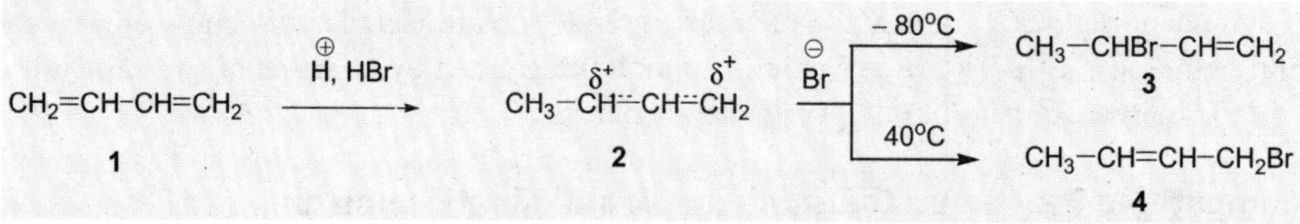

At low temperatures 1,2-addition product **3** predominates, but equilibrium sets in at higher temperatures in the presence of HBr, and 1,4-addition product **4** becomes the major product. Thus, at lower temperatures the reaction is *kinetically controlled* whereas it becomes *thermodynamically controlled* at elevated temperatures.

The situation is best illustrated by the energy diagram (**Fig. 5.6**).

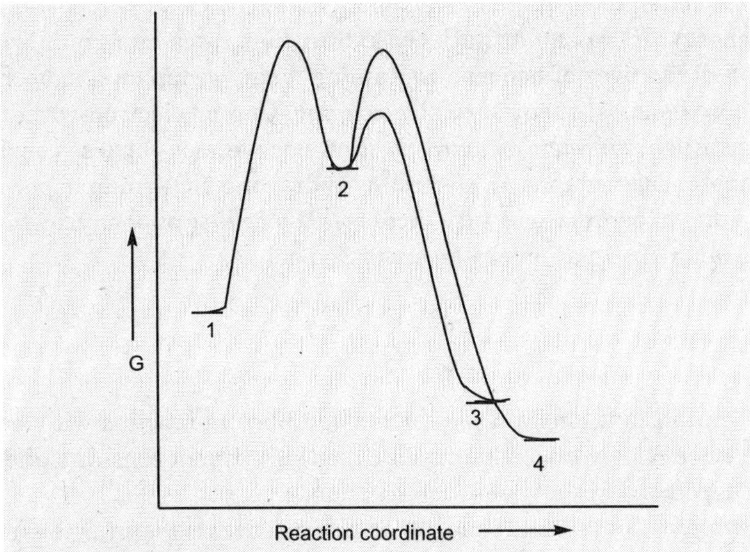

Fig.5.6: Energy diagram for kinetically and thermodynamically controlled reactions

However, it must be remembered that if the conditions of the reaction do not permit interconversion, then the kinetically controlled product will invariably predominate.

In another example, products obtained on sulphonation of naphthalene have been shown to be kinetically as well as thermodynamically controlled. At lower temperature, kinetically controlled product is obtained, whereas the thermodynamically controlled product is obtained as the major product on increasing the temperature.

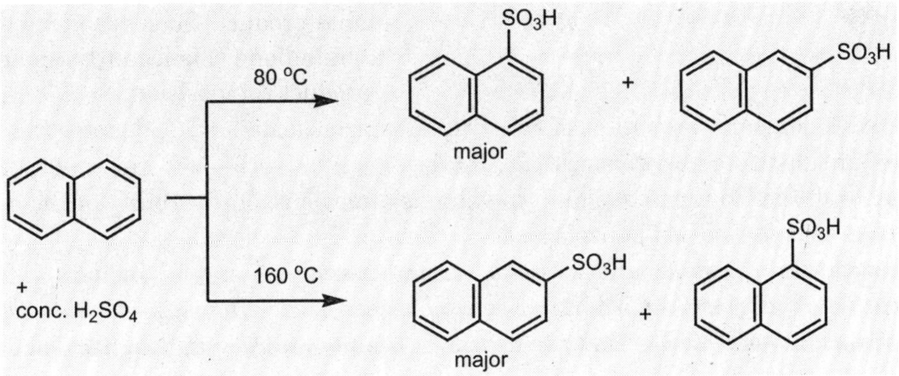

E. Hammett Equation

Hammett equation describes a *linear free energy relationship* (LFER) between reaction rates and equilibrium constants for many reactions involving benzoic acid derivatives with *meta* and *para*-substituents with two parameters, a substituent constant (σ) and a reaction constant (ρ).

Hammett stated that for any two reactions with two aromatic reactants only differing in the type of substituents, the change in free energy of activation is proportional to the change in Gibb's free energy. It was quantitatively known that rates and equilibrium of *meta* and *para*-substituted derivatives of benzene containing a ring group on a side-chain are affected similarly whether the substituents are electron-donating or electron-withdrawing. In other words, such substituents in *meta* or *para*-position behave more or less consistently in every situation. Hammett observed that an electron-withdrawing NO_2 group in *para* position, which increases the acidity of benzoic acid, also increases the acidity of phenylacetic acid.

The basic Hammett equation can be expressed as:

$$\log \frac{K}{K_o} = \sigma \rho$$

where, K is the equilibrium constant for a given equilibrium reaction with a substituent, and the reference K_o when R is hydrogen atom; σ is called substituent constant, and ρ is the reaction constant which depends only on type of the reaction.

Substituent constant (σ): Ionization of benzoic acid derivatives is taken as the reference reaction, and for the process at 25°C in water, the reaction constant (ρ) is taken as unity.

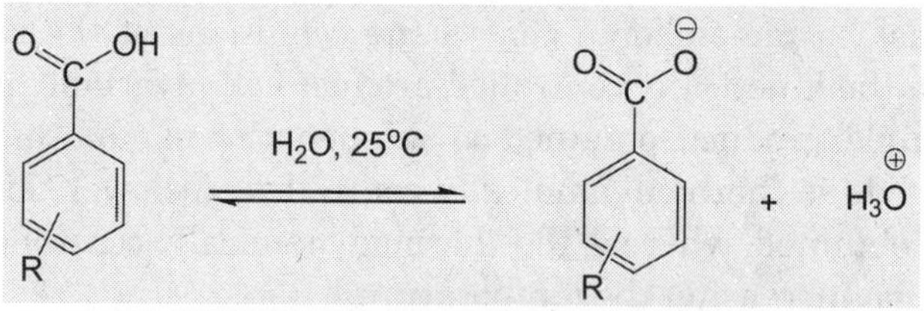

$$\log \frac{K}{K_o} = \sigma \quad (1)$$

$$\log K - \log K_o \quad (2)$$

where, K is the ionization constant of substituted benzoic acid, and K_o denotes the ionization constant of unsubstituted benzoic acid, *i.e.* R = H.

In general, the Hammett's sigma (σ) value is a measure of the ability of the substituent to modify the acid strength of benzoic acid.

The σ values are small and actually range between +1 and –1 including '0'. For the substituents that increase the acidity of benzoic acid, the σ value is positive, and those that decrease the acidity of benzoic acid have negative σ values. Moreover, the effect of substituents on the acidity of benzoic acid in *meta* and *para* positions are different; each substituent has Hammett's sigma values, σ_m and σ_p. For example, σ_m for –OH group is + 0.12, and σ_p value for the same group is – 0.37. This difference in σ values can be attributed to the fact that –OH group plays two opposing types of electronic effects at *meta* and *para*-positions.

When the substituent is in *meta*-position, its +R effect is not reflected on the ionization, but –I effect operates to withdraw electron density. However, in *p*-isomer, +R effect is strongly felt at the –COOH group, thereby showing a negative σ value.

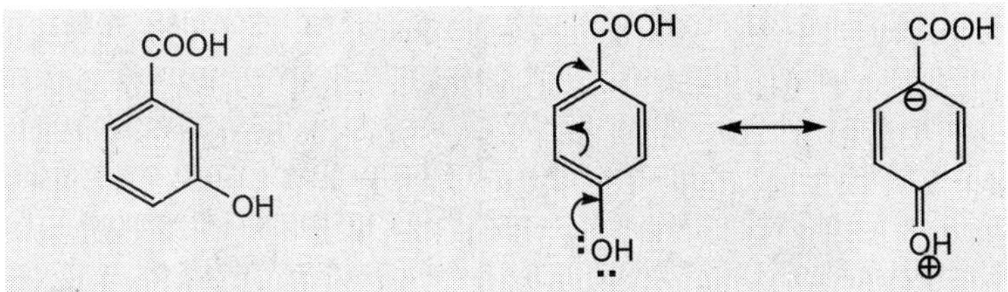

(– I effect of *m*-hydroxy group) (+ R effect of *p*-hydroxy group)

This treatment does not include *ortho*-substituents, which would introduce the steric effects. In case of polysubstituted benzoic acids, the σ values of substituents are additive in nature.

Hammett reaction constant (ρ)

Hammett reaction constant, denoted by the symbol ρ, is a measure of the sensitivity of a reaction towards the polar and resonance effects of the substituents. With the knowledge of substituent constant, it is now possible to obtain reaction constants for a wide range of organic reactions. A typical reaction is the alkaline hydrolysis of ethyl benzoates in a water/ethanol mixture at 30°C. Measurement of the reaction rate (K_o) of many substituted ethyl benzoates and their comparison ultimately results in a reaction constant of + 2.498.

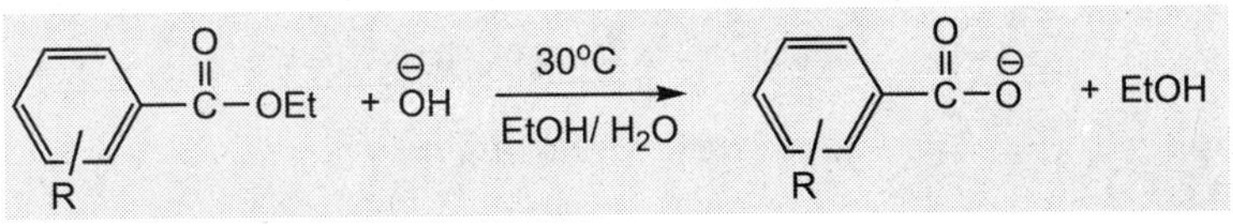

A plot of log (K/K_o) for a given reaction *versus* log (K/K_o) or σ_R with many differently substituted reactants gives a straight line. The graph of log (K/K_o) versus σ_R is linear, and the gradient is +2.498 (**Fig. 5.7**). The reaction constant or sensitivity constant ρ is equivalent to the shape of the Hammett plot, and information on the reaction and the associated mechanism can be obtained based on the values of ρ.

Now, the mathematical expression can be written as:

$$\log \frac{K}{K_o} = m \log \frac{K}{K_o}$$

$$\left[\left(\frac{-\Delta G}{2.303\ RT}\right) - \left(\frac{-\Delta G_o}{2.303\ RT}\right)\right] = m\left[\left(\frac{-\Delta G^{\#}}{2.303\ RT}\right) - \left(\frac{-\Delta G_o^{\#}}{2.303\ RT}\right)\right]$$

$$\text{or } (-\Delta G + \Delta G_o) = m(\Delta G^{\#} + \Delta G_o^{\#})$$

$$\Delta\Delta G = m\ \Delta\Delta G^{\#}$$

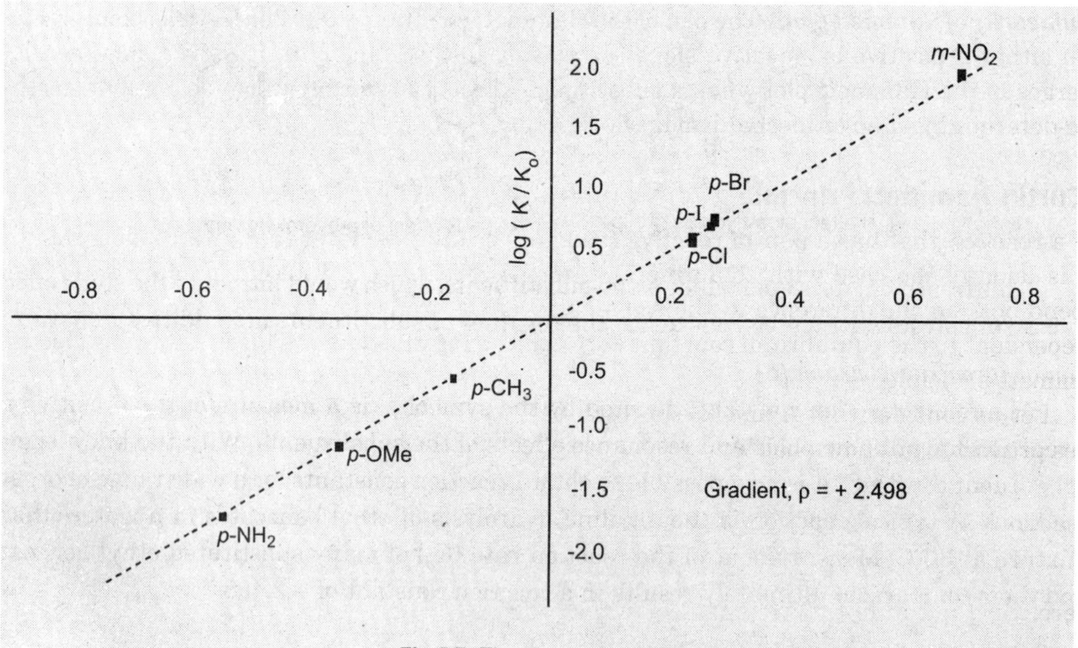

Fig 5.7: The graph of log (K/K_o) versus σ_R

This equation is called *linear free energy relationship*.

If the value of:

(i) $\rho > 1$, the reaction is more sensitive to substituents than benzoic acid, and negative charge is built during the reaction (or positive charge is lost).

(ii) $0 < \rho < 1$, the reaction is less sensitive to substituents than benzoic acid. and negative charge is built during the reaction (or positive charge is lost).

(iii) $\rho = 0$, no sensitivity to substituents, and no charge is built or lost.

(iv) $\rho < 0$, the reaction builds positive charge (or loses negative charge).

Some classical examples with ρ values are given below:

(a) Nucleophile attack on a carbonyl group as the rate-determining step is going to have a ρ value of +2 to +3, because in these reactions there is increase in negative charge in the transition state.

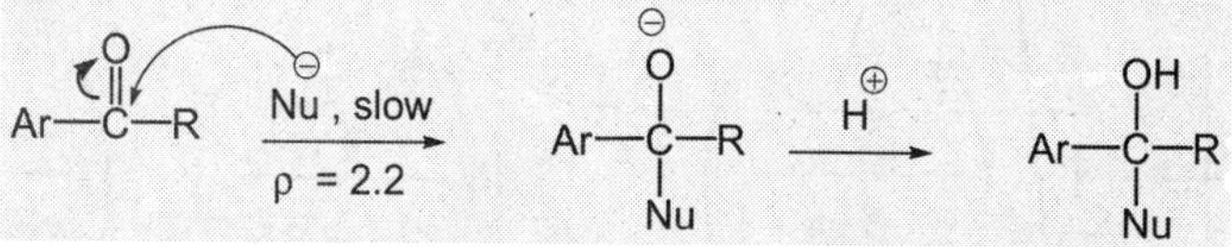

(b) An S_N1 reaction on the benzylic carbon has a large negative ρ value. The large negative ρ value is due to the decrease in electron-density at the site of the chemical reaction.

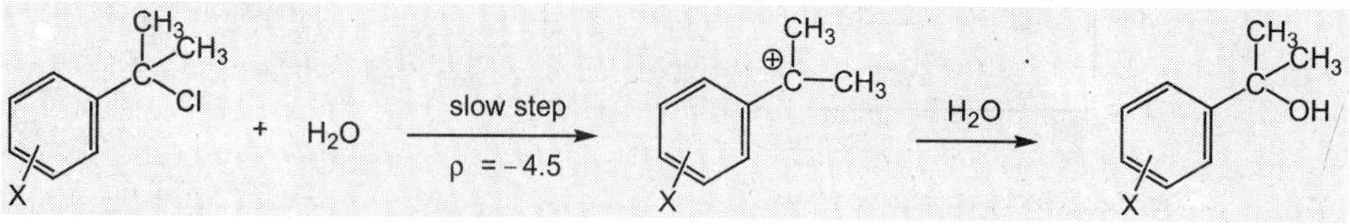

Nonlinearity of Hammett plots: The plot of the Hammett equation is typically seen as being linear, with either a positive or negative slope correlating to the value of ρ. However, nonlinearity emerges in the Hammett plot when a substituent affects the rate of a reaction, or changes the rate-determining step or mechanism of the reaction.

F. Curtin-Hammett Principle

For a reaction that has a pair of reactive reactants or intermediates that interconvert rapidly (as is usually the case with conformers), giving *irreversibly* a different product ratio, will depend only on the difference in the free energy of the transition state for each product and independent to the equilibrium constant between the intermediates. This is known as Curtin-Hammett principle.

Let us consider that A and B equilibrate rapidly, while A leads to the product C, and B gives product D as shown below:

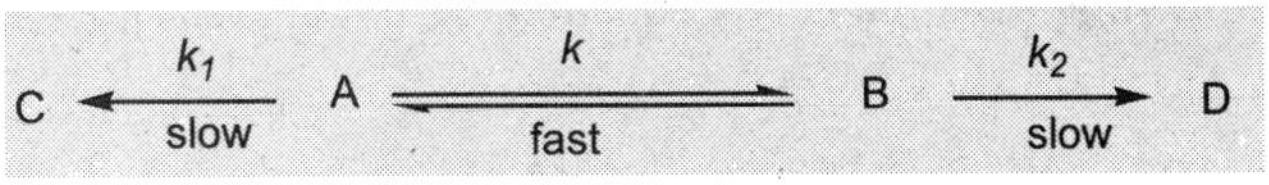

where, k is the equilibrium constant between A and B

k_1 = Rate constant for the formation of C

k_2 = Rate constant for the formation of D

Reaction coordinate free energy can be shown diagrammatically in **Fig. 5.8**.

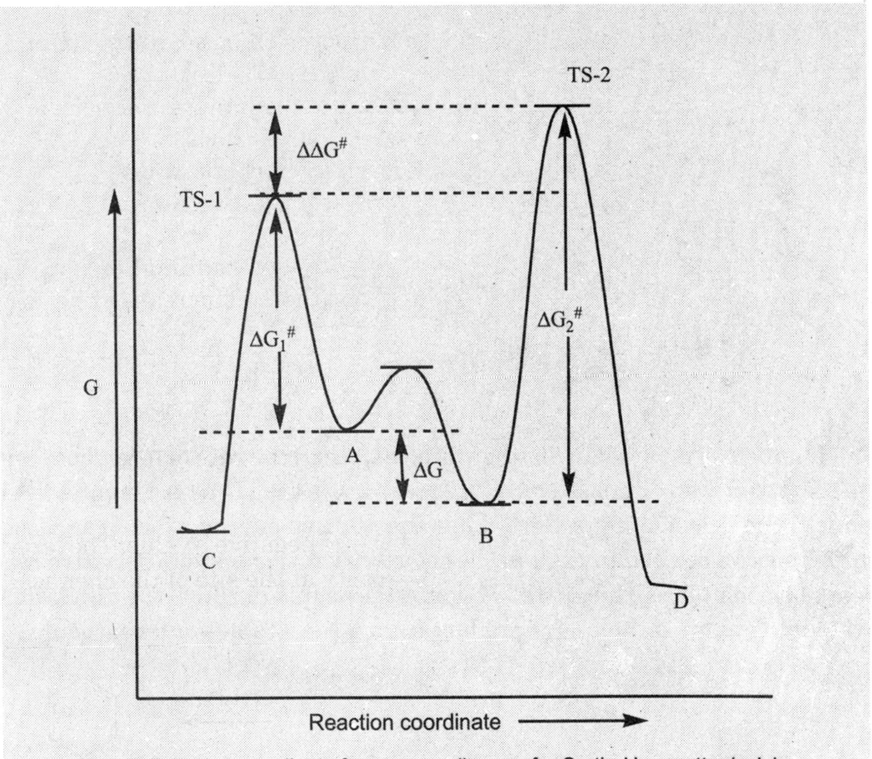

Fig. 5.8: Reaction coordinate free energy diagram for Curtin-Hammett principle

Mathematically, Curtin-Hammett Principle can be represented as,

$$\frac{[D]}{[C]} = e^{-\frac{\Delta\Delta G^{\#}}{RT}}$$

Rate of formation for compound C from A is given as,

$$\frac{d[C]}{dt} = k_1[A] \qquad (1)$$

Similarly, $$\frac{d[D]}{dt} = k_2[B] \qquad (2)$$

Dividing equation (2) by equation (1),

$$\frac{d[D]}{d[C]} = \frac{k_2[B]}{k_1[A]} = \frac{k_2 k[A]}{k_1[A]} = \frac{k_2 k}{k_1} \qquad (3)$$

$$= \frac{e^{-\Delta G_2^{\#}/RT} \cdot e^{-\Delta G/RT}}{e^{-\Delta G_1^{\#}/RT}}$$

$$= e^{-(\Delta G_2^{\#} + \Delta G - \Delta G_1^{\#})/RT}$$

$$= e^{-\Delta\Delta G^{\#}/RT} \qquad \text{if } K_{eq} = 1,\ \Delta G = 0$$

So the product ratio can be written as,

$$\frac{d[D]}{d[C]} = e^{\Delta\Delta G^{\#}/RT} \qquad (4)$$

Curtin-Hammett principle can also be stated as, when the rate of interconversion between A and B is much faster than either k_1 or k_2 (*i.e.* $k >> k_1$ or k_2), then Curtin-Hammett Principle tells us that product ratio C:D will not reflect k, but the relative energy of the transition states.

This principle can be used to explain the selectivity ratio for stereoselective reactions. The following reaction exemplifies the Curtin-Hammett principle in which the alkylation of tropane with methyl iodide produces the major product from a less stable conformation:

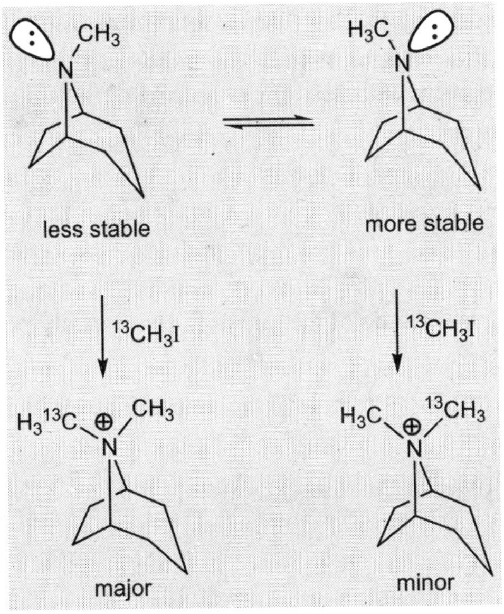

G. Taft Equation

As a modified Hammett equation, Taft developed an equation known as Taft equation. It is a linear free energy relationship (LFER) used in physical organic chemistry in the study of reaction mechanisms, and in the development of quantitative structure activity relationship for organic compounds. Hammett equation accounts for the influence of only field, inductive, and resonance effects on the reaction rates, whereas the Taft equation also concerns the steric effects of the substituent. The Taft equation is written as:

$$\log\left(\frac{k}{k_o}\right) = \rho^*\sigma^* + \delta\, Es$$

where $\log\left(\frac{k}{k_o}\right)$ = ratio of the rate of the substituted reaction compared to the reference reaction,

σ^* = polar substituent constant that describes the field and inductive effects of the substituent,

Es = steric substituent constant,

σ^* = sensitivity factor for the reaction to polar effects, and sensitivity factor for the reaction to steric effects.

To determine σ^*, Taft studied the hydrolysis of methyl esters (RCOOMe) under acid- as well as base-catalyzed conditions. Due to similar tetrahedral intermediates in both cases, it was proposed that under identical conditions, any steric factors should be nearly the same for the two mechanisms, and therefore, would not influence the ratio of the rates. However, because of the difference in charge built-up in the rate-determining steps, it was proposed that polar effects would only influence the reaction rate of the base-catalyzed reaction, since a new charge was formed. Taft defined the polar substituent constant σ^* as:

$$\sigma^* = \left(\frac{1}{2.48\rho^*}\right)\left[\log\left(\frac{k}{k_o}\right)_B - \log\left(\frac{k}{k_o}\right)_A\right]$$

where $\log\left(\frac{k}{k_o}\right)_B$ is the ratio of the rate of the base-catalyzed reaction compared to the reference reaction, $\log\left(\frac{k}{k_o}\right)_A$ is the ratio of a rate of acid-catalyzed reaction compared to the sensitivity of the reaction series.

Moreover,

$$Es = \frac{1}{\delta}\log\left(\frac{k s}{k_{CH_3}}\right)$$

where, *ks* is the rate of the studied reaction, and k_o is the rate of the reference reaction (R = methyl), and δ is the reaction constant that describes the susceptibility of a reaction series to steric effects. For the definite reaction series, δ was set to 1, and Es for the reference reaction was set to zero. This equation is combined with the equation for σ^* to give the complete Taft equation.

The polar sensitivity factor σ^* for Taft plots will describe the susceptibility of a reaction series to polar effects. When the steric effects of substituents do not significantly influence the reaction rate, Taft equation simplifies to a form of the Hammett equation.

$$\log\left(\frac{k}{k_o}\right) = \rho^* \sigma^*$$

- If $\rho^* > 1$, the reaction accumulates negative charge in the transition state, and is accelerated by electron-withdrawing groups.
- If $1 > \rho^* > 0$, negative charge is built up, and the reaction is mildly sensitive to polar effects.
- If $\rho^* = 0$, the reaction is not influenced by polar effects.
  If $0 > \rho^* > -1$, positive charge is built up, and the reaction is mildly sensitive to polar effects.
- If $-1 > \rho^*$, the reaction accumulates positive charge, and is accelerated by electron-donating groups.

5.2 METHODS OF DETERMINATION OF REACTION MECHANISMS

The mechanism of a reaction tells us the precise position of the atoms as a function of time in the reactant molecules as they are converted into products. Since many chemical reactions proceed very rapidly ($\sim 10^{-14}$s), it is not possible to follow the reaction path directly, and hence our knowledge of reaction mechanisms is mostly based on indirect evidence. After characterizing the product (s), the mechanism of a reaction is determined by considering all the possibilities that are consistent with the experimental observations. In most of the cases, new experiments are to be performed to exclude a good number of the proposed mechanisms. If such eliminations lead to only one mechanism, then it is said to be the *established mechanism* for that reaction.

Study of reaction mechanisms is of paramount importance in organic chemistry as thousands of, apparently diverse, reactions proceed by a limited number of common steps. Methyl bromide, for instance, reacts with hydroxide ion to produce methyl alcohol. Mechanism of this reaction involves a *transition state* in which there is simultaneous breaking of C–Br, and formation of C–OH bonds. Reactions of this type are commonly referred to as S_N2 (substitution nucleophilic bimolecular) reactions. Subsequently, it was shown that this type of mechanism operates not only in the reaction between methyl bromide and hydroxide ion, but in the reactions of all primary (and many secondary) alkyl halides with other nucleophiles such as HS and CN . Hence, an understanding of the S_N2 mechanism enables us to appreciate most of the chemical reactions of alkyl halides.

The proposed mechanism of a reaction should (a) be energetically reasonable, (b) account for the experimental observations, and (c) be in agreement with what is known about analogous reactions. There are a number of commonly used methods for determining the mechanism of a reaction. These include product analysis, determination of the presence of intermediates, isotopic labelling, isotope effects, stereochemical studies, and finally kinetic evidence. It should be noted that the above order does not necessarily reflect the order of experimentation. Further, it may not be necessary to apply all the methods of determine the mechanism of a particular reaction.

A. Product Analysis

Identification of products, byproducts if any, and determining their relative quantities constitute the first step in the investigation of a mechanism. Hydrolysis of isomeric allylic chlorides, for instance, yields a mixture of 85% of a tertiary alcohol, and 15% of a primary alcohol. Since product composition remains the same in either case, a common intermediate is suggested.

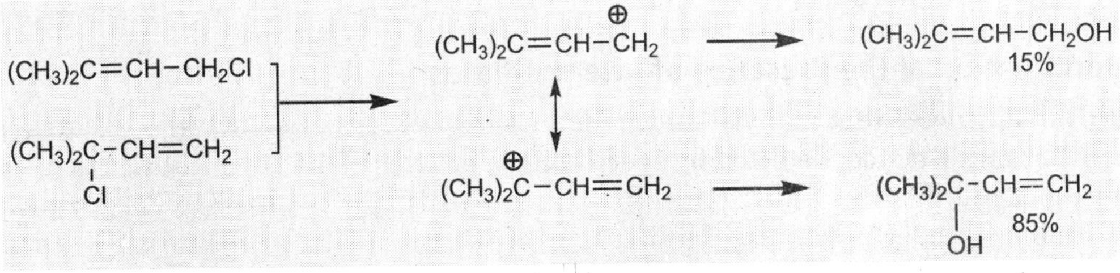

Product identification has helped in establishing the correct mechanism for the Beckmann rearrangement. Only one of the two geometrically isomeric oximes can furnish the substituted amide on treatment with phosphorus pentachloride. Earlier, chemists believed that the oxime which gives this amide has the migrating group (R′) nearer the leaving group (-OH).

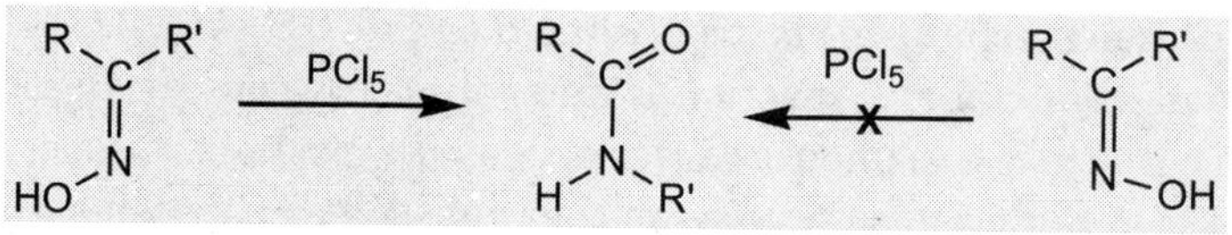

However, it was subsequently discovered that one of the oximes of 2-bromo-5-nitroacetophenone readily undergoes ring closure to benzisoxazole, thus establishing its structure as **12.** The other isomer (**14**) does not undergo ring closure, but readily gives the Beckmann rearrangement product, the amide **15.** The identification of **15** thus leads to the conclusion that the rearrangement involves a *trans* migration of the substituted phenyl group.

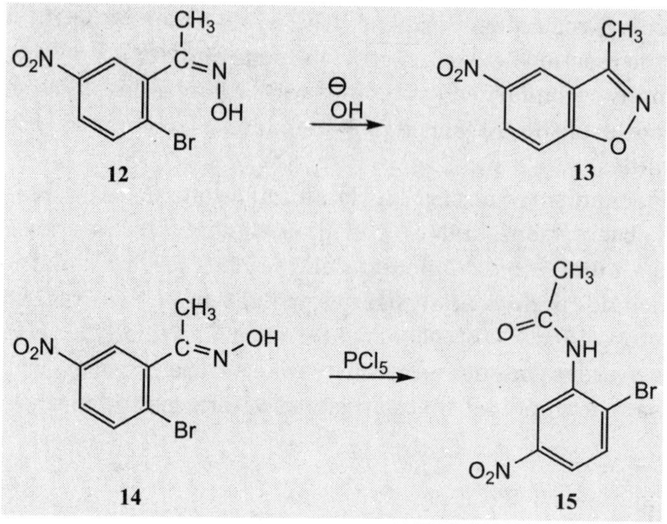

Another interesting example is the acid-catalyzed rearrangement of ethyl β-methyl-β-phenylglycidate (**16**). The product of the rearrangement was formulated as pyruvate (**17**), which could arise by the intermediacy of a carbocation (**18**), followed by hydride shift. A reinvestigation of the reaction, however, established the structure of the product as **19** which involves the migration of carbethoxy group.

B. Determination of the Presence of Intermediates

We have already noted that many organic reactions involve formation of intermediates, which can be sometimes isolated, detected by spectroscopy, or trapped by chemical reaction with an added compound.

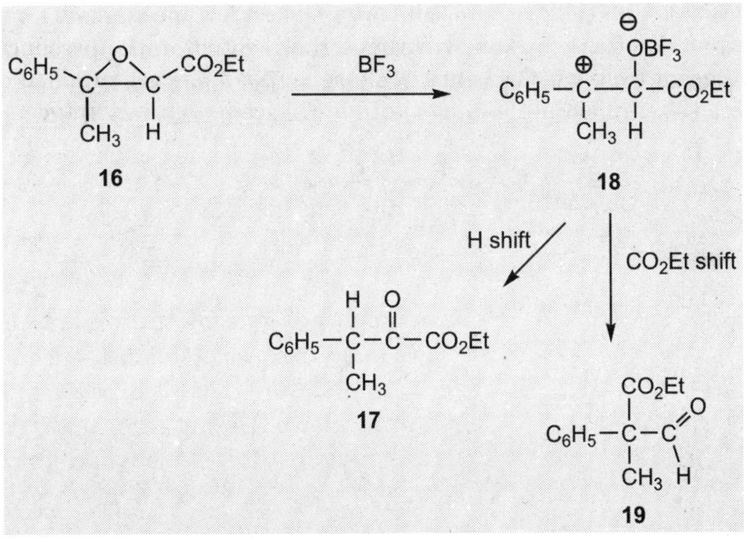

Isolation

Occasionally, it is possible to isolate stable intermediates by stopping the reaction before it is completed, or by using mild conditions. Three intermediates, for example, an N-bromoamide (RCONHBr), its anion ($RCON^{-}Br$), and an isocyanate (RNCO) have been isolated in the Hofmann rearrangement of an amide to an amine.

$$RCONH_2 \xrightarrow{NaOBr} RNH_2$$

Thus, any proposed mechanism for this rearrangement must account for the formation of all these intermediates.

Detection

Unstable intermediates are often detected by spectroscopic methods. The intermediacy of carbocation in S_N1 and $E1$ reactions has been established by NMR spectroscopy. Similarly, the intermediates of electrophilic substitution reactions, such as **20** have been characterized by their NMR and electronic spectra.

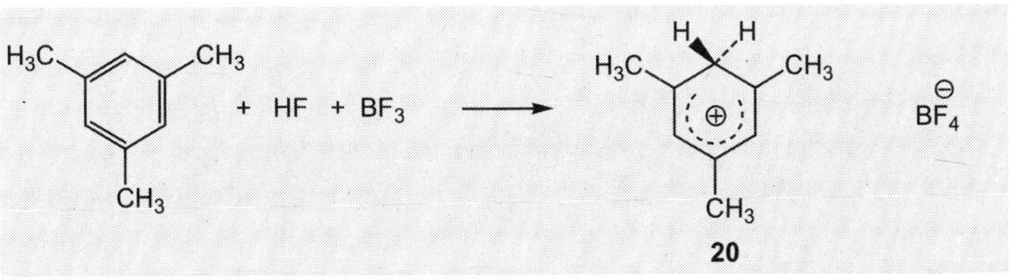

Radical intermediates are detected by the use of electron spin resonance (ESR) spectroscopy.

Trapping

An intermediate is often trapped by running the reaction in the presence of a compound with which it is known to react. Carbenes (**21**) are trapped as cyclopropane derivatives (**22**), whereas a benzyne (**23**) immediately reacts with anthracene to give triptycene (**24**).

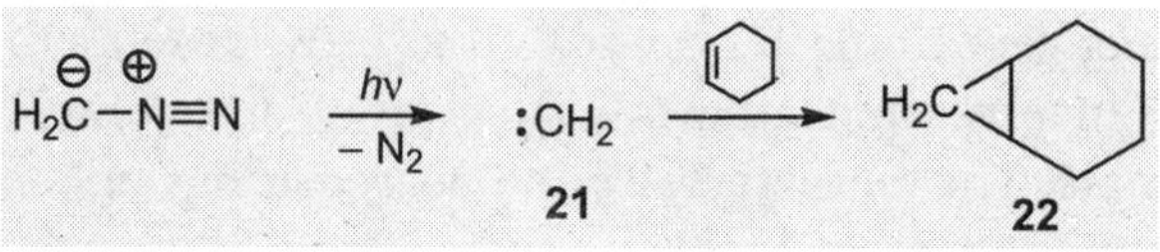

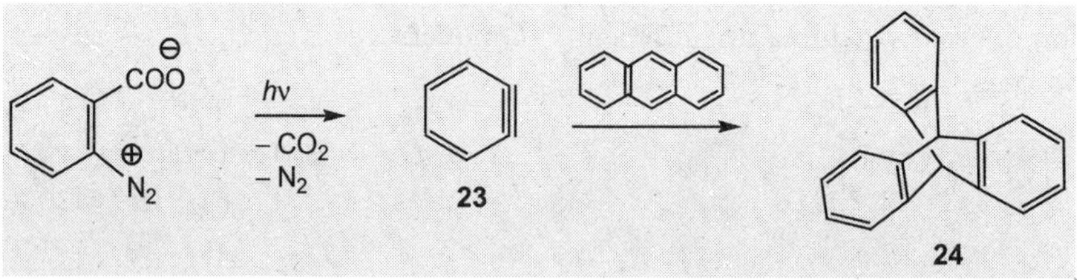

Nucleophiles are commonly employed to trap carbocations. Direct evidence for the intermediacy of a carbocation in the addition of a halogen to alkenes is provided by the observation that when the reaction is carried out in the presence of other nucleophiles, a mixture of products is obtained.

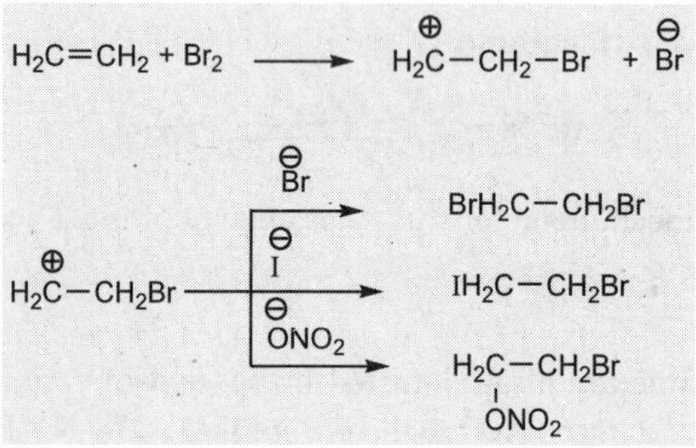

Free radicals are trapped by adding to the reaction system, bromine or nitric oxide, which readily reacts with them. Thus, a hydroxylamine derivative (**26**) is formed on heating azoisobutyronitrile (**25**) in the presence of nitric oxide.

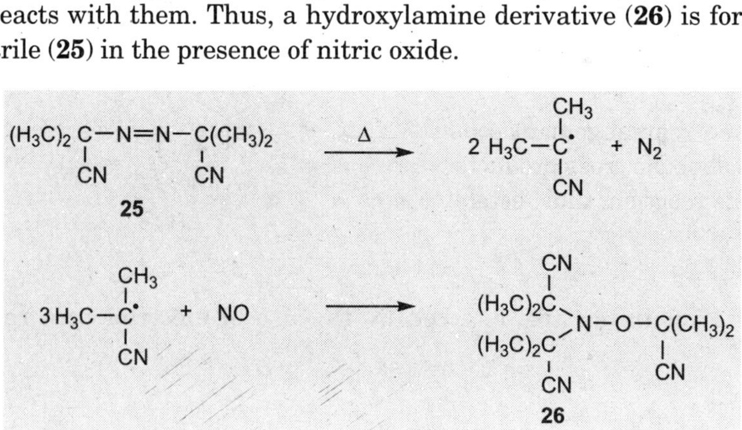

C. Crossover Experiments

Whether a rearrangement is a one-step or a two-step process, can be decided by carrying out the reaction with a mixture of two similarly constituted but non-identical reactants, and then analyzing the products. Since the migrating group must become free in a two-step process, we should get a product containing fragments of both the reactants in such a process. Fries rearrangement, which involves migration of an acyl group of a phenyl ester to the benzene ring, provides such an example. The two esters, **27** and **28**, rearrange under identical conditions and at approximately the same rate to give the products **29** and **30**, respectively. Rearrangement of a mixture of **27** and **28**, however, not only gives **29** and **30** but also the *cross products* **31** and **32**. Isolation of cross products unequivocally demonstrates that the Fries rearrangement is a two-step process.

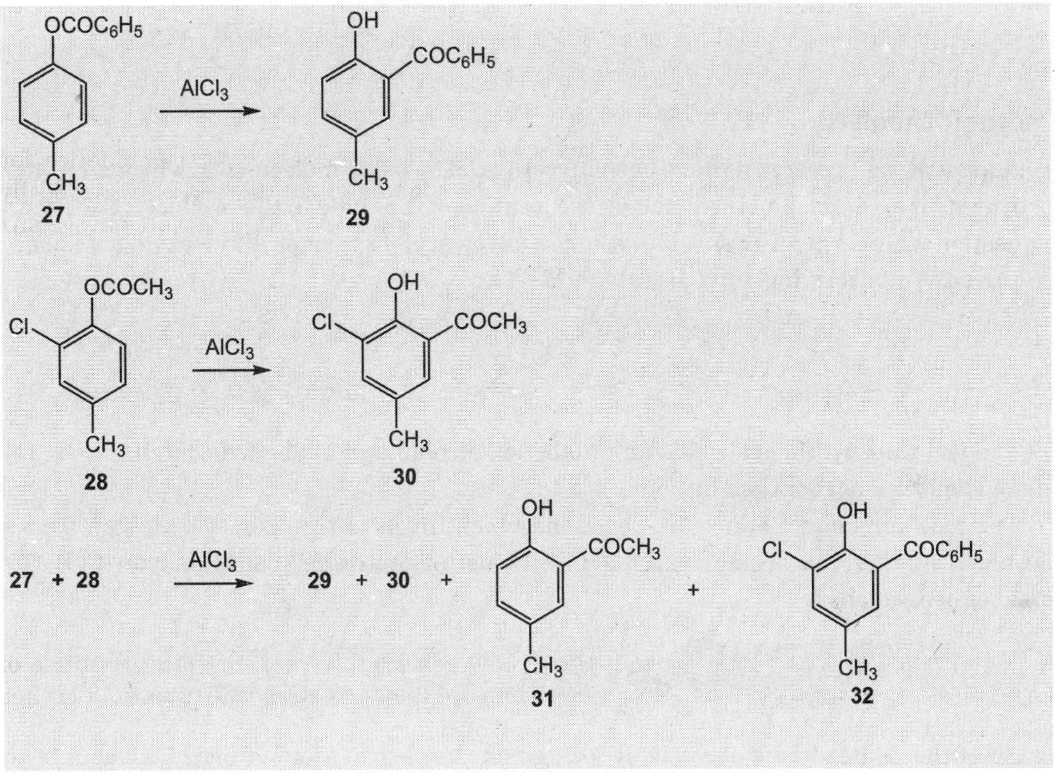

The Claisen rearrangement consists of the conversion of an allyl phenyl ether into an allyl phenyl alcohol. No cross products were obtained when a similar experiment as above was designed for this reaction, thus, establishing its intramolecular nature.

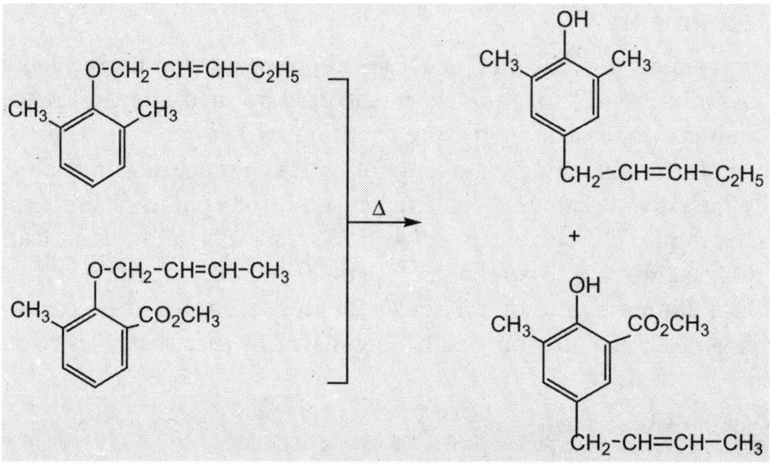

D. Isotopic Labelling

Mechanisms of many reactions have been investigated by using molecules in which an isotope is substituted for an atom, and then tracing the location of the same in the products. For example, the question whether hydrolysis of an ester involves acyl–oxygen or alkyl–oxygen fission, has been solved by hydrolyzing amyl acetate in $H_2{}^{18}O$.

$$H_3C-\overset{\overset{\displaystyle O}{\|}}{C}-O-C_5H_{11} + H_2{}^{18}O \xrightarrow{\ominus OH} CH_3-\overset{\overset{\displaystyle O}{\|}}{C}-{}^{18}\overset{\ominus}{O} + C_5H_{11}OH$$

The fact that hydrolysis yields an unlabelled alcohol and a labelled acid, has been taken as an evidence for acyl-oxygen fission.

The mechanism of Cannizzaro reaction, which involves transfer of hydrogen from one molecule to another, has been investigated by the use of deuterated benzaldehyde (C_6H_5CDO), instead of benzaldehyde.

$$2\,C_6H_5CDO + \overset{\ominus}{O}H \xrightarrow{H_2O} C_6H_5-\overset{\overset{\displaystyle O}{\|}}{C}-\overset{\ominus}{O} + C_6H_5CD_2OH$$

Since the deuterium is exclusively located on the carbon atom of benzyl alcohol bearing the hydroxyl group, the possibility of exchange of hydrogen with hydrogen atoms in the solvent is ruled out. Thus, there is a transfer of hydrogen (or deuterium) between the carbonyl carbons of two aldehyde molecules, and in the course of the reaction, hydrogen (or deuterium) never gets attached to an oxygen atom. A plausible mechanism for Cannizzaro reaction can thus be *visualized* as follows:

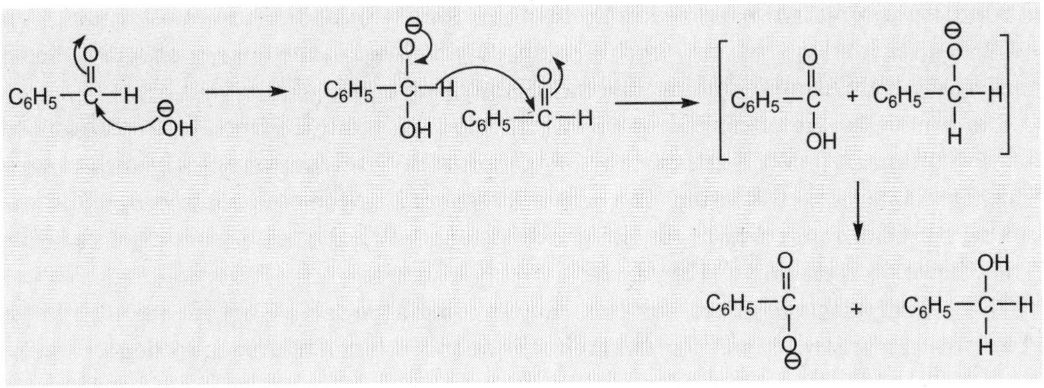

Reaction of chlorobenzene with potassium amide in liquid ammonia to yield aniline has been investigated by the use of labelled carbon atom. Direct substitution of chlorine with an amide ion appears to be an attractive possibility, but experiments performed with 1-^{14}C-chlorobenzene provided a sample of aniline, about half of which was labelled in the 1-position, and remaining half in the 2-position. A two-step mechanism involving the formation of an intermediate benzyne is consistent with the experimental observations.

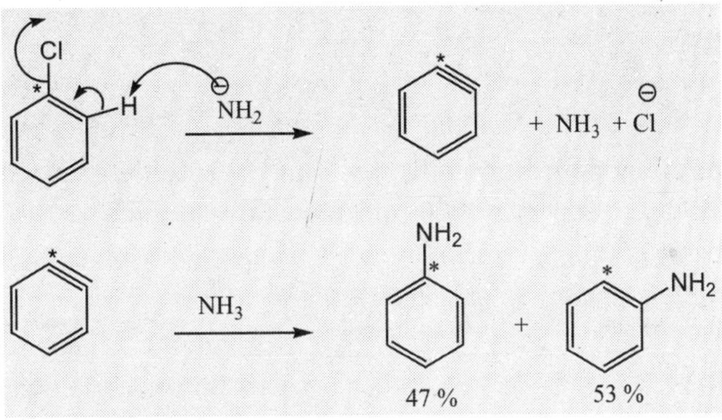

Isotope Effects

In the above discussion of isotopic labelling, we have made an assumption that there is no difference in the *chemical behaviour* of different isotopes, and hence the chemical properties of the compounds are not qualitatively changed when an isotope is substituted for an atom. In spite of their chemical similarity, the isotopes are not exactly identical, and often, the rate at which chemical reactions occur may vary with the isotope. This variation in the rate of a reaction due to a difference in the isotope present in the molecule is called the *isotope effect,* and it is usually small.

However, in a reaction in which the cleavage of a bond binding the particular atom (or its isotope) is the rate-determining step, the isotope effect can be of relatively large magnitude, and is called the *primary isotope effect*. This effect is greatest on replacing a hydrogen atom by a deuterium or tritium atom, and is expressed in terms of the rate ratio k_H / k_D or k_H / k_T, respectively. The value for this ratio is always greater than one, and this indicates that the compound containing hydrogen reacts faster than that with deuterium and tritium. Typical values for k_H / k_D and k_H / k_T are 7 and 16, respectively. Most of the isotope effect studies have been carried out by using the more easily available deuterium.

The reason for the primary isotope effect, when C–D and C–H bonds are broken in the rate-determining step of a reaction, may be traced to the greater *bond dissociation energy* of the former compared to the latter. The activation energy for the deuterium compound would, therefore, be greater in reactions involving the cleavage of C–D bond, and as expected the rate will slow down on replacing C–H by C–D.

The primary isotope effect has been found to be a powerful tool in the elucidation of mechanisms of many reactions. For example, the substitution of hydrogen by deuterium in the methyl groups of isopropyl bromide considerably slows down the rate of its reaction with the ethoxide ion.

	Relative rate
$(CH_3)_2CHBr + C_2H_5\overset{\ominus}{O} \longrightarrow H_2C{=}CH{-}CH_3 + C_2H_5OH + \overset{\ominus}{Br}$	1.0
$(CD_3)_2CHBr + C_2H_5\overset{\ominus}{O} \longrightarrow D_2C{=}CH{-}CD_3 + C_2H_5OD + \overset{\ominus}{Br}$	0.15

$$k_H/k_D = \frac{1}{0.15} = 6.66$$

The large isotope effect definitely indicates that the rate-determining step of the reaction involves breaking of a C–H bond, and rules out mechanisms such as the one given.

$$(CH_3)_2CHBr \xrightarrow{\text{slow}} (CH_3)_2\overset{\oplus}{C}H \xrightarrow[\text{fast}]{C_2H_5\overset{\ominus}{O}} H_2C{=}CH{-}CH_3 + C_2H_5OH$$

The correct mechanism for this reaction has been formulated as follows:

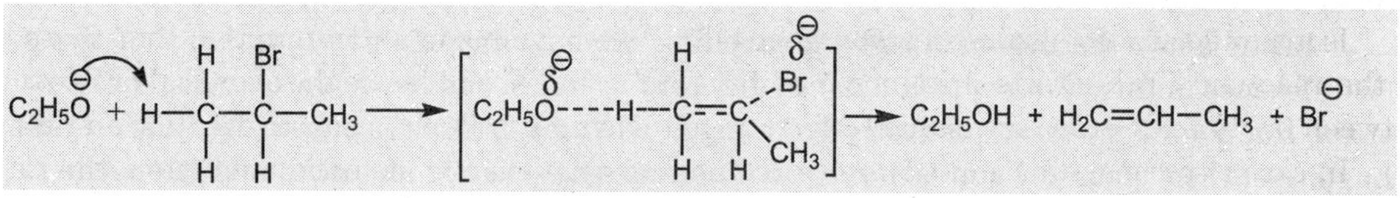

Nitration of benzene, on the other hand, does not show any isotope effect. In the nitration of tritiated sample of benzene, tritium atoms are replaced at the same rate as hydrogen atoms. Thus, the rate-determining step of the reaction does not involve breaking of the C–H (or C–T) bond. A mechanism may be visualized in which initial attack by the nitronium ion is the *slow step,* followed by the loss of a proton from the intermediate in a *fast step*.

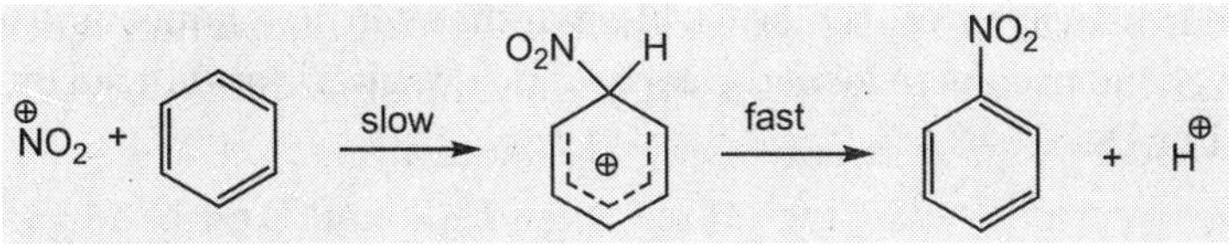

The kinetic isotope effect also tells us that in some cases C–H (or C–D or C–T) bond is broken during the rate-determining step. Vibrational frequency of a bond (C–H) is given by

$$\nu = 1/2\pi c\,(K(m_1 + m_2)/m_1 m_2)^{1/2}$$

where, K denotes constant force for the bond C–H (or C–D or C–T)

m_1 = mass of carbon atom

m_2 = mass of hydrogen atom

ν = vibrational frequency of the bond.

The bond energy, therefore, depends upon the frequency of the vibration, which depends on the reduced mass of the two atoms forming the bond. Due to greater mass of the deuterium, the vibrational energy of C–H bond is greater than that of a C–D bond. Hence, less energy is required for cleavage of C–H bond than that of C–D bond. For example, base-catalyzed bromination of acetone in aqueous medium is a second-order reaction, first-order with respect to acetone and first-order with respect to –OH, but independent of the concentration of bromine. A primary kinetic isotope effect of about 7 has been observed in the bromination. It shows that the rate-determining step involves removal of hydrogen by the catalyst base leading to the formation of the conjugate base of acetone.

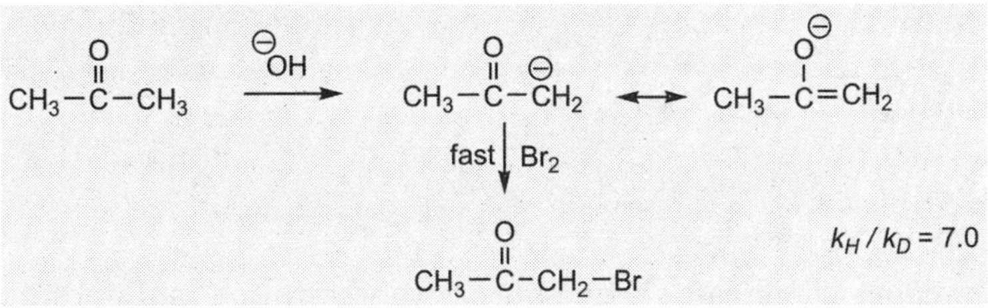

Secondary Isotope Effect

Isotope effects may also be observed when substituted hydrogen atom is not directly involved in the reaction. Such effects are known as *secondary isotope effects,* and their value ranges between 0.7 to 1.5. Secondary isotope effects may be normal ($k_H/k_D > 1$) or inverse ($k_H/k_D < 1$).

In case of nucleophilic addition reaction of carbonyl compounds, secondary isotope effect is observed. This is due to the conversion of sp^2 hybridized carbon atom into an sp^3 carbon atom.

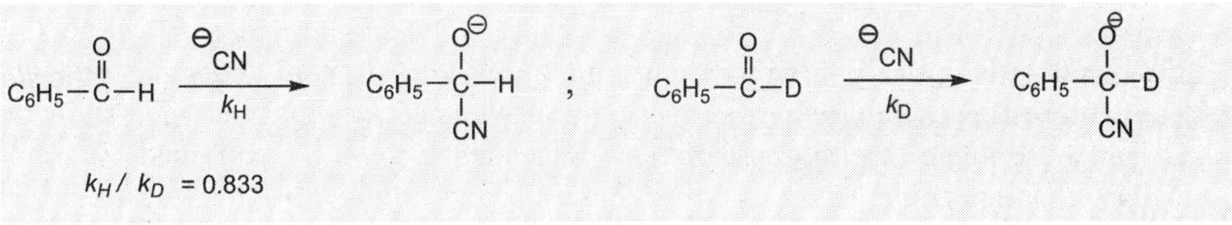

For example, in solvolysis of 1-chloro-1-phenylethane in 50% aqueous ethanol at 25°C, k_H / k_D = 1.15, the C–Cl bond breaks to form the carbocation, which is stabilized by hyperconjugation effect shown by H or D.

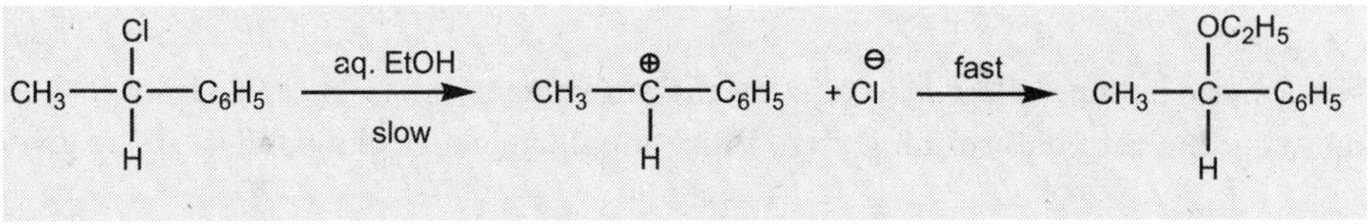

E. Stereochemical Evidence

Sometimes, the mechanistic course of a reaction can be delineated by following its *stereochemistry*. Mechanisms of nucleophilic substitution reactions, for instance, have been investigated by performing experiments with optically active compounds, and then checking optical activity of the products. Racemization of the product indicates participation by a planar intermediate, such as a carbocation, which is equally likely to be attacked by the nucleophile on either side, producing a mixture of equal number of molecules with *retention* and *inversion* of configuration. Thus, stereochemical (and kinetic) evidence suggests following mechanism for S_N1 reactions of alkyl halides:

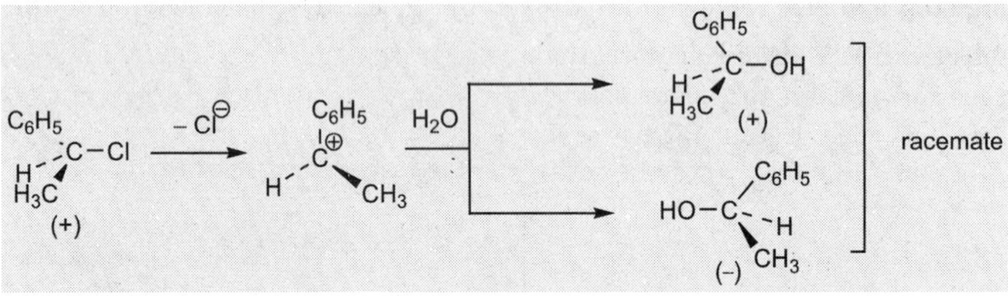

When the product is optically active but its configuration is different from that of the reactant, the process is known as *inversion of configuration*. For example, in the basic hydrolysis (S_N2 reaction) of 2-bromooctane, the nucleophile (^-OH) attacks the side of the carbon atom opposite to that of the leaving group (Br^-), and the reaction proceeds with stereochemical inversion, which is usually referred to as *Walden inversion*.

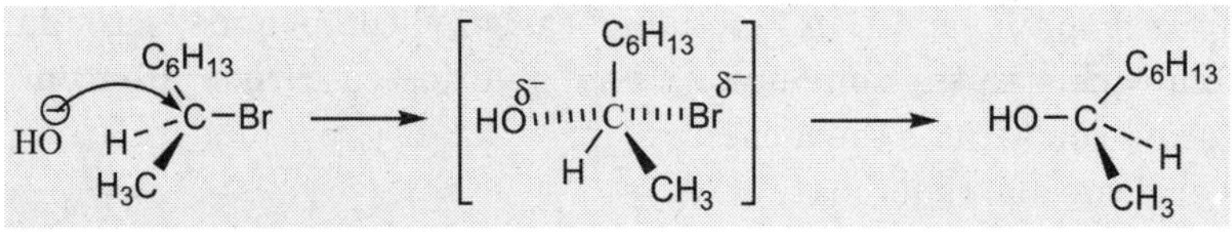

Racemization of optically active ketones, having an α-hydrogen, under basic conditions provides an interesting example where stereochemical evidence is complemented by the evidence gained from the isotopic labelling experiment. Racemization can be explained by assuming intermediacy of a carbanion, which is either planar, or pyramidal, and inverts rapidly. When the reaction is performed in presence of D_2O, the rate of racemization is found to be equal to the rate of deuteration. The following mechanism is in conformity with the experimental observations:

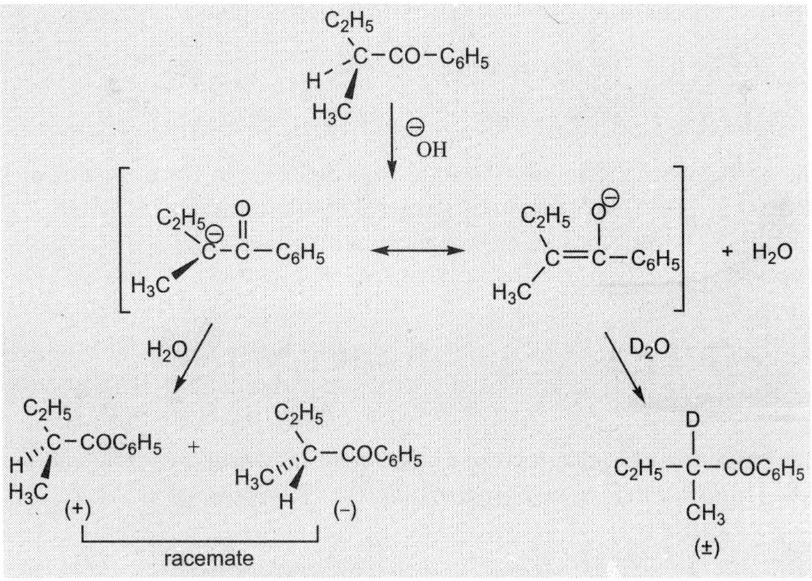

The observation that *cis*-dichloroethylene undergoes transformation to chloroacetylene twenty times faster than the corresponding *trans* isomer confirms the conclusions that *E2* reactions proceed with greater facility when the departing groups are *trans* than when they are *cis*.

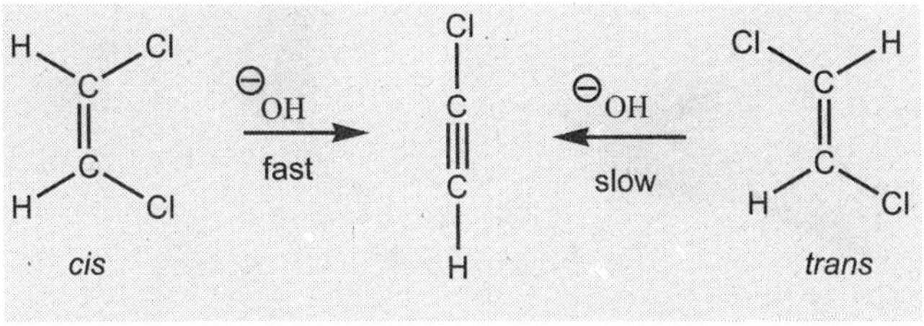

F. Kinetic Evidence

Although a reaction mechanism cannot be deduced unequivocally from kinetic studies, knowledge of the reaction order constitutes the most important step of the investigation. It is often possible to rule out several mechanisms merely by an inspection of the rate law. Perhaps, the most celebrated example of the use of kinetics in the investigation of mechanism is the hydrolysis of alkyl halides. The observation that the rate expression for reaction of methyl bromide with hydroxide ion is

$$\frac{-d\,[CH_3Br]}{dt} = k\,[CH_3Br]\;[\overset{\ominus}{O}H]$$

while that for the corresponding reaction of *tert*-butyl bromide is

$$\frac{-d\,[(CH_3)_3CBr]}{dt} = k\,[(CH_3)_3CBr]$$

in conjunction with stereochemical evidence, has led to the formulation of two different mechanistic paths (S_N1 and S_N2) for the hydrolysis of alkyl halides.

Another example is the Cannizzaro reaction, which is a *third-order reaction* at moderate base concentrations.

$$\frac{-d\,[HCHO]}{dt} = k_1\,[HCHO]^2\;[\overset{\ominus}{O}H]$$

However, at higher base concentrations it becomes an overall *fourth-order reaction*, second-order in aldehyde, and second-order in hydroxide.

$$\frac{-d\,[HCHO]}{dt} = k_2\,[HCHO]^2\;[\overset{\ominus}{O}H]^2$$

The observed kinetic data (and isotopic labelling experiments) helped in the formulation of the following two mechanisms:

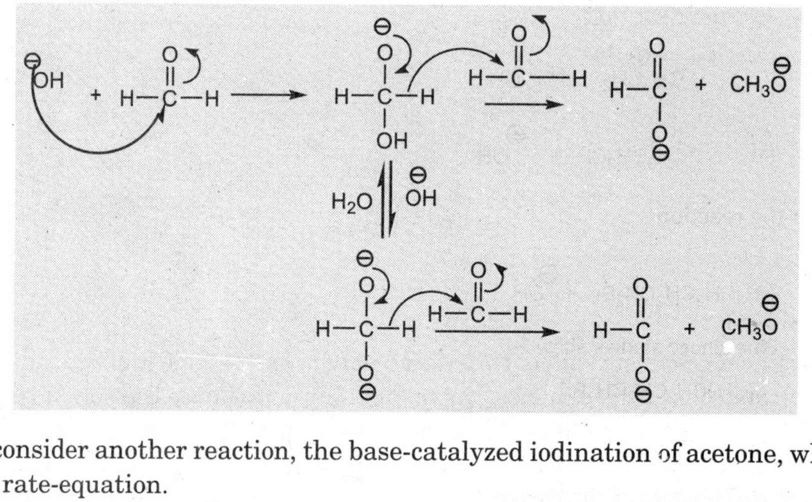

Let us consider another reaction, the base-catalyzed iodination of acetone, which is found to follow the rate-equation.

$$CH_3—CO—CH_3 \;+\; I_2 \xrightarrow{\overset{\ominus}{O}H} CH_3—CO—CH_2—I \;+\; H—I$$

$$\frac{-d\,[CH_3COCH_3]}{dt} = k\,[CH_3COCH_3]\,[\overset{\ominus}{O}H]$$

The observed kinetics suggests that the rate-determining step of the reaction is the removal of a proton by the base followed by a faster step involving the participation of iodine.

PROBLEMS AND REVIEW QUESTIONS

1. The pk_a value of *p*-chlorobenzoic acid is 3.98, while that of benzoic acid is 4.19. Calculate σ for *p*-chlorobenzoic acid.
2. The σ value for alkaline hydrolysis of methyl esters of substituted benzoic acids is 2.38, and the rate constant for alkaline hydrolysis of methyl benzoate under similar conditions is $2 \times 10^{-4}\,m^{-1}s^{-1}$. Calculate the rate constant for the hydrolysis of methyl *m*-nitrobenzoate ($\sigma_m = 0.71$)
3. Base-promoted hydrolysis of methyl mesitoate occurs through an attack on the alcohol carbon instead of the acyl carbon. Suggest an experiment that would confirm the mode of attack.
4. 1,3-Butadiene reacts with bromine to give 3,4-dibromo-1-butene as the major product at lower temperature, and 1,4-dibromo-2-butene at higher temperature. Explain.
5. On the basis of the given data, calculate how much faster *p*-bromobenzyl chloride will hydrolyze than *p*-nitrobenzyl chloride.
   σ *p*-Br = 0.26 , σ *p*-NO_2 = 0.81, $\rho = -1.31$
6. For which of the following mechanistic schemes is the rate law applicable:

$$\frac{d[CH_3CH_2Cl]}{dt} = k\,[CH_3CH_2Cl]\,[\overset{\ominus}{O}H]$$

(i) $CH_3CH_2Cl \xrightarrow{\text{slow}} CH_3\overset{\oplus}{C}H_2 + \overset{\ominus}{Cl}$

$CH_3\overset{\oplus}{C}H_2 + \overset{\ominus}{O}H \xrightarrow{\text{fast}} CH_3CH_2OH$

(ii) $CH_3CH_2Cl + \overset{\ominus}{O}H \xrightarrow{\text{slow}} CH_3CH_2OH + \overset{\ominus}{Cl}$

7. For the reaction

$$CH_3CH_2CH_2CH_2Br + \overset{\ominus}{O}H \longrightarrow CH_3CH_2CH_2CH_2OH + \overset{\ominus}{Br}$$

The kinetic studies show

$$\frac{d[CH_3CH_2CH_2CH_2Br]}{dt} = k[CH_3CH_2CH_2CH_2Br]\,[\overset{\ominus}{O}H]$$

Predict the effect of changes on the rate and rate constant *k* of the following: reactions:

(i) Concentration of substrate is doubled.
(ii) Concentration of hydroxide ions is halved.
(iii) The temperature is raised.

8. Explain the experimental observation that one of the compounds (**A** or **B**) forms intermolecular hydrogen bonds, whereas the other forms only intramolecular hydrogen bonds.

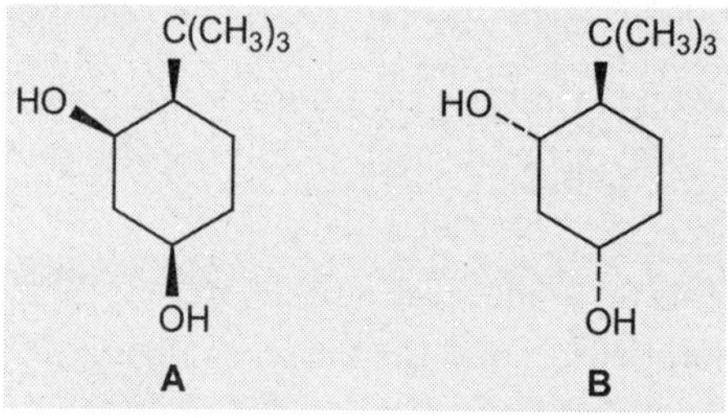

9. What are the requirements for a reaction to proceed in the forward direction? Discuss the role of kinetics in gaining an insight into the mechanism of a reaction.
10. A chemical reaction proceeds when the rate constant (k) is favourable. Discuss the factors that make one rate constant larger than another.
11. Define an intermediate. Draw energy profile diagrams for the two-step reactions with (a) Stable, intermediate and, (b) Unstable intermediate.
12. With the help of energy profile diagrams, demonstrate the role of catalyst in increasing the rate of a reaction.
13. With chosen examples, illustrate the terms– kinetic and thermodynamic control of a reaction.
14. (i) Differentiate between intermediate and transition state.
    (ii) State Hammond's postulate in determining the shape and geometry of transition state.
15. How does product analysis help in establishing the mechanism of a reaction? Illustrate your answer with suitable examples.
16. Discuss various methods for establishing the presence of an intermediate in a reaction.
17. What is isotope effect? How does isotopic labelling help in establishing the mechanism of a reaction?
18. Treatment of an optically active ketone with a base results in racemization. How do you explain this observation?
19. Discuss the significance of kinetic evidence in establishing the mechanism of a reaction?
20. Explain the following observations:
    (i) Naphthalene on treatment with conc. H_2SO_4 gives 1-naphthalene sulphonic acid and 2-naphthalenesulphonic acid as major products at 80°C and 160°C, respectively.
    (ii) A hydroxylamine derivative, I is formed when azoisobutyrometrile (AIBN) is heated in the presence of nitrous oxide.

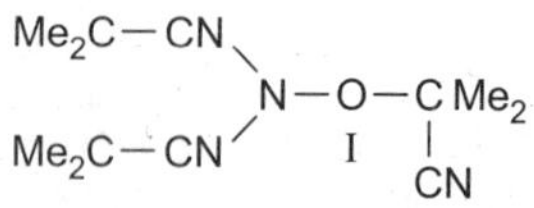

21. Define and illustrate the term 'isotope effect'. In what way is the knowledge of isotope effect useful in ascertaining the mechanism of a reaction?

22. Draw energy profile diagrams for the following:
    (i) A one-step reaction
    (ii) A two-step reaction in which the first step is rate-determining
    (iii) A three-step reaction in which the second step is rate-determining
    (iv) A catalyzed and an uncatalyzed reaction
23. Illustrate the use of the following techniques in the determination of the mechanism of reactions:
    (i) Crossover experiments (ii) Product analysis (iii) Isotope effect

SOLUTIONS

1. $\sigma_{x} = pKa\ (C_6H_5COOH) - pKa$ (*p*-chlorobenzoic acid).
2. According to Hammett's equation,

$$\log \frac{Ka\,(m\text{-nitro})}{Ka(C_6H_5COOH)} = \sigma m\text{-}NO_2 \times \rho$$

$$\frac{Ka\,(m\text{-}NO_2)}{Ka(C_6H_5COOH)} = 49$$

Structure of methyl mesitoate

3. 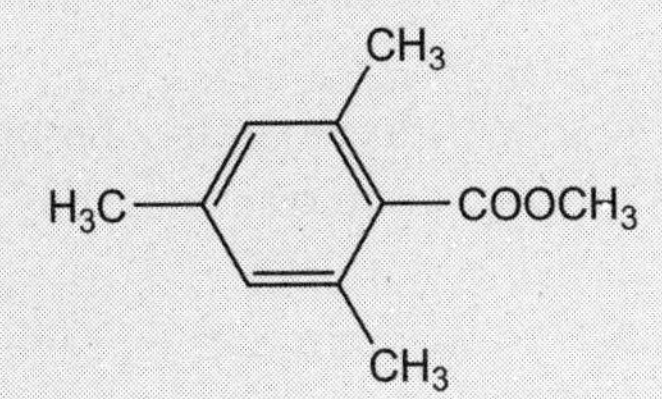

Labelled compound experiment can be used to predict the attack of $\overset{\ominus}{O}H$.

4. Hint: Thermodynamic and kinetic controlled reactions.
5. Hint: See problem 2.
6. For (ii), the rate law is applicable.
7. (i) The rate will become double,
   (ii) The rate will become half,
   (iii) The rate will increase.
8. 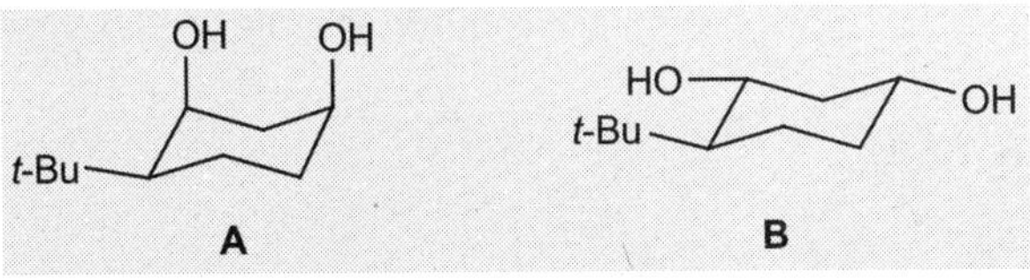

A forms intramolecular hydrogen bond while **B** will form intermolecular hydrogen bonds.

Chapter 6

Aliphatic Nucleophilic Substitution

Chapter Outline

6.1 DIRECT DISPLACEMENT PROCESS: S_N2 REACTIONS

6.2 CARBOCATION PROCESS: S_N1 REACTIONS

6.3 STEREOCHEMISTRY OF SUBSTITUTION REACTIONS

A. Steric Orientation in S_N2 Mechanism

B. Steric Orientation in S_N1 Mechanism

6.4 NEIGHBOURING GROUP PARTICIPATION

A. Lone Pair as Neighbouring Group

B. Double Bond as Neighbouring Group

C. Phenyl Group as Neighbouring Group: Phenonium Ions

D. Sigma Bond Participation

E. Classical vs Nonclassical Cations

6.5 S_Ni AND SET MECHANISMS

A. S_Ni Mechanism

B. SET Mechanism

6.6 FACTORS AFFECTING REACTIVITY IN SUBSTITUTION REACTIONS

A. Structure of Alkyl Group

B. Nature of Solvent

C. Nature of Leaving Group

D. Nature of Nucleophilic Reagent

6.7 AMBIDENT NUCLEOPHILES

A major advance in the study of reaction mechanisms was the investigation of the mechanism of nucleophilic substitution at a saturated carbon atoms. The reaction has great synthetic importance as it is utilized in a wide variety of interconversions involving many types of organic molecules. In a broad sense, the reaction may be depicted as follows:

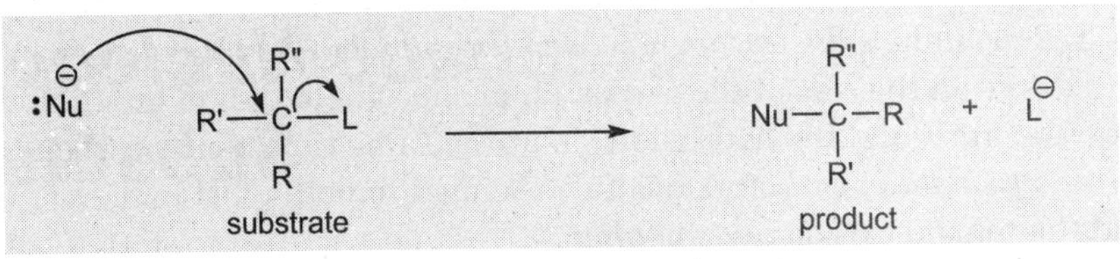

An inspection of the above equation reveals that a nucleophile (:Nu⁻) attacks on the carbon and displaces the leaving group (L). Also, in the process, the nucleophile furnishes both the electrons of the new covalent bond, and the leaving group departs with the electron pair of the bond undergoing cleavage.

Though the nucleophiles must contain an unshared electron pair, they may be neutral or negatively charged, whereas a substrate may be neutral or positively charged. Four possibilities may thus be visualized as below:

(i) $^{\ominus}OH + CH_3{-}Br \longrightarrow CH_3{-}OH + Br^{\ominus}$

(ii) $NH_3 + CH_3{-}Br \longrightarrow CH_3{-}\overset{\oplus}{N}H_3 + Br^{\ominus}$

(iii) $Cl^{\ominus} + CH_3{-}\overset{\oplus}{O}H_2 \longrightarrow CH_3{-}Cl + H_2O$

(iv) $H_2S + R{-}\overset{\oplus}{N}(CH_3)_3 \longrightarrow R{-}\overset{\oplus}{S}H_2 + N(CH_3)_3$

A nucleophilic substitution reaction may proceed either by a *direct displacement mechanism* in which bond making and bond breaking are simultaneous, or by a *two-step mechanism* in which C–L bond is broken first, followed by a reaction between the resultant carbocation and the nucleophile.

Direct Displacement Process

$$:Nu^{\ominus} + R{-}L \longrightarrow \underset{\text{transition state}}{[Nu\text{--}R\text{---}L]^{\ominus}} \longrightarrow Nu{-}R + :L^{\ominus}$$

Carbocation Process

$$R{-}L \longrightarrow R^{\oplus} + :L^{\ominus} \; ; \quad :Nu^{\ominus} + R^{\oplus} \longrightarrow Nu{-}R$$

6.1 DIRECT DISPLACEMENT PROCESS: S_N2 REACTIONS

Let us consider the reaction of methyl bromide with sodium hydroxide to yield methanol.

$$^{\ominus}OH + CH_3{-}Br \longrightarrow CH_3{-}OH + Br^{\ominus}$$

The kinetic data tell us that the rate of the reaction is dependent upon the concentration of both these reactants.

$$\text{Rate} = k\,[CH_3Br]\,[OH^-]$$

In other words, if the concentration of either of these reactants is cut to half, the rate should be halved, and doubling the concentration of either of them would result in doubling the reaction rate. This reaction proceeds by a direct displacement mechanism in which both the reactants are present in the rate-determining step. Ingold has described this mechanism by the symbol S_N2 meaning *substitution, nucleophilic, bimolecular.* It has been shown that the hydroxide ion attacks on the side of the carbon atom opposite to that of bromine. A bond forms between the carbon atom and the hydroxide ion at the same time as the carbon-bromine bond weakens. In *transition state,* carbon is partially bonded to both –OH and –Br, and the three hydrogens and the central carbon are *coplanar.*

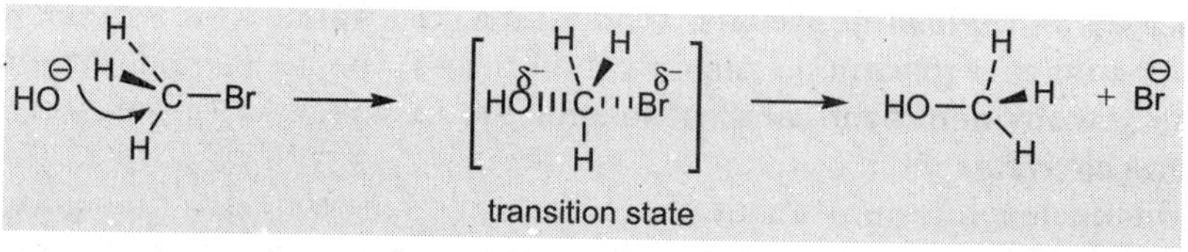

In transition state, hydroxide has a diminished negative charge since it is beginning to share its electrons with carbon, and at the same time bromine has developed a partial negative charge as it is tending to depart with the bonding electrons. When the carbon-oxygen bond is completely formed, the carbon-bromine bond is completely broken. The energy necessary to break a bond is compensated by the formation of another bond. The overall reaction is thus a concerted process occurring in one-step without any intermediates. **Fig. 6.1** represents the free energy *vs* reaction coordinate diagram for an S_N2 reaction.

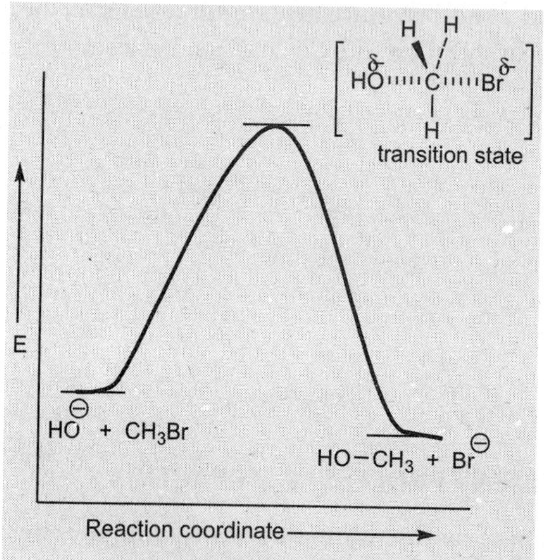

Fig.6.1: Energy diagram for S_N2 reactions

6.2 CARBOCATION PROCESS: S_N1 REACTIONS

In contrast to methyl bromide hydrolysis, the rate for *tert*-butyl bromide is not increased by the addition of sodium hydroxide even though hydroxide ions are consumed in the reaction. The rate of the reaction depends only on the concentration of *tert*-butyl bromide.

$$\overset{\ominus}{O}H + (CH_3)_3C{-}Br \longrightarrow (CH_3)_3C{-}OH + \overset{\ominus}{Br}$$

$$\text{Rate} = k\,[(CH_3)_3C{-}Br]$$

The mechanism of this reaction is called S_N1, which means *substitution, nucleophilic, unimolecular,* and is a two-step process. In the first step, C–Br bond ionizes to bromide ion (leaving group) and a carbocation, and in the second step the carbocation reacts with the hydroxide ion. Carbon atom of the carbocation has only six electrons in the outermost shell and, therefore, shows a strong tendency to react with any substance possessing an unshared electron pair. The rate-determining step of the reaction is the slow ionization of *tert*-butyl bromide and, as a result, the nature and concentration of the nucleophiles have no influence on the rate.

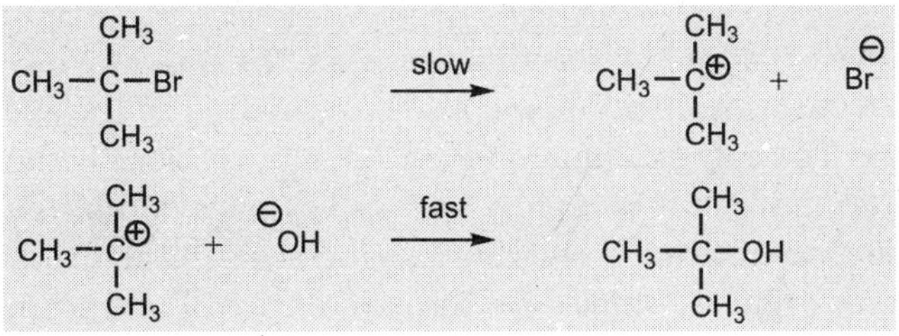

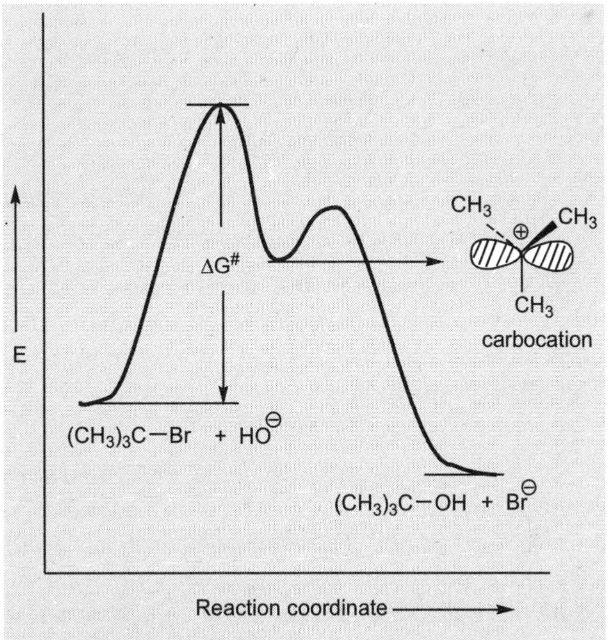

Fig. 6.2: Energy diagram for S_N1 reactions

Solvents play a very important role in S_N1 reactions as ionization is facilitated by the solvation of the carbocation and halide ions. The required energy for ionization is supplied, at least in part, by the formation of many ion-dipole bonds between the ions produced and the solvent molecules.

Ion Effects on S_N1 Reactions

An elegant proof for the intermediacy of carbocations in S_N1 reactions is provided by the study of substitution reaction of benzhydryl chloride (**1**), which generates an unusually stable carbocation (**2**).Treatment of **1** with tetramethylammonium fluoride in liquid SO_2 results in the substitution of chloride by a fluoride ion. Since the rate of the reaction is independent of the concentration of fluoride ions, it is an example of an S_N1 reaction.

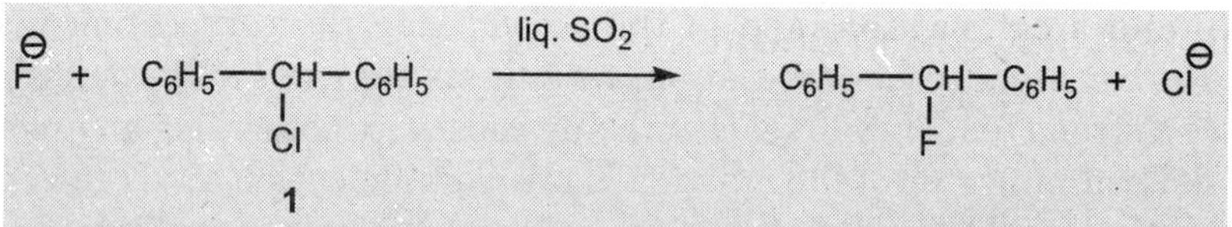

Rate = $k\,[(C_6H_5)_2CHCl]$

It has been observed that as the reaction proceeds, the rate at which substitution occurs starts decreasing. The rate of substitution can also be decreased by the addition of chloride ions in the reaction mixture. The only explanation for these observations is that the carbocation (**2**) formed in the rate-determining step is rapidly attacked by the available nucleophiles (Cl^- or F^-) with equal ease. As the concentration of chloride ions increases, either by the initial ionization or by the extra supply, they begin to compete for the carbocation. This results in the regeneration of the starting material, and, thus, the rate of substitution is retarded.

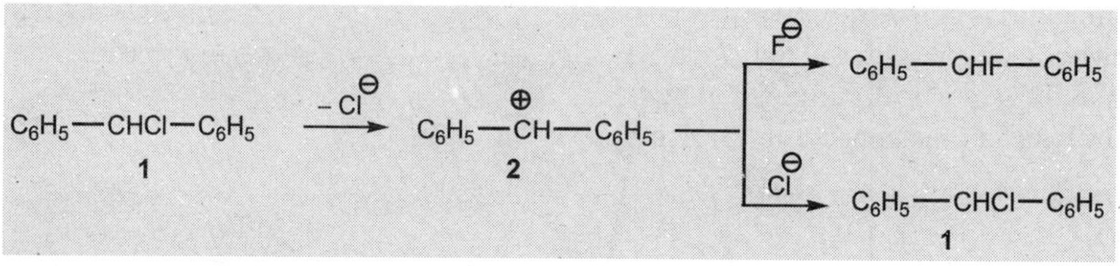

However, this ion effect, *i.e.* decrease in the reaction rate with time, is not observed in those S_N1 reactions which involve unstable carbocation intermediates. Because of their extreme reactivity, the carbocations have little chance of existing long enough to encounter the leaving nucleophiles present in small concentrations.

Solvolysis Reactions

Many solvents may be used as nucleophilic reagents, and when substitution reactions are carried out in these solvents, they are commonly referred to as *solvolysis* reactions. Common solvents are water, acetic acid and ethanol. These cause hydrolysis, acetolysis and ethanolysis, respectively. Since the solvent is present in excess, its concentration will not vary materially during the reaction, and as a result, kinetic data alone cannot decide between S_N1 and S_N2 mechanisms. More evidence is needed for this purpose.

Thus, the mechanism postulated for S_N1 reaction nicely explains the three experimental observations. First, the rate of reaction depends on the concentration of the alkyl halide and does not depend on the concentration of the nucleophile. Secondly, the mechanism shows the formation of a carbocation intermediate in the rate-determining step. Thirdly, as we will see in **Fig. 6.3**, in case of an alkyl halide in which the halogen (or the other leaving group) is attached to an asymmetric carbon, two stereoisomers are formed. This is because attack of the nucleophile on one side of the planar carbocation forms one stereoisomer, and attack on the other side generates the other stereoisomer.

6.3 STEREOCHEMISTRY OF SUBSTITUTION REACTIONS

A. Steric Orientation in S_N2 Mechanism

While considering S_N2 reactions, we assumed that the nucleophile attacks the side of the carbon atom opposite to that of the leaving group. But in case of hydrolysis of methyl bromide, we cannot distinguish this process from the one in which the hydroxide ion could approach from the same side of the bromine atom, as the carbon atom is not asymmetric, and **3** and **4** are identical.

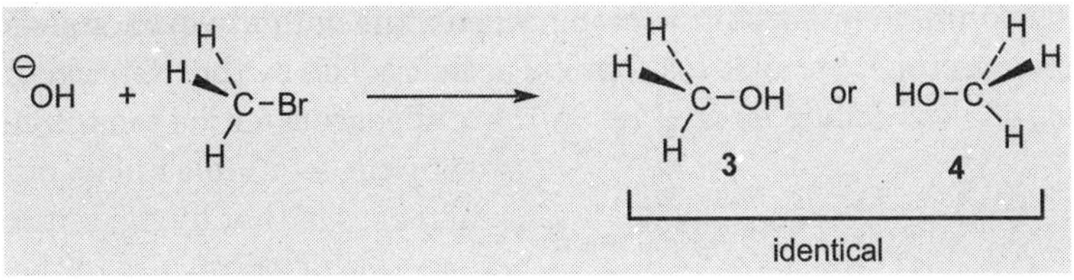

However, when the carbon atom on which substitution occurs is asymmetrically substituted, and the compound is *optically active*, it is possible to investigate the steric course of the reaction. Now, the two possible products **6** and **7** would no longer be identical, but would be mirror images of each other, and may be distinguished by their absolute configuration. This is well illustrated by the hydrolysis of 2-bromooctane (**5**), which exclusively gives **7.**

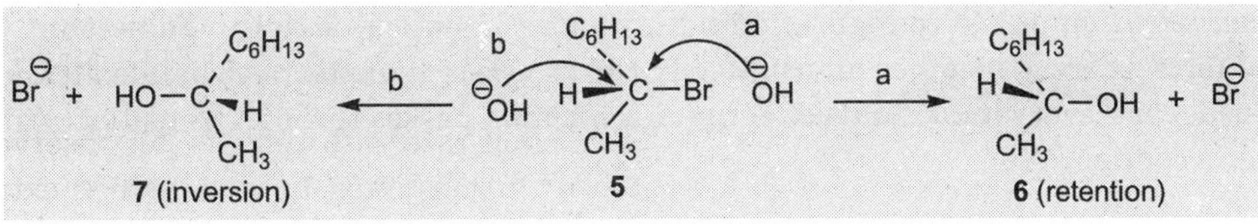

From this and many other substitution reactions, it is now established that the S_N2 reaction yields a product whose configuration is opposite to that of the substrate. In other words, the S_N2 reaction proceeds with *stereochemical inversion*, usually referred to as the *Walden inversion*.

We are now in a better position to describe the S_N2 reaction in which the backside attack by the nucleophile prompts the tetrahedral arrangement of the substrate to assume a *transition state* forcing the three atoms or groups attached to the carbon to a planar configuration perpendicular to the line of attack. After the expulsion of the leaving group, tetrahedral arrangement is regained, but *inverted*. The process may be likened to the turning inside-out of an umbrella in strong wind.

There still remains a problem. We need to relate the configuration of the reactant and product. We know that the direction of optical rotation and the configuration of two different compounds are not necessarily related, for they may have the same sign of rotation, but different configurations. The experimental proof of *Walden inversion* was obtained by a series of reactions in which an optically active alcohol was converted into its enantiomer having a rotation opposite in sign, but nearly equal in magnitude to that of the original alcohol.

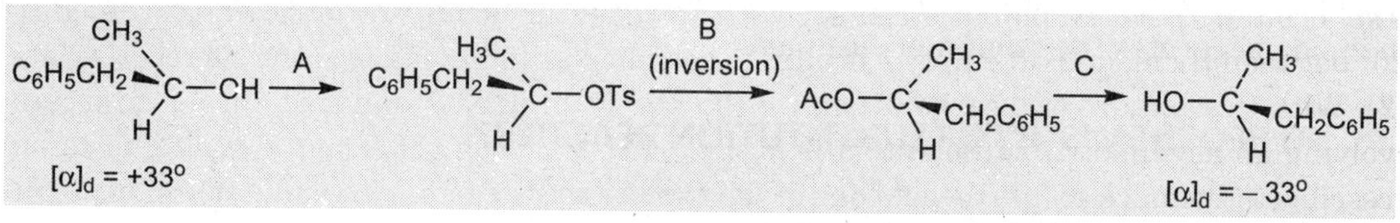

Since the asymmetric carbon–oxygen bond remains unchanged in reactions A (tosylation) and C (hydrolysis), an inversion of configuration must have occurred in the displacement of tosylate by acetate (B).

An ingenious way to establish the inversion of configuration in the S_N2 reactions has been described by Hughes and Ingold. They found that when optically active 2-iodooctane **(8)** was treated with sodium iodide in dry acetone solution, it not only exchanged ordinary iodide with the radioactive isotope, but also lost optical activity. The rate of the reaction depended on the concentration of alkyl iodide as well as on the radioactive iodide ions, confirming its S_N2 character.

$$\underset{\mathbf{8}}{C_6H_{13}CHICH_3} + I^{*\ominus} \longrightarrow C_6H_{13}CHI^*CH_3 + I^{\ominus}$$

$$\text{Rate} = k\,[C_6H_{13}CHICH_3]\,[I^{*\ominus}]$$

It was also observed that the rate of loss of optical activity was *twice* the rate of isotopic exchange. This experiment established the stereochemistry of the S_N2 reaction, for had the nucleophile attacked from the front side passing through the transition state (**9**), there would have been *retention* of the configuration with no loss in optical activity. To account for the observed facts, there is no other alternative but to conclude that the nucleophile attacks from the backside, and every individual act of the S_N2 reaction passes through transition state (**10**), and inverts configuration.

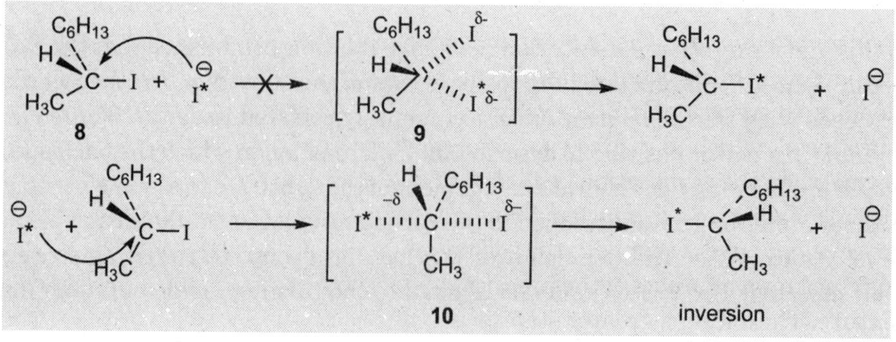

Since the rotation of inverted molecule exactly cancels the rotation of unreacted molecule, the rate of loss of optical activity is *twice* the rate of isotopic exchange.

B. Steric Orientation in S_N1 Mechanism

We have assumed that the first step in S_N1 reaction is the slow ionization of the substrate into a carbocation. The carbon atom in a carbocation is bonded to three other atoms or groups through its sp^2 orbitals. This is a planar structure with all the three bonds in a plane, making angles of 120° between them, and an empty *p* orbital perpendicular to the plane of the three sp^2 orbitals (**12**). With this picture of a carbocation, we can very well understand that once a carbocation involving an asymmetric carbon atom is formed, either lobe of the empty *p* orbital is equally accessible to the nucleophile to yield one of the two optical isomers. We would, therefore, expect equal number of molecules with *retention* and *inversion* of configuration to be formed in the reaction, *i.e.* the product should be racemic. The process may be illustrated by the hydrolysis of optically active 1-phenylethyl chloride (**11**).

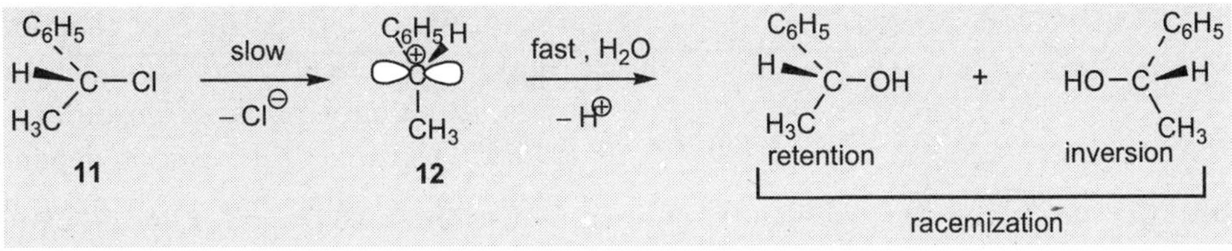

Usually, there is a larger proportion of molecules with inverted configuration than those of the same configuration, so that the product as a whole is not racemic. This may be explained by the fact that the ionization of an alkyl halide leads first to an ion pair **13** in which the departing halide ion by its close proximity to the carbocation shields its own side, thus helping the attack from the other side by the nucleophilic reagent or the solvent molecule. The situation can be represented as follows.

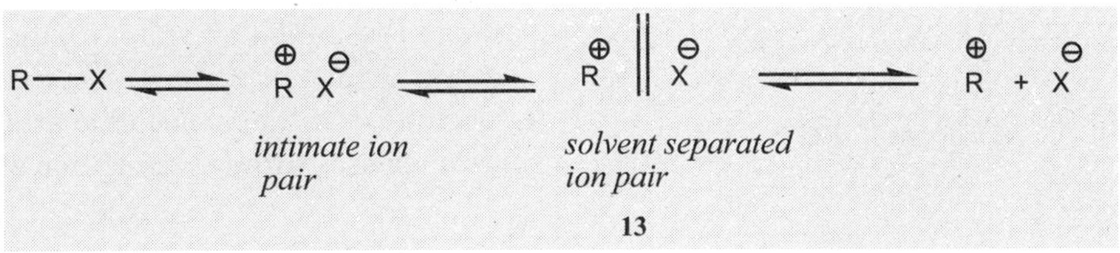

In the *intimate ion pair*, the leaving group (X^-) solvates the carbocation on the side from which it is departing, while solvent molecules are able to solvate the cation only from the opposite side, thus leading to inversion. In other words, in case of solvolysis, an *unsymmetrically solvated ion* (**14**) is produced, which may give rise to a product of inverted configuration.

However, when the carbocation lasts long enough, the leaving group is replaced by another solvent molecule leading to a *symmetrical* and *planar solvated ion* (**15**). Being symmetrical, this ion can form covalent bond either with the solvent molecule or the nucleophile on either side with equal ease giving equal amounts of the products with inverted and retained configurations, *i.e.* a racemic mixture. Therefore, a greater stability of the carbocation leads to greater racemization because of the ion's longer life, which enables it to become symmetrically solvated. All these processes are shown with water as the solvent in Fig. 6.3.

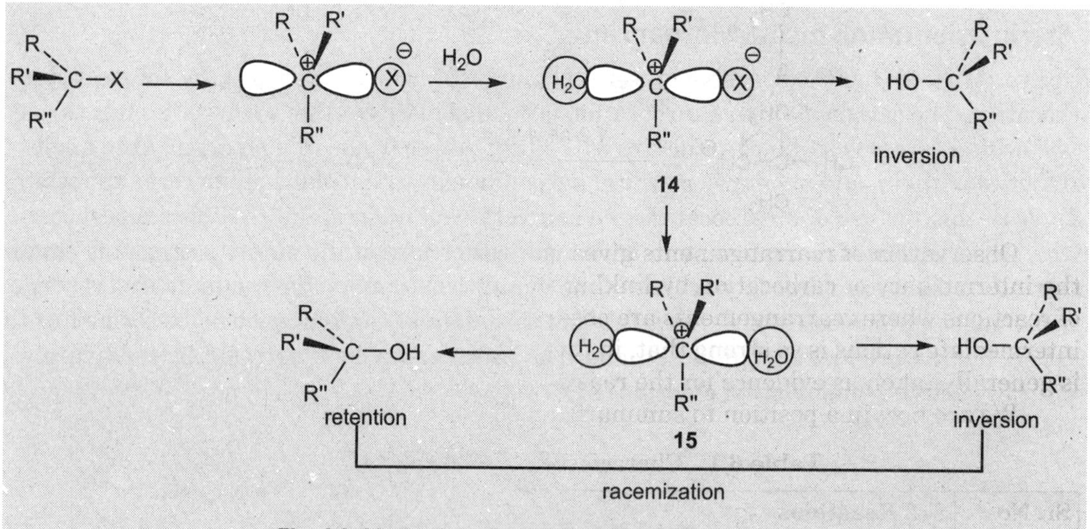

Fig. 6.3: Mechanism of an S_N1 reaction

The difference in the stereochemistry of the products obtained from S_N1 and S_N2 reactions can be easily visualized in the case of cyclic compounds. *Cis*-1-bromo-4-methylcyclohexane, for example, undergoes an S_N2 reaction generating a product with *trans* stereochemistry. However, when this compound undergoes an S_N1 reaction, both the *cis* and *trans* products are obtained due to planarity of the intermediate carbocation.

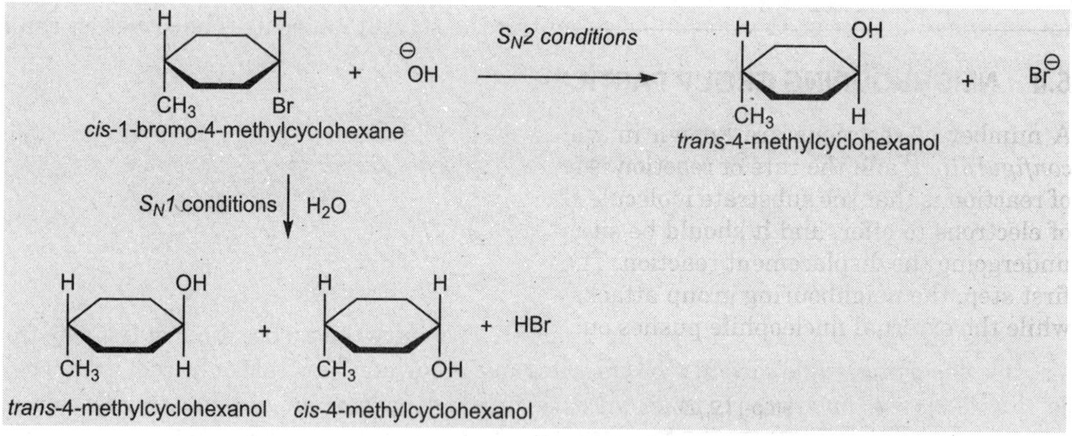

Rearrangement Reactions

The most striking feature in S_N1 reactions is the occurrence of the rearranged products. For example, in a solution of weakly nucleophilic ethanol, *neo*pentyl bromide undergoes a slow first order reaction to yield ethyl *tert*-pentylether. The process involves 1,2-shift of methyl group to generate a *tertiary* carbocation.

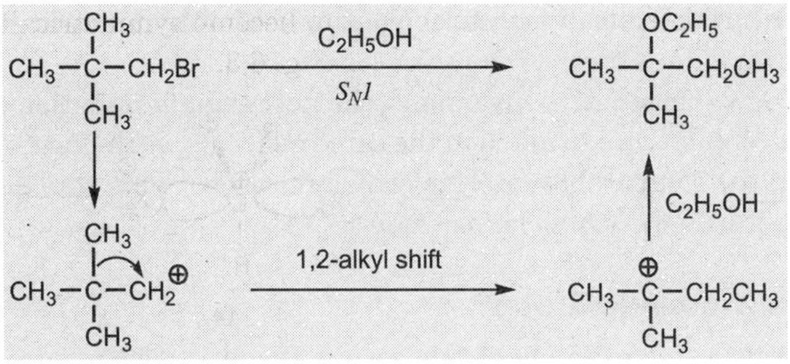

Observation of rearrangements gives powerful support to the S_N1 mechanism involving the intermediacy of carbocation by linking this mechanism to the mechanism of other kinds of reactions where rearrangements are observed. The correlation between rearrangement and intermediate cations is so strong that, in the absence of any other information, rearrangement is generally taken as evidence for the reaction to proceed through an S_N1 mechanism.

We are now in a position to summarize the characteristics of S_N2 and S_N1 reactions.

Table 6.1 Characteristics of S_N2 and S_N1 reactions

Sr. No	S_N2 Reactions	S_N1 Reactions
(i)	one-step mechanism	two-step mechanism
(ii)	bimolecular rate-determining step	unimolecular rate-determining step
(iii)	no carbocation intermediates, and hence no rearrangement	carbocation intermediates, and their rearranagements
(iv)	product has inversion of configuration	products are largely racemic
(v)	reactivity of alkyl halides follows the order: $CH_3 > 1° > 2° > 3°$	reactivity of alkyl halides follows the order: $3° > 2° > 1° > CH_3$

6.4 NEIGHBOURING GROUP PARTICIPATION

A number of reactions are known in which substitution occurs with complete *retention of configuration,* and the rate of reaction is *greater* than expected. The requirement for this type of reaction is that the substrate molecule should have a second functional group (Z) with a pair of electrons to offer, and it should be situated at a favourable distance from the carbon atom undergoing the displacement reaction. These reactions consist of two S_N2 substitutions; in the first step, the neighbouring group attacks as a nucleophile and pushes out the leaving group, while the external nucleophile pushes out the neighbouring group in the subsequent step.

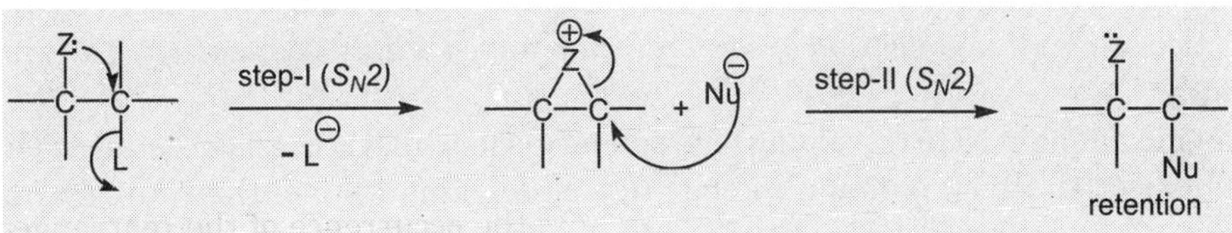

This process is often referred to as *neighbouring group participation* (NGP) where the substituent group acts as a nucleophile and the process affects the reaction rates and stereochemistry of the product. The requirement for NGP is that the nucleophilic substituent

located on the other carbon atom of the molecule should be able to attack from the backside of the leaving group.

Since there are two consecutive inversions, the net result is retention of configuration. In addition, as the nucleophile (Z) is situated in the same molecule, such reactions occur thousands of times faster than the comparable intermolecular reactions. This rate enhancement by the neighbouring group is called *anchimeric assistance.*

A. Lone Pair as Neighbouring Group

Perhaps, the most celebrated example of this type is the neighbouring group participation of the carboxylate anion in the conversion of 2-bromopropanoate anion (**16**) to the lactate anion (**18**). Treatment of 2-bromopropanoic acid with dilute alkali results in a substitution product with complete *retention of configuration*. This is explained by the consideration that before the leaving group (Br) has receded to a great distance, the pair of electrons on the carboxylate anion forms a weak bond with the carbocationic centre. This bond is apparently formed on the *side opposite* to that of the leaving group, thus protecting that side from attack by the nucleophile (hydroxide ion). The nucleophile then attacks the intermediate α-lactone (**17**) from the *frontside*, liberating the carboxylate anion, and effecting substitution with *retention of configuration*.

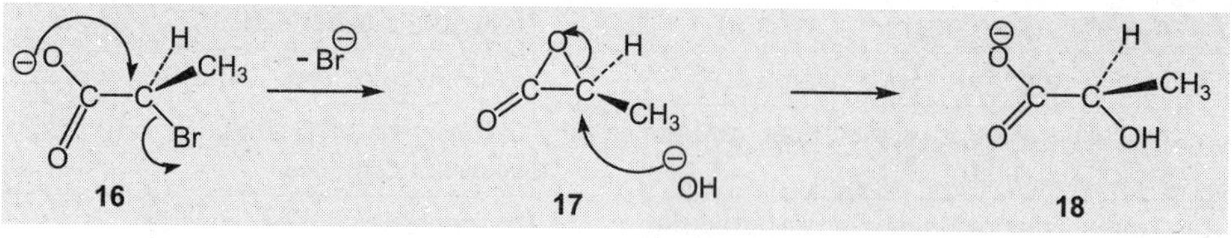

Kinetic data reveals that the formation of **17** is the rate-determining step, since once formed it rapidly reacts with the hydroxide ions because of the high strain associated with a three-membered ring. However, in the presence of a strong alkali solution, the hydroxide ions compete with the carboxylate ion, and **16** undergoes a second-order reaction to give the product having the inverted configuration as expected for an S_N2 reaction.

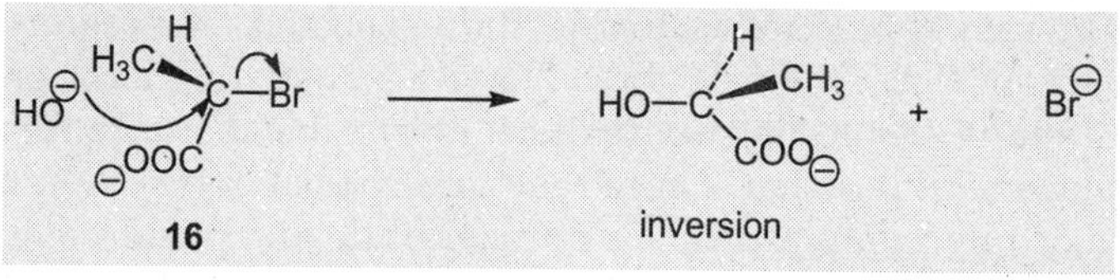

This inverted product racemizes with equal number of molecules of retained configuration, and in practice, we may get *retention of configuration mixed with racemization in any proportion to both limits.*

Another example of NGP (involving lone pair as nucleophile) is the base-catalyzed hydrolysis of mustard gas (**19**), which is found to obey first-order kinetics, *i.e.* the rate of hydrolysis is independent of the concentration of added alkali. The reaction proceeds *thousands of times* faster than the reactions involving analogous alkyl halides. Addition of other nucleophiles to the reaction mixture yields mixed products but the reaction rate remains unchanged. Moreover, the reaction rate begins to decrease with time, thereby, suggesting that the chloride ions regenerate the starting material. It is, therefore, suggested that the rate-determining step of the reaction is the reversible formation of a cyclic sulphonium ion (**20**), which being highly strained undergoes rapid hydrolysis.

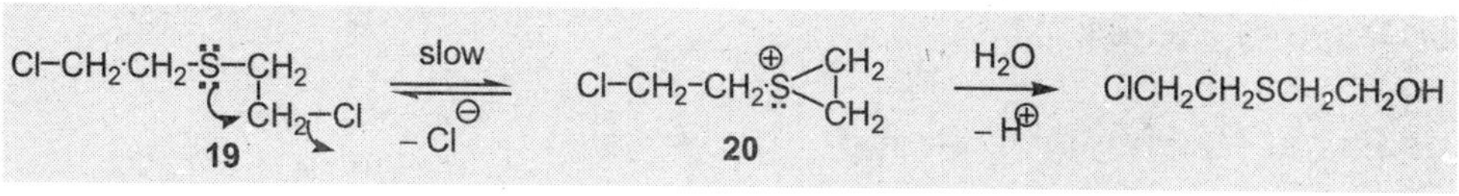

Properly situated halogen atoms also offer anchimeric assistance, *e.g.* the rate of acetolysis of *trans*-2-iodocyclohexyl brosylate (**21**)[1] is many times faster than the acetolysis of the *cis* isomer (**22**).

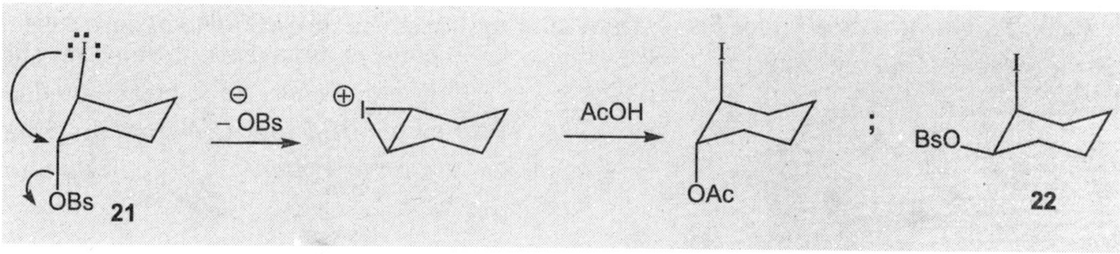

It is interesting to note that in many instances a stable bond is formed between a neighbouring group, and the carbon atom undergoing displacement reaction resulting in the isolation of cyclic products. This is exemplified in the formation of epoxides from chlorohydrins in the presence of a base. Reactions of this type may be regarded as internal S_N2 reactions requiring a backside attack by the internal nucleophile.

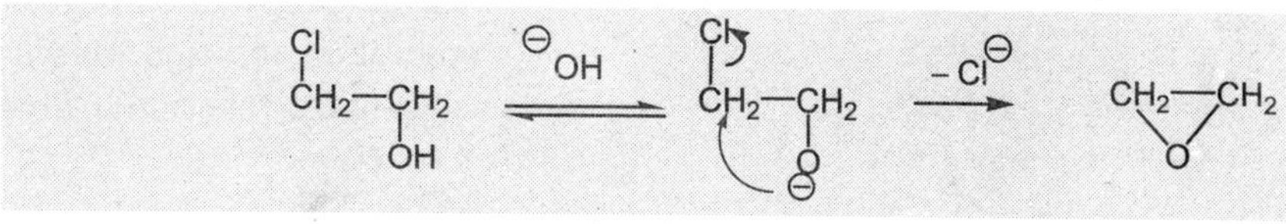

This geometric requirement is well illustrated by the observation that *cis*-2-chlorocyclohexanol (**24**) forms epoxide far more slowly than its *trans* analog (**23**).

---

[1]The OH can be made a better leaving group by its conversion to groups such as sulphonic esters. Two such moieties are brosylates and tosylates, which are commonly used in the study of nucleophilic substitution reactions.

Brosylates (OBs)
(*p*-Bromobenzenesulphonates)

Tosylates (OTs)
(*p*-Toluenesulphonates)

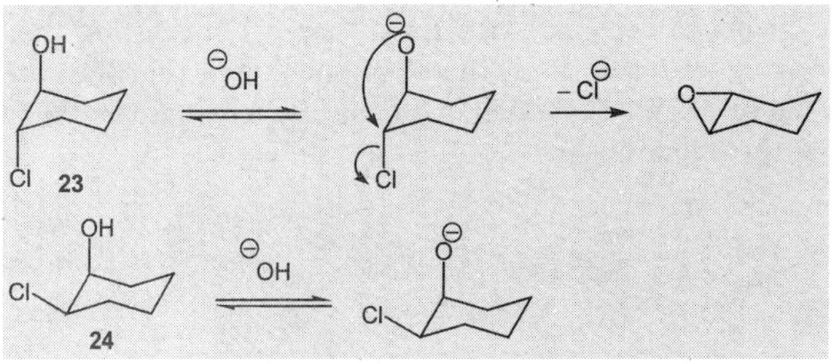

I. Oxonium Ions

An illustrative example of neighbouring group participation by lone pairs is provided by acetolysis of *trans* (**25**) and *cis* (**26**) diastereomers of 2-acetoxycyclohexyl *p*-toluenesulphonate. **25** reacts with acetic acid hundreds of times faster than **26**. Furthermore, the products in both cases have the same stereochemistry (**27**).

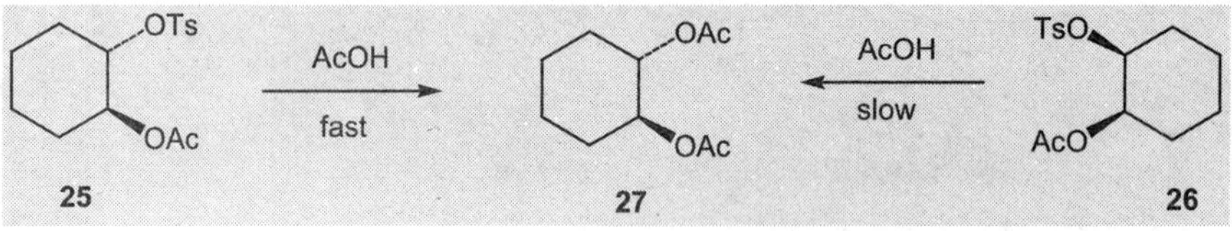

It is thus clear that solvolysis of **25** proceeds with retention, and that of **26** with inversion of configuration.

To explain these experimental observations, the planar structures are to be drawn in the chair conformation.

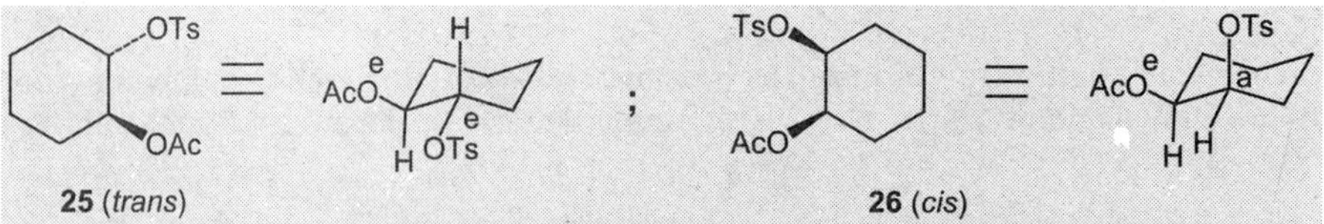

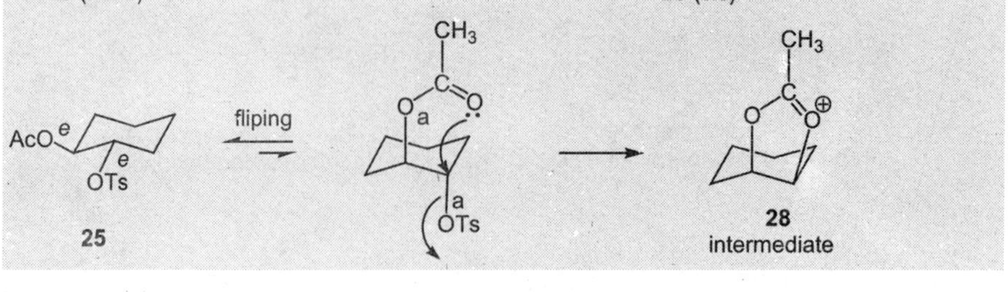

The intermediate **28** is symmetrical as the positive charge is delocalized on both the oxygens.

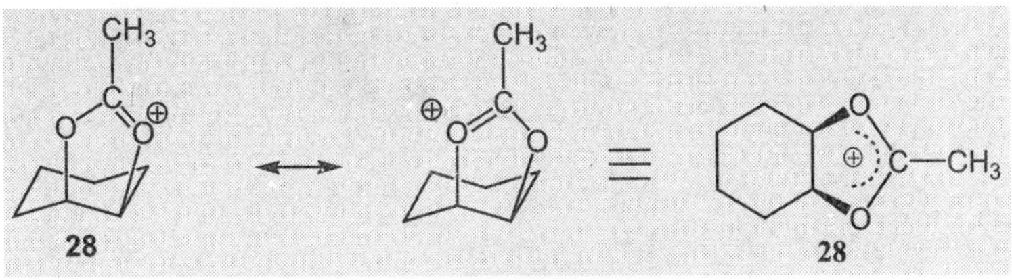

When the reaction is performed in ethanol, the intermediate **28** has indeed been trapped.

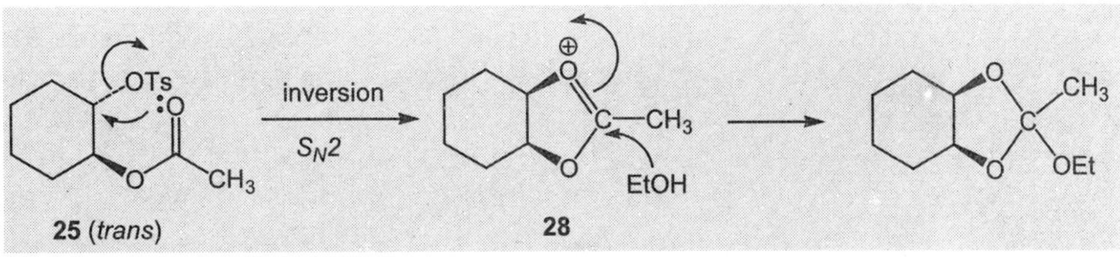

The intermediate **28** is then attacked by acetic acid in another S_N2 reaction resulting in the second inversion. As there are two inversions, it results in the retention of configuration in the product **27**.

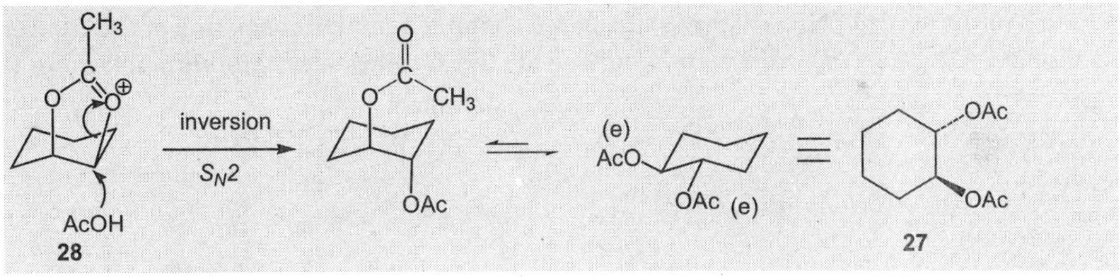

The observed stereochemistry, *i.e.* retention of configuration can only be explained by two consecutive S_N2 reactions, which are possible by *neighbouring group participation*.

The other diastereomer **26** has no option but to react by S_N2 reaction resulting in the inversion of configuration. It is to be noted that there cannot be intramolecular displacement of the tosyl group by placing acetyl group in either the axial or equatorial position, as the lone pair of oxygen cannot reach at the back of the –OTs group.

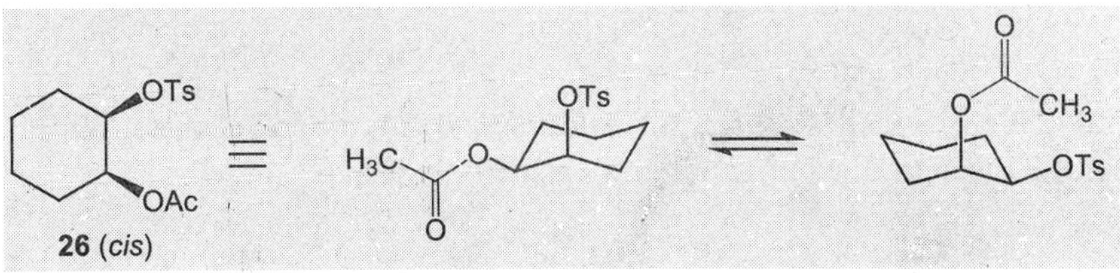

The solvolysis of **26** thus proceeds as follows:

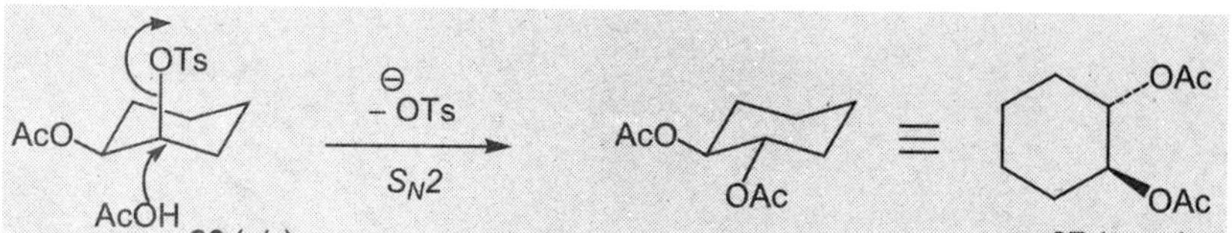

26 (*cis*) 27 (*trans*)

As the intermediate oxonium ion **28** is delocalized and achiral, a pure enantiomer of the starting material generates the racemic product. It is understandable as the attack on one carbon atom of **28** gives one enantiomer, while on the other gives the mirror image.

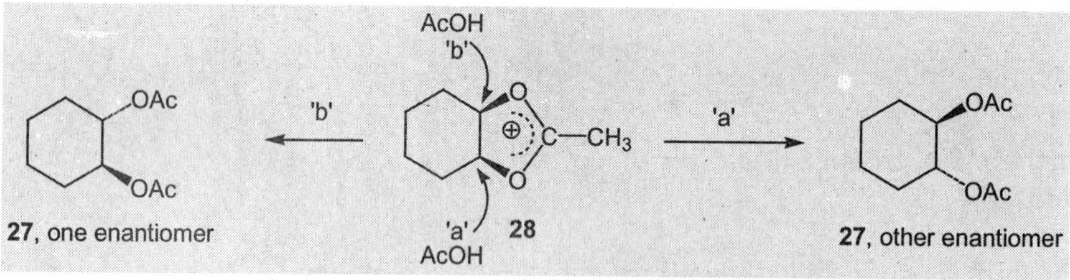

27, one enantiomer 28 27, other enantiomer

II. Bromonium Ions

In 1939, Winstein and Lucas performed several experiments, which are considered as landmark in the study of NPG. They investigated a simple reaction consisting of the treatment of 3-bromo-2-butanol (**29**) with HBr resulting in the formation of 2,3-dibromobutane (**30**).

29 30

Compound **29** exists as a pair of diastereomers, *threo* (**31**) and *erythro* (**32**), which in turn exist as pairs of two enantiomers.

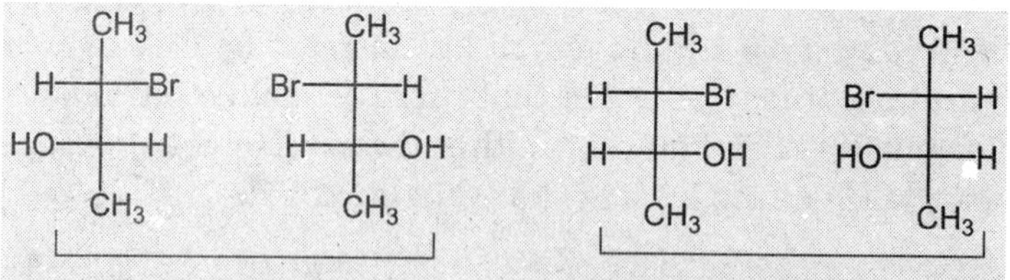

threo (racemic) **31** *erythro* (racemic) **32**

Compound **30** also exists as *threo* (**33**) and *erythro* (**34**) diastereomers. While **33** has a pair of enantiomers, **34** is a *meso* compound, and optically inactive.

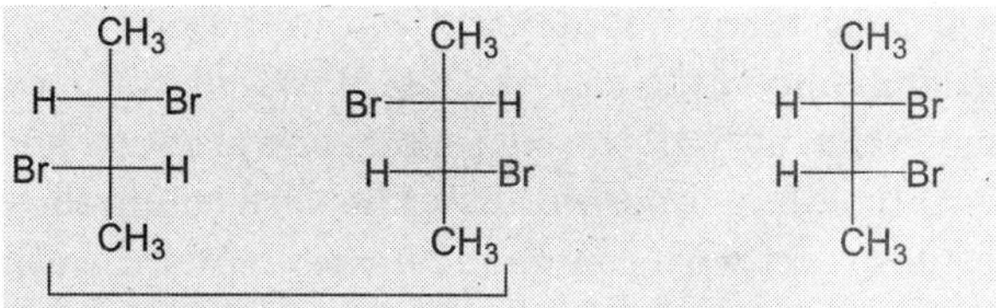

threo (racemic) **33** *erythro* (*meso*) **34**

Treatment of *erythro* dl pair (**32**) (or an enantiomer) invariably gives only *meso* dibromide (**34**).

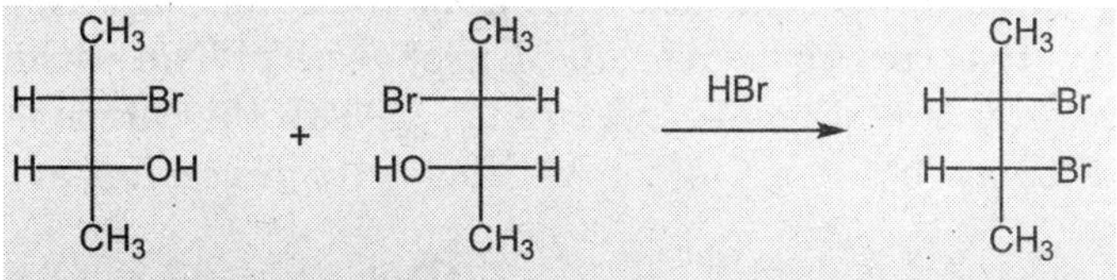

erythro (racemic) **32** *meso* **34**

Furthermore, treatment of *threo* dl pair (**31**) (or an enantiomer) gives the dl pair of the dibromide (**33**).

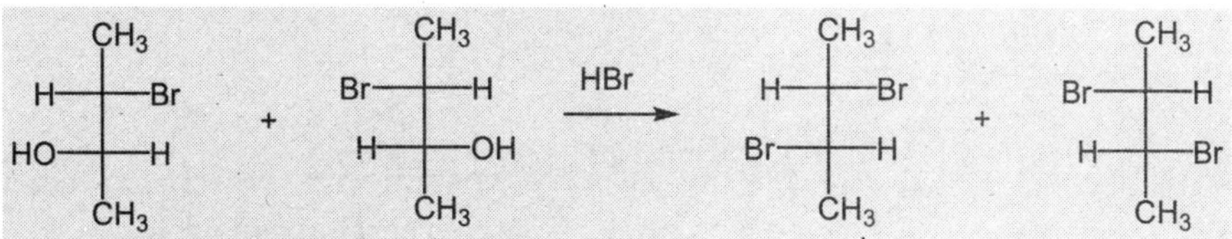

threo (racemic) **31** *threo* (racemic) **33**

Similar results have been obtained with optically active substrates. Optically active *erythro* (**32**) gave *meso* dibromide (**34**).

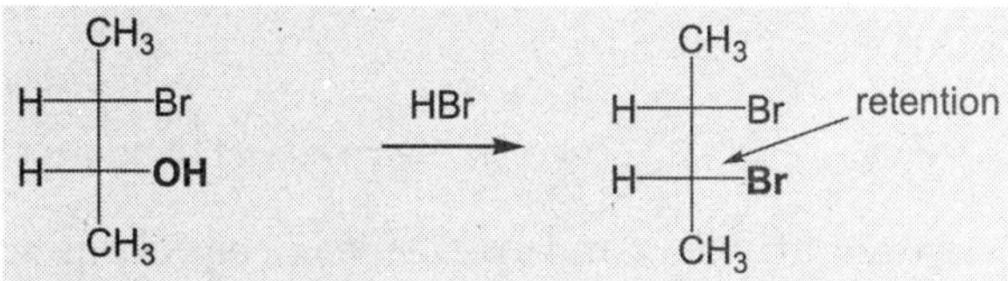

erythro (optically active) **32** *meso* **34**

Racemic *threo* dibromide (**33**) was similarly obtained by treatment of an enantiomer of *threo* 3-bromo-2-butanol (**31**).

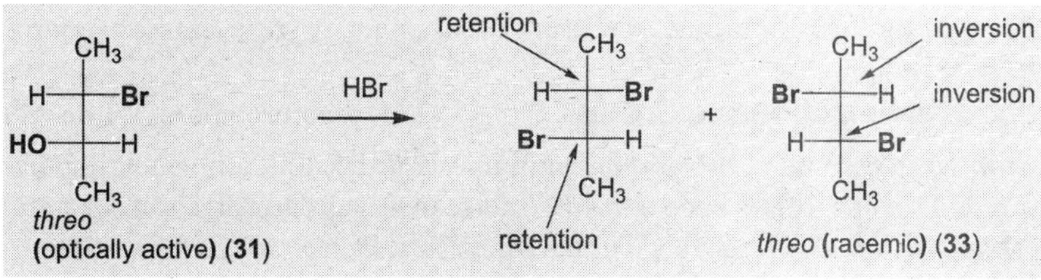

threo (optically active) (**31**) *threo* (racemic) (**33**)

On careful analysis of these results, it appears that in the case of *erythro* compound, there is retention of configuration at one of the carbons (bold). While in the case of *threo* compound, there is *retention* of configuration at both the carbons in one product, and *inversion* at both the carbons in the other product (bold). In other words, while performing the reaction with the *threo* compound, half the molecules react with complete *retention*, and the other half with *inversion* of configuration at both the carbons, resulting in the formation of a racemic product **(33)**.

To explain these observations, formation of a bridged bromonium ion was proposed, which is generated after the departure of the leaving group (H_2O). In the second step of the reaction, the nucleophile (Br^-) attacks the bromonium ion only from the unhindered side on either of the two carbons with equal probability. The attack on one carbon (pathway 'a') results in the retention of configuration at both the carbons. When Br^- attacks the other carbon (pathway 'b'), it gives the product with inversion at both the centres.

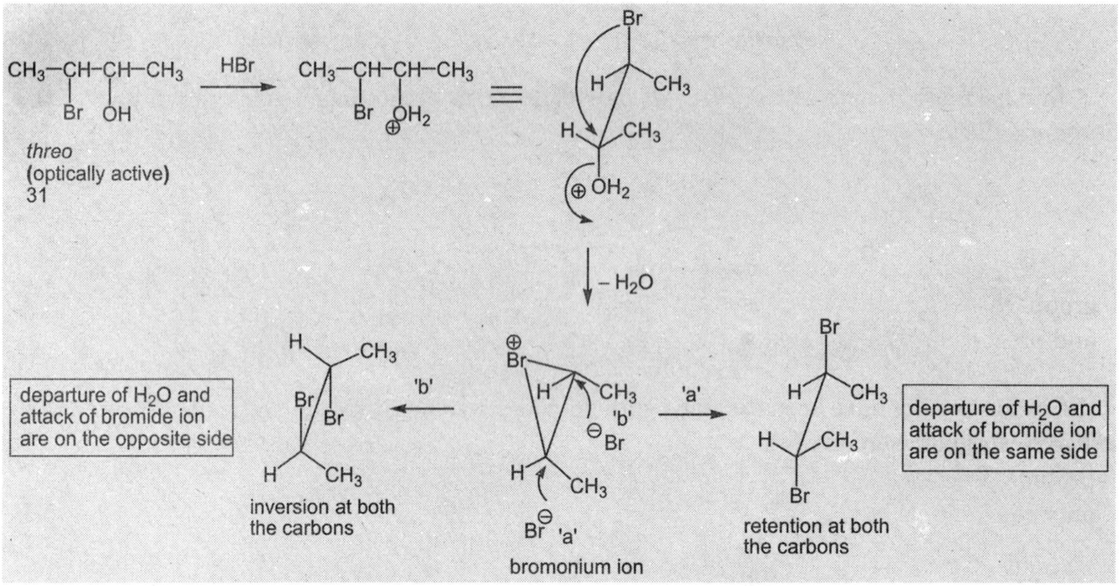

As both the products are enantiomers and are formed in equal amounts, a racemic mixture is obtained.

We can thus conclude that the transformation of 3-bromo-2-butanol (**29**) to 2,3-dibromobutane (**30**) consists of two successive nucleophilic substitutions of S_N2-type. In the first step, the nucleophile is the neighbouring bromine in the molecule, and in the second step it is the bromide ion (Br^-) from outside the molecule. Had there been any formation of a carbocation intermediate, we would have obtained a mixture of *threo* and *erythro* diastereomers.

B. Double Bond as Neighbouring Group

If properly situated, electrons of the double bond may also act like other neighbouring groups having unshared electrons. It has been shown, for instance, that acetolysis of tosylate **35** is 10^{11} times faster than that of **36**. Moreover, it proceeds with *retention of configuration*.

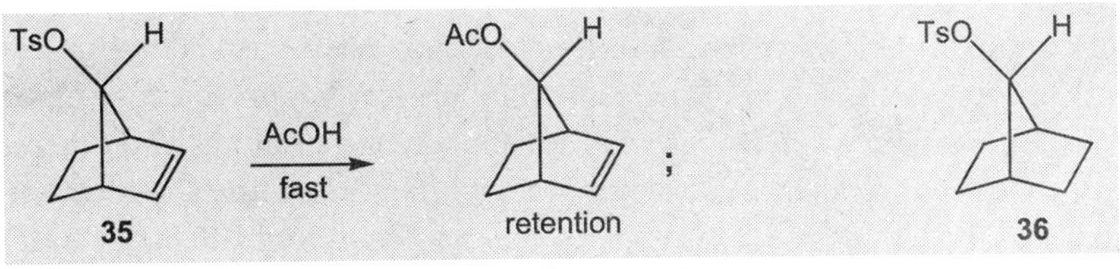

A norbornenyl cation (**37**) in which the positive charge is delocalized by a double bond has been invoked to explain these experimental observations.

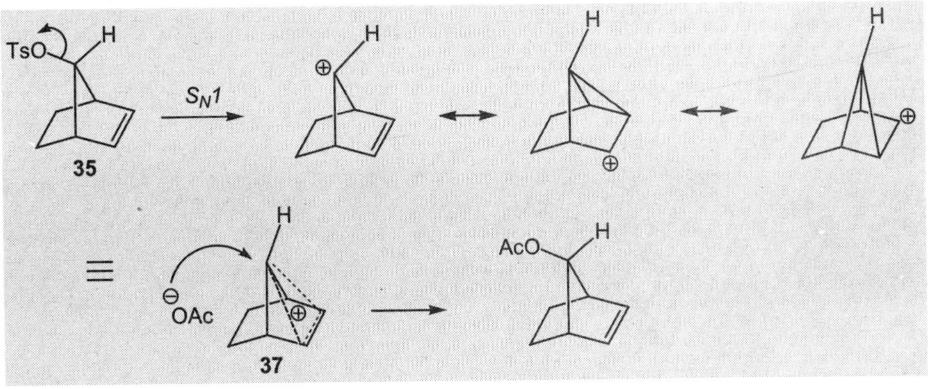

The rate data strongly suggest that the double bond assists in the departure of the tosyloxy group. No such anchimeric assistance is available to **36** and hence its rate of solvolysis is normal and proceeds with *inversion of configuration*.

The evidence for the intermediacy of a nonclassical cation (**37**) comes from a study of the NMR spectrum of a related cation, norbornadienyl (**38**), which is somewhat more stable. The NMR spectrum of this cation indicated that 2 and 3 protons are not *equivalent* to 5 and 6 protons, suggesting, thereby, that there is an interaction of the positively charged carbon with only *one double bond*.

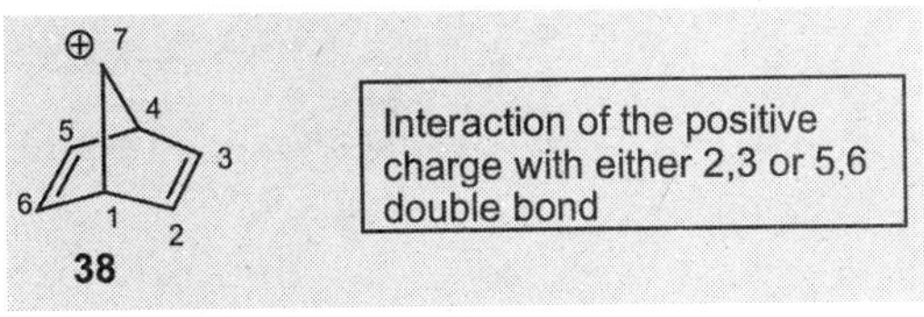

It may be noted that in the case of **35**, the double bond is located at a favourable position for backside attack on the carbon bearing the tosyloxy group. This explains the observed, very large, rate enhancement. However, this rate enhancement can be reduced by the presence of an electron-donating group at the 7-position of **35** and **36**. Thus, the difference in the rate of solvolysis of **39** and **40** is *far less* than what was observed between **35** and **36**.

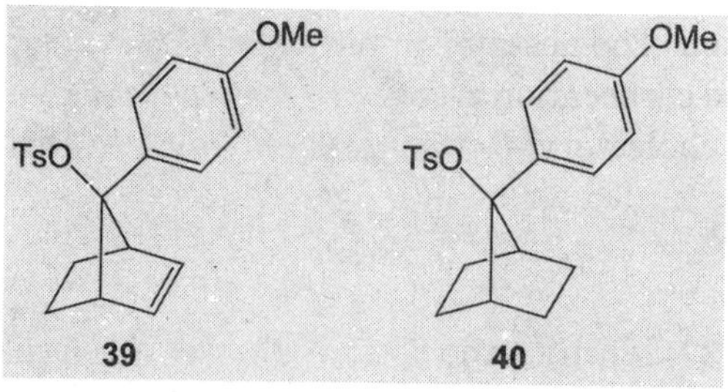

This observation suggests that NGP is considerably *reduced* if the stability of the potential carbocation is increased by introducing an appropriate substituent. In the case of **39**, the incipient positive charge developed at position 7 in the transition state is greatly stabilized by *p*-anisyl group, and further stabilization of charge by the participation of double bond is really not required.

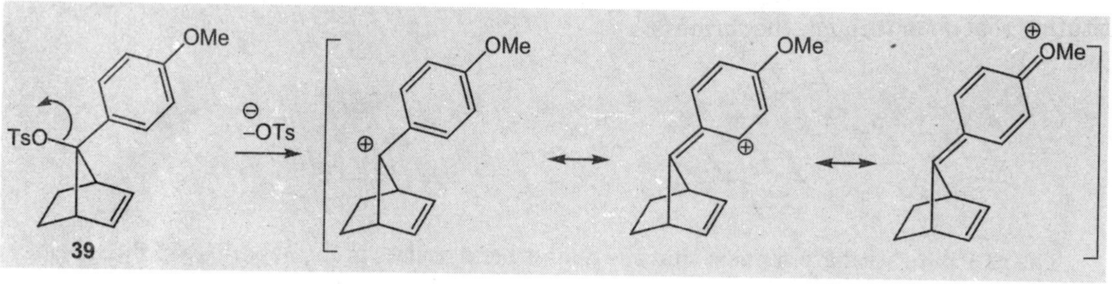

Similar participation of a double bond has been invoked to explain the acetolysis of cholesteryl chloride (**41**) to give cholesteryl acetate (**42**) with *retention of configuration*, while its dihydro derivative cholestanyl chloride (**43**) undergoes normal S_N2 reaction with *inversion of configuration* (**44**).

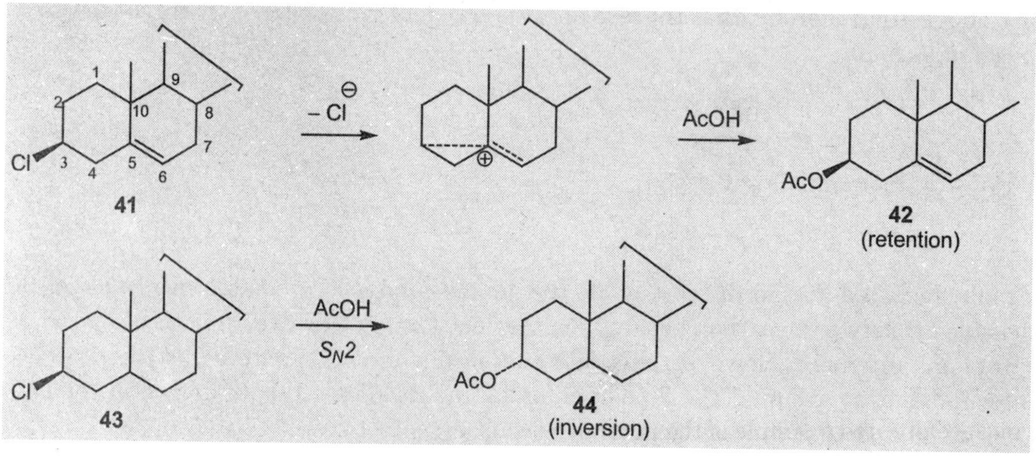

It has been suggested that the π electrons of the 5,6-double bond in **41** have configuration holding properties for the 3-position, because the positive charge initially formed on the 3-position gets stabilized through delocalization between 3,5- and 6-positions. The conformational structure of **41** permits such delocalization. The acetate ion has only β face to attack as α face is blocked. The formation of the delocalized carbocation is supported by the isolation of *i*-cholesteryl methyl ether (**46**) on the treatment of cholesteryl-3-β-tosylate (**45**) with KOAc in CH_3OH.

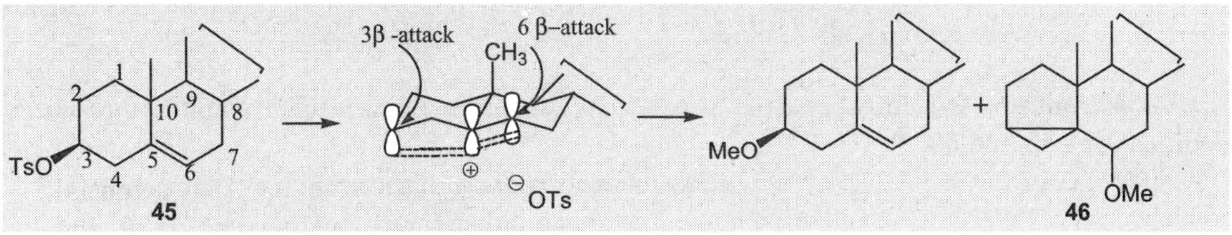

(C) Phenyl Group as Neighbouring Group: Phenonium Ions

There is an increasing evidence that neighbouring phenyl group participates in nucleophilic substitution reactions through the formation of a resonance-stabilized phenonium ion.

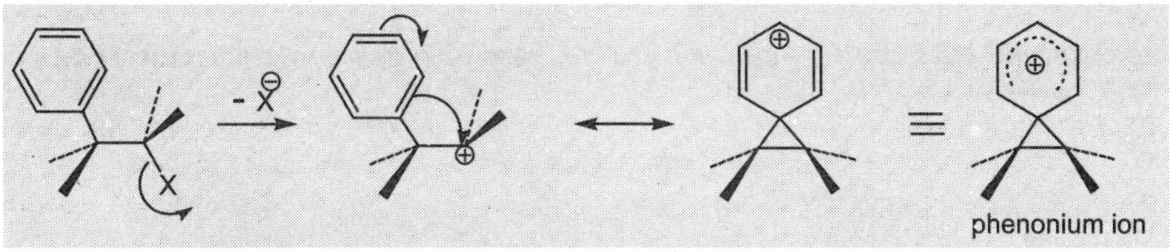

As expected, substitution takes place with retention of configuration (path 'a'), and is accompanied by a rearranged product (path 'b').

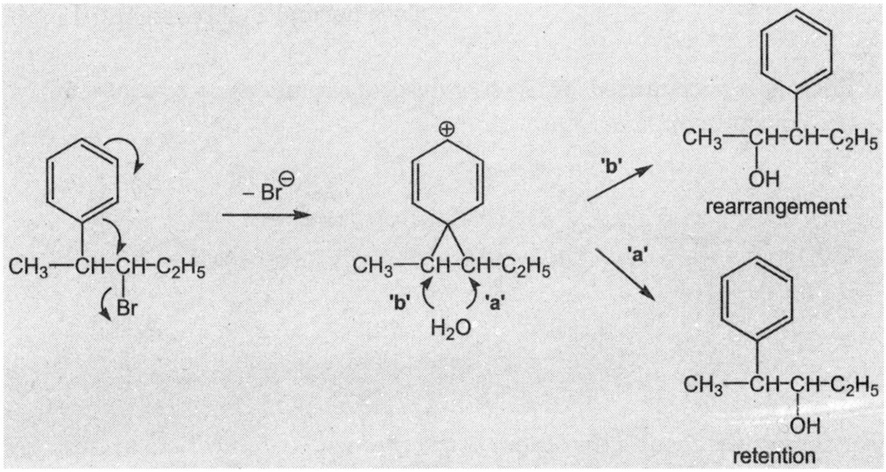

The most celebrated example of the participation of phenyl group resulting in the formation of phenonium ions is provided by a study of the acetolysis of 3-phenyl-2-butyl tosylate (**47**). Stereochemistry of the product acetate **48** was thoroughly investigated by Cram in 1949, and the findings generated a heated debate about the existence of bridged carbocations lasting more than 20 years.

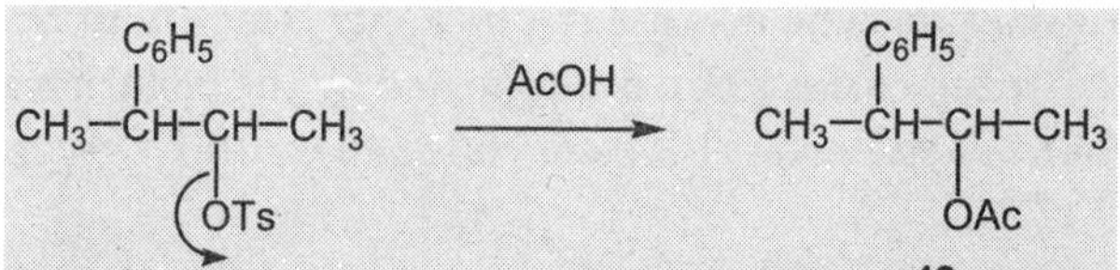

47 **48**

As **47** contains two chiral centres, it exists as two pairs of enantiomers and two racemic modifications **49** and **50**.

erythro tosylate (racemic) **49**

threo tosylate (racemic) **50**

It was observed that the racemic *erythro* tosylate (**49**) gave only racemic *erythro* acetate (**51**).

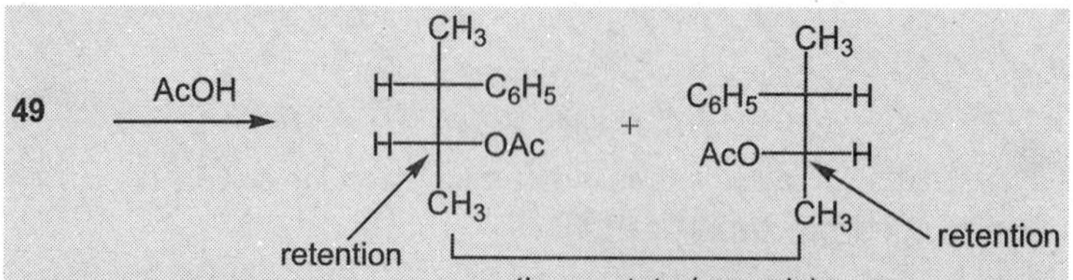

erythro acetate (racemic) **51**

Similar findings were reported for the acetolysis of racemic *threo* tosylate (**50**), which gave only racemic *threo* acetate (**52**).

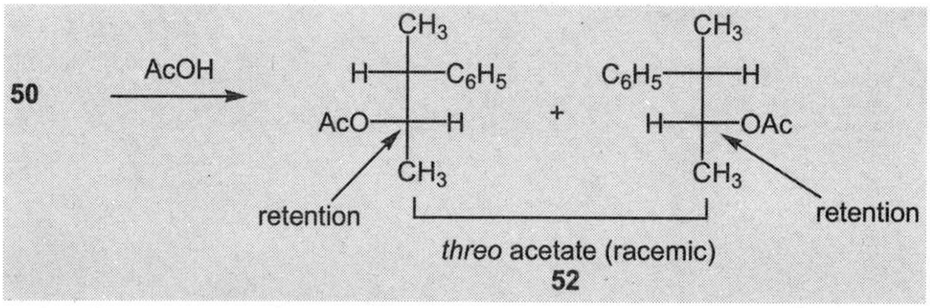

threo acetate (racemic) **52**

It is clear from these results that there is retention of configuration; acetoxy group (OAc) has taken the place vacated by tosyloxy group (OTs).

However, when the reaction was performed using optically active *threo* tosylate (**50**, one enantiomer), surprisingly the product was found to be the same as obtained in the acetolysis of 50, ie. racemic three acetate (**52**).

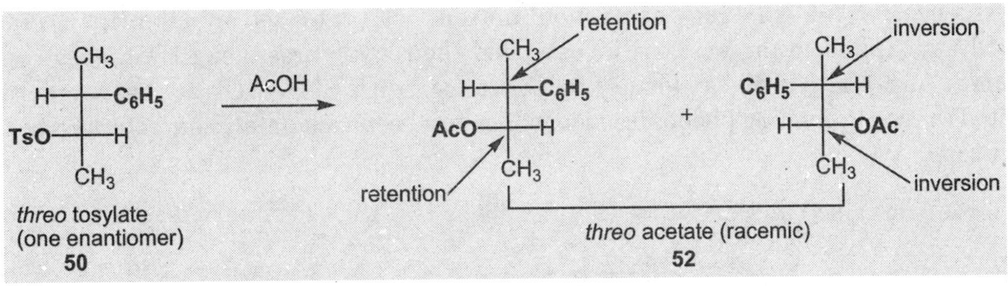

There is retention of configuration at *both* the carbons in one product, and inversion at both the carbons in the other product (bold). In other words, half the molecules react with complete retention, and the other half with inversion of configuration at both the carbons. Intermediacy of a phenonium ion neatly explains these experimental observations.

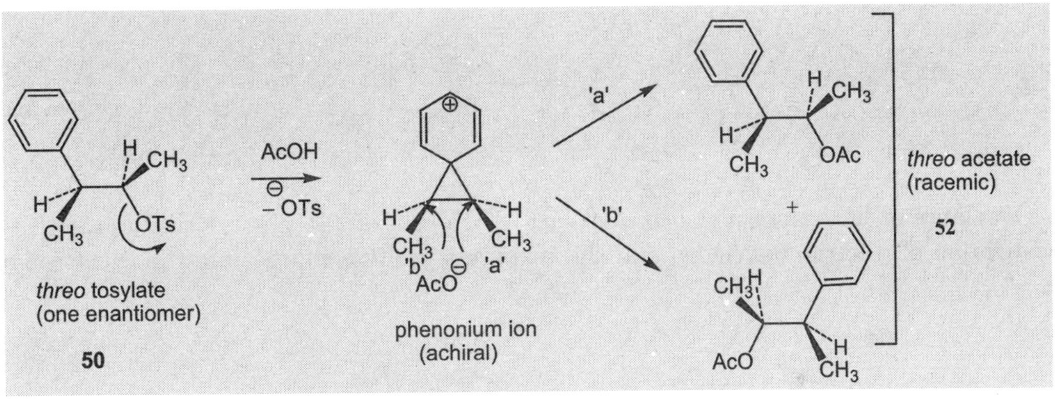

There are equal chances for pathways 'a' and 'b' to operate and thus the product is an equal mixture consisting of a racemic product.

As the phenonium ion obtained from optically pure *erythro* tosylate is chiral, the product acetate (**51**) is also enantiomerically pure.

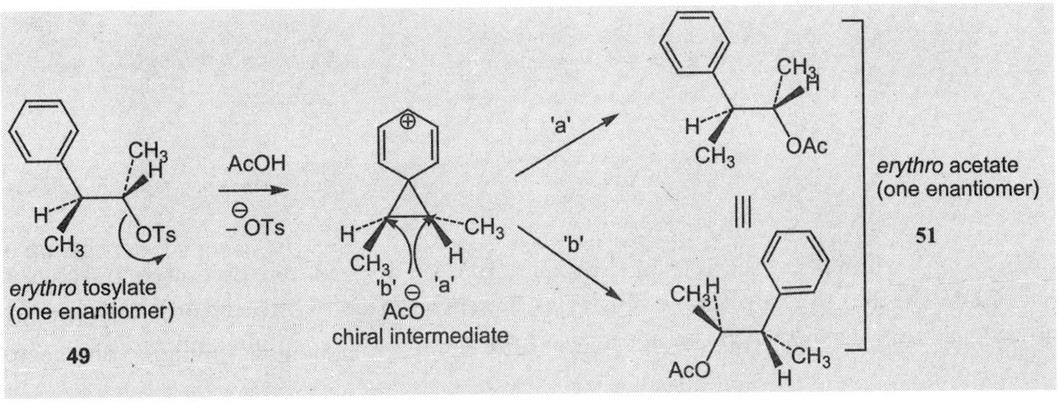

Evidence for the existence of phenonium ions was provided by Olah in 1970s who prepared many phenonium ions and investigated their structures using NMR spectroscopy. Treatment of β-phenylethyl chloride, for instance, with SbF_5–SO_2 at low temperature generated the corresponding phenonium ion, which has been adequately characterized by NMR spectroscopy.

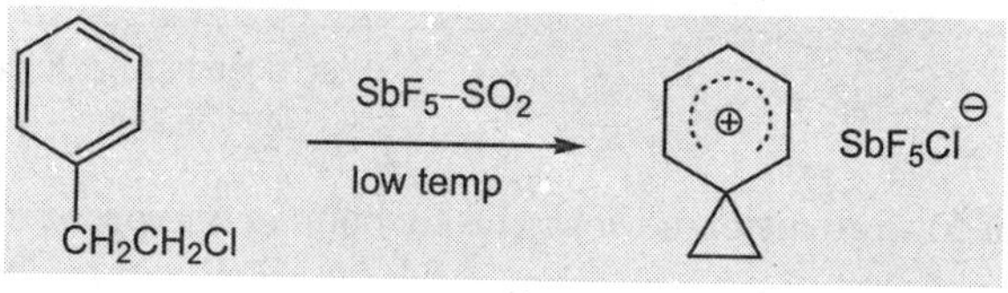

Resonance in the phenyl ring has also been demonstrated by a combined application of ^1H and ^{13}C NMR spectroscopy.

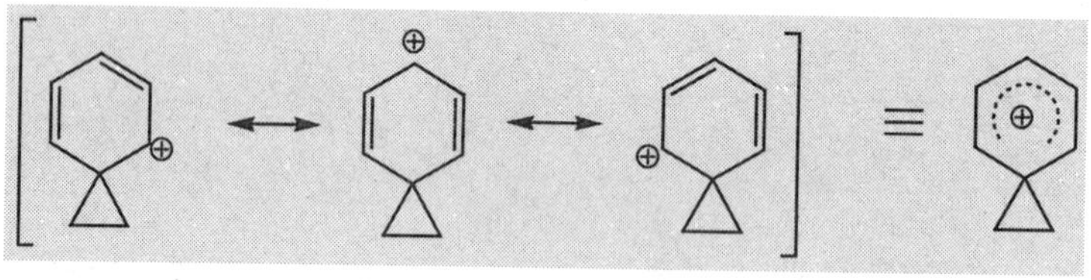

Overlapping of cyclopropyl orbitals with π-orbitals of the phenyl ring helps in the delocalization of electron-deficiency and the resultant positive charge into the cyclopropane ring.

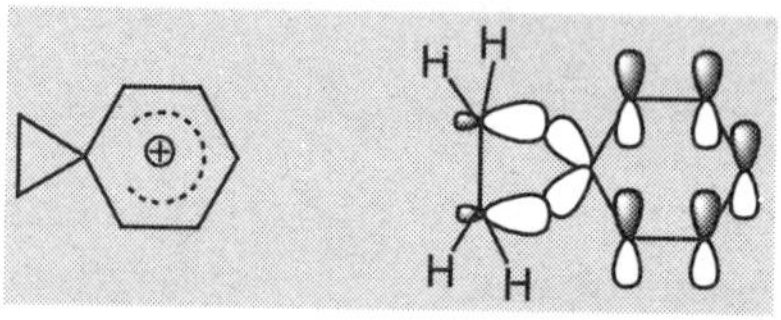

It has also been shown by NMR spectroscopy that *tertiary* phenonium ions exist as open-chain ions to some extent.

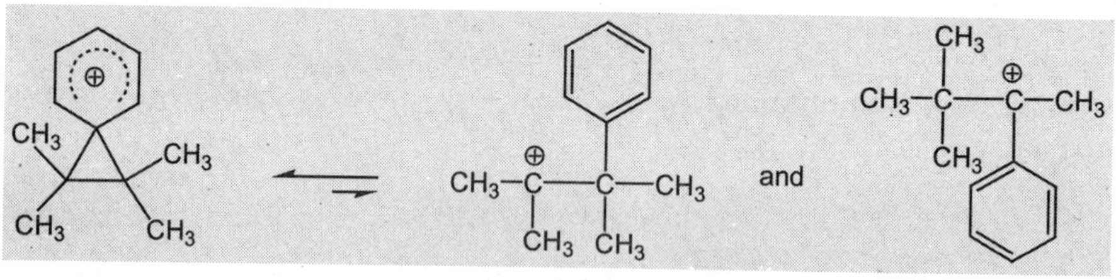

Such equilibrium is not possible for the phenonium ion derived from β-phenylethyl chloride ($C_6H_5CH_2CH_2Cl$) due to the primary nature of the carbocation, which is very unstable. The NMR studies have also revealed the absence of aromatic character in phenonium ions.

D. Sigma Bond Participation

The conversion of camphene hydrochloride (**53**) into *iso*bornyl chloride (**54**) was reported by Meerwein in 1922.

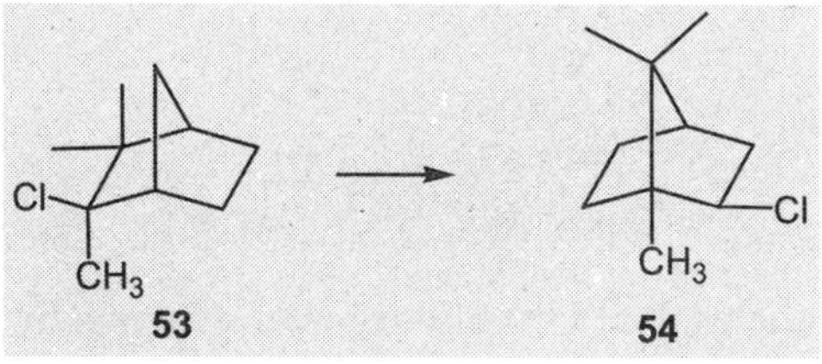

Mechanism of this reaction was investigated for a long time, and is now believed to proceed in the following way.

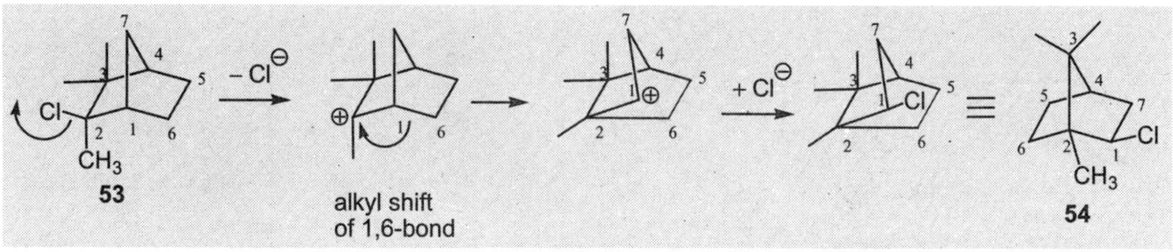

There are two other observations: the rate of solvolysis of **53** is thousands times faster than *tert*-butyl fluoride, and then there is no trace of the *endo* isomer of **54**.

To explain these observations, Winstein carried out experiments with simpler systems. Acetolysis of diastereomeric pair (*exo* or *endo*) norbornyl brosylate (**55** and **56**) provided an enantiomeric pair of *exo*-norbornyl acetate (**57**).

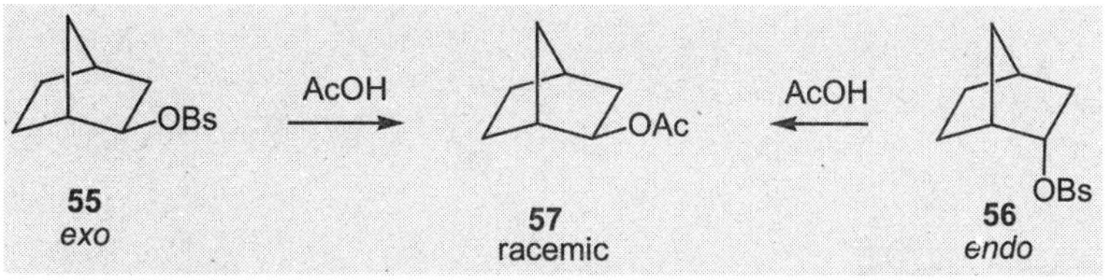

The same product was obtained by taking either of the enantiomers of *exo* and *endo* norbornyl brosylate.

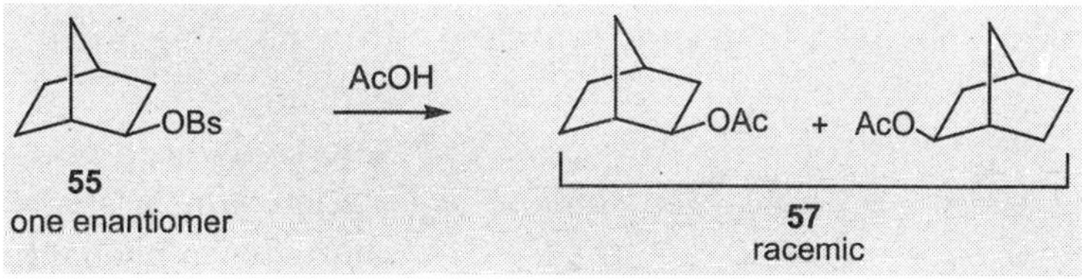

Further, it was found that the rate of acetolysis of *exo* isomer (**55**) is 350 times more than that of *endo* brosylate (**56**).

All these experimental observations were explained by the intermediacy of a nonclassical ion, which is formed by the loss of brosylate anion with anchimeric assistance from C6,C1 bond (rate -acceleration).

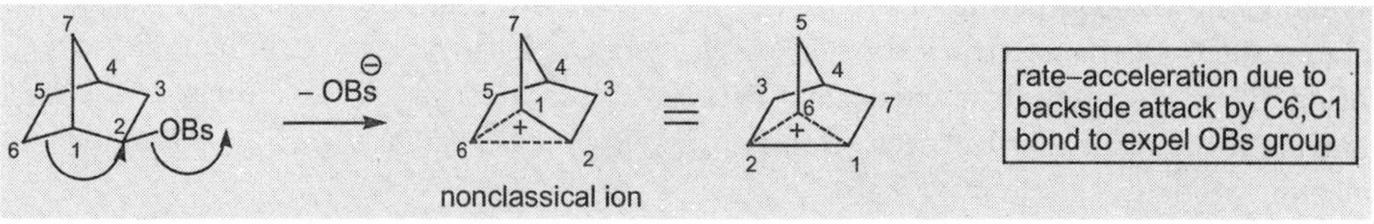

The bridged cation (nonclassical ion) undergoes solvolysis by attack at either C2 or C1 to yield the observed product.

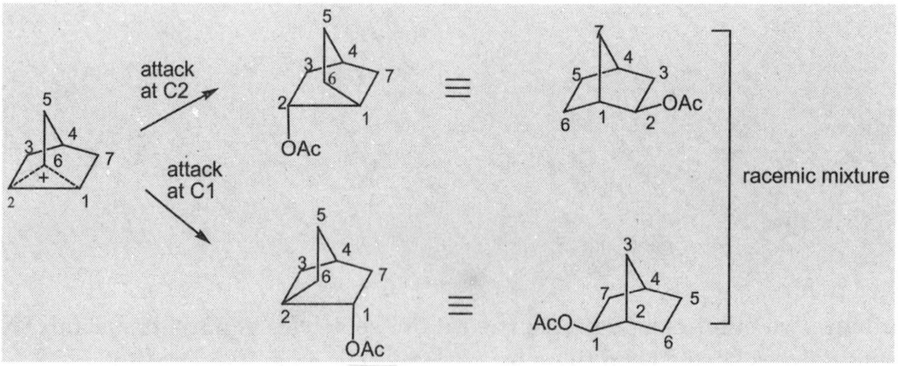

However, in the case of *endo* brosylate, C6, C1 bond is not in a position to help in the expulsion of brosylate ion, and hence it undergoes normal S_N1 reaction resulting in the formation of classical cation, which is then stabilized to the nonclassical structure.

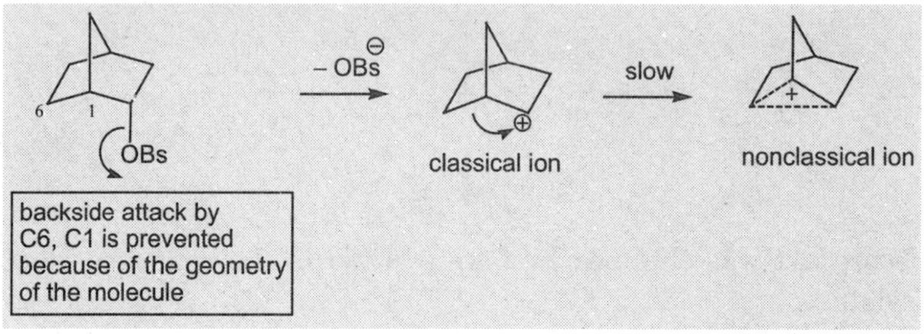

Nonclassical ion, obtained from *endo* brosylate, gives the same product as expected, *exo* and racemic.

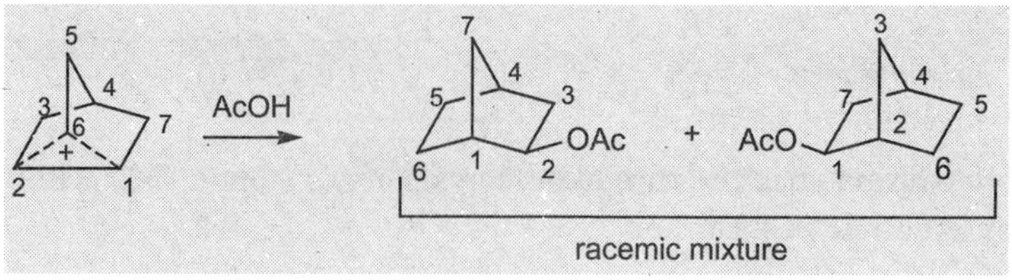

Another example is solvolysis of 2-*endo* norbornyl chloride, which takes place at the same rate as the corresponding cyclopentyl and cyclohexyl derivatives. But the *exo*-isomer undergoes solvolysis 80 times faster than the *endo* isomer indicating anchimeric assistance during the solvolysis of the *exo* compound. The C6–C1 bond is situated at the rearside of the ionizable group, and the σ-electrons attack the carbon bearing the ionizable group, thus facilitating the ionization. The intermediate nonclassical carbocation is attacked by the nucleophile (in this case solvolent molecules) with equal probability at either C1 or C2 resulting in the formation of a racemic product.

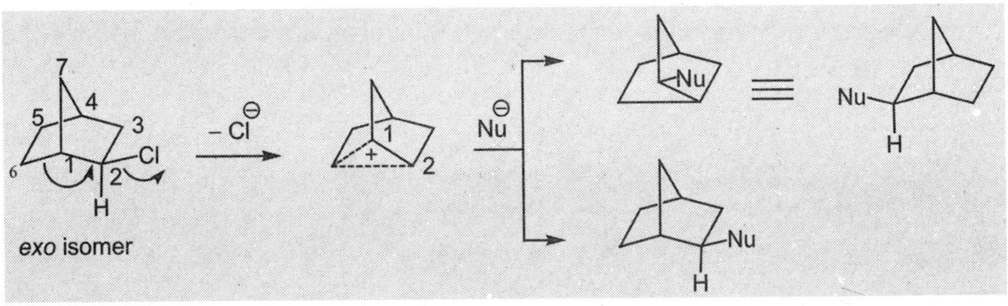

The *endo* isomer does not get such anchimeric assistance, since the C6, C1 bond is situated on the same side of the ionizable group, and consequently, its rate of solvolysis is normal.

Another interesting example is the acetolysis of *exo*-2,2,1-bicycloheptyl bromobenzene-*p*-sulphonate which is 350 times faster than acetolysis of its *endo* isomer. This has again been explained by rearside attack of σ-electrons on the back of the carbon, precisely at the same time as the ionizable group starts leaving the carbon. This leads to the formation of a nonclassical carbocation, which reacts with acetic acid to yield the racemic mixture of acetates. No such anchimeric assistance is available to the *endo* isomer, which undergoes slow acetolysis through classical carbocations.

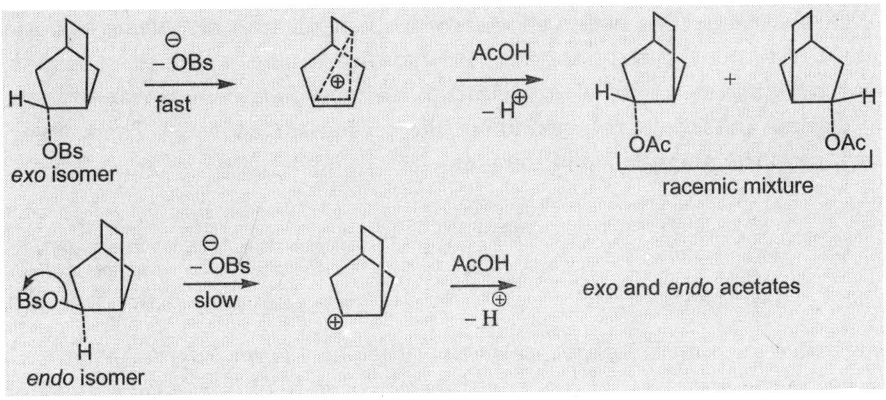

E. Classical *vs* Nonclassical Cations

We have referred to the formation of carbocations in which the positive charge is delocalized by a double bond (not in allylic position) or by a single bond. Such cations are called *nonclassical carbocations* as opposed to the *classical carbocations* in which the positive charge is localized on one carbon atom or delocalized by resonance involving an unshared pair of electrons or a double bond in the allylic position.

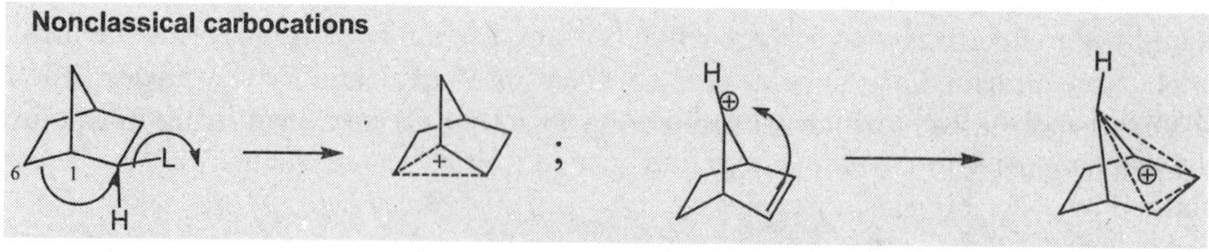

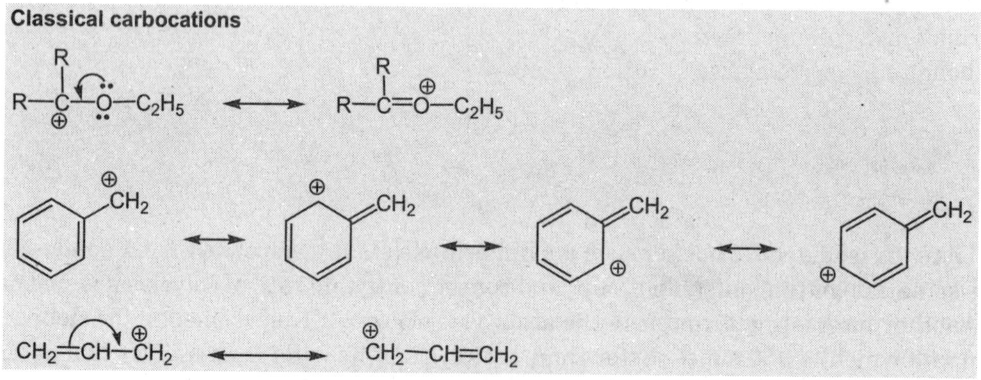

It may also be noted that participation of a double or a single bond in the departure of the leaving group may or may not lead to the formation of a nonclassical carbocation, which is a discrete reaction intermediate in many reactions.

Brown has taken a different view regarding nonclassical or bridged ions as intermediates in solvolysis reactions, and has explained the kinetic and stereochemical results on the basis of classical carbocations. The observed stereospecificity in the acetolysis of optically active *erythro*-and *threo*-3-phenyl-2-butyl tosylate, for instance, can be explained by assuming that the solvolysis intermediate is a pair of *equilibrating classical ions*, whose rate of interconversion is much faster than the rate of rotation about the carbon-carbon bond. The solvent, therefore, cannot attack from the phenyl side of the ions.

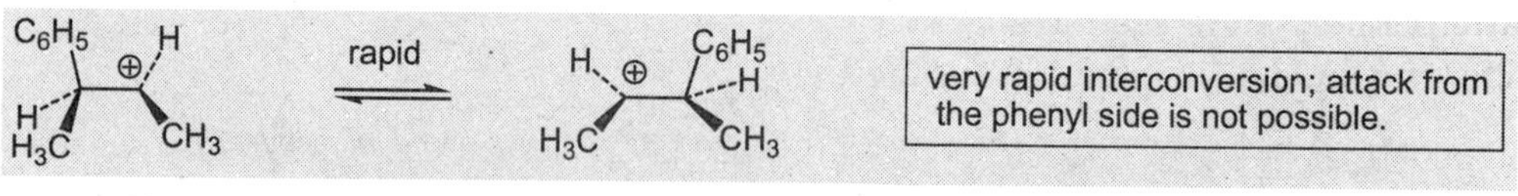

The suggested anchimeric assistance given by the phenyl group in the formation of a bridged phenonium ion in the solvolysis of 3-phenyl-2-butyl tosylate (**47**) was thus questioned. It was shown that there is not much difference in the reaction rates between **47** and unsubstituted *sec*-butyl tosylate. Why is there no rate acceleration in the solvolysis of **47**? It was suggested that the formation of bridged ions need not be invoked in such cases, as a pair of *rapidly equilibrating* open carbocations with phenyl on one carbon and then on the other carbon, shields backside attack by the solvent, explaining the observed stereochemistry.

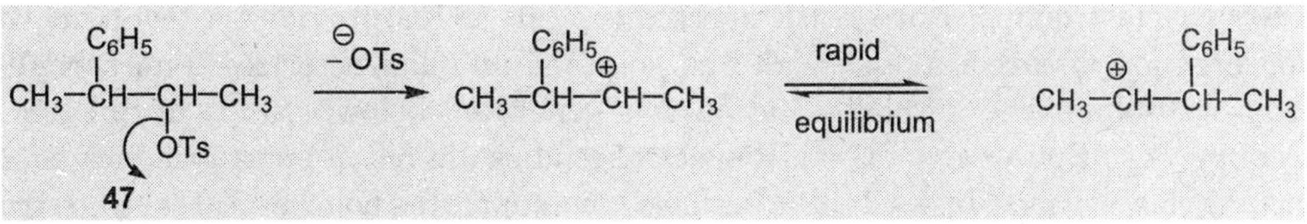

Brown has also challenged the formation of nonclassical cations in the reactions involving the solvolysis of norbornyl systems, and has argued that all the experimental observations can be explained by describing the intermediate as *rapidly equilibrating classical cations*. The main arguments are: (1) The rate of hydrolysis of the *exo* isomer is normal, and of the *endo* isomer is abnormally low because of steric hindrance to the removal of the leaving group in that direction. (2) Formation of racemic mixture can be explained by assuming solvolysis *without* the assistance of 1,6 bond to generate classical cation, which is in rapid equilibrium with another cation.

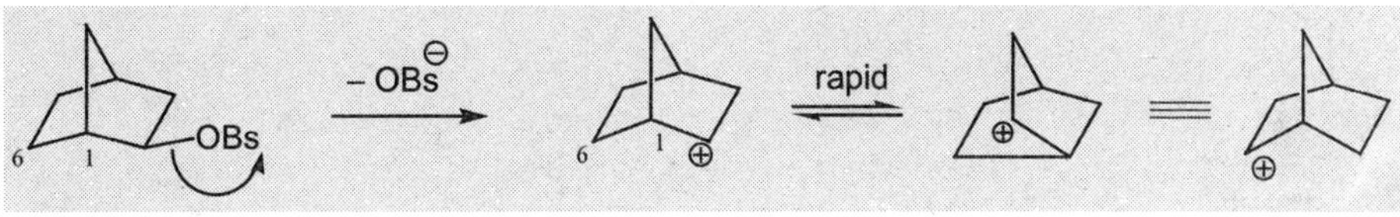

There is rapid interconversion, like the action of a windshield wiper, resulting in the attack of nucleophile from the same side as was proposed for a nonclassical cation.

We have to understand that when a nonclassical ion is lower in energy than the classical ion, it will be an intermediate (not the transition state), and the bridged ion will be the most stable structure. On the other hand, if the nonclassical ion is higher in energy than the classical structure, it will be a transition state. Thus, what was suggested is that a nonclassical structure is not an *intermediate,* but a *transition state*. The two alternative structures for norbornyl cations received extensive attention, and a lot of work was done between 1950 and 1980.

Anchimeric assistance by C6–C1 bond has been invoked to explain the solvolysis of camphene hydrochloride (**53**), which was found to be thousands of times faster than *tert*-butyl chloride. It was shown that many *tert*-cyclopentyl chlorides also solvolyze much faster than *tert*-butyl chloride. It was suggested that faster solvolysis of camphene hydrochloride (**53**) (and other *tert*-butyl chlorides) is due to the relief in steric strain associated with solvolysis. For example, with the loss of chloride ion in **53**, sp^2-hybridized carbon moves in the plane of the ring. Thus, the cation is now more comfortable sterically, as there is disappearance of four non-bonded contacts (two for chlorine and two for methyl).

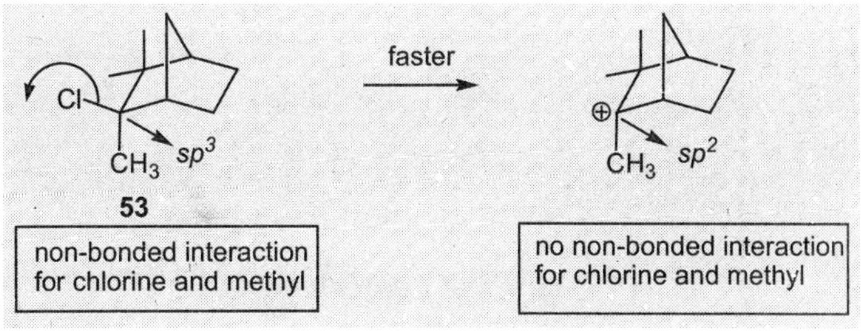

Advent of low temperature NMR spectroscopy in 1970s provided a technique for the direct observation of carbocations. Olah has prepared and characterized 2-norbornyl cation using this technique. The norbornyl cation was generated below –150°C in SbF_5–SO_2 and FSO_3H–SbF_5–SO_2. Proton and ^{13}C NMR spectral studies did not support existence of rapidly equilibrating classical ions, instead these findings have supported nonclassical bridged structure. Theoretical calculations suggested stabilization energy of about 6-10 kcal (~25-42 kJ)/mol for the nonclassical bridged ion. Further studies have revealed that for the norbornyl cation, bridged ion is more stable than the classical ion, even in solution.

The low temperature NMR studies have also revealed that, in case, departure of the leaving group leads to the formation of a stable carbocation, there may not be any participation of σ-bond. Spectral study of 2-methyl- and 2-ethylnorbornyl cations shows less delocalization, whereas 2-phenyl norbornyl cation has a classical structure without any delocalization.

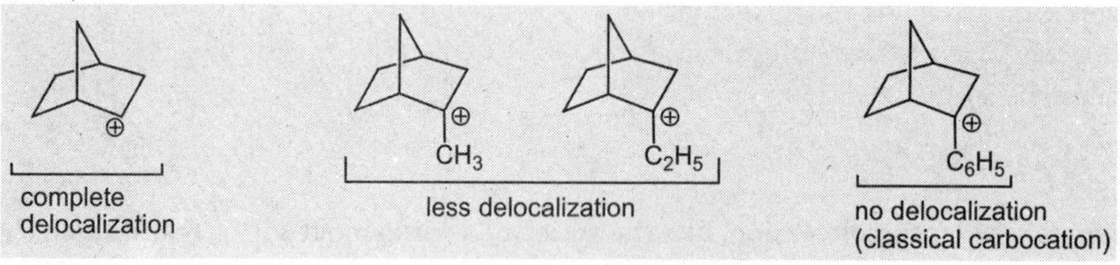

As methoxy and halo groups also stabilize the positive charge, 2-methoxy and 2-chlorobornyl cations show no delocalization, and hence have a classical structure.

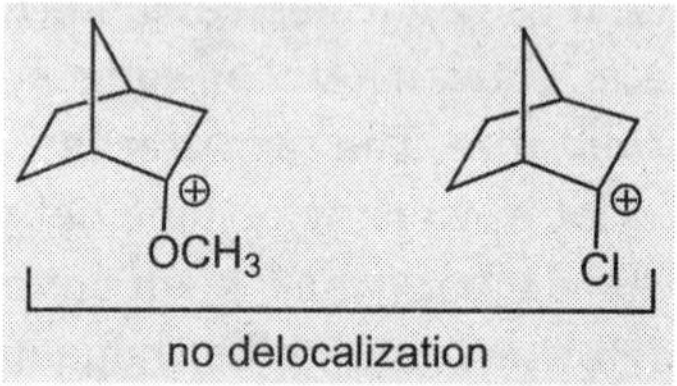

6.5 S_Ni AND SET MECHANISMS

A. S_Ni Mechanism

Alcohols react with thionyl chloride to give chlorosulphites, which subsequently decompose to alkyl halides.

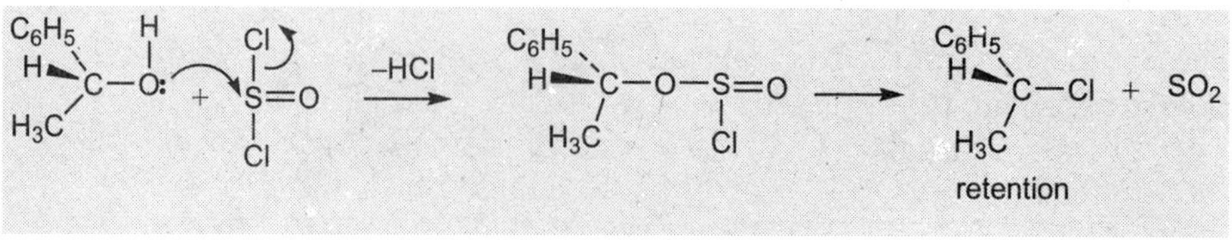

$$\text{Rate} = k\,[\text{ROH}]\,[\text{SOCl}_2]$$

Unexpectedly, the alkyl chlorides have been found to have the same configuration as the starting alcohols. The stereochemistry of the reaction is thus inconsistent with the rules formulated for either S_N2 or S_N1 reactions, and a different reaction path, termed the S_Ni mechanism (*substitution, nucleophilc, internal*) has been proposed. This consists in internal transformation through a cyclic transition state in which chlorine can attack only from the frontside, leading to the retention of configuration.

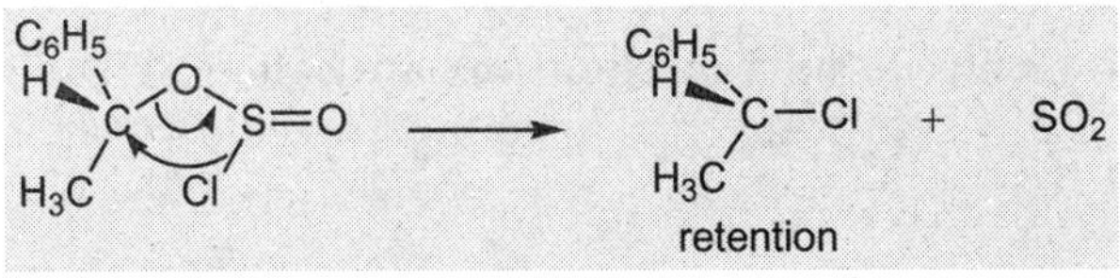

Support for S_Ni mechanism comes from the observation that the addition of pyridine to the reaction mixture leads to the formation of alkyl chlorides with *inverted configuration*. This is to be expected as the free chloride ions attack from the rear in the normal S_N2 fashion, leading to inversion of configuration.

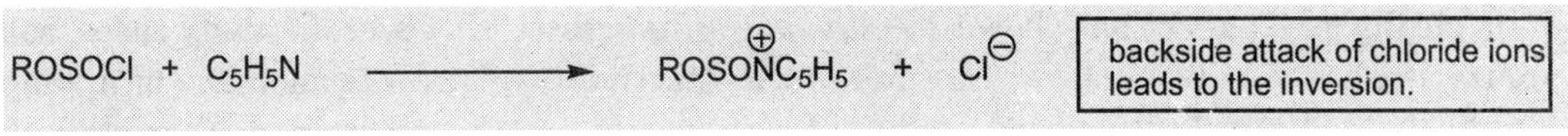

Recent evidence, however, points out that the decomposition of chlorosulphite begins, like a typical S_N1 reaction, with the formation of an ion pair. A part of the leaving group (chloride ion) then attacks from the front, since it is unable to get to the rear.

Rearrangement in the carbocation part, before the nucleophilic attack, may lead to the rearranged product, as seen in the following example:

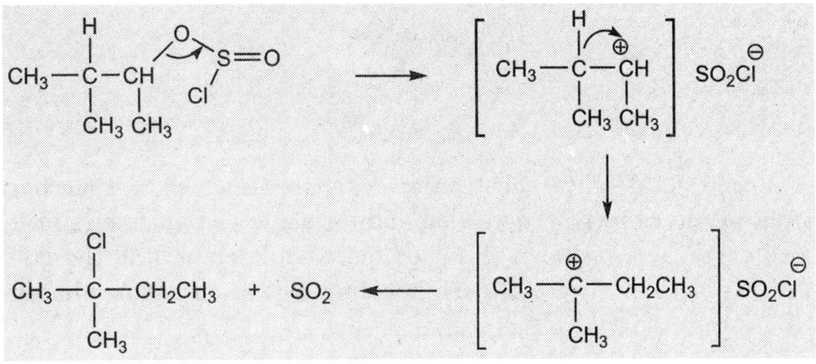

The S_Ni mechanism is rarely encountered, another example being the reaction of 2-octyl alcohol with phosgene that proceeds with retention of configuration.

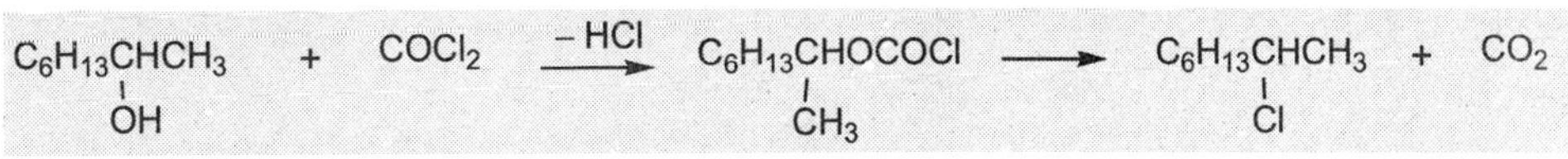

B. SET Mechanism

Involvement of free radicals (or radical ions) has been shown in many aliphatic nucleophilic substitution reactions. Such radicals have indeed been trapped by ESR spectroscopy. In the reaction of the type,

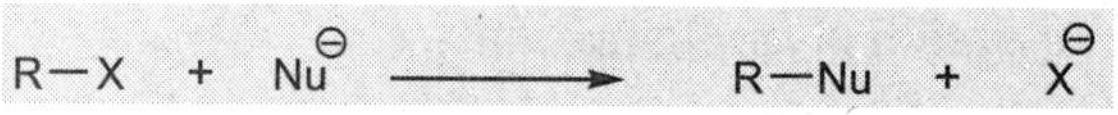

following steps may be involved. Such mechanisms are called SET (single electron transfer) mechanisms.

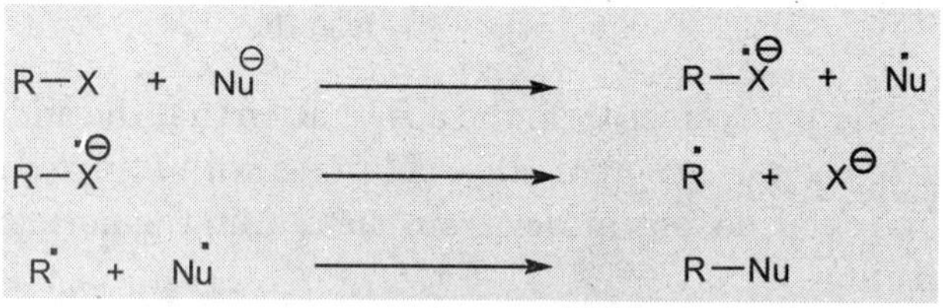

Typically, these reactions show *inversion of configuration.* However, in many cases, some racemization has also been observed. In case the reaction involves intermediacy of a simple free radical, a completely racemized product (R–Nu) would have been obtained. To explain this observation, it has been suggested that Nu^- approaches R–X from the backside, and the radical $R^\bullet$ is tied up in a solvent cage having the nucleophile on the back.

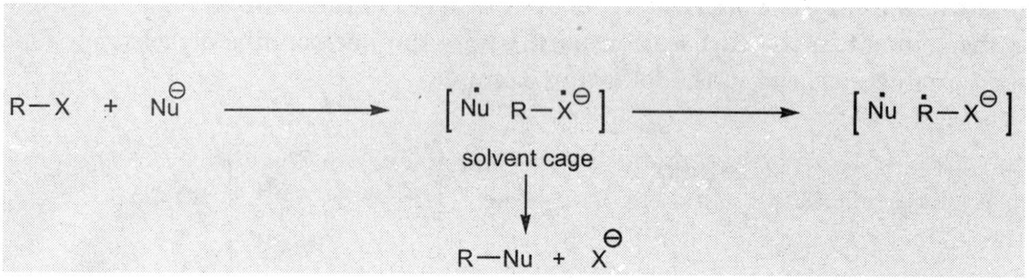

In case of certain nucleophilic substitution reactions in which S_N1 mechanism is highly probable, it has been shown by ESR detection of intermediates that free radicals are actually involved. In such cases, a carbocation is a good electron acceptor, and the nucleophile is an electron donor. The reaction between the triphenylmethyl cation and *tert*-butoxide ion can be explained by SET mechanism.

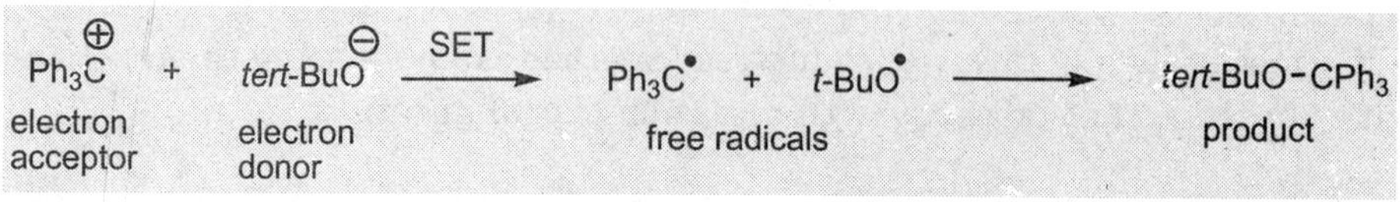

6.6 FACTORS AFFECTING REACTIVITY IN SUBSTITUTION REACTIONS

We have seen that the hydrolysis of methyl bromide takes place by the S_N2 mechanism, whereas *tert*-butyl bromide hydrolyzes by the S_N1 mechanism. Other alkyl halides undergo hydrolysis by either mechanisms, depending on their structures and the experimental conditions employed. How can we predict which of the two mechanisms apply in a given reaction of a particular reactant? It has been found that at least four factors influence the relative reactivity of aliphatic compounds in substitution reactions.

A. Structure of Alkyl Group

In S_N2 reactions, the order of reactivity of various alkyl halides follows the sequence, CH_3X > primary > secondary > tertiary. Electron withdrawal or electron release by substituents does not affect the rate of the S_N2 reactions. However, it is influenced by *steric hindrance* around the carbon site undergoing displacement, since the nucleophile has to push in among the three carbon substituents to reach the *transition state*. As hydrogen atoms are replaced by the larger alkyl groups, nucleophile finds it increasingly difficult to approach the carbon site. Also, if the substituents are bulky, the transition state would be heavily crowded, and the non-bonding interaction between various substituents would raise its energy.

While considering the S_N2 reaction of alkyl bromides with iodide ions, it was observed that if we presume the relative rate of displacement of bromide to be 1 for ethyl bromide (C_2H_5Br), it is 150 for methyl bromide (CH_3Br), and 0.008 and 0.001, respectively, for *iso*propyl [$CH_3)_2CHBr$] and *tert*-butyl bromide [$(CH_3)_3CBr$].

$$\text{R—Br} + \overset{\ominus}{\text{I}} \xrightarrow{S_N2} \text{R—I} + \overset{\ominus}{\text{Br}}$$

The relative rate of displacement for bromide ion in *neo*pentyl bromide ($(CH_3)_3CCH_2Br$) has been found to be 0.00001. This last example, *neo*pentyl bromide, is a primary alkyl halide, but even then it is very unreactive in S_N2 reactions because of steric hindrance in the transition state by methyl groups on the β-carbon.

In contrast to S_N2 reactions, the sequence of reactivity of alkyl halides has the following order in S_N1 reaction: tertiary > secondary > primary > CH_3X. The sequence is so pronounced that it is almost impossible to avoid S_N1 reactions with tertiary halides.

As we know, carbocations are intermediates in S_N1 reactions, and therefore, the factors favouring formation and stability of these ions would increase the reactivity in S_N1 reactions. The alkyl groups are electron-releasing, and so when attached to a carbocation, they tend to release electrons to that carbon, and thus reduce the positive charge by acquiring a little positive charge themselves. Because of the dispersal of the charge, carbocations tend to become more stable. Thus, carbocation stability runs in the same order as alkyl halides reactivity in S_N1 reactions. The order of stability of these carbocations is also explained by hyperconjugative effects.

Interesting examples of the effect of structure on reactivity are provided by compounds with potential leaving groups at *bridgehead carbons*. Thus, 1-chloroapocamphane **(56)** does not hydrolyze upon being refluxed with alcoholic potassium hydroxide for 21 hours. The low S_N2 reactivity of **56** is attributed to the fact that the nucleophile cannot approach from the rear. Moreover, the bridgehead carbon atoms cannot assume coplanarity because of the steric requirements of the bridged ring structure; ionization leading to the formation of a carbocation is also prevented. Similar arguments explain the failure to replace bromine in 1-bromotriptycene **(57)**.

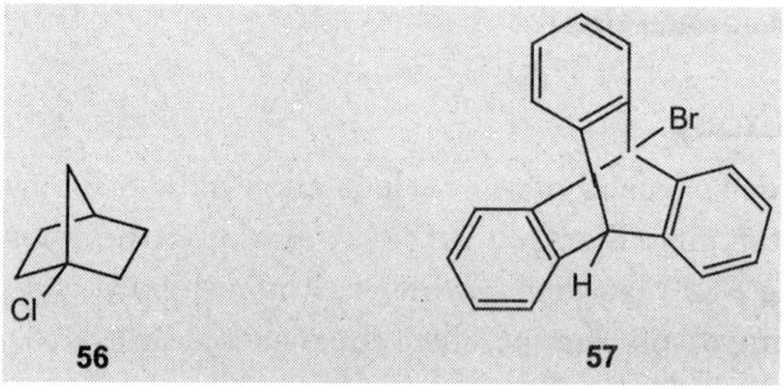

Steric Acceleration

Another factor that favours S_N1 substitution is steric. If the alkyl groups attached to the carbon undergoing substitution are large, they facilitate the attainment of coplanarity of the carbocation, because in the process the bond angle is necessarily increased from 109.5° to 120°, thereby, reducing the non-bonded interactions amongst the alkyl groups.

As a consequence of these structural features, primary halides mostly react by the S_N2 mechanism and tertiary halides mostly by the S_N1 mechanism. Secondary halides may react by either, depending upon the conditions of the reaction, notably the solvent and nucleophile.

Effect of Unsaturation at α- and β- Carbons

Vinylic halides and aryl halides do not undergo S_N1 or S_N2 reactions. There are two reasons for the failure of S_N1 reactions with these substrates. First, vinylic and aryl cations are very unstable, in fact, more unstable than primary carbocations because the positive charge is located on an sp^2 hybridized carbon, which is very resistant to becoming positively charged. Second, sp^2 carbon forms much stronger bond with the halogen, and it is very hard to break the carbon-halogen bond.

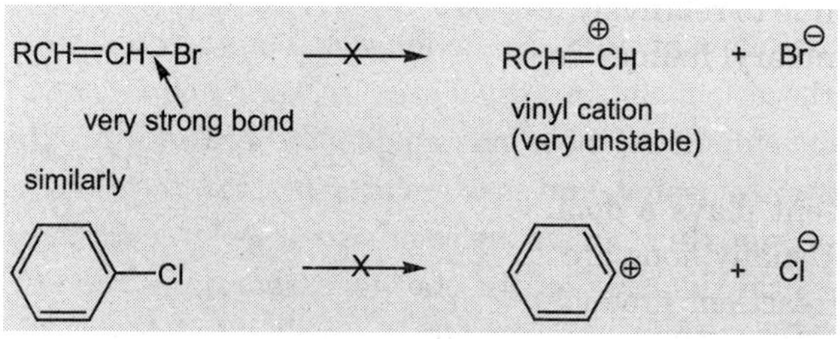

These compounds do not undergo S_N2 reactions, because when the nucleophile approaches the backside of the sp^2 carbon, it is repelled by the π-electron cloud of the double bond.

On the other hand, allylic and benzylic halides readily undergo S_N1 reactions because of the ease of formation of resonance-stabilized carbocations.

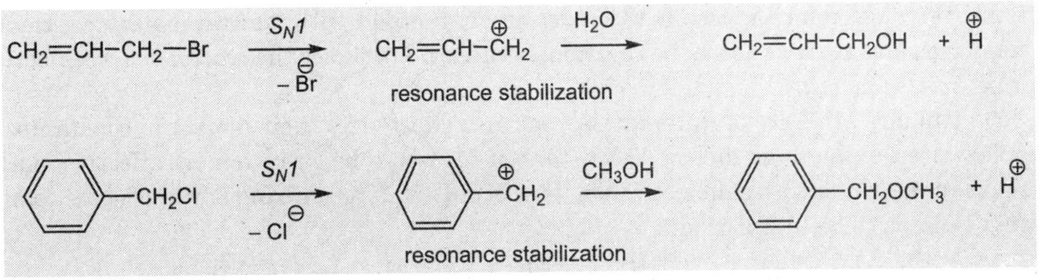

When two different groups are bonded to the sp^2 carbons of an allylic halide, two products are obtained.

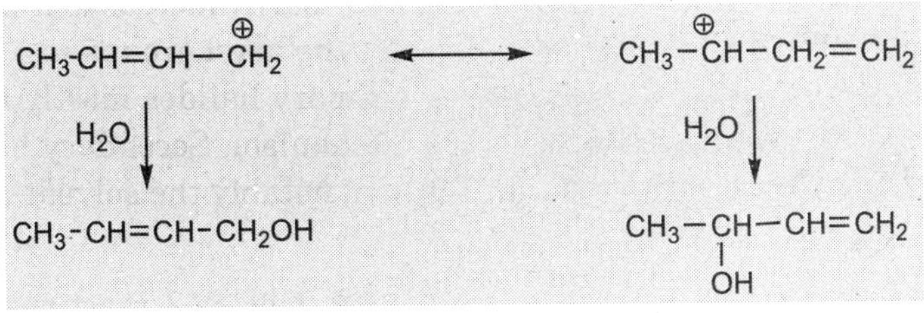

In summary, we can say that methyl halides and primary alkyl halides undergo only S_N2 reactions, as an S_N1 reaction involves the intermediacy of *highly unstable carbocations.* Tertiary alkyl halides, on the other hand, undergo only S_N1 reaction because of the ease of formation of a *stable carbocation,* and also due to steric hindrance making them unreactive in an S_N2 reaction. Secondary alkyl halides as well as benzylic and allylic halides undergo both S_N2 and S_N1 reactions due to relatively less steric hindrance and formation of relatively stable carbocations. Vinylic and aryl halides do not undergo S_N2 or S_N1 reactions.

B. Nature of Solvent

The nature of the solvent plays a dominant role in deciding the mechanism and rate of a substitution reaction. Usually, ions are stabilized by solvation in good ionizing solvents, and therefore, substitution reactions producing ions from neutral reactants are accelerated by good ionizing solvents.

The S_N1 reaction begins with the initial ionization of a neutral molecule, and thus, it is facilitated by good ionizing solvents. Cations are solvated by solvent molecules through coordination with electron-pairs, whereas anions are solvated by solvent molecules through hydrogen-bonding.

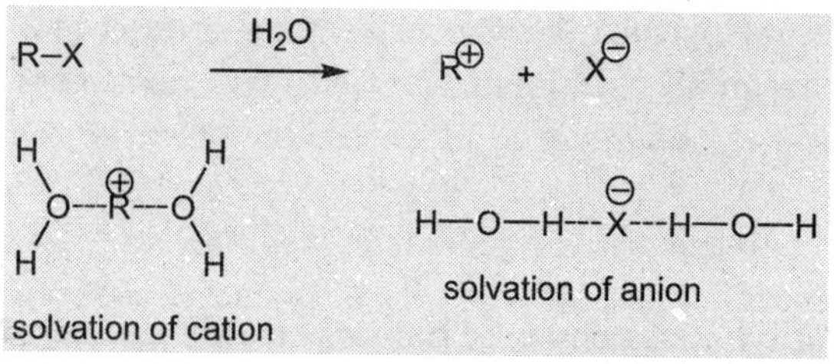

Because of these bonds, the energy of transition state of an S_N1 reaction is lowered. In S_N2 reactions, the most common case is the reaction of an anion with neutral molecules. Hence a good ionizing solvent slows down the reaction, because it stabilizes the *reactant* more effectively than the *transition state*.

The ionizing property of different solvents is reflected in their dielectric constants. For example, water, with a high dielectric constant of 81, would be a much more effective ionizing solvent than methanol with a dielectric constant of only 33. The ionizing power of some common solvents is found to be in the order:

$$H_2O > HCOOH > CH_3OH > CH_3CH_2OH > CH_3COOH$$

It is sometimes possible to change the mechanistic course of a substitution reaction by changing the polarity of solvent. Methyl bromide, which normally undergoes hydrolysis by the S_N2 mechanism in aqueous alkaline ethanol, hydrolyzes by the S_N1 mechanism in formic acid containing a small amount of water.

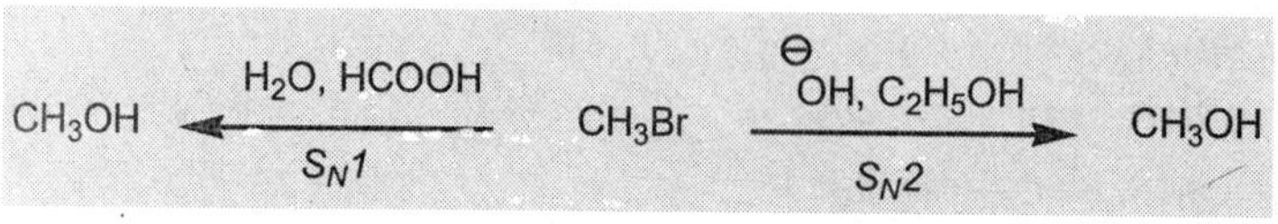

C. Nature of Leaving Group

The reactivity of a substrate in substitution reactions is partly determined by the nature of the leaving group. Since the leaving group breaks away as a base, weak bases are easier to displace than strong bases. And thus, we may generalize that the less basic the substituent, the more easily is it displaced by the solvent in S_N1 reactions, or by an attacking nucleophilic reagent in S_N2 reactions. For example, alcohols are resistant to substitution reaction in non-acidic media, because of the strong basicity of the (HO^-) group which makes it a poor leaving group. However, in presence of acids, the hydroxyl function is displaced as H_2O, which being a weak base, is a good leaving group. Consequently, alcohols undergo substitution reactions only in acidic media.

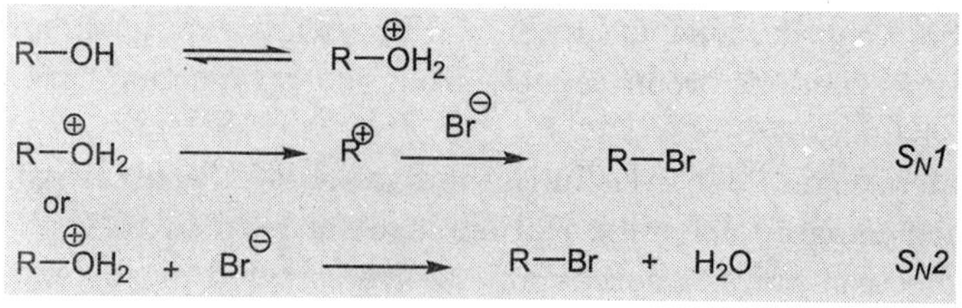

Other strong bases, which are very poor leaving groups in substitution reactions, are: H^-, RO^-, and R_2N^-. In case of alkyl halides, the ease of separation of halide ions is in the order, $I^- > Br^- > Cl^- > F^-$. This is the reverse of the order of carbon–halogen bond strength, as expected.

In general, we can say the *weaker* the basicity of a group, the *better* is its leaving ability. Weak bases are always stable bases as they are not bonded as strongly to the carbon as strong bases. A weaker bond is thus easier to break. As larger atoms are better able to stabilize their negative charge, the relative basicities of the halide ions are:

(weakest base) $I^- < Br^- < Cl^- < F^-$ (strongest base)

As weak bases are better leaving groups, relative leaving abilities of halide ions follow the order: (best leaving group) $I^- > Br^- > Cl^- > F^-$ (worst leaving group). Relation of alkyl halides in S_N2 reactions thus follows the order:

(most reactive) R—I > R^-Br > R^-Cl > R^-F (least reactive)

R–Br > R–Cl > R–F

D. Nature of Nucleophilic Reagent

Since the transition state for the S_N2 reaction involves a nucleophile, the nature of nucleophilic reagent plays an important role in determining the rate of the reaction. All nucleophiles are bases but the nucleophilicity of a reagent, that is its ability to donate an electron pair to an electron deficient carbon, may or may not parallel its basicity, which is defined as its ability to donate an electron pair to a proton. Hydroxide ion (HO^-) is a strong base, and also happens to be a good nucleophile. On the other hand, iodide ion is a very weak base, but still a good nucleophile. Relative nucleophilicities of some common nucleophiles in S_N2 reactions are as follows:

$C_6H_5S^-$(4,70,000), CN^- (5,000), I^- (3700), OH^-, $C_2H_5O^-$ (1,000), Br^- (500), Cl^- (80), NO_3^- (1).

The above relative data on the reactivities of the halide ions reveals that larger atoms of the same family of elements are more nucleophilic than smaller ones. This size effect may be explained in two ways: firstly, in the larger atoms, the external electrons are less tightly bound by the nucleus, and consequently, they are more easily available to the electron-deficient carbon in the *S_N2 transition state* allowing more effective bonding. In other words, because of their big size, the atoms become more *polarizable,* and hence, better nucleophiles. The second explanation originates from the fact that a nucleophile is surrounded by solvent molecules through hydrogen bonding, and before attacking a carbon atom the solvated nucleophile must break away from some of these bonds. It is well known that the solvation energies of smaller ions are always greater than those of larger ions, and so the latter are better nucleophiles. This has been confirmed by the observation that in the non-hydroxylic solvents, such as acetone, iodide ion is less nucleophilic than bromide ion. Thus, basicity, polarizability, and solvation are all involved in determining the nucleophilicity of a given reagent.

Hard and Soft Nucleophiles: Many nucleophiles, such as RS^-, I^-, and R_3P, attack on saturated carbon to form a covalent bond. These nucleophiles which are not basic, large, neutral, and normally having high-energy electrons are referred to as *soft nucleophiles*. On the other hand, there are *hard nucleophiles,* such as RO^- and NH_2^-, which attack on the carbonyl carbon. These species are normally small with closely held electrons, basic, and have high charge density.

Steric factors in the attacking nucleophile may sometimes assume significance in S_N2 reactions. Quinuclidine (**58**), for instance, reacts hundreds of times faster than triethylamine

(**59**) with *iso*propyl iodide. This should not appear surprising if we consider that the substituents in **58** are tied back by the ring system, whereas freely rotating ethyl groups of **59** partly cover the unshared pair of electrons on nitrogen.

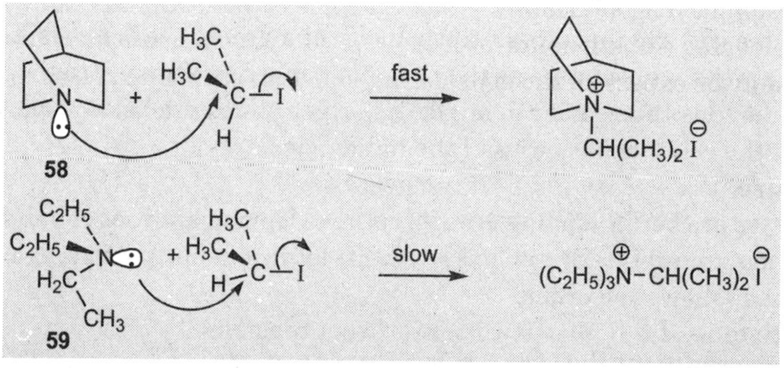

In an S_N1 reaction, the rate-determining step is the ionization of the substrate. As a result, nucleophilicities of nucleophiles do not change the rate of the reaction. However, if there are several nucleophiles present in the reaction mixture, the intermediate carbocation will form a product with the better nucleophile. In this way, depending upon their relative nucleophilicities, various nucleophiles may change the *product composition,* but the rate of reaction remains *unchanged.*

Synthetic Utility of Nucleophilic Substitution Reactions

Nucleophilic substitution is not only very important because of mechanistic considerations, but it also has considerable practical utility. Many useful synthetic procedures involve the nucleophilic substitution of one group by another; some typical examples are: preparation of alcohols by hydrolysis of alkyl halides, preparation of alkyl halides by treating alcohols with halogen acids, synthesis of ethers by Williamson's method, cleavage of ethers by hydriodic acid, synthesis of amines by the action of ammonia on alkyl halides, synthesis of higher acetylenes by the action of acetylides on alkyl halides, and alkylation of active methylene compounds with alkyl halides.

An interesting example of substitution reactions is provided by epoxides, which are usually prepared by an internal displacement reaction. Unlike most ethers, epoxides react easily with many nucleophilic reagents due to the ease of opening the highly strained three-membered ring. These substitution reactions resemble other nucleophilic displacements except that the leaving group, which is the oxygen of the oxide ring, remains a part of the product molecule. The epoxides may be cleaved either under acidic or alkaline conditions.

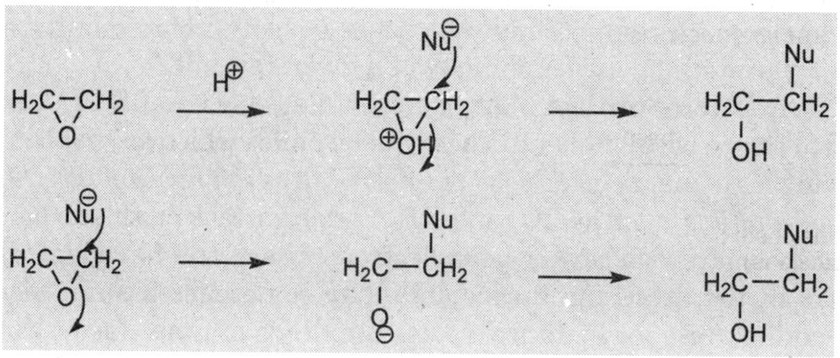

Under acidic conditions, –OH leaves as a neutral species, and therefore, an epoxide ring can be opened up even by relatively weak nucleophiles.

Under alkaline conditions, lower reactivity of the non-protonated epoxide is compensated by more strongly nucleophilic reagent. However, acid-catalyzed reactions of unsymmetrical epoxides having electron-releasing substituents proceed *via* S_N1 mechanism involving carbocation intermediates. As may be expected, the epoxide ring opens in such a way so as to give the more stable carbocation.

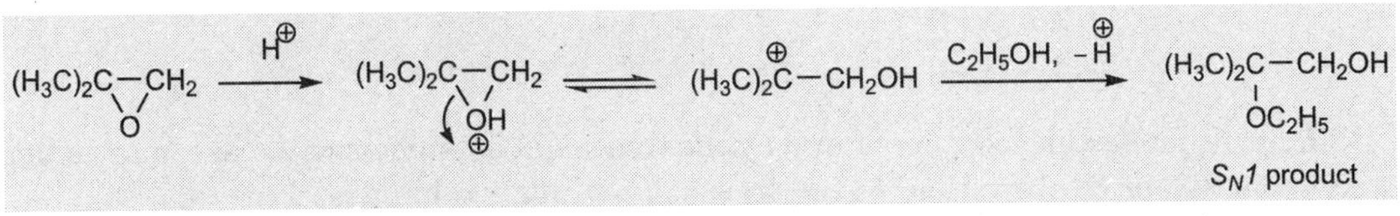

Under alkaline conditions, however, the nucleophile attacks the less substituted carbon of the epoxide ring because of steric factors, and normal S_N2 product is obtained.

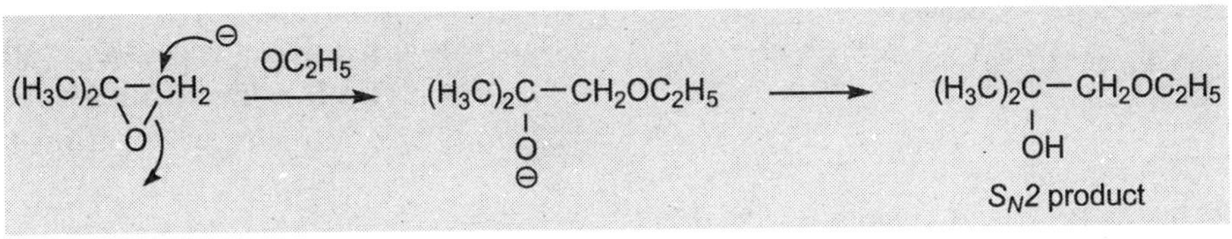

6.7 AMBIDENT NUCLEOPHILES

Many nucleophiles may be represented by two or more resonance forms in which the unshared pair of electrons may reside on two or more atoms. In these cases, the nucleophiles may attack in two or more different ways to give different products.

Removal of a proton from ethyl acetoacetate (**60**), for instance, gives an ion which is a resonance hybrid of **61** and **62**.

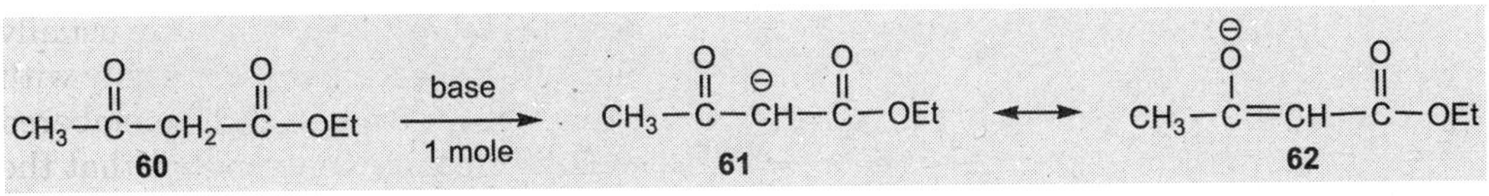

Such ions are called *ambident anions*, since they have two possible attacking atoms leading to the formation of two compounds. Similar type of anions are derived from compounds such as ethyl malonate and acetylacetone.

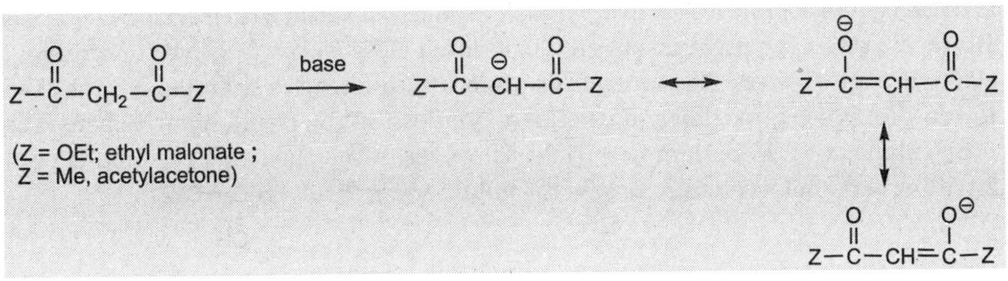

On treatment with alkyl halides, these ions attack the carbon (leading to C–alkylation) or oxygen (leading to O–alkylation).

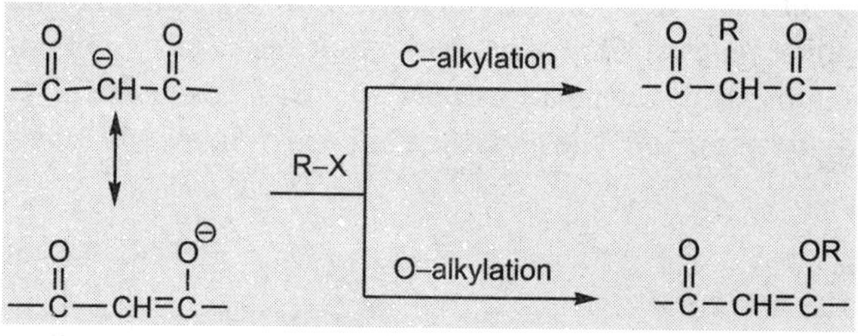

Similarly, phenoxide ions, generated by the treatment of phenols with base, may undergo C–alkylation (**63**) or O–alkylation (**64**) on treatment with alkyl halides.

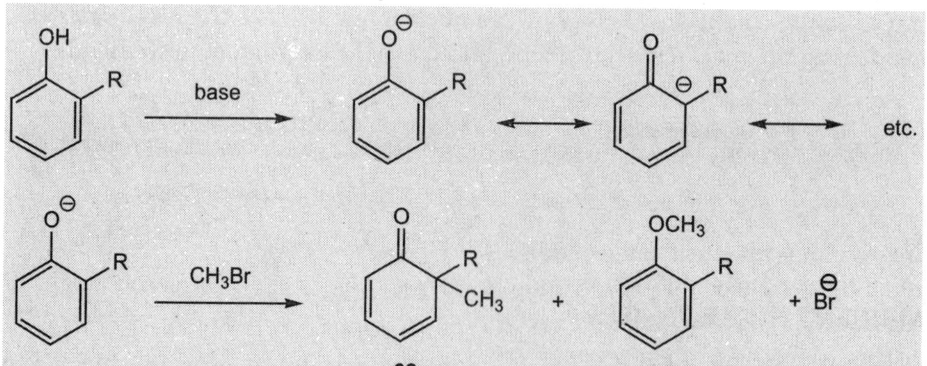

63 **64**

The positive counterion plays an important role in deciding the atom which acts as a nucleophile. For example, in case of NaCN, the attack of cyanide ion takes place from the carbon atom, whereas in AgCN, attack takes place from the nitrogen atom.

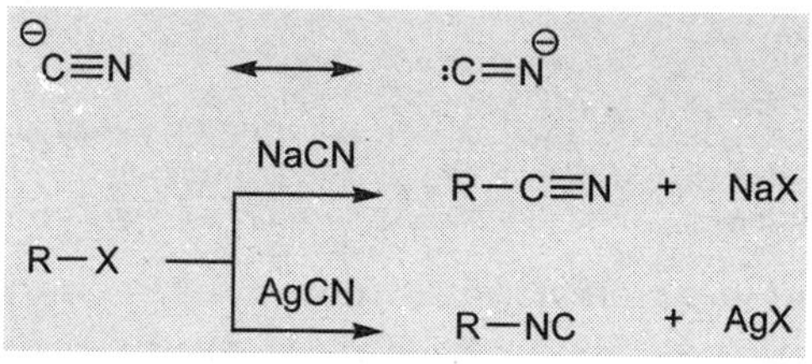

In many cases, it has been observed that the nature of solvent influences the course of reaction. When a nucleophile is relatively free, it normally attacks with its more electronegative atom. However, when the atom is solvated or located close to the positive counterion, attack takes place through the less electronegative atom. An illustrative example is that of sodium-naphthoxide (**65**), which gives the product of O–alkylation (**66**), on treatment with benzyl bromide in dimethylsulphoxide (DMSO), in 95% yield. However, when the same reaction is performed in 2,2,2-trifluoroethanol, the major product is obtained by C–alkylation (**67**).

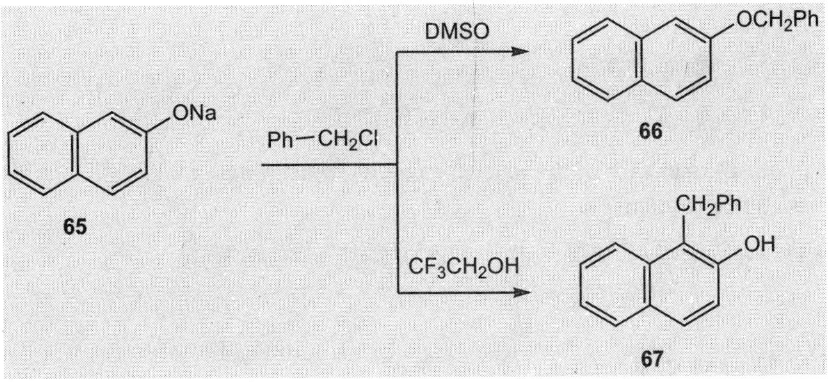

Ambident dianions (**68**) can be generated by treating compounds such as ethyl acetoacetate with two moles of a strong base.

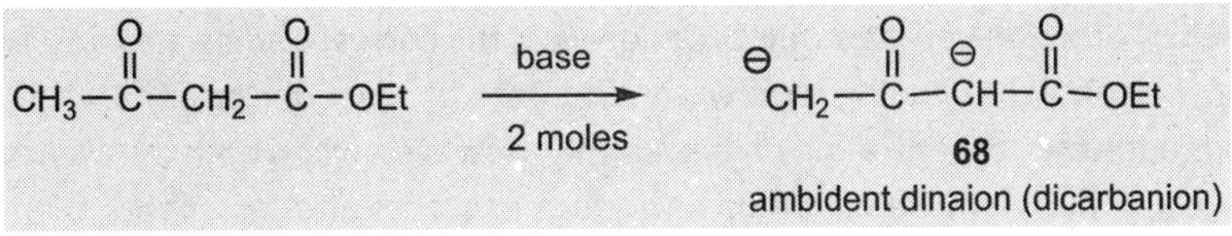

Hydrogen atoms attached to the middle carbon atom of compounds such as ethyl acetoacetate, being flanked by two carbonyl groups (one keto carbonyl and the other ester carbonyl), would be more acidic than those on the end carbon being attached to a single carbonyl group. The strength of the respective conjugate bases will be in the reverse order. Taking advantage of this, the enol carbon atom in the ambident dianion (**68**) can be preferentially alkylated by reacting it with one molecular proportion of an alkyl halide.

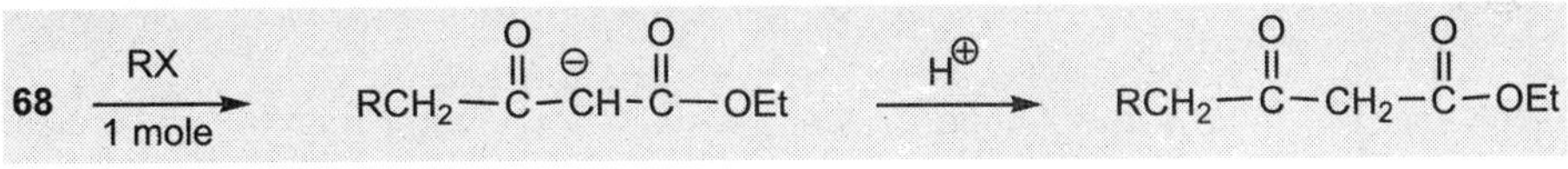

In case one mole of base is used, the alkylation at methylene can be carried out.

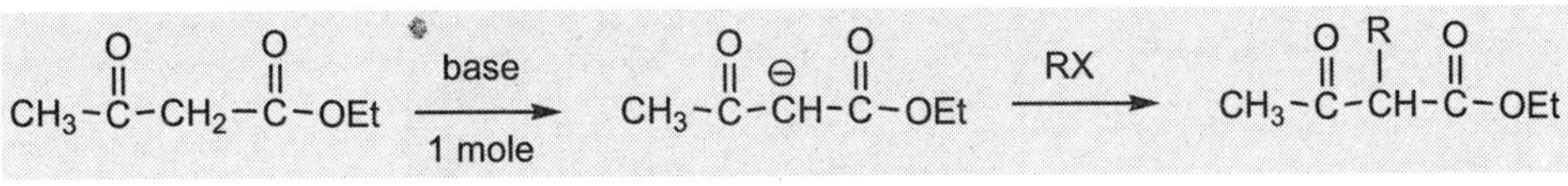

PROBLEMS AND REVIEW QUESTIONS

1. Which of the two alkyl chlorides in each set will undergo faster hydrolysis in an S_N1 process? Suggest reasons.

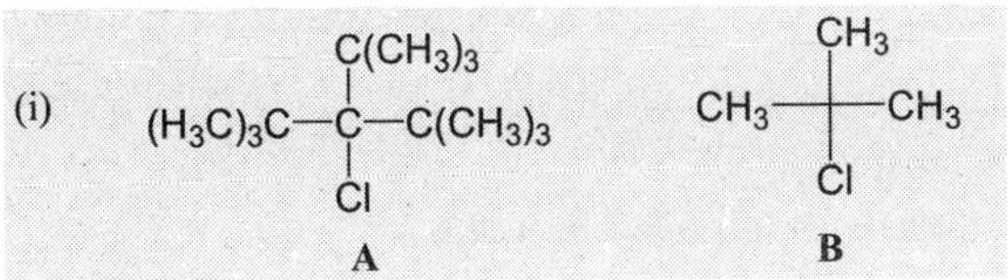

(ii) 

A B

2. Which substrate in each of the following pair will react more rapidly with hydroxide ions under S_N2 reaction conditions?

(i) CH_3CH_2I and CH_3CH_2Cl (A, B)

(ii) p-$BrC_6H_4CH_3$ and $C_6H_5CH_2Br$ (A, B)

(iii) CH_3Cl and CH_2Cl_2 (A, B)

(iv) 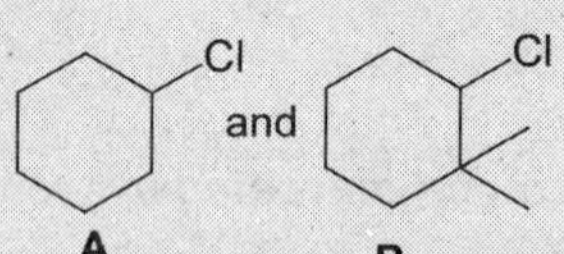

A B

3. Both benzyl bromide and *iso*propyl bromide yield the corresponding alcohols, benzyl alcohol and *iso*propyl alcohol, on heating with water. The rate of reaction for either of the alkyl bromides depends upon the concentration of halides only. However, when both the alkyl halides are treated separately under identical conditions, the former reacts at a faster rate than the latter. What inference do you draw from these observations?
4. Sketch a suitable mechanistic scheme for each of the following transformations.

(i) 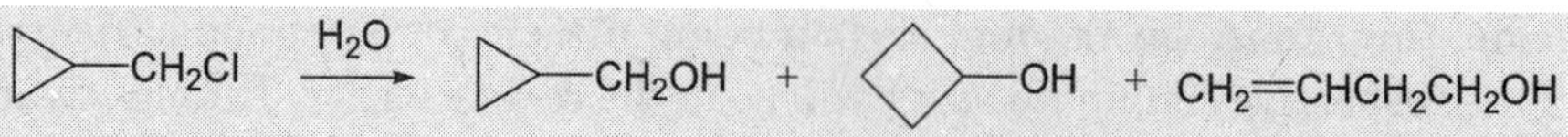

(ii) 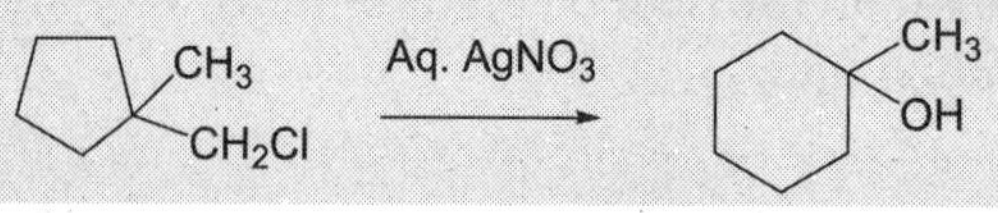

5. Why does the *trans* isomer **I** undergo acetolysis 670 times faster than the *cis* isomer **II**? Discuss the stereochemistry of products also.

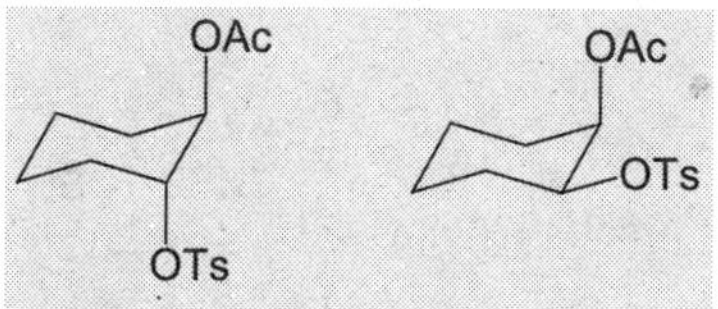

I II

6. *Trans*-2-chlorocyclohexanol gives epoxycyclohexane in high yield on treatment with a base, whereas the *cis* isomer does not react in this way. Explain why?
7. Offer suitable explanation for the following observations:
   (i) Chloromethyl methyl ether ($ClCH_2OCH_3$) undergoes S_N1 reaction.
   (ii) The major products in the reaction of alkyl bromides with $NaNO_2$ and $AgNO_2$ are corresponding nitroalkanes and alkyl nitriles, respectively.
   (iii) Halocyclopropanes are reluctant to undergo S_N2 reaction.

8. Predict the product(s), and write the mechanism for each of the following reactions:

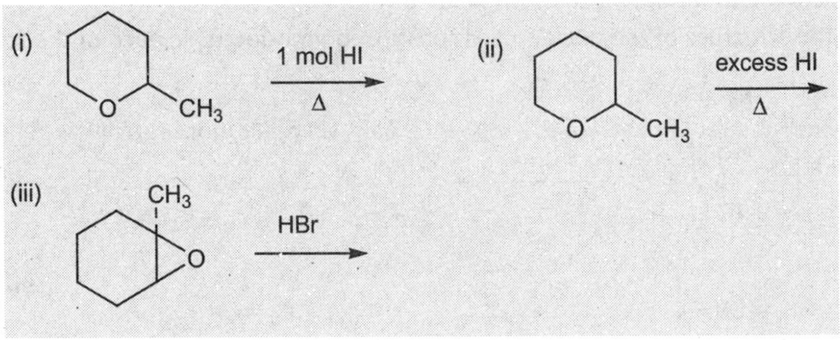

9. Briefly account for the stereochemistry of the following reaction sequences:

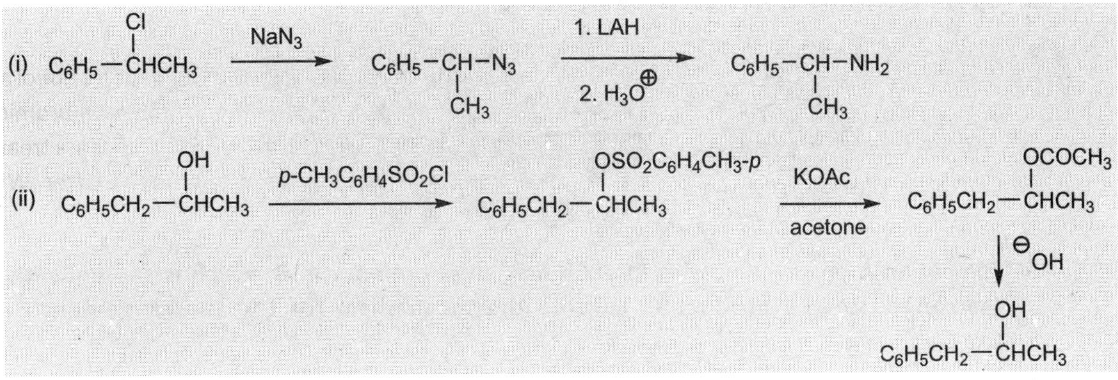

10. Outline the synthesis of the following compounds from suitable nucleophiles and an appropriate halide.

(i) $CH_3C{\equiv}CCH_2CH_2CH_3$ (ii) $(CH_3)_2CH{-}CH(NH_2){-}COOH$

(iii) $CH_3CH_2CH_2OCH_3$ (iv) $(CH_3)_2CHCH_2CH_2OCOCH_2CH(CH_3)_2$

11. (i) It is proposed to carry out a number of nucleophilic displacement reactions of ethyl bromide. Tell which of the following reagents would yield the product(s):
   (a) C_6H_5ONa (b) CH_3SNa
   (c) KI (d) KCl
   (ii) In the reaction, $CH_3I + {}^{-}OCH_3 \rightarrow CH_3OCH_3 + I^{-}$, rate is $k\,[CH_3I]\,[{}^{-}OCH_3]$. What shall be the effect of the following on the rate of the reaction?
   (a) The concentration of both the reactants is doubled.
   (b) The concentration of CH_3O^{-} is increased four times, and that of CH_3I is tripled.
   (c) The concentration of CH_3I is halved.
   (d) The temperature is raised.

12. Warming 2-bromo-3-methylbutane in moist acetone gives a product **A**. The IR spectrum of **A** displays a broad peak at about 3400-3200 cm^{-1}, whereas its NMR spectrum shows signals at δ

0.9 (*d*, 6H), 1.1(*d*, 3H), 1.5 (*m*, 1H), 2.6 (*s*, 1H), exchangeable with D_2O and 3.5 (*m*, 1H). Identify **A**, and comment on the mechanism of its formation.

13. Write the structure of the nucleophilic substitution product(s) in each of the following:

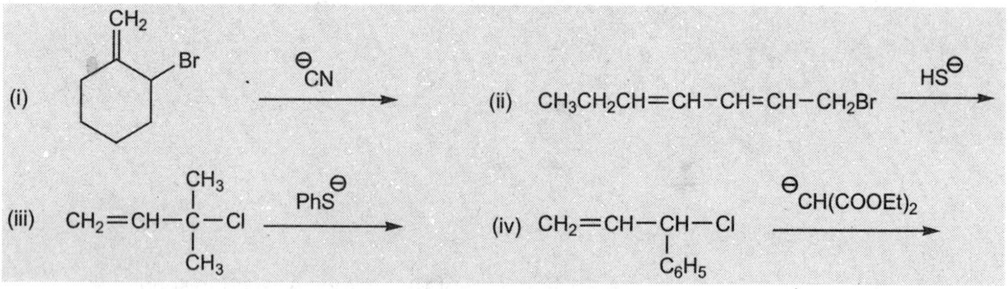

14, Outline the mechanism for the following transformation:

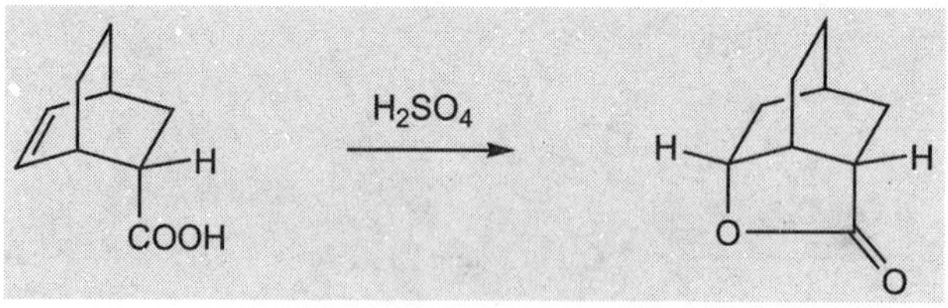

15. Compound **A** on reaction with Br_2/PCl_3 first gives bromo acid **B**, which is subsequently rearranged to the product **C**. Outline the mechanism for the rearrangement of **B** to **C**.

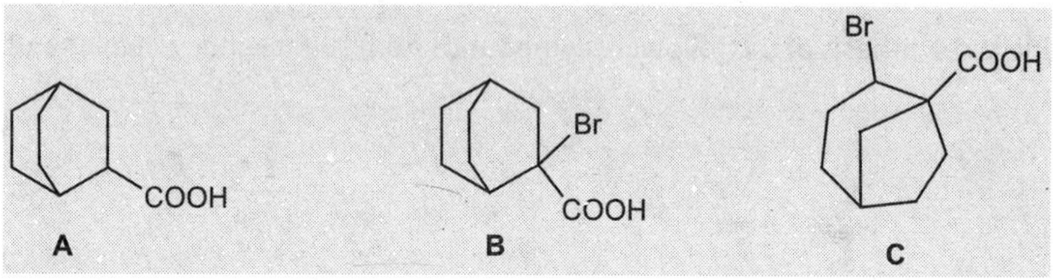

16. Explain the following observations:
   (i) Optically active *sec*-butyl chlorosulphite on warming in dioxan affords *sec*-butyl chloride with retention of configuration, while the same reaction when carried out in pyridine leads to the product with inversion.
   (ii) Solvolysis of compound **A** in methanol gives two ethers, while reaction of compound **B** with cyanide ion in acetone leads to the substitution product with inversion of configuration.

   (iii) The rate of solvolysis of compound **A** in ethanol does not change on addition of ethoxide ion, and it proceeds faster than compound **B**.

A B

17. The acetolysis of compound **X** gives a mixture of two products **Y** and **Z**. Give the structures of **Y** and **Z**, and explain their formation.

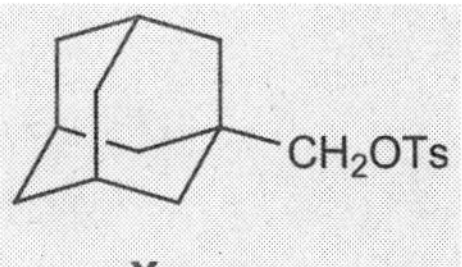

X

18. Hydrolysis of $ClCH_2CH_2CH_2CH_2SH$ with dilute alkali yields a compound whose NMR spectrum displays two sets of closely spaced multiplets of four proton intensity each. Identify the product, and account for its formation.
19. The kinetic rate law for the reaction

$$\underset{\textbf{A}}{C_6H_5-\overset{\displaystyle Cl}{\overset{|}{C}H}-C_6H_5} + F^{\ominus} \xrightarrow{\text{liq. } SO_2} C_6H_5-\overset{\displaystyle F}{\overset{|}{C}H}-C_6H_5 + Cl^{\ominus}$$

is Rate = k[**A**]

It was observed that:

(i) the rate of the reaction showed a gradual fall with the progress of the reaction.

(ii) the rate of the reaction also slowed down by the addition of chloride ion from outside.

Explain these observations.

20. Why does quinuclidine react faster than triethylamine with *iso*propyl iodide in an S_N2 reaction?
21. Mustard gas gets easily hydrolyzed in aqueous NaOH. The rate of hydrolysis is independent of hydroxide ions, and is decreased by chloride ion. Suggest a mechanism consistent with this data.
22. What are ambident nucleophiles? Explain their regioselectivity with suitable examples.
23. Offer suitable explanations to the following observations:
    (i) (+) 2-Iodobutane undergoes racemization when it is allowed to react with iodide ion.
    (ii) Ethanolysis of 2-(*p*-hydroxyphenyl)ethyl bromide in presence of ethoxide ion is 10^6 times faster than that of 2-(*p*-methoxyphenyl)ethyl bromide.
    (iii) The solvolysis of phenylethyl tosylate in CF_3CO_2H is about 3040 times faster than that of ethyl tosylate.
24. Acetolysis of optically active *erythro*- and *threo*-3-phenyl-2-butyl tosylates proceeds with retention of configuration. Does this experiment provide an evidence for the existence of phenonium ions?
25. Comment upon the fact that iodide ion is both a good leaving group and a good nucleophile.
26. Predict the effect of solvent polarity on the rate of an S_N1 reaction.
27. What are the evidences for the intermediacy of carbocations in S_N1 reactions?
28. Comment upon the nucleophilic substitution reactions of allylic halides.
29. What is anchimeric assistance? How do you distinguish phenonium ions from nonclassical carbocations? Discuss giving appropriate examples.

30. Predict the product(s) in the following reactions:

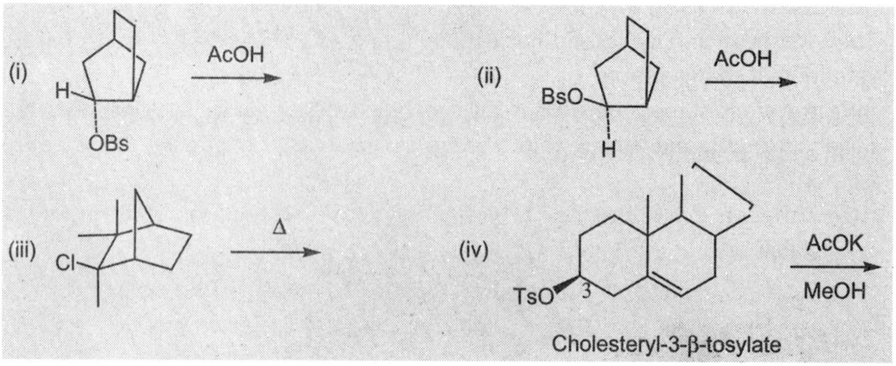

SOLUTIONS

1. (i) **A** reacts faster than **B** since there is more relief of steric strain by expanding its angles from 109.5° to 120° in transition state leading to the formation of the intermediate carbocation.
   (ii) **A**. The ease of S_N1 reaction goes parallel to the ease of the formation of carbocations. The halide **A** is expected to attain planarity (an essential condition for the formation of carbocations) more easily than **B** because of an extra CH_2 group in the bridge.
2. (i) **A** (I^- is a good leaving group), (ii) **B** (**A** will not react), (iii) **A**, (iv) **A**.
3. Both the reactions are S_N1 displacements involving benzyl and *iso*propyl cations. Since benzyl cation is more stable than *iso*propyl cation, it is formed at the faster rate than *iso*propyl cation. This is the reason why under identical conditions, benzyl bromide reacts at a faster rate than *iso*propyl bromide.

4.

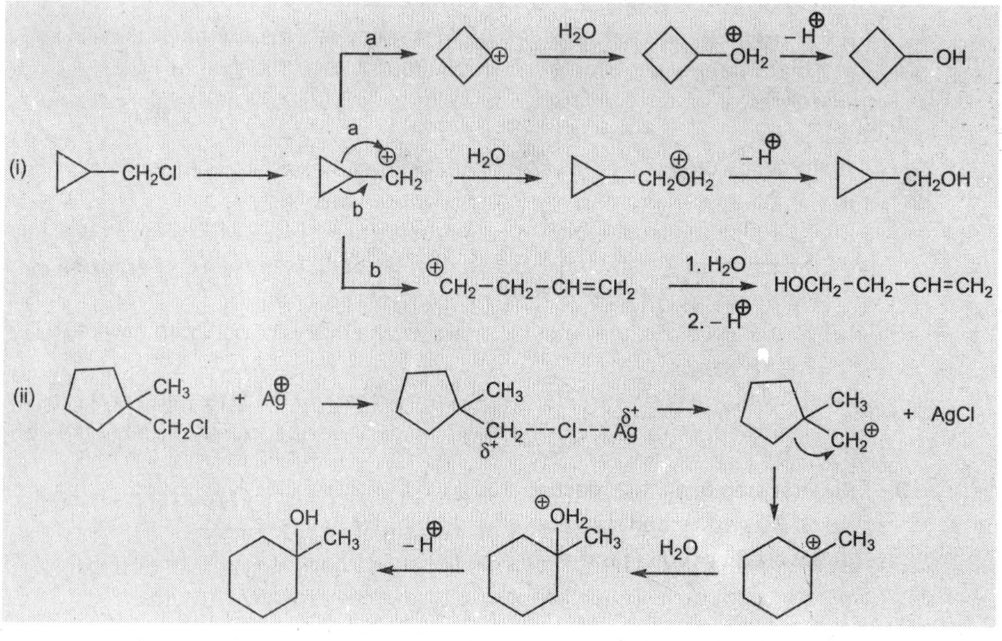

5. In *trans* isomer **I**, both acetoxy and tosyloxy are axial. So, acetoxy group can provide an anchimeric assistance by attacking from the backside of the leaving tosyloxy group to give a cyclic intermediate. Then, ring opening by second nucleophlic attack of acetic acid will lead to the product with retention of configuration.
6. In case of *trans* isomer, there is possibility of neighbouring group participation by alcoholic group (alcoholic group and leaving chloro group can have axial positions). This is not possible in *cis* isomer.
7. (i) Although it is a primary alkyl halide, the corresponding carbocation is resonance-stabilized.

$$CH_3{-}O{-}\overset{\oplus}{C}H_2 \longleftrightarrow CH_3{-}\overset{\oplus}{O}{=}CH_2$$

(ii) $NaNO_2$ and $AgNO_2$ are ambident nucleophiles and, therefore, can lead to the formation of two isomeric products. It is known that if a particular substitution reaction attains an increasing degree of S_N1 nature, the incoming nucleophile tends to attack with its most electronegative atom, but if the reaction is of more S_N2 nature, the nucleophile attacks with a less electronegative atom. $AgNO_2$ (Ag^+) promotes the formation of carbocation giving S_N1 reaction and therefore will react with alkyl bromide involving oxygen atom of the nucleophile NO_2^-. When $NaNO_2$ is used as nucleophile, the reaction primarily takes place through S_N2 and involves less electronegative (nitrogen) of nucleophile in the formation of the product.

$$RBr + AgNO_2 \longrightarrow \underset{(major)}{RONO} + \underset{(minor)}{RNO_2}$$

$$RBr + NaNO_2 \longrightarrow \underset{(major)}{RNO_2} + \underset{(minor)}{RONO}$$

7.

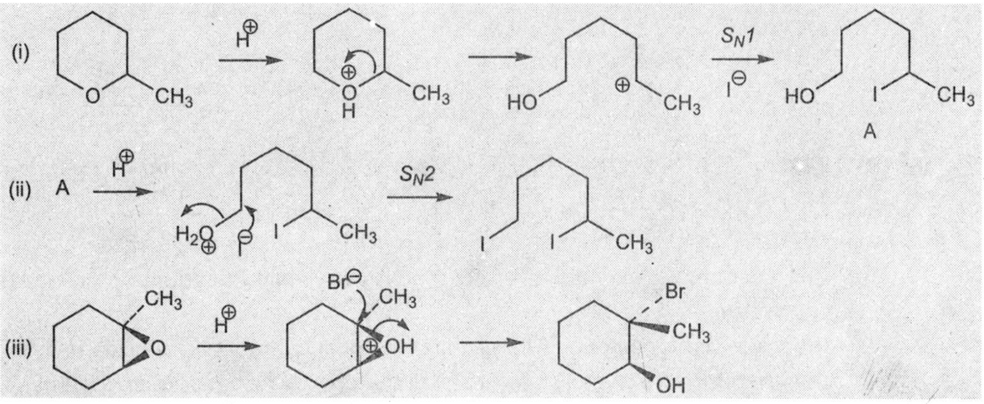

9. (i) First step is an S_N2 reaction and hence inversion of configuration has taken place at this stage. The second step does not involve any change in stereochemistry.
(ii) Inversion of configuration occurs in displacement of the tosylate by acetate (second step).

10. (i) $CH_3C{\equiv}C^{\ominus}$ and $CH_3CH_2CH_2Br$ (ii) $(CH_3)_2CH{-}CH(Br){-}COOH$ + NH_3

(iii) $CH_3CH_2CH_2O^{\ominus}$ + CH_3I (iv) $(CH_3)_2CHCH_2COO^{\ominus}$ + $(CH_3)_2CHCH_2CH_2Br$

11. (i) a, b and c will yield the products. No product will be obtained from d.
   (ii) (a) the rate becomes $2 \times 2 = 4$ times faster.
   (b) the rate becomes $3 \times 4 = 12$ times faster.
   (c) the rate is halved.
   (d) the rate of the reaction increases.

12. $CH_3{-}CH(OH){-}CH(CH_3){-}CH_3$

13. The major products obtained in these reactions are formed by S_N2' mechanism.

(i) 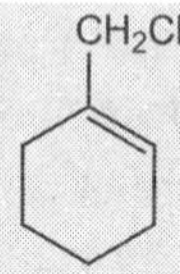

(ii) $CH_3CH_2CH(SH){-}CH{=}CH{-}CH{=}CH_2$ + $CH_3CH_2{-}CH{=}CH{-}CH{=}CH{-}CH_2SH$ + $CH_3CH_2{-}CH{=}CH{-}CH(SH){-}CH{=}CH_2$

(iii) $PhS{-}CH_2{-}CH{=}C(CH_3)_2$ (iv) $CH_2{=}CH{-}CH(C_6H_5){-}CH(COOEt)_2$ + $(EtOOC)_2CH{-}CH_2{-}CH{=}CH(C_6H_5)$

14. 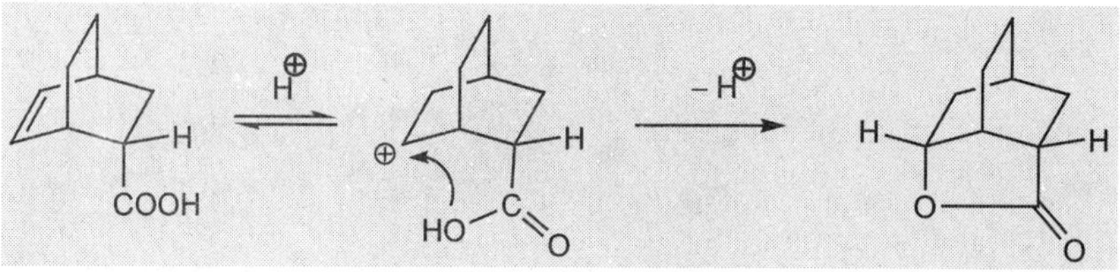

15. B ≡ 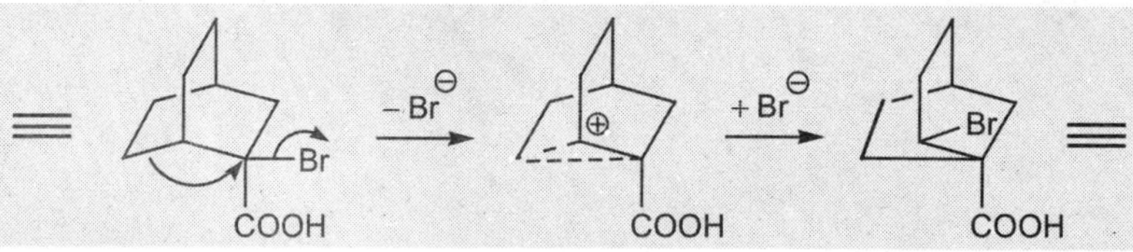 ≡ C

16. (i) The reaction in dioxan involves S_Ni mechanism while in pyridine proceeds through S_N2 mechanism.
   (ii) The solvolysis of compound **A** can generate resonance-stabilized secondary allylic carbocation and, therefore, involves S_N1 mechanism thereby giving a pair of enantiomeric ethers. On the other hand, reaction of compound **B** with strong nucleophile (CN^-) in acetone occurs *via* S_N2 mechanism and gives inverted product.

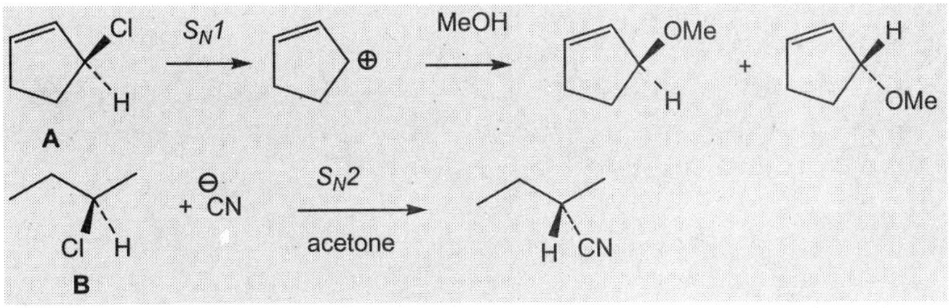

17.

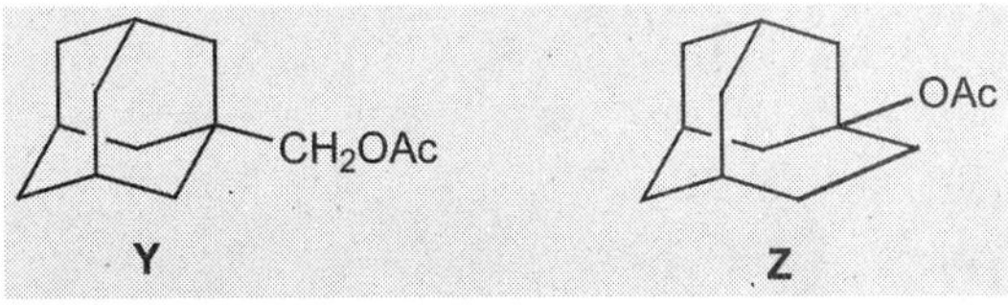

Chapter 7

Addition to Carbon-Carbon Multiple Bond

Chapter Outline

7.1 ELECTROPHILIC ADDITIONS

- A. Addition of Halogens and Hydrogen Halides
- B. Hydroboration Reaction
- C. Epoxidation and Hydroxylation
- D. Addition of Water and Oxymercuration-Reduction
- E. Hydrogenation of Alkenes and Alkynes
- F. Hydrogenation of Aromatic Rings
- G. Ozonolysis
- H. Addition to Conjugated Dienes

7.2 NUCLEOPHILIC ADDITIONS

- A. Cyanoethylation
- B. Michael Addition

Addition to a carbon-carbon multiple-bond may be initiated by an electrophile, a nucleophile, or a free radical. All the three reactions take place by a two-step mechanism, the second step consisting of the combination of the resultant intermediate with a negative species, a positive species, or a neutral entity (a free radical).

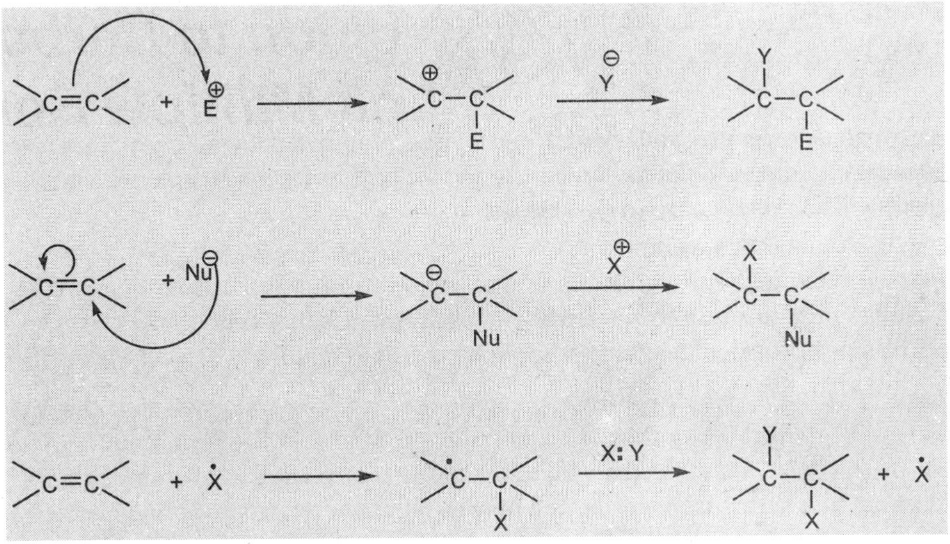

Since a multiple-bonded carbon atom can readily supply the comparatively loosely held π electrons to an electrophile, electrophilic addition is the characteristic reaction of the alkenes. In these reactions, π electrons of the double bond are attracted to an electrophile resulting in the addition of the electrophile to one of the sp^2 carbons of the alkene. The reaction concludes with the addition of a nucleophile to the other sp^2 carbon.

In contrast, the π electrons cloud tends to shield the molecule from the attack of nucleophilic reagents. Such reactions, however, may be initiated in the presence of substituents in the molecule which have a high electron-attracting power, and thus, are able to polarize the π electrons in such a manner that the β-carbon becomes electron-deficient, and hence, enters into a bonding interaction with an approaching nucleophile.

We shall discuss the detailed mechanisms of these two types of reactions in this Chapter, in addition to some special cases such as epoxidation and hydroxylation where the attack on the two carbons of the double bond is simultaneous. A free radical addition to alkenes is quite general, and reactions of this type will be examined in detail in Chapter 13.

7.1 ELECTROPHILIC ADDITIONS

A. Addition of Halogens and Hydrogen Halides

I. Addition of Halogens

Addition of halogen to a carbon-carbon double bond provides a typical example of an electrophilic addition reaction involving a two-step mechanism with a carbocation intermediate. The best evidence for the mechanism emerges from the observation that when ethylene is treated with bromine in the presence of other nucleophiles, a mixture of products is obtained.

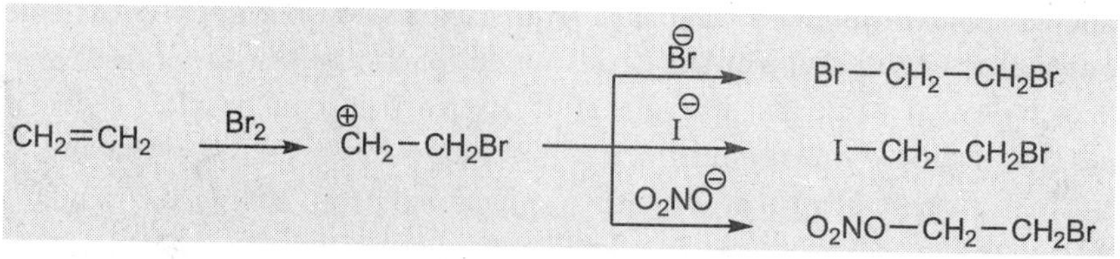

Since various reagents undergoing electrophilic addition to alkene follow a common mechanistic path, a general discussion on the addition of halogens to alkenes will be typical. Three aspects of the reaction are to be considered:

- Formation of the π-complex
- Stereochemical consequences of the reaction and the choice of addition of the electrophile to an unsymmetrically substituted alkene, and
- The side of the double bond attacked by the electrophile in case of cyclic alkenes.

(i) Formation of the π-Complex: Spectroscopic and other evidences suggest that the initial step of an electrophilic addition, say that of bromine to an alkene, is the formation of a π-complex **(1)** formed by the attack of π electron cloud on the vacant orbital of the electrophile. Such a complex does not involve actual bonding, and may be regarded as an association in which halogen molecule is embedded in one lobe of the π orbital of the double bond. This, in most cases, leads to the formation of a cyclic intermediate known as halonium ion **(1a)** (bromonium ion in this particular case).

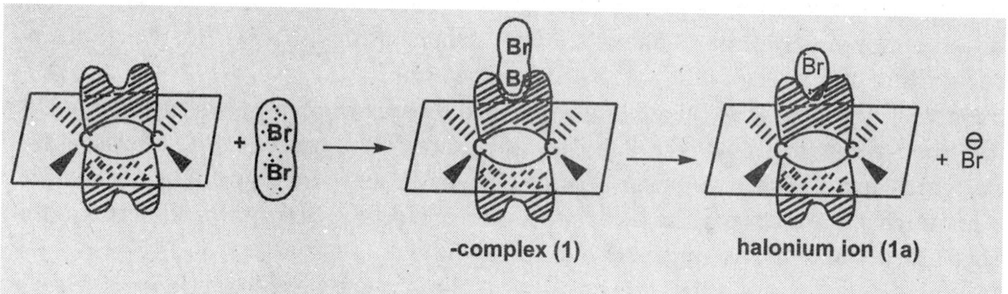

(ii) Stereochemical Consequences: It has been found that in cases where stereochemistry of polar addition can be detected, it invariably occurs across the double bond in a *trans* and *antiparallel* manner. Addition of bromine to cyclopentene, for instance, yields *trans* dibromocyclopentane, rather than the *cis* isomer.

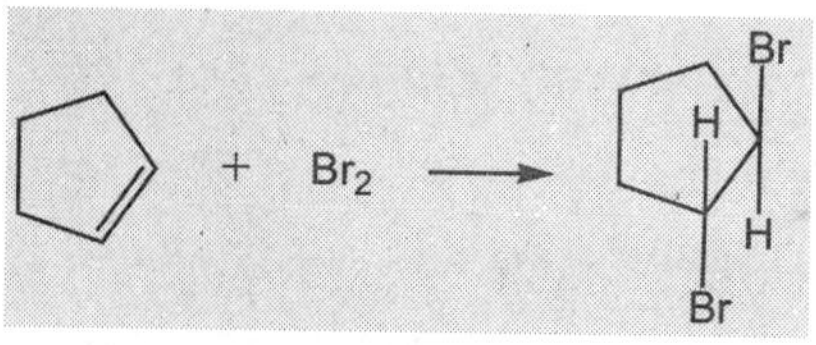

How do we explain the exclusive formation of *trans* product? Carbocation is a planar structure, and if it is an intermediate in the reaction, we should expect the bromide ion to attack from either side of the carbocation to give a mixture of *cis* and *trans* products. Apparently, it is not so as the first and second steps of the bromine addition are taking place from the opposite sides of the molecules. It appears, therefore, that the free carbocation is not formed, and instead, the participation of a bromonium ion has been envisaged. Bromonium ion **(1b)** is a cyclic intermediate in which bromine is bonded to both carbons of the double bond, and carries the positive charge as the hybrid of two localized carbocation structures. Attack of bromide ion on this intermediate can only occur from the opposite side resulting in the exclusive formation of the *trans* addition product.

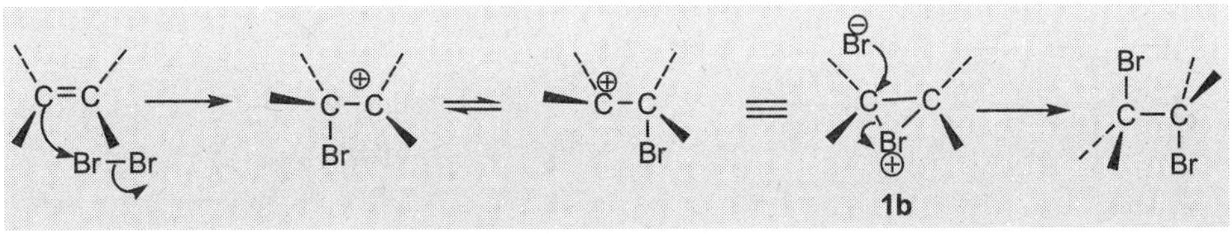

If **1** is symmetrically substituted, attack by a nucleophile could occur at either carbon. But in case of **1** derived from an unsymmetrical alkene, the nucleophile preferably attacks the more substituted carbon.

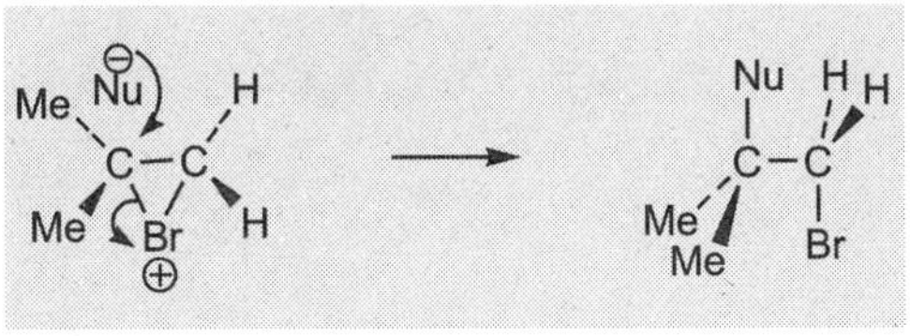

Because of this observed selectivity of the nucleophile, the bromonium ion may be regarded as a resonance hybrid of the following structures.

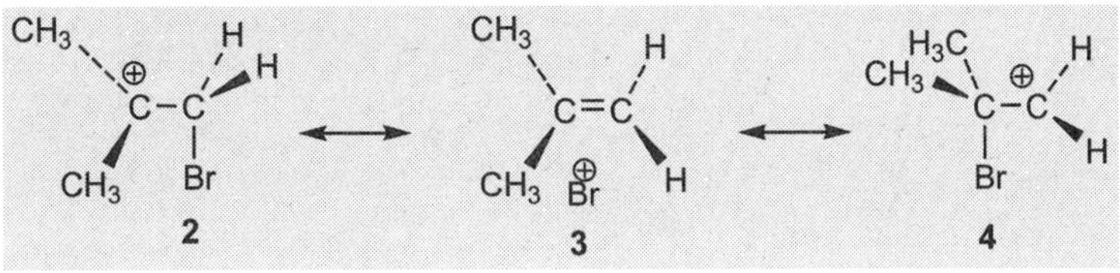

Clearly, **2** being a tertiary carbocation, will have greater contribution to the hybrid structure, and consequently, bond cleavage of the bromonium ion occurs at the carbon which produces more stable carbocation followed by the attack of a nucleophile from the backside. Thus, the mechanistic course of reaction of the intermediate bromonium ion with a nucleophile resembles rather closely the acid-catalyzed ring-opening of epoxides.

Direct evidence for the intermediacy of bromonium ion has been provided by low-temperature NMR studies of the product obtained on treatment of 1-bromo-2-fluoropropane **(5)** with antimony pentafluoride in sulphur dioxide. The NMR spectrum of the product was found to be consistent with the structure **6.**

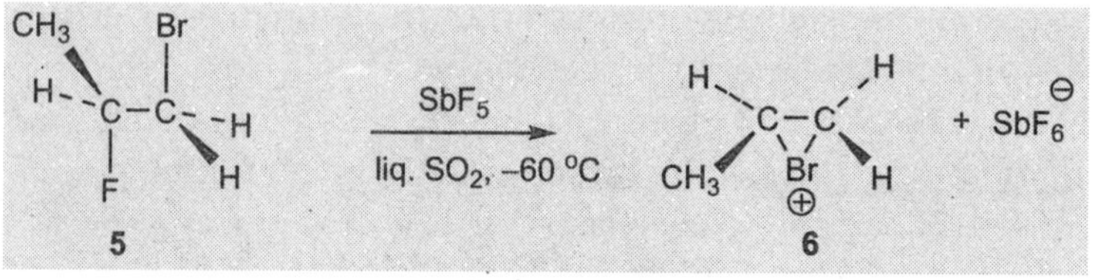

Bromination of alkenes is a stereospecific process as the geometry of the starting alkenes determines which diastereoisomer is obtained. With the intervention of bromonium ion intermediates, we can readily account for the formation of *dl*-2, 3-dibromobutane (**7**) and *meso*-2, 3-dibromobutane (**8**) from the bromination of *cis* and *trans*-2-butenes, respectively.

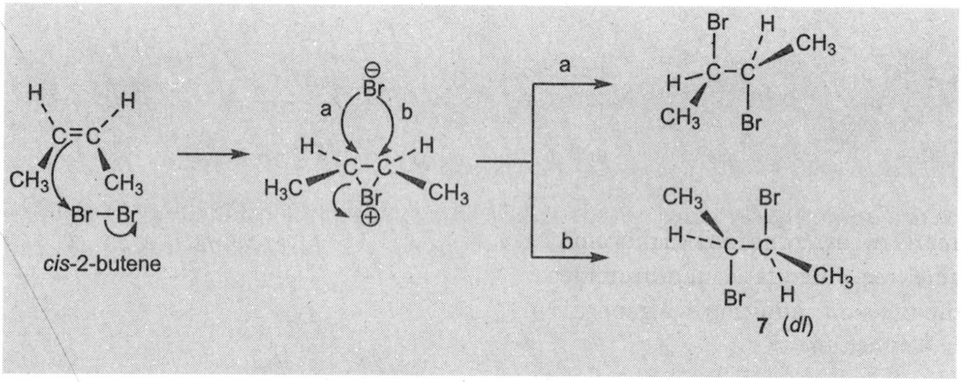

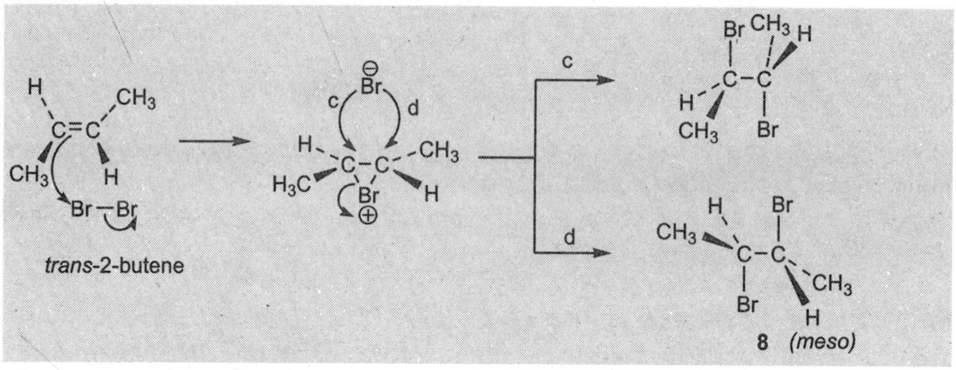

(iii) Addition to Cyclic Alkenes: Since all the four substituents attached to the two carbon atoms of a double bond are coplanar, the π electrons do not discriminate between the electrophile's approach from above or below the plane in an *acyclic* system. *Cyclic* alkenes do not, however, have this option in the presence of a large *quasi-axial* group, which sterically hinders the approach of the electrophile on its side of the ring, thus permitting its approach from the other side. Because of the intervention of a halonium ion intermediate, the nucleophile then attacks from the backside of the ring producing a *trans diaxial* product. We can similarly explain the formation of the racemic *anti*-1,2-dibromocyclohexane with the intermediacy of a bromonium ion.

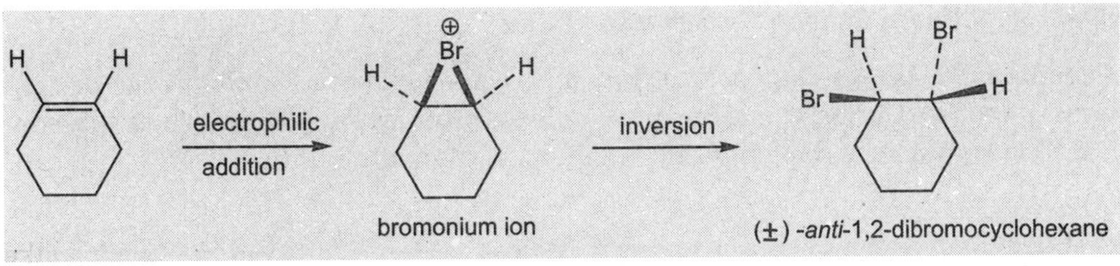

The process can also be well-illustrated in the bromination of 2-cholestene **(9)** to give a *trans* diaxial dibromide, the net result being an antiparallel addition of Br_2.

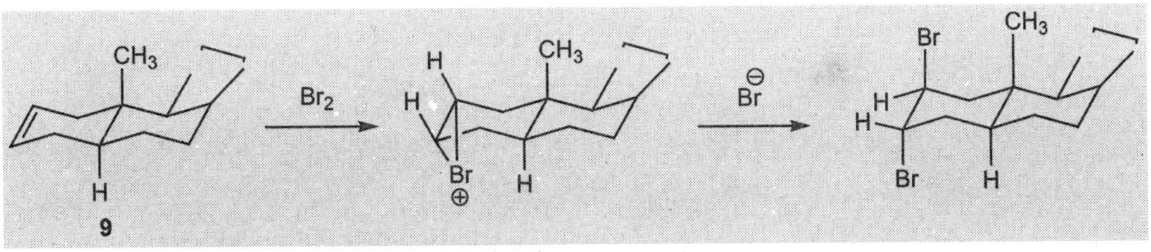

In many cases, *trans* diaxial dibromides undergo rearrangement to the thermodynamically more stable *trans* diequatorial dibromides.

Although *trans* addition is a general rule, addition of chlorine to acenaphthylene (**A**) takes place in a *cis* fashion.

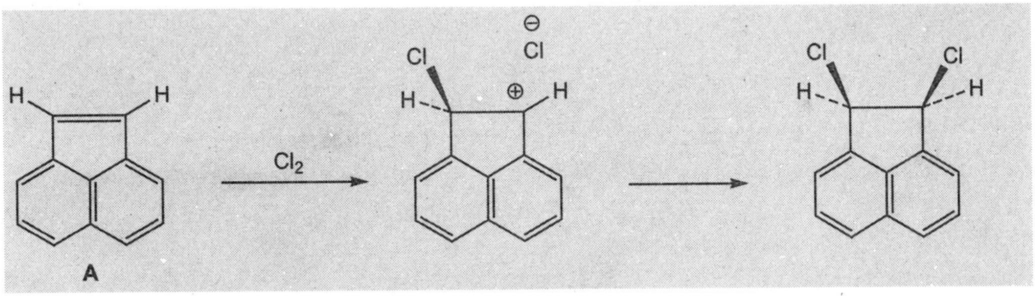

(iv) Effect of Substituents on the Rate of Addition Reactions: Since electrophilic attack depends on the degree of accessibility to the π electrons, any effect that increases the electron density of the double bond will enhance its reactivity towards electrophilic reagents. These substituents increase the rate of addition by stabilizing the intermediate cationoid species, and thus, lowering the activation energy. A typical sequence of reactivity of various alkenes is in the order:

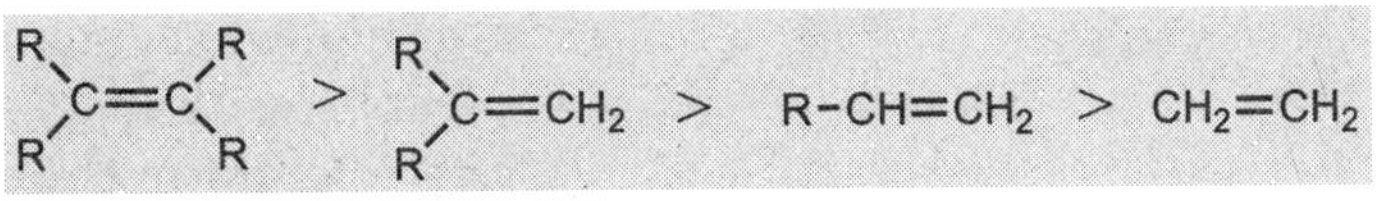

Conversely, if the electron density is reduced by an electron-withdrawing group, such as $-NO_2$, –CN, or a halogen, the developing cationoid species will be destabilized, and the rate of addition will be retarded.

II. Addition of Hydrogen Halides: Markownikoff's Rule

In the addition of hydrogen halides to a carbon-carbon double bond, *proton is the electrophile,* and forms a hydrogen-bridged intermediate called a protium ion **(10)**, which is subsequently attacked from backside by the halide ion resulting in the *trans* addition.

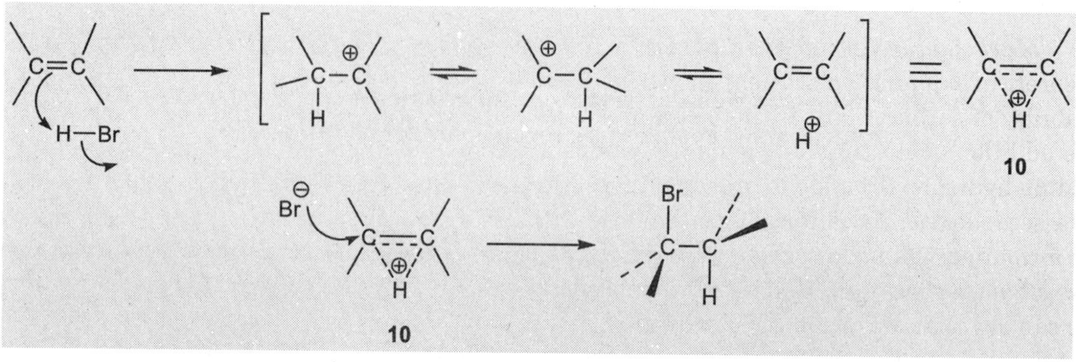

In the case of unsymmetrical alkenes, the position of attachment of the nucleophile will be governed by the nature of the substituent. In propene, for instance, the electron-donating inductive effect of the methyl group will stabilize the positive ionic charge on the carbon atom to which it is attached, so that the nucleophile will preferably attack on that very carbon. This explains what is known as Markownikoff's rule. This rule can be broadly defined as follows: The hydrogen (or an electrophile) adds to the sp^2 carbon that is bonded to greater number of hydrogens.

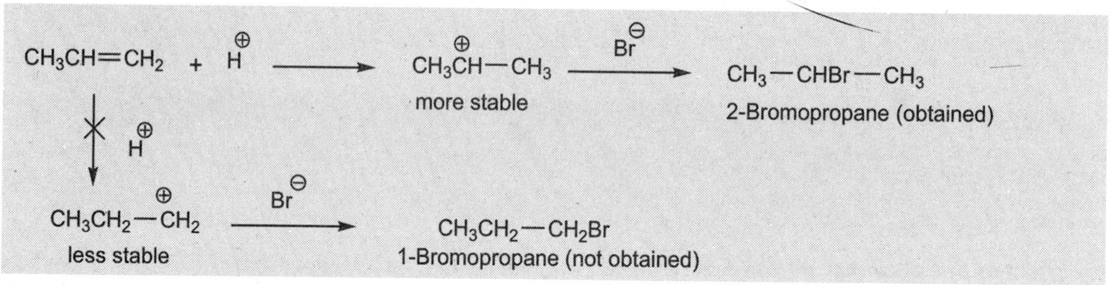

We can similarly explain the preferred formation of 2-iodo-2-methylbutane from the reaction of 2-methyl-2-butene and HI.

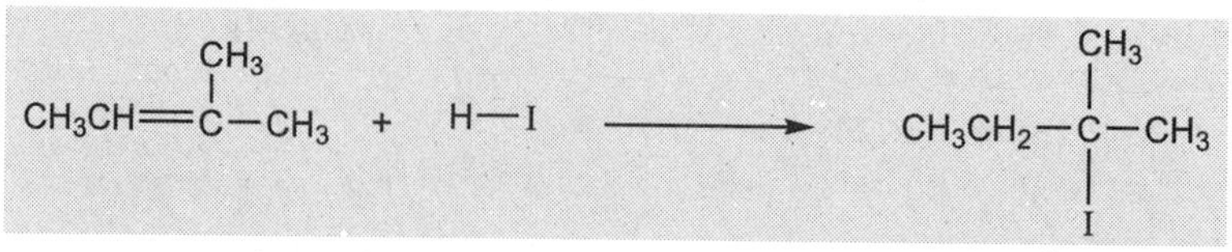

1-Bromo-1-methylcyclohexane is obtained in the addition reaction of 1-methylcyclohexene and HBr.

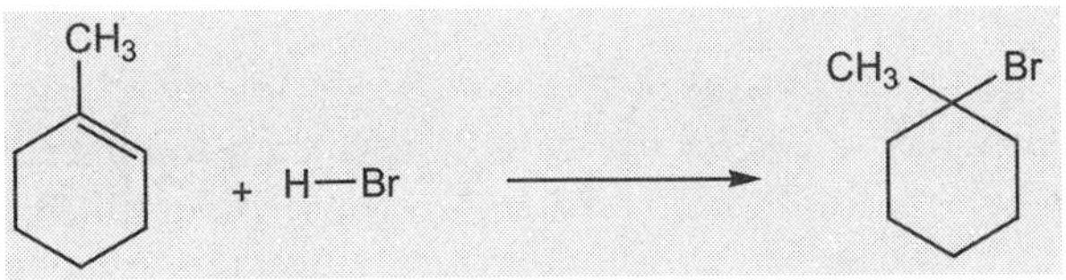

Markownikoff's rule was purely empirical when it was formulated. However, now a theoretical explanation of orientation of addition to olefins can be offered on the basis of relative stability of carbocations. If there is a probability of formation of more than one carbocations, the addition of electrophile yields the more stable one. For example, as we have seen, that in adding hydrogen bromide to propene, there is a possibility of the formation of either a primary or a secondary carbocation. Since the latter is *more stable*, the activation energy required for its formation will be lower as compared to the activation energy required for the formation of *less stable* carbocation. Hence, 2-bromopropane is formed as an exclusive product. Therefore, we can say that *Markownikoff's addition occurs through a more stable carbocation* (**Fig. 7.1**).

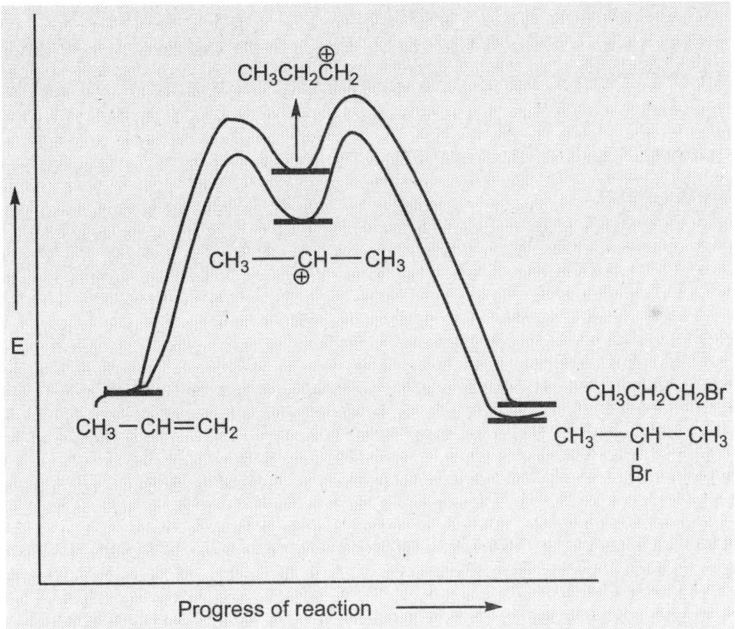

Fig.7.1: Potential energy diagram for the addition of HBr to propene

We can now conclude that *validity of Markownikoff's rule depends upon the stabilities of the intermediate carbocations, i.e. the energies of the transition states involved.*

Markownikoff's rule can be extended to cover additions of highly polarized reagents such as $X^{\delta+}Y^{\delta-}$ to an unsymmetrical alkene. The product is formed by the initial addition of X^+ to give the more stable carbocation, regardless of whether the true intermediate is an open carbocation or a bridged ion. For example, it is known that chlorine is the cationoid part of the hypochlorous acid, and on this basis, its addition to *iso*butylene can be rationalized.

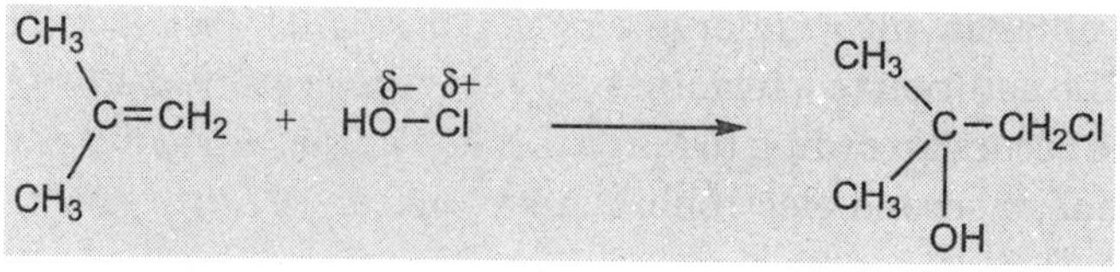

Because of the intermediacy of a bridged halonium ion, additions of hypochlorous acid to alkenes take place in a *trans* manner.

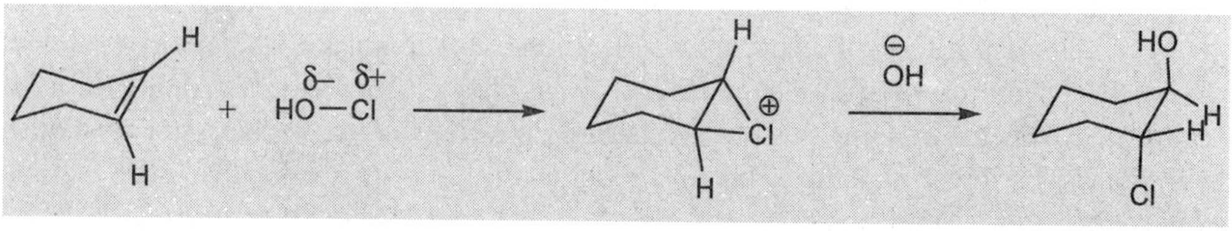

In contrast, addition of DBr to acenaphthylene (**A**) gives the *cis* isomer, as evident from the NMR coupling constants of vicinal protons. It has been suggested that there is an initial formation of a tight ion pair (in which the carbocation is stabilized by *p-pi* conjugation with the benzene rings, and the bromide ion is attached on the same side of the carbon-carbon double bond as the entering electrophile) that collapses to the product faster than rotation or attack by another bromide ion. The addition appears to be *cis*-wise only where an aromatic ring is attached to the double bond.

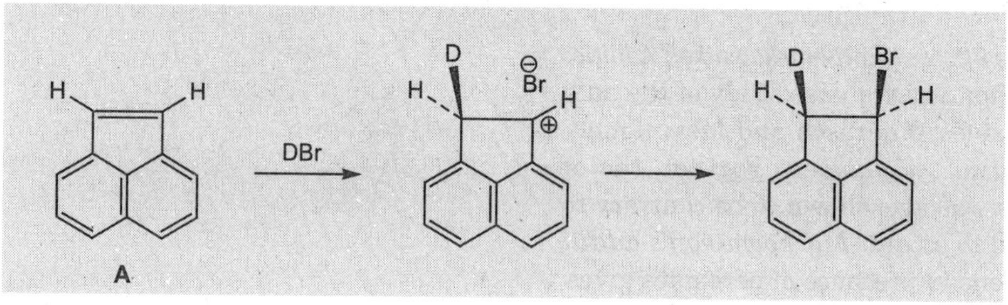

Addition of HCl to vinyl chloride yields ethylidene chloride rather than ethylene chloride, mainly because of electron delocalization of the intermediate carbocation. This conjugative stabilization more than compensates for the electron-attracting inductive effect of chlorine.

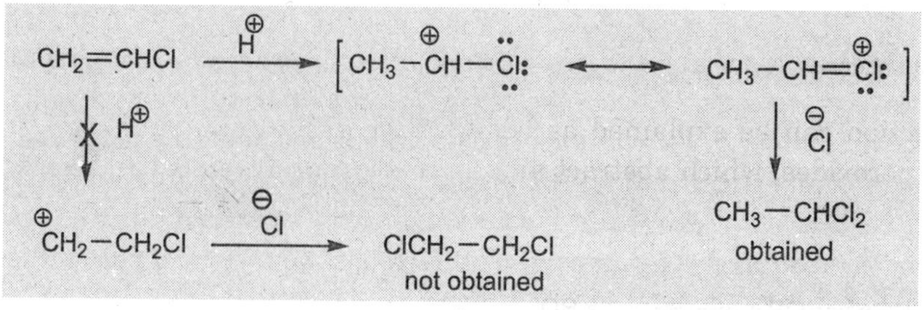

Alkynes, like alkenes, also undergo electrophilic addition reactions with halogens and hydrogen halides. The addition to a terminal alkyne with electrophiles such as hydrogen halides is regioselective, and occurs according to Markownikoff's rule. Since the product of such additions to alkyne is an alkene, a second electrophilic reaction can occur to give a saturated product.

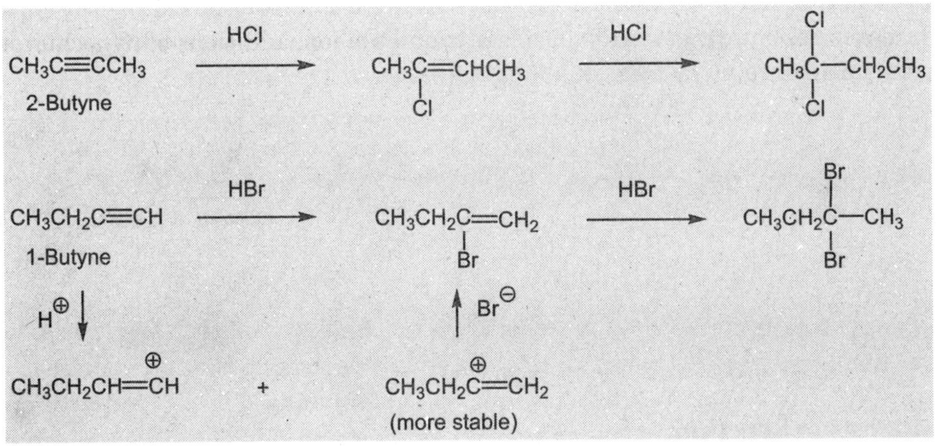

As would be expected, the reactivity of alkynes towards electrophiles is increased by electron-repelling substituents, and retarded by electron-withdrawing substituents. In spite of this close similarity, the reaction rates of alkynes with electrophilic reagents are rather slower than that of alkanes.

Peroxide Effect or Anti-Markownikoff Addition

After an extensive study of the addition of hydrogen bromide to alkenes in the presence of peroxides, Kharasch and Mayo found that under these conditions the reaction involves a *free radical mechanism*. Further, the orientation of addition in the case of unsymmetrical alkenes was also shown to be contrary to Markownikoff's rule. Such additions are sometimes referred to as *anti-Markownikoff's additions.* For instance, the addition of hydrogen bromide to propene in presence of peroxides gives 1-bromopropane rather than 2-bromopropane. Since the orientation of addition is reversed due to peculiar effect of peroxides, this is also termed as *peroxide* or *Kharasch effect.*

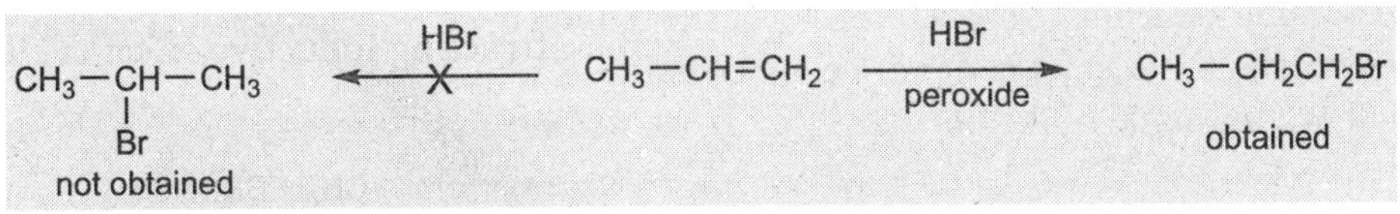

The reaction can be explained as being initiated by the alkoxy radicals, produced by homolysis of peroxides, which abstract an atom of hydrogen from HBr generating a bromine atom.

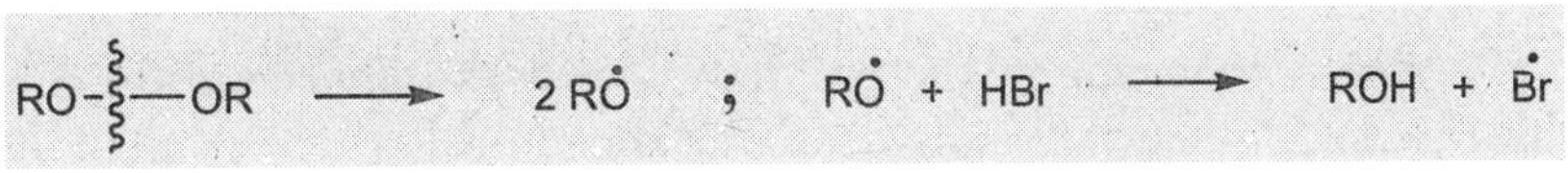

The bromine atom can add to either of the two carbon atoms of the double bond in an unsymmetrical alkene, such as propene, producing either a primary or secondary radical. The preferential formation of the latter is in accordance with the order of stability of radicals: tertiary > secondary > primary.

The secondary radical abstracts a hydrogen atom from HBr, thus producing 1-bromopropane, and the bromine atom propagates the chain further. In addition to peroxides, such free radical additions are also initiated by light, high temperature, or even oxygen.

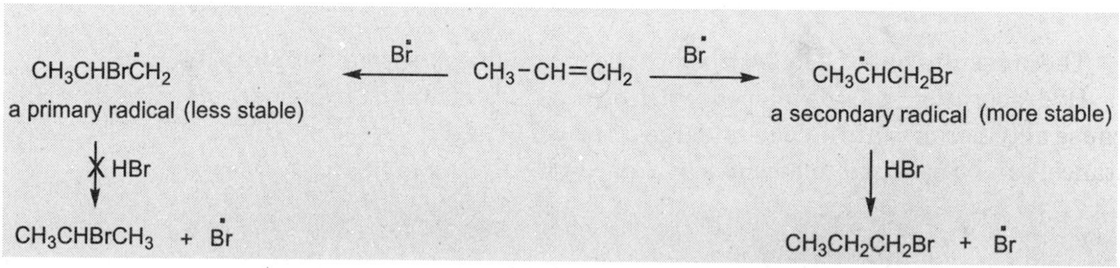

B. Hydroboration Reaction

Borane (BH_3) itself is unknown, but its dimer, diborane (B_2H_6), functions as if it were the hypothetical monomer borane. Because of the vacant orbital on boron, borane may be regarded as an electrophile, and attacks the π electrons of an alkene. In accordance with the Markownikoff's rule, the positive charge develops on that carbon of the double bond which is best able to support it. Intramolecular hydride shift then completes the hydroboration.

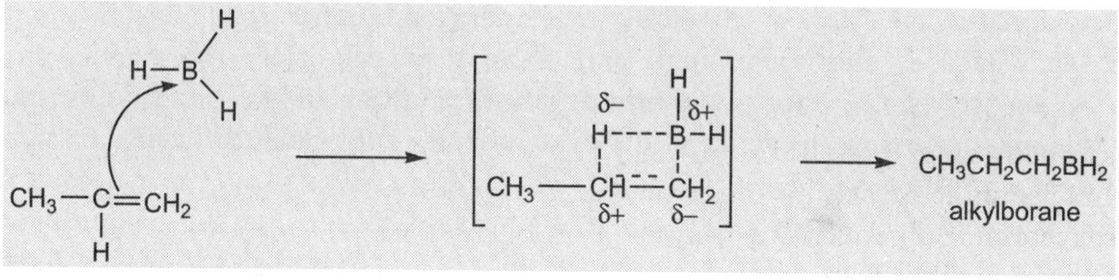

Because of the one-step four-centre process, hydroboration is *cis*-addition; the hydrogen and the boron become attached to the same side of the double bond. As there is no free carbocation, rearrangement in the alkene is not to be expected.

Alkylboranes behave like boranes, and react further with excess alkene to form di- and trialkylboranes.

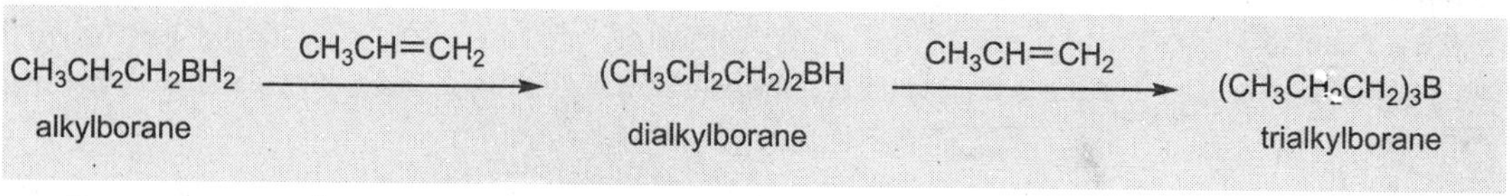

The rate of addition of borane to alkenes decreases with increase in the number of alkyl substituents on the double bond. Trisubstituted alkenes normally give dialkylboranes, and tetrasubstituted alkenes form only monoalkylboranes.

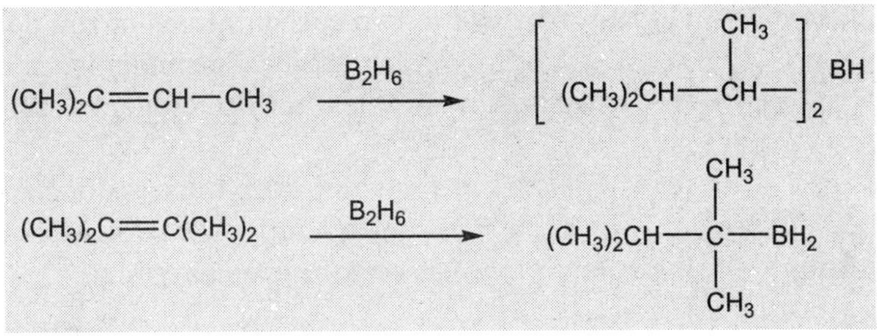

These results have been ascribed to the increasing steric bulk of substituents.

Hydroboration was developed by Brown as a reaction of tremendous synthetic utility, because alkylboranes are able to undergo a variety of useful transformations. Alkylboranes, for instance, can be oxidized to alcohols by treatment with hydrogen peroxide.

$$(CH_3CH_2CH_2)_3B \xrightarrow[NaOH]{H_2O_2} 3\ CH_3CH_2CH_2OH$$

The overall result of these two-step reactions with diborane followed by the oxidation of alkylborane appears to be *anti-Markownikoff* addition of water to a carbon-carbon double bond, although *mechanistically, hydroboration does not violate Markownikoff's rule*. Another significant feature of the whole process is the stereospecificity of both the steps, which results in *cis* addition of water across a double bond.

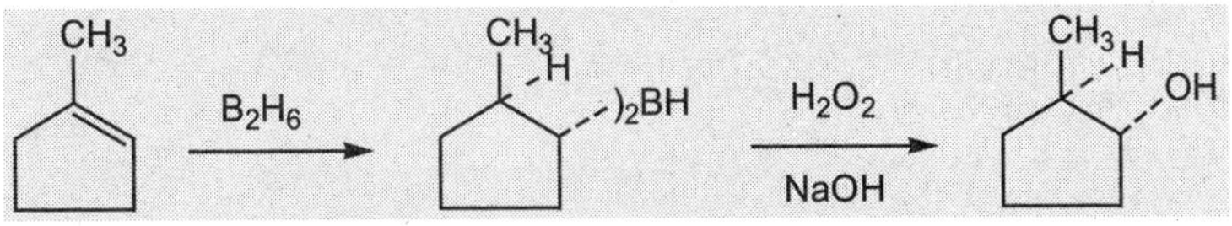

The mechanism for the alkaline hydrogen peroxide oxidation of alkylboranes involves nucleophilic attack by the hydroperoxide anion followed by the rearrangement of hydroperoxyboron. Subsequent hydrolysis of the borates yields the alcohol. The oxidation of alkylboranes with alkaline hydrogen peroxide proceeds in the following steps:

Step I: Formation of hydroperoxide ion

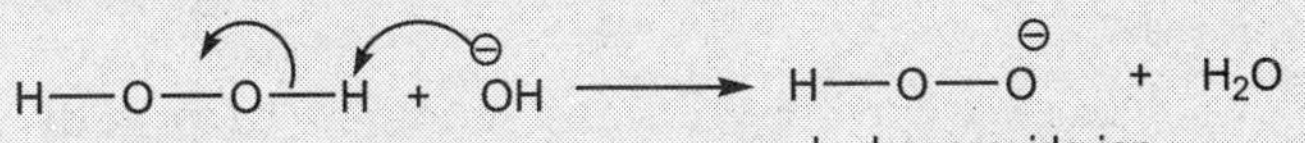

Step II: Nucleophilic addition of hydroperoxide ion on boron atom of trialkylborane.

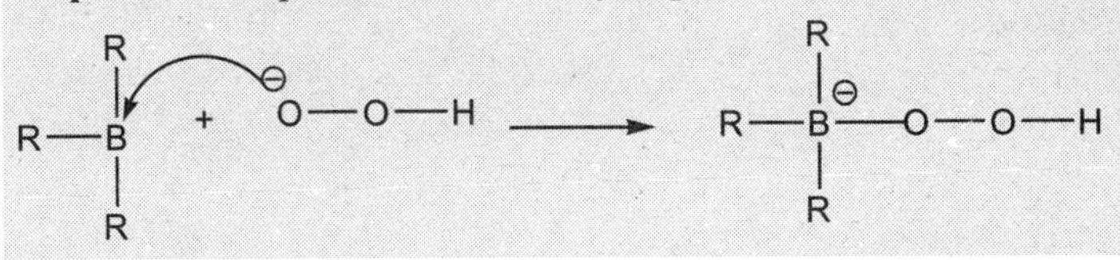

Step III: Rearrangement of hydroperoxyboron

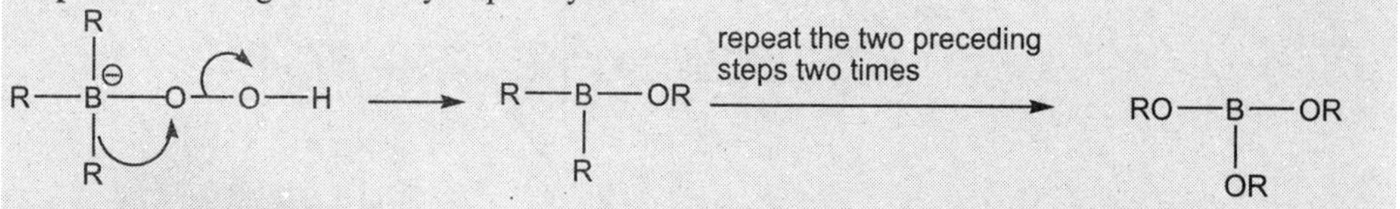

Step IV: Hydrolysis of borate

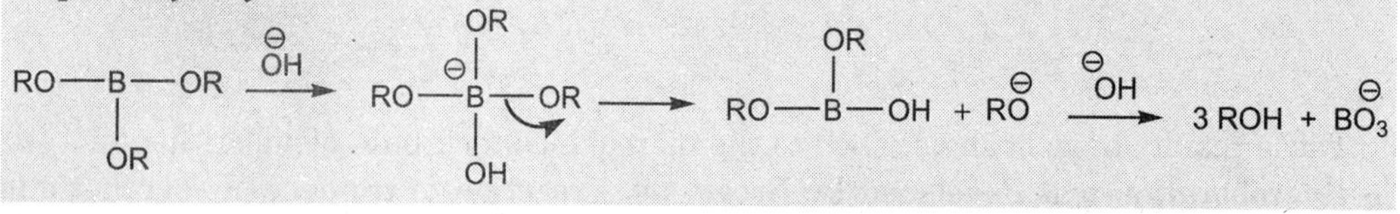

It is interesting to see that in the overall process of hydroboration-oxidation, one mole of BH_3 reacts with three moles of an alkene producing three moles of alcohol.

Alkylboranes are readily cleaved to alkanes on boiling with acetic or propionic acid. Thus, the hydroboration-acid hydrolysis sequence provides a valuable method of effecting the reduction of a double bond.

$$n\text{-}C_4H_9CH{=}CH_2 \xrightarrow{B_2H_6} (n\text{-}C_4H_9CH_2CH_2)_3B \xrightarrow{CH_3CH_2COOH} n\text{-}C_4H_9CH_2CH_3$$

That the reduction product is invariably *cis,* is elegantly demonstrated in the reaction of norbornene with deuterated diborane, followed by acid hydrolysis using deuterated acid.

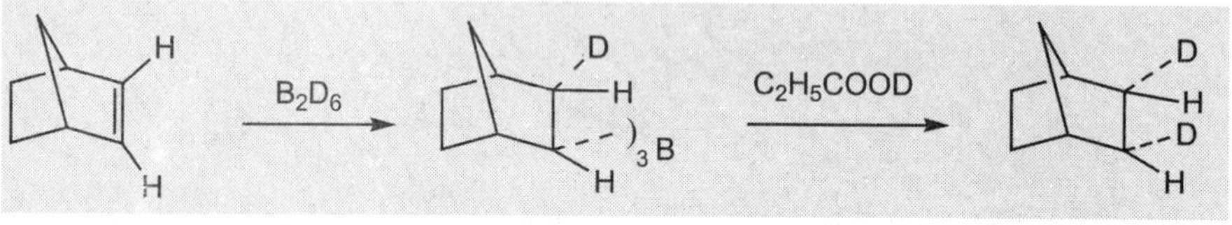

Alkylboranes can also be oxidized to carbonyl compounds on treatment with chromic acid.

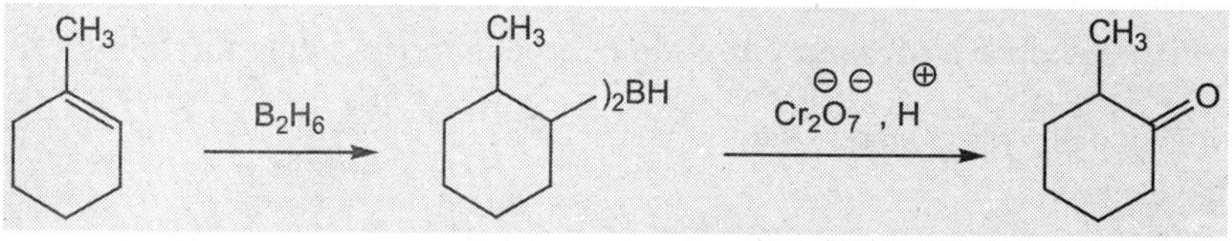

Hydroboration proceeds faster with acetylenes than alkenes, and treatment of the products (vinylboranes) with acetic acid yields *cis* olefins of high purity, whereas peroxide oxidation of vinylboranes gives carbonyl compounds.

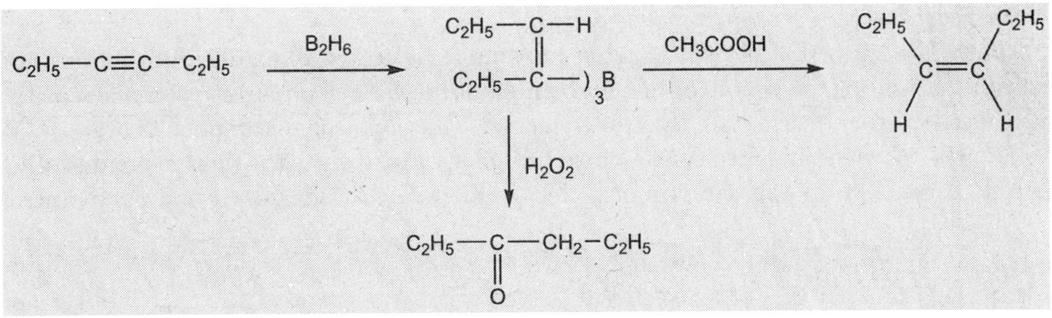

Further Applications of Hydroboration Reaction: Reduction of Carbonyl Compounds

Aldehydes and ketones can be reduced to primary and secondary alcohols, respectively, by boranes. All the three hydrogens can be transferred from BH_3, so that three molecules of an aldehyde or ketone can be reduced by one molecule of borane.

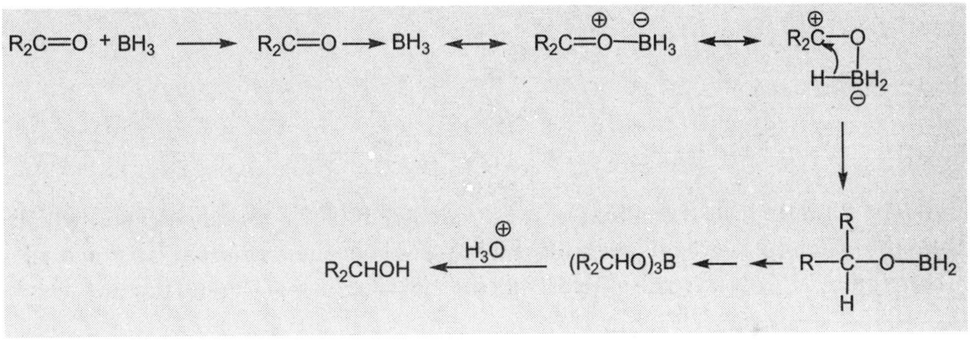

In diarylketones the reduction can proceed to the hydrocarbon stage. This is explained by the consideration that the borate anion being a good leaving group, a resonance-stabilized carbocation is generated, which is reduced to the hydrocarbon by the uptake of a hydride ion.

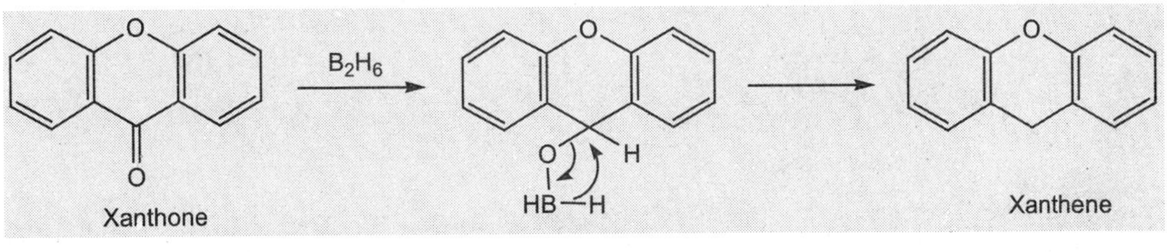

Reduction of an aralkyl ketone to the hydrocarbon stage can be effected by boranes by the addition of boron trifluoride, which facilitates the departure of borate anion.

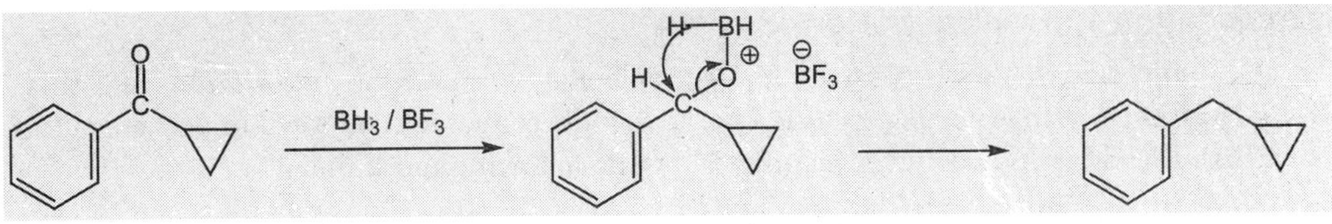

Stereochemical Aspects of Hydroboration Reaction

Dialkylboranes, though less reactive, are more selective reagents than borane itself. Further, by increasing the bulk of the alkyl substituents, the dialkylboranes can be made sensitive to steric control in order to give stereospecific reduction. Thus, *bis*-isopinocamphenylborane (R_2BH), obtained from α-pinene and diborane, reduces camphor exclusively to the thermodynamically less stable *iso*-borneol by an approach from the less hindered *endo* side (steric approach control).

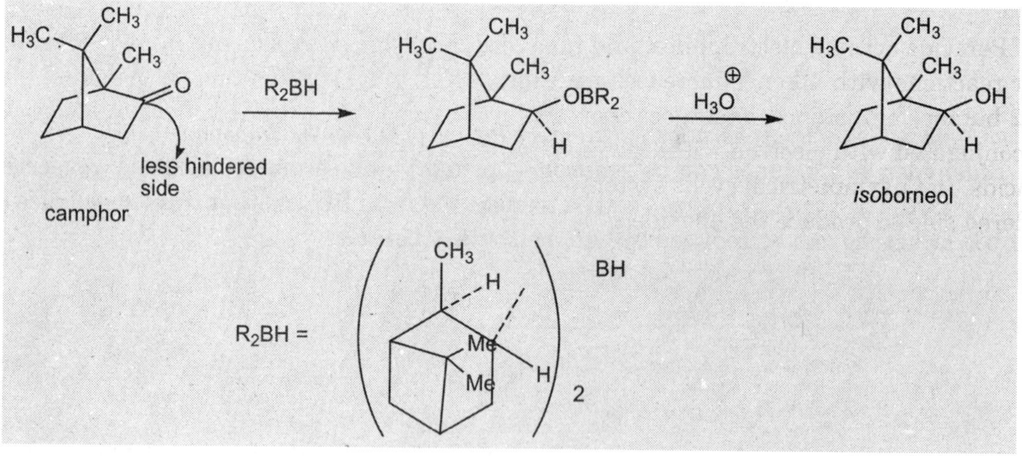

Similarly, 2-methylcyclohexanone can be reduced 100% to the *cis* alcohol, which is less stable than the other epimer by using bulky substituted boranes, whereas borane itself gives a mixture containing 75% of the more stable *trans* alcohol, and 25% of the *cis* (product development control).

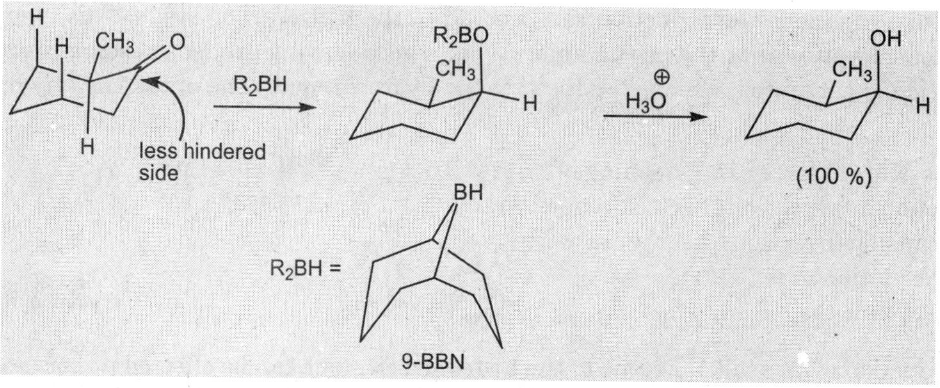

C. Epoxidation and Hydroxylation

The reaction of alkenes with peracids to form epoxides is known as *epoxidation,* and this is effected best by treating the alkene with a peracid, most conveniently, *m*-chloroperbenzoic acid (*m*-CPBA). Epoxidation is believed to proceed by the following mechanism.

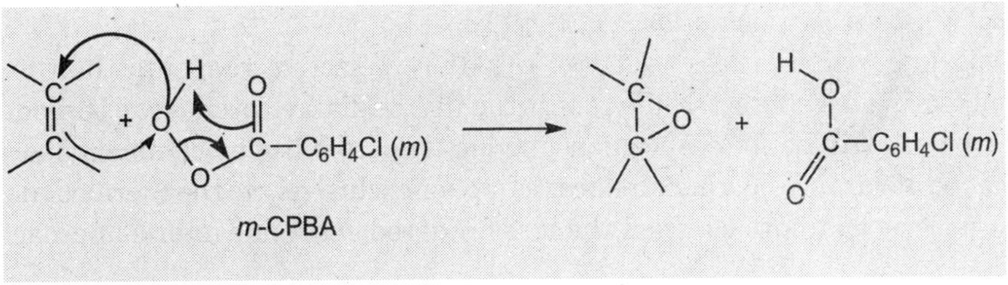

Peracids act like electrophiles, and in accordance with the above mechanism, the rate of their reactions with alkenes increases with increasing alkyl substitution at the double bond. *Cis*-2-butene, for instance, undergoes epoxidation 500 times faster than ethylene. Alkenes which are conjugated with electron-withdrawing groups, such as a carbonyl group, do not react with peracids. In unsymmetrical cyclic systems, the peracids usually attack the alkene from its less hindered side to produce the epoxide.

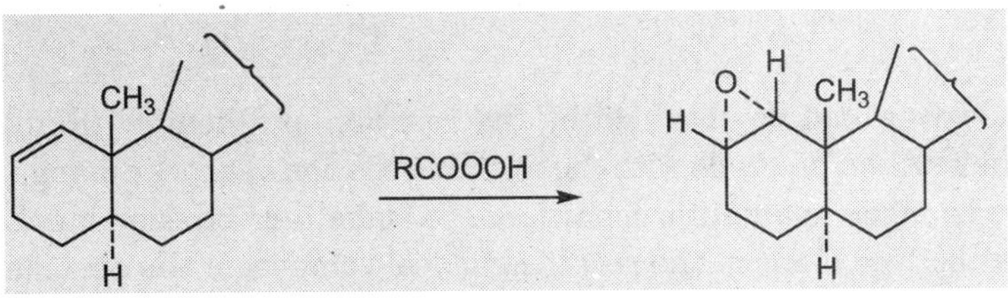

Epoxidation is a stereospecific process, as the formation of two C–O bonds occurs on the same face of the π-bond of alkene. As expected, *cis*-alkene gives *cis*-epoxide, and *trans*-alkene gives *trans*-epoxide.

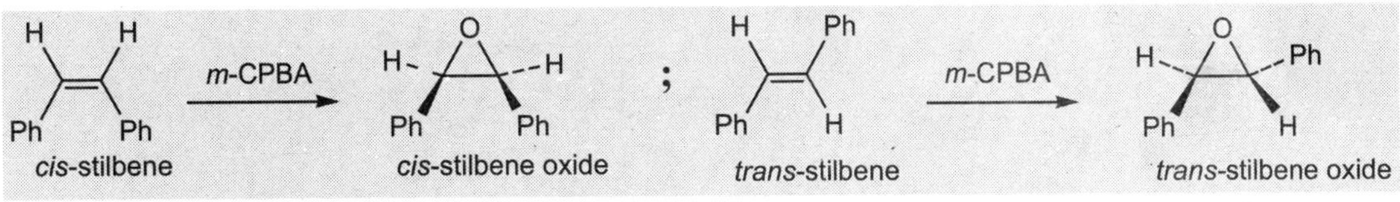

The acid-catalyzed ring-opening of the epoxides leads to the formation of diols. This is an S_N2 reaction in which water acts as a nucleophile, and attacks the carbon on the side opposite to oxygen causing inversion of configuration at one of the carbon atoms. This is one of the methods for hydroxylation of an alkene.

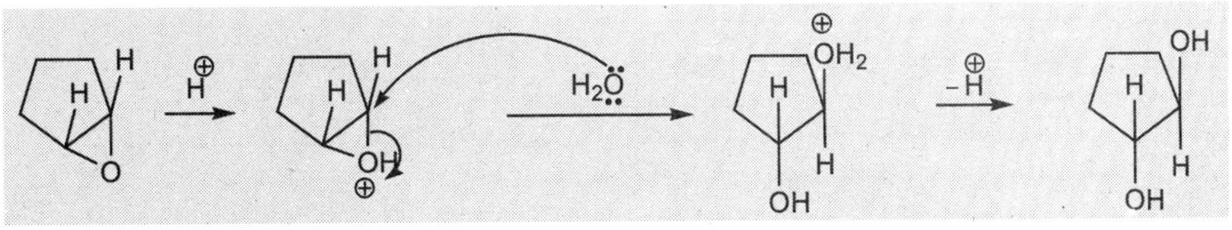

In the presence of strong peracids, epoxides cannot be isolated, rather they undergo ring-opening to yield a *trans*-1,2-hydroxy ester **(11)**. Further, hydrolysis of this ester yields a vicinal *trans* glycol **(12)**. Thus, there is an overall *trans* addition of two hydroxyl groups to the double bond.

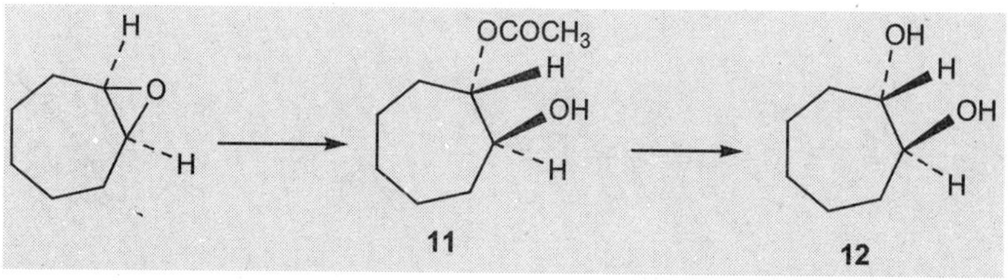

Ring-opening of a cyclohexene epoxide **(13)** yields a glycol **(14)** having the two hydroxyl groups *diaxial,* rather than *diequatorial.*

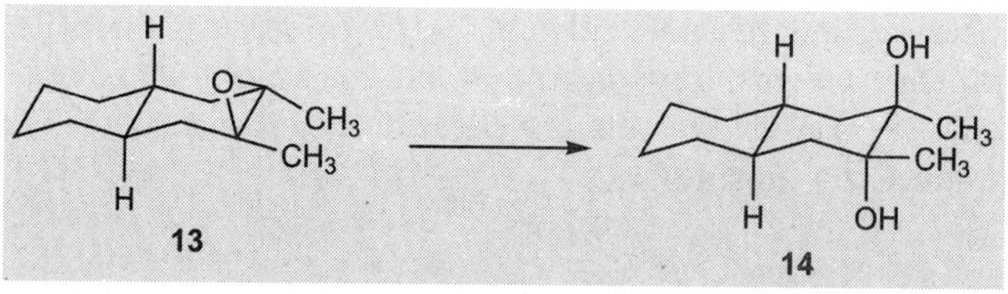

A *cis*-diol, however, can be obtained by the reaction of alkene with alkaline potassium permanganate. The reaction proceeds with the initial formation of a cyclic manganese ester, which undergoes further changes in alkaline conditions. ^{18}O labelling experiments have conclusively demonstrated that both oxygens of the resulting glycol come from the permanganate ion.

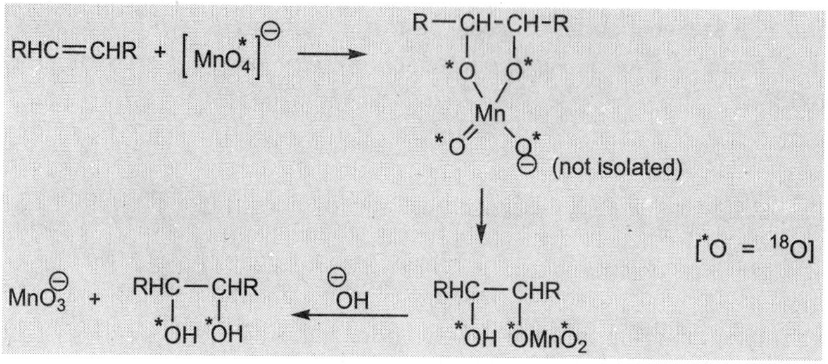

As there is splitting of the manganese–oxygen bond in the reaction, there is no inversion of configuration at the carbon atoms, and this results in a stereospecific *cis* addition.

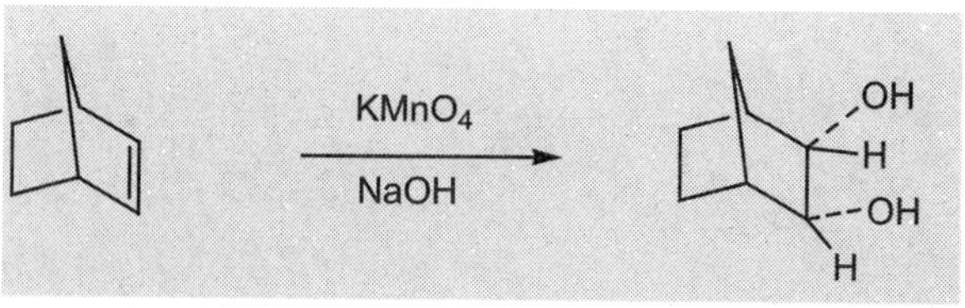

Cis-hydroxylation can be better achieved by the reaction of alkene with osmium tetroxide.

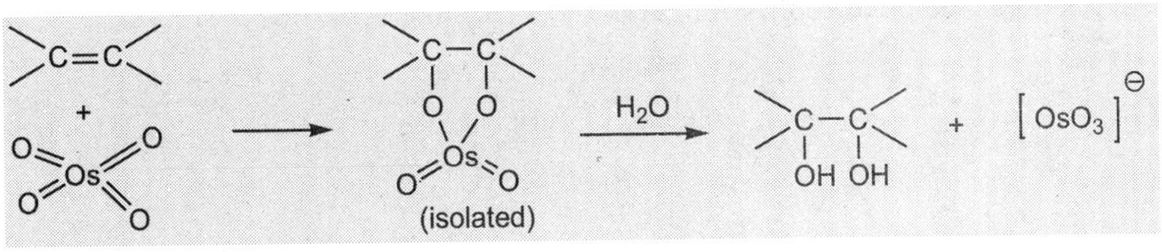

Osmium tetroxide is a highly specific reagent for the *cis*-hydroxylation on the less hindered side of the alkene.

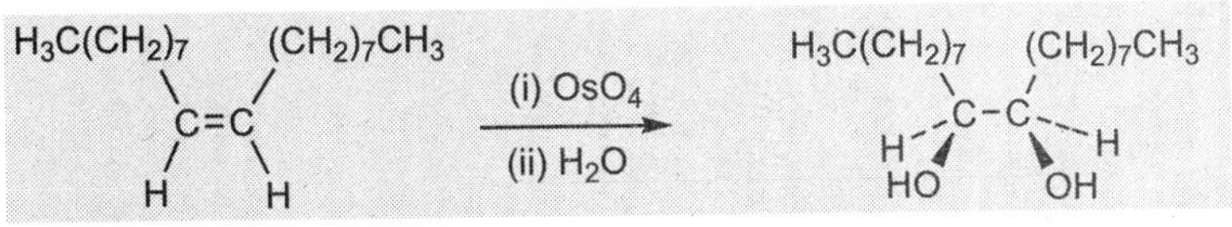

Thus, depending upon the choice of reagents, *cis* or *trans*-diols can be obtained by the hydroxylation of alkenes.

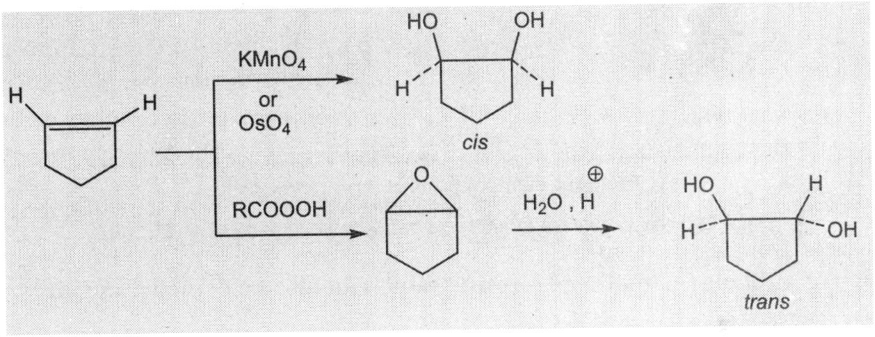

D. Addition of Water and Oxymercuration-Reduction

I. Acid-Catalyzed Addition of Water

Addition of water to an alkene (hydration) can be carried out using aqueous acid. There is an initial formation of carbocation by the attack of the electrophile (H^+) to the sp^2 carbon that is bonded to the greater number of hydrogens. In the second step, the nucleophile (H_2O) adds to the other sp^2 carbon in a fast step. The product, a protonated alcohol, rapidly loses a proton to give the alcohol.

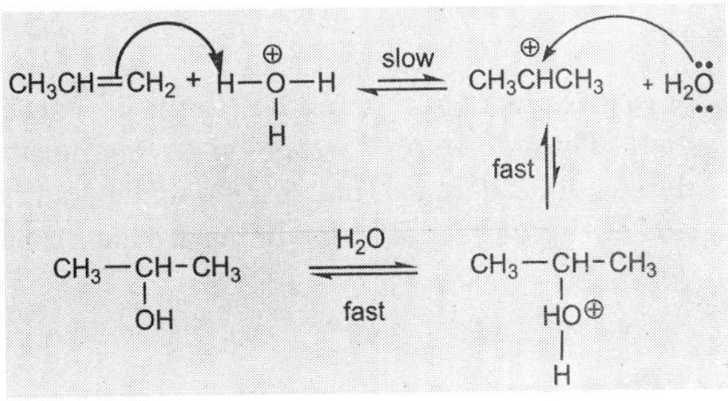

The reaction can be outlined by a reaction coordinate diagram (**Fig. 7.2**).

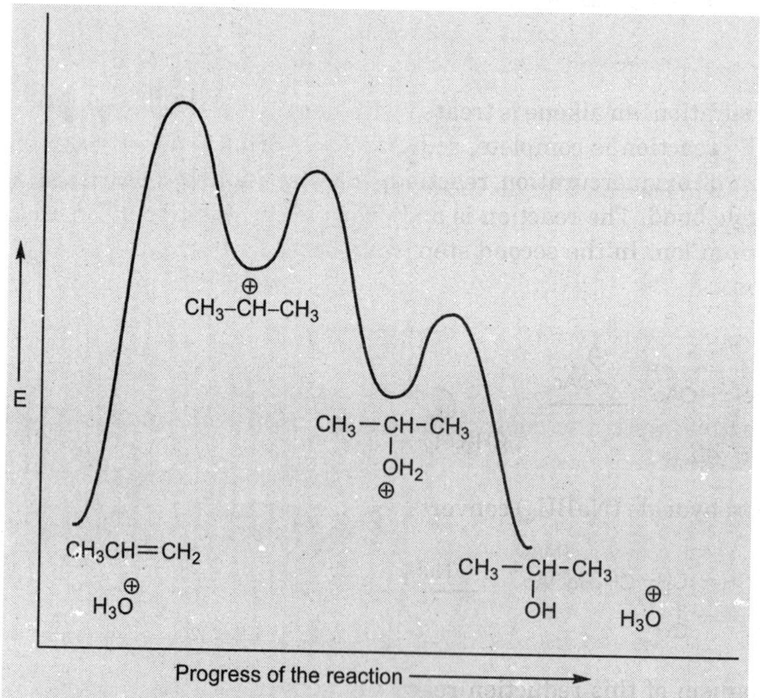

Fig.7.2: Energy diagram for the acid-catalyzed hydration of an alkene

Addition of alcohols to an alkene forming ether can also be depicted in similar steps.

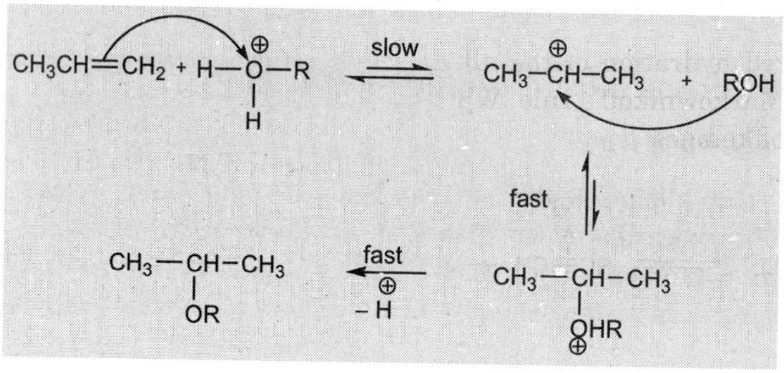

II. Oxymercuration-Reduction

Under normal laboratory conditions, water is added to an alkene by a procedure known as *oxymercuration-reduction*. The addition of water by oxymercuration-reduction has two advantages over acid-catalyzed addition. It does not require acidic conditions that are harmful to many organic molecules, and as carbocation intermediates are not formed, the rearrangements do not occur.

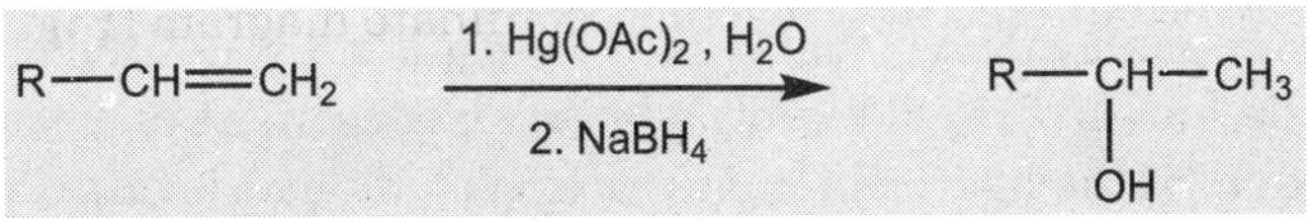

In oxymercuration, an alkene is treated with mercuric acetate in aqueous tetrahydrofuran (THF). When the reaction is complete, sodium borohydride is added to the reaction mixture. In the first step of oxymercuration reaction, the electrophilic mercury of mercuric acetate adds to the double bond. The reaction is analogous to the addition of Br_2 to an alkene to form a cyclic bromonium ion. In the second step, water attacks the more substituted carbon of the mercurinium ion.

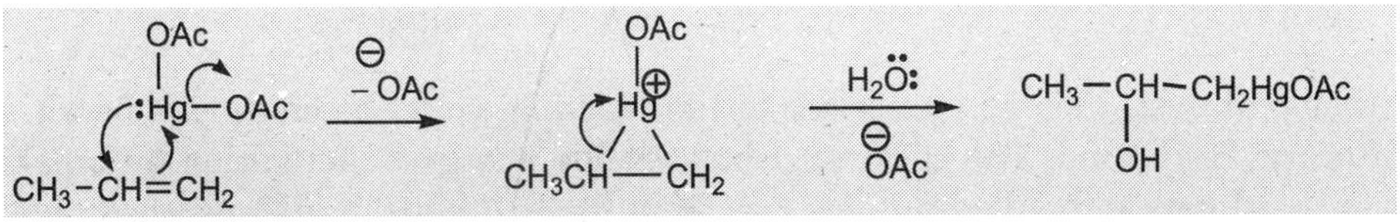

Sodium borohydride ($NaBH_4$) converts the C–Hg bond into a C–H bond.

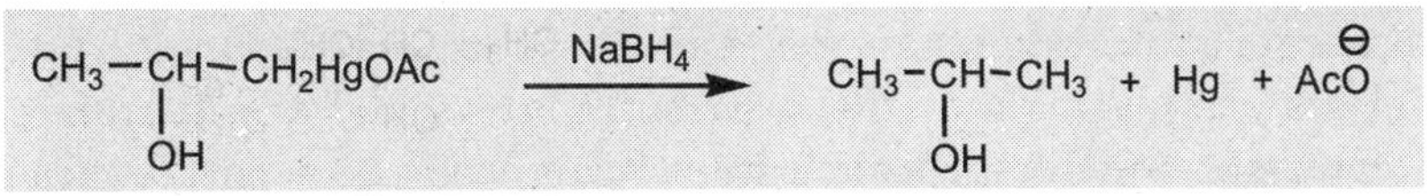

The mechanism of this reduction reaction is not fully understood, although it is known that the intermediate is a radical. The overall reaction (oxymercuration-reduction) gives the same product that would be formed from the acid-catalyzed addition of water; the hydrogen adds to the sp^2 carbon bonded to the greater number of hydrogens, and OH adds to the other sp^2 carbon.

Acid-catalyzed hydration of the alkynes also requires mercuric ions, and proceeds in accordance with Markownikoff's rule. While acetylene gives acetaldehyde, other alkynes lead to the formation of ketones.

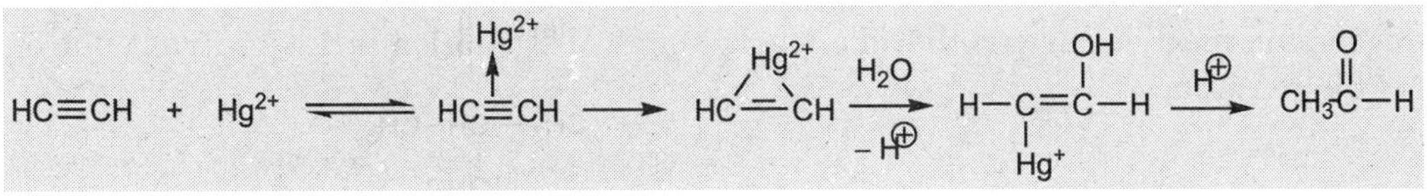

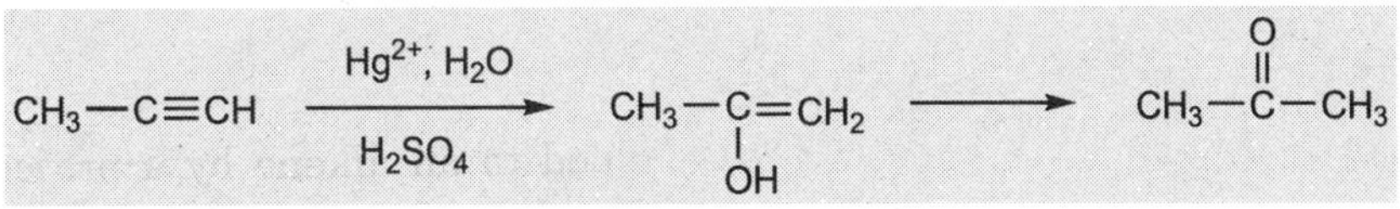

E. Hydrogenation of Alkenes and Alkynes

In the presence of a metal catalyst such as platinum, palladium, or nickel, hydrogen (H_2) adds to the double and triple bonds of alkenes and alkynes, respectively, to form an alkane. Hydrogen adds to an alkyne in the same manner that it adds to an alkene. It is difficult to stop the reaction at the alkene stage, because hydrogen readily adds to alkenes in the presence of these catalysts. Eventually, the product of the hydrogenation reaction of alkynes is an alkane.

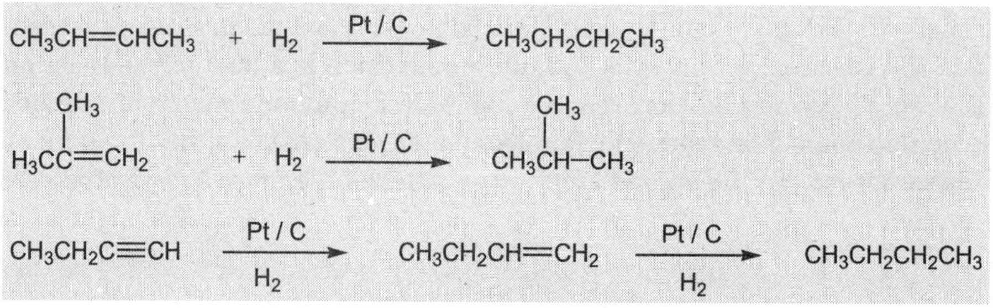

Without catalyst, the energy barrier to the reaction would be enormous because of the strength of the H–H bond. The catalyst decreases the energy of activation by breaking the H–H bond. Platinum and palladium are used in a finely divided state adsorbed on charcoal (Pt/C, Pd/C). The platinum catalyst is frequently used in the form of PtO_2, which is known as *Adam's catalyst*.

Due to presence of a catalyst in these reactions, the process is called *catalytic hydrogenation*. The metal catalysts are insoluble in the reaction mixture, and are, therefore, classified as heterogeneous catalysts. A heterogeneous catalyst can easily be separated from the reaction mixture by filtration. It can then be recycled, which is an important property, since metal catalysts are expensive. The reaction can be stopped at the alkene stage if a poisoned (partially deactivated) metal catalyst is used. The most commonly used partially deactivated metal catalyst is Lindlar catalyst, which is prepared by precipitating palladium on calcium carbonate, and treating it with lead(II) acetate and quinoline. This treatment modifies the surface of palladium, making it much more effective at catalyzing the addition of hydrogen to a triple bond than to a double bond.

Heat of Hydrogenation and Relative Stability

The heat released in a hydrogenation reaction is called the heat of hydrogenation. As hydrogenation reactions are exothermic, the heat of hydrogenation is the positive value of the ΔH^o of the reaction.

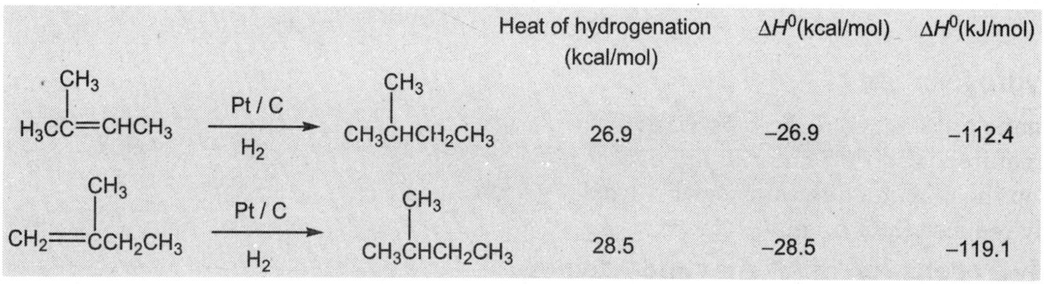

Since we do not know the precise mechanism of hydrogenation reaction, we cannot draw a reaction coordinate diagram for it. We can, however, draw relative energy levels of reactants (alkenes) and the products. Three catalytic hydrogenation reactions shown in **Fig.7.3** give the same product, and therefore, the energy of the product is same for each reaction. The three reactions, however, have different heats of hydrogenation, reflecting that these three reactants must have different energies. Most stable alkene has the smallest heat of hydrogenation. For example, 3-methyl-1-butene releases the most heat, so it must be the least stable of the three alkenes. This is in agreement with the observation that greater number of alkyl substituents on the sp^2carbons increases the stability of an alkene.

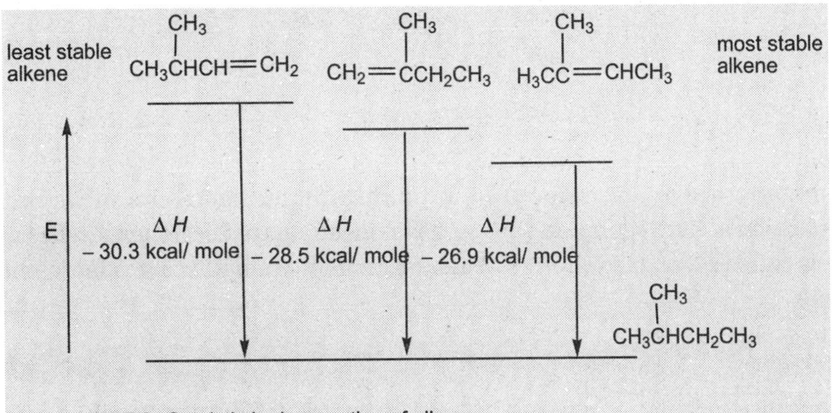

Fig.7.3: Catalytic hydrogenation of alkenes

Both *trans*- and *cis*-2-butenes have two alkyl groups bonded to the sp^2 carbons, but *trans*-2-butene has less heat of hydrogenation. This means that the *trans* isomer, in which the large substituents are apart, is more stable than the *cis* isomer, where the large substituents are closer.

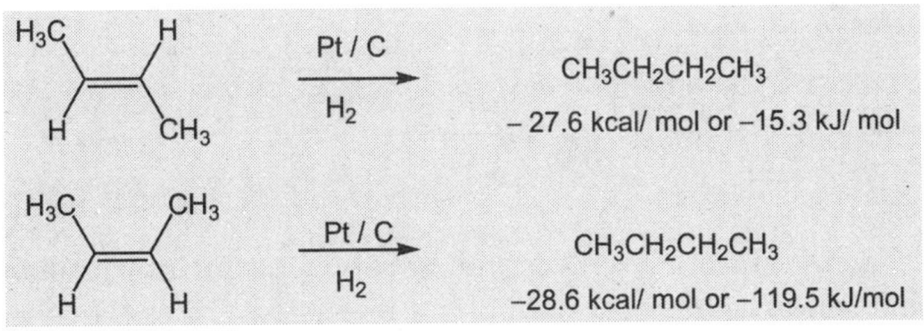

F. Hydrogenation of Aromatic Rings

Aromatic rings can be reduced by catalytic hydrogenation, but higher temperatures (100 to 200°C) are required than that for ordinary double bonds. With benzene rings, it is usually impossible to stop the reaction after only one or two double bonds have been reduced, as olefins are more easily reduced than aromatic rings.

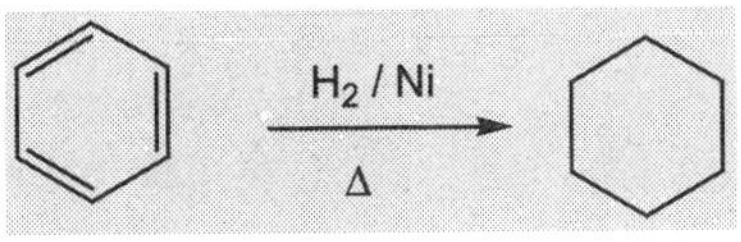

Thus, 1 mole of benzene when treated with 1 mole of hydrogen gives no cyclohexadienes or cyclohexene, but only 1/3 mol of cyclohexane and 2/3 mol of the recovered benzene. However, this is not true for all aromatic systems, for example, with phenanthrene it is easy to stop the hydrogenation after only the 9,10-bond has been reduced.

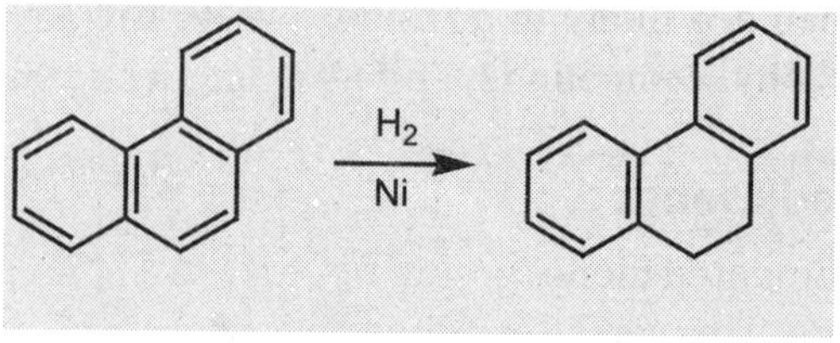

Birch Reduction

When aromatic rings are reduced by lithium in liquid ammonia in the presence of an alcohol, 1,4-addition of hydrogen takes place and nonconjugated cyclohexadienes are produced. This reaction is called *Birch reduction*. By this reaction an aromatic ring is more readily reduced than an alkene.

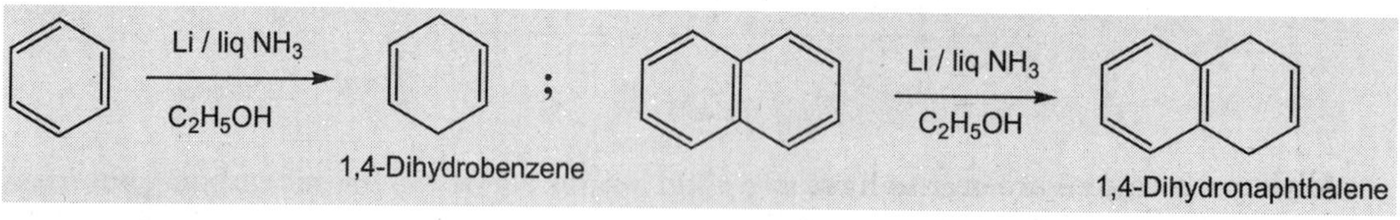

G. Ozonolysis

Each of the three oxygen atoms of ozone has electrophilic character as evident by the following resonating structures.

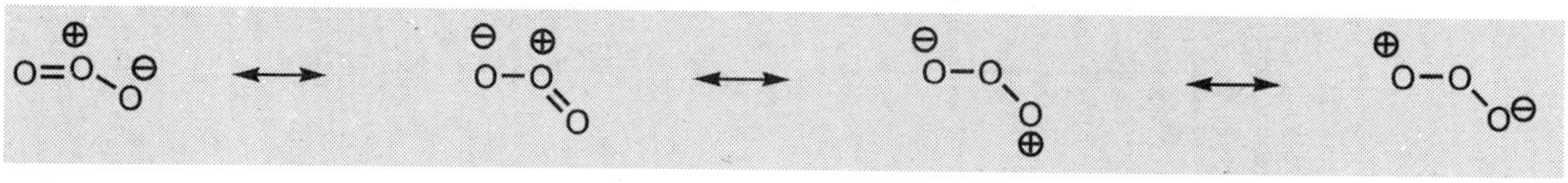

The alkenes add ozone to form ozonides which are very useful intermediates in degradative studies. The mechanism of ozonide formation is a multistep process. In the first step, cyclic molozonide **(15)** is formed by the electrophilic addition of ozone to an alkene. The molozonide, being unstable, subsequently decomposes into two fragments. Recombination of these fragments in an alternative way yields the ozonide **(16)**.

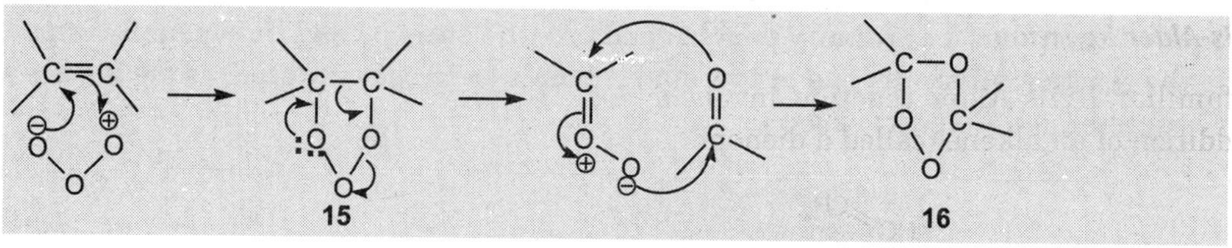

Participation of molozonide has been demonstrated by the isolation of molozonide of *trans*-2, 5-dimethylhex-3-ene at – 80°C, which rearranges to the ozonide on warming up to – 60°C. The ozonide undergoes reductive decomposition with zinc and acetic acid or hydrogen in presence of a catalyst such as palladium. Hydrolytic cleavage of the ozonide, on the other hand, gives rise to hydroperoxides, which are likely to adversely affect the advantages of ozonolysis as a preparative method for carbonyl compounds, and as a diagnostic one for the ethylenic bond.

H. Addition to Conjugated Dienes

Electrophilic addition to conjugated dienes occurs not only to a pair of adjacent carbons, but also to the carbons at the two ends. This can be explained by assuming that in the initial electrophilic attack, addition occurs in the direction which leads to a resonance stabilized allylic cation. The positive charge of this intermediate resides primarily on two carbons and either of these positions may be attacked by the nucleophile yielding the 1,2- or the 1,4-addition product.

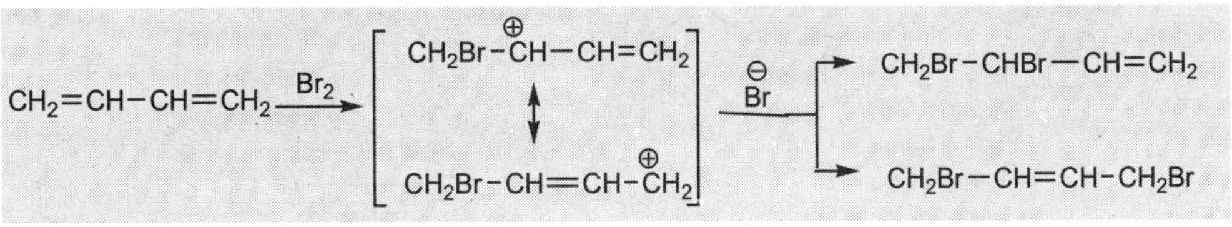

Allylic carbocations are unusually stable, and hence, are formed preferentially. For example, treatment of 2,4-hexadiene with HCl yields products derived from the allylic carbocation (**17**), rather than the alternative carbocation **18** although in both the structures, the positive charge is located on a secondary carbon atom.

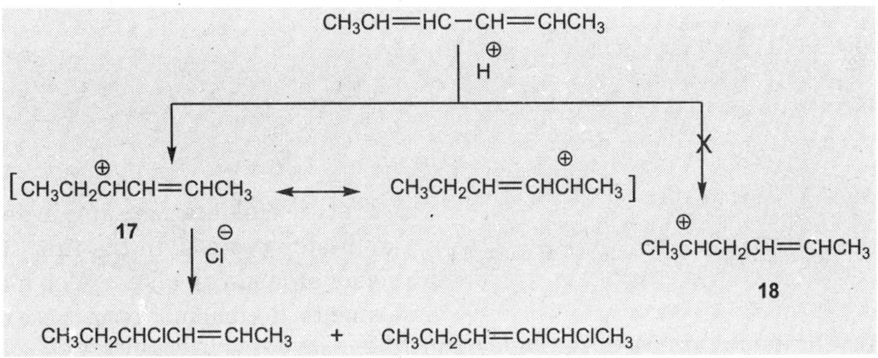

I. Diels-Alder Reaction

The familiar Diels-Alder reaction involves the formation of a six-membered ring by the 1,4-addition of an alkene (called a dienophile) to a conjugated diene.

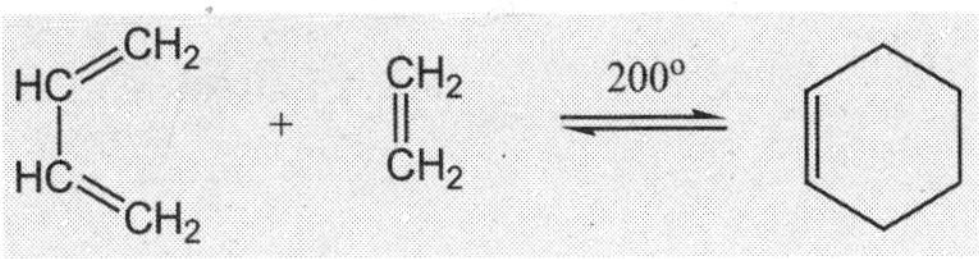

The reaction proceeds faster when the dienophile has electron-withdrawing substituents such as –COOH, –COOR, –CHO, $–NO_2$ and –CN. On the other hand, reaction rates rise with increasing electron supply in the diene; 1,3-butadiene, for instance, is less reactive than its mono-, di-, and trimethyl derivatives. Many such reactions take place under relatively milder conditions.

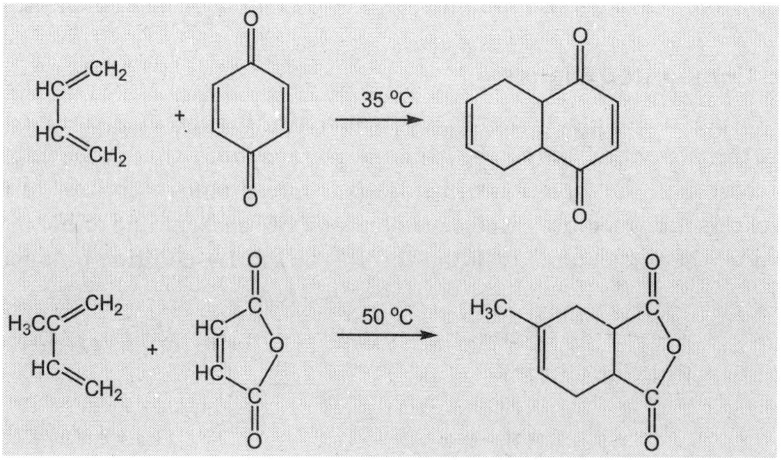

These reactions show an extremely high degree of stereospecificity, for example, maleic acid gives *cis* acid **(19)** and fumaric acid yields the *trans* isomer **(20)**, both by *cis* addition.

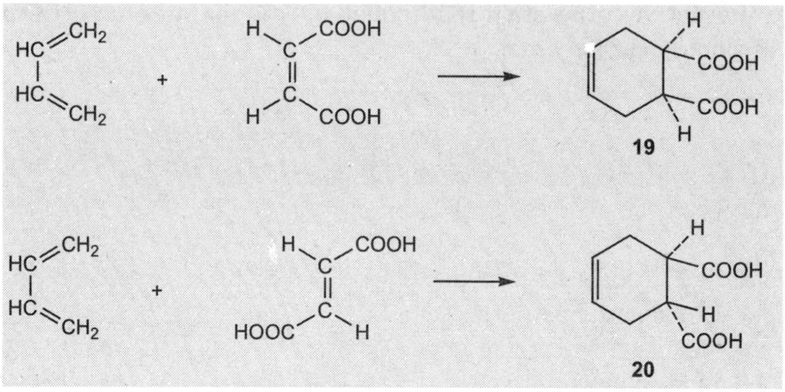

When cyclic dienes are used, the *endo* adduct **21** is favoured over the *exo* **22.**

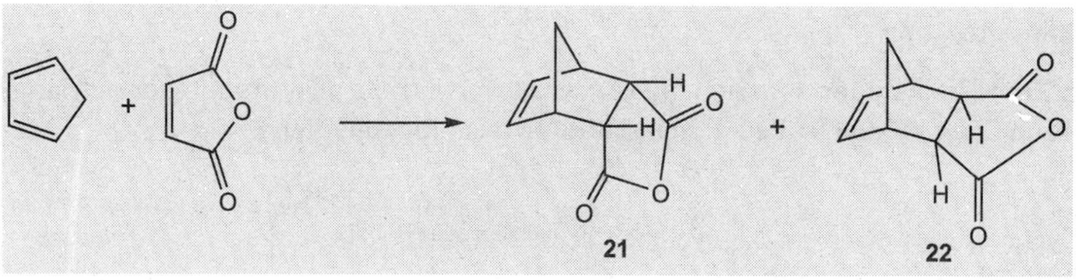

A carbon-carbon triple bond may also participate in the reaction.

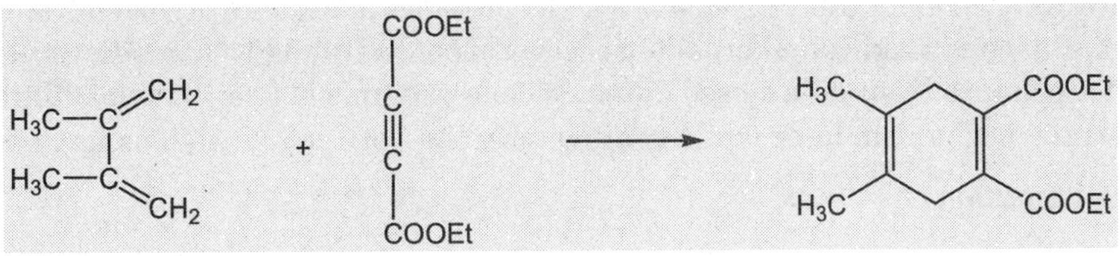

This reaction has been employed in determining the structures of substituted homoannular cyclohexadienes. For example, the structures of the following isomers of methylcyclohexadienes can be determined.

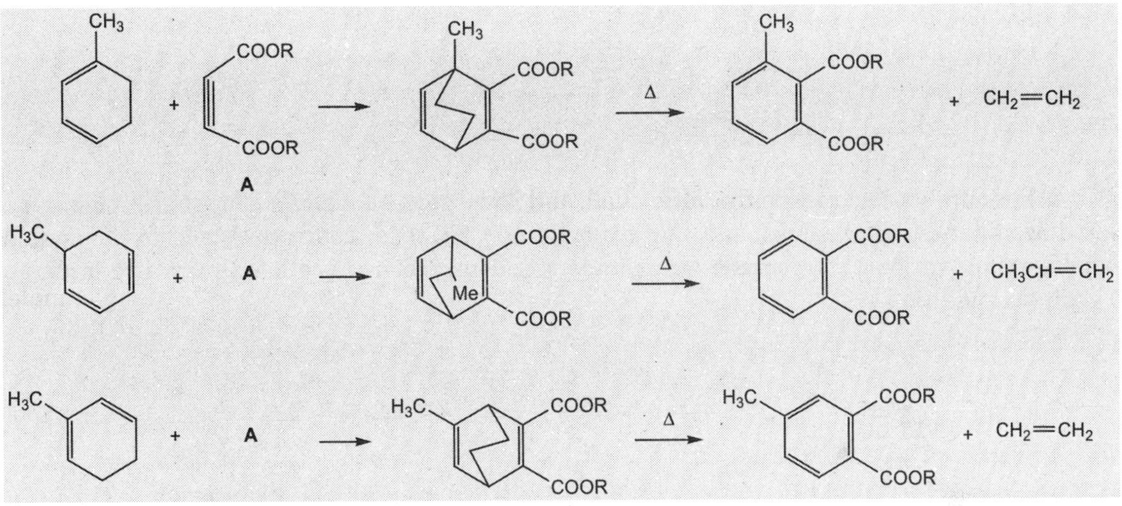

By identifying phthalic esters through hydrolysis and pyrolysis to the corresponding phthalic anhydride, the structures of the homoannular cyclohexadienes are very conveniently determined.

Success of the Diels-Alder reaction depends upon the geometric arrangement about a single bond of the diene component; in other words, the diene must have *s-cis* conformation. Since the energy barrier for the interconversion of *s-cis* **(23)** and *s-trans* **(24)** acyclic dienes is very low (5 kcal/mole), the reaction proceeds well with the latter dienes as well because of the equilibrium between the two conformational structures, even though *s-trans* is thermodynamically more stable.

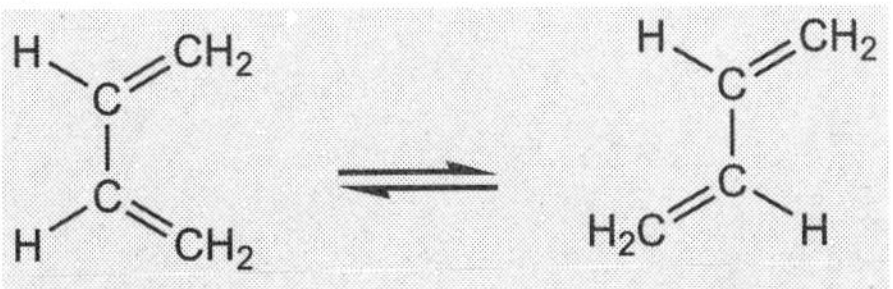

23 **24**

Cyclic dienes, such as 1,3-cyclopentadienes are fixed in the *s-cis* conformation, and are especially reactive.

The Diels-Alder reaction usually does not appear to involve ionic or free radical intermediates, and is remarkably insensitive to the presence or absence of solvents and catalysts. There is simultaneous making and breaking of several bonds through a transition state in which all first order changes occur on a closed curve. It is an example of (4 + 2) cycloaddition, and we shall discuss its mechanism later while dealing with the pericyclic reactions in Chapter 15.

II. 1,3-Dipolar Addition

This type of addition may be regarded as a (2 + 3) cycloaddition, which provides a versatile method for synthesizing five-membered heterocyclic compounds by the addition of a 1,3-dipolar compound to a double bond. The generalized equation may be written as follows:

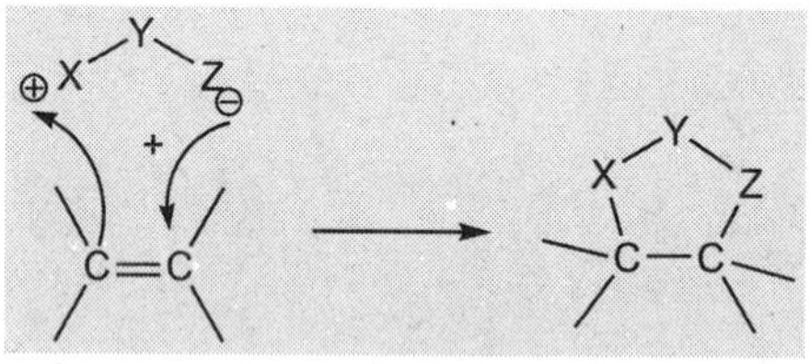

X has only six electrons in the outer shell, and Z has at least one pair of unshared electrons, *e.g.* diazoalkanes, alkyl azides, and azoxy compounds. All these compounds add to the double bond forming various heterocyclic compounds. Carbon-carbon triple bond may also undergo 1,3-dipolar addition.

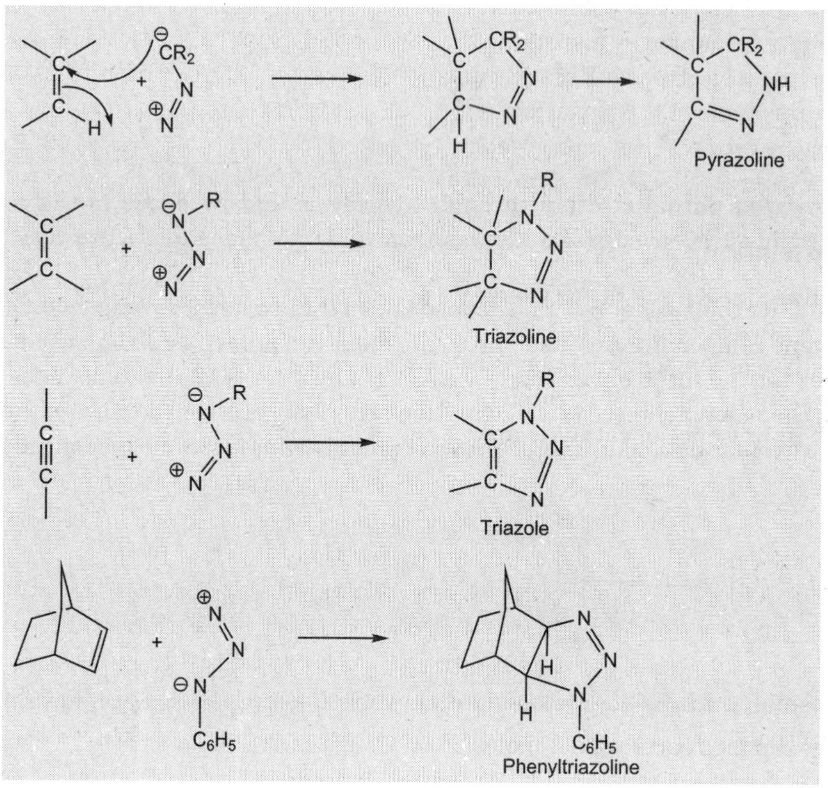

Formation of molozonide **(15)**, as discussed earlier, may be regarded as an example of 1,3-dipolar addition.

A convenient method for lengthening a carbon chain by three is based on 1,3-dipolar addition of a keto carbene to a ketene.

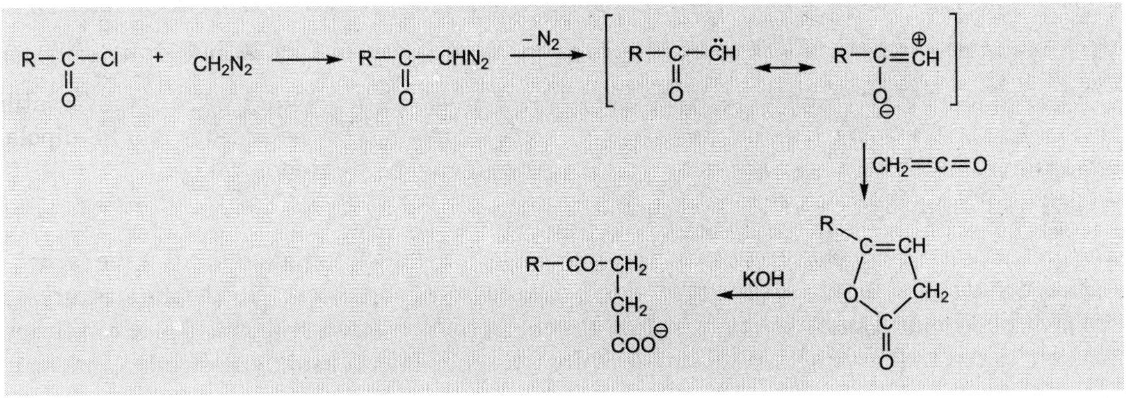

7.2 NUCLEOPHILIC ADDITIONS

We have already noted that an electron-withdrawing group deactivates a carbon–carbon double bond towards electrophilic addition, but at the same time its presence makes the alkene more susceptible to attack by a nucleophile. The role of electron-withdrawing substituents is to stabilize the intermediate carbanion by dispersing the negative charge. Thus, a carbon–carbon double bond conjugated with an electron sink readily undergoes nucleophilic addition. Common examples are the addition reactions of α,β-unsaturated nitriles, aldehyde, ketones, and esters. Nucleophilic addition is the reverse of elimination reaction involving carbanion (*E1cB* mechanism), and both reactions are interrelated by the principle of *microscopic reversibility*.

A. Cyanoethylation

The 1,4-addition reactions of acrylonitrile are known as *cyanoethylations*. The β-carbon is attacked by a nucleophilic reagent, and the negative charge is stabilized by the cyano group in the intermediate anion. The reaction is then completed by the abstraction of a proton from the solvent. These steps can be illustrated by taking a simple example of the addition of ethanol to acrylonitrile in presence of catalytic amounts of a strong base.

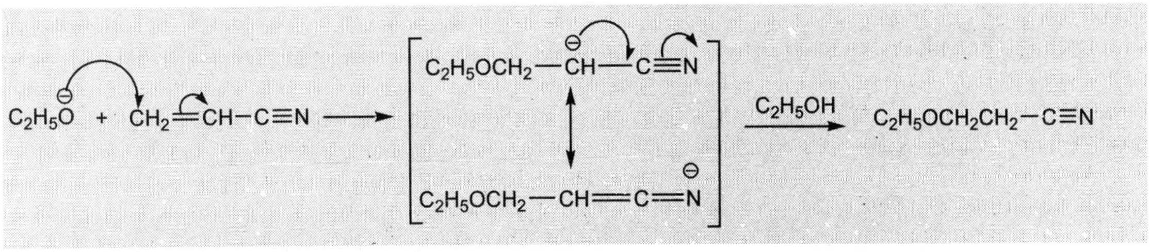

Cyclohexanone reacts with 4 moles of acrylonitrile in the presence of a base.

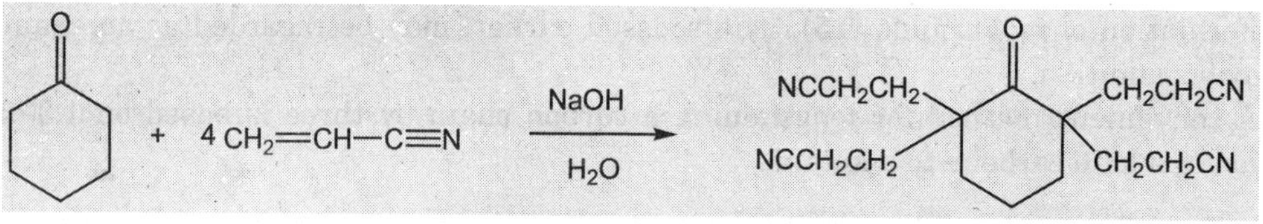

Ammonia and amines do not require the presence of a base, but act as their own catalysts.

$$(CH_3)_2\ddot{N}H \;+\; CH_2{=}CHCN \longrightarrow (CH_3)_2NCH_2CH_2CN$$

B. Michael Addition

The presence of the carbonyl group is responsible for the facile nucleophilic addition at the carbon-carbon double bond in α,β-unsaturated aldehydes, ketones, and esters. Nucleophilic attack at the β-carbon yields a stable anion, which abstracts a proton from the solvent to produce either the keto or the enol form of the addition product. The enol form is usually less stable, and so it is rapidly equilibrated to the more stable keto form.

Probably, the most important addition reaction of α,β-unsaturated carbonyl compounds is the nucleophilic addition of enolate anions to the carbon–carbon double bond. This process in which a new carbon–carbon bond is produced is called *Michael addition*. The α,β-unsaturated compounds used in this reaction include any unsaturated system having an electron-withdrawing group capable of stabilizing the intermediate anion. The enolate anions are commonly produced from malonic ester, ethyl cyanoacetate, and ethyl acetoacetate. The mechanism of Michael addition, with ethyl malonate as the source of the nucleophile, may be depicted with the nucleophile = $^{-}CH(COOEt)_2$.

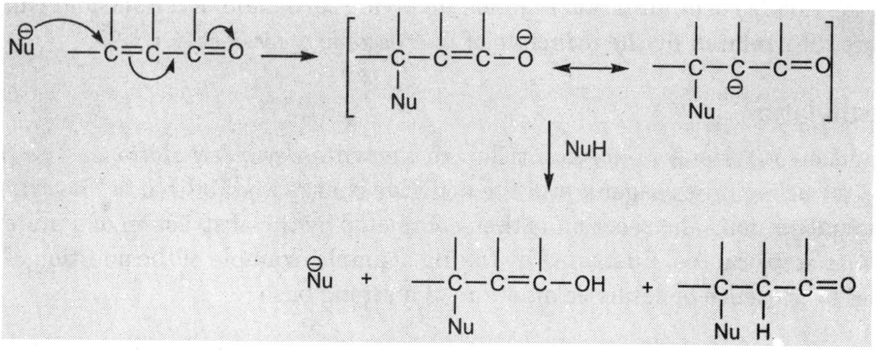

Double Michael additions are often employed for synthesizing ring compounds.

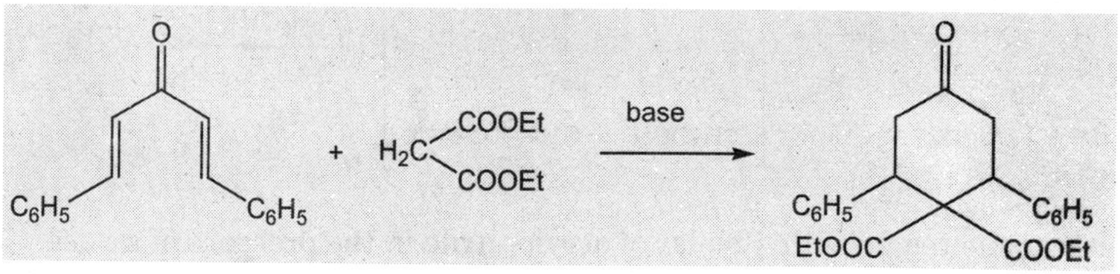

PROBLEMS AND REVIEW QUESTIONS

1. Which of the following sequence would be best for the preparation of 2-bromobutane:

(i) $CH_3CH{=}CHCH_3 \xrightarrow{Br_2}$

(ii) $CH_3CH_2CH{=}CH_2 \xrightarrow{HBr}$

(iii) $CH_3CH_2CH{=}CH_2 \xrightarrow[\text{peroxide}]{HBr}$

2. On heating *cis*-2-butene to a temperature greater than 300°C, a mixture of two isomeric 2-butenes results.
   (i) What is the mechanism of this reaction?
   (ii) Which butene isomer would you expect to predominate in the mixture when equilibrium is established between them?
3. On carrying out reaction of 2-butene with HCl in the presence of ethanol, one of the products formed has the molecular formula $C_6H_{14}O$. Assign a structure to this product, and discuss the mechanism leading to its formulation.
4. Addition of an electrophile to 2,4-hexadiene generates two intermediates. Write their structures, and show which one is more stable.
5. Write structures for the products expected for the following two reactions indicating the most stable conformations:

6. How many products do you expect to be formed by the addition of 1 mole of bromine to 1-phenyl-1,3-butadiene? Write their structures, and state which one is the most stable.
7. How many products do you expect when 2-methyl-2-butene is treated with bromine in the presence of methyl alcohol containing lithium chloride? Discuss mechanisms of their formation.
8. Predict the stereochemistry of the product formed by the addition of bromine on- (i) *trans*-1,2-dideuterioethene and (ii) *cis*-1,2-dideuterioethene.
9. Propose a mechanism for the following reaction:

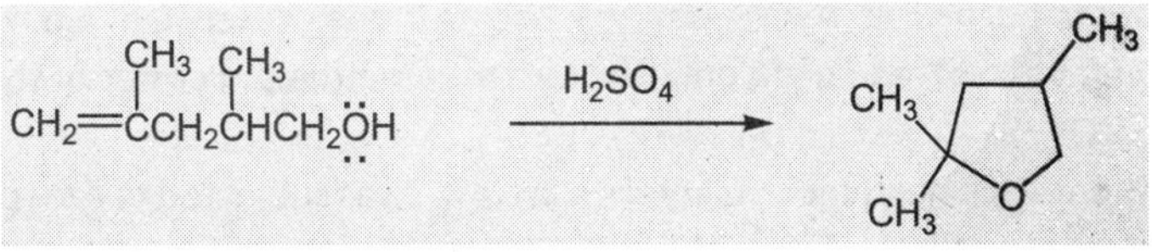

10. Write a feasible mechanism for the acid-catalyzed isomerization of *cis*-2-butene to *trans*-2-butene.

11. What would be the major product obtained by the addition of HBr to each of the following compounds?

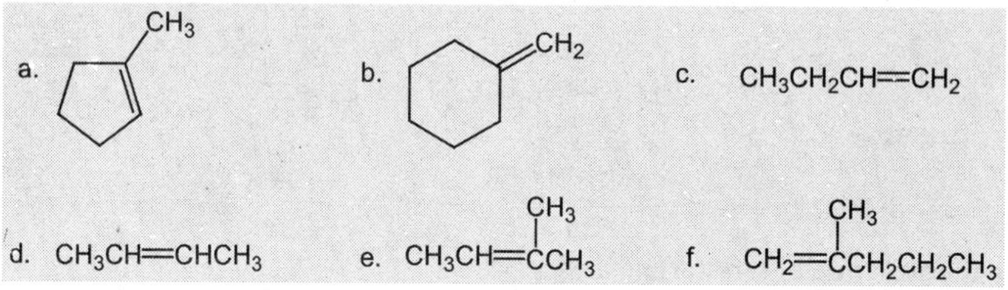

12. How many moles BH_3 are needed to react with 2 moles of 1-pentene?
13. What will be the major product of reaction of 2-methyl-2-butene with each of the following reagents?
    (a) HBr
    (b) $Hg(OAc)_2$, H_2O followed by $NaBH_4$
    (c) HBr + Peroxide
    (d) BH_3 / THF, followed by H_2O_2 / hydroxide ion.
14. On the basis of the fact that the π bond of an alkyne is weaker than the π-bond of an alkene, it is expected that an alkyne might be more reactive than an alkene. However, electrophilic additions to an alkyne occur more slowly than an alkene. Explain.
15. Complete the following reactions, and sketch a plausible mechanism for each of them.

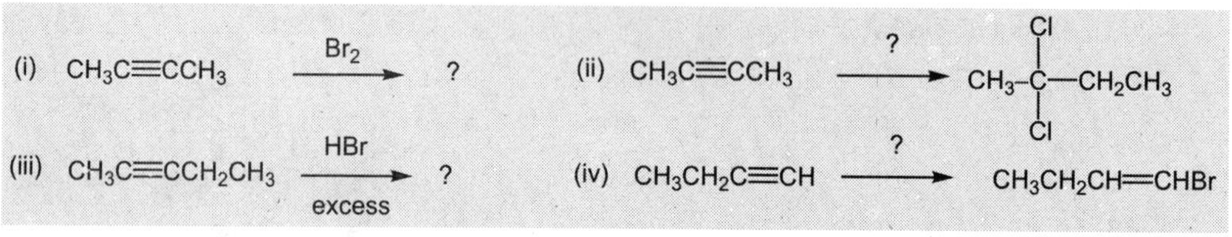

16. It has been observed in hydroboration of alkynes that one equivalent of BH_3 adds to the internal alkynes, and reaction can be stopped at the alkene stage, whereas in the case of terminal alkynes, it is harder to stop the addition reaction at the alkene stage.
    (i) Giving suitable examples, offer explanation for such observations.
    (ii) Which is the suitable reagent to stop the reaction at alkene stage in the case of terminal alkynes?
17. 2-Pentyne undergoes hydration in the presence of H_2SO_4 to give a mixture of two isomeric ketones. On the other hand, hydration of 1-pentyne occurs in the presence of a mixture of sulphuric acid and mercuric sulphate, and gives only one ketone. Give the structures of the products along with mechanisms.
18. A hydrocarbon which has a molecular formula C_6H_{12}, was subjected to ozonolysis giving equimolar amounts of ethyl methyl ketone ($CH_3CH_2COCH_3$) and acetaldehyde (CH_3CHO). Assign structure to it.
19. Ozonolysis of alkenes (A), (B), (C), (D), and (E) gave the following products:

(A) $CH_3-\overset{O}{\overset{\|}{C}}-CH_2CH_2CH_2CH_2-\overset{O}{\overset{\|}{C}}-H$ (only) (B) $CH_3-\overset{O}{\overset{\|}{C}}-CH_3 + CH_3-\overset{O}{\overset{\|}{C}}-H$

(C) $CH_3CH_2-\overset{O}{\overset{\|}{C}}-H$ (only) (D) $CH_3-\overset{O}{\overset{\|}{C}}-\overset{O}{\overset{\|}{C}}-H$ + 2 moles of $CH_3-\overset{O}{\overset{\|}{C}}-CH_3$

(E) 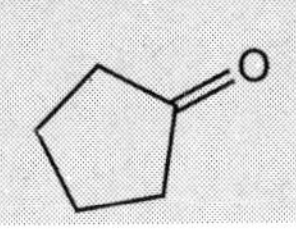 + $H-\overset{O}{\overset{\|}{C}}-H$

Give the structure of the starting alkenes.

20. Which alkyne and other reagents/conditions would be suitable for the synthesis of
    (i) cyclohexyl methyl ketone, and
    (ii) 3-hexanone? Sketch the mechanisms.
21. Which of the following compounds has the least heat of hydrogenation?

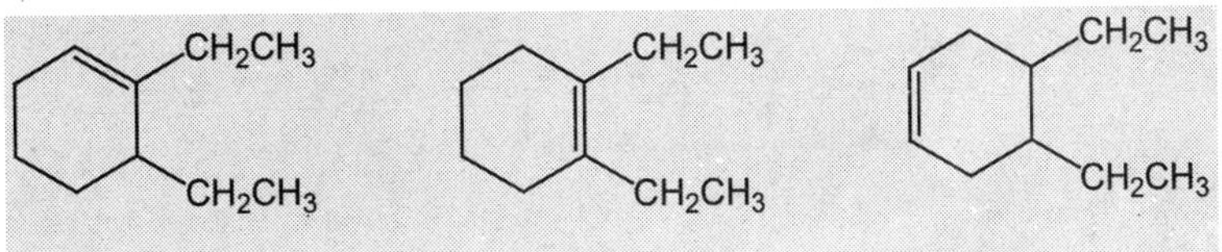

A B C

22. The addition of HBr to which of the following alkenes is more highly regioselective?

a. 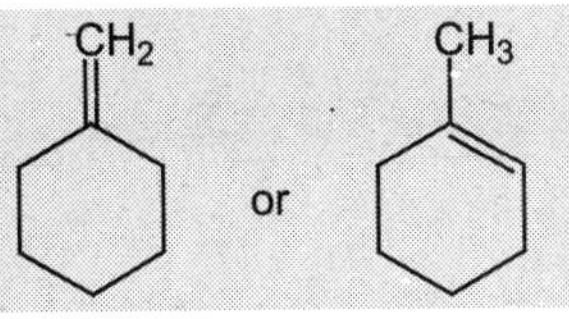

b. $CH_3\overset{CH_3}{\overset{|}{C}}=CHCH_3$ or $CH_3CH_2\overset{CH_3}{\overset{|}{C}}=CH_2$

23. Write the structures of all the products that could possibly be formed when 1-pentene is treated with bromine water containing a little NaCl. Sketch a suitable mechanistic scheme that explains the formation of these products.
24. What product would be obtained from hydroboration-oxidation of 1-methylcyclohexene?
25. What are the conditions required for the nucleophilic addition to a carbon-carbon double bond? Discuss the mechanism of cyanoethylation and Michael addition.
26. Complete the following equations showing the stereochemistry of the products wherever possible:

(i) *trans*-2-Butene $\xrightarrow{Br_2/CCl_4}$

(ii) *cis*-2-Butene $\xrightarrow{HCOOOH}$

(iii) 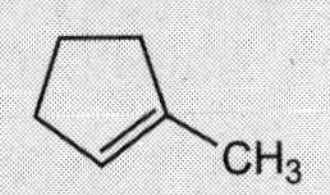 $\xrightarrow[2.\ H_2O_2\ /\ H\overset{\ominus}{O}]{1.\ B_2H_6}$

Discuss the mechanisms involved in these processes.

27. Predict the products of the following reactions:

(i) $CH_3CH{=}CH_2 \xrightarrow{HBr}$ (ii) $3\ CH_3CH{=}CH_2 \xrightarrow{BH_3}$

(iii) $RCH{=}CHR \xrightarrow[\ominus OH]{KMnO_4}$ (iv) $CH_2{=}C{=}CH_2 \xrightarrow{HCl}$

(v) $CH_2{=}CH{-}CH{=}CH_2 \xrightarrow{Br_2}$

28. Discuss the mechanism of ozonide formation.
29. Predict the product(s) in the following reactions:

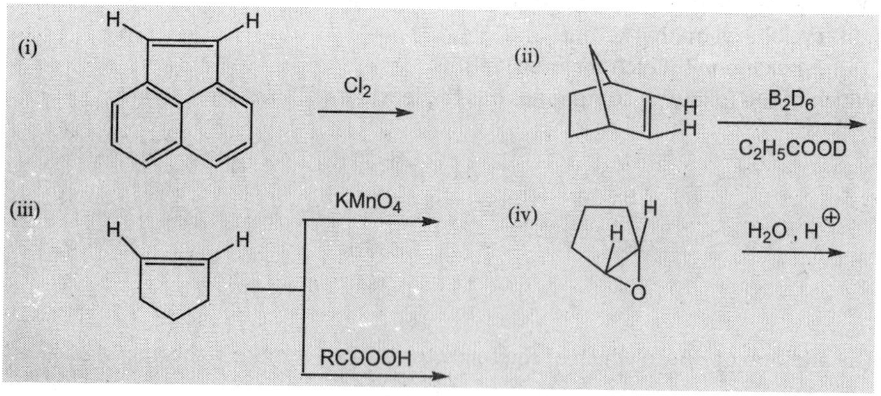

SOLUTIONS

1. (ii) The addition takes place according to Markownikoff's rule to give the desired product.
2. (i) Free-radical *cis-trans* isomerization;
   (ii) *trans*-2-butene, more stable is the favoured isomer.
3. $$CH_3CH{=}CHCH_3 \xrightarrow{H^{\oplus}} CH_3\overset{\oplus}{C}H{-}CH_2CH_3 \xrightarrow[-H^{\oplus}]{C_2H_5OH} CH_3\underset{OC_2H_5}{\underset{|}{C}H}{-}CH_2CH_3$$
4. $$CH_3CH{=}CH{-}CH{=}CHCH_3 \xrightarrow{E^{\oplus}} CH_3\overset{\oplus}{C}H{-}\underset{E}{\underset{|}{C}H}{-}CH{=}CHCH_3 \text{ and } CH_3\underset{E}{\underset{|}{C}H}{-}\overset{\oplus}{C}H{-}CH{=}CHCH_3$$
(more stable, allylic type)

5. (i)

(ii)

6. 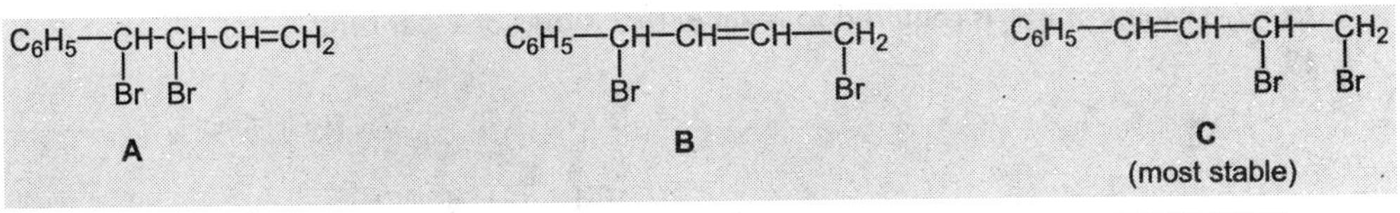

7. 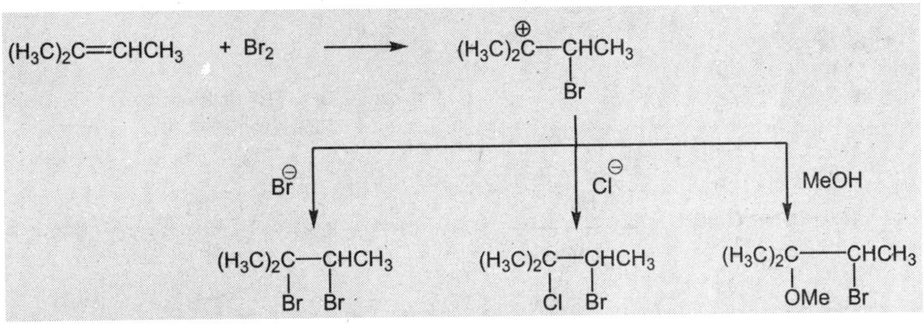

8.

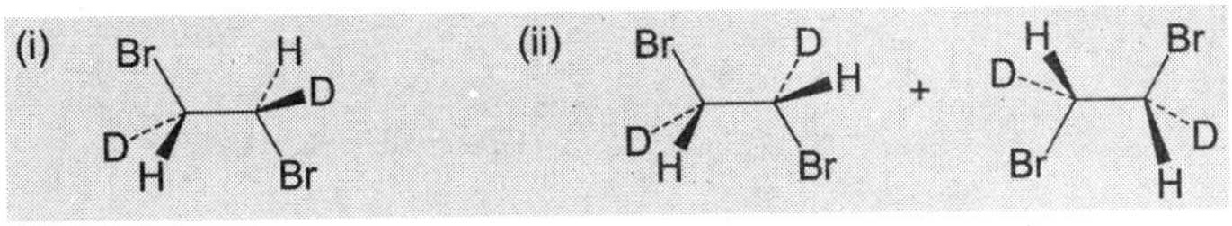

9.

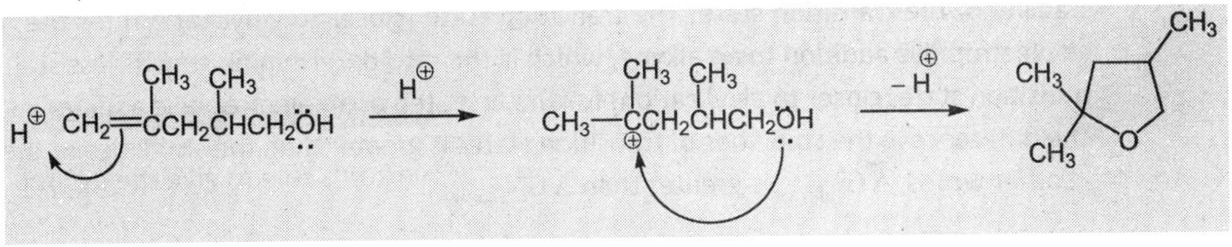

Intramolecular participation by the alcoholic group is favoured.

10.

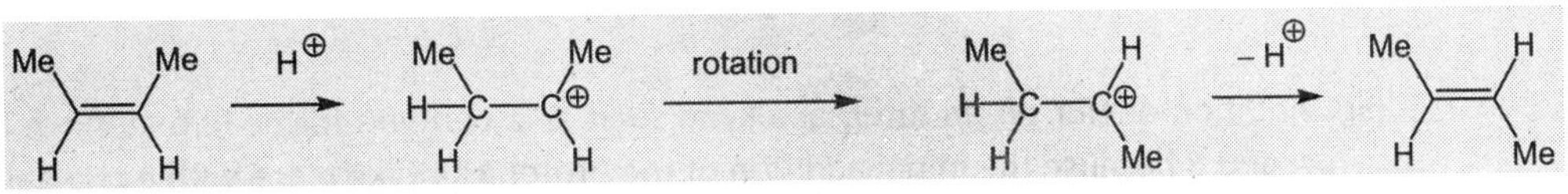

11.

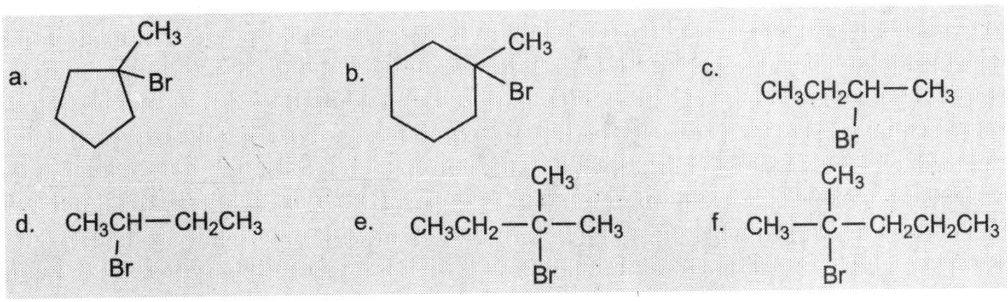

12. 2/3 Moles of BH_3 is required to react with 2 moles of 1-pentene.

13.

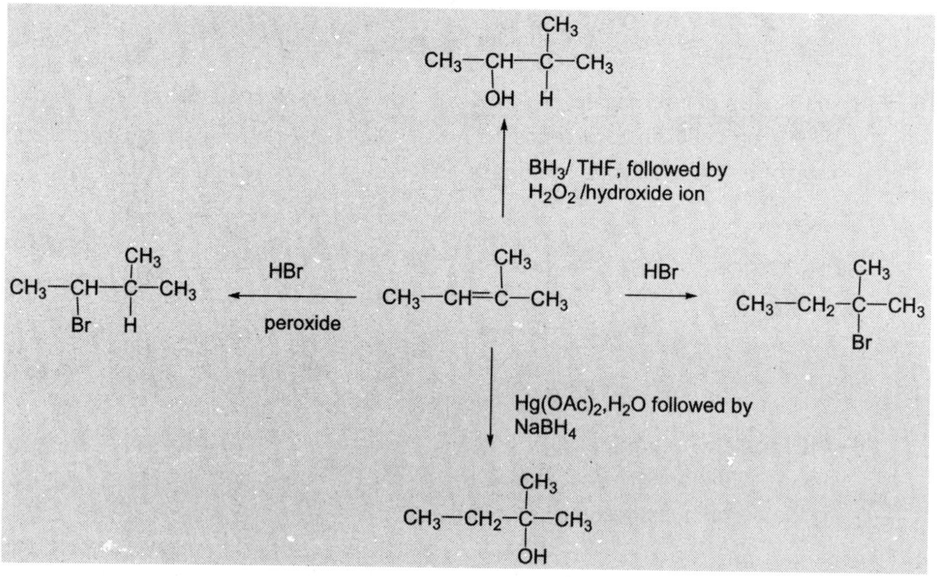

14. The reactivity depends on $\Delta G^{\#}$, which in turn depends on the stability of the reactant and the stability of the transition state. The transition state (closer to vinyl cation) for the first step of an electrophilic addition to an alkyne, which is the rate-determining step, is less stable than the transition state (closer to alkyl cation) for the first step of an electrophilic addition to an alkene. The difference in the stabilities of transition states is greater than the stabilities of the reactants. In other words, $\Delta G^{\#}_{alkyne}$ is greater than $\Delta G^{\#}_{alkene}$.

15.

(i)

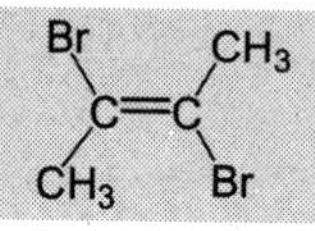

(ii) HCl (excess)

(iii) Addition of HBr on an internal alkyne such as 2-pentyne forms two geminal dibromides *A* and *B* because the initial addition of the proton can take place with equal ease to either of the *sp* carbons.

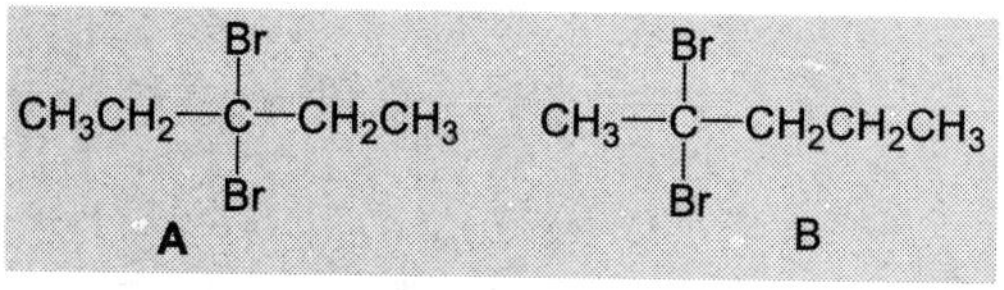

(iv) Addition of HBr in the presence of peroxide follows *anti*-Markownikoff's rule.

$$CH_3CH_2C\equiv CH \xrightarrow[\text{peroxide}]{HBr} CH_3CH_2CH=CHBr$$

16. (i) In the case of internal alkynes such as 2-butyne, the bulky groups on the boron-substituted alkene prevent the occurrence of second addition. However, there is less hindrance in the case of terminal alkynes and hence they undergo second addition with BH_3.

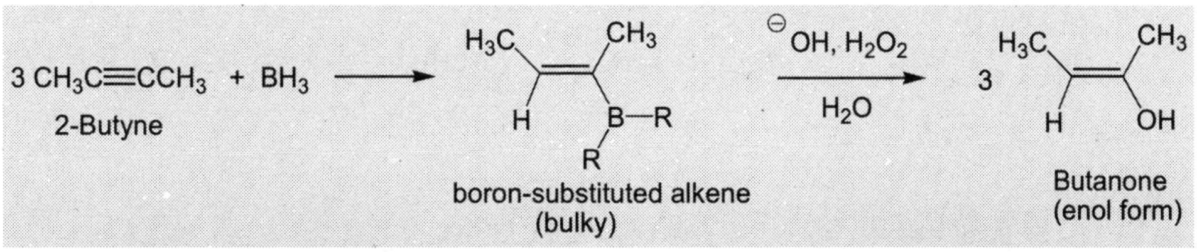

(ii) The problem in terminal alkynes can be solved by using a special reagent called disisoamylborane. The bulky alkyl groups of disisoamylborane prevents the second addition to the boron-substituted alkene formed in the first stage.

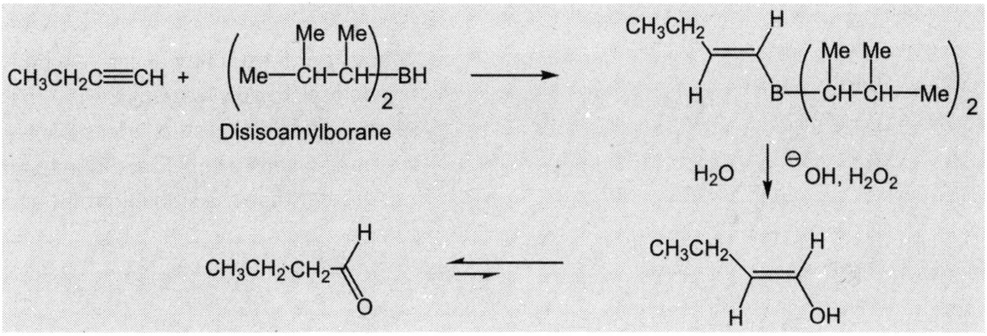

17. Hydration of 2-pentyne using sulphuric acid gives two isomeric ketones **A** and **B**. On the other hand hydration of 1-pentyne in the presence of H_2SO_4 and $HgSO_4$ will give only one ketone **A**.

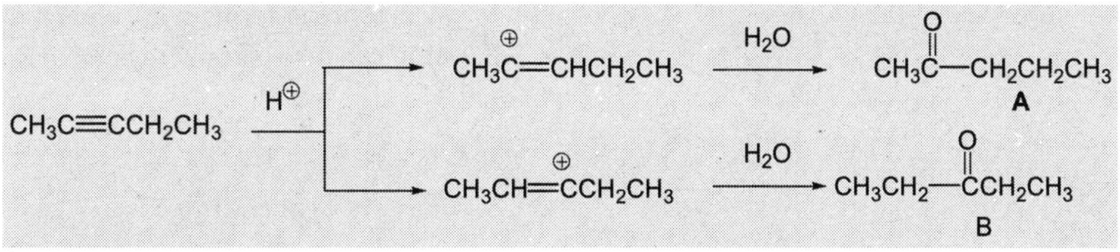

18. 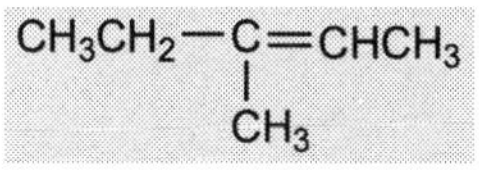

19. (A) 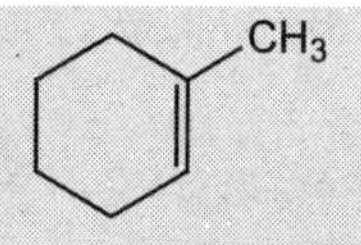 (B) 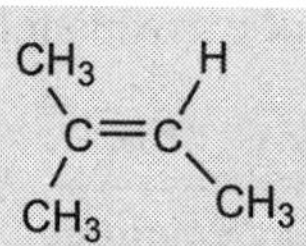 (C) $CH_3CH_2CH{=}CHCH_2CH_3$

(D) 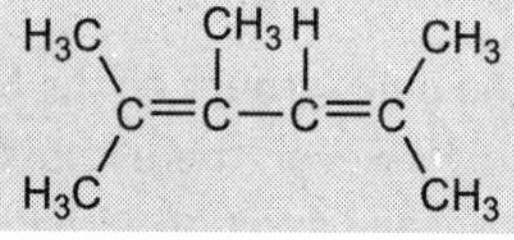 (E) 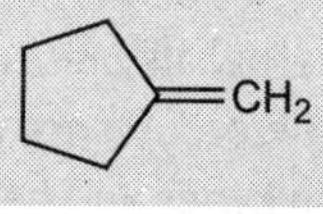

20. (i) 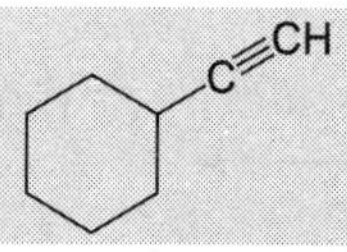 $+ H_2SO_4 + HgSO_4$ (ii) $CH_3CH_2C{\equiv}CCH_2CH_3 + H_2SO_4$

21. **B**

22. a. 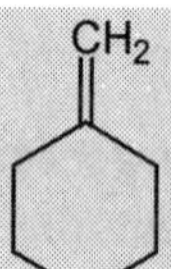 b. 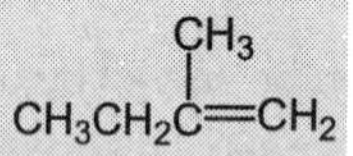

Chapter 8 Elimination Reactions

Chapter Outline

8.1 β-ELIMINATION

- **A.** *E2* and *E1cB* Mechanisms
- **B.** *E1* Reactions: The Unimolecular Mechanism
- **C.** Competition between *E2* and *E1* Reactions
- **D.** Regioselectivity of *E2* and *E1* Reactions
- **E.** Orientation in Elimination Reactions
- **F.** Stereochemical Preferences
- **G.** Cis-elimination in *E2* Reactions
- **H.** Elimination vs Substitution
- **I.** Thermal Elimination
- **J.** Dehalogenation

8.2 α-ELIMINATION: CARBENES

- **A.** Generation of Carbenes
- **B.** Reactions of Carbenes
- **C.** Carbenoids

Elimination reactions, in which two groups are removed from a molecule, neither being replaced by another group, are the reverse of addition reactions. Usually, they involve the loss of two substituents from vicinal atoms resulting in the formation of a double or triple bond. Most commonly, a proton is lost from one carbon, whereas a nucleophile is lost from the adjacent carbon; these two carbon atoms are usually referred to as β- and α- carbons, respectively.

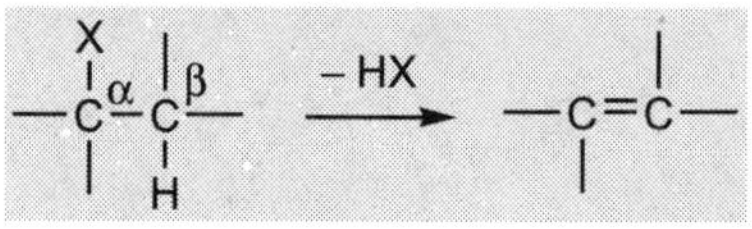

This type of elimination reaction is known as β-elimination or a 1,2-elimination reaction. The most familiar examples of β-elimination reactions include dehydrohalogenation of alkyl halides, dehydration of alcohols, pyrolysis of esters, and the Hofmann degradation of quaternary ammonium hydroxides. There are a few reactions in which both the groups are lost from the same carbon atom. These are called α-elimination reactions, and the most common example of this type is the generation of dichlorocarbene from chloroform.

8.1 β-ELIMINATION

In analogy with substitution reactions, β-elimination reactions are divided into *E1* (*Elimination, unimolecular*) and *E2* (*Elimination, bimolecular*) reactions. In some cases, *E1cB (Elimination, unimolecular, conjugate base)* mechanism is also operative.

A. *E2 and E1cB* Mechanisms

Many elimination reactions are successful only when carried out in the presence of a strong base. The formation of ethylene on the treatment of ethyl bromide with sodium ethoxide is an example of this type. The rate of alkene formation is proportional to the concentration of ethyl bromide as well as that of sodium ethoxide.

$$CH_3CH_2Br + C_2H_5O^{\ominus} \longrightarrow CH_2{=}CH_2 + C_2H_5OH + Br^{\ominus}$$

$$\text{Rate} = k\,[CH_3CH_2Br]\,[C_2H_5O^{\ominus}]$$

Two types of mechanisms can be written, both of which are in agreement with the kinetic data. In the first mechanism, the base abstracts a proton from the β-carbon, and simultaneously the leaving group departs from the α-carbon along with the pair of bonding electrons.

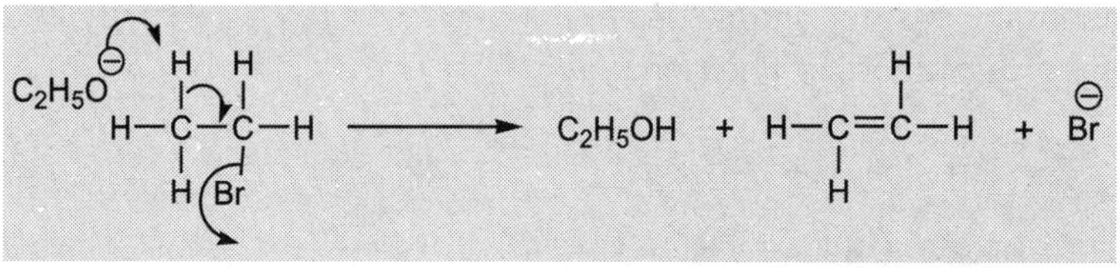

That the rate-determining step involves breaking of the C–H bond has been shown by kinetic isotope effects. Changing H for D can affect the rate of the reaction, only if that H (or D) is involved in the rate-determining step. The process is illustrated when 2-phenylethyl bromide **(1)** is treated with ethoxide ion to give styrene **(2)**.

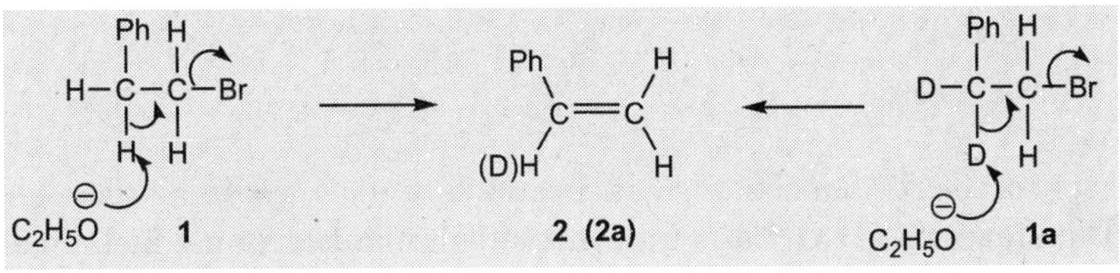

In *E2* mechanism, a hydrogen or deuterium has to be abstracted from the substrate (**1** or **1a**). It is known that the C–D bond is stronger than the C–H bond, and thus requires more energy to be broken. It may then be inferred that the rate of elimination in **1** should be much faster than **1a** which has indeed been found to be the case. It has been shown that **1** *reacts 7 times faster* than **1a**. Such reaction pathways are known as *E2* mechanism.

In the second mechanism (*E1cB*), the base rapidly removes proton from the β-carbon resulting in the formation of carbanion, which loses the leaving group in a rate-determining step. Since the conditions of base-catalysed elimination reactions do not allow the formation of an unstabilized carbanion, it is reasonable to presume that if formed, they must be either rapidly reconverted to the substrate or converted to the alkene.

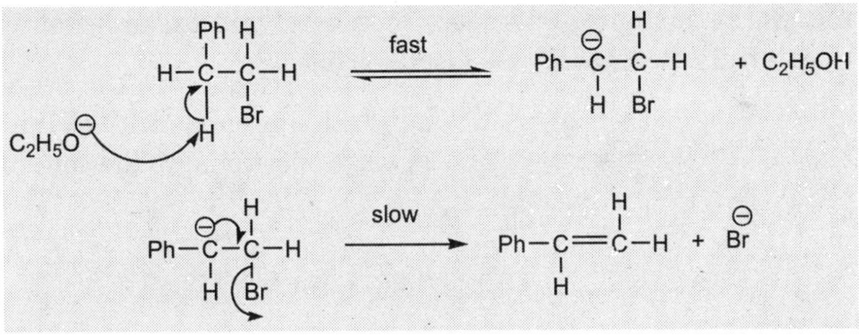

In this mechanism, the overall rate is limited to that of the slower second stage, which *depends* only on the concentration of the conjugate base of the reactant. This mechanism is called as *E1cB* as the leaving group is lost from the conjugate base of the starting molecule, and the reaction is unimolecular.

The distinction between *E2* and *E1cB* mechanism can also be made very easily by means of labelling experiments. Consider again the reaction of **1** and ethoxide to yield styrene (**2**) which follows the second order kinetics. The reaction may follow either mechanism.

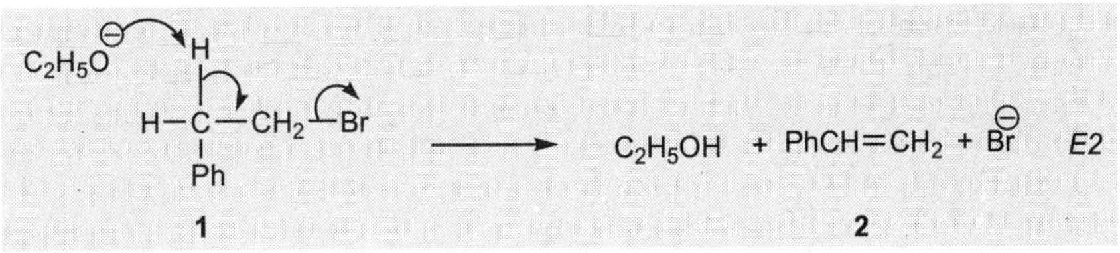

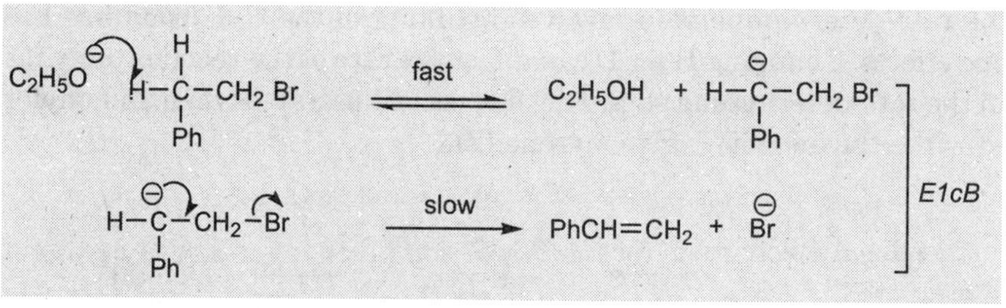

The first step of the *E1cB* mechanism is reversible, and hence, when the reaction is carried out in C_2H_5OD instead of C_2H_5OH, the intermediate carbanion should pick up deuterium.

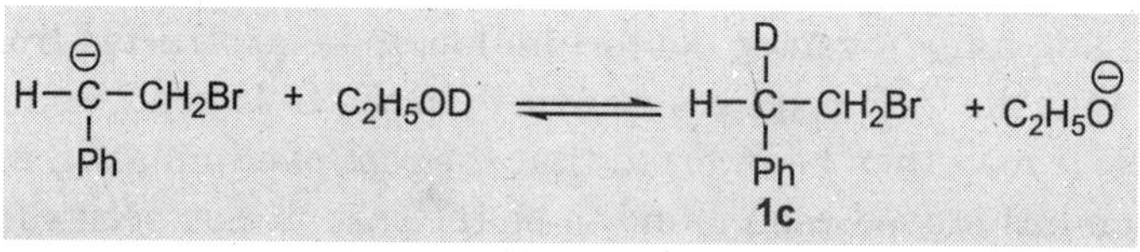

If the *E1cB* mechanism is correct, we should be able to recover labelled 2-phenylethyl bromide (**1c**) after a partial transformation to styrene. On the other hand, there should be no incorporation of deuterium if the *E2* mechanism is operative. Actual experiments have shown that *there is no deuterium incorporation,* and hence, the *E1cB* mechanism does not operate in this case.

However, this mechanism does operate under special circumstances. 1,1,1-Trifluoro-2,2-dichloroethane (**3**), for instance, undergoes base-catalyzed exchange of β-hydrogen atom with the solvent deuterium faster than dehydrofluorination.

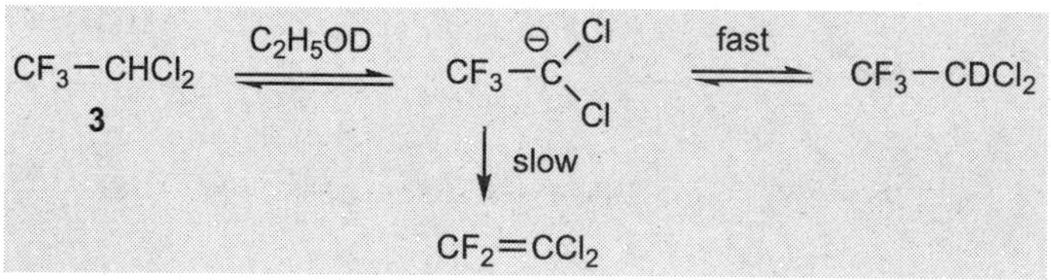

A strong carbon-fluorine bond (and consequent poor leaving ability of fluoride ion) coupled with the electron-withdrawing effect of halogens explains the formation of carbanion before elimination.

E1cB reactions are quite pronounced when proton that is to be removed is located α to the carbonyl group. The rather acidic proton is easily removed by the base without leaving group departing at the same time. The resultant anion is stabilized due to delocalization on the carbonyl group.

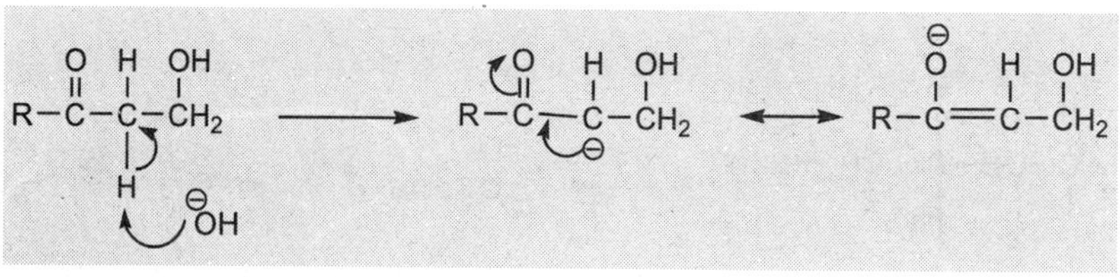

In the next step, the anion loses a leaving group to generate an alkenone.

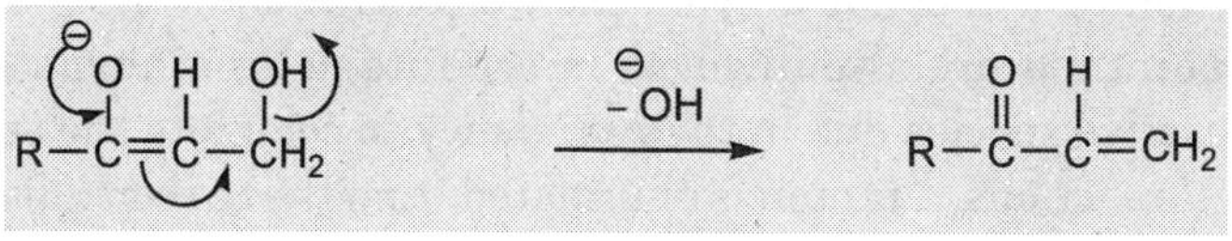

Thus, a generalized mechanism for *E1cB* reaction involving a carbonyl compound can be depicted as follows:

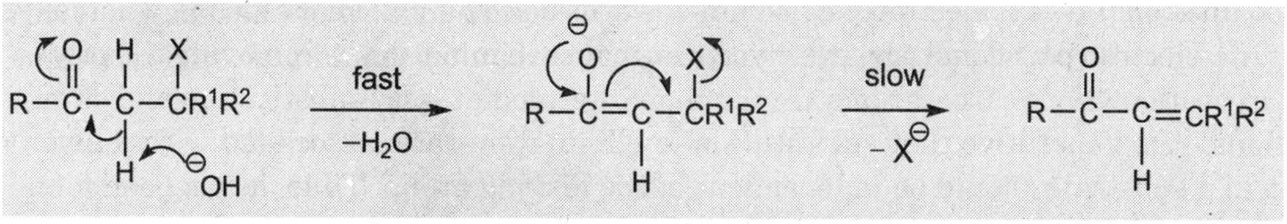

In summary, removal of a proton and loss of the leaving group occurs *simultaneously* in *E2* elimination, whereas removal of a proton is the *first step* in *E1cB* reaction. In the subsequent rate-determining step of *E1cB* reaction, the leaving group departs from the conjugate base of the substrate.

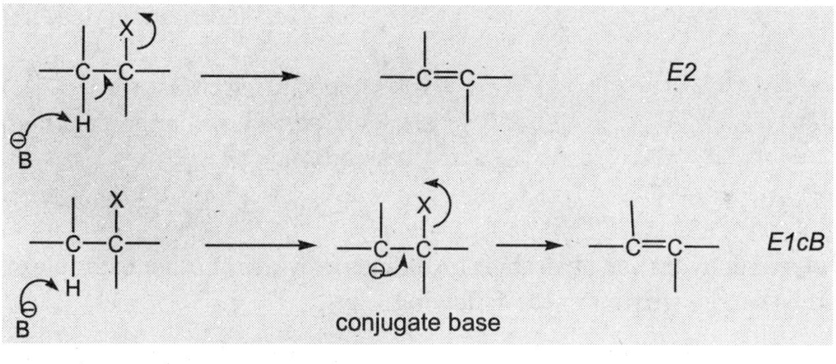

B. *E1* Reactions: The Unimolecular Mechanism

The main feature of this mechanism is that under the influence of solvation forces, the electron-attracting group ('leaving group') breaks away along with the bonding electrons. The resultant *carbocation* subsequently loses a proton to the solvent or to some other proton acceptor.

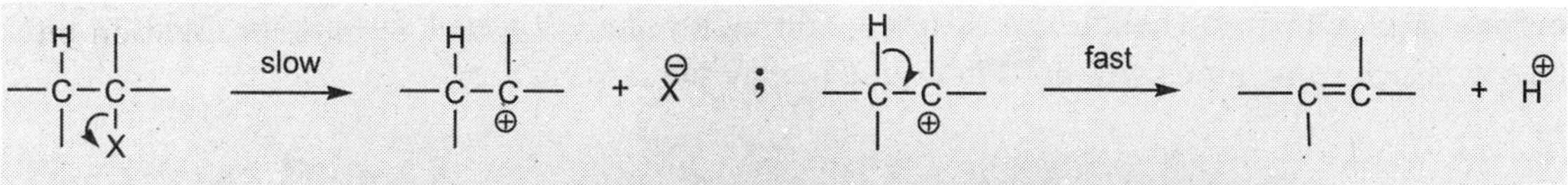

Rate = k 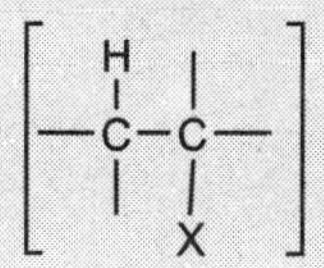

The reaction, thus, has two stages, of which the *first is the rate-determining step,* and as a result, the reaction rate depends only on the concentration of the first reactant. We may recall here that carbocations are also involved in the rate-determining step of S_N1 reactions, and they are very reactive species due to their tendency to complete the octet of the positively charged carbon. In *E1* reactions, a proton is eliminated from the carbon adjacent to the positive, electron-deficient carbon and the pair of electrons formerly shared by this hydrogen is available for the formation of a π-bond.

E1 and S_N1 reactions usually occur simultaneously. Substitution is favoured by the presence of good nucleophiles (*i.e.* electron pair donors towards electron-deficient carbon), whereas strong bases (*i.e.* electron pair donor towards hydrogen) cause elimination by removing the proton from the carbocation. Since carbocations are common intermediates for substitution and elimination reactions, it is imperative that substitution to elimination ratio associated with a given alkyl group in a substrate should be independent of the leaving group. Thus, it has been found that the mixture containing substitution and elimination products is the same whether the starting material is *tert*-butyl chloride, *tert*-butyl bromide, or *tert*-butyl iodide.

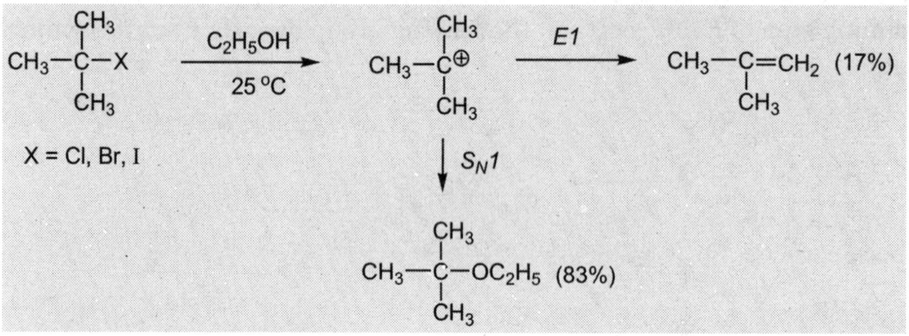

Acid-catalyzed dehydration of alcohols to alkenes provides another example of *E1* reactions. The various steps can be written in the following way.

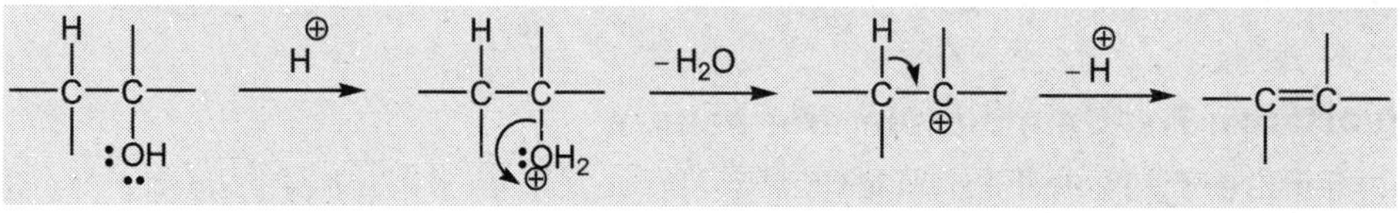

As we know, the carbocations are planar species and their formation at the bridgehead position will be a difficult process. Thus, bicyclic structures prevent the bridgehead carbon becoming planar even though the cation would be tertiary. It is not formed due to very high energy. Such compounds will not undergo *E1* reactions.

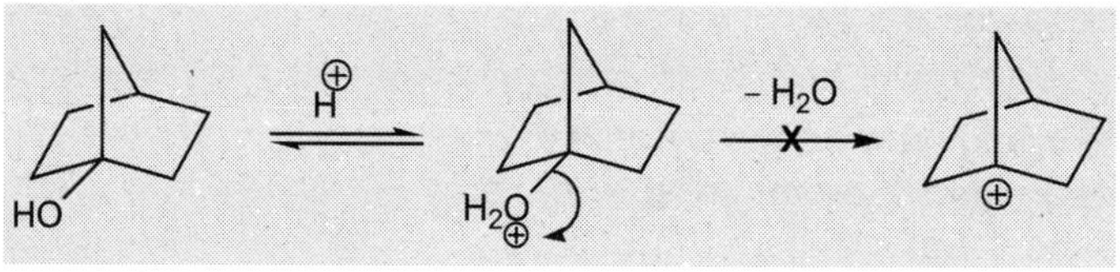

C. Competition between *E2* and *E1* Reactions

We have seen that in an *E1* reaction, the rate-determining step is unimolecular, and *does not involve the base*. The leaving group departs in the first step, and the proton is removed in the second step.

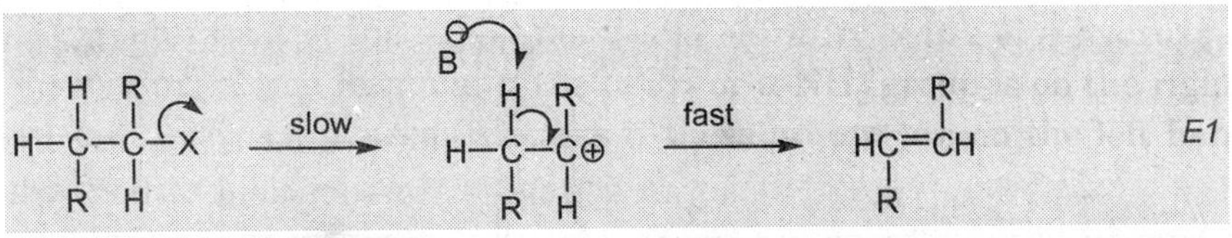

Rate = *k* [alkyl halide]

In contrast, *E2* reaction has a bimolecular rate-determining step *requiring the base*. Loss of the leaving group is simultaneous with the removal of proton by the base.

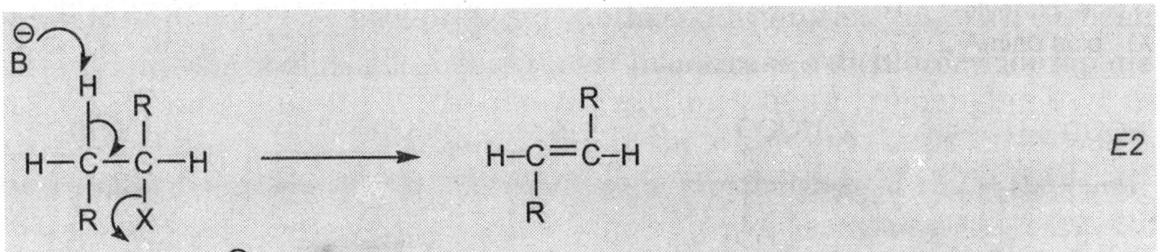

Rate = $k\,[\overset{\ominus}{B}]$ [alkyl halide]

Primary alkyl halides undergo only *E2* elimination reactions. They cannot undergo *E1* reactions because of the difficulty encountered in forming primary carbocations. Secondary and tertiary alkyl halides undergo both *E2* and *E1* reactions.

For those alkyl halides that can undergo both *E2* and *E1* reactions, the *E2* reaction is favoured by the same factors that favour an S_N2 reaction, and *E1* reaction is favoured by the same factors that favour an S_N1 reaction. Thus, an *E2* reaction is favoured by a high concentration of a strong base and an *aprotic polar solvent* (*e.g.* DMSO or DMF), whereas an *E1* reaction is favoured by a weak base and a *protic polar solvent* (*e.g.* H_2O or ROH).

D. Regioselectivity of *E2* and *E1* Reactions

An alkyl halide, such as 2-bromopropane, has two β-carbons from which a proton can be removed in an *E2* reaction. Because two β-carbons are identical, the proton can be removed with equal ease from either one of them. The product of this elimination reaction is propene.

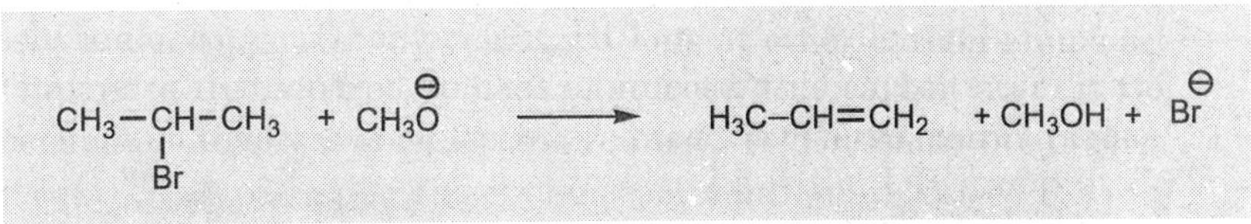

In contrast, 2-bromobutane has two structurally different β-carbons from which a proton can be removed. So, when 2-bromobutane reacts with a base, two elimination products are formed: 2-butene and 1-butene. Thus, *E2* reaction is regioselective, because of the preferred formation of 2-butene (80%).

$$\underset{\text{2-Bromobutane}}{CH_3-\underset{|}{\overset{}{CH}}(Br)-CH_2CH_3} + CH_3O^{\ominus} \longrightarrow \underset{\substack{\text{2-Butene}\\80\%}}{CH_3CH=CHCH_3} + \underset{\substack{\text{1-Butene}\\20\%}}{CH_2=CHCH_2CH_3}$$

Regioselectivity of an *E2* reaction is determined by the alkene which is formed more easily *i.e.* which is formed faster. The reaction co-ordinate diagram for *E2* reaction of 2-bromobutane is shown in **Fig. 8.1.**

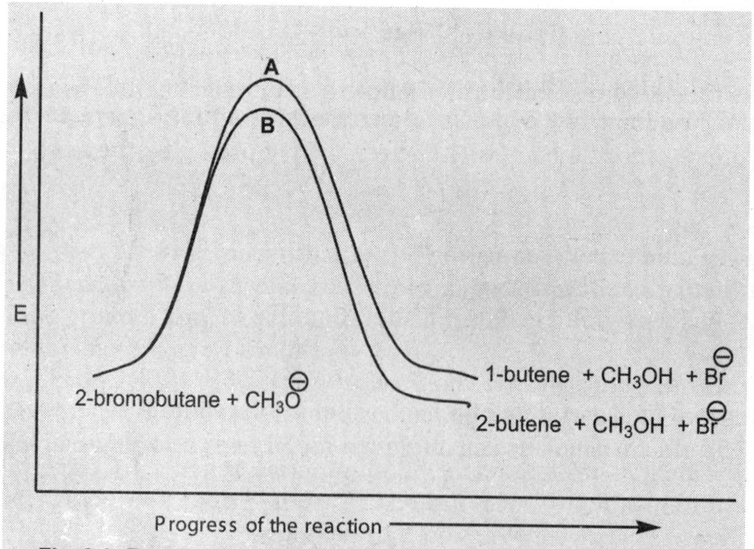

Fig. 8.1: Energy diagram for the *E2* reaction of 2-bromobutane

Transition states leading to the formation of 1- and 2-butenes may be depicted as:

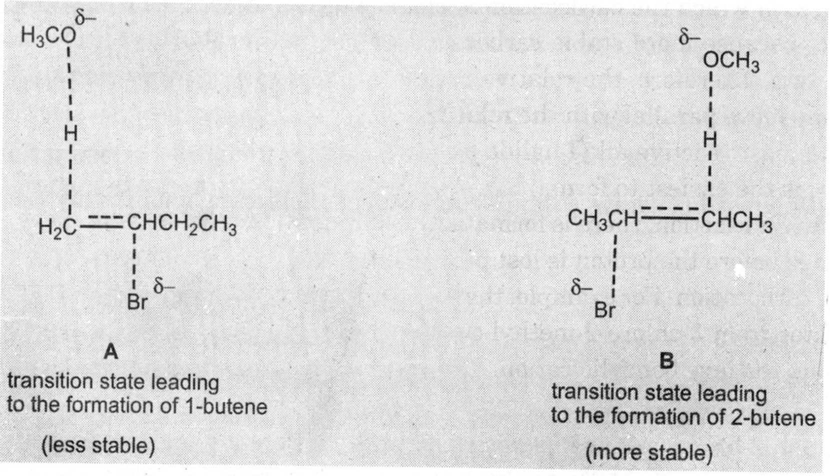

A
transition state leading to the formation of 1-butene
(less stable)

B
transition state leading to the formation of 2-butene
(more stable)

It may be noted that in these transition-states, the C–H and C–Br bonds are partially broken to generate an alkene-like structure. It is reasonable to assume that factors stabilizing an alkene will also stabilize the transition state leading to its formation. It is known that greater number of alkyl substituents bonded to the sp^2 carbons of an alkene increase its stability. This explains the greater stability of 2-butene (two methyl substituents bonded to its sp^2 carbons) as compared to 1-butene (one ethyl substituent). Thus, the more stable of the two alkenes is found to be a major product of the reaction.

Preferential formation of the more highly substituted alkene in *E1* elimination reaction is based on the assumption that the entropies of the product-determining transition states parallel those of the isomeric alkenes.

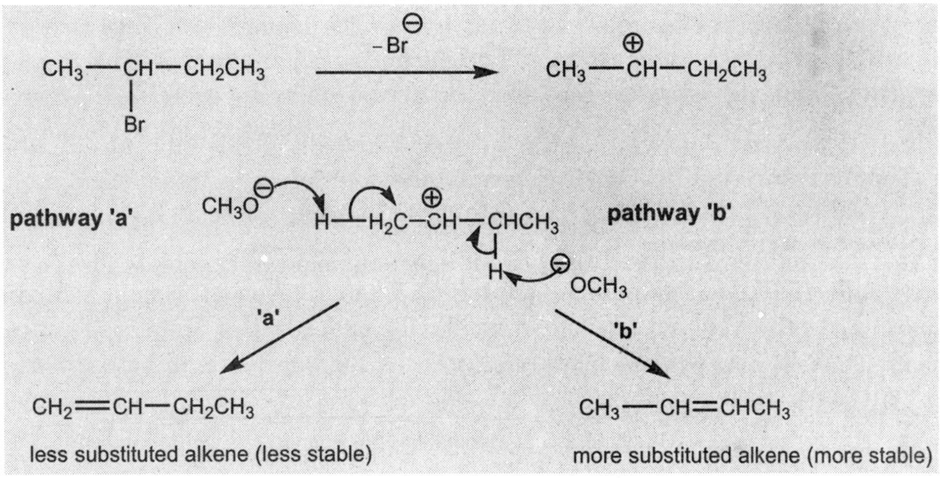

Because the first step is the rate-determining step, the rate of an *E1* reaction depends on the ease with which the carbocation is formed. The *more stable* the carbocation, the *easier* it is formed, because more stable carbocations have more stable transition states leading to their formation. Therefore, the relative reactivities of a series of alkyl halides with the same leaving group move parallel with the relative stabilities of the carbocations. A tertiary benzylic halide is the most reactive alkyl halide because a tertiary benzylic cation, being a very stable carbocation, is the easiest to form.

As in an *E1* reaction, there is formation of a carbocation intermediate, the carbon skeleton can rearrange before the proton is lost provided the rearrangement leads to the formation of a more stable carbocation. For example, the secondary carbocation, that is formed when a chloride ion dissociates from 3-chloro-2-methyl-2-phenylbutane, undergoes a 1,2-methyl shift to form a more stable tertiary benzylic cation. This cation then undergoes deprotonation to form the alkene.

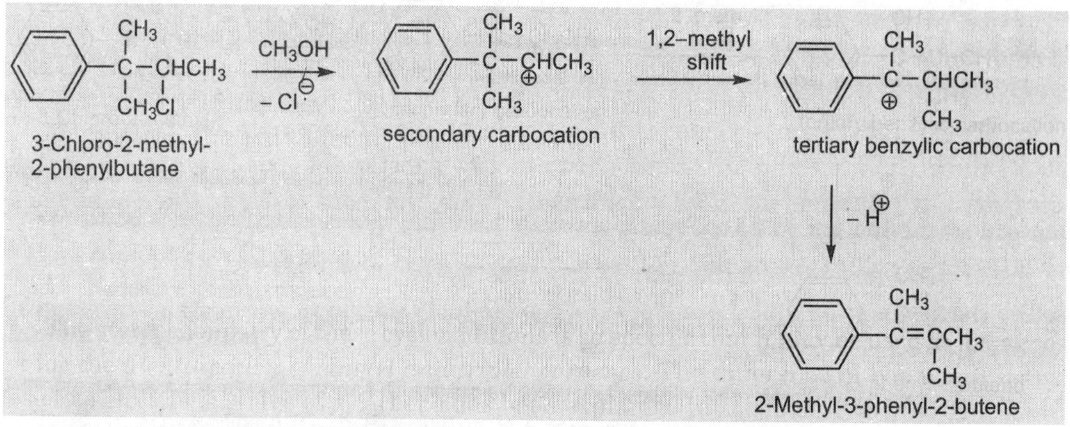

E. Orientation in Elimination Reactions

The elimination reaction of unsymmetrical substrates usually yields mixtures of all possible products. There are two empirical rules governing the orientation in these reactions.

The Saytzeff Rule

It states that neutral substrates (alkyl halides or sulphonates) capable of forming a double bond in either direction of the chain preferably yield that alkene in which there is *greater number* of alkyl groups attached to the double bond. This rule applies to *E1* reactions and to most of the *E2* reactions. The following examples are typical:

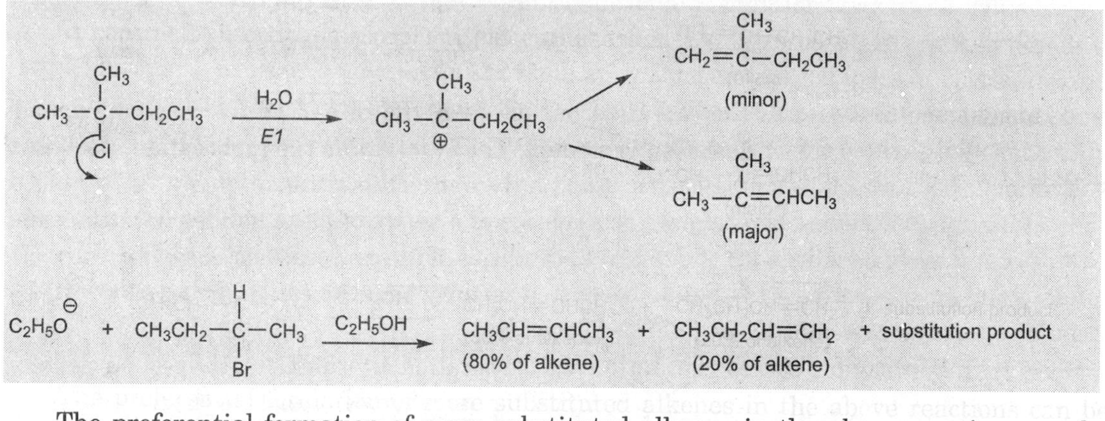

The preferential formation of more substituted alkenes in the above reactions can be correlated with the relative stability of various alkenes. Calculations from heats of combustion and hydrogenation establish that the stability of a double bond is increased by alkyl substitution on the double-bonded carbon atoms. Thus, the stabilities of alkenes follow the order:

$$R_2C{=}CR_2 > RCH{=}CR_2 > RCH{=}CHR > RCH{=}CH_2 > CH_2{=}CH_2$$

The greater stability of more substituted alkene is sometimes explained on the basis of hyperconjugation.

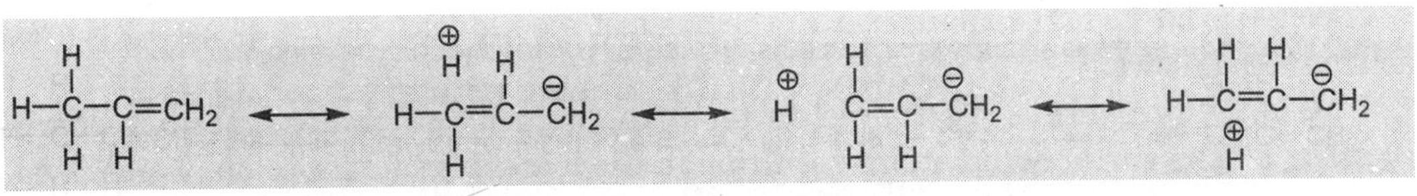

The number of resonance structures of this hybrid increases with the increase in the number of alkyl substituents, and this consequently increases the stabilization. However, it should be remembered that actually it is the relative stability of the *transition state* leading to the alkene, rather than that of the alkene itself which is responsible for Saytzeff type of elimination. In the *transition state*, sigma electrons of γ-carbon-hydrogen bond become hyperconjugated with the partly formed α,β-double bond, thereby, reducing its energy and facilitating the reaction.

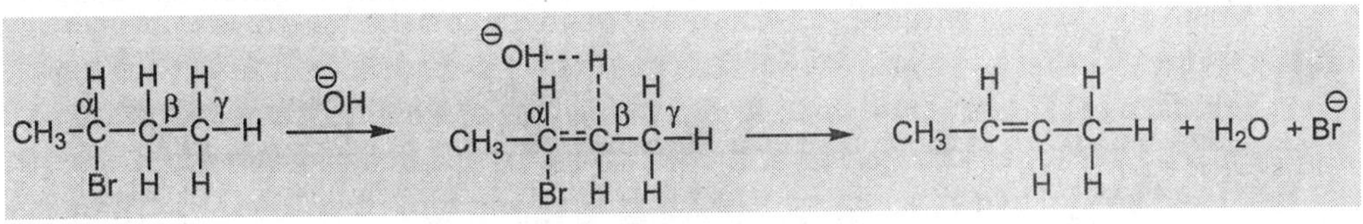

I. The Hofmann Rule

It states that charged substrates (quaternary ammonium and sulphonium salts) yield predominantly the least substituted alkene.

Let us first examine the mechanism of elimination reaction of a quaternary ammonium hydroxide. On heating, such compounds yield an alkene and a tertiary base. For example,

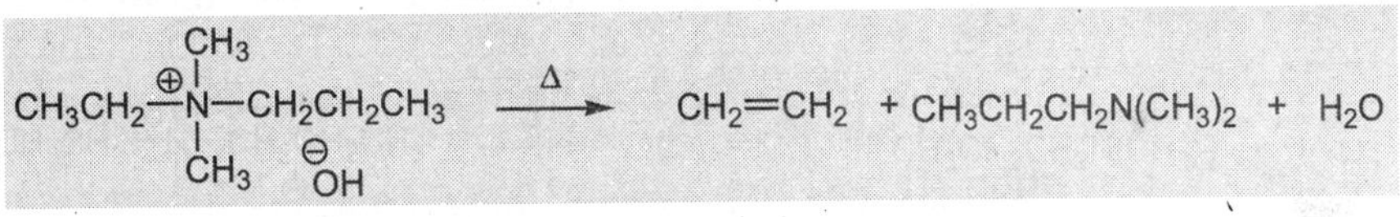

The hydroxide ion abstracts a proton from the β-carbon resulting in the formation of a double bond, and expulsion of a tertiary amine.

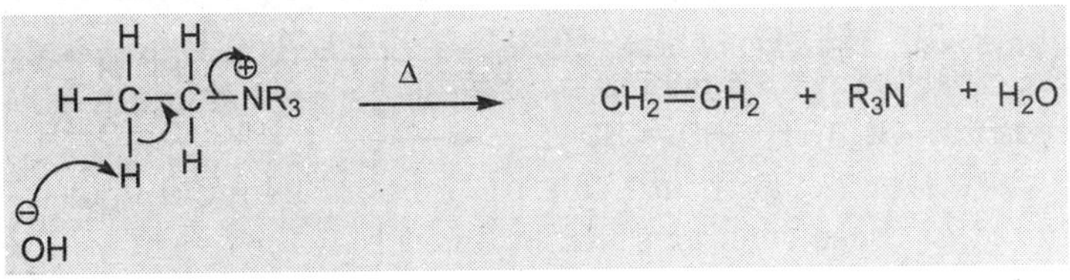

It has been observed that quaternary ammonium hydroxides as well as sulphonium salts preferably yield *less substituted alkenes*.

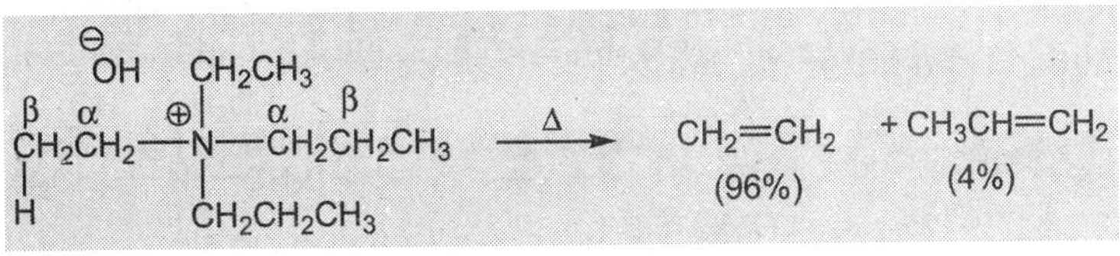

Ingold has offered an explanation on the basis of the inductive effect of the alkyl groups, which runs as follows:

Due to the presence of a strong electron-attracting leaving group, a positive charge is induced on all the surrounding atoms resulting in the loosening of a β-proton, which is easily abstracted by a base, thus causing an *E2* reaction. However, if an alkyl group is attached to the β-carbon, it tends to neutralize the induced positive charge by releasing electrons due to its inductive effect. The loosening of the β-proton is then somewhat arrested, and hence, the reaction is inhibited. It follows that if a salt has ethyl as well as *n*-propyl groups attached to the positively charged nitrogen, the *n*-propyl group will show relatively little tendency to lose a hydrogen.

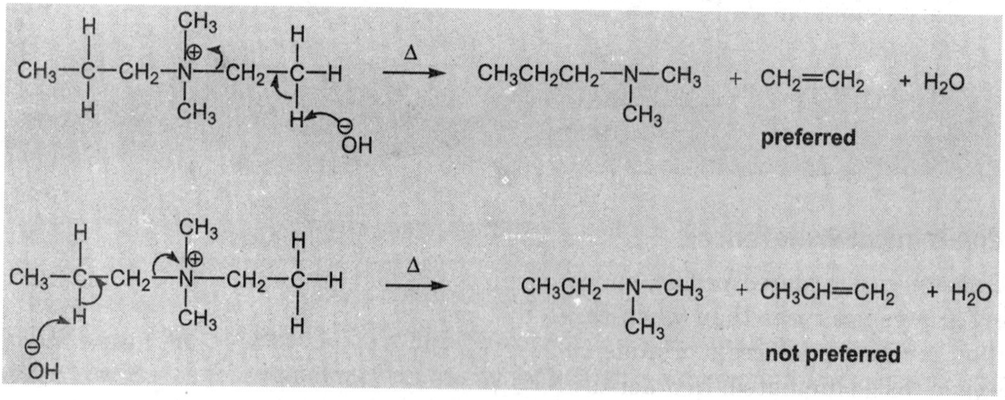

For the same reason an *iso*butyl group shows still less tendency to lose a proton, if either an *n*-propyl or an ethyl group is available in the molecule.

Another way to look at the role played by the inductive effect of alkyl groups in Hofmann elimination is that the transition state of the reaction has considerable *carbanion character*. In other words, the proton removal runs slightly ahead of the loss of the leaving group. It appears quite reasonable as carbon-nitrogen and carbon-sulphur bonds are more difficult to break than carbon-halogen bonds of the alkyl halides. The alkyl groups are electron-donating, and so, primary carbanions are more stable than secondary or tertiary carbanions. The proton loss occurs from the β-carbon, which generates the more stable carbanion, and hence, the less substituted alkene is formed.

Steric factors also play a significant role in Hofmann eliminations. It is apparent that the groups eliminated in these reactions are bulkier than those lost in the Saytzeff type of eliminations. Indeed, it has been found that *E2* reactions of alkyl halides can be forced to follow Hofmann type of elimination by sufficiently increasing the bulk of the attacking base, or by increasing the degree of branching in the substrate. Treatment of **4** with base, for instance, yields **5** in preference to **6** mainly due to steric strain resulting from the interaction of methyl and *tert*-butyl groups in the latter alkene.

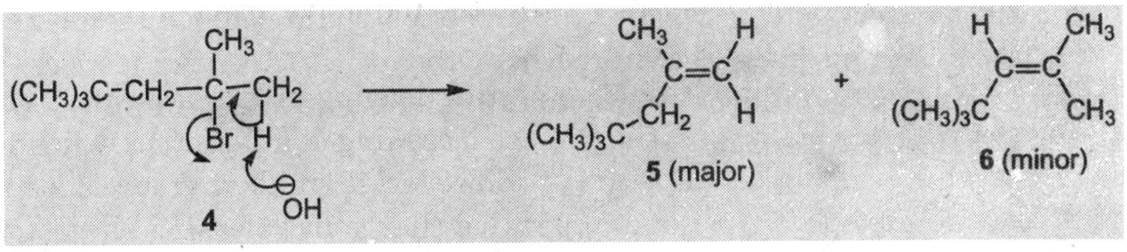

A similar steric effect seems to operate when the size of attacking base is increased, *e.g.*

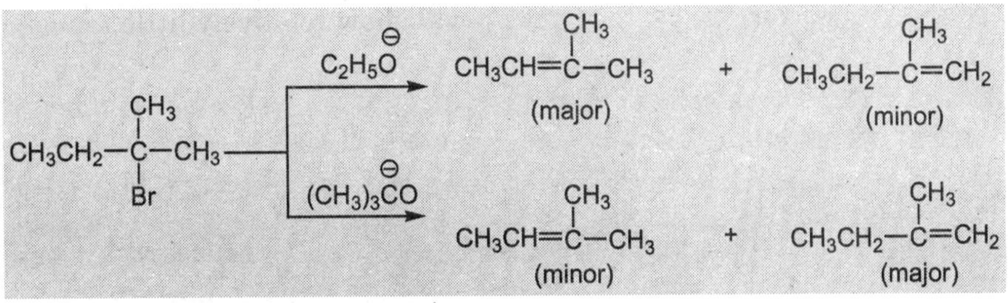

F. Stereochemical Preferences

Most *E2* reactions are stereospecific, and proceed with considerably greater facility when the departing groups are *trans* than when they are *cis*. A plausible explanation may be that an elimination reaction produces *p* orbitals from what were the bonds to the leaving group and the hydrogen to be eliminated. In order to achieve maximum overlap of these *p* orbitals in the transition state they should be coplanar, and for minimum energy of the transition state the two groups to be eliminated must be anti-parallel to each other across the central bond, which becomes the π-bond.

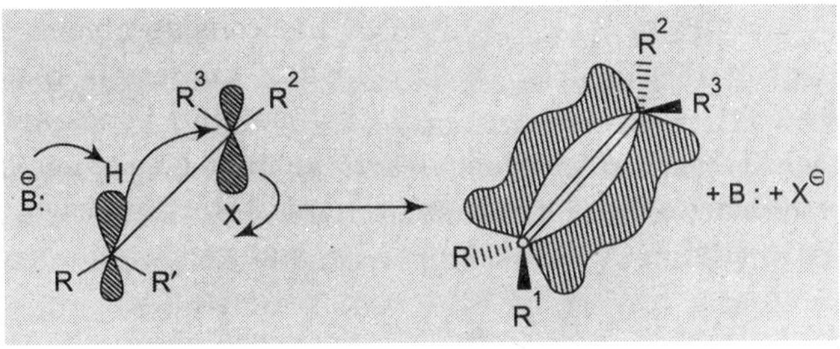

In *E2* as well as in S_N2 reactions, the leaving group is displaced by a backside attack, but in the former case, the attacking species is a base, whereas in the latter it is a nucleophile. Many examples of elimination reactions leading to acetylenes and alkenes can be cited to demonstrate the stereospecificity of these reactions. For instance, **7** is converted by a base to 2-butyne at a faster rate than the *cis* isomer (**8**).

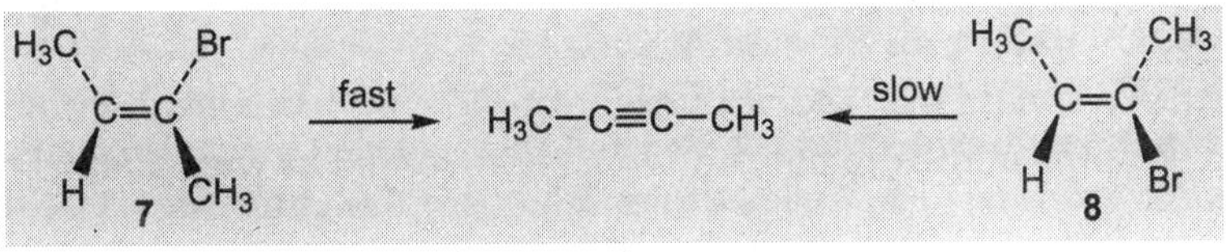

Similarly, the products of *E2* reactions of the open-chain compounds possessing two asymmetric carbons depend upon the stereochemistry of the starting isomer. Thus, the dehydrobromination of *meso* stilbene dibromide (**9**) yields the *cis* olefin (**10**), whereas the *dl*-bromide (**11**) produces the *trans* olefin (**12**).

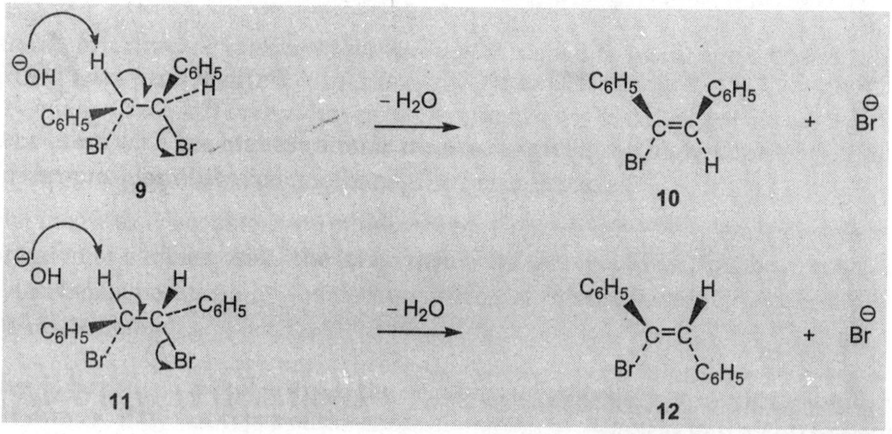

Elimination reactions of cyclic compounds have been studied in considerable detail to shed more light on the stereochemistry of *E2* reactions. In these compounds, rotation about the carbon–carbon single bonds is restricted by the ring structure. Let us take the case of unsubstituted cyclohexyl halides. There exists an equilibrium in which two conformations are present, one with the axial halide, and the other in which it is equatorial. We know that the equilibrium will have a predominance of the equatorial conformer.

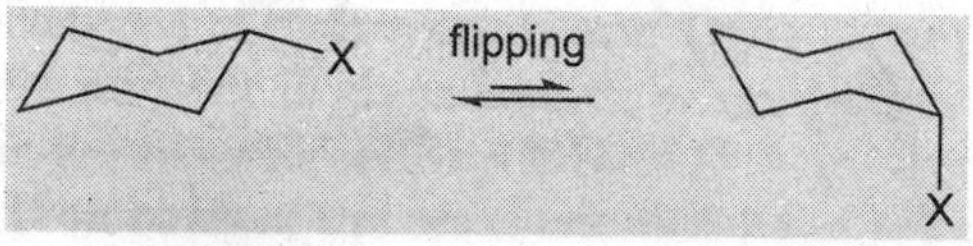

However, it is the axial conformer which has the right geometry for *E2* elimination, as the axial group is *antiperiplanar* to the C–H bonds.

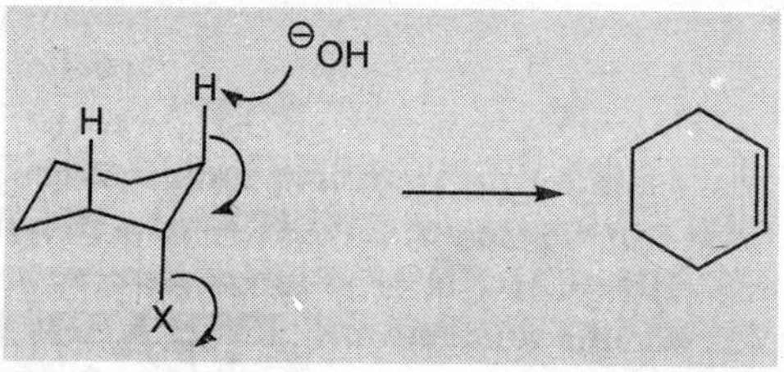

In the more stable equatorial conformer, the X is antiperiplanar only to C–C bonds, and thus, cannot be eliminated by an *E2* mechanism.

It has indeed been found that the *cis* isomer of 4-*tert*-butylcyclohexyl-*p*-toluene sulphonate (**13**) in which the flipping of the cyclohexane ring is prevented by the equatorially bound bulky *tert*-butyl group, readily undergoes *E2* reaction as the leaving group (OTs) is situated anti-parallel (antiperiplanar) to the hydrogen. On the other hand, the *trans* isomer (**14**) lacks this geometric requirement, and hence, does not undergo *E2* reaction under the same conditions.

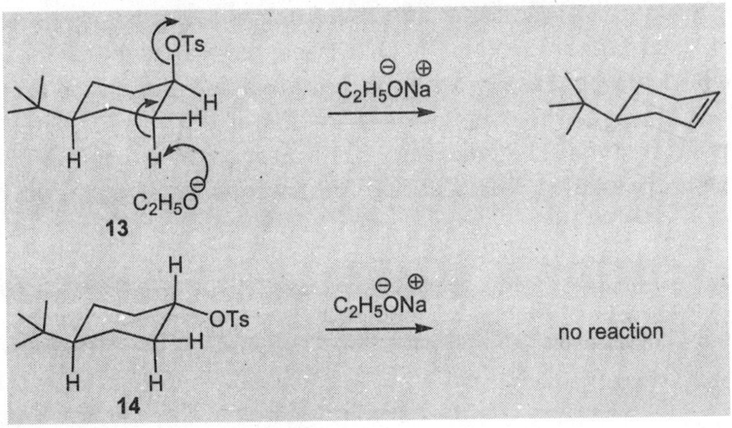

A dramatic example of the requirement of *trans* diaxial orientation in the elimination reactions of cyclohexane derivatives is provided by menthyl chlorides. Although menthyl chloride exists predominantly in the more stable conformation in which all the three substituents are in the equatorial position, a conformational flip may make these bonds axial (**15**). It undergoes a facile elimination reaction with sodium ethoxide in ethanol to yield 2-menthene (**16**) which is an *anti*-Saytzeff product, thus illustrating the importance of the *trans* arrangement of hydrogen and chlorine situated on carbon atoms 2 and 3, respectively. A similar reaction with neomenthyl chloride (**17**) yields 3-menthene (**18**) in addition to 2-menthene (**16**), because a *trans* coplanar arrangement is possible for axial chlorine with the axial hydrogens situated on both C-2 and C-4. As would be expected, the Saytzeff product (**18**) predominates over *anti*-Saytzeff product (**16**).

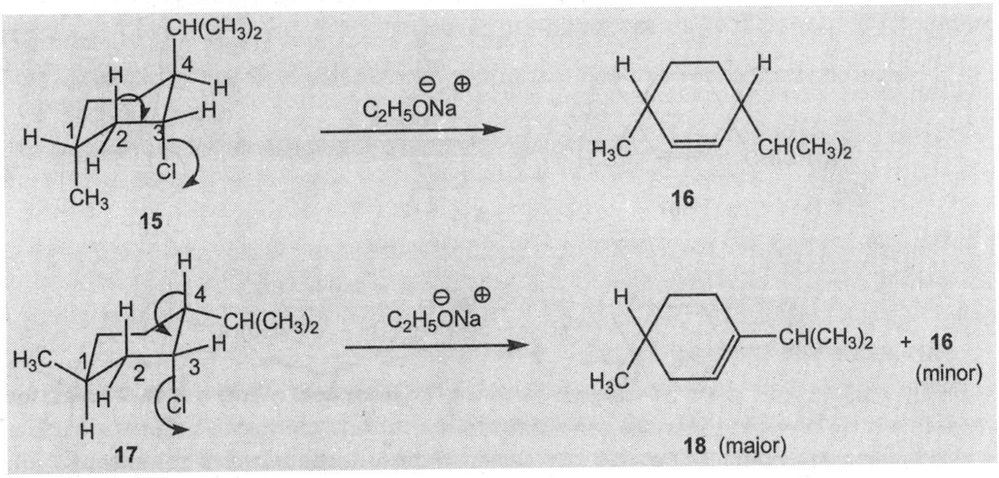

G. Cis-elimination in E2 Reactions

Although in a vast majority of cases involving open-chain systems, *trans* elimination is a general rule due to the ease with which free rotation around carbon-carbon single bond produces the *trans* periplanar (antiparallel) conformation, nevertheless, it has been pointed out that certain cyclic compounds violate the *trans* rule. In such cases, the necessary orbital overlap for an *E2* reaction can also be achieved if the two groups that are being eliminated are *cis* and *coplanar.* A dihedral angle of 0° serves the same purpose as that of 180°, and is actually preferred to a *trans* elimination with a dihedral angle of 120°. The most striking example of this type is found in 2,3-dichloronorbornane series, where *cis* elimination from **19** is faster than *trans* elimination from **20.**

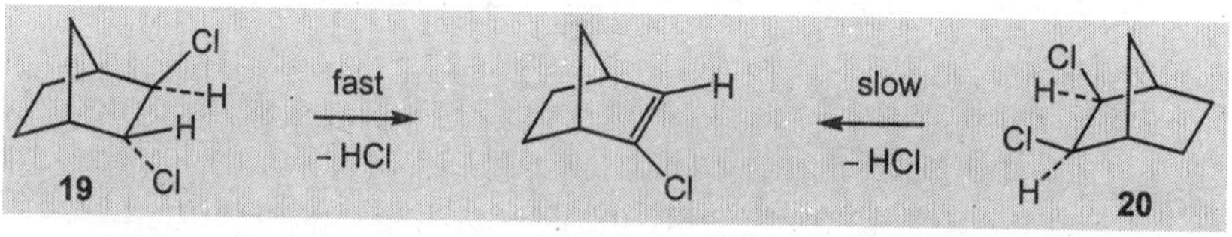

Moreover, **21** undergoes a facile *cis E2* reaction yielding a product completely free from deuterium.

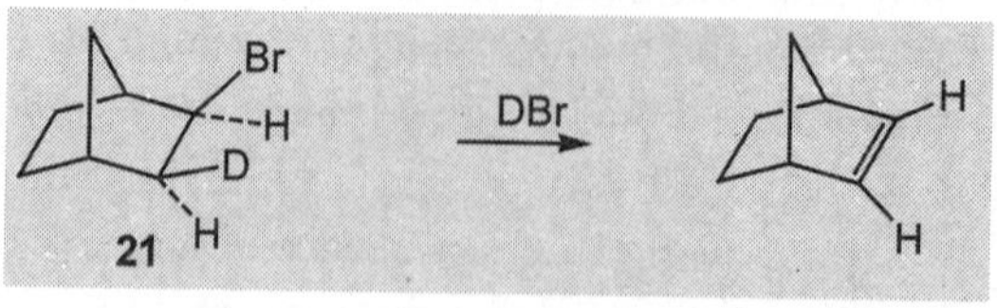

It has been suggested that due to geometric rigidity of the norbornane system, coplanarity in these cases can be achieved in the transition state leading to *cis* rather than *trans* elimination.

In contrast to *E2* reactions, *E1* reactions are not *stereospecific*. The reason is not far to seek if we consider the planarity of the intermediate carbocation. *Cis* and *trans* isomers produce the same carbocation in the initial step of the *E1* reaction, and therefore, the *product composition* remains the same with either isomer. This is well illustrated in the following example.

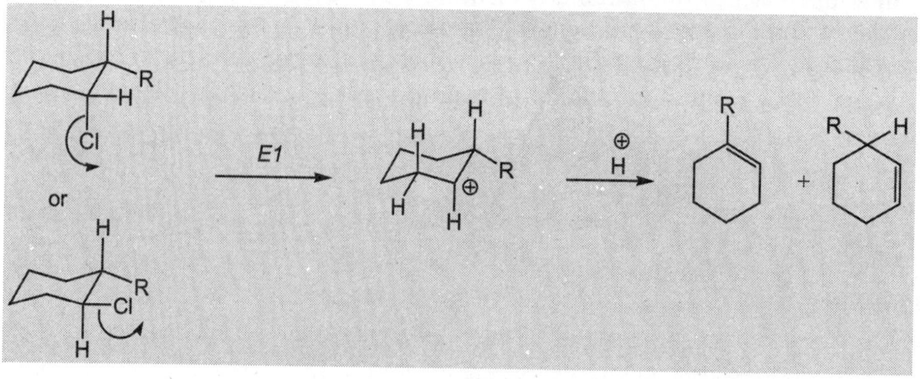

H. Elimination *vs* Substitution

Like S_N1 and $E1$ reactions, S_N2 and $E2$ reactions also occur simultaneously. If the attacking reagent is a good base, it abstracts the β-proton and causes elimination, whereas a good nucleophile attacks the α-carbon causing substitution.

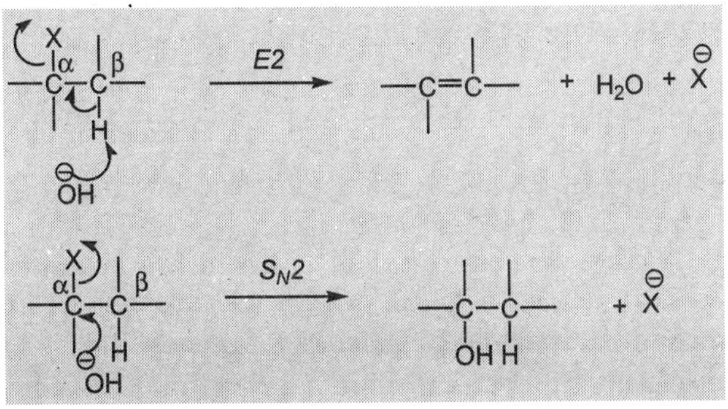

The following factors influence the extent of elimination and substitution:

(I) Structure of the reactant (II) Nature of the base

(III) Nature of the solvent (IV) Effect of temperature

I. Structure of the Reactant

Generally where elimination and substitution are competing reactions, the proportion of elimination increases with increased chain branching of the reactant. One reason for this preference may be ascribed to greater stability of highly substituted alkenes formed from substrates such as alkyl halides. The other reason is purely steric, and is applicable only to unimolecular reactions where carbocations are the common intermediates. As we know, carbocations are planar structures with bond angles of 120°, and with the increase in branching, they will show more resistance to the decrease in bond angle (120° to 109.28°) brought about by the substitution. On the other hand, elimination will be preferred as it does not involve any change in bond angles.

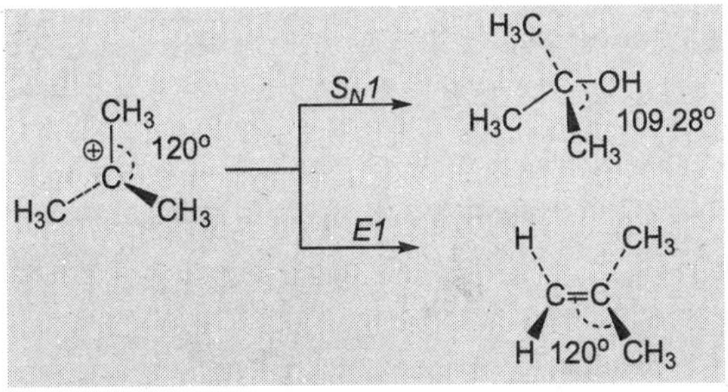

II. Nature of the Base

A good nucleophile may not necessarily be a good base or *vice versa*. This can be demonstrated by comparing the nucleophilic and basic properties of some common reagents. The basicity of these reagents are found to be in the order

$$H_2O,\ I^{\ominus},\ Cl^{\ominus} < CH_3COO^{\ominus} < SH^{\ominus} < OH^{\ominus}$$

whereas, the nucleophilicities of the same reagents follow different order,

$$H_2O < CH_3COO^{\ominus} < Cl^{\ominus} < OH^{\ominus} < I^{\ominus} < SH^{\ominus}$$

To some extent, the elimination/ substitution ratio depends upon the nature of the base involved. In bimolecular reactions (*E2* and S_N2), for example, if we have a weak base but a strong nucleophile, *e.g.* iodide, there will be more substitution than elimination. On increasing the concentration and strength of the base, unimolecular reaction tends to bimolecularity. This change in molecularity is more pronounced in *E1* than in S_N1 reactions, leading to a higher proportion of the elimination product. This is precisely the reason for employing strong bases at high concentrations in the preparative methods for alkenes.

III. Nature of the Solvent

A change in the polarity of the solvent also affects the course of a reaction. In bimolecular reactions, for example, the $E2/S_N2$ ratio depends on how well the solvent is able to solvate the two transition states of the same reactant. Usually, bimolecular reactions are favoured by a decrease in the polarity of solvent, but this effect is more pronounced in *E2* reactions. Thus, a less polar solvent not only favours bimolecular reactions, but also *E2* reactions over S_N2 reactions.

IV. Effect of Temperature

Elimination reactions usually have higher activation energy than the accompanying substitution reactions. Therefore, increase in the temperature of the reaction mixture increases the extent of elimination.

I. Thermal Elimination

A number of compounds undergo *pyrolytic eliminations* reacting unimolecularly through cyclic mechanisms. These reactions are carried out in the gas phase, and proceed in a concerted fashion yielding the products of *cis* eliminations. A common example is the pyrolysis of acetate esters resulting in the formation of alkenes.

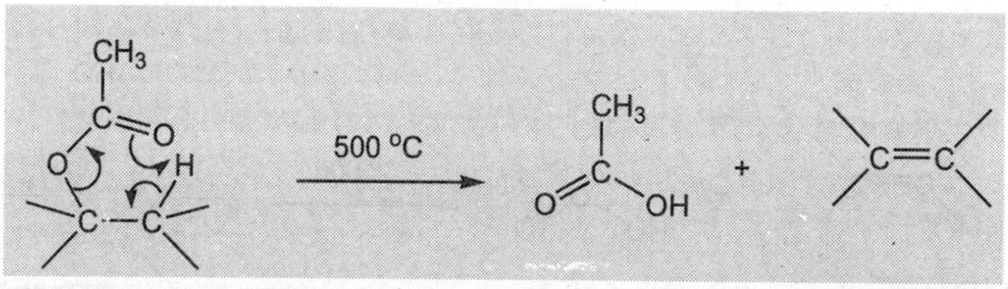

The existence of a cyclic, concerted mechanism has been demonstrated by the pyrolysis of esters **22** and **23**. As expected, there is complete retention of deuterium in the alkene obtained from **22** whereas no deuterium was found in the product obtained from the pyrolysis of **23**.

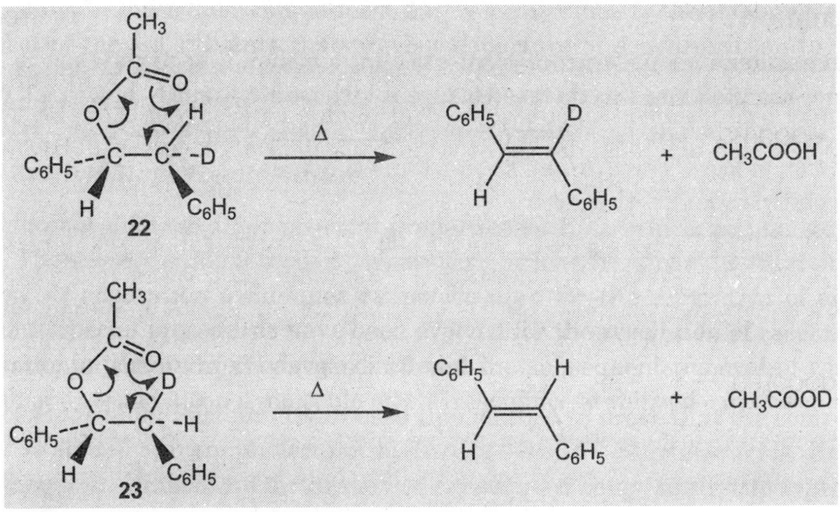

The synthetically important Tschugaev reaction involves pyrolysis of a xanthate at relatively low temperatures.

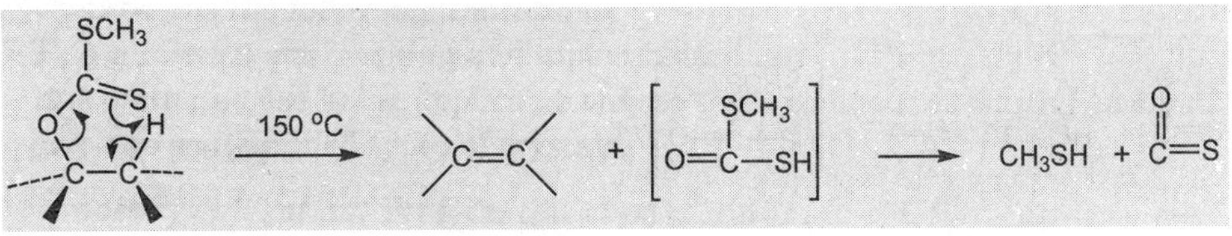

Unsymmetrical acetate and xanthate esters yield mixture of all the possible alkenes, but there is usually a predominance of the more highly substituted alkene. Another analogous reaction of preparative value is the pyrolysis of amine oxides to yield alkenes. This reaction is also a cyclic process resulting in *cis* elimination.

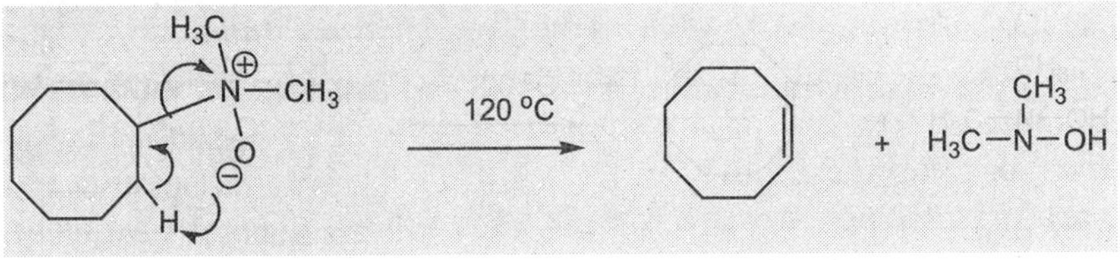

Alkyl halides also undergo pyrolytic dehydrohalogenation *via* a four-membered cyclic transition state.

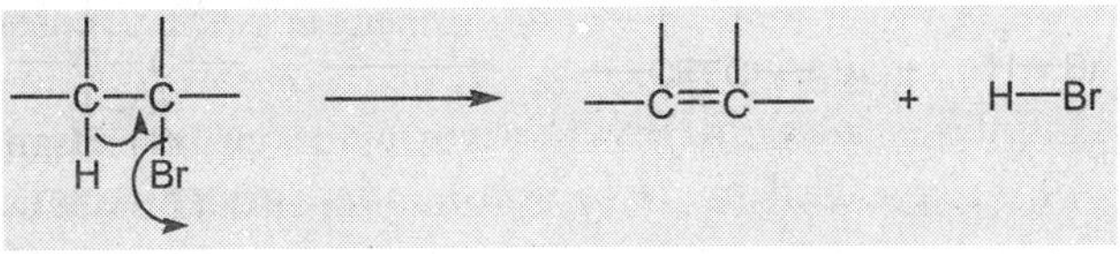

J. Dehalogenation

Elimination reactions are not confined only to those reactions in which one of the leaving groups is hydrogen. Reactive metals such as Zn and Mg are capable of removing halogens from 1,2-dihalides to yield alkenes. Like other *E2* reactions, these reactions also proceed by a concerted *trans* elimination process. Thus, *dl*-and *meso*-2,3-dibromobutanes yield *cis* and *trans*-2-butenes, respectively, on treatment with zinc.

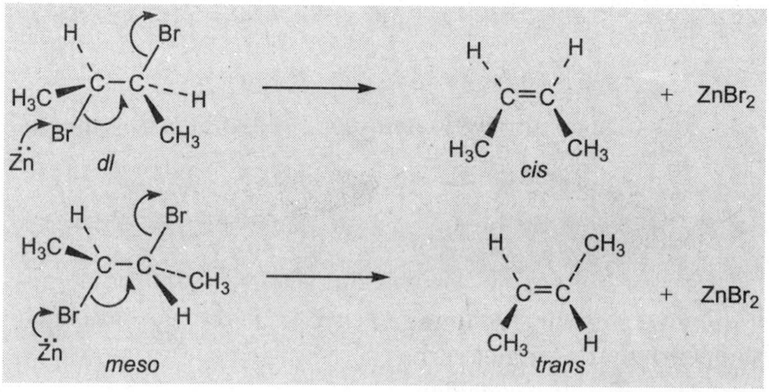

A similar treatment of *trans*-2,3-dibromo-2-butene with zinc results in the formation of dimethylacetylene.

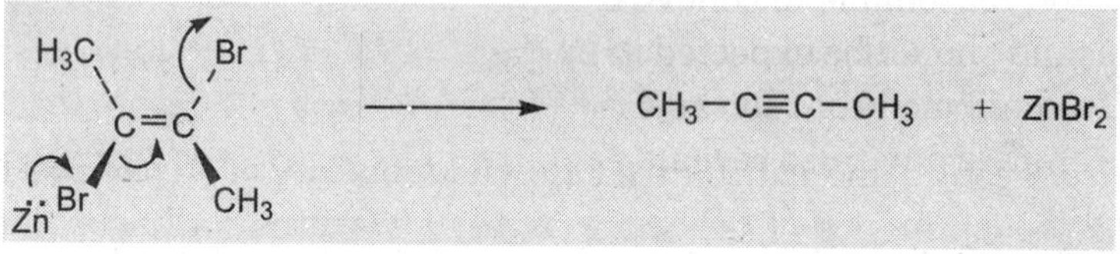

8.2 α-ELIMINATION: CARBENES

A reaction in which both the groups and atoms are lost from the same carbon is called an α-elimination reaction, and results in the formation of *carbenes*. In view of recent interest in the chemistry of these reactive species, we shall discuss their structure, methods of generation, and chemical reactions in some detail.

Carbenes are key intermediates in a wide variety of chemical and photochemical reactions. Historically, as early as in 1862, the intermediacy of carbon dichloride was predicted in the alkaline hydrolysis of chloroform. But it was only in 1950 that the detailed mechanism of their formation was clearly established by the pioneering studies of Jack Hine and his associates. Another important landmark in the study of carbene was the discovery that they can easily be trapped by an alkene to yield cyclopropanes.

Carbenes can be defined as neutral, divalent carbon intermediates in which a carbon is covalently bonded to two atoms, and has two non-bonded orbitals containing two electrons between them. Theoretical considerations suggest that there are two possible kinds of carbenes, *singlet* and *triplet* carbenes. In the *singlet* state, a carbon atom is presumed to approximate sp^2

hybridization. Two of the three sp^2 hybrid orbitals are utilized in forming two covalent bonds, whereas the third hybrid orbital contains the unshared pair of electrons. The remaining *p* orbital remains vacant. Thus, singlet carbene (**24**) resembles a carbocation (**25**) very closely.

On the other hand, carbon atom of a triplet carbene is *sp* hybridized, and is a linear or non-linear species. These two hybrid orbitals are involved in bond formation with two groups and the remaining two electrons are placed, one each, in the equivalent, mutually perpendicular orbitals. These electrons have parallel spins, and a carbene with this structure is said to be in a *triplet* state (**26**).

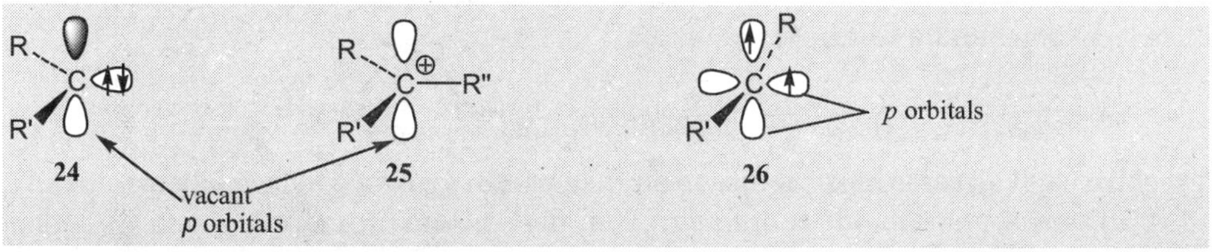

Stability of Carbenes

At first sight, it appears that a singlet carbene has lower energy as the unshared electron pair is in sp^2 hybrid orbitals. But, the consideration of the electron repulsion energy that must be overcome to pair two electrons in a single orbital places it at a higher energy level than a *triplet structure*.

It is, therefore, reasonable to believe that the triplet state of a carbene is more stable than the singlet state, and should be expected to be the ground state. Electron spin resonance (ESR) studies have also confirmed that for aryl and diaryl carbenes the ground state is their *triplet structure*. These studies have also revealed that triplet carbenes are not linear, but have a bond angle of 140° to 150°.

A. Generation of Carbenes

- From cycloelimination
- From diazoalkanes and ketenes
- From carbanions

From Cycloelimination

Cycloelimination is the reverse of cycloaddition reaction, and may be defined as the cleavage of a cyclic compound into various fragments. Cyclopropanes are easily synthesized by cycloaddition of carbenes to alkene, whereas their photochemical or thermal cleavage offers a method for generating carbenes.

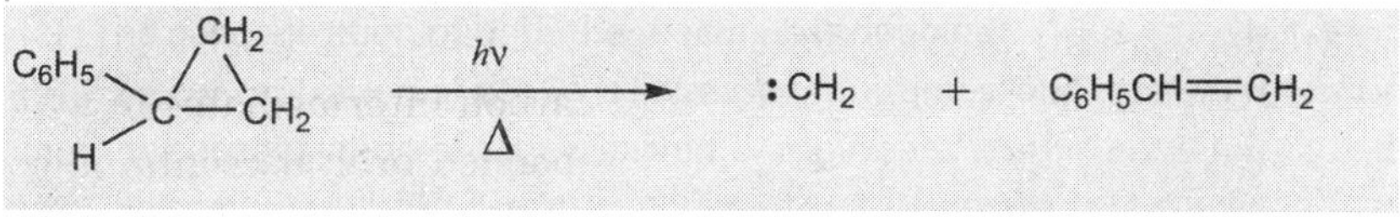

7-Norbornadienyl ketals such as **27** are good source of dialkoxycarbenes (**28**).

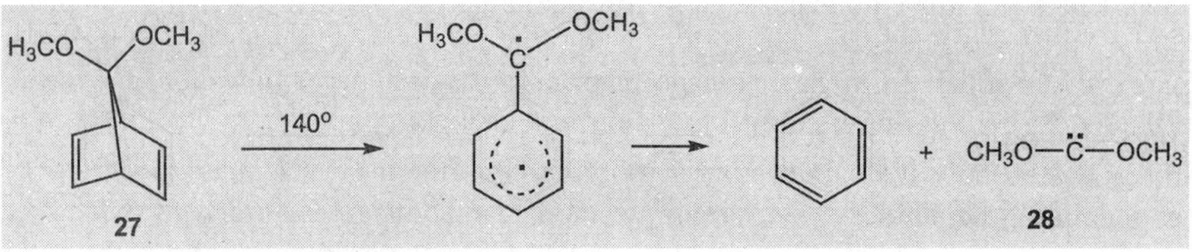

Oxiranes are related to cyclopropanes in being three-membered ring compounds, and because of the steric strain associated with a three-membered ring, they also decompose easily to produce a carbene and a carbonyl compound.

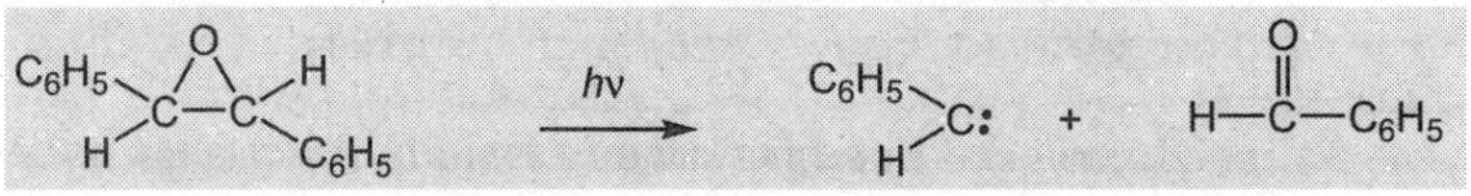

Photochemistry of several substituted oxiranes was investigated, and it was found that their principal mode of decomposition involves formation of carbenes under these conditions. The most probable mechanistic pathway can be written as follows:

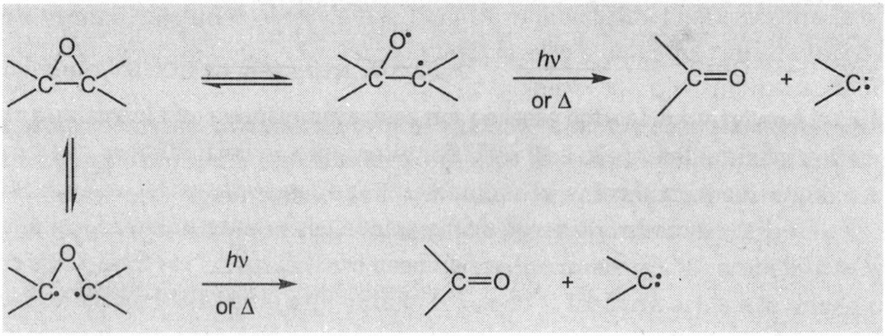

From Diazoalkanes and Ketenes

Diazo compounds are one of the best known sources of many carbenes. Photolysis or thermolysis of these compounds generally provides the singlet state carbene, which in the presence of alkenes rapidly reacts to form cyclopropanes. It has been found that this reaction is faster than the conversion of a singlet carbene to the more stable triplet state. However, when the decomposition is carried out in the gas phase in high dilution, the singlet carbene gets sufficient time for decaying to a triplet state, which then reacts with alkenes to produce cyclopropanes.

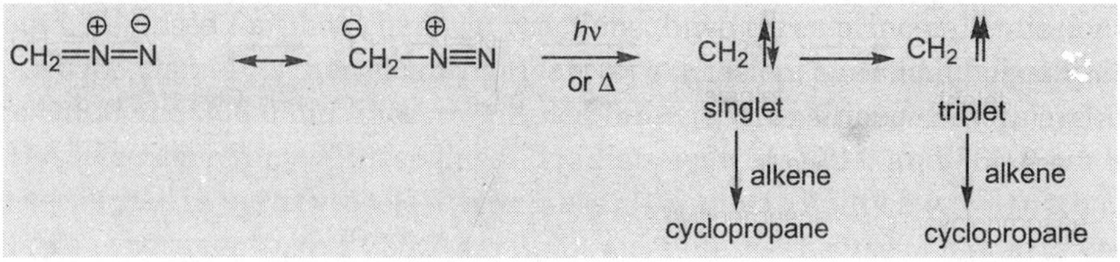

As we shall examine shortly, it is possible to distinguish the singlet from the triplet carbene by studying the stereochemistry of the cyclopropane formed in the reaction.

Ketenes, which are structurally similar to diazoalkanes, also produce carbene on photochemical decomposition.

$$R_2C{=}C{=}O \xrightarrow{h\nu} R_2C\!: \;+\; CO$$

From Carbanions

The detailed mechanism of the hydrolysis of chloroform to generate dichlorocarbene **(29)** has been worked out. In the first step of the reaction, the conjugate base of chloroform is formed, which then undergoes unimolecular reaction to give dichlorocarbene.

$$CHCl_3 + \overset{\ominus}{O}H \rightleftharpoons \overset{\ominus}{C}Cl_3 + H_2O$$

$$\overset{\ominus}{C}Cl_3 \rightleftharpoons \overset{\ominus}{Cl} + \underset{\mathbf{29}}{:CCl_2}$$

The best evidence for this mechanism comes from the observation that the rate of hydrolysis is slowed down on addition of chloride ions as they react with dichlorocarbene to regenerate the anions. Dichlorocarbene rapidly reacts with water and hydroxide ions to give a mixture of products.

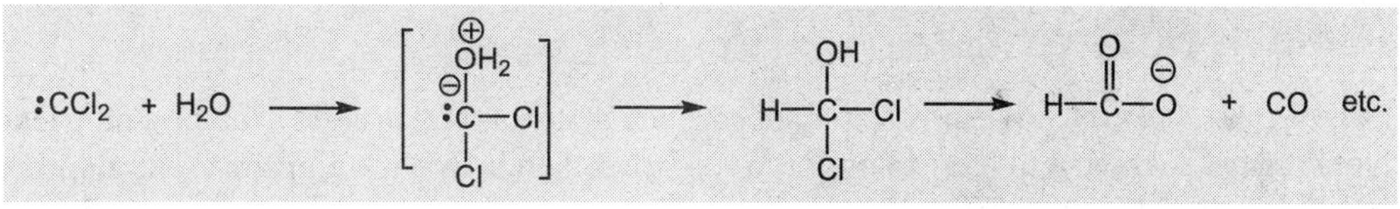

Thus, the whole process is an α-elimination by a unimolecular reaction of the conjugate base of chloroform, and may be symbolized by the term 1,1-*E1cB* process. This method of generating dihalocarbenes has been extended to a variety of haloforms and bases. In non-aqueous media, carbenes can be trapped as cyclopropanes.

B. Reactions of Carbenes

- Cycloaddition reactions
- Insertion reactions
- Reactions with nucleophiles

Cycloaddition Reactions

Singlet carbenes add to alkenes in a stereospecific fashion, i.e. stereochemistry of the alkene is retained in cyclopropane.

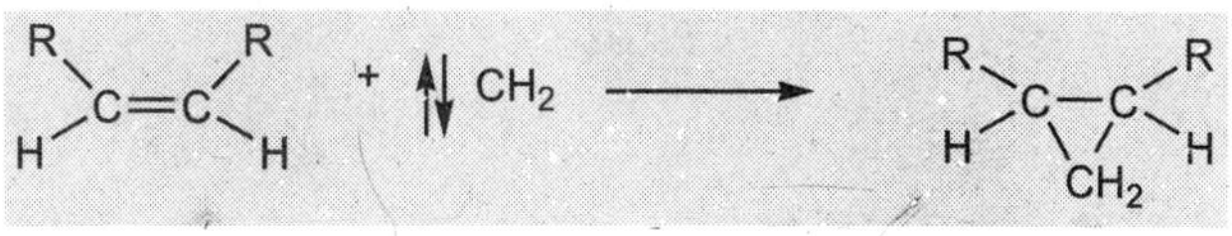

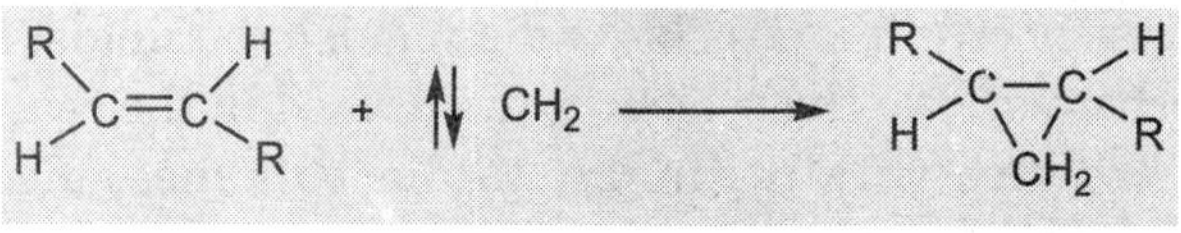

In contrast, the addition of triplet carbenes to alkenes takes place non-stereospecifically.

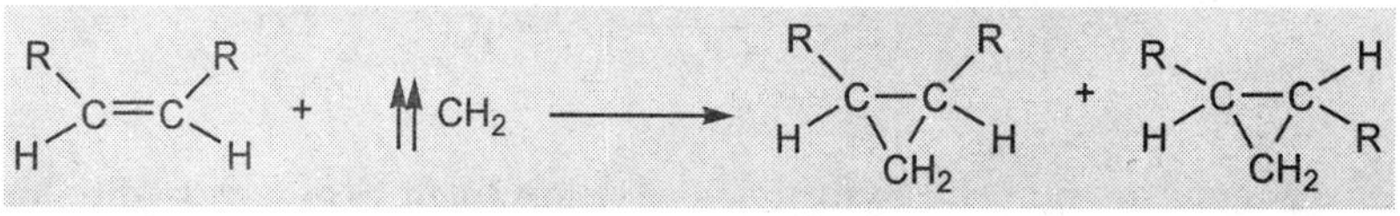

The stereochemistry of these cycloadditions is so specific that it may be used as a diagnostic test for the *identification of singlet and triplet carbenes*. The difference in behaviour between these two electronic states of a carbene has been explained in the following way. The addition of a singlet carbene to an alkene is a *concerted* process as there is no spin restriction on the simultaneous formation of two new σ bonds of cyclopropane, and hence, it is a *stereospecific* process.

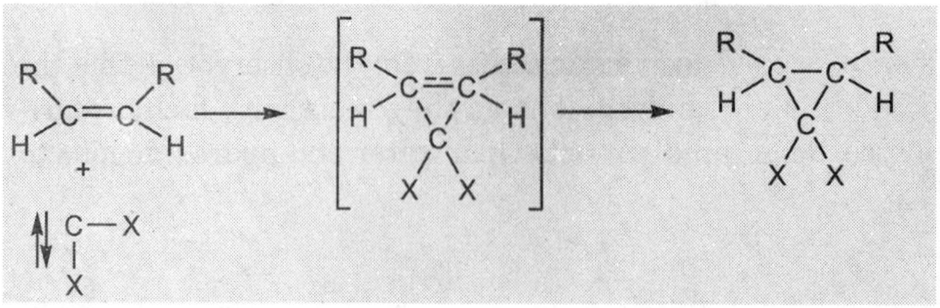

In contrast, the cycloaddition of a triplet carbene is a *step-wise process*. The first step leads to a triplet diradical, which will have to undergo spin inversion before it collapses to a cyclopropane. Since this diradical has a C–C single bond, there is a possibility of free rotation about that bond. If this rotation is faster than spin inversion, the reaction becomes *non-stereospecific*.

These steps can be visualized as follows:

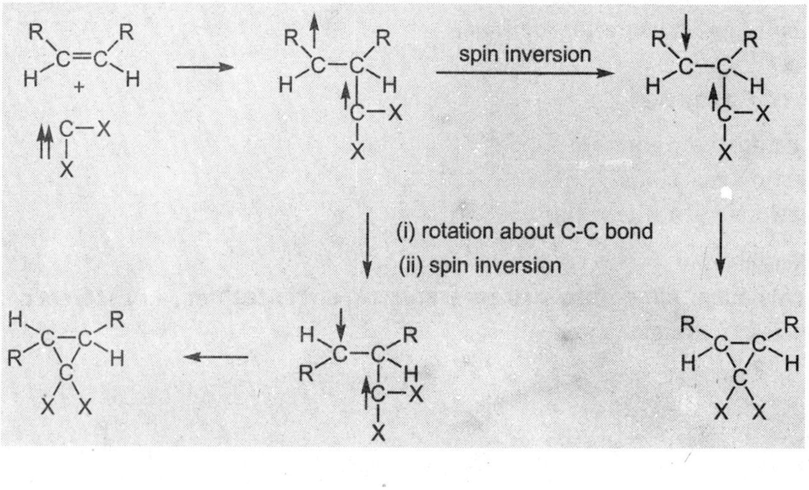

Carbenes are also capable of forming addition products with the aromatic system. One such example is the addition of carbalkoxy carbene **(30)** to benzene yielding the cycloheptatriene esters, which are commonly referred to as 'Buchner acid esters' **(31)** in recognition of Buchner who discovered this reaction in 1885.

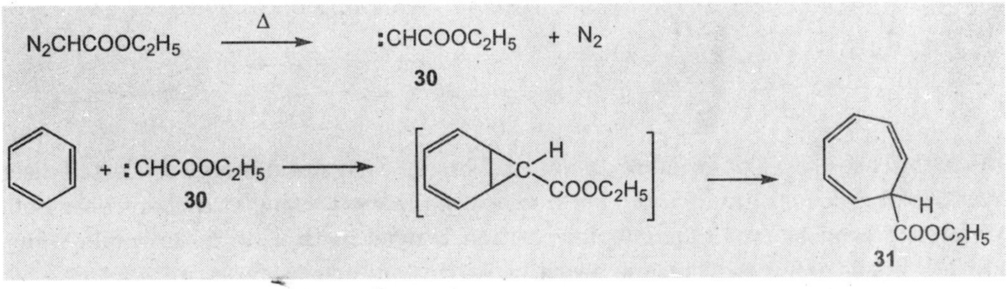

Many benzene derivatives also undergo similar addition reaction with carbenes produced by pyrolysis or photolysis of diazoalkanes. At least in one case a dichlorocyclopropane compound has been isolated.

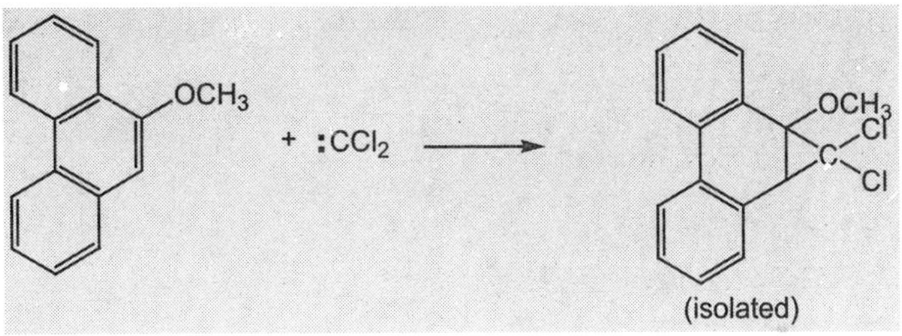

Insertion Reactions

Frequently, carbenes can be inserted into a C–H bond; the reaction may be inter or intramolecular.

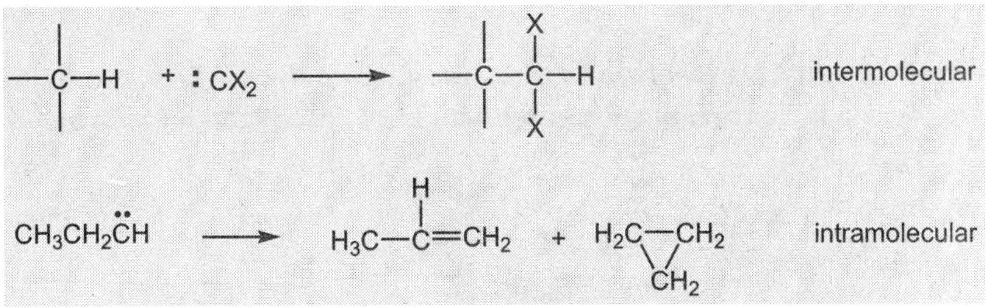

Intermolecular insertion of carbenes into a C–H bond may occur either by a one-step process, or it may be a two-step abstraction-recombination process.

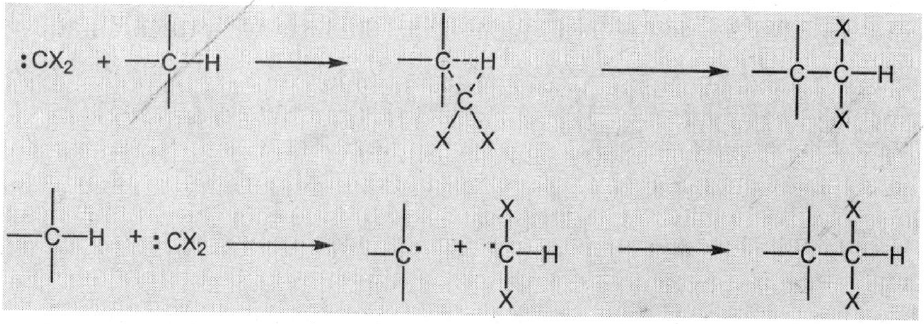

As in the case of cycloaddition reactions, the singlet carbenes undergo concerted insertion, whereas triplet carbenes are inserted *via* a two-step process. A singlet carbene does not show any preference for insertion to a particular position. Singlet methylene, for example, is inserted into all C–H bonds of 2-methylbutane, practically, without any preference.

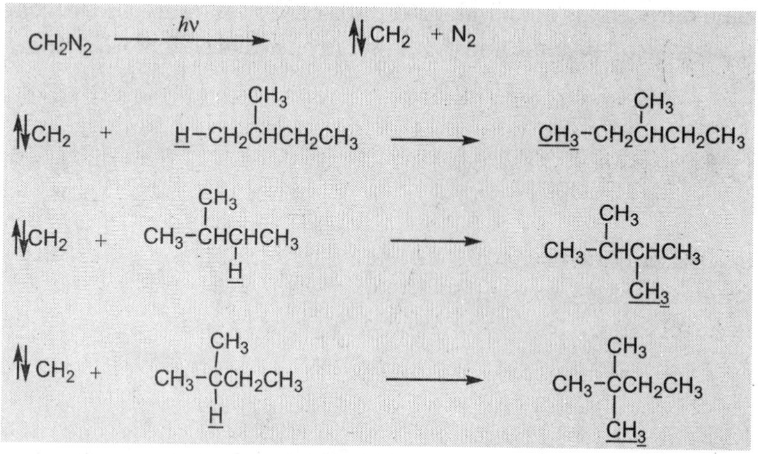

Triplet carbenes usually display more selectivity for a particular position in insertion reactions. Triplet methylene shows primary : secondary : tertiary selectivity in the ratio 4.4 : 1.6 : 1.0.

Alkylcarbenes, usually, are inserted intramolecularly producing an alkene or a cyclopropane. The mechanism for the formation of these products may be written as follows.

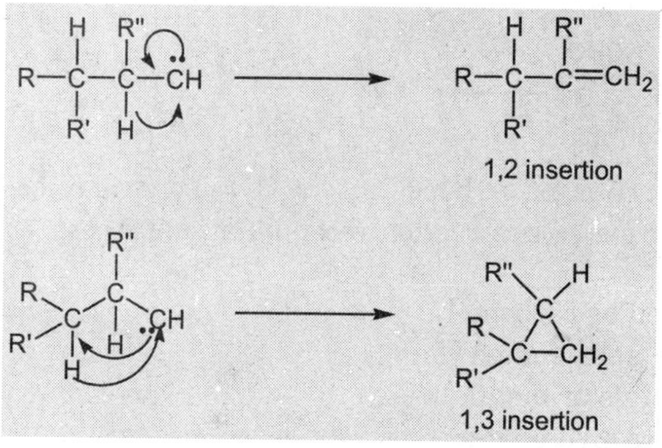

Thus, arylcarbenes can be inserted intramolecularly into aromatic C–H bonds.

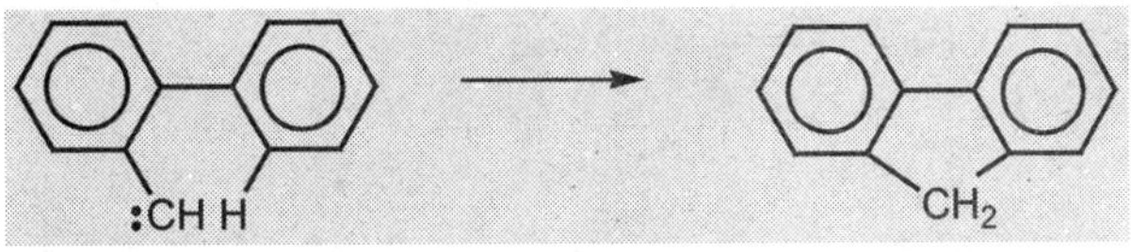

Carbenes are intermediates in the Wolff rearrangement, which involves the 1,2-shift of an alkyl group.

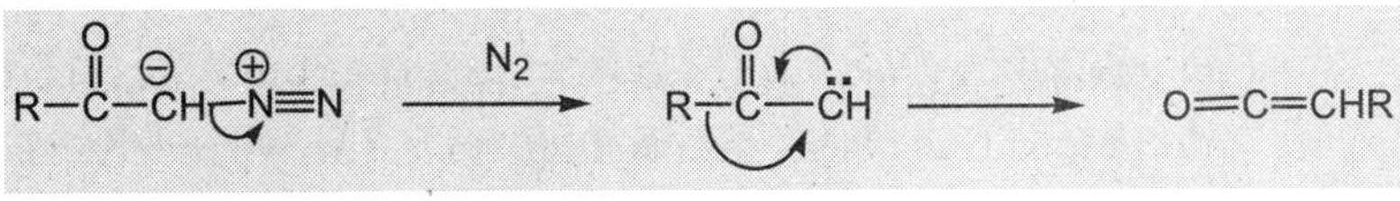

Reactions with Nucleophiles

Carbenes react with many nucleophiles. A common method of trapping carbenes involves insertion into the –OH bond of an alcohol.

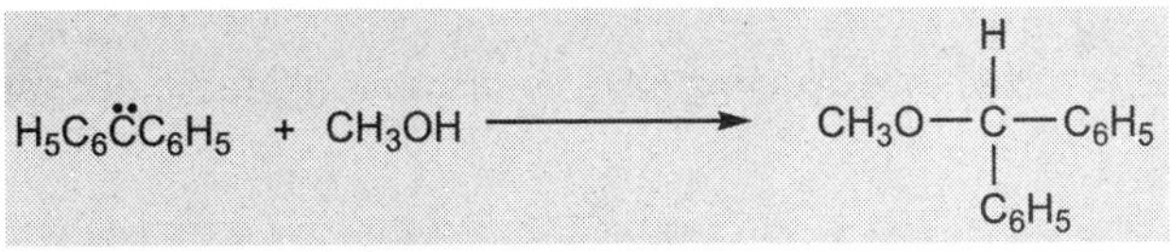

The Reimer-Tiemann conversion of phenoxide to *o*-hydroxybenzaldehyde (salicylaldehyde) involves an electrophilic attack by dichlorocarbene.

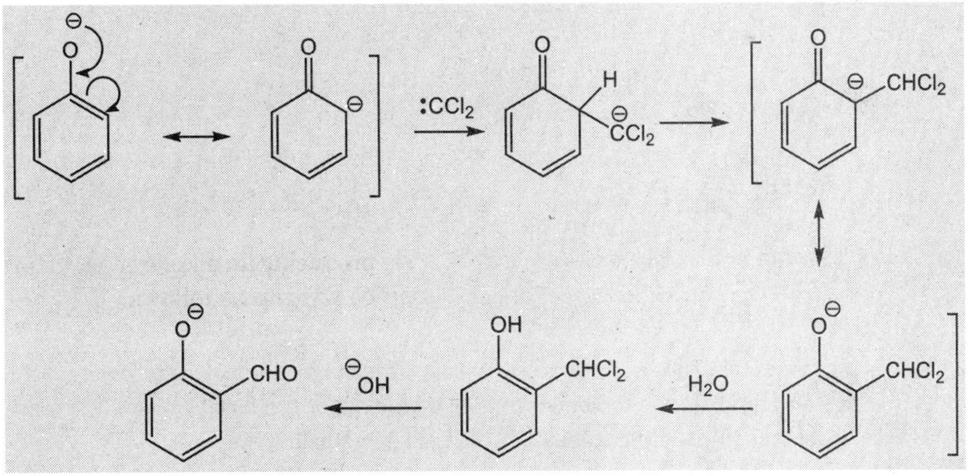

C. Carbenoids

Carbenoids can be defined as carbene-like organometallic intermediates, which are produced by the action of metal on halogenated compounds. Typically, when *gem*-dihalogen compounds **(32)** are treated with metals or metalalkyls, unstable intermediates **(33)** are obtained, which decompose with alkenes to give products appearing to originate from carbenes.

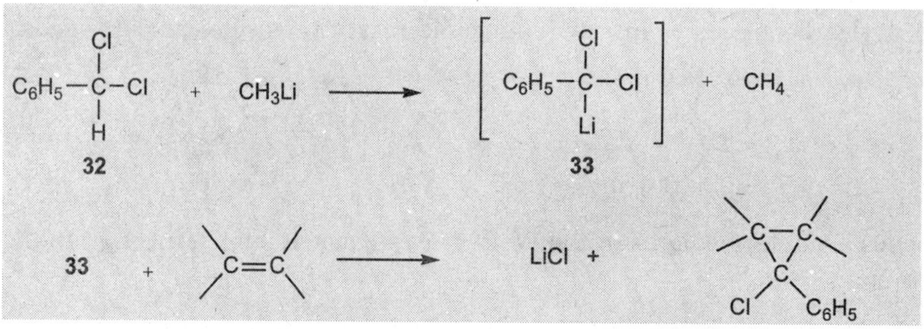

Most of the available evidence is against the participation of free carbenes in these reactions. For example, the rate of decomposition of the intermediate in the above reaction is dependent on the concentration of alkenes. Also, the 'phenylcarbene' generated in the above reaction differs from phenylcarbenes produced by the decomposition of phenyldiazomethane in its selectivity towards various alkenes.

PROBLEMS AND REVIEW QUESTIONS

1. Which alcohol in each of the following pairs would undergo facile dehydration? Give reasons for your answer.

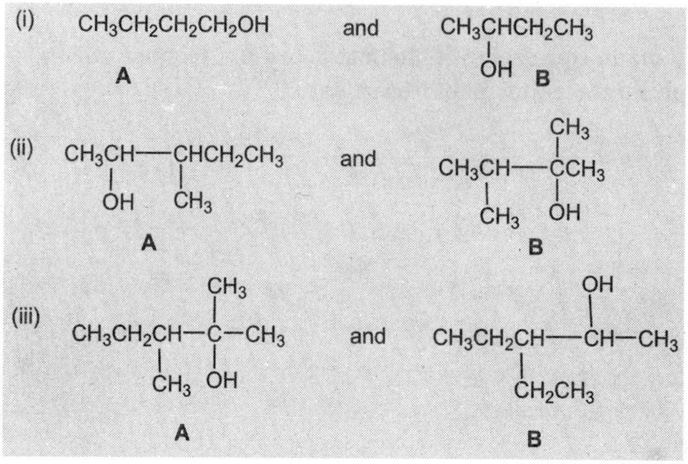

2. Predict the structures of alkenes expected from dehydrohalogenation of the following alkyl halides. In case of isomeric alkenes, predict which shall be the major product?
   (i) 2-Chlorobutane (ii) 1-Chlorobutane (iii) 2-Chloro-3-methylbutane
   (iv) 3-Chloro-2-methylpentane (v) 2-Chloro-3-methylpentane, and (vi) 2-Chloro-3,3-dimethylpentane
3. Acid-catalyzed dehydration of *neo*pentyl alcohol produces two isomeric alkenes. Predict their structures, and discuss a suitable mechanistic pathway for the reaction. Which alkene do you think shall be the major product? Give reasons.

4. When *cis*-2-chlorocyclohexanol is treated with a base, the product is cyclohexanone. Outline the mechanism.
5. Which alkyl halide do you expect to be more reactive in an *E2* reaction?

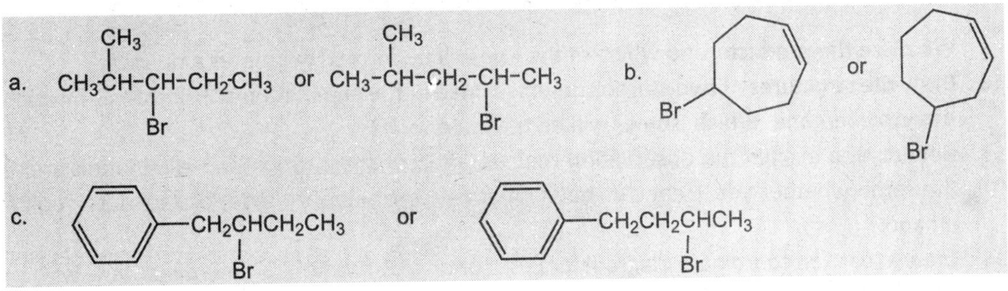

6. For each of the following reactions:
   (i) Indicate whether elimination will occur *via* an *E2* or an *E1* reaction, and
   (ii) Give the major elimination product of each reaction. Also provide reason(s).

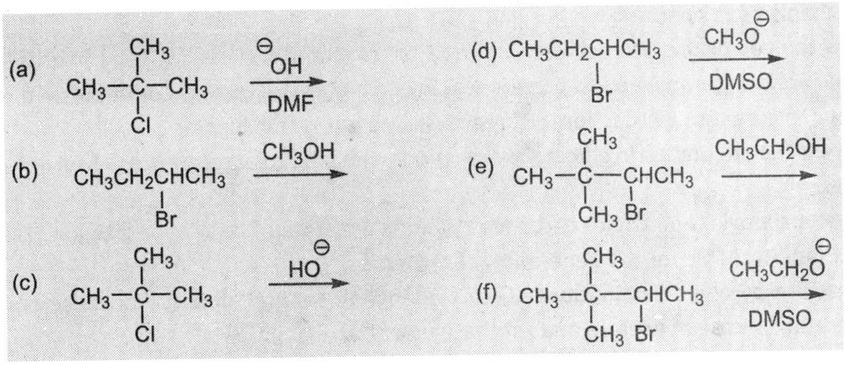

7. The reaction of ethyl bromide with *tert*-butoxide ion gives *tert*-butyl ethyl ether (substitution product), whereas *tert*-butyl bromide on treatment with ethoxide gives 2-methylpropene (elimination product). Explain.
8. Indicate which of the compounds in each pair will give a higher substitution/elimination ratio when it reacts with *iso*propyl bromide:
   (a) Ethoxide or *tert*-butoxide ion
   (b) ^{-}OCN or ^{-}SCN
   (c) Cl^- or Br^-
   (d) CH_3S^- or CH_3O^-
9. How many alkenes are formed from *E1* reaction of 3-bromo-2,3-dimethylpentane? Draw their structures, and rank them according to the amount that would be formed.
10. Give the products obtained from *E1cB* reaction, and outline their mechanism also.

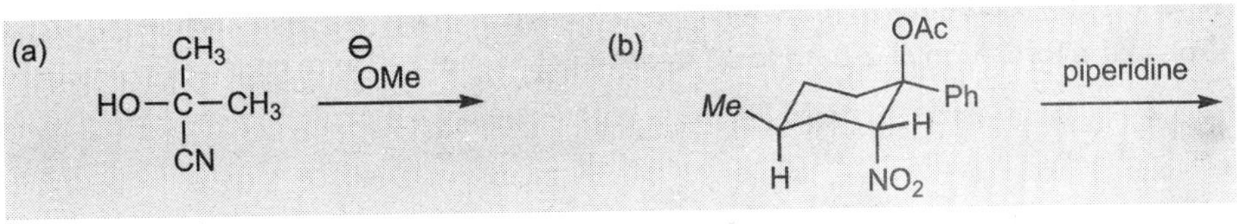

11. Two elimination products are obtained in the following *E2* reaction:

$$CH_3CH_2CHDCH_2Br \xrightarrow{\ominus OH}$$

What are the products, and which of the two will be formed in greater amount?
12. Draw the structures of two stereoisomers formed in the elimination of HBr from 3-bromo-2,2,3-trimethylpentane. Which isomer will be the major one?
13. How do you explain the observation that the ratio of substitution and elimination products is the same whether you treat *tert*-butyl chloride, *tert*-butyl bromide, or *tert*-butyl iodide with ethanol?
14. Draw a reaction coordinate diagram for *E2* reaction of 2-bromo-2,3-dimethylbutane with sodium *tert*-butoxide.
15. Which isomer does react more rapidly in an *E2* reaction, *cis*-1-bromo-4-*tert*-butylcyclohexane or *trans*-1-bromo-4-*tert*-butylcyclohexane? Explain your choice.
16. With suitable examples, illustrate *E1, E2,* and *E1cB* reactions. Outline conditions for each of these mechanisms to operate.
17. How can you distinguish between *E1* and *E1cB* reactions by labelling experiments?
18. What are the stereochemical preferences in elimination reactions? Comment on the observation that in contrast to *E2* reactions, *E1* reactions are not stereospecific.
19. Discuss the orientation in elimination reactions with particular reference to Saytzeff and Hofmann rules.
20. What are singlet and triplet carbenes? What is the state of their hybridization? Discuss various methods available for the generation of carbenes.
21. What are pyrolytic eliminations? Give evidence in support of a cyclic, concerted mechanism operating in case of acetate and xanthate esters?
22. Discuss the factors which influence the extent of elimination and substitution.
23. Comment on the following:
    (i) Carbenoids
    (ii) Tschvgaev reaction
    (iii) Stereoselectivity in *E2* reactions
    (iv) Saytzeff and Hofmann rules in elimination reactions.
24. How do you account for the formation of dichlorocarbene in the alkaline hydrolysis of chloroform? Comment on the subsequent conversion of dichlorocarbene to the observed products.
25. Indicate how each of the following factors affects an *E1* reaction:
    (i) Structure of alkyl halide, (ii) Concentration of the base, (iii) Solvent, and (iv) Strength of the base.
26. Explain why *E2* reaction is more common than *E1* and *E1cB* reactions?
27. How does conformation of a molecule affect its reactivity in *E2* reaction?
28. What are the conditions that favour *E1cB* mechanism? Illustrate with examples.
29. Explain the formation of 2-butene on treatment of 1-butanol with conc. H_2SO_4.

30. Predict the product(s) in the following reactions:

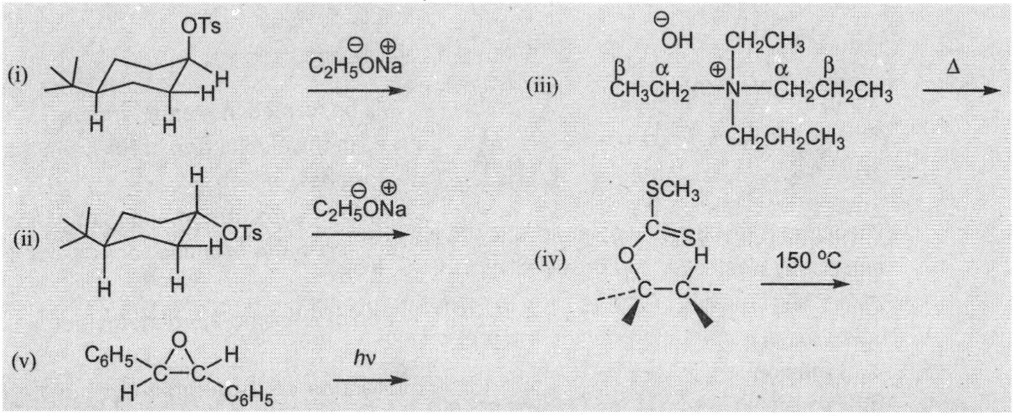

SOLUTIONS

1. (i) **B** (ii) **B** (iii) **A**
2. (i) 1-Butene (minor), 2-butene (major) (ii) 1-Butene (iii) 3-Methyl-1-butene (minor), 3-Methyl-2-butene (major); (iv) 4-Methyl-2-pentene (minor), 2-methyl-2-pentene (major); (v) 2-Methyl-1-pentene (minor), 2-methyl-2-pentene (major); (vi) 3,3-Dimethyl-1-pentene.
3. 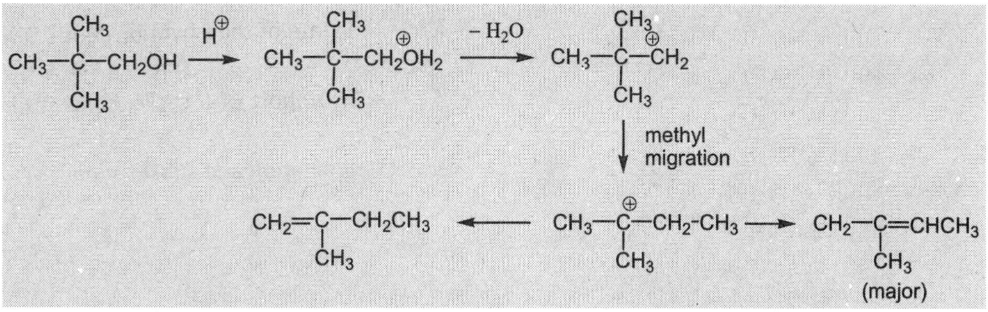

4. 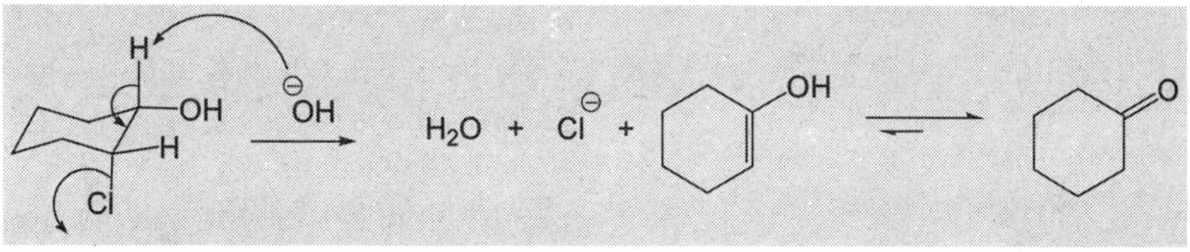

5. 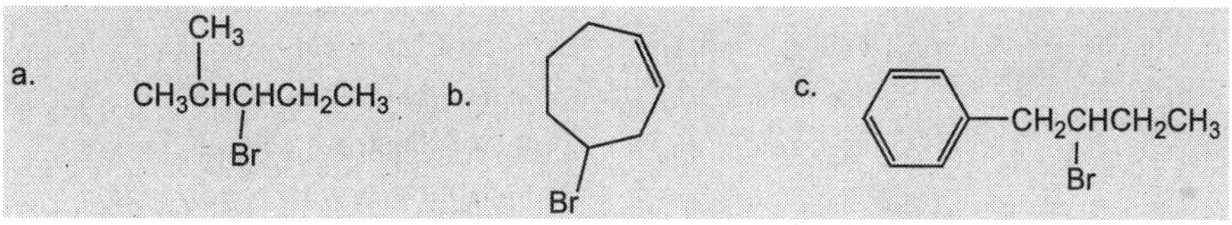

6. 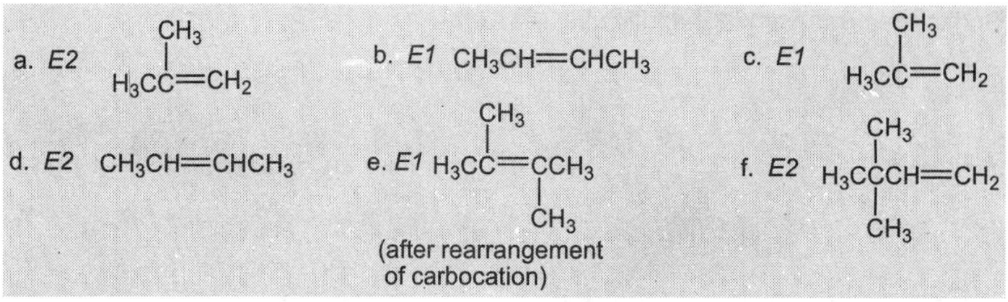

E2 reaction is favoured by polar aprotic solvents (DMF, DMSO), on the other hand *E1* reaction is favoured by weak base and protic solvents (H_2O, ROH).

7. Primary alkyl halides primarily form substitution product, ether, whereas tertiary alkyl halides lead to the elimination product in the presence of a strong base.

8. a. Ethoxide b. ^{-}SCN c. Br^{-} d. CH_3S^{-}

9. Highly substituted alkene (**1**) is most stable. Alkene (**2**) is more stable than *cis*-alkene (**3**).

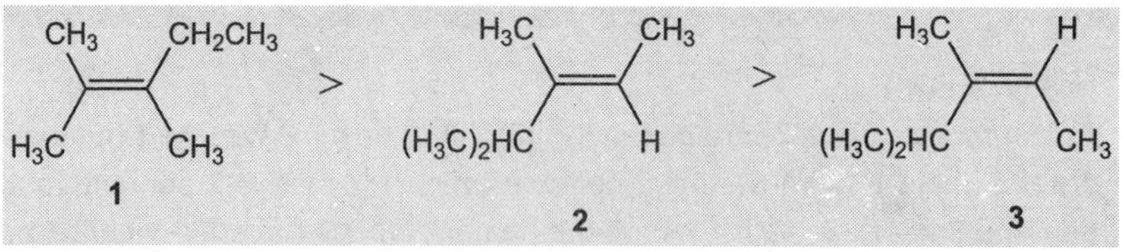

10. 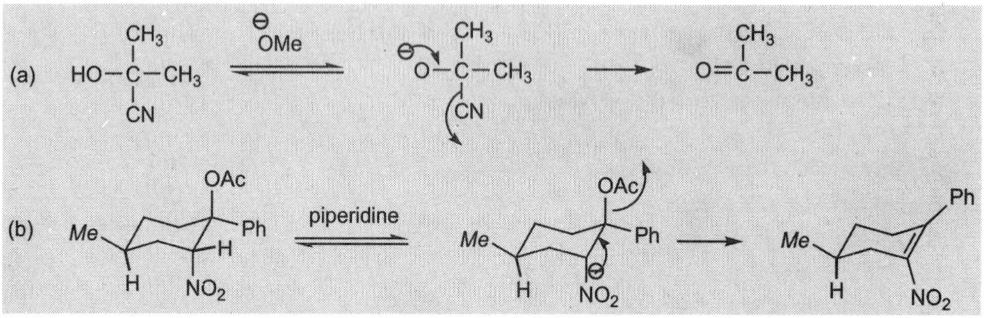

11.
$$CH_3CH_2CHDCH_2Br \xrightarrow{\ominus OH} \underset{\text{(major)}}{CH_3CH_2CD{=}CH_2} + \underset{\text{(minor)}}{CH_3CH_2CH{=}CH_2}$$

12.. 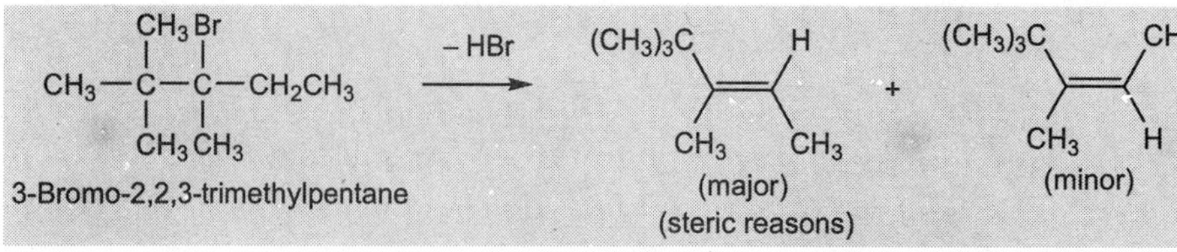

Chapter 9

Reactions of the Carbonyl Group

Chapter Outline

9.1 REACTIONS OF ALDEHYDES AND KETONES

- A. Reaction with Water and Alcohols
- B. Condensation with Derivatives of Ammonia
- C. Reaction with Hydride Donors
- D. Addition of Carbanions
- E. Halogenation of Ketones

9.2 REACTIONS OF CARBOXYLIC ACID DERIVATIVES

- A. Esterification and Hydrolysis
- B. Claisen Condensation
- C. Acyloin Condensation
- D. Hydrolysis of Amides
- E. Hydrolysis of Acid Halides and Anhydrides
- F. Decarboxylation of Carboxylic Acids

The carbonyl group may be regarded as the backbone of organic chemistry. In this group, carbon and oxygen are joined by a double bond, which is polarized by the strongly electronegative oxygen atom. This electron imbalance in the π bond renders the carbon atom *electron-deficient,* and causes the carbonyl group as a whole to exercise an electron-withdrawing effect.

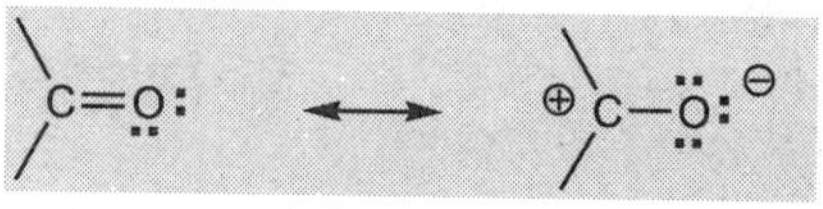

It is this electron imbalance that dominates the chemistry of carbonyl compounds in the following two ways: (a) carbonyl carbon can be attacked directly by a nucleophile, and

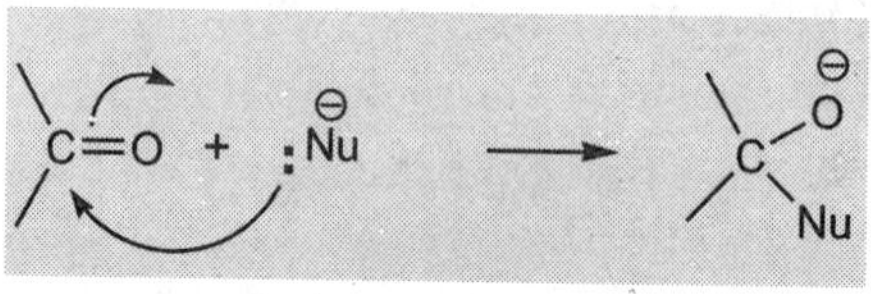

(b) The electron deficiency on the carbonyl carbon renders the α-hydrogen more acidic, which gives rise to mesomeric anion in the presence of a base.

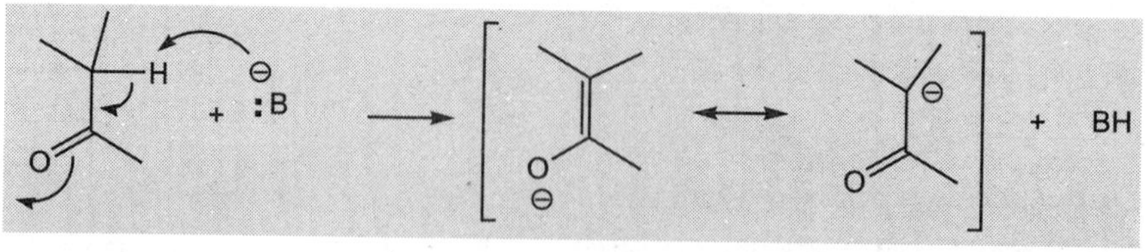

Compounds containing the carbonyl group can broadly be classified into the following two categories: (i) aldehydes and ketones, (ii) carboxylic acids and their derivatives, such as esters and amides. Compounds of the first category display characteristic behaviour of carbonyl compounds, whereas those of the second category do not exhibit many of these reactions, because of the conjugation of carbonyl group with the lone pair of electrons situated on the adjacent heteroatom.

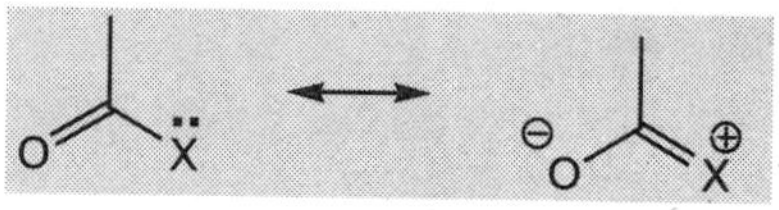

9.1 REACTIONS OF ALDEHYDES AND KETONES

Structure of the Carbonyl Group

The carbon atom of the carbonyl group is sp^2 hybridized. The three sp^2 orbitals form σ bonds to the carbonyl oxygen and two substituents. The carbonyl oxygen is also sp^2 hybridized using one of its sp^2 orbitals to form a σ-bond with the carbonyl carbon. The other two sp^2 orbitals contain a lone pair. The π-bond between carbon and oxygen is formed by the overlap of the remaining p orbital of the carbonyl oxygen with the remaining p orbital of the carbonyl carbon. The sp^2 hybridization implies that the carbonyl group is planar, and the angle between the substituents is approximately 120°.

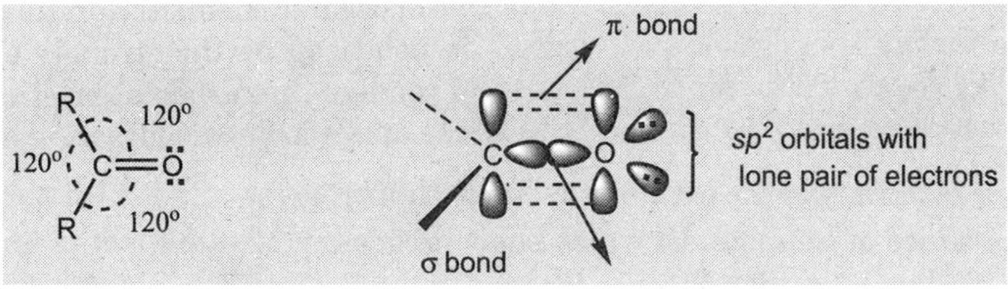

Effect of Structure on Reactivity

The groups attached to the carbonyl carbon greatly influence the reactivity of aldehydes and ketones towards nucleophiles in the following two ways. First, an alkyl group being electron-donating would decrease the positive character of the carbonyl carbon, and thus, makes it less susceptible to nucleophilic attack, and second, the sheer bulk of an alkyl group presents greater steric hindrance than a smaller hydrogen atom to the approaching nucleophile. On these considerations, we anticipate decreasing reactivity with increasing alkyl substitution as displayed by the following series:

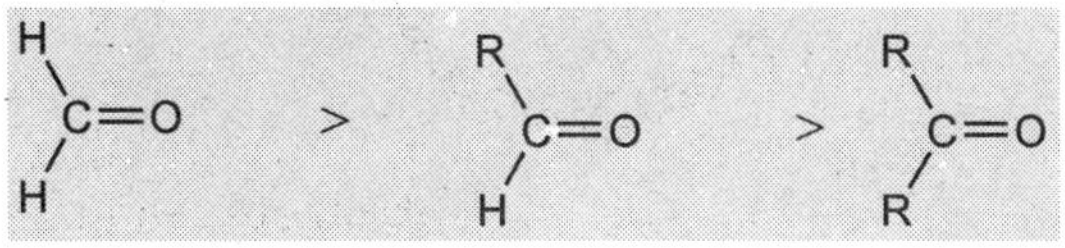

Similar influences are exerted by an aryl substituent as delocalized π orbitals of the ring act as an electron source. As may be expected, electron-withdrawing α-substituents such as $-NO_2$ or halogen would increase the positive character of the carbonyl carbon, and thus facilitate the attack of nucleophiles. It is, therefore, not surprising that compounds such as **1** and **2** add nucleophilic reagents more readily than acetaldehyde.

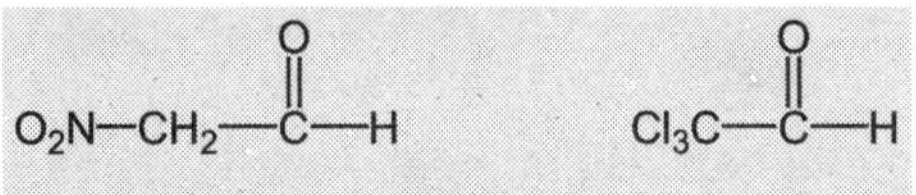

1 **2**

A. Reaction with Water and Alcohols

The carbonyl group of aldehydes and ketones reacts with water producing *gem*-diols which are generally too unstable to be isolated.

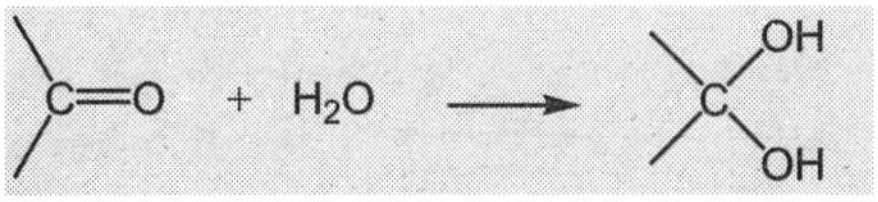

Their existence has been confirmed by UV and IR spectroscopy. Exchange with water containing excess ^{18}O provides a convenient method of deciding the amount of the hydrate formed as the hydrate once formed, will lose $H_2{}^{18}O$ and H_2O with equal ease.

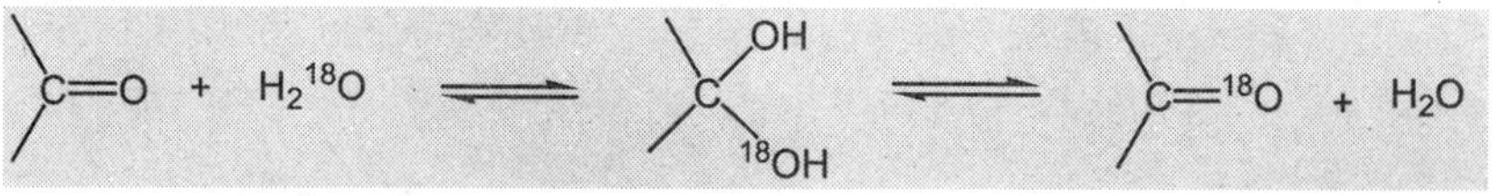

Decreasing reactivity of the carbonyl group with increasing alkyl substitution is reflected in the following series of compounds, which show progressively diminishing tendency to form hydrates when treated under similar conditions.

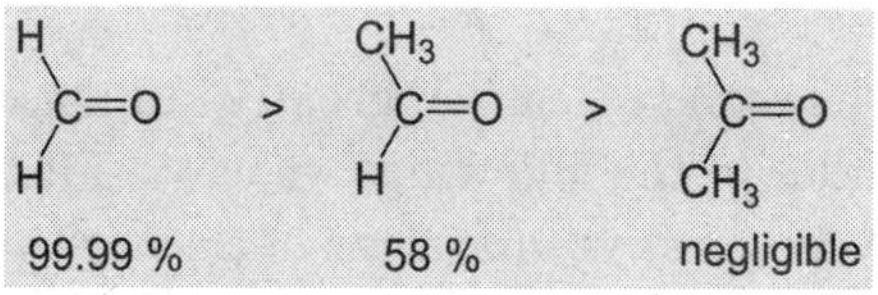

The hydration reaction is subject to both acid and base catalysis.

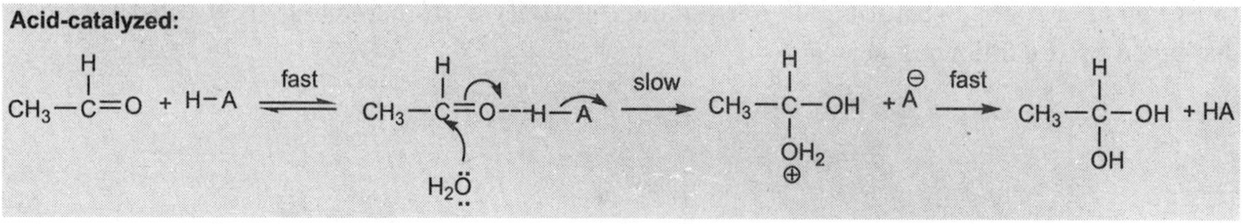

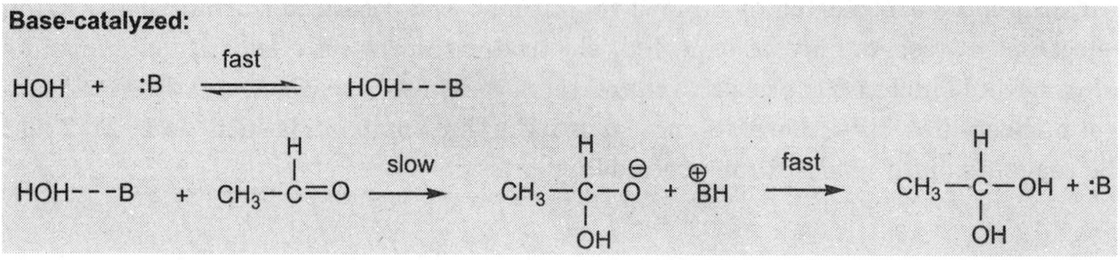

In some cases, the *gem*-diols are isolable, especially when the carbonyl compounds are heavily substituted with the electron-withdrawing substituents. The common examples of stable *gem*-diols are chloral hydrate (**3**) and ninhydrin (**4**).

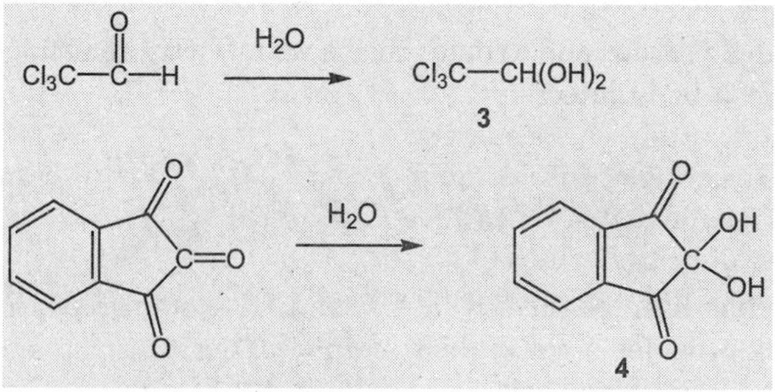

3

4

The stability of these diols can, at least partly, be ascribed to the combined electron-with drawing effect of the substituents that prevent departure of a hydroxyl ion to generate the carbonyl compound.

Alcohols add to aldehydes in a similar fashion producing hemiacetals **5** which are subsequently converted to acetals **6**. In acidic medium the reaction takes place in the following steps:

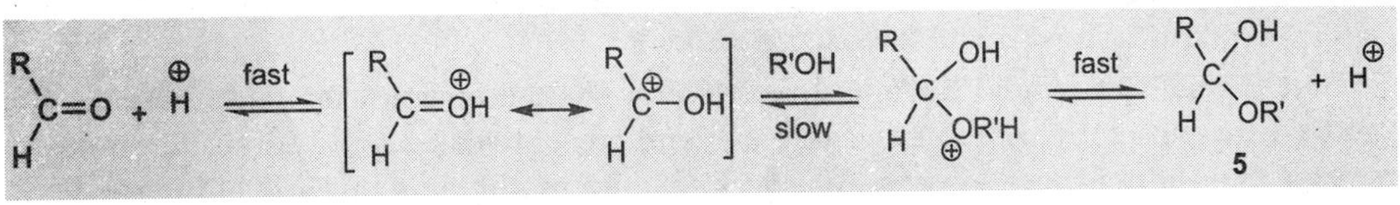

Protonation of the hydroxyl oxygen in **5** and the subsequent removal of water followed by the attack of a second alcohol molecule on the resonance-stabilized carbocation leads to the formation of an acetal **6**.

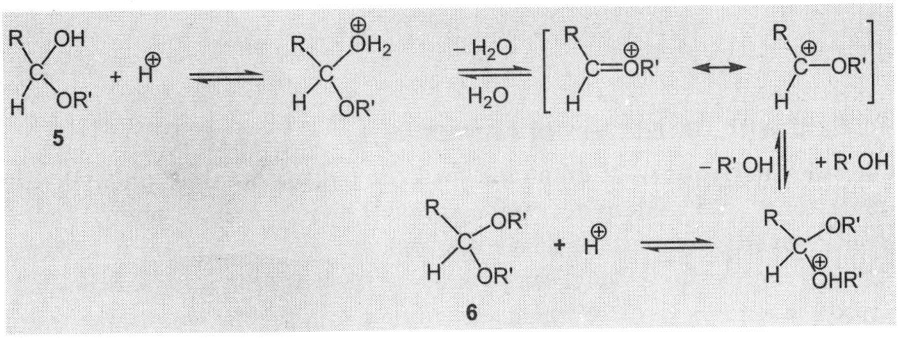

There is an equal probability of protonation of the more basic alkoxy oxygen in the hemiacetal molecule. This, indeed, does take place, and results in the regeneration of conjugate acid of the aldehyde by strong acid.

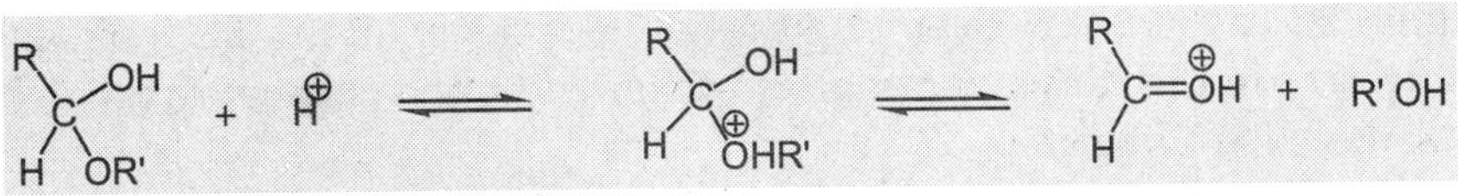

Intramolecular hemiacetal formation resulting in a cyclic structure is of great importance in sugar chemistry. These cyclic structures are interconvertible in solution, and give rise to a phenomenon termed as *mutarotation,* which can be defined as a change in the optical rotation of the solution till it attains a constant value. For instance, either form of glucose establishes an equilibrium consisting of 64% of the β-anomer and 36% of the α-anomer with a very small amount of free aldehyde (0.024%). As a result, optical rotation of any isomer of glucose becomes constant at the equilibrium value of +52.50°.

Mutarotation of glucose is subject to both general acid and general base catalysis. The opening of the cyclic hemiacetal appears to be a concerted process involving the simultaneous removal of a proton from the acetal –OH group and donation of the proton to the ethereal oxygen atom. Since water has both acidic and basic properties, mutarotation proceeds more readily in aqueous solution.

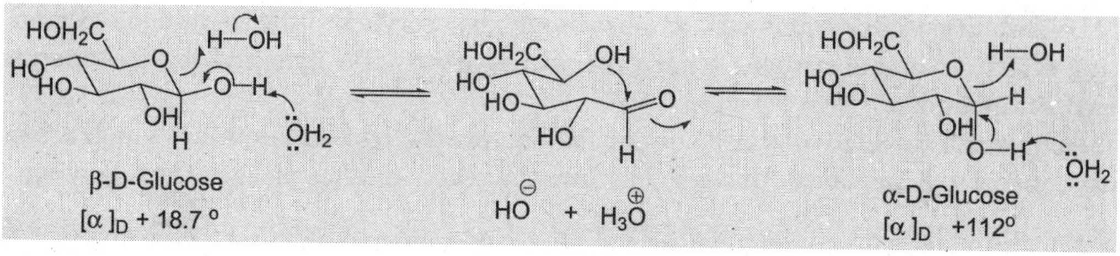

That the solvent should have both acidic and basic properties for mutarotaion is most elegantly exemplified by pyridine (weak base) and cresol (weak acid). Mutarotation does not take place in either of these solvents, but does so quite readily in a mixture of the two.

Similarly, ketones on reaction with alcohols lead to the formation of corresponding ketals.

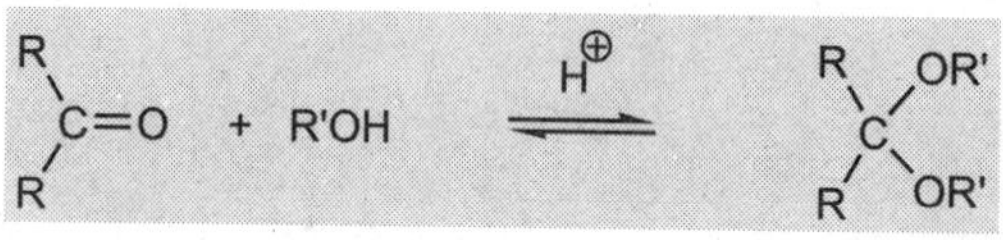

B. Condensation with Derivatives of Ammonia

Amines and other derivatives of ammonia such as hydroxylamine, phenylhydrazine and semicarbazide react with aldehydes and ketones to yield oxime, phenylhydrazone and semicarbazone, respectively. Most of these reactions are subject to general acid catalysis and are found to be first order in carbonyl compound and first order in ammonia derivatives.

Spectroscopic studies have revealed that under mild acidic conditions there is a *fast* disappearance of the carbonyl function followed by a *slow* appearance of the product. The rate of reaction is increased by an increase in acidity but beyond certain limits the rate of reaction falls rapidly with further increase of acidity. It has been found that the rate of condensation of an ammonia derivative with an aldehyde or ketone is maximum at about *pH* 5. Under mild acidic conditions these data have been interpreted in terms of an initial, rapid addition of an ammonia derivative to the carbonyl compound followed by the acid-catalyzed dehydration of adduct in the rate-controlling step.

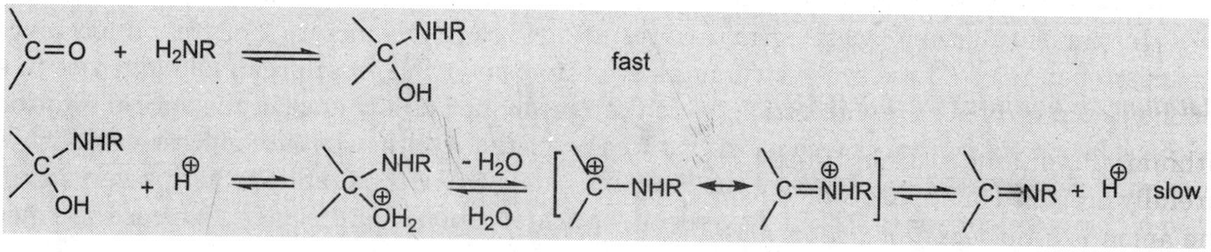

Obviously, an increase in acidity would promote the reaction because it increases the rate of the dehydration step. But with a further increase in acidity, the nucleophile (RNH_2) may get protonated (RN^+H_3) and, thus deprived of its lone pair of electrons, it is no longer in a position to attack the positive carbonyl carbon. Thus the rate of the reaction diaplays dependence on the *pH* of the reaction medium as shown in **Fig. 9.1.**

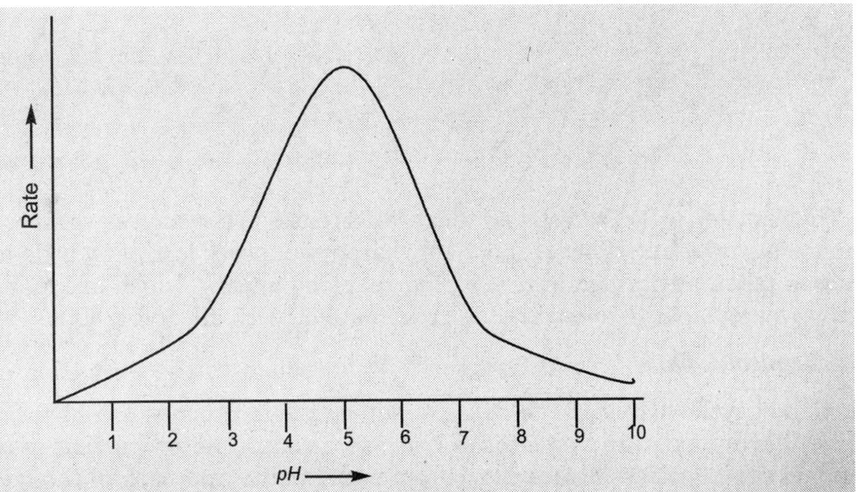

Fig. 9. 1 Rate of condensation of an ammonia derivative with an aldehyde or a ketone as function of *pH*

The reaction of carbonyl compounds with secondary amines such as pyrrolidine, piperidine, and morpholine gives enamines which have found synthetic use in the introduction of a primary alkyl group at the α-position to the carbonyl. For example,

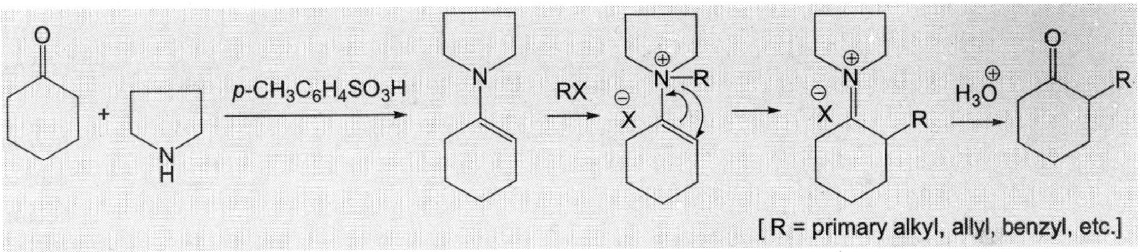

C. Reaction with Hydride Donors

I. Lithium Aluminium Hydride (LAH)
II. Sodium Borohydride ($NaBH_4$)
III. Meerwein-Ponndorf-Verley Reduction
IV. Cannizzaro Reaction

I. Lithium Aluminium Hydride (LAH)

Lithium aluminium hydride ($LiAlH_4$) is sensitive to protic solvents and, therefore, used in carefully prepared nonhydroxylic solvents. It may be regarded as derived from a metal hydride. The anion is a nucleophilic agent, and normally attacks the polarized multiple bond such as C=O, C=N, C≡N, and N=O at the more positive atom. All the four hydrogen atoms in the anion may be successively used in reductions as shown in the following scheme.

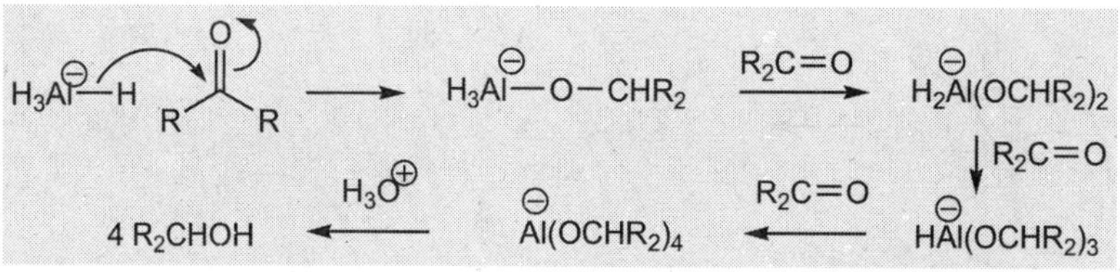

The first step is the fastest, and the rate decreases in the successive steps since the electron-withdrawing inductive effect of the alkoxyl groups opposes loss of hydride ion. This reactivity order was taken advantage in developing reducing agents, which are less reactive but more selective, by replacing two or three hydrogen atoms with alkoxyl groups.

Stereochemical Aspects

Reduction of cyclic ketones such as substituted cyclohexanones can give rise to two epimeric alcohols, having axial and equatorial hydroxyl group. The proportion of these epimers will depend on two opposing factors: steric hindrance to the approach of the metal hydride ion to the carbonyl group, called *steric approach control* (*S.A. control*), and the relative stability of the final product, called *product development control* (*P.D. control*). A rigid system like the *trans*-β-decalone derivative, as shown below, gives predominantly the equatorial (*P.D. control*) or the axial (*S.A. control*) epimer, depending on the size of the groups R^1 and R^2, responsible for 1,3-diaxial steric interaction in the event of axial approach of the reagent.

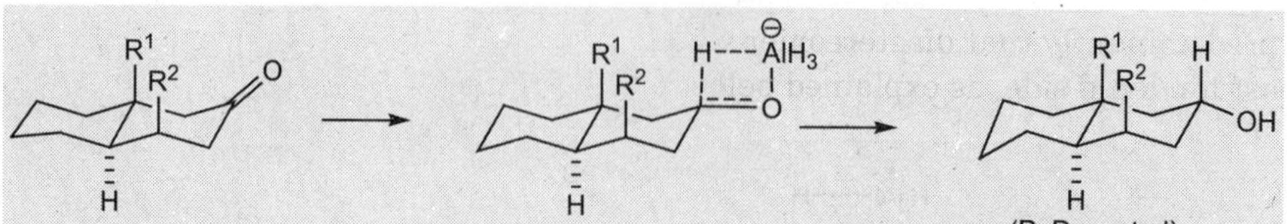

(P. D. control)

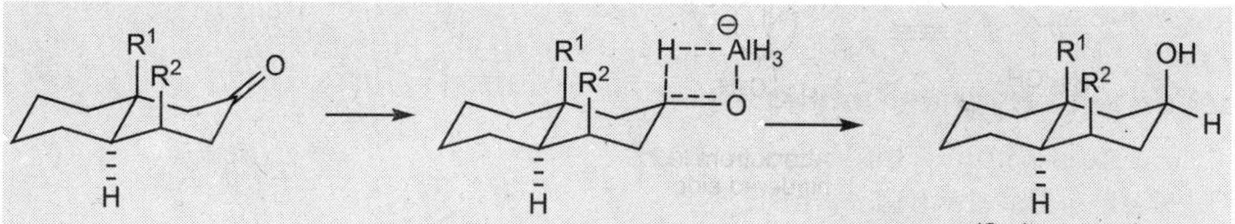

(S. A. control)

The predominant reduction of camphor to the less stable isoborneol is clearly due to steric hindrance to the approach of reagent to hindered *exo* side of carbonyl group.

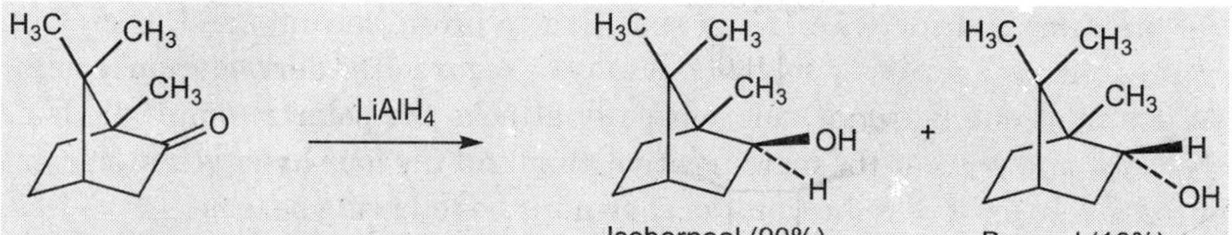

Isoborneol (90%)
(S. A. control)

Borneol (10%)

But, normally, LAH reduction gives a mixture of alcohols in which the more stable alcohol predominates as in the reduction of 4-*tert*-butylcyclohexanone.

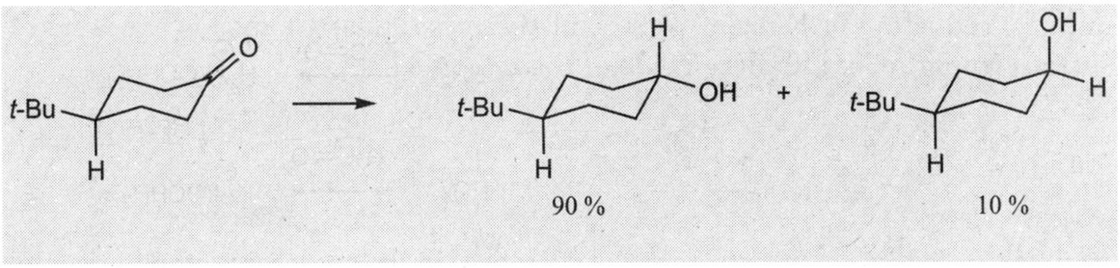

The proportion of the more stable epimer may be increased if $AlCl_3$ and an excess of the ketone to be reduced are present. Thus, equilibration to the more stable epimer takes place through a step very much like Ponndorf reduction. $AlCl_3$ possibly polarizes the carbonyl group for efficient hydride transfer.

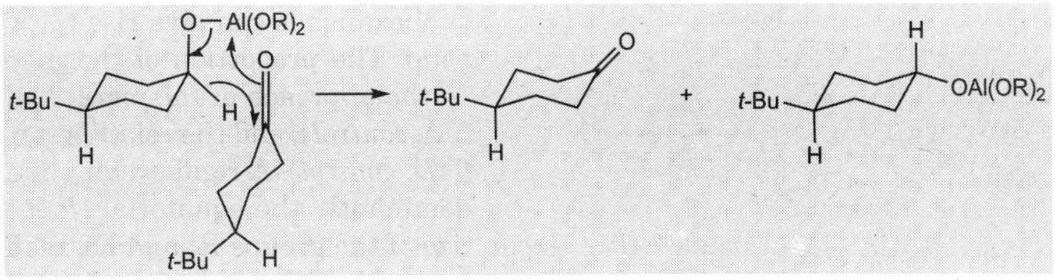

The asymmetric reduction of an optically active ketone having three different α-substituents gives predominantly that diastereomer which is formed by the attack of hydride reagent from the least hindered side, as explained below:

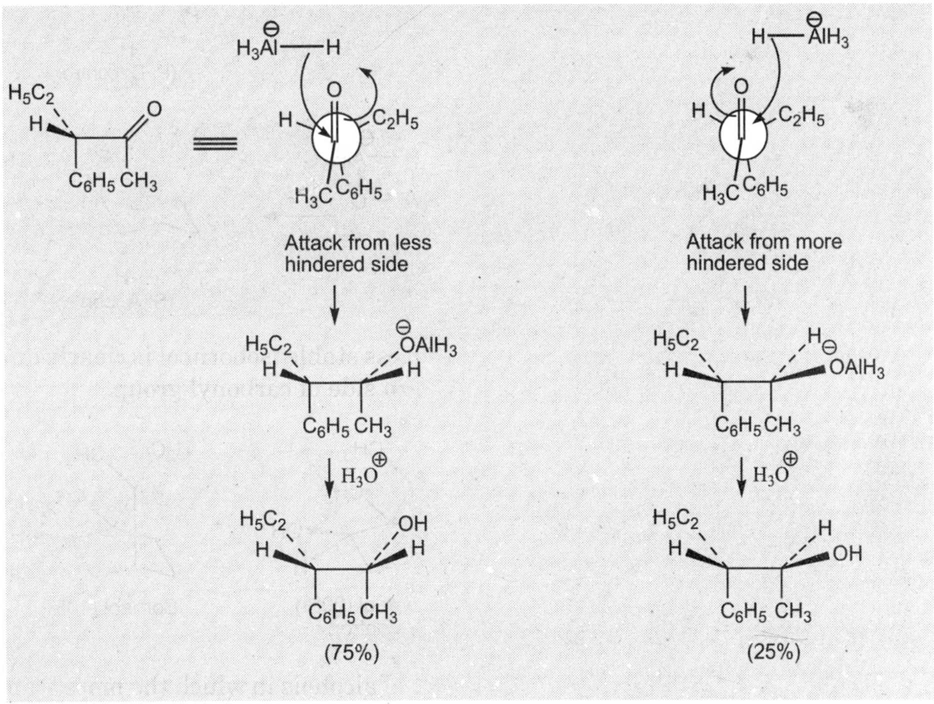

Similarly, reduction of benzoin to 90% of the *meso* diastereomer is explained by metal hydride attack from the less hindered side of the intermediate.

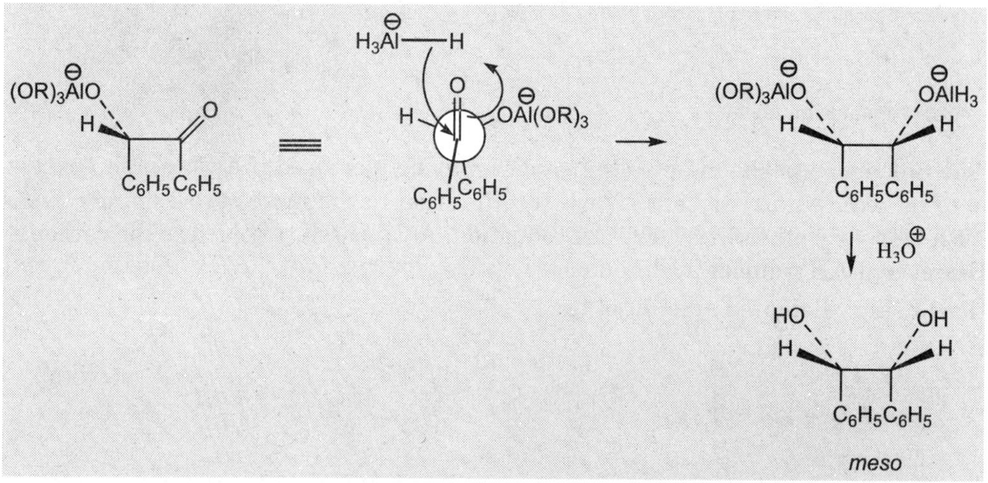

LAH is a versatile reagent capable of reducing a variety of functional groups, as illustrated below.

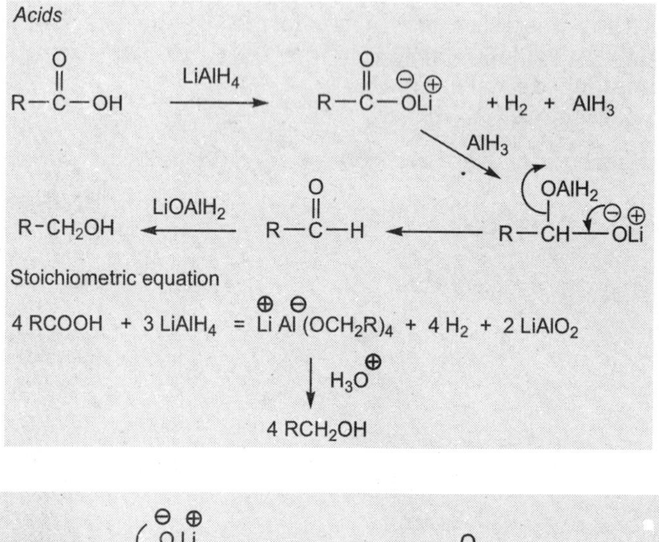

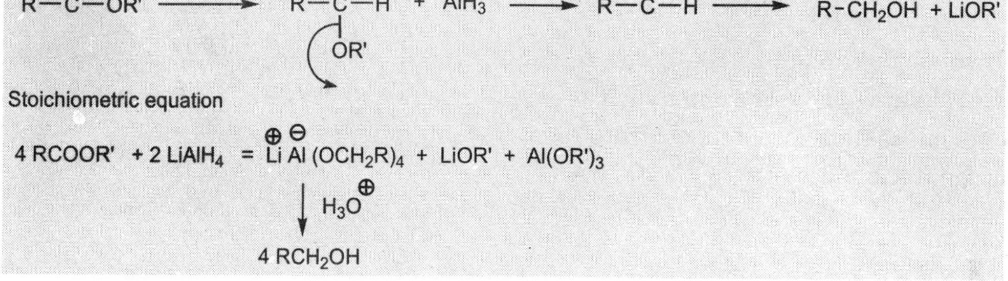

Amides

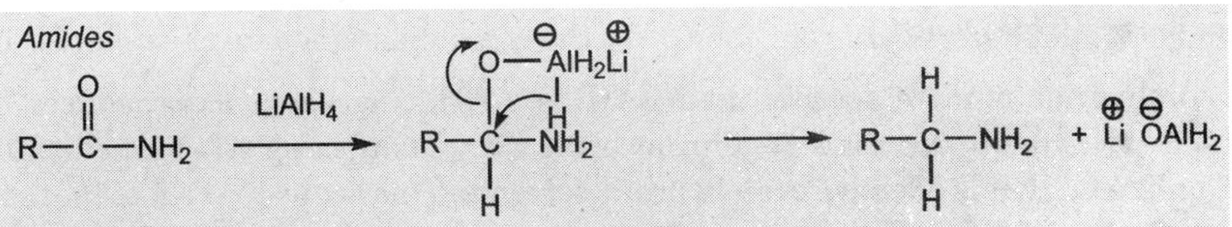

Stoichiometric equation

$$2\,RCONH_2 + LiAlH_4 = 2\,RCH_2NH_2 + LiAlO_2$$

Similarly, nitriles are also easily reduced by LAH to corresponding amines.

However, LAH reduces acylimidazoles and acylaziridines to aldehydes under controlled conditions using calculated amount of the reagent.

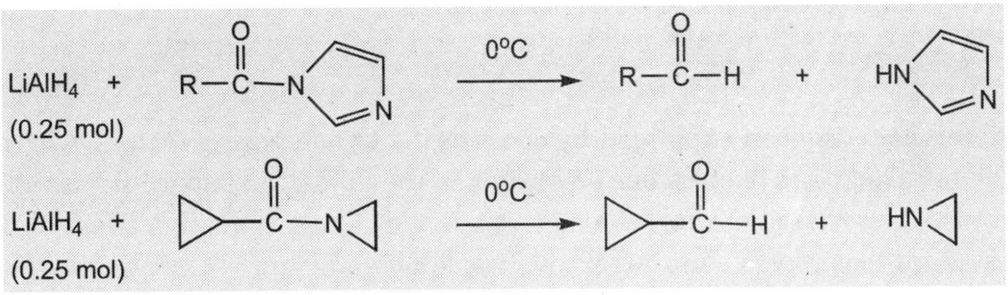

LAH normally does not reduce olefinic double bond, although exceptions are known. For instance, cinnamaldehyde when treated with excess of LAH gives an excellent yield of 3-phenylpropanol. This has been explained by assuming the following sequence of reactions involving a cyclic intermediate after the second stage of reduction.

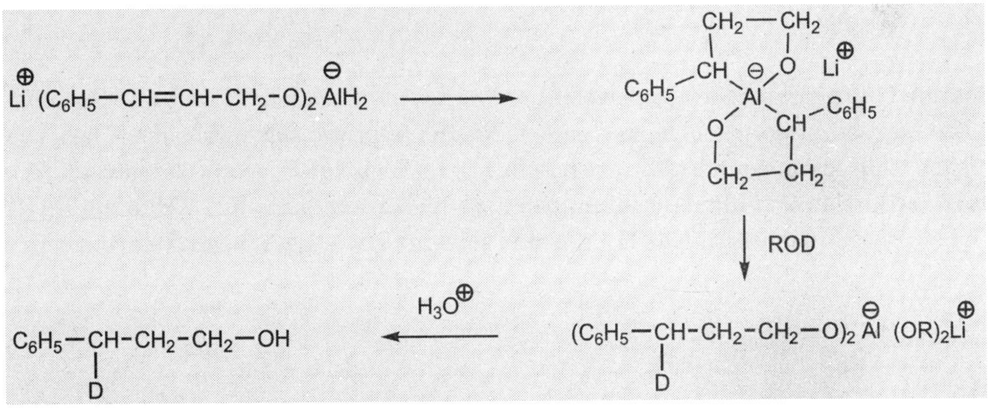

Inverse addition of LAH at lower temperature can avoid such reduction of C=C bonds.

LAH hydrogenolyzes groups such as oxides, sulphonates of alcohols and alkyl bromides by an S_N2 displacement mechanism. Thus, oxides are opened by the attack on the less substituted carbon for steric reasons.

II. Sodium Borohydride ($NaBH_4$)

Sodium borohydride is more specific than LAH as a reducing agent as it reduces only the ketones and aldehydes to the corresponding alcohols without affecting other functional groups. Further, unlike LAH, it is effective even in protic solvents. This is due to the fact that reduction of carbonyl group, which is fast, can be effected in protic solvents. Sodium borohydride also reduces lactones and acid chlorides, but not esters.

The mechanism of reduction by $NaBH_4$ runs parallel to that of LAH, with the difference that the first step in the former is the slowest and rate determining.

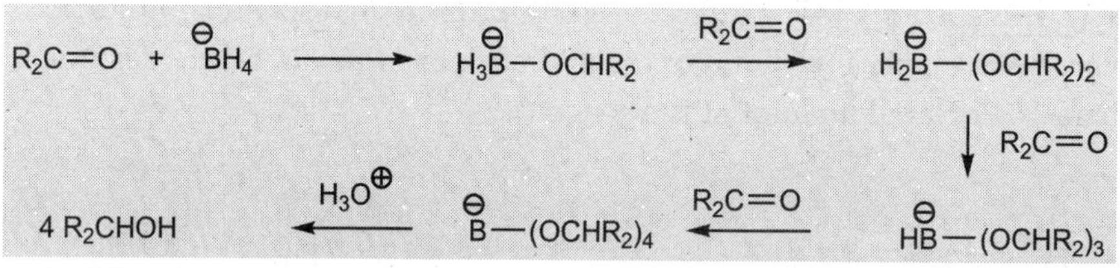

This difference has been explained by considering two opposing factors. The mesomeric effect of two unshared pairs of electrons on oxygen of the alkoxyl group would assist the loss of hydride ion, while, as stated earlier, the electron-withdrawing inductive effect of the alkoxyl group would retard the process. For the relatively small boron atom substantial *p* orbital overlap of the type depicted below is possible, whereas in the case of larger aluminium atoms, this type of overlap is less effective. So, the rate-retarding inductive effect of the alkoxyl group becomes more dominant in the latter case.

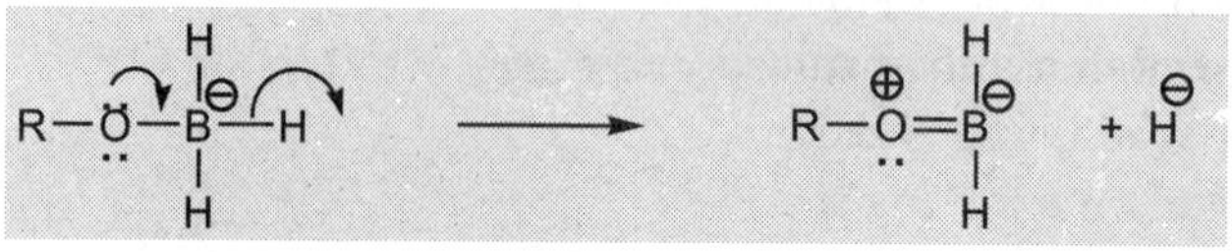

Sodium borohydride is a less powerful reducing agent than LAH, but has the advantage that it can be used in protic solvents. Indeed, $NaBH_4$ in protic solvents, sometimes, produces more highly reduced compounds than are possible by LAH or $NaBH_4$ in aprotic solvents. A case in point is the reduction of *N*-alkyl pyridinium salts, which are reduced by LAH to the 1,2-dihydro stage, whereas reduction with $NaBH_4$ in protic solvents gives 1,2,5,6-tetrahydropyridines.

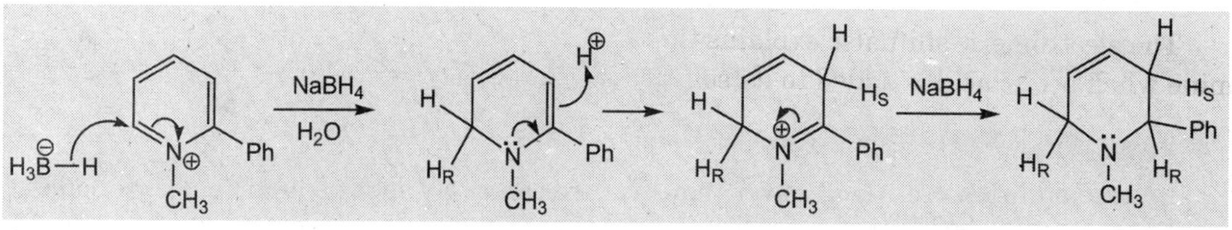

(H_R signifies hydrogen from the reagent borohydride, H_S that from the solvent, and H the original aromatic hydrogen).

The uptake of proton from the solvent is made possible by the *electromeric* shift.

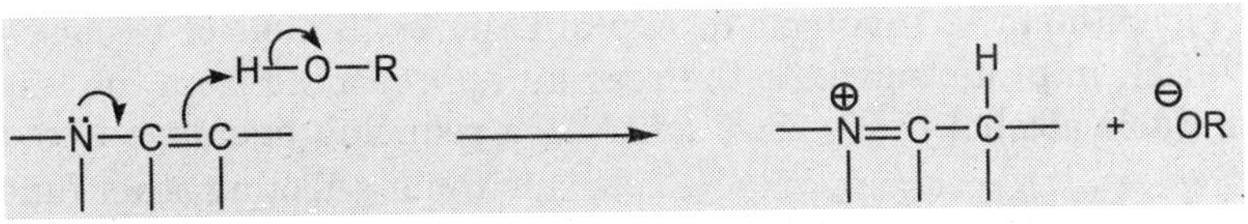

The pyridine ring can be completely reduced if the susbstituent on the nitrogen atom is large, such as *iso*propyl in the following example.

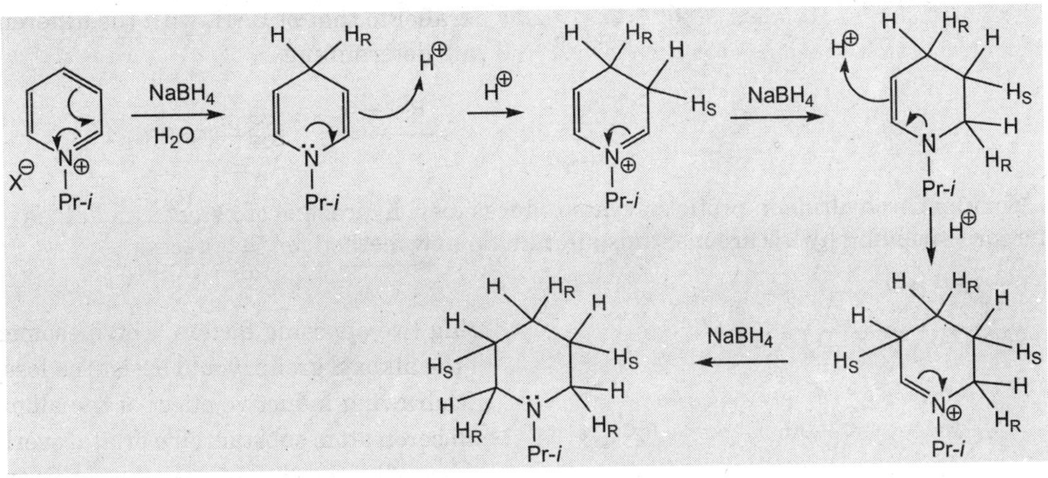

It appears that the large group on the nitrogen hinders the transfer of hydride ion to the 2-position. The potential positive charge at the 2-position, therefore, undergoes mesomeric shift to the 4-position to combine with hydride ion. Then the fully reduced pyridine is obtained by a double sequence of proton transfer and hydride attack. Thus, 2,4,6-trideuterated piperidine derivative can be prepared using $NaBD_4$.

The above mechanism explains the following reduction by formic acid-*d,* which is thus a hydride transfer reagent.

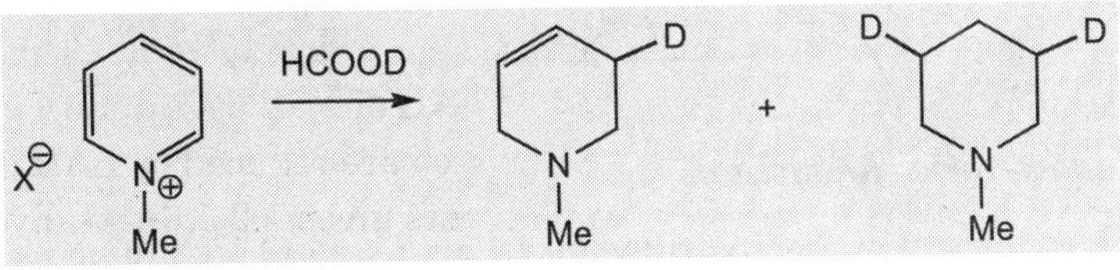

The electromeric shift also explains the reduction of an α,β-unsaturated steroidal tertiary amine when acetic acid is added to its solution containing sodium borohydride.

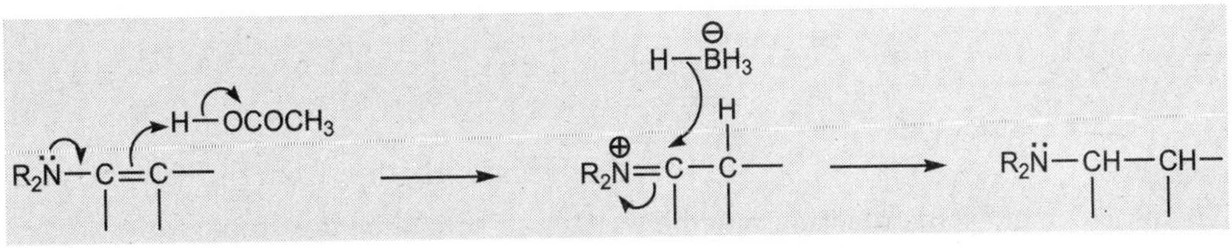

A similar mechanism is involved to explain the reduction of cholest-4-ene-3-one to cholesterol by $NaBH_4$ in protic solvents, the reagent approaching from the less hindered side to give the more stable alcohol.

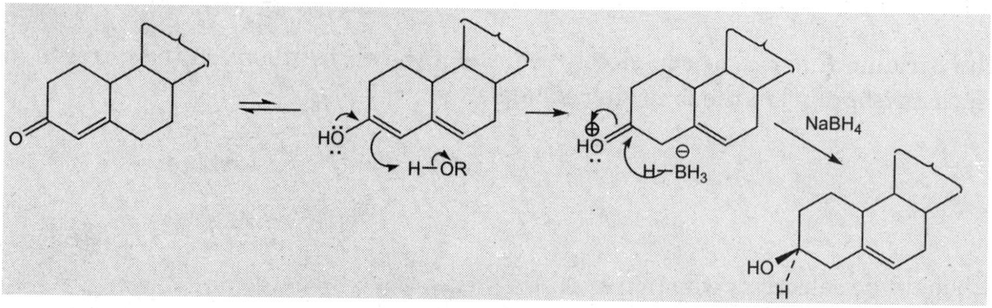

Sodium borohydride in protic solvents reduces tosyl hydrazone of ketones and aldehydes to the corresponding hydrocarbons, thus affording a new method for the process,

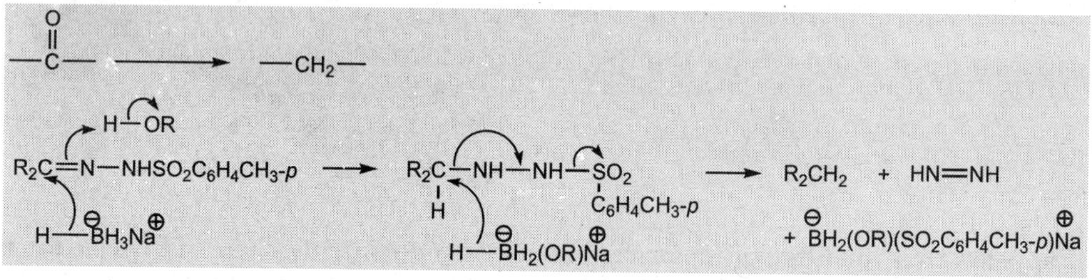

Stereochemical Aspects: Sodium borohydride, like LAH, gives the more thermodynamically stable alcohol in its reduction of acyclic and cyclic ketones, unless steric factors in the approach of the borohydride ion to the carbonyl function become dominant. For instance, norbornanone gives the less stable *endo* alcohol by borohydride attack from the less hindered *exo* side (*SA* control).

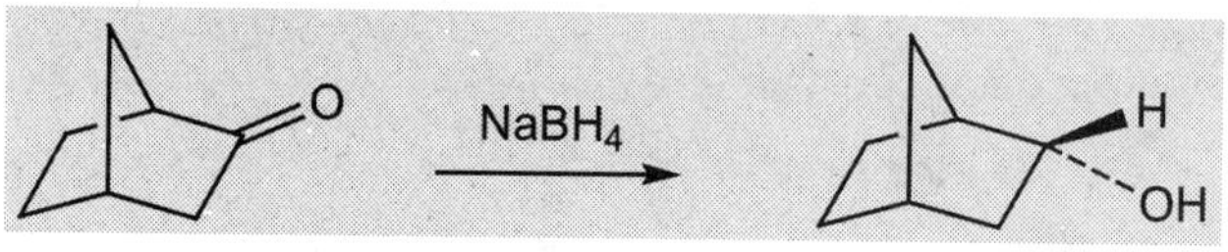

III. Meerwein-Ponndorf-Verley Reduction

This reaction involves the transfer of hydride ions from the metal salts of primary and secondary alcohols to the carbonyl carbon. A common procedure for the reduction of ketones to secondary alcohols consists in heating them in *iso*propyl alcohol in the presence of aluminium *iso*propoxide, and distilling off the acetone from the reaction mixture.

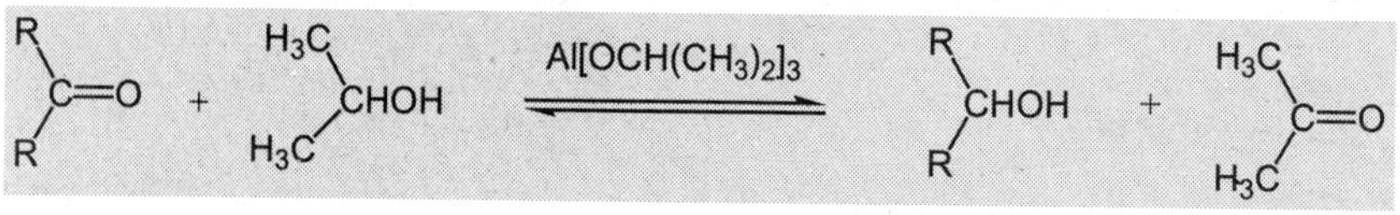

The mechanism of this reaction may be written in the following steps:

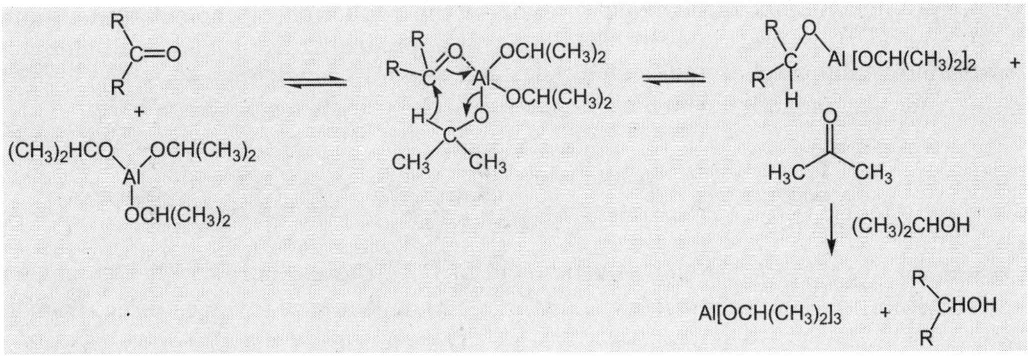

Although the selective reduction of an aldehyde or ketone to the alcohol stage is the rule, reduction of a diaryl ketone system may be pushed to the hydrocarbon stage by using excess of aluminium *iso*propoxide at elevated temperature (compare hydroboration). For instance, when 9,9-dimethyl-10 anthrone is reduced with excess of aluminium *iso*propoxide in xylene, the hydrocarbon product is obtained in good yield. This is explained by the following mechanism.

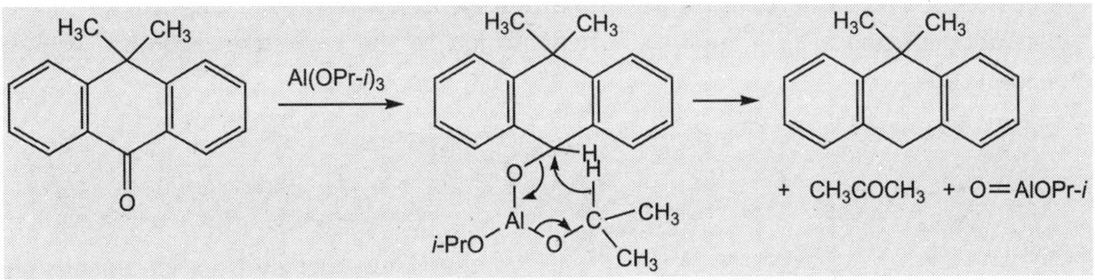

Reduction of an optically active ketone by aluminium *iso*propoxide produces a new chiral centre. In most cases, both diastereomers are formed, often in comparable amounts. For instance, camphor gives a mixture of borneol and isoborneol. But benzoin or benzyl gives 90 per cent of *meso*-dihydrobenzoin as in the case of LAH reduction. This shows that aluminium *iso*propoxide is also sensitive to steric approach control, and preferably approaches the substrate from the less hindered side.

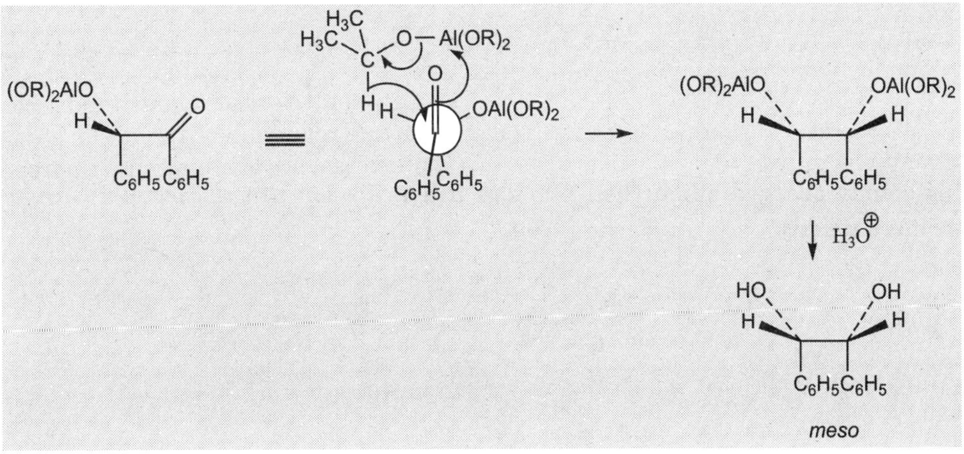

VI. Cannizzaro Reaction

A characteristic reaction of aromatic aldehydes and aliphatic aldehydes without α-hydrogens is the self oxidation-reduction which they undergo in the presence of a strong base.

$$2\,(CH_3)_3CCHO \xrightarrow{\overset{\ominus}{O}H} (CH_3)_3C{-}CO\overset{\ominus}{O} + (CH_3)_3C{-}CH_2OH$$

When the *Cannizzaro reaction* is conducted in D_2O, the alcohol formed has no carbon-bound deuterium, thereby establishing that the reaction sequence involves direct transfer of a hydrogen from one aldehyde molecule to other. The kinetics of the *Cannizzaro reaction* of benzaldehyde in aqueous methanol follows the third order.

$$2\,C_6H_5CHO + \overset{\ominus}{O}H \longrightarrow C_6H_5CO\overset{\ominus}{O} + C_6H_5CH_2OH$$

$$\text{Rate} = k\,[C_6H_5CHO]^2\,[\overset{\ominus}{O}H]$$

Here, an aldehyde molecule produces an anion by interaction with a hydroxide ion which, in the rate-determining step, transfers a hydride ion to the carbonyl carbon of the second aldehyde molecule.

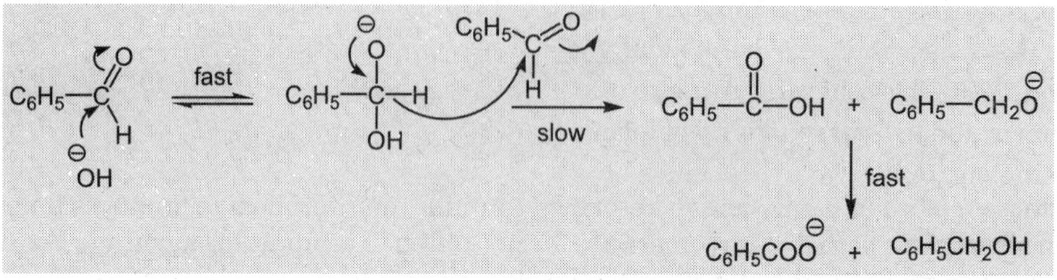

However, *Cannizzaro reaction* of furfural and formaldehyde has been found to be fourth order reaction; rate = k [aldehyde]^2 [$^-$OH]^2 suggesting that a second anion is formed from the first by the transfer of its proton to another hydroxide ion.

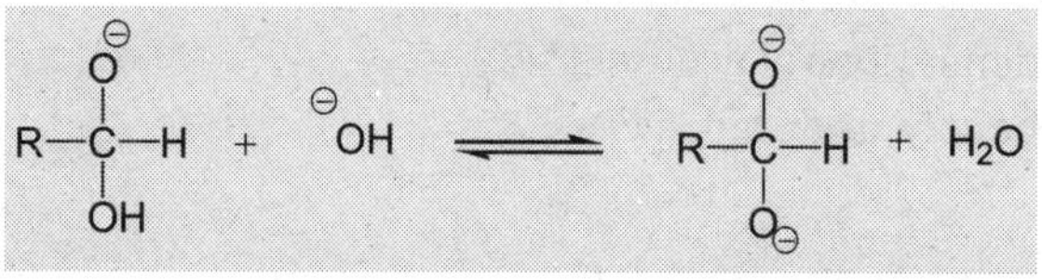

This doubly charged anion then provides a hydride ion to the carbonyl carbon of the aldehyde more readily.

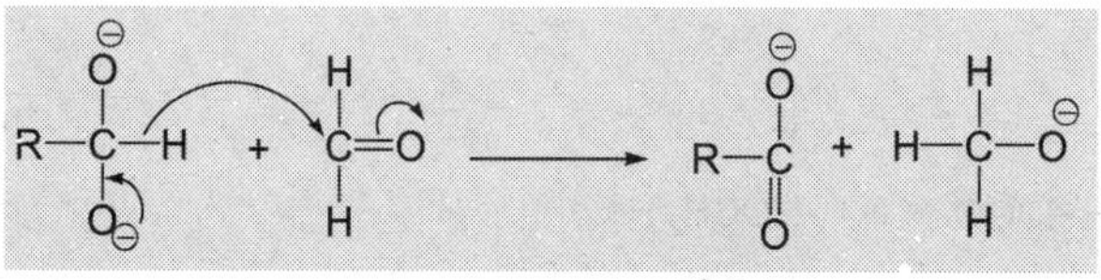

A study of the rate of Cannizzaro reaction of substituted benzaldeydes has revealed that electron-withdrawing substituents increase the reactivity of the aldehydes. This is to be expected since both the molecules of the aldehyde acquire some negative charge in the transition state.

Intramolecular Cannizzaro Reaction

Carbonyl compounds such as 1,2-dialdehydes and α-ketoaldehydes are capable of transferring a hydride ion, and hence, they may undergo internal Cannizzaro reaction. A thoroughly studied example is the conversion of phenyglyoxal **(7)** to the salt of mandelic acid **(8)** under the influence of alkali. It is a second order reaction, first order in hydroxide and second order in phenylglyoxal, and there is no incorporation of deuterium in the product when the reaction is carried out in D_2O. The following mechanism has been suggested in accordance with the above facts.

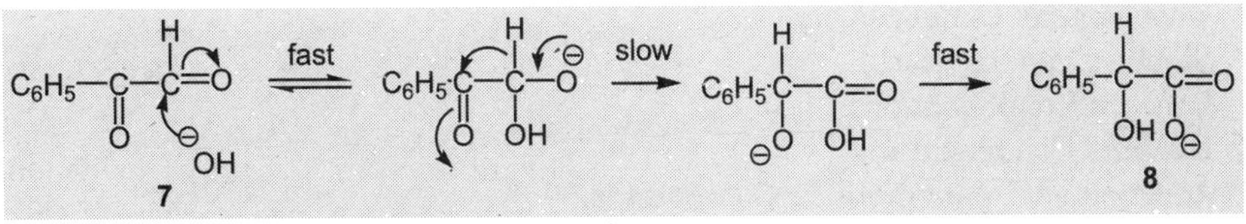

Crossed Cannizzaro Reaction

The Cannizzaro reaction between two dissimilar aldehyde molecules may yield all possible products. However, when one of the aldehydes is formaldehyde, the formate ion and the alcohol corresponding to the other aldehyde are formed exclusively. The exclusive formation of the formate ion is due to high reactivity of formaldehyde, which preferentially adds a hydroxide ion before losing one of its two hydrogens as hydride ion to the less reactive aldehyde.

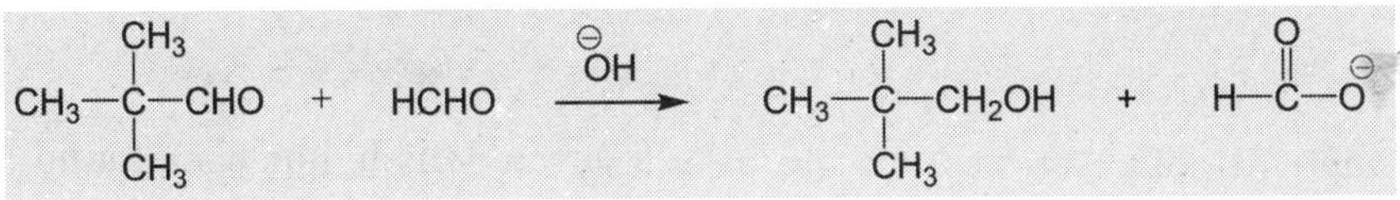

An industrial application of the crossed Cannizzaro reaction, coupled with aldol condensation, is in the synthesis of pentaerythritol **(9)** from formaldehyde and acetaldehyde. As a result of three aldol condensations of acetaldehyde with formaldehyde, trihydroxypivalaldehyde is formed, which subsequently undergoes crossed Cannizzaro reaction with formaldehyde.

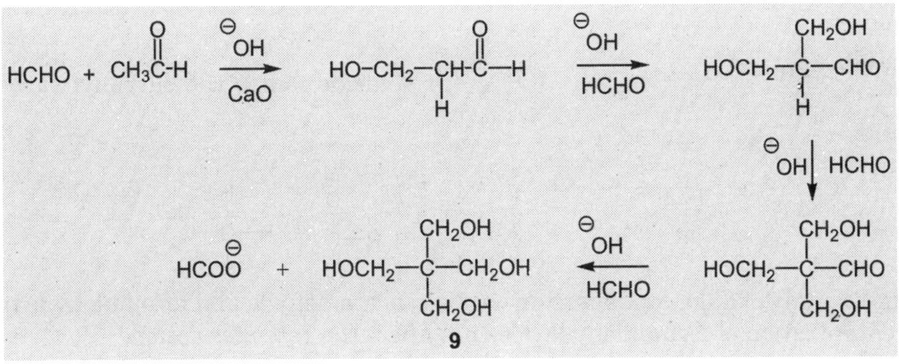

D. Addition of Carbanions

One of the characteristic properties of carbonyl group is to increase the acidity of α-hydrogens by stabilizing the negative charge in the conjugate base through resonance.

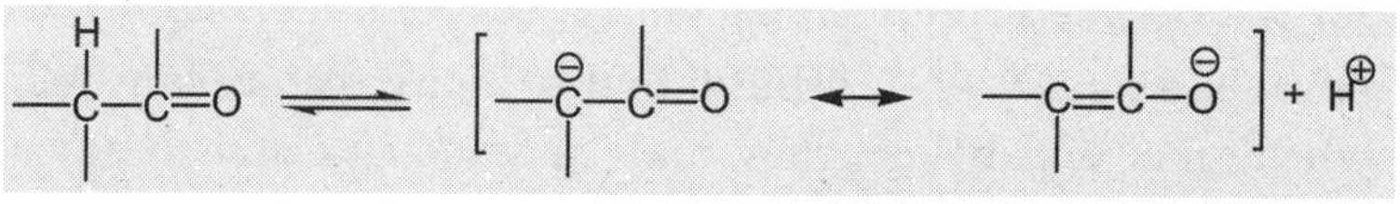

Since part of the charge is carried by carbon, the resonance hybrid is termed as carbanion. It behaves as a nucleophile, and readily attacks the carbonyl carbon of the second molecule. A wide variety of synthetically important reactions involve these intermediates.

I Aldol Condensation
II Perkin Condensation
III Knoevenagel Condensation
IV Stobbe Condensation
V Benzoin Condensation
VI Darzens Condensation
VII Addition of Organometallic Compounds
VIII Wittig Reaction
IX Mannich Reaction

I. Aldol Condensation

Aldol condensation is a condensation between two molecules of an aldehyde or a ketone in the presence of a base to yield a β-hydroxyaldehyde or β-hydroxyketone.

$$2\,CH_3CHO \xrightarrow{\overset{\ominus}{O}H} CH_3CHOHCH_2CHO$$

aldol (ald = aldehyde; ol = alcohol)

Since aldehydes and ketones lacking α-hydrogens do not undergo aldol condensation, a key step of the reaction is the formation of the enolate anion (carbanion) by removal of an α-hydrogen by a base. Like other nucleophiles, the carbanion then attacks the carbonyl carbon of the second molecule, and the anion so formed accepts hydrogen from the solvent molecule to yield the product. In other words, one molecule is converted into a nucleophile by forming its enolate in basic conditions, and the second molecule acts as an electrophile.

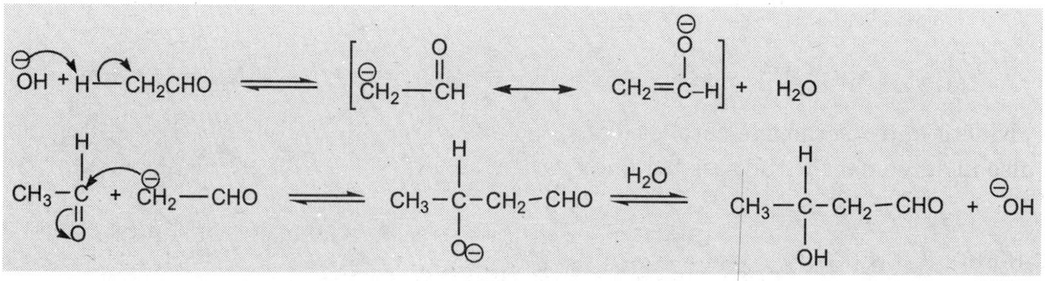

A kinetic study of aldol condensation between acetaldehyde and formaldehyde has shown that the concentration of formaldehyde does not affect the rate of reaction.

$$HCHO + CH_3CHO \xrightleftharpoons{\overset{\ominus}{O}H} HOCH_2CH_2CHO \; ; \; \text{Rate} = k\,[CH_3CHO]\,[\overset{\ominus}{O}H]$$

It follows that the rate-determining step of the reaction involves only acetaldehyde, and as soon as its anion is formed it rapidly attacks the carbonyl carbon of formaldehyde.

$$CH_3CHO + \overset{\ominus}{O}H \xrightleftharpoons{\text{slow}} \overset{\ominus}{C}H_2CHO + H_2O$$

$$\overset{\ominus}{C}H_2CHO + HCHO \xrightleftharpoons{\text{fast}} \overset{\ominus}{O}{-}CH_2CH_2CHO$$

$$\overset{\ominus}{O}{-}CH_2CH_2CHO + H_2O \xrightleftharpoons{\text{fast}} HO{-}CH_2CH_2CHO + \overset{\ominus}{O}H$$

When aldol condensation of acetaldehyde is carried out in D_2O, neither the aldol nor the recovered acetaldehyde was found to contain any deuterium bound to carbon. Thus, the formation of the carbanion is not only the slow step, but is also not reversible; had it been reversible the carbanion would have reacted with D_2O to regenerate a deuterated acetaldehyde that would have yielded a deuterated aldol. However, there is considerable deuterium exchange of acetaldehyde when the reaction is run in a very dilute solution, as under these conditions the carbanion does not easily encounter the carbonyl compound, and hence, reacts with D_2O to generate deuterated acetaldehyde.

In contrast, aldol condensation of acetone follows third order rate law, rate = $k\,[CH_3COCH_3]^2\,[OH^-]$.

This suggests that the attack of anion on acetone is the rate-determining step of the reaction. This is to be expected as carbonyl carbon of acetone is less reactive than that of an aldehyde. The mechanism may be formulated in the following steps.

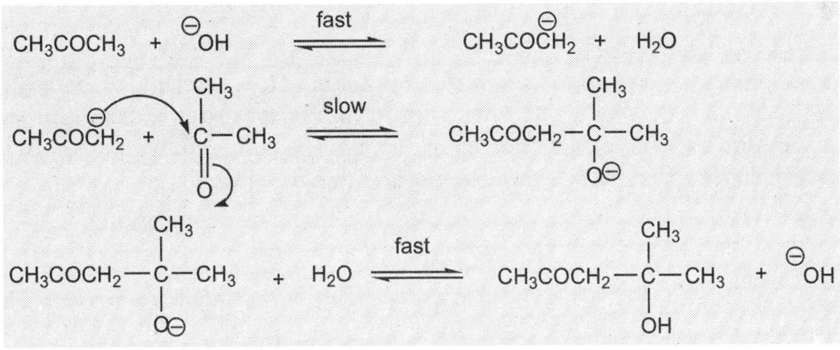

Kinetic data of ethoxide-catalyzed aldol condensation of benzaldehyde with acetophenone are also in agreement with those obtained for acetone.

$$C_6H_5CHO + CH_3COC_6H_5 \xrightarrow{C_2H_5\overset{\ominus}{O}} C_6H_5CHOHCH_2COC_6H_5 \xrightarrow{-H_2O} C_6H_5CH{=}CHCOC_6H_5$$

Rate = $k\,[C_6H_5CHO]\,[C_6H_5COCH_3]\,[C_2H_5O^-]$.

Since dehydration of a β-hydroxy ketone is a fast reaction, the above data indicate that the anion of acetophenone ($^{-}CH_2COC_6H_5$) is formed in a *fast* step, which subsequently attacks on the carbonyl carbon of benzaldehyde in a rate-determining step.

The appearance of a dehydrated product in the above reaction is not unique, since many aldol condensations employing strong bases such as alkoxide or hydroxide ions do yield the dehydrated products. Their mechanism of formation involves an enolate ion, which subsequently eliminates a hydroxide ion.

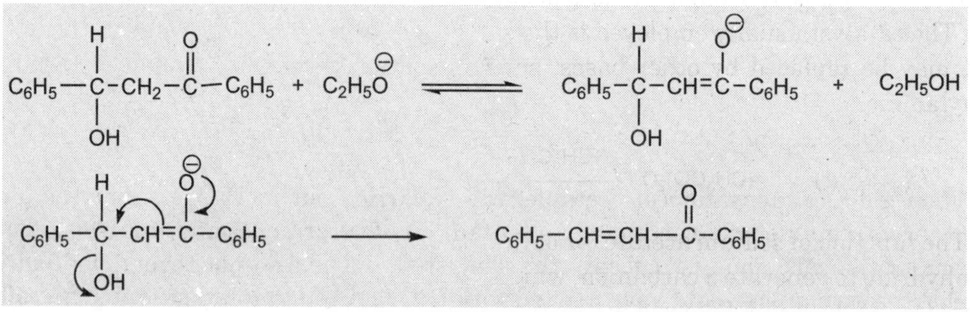

This forms the basis of Claisen-Schmidt reaction, which may be defined as a base-catalyzed condensation of an aldehyde or ketone to form an α,β-unsturated aldehyde or ketone.

Crossed Aldol Condensation

Crossed aldol condensation between two different carbonyl compounds, both having α-hydrogens, yields a mixture of four possible products. However, if one of the reactants has no α-hydrogen, it may function as a substrate for the other enolizable carbonyl compound.

$$C_6H_5CHO + CH_3CHO \xrightleftharpoons{^{\ominus}OH} C_6H_5CH{=}CHCHO + H_2O$$

Acid-catalyzed Aldol Condensation

Aldol condensations are also catalyzed by acids. The reaction of benzaldehyde with acetophenone in the presence of an acid was found to be first order in both the reactants, and it appears reasonable to believe that the acid catalyzes the conversion of acetophenone into the enol form, which then reacts with the protonated benzaldehyde.

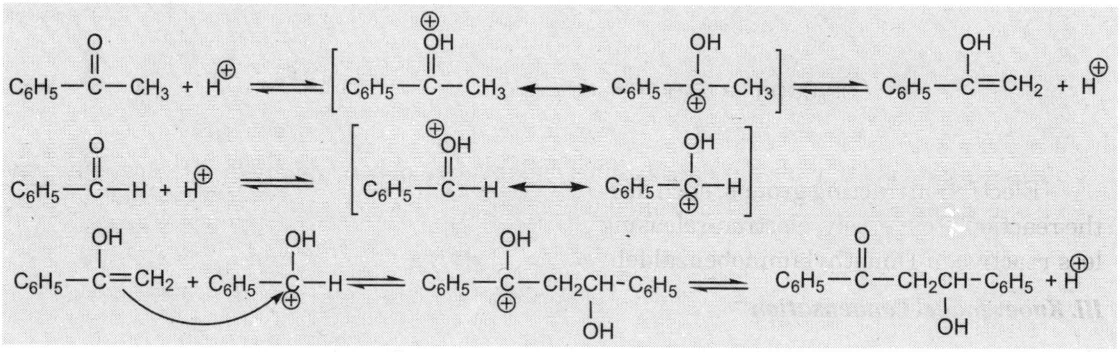

Usually, the aldol products are easily dehydrated under acidic conditions.

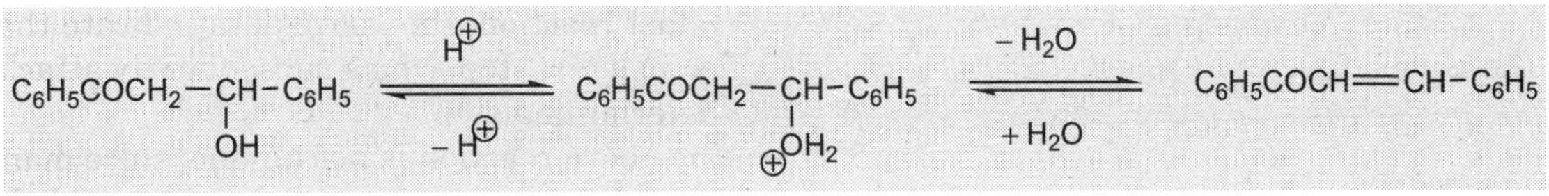

II. Perkin Condensation

Perkin discovered that aromatic aldehydes condense with an acid anhydride to yield cinnamic acids. The catalyst usually employed is the salt of the carboxylic acid related to the anhydride, but it may be replaced by other bases, such as sodium carbonate, quinoline, pyridine, and triethylamine.

$$C_6H_5CHO + (CH_3CO)_2O \xrightarrow{CH_3COONa} C_6H_5CH{=}CHCOOH + CH_3COOH$$

The function of sodium acetate (or of the other bases) is to abstract an α-hydrogen atom of the anhydride to generate a carbanion, which then attacks the carbonyl carbon of the aldehyde. Protonation of the anion followed by the removal of water produces an unsaturated anhydride, which is finally hydrolyzed to form the component acids on being poured into water at the end of the reaction.

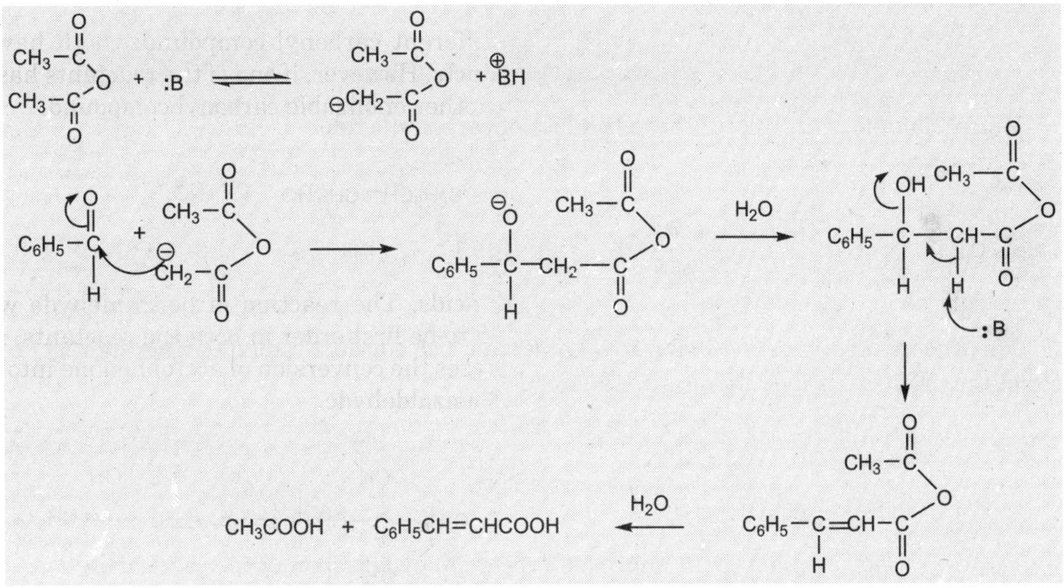

Electron-attracting groups, such as *p*-nitro on the aromatic ring, increase the efficiency of the reaction. Conversely, electron-releasing groups on the aromatic ring make the benzaldehydes less reactive. *p*-Dimethylaminobenzaldehyde, for instance, does not undergo Perkin reaction.

III. Knoevenagel Condensation

Base-catalyzed condensations between an aldehyde or ketone and any compound having an active methylene group are usually called Knoevenagel condensation.

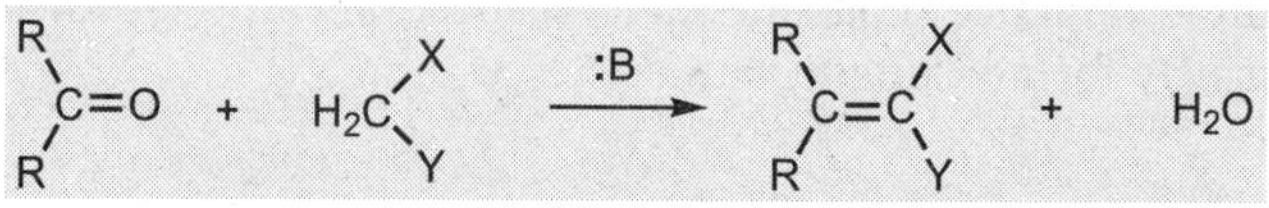

The effectiveness of various activating groups in the active methylene compound is found to be in the order $NO_2 > CN > COCH_3 > COC_6H_5 > COOC_2H_5$. The basic catalyst employed is usually an amine. Since the rate of reaction decreases appreciably on adding acids, it appears certain that the function of the base is to remove a proton from the active methylene compound. The resonance-stabilized carbanion then attacks the carbonyl carbon of the aldehyde or ketone. Subsequent protonation of the anion followed by dehydration yields the product.

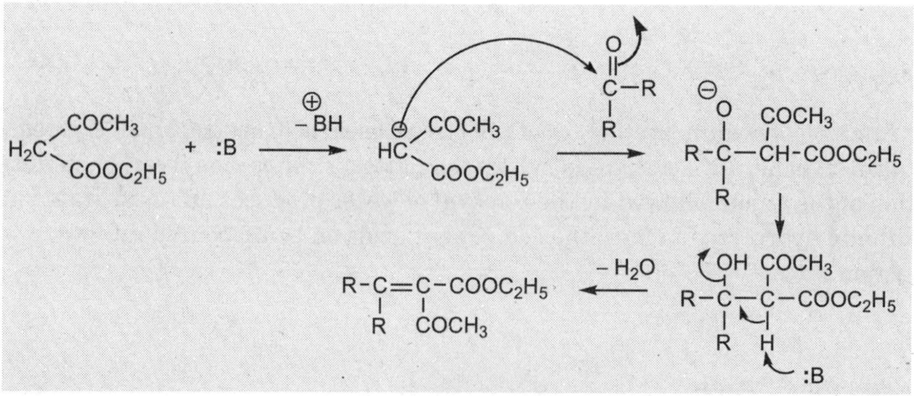

When malonic acid is used, there is a loss of carbon dioxide.

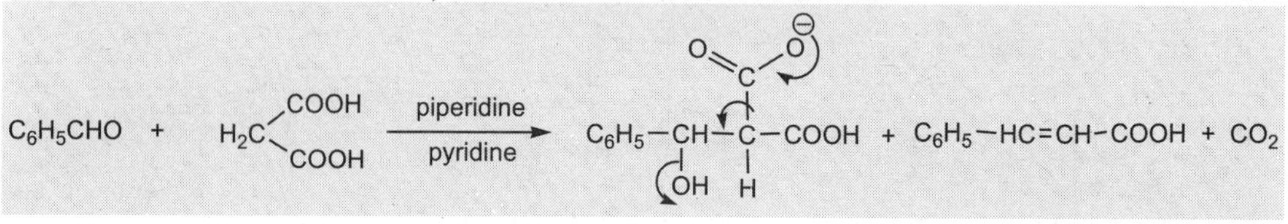

The *Knoevenagel reaction* being reversible, and the equilibrium favouring the reactants, yield of the products is usually not satisfactory.

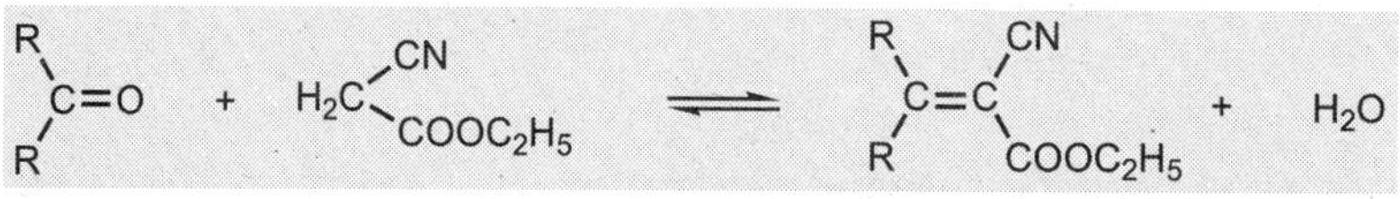

However, Cope introduced an improved procedure by performing the reaction in benzene, and removing the water produced as an azeotropic mixture in the Dean and Stark apparatus, thus, shifting the equilibrium completely to the right. It is, therefore, now known as the *Cope-Knoevenagel reaction.*

Carbonyl compounds containing α-hydrogens can also be condensed successfully, since basic conditions are so mild that self-condensation is rarely a hazardous side reaction. Knoevenagel condensations are sometimes accompanied by the formation of *bis*(active methylene) adducts from one molecule of the carbonyl compound and two molecules of the active methylene component. The available evidence suggests Michael addition of the second molecule of methylene component to the initially formed unsaturated compound.

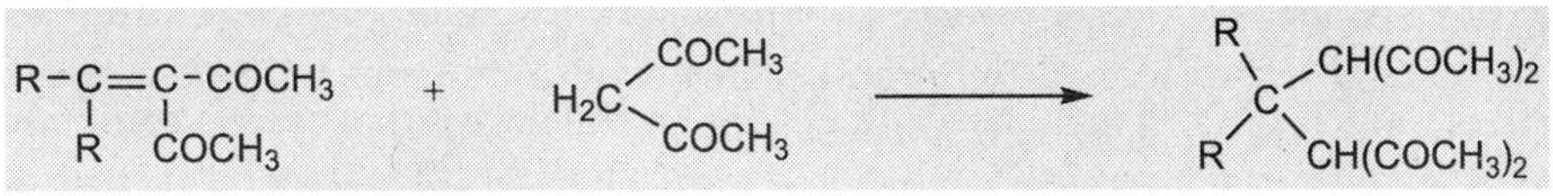

IV. Stobbe Condensation

In the presence of bases such as sodium ethoxide, diethyl succinate condenses with aldehydes and ketones in such a manner that enolate of the ester adds to carbonyl group of the reactant molecule.

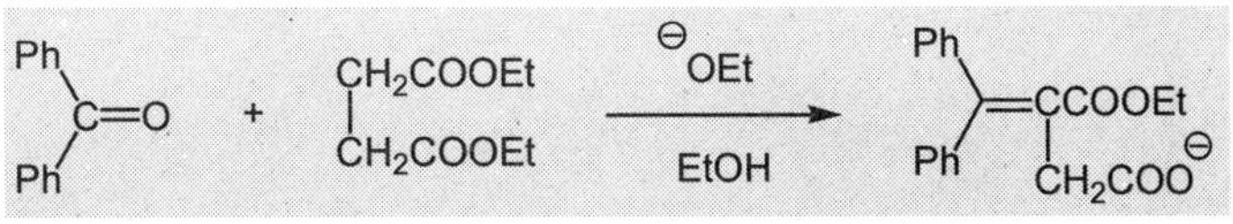

A generalized mechanism for the reaction, known as Stobbe condensation, can be written in the following steps:

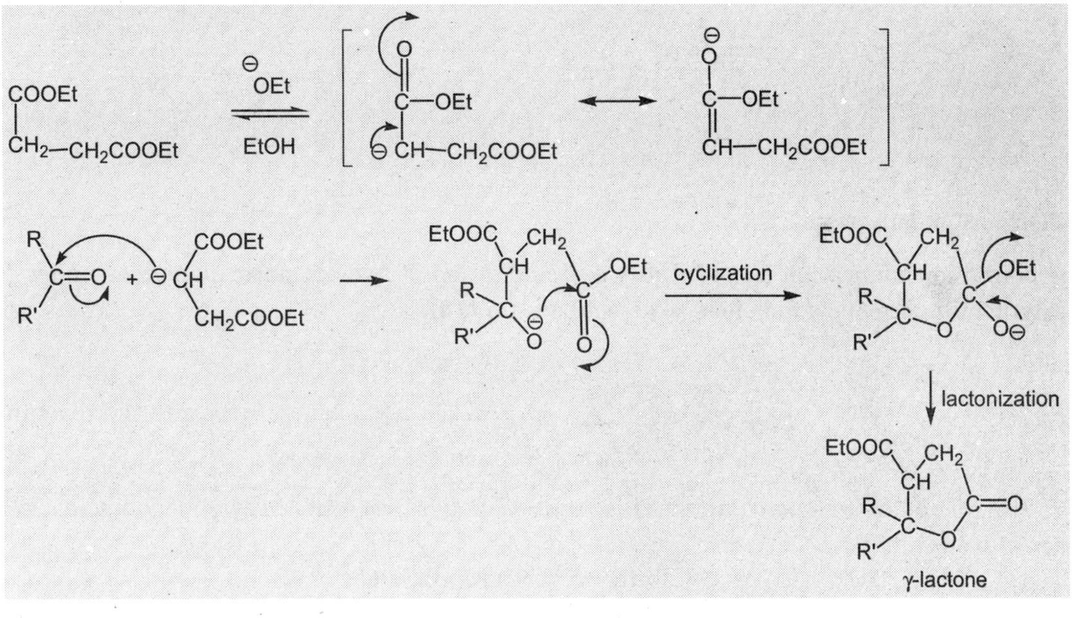

The intermediate γ-lactones have, indeed, been isolated, and they readily undergo base-catalyzed ring-opening to give the corresponding salts of the carboxylic acids.

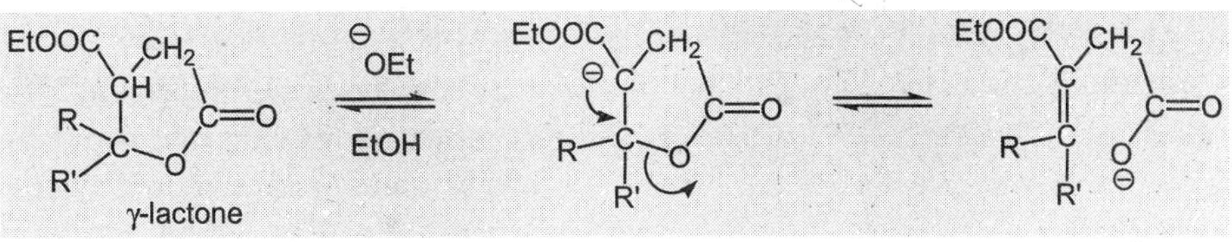

The fact that succinic esters react better than other esters, and one ester group is always cleaved, provides good evidence for the mechanism outlined above.

As expected, Stobbe condensation does not proceed with aldehydes containing α-hydrogens due to the competing self-condensation process. However, aldehydes with no α-hydrogen, such as benzaldehyde, have been successfully employed.

An interesting application of Stobbe condensation is the ring *extension* of aromatic compounds:

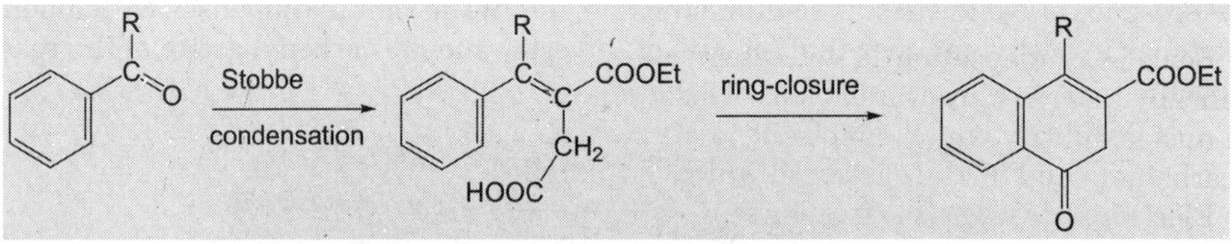

The ketone on reduction gives an alcohol, which readily undergoes loss of water due to the formation of aromatic system.

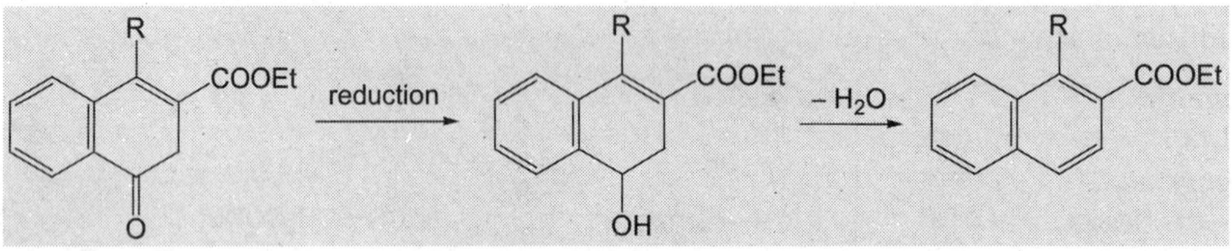

V. Benzoin Condensation

The benzoin condensation is essentially a dimerization of two aromatic aldehydes under the catalytic influence of cyanide ions to give a benzoin **(10)**.

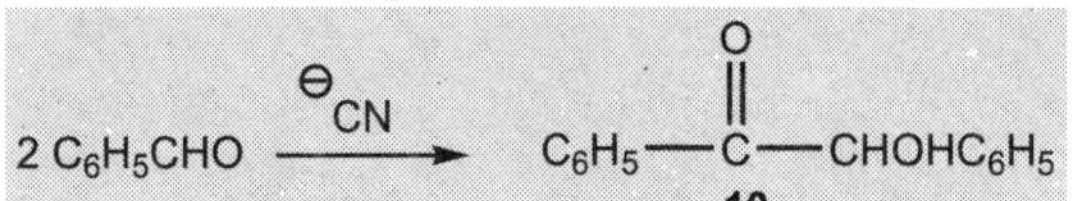

10

Kinetics of the reaction indicates that the reaction is first order in cyanide ion, and second order in benzaldehyde.

$$\text{Rate} = k\,[\text{CN}^-]\,[\text{C}_6\text{H}_5\text{CHO}]^2$$

The hydrogen atom attached to the carbonyl group of aldehyde is not active enough to be removed easily, but the addition of cyanide ion to the carbonyl carbon places this hydrogen in the α-position of the nitrile, thus rendering it relatively acidic. The carbanion, thus generated, attacks the carbonyl carbon of the second aldehyde molecule in a rate-determining step forming an unstable cyanohydrin of benzoin, which immediately breaks down into benzoin and hydrogen cyanide.

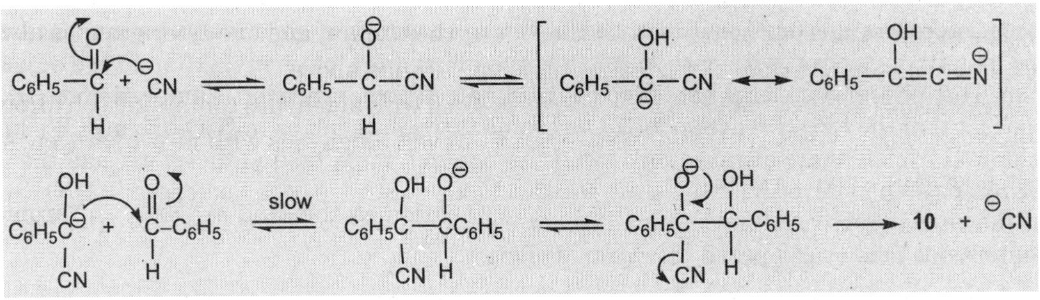

Benzoin condensation is not catalyzed by hydroxide ion or by bases in general, but is specifically catalyzed by cyanide ions. This unique catalytic property of cyanide is due to (a) its good nucleophilicity which enables it to attack the carbonyl carbon, (b) its ability to stabilize the carbanion, and (c) the ease with which cyanide ion departs in the last step.

Electron-attracting substituents are expected to inhibit benzoin condensation as electron-withdrawal from the intermediate carbanion would decrease its nucleophilic chararacter, and thus, subsequent attack on carbonyl carbon would be slow. Accordingly, *p*-nitrobenzaldehyde does not undergo benzoin condensation. Electron-donating substituents such as *p*-$N(CH_3)_2$ and *p*-OCH_3, also retard the reaction, since increase in electron density at the carbonyl carbon makes the addition of anions (^-CN and the carbanion) somewhat more difficult.

(i) Benzilic Acid Rearrangement: Benzoin (**10**) is oxidized by nitric acid to α-diketone, benzil (**11**), which undergoes an interesting base-catalyzed reaction known as *benzilic acid rearrangement*.

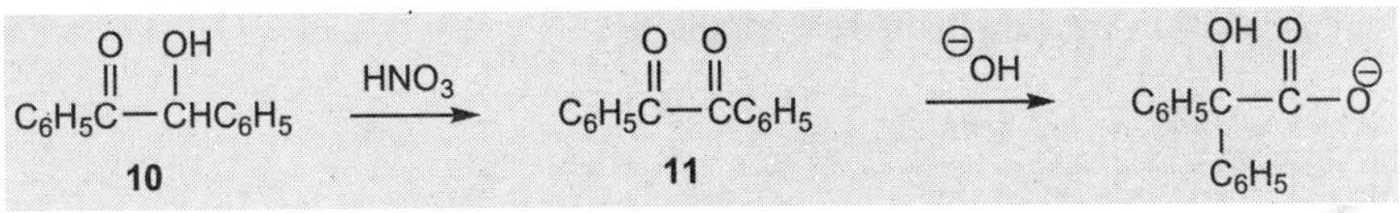

$$\text{Rate} = k\,[C_6H_5COCOC_6H_5]\,[OH^-]$$

If methoxide ion rather than a hydroxide ion is used, methyl ester of benzilic acid $[(C_6H_5)_2C(OH)COOCH_3]$ is obtained. The reaction is not subject to general base catalysis, and when it is carried out in $H_2{}^{18}O$, benzil exchanges ^{18}O at a faster rate than it undergoes the rearrangement. All these observations are in agreement with a mechanism that involves a nucleophilic attack at a carbonyl group by the hydroxide ion (or methoxide ion), which is a reversible and fast step. This intermediate then undergoes a rate-determining rearrangement, which is actually a 1,2-shift of phenyl group to the adjacent carbonyl carbon.

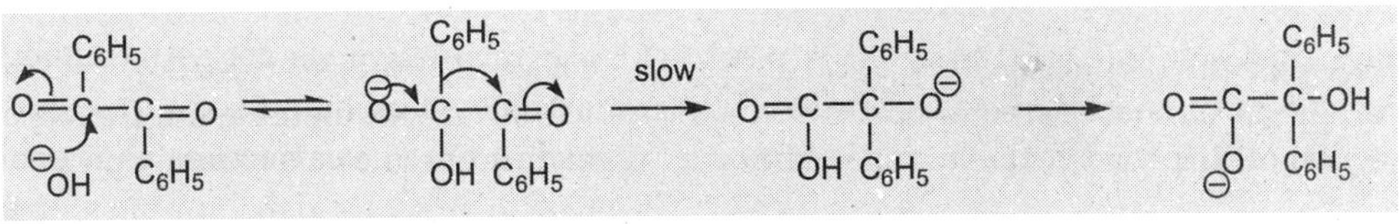

The reaction is similar to intramolecular Cannizzaro reaction of the phenylglyoxal except that phenyl has taken the place of hydrogen. The driving force(*s*) for phenyl migration may be either the partially positive carbonyl carbon, which 'pulls' the phenyl group analogous to other carbocation rearrangement, or it may be the strong tendency of negatively charged oxygen to share its pair of electron with carbon that results in 'pushing' of phenyl group to the adjacent carbon. That proton transfer is not involved in the rate determining step is amply demonstrated by the fact that the rate of reaction in D_2O-dioxane is twice as fast as in H_2O-dioxane. We know that proton transfers are always more rapid than deuteron shifts, and therefore, had the rate determining step involved a proton shift, the reaction would have been decreased in D_2O rather than showing an increase. The observed increase in the reaction rate is due to greater basicity of deuteroxide ions as compared to hydroxide ions.

VI. Darzens Condensation

The Darzens condensation involves the addition of a carbanionic ester, derived from an α-chloro or α-bromo ester by the action of a strong base, to the carbonyl carbon followed by the expulsion of halide ion. Glycidic esters (α,β-epoxy esters) are thus obtained.

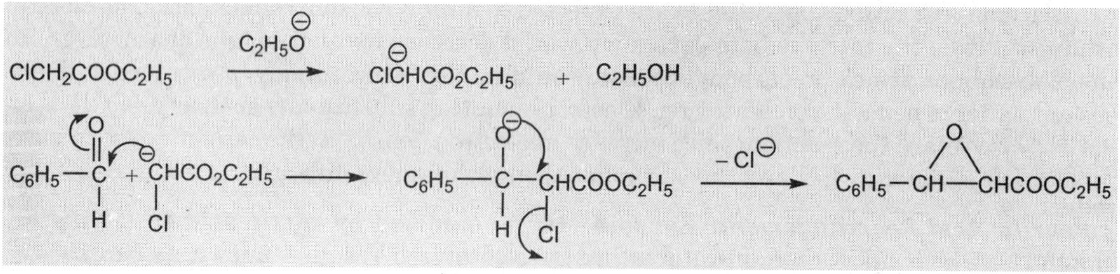

Glycidic esters can be saponified to salts of the corresponding glycidic acids, which readily undergo decarboxylation in the presence of acids.

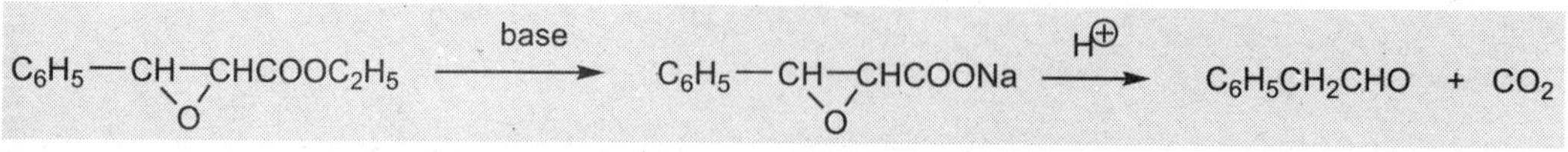

The overall process thus consists of the addition of one carbon to an aldehyde ($C_6H_5CH_2CHO$ from C_6H_5CHO).

VII. Addition of Organometallic Compounds

(i) Grignard Reagents: In ether solutions, Grignard reagents exist in an equilibrium, which may be formulated as

$$2\,RMgX \rightleftharpoons R_2Mg + MgX_2$$

Although these species, RMgX and R_2Mg, are not really ionized; the alkyl groups have considerable carbanion character due to the polar nature of carbon–metal bond. Hence, most of the reaction of Grignard reagents is explained by regarding them as carbanion donors.

$$RMgX \rightleftharpoons \overset{\ominus}{R} + \overset{\oplus}{Mg}X$$

$$R_2Mg \rightleftharpoons \overset{\ominus}{R} + \overset{\oplus}{Mg}R$$

Thus, a straightforward mechanism for the addition of Grignard reagents to a carbonyl compound may be represented as follows:

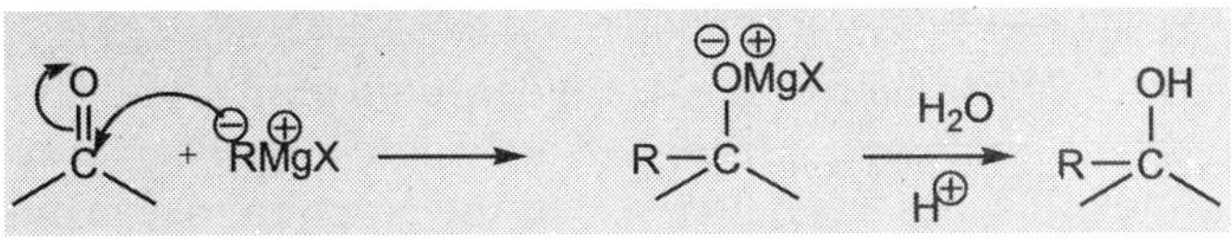

However, it has been shown that in ether solution, ketones and Grignard reagent are associated into a complex that makes the carbonyl carbon more positive, and therefore, more susceptible to attack by a second molecule of the Grignard reagent.

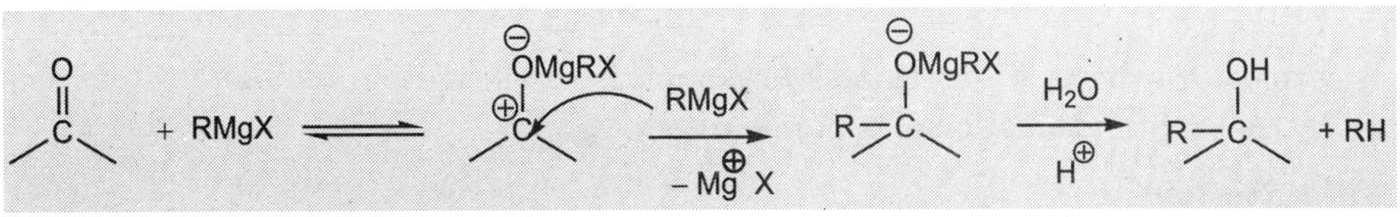

Being a Lewis acid, magnesium bromide coordinates with a carbonyl group better than a Grignard reagent, and so, its addition to the reaction mixture increases the yield of the desired product.

(ii) Reformatsky Reaction: Organozinc compounds are less reactive, and so, unlike Grignard reagents, they do not react with ester function. It is thus possible to prepare an ester enolate anion by reaction of an α-halo ester with zinc. The addition of such enolate anion to the carbonyl carbon of an aldehyde or ketone is known as the *Reformatsky reaction*.

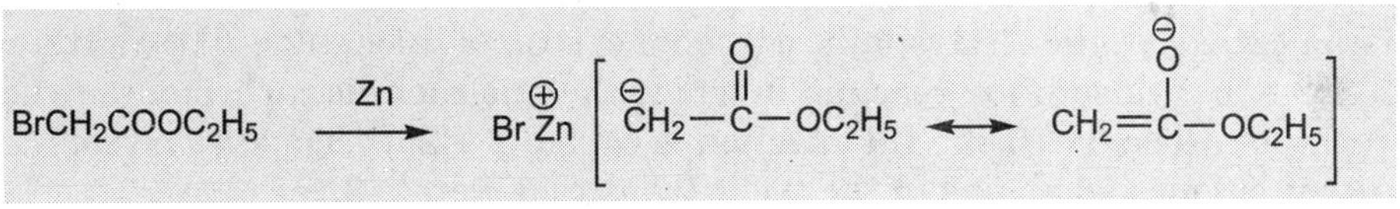

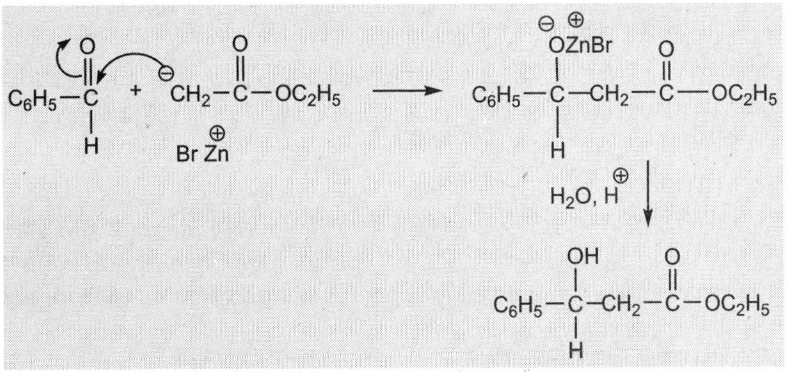

The β-hydroxy esters, so obtained, readily undergo acid-catalyzed dehydration to α,β-unsaturated esters.

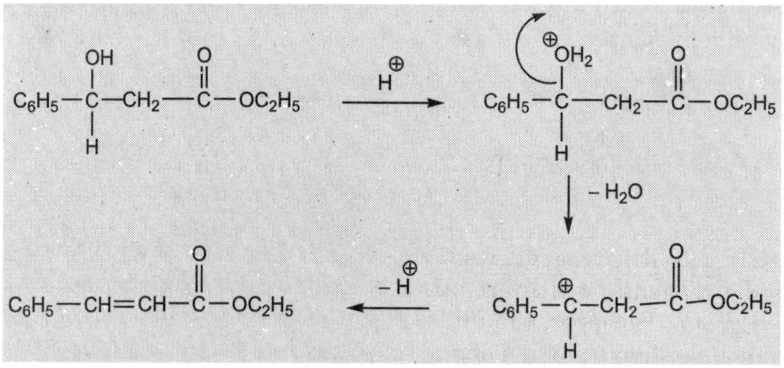

A serious drawback of Reformatsky reaction is the formation of β-keto esters as a side product by the self-condensation of α-bromo esters.

VIII. Wittig Reaction

The reaction of an aldehyde or ketone with alkylidenetriphenyl phosphoranes (Wittig reagents; phosphonium ylide **12**) to form alkenes is known as Wittig reaction.

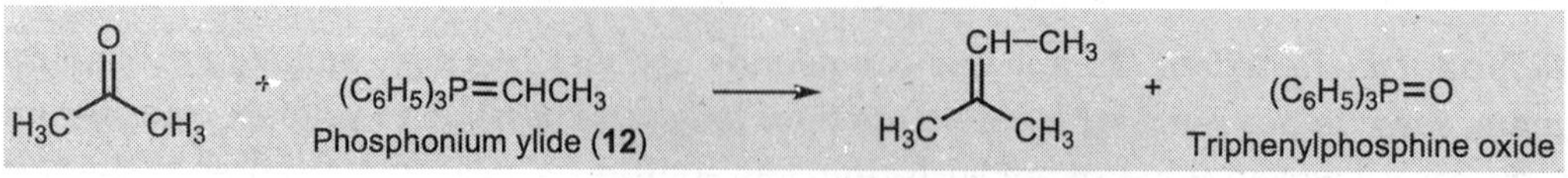

The key step of the reaction is the nucleophilic attack by alkylidenetriphenyl phosphoranes, which is actually a carbanion stabilized by resonance with the *d* orbitals of an adjacent phosphonium cation, on the electrophilic carbonyl carbon of the ketone. The Wittig reaction is a concerted [2 + 2] cycloaddition reaction in which the nucleophilic carbon of the ylide attacks the electrophilic carbonyl carbon. The reaction is called [2 + 2] cycloaddition reaction as there is involvement of four π electrons in the cyclic transition state (**13**).

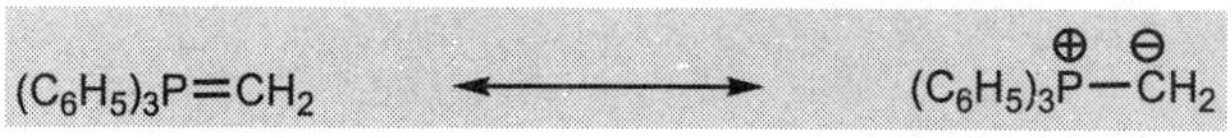

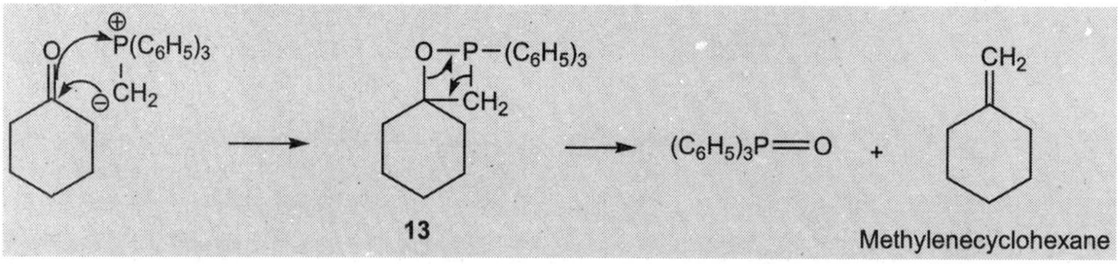

The driving force for the reaction is believed to be the formation of a phosphorus–oxygen bond at the expense of a phosphorus–carbon bond. As expected, Wittig reaction is accelerated by the presence of electron-withdrawing groups in carbonyl compounds. Alkylidenetriphenyl phosphoranes, also known as ylides because of the presence of opposite formal charges on the two adjacent atoms in one of the resonance forms, are usually prepared by treating triphenylphosphine with an alkyl halide, and then removing alpha hydrogen from the resultant phosphonium salt by the action of strong bases such as sodium amide, sodium hydride, or organolithium compounds.

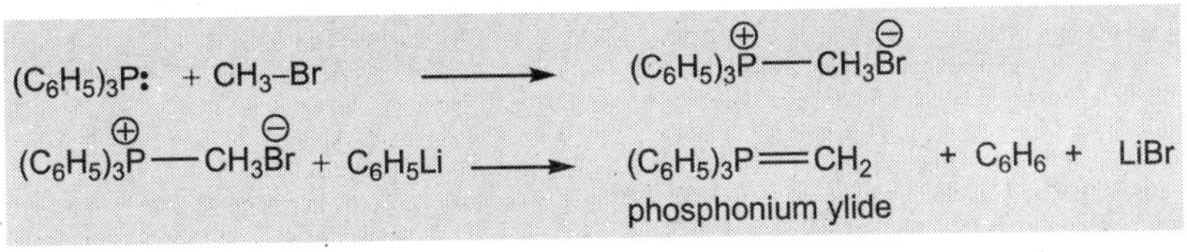

The reactivity of Wittig reagents vary according to their structures. Usually, they are highly reactive, unstable reagents, which are strongly nucleophilic, and consequently react readily with a carbonyl group. However, the introduction of electron-withdrawing substituents such as carbonyl in the alkylidene portion of an ylide decreases its nucleophilicity, and the reaction with carbonyl compounds is somewhat hindered. An example of ylide of this type is **14**, which is obtained by the reaction of α-bromo acetophenone and triphenylphosphine. It is a stable solid, and fails to react with cyclohexanone. Its unusual stability has been ascribed to the following resonance structures (**14**,**14a** and **14b**).

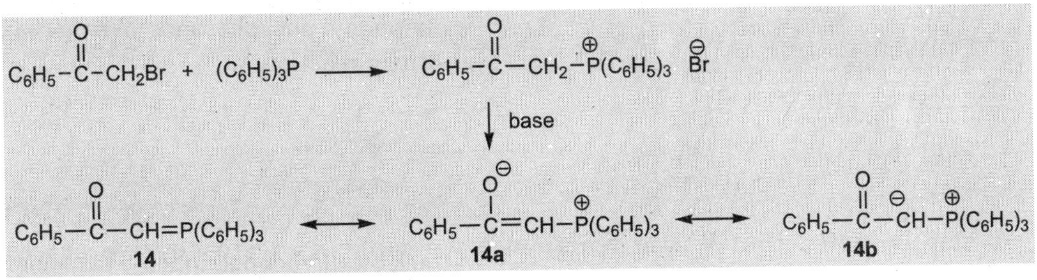

The difficulty in preparing alkenes having electron-withdrawing group in the α-position has been overcome by the introduction of a modified Wittig reaction in which a phosphonate ylide **(15)** attacks on the carbonyl carbon as carbanion.

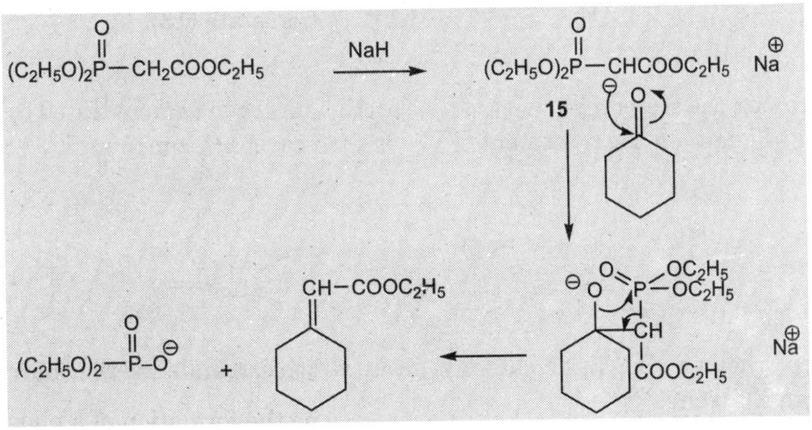

The Wittig reaction has great synthetic utility, because the position of the carbon-carbon double bond is unambiguous in the product. The following examples illustrate the versatility of the reaction.

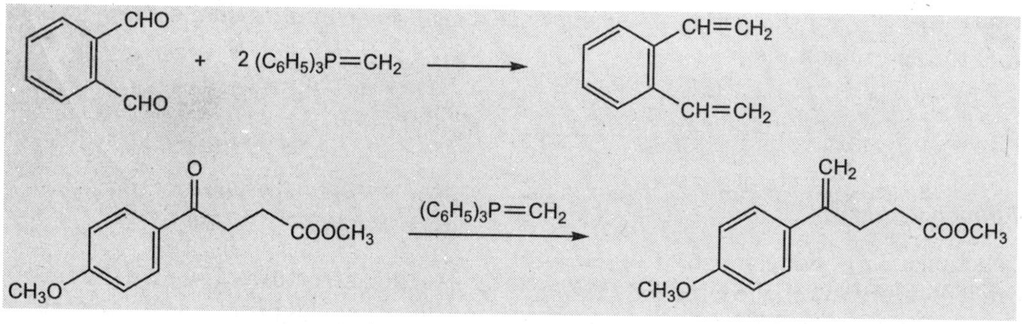

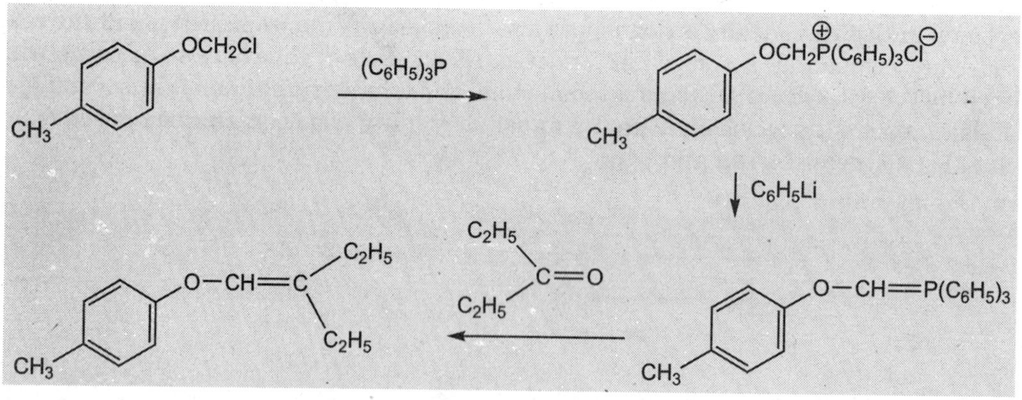

It may be noted that Wittig reaction is a completely regioselective process as methylenecyclohexane was found to be the only product from cyclohexanone.

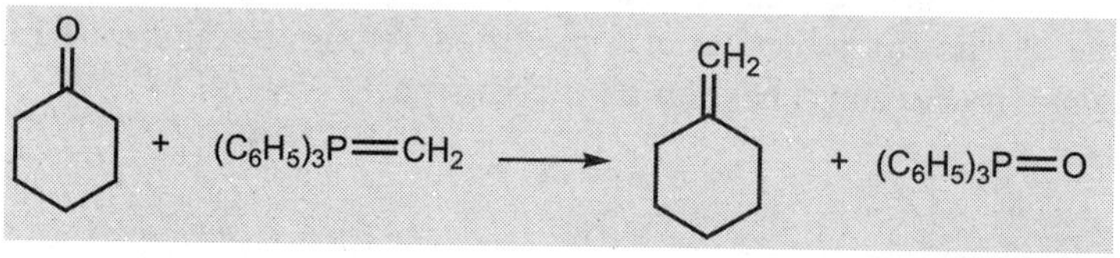

Furthermore, the stereoselectivity of the reaction has been demonstrated by the formation of primarily *Z*-alkenes on the treatment of ylides with carbonyl compounds.

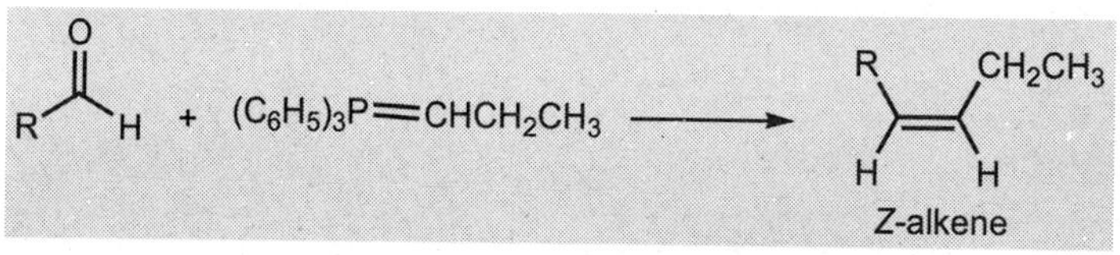

However, when an ylide has a carbonyl group which can stabilize the charge, the product is primarily an *E*-alkene.

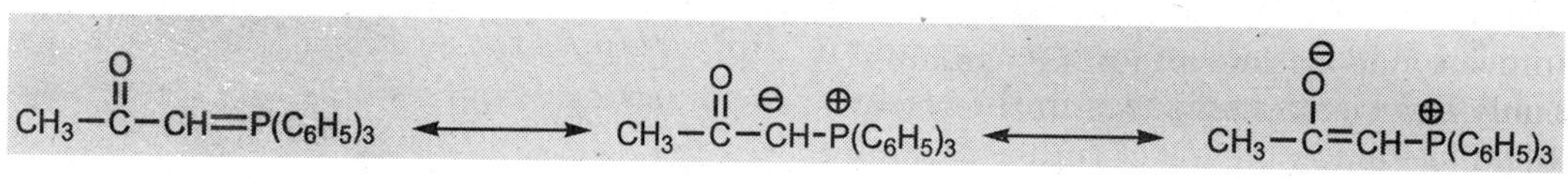

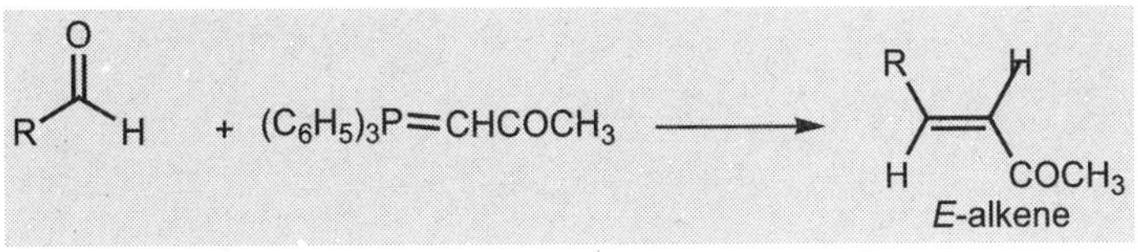

IX. Mannich Reaction

The reaction of an enolic compound (active methylene compound) with formaldehyde and amine to form a β-aminocarbonyl compound (Mannich base) is known as the *Mannich reaction*. It is a very convenient method for introducing a carbon atom in enolizable ketones.

$$C_6H_5COCH_3 + HCHO + (CH_3)_2NH \longrightarrow C_6H_5COCH_2CH_2N(CH_3)_2$$

Mannich reaction has been found to be a third order reaction, first order in each reactant.

$$\text{Rate} = k\,[C_6H_5COCH_3]\,[HCHO]\,[(CH_3)_2NH]$$

It is believed that in the first step of the reaction the amine and formaldehyde condense to form the imminium salt **(16)**.

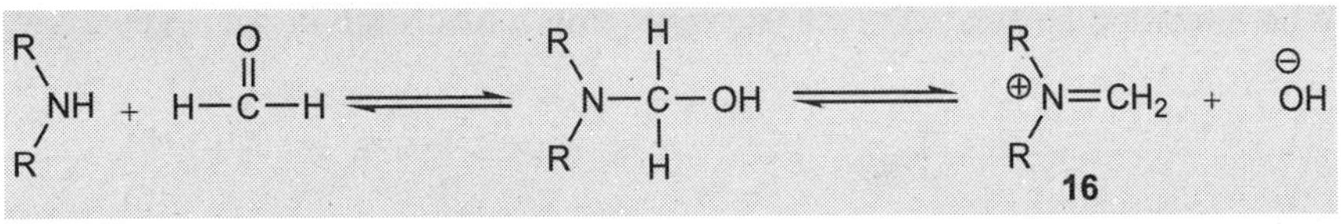

The imminium salt is then attacked by enolate anion of the active methylene compound in a rate controlling step.

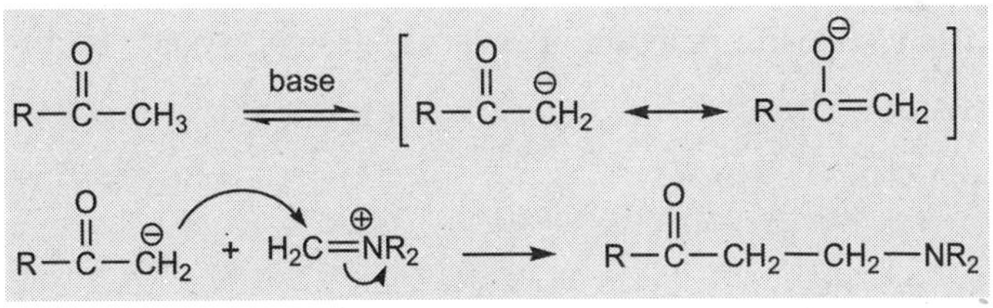

Since the ease of enol formation is increased by the presence of α-alkyl substituents, unsymmetrical ketones such as **17** react predominantly at more highly substituted α-position.

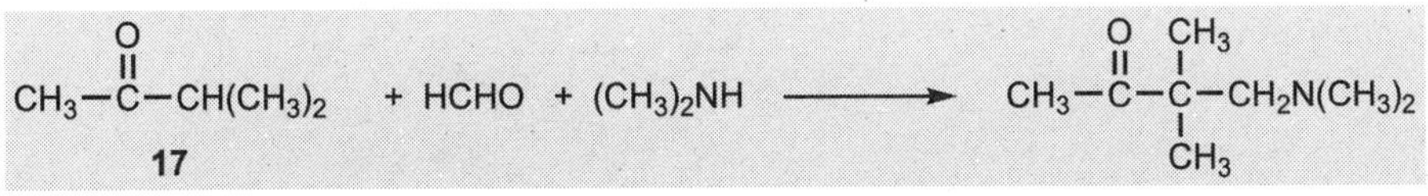

The use of ammonia and primary amines is undesirable in Mannich reaction as the first formed Mannich base will still carry hydrogen(s) on the nitrogen atom, and thus, further participate in the reaction. Interestingly, a number of cyclic products have been prepared *via* double Mannich condensation. A familiar example is the Robinson's synthesis of tropinone (**18**), which not only involves a double Mannich condensation, but at the same time illustrates the use of an aldehyde (succinaldehyde) other than formaldehyde.

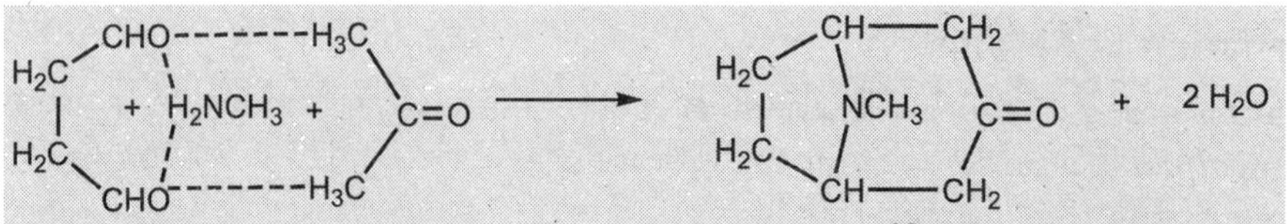

18

Mannich reaction is not confined to enolizable ketones only. Phenols, nitroalkanes and indoles are also attacked by imminium salts, which are very strong electrophilic agents.

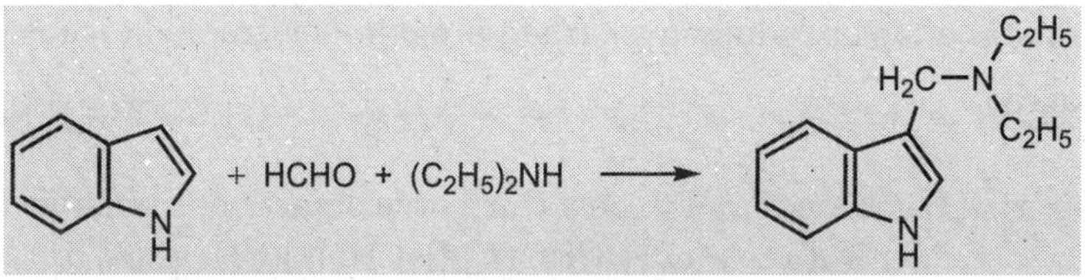

Synthetic importance of Mannich reaction lies in the facile conversion of a Mannich base into various products of interest. For instance, an α,β-unsaturated ketone or a saturated ketone may readily be obtained by heating or hydrogenating a Mannich base, respectively.

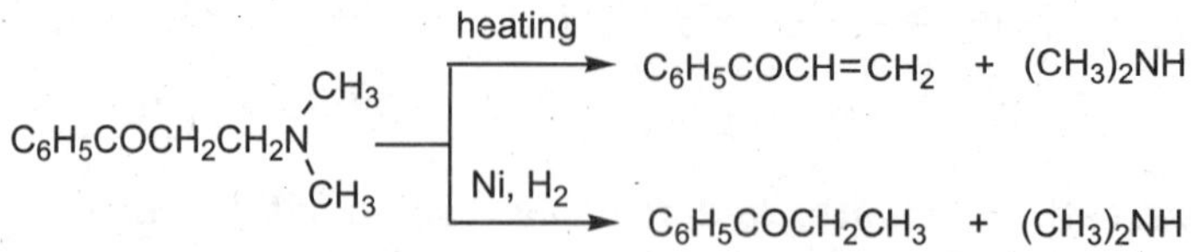

An important application of Mannich reaction is found in the *Robinson-Mannich ring annelation,* which is a composite reaction of several base-catalyzed reactions.

$$CH_3-\overset{O}{\overset{\|}{C}}-CH_3 + HCHO + (CH_3)_2NH \xrightarrow[\text{reaction}]{\text{Mannich}} CH_3COCH_2CH_2N(CH_3)_2$$

$$CH_3COCH_2CH_2N(CH_3)_2 + CH_3\text{–}I \longrightarrow CH_3COCH_2CH_2\overset{\oplus}{N}(CH_3)_3\,\overset{\ominus}{I} \xrightarrow[\text{elimination}]{\text{Hofmann}} CH_3COCH{=}CH_2$$

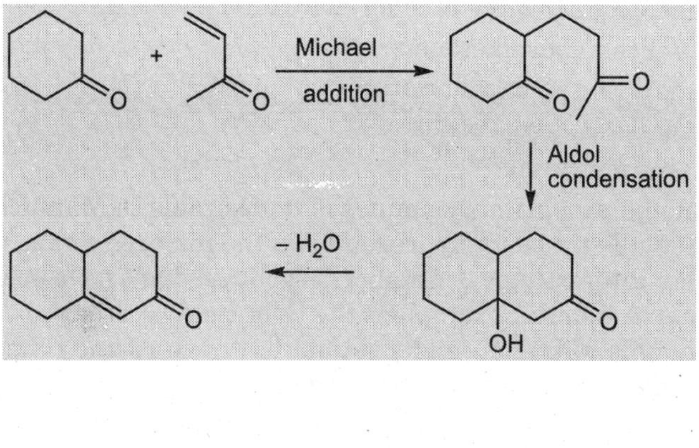

E. Halogenation of Ketones

The formation and subsequent reactions of anions derived by removal of a proton from the carbon atom *alpha* to a carbonyl group are extremely important. These anions are usually called enolate anions, and they are intermediates in a wide variety of reactions. The base-catalyzed halogenation of ketones demonstrates the existence of carbanions, and at the same time illustrates the role played by *keto-enol tautomerism* in the chemistry of carbonyl compounds.

I. Base-Catalyzed Halogenation

Ketones having α-hydrogens react with bromine to form bromoketones; the reaction is accelerated by addition of bases.

$$CH_3COCH_3 + Br_2 + :B \longrightarrow CH_3COCH_2Br + \overset{\ominus}{Br} + \overset{\oplus}{\mathbf{HB}}$$

The rate of reaction is proportional to the concentration of ketone and the base, but is independent of the concentration of bromine, thus indicating that the rate-determining step of the reaction does not involve bromine. Iodination of acetone in the presence of base also obeys the same kinetics, and moreover, under the given conditions, bromination and iodination proceed at identical rates. The simplest mechanism that is consistent with these data can be written in the following two steps:

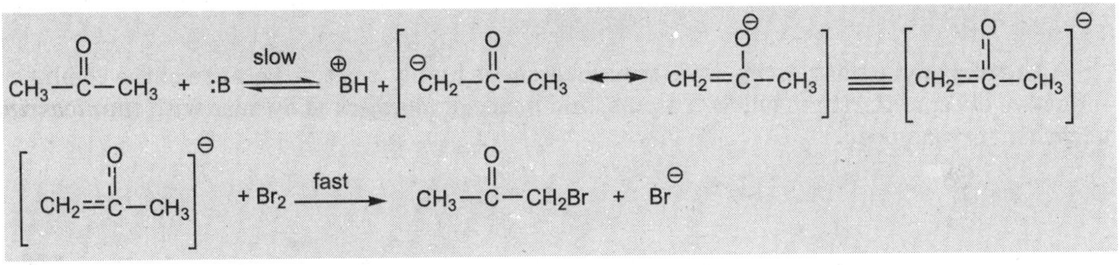

Another evidence for the existence of carbanion is provided by the base-catalyzed racemization of (+)–phenyl *sec*-butyl ketone (**19**). Here, the abstraction of proton by base yields the carbanion, and since it has a flat structure and the reaction is reversible, the proton may get attached to either face, producing the racemic mixture. Also, the rate constants for racemization and for bromination of the above ketone are found to be identical. The rate of racemization is also found to be equal to the rate of deuteration when the reaction is carried out in presence of a base in D_2O. All these observations strongly support the formation of planar enolate anion (**20**) in the rate-determining step.

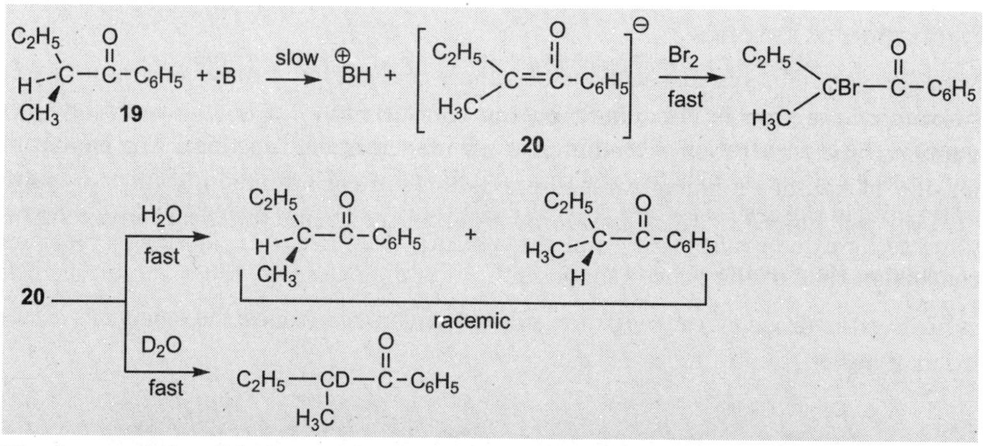

The base-catalyzed halogenation of ketones has limited synthetic utility as the reaction yields polyhalogenated compound. The halogens are electron-withdrawing, and so, the hydrogens situated on α-carbon become more acidic in the first formed α-halo ketone, and are easily replaced by halogens. Base-catalyzed halogenation of a methyl ketone thus results in the formation of trihaloketone.

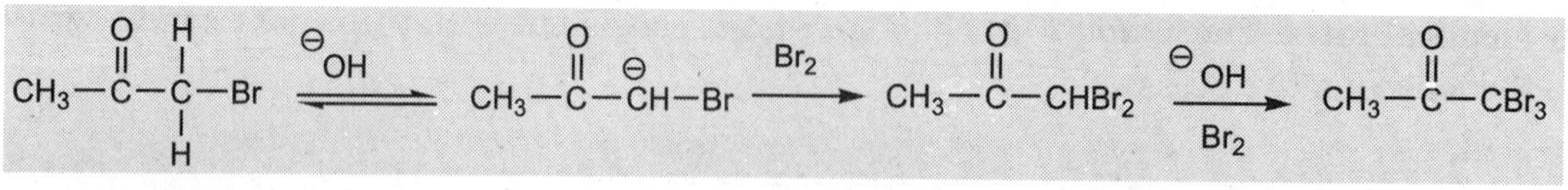

The carbonyl carbon atom of the trihaloketone is highly positive because of the combined inductive effect of the three halogen atoms, and hence, it is attacked by base with the cleavage of carbon–carbon bond.

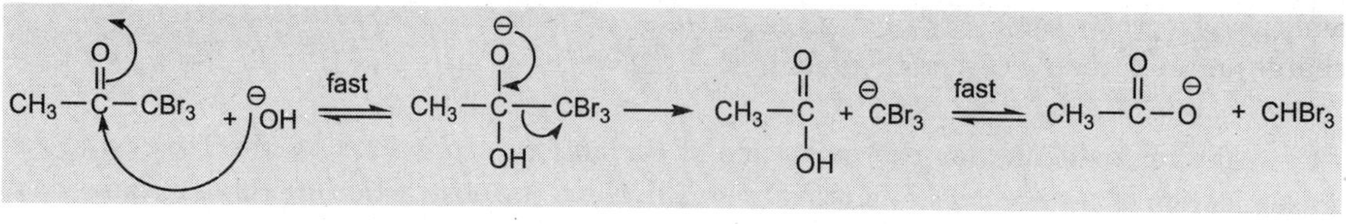

This reaction is known as the *haloform reaction.*

Since alkyl groups are electron-repelling, α-alkyl substituents are expected to retard the base-catalyzed halogenation of ketones. Thus, halogenation of an unsymmetrical ketone takes place on that carbon which bears fewer alkyl groups.

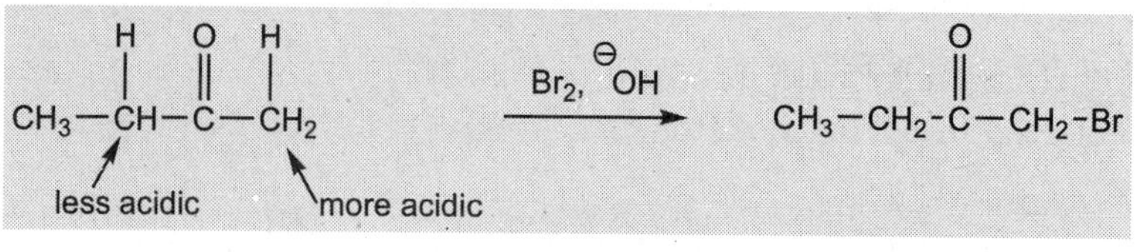

II. Acid-Catalyzed Halogenation

The halogenation of ketones is also accelerated by addition of acids.

$$CH_3COCH_3 + Br_2 \xrightarrow{H^{\oplus}} CH_3COCH_2Br + HBr$$

The rate of reaction is dependent on the concentration of ketone and the acid, but independent of the concentration of bromine. As in base-catalyzed halogenation, bromination and iodination proceed at identical rates; the rates of halogenation and deuteration are the same in heavy water, and in the case of an optically active ketone, the rate of acid-catalyzed halogenation is the same as its rate of racemization in the absence of halogen. All these observations lead to the conclusion that in the above reactions the rate-determining step involves formation of an enol (**21**).

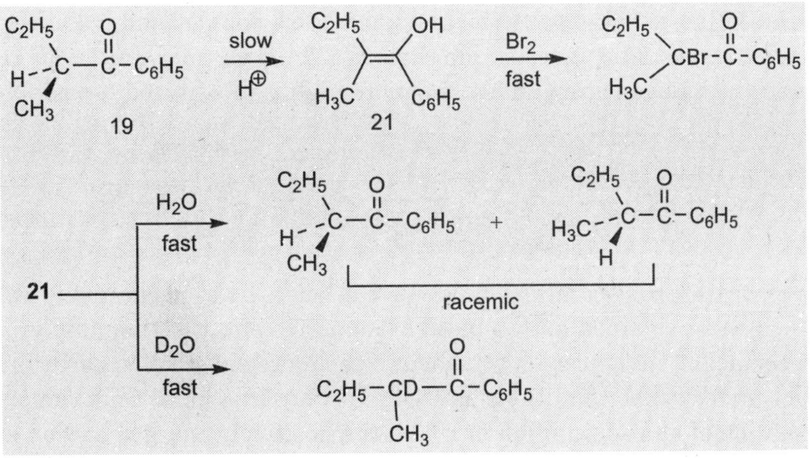

Since enolization (as measured by the rate of halogenation) is accelerated by acids, it is not a simple direct transfer of a proton from a carbon to oxygen. The first step is the reversible addition of a proton to the carbonyl oxygen yielding an oxonium ion (**22**); the enhanced positive charge on the carbonyl carbon then facilitates the removal of a proton from α-carbon.

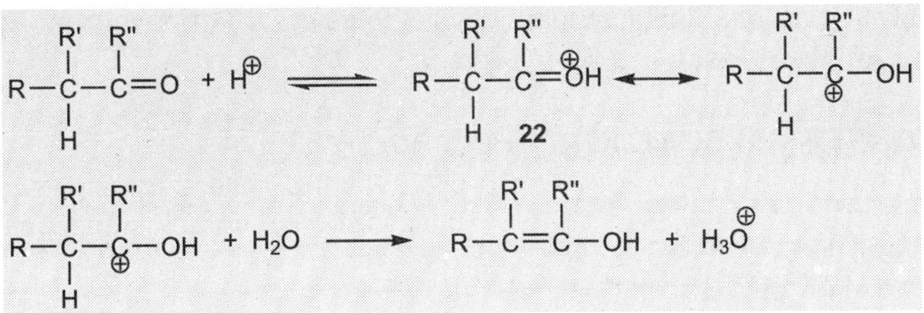

The enol reacts rapidly with bromine to yield a protonated bromoketone, which is subsequently converted into a bromoketone by the loss of proton.

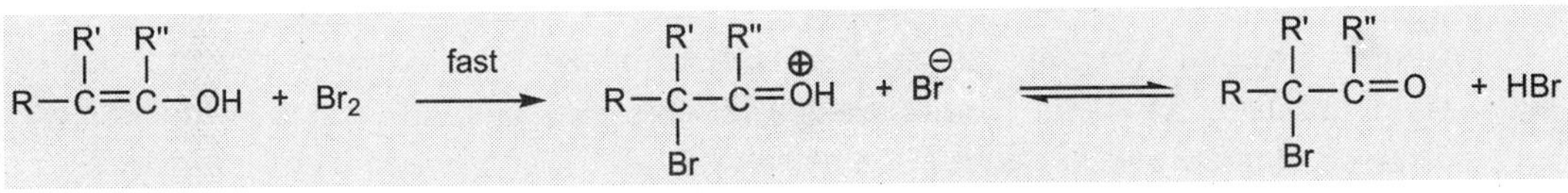

Since the effect of halogen is to withdraw electrons, monohaloketones show much less tendency to protonation and subsequent enolization, and hence it is possible to isolate them under these conditions, unlike in base-catalyzed halogenation. Acid-catalyzed halogenation depends upon the extent of enolization, and it is, therefore, logical to expect that in an unsymmetrical ketone (such as $CH_3CH_2COCH_3$) the position taken by the halogen should depend on the ease of formation of two isomeric enols, **23** and **24**.

23 (more stable) **24** (less stable)

We may recall that an alkyl substituent adjacent to a double bond stabilizes it by 2.5 kcal (10.4 kj) per mole, and so **23** is more stable than **24**. Thus, an unsymmetrical ketone yields a product in which the halogen atom enters the more highly substituted α-position.

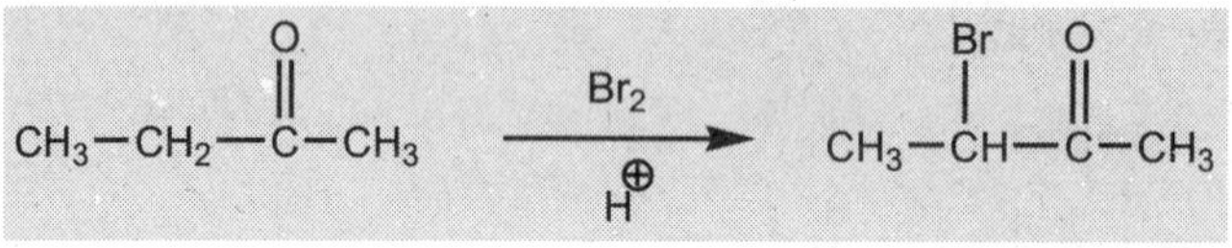

This is in contrast to base-catalyzed halogenation where halogen enters the least substituted α-position, which was explained on the basis of inductive effect of the alkyl groups resulting in decrease in the acidity of hydrogen attached to the more substituted carbon. The inductive effect does exist under acidic conditions as well, but here it is more than compensated for by the initial formation of an oxonium ion in which the positively charged carbon atom pulls the electron density away from the α-C–H bonds, making easier the departure of that α-proton which leads to the more stable enol.

Thus, the reactions of carbonyl compounds in the presence of a base usually involve *enolate anions*, whereas acid-catalyzed reactions go through *enol intermediates*. In fact, enols are present in equilibrium with the carbonyl compounds even in the absence of added acidic or basic catalyst, but the equilibria appears always to be greatly in favour of the keto forms in compounds containing an isolated carbonyl group.

9.2 REACTIONS OF CARBOXYLIC ACID DERIVATIVES

These compounds have the general formula R–COX where X may be a halogen, OR, NH_2, or O^-. The non-bonding electrons on X participate in resonance with the carbonyl double bond, and thus confer extra stability to the carbonyl grouping.

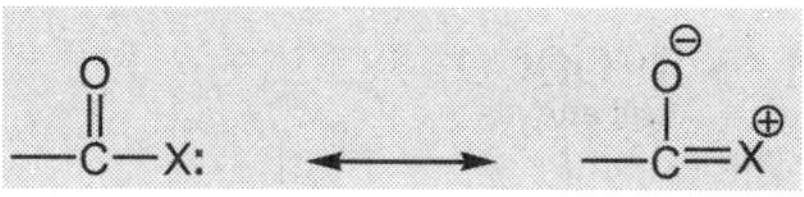

The order of increasing interaction of these electrons with the carbonyl π-orbital is X < OR < NH_2 < O^-. The reason for the maximum delocalization in carboxylate anion is that the excess charge is shared equally by the two oxygen atoms without involving charge separation.

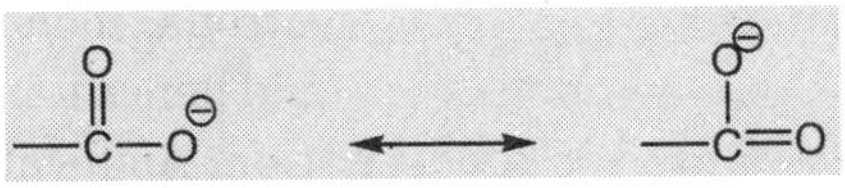

The evidence for the participation of lone pair of electrons situated on the heteroatom is provided by the calculation of bond lengths in acid derivatives. Both the C–O bonds of a carboxylate anion, for instance, are of equal lengths with a value somewhat intermediate between a normal C=O and a carbon-oxygen single bond. Similarly, in formamide the carbon-nitrogen bond length is only 1.34 Å as against the normal carbon–nitrogen single bond length of 1·47 Å. The NMR studies of protonated acetamide show that it has structure **25**, and since protonation takes place only at the oxygen atom, it is, therefore, reasonable to believe that the basic properties of nitrogen are transferred to oxygen due to the following resonance structures:

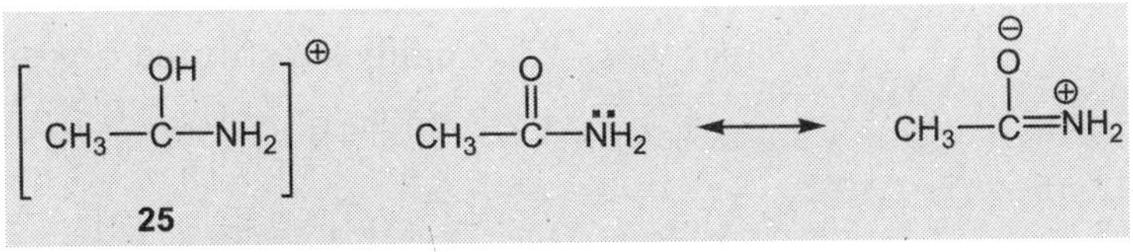

An important consequence of this delocalization is the reduction in partial positive charge of the carbonyl carbon, which would now be attacked by nucleophiles less readily. It is, therefore, not surprising that carboxylic acid derivatives do not undergo usual nucleophilic addition reactions so typical of aldehydes and ketones. Viewing alternatively, bonding between the carbonyl carbon of an acid derivative and a nucleophile destroys the conjugative stabilization of the system, and hence, it is resisted.

However, the carbonyl carbon continues to remain an electrophile, and hence, will be attacked by a nucleophile in carboxylic acid derivatives such as acyl chlorides, acid anhydrides, esters, carboxylic acids, and amides. As we shall see soon that when a nucleophile attacks the carbonyl carbon of a carboxylic acid derivative, the π-bond breaks and a tetrahedral intermediate is formed. Because of the presence of two electronegative atoms, the tetrahedral intermediate is unstable, and expels one of the groups to reform the carbon-oxygen double bond.

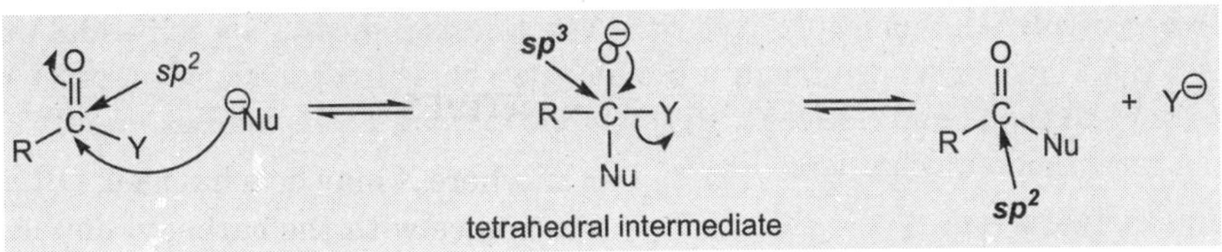

The relative reactivities of carboxylic acid derivatives follow the order:

RCOCl	> RCOOCOR >	RCOOR,	RCOOH >	$RCONH_2$
acyl chloride (most reactive)	acid anhydride	ester,	acid	amide (*least reactive*)

Acyl chlorides are most reactive, whereas amides least reactive. The reactivity of carboxylic acid derivatives thus depends on the basicity of the substituent attached to the acyl group. In general, *less basic* is the substituent; *more reactive* will be the carboxylic acid derivative. The relative basicities of the leaving groups follow the order:

Cl^-	$<$ ^-OCOR	$<$ ^-OR, ^-OH	$<$ $^-NH_2$
(weakest base)			(strongest base)

Thus, acyl chlorides are most *reactive* because chloride ions are weakest bases. The strongest basicity of amide ions, on the other hand, make the amides least reactive.

A. Esterification and Hydrolysis

There are a number of distinct mechanisms by which esterification of an acid and hydrolysis of an ester may proceed. This plurality of mechanisms is due to the following three reasons:

(i) Esterification and hydrolysis may proceed either through an acyl–oxygen bond cleavage **(A)** or an alkyl–oxygen bond cleavage **(B)**.

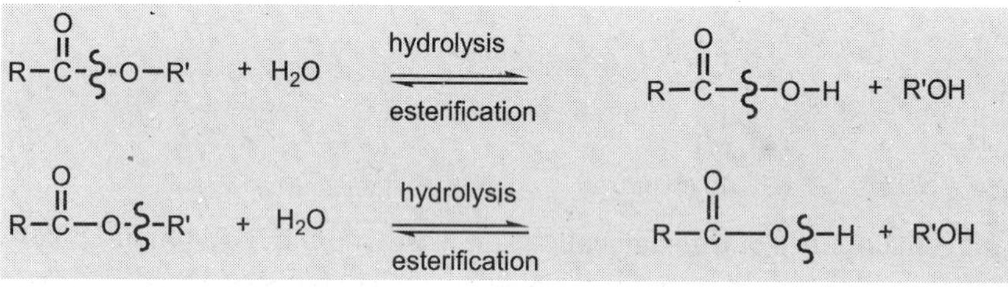

(A)

(B)

(ii) Hydrolysis of an ester may be accomplished either by acids or alkalis, and depending upon the reagent, the species undergoing hydrolysis may be a neutral ester molecule (R–COOR') or an ionic conjugate acid $(RCOOHR')^+$

(iii) The esterification and hydrolysis may proceed either by a unimolecular or a bimolecular mechanism.

Since esterification in basic medium is not possible because of the conversion of carboxylic acid to a resonance-stabilized carboxylate anion, depending upon the above three factors, there are four possible mechanisms by which esterification may proceed. Similar considerations visualize eight mechanisms by which hydrolysis may proceed. Ingold has suggested a shorthand notation to describe these bewildering number of mechanisms in which **A** or **B** refers to the acidic or basic (and neutral) medium; acyl–oxygen or alkyl–oxygen cleavage is symbolized by *AC* or *AL*, respectively, and molecularity or the reaction is indicated by the usual number 1 or 2. Thus, by a $B_{AC}2$ hydrolysis we mean a bimolecular basic hydrolysis of an ester proceeding through the acyl–oxygen bond cleavage. In the following pages, we shall exemplify a few of the relatively more common mechanisms like,

I Basic Hydrolysis
II Acidic Hydrolysis and Esterification
III The Acylium ion Mechanism
IV The Carbocation Mechanism

I. Basic Hydrolysis (The $B_{AC}2$ mechanism)

The hydrolysis of esters by hydroxide ions in aqueous solution is kinetically a second-order reaction, first order in ester, and first order in hydroxide ion. It is, therefore, a bimolecular reaction. The evidence for the acyl–oxygen bond fission is provided by carrying out the hydrolysis of *n*-amyl acetate in water enriched in ^{18}O, and then isolating *n*-amyl alcohol without ^{18}O.

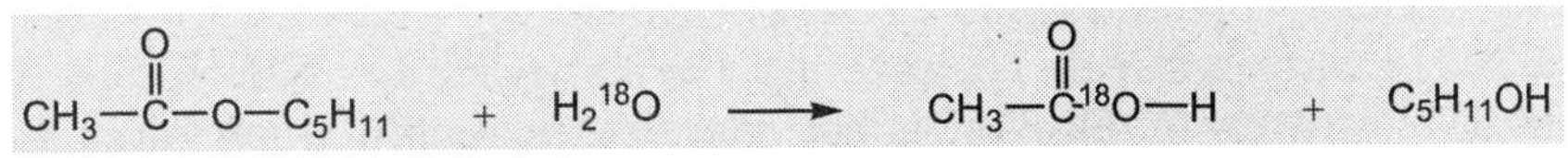

The hydrolysis can be thus depicted as involving a $B_{AC}2$ reaction for which two mechanisms are consistent with the above data.

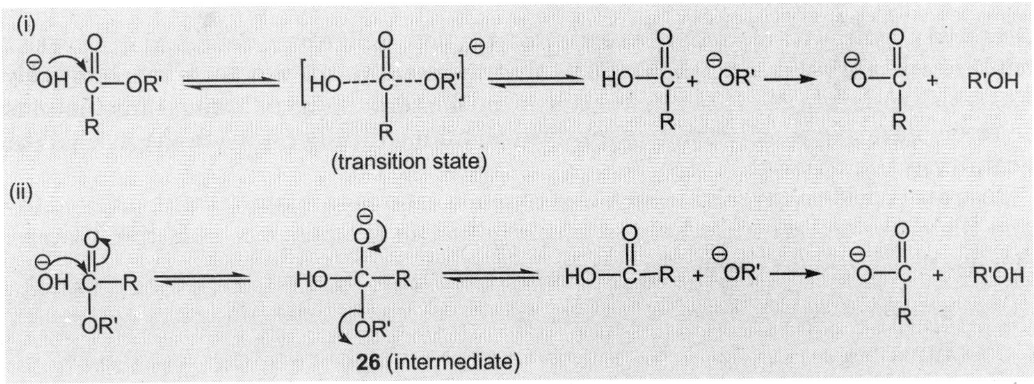

In the first of these two mechanisms, there is a *transition* state (*cf.* S_N2 reaction) with an *energy maximum* in the reaction coordinate, whereas the second mechanism postulates a *true intermediate* (**26**) with an *energy minimum*.

Bender solved the problem of deciding between these two mechanisms by hydrolyzing esters (ethyl, *iso*-propyl, and *tert*-butyl benzoates) labelled in the carbonyl group with ^{18}O in ordinary water containing hydroxide ions. When the hydrolysis was stopped before completion, unreacted esters were found to contain less ^{18}O content. This decrease in ^{18}O content of the ester cannot be explained by the first mechanism as it is a concerted process. But, if we assume that the proposed tetrahedral intermediate (**26a**) of the second mechanism lives long enough to survive a proton shift, and establish an equilibrium with another intermediate (**27**), then the two oxygen atoms become equivalent. Since the basicities of alkoxide and hydroxide ions are comparable, the two intermediates may lose these ions at comparable rates (paths *a* and *b*, respectively), and then it appears quite reasonable to expect unlabelled ester to be formed in the process.

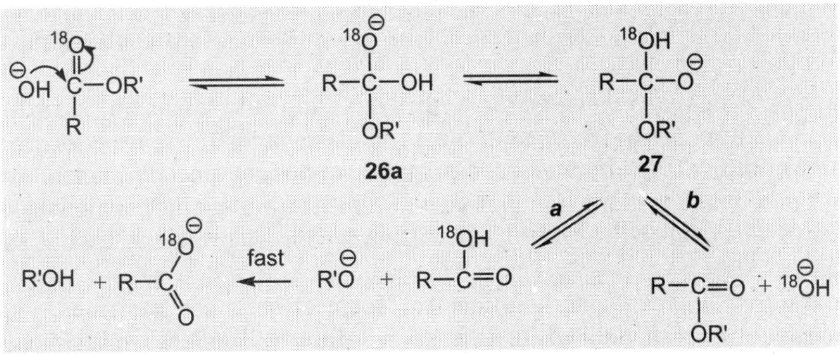

The effect of substituents on $B_{AC}2$ hydrolysis of aromatic esters also falls in line with the above conclusions. Electron-withdrawing substituents in either the acyl or alkyl part of the esters facilitate attack by hydroxide ions. It was also observed that while the subsituents such as Cl, Br, and NO_2 group accelerate the reaction, CH_3, p-OCH_3 and NH_2 groups retard it.

The rate of hydrolysis of an ester is considerably increased when carried out in basic conditions. Hydroxide ion increases the rate of formation of the tetrahedral intermediate, because OH^- is a better nucleophile than H_2O.

II. Acidic Hydrolysis and Esterification (The $A_{AC}2$ mechanism)

Acid-catalyzed hydrolysis of esters is a reversible reaction, unlike base-catalyzed hydrolysis. The principle of *microscopic reversibility* tells us that the mechanism of reverse reaction is known with as much certainty as is the mechanism of the forward reaction. This means that a study of the mechanism of acid-catalyzed hydrolysis will automatically establish the mechanism for acid-catalyzed esterification.

The rate of acid-catalyzed hydrolysis of common esters is found to be proportional to both H^+ and the ester, and when hydrolysis is carried out in the presence of $H_2{}^{18}O$, the product alcohol has no ^{18}O content. Thus, it is an $A_{AC}2$ reaction.

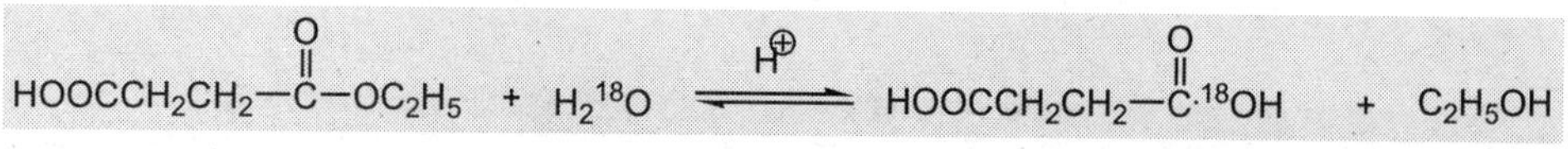

Again, Bender observed the loss of ^{18}O content in the unreacted ethyl benzoate when the hydrolysis was stopped before completion. This loss can only be explained by assuming two intermediates **28** and **29** whose life is long enough to permit proton exchange, thus, making the two oxygen atoms equivalent.

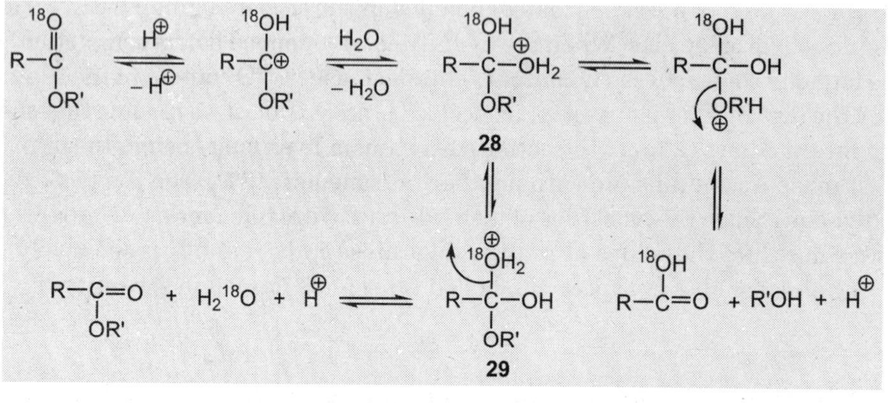

In contrast to basic hydrolysis, acidic hydrolysis (and esterification) is rather insensitive to polar effects. This appears quite reasonable as the two parts of the process, protonation and subsequent nucleophilic attack by water, require two opposing electronic influences. Electron-donating groups accelerate protonation, but they inhibit the subsequent attack by a nucleophile. On the other hand, electron-withdrawing substituents repress protonation, but they accelerate the nucleophilic attack.

These reactions, however, are subject to steric effects; for instance, esterfication of 2-methylbenzoic acid is retarded while that of 2,6-dimethylbenzoic acid is completely prevented.

III. The Acylium ion Mechanism (The $A_{AC}1$ mechanism)

Hydrolysis and esterification of sterically hindered compounds, such as derivatives of trialkylacetic acids and *ortho* disubstituted benzoic acids, are carried out by dissolving them in concentrated sulphuric acid, and then pouring this solution into cold water (for hydrolysis) or into an alcohol (for esterification). Mesitoic acid (2,4,6-trimethylbenzoic acid; **30**) provides a good example of this type, whose solution in conc. sulphuric acid indicates the presence of four particles.

$$\underset{\mathbf{30}}{\text{Ar-COOH}} + 2H_2SO_4 \longrightarrow \underset{\mathbf{31}}{\text{Ar}\overset{\oplus}{\text{CO}}} + H_3\overset{\oplus}{O} + 2\,HSO_4^{\ominus}$$

Ar = 2,4,6-trimethylphenyl

The available evidence indicates that like $A_{AC}2$ mechanism, there is an initial formation of an oxonium ion by the addition of a proton to the substrate, which then undergoes a rate-controlling heterolytic fission to yield an acylium ion **(31).** The acylium ions are linear, and their formation is favoured primarily because of relief in the steric strains of the sterically hindered compounds. Another factor that favours their formation is the conjugation of carbonyl group with the unsaturated system, thus delocalizing the positive charge over the whole ion **31**. A rapid attack by water or alcohol on this ion results in the formation of acid or ester, respectively. Thus, esterification (or hydrolysis of methyl ester) of mesitoic acid proceeds as follows:

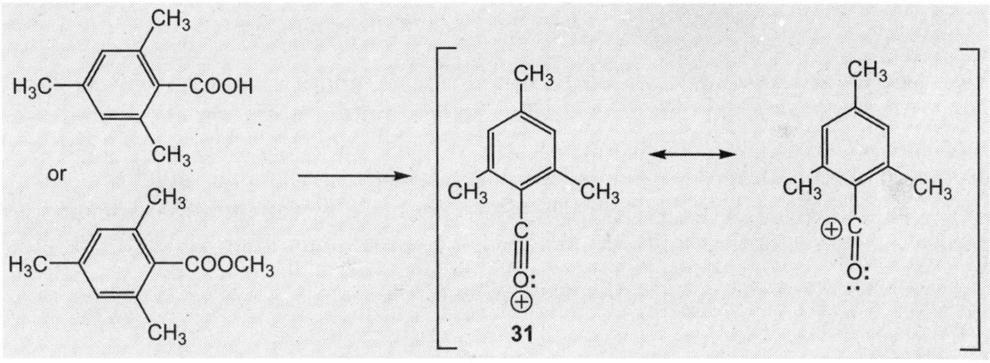

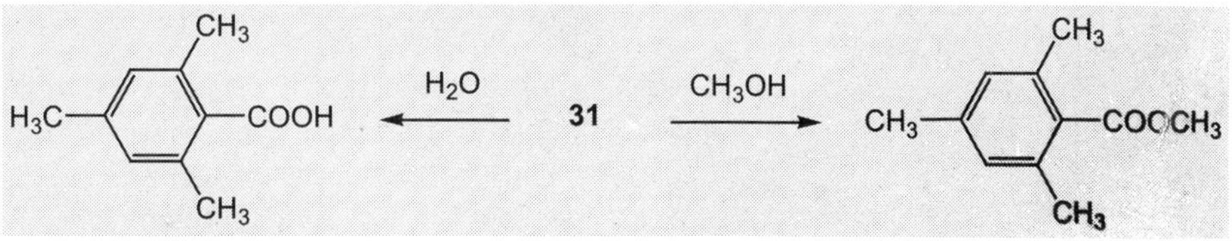

IV. The Carbocation Mechanism (The $A_{AL}1$ mechanism)

When groups such as *tert*-alkyl and benzyl, capable of forming stable carbocations, are the alkyl moiety in the substrate, hydrolysis and esterification may take place by an alkyl-oxygen bond cleavage. A mechanism of this type occurs in the solvolysis of *tert*-butyl benzoate, which on treatment with methanol and acid gives *tert*-butylmethyl ether and benzoic acid, rather than *tert*-butyl alcohol and methyl benzoate, the products of normal solvolysis. The following mechanism is consistent with the product analysis.

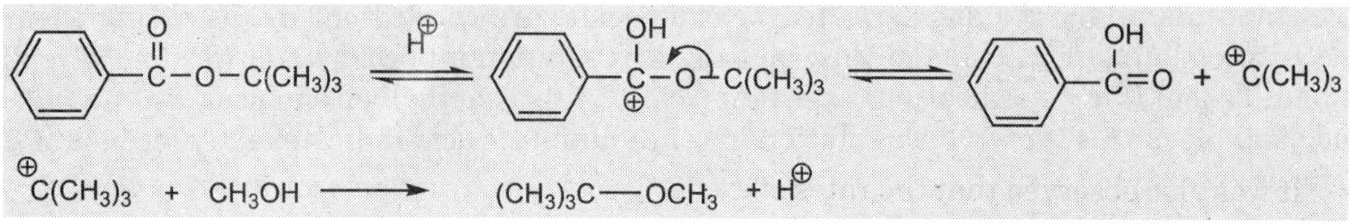

Confirmation of this mechanism is provided by carrying out the hydrolysis in $H_2{}^{18}O$.

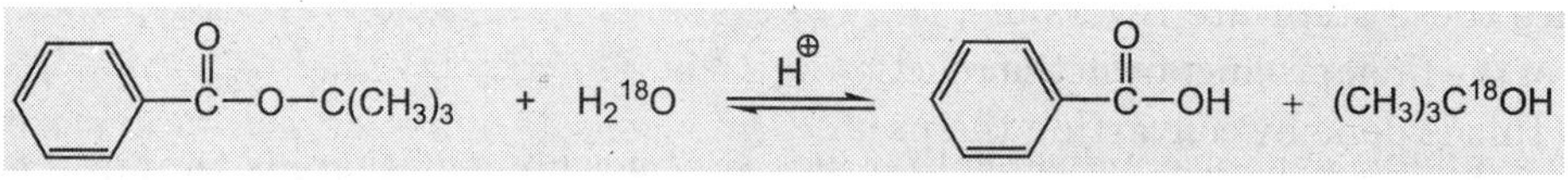

This type of mechanism has been labelled as $A_{AL}1$ mechanism. The intermediate has features characteristic of a free carbocation, *i.e.* its racemization and rearrangement, if the structure permits.

B. Claisen Condensation

Ethyl acetate can be converted to ethyl acetoacetate.

$$2\ CH_3COOC_2H_5 \longrightarrow CH_3COCH_2COOC_2H_5$$

This reaction is known as the Claisen condensation, and it covers various condensations between a carboxylic ester and an *alpha* hydrogen containing ester, ketone or nitrile, yielding a β-keto ester, ketone or nitrile, respectively. These reactions are usually carried out by means of sodium ethoxide in ethanol solution. Claisen condensation resembles aldol condensation in the sense that it involves the attack of carbanion, generated by the removal of acidic hydrogen on the carbonyl carbon of the ester. Expulsion of the ethoxide ion then yields the product.

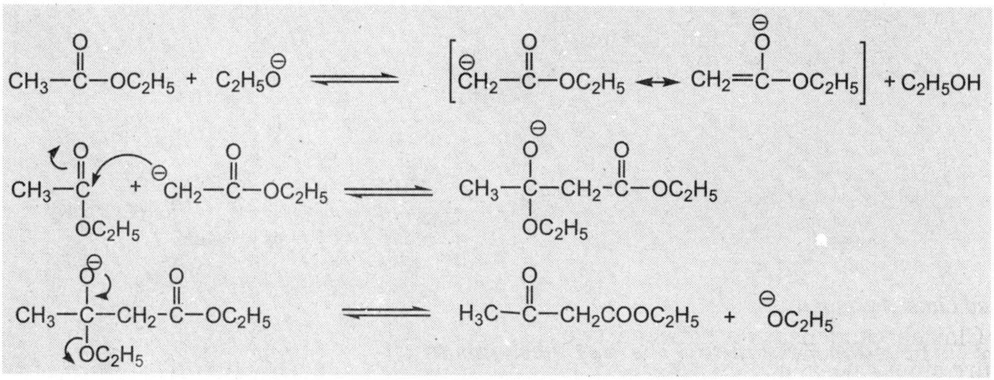

The reversible formation of the carbanion in the above sequence has been demonstrated by treatment with C_2H_5OD containing sodium ethoxide, whereby, these esters underwent hydrogen exchange at the *alpha* position.

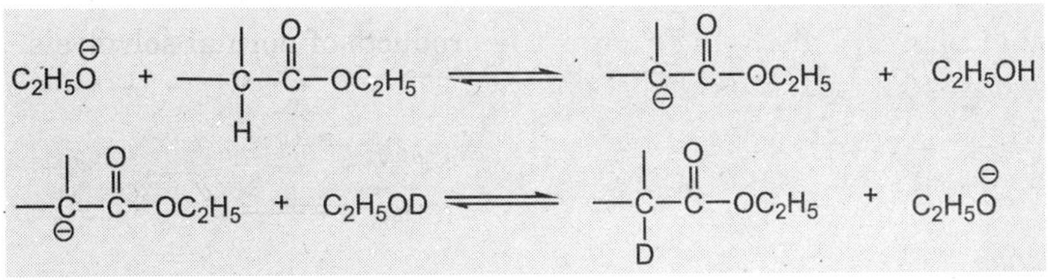

It was also observed that the rates of exchange of these hydrogens parallel the reactivity of these esters in Claisen condensation. The same conclusion was reached by the observation that optically active esters containing an α-hydrogen racemize in the presence of sodium ethoxide.

Since these steps are reversible, the success of Claisen condensation depends upon the removal of the final product from the reaction site, and thus, forcing the equilibrium to the right. Usually, this is done by converting the product ester into an enolate anion, which is stabilized by resonance.

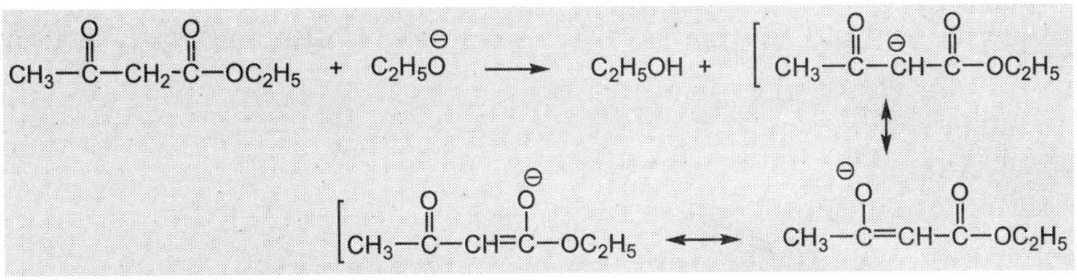

It follows that when the product β-keto ester does not have an α-hydrogen, the equilibrium of the condensation cannot be shifted to the right, and hence, esters without *two* α-hydrogens do not undergo Claisen condensation in the presence of sodium ethoxide. The equilibrium in such reactions, however, may be forced to the right by use of more strongly basic catalysts, such as sodium triphenylmethide and sodium amide, which are capable of bringing about the formation of β-keto esters that contain no α-hydrogen atoms.

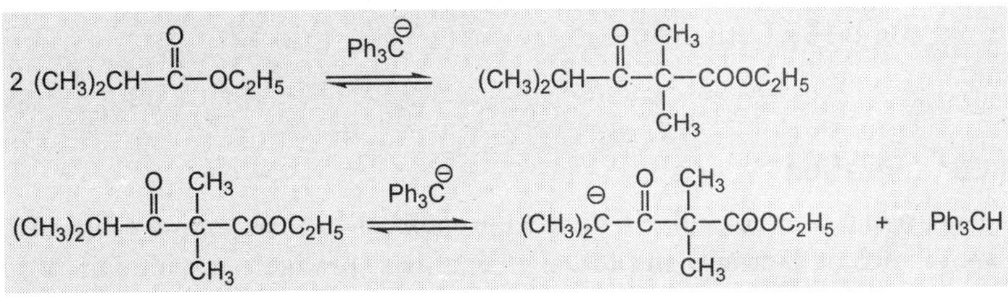

Crossed Claisen Condensation

Claisen condensation between two different esters, both having α-hydrogens, yields a mixture of four possible keto esters. Hence, it has little synthetic importance. However, esters such as those of formic, oxalic, and benzoic acids, which do not have α-hydrogens, but possess high carbonyl reactivity, can be successfully condensed with a second ester to offer a single product. For instance,

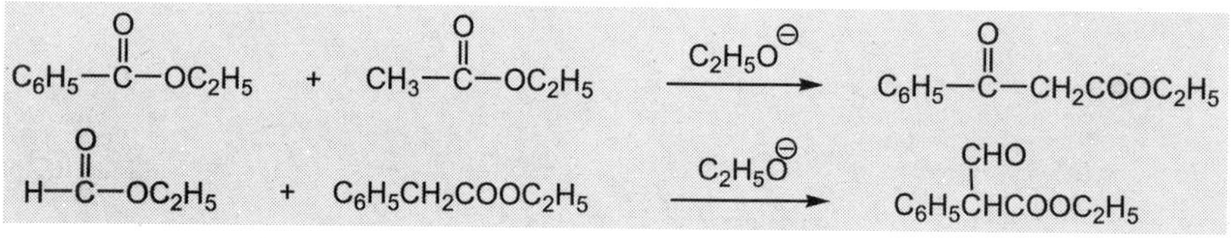

Dieckmann Reaction

Dieckmann synthesized cyclic β-keto esters by an intramolecular Claisen condensation of esters of dicarboxylic acids. In this reaction, the carbanion attacks the carbonyl carbon of the second ester group in the same molecule. Expulsion of the ethoxide ion leads to the product.

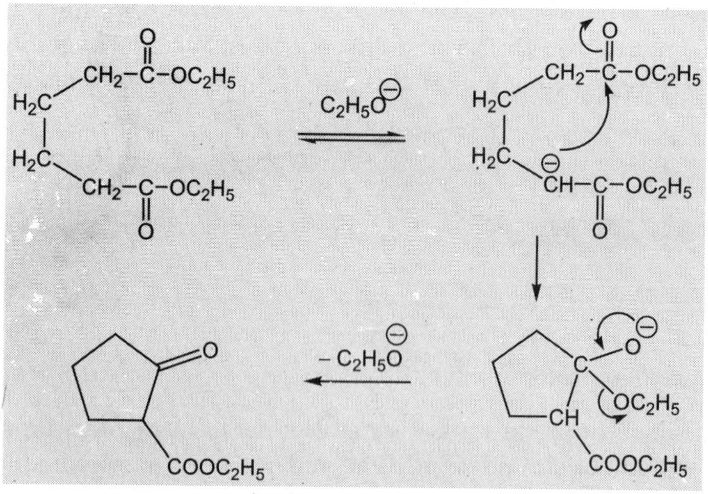

A similar mechanism can be written for the formation of a β-keto ester from 1,7-diester.

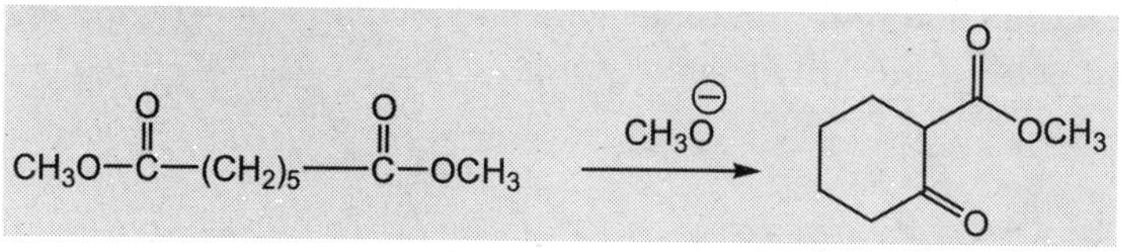

C. Acyloin Condensation

Acyloin condensation is a bimolecular reduction brought about by treating esters with sodium in inert solvents such as benzene and toluene. The most probable mechanism of reaction is the one in which sodium donates an electron to the ester molecule leading to the production of anion radicals. Subsequently, the two anion radicals couple to yield the dianion, which forms an acyloin (α-hydroxy ketone; **32**) by further reduction and acidification.

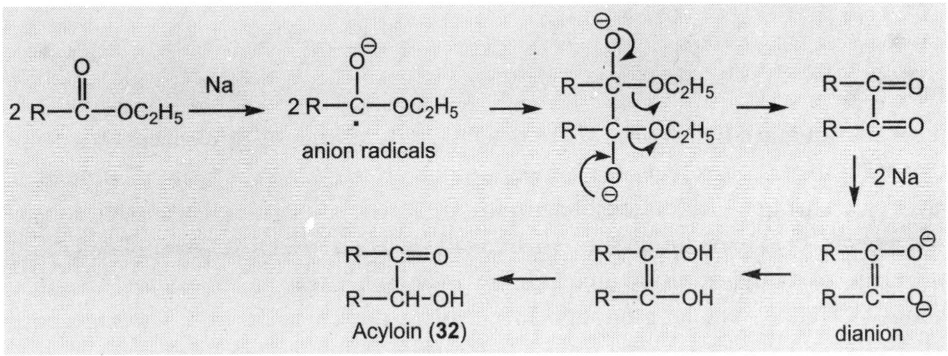

If the ester of a dibasic acid is used, a cyclic acyloin is formed, and thus, this method offers a convenient method for the preparation of large rings.

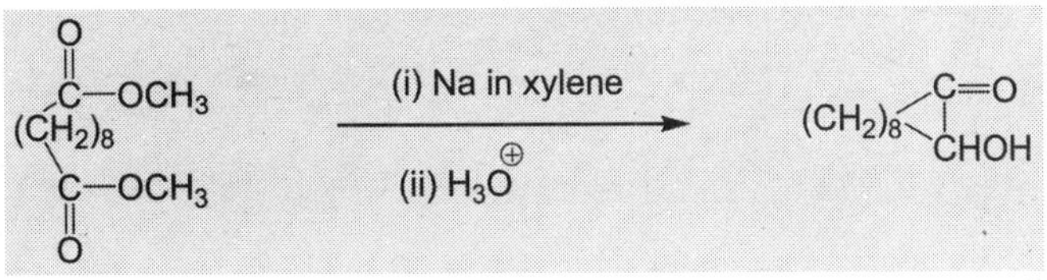

Bouveault-Blanc Reduction of Esters

In the presence of a protonic solvent such as ethanol, the initially formed anion radical reacts further with sodium to yield a divalent anion. Protonation, followed by explusion of the ethoxide ion, converts this divalent anion into an aldehyde. Repetition of the above sequence produces an alkoxide, which finally yields a primary alcohol on acidification.

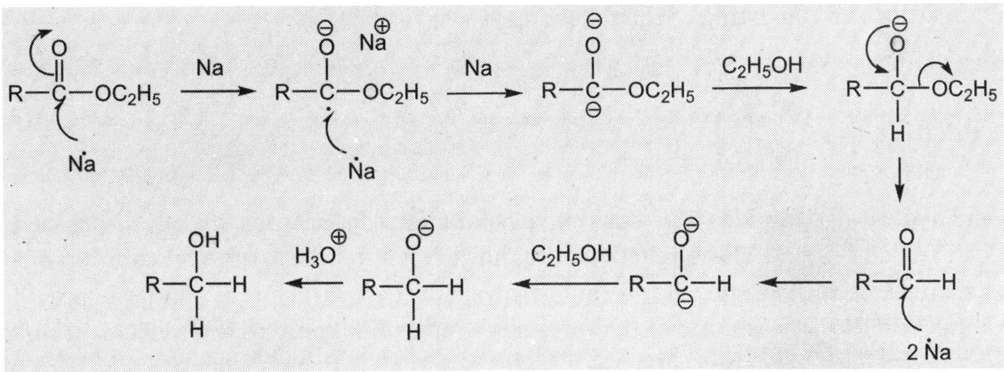

D. Hydrolysis of Amides

The hydrolysis of amides, like that of esters, is catalyzed by both acids and bases. The rates of hydrolysis have been found to be proportional to [amide] [$^{-}$OH] and [amide] [H^+] in basic and acidic media, respectively. Hydrolysis of *N*-substituted amides yields amines and carboxylic acids, and therefore, involves fission of the acyl–nitrogen bond rather than the alkyl–nitrogen bond.

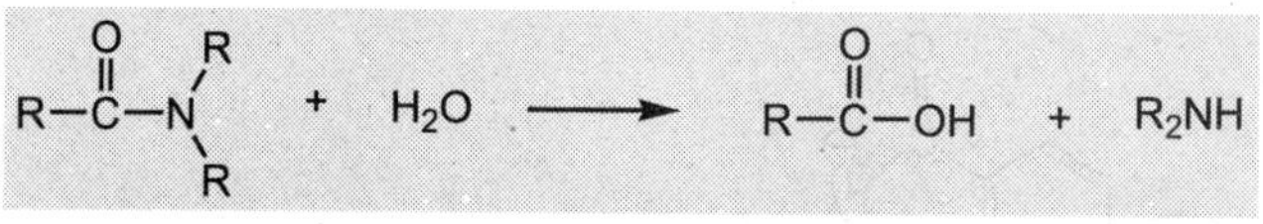

Thus, basic and acid hydrolyses of amides follow $B_{AC}2$ and $A_{AC}2$ mechanisms, respectively, in analogy with ester hydrolysis. The mechanism for $B_{AC}2$ hydrolysis of amides has been established by treating ^{18}O labelled benzamide in water containing hydroxide ions. It led to more exchange of ^{18}O in benzamide than its hydrolysis. It is, therefore, evident that the reaction proceeds by the carbonyl-addition mechanism in which there is reversible formation of an intermediate (**33**) that lives long enough to permit a proton shift, and establish equilibrium with another intermediate (**34**). The two oxygen atoms thus become equivalent.

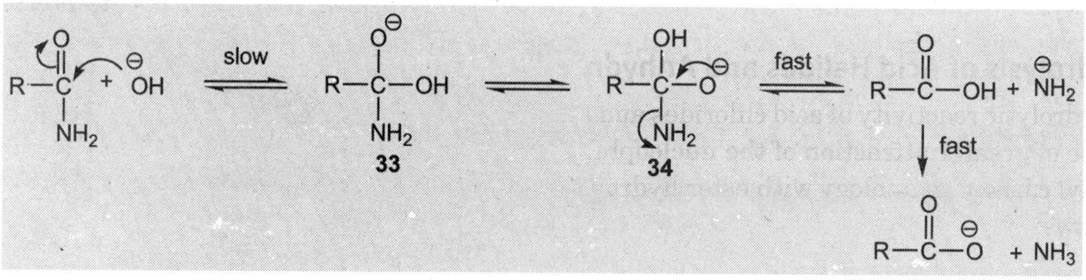

Basic hydrolysis of amides is accelerated by the presence of electron-withdrawing groups, and retarded by electron-donating groups. This is to be expected as the rate-determining step of the reaction involves an attack of nucleophilic ion on the carbonyl carbon. The hydrolysis is also subject to steric retardation by bulky groups.

Based on similar evidence, it appears almost certain that acid hydrolysis of amides follows an $A_{AC}2$ mechanism consisting of the following steps:

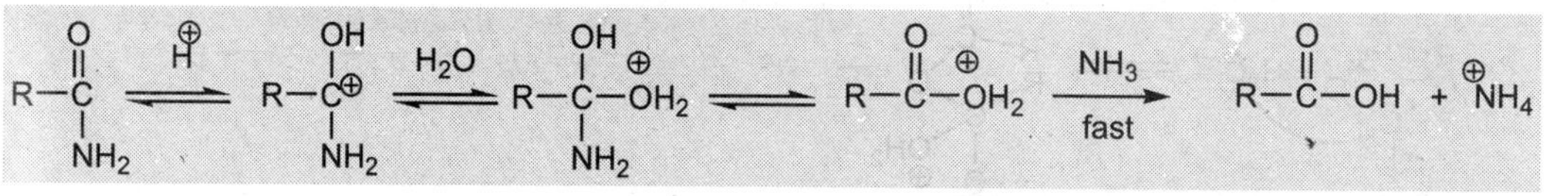

Acid hydrolysis of amides is insensitive to polar effects for reasons already discussed under $A_{AC}2$ hydrolysis of esters. These reactions are, however, subject to steric retardation by bulky groups. An interesting example of internal catalysis in the hydrolysis of amides is provided by phthalamic acid (**35**), whose rate of hydrolysis is proportional only to the concentration of the undissociated acid. It hydrolyzes at a much faster rate than benzamide or *p*-carboxybenzamide. Labelling experiments proved the intermediacy of phthalic anhydride (**36**), and on this basis a mechanism involving an internal nucleophilic attack by the carboxylate anion on the protonated amide group has been proposed.

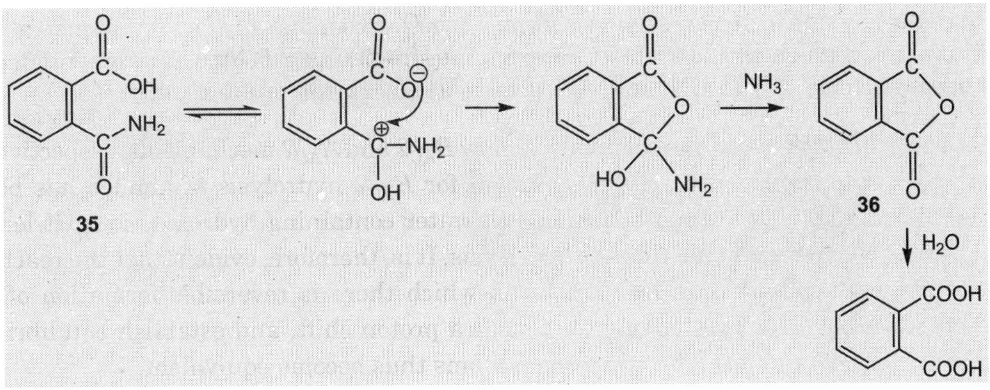

E. Hydrolysis of Acid Halides and Anhydrides

The hydrolytic reactivity of acid chlorides and anhydrides is more than that of carboxylic esters, because of greater attraction of the nucleophilic water molecule to the relatively more positive carbonyl carbon. In anology with ester hydrolysis, the mechanism of hydrolysis may be written as follows:

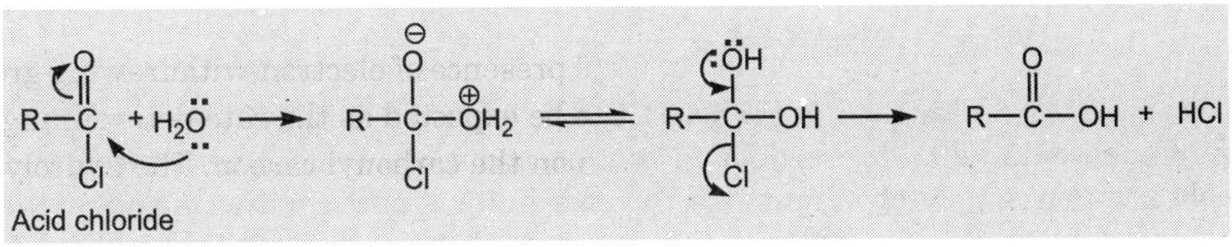

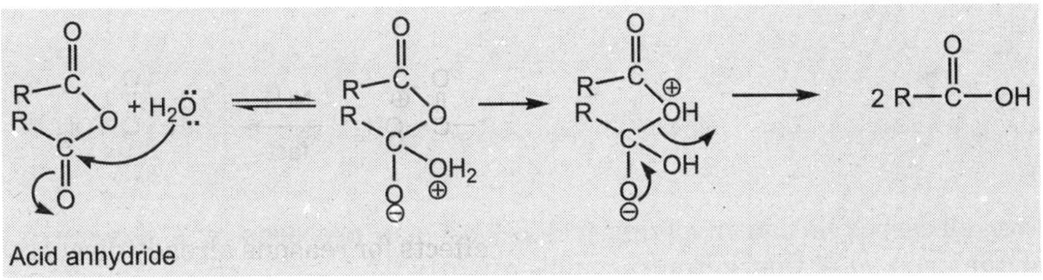

F. Decarboxylation of Carboxylic Acids

Typically, decarboxylation of a carboxylic acid takes place through a carboxylate anion from which the group R departs along with its bonding electron pair. Once formed, the carbanions react with a proton source (in solution, usually water) to give a new carbon–hydrogen bond.

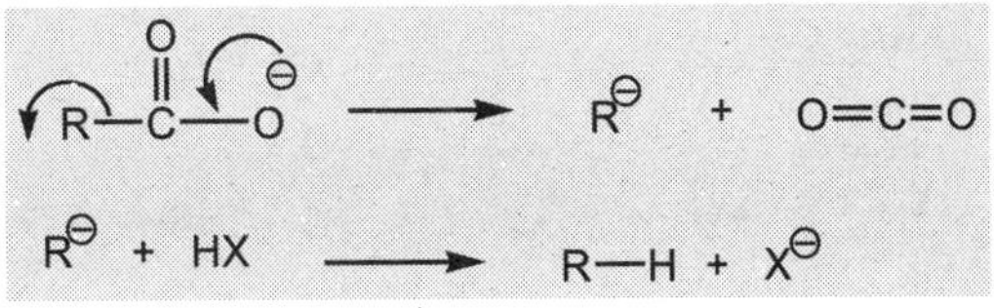

Since the key step in above reaction is formation of a carbanion, any factor that increases the stability of the carbanion should promote decarboxylation. Thus, acids with electron-withdrawing substituents such as $-NO_2$, –CN, and –CO groups decarboxylate quite readily.

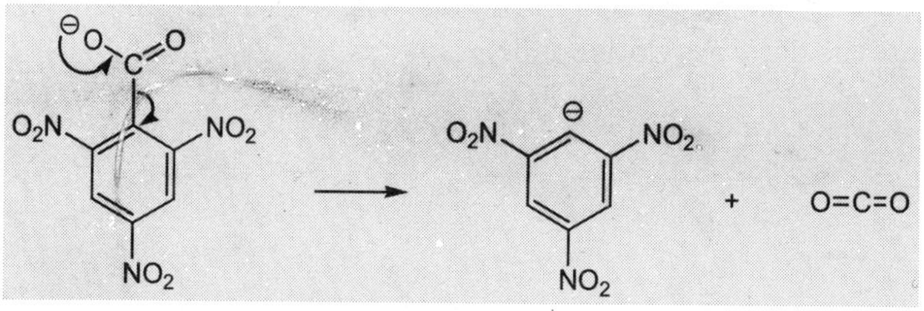

When the acid does not have carbanion stabilizing group, decarboxylation is difficult, and is usually accompanied by decomposition of the R group.

I. Decarboxylation of β-Keto Acids

A kinetic investigation of decarboxylation has revealed that decarboxylation of acetoacetic acid in aqueous solution follows the rate:

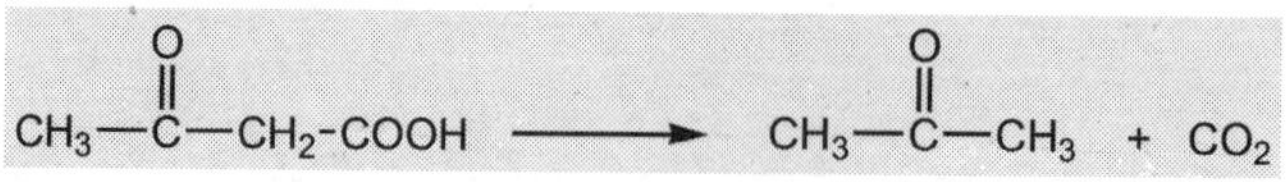

$$\text{Rate} = k\,[CH_3COCH_2COOH\,]\,[CH_3COCH_2COO\,]$$

A similar rate-law was observed for the decarboxylation of α,α-dimethylacetoacetic acid, and this observation completely eliminated the possibility of participation of enol in the decarboxylative mechanism. It, thus, appears reasonable that the two terms in the above rate-law represent the unimolecular decarboxylation of the keto acid and its anion.

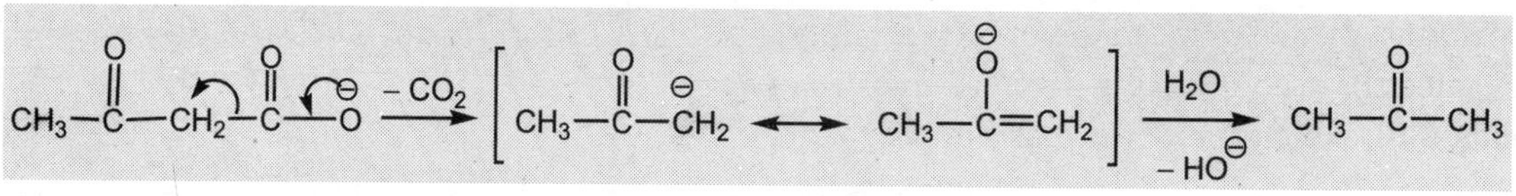

However, decarboxylation involving free β-keto acid proceeds by a six-centre transition state, which yields an *enol* that ketonizes subsequently.

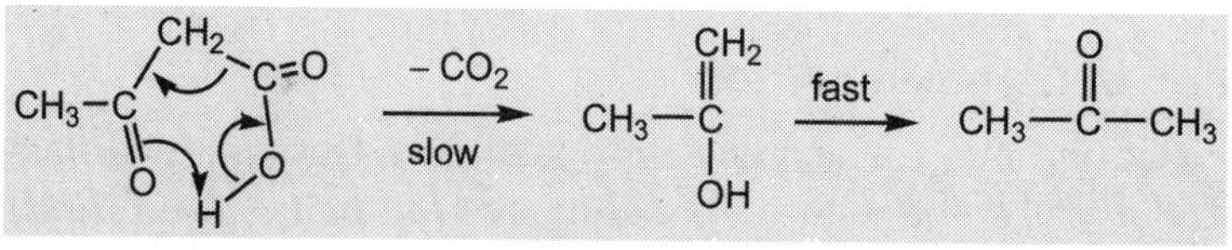

The intermediacy of the enol form in the above reaction is demonstrated by trapping it with halogens (bromine or iodine).

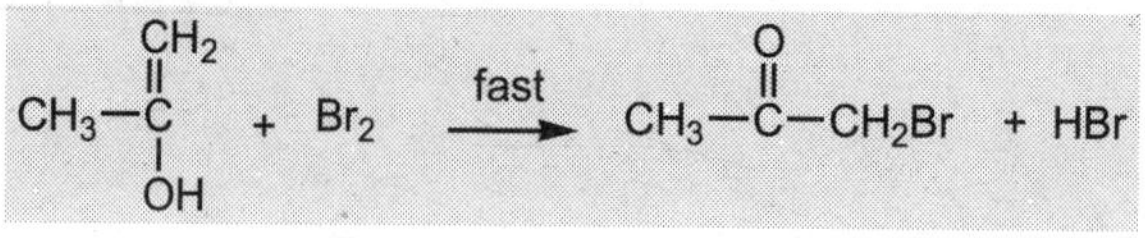

It has already been shown that acetone does not react significantly with bromine, and thus, the isolation of bromoacetone from the reaction mixture is a clear evidence for the existence of enol form, which forms in the rate-determining step. Accordingly, if the structure of the keto acid is such that the four atoms attached to the double bond of the intermediate enol are incapable of lying in a plane, the decarboxylation fails to take place.

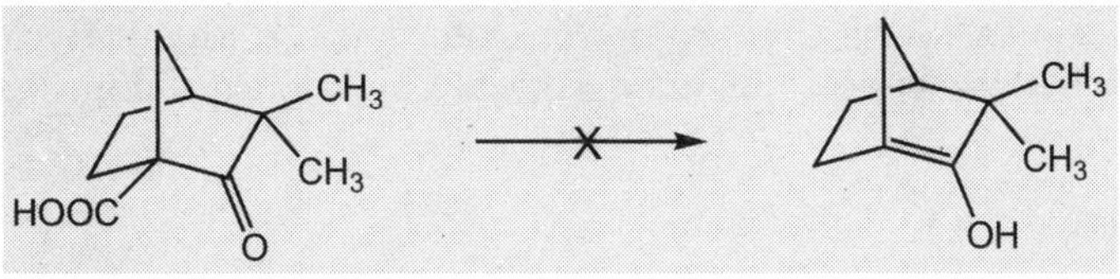

Mechanism of decarboxylation of malonic acid **(37)** and β,γ-unsaturated acids **(38)** may also be visualized in an analogous way.

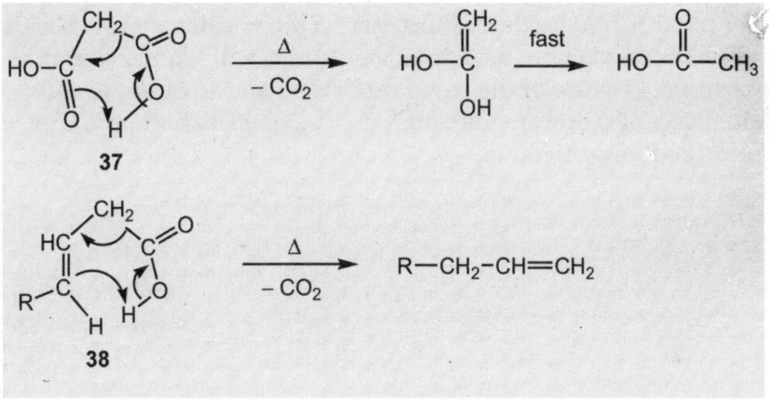

A study of the mechanism of thermal decarboxylation of α,β-unsaturated acids **(39)** has revealed that the first step involves their rearrangement to the corresponding β,γ-unsaturated acids **(38)**, which then decarboxylate through the above mechanism.

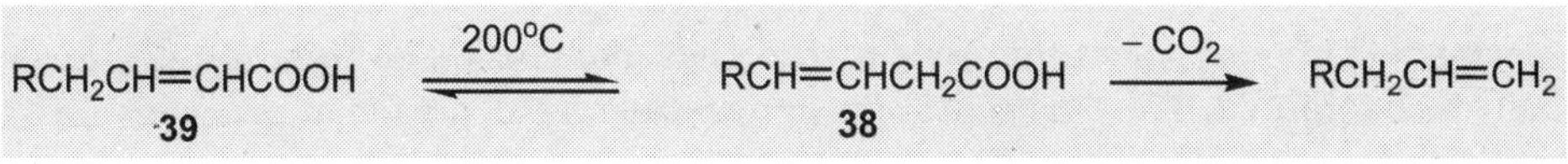

α,β-Unsaturated acids of the type $R_3CCH{=}CHCOOH$, in which the double bond is unable to undergo isomerization to β,γ, fail to decarboxylate under the above conditions.

II. Decarboxylation of Glycidic Acids

The acid-catalyzed decarboxylation of glycidic acids (**40**) presents an interesting example of a mechanism proceeding through a carbocation. For quite some time, it was believed that the reaction involves a concerted process in which decarboxylation and epoxide ring opening occur simultaneously yielding an *enol*, which finally tautomerises to the carbonyl compound.

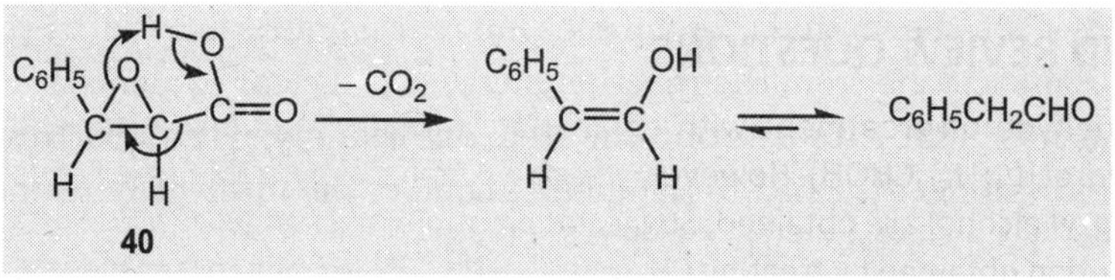

However, it has been shown that the decarboxylation of α-phenylglycidic acid **(41)** yields phenylacetaldehyde instead of acetophenone expected by a concerted mechanism.

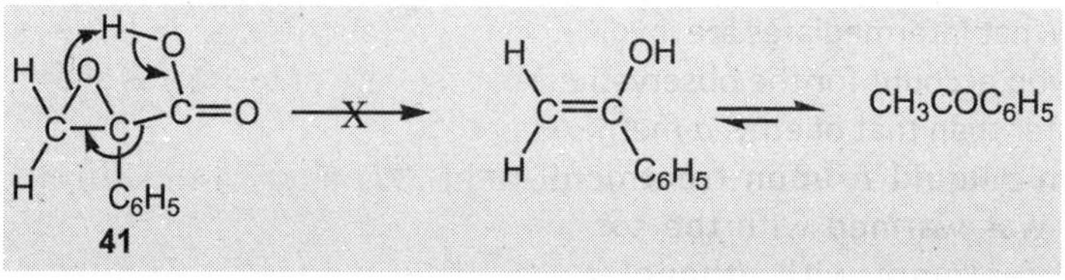

To account for the observed product, a carbocation intermediate formed by the isomerization of the initially protonated glycidic acid has been proposed. The carbocation is exclusively formed at the α-position, because of the resonance stabilization of the positive charge over the phenyl ring. Elimination of a proton and subsequent isomerization produces a species, which decarboxylate *via* an *enol* intermediate.

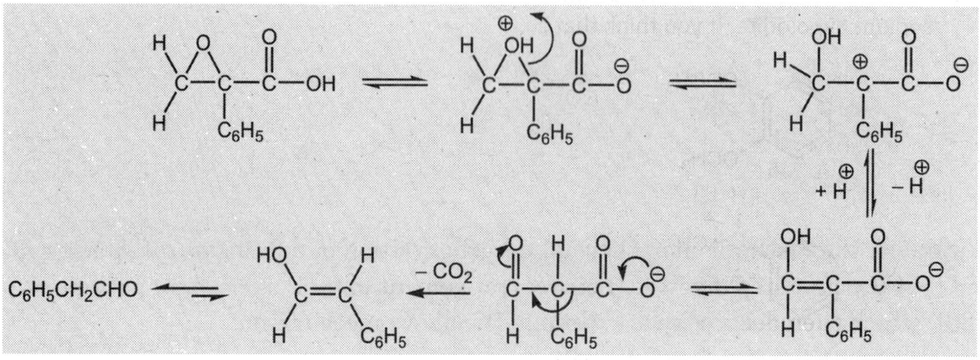

The formation of phenylacetaldehyde by the decarboxylation of β-phenylglycidic acid **(40)** can similarly be explained. The carbocation is generated on the β-carbon precisely for the same reason. Facile decarboxylation then yields the product.

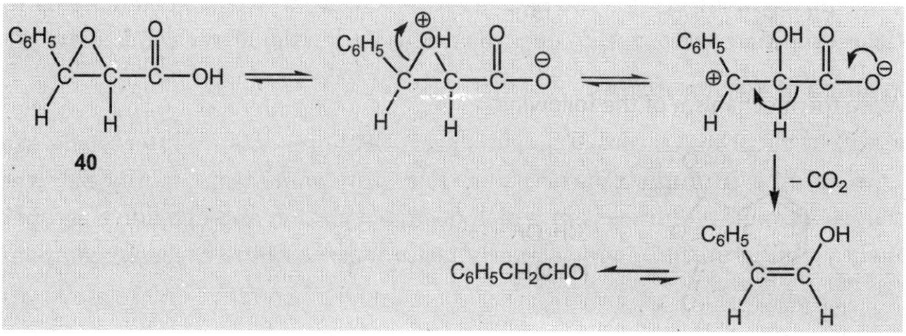

PROBLEMS AND REVIEW QUESTIONS

1. Benzaldehyde on treatment with equimolar amounts of phenyl magnesium bromide yields benzhydrol [$(C_6H_5)_2CHOH$]. However, when two moles of benzaldehyde are used, benzophenone and benzyl alcohol are obtained. Suggest a plausible mechanism.
2. The solution obtained by treating CH_3–CH=CH–CH_2Cl with magnesium was treated with acetone. After usual work-up, it was discovered that the product contained a mixture of two alkanols. Identify the alkanols, and suggest a suitable mechanism for their formation.
3. Cyanohydrins are often prepared by treating the aldehyde with acetone cyanohydrin and a trace of base. What intermediates are involved in this reaction? Why does the reaction proceed?
4. How do you account for the observation that the rate of hydrolysis of ethyl *p*-nitrobenzoate is much faster than that of ethyl *p*-methoxybenzoate?
5. An organic liquid **A** upon treatment with thionyl chloride yielded another liquid **B**. When **B** was warmed with the sodium salt of **A**, a third liquid **C** was obtained. Liquid **C** upon treatment with ethanol gave fourth liquid **D,** which showed the following NMR spectral characteristics:
   δ 1.2 (*t*, 3H), 2.0 (*s*, 3 H) and 4.1 (*q*, 2H).
   Identify **A**, **B**, **C,** and **D,** and also comment upon their mechanisms of formation.
6. Cyclohexanone forms an oxime which is tautomeric. Write the structures for both the tautomers.
7. Give the structure of the product(s) obtained when each one of the following is treated with sodium hypoiodite. If you think that a certain compound would not react, write 'No reaction'.

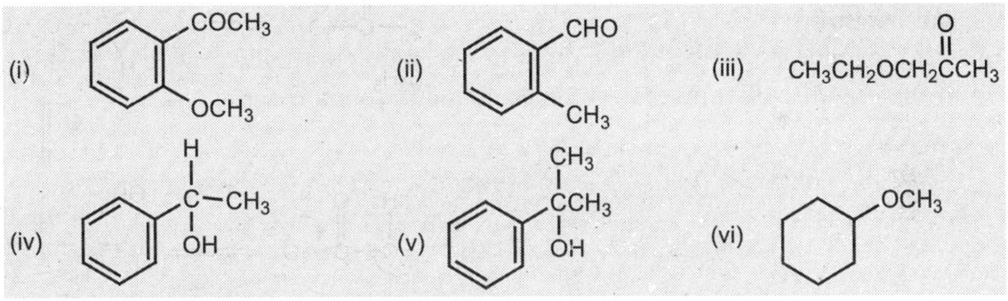

8. Using $H_2{}^{18}O$ as the source of labelled oxygen, outline the synthesis of various methyl benzoates, **A**, **B**, and **C**.

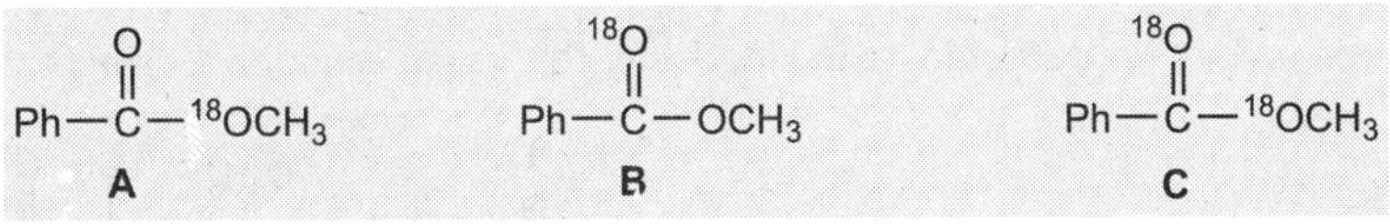

9. Write the mechanism of the following reaction:

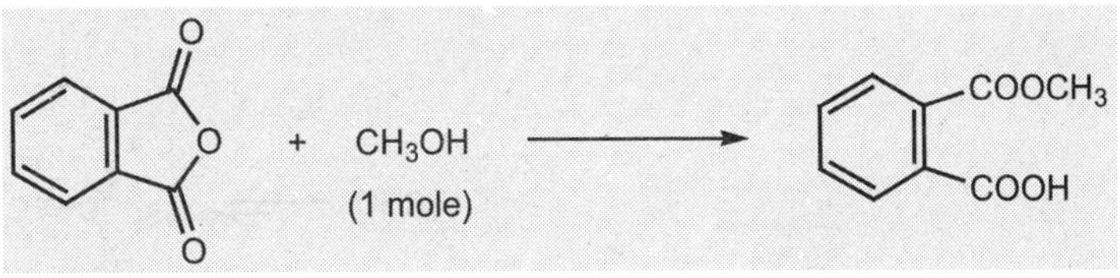

10. Suggest an experiment to show that in Cannizzaro reaction, transfer of hydride to the carbonyl carbon takes place from the substrate, and not from the solvent.
11. Illustrate the mechanism of the formation of semicarbazone. How do you account for the observation that only semicarbazone of acetone is isolated when a molecule of semicarbazide is treated with a mixture of acetone (1 mole) and benzaldehyde (1 mole), whereas only semicarbazone of benzaldehyde is isolated if the reaction mixture is allowed to stay for several hours?
12. Predict the product(s), and propose a suitable mechanism for each of the following transformations:

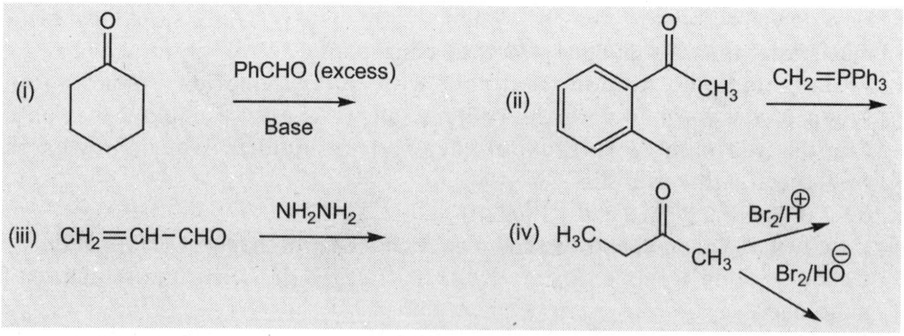

13 Hydroxide-ion promoted hydrolysis of *S*-2-butyl acetate gives *S*-2-butanol and acetate ion. Outline a suitable mechanism based on this observation. Also explain how you rule out other possible mechanisms.
14. List the following amides in the order of decreasing reactivity toward hydroxide ion promoted hydrolysis. Also offer mechanistic explanation to your answer.

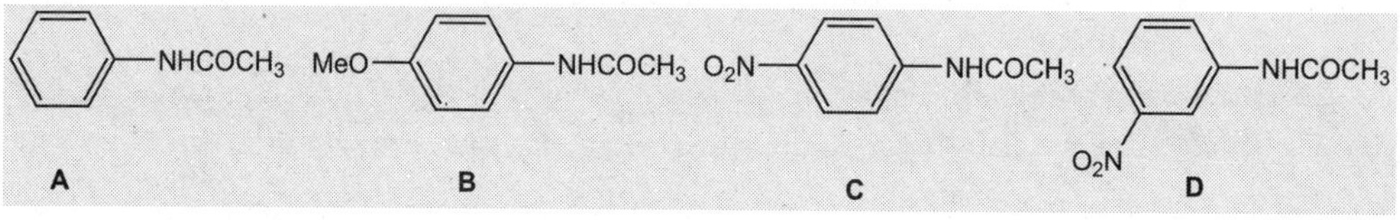

15. Define 'transesterification'. What are the suitable conditions to bring about the desired transesterification?
16. A pair of alcohols is formed in the LAH reduction of ketones **A** and **B**. Which alcohol in each pair would you expect to predominate in kinetically controlled reactions? Give reasons.

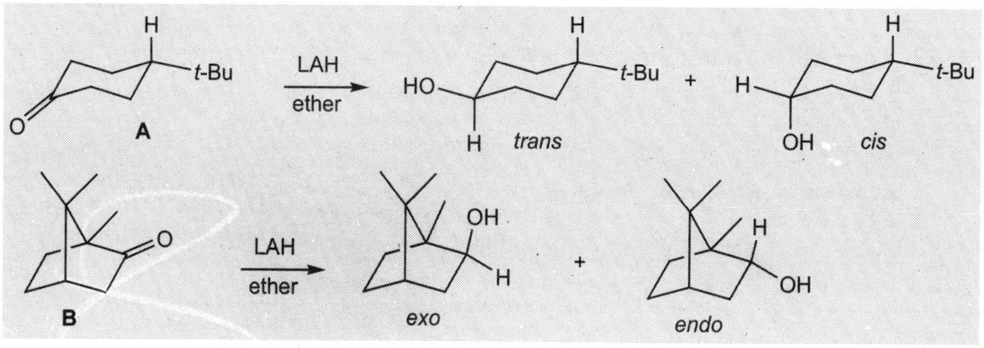

17. Write the structure of the product obtained on treating compound **A** with LAH.

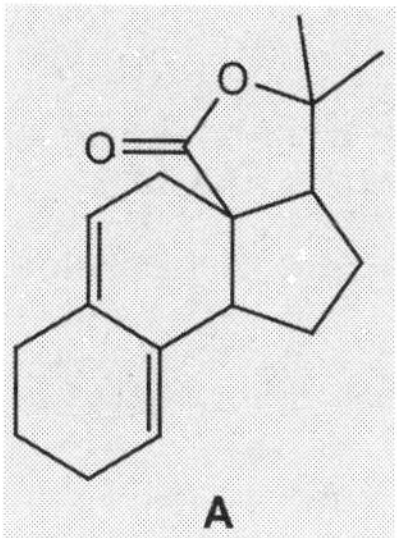

A

18. What reagents would be required to bring about the following transformations?

(i) $CH_2{=}CH(CH_2)_3COCH_3 \xrightarrow{?} CH_2{=}CH(CH_2)_3CHOHCH_3$

(ii) $CH_3CO(CH_2)_3COOCH_3 \xrightarrow{?} CH_3CHOH(CH_2)_3COOCH_3$

(iii) $CH_3COCH_2CH_3 + C_6H_5COOC_2H_5 \xrightarrow{?} C_6H_5COCH_2COCH_2CH_3 + C_6H_5COCH(CH_3)COCH_3$

19. A dicarboxylic acid **A** ($C_4H_6O_4$), gave product **B** ($C_6H_{10}O_4$) upon treatment with excess of methanol and a trace of H_2SO_4. Subsequent treatment of **B** with LAH followed by usual work-up afforded **C** ($C_4H_{10}O_2$). Assign structures to **A, B,** and **C**. Also outline the mechanism for each step.
20. Account for the following observations:
    (i) Phenyl esters are more reactive than alkyl esters towards nucleophiles.
    (ii) The rate of aminolysis cannot be increased by H^+, HO^- or RO^-.
    (iii) Conversion of an ester such as methyl butanoate to butanal is only possible by using diisobutylaluminium hydride at −78°C, but not with LAH or $NaBH_4$.
21. How will you convert 2-carbethoxycyclohexanone (**A**) into 2-hydroxymethylcyclohexanone (**B**)? Outline the mechanism for all the steps involved in this conversion.

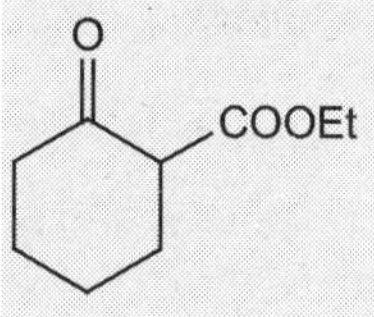

A

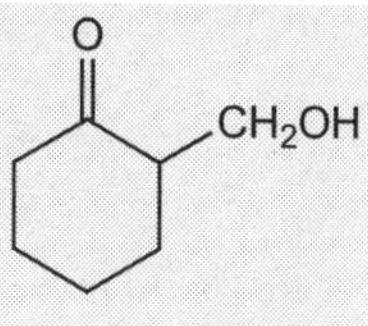

B

22. Complete the transformations, and outline the mechanism for each of the reaction.

(i) $PhCHO + Ph_3P{=}CHPh \longrightarrow$ **A** (major) + **B** (minor)

(ii) $Ph{-}CH{=}CH{-}COPh \xrightarrow{\textbf{C}} Ph{-}CH(Ph){-}CH_2COPh$

(iii) **D** $\xrightarrow[2.\ H^{\oplus},\ H_2O]{1.\ Me_2CuLi}$ 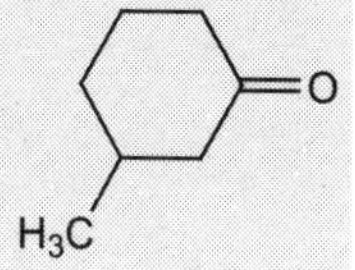

23. List the following esters in order of increasing reactivity toward hydroxide ion promoted hydrolysis.

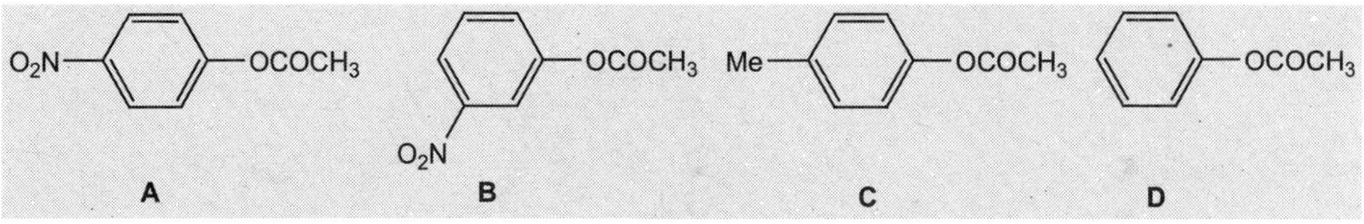

24. Base-catalyzed condensation of diesters **A** (M. F. = $C_7H_{12}O_4$) and **B** (M. F. = $C_8H_{14}O_4$) with a ketone RCOR yields the products **C** and **D**, respectively, give the
   (i) structures of **A** and **B**,
   (ii) reagents/conditions required for the formation of **C** and **D**
   (iii) names
   (iv) mechanisms for these reactions.

25. Ethylene glycol is an extremely useful reagent for the protection of aldehydes and ketones. Write a mechanism for the following reaction.

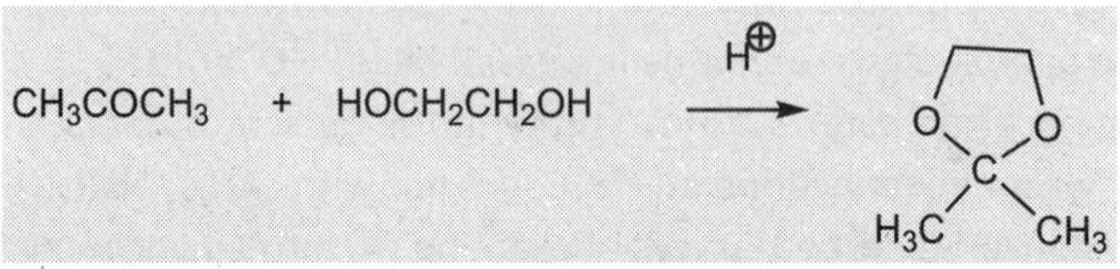

26. α-Hydroxy aldehydes can exist as an equilibrium mixture of open-chain hydroxyaldehyde and cyclic hemiacetal form. Write a mechanism for the following case.

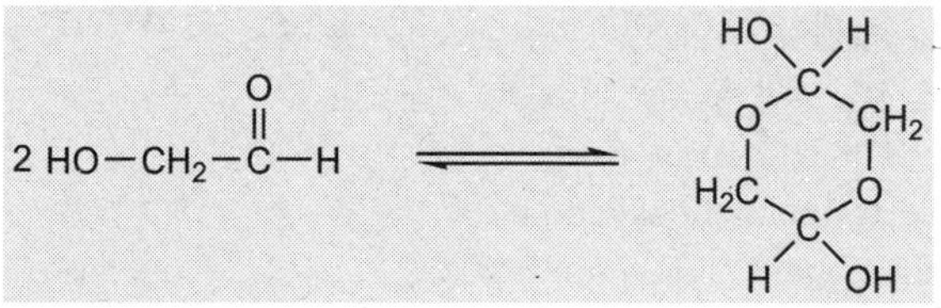

27. Complete the following reactions, and outline the mechanism for each of them.

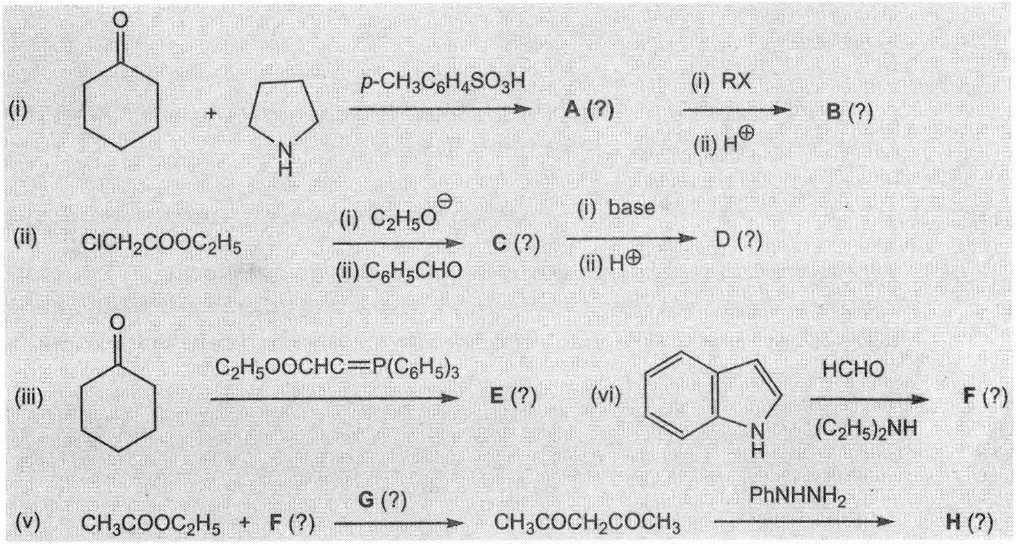

28. What is the effect of structure on the reactivity of carbonyl group? Discuss an experiment which demonstrates the formation of *gem*-diols on the treatment of carbonyl compounds with water. Is it possible to isolate the *gem*-diols?
29. Discuss the mechanism of formation of hemiacetals and acetals. How is it related with the phenomenon of mutarotation?
30. Discuss the mechanism of reactions involving transfer of a hydride ion from a donor molecule to the carbonyl carbon.
31. How do you explain the acidity of α-hydrogens in carbonyl compounds? Discuss the mechanism of aldol condensation. Could this reaction be catalyzed by acids as well?
32. Discuss the mechanisms of the following reactions:
    (i) Perkin condensation
    (ii) Benzoin condensation
    (iii) Mannich reaction
33. In the presence of a strong base, an α-bromo ester reacts with a carbonyl compound resulting in the formation of an α,β-epoxy ester, commonly referred to as a glycidic ester. How can glycidic esters be converted to carbonyl compounds? Discuss the mechanisms of the process involved.
34. What are Wittig reagents? How do they react with aldehydes and ketones?
35. Classify and discuss the various mechanisms by which esterification of an acid and the hydrolysis of an ester may proceed.
36. Discuss the mechanism of Claisen condensation. How will you synthesize an α-keto ester without α-hydrogen atoms?
37. Comment on the mechanisms involved in the hydrolysis of amides, acid halides, and anhydrides. Discuss their comparative reactivity.
38. What is the driving force in the decarboxylation of carboxylic acid? Discuss the mechanism of decarboxylation of β-keto acid.

39. Discuss the mechanism of base-catalyzed exchange of oxygen of acetone with $H_2{}^{18}O$.
40. Suggest a plausible mechanism for the formation of a Grignard reagent from an alkyl halide and magnesium.
41. Discuss the mechanism of Cannizzaro reaction. Illustrate by choosing suitable examples the mechanistic course of intramolecular and crossed Cannizzaro reaction. What is the role of deuterium in establishing the mechanism of this reaction?

SOLUTIONS

1. With equimolar reactants, we get benzhydrol as the normal product of a Grignard reaction. However, with two moles, the salt of benzhydrol is oxidized to benzophenone by a hydride transfer to the excess benzaldehyde. The driving force for this transfer may be the presence of a negative charge on alkoxide oxygen.

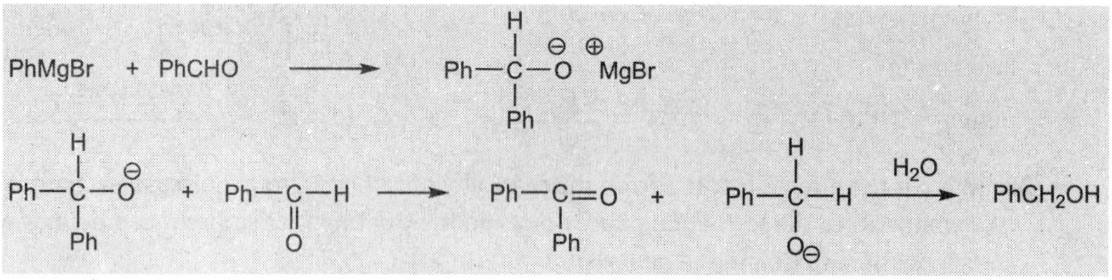

2. In the first step, there is the formation of a resonance-stabilized free allylic radical. Coupling of such radicals with ·MgX can give rise to two Grignard reagents and hence two products.

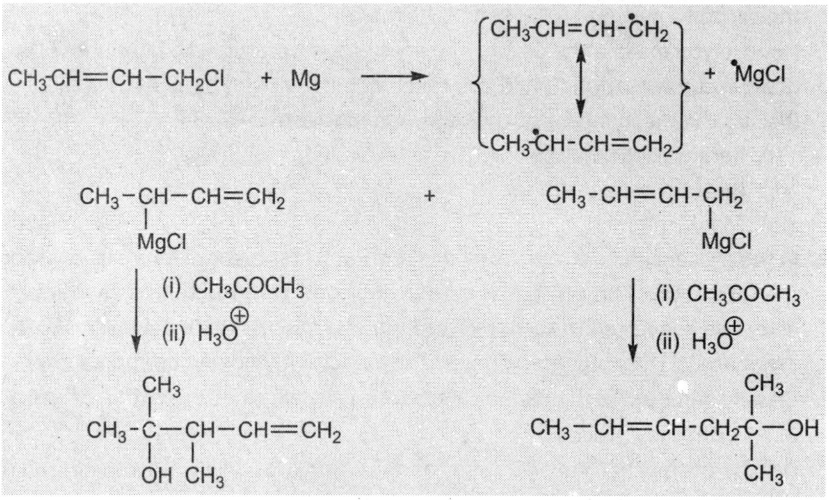

3. Being reversible reaction, cyanohydrin formation is more favourable for aldehydes than ketones. Acetone cyanohydrin decomposes to acetone and HCN, which reacts with an aldehyde to form more stable cyanohydrin.

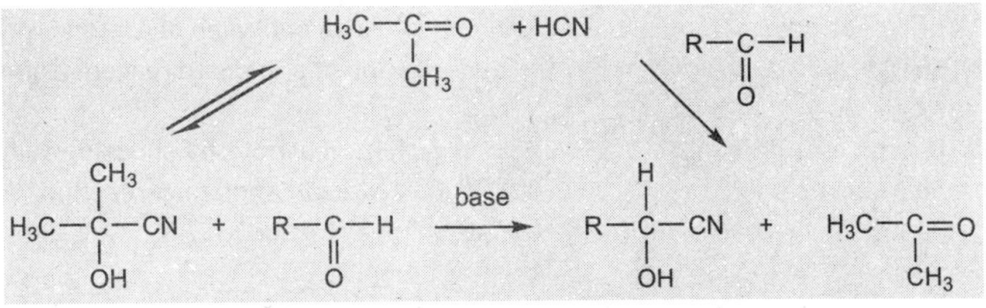

4. Electron-withdrawal by nitro group accelerates the attack of hydroxide ion on the carbonyl carbon of the ester group by effectively dispersing the negative charge. Reverse is true for the methoxy substituent.
5. **A**, CH_3COOH; **B**, CH_3COCl; **C**, $(CH_3CO)_2O$; **D**, $CH_3COOC_2H_5$.

6.

7. It is iodoform reaction. Only those compounds will react which have $-COCH_3$ or $-CH(OH)CH_3$, to give a yellow precipitate of iodoform.

8.

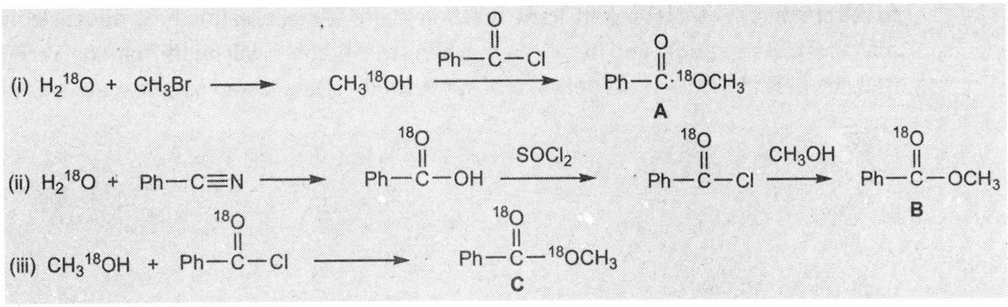

9.

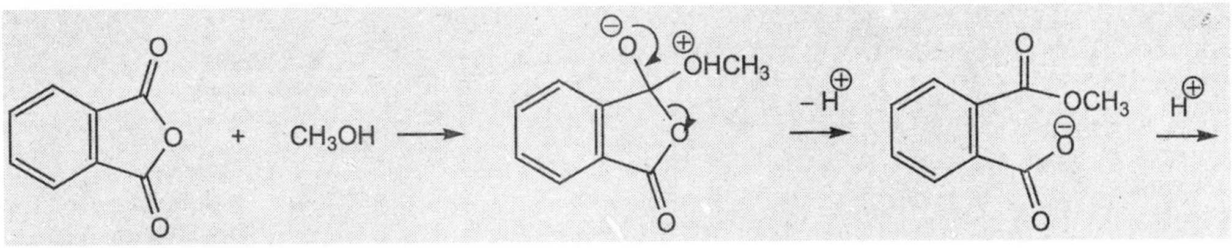

10. Use of RCDO instead of RCHO and look for RCD_2OH.
11. The reaction is reversible. Because of high reactivity acetone reacts faster than benzaldehyde, but latter gives the more stable product. We can, therefore, say that formation of semicarbazone of acetone is kinetically controlled, whereas the other product arises as a result of thermodynamic control.

12.

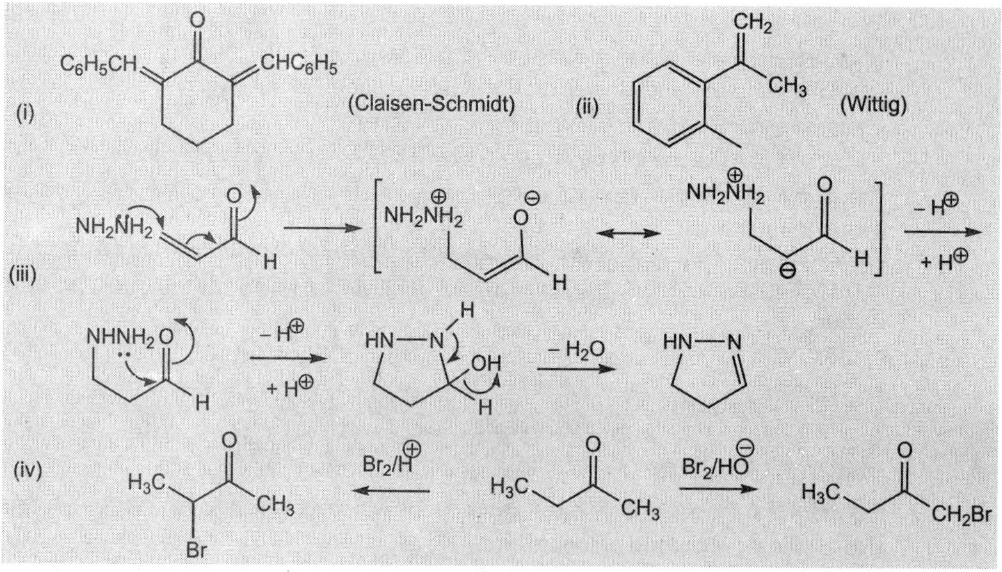

13. There are three possible mechanisms for hydroxide-ion promoted hydrolysis of esters: (i) nucleophilic acyl substitution, (ii) S_N2 mechanism and (iii) S_N1 mechanism. If the reaction involves mechanism (i), the product alcohol will have the same configuration (as the reactant) as no bond to the chirality centre is broken. If the reaction is an S_N2 reaction (ii), the product alcohol will have opposite configuration, and in case of mechanism (iii), there will be formation a racemic mixture (may be partial). Thus, nucleophilic acyl substitution is suggested for the mechanism.

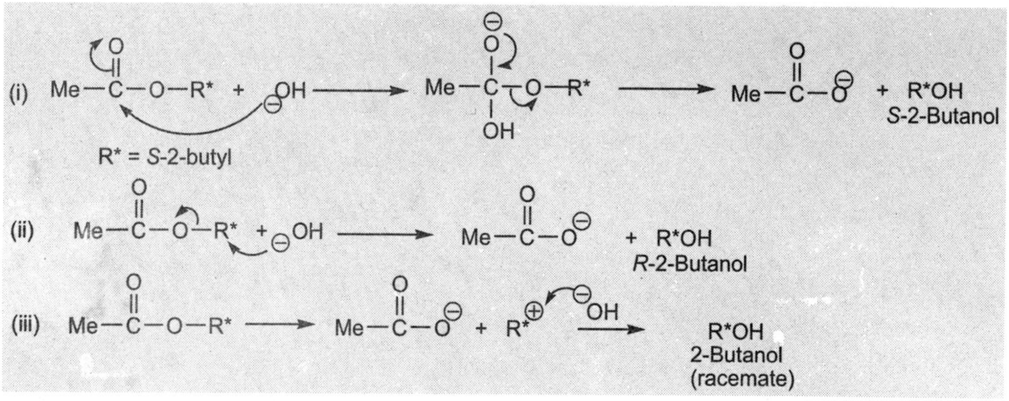

14. **C > D > A > B**
15. The reaction of an ester with an alcohol to form a new ester and new alcohol is known as 'transesterification'. The reaction is reversible and catalyzed either by acid or conjugate base (R^3O^-) of the reactant alcohol.

$$R^1COOR^2 + R^3OH \rightleftharpoons R^1COOR^3 + R^2OH$$

Two conditions are significant for effecting to the reaction to drive the equilibrium to the right: (i) to use an excess of the reactant alcohol (R^3OH), (ii) the boiling point of the product alcohol (R^2OH) should be much lower than the boiling point of the reactant alcohol so that the same could be separated by distillation.

16. Reduction of **A**: As there is no favoured site for the transfer of hydride from aluminium to carbon, the favoured product is expected to be more stable. This is the *trans* product in which the OH group is equatorial.

    Reduction of **B**: The transfer of hydride ion to the carbonyl carbon of **B** is apparently more hindered in one direction than the other. Approach from the less-hindered side is predicted to be more favourable and will lead to the *exo*-alcohol.

17.

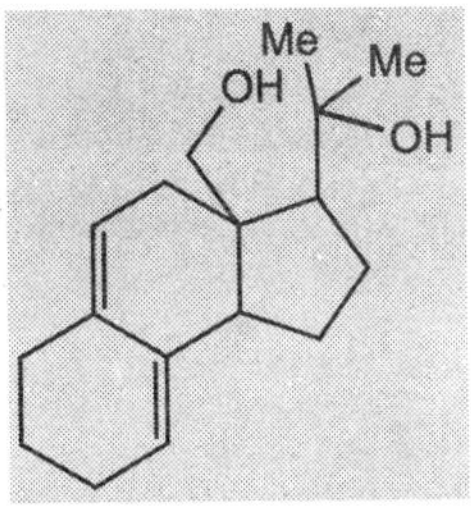

18. (i) $NaBH_4$, (ii) $NaBH_4$, (iii) C_2H_5ONa.

19.

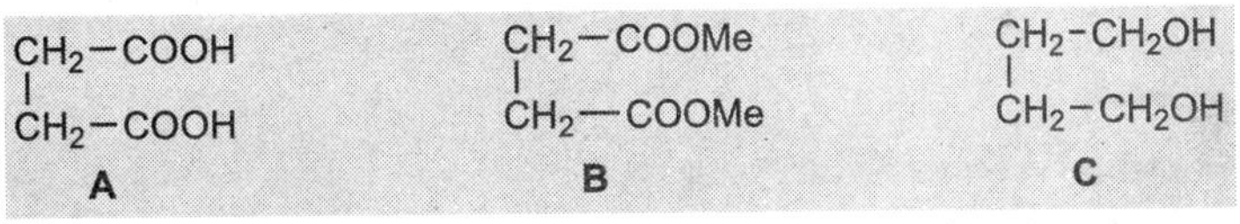

21, Ketonic group of starting β-keto ester **A** will be protected by forming ketal and then reduction of ester group will be effected by using LAH. Finally, desired product will be obtained by acid hydrolysis of the ketal.

22.

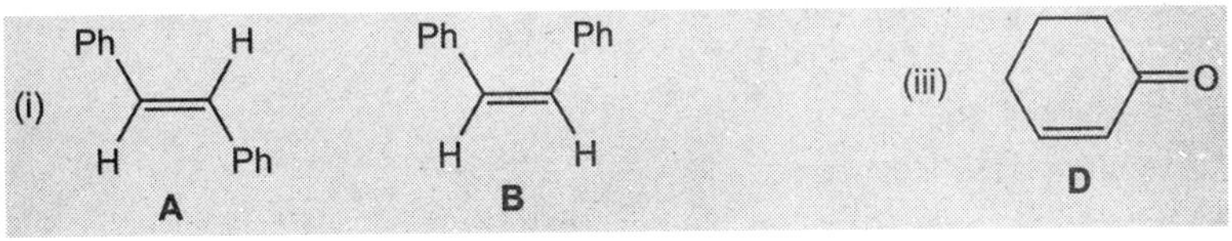

Chapter **10**

Aromatic Electrophilic Substitution

Chapter Outline

10.1 ELECTROPHILIC SUBSTITUTION REACTIONS OF BENZENE

- **A.** Nitration
- **B.** Halogenation
- **C.** Sulphonation
- **D.** Mercuration
- **E.** Friedel-Crafts Reaction

10.2 REACTIVITY AND ORIENTATION

- **A.** Reactivity
- **B.** Orientation
- **C.** Effect of Substituents
- **D.** Orientation in Other Ring Systems
- **E.** The *ortho-para* Ratio
- **F.** Competition between the Substituents
- **G.** *Ipso* Substitution

10.3 QUANTITATIVE TREATMENT

10.4 SOME SYNTHETICALLY USEFUL AROMATIC ELECTROPHILIC SUBSTITUTION REACTIONS

- **A.** Diazonium Coupling
- **B.** Vilsmeir-Haack Reaction
- **C.** Gattermann-Koch Reaction
- **D.** Bischler-Napieralski Reaction
- **E.** Houben-Hoesch Reaction

Resonance confers extra stability to the aromatic rings, and such systems undergo substitution rather than addition reactions. Further, these substitutions are normally brought about by the reagents, which are electrophilic in character. A look at the structure of benzene reveals that the σ skeleton of benzene is planar, and is sandwiched between the two lobes of π-electron cloud, one above and the other below this plane. This delocalization of electrons in the aromatic system is responsible for holding the carbons together with a greater strength than in the case of alkenes. But still π-electrons are loosely held as compared to σ electrons. Hence, these electrons are easily available to an electrophile, which is a positive ion or a dipole or induced dipole. The group that departs must, therefore, leave without its electron pair.

Thus, the typical reactions of the benzene ring are *electrophilic substitution reactions* just as the typical reactions of alkenes are *electrophilic addition reactions*. Further, these reactions are characteristic of benzene ring, and also of many other aromatic systems (benzenoid or non-benzenoid). These reactions are very important synthetically as well as mechanistically.

In an electrophilic aromatic substitution reaction, a covalent bond between the aromatic compound and the electrophilic reagent is formed for which the necessary electron pair is supplied by the former. The reaction can be visualized as taking place through the following two different pathways – (i) concerted process, and (ii) two-step process.

In the concerted mechanism, the covalent bond formation and C–H bond breaking are *simultaneous*, and the reaction passes through a *transition state* (**1**), which resembles the *transition state* of the S_N2 reaction of aliphatic compounds.

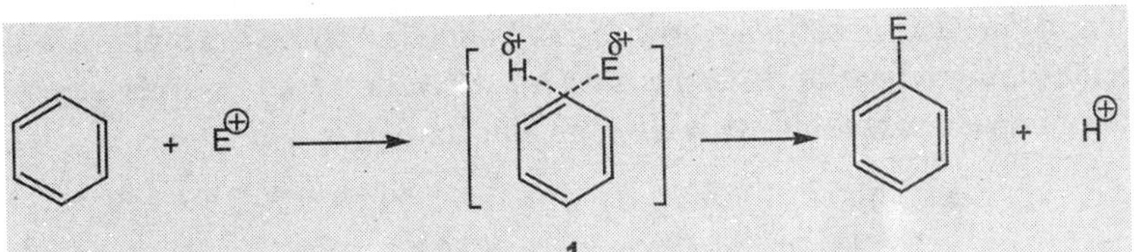

1

Alternatively, it may be a two-step process initiated by the addition of electrophile to the nucleus, forming an *intermediate* (**2**), followed by the loss of a proton. Either of them can be the rate-determining step of the reaction. This mechanism is also known as *arenium-ion mechanism*.

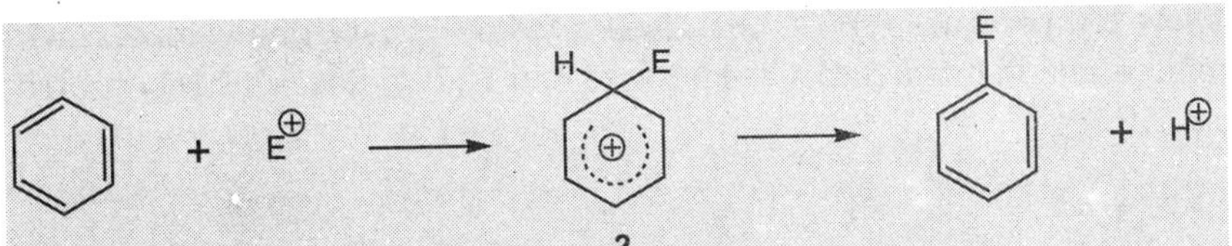

2

Now the question arises: which of the two processes represents the correct mechanism? On the basis of experimental observations, it is now certain that the second mechanism operates in aromatic electrophilic substitution reactions. In this mechanism, the first step involves the addition of the electrophile to the aromatic nucleus forming a resonance-stabilized intermediate called cyclohexadienyl cation or arenium-ion. Intermediates of this type, which correspond to the energy minimum on the reaction coordinate diagram, were first described by Wheland, and are sometimes referred to as *Wheland intermediates.* The Wheland intermediate loses a proton in the second step producing the substituted benzene derivative. Either of these steps can be the rate-determining step of the reaction.

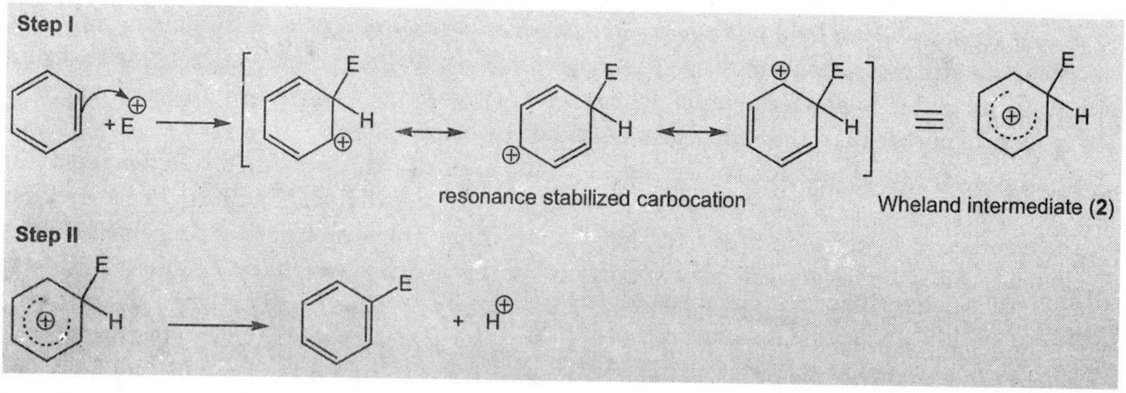

On the basis of the following two observations, it is now certain that the second mechanism operates in electrophilic aromatic substitutions.

(a) Isolation of Intermediates

The intermediates of aromatic substitutions have not only been detected, but have also been isolated and characterized in several cases. Following three examples are illustrative:

(i) Cyclohexadienyl cation intermediate has been detected by NMR spectroscopy when fluorobenzene is treated with HF-SbF_5 in liq. SO_2.

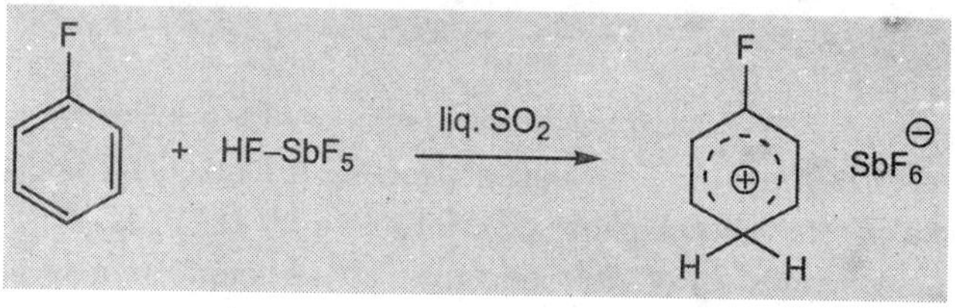

(ii) When benzotrifluoride is treated with nitryl fluoride in the presence of boron trifluoride at low temperature (–80°C), it gives a yellow crystalline complex, which has been isolated and identified. On warming to –50°C, it yields *m*-nitrobenzotrifluoride, BF_3, and HF.

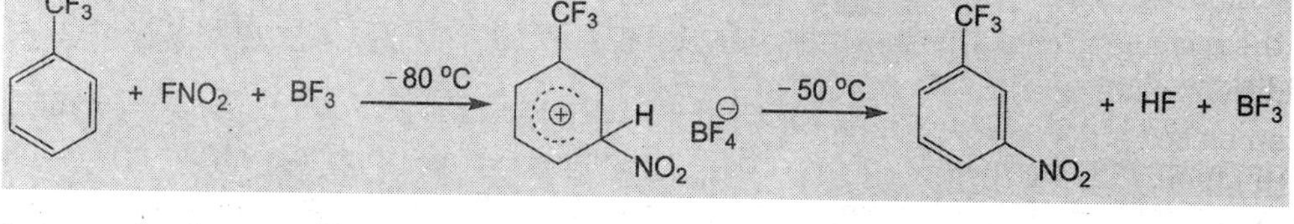

(iii) A σ-complex intermediate has also been isolated by the treatment of mesitylene with ethyl fluoride and BF_3 at –80°C. The normal substitution product was obtained on warming this intermediate.

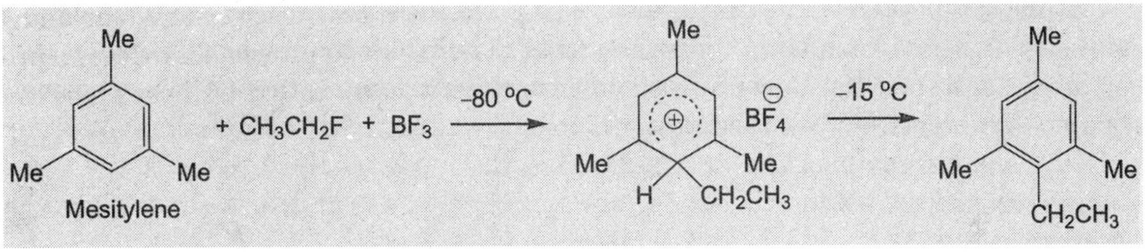

(b) Isotope Effects

A difference in the rate of reaction due to a difference in the isotope present is called the *isotope effect*. We know that a carbon–hydrogen bond is broken more readily than a bond between carbon and deuterium, which in turn breaks more easily than a carbon–tritium bond. It, therefore, follows that whenever hydrogen is lost in the rate-determining step of a reaction, it will show an isotope effect on replacing the hydrogen by deuterium. In other words, a deuterated substance should undergo substitution more slowly than a non-deuterated compound. Such isotope effects have not been observed in most aromatic substitutions. For example, the rates of nitration of *deuterio* and *tritio*-benzenes are the same as that for benzene. Thus, it is reasonable to conclude that proton elimination is not the rate-determining step of the reaction.

Substitution versus Addition Reactions: The tendency of aromatic compounds to undergo substitution rather than addition is associated with *aromaticity,* which is the stability conferred upon a cyclic planar conjugated polyene having $(4q + 2)$ π-electrons. The addition product will no longer retain this special feature, and will, therefore, be far less stable than the starting aromatic compound. Hence, it will not be formed. Completely delocalized π orbitals are regained by the expulsion of a proton from the *intermediate* **(2)**, thus recovering aromatic stability.

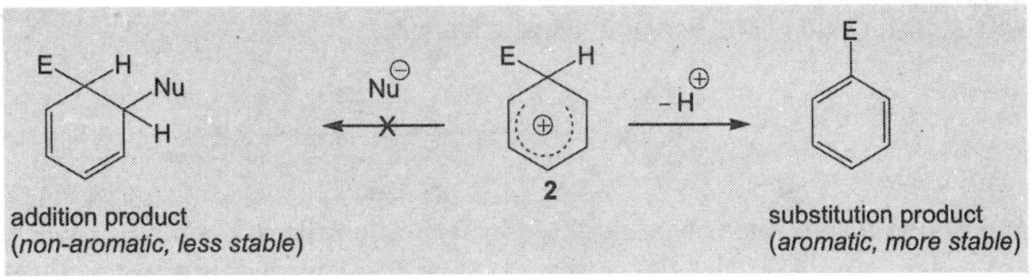

Thus, benzene and other aromatic compounds do not undergo electrophilic addition reactions since the second step is endothermic at least by 150.48 kJ/mol or 35.82 kcal/mol (equal to the resonance energy of benzene). On the other hand, alkenes prefer to undergo electrophilic addition reaction since no such effects are involved. The first step involving the initial attack of an electrophile on the double bond to produce a carbocation as intermediate is the same for both alkenes as well as benzene. But, the rate of addition to benzene is much slower than that of alkenes. The reason is that the carbocation intermediate from benzene is much less stable than benzene itself, because it has 150.48 kJ/mol or 35.82 kcal/mol higher energy (resonance energy) than benzene. In other words, E_{act} for the formation of carbocation intermediate from benzene is much higher than that of alkenes.

Role of σ- and π-Complexes: The intermediates in aromatic substitutions are commonly called σ-complexes. In these complexes, there is actual bonding of the reagent to one of the ring carbon atoms, and since the π-orbitals now involve only five atoms, there is loss of aromaticity. Experimental evidence, however, suggests that the formation of a σ-complex follows the initial formation of another complex known as π-complex. In contrast to the σ-complex, the π-complex does not involve actual bonding. Actually, the electrophile is held near the π-electron cloud of the aromatic ring.

The following two experiments clearly support the formation of σ- and π-complexes:

(i) In the presence of $AlCl_3$, benzene forms a 1 : 1 complex with HCl. The complex is coloured, and its solution conducts electricity, signifying its ionic character. Moreover, when DCl is taken in place of HCl, the benzene protons are exchanged with deuterium. This can be explained by assuming that the deuterium cation attaches itself to the ring carbon, and thus, becomes equivalent to the proton on the same carbon. Either of them can be lost on decomposition of the complex to yield the products. All these properties are explained by assigning structure **3** to the complex.

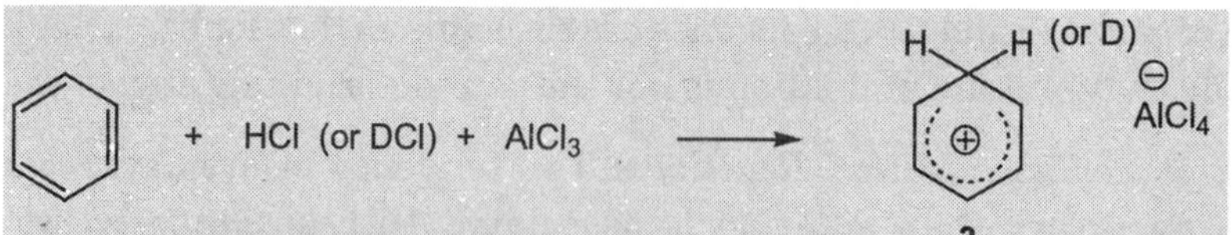

3

(ii) In the absence of $AlCl_3$, however, benzene still forms a 1 : 1 complex with HCl. But, in contrast to the σ-complex, the solution of this complex is colourless, and does not conduct electricity. Finally, when HCl is substituted for DCl, the recovered benzene is not found to contain any deuterium. This type of complex is called a π-complex. It is formed by the loose association of a proton (or any other electrophile) with the π-electron cloud of the benzene ring as shown.

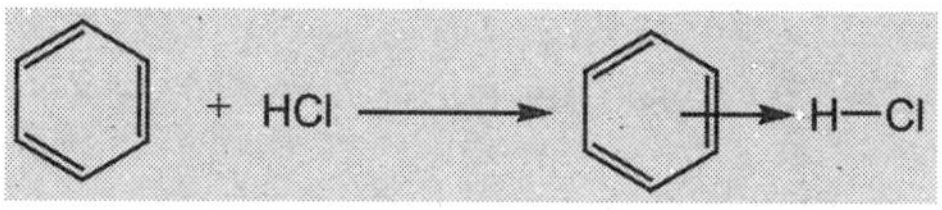

As mentioned above, the initial step of electrophilic aromatic substitution involves the attack of the electrophile on the π-electron cloud of the ring leading to the formation of a π-complex. Now, in the π-complex, the electrophile removes a pair of electrons from the ring and enters into covalent bond formation with any one of the ring carbons to yield the σ-complex. This σ-complex loses a proton, which before departing from the substrate again forms a π-complex. Thus, a complete picture of the mechanism of electrophilic substitution would be as follows:

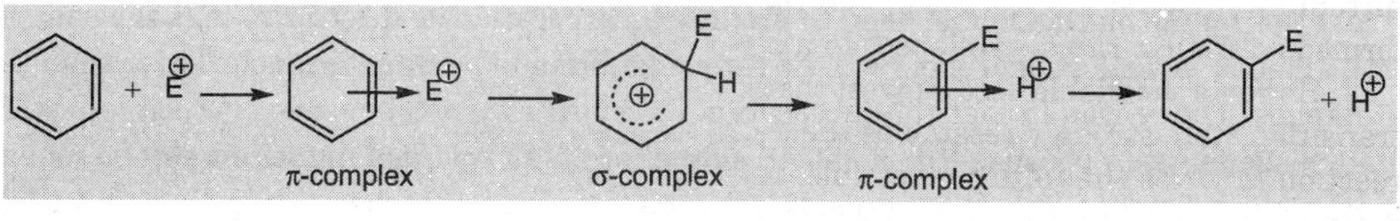

The question now arises: which of the two complexes, σ or π, is involved in the rate-determining step of the reaction? In case the π-complex is formed in the rate-determining step, the energy profile diagram of substitution would be as shown in **Fig. 10.1**.

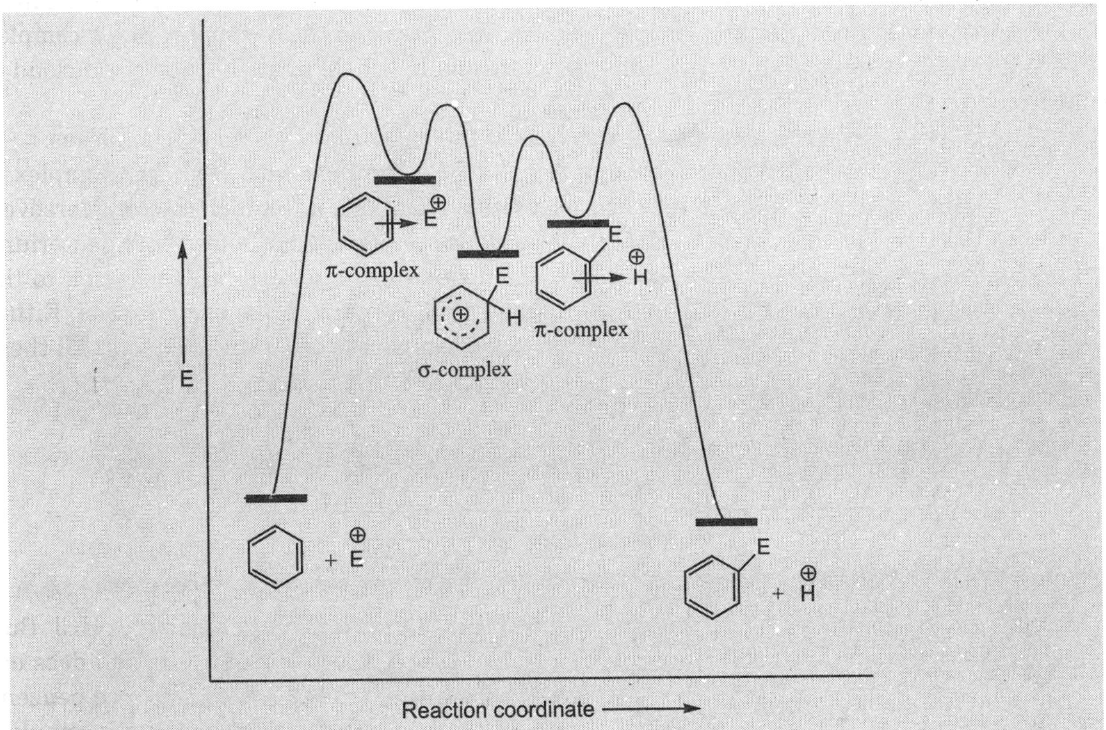

Fig. 10.1: Energy profile diagram of aromatic substitution having formation of π-complex as rate-determining step

With σ-complex formation being the rate-determining step, the energy profile diagram of the reaction would be as shown in **Fig. 10.1.**

This problem has been solved by comparing the relative rates of substitution with the relative stabilities of the σ- and π-complexes. Bromination studies provide a good example. The rate of bromination of a number of alkylbenzenes has been compared with the relative stabilities of σ- as well as π-complexes. It has been shown that the rate of bromination parallels the stabilities of the corresponding σ-complexes more closely than those of π-complexes. Thus, it can be said that in this reaction the first step is rapid, reversible formation of a π-complex and a high-energy transition state is involved in the conversion of the π-complex to the σ-complex.

Olah, on the other hand, has provided evidence that it is the formation of π-complex, which constitutes the rate-determining step in nitration of a series of aromatic hydrocarbons using nitronium fluoroborate ($N^+O_2BF_4^-$) in tetramethyl sulphone solvent (**Fig. 10.1**). The initially formed π-complex rapidly collapses to a σ-complex, subsequently yielding the product.

However, available data suggest that in most of the aromatic substitutions the high-energy transition state strongly resembles the σ-complex, and thus, the more probable course for the reaction involves the initial reversible, rapid formation of a π-complex, followed by the rate-determining formation of a σ-complex (**Fig. 10.2**).

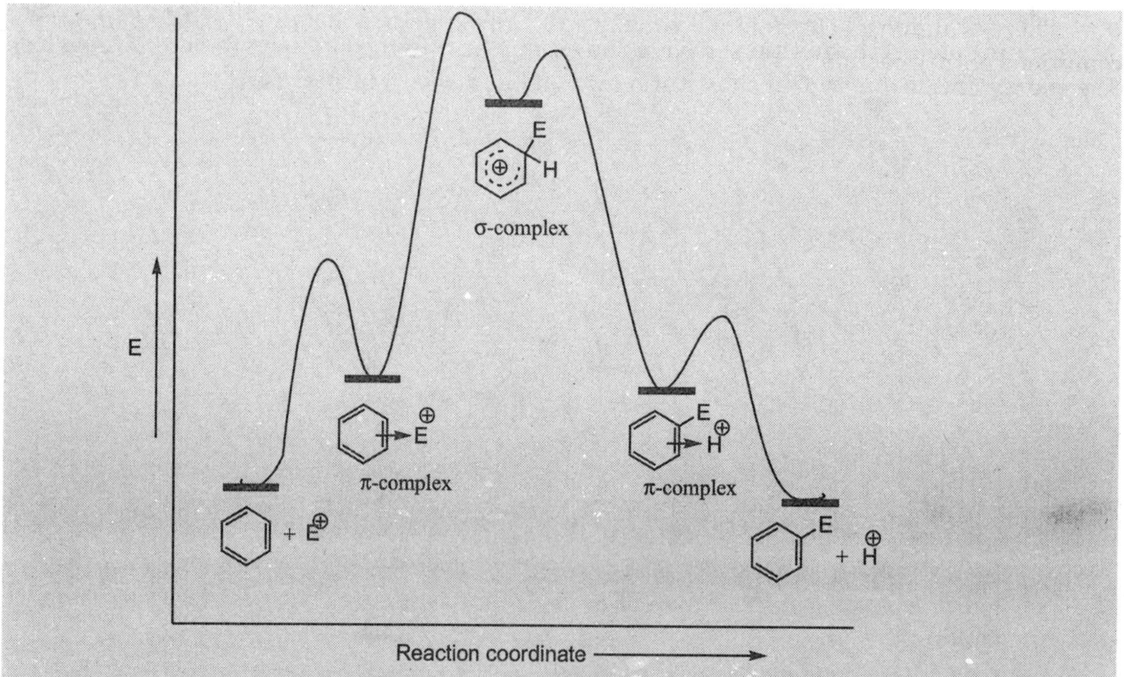

Fig. 10.2: Energy profile diagram of aromatic substitution having formation of σ-complex as the rate-determining step

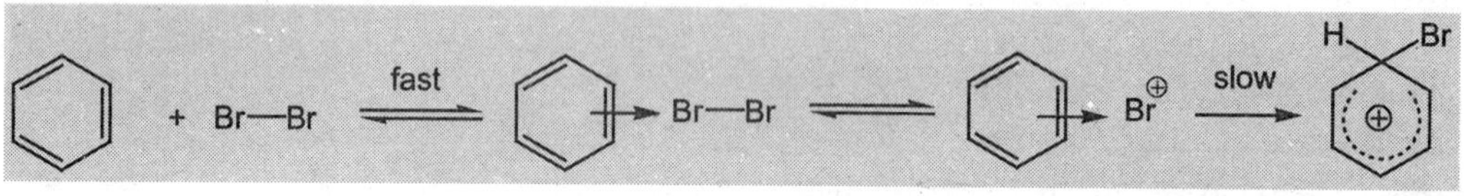

10.1 ELECTROPHILIC SUBSTITUTION REACTIONS OF BENZENE

We are now in a position to discuss in detail the mechanisms of some common electrophilic substitution reactions of benzene.

A. Nitration

Nitration is usually brought about by treating an aromatic compound with a mixture of nitric acid and sulphuric acid, which is called the *nitrating mixture*.

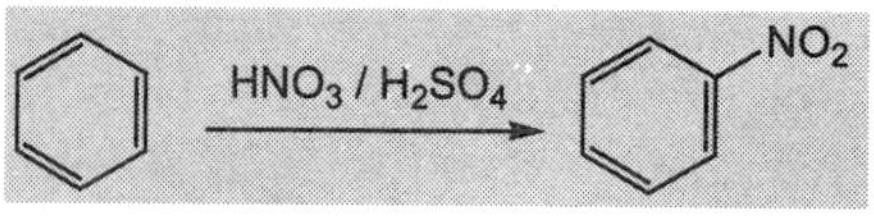

This reaction involves three steps:

(i) generation of the electrophile

(ii) formation of σ-complex

(iii) loss of proton from the σ-complex.

There is an initial electrophilic attack by the nitronium ion, which is produced under the conditions of the reaction as shown by the following equilibria:

$$H_2SO_4 + HNO_3 \rightleftharpoons HSO_4^{\ominus} + H_2\overset{\oplus}{O}{-}NO_2$$

$$H_2\overset{\oplus}{O}{-}NO_2 \rightleftharpoons H_2O + \overset{\oplus}{N}O_2 \quad \text{(nitronium ion)}$$

$$H_2SO_4 + H_2O \rightleftharpoons HSO_4^{\ominus} + H_3O^{\oplus}$$

$$\text{overall} \quad 2\,H_2SO_4 + HNO_3 \rightleftharpoons 2\,HSO_4^{\ominus} + H_3\overset{\oplus}{O} + \underset{\text{electrophile}}{NO_2^{\oplus}} \qquad \text{(step 1)}$$

The existence of nitronium ion in the acid mixture is evident from the following study:

(i) The best evidence for its existence comes from cryoscopic studies. The *vant Hoff* factor for the reaction is 4.0 which is in complete agreement with the formation of nitronium ion. The results of these studies are consistent with the establishment of the above equilibria.

(ii) Other species capable of producing nitronium ions, such as $NO_2^+BF_4^-$, $NO_2^+NO_3^-$, and $NO_2^+ClO_4^-$ also nitrate aromatic hydrocarbons.

(iii) The relative reactivities of different aromatics towards these nitronium salts are the same as their relative reactivities towards the nitric acid-sulphuric acid mixtures.

This study confirms the participation of nitronium ions. The isolation of intermediates and the absence of any isotope effect support the view that addition of the nitronium ion to the aromatic nucleus is the *rate-determining* step.

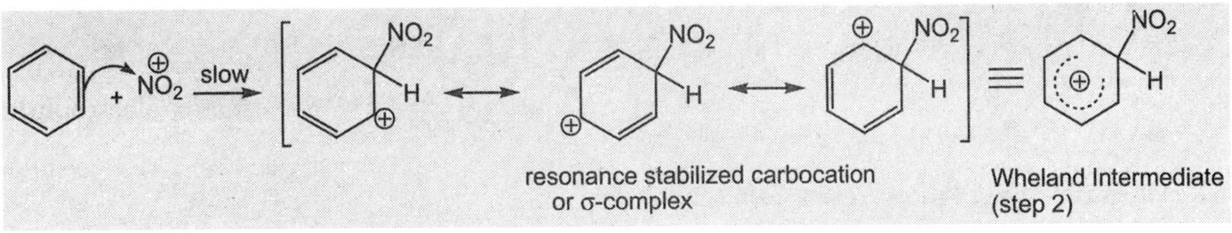

In the last step, the σ-complex or the carbocation intermediate loses a proton to the base to give nitrobenzene.

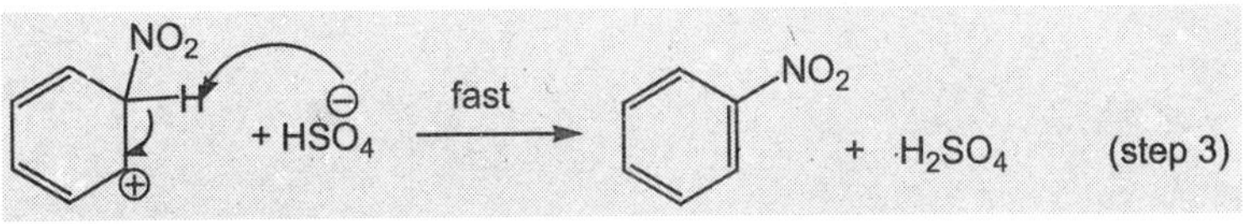

It has been found that the rate of nitration of hexadeuterobenzene (C_6D_6) and benzene (C_6H_6) are the same. Hence, the reaction does not show any isotope effect. Had it been the case then the rate of reaction in case of C_6D_6 should have been slower than in the case of C_6H_6. Therefore, the loss of a proton is not the *rate-determining* step.

B. Halogenation

I. Halogenation with Halogens

Although in many cases aromatic compounds can be halogenated by direct treatment with molecular halogen, customarily halogenation is brought about in presence of catalysts such as metal halides. The main function of the catalyst is to partially or completely polarize the halogen–halogen bond.

$$FeCl_3 + Cl_2 \rightleftharpoons \overset{\delta^+}{Cl}—\overset{\delta^-}{Cl}\text{----}FeCl_3 \quad \text{or} \quad \overset{\oplus}{Cl} + \overset{\ominus}{FeCl_4}$$

The positive end of the halogen dipole (or the free halogen cation, itself) attacks the aromatic ring resulting in the formation of a σ-complex. The carbocation intermediate or σ-complex loses a proton to the base, $FeCl_4^-$, to give chlorobenzene. This step is fast, and thus, does not affect the rate of the reaction.

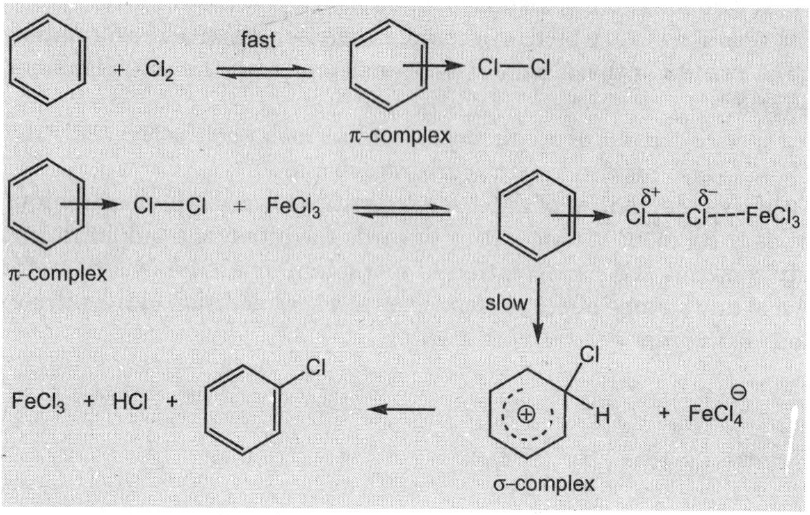

II. Halogenation with Interhalogen Compounds

Aromatic compounds undergo halogenation on treatment with some interhalogen compounds in the presence of Lewis acid catalysts. More electropositive atom acts as an electrophile. For example, with iodine monochloride (ICl), benzene gives iodobenzene (and not chlorobenzene).

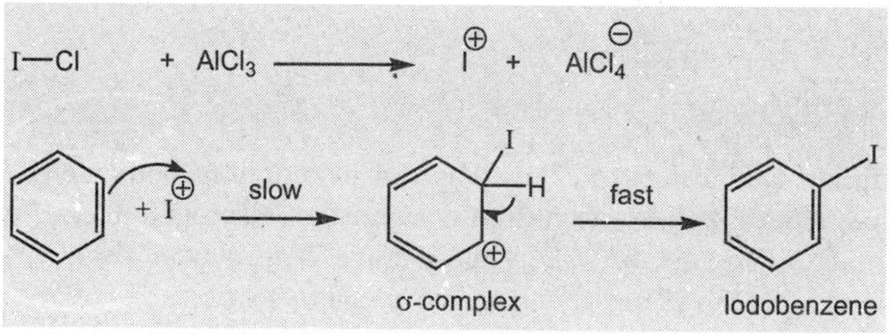

III. Halogenation with Hypohalous Acids

Halogenation can also be effected by using hypochlorous or hypobromous acids in presence of strong acids. The function of strong acid catalyst is to protonate the hypohalous acids, thus making them powerful electrophilic species.

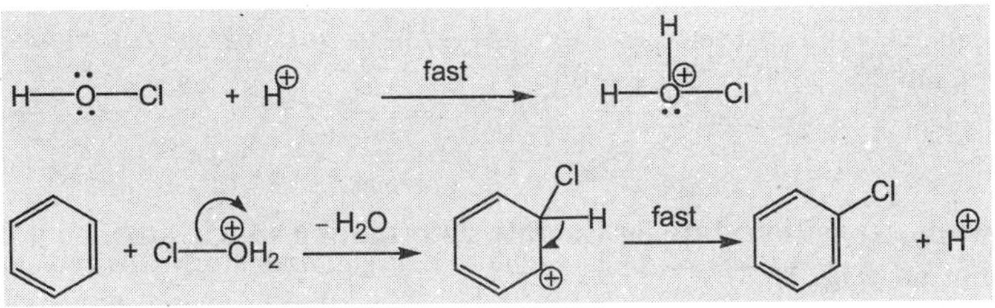

Instead of a protonated hypohalous acid, the attacking species can also be a positive halogen ion without affecting the course of the reaction.

$$H_2\overset{\oplus}{O}-Cl \rightleftharpoons H_2O + \overset{\oplus}{Cl}$$

C. Sulphonation

Sulphonation is brought about by treating benzene or other aromatic compounds with conc. H_2SO_4 or oleum $H_2S_2O_7$ ($H_2SO_4 + SO_3$). In either case, the sulphonating agent is protonated SO_3 ($^+SO_3H$) having the structure

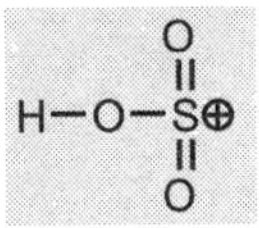

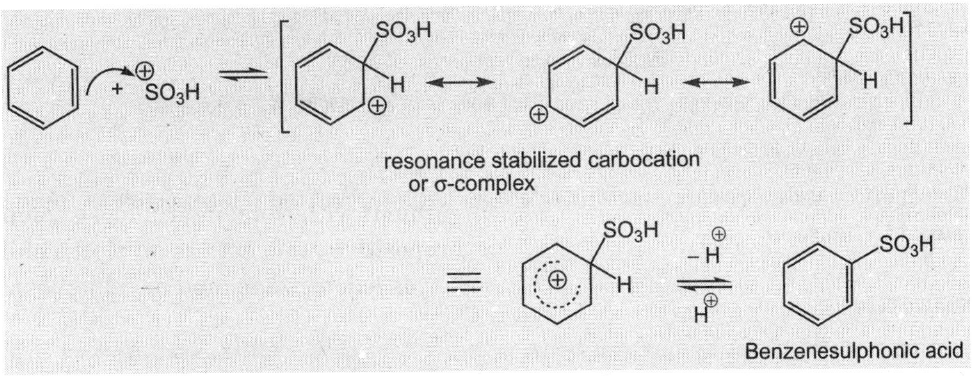

Sulphonation of benzene is a reversible reaction as heating of benzenesuphonic acid in dilute acids regenerates benzene.

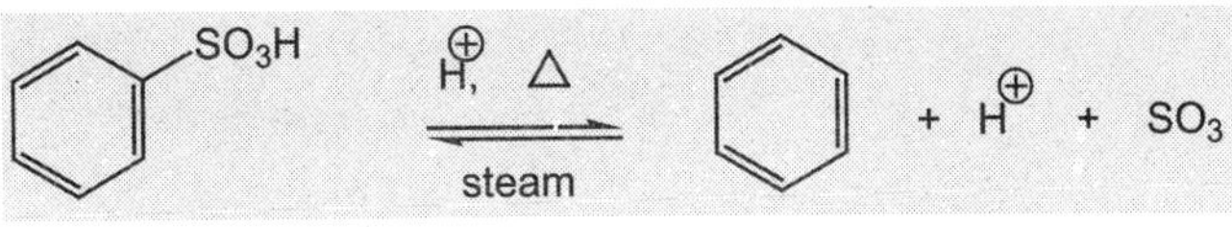

According to *the principle of microscopic reversibility*, mechanism of a reaction in the reverse direction must have the same intermediates as in the forward direction, and that the rate-determining step must be the same in both directions. In sulphonation reaction, the rate-determining step is the nucleophilic attack of benzene on the $^{+}SO_3H$ ion. In desulphonation, the rate-determining step is the loss of the $^{+}SO_3H$ ion from the benzene ring.

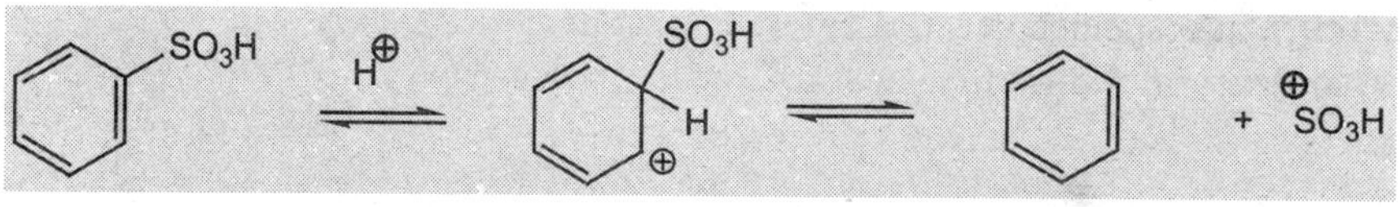

In other words, transition state for the rate-determining step in sulphonation is the same as in desulphonation (**Fig. 10.3**).

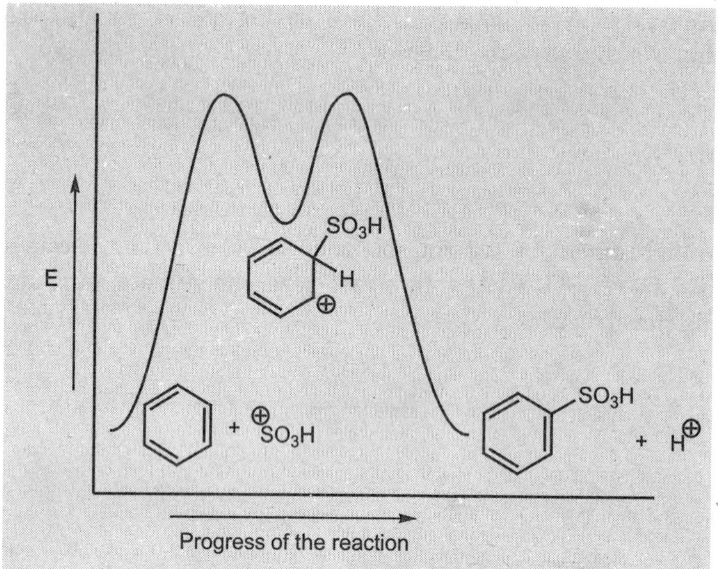

Fig.10.3: Energy profile diagram of reversible sulphonation of benzene

The reaction shows *positive isotope effect* as hexadeuterated benzene reacts more slowly as compared to benzene.

D. Mercuration

This reaction is carried out by heating benzene with mercuric acetate. The product is phenylmercuric acetate.

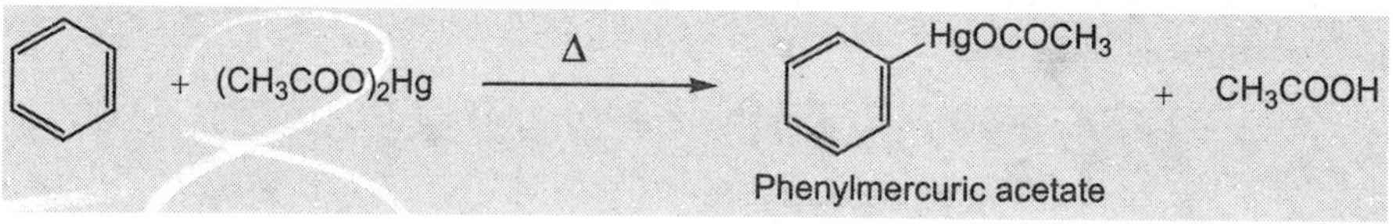

It provides another example of an electrophilic aromatic substitution exhibiting a *positive isotope effect*. It has been observed that benzene reacts six times faster than perdeuterobenzene under similar conditions, thereby demonstrating that like sulphonation, this reaction also involves breaking of a C–H bond in the rate-determining step. An inspection of the energy profile diagram shown in **Fig. 10.4** makes it clear. Like other electrophilic aromatic substitutions, this reaction also takes place in three steps.

The mercurating species is a positively charged acetomercuric ion, $^{+}HgOCOCH_3$, **formed** by the initial ionization of mercuric acetate.

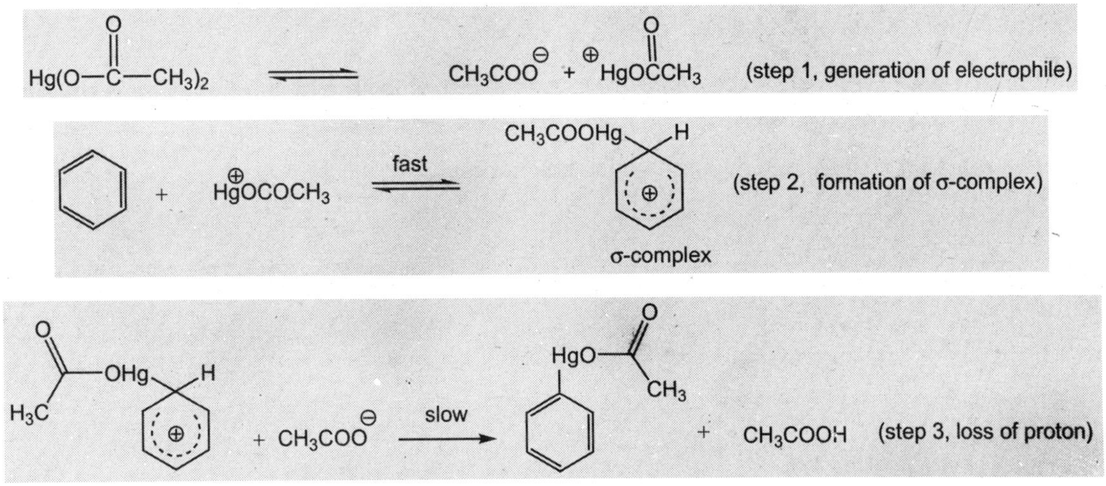

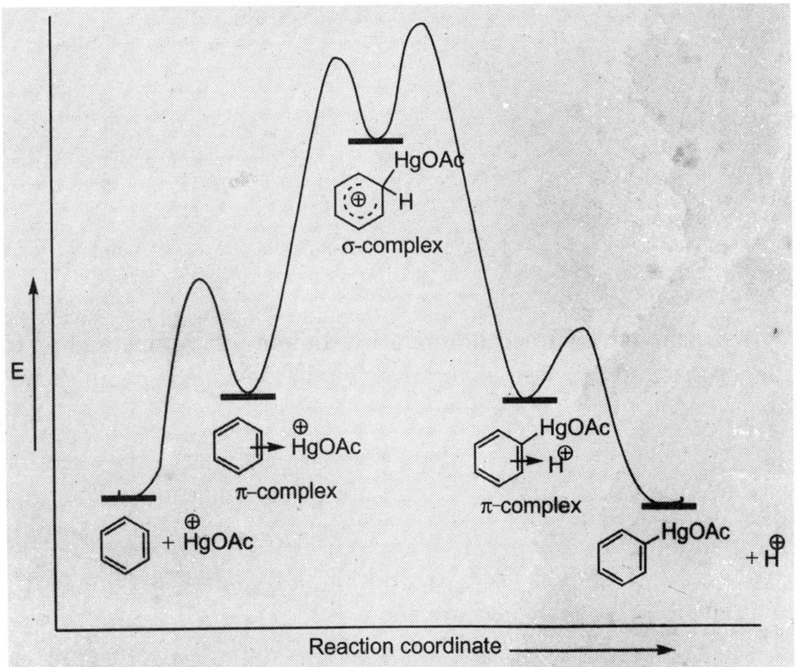

Fig. 10.4: Energy profile diagram of mercuration **of benzene**

F. Friedel-Crafts Reaction

Friedel-Crafts reaction is used to describe two different processes taking place by the same mechanism. These processes are called (i) Friedel-Crafts alkylation, and (ii) Friedel-Crafts acylation.

I. Friedel-Crafts Alkylation

Aromatic compounds react with alkyl halides in presence of Lewis acid catalysts such as $AlCl_3$, BF_3, $GaCl_3$, and $AlBr_3$ to yield alkylated products.

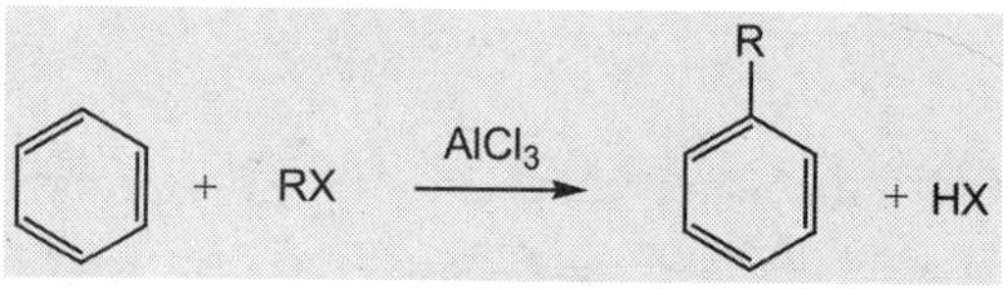

The reaction takes place by a step-wise mechanism.

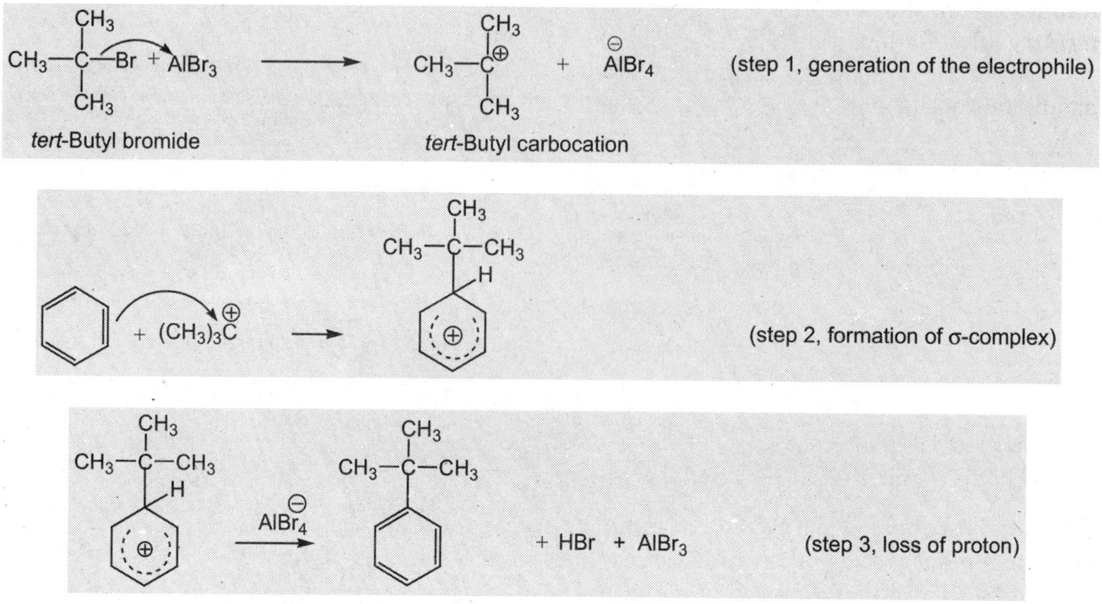

However, with primary and secondary alkyl halides, it appears that instead of free carbocations, the electrophilic species is an alkyl halide-Lewis acid complex with a positively polarized carbon.

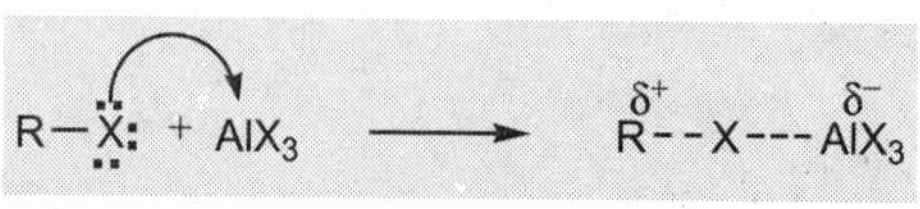

The principal evidence against the participation of free carbocations with primary alkyl halides is based on the following two observations:

1. These reactions are first-order with respect to the aromatic compound, the Lewis acid, and the alkyl halide.

2. When toluene is methylated using methyl bromide and methyl iodide, the isomeric xylenes are formed in different ratios.

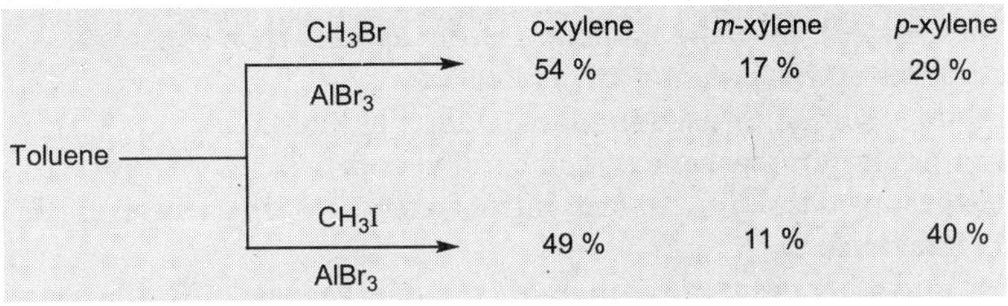

Free carbocation (CH_3^+ in either case) should have reacted identically with toluene irrespective of their source of generation.

In the case of primary alkyl halides, the reason for the alkylation proceeding *via* polarized alkyl halides is the high energy required to generate a *primary* carbocation. *Secondary* alkyl halides are on the border line, and alkylation may take place through either mechanism. With *tertiary* alkyl halides, the alkylation involves a free carbocation.

Aromatic compounds can be alkylated with carbocations obtained from other sources such as alkenes or alcohols.

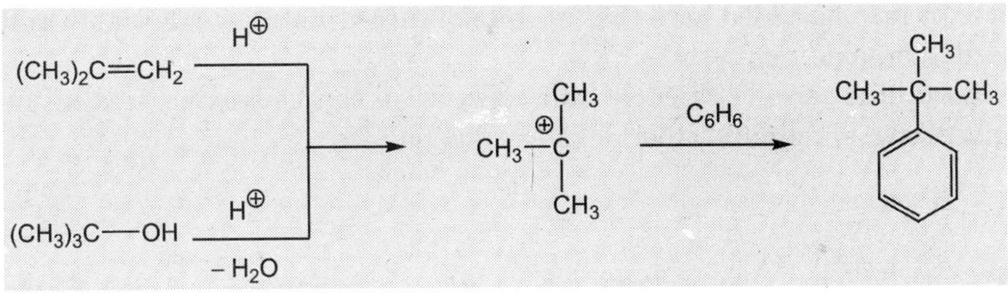

Limitations of Friedel-Crafts

Though quite versatile, Friedel-Crafts alkylation has the following two main limitations:

(i) Rearrangement of Carbocations: Friedel-Crafts alkylation is often accompanied by rearrangements in the alkylating groups. For example, the alkylation of benzene with *n*-propyl chloride leads to the formation of a mixture of *n*- and *iso*propylbenzenes. The formation of the rearranged product can be rationalized on the basis of the relative carbocation stability, although free carbocations may not be present in the reaction.

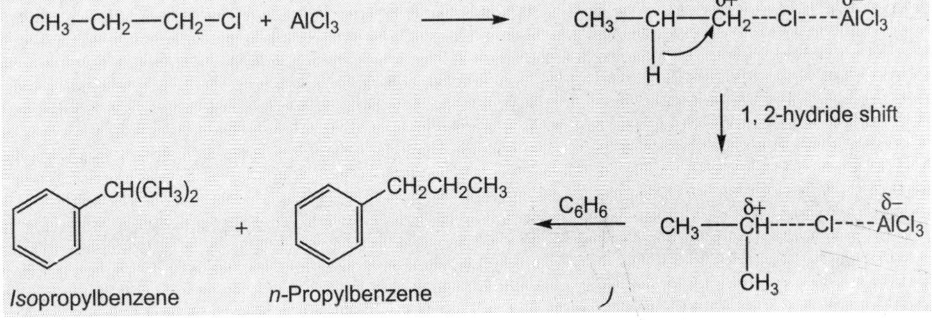

(ii) Formation of Polysubstituted Products: Another experimental limitation of the Friedel–Crafts alkylations is the formation of polysubstituted products. The reason for this is that the introduction of one alkyl substituent activates the ring towards a second substitution. This can be avoided, to some extent, by using an excess of the hydrocarbon substrate.

Other limitations of this reaction are as follows:

(a) Aryl halides cannot be used in place of alkyl halides.

(b) Substrates like nitrobenzene cannot be alkylated, because of the deactivating influence of nitro group on the ring. In fact, nitrobenzene is used as a solvent in many Friedel-Crafts reactions.

(c) Aniline and other aromatic amines fail to undergo Friedel-Crafts reaction, because of the formation of a complex between the base and Lewis acid. Since this complex has a positive charge on nitrogen, the ring is deactivated.

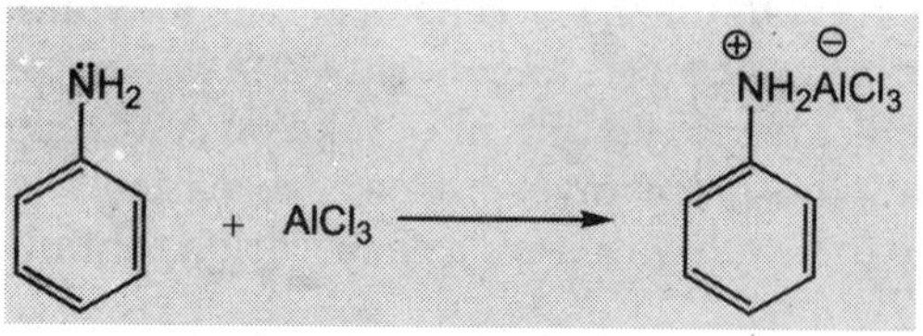

However, it may be added here that in spite of these limitations, Friedel-Crafts reaction is widely used in organic synthesis.

The important aspects of Friedel-Crafts alkylation can be summarized in the following chart using benzene, aluminium chloride, and alkyl halides.

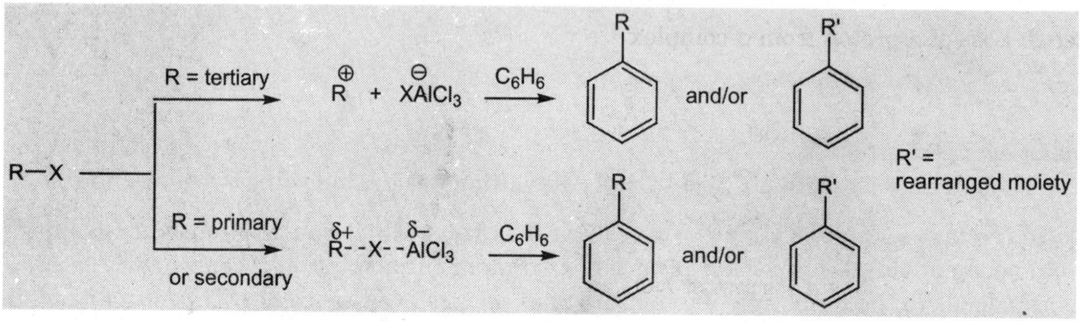

II. Friedel-Crafts Acylation

Benzene and other aromatic compounds react with acid chlorides or acid anhydrides in presence of anhydrous $AlCl_3$ (or other Lewis acid catalysts) to provide acylated products (ketones).

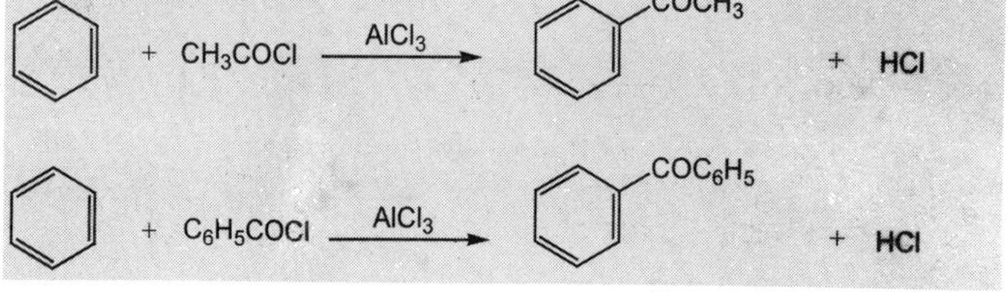

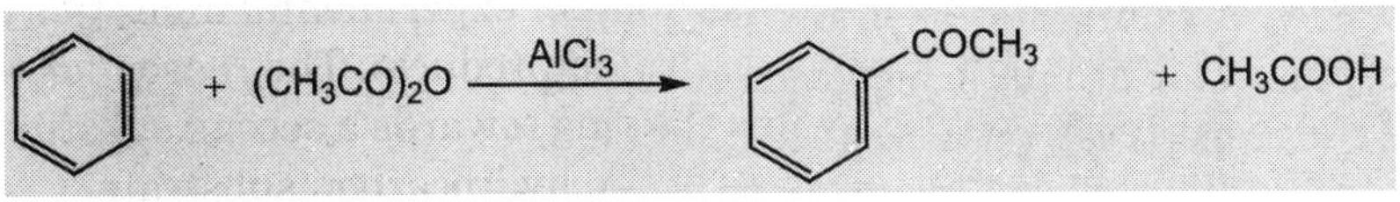

Like Friedel-Crafts alkylation, the acylation reaction takes place by a step-wise mechanism. Various steps involved in the reaction are as follows:

Step 1: Generation of an electrophile
Acylium ion (RCO^+), which is the electrophile in the reaction, is generated by the reaction of acyl chloride or acid anhydride with the Lewis acid (like $AlCl_3$).

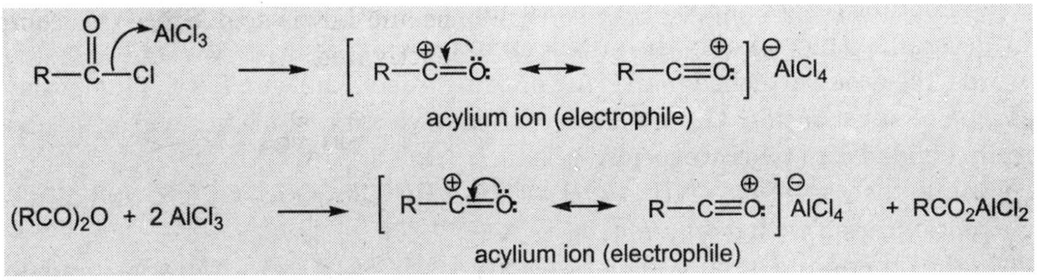

Step 2: Formation of resonance stabilized σ-complex

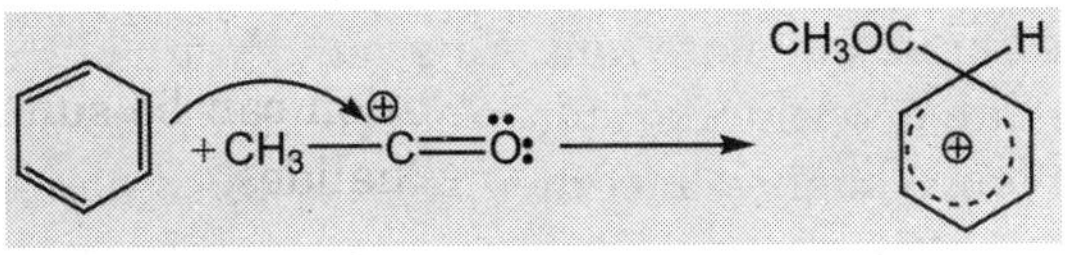

Step 3: Loss of a proton from σ-complex

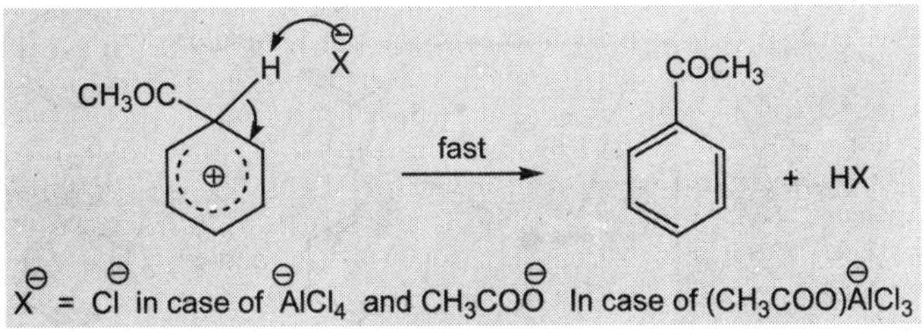

Thus, a generalized picture of Friedel-Crafts acylation can be written in the following three steps.

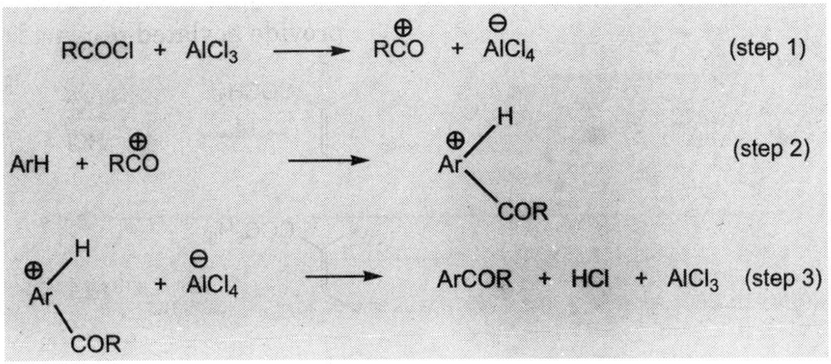

Olah prepared a number of acylium salts such as $CH_3CO^+BF_4^-$ and $CH_3CO^+ SbCl_6^-$, which have been used for aromatic acylation reactions. The structure of these salts has been established by NMR spectroscopy. Unlike alkylation, acylation reactions are not accompanied by rearrangement within the acyl group, and also, there is no polysubstitution as the aromatic ring is deactivated after the introduction of the first acyl group.

10.2 REACTIVITY AND ORIENTATION

A. Reactivity

When an electrophilic reagent attacks a substituted aromatic ring, the group already attached to the ring determines how readily the attack occurs and at what position of the ring it occurs. In other words, the group attached to the ring not only affects the *reactivity,* but also determines the *orientation* of substitution. On the basis of these two characteristics, various substituents are generally divided into two categories.

1. Groups such as $–NR_2$, $–NH_2$, –OH, $–O^-$, –NHCOR, –OR, –OCOR, and alkyls lead to *o,p*-orientation with *activation*.
2. On the other hand, groups such as $–N^+R_3$, $–CX_3$, –C(R)=O, –CN, $–SO_2$, and $–NO_2$ lead to *meta* orientation with *deactivation.*

The rate of an electrophilic substitution, like any other reaction, depends on the difference between the energy of the transition state and the ground state of the reactants, *i.e. energy of activation* (E_{act}). Since the rate-determining step of electrophilic aromatic substitution is usually the formation of a σ-complex, a slightly modified form of the reaction coordinate diagram for an aromatic electrophilic substitution on benzene, ignoring the π-complex formation, can be constructed as shown in **Fig. 10.5**.

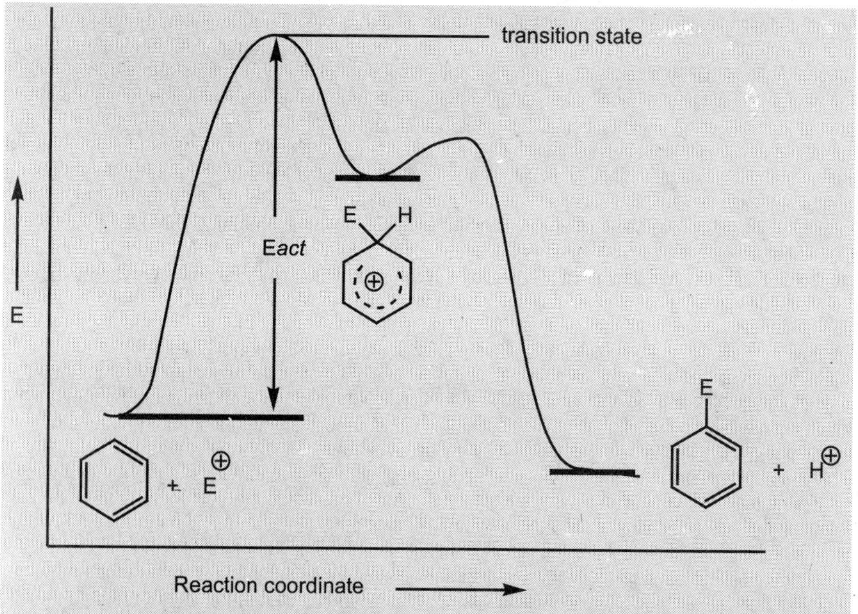

Fig. 10.5: Energy profile diagram for a typical electrophilic aromatic substitution

Similar energy profile diagrams can be constructed for substituted aromatic rings. The rate of substitution at *o*-, *p*-, or *m*- positions then depends on the height of the energy barrier between the reactant and the transition state (E_{act}). If any effect due to the substituent lowers E_{act}, the rate of substitution increases, and the favoured orientation of attack is the one with the lowest E_{act}.

Since it is difficult to assess the energy of transition state directly, the energy of the σ-complex is often assessed, as the transition state closely resembles it in energy and geometry. Thus, any effect of the substituent, which stabilizes the σ-complex, will enhance the rate of substitution.

The substituents of category (i) are electron-donating (either by inductive or by resonance effect), and exert an electrostatic effect, which stabilizes the positively charged σ-complex, and thus, increases the rate of the reaction. On the other hand, substituents of category (ii) are electron-withdrawing, which destabilize the σ-complex, *i.e.* the positive charge is intensified due to which the rate of the reaction decreases.

It is worthwhile to mention here that the release or withdrawal of electrons may occur due to inductive effect alone or due to a combination of both inductive and resonance effects. The two effects may reinforce or oppose each other.

B. Orientation

The orientation effect of the substituents can best be explained by writing all the possible resonance forms of the charged intermediate for each of the three possible reaction courses.

In toluene where the methyl group is electron-donating (by inductive effect), the following resonance forms can be written for attack of electrophile at *para*, *meta,* and *ortho* positions.

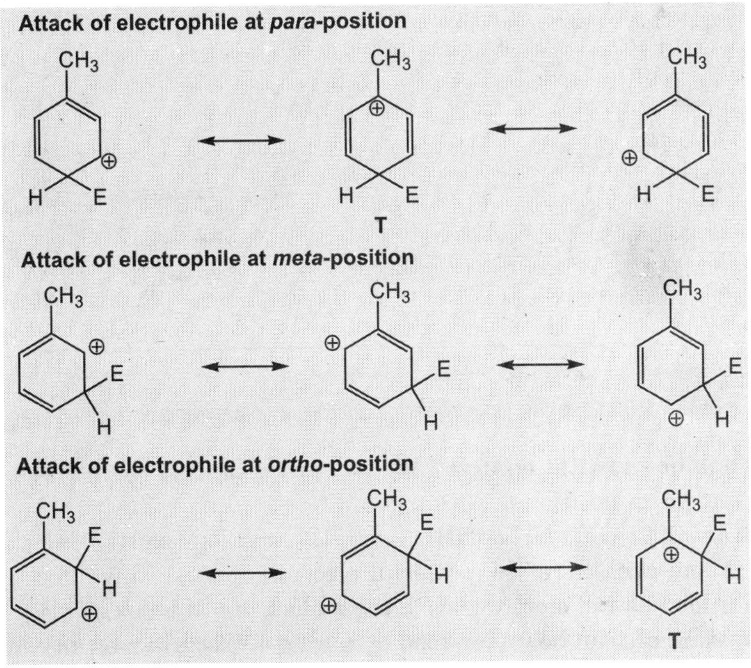

In case of *ortho* and *para* attacks, there in one structure (marked **T**) in each case where the positive charge is located on the carbon atom to which the methyl group is attached, whereas no such structure can be written when the electrophile attacks at the *meta* position. As structures marked **T** are *tertiary carbocations*, they are more stable than rest of the structures, which are all *secondary carbocations.* Thus, in this case, substitution at *o*- and *p*- is preferred to *m*-substitution.

In other words, the energy of *o*- and *p*-substituted Wheland intermediates (σ-complexes) is considerably lower than that of *m*-substituted intermediates. However, it may be noted that its energy is lower than the corresponding intermediate in benzene. This can be easily understood by an inspection of the energy profile diagram depicting substitution in benzene, and *o*-, *m*-, and *p* substitutions in toluene (**Fig. 10.6**).

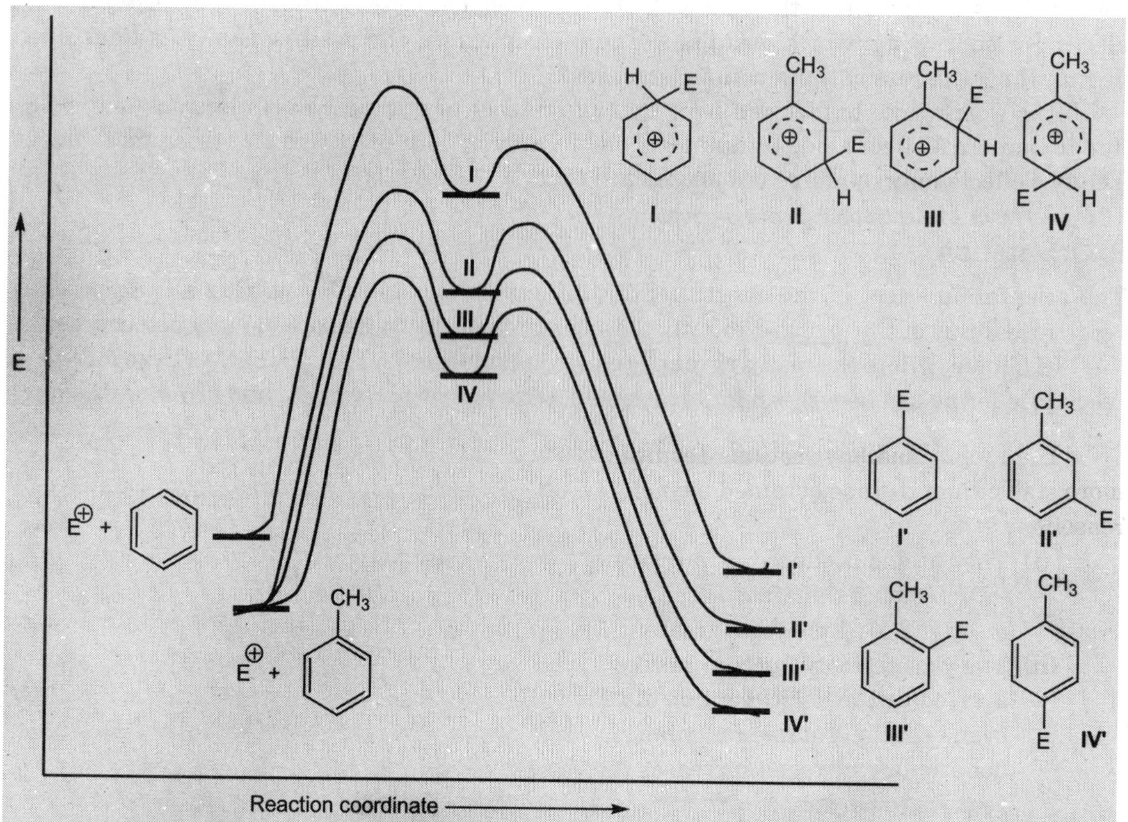

Fig. 10.6: Energy profile diagram depicting electrophilic substitution in benzene and toluene

Thus, electrophilic substitution in toluene is a faster process than in benzene and occurs mainly at *ortho*- and *para*- positions.

Groups such as –OH, –OR, $–NH_2$, –NHR, and $–NR_2$ have lone pair(s) of electrons on nitrogen and oxygen atoms and are, therefore, powerful electron donors. This effect is due to strong interaction of their lone pair of electrons with the π-electrons of benzene ring. The orientation effect of these substituents can be understood by taking $–\ddot{N}H_2$ group as an example. As in the case of toluene, the following contributing resonance forms can be written for the σ-complex formed as a result of attack of an electrophile on *para*-, *meta*-, and *ortho*- positions.

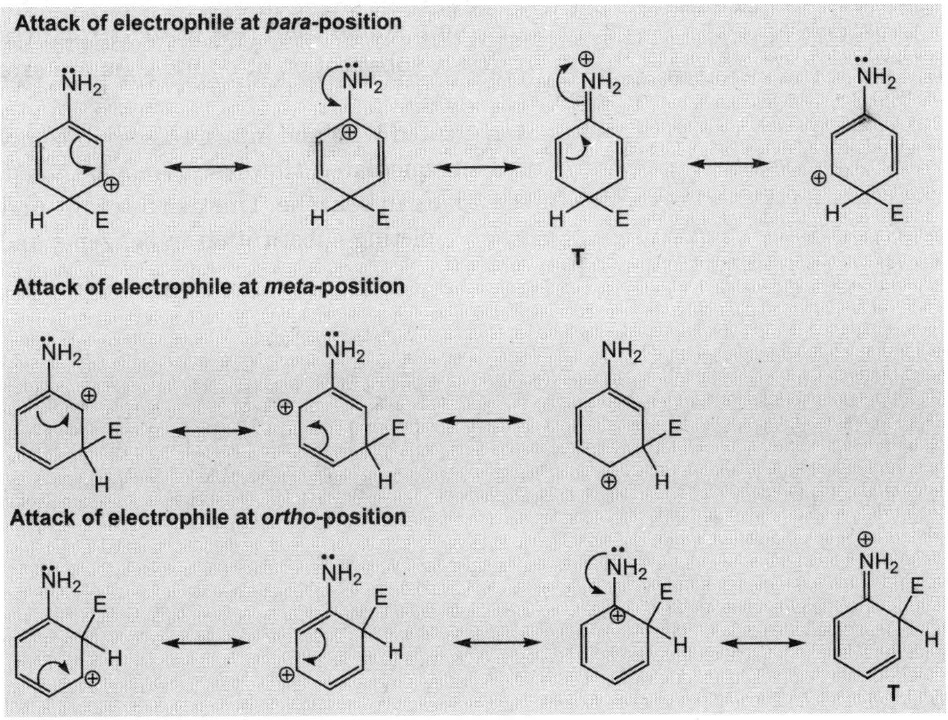

The cyclohexadienyl cations resulting from *ortho*- and *para*-attack of electrophile are more stable than the one obtained by *meta*-attack of electrophile. It is due to the following two reasons:

(i) The cations produced as a result of *o*- and *p*-attack are hybrids of four contributing forms each as against three such forms of a similar intermediate resulting from *m*-attack. We know that greater the number of contributing forms, stabler the species.

(ii) The resonating structure marked **T** of the cations derived from *o*- and *p*-attack is derived by the delocalization of nitrogen lone pair into the π-system of the ring. This is an especially stable contributing resonance form. It may also be noted that electron donation by nitrogen increases the rate of substitution at all the positions of aniline relative to benzene.

Electron-withdrawing groups such as —NO_2, —CN, —COOH, —CHO, and —SO_3H deactivate all ring positions relative to benzene due to their electron-withdrawing inductive (–I) and resonance (–R) or mesomeric (–M) effects. In fact, they are more deactivating for *o*- and *p*- substitutions than *m*-substitution. Hence, in the presence of such substituents, further substitution is slower, and takes place only at *m*-position. The reason for this is not far to seek if we compare the resonance forms of σ-complexes resulting from an electrophilic attack on benzene, and at *o* –, *m*–, and *p*-positions of nitrobenzene, which provides an example of aromatic substrate having a substituent belonging to this category.

The σ-complex resulting from the addition of electrophile to benzene has only one positive charge in all its contributing forms, while the σ-complex obtained from similar addition of electrophile to nitrobenzene has two positive charges, one on the ring carbon and the other on the nitrogen atom of the nitro group. This results in electrostatic repulsions rendering the transition states (leading to the intermediates) less stable in nitrobenzene as compared to benzene.

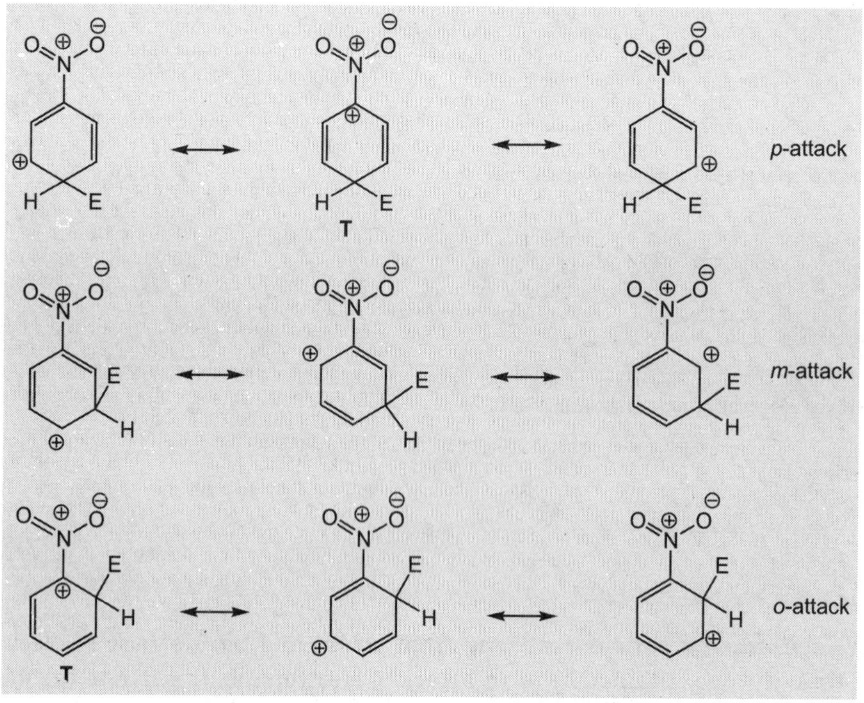

Since the attack at *ortho-* and *para-* positions places a positive charge adjacent to nitro group already having a positively charged nitrogen atom, structures marked **T** are strongly *destabilized,* because of greater electrostatic repulsion. The intermediate from *meta*-attack has no such form with similar charges on adjacent atoms. Therefore, its transition state is most stable, and attack at *meta* position is preferred. The situation can best be illustrated by an energy profile depicting electrophilic substitution of nitrobenzene and benzene (**Fig. 10.7**).

The Case of Halogens

Halobenzenes are less reactive than benzene, but yield *ortho-* and *para-* substitution products. This apparent anomalous phenomenon can be explained by a consideration of the inductive and resonance effects of the halogens. Halogens withdraw electrons through inductive effect (–I), thereby deactivating the ring towards electrophilic substitution. On the other hand, halogens are capable of sharing their pairs of electrons with the aromatic ring, and can accommodate a positive charge. Writing various resonating structures of the positively charged intermediate for *o*-, *p*-, and *m*-attacks makes it clear that the attack at *o*- and *p*- positions places the positive charge on the carbon adjacent to halogens.

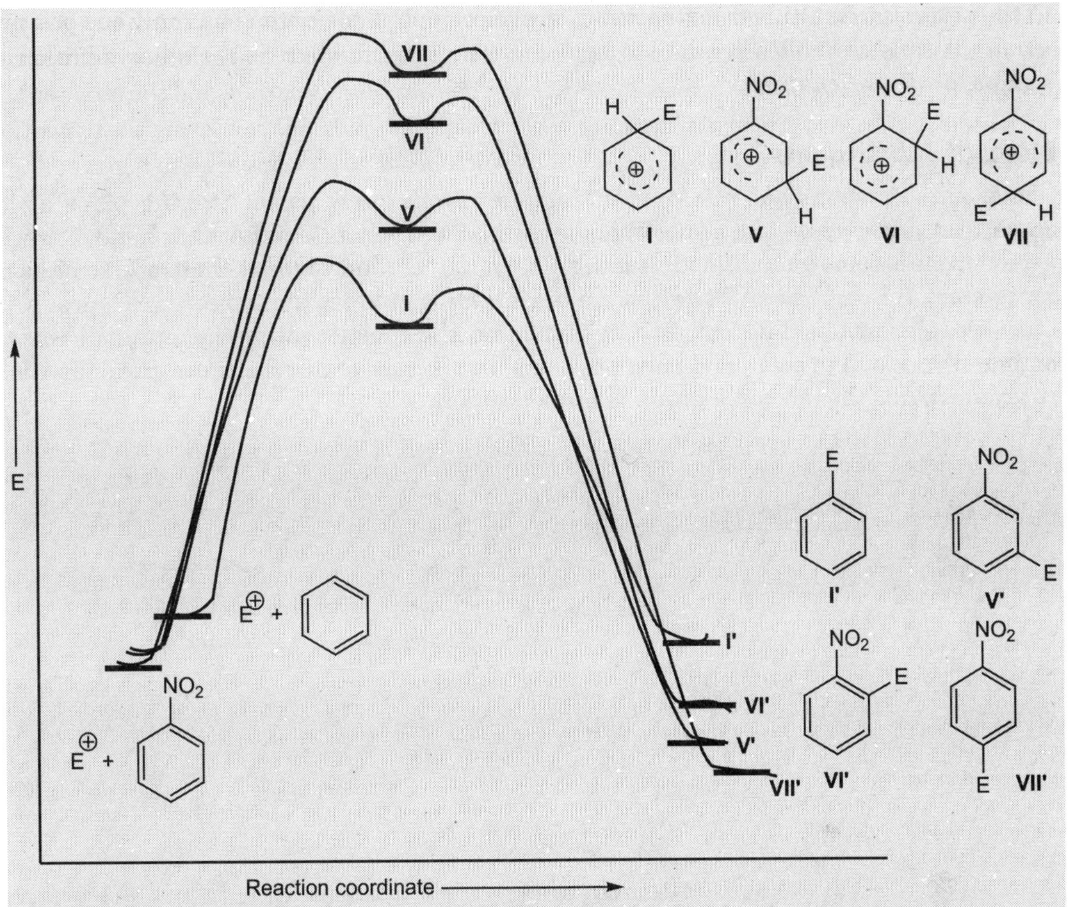

Fig. 10.7: Energy profile diagram depicting electrophilic substitution of benzene and nitrobenzene

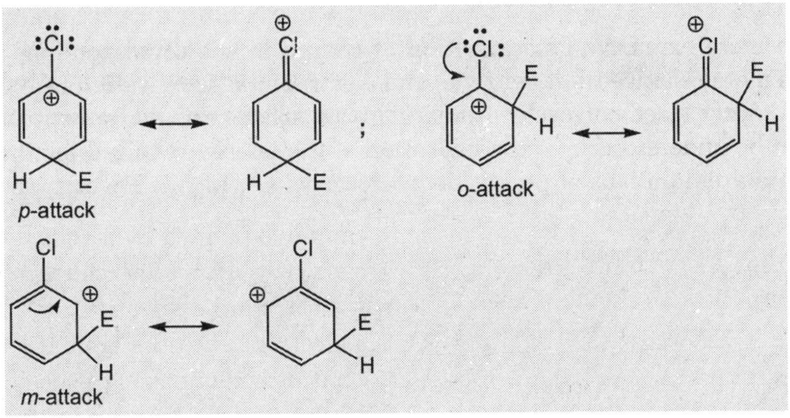

Although *o*- and *p*-positions are activated to some extent because of the resonance effect, the inductive effect overshadows it, and causes net electron-withdrawal leading to deactivation

at all the positions. But, this deactivation is less for *o*- and *p*- positions than for the *m*-position. *Reactivity* is thus controlled by the stronger inductive effect, whereas the weaker resonance effect controls the *orientation*.

C. Effect of Substituents

We have already seen that nitration of toluene is relatively easier (taking place at low temperature), and gives *o*- and *p*-nitrotoluenes as major products. On the other hand, nitration of nitrobenzene is somewhat difficult (taking place at 90 °C), and yields *m*-dinitrobenzene as the major product. It is thus clear that although the incoming group is the same, *i.e.* NO_2 group, the position occupied by it is different. In case of toluene, the substituent methyl group directs the incoming nitro group to *ortho* and *para* positions, but in case of nitrobenzene, the substituent nitro group *directs* the incoming group to *meta*-position

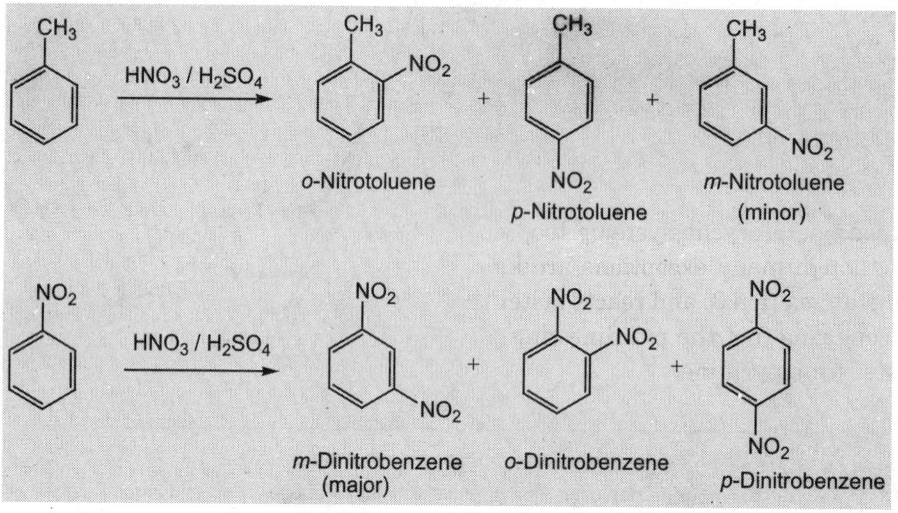

D. Orientation in Other Ring Systems

Because of the more extensive delocalization of charge in the corresponding arenium ions, naphthalene is more reactive than benzene, and electrophilic substitution is faster. Similarly, anthracene and other fused polycyclic aromatic hydrocarbons are also substituted faster than benzene. It can be understood by a consideration of the energy required for the formation of the intermediate arenium ions of polycyclic hydrocarbons, which is lower than that required for benzene.

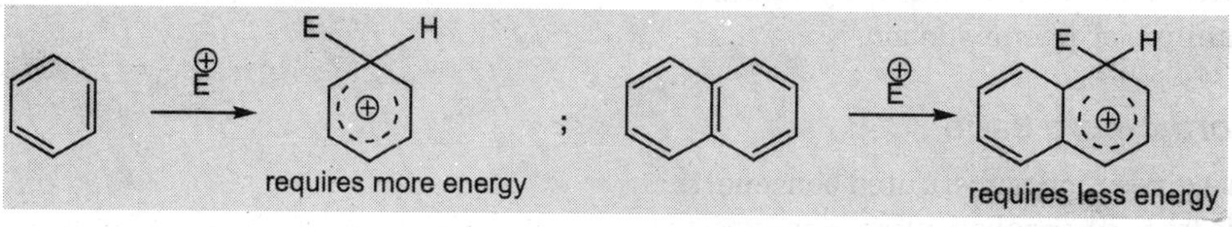

In the fused ring systems, the positions are not equivalent, and there is usually a preferred orientation even in the unsubstituted ring. Electrophilic substitution in naphthalene takes

place at 1-position as the σ-complex is more stable than the one formed as a result of attack at 2-position.

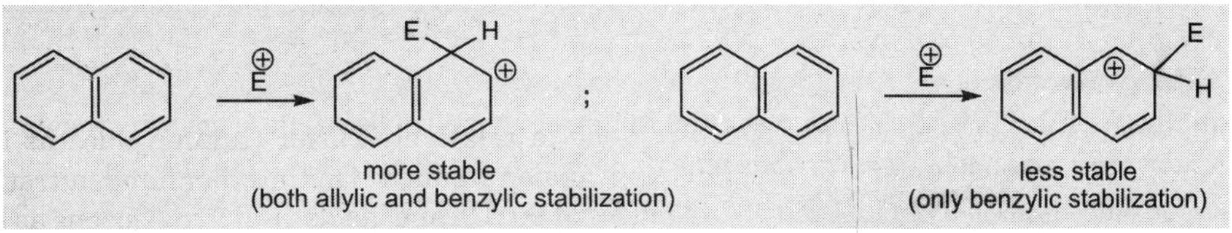

When fused ring systems have substituents, successful predictions can often be made by using a combination of above principles. Thus, ring A of 2-methylnaphthalene is activated by the methyl group, ring B is not. We, therefore, expect substitution in ring A. The methyl group activates the positions *ortho* to itself (1 and 3), but not position 4, which is *meta* to it.

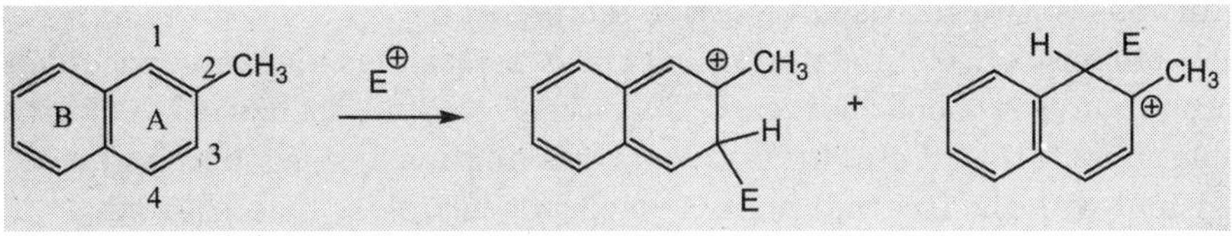

For fused heterocyclic systems too, we can often make predictions based on the above principles, though many exceptions are known. Thus, indole is primarily substituted in the pyrrole ring (at position 3) and reacts faster than the benzene, while quinoline generally reacts in the benzene ring (not the pyridine ring) at the 5 and 8 positions and slower than benzene, though faster than pyridine.

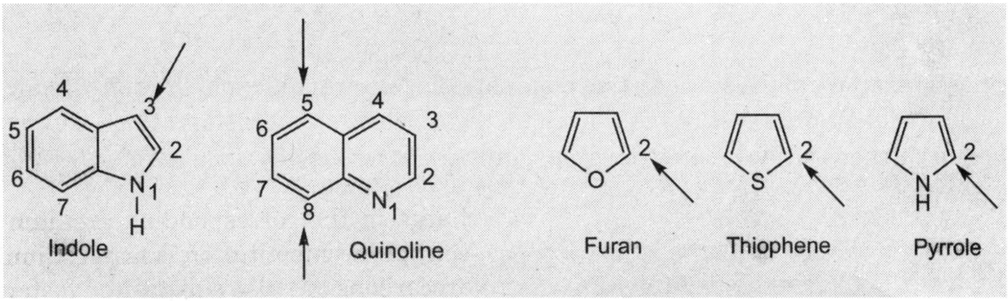

Heterocyclic compounds too have non-equivalent positions, and the principles are similar. Furan, thiophene, and pyrrole are preferentially substituted at the 2-position, and all are substituted faster than benzene. Pyrrole is particularly reactive, with a reactivity approximating that of aniline or phenoxide ion.

E. The *ortho-para* Ratio

Since in a given monosubstituted benzene, there are two *ortho* and one *para* positions, which are available for further substitution, it is expected that if the substituent is *ortho, para* directing, the *ortho* to *para* ratio should be 2 : 1. But, usually in *o*-, *p*-directing systems, the *ortho*-substitution occurs to a lesser extent than expected. This difference may be due to many factors, *i.e.* polar effects of the substituent, effect of solvents, temperature and size of the substituents, and the entering groups (steric factors). It has been shown that steric factors play dominant role in determining the *ortho* to *para* ratio.

Thus, nitration of various alkylbenzenes has revealed that *ortho* to *para* ratio decreases gradually (from 1.5 to 0.25) as one goes from toluene to *tert*-butylbenzene. This is in accordance with the order: toluene (1.5) > ethylbenzene (0.95) > *iso*propylbenzene (0.5) > *tert*-butylbenzene (0.25). This decrease in *o*/*p* ratio has been explained on the basis of steric effects of the substituents already present. In case of *o*-substitution, the steric interaction between the electrophile and the group already present on the ring raises the energy of the transition state, thus making it less stable. Probably, this is the reason why the proportion of *o*-isomer in the *o*-, *p*-mixture increases if the reaction is carried out at higher temperatures. For instance, nitration of acetanilide at 0°C gives *p*-nitroacetanilide almost exclusively. However, the amount of *o*-isomer increases with the rise in reaction temperature.

The role of steric effects in determining the *o*/*p* ratio has elegantly been demonstrated in the halogenation of *tert*-butylbenzene. Thus, when *tert*-butylbenzene is chlorinated, there is appreciable substitution at *o*-position, but its bromination at *o*-position is almost negligible. This is consistent with the fact that bromine molecule is bulkier than chlorine.

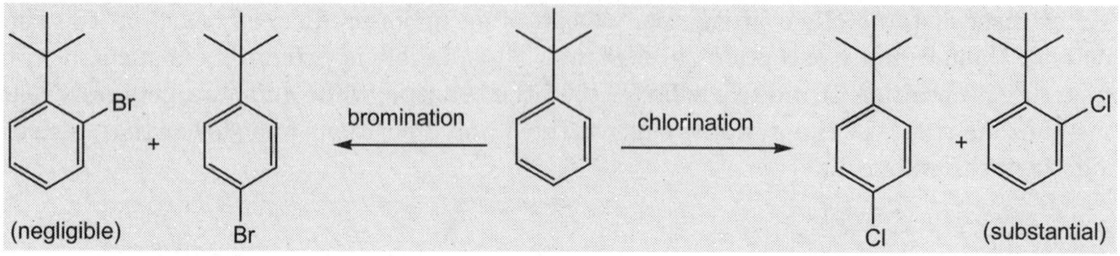

A classic example of the steric factors dominating in electrophilic substitution (bromination, nitration, Friedel-Crafts reaction) is *p*-cymene. From electronic considerations, substitution should occur at the position *ortho* to *iso*propyl group, but invariably, it takes place *ortho* to the methyl group.

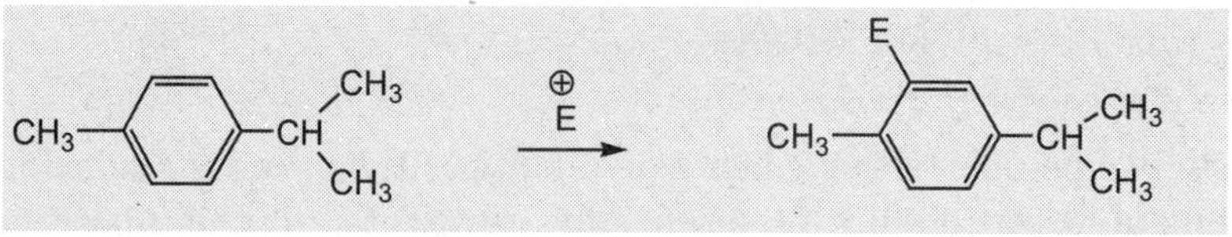

However, a different situation arises when the substituent already present has a pair of non-bonding electrons. Electronic factors now dominate, and result in increase in the amount of *p*-isomer. This is due to the fact that one of the resonance forms of the σ-complex assumes *p*-quinonoid structure in case of *p*-attack, while that obtained from *ortho*-attack assumes an *ortho*-quinonoid structure. As the *para* quinonoid structures are more stable than *ortho*-quinonoids, it can be assumed that the form **T'** would be more stable than the form **T**, and hence, substitution is preferred at *para*-position.

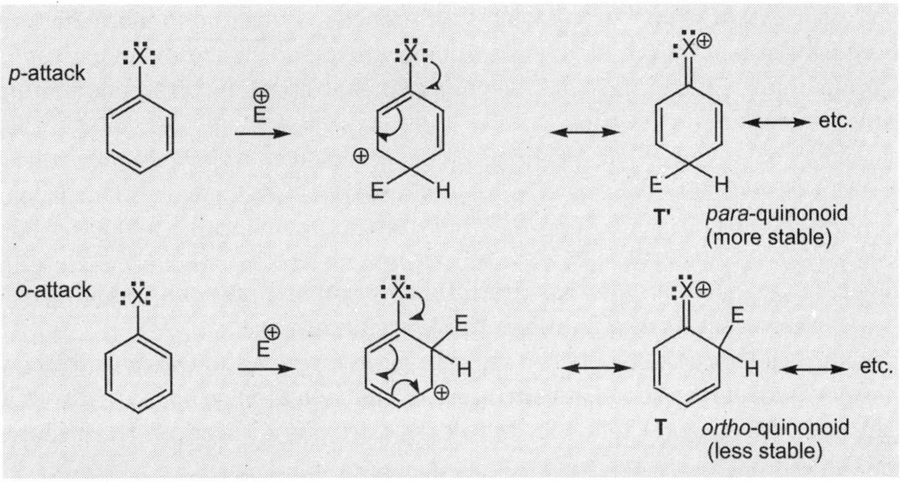

F. Competition between the Substituents

When two or more groups are present in the ring, the orientation and reactivity effects as described above still hold good. Both the groups can direct further substitution.

(a) Usually, the effects of these groups are additive provided they are suitably situated. Thus, for instance, the position to be occupied by the new group (indicated by arrows) is in accordance with this prediction.

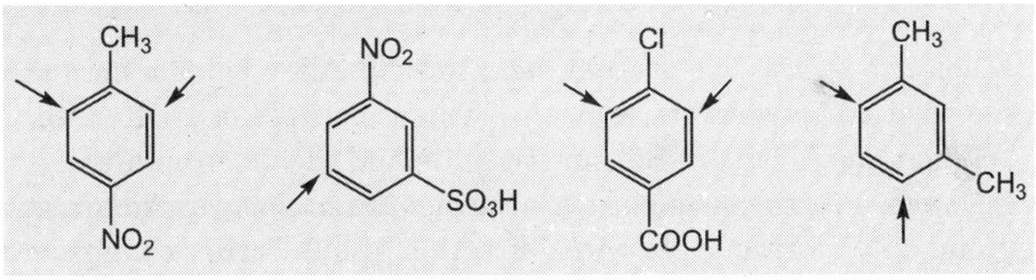

(b) However, when the two groups are located at such positions where they oppose each other, predictions regarding the position taken by the electrophile may become somewhat difficult. For instance, when two groups belonging to the same category are present in *competing positions*, it is hard to predict which will control the orientation. Thus, all four isomers are possible in case of *o*-methoxyacetanilide. However, the amount of isomer having the incoming group *ortho* to anilide group would be less due to steric effect.

In some cases, it is possible to predict the product composition by invoking the resonance and inductive effects of the substituents already present in the molecule. But, in practice, mixtures having all possible isomers in nearly all proportions are usually obtained. In many cases, it is hard to explain the experimental results. For instance, in case of *ortho*-bromofluorobenzene, substitution takes place *para* to fluorine, while in case of *para*-bromofluorobenzene, electrophile enters *ortho* to bromine.

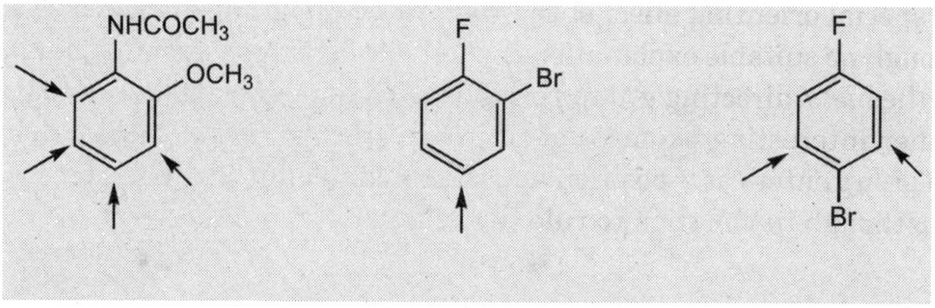

However, it is possible to make the following generalizations, which are based on experimental observations as well as theoretical considerations:

(i) If there is a competition between an *ortho, para* directing and *meta* directing group, the former controls the site of substitution. Even among the *ortho, para* directing groups, the one with stronger tendency to release electrons will control the orientation.

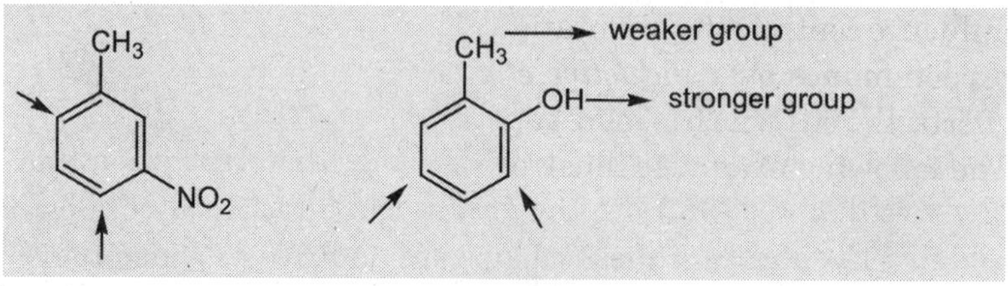

Thus, various groups can be arranged in the decreasing activation of the ring:

$$-OH, -NR_2, -O^-, -NH_2 > -OR > -OCOR > -NHCOCH_3 > C_6H_5 > CH_3$$

(ii) When two groups are present *meta* to each other, seldom does the substituent enter the position between these groups. This is because of greater steric hindrance at this position due to bulk of the substituents and the electrophile.

(iii) If a *meta*-directing group is situated *meta* to an *ortho/para* directing group, the third group enters *ortho* to the *meta*-directing group, rather than *para* to it. For instance, 3-chloronitrobenzene on chlorination yields mainly 3,6-dichloronitrobenzene with a little amount of 2,3-dichloronitrobenzene. The latter compound is formed in violation of rule (ii), but 3,4- dichloronitrobenzene is not formed at all.

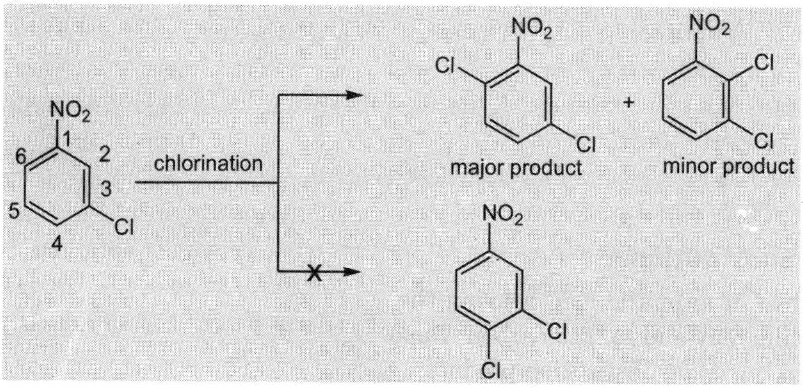

This special orienting effect of the *meta*-directing group is known as the *ortho* effect. Although no suitable explanation is available for this observation, it may be presumed that the *meta*-directing group possibly gives some intramolecular assistance. Another interesting example of this is the dinitration of 4-bromotoluene. The first nitro group enters at 2-position and the second enters at 3-position (*ortho* to first NO_2 group though in violation to rule (ii)) and not at 5-or 6-position.

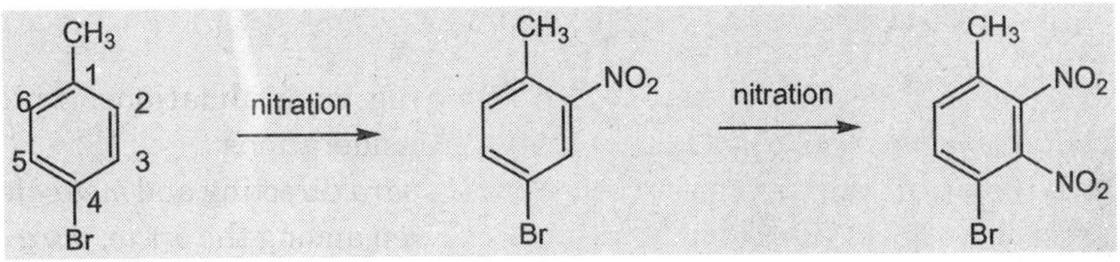

(iv) In case where both the substituents are *o*/*p* directing, the one which exerts stronger influence controls the orientation of the third group. Generally, resonance effects take predominance over *inductive effects,* and if both the groups operate *via* resonance effects, the one which is more capable of sharing electron pair exerts stronger influence. The following examples illustrate this rule:

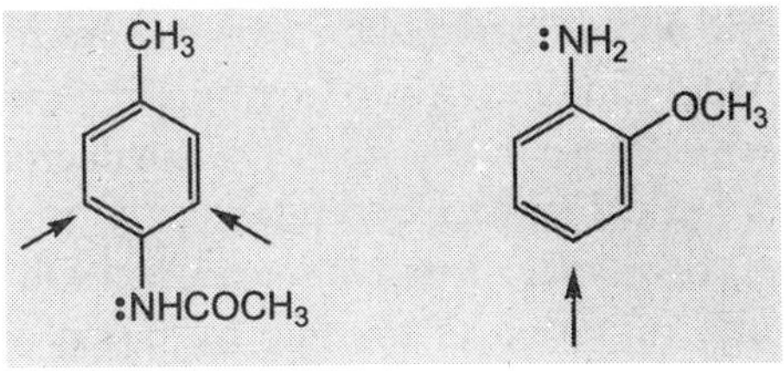

(v) When both the groups already present on the ring are *meta* directing, further substitution in the ring becomes more difficult. However, if the third group is actually introduced, its orientation is governed by the less powerful of the two *meta*-directing groups.

(vi) If the two *ortho*/*para* directing groups are *meta* to each other, little substitution occurs at the position which is *ortho* to both.

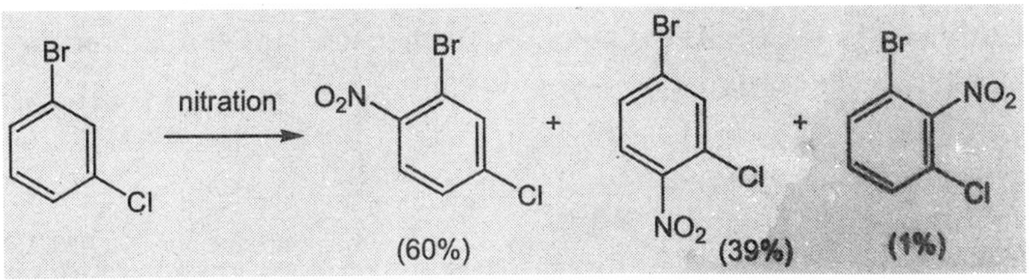

G. *Ipso* Substitution

The carbon of aromatic ring bearing the substituent is the ***ipso*** **carbon**. In some cases, the electrophile may add to *ipso* carbon. Departure of the substituent which was already present results in the *ipso*-substitution product.

The silicon atom has the ability to stabilize a positive charge on a β-carbon atom, because it is relatively electropositive, and also partly due to carbon-silicon hyperconjugation. Due to this effect, arylsilanes undergo electrophilic substitution at the site of trimethylsilyl group. Thus, the incoming electrophile is directed by the silyl group to the carbon atom carrying the trimethylsilyl group.

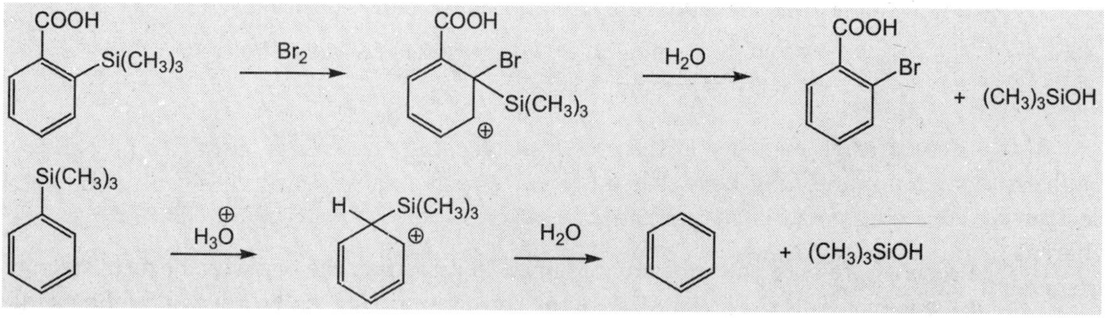

10.3 QUANTITATIVE TREATMENT

A convenient method for the quantitative expression of orientation and reactivity in aromatic substitution is through the *partial rate factor,* which is represented by the symbol *f*. It may be defined as the rate of substitution at a given position in *substituted benzene* relative to that of any one position in benzene.

Partial rate factors are calculated by knowing the proportions of the isomers from the substituted benzene, and the relative rates of reaction of the substituted benzene and benzene itself. For example, in the nitration of toluene by nitric acid in acetic acid at 45°C, *para*, *meta* and *ortho*-nitrotoluenes were obtained in 40, 3.2, and 57% yields, respectively, and the relative reactivity of toluene to benzene was found to be 24.5. With the help of the data, we can conveniently calculate partial rate factors for all the three positions of toluene. The reactivity of the *para* position of toluene to benzene may then be expressed as

$$\frac{40}{100} \times 24.5$$

But, since there are six positions available in benzene in comparison to only one *para* position in toluene, the correct expression should be

$$p_f Me = 6 \times \left(\frac{40}{100}\right) \times 24.5 = 59$$

($p_f Me$ = partial rate factor for the reactivity of *para* position of toluene)

Partial rate factors for *ortho* and *meta* positions are similarly calculated, and have been found to be 4.2 and 2.4, respectively (remembering that there are two *ortho* and two *meta* positions in toluene). Consideration of partial rate factors is very useful since one can tell at a glance whether a particular position is activated ($f > 1$) or deactivated ($f < 1$) in relation to benzene.

Brown has applied the modified Hammett equation for the quantitative prediction of partial rate factors, and has obtained good results for the relative reactivities of *meta* and *para* positions. However, this treatment does not apply for *ortho* substitution because of steric interactions. The modified Hammett equation is written as

$$\log f = \sigma^+ \rho$$

where σ^+ (sigma plus) = substitutent constant

ρ (rho) = reaction constant

σ^+ is a quantitative measure of the electron-releasing or electron-withdrawing power of a substituent group, and may have a negative or positive value. Positive σ^+ value indicates electron-withdrawing nature of the substituent resulting in the deactivation of aromatic ring, whereas negative value suggests electron donation by the substituent. Some typical σ^+ values are given in **Table 10.1**

Table 10.1 Substituent constants for electrophilic substitution

Substituents	***meta*** **(σ_m+)**	***para*** **(σ_p+)**
Methoxy	0.047	–0.77
Carboxylate	–0.028	–0.023
Methyl	–0.066	–0.31
Ethyl	–0.064	–0.29
Hydrogen	0	0
Carboxyl	0.322	0.421
Bromo	0.405	0.150
Cyano	0.562	0.659
Nitro	0.674	0.790

ρ is a measure of the sensitivities of various reaction series to changes in the polarity of the substituents. It usually has a negative value, and lies between –4 and –10. Some typical ρ values for electrophilic substitution (at 25°C) are given in **Table 10.2**.

Table 10.2 Reaction constants for electrophilic substitution

Reaction	ρ
Ethylation (C_2H_5Br-$GaBr_3$)	–2.4
Mercuration (Hg acetate / HOAc)	–4.0
Positive bromination (HOBr-dioxane)	–6.2
Acylation (CH_3COCl-$AlCl_3$)	–9.1
Molecular chlorination (HOAc)	–10.0
Molecular bromination (HOAc-H_2O)	–12.1

Larger negative values of ρ are suggestive of the strong influence of the substituents on the rate of substitution, whereas smaller values indicate lesser influence. By substituting the values of σ^+ and ρ in the modified Hammett equation, partial rate factors may be calculated. For example, when anisole is ethylated by Friedal-Crafts ethylation, partial rate factor for the *para* position can be calculated in the following way:

$$\log f = \sigma^+ \rho$$
$$\log p_f^{OMe} = (-0.778) \times (-2.4)$$
$$= 1.867$$
$$p_f^{OMe} = 74$$

10.4 SOME SYNTHETICALLY USEFUL AROMATIC ELECTROPHILIC SUBSTITUTION REACTIONS

A. Diazonium Coupling

Primary amines on reaction with nitrous acid (generated *in situ* by the action of sodium nitrite with acids such as HCl and H_2SO_4) give diazonium ions. This process is called *diazotization*. Aliphatic diazonium ions are too unstable to be isolated, whereas aromatic diazonium ions are moderately stable in aqueous solution at low temperatures. When present with anions of low nucleophilic power such as Cl^- and HSO_4^-, aryldiazonium salts may be isolated as solids.

$$ArNH_2 \xrightarrow[\text{0- 5 °C (diazotization)}]{NaNO_2\text{-}HCl} Ar{-}\overset{\oplus}{N}{\equiv}N\ \overset{\ominus}{Cl}$$

Aromatic diazonium ions react with active aromatic substrates such as amines and phenols to give azo compounds. Such reactions are known as azo *coupling reactions*. Presumably because of the size of the attacking species, substitution is mostly *para* to the activating group, unless that position is already occupied (*e.g.* β-naphthol), in which case *ortho* substitution takes place.

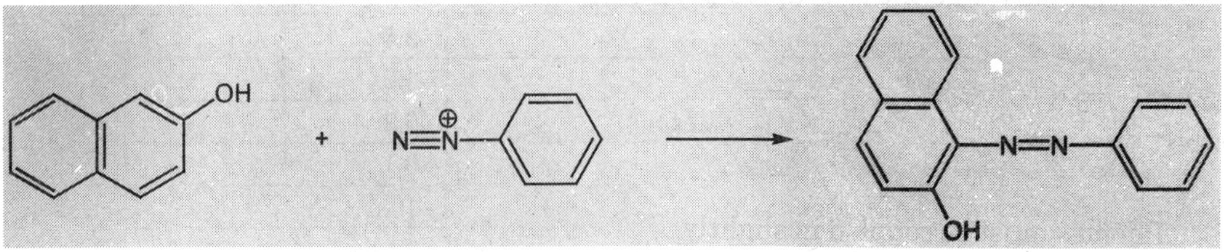

The mechanism of azo coupling reaction is analogous to other electrophilic aromatic substitutions, and involves two steps. In the first step, diazonium ion adds to the other aromatic compound, *e.g.* *N,N*-dimethylaniline. Then loss of a proton produces aromatic azo compound.

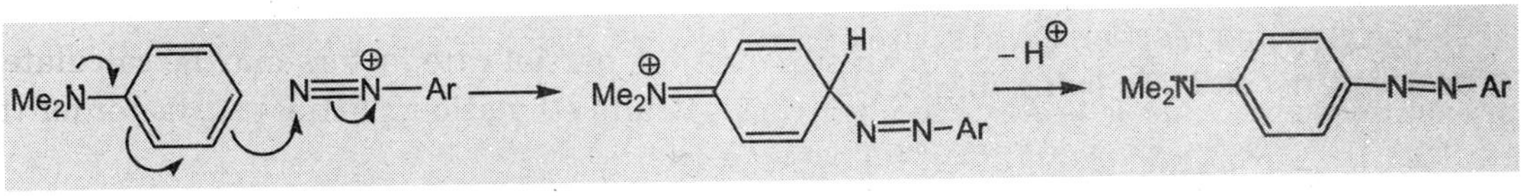

The *pH* of the solution is important both for amines and phenols. For amines, the solutions may be mildly acidic. If the acidity is too high, the reaction does not occur because the concentration of free amine becomes too small. Among the amines, tertiary amines on coupling with diazonium ions give *C*–azo compounds (*C*-coupled products), whereas primary and secondary amines show preference for the formation of *N*–azo compounds (*N*-coupled products). For example, when aniline (a primary amine) is partially diazotized with sodium nitrite and hydrochloric acid, the addition of sodium acetate liberates the unreacted aniline (free base) from its conjugate acid. Finally, coupling reaction between aniline and diazoniun ions produces diazoaminobenzene.

$$PhNH_3^{\oplus}Cl^{\ominus} + AcONa \longrightarrow PhNH_2 + AcOH + NaCl$$

$$PhNH_2 + PhN_2^{\oplus} \xrightarrow{-H^{\oplus}} PhNHN{=}NPh$$

Diazoaminobenzene
(*N*-azo coupled product)

The *N*-azo compounds can be converted to *C*-azo compounds by rearrangement using a mineral acid. The rearrangement occurs by uncoupling followed by *C*-coupling.

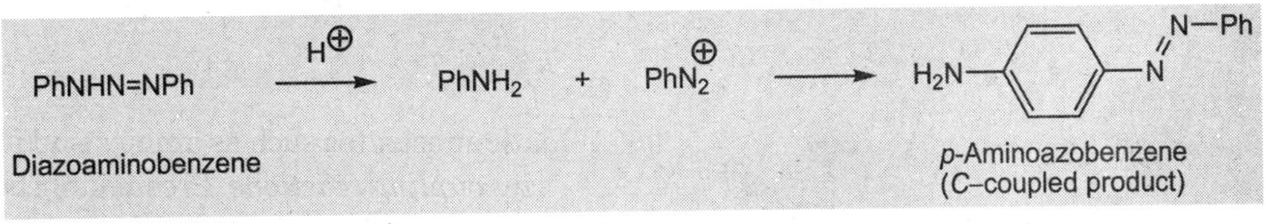

Primary and secondary aromatic amines can undergo direct C-coupling under two circumstances:

(i) When strong electron-withdrawing group such as NO_2 is present at *para* position of the diazonium ion, which makes the diazonium ion more reactive
(ii) When coupling reaction is carried out in aqueous formic acid, which is acidic enough to rearrange the *N*-coupled product to the *C*-coupled product.

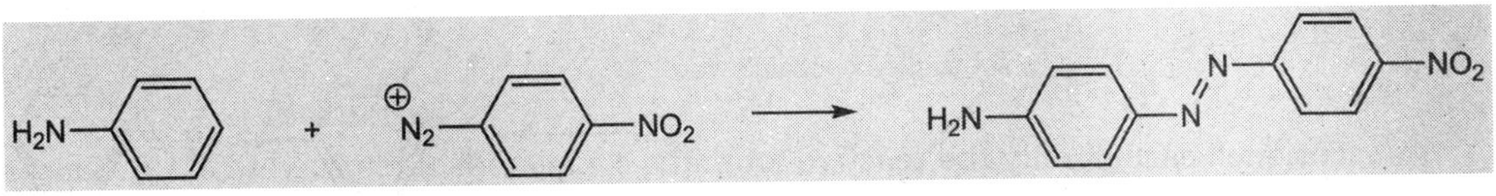

Phenols must be coupled in slightly alkaline solution where they are converted to the more reactive phenoxide ions as phenols themselves are not active enough for the reaction. However, neither phenols nor amines react in strongly alkaline solution as the corresponding diazonium ions are converted to a diazo hydroxide (ArN=N—OH).

Some coupling reactions, which are otherwise very slow (in cases where the coupling site is crowded), are catalyzed by pyridine.

B. Vilsmeir-Haack Reaction

A combination of disubstituted formamides (*e.g.* *N,N*-dimethylformamide; DMF) and phosphorus oxychloride is called the Vilsmeir-Haack reagent (or Vilsmeir reagent). This reagent is most commonly used for the formylation of active aromatic compounds such as phenol ethers and *N,N*-dialkylanilines. Such formylation is known as *Vilsmeir-Haack reaction* (or Vilsmeir reaction).

$$ArH + \underbrace{HCONR_2 + POCl_3}_{\text{Vilsmeir-Haack reagent}} \longrightarrow ArCHO$$

The mechanism of formation of an *ortho*-substituted benzaldehyde is outlined in the following scheme. *N,N*-Disubstituted formamide reacts with phosphorus oxychloride to generate a resonance stabilized electrophilic species, which subsequently attacks *ortho* position of the aromatic substrate. The resulting intermediate is aromatized by the loss of proton to give unstable iminium salt, which is easily hydrolyzed to the product. Similar steps are followed for the formation of a *para*-substituted benzaldehyde.

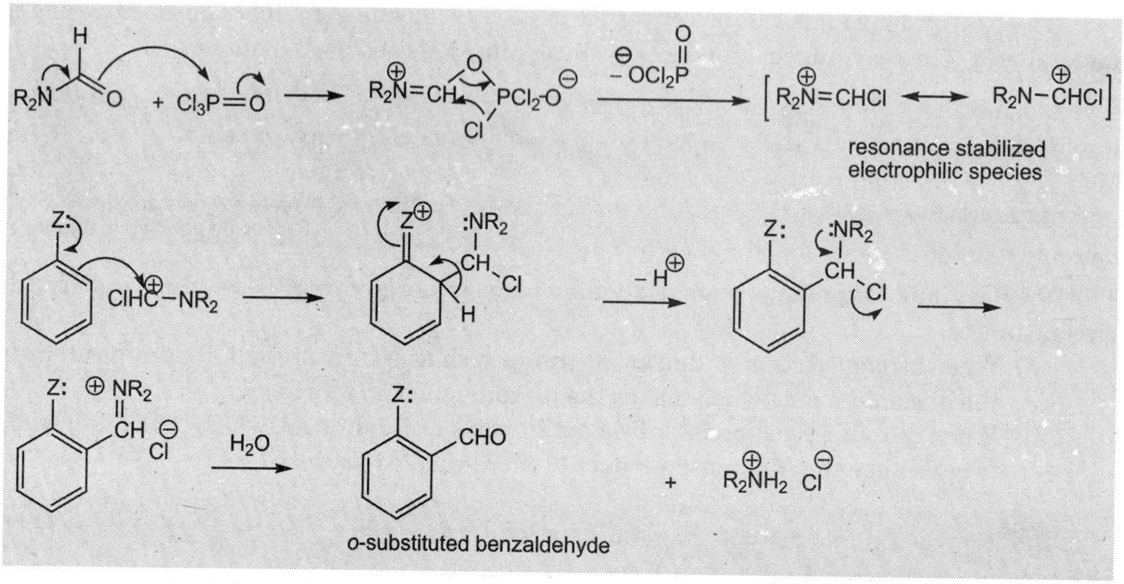

As mentioned earlier, only the reactive aromatic compounds are formylated, *e.g.* benzene and naphthalene are unreactive, but anthracene gives 9-formylanthracene in high yield. Similarly, azulenes and ferrocenes are smoothly formylated using $HCONR_2$ and $POCl_3$. Heterocycles can also be formylated, in case they are more active than benzene. The method is particularly effective for compounds such as pyrrole and indole, which cannot be formylated by using other standard procedures.

Although DMF and *N*-phenyl-*N*-methylformamide are common amides, other arylalkyl amides and dialkyl amides are also used. $COCl_2$ has been used in place of $POCl_3$. The scope of this reaction has also been extended for preparation of ketones. For example, use of *N,N*-dimethylacetamide in place of DMF offers a convenient method for aryl methyl ketones.

$$\underset{}{ArH} + \underset{N,N\text{-Dimethylacetamide}}{CH_3CONMe_2} + POCl_3 \longrightarrow \underset{\text{Aryl methyl ketone}}{ArCOCH_3}$$

C. Gattermann-Koch Reaction

Certain aromatic compounds can be formylated with a mixture of CO and HCl in the presence of $AlCl_3$. The method is known as the *Gattermann-Koch reaction*. An easy way to prepare the reagent mixture is to add chlorosulphuric acid, $ClSO_3H$, on formic acid, which generates CO, HCl, and H_2SO_4. Cuprous chloride is required, if the reaction is to be carried out at atmospheric pressure. In absence of cuprous chloride, pressure of 100 to 250 atm is required.

$$HCOOH + ClSO_3H \longrightarrow CO + HCl + H_2SO_4$$

$$ArH + CO + HCl \xrightarrow[CuCl]{AlCl_3} ArCHO \quad \text{(Gattermann-Koch reaction)}$$

Earlier, it was thought that the mixture of CO and HCl generated formyl chloride (HCOCl), which acted as an elctrophile in the formylation of aromatic substrate. Since formyl chloride has never been isolated, it is now considered that probably the attacking species is the resonance stabilized formyl cation (HCO^+), formed by transfer of a proton from the initially formed H^+ $AlCl_4^-$ (by the reaction between HCl and $AlCl_3$) to carbon monoxide. The function of cuprous chloride is to coordinate with CO, providing a higher local concentration. At high pressure, the concentration of HCO^+ is high enough for the success of the reaction, and therefore, CuCl is unnecessary.

$$HCl + CO + AlCl_3 \longrightarrow \underset{\text{resonance stabilized formyl cation}}{[\,HC\equiv\overset{\oplus}{O} \longleftrightarrow H\overset{\oplus}{C}=O\,]} + AlCl_4^{\ominus}$$

$$ArH + H\overset{\oplus}{CO} \xrightarrow{-H^{\oplus}} ArCHO$$

The formylation reaction has been largely limited to benzene and alkylbenzenes. It fails with amines, phenols, and phenolic ethers because of the formation of complexes with Lewis acids. It is also unsuccessful with aromatic compounds of lower nuclear reactivity than the halobenzenes. Therefore, nitrobenzene may be used as a solvent. This formylation process largely gives *para* substitution.

The reaction suffers from the drawback that rearrangement and disproportionation also occur in case of polyalkylated benzenes such as *p*-xylene, which gives 2,4-dimethylbenzaldehyde, obviously involving rearrangement.

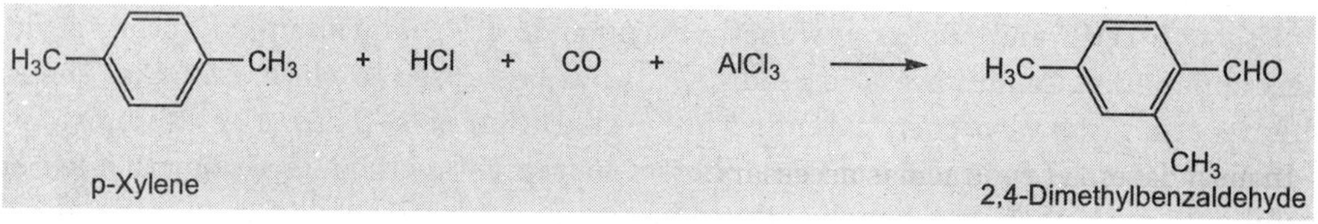

D. Bischler-Napieralski Reaction

In this reaction, amides of the type **4** are cyclodehydrated with an acidic reagent such as phosphorus oxychloride to give 1-substituted-3,4-dihydroisoquinolines **(5)**. The latter undergo dehydrogenation with palladium-carbon to yield corresponding isoquinolines **(6)**. The amides **(4)** are obtained by acylation of β-phenylethylamine.

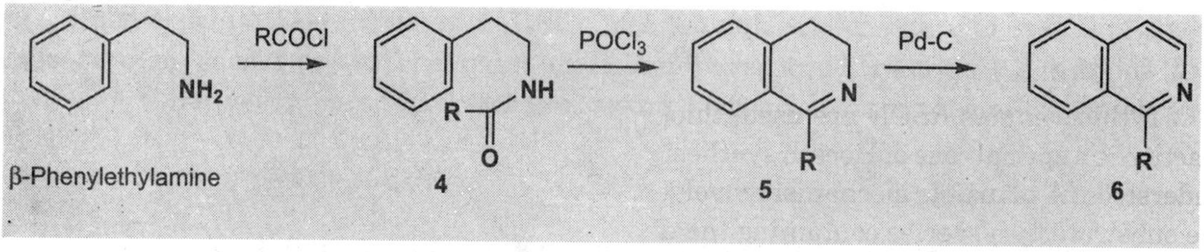

Cyclodehydration of **4** to **5** probably involves the following mechanistic pathway:

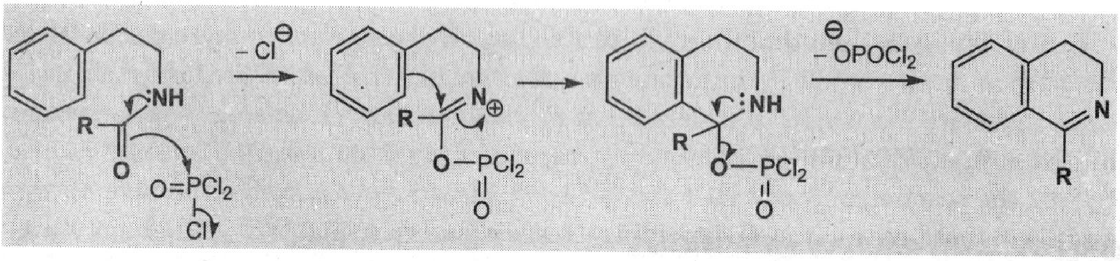

Higher yields can be obtained if the amide is treated with PCl_5 to give an imiodoyl chloride, $ArCH_2CH_2N–CR–Cl$, which is isolated and then cyclized by heating.

An important application of the *Bischler-Napieralski reaction* is the synthesis of an alkaloid, papaverine:

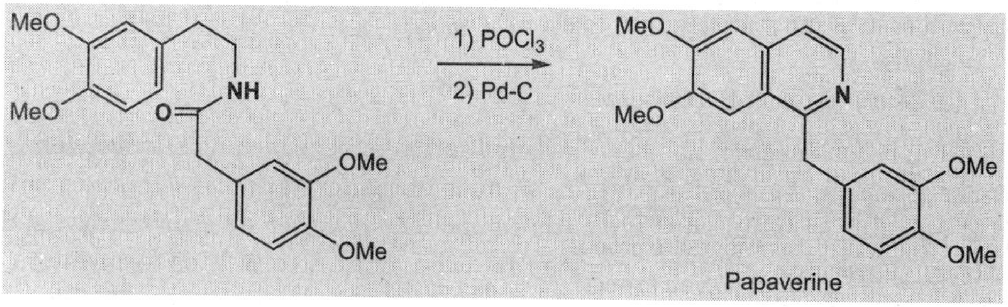

E. Houben-Hoesch Reaction

Friedel-Crafts acylation with nitriles and HCl is called the *Hoesch* or *Houben-Hoesch reaction.*

$$ArH + RCN \xrightarrow[ZnCl_2]{HCl} ArCOR$$

In most cases, a Lewis acid is necessary, zinc chloride being the most common. The reaction is useful only with phenols, phenolic ethers, and some reactive heterocyclic compounds, *e.g.* pyrrole. Monohydric phenols, however, generally do not give ketones, but are attacked at the oxygen to produce imino esters. Several nitriles including aryl nitriles have been used. In case of aryl nitriles, it is necessary first to treat them with HCl and $ZnCl_2$, and then add the substrate at 0°C.

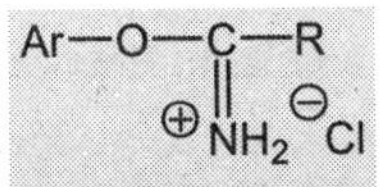

an amino ester

If thiocyanates RSCN are used, thiol esters ArCOSR can be obtained. The Gattermann reaction is a special case of Hoesch synthesis. The reaction course is complex, and not completely understood. A plausible mechanism involves two stages. The first stage consists of an attack on the substrate by a species containing the nitrile and HCl to give an imine salt **7**. In the second stage, the salts are hydrolysed to products.

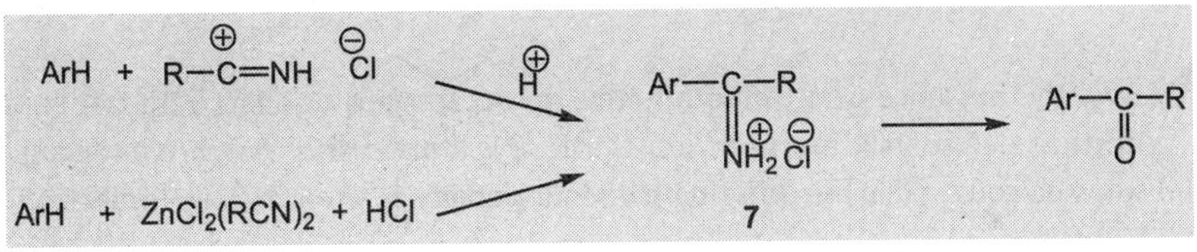

PROBLEMS AND REVIEW QUESTIONS

1. It has been shown that chlorination of aromatic ring can be effected with hypochlorous acid only in the presence of a proton catalyst. Explain.
2. Explain why compared to aniline ($C_6H_5NH_2$), acetanilide ($C_6H_5NHCOCH_3$) is somewhat deactivated towards electrophilic aromatic substitution.
3. From each of the following sets, select the most reactive and least reactive substrate towards ring nitration:
   (i) Benzene, toluene, nitrobenzene, and bromobenzene
   (ii) *m*-Dinitrobenzene, *m*-nitrotoluene and toluene
   (iii) Benzene, mesitylene, *m*-xylene and toluene.
   (iv) Chlorobenzene, 2,4-dinitrochlorobenzene and *p*-chloronitrobenzene.
4. The following experiments are proposed in order to decide whether aromatic sulphonation exhibits isotope effect. What do you expect to observe in each case, if this is so?
   (i) Measurement of relatives rates of sulphonation of benzene and benzene-D_6.
   (ii) Sulphonation of a 1 : 1 mixture of benzene and benzene-D_6 with one mole of sulphonating agents.

(iii) Determination of *o/p* ratio of the sulphonation products in case of toluene and toluene-4-*D*.

5. Predict the product(s), and outline the mechanism for the process when benzene is treated with (i) $(CH_3)_3CCl$ in presence of $AlCl_3$, (ii) $(CH_3)_2C{=}CH_2$ in presence of H_2SO_4, (iii) $(CH_3)_2CHCH_2Cl$ in presence of $AlCl_3$, and (iv) $(CH_3)_3COH$ in presence of H_2SO_4.
6. Write the mechanisms for the reaction of (i) *tert*-butyl bromide, (ii) *iso*butyl bromide, (iii) *sec*-butyl bromide, and (iv) *n*-butyl bromide with benzene in the presence of anhydrous $AlCl_3$.
7. (i) In Friedel-Crafts reactions, nitrobenzene is often used as a solvent. Why does it not compete with aromatic substrates for this reaction?

   (ii) More than one equivalent of $AlCl_3$ is used in Friedel-Crafts acylation reaction in order to obtain a good yield. Suggest a reason. How does your answer help to explain why multiple acylations are not observed?
8. Predict the product(s), and sketch a suitable mechanism for the following reactions:

$$\text{(i) } C_6H_6 + ICl \xrightarrow{FeX_3} \qquad \text{(ii) } C_6H_6 + FCHO \xrightarrow{BF_3}$$

9. Sketch a suitable mechanism for the following transformation:

$$C_6H_5C(CH_3)_3 + Br_2 \xrightarrow{AlBr_3} C_6H_5Br + (CH_3)_2C{=}CH_2$$

10. (i) Will the presence of a trifluoromethyl group activate or deactivate the benzene ring for further electrophilic aromatic substitution? Is it *meta* or *ortho-para* directing group?

    (ii) How do you explain the *ortho-para* directing ability of an alkyl group in electrophilic aromatic substitution?
11. The NMR spectrum of mesitylene ($C_6H_3(CH_3)_3$), exhibits two sharp singlets at δ 2.25 (9H) and 6.75 (3H). When treated with HF/SbF_5 in liquid SO_2, it gave a coloured crystalline solid, which was polar in nature. A look at the NMR spectrum of the solid revealed four sharp singlets at δ 2.8 (6H) and 2.9 (3H), 4.6 (2H) and 7.7 (2H). Suggest a suitable structure of the product.
12. Nitration of chlorobenzene gave rise to a mixture of two products, **A** and **B**. Upon reduction, **A** gave **C** (C_6H_6NCl). The NMR spectrum of **C** showed signals at δ 3.6 (*s*, 2H, disappears on shaking with D_2O), 6.6 (*d*, 2H) and 7.2 (*d*, 2H). Assign structures of **A**, **B** and **C**.
13. The insecticide DDT is prepared from chloral and chlorobenzene. Suggest a plausible mechanism.

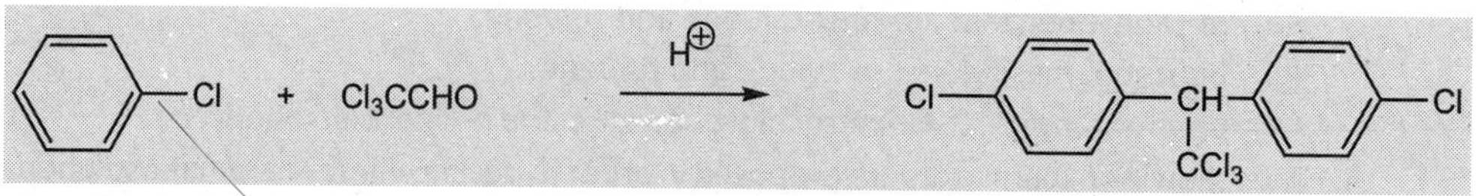

14. Give the structure of the product(s), and outline the mechanism involved in each of the following reactions.

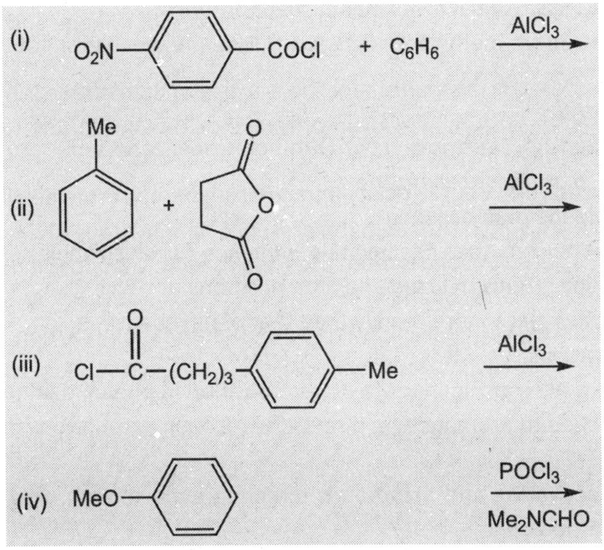

15. The Gattermann-Koch reaction for the formylation of an aromatic ring is regarded as aromatic electrophilic substitution reaction. Consider the following reaction sequence for the synthesis of benzaldehyde.

$$C_6H_6 + CO + HCl \xrightarrow[100\ atm]{AlCl_3} C_6H_5CHO + HCl$$

If we use DCl in place of HCl, shall we get C_6H_5CHO or C_6H_5CDO?

16. Predict the major product in each of the following reactions:
(i) Monochlorination of *p*-cumene, (ii) Monobromination of *o*-cresol, (iii) Mononitration of *p*-fluorobromobenzene, (iv) Mononitration of benzotrichloride, (v) Mononitration of *m*-bromochlorobenzene.

17. Comparative rate constants for nitration and sulphonation of benzene are given in the following equations:

$$C_6H_6 + \overset{\oplus}{N}O_2 \underset{k_{-1}}{\overset{k_1}{\rightleftharpoons}} \overset{\oplus}{C_6H_5}\begin{smallmatrix}H \\ NO_2\end{smallmatrix} \xrightarrow{k_2} C_6H_5NO_2 + \overset{\oplus}{H}\ ;\ k_2 >> k_{-1}$$

$$C_6H_6 + SO_3 \underset{k_{-1}}{\overset{k_1}{\rightleftharpoons}} \overset{\oplus}{C_6H_5}\begin{smallmatrix}H \\ SO_3^{\ominus}\end{smallmatrix} \xrightarrow{k_2} C_6H_5SO_3^{\ominus} + \overset{\oplus}{H}\ ;\ k_2 \sim k_{-1}$$

(i) Draw energy profile diagrams for these processes.
(ii) Account for the fact that isotope effect is observed in sulphonation reaction, but not in nitration.

18. Explain the following observations:
(i) Phenol on treatment with bromine-water and bromine in CS_2 or CCl_4 forms 2,4,6-tribromo derivative and monobromo derivative (*ortho* and *para*), respectively.

(ii) Nitration of aniline is sluggish, and it gives mainly *meta* product.

19. How will you convert toluene into *p*-methylacetophenone? Give the mechanism of the reaction involved.
20. Justify the following statements:
    (a) Nitration of bromobenzene gives *ortho* and *para* bromonitrobenzene, while bromination of nitrobenzene gives *meta* bromonitrobenzene
    (b) It is easier to nitrate toluene than benzene.
    (c) Chlorine atom in chlorobenzene has deactivating influence for electrophilic substitution, yet it directs the incoming groups to *ortho* and *para* positions.
    (d) Nitration of benzene takes place more readily than that of nitrobenzene.
21. Explain the following observations:
    (i) It is easier to introduce a nitro group in *ortho* or *para* position in phenol than in toluene.
    (ii) $-NH_2$ group when attached to aromatic ring directs the incoming group to *ortho* and *para* position.
    (iii) Alkoxy group acts as activating and *ortho-para* directing even though oxygen is more electronegative than carbon.
    (iv) In case of *o*- and *p*-substitution, the *p*-isomer generally predominates.
22. Arrange the following compounds in increasing order of reactivity towards electrophilic substitution: benzene, aniline, acetophenone, toluene, and acetanilide.
23. What is the essential difference between Friedel-Crafts alkylation and Friedel-Crafts acylation?
24. Polysubstitution products are often obtained in Friedel-Crafts alkylation when the alkylating agent is present in excess. However, no polysubstitution has been observed in Friedel-Crafts acylation reaction. How do you explain this observation?
25. Discuss the mechanism of sulphonation of benzene. In what respect does it differ from nitration?
26. Write the important resonance contributing forms for σ-complexes for the *meta* chlorination of aniline, and then show why does the reaction take place at *ortho*- and *para*-positions.
27. Explain why it is necessary to convert amino group into acylamino group for effecting monobromination or mononitration of aniline.
28. (i) What is the role of Lewis acid catalyst in electrophilic reactions of benzene?
    (ii) Can a proton act as an electrophile? Predict the product obtained from toluene upon treatment with D_2SO_4.
29. What are σ- and π-complexes? Illustrate their role in aromatic electrophilic substitution reactions with the help of an energy profile diagram.
30. It is well known that aromatic electrophilic substitution is a two-step process. Describe evidence for stepwise mechanism of electrophilic substitution in benzene.
31 Draw σ-complex for the electrophilic nitration of styrene ($C_6H_5CH{=}CH_2$) at *o, m*-, and *p*-positions. On the basis of stability of these complexes, predict whether $-CH{=}CH_2$ group is *o/p* or *m*-directing.
32. How do you account for the observation that in *o,p*-directing systems, usually *ortho* substitution occurs to a lesser extent than expected? Are there exceptions to this general rule?
33. In the presence of two groups: an *ortho/para* and *meta*-directing, what will be the position for the third substituent on the phenyl ring?

34. Comment upon the statement: In case of halogens, reactivity is determined by stronger inductive effects, while orientation is largely controlled by resonance.
35. (i) Methyl group attached to the benzene ring has no lone pairs of electrons, but is still *ortho/para-* directing in nature. Explain.
    (ii) Why is Friedel-Crafts acylation preferred over Friedel-Crafts alkylation in organic synthesis?
36. Giving suitable examples, discuss *ipso* substitution.

SOLUTIONS

1. The electrophile, $H_2O^+–Cl$ or probably Cl^+ may be generated as follows:
   $Cl–OH + H^+ \longrightarrow H_2O^+–Cl \longrightarrow Cl^+ + H_2O$
2. Because of the presence of carbonyl group, the nonbonding electrons on nitrogen are less readily available for donation to the ring. Furthermore, this places partial positive charge on nitrogen which exerts a greater electron withdrawal than amino nitrogen.

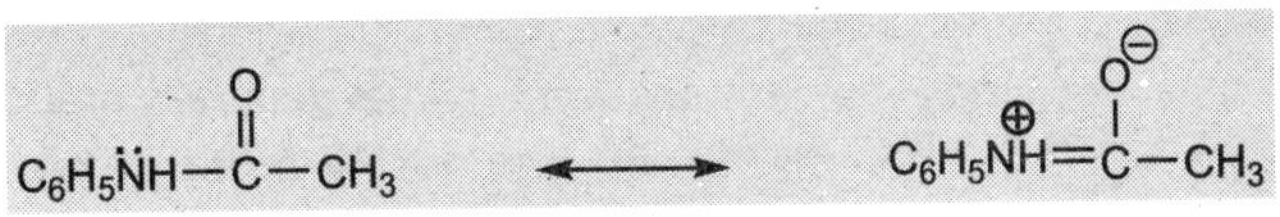

3. *Most reactive*: (i) Toluene (ii) Toluene (iii) Mesitylene (iv) Chlorobenzene
   Least reactive: (i) Nitrobenzene (ii) *m*-Dinitrobenzene (iii) Benzene (iv) 2,4-Dinitrochlorobenzene
4. (i) The rate of sulphonation of C_6D_6 will be slower than that of C_6H_6.
   (ii) The mixture will contain more of $C_6H_5SO_3H$, than $C_6D_5SO_3H$ (D).
   (iii) The *o/p* ratio in the case of toluene-4-D would be larger than that in the case of toluene.
5. *Tert*-butylbenzene.
6. The product in all the three cases is *tert*-butylbenzene. In the first case, the *tert*-butyl cation is formed directly, whereas in other cases the same cation is formed by the rearrangement of initially formed primary or secondary cation (the *tert*-butyl cation is most stable).
7. (i) Due to the presence of a strong electron-withdrawing nitro group, the rate of reaction of nitrobenzene with an electrophile is much slower than that of benzene. The difference between the two rates is so much that nitrobenzene remains practically unaffected.
   (ii) The product of Friedel-Crafts acylation complexes with $AlCl_3$. The result is a deactivated benzene ring which is unable to acylate further.

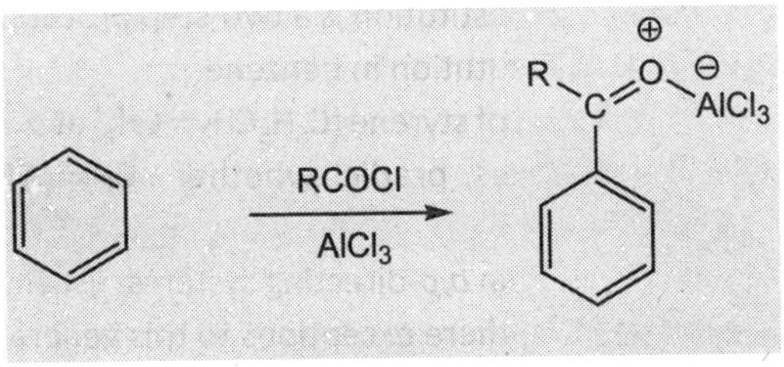

8. 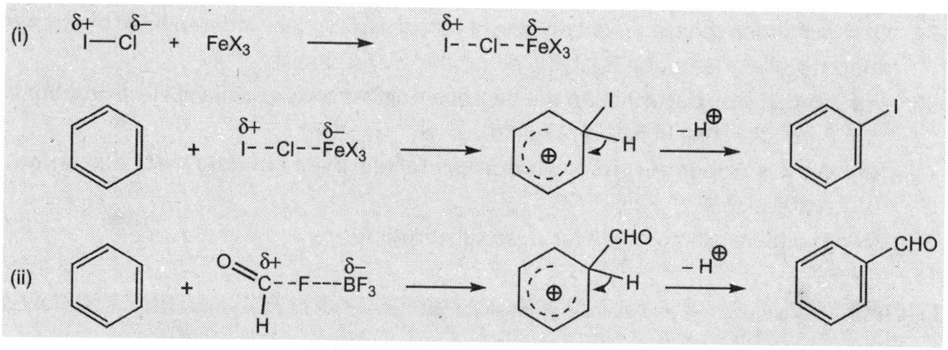

9. 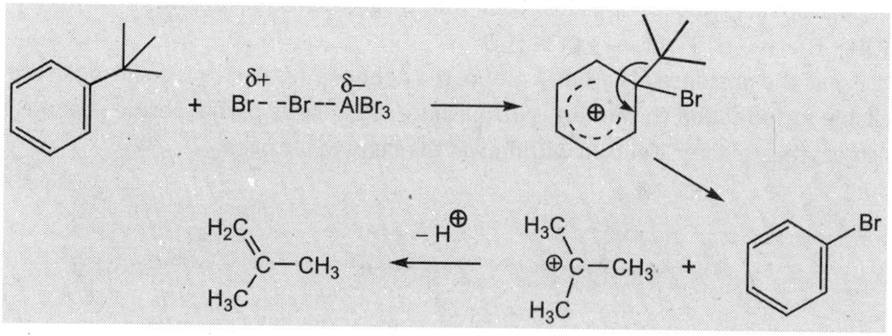

10. (i) The ring is deactivated due to electron-withdrawing effect of three fluorine atoms and CF_3 group would be a *meta* director.

11. 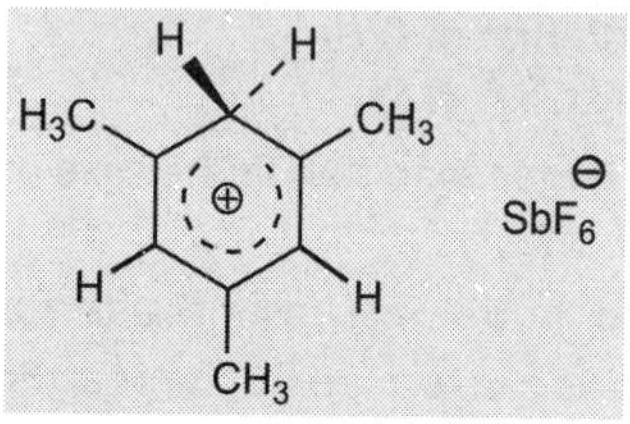

12. 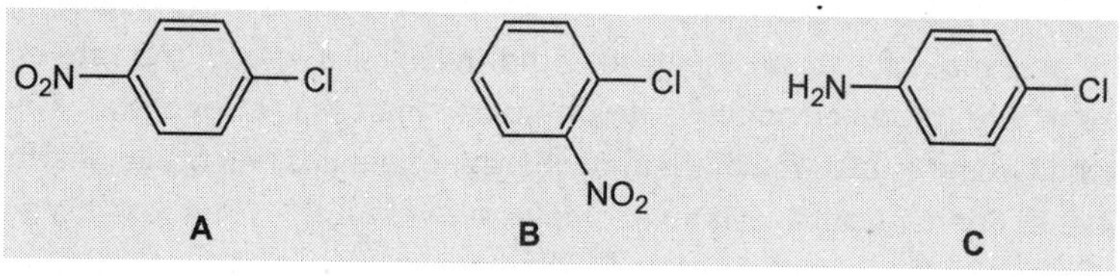

13. 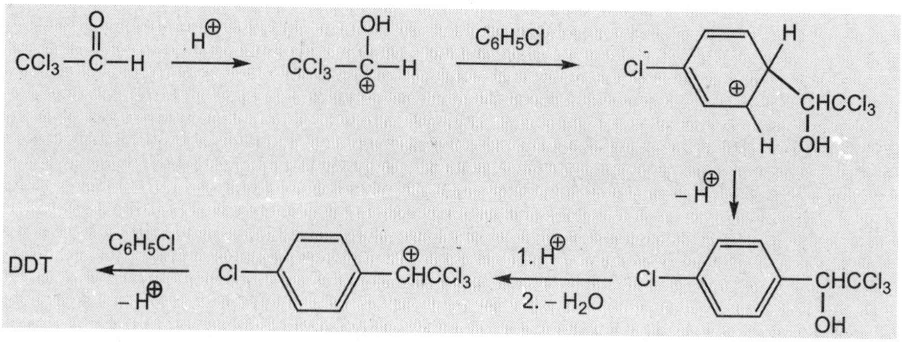

14. (i) 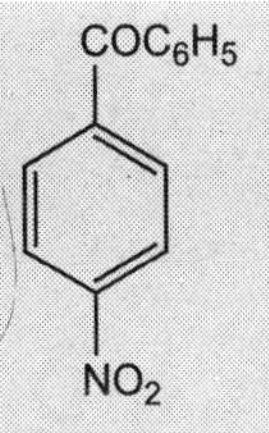 (ii) 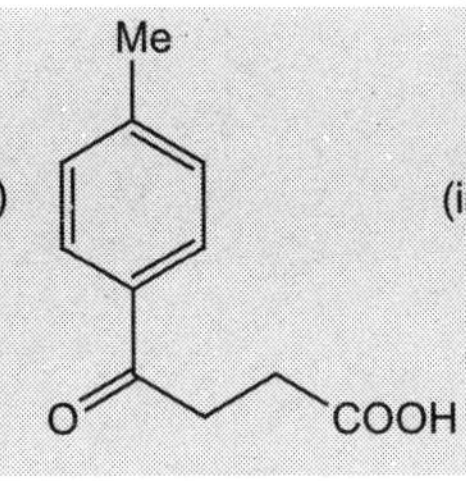 (iii) 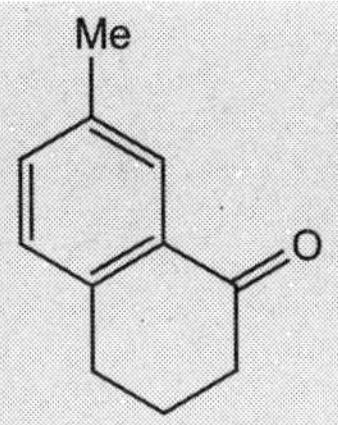 (iv) 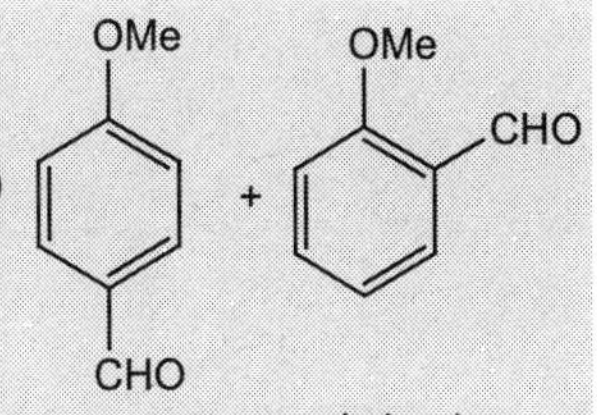

(major) (minor)

15. C_6H_5CDO

16. (i) 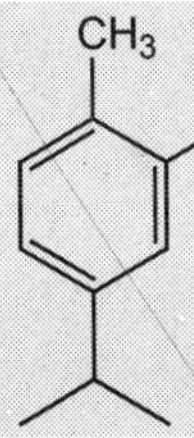 (ii) 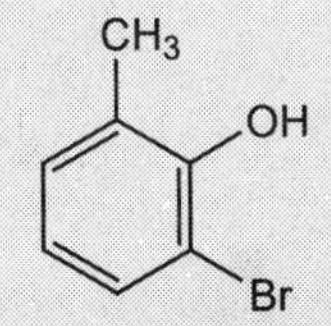 and 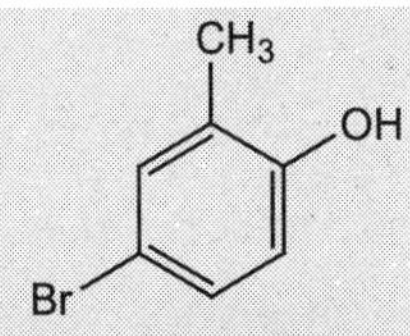

(major)

(iii) 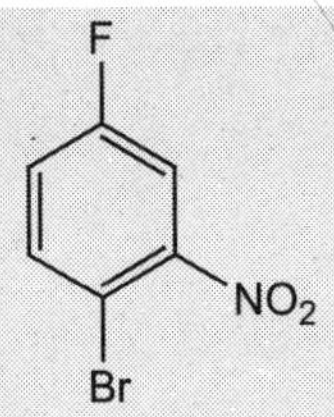 (iv) 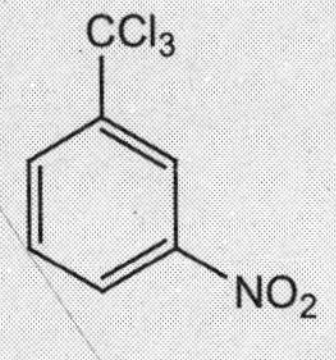 (v) 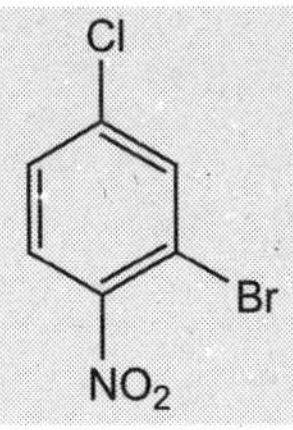 and 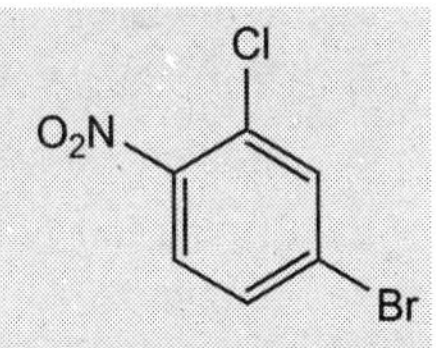

Chapter 11

Aromatic Nucleophilic Substitution

Chapter Outline

11.1 MECHANISMS

- **A.** Unimolecular Mechanism
- **B.** Bimolecular Mechanism
- **C.** Elimination-Addition Mechanism: Benzynes

11.2 AROMATIC NUCLEOPHILIC REARRANGEMENTS

- **A.** von-Richter Rearrangement
- **B.** Sommelet-Hauser Rearrangement
- **C.** Stevens Rearrangement
- **D.** Smiles Rearrangement
- **E.** Bucherer Reaction

11.1 MECHANISMS

Aromatic nucleophilic substitution may be described as a process in which a bond with an aromatic ring carbon is formed by a nucleophilic reagent (Nu^-) with the corresponding displacement of a group, X, along with its bonding electrons.

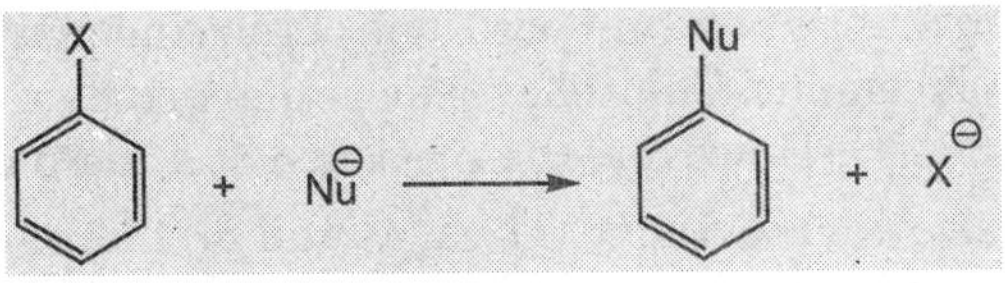

Nucleophilic substitution is not a typical reaction of aromatic compounds as the presence of the π-electron system in their σ-bond framework repels the unshared electron pair of the nucleophile. In addition, many potentially displaceable groups have unshared electron pairs, which are conjugated with the ring, and thus, their displacement is even more difficult. However, appropriately substituted aromatic compounds do undergo nucleophilic substitutions. As in aliphatic compounds, nucelophilic substitution in aromatic compounds may be either an unimolecular (S_N1) or a bimolecular (S_N2) process.

A. Unimolecular Mechanism

This involves an initial rate-determining heterolysis of the bond attached to the displaced group followed by reaction of the phenyl cation with the nucleophile. Substitution by this process is far less common in aromatic compounds than in the aliphatic series mainly because of the lack of internal conjugation to stabilize the phenyl cation formed after heterolysis. Probably, best known example of S_N1 reactions is spontaneous decomposition of aryldiazonium salts, which yield different products in the presence of various nucleophiles.

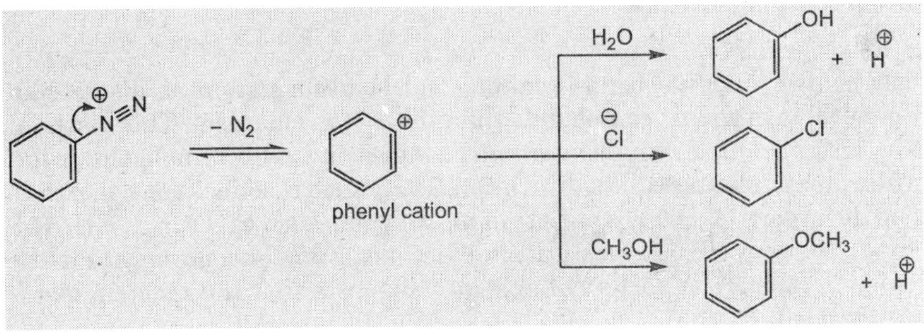

The main evidence for the intermediacy of the aryl cation comes from the observation that the rate of decomposition follows first order kinetics, and is independent of the nature and concentration of anions present. It is thus presumed that the rate-determining step in the reaction is the formation of the aryl cation, which being very reactive, rapidly combines with the anions present in the reaction mixture. That the cleavage of carbon-nitrogen bond is a reversible step has been established by the observation that when $Ar^{15}N^{+}{\equiv}N$ is allowed to decompose, the recovered starting material also contains $ArN^{+}{\equiv}{}^{15}N$.

In conformity with other S_N1 reactions, we should expect rate-acceleration by electron-releasing substituents, and rate-retardation in the presence of electron-withdrawing substituents. However, these expectations are fulfilled only in case of *meta*-substituents where substituents such as –OH, –OMe, and –Me accelerate the rate, and groups such as –COOH, –Cl, and $–NO_2$ retard the rate of reaction.

B. Bimolecular Mechanism

Normally, aryl halides are characterized by very low reactivity towards the nucleophilic reagents. However, presence of some groups at certain positions of the ring markedly activates the halogen of aryl halides towards displacement. Chlorobenzene is converted into phenol by aqueous NaOH only at temperature over 300°C. However, the presence of a nitro group, *ortho* or *para* to the chloro group greatly increases its reactivity. Thus, *o*- and *p*-chloronitrobenzenes are converted into corresponding nitrophenols by treatment with aqueous NaOH at 160°C.

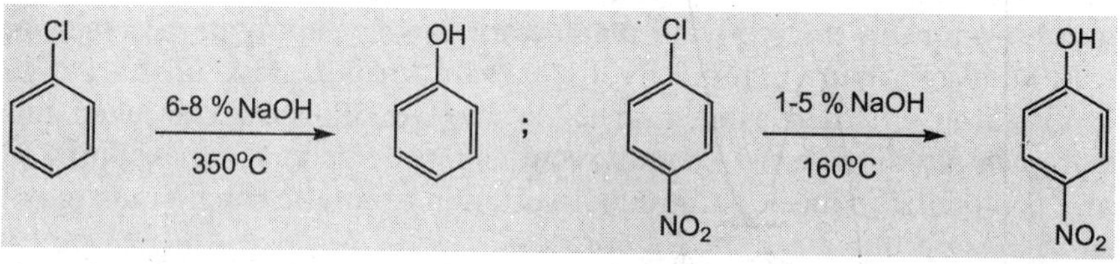

As the number of *ortho* and *para* nitro groups on the ring is increased, the reactivity increases. The corresponding phenols are obtained from 2,4-dinitrochlorobenzene by treatment with hot aqueous sodium carbonate, and from 2,4,6-trinitrochlorobenzene by simple treatment with water.

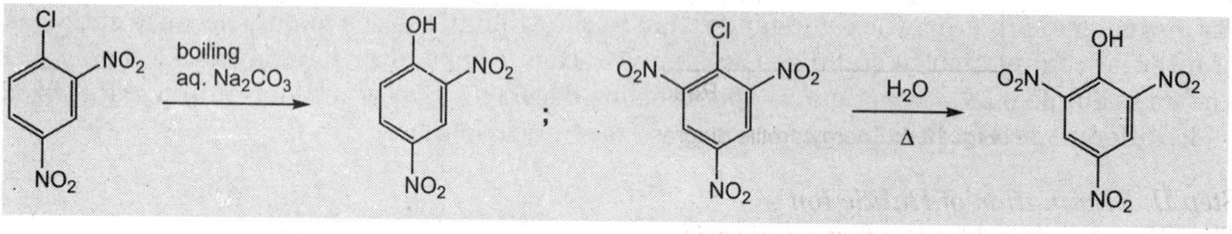

This mechanism involves bond-breaking and bond-formation as discrete steps, and is much more general in aromatic compounds than the S_N1 mechanism. This mechanism is often referred to as *intermediate complex mechanism* indicating its resemblance to the σ-complex mechanism of aromatic electrophilic substitution. Many such reactions consist of the expulsion of halogens from the aromatic ring by a variety of the reagents such as $^-OCH_3$, ^-OH, and piperidine. Like other S_N2 reactions, these substitution reactions follow second order kinetics (*i.e.* they are bimolecular in nature). A variety of evidence suggests that the reaction goes through an *addition-elimination process*.

Step I: Addition of the Nucleophile to Produce Cyclohexadienylide Anion

This step involves the addition of the nucleophile to the carbon carrying the halogen to produce an intermediate called cyclohexadienylide anion or *Meisenheimer complex*, which is stabilized by resonance.

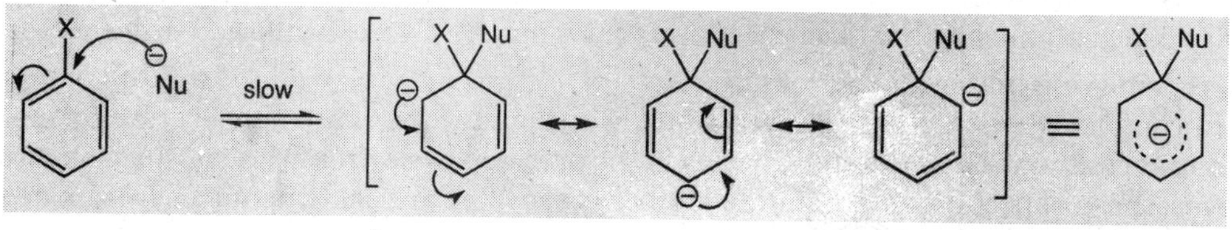

In many cases, this step is rate-determining. The reason is that there is rehybridization of carbon undergoing substitution from sp^2 to sp^3, thereby destroying the aromaticity of the ring. This raises the E_{act} for this step (**Fig. 11.1**).

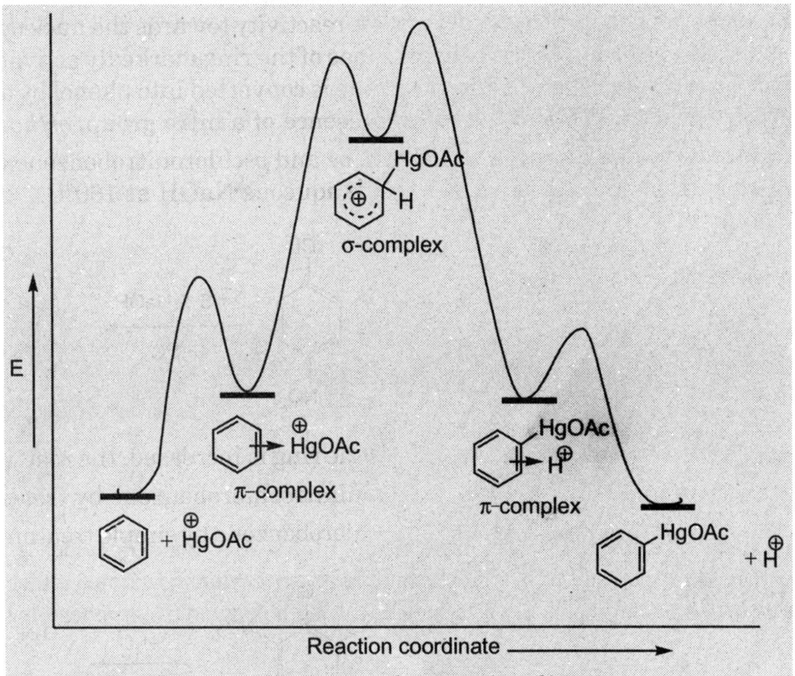

Fig. 10.4: Energy profile diagram of mercuration of benzene

Step II: Elimination of Halide Ion

The other step involves the fast elimination of halide anion from the intermediate. The driving force for the elimination of halide ion is that the ring regains its aromatic character.

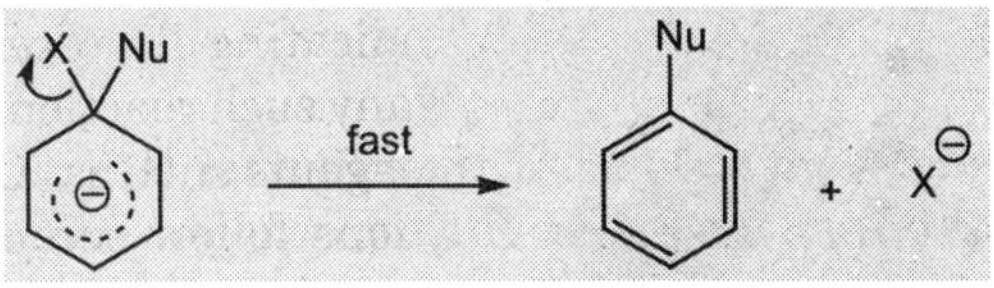

It can be seen that this mechanism greatly resembles the arenium-ion mechanism of electrophilic aromatic substitution. It may be noted here that formation of the cyclohexadienylide adduct is not possible unless some electron-attracting groups such as $-NO_2$, and $-CN$, are present at the *ortho* and *para* positions. The presence of these groups stabilizes the intermediate by dispersing the negative charge.

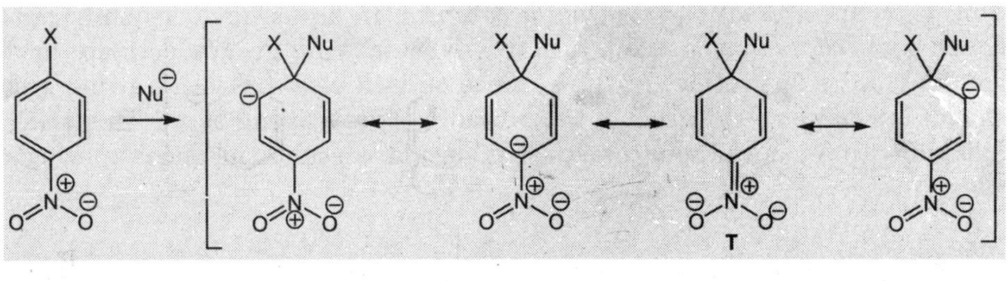

Such delocalization of the negative charge *via* structure marked as **T** is not possible with *meta* nitro derivatives, and hence, they do not undergo nucleophilic substitution.

A significant example of aromatic nucleophilic substitution is the reaction of an amine with 2,4- dinitrofluorobenzene. This reaction was used by Sanger to develop a method for identification of *N*-terminal amino acid in proteins. 2,4-Dinitrofluorobenzene is also known as *Sanger's reagent*.

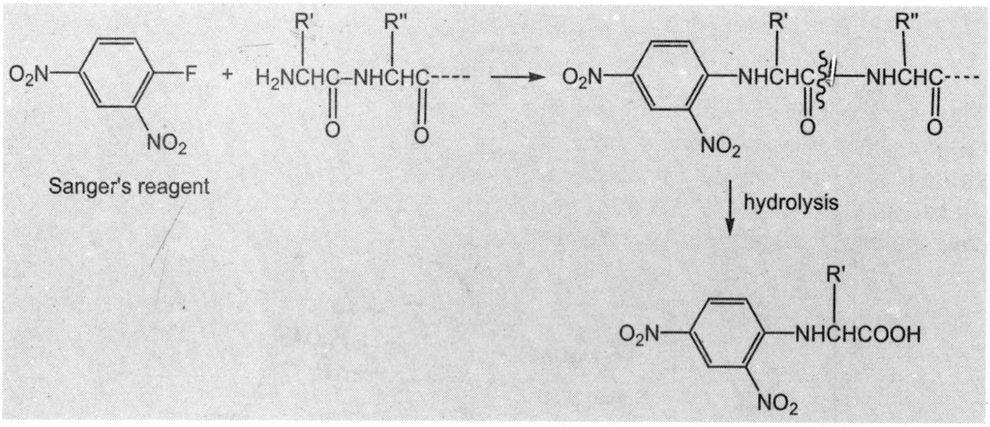

There is a variety of evidence for this mechanism. The most convincing evidence was isolation of the intermediates (Meisenheimer salts). A typical example is salt **1** which is formed either by the action of potassium ethoxide on 2,4,6-trinitroanisole (**2**) or potassium methoxide on trinitrophentole (**3**).

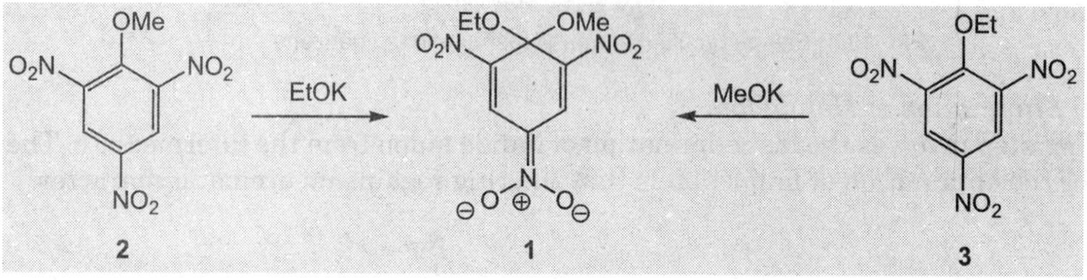

Further evidence comes from studies of the effect of leaving group on reaction. If the mechanisms were similar to either the S_N1 or S_N2 mechanism, the Ar–X bond would be broken in the rate-determining step. But, in S_NAr mechanism, this bond is not broken until after the rate-determining step. If S_NAr mechanism is operating, a change in leaving group should not have much effect on the rate of the reaction.

C. Elimination-Addition Mechanism: Benzynes

In contrast to the addition-elimination mechanism, the nucleophilic aromatic substitution may also take place by an *elimination-addition mechanism*. In aromatic nucleophilic substitution involving the bimolecular mechanism, electron-withdrawing groups activate aryl halides towards nucleophilic substitution. In the absence of such electron-withdrawing groups, aryl halides can still undergo substitution in the presence of very strong bases. This mechanism of nucleophilic substitution is different than the bimolecular mechanism, and is known as *benzyne mechanism*.

When an aryl halide such as chlorobenzene is treated with a strong base, *e.g.* potassium amide in liquid ammonia, it is converted into aniline. This is not a simple displacement. Instead, the reaction involves two stages, elimination (stage I), and addition (stage II). The intermediate is called benzyne (or dehydrobenzene).

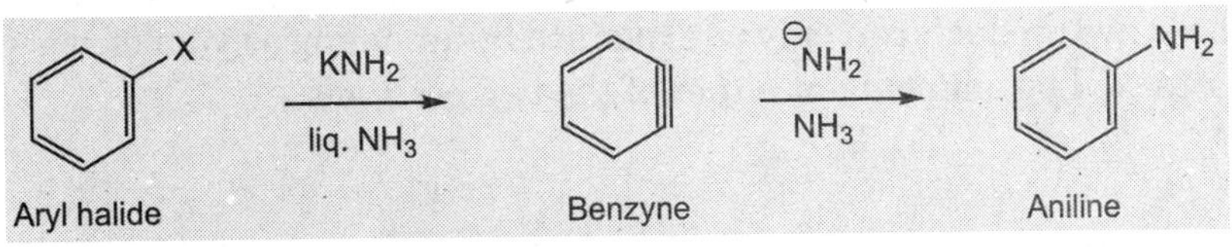

The overall process is an elimination-addition reaction, and hence is also known as the *elimination-addition mechanism* of aromatic nucleophilic substitution. Since the reaction may be regarded as going through a vicinal rearrangement in which the entering group comes in *ortho* to the leaving group, it is sometimes referred to as *cine*-substitution. In the presence of other nucleophiles, mixed products are obtained.

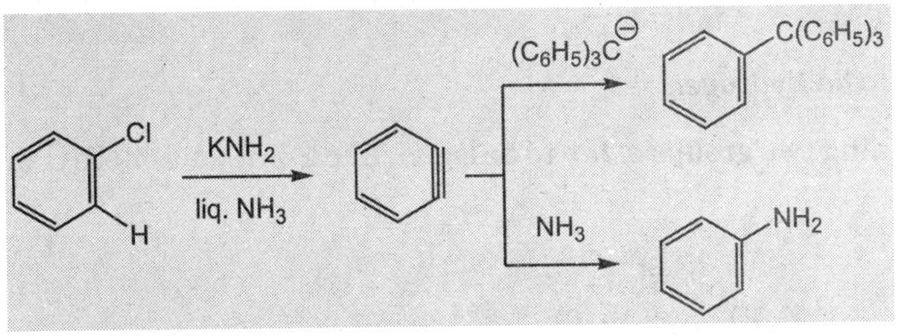

Each stage (I and II) of the reaction involves two steps:

Stage I: Two-Step Elimination of HX

Step I: Abstraction of a proton from *o*-position by a strongly basic amide anion to produce a carbanion, and

Step II: Loss of bromide ion from carbanion to produce benzyne.

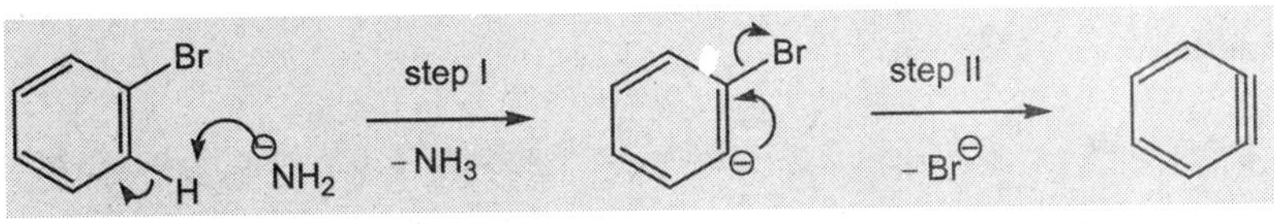

Stage II: Two-Step Addition of the Nucleophile

Step I: Addition of amide ion to benzyne to produce another carbanion,

Step II: Abstraction of a proton by carbanion from ammonia.

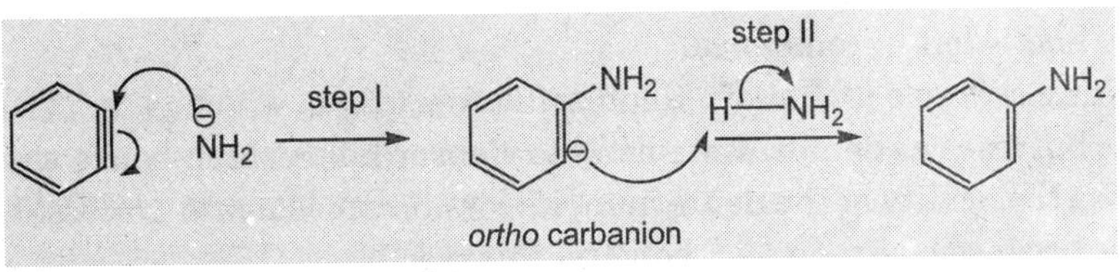

The following facts are in agreement with the proposed mechanism.I

I. Tracer Technique

The most convincing evidence for the intermediacy of benzynes in these types of substitution reactions has been provided by the fact that halobenzene-1-^{14}C yields almost equal amounts of aniline-1-^{14}C and anilne-2-^{14} C on treatment with potassium amide in liquid ammonia. The formation of these two products can only be explained by assuming a symmetrical intermediate, which can be attacked by ammonia at either of the two positions.

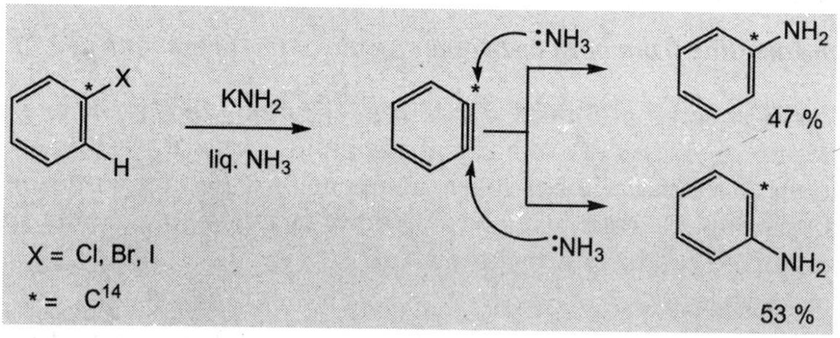

II. Condition of an **ortho** Hydrogen

Compounds containing two groups *ortho* to halogen, like 2-bromo-3-methylanisole, do not react at all.

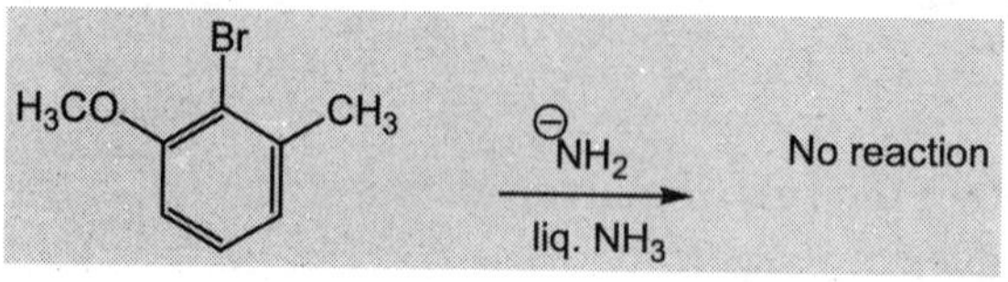

III. Isotope Effect

When a 50 : 50 mixture of bromobenzene and *o*-deuterobromobenzene is allowed to react with a limited amount of amide ion, recovered unreacted material contains more of deuterobromobenzene than bromobenzene.

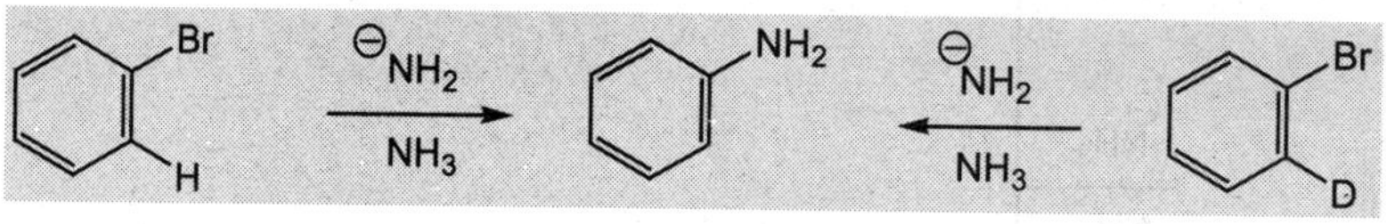

This can be explained by isotope effect. The isotope effect shows that not only *ortho* hydrogen is involved, but it is involved in the *rate-determining step*. Deuterium is abstracted more slowly in the first step, and is left unconsumed.

An interesting example is that of *o*-deuterofluorobenzene, which is converted into aniline only very *slowly*, but loses its deuterium *rapidly* to yield ordinary fluorobenzene (*pathway a*). As fluorine increases the acidity of the deuterium, its loss becomes a fast process. Loss of fluorine is a slow process because of strong C–F bond (*pathway b*).

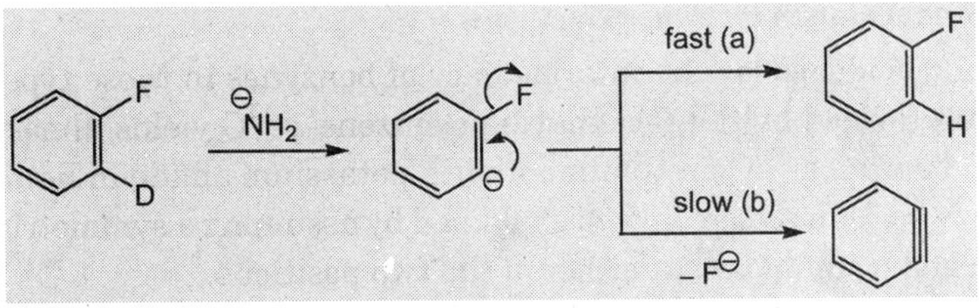

IV. Formation of m-Anisidine from both m-Bromoanisole and o-Bromoanisole:

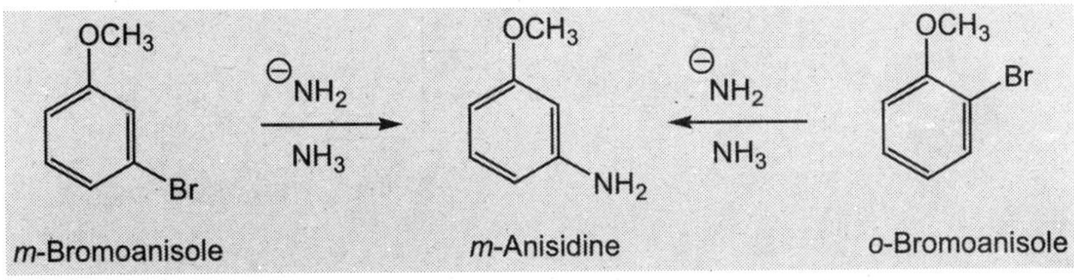

The reason for the same product is formation of the same intermediate benzyne **3**.

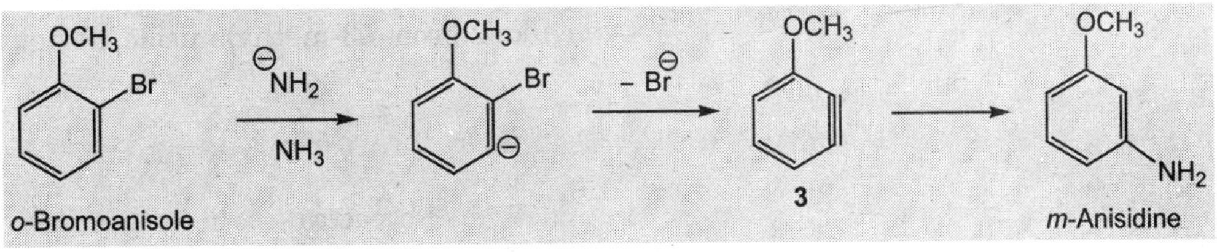

Since in *o*-bromoanisole only one *ortho* hydrogen is present, there is possibility of formation of only one benzyne **3**. On the other hand, *m*-bromoanisole could, in principle, give rise to two isomeric benzynes **3** and **4**.

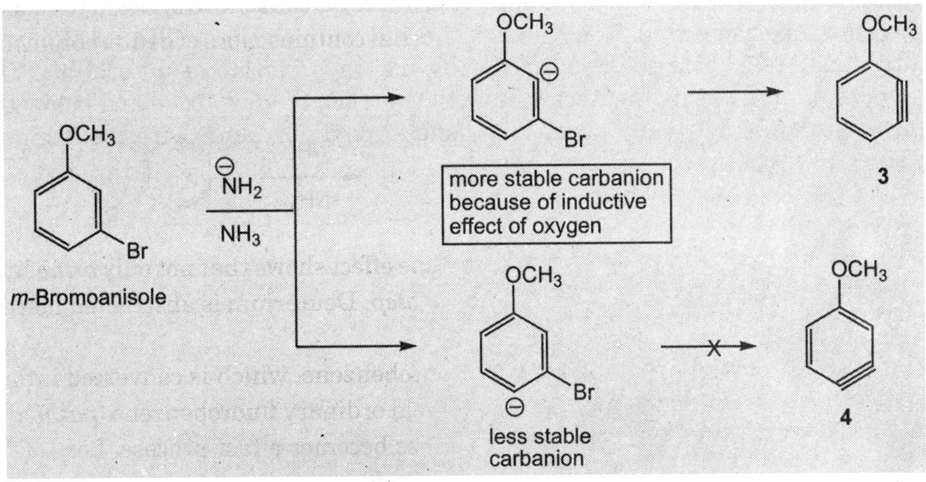

However, it generates only benzyne **3** because in the first step, the negative charge develops on the carbon that can best accommodate it. Thus, **3** yields *m*-anisidine because the negative charge generated in this way could be stabilized by methoxyl group.

V. Trapping Experiments for Detection of Benzynes

Benzynes have been detected either through spectroscopy or by their participation in such reactions as dimerization, and by trapping through cycloaddition to substances like furan or anthracene.

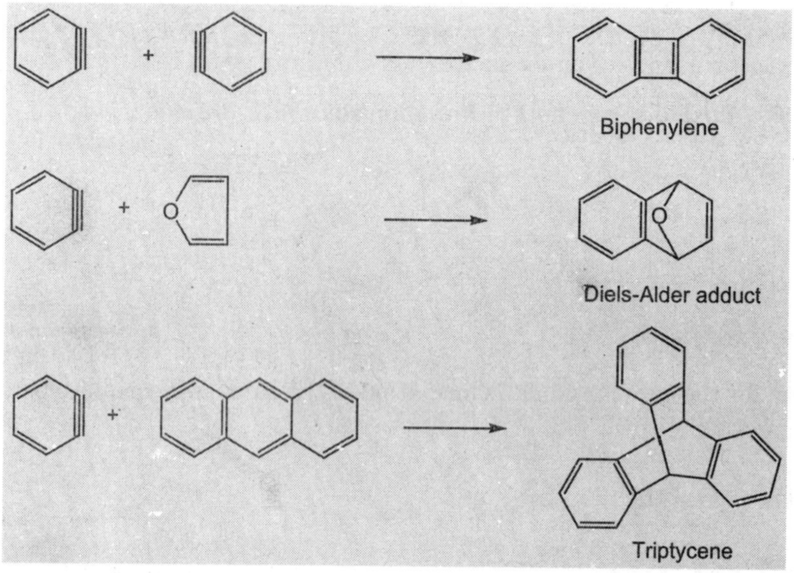

Thus, benzyne mechanism operates only in unactivated structures where substitution by addition-elimination mechanism is not possible. Other requirements for this mechanism are the presence of hydrogen atom *ortho* to the halogen, and a very strong base.

Structure of Benzyne: The benzyne bond is not like the triple bond of acetylene where the two carbons form a σ-bond using *sp* orbitals, and the remaining *p* orbitals are used to form two π-bonds. Such a structure is not possible in benzyne because of the hexagonal geometry associated with the benzene ring. Benzyne has the structure shown in **Fig. 11.2** in which an additional bond is formed between two carbons by sideways overlap of sp^2 orbitals. This new π-bond is *abnormal,* and has little interaction with the π cloud lying above and below the ring. The sideways overlap is not really good. This bond is unstable, and therefore, benzynes are extremely reactive chemical species.

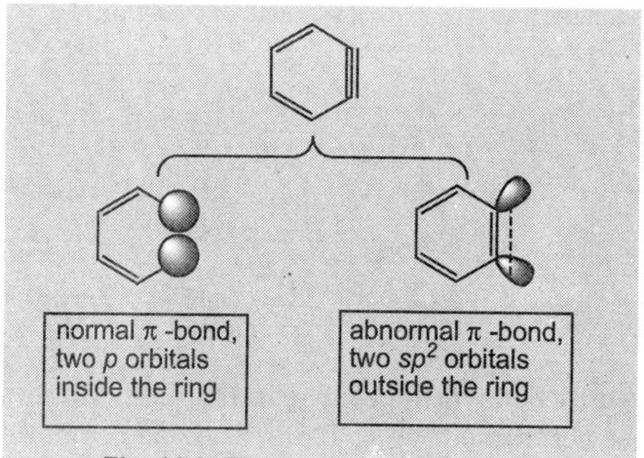

Fig. 11.2: The *p* orbitals of benzyne

11.2 AROMATIC NUCLEOPHILIC REARRANGEMENTS

A. von-Richter Rearrangement

In addition to the nucleophilic displacements discussed in **Section 11.1**, there is yet another reaction in which the nucleophile enters the aromatic ring *ortho* to the position occupied by the leaving group. For instance, when halogenated aromatic nitro compounds (**5**) are heated with cyanide ions in the presence of aqueous ethanol, the nitro group is displaced, and a carboxyl group enters the ring *ortho* to the position occupied by the leaving group. Thus, this is a *cine*-substitution, and is called *von-Richter rearrangement*.

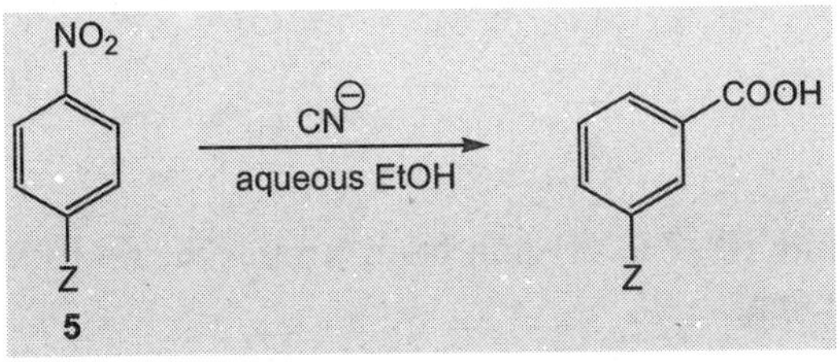

The reaction gives best results when electron-withdrawing groups are located in *ortho* and *para* positions, but yields are low usually ranging between 20–50%. A plausible mechanism for this rearrangement is outlined in the following scheme.

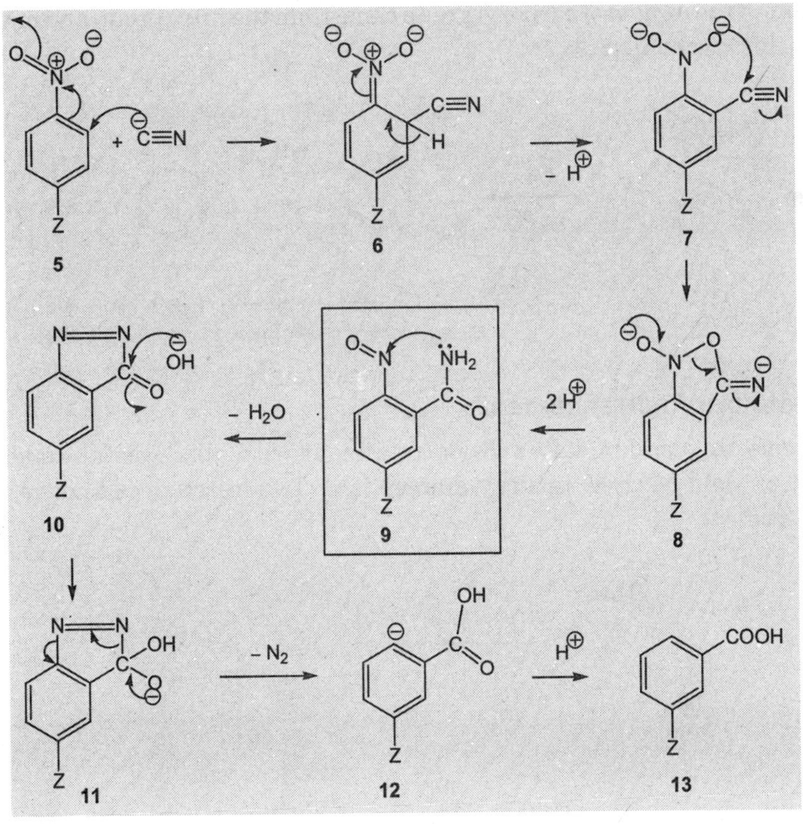

It may be noted that this mechanism involves at least one stable compound, **9**. Since the mechanism postulates the existence of **7**, which has the unusual functional group $-NO_2^-$, it is possible that **9** could be formed without going through **7**.

The mechanism is supported by the fact that among the side products isolated from reaction of **5** (Z = Cl) was **14**, which could arise by the reduction of **9**.

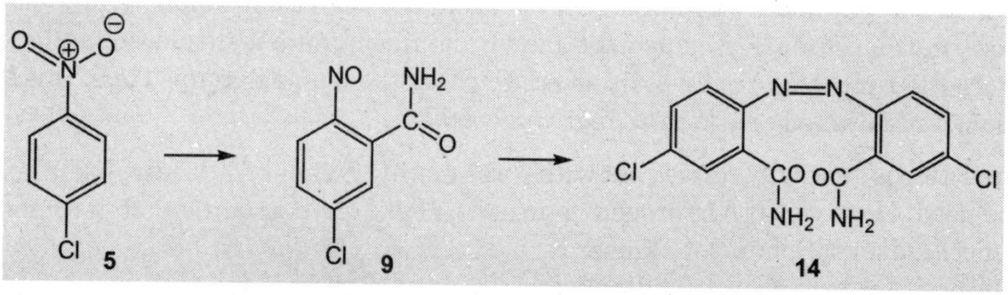

Further, the compound **10** (Z = H) has been prepared, and its treatment with cyanide gives nitrogen and benzoic acid.

$$\mathbf{10} \xrightarrow{CN^-} \mathbf{11} \xrightarrow{-N_2} \mathbf{12} \longrightarrow \mathbf{13}$$

When **5** is treated with cyanide ions in $H_2{}^{18}O$, one-half of oxygen in the product is labelled showing that one of the oxygen of carboxyl group came from the nitro group and one from solvent as required by this mechanism.

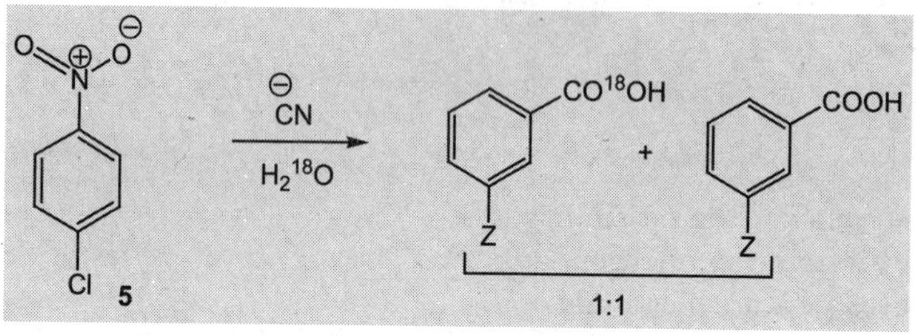

B. Sommelet-Hauser Rearrangement

Benzyl quaternary ammonium salts (**15**) on treatment with alkali-metal amides undergo a rearrangement to yield benzylic tertiary amines (**16**). This reaction is known as *Sommelet-Hauser rearrangement*.

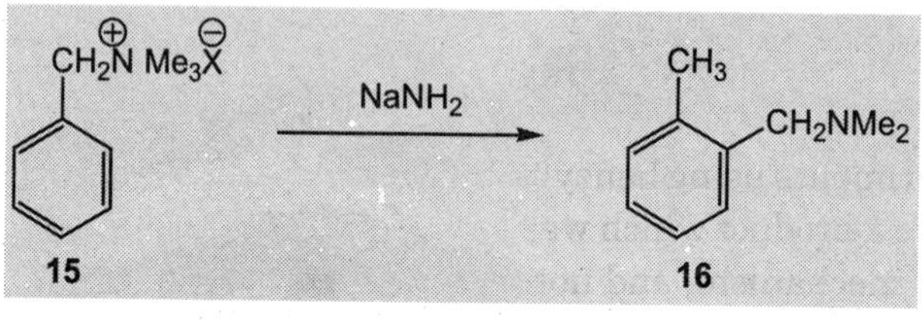

Since the product is a benzylic tertiary amine, it can be further alkylated, and the resulting quaternary ammonium salt again can undergo rearrangement. This process can be repeated around the ring until all the *ortho* positions are occupied.

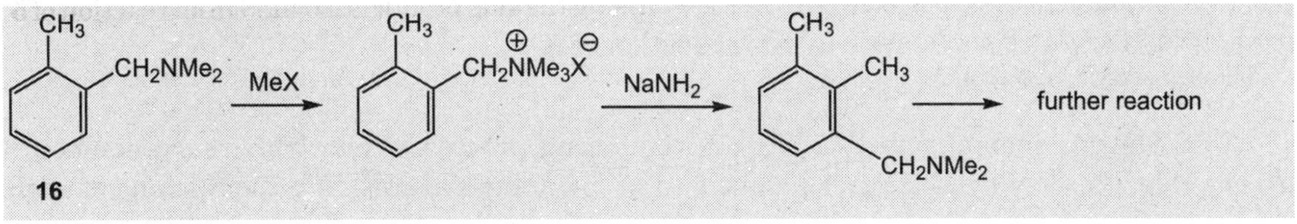

The reaction is generally carried out with three methyl groups on nitrogen, but other groups can also be used. However, if β hydrogen is present, *Hofmann elimination* often competes.

The mechanism of Sommelet-Hauser rearrangement involves the initial deprotonation of the benzylic position, as it is most acidic, to give ylide **A**, which is in equilibrium with the second ylide **B**. Although ylide **B** is present in small amount, it undergoes rearrangement and shifts the equilibrium in its favour. It finally undergoes aromatization to generate the final product.

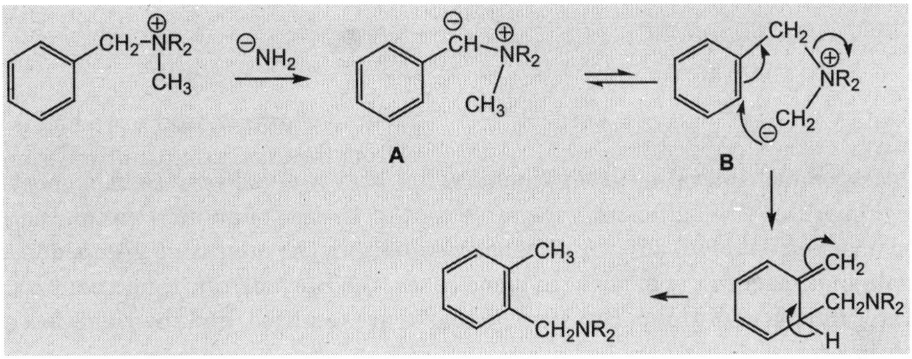

One may also consider the possibility of a *second* mechanism in which a methyl group actually breaks away from the nitrogen, and then attaches itself to the ring. This is not the case, and is supported by a product study and labelling experiments. If the second mechanism operates, **17** should give **18**. But the first mechanism predicts the formation of **19**, which is what was actually obtained.

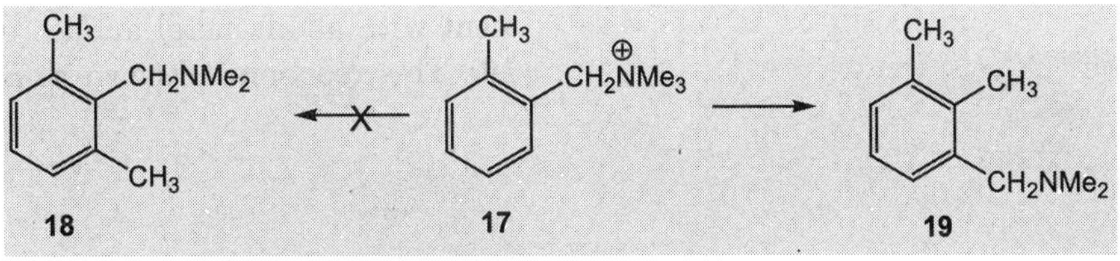

The labelling experiments using benzyltrimethylamine, labelled with ^{14}C in the α-position of the benzyl group, gave a product which was labelled in the ring methyl group. Such a product is predicted by the first mechanism, and not by the second.

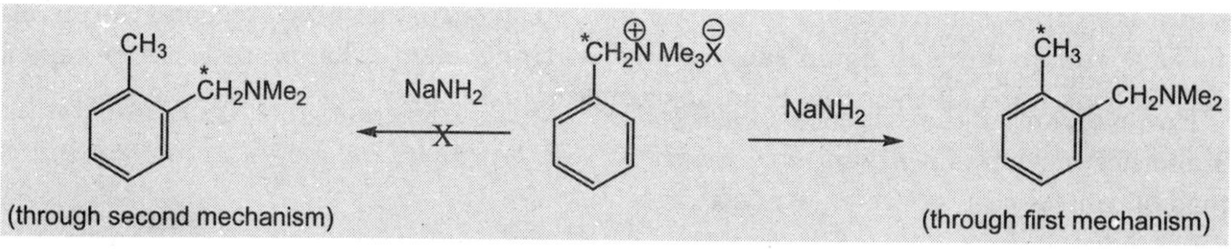

The Stevens rearrangement is also a competing process. When both rearrangements are possible, the Sommelet-Hauser is favoured at low temperatures, and the Stevens at high temperatures.

C. Stevens Rearrangement

When a quaternary ammonium salt containing an electron-withdrawing group Z (such as RCO, ROOC, and phenyl) on one of the carbons attached to the nitrogen is treated with a strong base (such as NaOR, or $NaNH_2$), the product obtained is a rearranged tertiary amine. This reaction is called *Stevens rearrangement*.

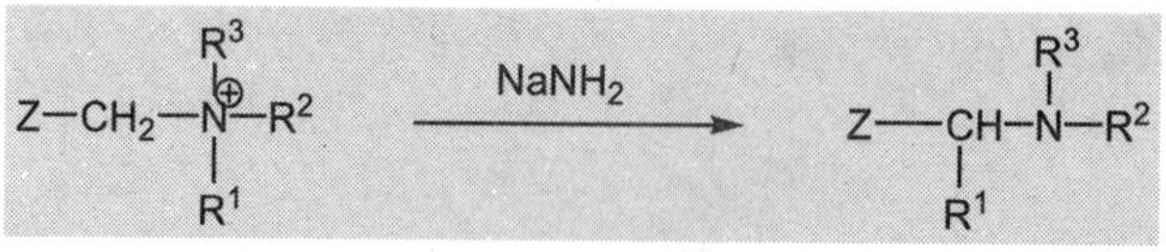

The most common migrating groups include allyl, benzyl, benzhydryl, and phenacyl, though even methyl migrates to a sufficiently negative center. In case of an allyl group, there may or may not be involvement of an allylic rearrangement within the migrating group, depending on the substrate and reaction conditions. In some cases, the Stevens rearrangement occurs with the substrates having no Z group. But stronger bases are required, and the yields are generally low in such cases.

Two mechanisms (mechanism-I and mechanism-II) suggested for the Stevens rearrangement show its intramolecular nature as evidenced by (i) crossover experiments, (ii) ^{14}C labelling experiments, and by (iii) the fact that retention of configuration is found at R. The first step is loss of the acidic proton to give the ylide **20**. The rearrangement of **20** to tertiary amine may involve either radical pair (mechanism-I) in a solvent cage or ion pairs in solvent cage (mechanism-II).

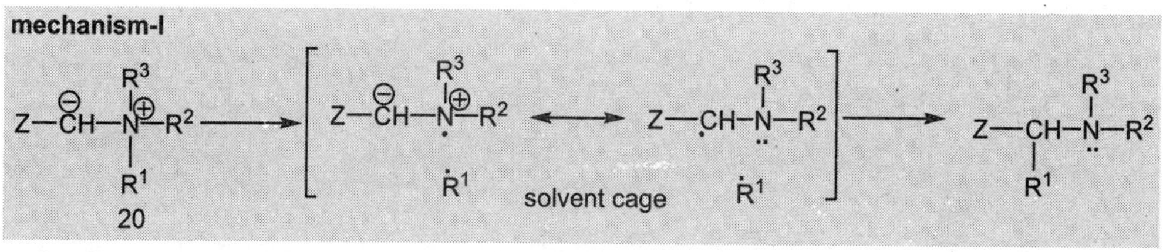

Mechanism-I is supported by the following observations:

(i) The radicals recombine rapidly in order to account for the fact that R^1 does not racemize (since radicals are held together in a solvent cage, they do not drift).

(ii) In some cases, small amounts of coupling products (R^1R^1) have been isolated. This would be expected if R^1 radical leaked from the solvent cage.

However, not all the evidence is easily compatible with mechanism-I. Therefore, it is likely that another mechanism-II similar to mechanism-I, but involving ion pairs in a solvent cage instead of radical pair, operates in some cases.

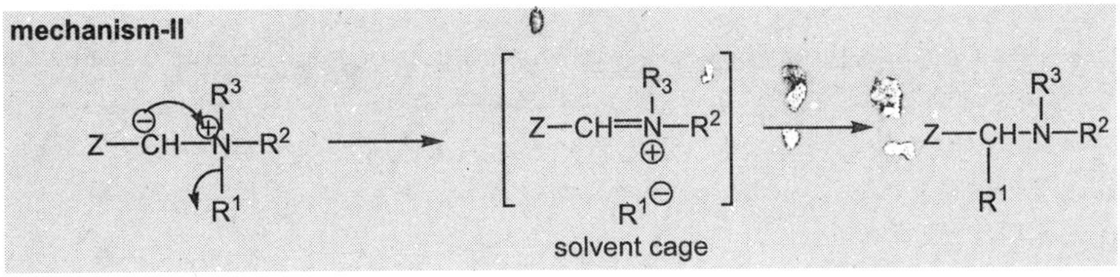

D. Smiles Rearrangement

Smiles rearrangements are simply intramolecular nucleophilic substitutions as shown in the following equation.

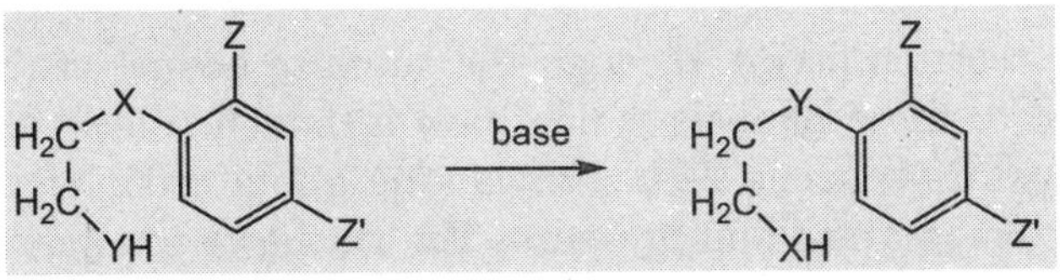

Z and Z' are the activating groups for a nucleophilic aromatic substitution reaction, which are electron-withdrawing substituents, to stablize the cyclohexadienylide anion generated in the reaction. In this rearrangement, the chain linking X and Y can be aromatic as well as aliphatic. For example, *o*-(*o*-nitrosulphonyl) phenol undergoes Smiles rearrangement to *o*-(*o*-nitrophenoxy) sulphinic acid in the presence of hydroxide ions.

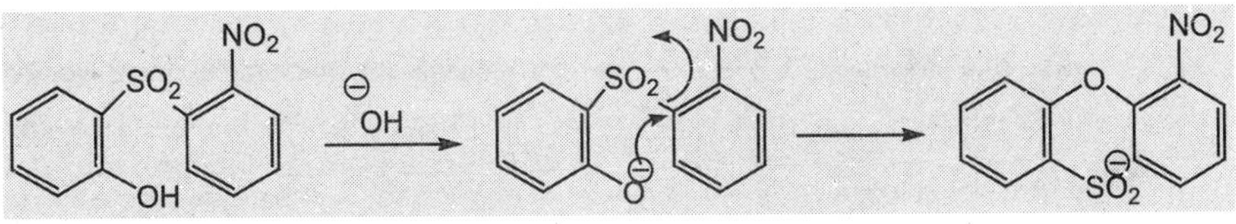

In this case, SO_2Ar is the leaving group, and ArO^- the nucleophile. The nitro group serves to activate the *ortho* position of the ring. Other examples of the substrates that can undergo Smiles rearrangement are:

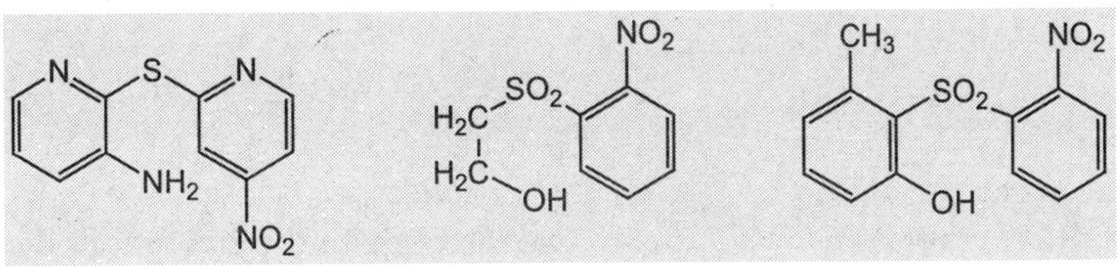

E. Bucherer Reaction

Certain phenols such as 2-naphthol react with aqueous ammonium sulphite (or ammonia and sodium bisulpite) to give aromatic amines. This reaction is known as *Bucherer reaction*, and occurs through the sulphite adduct of the keto-tautomer as shown in the following scheme.

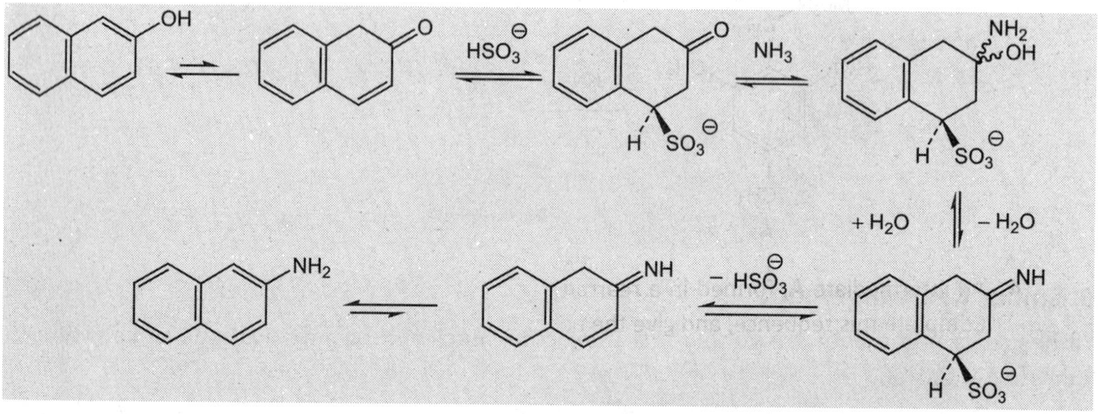

Since Bucherer reaction occurs through the sulphite adduct of the keto-tautomer, its application is limited to those phenols which have a tendency to ketonize, such as 1- and 2-naphthol and *m*-dihydroxybenzene. 2-Naphthylamine is obtained in high yields by performing the reaction at high temperature and pressure. The Bucherer reaction is reversible, and by careful control of the relative amounts of ammonia and water, it may be used to convert amines into phenols.

PROBLEMS AND REVIEW QUESTIONS

1. When the three isomeric chlorotoluenes are treated with KNH_2 in liquid NH_3, they yield different toluidines as shown below:

(i) *o*-Chlorotoluene ⟶	*o*-toluidine + *m*-toluidine (no *p*-isomer)
(ii) *m*-Chlorotoluene ⟶	*o*-toluidine + *m*-toluidine + *p*-toluidine
(iii) *p*-Chlorotoluene ⟶	*m*-toluidine + *p*-toluidine (no *o*-isomer)

Explain these observations.

2. Complete the following reactions:

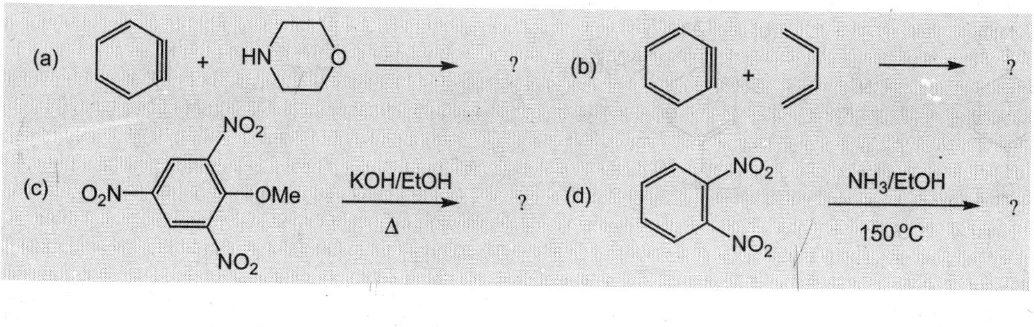

3. Benzyl quaternary ammonium salt **A** on reaction with KNH_2 gives product **B** rather than normal product of Sommelet-Hauser rearrangement. Explain the significance of this observation in establishing the mechanism of Sommelet-Hauser rearrangement.

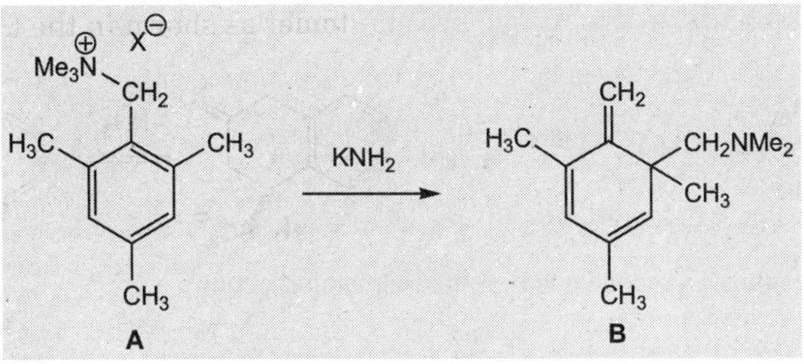

4. An intermediate **A**, formed in a rearrangement process, can be trapped by treatment with **B**. Complete this sequence, and give the name of this rearrangement.

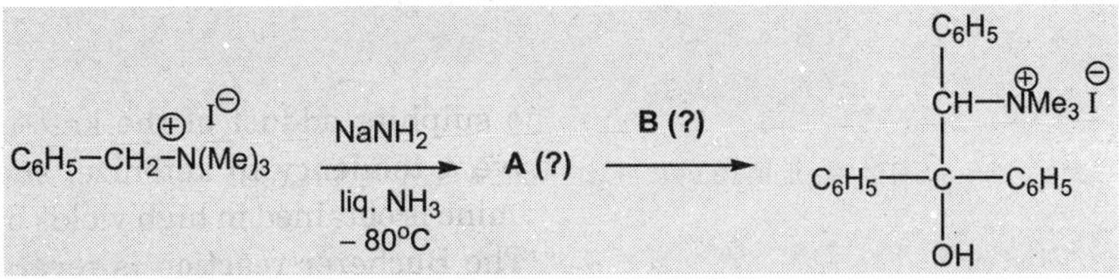

5. Outline the mechanism for the synthesis of the following two heterocyclic compounds:

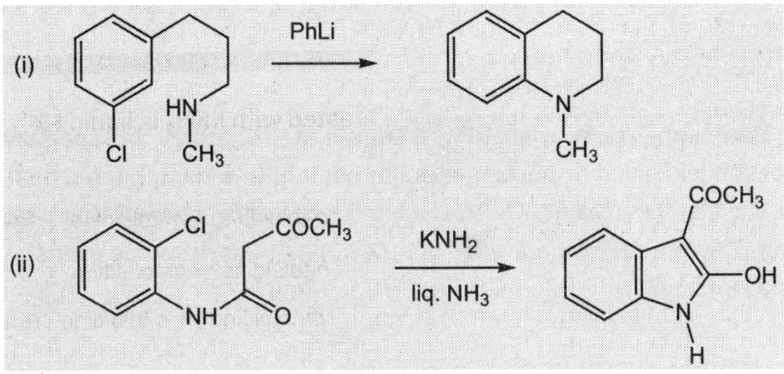

6. Predict the product in the following reactions, and outline the mechanism of each reaction.

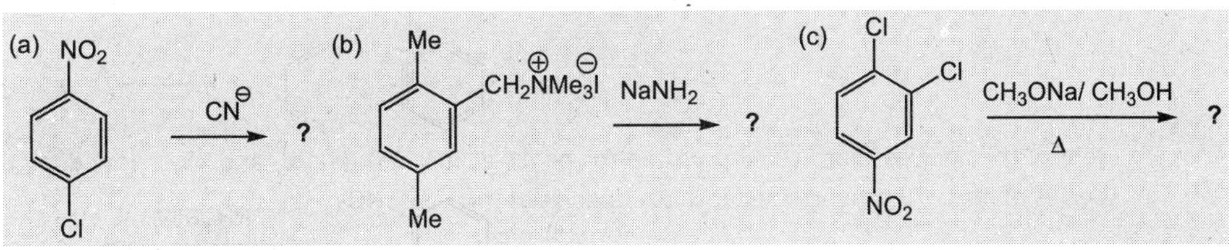

7. Complete the following reactions, and give their name and mechanism:

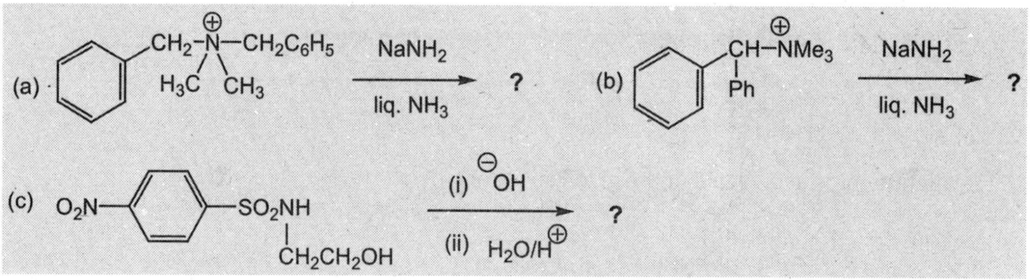

8. Give the product and mechanism of the following reaction:

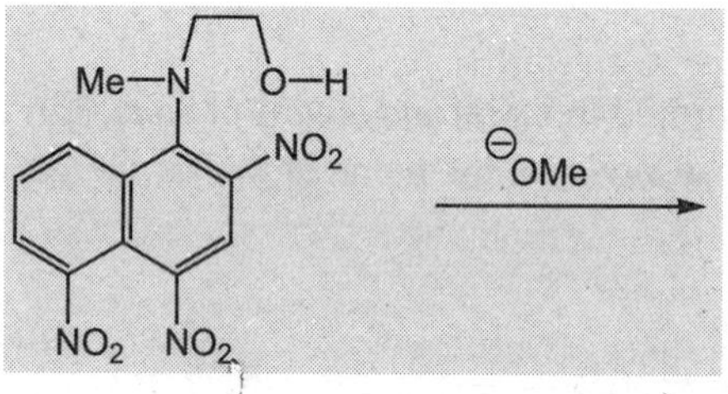

9. Give the mechanistic steps involved in the following conversions:

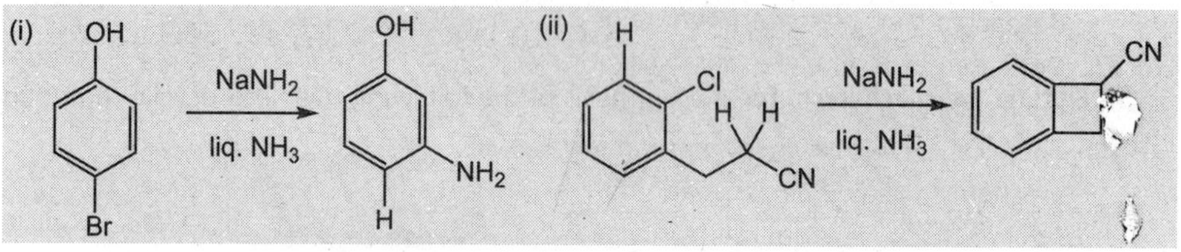

10. Benzene diazonium-2-carboxylate (**A**) was subjected to photolysis in a specially designed apparatus in conjunction with a mass spectrometer. Immediately, there appeared peaks at m/z 28, 44, and 76. After a while, the peak at m/z 76 started disappearing, and another peak at m/z 152 started appearing. The rate of disappearance of m/z 76 peak was the same as that of the appearance of the peak at m/z 152.

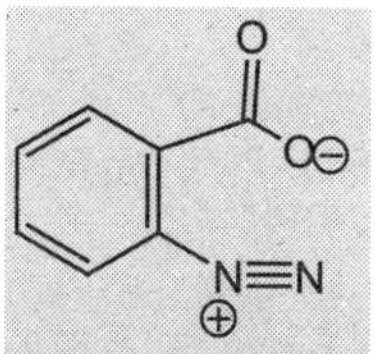

A

Identify the species that are responsible for peaks at m/z 28, 44, 76, and 152. How does the species of mass 76 convert itself into the species at m/z 152?

11. Complete the following reaction, and outline its mechanism.

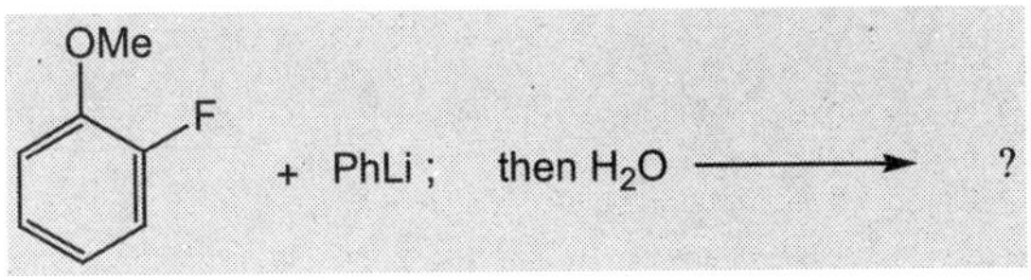

12. Identify products **A** to **E** in the given sequence of reactions.

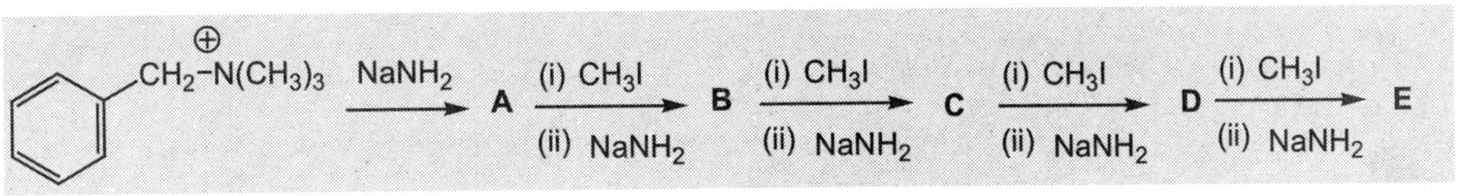

13. 2-Bromoanisole reacts with $NaNH_2$ in liquid ammonia to give *m*-anisidine, whereas 2-bromo-3-methylanisole does not. What inference can be drawn regarding the condition for intermediacy of benzynes in such reactions?
14. Discuss the role of tracer technique in establishing the intermediacy of benzynes in aromatic nucleophilic substitutions.
15. Explain why in the presence of $NaNH_2$ in liquid ammonia, both *m*-bromoanisole and *o*-bromoanisole yield the same product, *m*-anisidine.
16. After the reaction of fluorobenzene with deuterated amide ion in deuterated ammonia, the unreacted starting material contains deuterium. Explain the importance of this observation.
17. Design an experiment to show the intermediacy of cyclohexadienylide in bimolecular aromatic nucleophilic displacement reactions.
18. There can be two mechanisms for nucleophilic aromatic substitution, namely addition-elimination mechanism and elimination-addition (benzyne) mechanism. How will you establish which of the two mechanisms is operating in a particular case?
19. What is Sommelet-Hauser rearrangement? What are the competing processes in this reaction? Discuss how these competing processes can be controlled.
20. Discuss synthetic applications and mechanism of the following reactions:
    (i) Stevens rearrangement
    (ii) Smiles rearrangement
    (iii) Bucherer reaction

SOLUTIONS

1. These reactions involve aryne intermediates. The formation of different toluidines can be rationalized by the following considerations. In *o*-chlorotoluene, only one *o*-hydrogen is available and so its elimination generates only one aryne, whereas the availability of two *o*-hydrogens in *m*-chlorotoluene generates two arynes. Although two *o*-hydrogens are available in *p*-chlorotoluene, their elimination generates only one aryne.

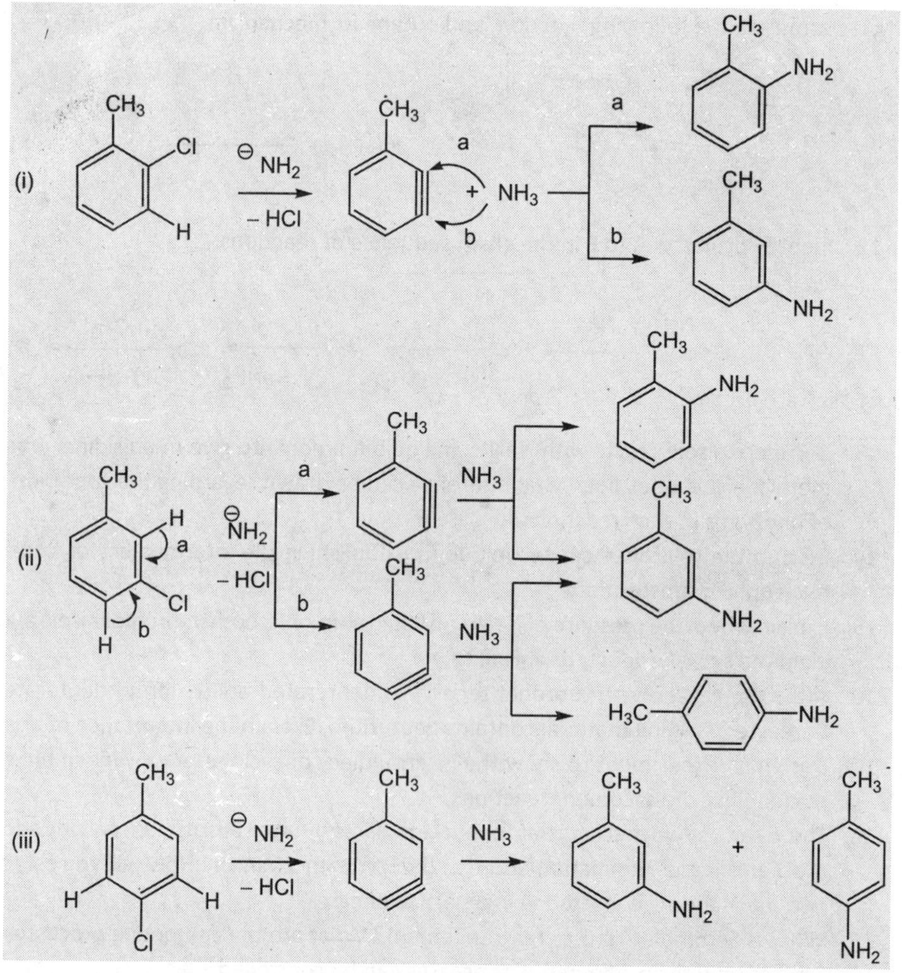

2. 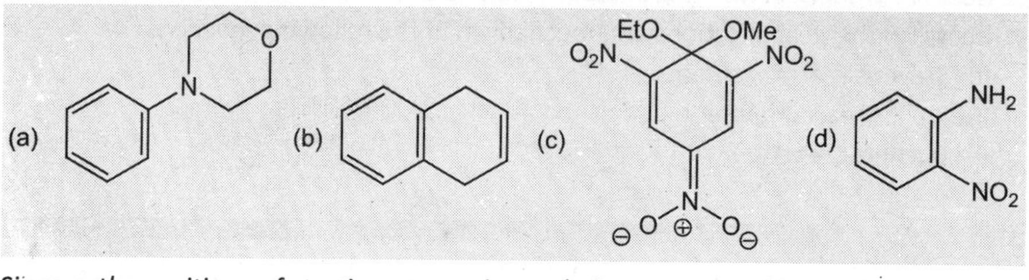

3. Since *ortho* positions of starting ammonium salt **A** are occupied by methyl groups, it could not give normal product as the reaction takes place *via* a cyclic mechanism (involving *o*-vacant position).

4. **A** is $C_6H_5-\overset{\ominus}{C}H-\overset{\oplus}{N}(CH_3)_3$; **B** is $C_6H_5COC_6H_5$

The name of the rearrangement is Sommelet-Hauser.

5. Both the syntheses involve generation of aryne intermediates which undergo cyclization by participation of internal nucleophile.

5.

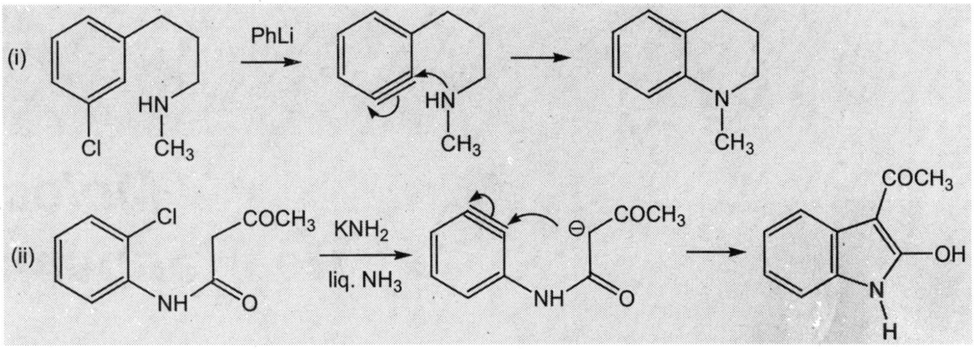

6.

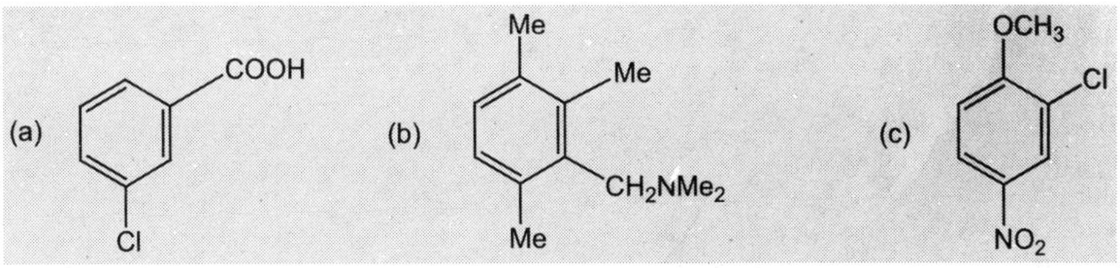

7.

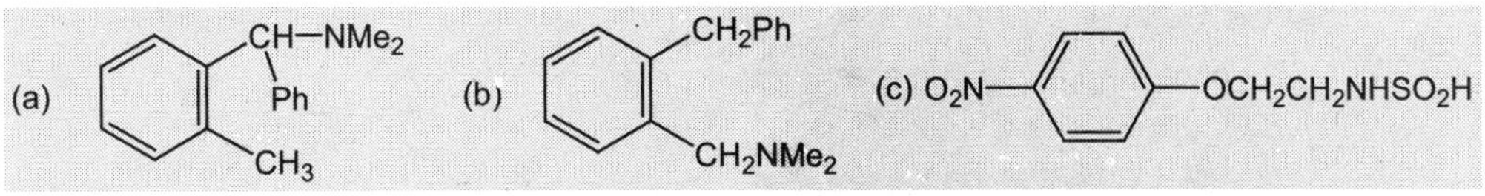

8.

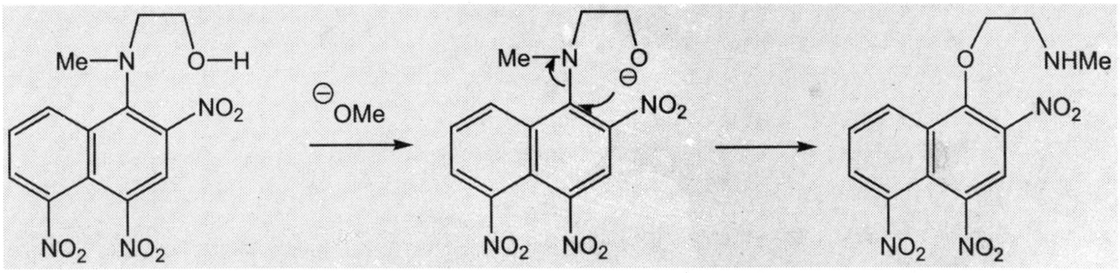

9.

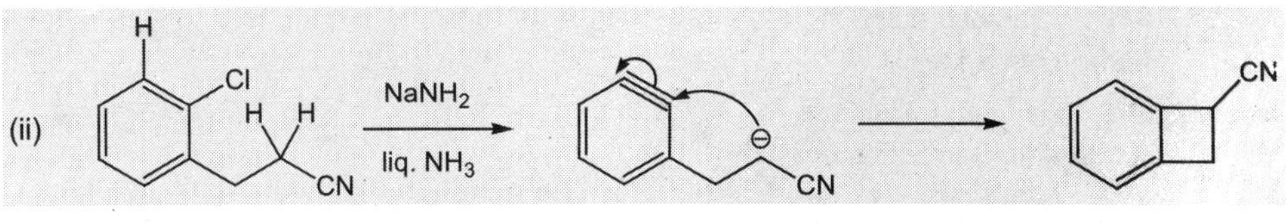

10. **A** undergoes photochemical decomposition to give benzyne (m/z 76), CO_2 (m/z 44) and N_2 (m/z 28). Subsequent dimerization of benzyne then leads to the formation of biphenylene.

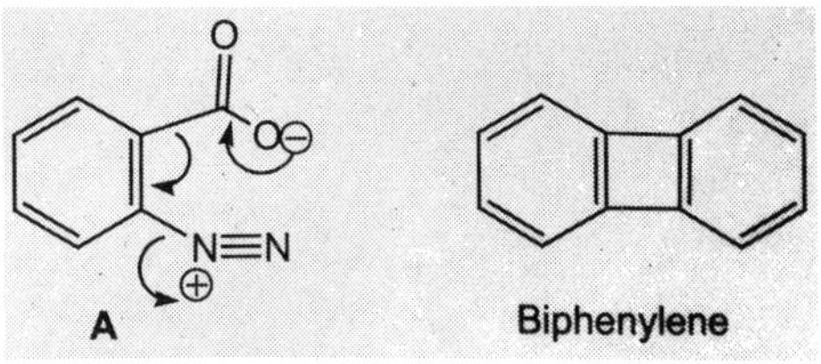

Chapter 12

Molecular Rearrangements

Chapter Outline

12.1 MOLECULAR REARRANGEMENTS INVOLVING ELECTRON-DEFICIENT SPECIES

A. Carbocation Rearrangements

I Rearrangements of 1,2-Glycols

II Acid-Catalyzed Rearrangements of Carbonyl Compounds

III Rearrangement of Epoxides

IV Demjanov Rearrangement

V Transannular Rearrangement

VI Rearrangement in Bicyclic Compounds

B. Other Related Rearrangements

I Wolff Rearrangement

II Hofmann Rearrangement

III Curtius, Schmidt and Lossen Rearrangements

IV Baeyer-Villiger Oxidation

V Beckmann Rearrangement

12.2 FURTHER MOLECULAR REARRANGEMENTS

A Fries Rearrangement

B Favorskii Rearrangement

C Benzidine Rearrangement

D Claisen Rearrangement

E Cope Rearrangement

The term 'molecular rearrangement' refers to the migration of a group from one atom to another within the same molecule.

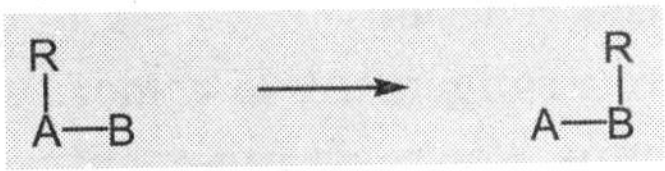

The group R may either completely detach itself from atom A before combining with atom B of the same molecule (two-step process) or, alternatively, it may remain partly attached to both the atoms A and B during the whole process of the rearrangement (one-step process). The rearrangements of the latter category occur more frequently.

12.1 MOLECULAR REARRANGEMENTS INVOLVING ELECTRON-DEFICIENT SPECIES

In the present section, we shall limit the discussion to those rearrangements in which the atom B has only six electrons in its valence shell and is, therefore, *electron-deficient*. In a great majority of such reactions, carbon atoms constitute the migration origin and terminus, *i.e.* a hydrogen or a group migrates with a pair of its bonding electrons from an adjacent carbon to the electron-deficient carbon atom in order to form a more stable carbocation. If R is a hydrogen, the process is called a *1,2-hydride shift* and in case it is an alkyl or aryl group, it is correspondingly termed as *1,2-shift* of an *alkyl* or *aryl* group. In carbocation rearrangements the reaction may proceed through a *classical* or *nonclassical* (bridged) carbocation intermediate or both. The whole process is summarized as below:

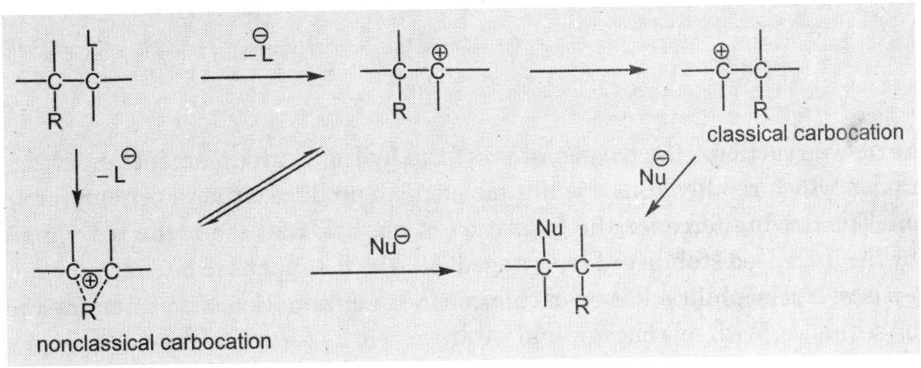

Carbocation rearrangements take place far more readily than those involving either a free radical or carbanion. A reasonable explanation for this observation depends upon the number of electrons in the transition states. In the carbocation transition state, there is a new molecular orbital which is formed as a result of the combination of an atomic orbital of the migrating group with the atomic orbitals of the other two carbon atoms. This molecular orbital can accommodate only two electrons. In contrast, free radical and carbanion rearrangements have to pass through cyclic transition states in which three and four electrons, respectively, are to be accommodated in the molecular orbital embracing the three carbon atoms and should, therefore, be less stable than carbocation transition states.

We have already seen that the two important reactions of carbocations are:

(i) Reaction with a nucleophile (S_N1 reaction), and

(ii) Elimination of a proton (*E1* reaction)

The third alternative is its rearrangement to a more stable carbocation by 1,2-shift of hydrogen atoms or alkyl or aryl groups. The rearrangements of alcohols under acidic conditions (commonly known as the Wagner-Meerwein rearrangement) provide typical examples.

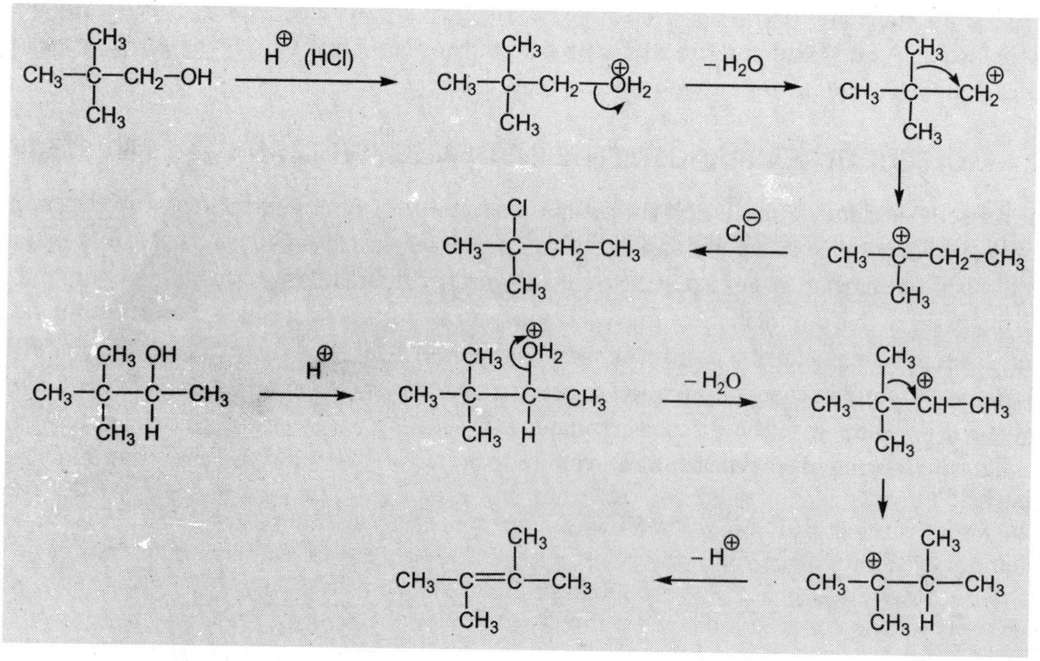

In the above reactions, the oxygen atom of the hydroxyl group accepts a proton yielding an oxonium ion which readily loses a water molecule to produce either a primary or secondary carbocation. The driving force for the migration of methyl, that is for the rearrangement, is provided by the increased stability of rearranged ion which happens to be a *tertiary* carbocation in *both the cases*. Nucleophilic addition of chloride ion or elimination of a proton then completes the recation sequence. With this background we are now in a position to appreciate some common carbocation rearrangements described in the following pages.

A. Carbocation Rearrangements

I. Rearrangements of 1,2-Glycols

Acid-catalyzed rearrangement of 1,2-glycols yields carbonyl compounds, the classical example being the acid-catalyzed rearrangement of pinacol (**1**) to pinacolone (**2**).

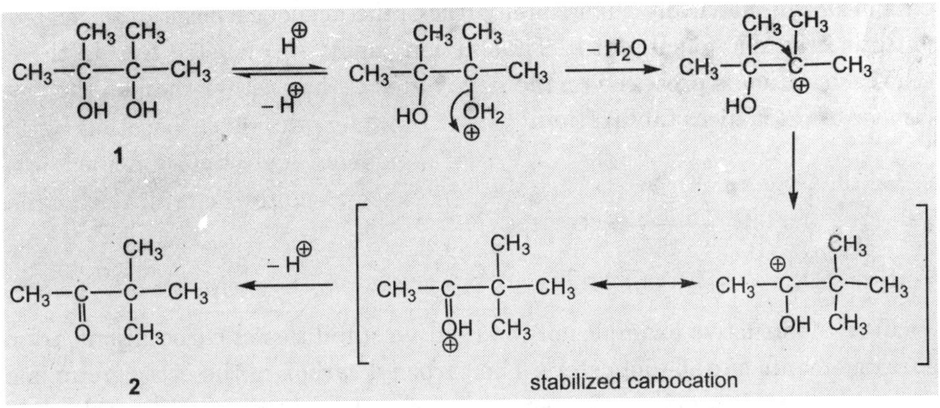

The driving force for the migration of the methyl group is provided by the resonance stabilization of the rearranged ion by the attached oxygen. The rearrangement is an intramolecular process in which the methyl group begins to attach itself to the positive carbon before separating from its original position. This is verified experimentally by rearranging two different pinacols together, which results in the formation of only two rearranged products. The absence of *mixed products* indicate that the migrating group of one compound never becomes sufficiently free of the initial carbocation so as to get itself attached to the carbocation produced from the other pinacol.

The pinacol rearrangement may be viewed as a rearrangement with a 'push' and a 'pull'. The carbocation formed by the departure of water *pulls* the migratory group across simultaneously as the lone pair of oxygen *pushes* it. The intermediacy of carbocation in this rearrangement can be demonstrated in a variety of ways.

First, when the pinacol is rearranged in $H_2{}^{18}O$ and the reaction is stopped before completion, the recovered pinacol is found to contain the ^{18}O isotope suggesting the reversible formation of a carbocation.

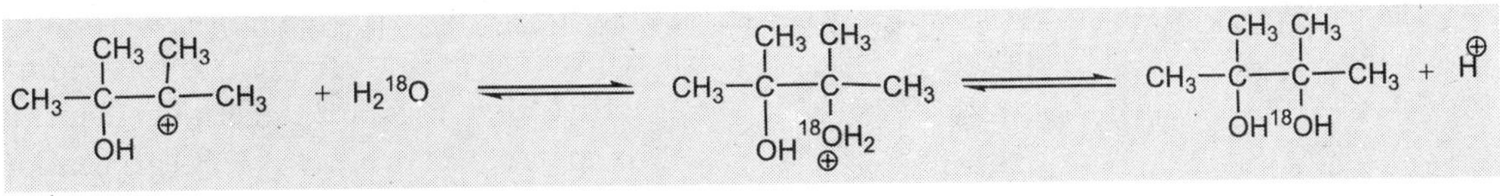

Second, any other reaction capable of producing the same carbocation results in the formation of pinacolone. For example, the reaction of the bromohydrin (**3**) with silver oxide, and the deamination of the aminohydrin (**4**) with nitrous acid yields pinacolone (**2**).

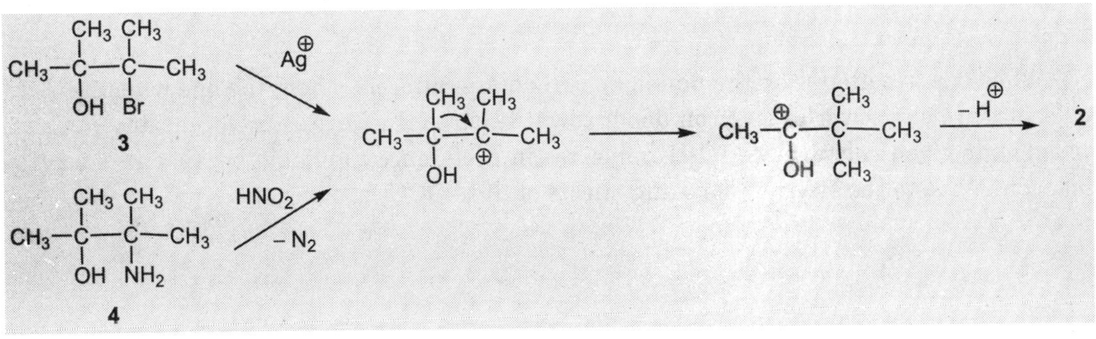

In case of unsymmetrically substituted glycols, the positive charge is generated on that carbon which is best able to support it. Thus, it can easily be predicted that in the following example the carbocation is produced on the phenylated carbon rather than on the methylated carbon because of resonance stabilization.

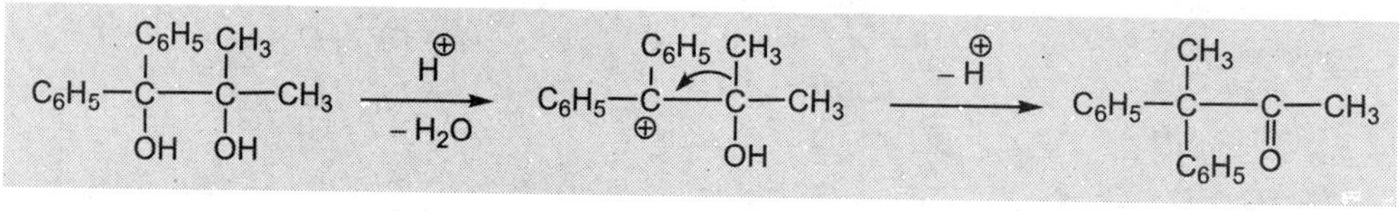

Particularly instructive examples of the relative stabilities of carbocations are provided by the rearrangements of α, β-diol esters. The carbonyl carbon of the ester group is partially positive, and therefore, it prevents the generation of a carbocation at the α-carbon because of mutual repulsion between two similar charges. However, the positive charge can be forced to develop at the α-carbon by stabilizing it over a phenyl substituent (**6**) together with increasing the energy of the alternate β-carbocation by making it a primary one (**5**).

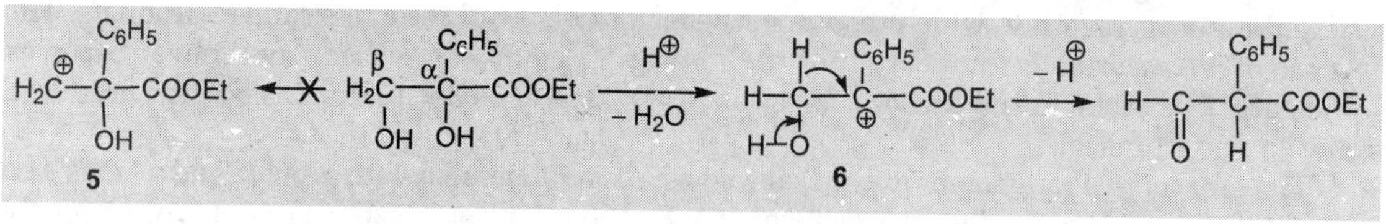

Carbocation is produced at the β-carbon when both the hydrogen atoms are substituted by methyl groups, thus indicating that a tertiary carbocation (**7**) is more stable than a benzylic carbocation adjacent to a carbethoxy group.

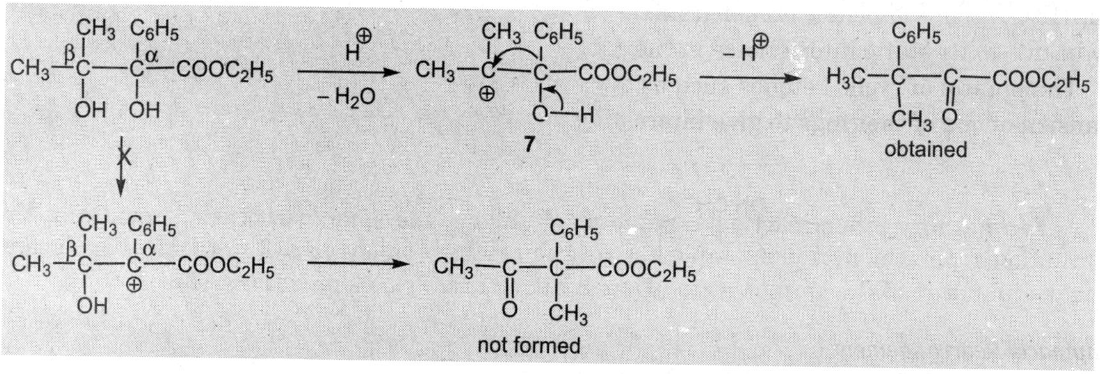

When the two groups on the non-ionic carbon are different, then, the one which is better nucleophile (more powerful electron donor towards carbon) migrates preferentially. Thus, in the example given below, p-$MeOC_6H_4$ migrates in preference to phenyl because it is a better nucleophile due to the electron-donating effects of the $-OCH_3$ group.

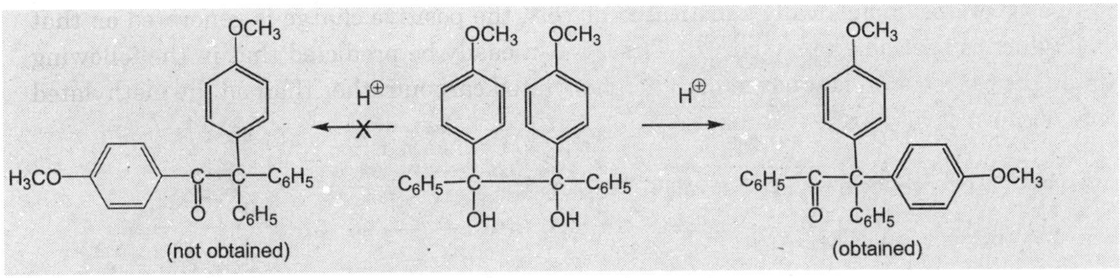

Usually, phenyl group migrates in preference to alkyl groups. The following values were established for the migratory aptitude of substituted phenyl groups with reference to phenyl which has been assigned the value of 1.

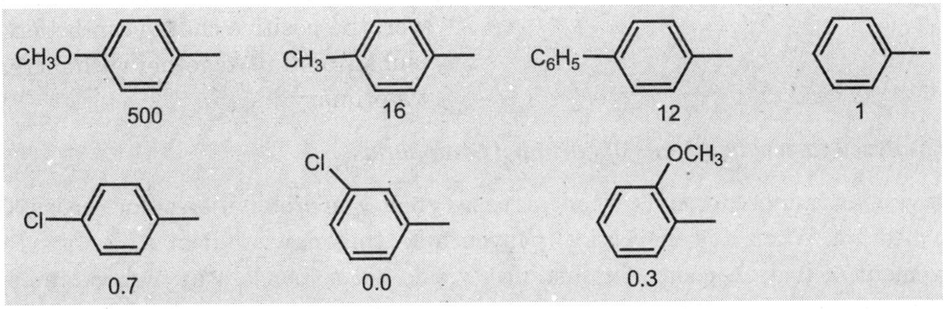

It is evident from this data that the migratory aptitude of a group is closely linked with its ability to release electrons and this can be taken as an indication that the migrating group starts forming a bond with the adjacent electron-deficient centre even though the departing group (H_2O) is not completely broken from the molecule. The anomalous behaviour of *o*-$MeOC_6H_4$ may be due to its steric interference in the transition states with the non-migrating groups.

The pinacol of cyclic ketones such as cyclopentanone undergoes rearrangement with the expansion of one of the rings to give interesting spirocyclic ring systems.

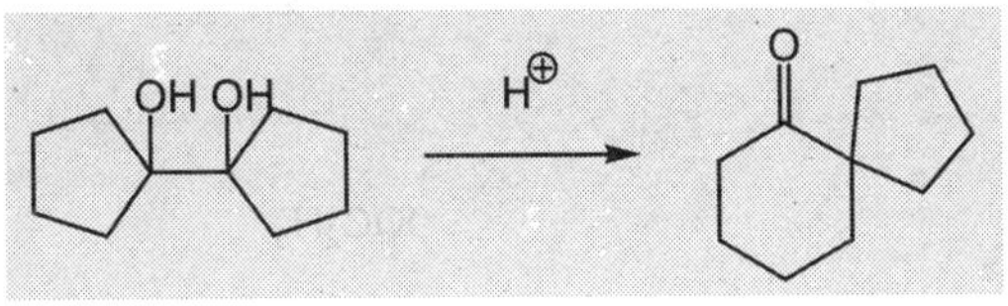

Semipinacol Rearrangement

An alternate product (in which regioselectivity of the reaction is in the opposite direction of the pinacol rearrangement) can be obtained by modifying the conditions. In such cases, a hydroxyl group provides the electrons to 'push' the migrating group across, but the 'pull' comes from the departure of leaving groups other than water such as tosylate, halide and nitrogen. These types of rearrangements are known as semipinacol rearrangements. An illustrative example of this is the conversion of 2-phenylpropane-1,2-diol to 3-phenyl-2-propanone *via* tosylate. Under acidic conditions normal pinacol rearrangements give 2-phenylpropanal.

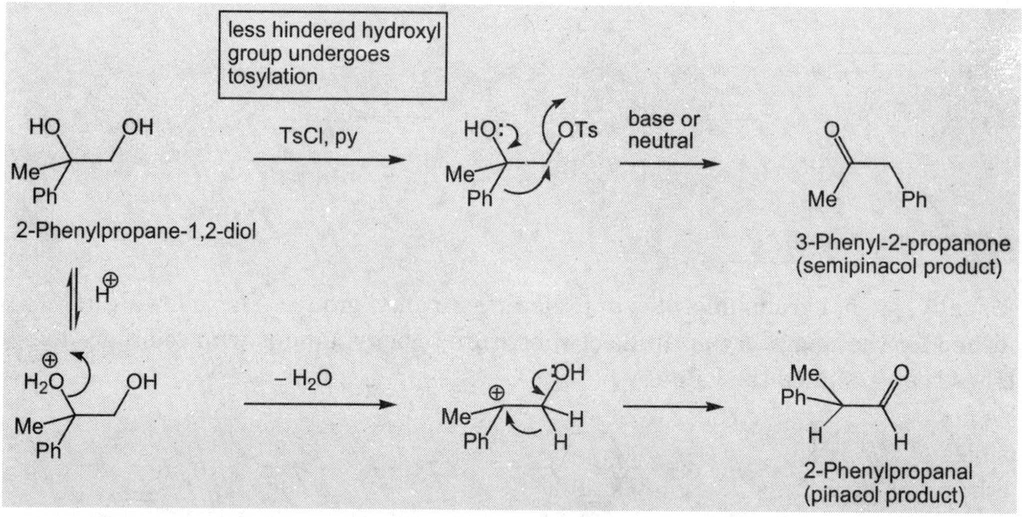

II. Acid-Catalyzed Rearrangements of Carbonyl Compounds

The positive character of carbonyl carbon increases greatly on protonation of the carbonyl oxygen in acidic medium. When it is energetically favourable, this may result in a 1,2-shift. Pinacolic rearrangement of hydrobenzoin, for instance, yields desoxybenzoin at higher temperatures and diphenylacetaldehyde at lower temperature. Obviously, diphenylacetaldehyde rearranges further to desoxybenzoin for which a possible mechanism may be written as follows:

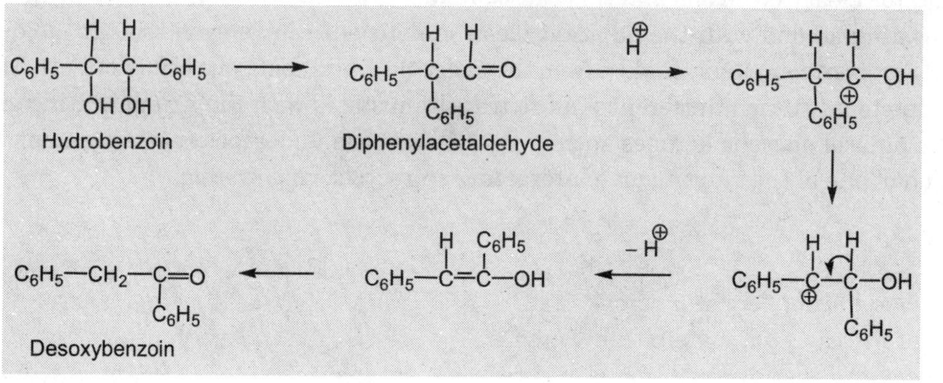

Generally speaking, when a secondary or tertiary aldehyde (RR'HCCHO or RR'R'CCHO) is treated with concentrated sulphuric acid, it undergoes rearrangement in the above fashion leading to an isomeric ketone. Rearrangements under relatively mild conditions, however, may be achieved in α-hydroxy ketones and α-hydroxy aldehydes for which the following mechanism appears quite reasonable.

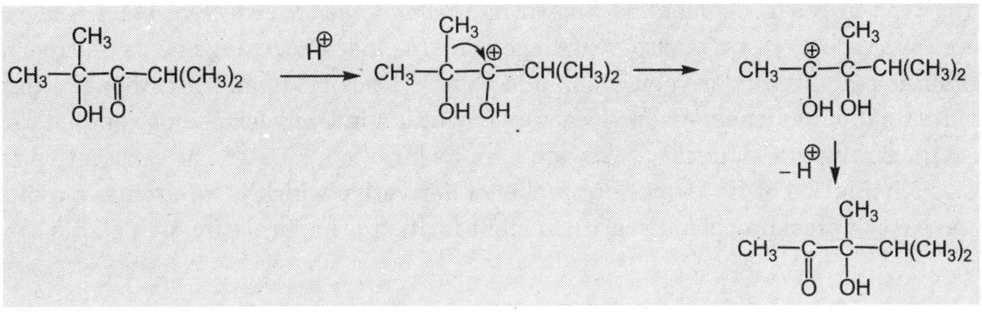

The steps involved in the conversion of an α-hydroxy aldehyde to an α-hydroxy ketone [$(C_6H_5)_2$–COH–CHO → C_6H_5CO–CHOH–C_6H_5] may be written similarly.

A novel rearrangement of an α-keto ester into two β-keto esters has been reported. The most plausible mechanism involves a sequence of two migrations (phenyl and ethoxycarbonyl; methyl and ethoxycarbonyl) from the protonated keto ester as shown in the following scheme.

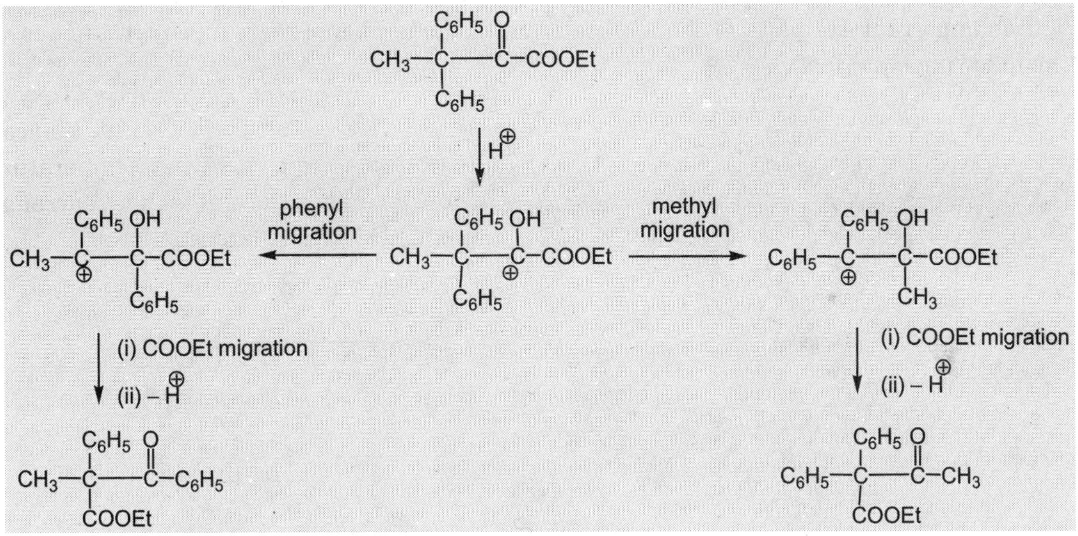

The Dienone-Phenol Rearrangement: 4,4-Dialkylcyclohexadienones and related compounds, on treatment with an acid, undergo alkyl migration with the formation of phenols. When a reaction is carried out in acetic anhydride in the presence of catalytic amount of sulphuric acid, it gives phenol acetate, which, on hydrolysis can be converted to the corresponding phenol.

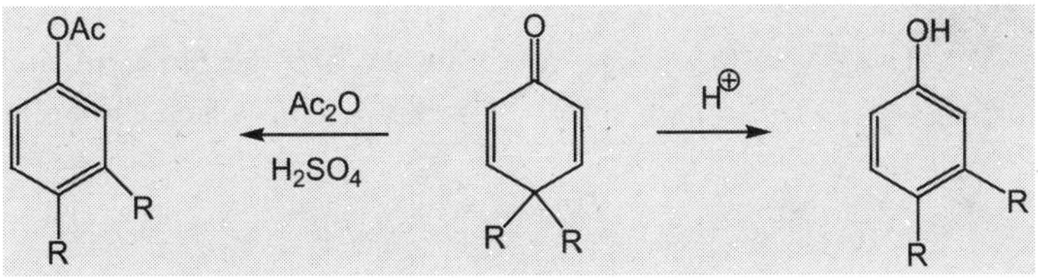

This type of rearrangement is known as *dienone-phenol rearrangement* and may be considered as a type of *reverse* pinacol rearrangement. In pinacol rearrangements, the rearranged carbocation is stabilized by oxygen atom, and rapidly loses H^+ to give a carbonyl compound. On the other hand, in dienone-phenol rearrangement, an initially formed protonated carbonyl compound rearranges to a tertiary carboaction in its key step. Finally, the carbocation rapidly undergoes elimination of H^+ to produce a phenol derivative which is an aromatic compound. Thus, the ease of dienone-phenol rearrangement is due to the formation of a stable aromatic product.

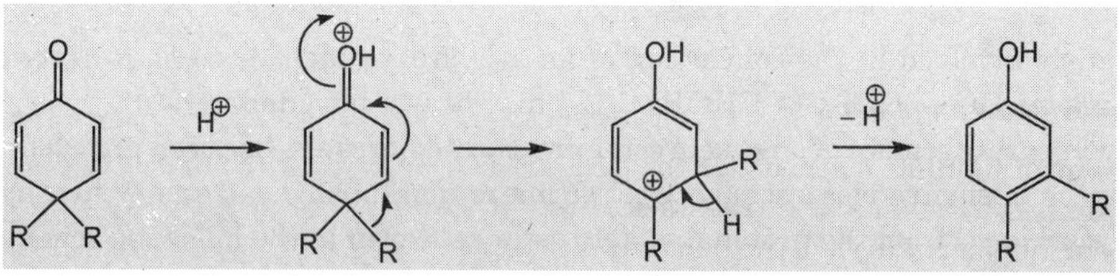

Two important examples of the application of dienone-phenol rearrangement are shown in the following equations.

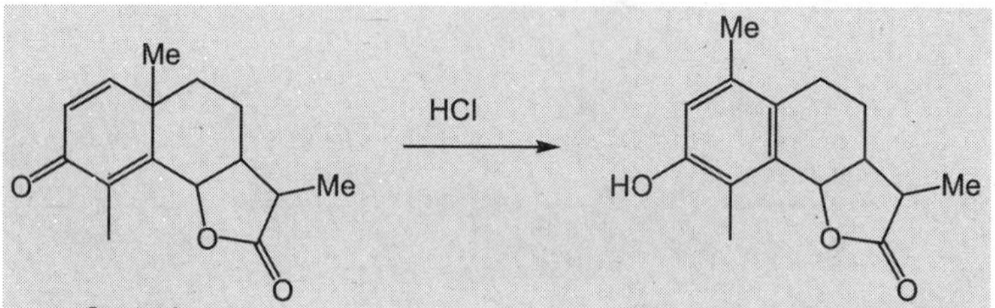

Santonin

Desmotroposantonin

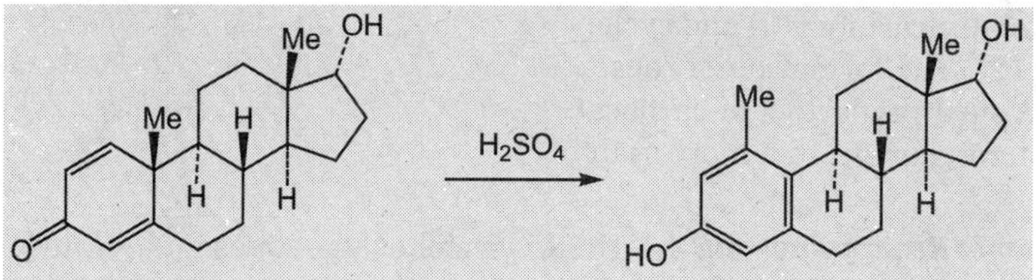

1-Methyloestradiol

In the above examples, there is migration of an alkyl group. The following example illustrates the migration of a ring system.

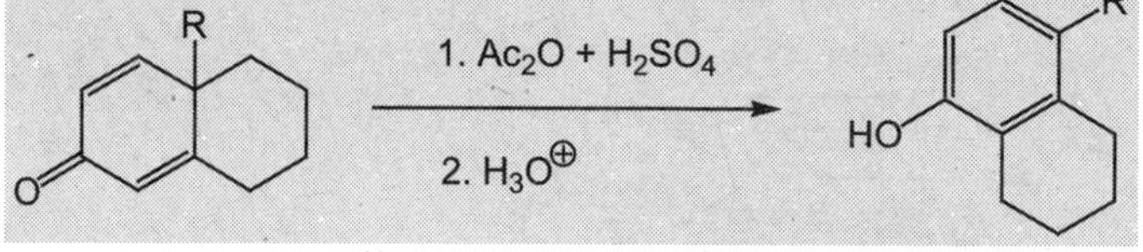

III. Rearrangement of Epoxides

Epoxides rearrange to carbonyl compounds under the influence of acids. Rearranged products are usually similar to those obtained from the acid-catalyzed rearrangement of the corresponding 1,2-glycols.

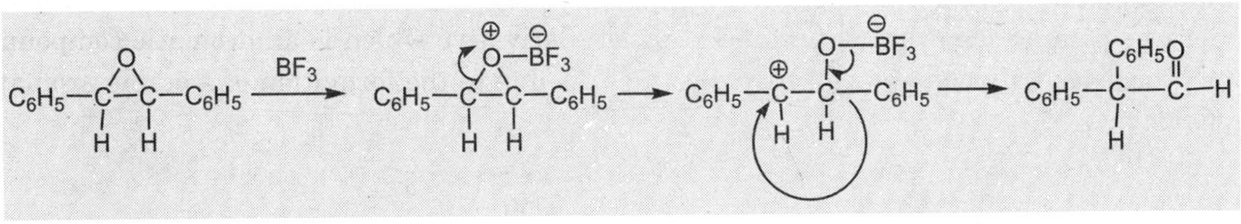

VI. Demjanov Rearrangement

One of the most familiar methods for the generation of carbocations is the deamination of the primary amines by means of nitrous acid. In many cases, the initially formed carbocations undergo rearrangement to yield the rearranged products. Thus, *n*-propylamine on treatment with nitrous acid yields some *n*-propyl alcohol together with a major proportion of *iso*propyl alcohol.

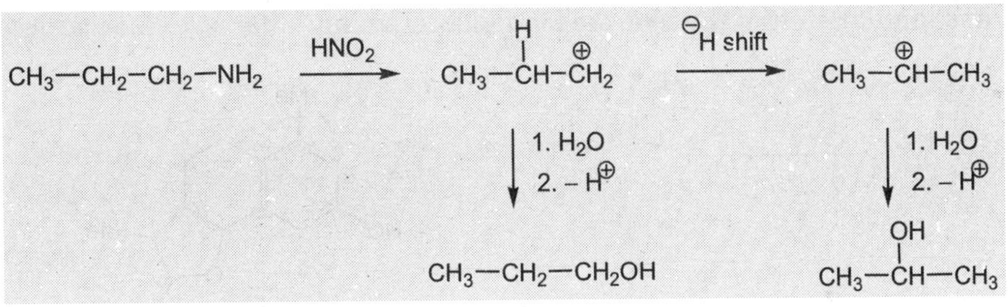

This reaction forms the basis of Demjanov's method of expanding and contracting alicyclic ring systems. For example, when cyclobutylmethylamine (**8**) is treated with nitrous acid it gives both cyclobutylcarbinol (**9**) and cyclopentanol (**10**) besides methylenecyclobutane (**11**) and cyclopentene (**12**). The formation of these products can be readily rationalized by considering all the three typical properties of carbocations, *viz*. addition of nucleophiles, elimination of a proton and rearrangement to a more stable ion.

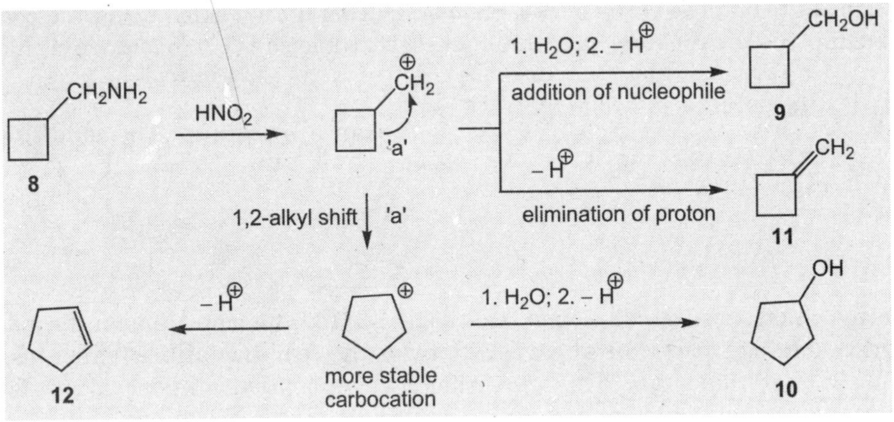

Cyclobutylamine (**13**) yields both cyclobutanol (**14**) and cyclopropyl carbinol (**15**). The mechanism of ring contraction in this case can also be written in a similar fashion.

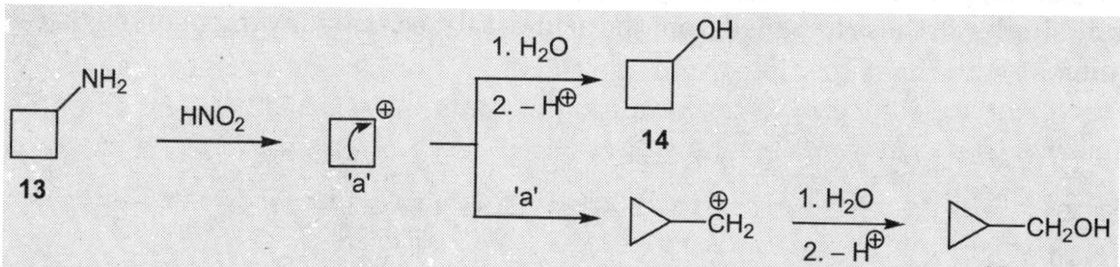

13 14 15

The formation of cyclopropyl methyl cation may be due to its relative stability by significant charge delocalization involving the ring bonds and the positive charge since the C–C bonds in cyclopropane have appreciable π-character.

V. Transannular Rearrangements

Many cases are known where the migration origin and migration terminus are separated by more than two or three bonds. An illustrative example is the formolysis of cyclooctene oxide in which a transannular shift of the hydride ion follows the initial formation of carbocation.

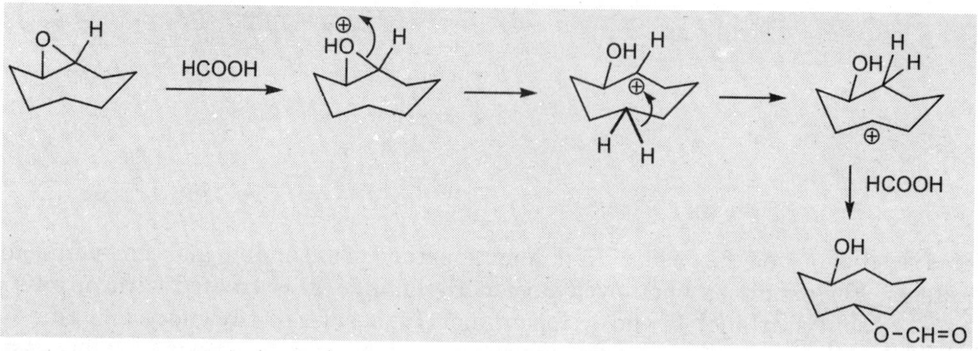

VI. Rearrangement in Bicyclic Compounds

Wagner and Meerwein discovered many rearrangements in the bicyclic terpene series involving not only alkyl shifts but also shifts of ring bonds resulting in ring expansions and contractions. A typical example is the conversion of camphene hydrochloride (16) to *iso*bornyl chloride (17).

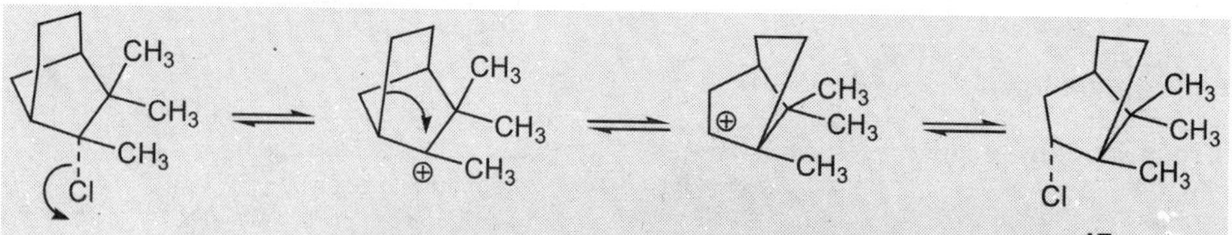

16 17

*Iso*bornyl chloride is the *exo* isomer, that is, it has the chlorine atom on the side opposite to the migrating bridge. It is thus apparent that it is the rear side attack which results in the *inversion at the migration origin*. This stereochemical implication is a general rule for Wagner-Meerwein rearrangements accompanying substitution and isomerization.

Another point of interest in the above sequence is the reversibility of the reaction. The migration *terminus* in the forward process becomes migration origin in the reverse process. Therefore, from the *principle of microscopic reversibility,* it can be derived that both the reactions should have a common mechanism and identical stereochemical rules. There is inversion of configuration at both the migration origin and the migration terminus. Subsequently, we shall examine the evidence that proves the retention of configuration in the migrating group. We may summarize these rules in the following way.

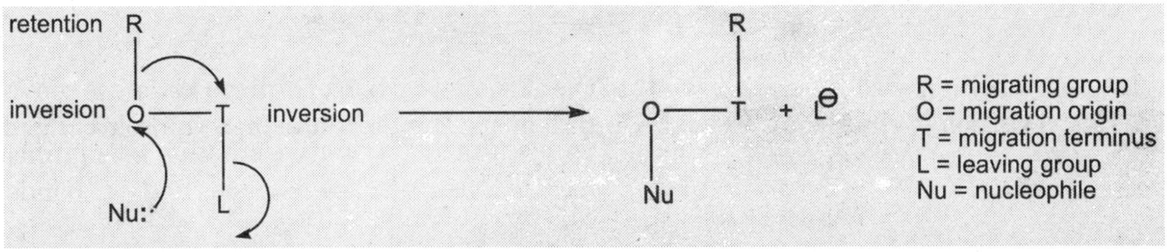

These stereochemical rules have a general applicability for all 1,2-shifts and can be depicted in terms of the following orbital diagram.

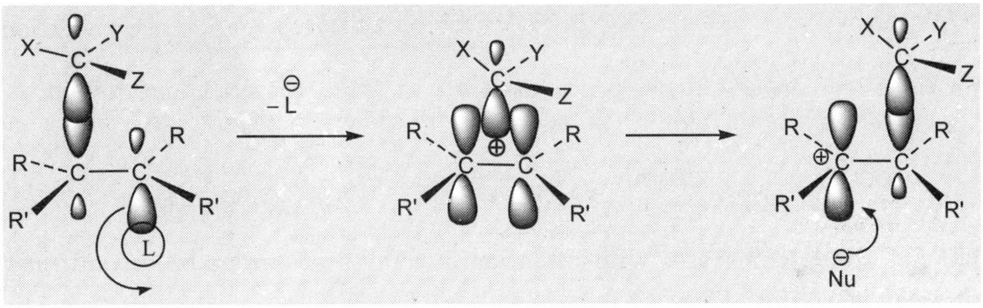

B. Other Related Rearrangements

A number of rearrangements are known in which not a carbocation but a related species is involved. In all these processes, there is 1,2-shift of a group together with its pair of bonding electrons. The requirement for these rearrangements to occur is that the initial bond fission must produce an atom with an incomplete octet, normally a sextet. The most common rearrangements of this type involve carbene, nitrene, positive nitrogen and positive oxygen as intermediates.

I. Wolff Rearrangement

The Wolff rearrangement converts an α-diazo ketone into a ketene and nitrogen.

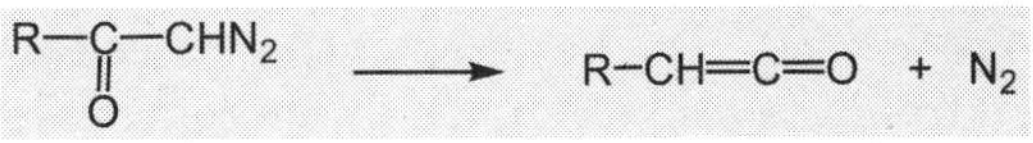

The rearrangement is carried out either by heating in solvents or with catalysts like silver oxide. In presence of water, alcohols, ammonia or amines, the ketenes are converted into the corresponding derivatives.

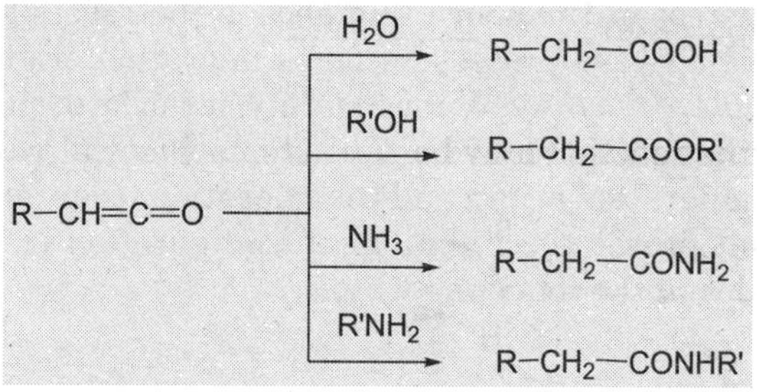

Isotopic labelling with ^{13}C has shown that the carbonyl carbon of α-diazo ketone is converted into carboxyl carbon of the resulting acid. On this basis, the mechanism of the rearrangement can be written in the following steps:

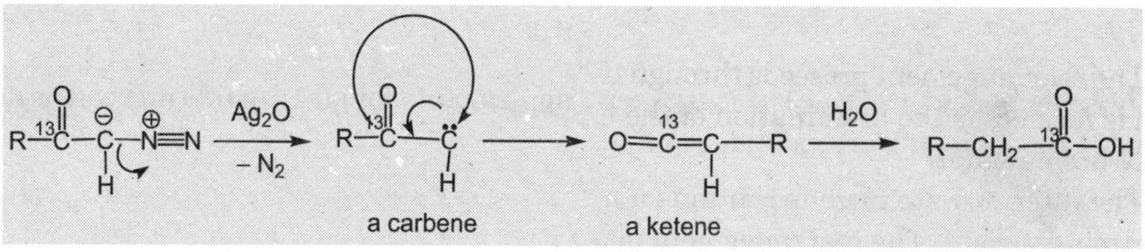

The Wolff rearrangement has practical utility in Arndt-Eistert homologation, in which, an acid is converted into its higher homologue by the following steps:

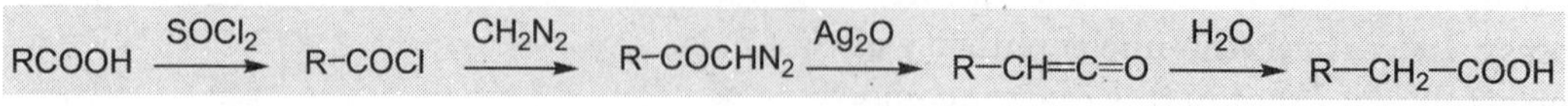

That the group R migrates with retention of configuration has been confirmed by the following transformations involving an optically active acid.

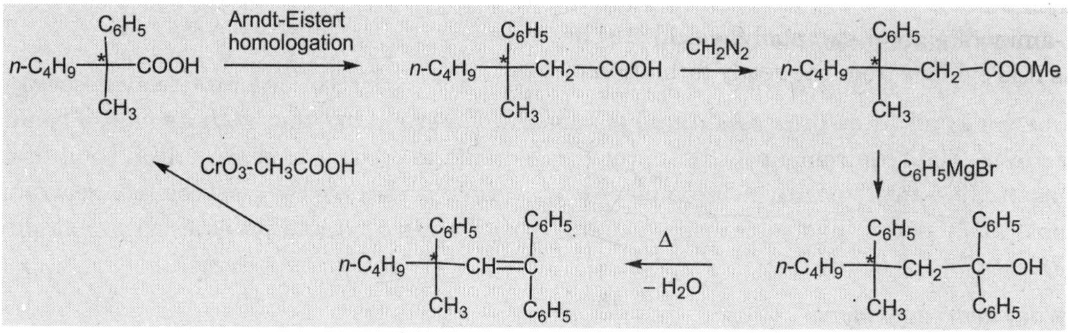

The acid obtained by the Arndt-Eistert synthesis was degraded to the original acid by Barbier-Wieland degradation (The Barbier-Wieland degradation is used for decreasing the length of an acid by one carbon atom. The key step of this procedure is the oxidative cleavage of alkene which is obtained from the acid as shown in the conversion). The acid thus obtained was found to have the same configuration as the original acid.

II. Hofmann Rearrangement

The Hofmann rearrangement converts an amide to a primary amine by the action of halogens and base.

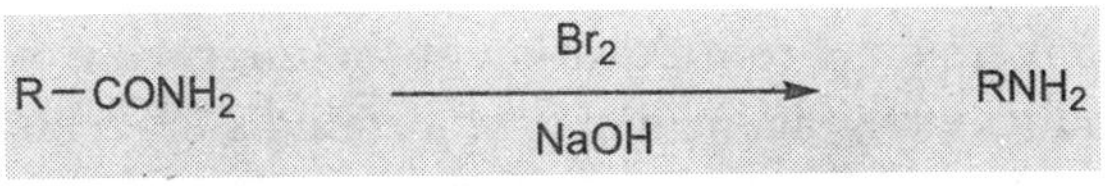

The mechanism of the reaction may be depicted as follows:

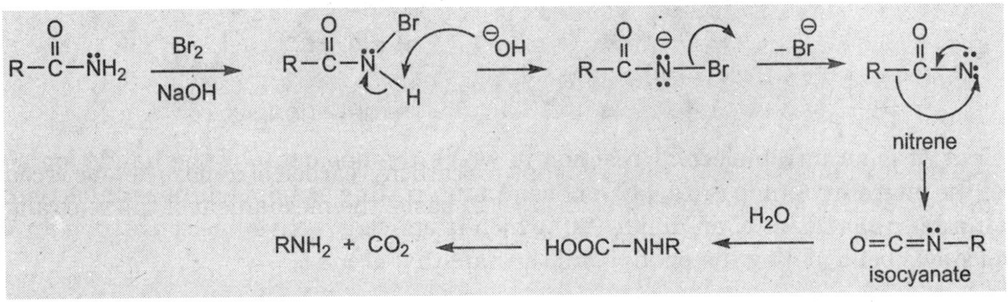

The rearrangement proceeds through the generation of a reactive acyl nitrene intermediate which is formed by the bromination of the amide followed by elimination of a molecule of HBr from *N*-bromamide.

The intermediate nitrene contains a nitrogen atom having a sextet of electrons and so it is electron-deficient. The rearrangement of nitrene to isocyanate is analogous to the conversion of an acylcarbene to a ketene. The isocyanates can be isolated as stable products under aprotic conditions. It has been established by isotopic labelling that when two amides are rearranged, no cross products are obtained and hence the group migration must be an *intramolecular process*. Increase in the nucleophilicity of the migrating group facilitates the reaction. It has been established that in this rearrangement, there is *retention of configuration* at the migrating atom. The most effective argument in its support has come from the observation that Hofmann rearrangement of β-camphoramic acid (**18**) in which COOH and $CONH_2$ groups lie *cis* to each other as it is formed from *cis* camphoric acid which readily yields an anhydride) yields 1-aminodihydro-α-campholytic acid (**19**) having COOH and NH_2 groups also *cis* to each other as it is evident from the ready formation of a lactam.

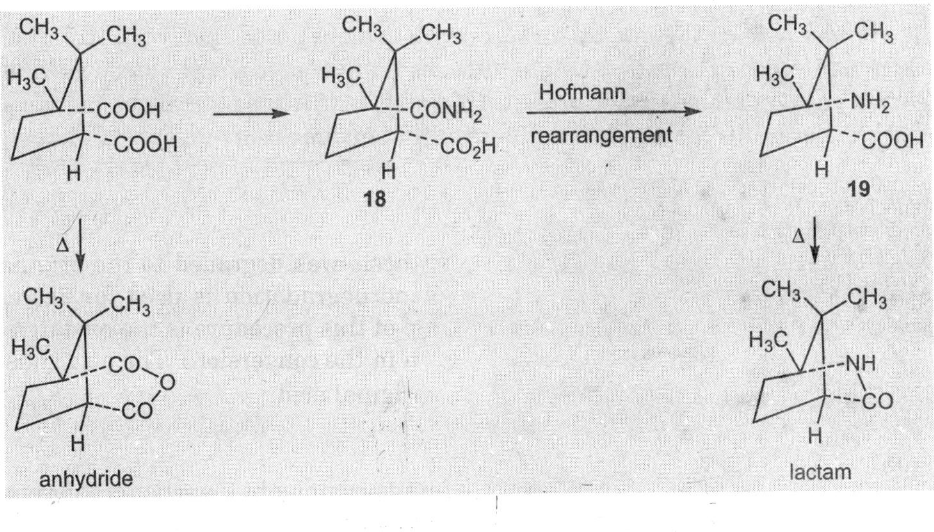

Had there been an inversion of configuration at the migrating carbon, a *trans* amino acid would have been the product which is incapable of forming such a lactam.

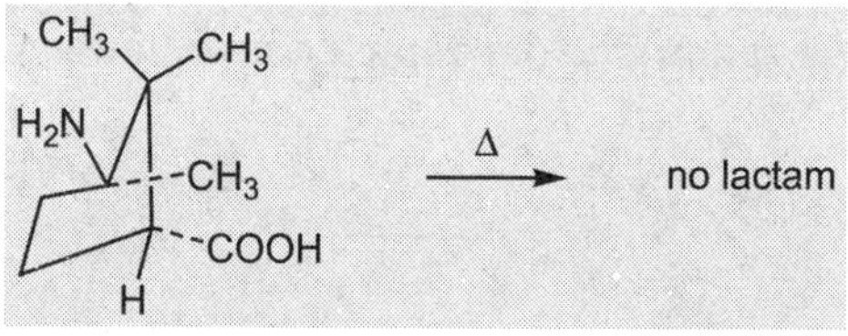

Thus, it is an intramolecular reaction in which the departure of the halide ion and the shift of the migratory group are synchronous. An interesting example in this connection is the Hofmann rearrangement of an amide (**20**) which is optically active due to restricted rotation around single bond joining the phenyl and the naphthyl groups.

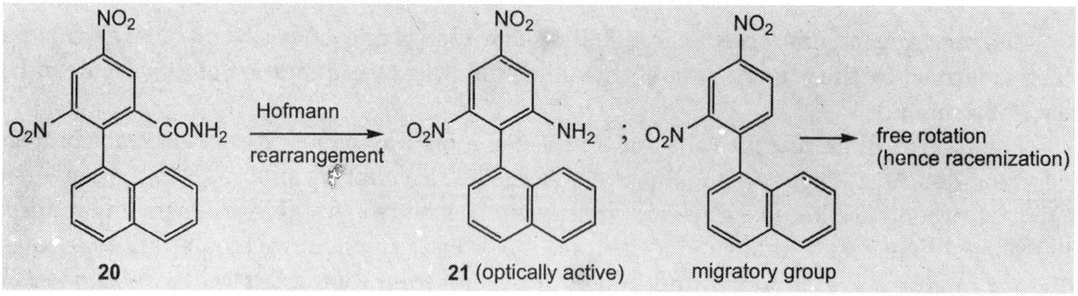

The restricted rotation and hence the optical activity of the amide is due to the presence of two bulky *meta* substituents (NO_2 and $CONH_2$). Had the migratory aryl group become free during the rearrangement, the rotation would not be restricted and would have resulted in racemization. Since the product, amine (**21**), has been found to be optically pure the rearrangement proceeds through a transition state in which the migratory group is partially bonded to both the migration origin and the terminus.

III. Curtius, Schmidt and Lossen Rearrangements

As in the Hofmann rearrangement, these rearrangements also involve shifting of a group from carbon to nitrogen. The intermediate nitrenes rearrange to isocyanates which are finally converted into primary amines in the presence of water. In Curtius rearrangement, an acyl azide is heated which loses nitrogen and then the resulting nitrene rearranges to an isocyanate.

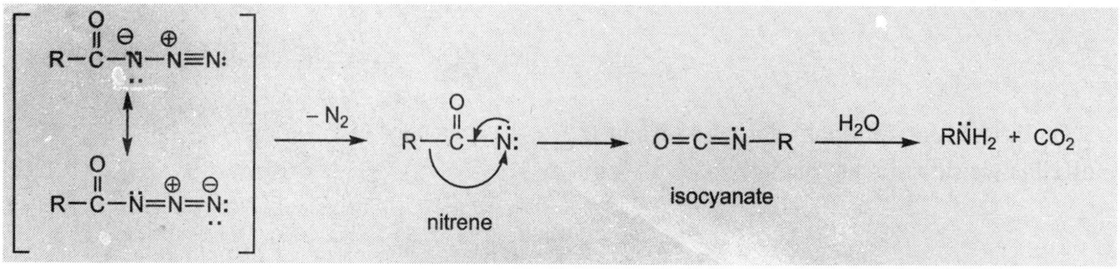

In the Schmidt modification of this reaction, the acyl azide is prepared by condensing a carboxylic acid and hydrazoic acid (HN_3) in the presence of sulphuric acid. The acyl azide is normally not isolated but allowed to decompose and rearrange in the reaction mixture itself. The following mechanism is consistent with the experimental observations.

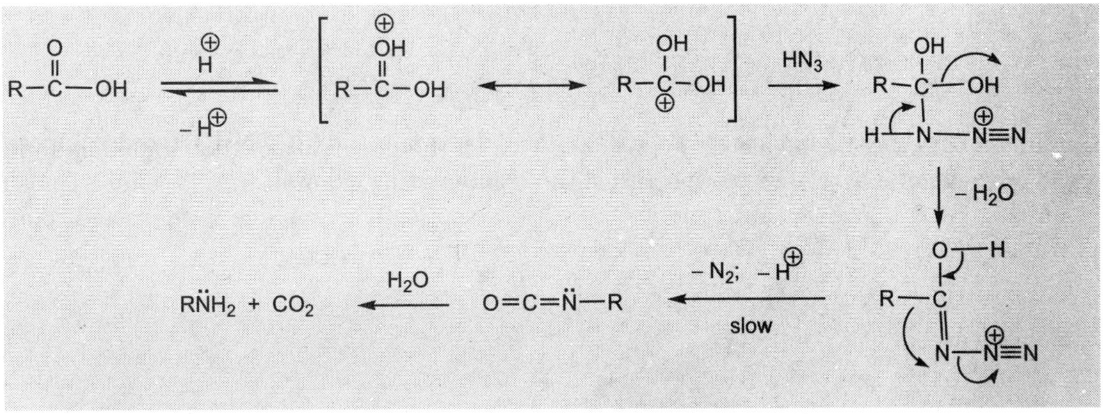

Since *N*-alkylated cations derived from the condensation of alkyl azides (RN_3) and carboxylic acids do not undergo Schmidt rearrangement, it appears that the removal of water molecule precedes the rate-determining step.

Carbonyl compounds also undergo Schmidt reaction when treated with hydrazoic acid in sulphuric acid. Ketones produce *N*-substituted amides, whereas aldehydes generally yield a mixture of the corresponding nitriles and *N*-formyl derivative. A probable mechanism of this rearrangement may be written as follows:

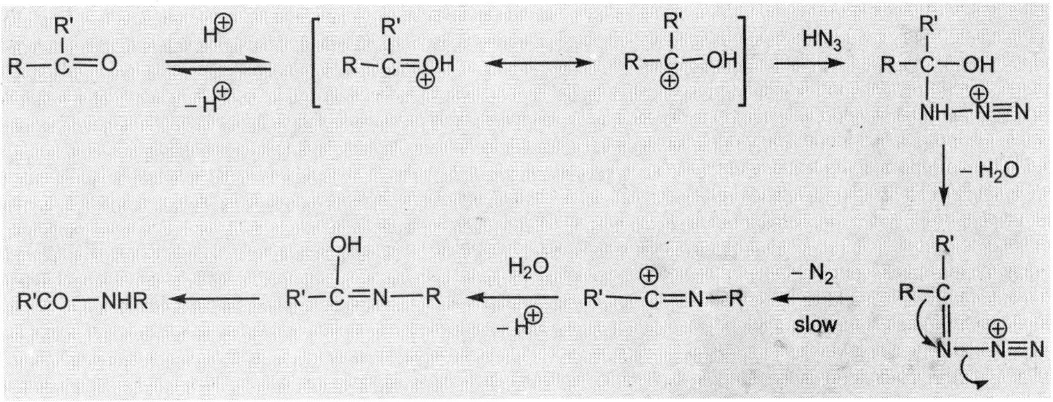

In the case of aldehydes, the loss of a proton from the carbocation may be preferred and nitriles would thus be formed.

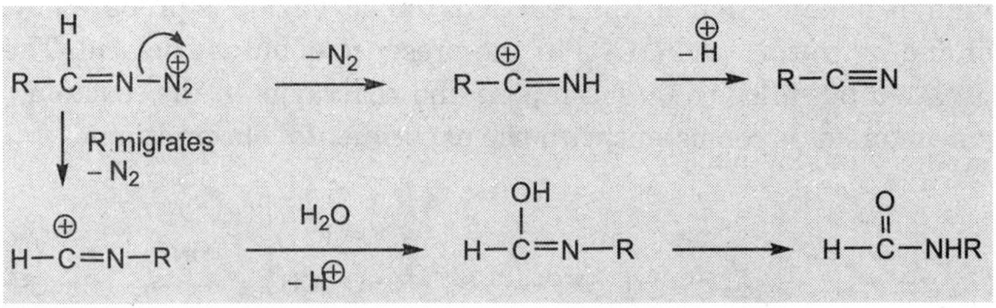

In the Lossen rearrangement, a derivative of hydroxamic acid (RCONHOH) is decomposed in the presence of a base. The mechanism of the reaction is as follows:

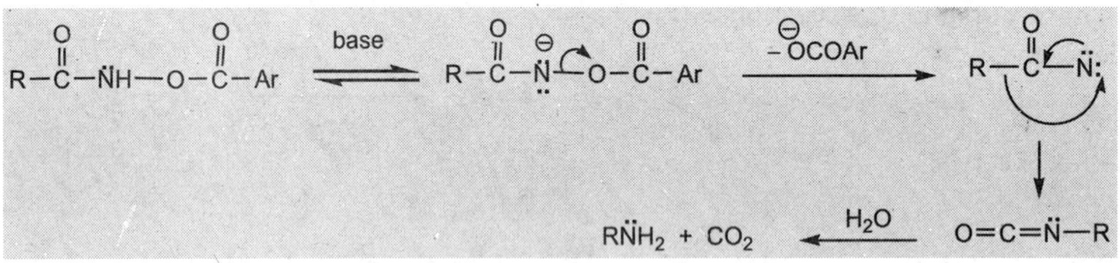

Since ArCOO leaves as Ar COO^-, its separation and hence the rearrangement is facilitated by the presence of electron-withdrawing substituents in *meta*- or *para*-positions.

All these reactions like Curtius, Schmidt and Lossen are intramolecular rearrangements, in which, the migrating group never becomes free. Also, the migrating group shifts from carbon to nitrogen with retention of configuration. These conclusions have been proved in an elegant way by rearranging the appropriate derivatives of α-benzylpropionic acid, which yielded α-benzylethylamine having the same configuration as the starting materials.

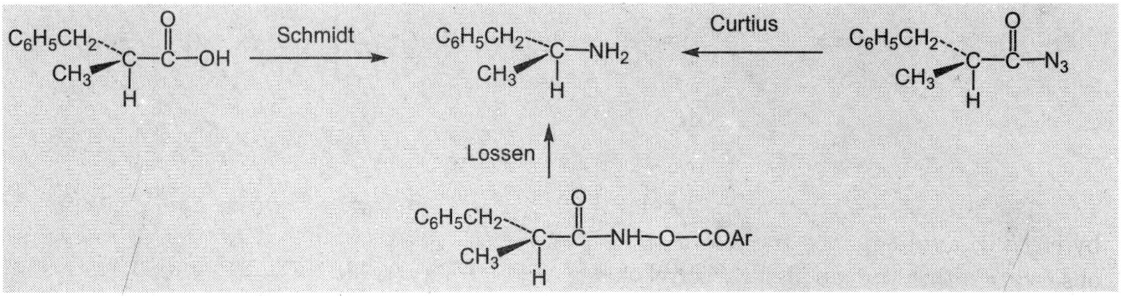

IV. Baeyer-Villiger Oxidation

There are several examples of rearrangement of groups from carbon to electron-deficient oxygen atom. Out of these, the Baeyer-Villiger reaction is the most interesting rearrangement of this type. It involves the oxidation of ketones to esters by means of peracids such as perbenzoic acid, peracetic acid, peroxytrifluoroacetic acid and permonosulphuric acid. An oxygen atom is inserted next to the carbonyl group.

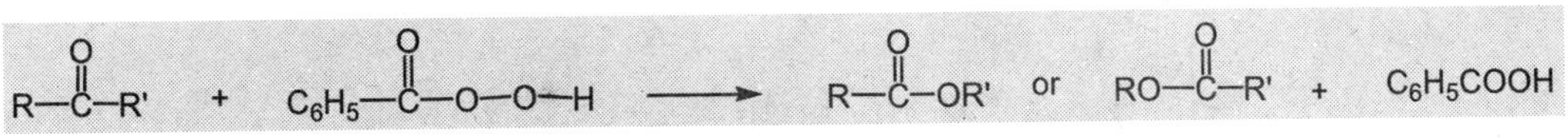

Cyclic ketones yield lactones.

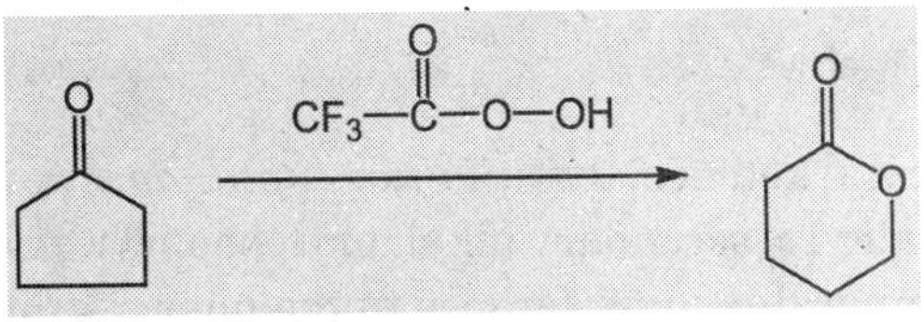

The key step in this reaction is the rearrangement of a group from carbon to the electron-deficient oxygen.

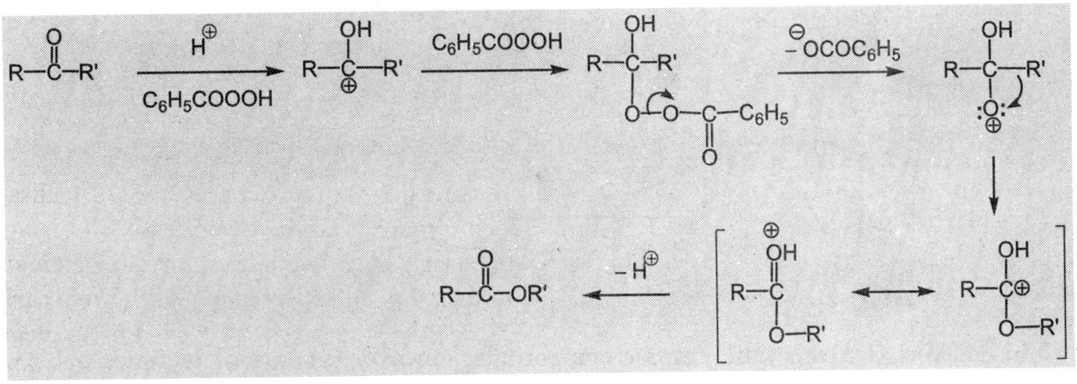

Most probably the loss of the carboxylate anion and the rearrangement are concerted.

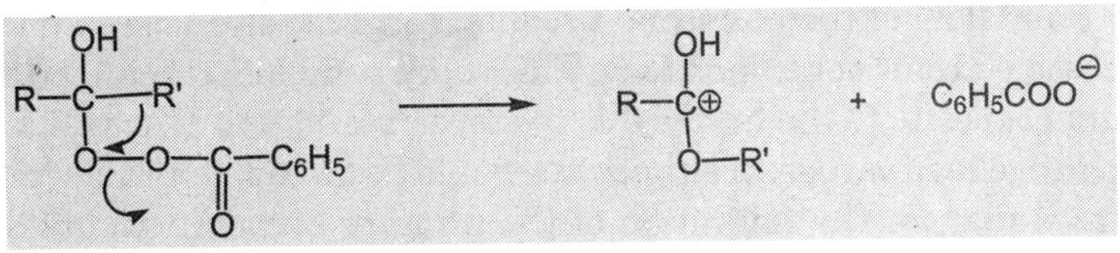

The rate of the reaction is accelerated by *electron-donating* groups in the ketone and by *electron-withdrawing* groups in the peracid. The above mechanism is supported by the observation that the labelled carbonyl oxygen atom of the ketone becomes the carbonyl oxygen atom of the ester.

Like other rearrangements involving electron-deficient species, the Baeyer-Villiger reaction also takes place with *the retention of configuration* in the migrating group.

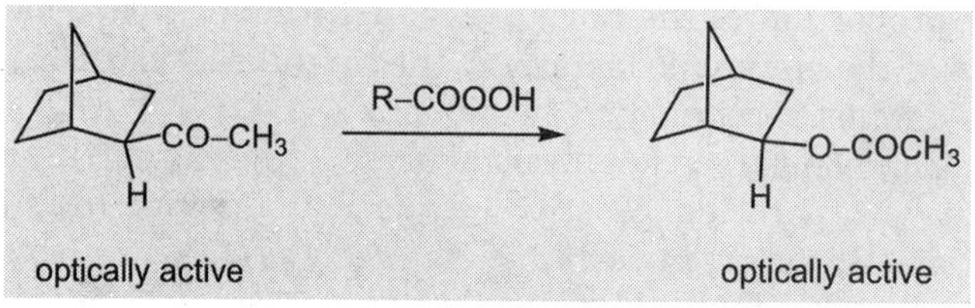

Finally, in unsymmetrical ketones, the *migratory aptitude* of various groups has been found to be in the order: tertiary alkyl > secondary alkyl, aryl, benzyl > primary alkyl > methyl. Even a bridged *tert*-alkyl group migrates in preference to the phenyl group.

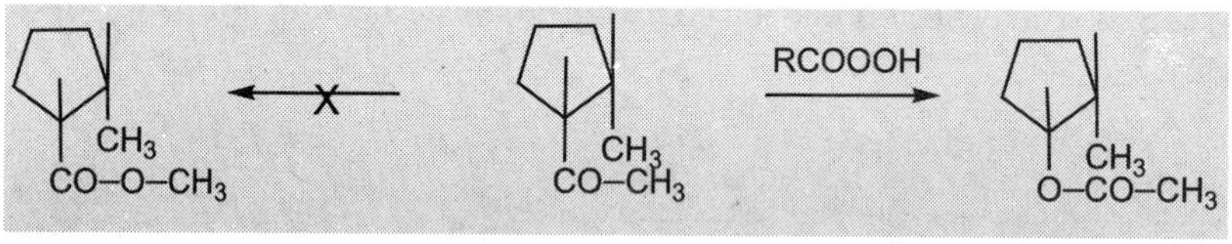

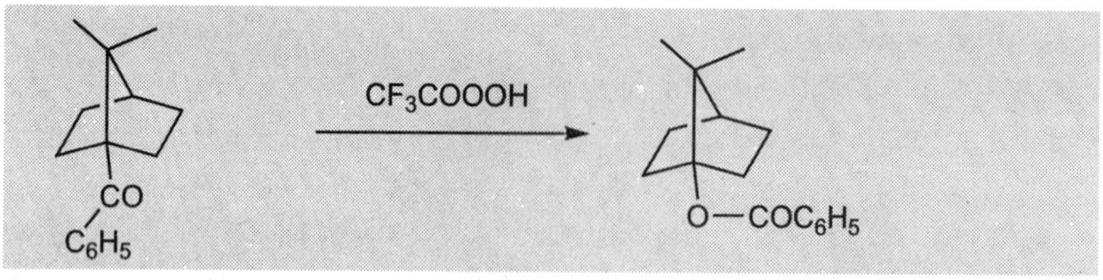

In condensed carbocyclic ketonic compounds, however, abnormal lactones are often formed in varying proportions by migration of the carbons having lesser migratory aptitude. For instance, the lactones obtained by the Baeyer-Villiger oxidation of camphor and apocamphor contain 25% and 94% of the abnormal product, respectively. These abnormalities can be explained on stereoelectronic considerations. The bulky peracid molecule attacks the carbonyl group from the side opposite to the bridge due to steric interaction. The transition state for the migration of the bridgehead carbon in either case would be an unfavourable twisted boat (atoms 1,2,3,4,5,6) whereas that for the migration of the primary carbon it would be the favourable chair shape.

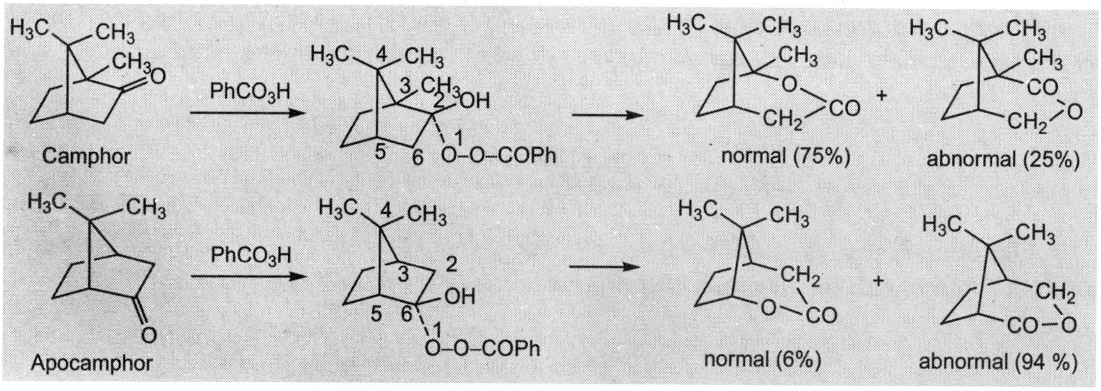

The much larger proportion of the normal product from camphor may be explained by the relative dominance of the electronic factor, *i.e.* the migrating carbon for normal product is tertiary in camphor as against secondary in apocamphor.

Ketones containing carbon-carbon double bond may undergo epoxidation or Baeyer-Villiger oxidation on treatment with peracids. Mostly epoxidation of double bond occurs faster than Baeyer-Villiger oxidation. However, in certain cases such as **22** where carbon-carbon double bond is less reactive, Baeyer-Villiger oxidation takes place leaving the double bond intact.

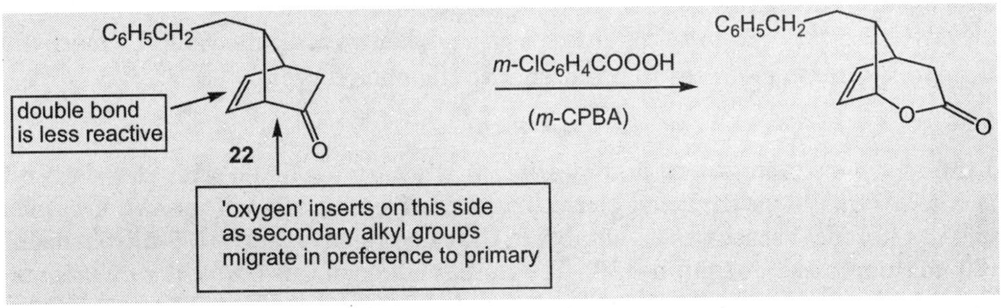

The Baeyer-Villiger rearrangement takes place exclusively in the presence of a reactive double bond when hydrogen peroxide is used in place of peracids (**23** → **24**).

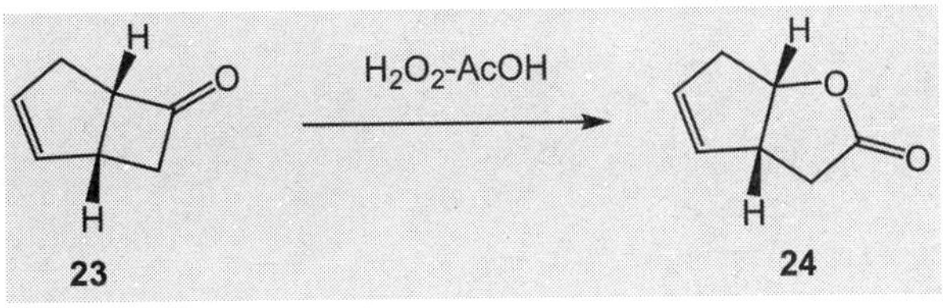

V. Beckmann Rearrangement

The acid-catalyzed transformation of a ketoxime to an *N*-substituted amide is known as the Beckmann rearrangement.

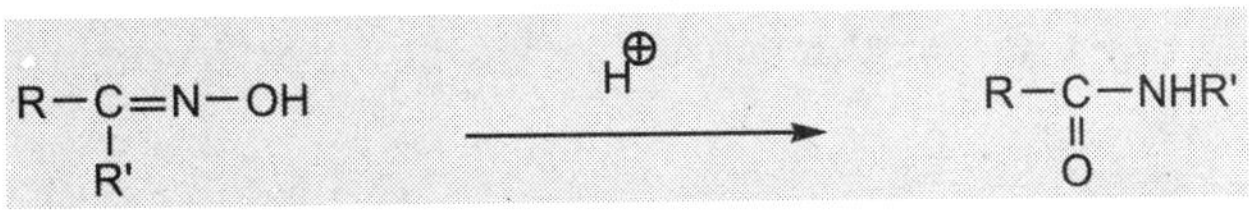

This rearrangement is catalyzed by a variety of acidic reagents such as H_3PO_2, H_2SO_4, P_2O_5, $SOCl_2$ and PCl_5, and a simplified mechanism can be written in the following steps.

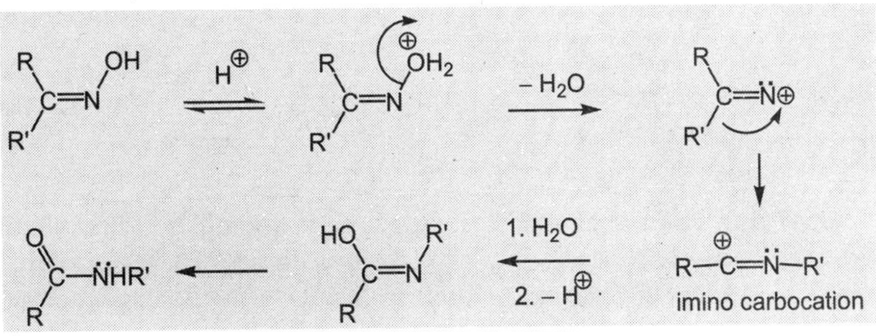

The reaction proceeds with the protonation of the OH of the oxime which is followed by the loss of water to generate an electron-deficient species. Subsequently, migration of a group from adjacent carbon to the electron-deficient nitrogen results in an imino carbocation which then reacts with water and finally gives the product amide after loss of a proton.

The rearrangement is highly *stereospecific*, wherein, the migrating group always approaches the nitrogen atom on the side opposite to the oxygen atom.

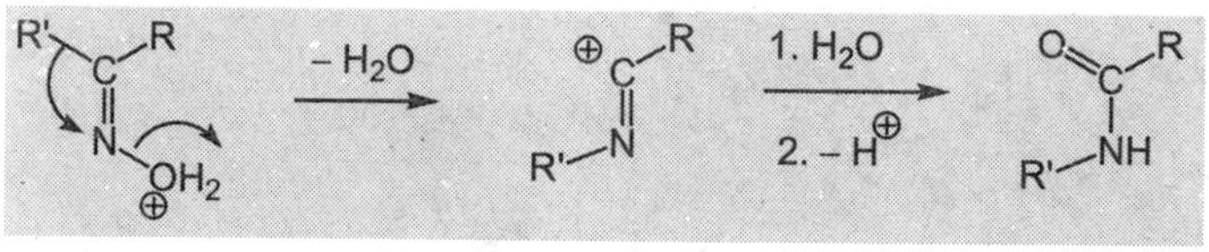

In the above example, there is migration of R' group to the electron-deficient nitrogen, which is *anti* or *trans* to the hydroxyl group. The exclusive migration of the *anti* group has been confirmed in a number of cases, most notably in the conversion of 2-chloro-5-nitrobenzophenone oxime (**25**) to chloronitrobenzanilide (**26**). The configuration of this particular oxime has earlier been established by its ready conversion to nitro-substituted phenylbenzisoxazole (**27**), showing that the nitrated benzene ring lies on the same side of the C=N linkage as the OH group. When this oxime undergoes Beckmann rearrangement, it is found that the phenyl group, rather than the nitrated benzene ring, migrates to the nitrogen.

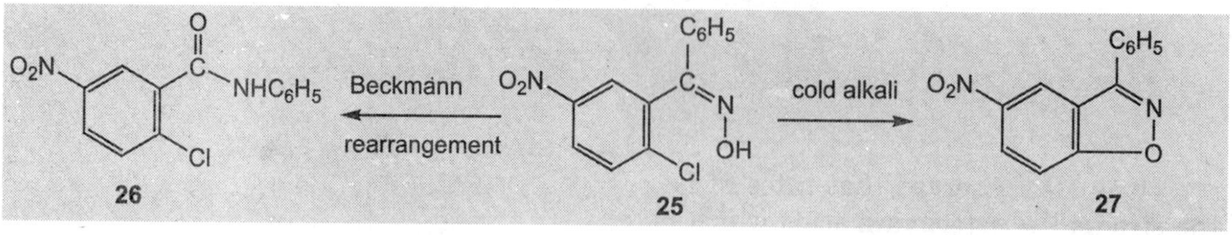

Since it is always the *anti* group that migrates, it appears that either the breakage of the N–O bond and the group migration are synchronous or these two steps follow each other quickly. The migration of *anti* group is so much certain in the Beckmann rearrangement that it is normally possible to establish the configuration of a particular oxime by identifying its rearrangement products.

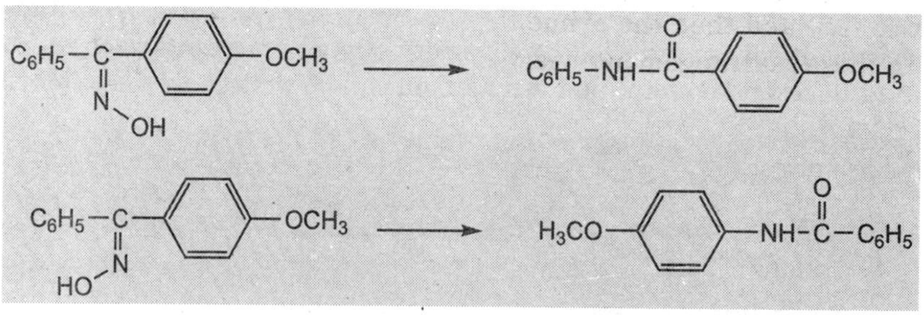

It is to be noted that in case of aliphatic ketones, if one of the alkyl chains is branched, more of the oxime with the hydroxyl group *anti* to that chain will be formed and accordingly more of the branched group will migrate. Such a situation is illustrated by considering the example of *iso*propyl methyl ketone.

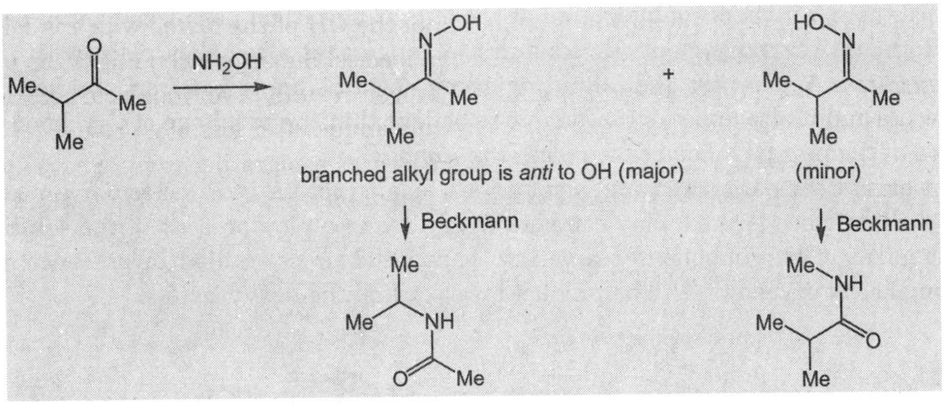

Another mechanistic feature of the rearrangement has been worked out by Kuhara and later by Chapman who established that the rearrangement does not take place in the oxime itself, but in its acyl derivative which ionizes and subsequently rearranges. The ionization of the acyl derivative has been shown to be the rate-controlling step of the rearrangement. Typically, the benzene sulphonic ester of benzophenone oxime **(28)** undergoes rearrangement without any acid catalyst.

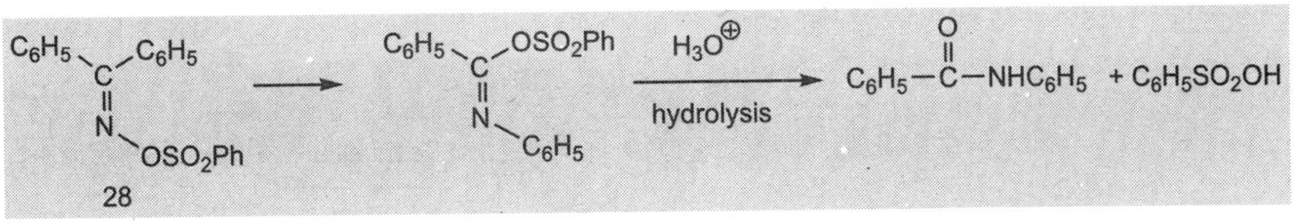

It is also observed that rates of isomerization of various oxime esters run parallel to the strength of esterifying acids which is found to be in the following order: $C_6H_5SO_2OH$ > C_6H_5COOH > CH_3COOH.

Finally, the carbon atom of the migrating group *retains its configuration* in the Beckmann rearrangement. By the following series of reactions, Kenyon and Young not only proved retention of configuration but also correlated Beckmann and Hofmann rearrangements. Optically active 3-heptylcarboxylic acid **(29)** was converted to its amide **30** and subjected to Hofmann rearrangement. In another set of experiment the same acid **(29)** was converted into 3-heptyl methyl ketone **(32)** and then the oxime **33** of this ketone was rearranged by the Beckmann method. Both these rearrangements yielded the same optical isomer of 3-heptylamine **(31)**.

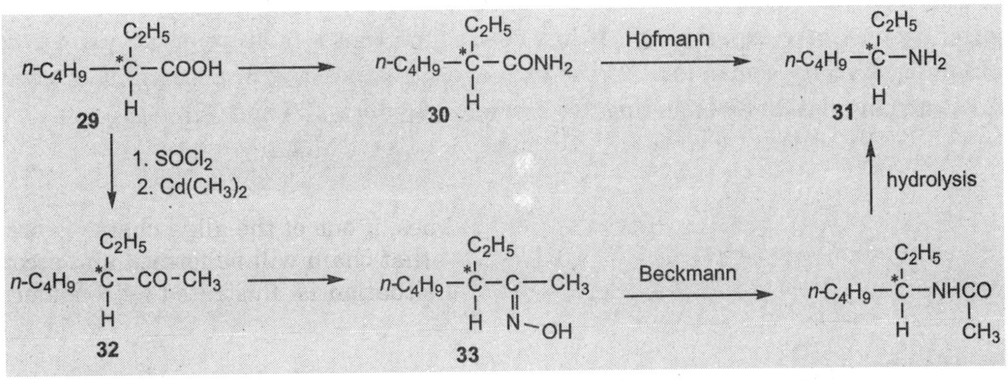

Since the retention of configuration at the migrating carbon has already been confirmed in the Hofmann rearrangement, Beckmann rearrangement also takes place with *retention of configuration*. Apparently, the migrating group never becomes free from the remainder of the molecule and it also appears reasonable to believe that the breakage of C–C bond and the formation of the new C–N bond take place *synchronously*.

However, Beckmann rearrangement as a whole is not an intramolecular process since the carbocation, once formed, may combine with a hydroxyl ion present in the solution. The rearrangement of benzophenone oxime into benzanilide was studied in presence of water containing heavy oxygen (^{18}O). The product incorporated the heavy isotope.

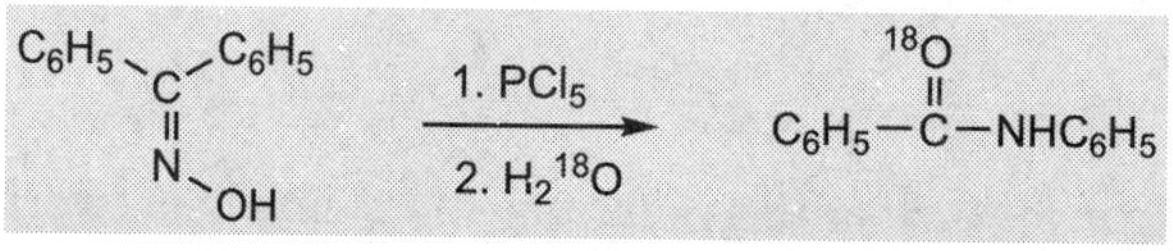

It is, therefore, evident that the Beckmann rearrangement does not proceed by intramolecular exchange of hydroxyl and the phenyl group, rather, it passes through a carbocation which accepts hydroxyl ion or a water molecule from the solvent. Thus, there is complete loss of the original oxime oxygen from the molecule during the reaction. A reasonable mechanism for the Beckmann rearrangement may be summed up as under:

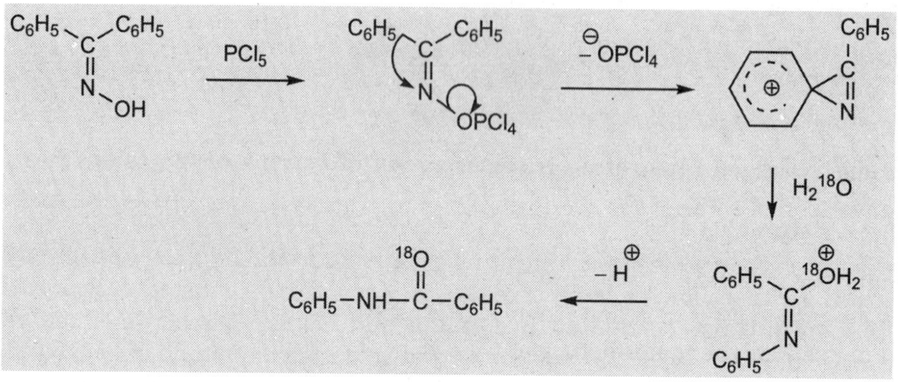

In the presence of strong acids, the reaction is initiated by protonation of the oxime followed by loss of water to yield the same species as obtained by the decomposition of ester.

Beckmann rearrangement is not an intramolecular process, this statement is further supported by *crossover* experiments. It has been found that a mixture of *two* oximes **34** and **35** containing tertiary centre next to the oxime, on treatment with a strong acid gives *four* *N*-substituted amides **36-39** including *two crossover* products (**38** and **39**).

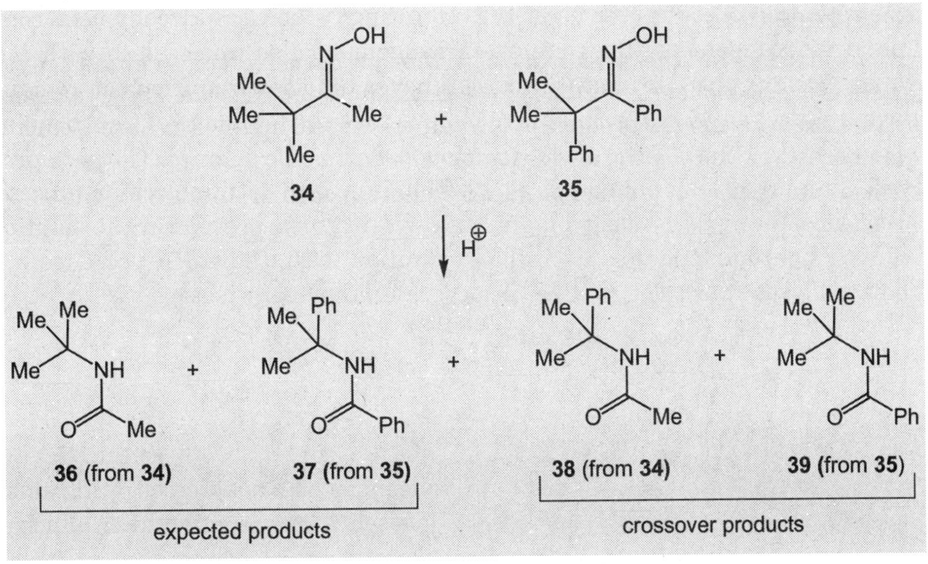

Beckmann rearrangement is not confined to ketoximes, and some aldoximes have also been shown to undergo this rearrangement. Thus, *syn* and *anti*-benzaldoximes undergo rearrangement under the influence of PPA.

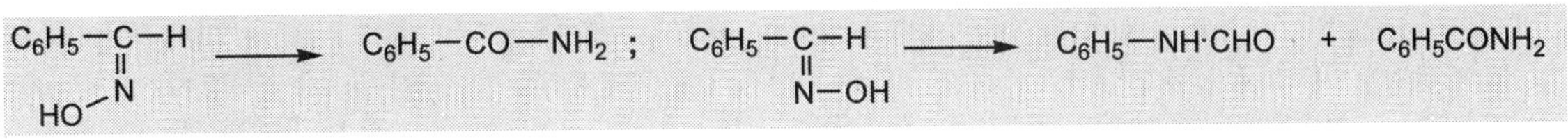

The formation of benzamide in the rearrangement of the *syn*-isomer has been explained by the ready conversion of some of the *syn*-form into *anti* which then rearranges with a hydrogen migration.

Beckmann rearrangement has also been used for the enlargement of rings. An example of considerable industrial importance involves the rearrangement of cyclohexanone oxime to ω-caprolactam used for the manufacture of Nylon-6.

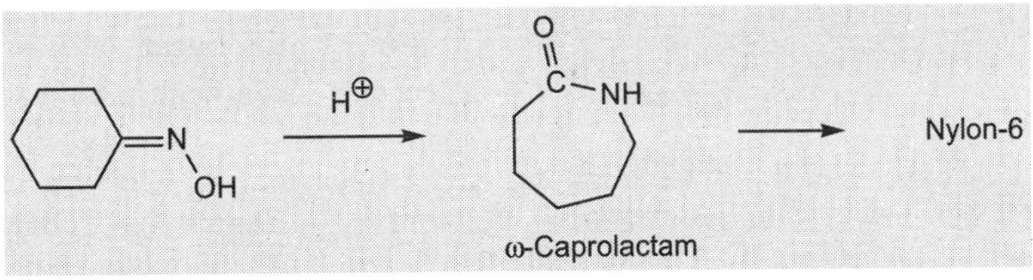

12.2 FURTHER MOLECULAR REARRANGEMENTS

In the previous Section (12.1), we had confined ourselves to the rearrangements involving electron-deficient species. Here we shall be concerned with such rearrangements These are: (A) Fries, (B) Favorskii, (C) Claisen and (D) Cope rearrangements which do not fall in that category.

A. Fries Rearrangement

A preparative method for the synthesis of *ortho*- or *para*-hydroxy ketones involves the rearrangement of the phenolic esters in the presence of aluminum chloride. These rearrangements, known as Fries rearrangements produce at low temperatures, mainly the *para* isomer (**40**), and at higher temperatures, mainly the *ortho* isomer (**41**) of phenolic ketones. The *para* product can be transformed into the *ortho* product on further heating with aluminum chloride.

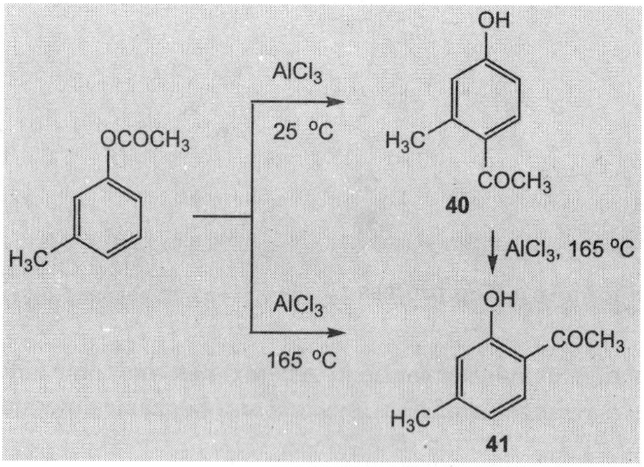

Although the *para* isomer is formed more rapidly, it appears that the *ortho* compound is the more stable of the two. Greater stability of the *ortho* product may be due to the existence of intramolecular hydrogen bonding (**42**). In the presence of aluminium chloride, however, aluminium complexes are formed rather than phenols, and again, the complex of *ortho* hydroxy ketone gets stabilized due to the union of the two oxygen atoms through aluminium (**43**).

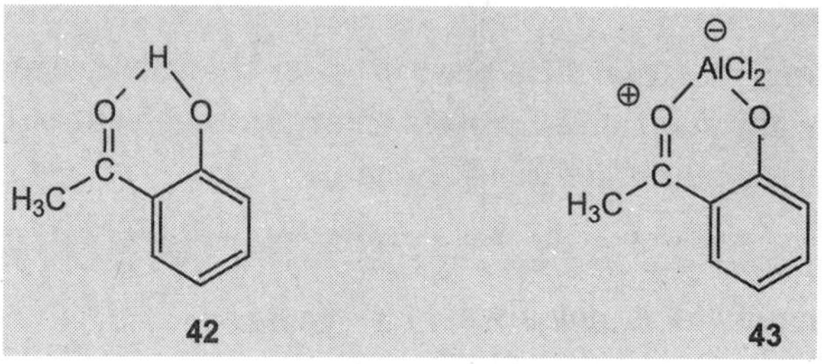

The exact mechanism of the Fries rearrangement is somewhat uncertain as it is sometimes a one-step and sometimes a two-step process. As we have already seen (Chapter 5), the two alternatives are often distinguished by running the reaction of a phenolic ester in the presence of another aromatic compound and then checking for *cross products*. The two esters, **44** and **45**, rearrange under identical conditions and at approximately the same rate to give the products **46** and **47**, respectively. Rearrangement of a mixture of **44** and **45**, however, not only gave **46** and **47** but also the cross products **48** and **49**. Isolation of cross products supports the two-step mechanism involving an initial dissociation and subsequent electrophilic attack of the acyl cation on the complex derived from the phenoxide and $AlCl_3$.

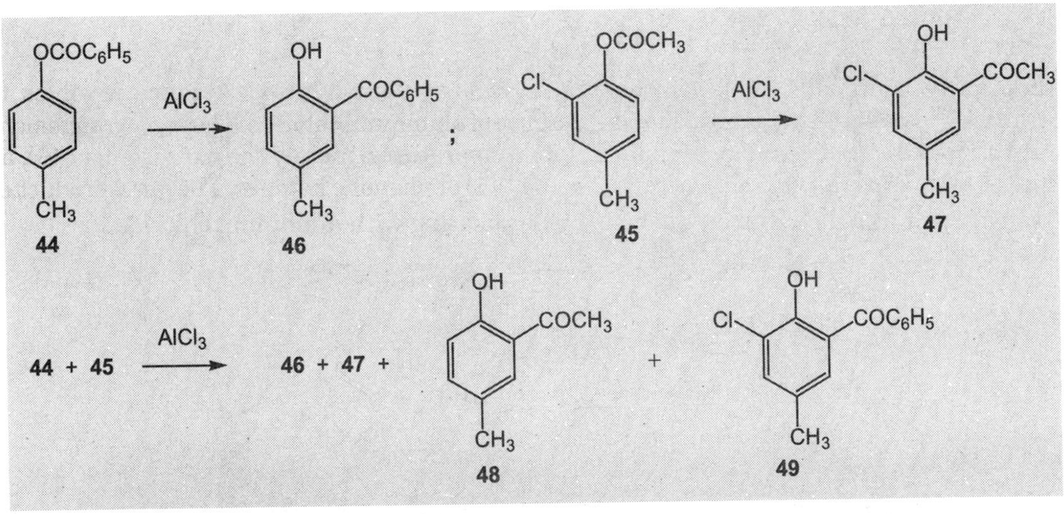

The mechanism for a two-step process may be written in the following steps.

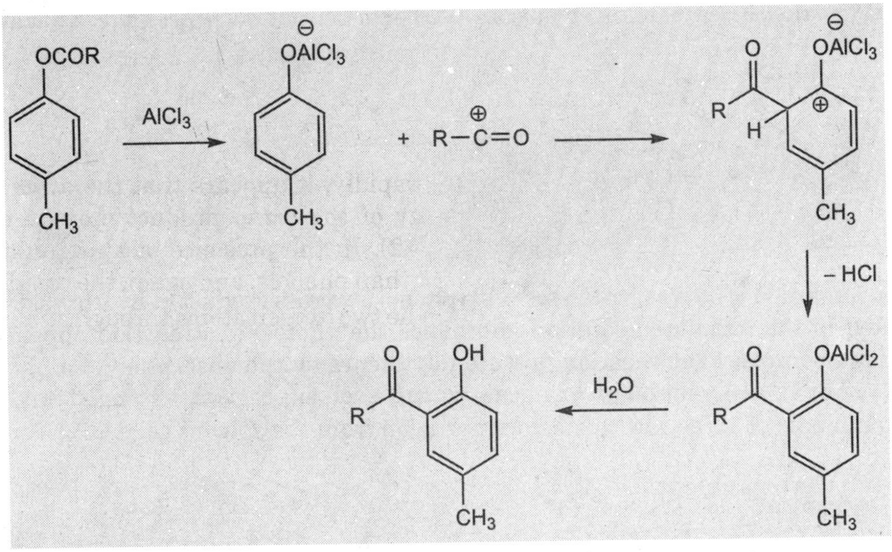

That the rearrangement is not always a two-step process has been demonstrated by rearranging two similarly constituted phenolic esters **50** and **51** in the presence of toluene. Acylated toluene (**52**, cross product) was indeed obtained from **50**, but no such cross acylation took place in the rearrangement of **51** indicating the one-step mechanism of the latter reaction.

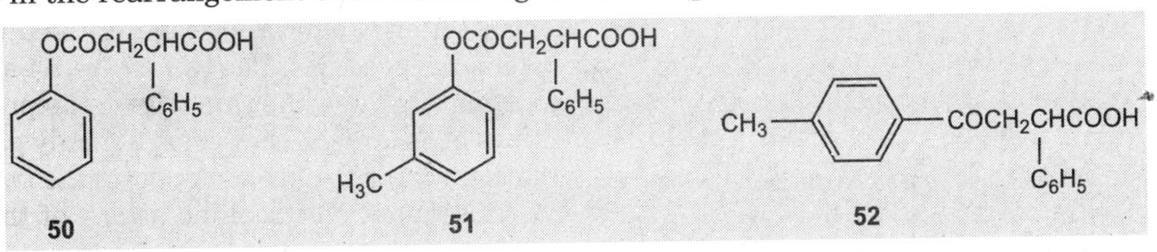

The one-step process is believed to proceed *via* the following mechanism.

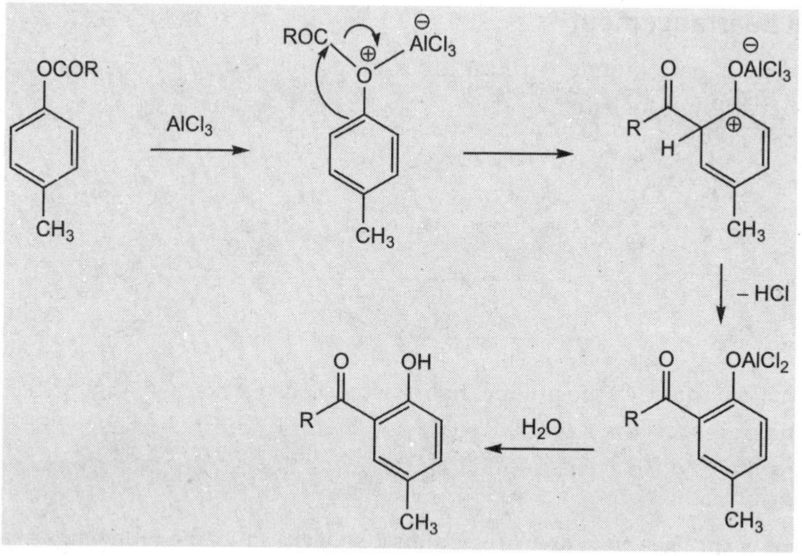

Photo-Fries Reaction: Fries rearrangement can also be carried out in the presence of ultraviolet light.

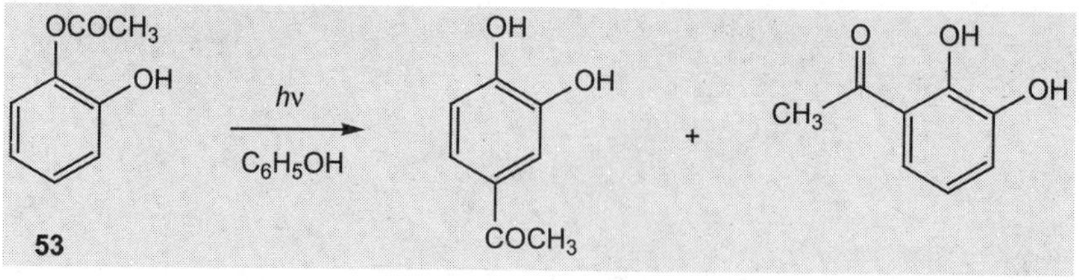

When a solution of **53** was irradiated with phenol, no cross product (*o*-hydroxyacetophenone) was detected in the irradiated solution and hence the photo-Fries reaction appears to be an intramolecular process. The reaction presumably occurs by the absorption of a photon by the ester followed by its dissociation into radicals within a solvent cage. Rearranged products result by the recombination of the radicals before diffusion from the solvent cage.

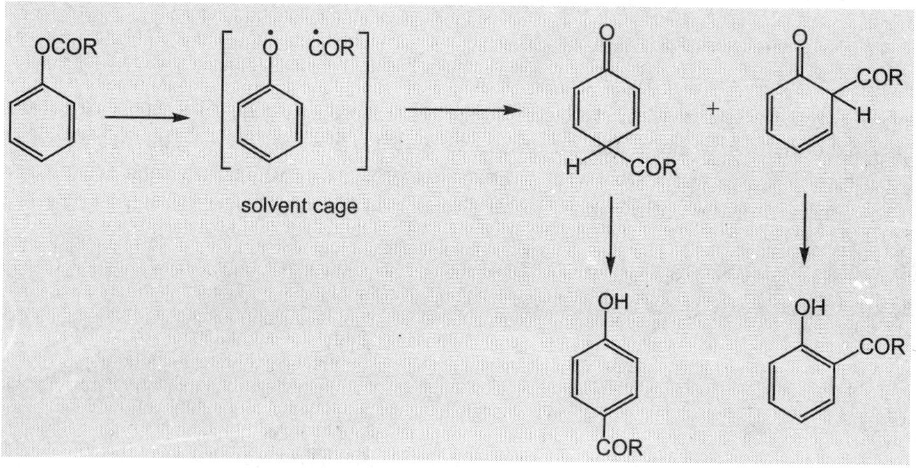

B. Favorskii Rearrangement

Treatment of an α-halo ketone with an alkoxide does not lead to elimination or substitution product, rather it gives an ester.

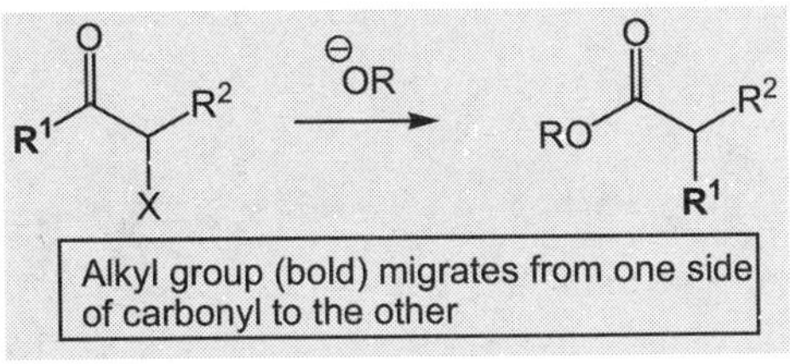

This reaction, in which the product has a rearranged carbon skeleton with the same number of carbon atoms, is called the Favorskii rearrangement. A similar treatment with hydroxide ions gives a carboxylic acid. In cyclic α-halo ketones the process leads to ring contraction.

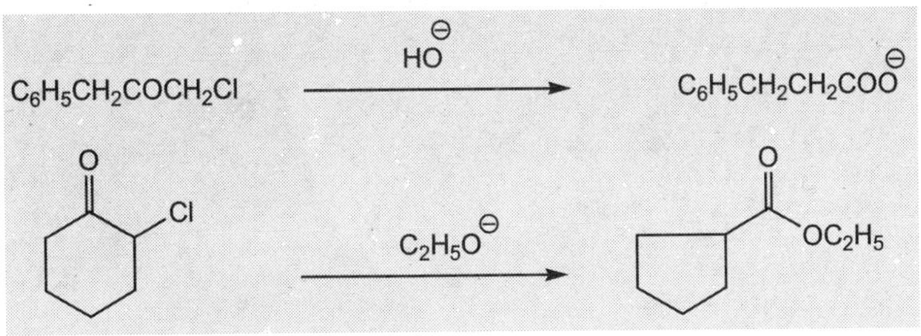

A reasonable mechanism for this rearrangement envisages the abstraction of an α-proton so as to form a carbanion. Nucleophilic attack by the carbanion on the carbon atom bearing the halogen results in the displacement of the halogen. Subsequent opening of the resultant transient cyclopropanone ring gives the product.

Evidence supporting the above mechanism has been obtained by treating 2-chlorocyclohexanone (**54**) labelled with ^{14}C at the chlorine bearing carbon. The product was found to contain half the ^{14}C at the α-carbon atom (**56**) and half at the β-carbon (**57**). This finding demonstrated that the two carbon atoms that are *alpha* to the carbonyl group (C-2 and C-6) in **54** become equivalent at some stage during the course of the rearrangement. Intermediacy of a symmetrical cyclopropanone (**55**) which opens up with equal probability on either side of the carbonyl group fits in best with the experimental observations.

However, in cases where the cyclopropanone intermediate is not symmetrically substituted, it opens up on that side which gives the more stable carbanion. Isomeric ketones **58** and **59**, for instance, yield the same ester **61** on treatment with alkoxide as the intermediate cyclopropanone **60** opens up in such a way as to produce a resonance-stabilized carbanion.

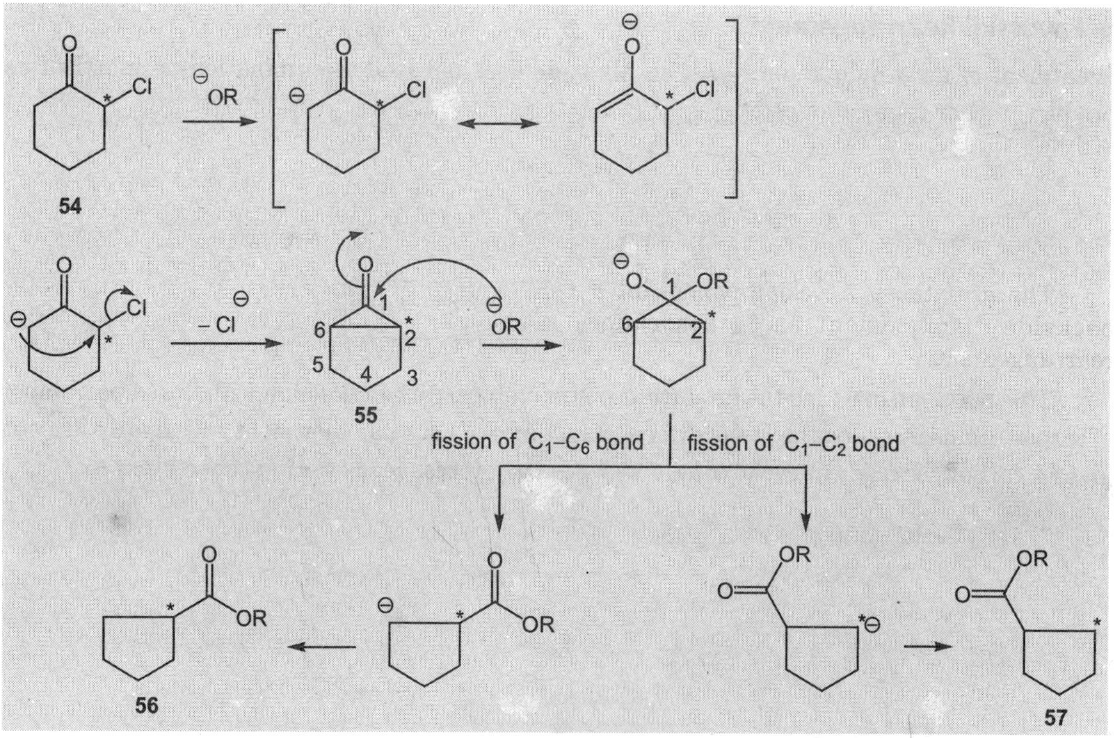

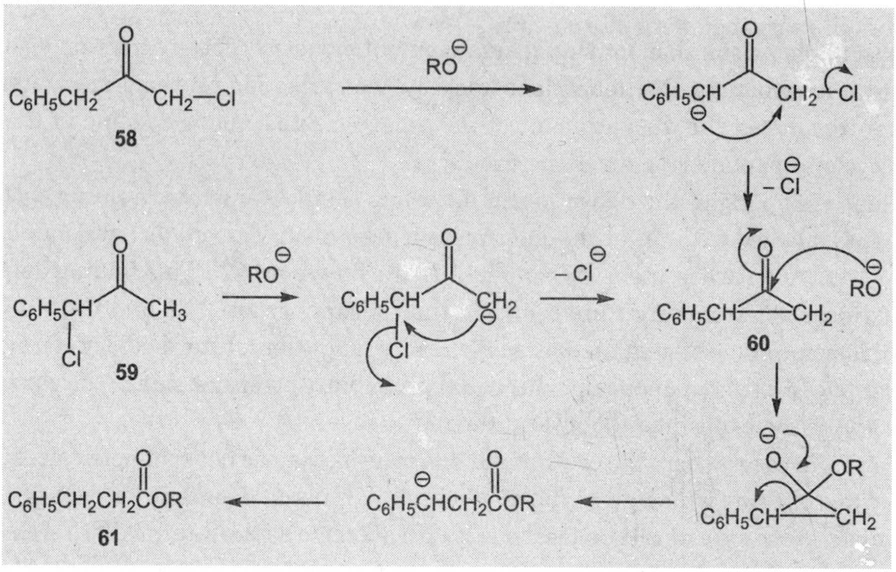

Although cyclopropanone intermediates have not been isolated, in at least one case, a cyclopropanone intermediate **62** has been trapped when furan was present in the reaction mixture, giving the adduct **63**.

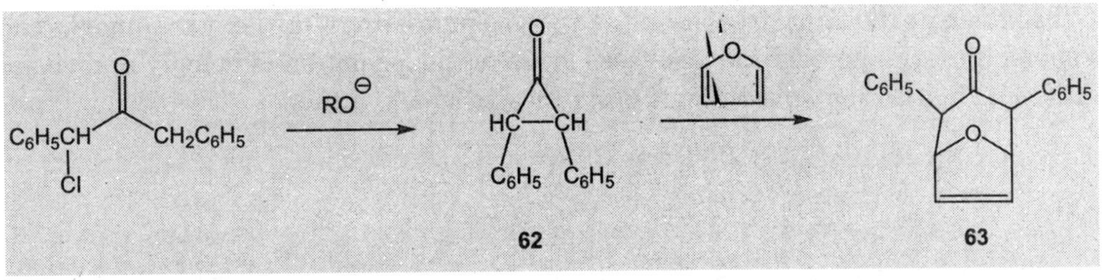

The formation of a cyclopropanone intermediate presumably proceeds by an intramolecular backside displacement as **64** rearranges in contrast to **65** which does not undergo rearrangement.

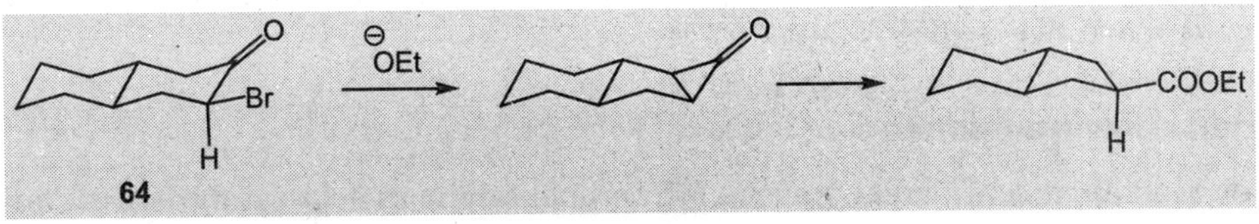

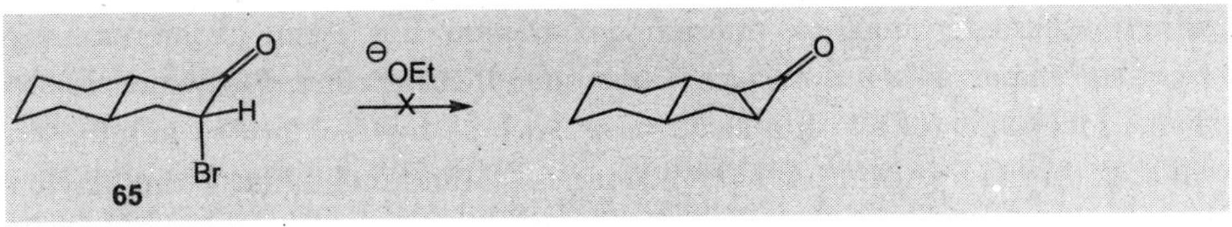

Favorskii rearrangement of *gem*- (**66**) and *vic*-dihalo ketones (**67**) yields α,β-unsaturated esters. The intermediate cyclopropanone, which is the same in both the cases, opens up with simultaneous expulsion of the halide ion.

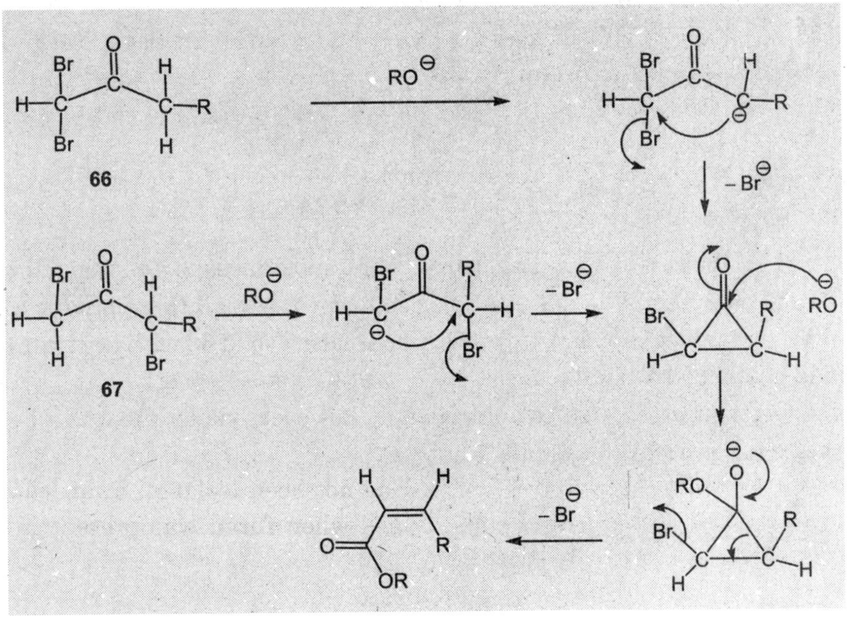

When there are no acidic protons in the starting halo ketone, the ketone cannot be enolized. The Favorskii rearrangement in such cases leads to the formation of heavily branched esters and acids which are otherwise difficult to synthesize by alkylation.

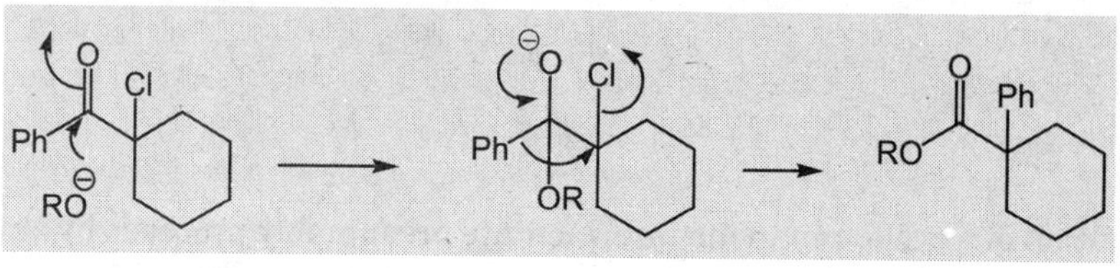

The mechanism of Favorskii rearrangement in such cases is rather similar to benzilic acid rearrangement. By analogy with the *semipinacol*, these reactions may be better called *semibenzilic* rearrangements.

C. Benzidine Rearrangement

Benzidine rearrangement, which includes the whole family of acid-catalyzed rearrangements of aromatic hydrazo compounds, is one of the most thoroughly investigated but still one of the most baffling mechanistic problems in organic chemistry. The name of the rearrangement is derived from the major product, 4,4'-diaminobiphenyl (benzidine, **69**) which is obtained by treating hydrazobenzene (**68**) with a strong acid such as hydrochloric or sulphuric acid. The other product of the rearrangement is the isomeric 2,4'-diaminobiphenyl (diphenyline, **70**).

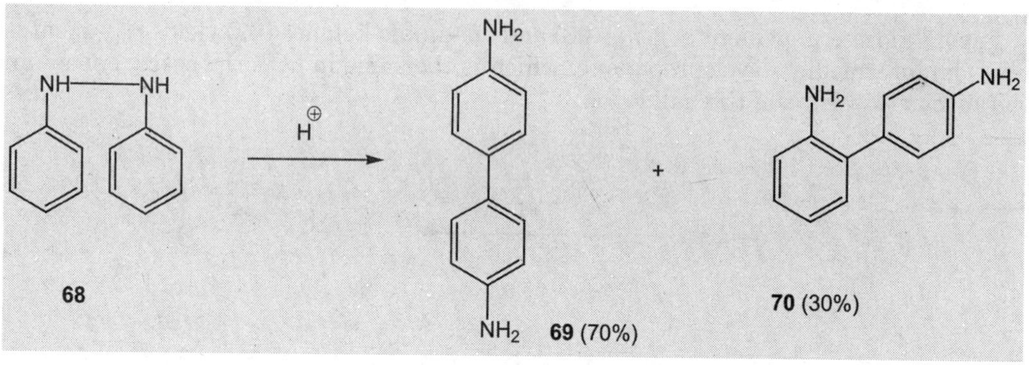

68 **69** (70%) **70** (30%)

In addition to these two products, three more substances, *ortho* benzidine (**71**), *ortho* semidine (**72**) and *para* semidine (**73**) may also be formed. Not all these products may arise in any one example, but under certain conditions, like treating solid **68** with dry hydrogen chloride, all the possible products are obtained.

The rearrangement of hydrazonaphthalenes, however, yields the *ortho* benzidine type compound as the major product.

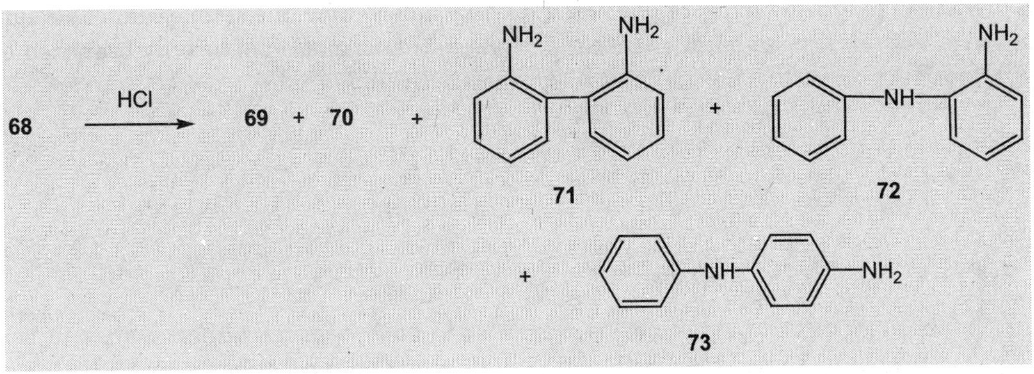

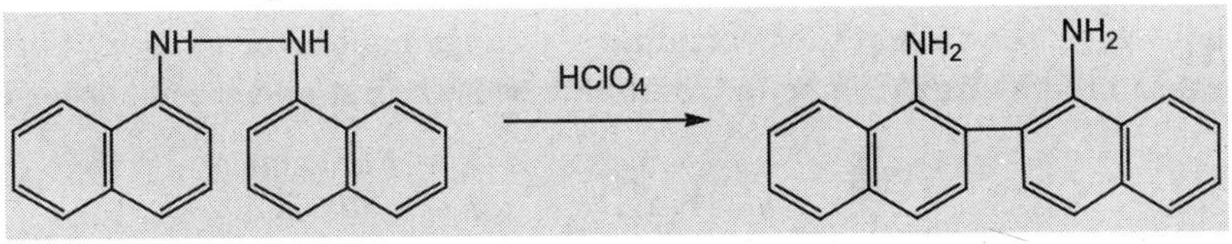

The question of the mechanism of the benizidine rearrangement has attracted much attention and several divergent mechanisms have been proposed. The *intramolecularity* of the reaction has been established by a number of experiments. Rearrangement of a mixture of 2,2'-dimethoxyhydrazobenzene (**74**) and 2,2'-diethoxyhydrazobenzidine (**75**) yields 3,3'-dimethoxy (**76**) and 3, 3'-diethoxybenzidine (**77**). No trace of cross product, *i.e.* 3-ethoxy-3'-methoxybenzidine (**78**) was found in the product mixture.

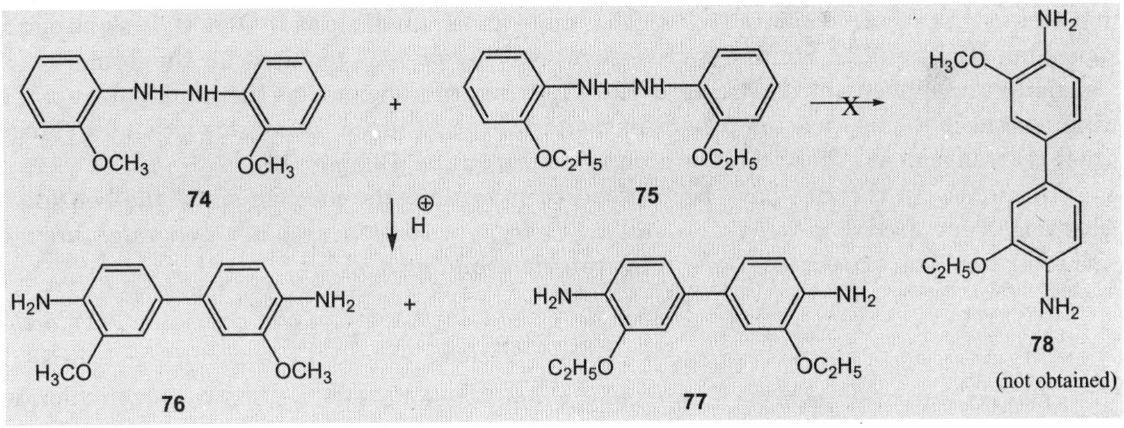

In a more conclusive experiment, a mixture of 2,2'-dimethylhydrazobenzene (**79**) and 2-methyl-^{14}C-hydrazobenzene (**80**) was rearranged. Again, no trace of the cross product, **81**, was found in the reaction mixture.

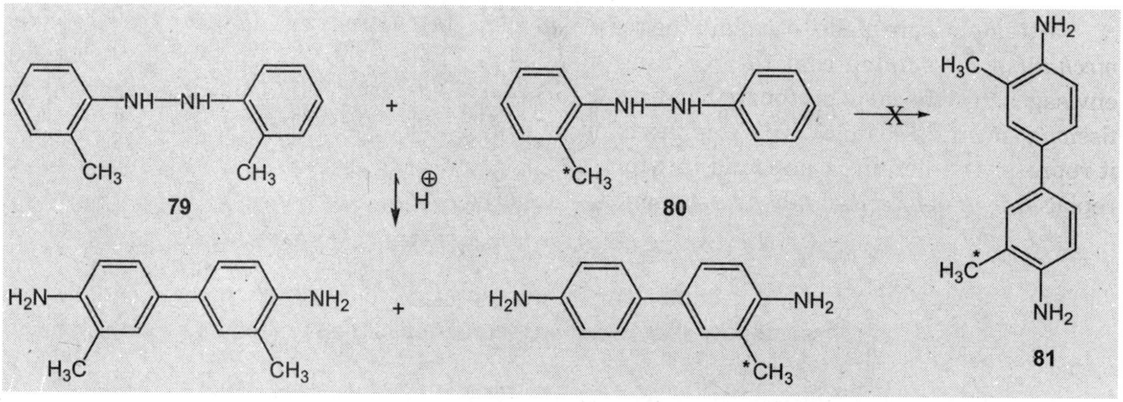

Further studies have shown that the benizidine rearrangement involves *specific acid catalysis*. It is first order in the hydrazo compound, and may be both first and second order in the acid.

$$\frac{-d\,[\mathrm{Hyd}]}{dt} = k_2\,[\mathrm{Hyd}]\,[\mathrm{H}^+] + k_3\,[\mathrm{Hyd}]\,[\mathrm{H}^+]^2$$

A number of *ortho* hydrazo compounds such as hydrazotoluene rearrange under a *mixed* first and second order catalysis; the overall result then is near the mean of 1.5. That the protonation at one or both of the nitrogen atoms is not a slow, rate-determining step has been proved by comparing the rate of rearrangements in H_2O (where the acid is H_3O^+), and D_2O (where the acid is D_3O^+). Had the protonation been the rate-determining step of the rearrangement, there should have been retardation in the rate of reaction in D_2O. Actually, there was a twofold increase in the rate of reaction in D_2O which is quite understandable as D_3O^+ in D_2O is a stronger acid than H_3O^+ in H_2O. Similarly, the loss of two ring protons required for the formation of benzidine and diphenyline is also not a slow, rate-determining step as the replacement of the ring protons in the hydrazo compounds by deuterium or tritium does not slow down the rate of their rearrangement (*cf* electrophilic aromatic substitution, Chapter 10).

A number of theories have been proposed to explain the mechanism of the benzidine rearrangement. According to the *caged radical theory*, a two proton mechanism operates through the formation of *two cation radicals* in the rate-determining step.

$$\mathrm{Ar}\overset{+}{\mathrm{N}}\mathrm{H_2}\!-\!\overset{+}{\mathrm{N}}\mathrm{H_2Ar} \longrightarrow \mathrm{Ar}\overset{+\bullet}{\mathrm{N}}\mathrm{H_2} \;/\; \mathrm{H_2}\overset{+\bullet}{\mathrm{N}}\mathrm{Ar}$$

Each of the cation radicals is resonance-stabilized and is prevented from diffusing away from the other by a cage of solvent molecules. All types of rearrangement products are formed by a recombination of fragments within the solvent cage. However, the theory appears less likely as no radical could be detected by ESR spectroscopy in the rearrangement of hydrazobenzene. Further, in one-proton rearrangements, the homolytic dissociation of the monocation into the unsymmetrical radical ion $\left(\mathrm{Ar}\overset{+\bullet}{\mathrm{N}}\mathrm{H_2} \;/\; \mathrm{H}\overset{\bullet}{\mathrm{N}}\mathrm{Ar}\right)$ is difficult to visualize in the *absence* of repulsion between like charges that operates in the dication.

Attempts were made to explain the rearrangement through what is known as the *π-complex mechanism*. Assuming that the rearrangement is kinetically first order in acid, the theory envisages that the monoprotonated hydrazo compound undergoes a rate-determining heterolytic fission to form a π-complex. This complex **82** is not to be confused with the *transition state*, rather it represents a definite compound in which the covalent bond between the nitrogen atoms is replaced by a delocalized π-bond between the aromatic rings.

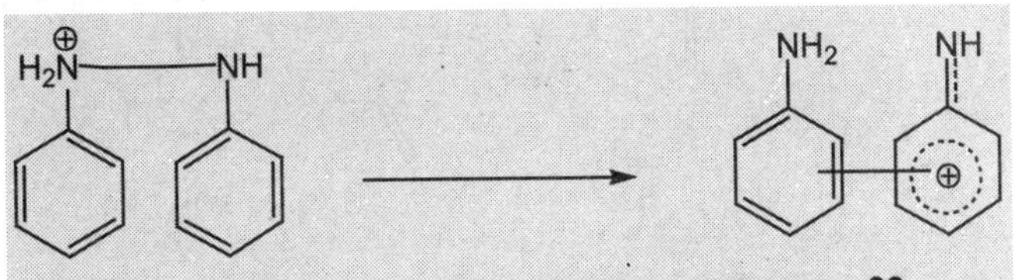

82

The aromatic rings in **82** are held in parallel planes by the π-bond and they can rotate and collapse to form the various products. With the discovery of second order acid kinetics it became necessary to modify the original theory and it was then proposed that the attack of second proton on **82** was rate-determining.

By far the most satisfactory theory to explain the rearrangement is due to Ingold and is known as the *polar transition state theory*. This theory stresses the development of a polar transition state in the rate-determining step of the rearrangement. It is illustrated here by a consideration of the rearrangement of **83** *via* a two-proton mechanism. Since there are two adjacent positively charged nitrogen atoms in the diprotonated molecule of **83**, repulsion between them facilitates the cleavage of the N–N bond. One of the loosely bound aromatic rings formed as a result of heterolysis carries two units of positive charge and thus behaves like a dication, whereas the second ring which does not possess any charge has the character of an arylamine and will have *fractional dipolar charge* distributed in the same manner as in an aniline molecule.

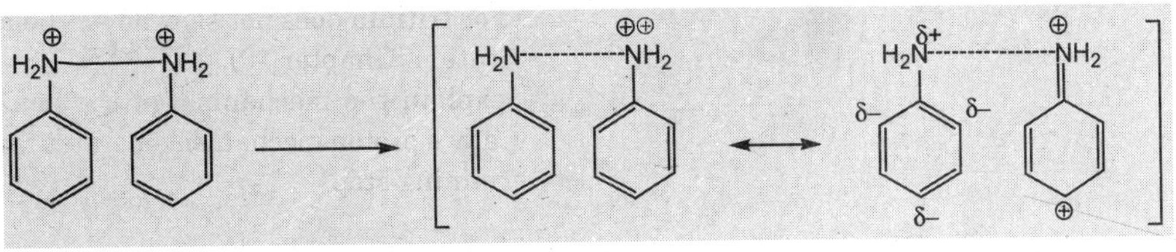

It becomes apparent immediately from an inspection of the charge distribution in the *transition state* that the bond formation can only take place at 4,4'-and 2,4'-positions. Thus the rearrangement of **83** involves a transition state in which the heterolysis of the N–N bond is concerted with the formation of new bonds at those positions which are most favourable in the transition state. Based on these assumptions a two-proton mechanism for the formation of **84** and **85** from the rearrangement of **83** can be written as follows.

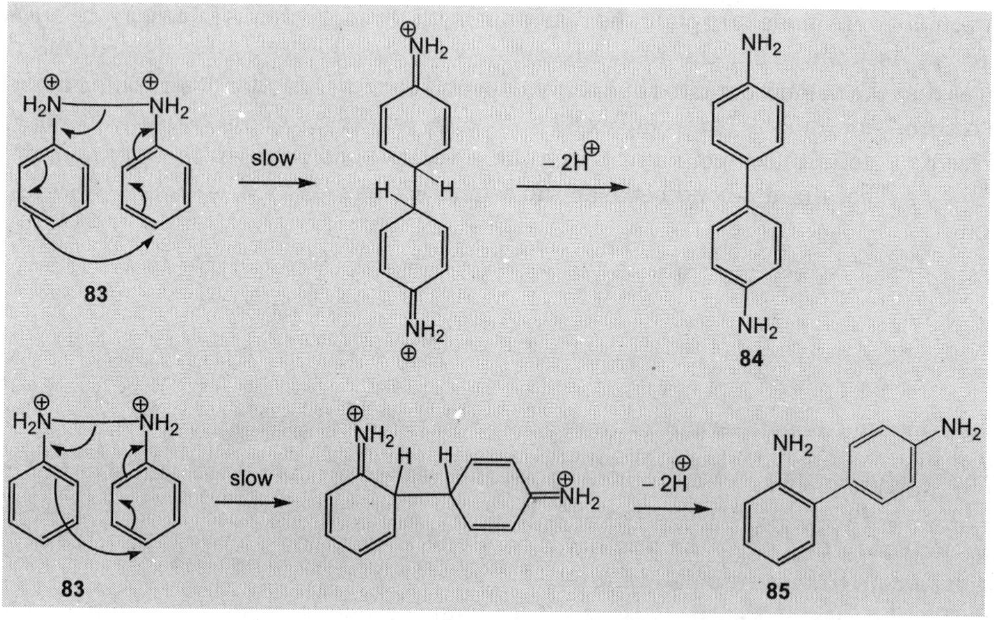

D. Claisen Rearrangement

Claisen discovered that when a simple or substituted allyl ether of a phenol is heated at about 200°C, a fascinating isomerization takes place in which the allyl group migrates from the ether oxygen atom to a carbon atom of the aromatic ring. The migrant allyl group moves to an *ortho* position when such a position is unsubsubstituted or to the *para* position if both the *ortho* positions are blocked. Migration to *a meta* position has not been observed.

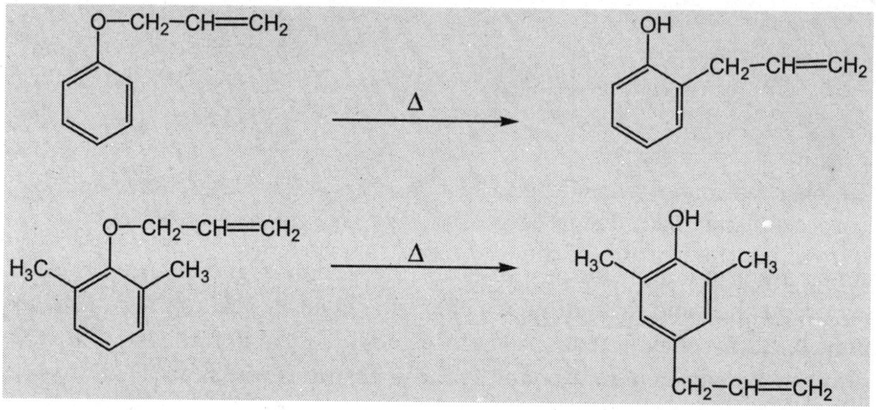

The reaction does not require a catalyst and follows first order kinetics thereby showing that each individual act of rearrangement involves only one molecule. Moreover, heating a mixture of two different ethers, such as **86** and **87**, in which the allylic groups are different, does not lead to the formation of *cross products*, **88** and **89**.

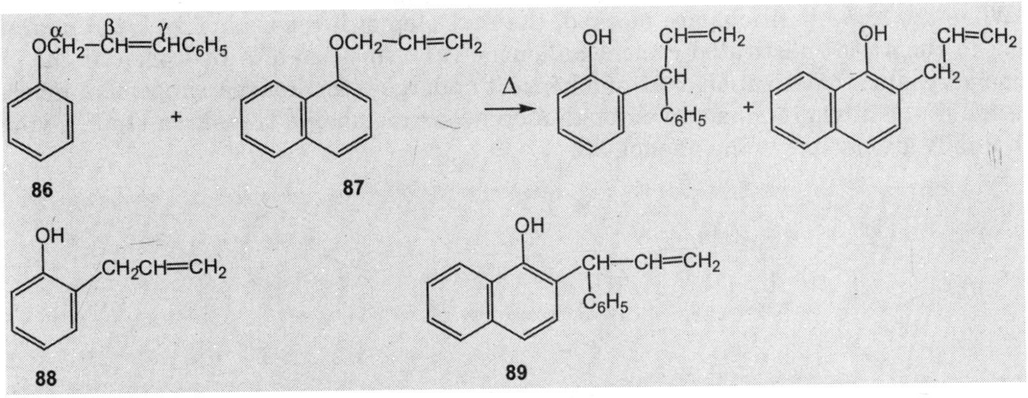

The rearrangement, therefore, is an *intramolecular,* one-step process. Another important feature of the rearrangement is that when the migration takes place to the *ortho* position, the carbon atom by which the migrating allyl group gets attached to the ring carbon is γ to the oxygen atom in the original ether. In other words, *ortho* isomerization is accompanied by an inversion of the position of substituent with respect to that of the starting compound. However, no such inversion is observed in *para* isomerization.

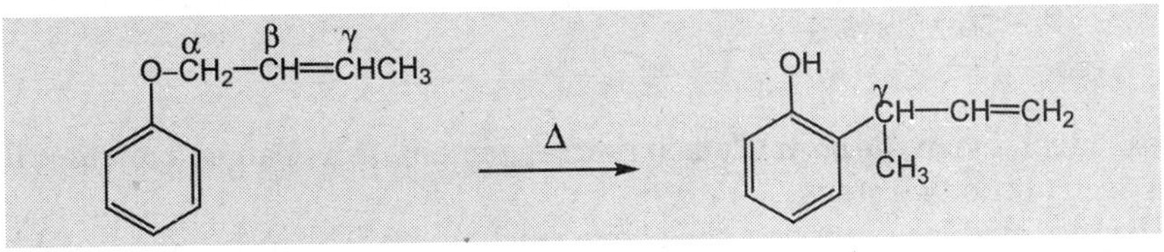

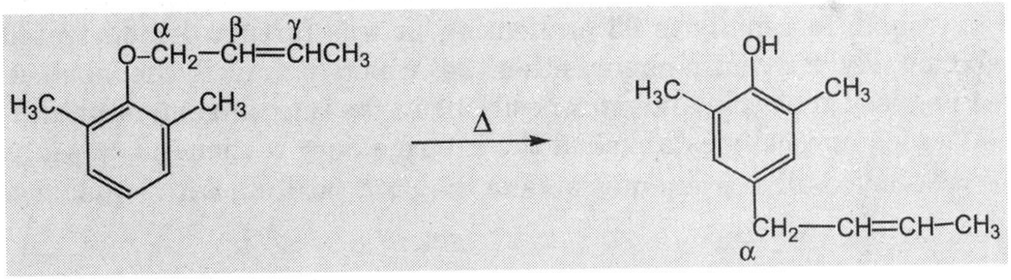

All these observations if considered together indicate that the *ortho* rearrangement is a *concerted* process. The reaction proceeds through a cyclic six-membered transition state (Chapter 15) in which the rupture of the oxygen-allyl bond is synchronous with the formation of a carbon-carbon bond at an *ortho* position. A cyclohexadienone intermediate (**90**) is thus formed which subsequently converts itself to the more stable aromatic compound, phenol, by a prototropic change.

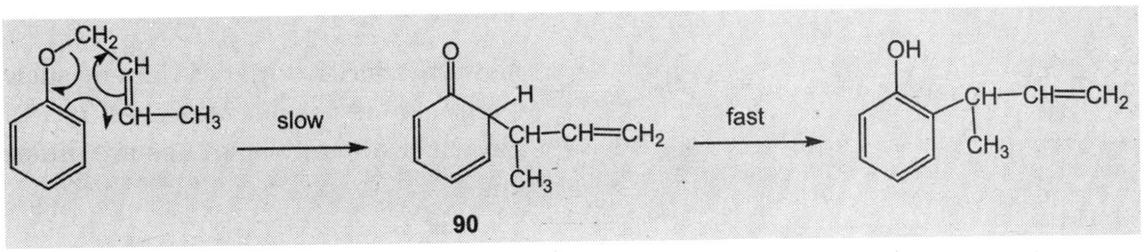

When the *ortho* positions are blocked, the first step still consists of an *ortho* migration leading to the *ortho*-substituted cyclohexadienone (**91**). Absence of a hydrogen at the *ortho* position prevents aromatization of **91** and hence it undergoes further rearrangement involving migration of the allyl group, again through a cyclic six-membered transition state, giving **92** which finally aromatizes to the phenol, **93**.

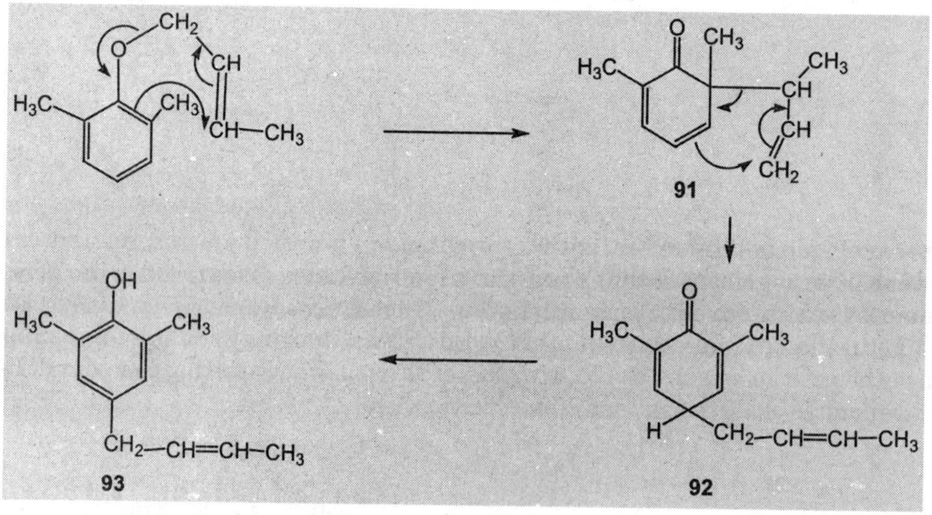

Absence of inversion in the *para* Claisen rearrangement, thus, can be explained by assuming inversions in each of the two steps.

A large body of evidence supports the above mechanism. The intermediate **91** has not only been trapped by means of a Diels-Alder reaction but has also been synthesized independently. It was found to transform rapidly to **93** on heating. It was further demonstrated that when *ortho*-α-^{14}C-labelled allyl 2,6-diallylphenyl ether (**94**) was rearranged, the labelled carbon was equally divided between the *ortho* and *para* positions in the rearrangement product (**96a/96b**). This experiment unequiovocally established the intermediacy of dienone (**95**) in which either the labelled or unlabelled allyl group migrates to the *para* position with equal ease.

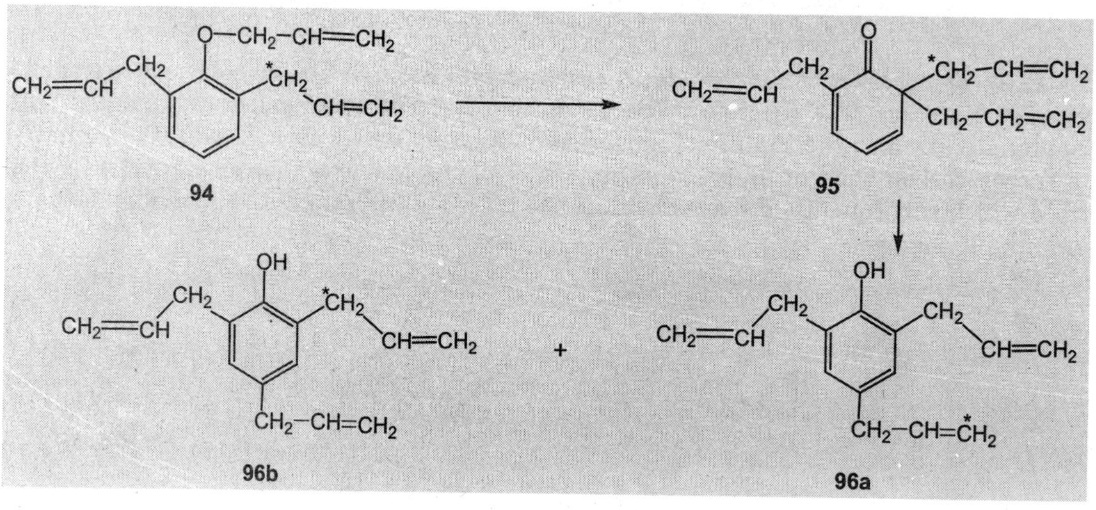

Reversibility of the cyclic mechanism operating in the Claisen rearrangement has been most elegantly demonstrated by heating labelled allyl 2,6-dimethyl-4-allylphenyl ether which resulted in the distribution of ^{14}C between its original position and the terminal position of the O-allyl group.

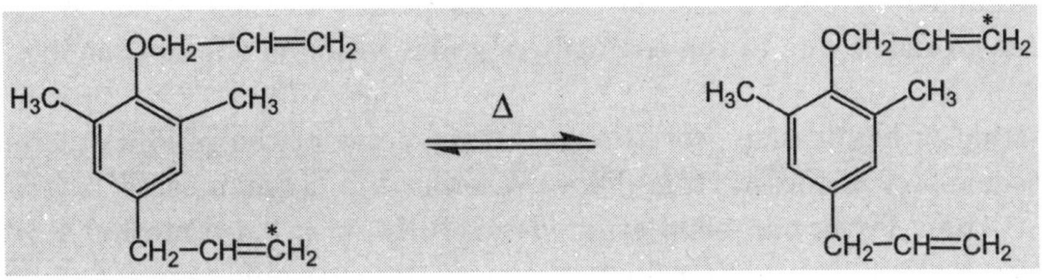

In accordance with the prediction of a cyclic mechanism, Claisen rearrangement does not greatly depend on the presence or absence of substituent groups on the ring. Claisen rearrangement is not confined to only aromatic systems. It can be generalized in the following way:

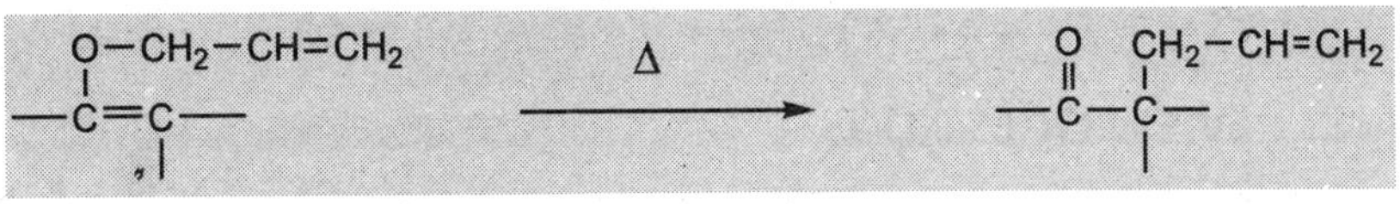

E. Cope Rearrangement

The Claisen rearrangement, as we have just seen, proceeds through a six-membered cyclic transition state consisting of one oxygen and five carbon atoms. When the cyclic transition state consists of *six carbon atoms*, the process is referred to as Cope rearrangement. The thermal conversion of 3,4-dimethyl-1,5-hexadiene to 1,6-dimethyl-1,5-hexadiene is a typical example. Substitution by phenyl or carbethoxy groups at 3- or 4-positions facilitates the reaction.

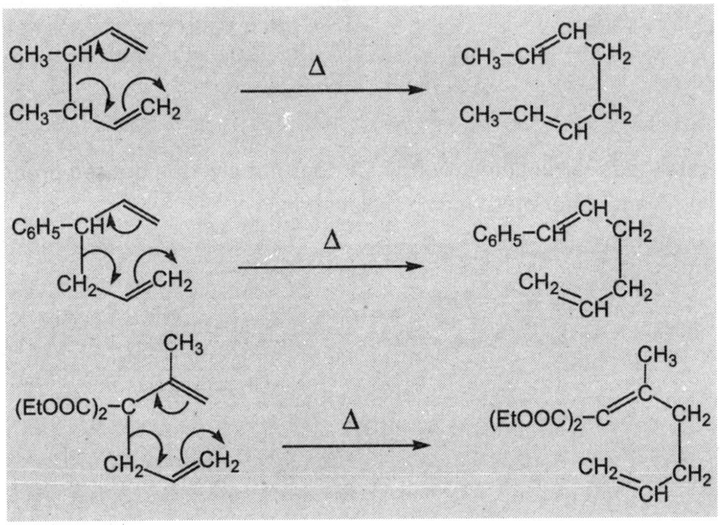

The Cope rearrangement is a reversible process; the equilibrium mixture has a great proportion of thermodynamically stable isomer. It follows first order kinetics. Like Claisen rearrangement, it is an intramolecular process because rearrangement of a mixture of two different 1,5-hexadienes does not lead to the formation of *cross products*. The reaction proceeds through a six-membered cyclic transition state and the acceleration by substitutents such as phenyl or ester groups is due to the conjugative stabilization of the new double bond in the product diene.

The reaction is highly *stereospecific* as the geometry of the product diene is dictated by the stereochemistry of the starting diene. Whether the six-membered transition state is in the boat or chair form, has been studied carefully with 3,4-dimethyl-1,5-hexadienes. If the transition state is in the chair form, the *meso* isomer would give the *cis, trans* isomer while a boat-type transition state would give either the *cis, cis* or the *trans, trans* product. The exclusive formation of *cis, trans* product establishes that a chair-type transition state is highly favoured over a boat-type. The above conclusion is further supported by the predominant formation of the *trans, trans* isomer from a racemic mixture of 3,4-dimethyl-1,5-hexadiene (Chapter 15).

PROBLEMS AND REVIEW QUESTIONS

1. A compound containing C, H and O displays a weak molecular ion at m/z 88. The NMR spectrum shows signals at about δ 1.0, 3.5 and 5.2 in the ratio of 9 : 2 : 1. On heating with conc. H_2SO_4, an alkene, showing molecular ion at m/z 70 is obtained, which on hydrogenation yields 2-methylbutane. Deduce the structure of the unknown compound and propose a mechanism for the transformation.
2. Draw the structure of the products and depict a reasonable mechanism for each of the following reactions.

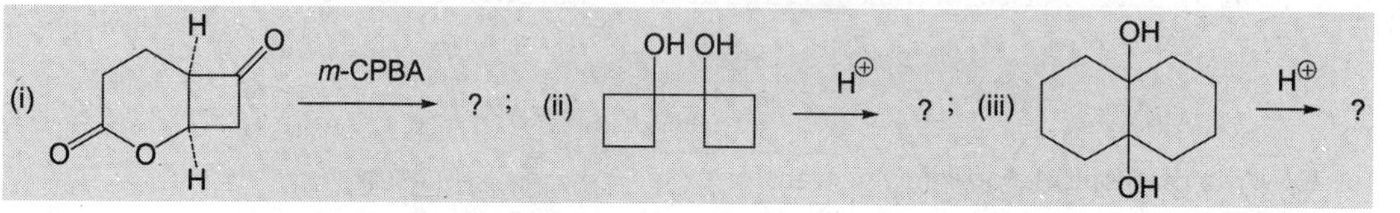

3. The acid-catalyzed rearrangement of diol **A** does not give the desired product **B**. How will you solve this problem? Offer mechanistic explanation.

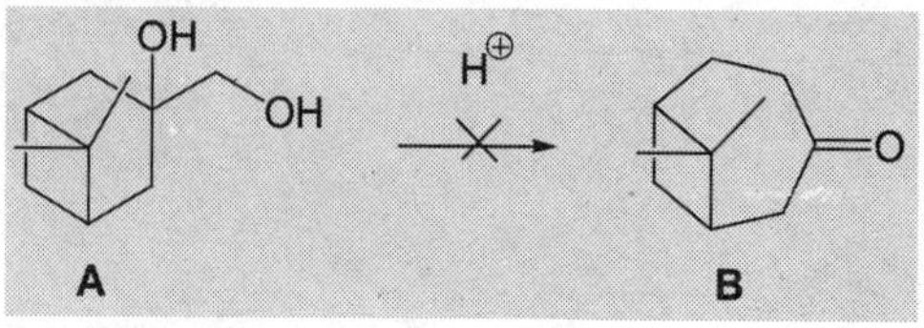

4. Sketch mechanisms for the following transformations:

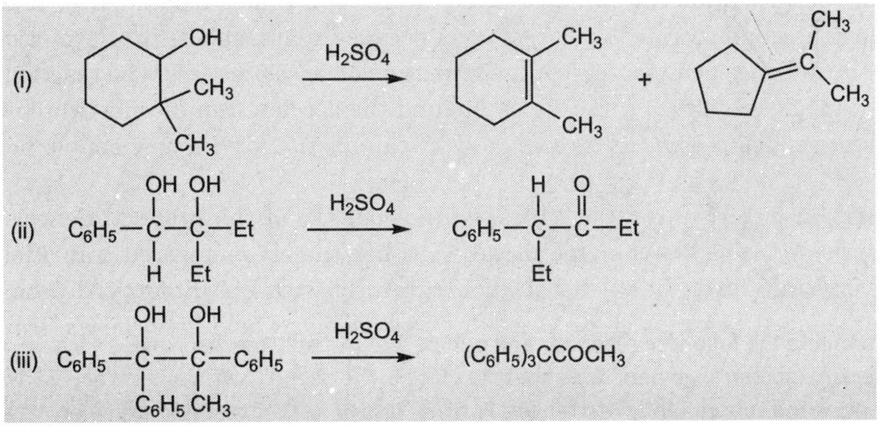

5. Treatment of 2,2,6,6-tetramethylcyclohexanol with acid leads to the formation of an alkene after rearrangement. Assign structure to the alkene and sketch mechanism leading to its formation.
6. A compound having C, H and O displayed weak molecular ion at m/z 114. Treatment with acid yielded an alkene showing molecular ion at m/z 96. Its NMR spectrum showed an olefinic proton, a methyl on double bond and 8 methylene protons. Predict the structure of the alkene and formulate a mechanism for this transformation.
7. When 1,3-propanediol is dissolved in SbF_5-SO_2, the NMR spectrum shows the presence of only one substance, propionaldehyde. How do you explain this observation?
8. Diol **A** undergoes rearrangement to ketones **B** and **C** on treatment with acid. Explain this observation giving mechanism.

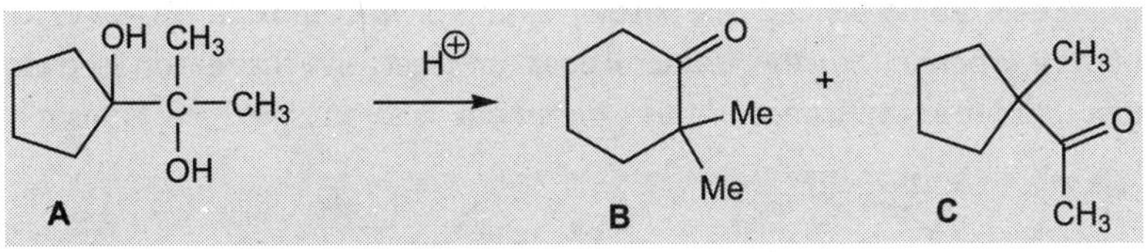

9. Write mechanisms for following transformations:

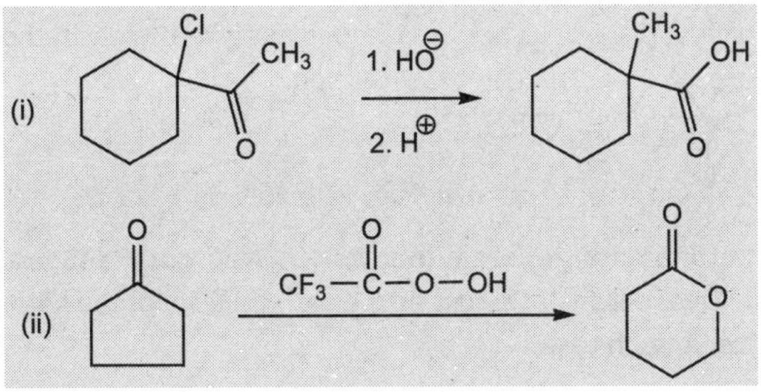

10. Formulate a reasonable mechanism for acid-catalyzed decomposition of triphenylmethyl hydroperoxide (Ph_3C–O–OH) to benzophenone and phenol.

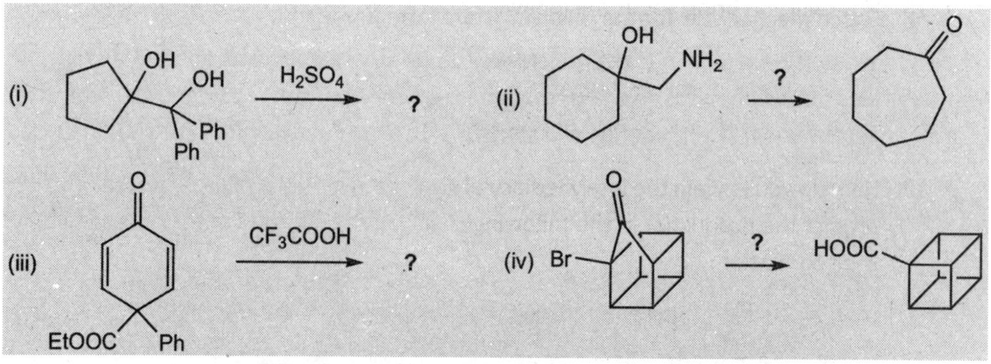

11. Complete the following reactions and outline a reasonable mechanism for each one of them.
12. Beckmann rearrangement does not take place in the oxime itself, but in its acyl derivative which ionizes and subsequently rearranges. Further, rate of isomerization of oxime esters are affected by the nature of esterifying acids. Tell what will be order of rates of isomerization on using the acids: $C_6H_5SO_2OH$, C_6H_5COOH, CH_3COOH.
13. Acid-catalyzed rearrangement of two isomeric diols **A** and **B** ($C_{10}H_{18}O_2$) gives the same spiroketone **C**. What are the structures of **A** and **B**? Outline the mechanism for their rearrangement to **C**.

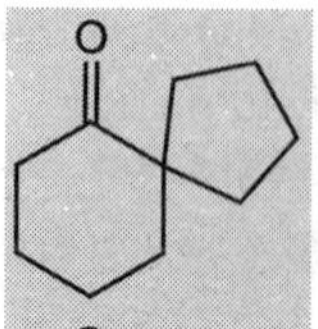

C

14. One isomer of 3-bromo-decal-2-one (**A**) on treatment with ethoxide ions leads to the formation of the product **P**, whereas the other isomer **B** does not react. Give the reason for this observation. Also comment upon the significance of this experiment in establishing mechanism of the reaction.

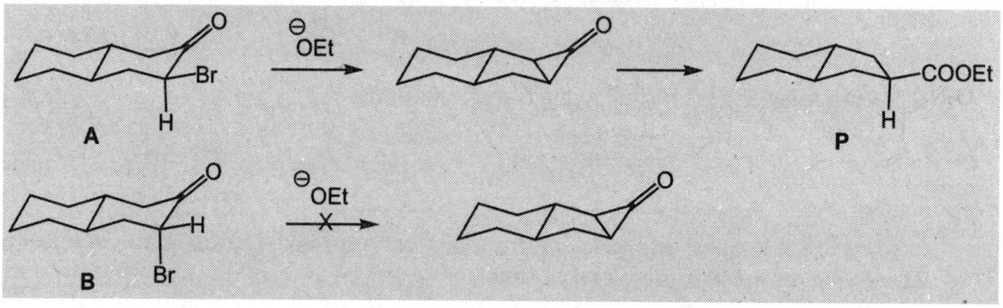

15. Unsaturated ketones are generally not considered good substrates for Baeyer-Villiger rearrangement. Give reasons ? Discuss how electrophilicity of the ketone and nucleophilicity of the alkene matter in such cases.

16. How will you bring about the following conversion? Outline the mechanism.

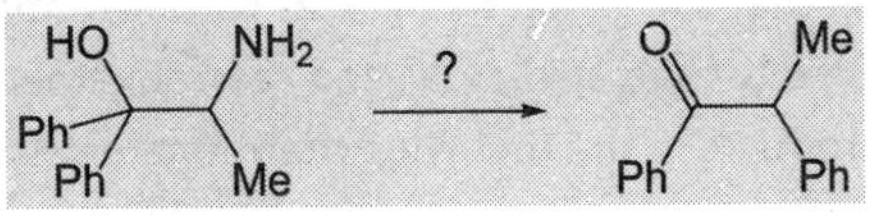

17. How do you explain the intermediacy of carbocations in the pinacol-pinacolone rearrangement? Predict the product(s) in the following rearrangements and suggest a plausible mechanism for their formation.

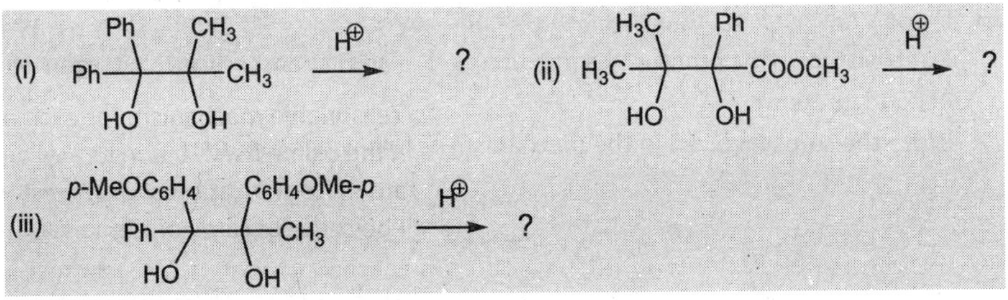

18. Write structure of the products (**A, B, C**) in the given reaction sequence and offer mechanistic explanation for their formation.

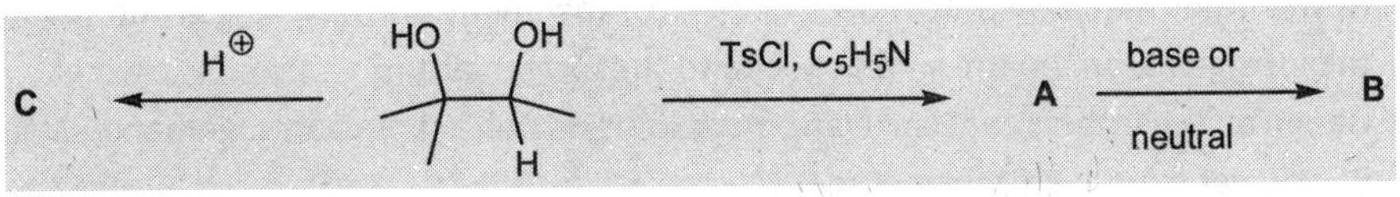

19. How can you convert cyclohexanone oxime into ω-caprolactam, which is used for the manufacture of Nylon-6? Outline the mechanism.
20. Compound **A** on Beckmann rearrangement gives anilide **B**. On the other hand, there is formation of benzisoxazole derivative **C** when **A** is treated with cold alkali. Comment on the significance of this experiment in establishing mechanism of Beckmann rearrangement.

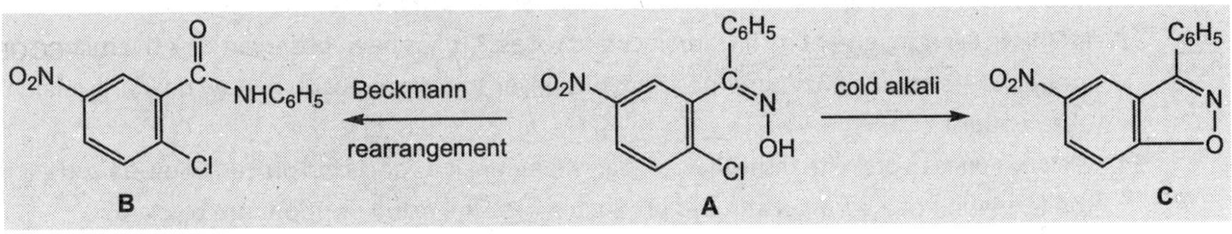

21. Giving mechanism, discuss row Favorskii rearrangement of *gem*- and *vic*-dihalo ketones leads to the formation of α,β-unsaturated esters.
22. How do you explain the formation of three products- cyclobutylcarbinol, cyclopentanol and methylenecyclobutane on the treatment of cyclobutylmethylamine with nitrous acid?
23. Is there inversion at the migration origin in a 1,2-shift? Explain this by considering the conversion of camphene hydrochloride to isobornyl chloride.

24. Discuss the mechanism of Wolff rearrangement. Demonstrate its utility in Arndt-Eistert homologation showing that the group R migrates with retention of configuration.
25. What is the driving force for the acid-catalyzed rearrangement of carbonyl compounds? Write the structure of the products in the following rearrangements:

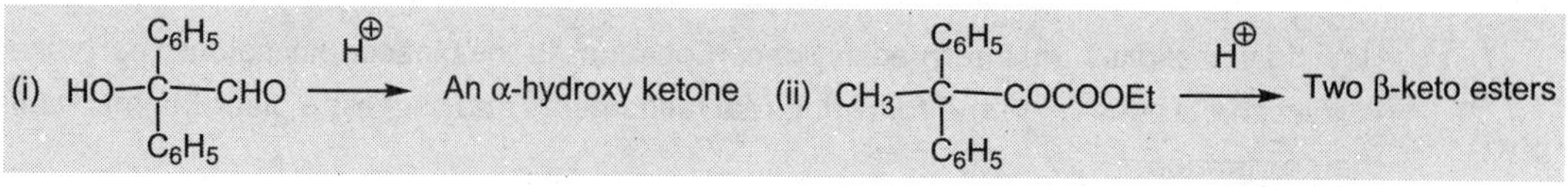

26. Discuss mechanism of Curtius, Schmidt and Lossen rearrangements. How will you establish that in the process, the migrating group shifts from carbon to nitrogen with the retention of configuration?
27. Write the steps involved in the mechanism of following reaction:

$$RCONH_2 \xrightarrow[H_2O]{Br_2/NaOH} RNH_2$$

How will you establish that in such a transformation there is retention of configuration at the migrating atom? Is it an inter- or intramolecular process?

28. How will you establish that in Beckmann rearrangement the migrating group always approaches the nitrogen atom on the side opposite to the oxygen atom?
29. Design an experiment that correlates Hofmann and Beckmann rearrangements. What information do you get from such an experiment?
30. Rearrangement of phenolic esters in presence of aluminium chloride yields *ortho* or *para*-hydroxy ketones. What is the mechanism of this reaction? Is it a one-step or two-step process?
31. How will you establish the intermediacy of a cyclopropanone in Favorskii rearrangement? What will be the direction of opening of such an intermediate in case it is unsymmetrically substituted?
32. How will you prove that benzidine rearrangement is an intramolecular process? Discuss its mechanism assuming the development of a polar transition state.
33. Is Claisen rearrangement an intramolecular process? It has been observed that the migrant allyl group moves to *ortho* (or *para*) position, but never to a *meta* position. How do you explain this observation?
34. Give evidence to prove that the Claisen rearrangement is a concerted process. How do you explain the formation of a *para*-substituted product in case the *ortho* positions are blocked?
35. Discuss the mechanism of Cope rearrangement. Is Cope rearrangement an intramolecular process?
36. Discuss mechanisms of the following transformations:

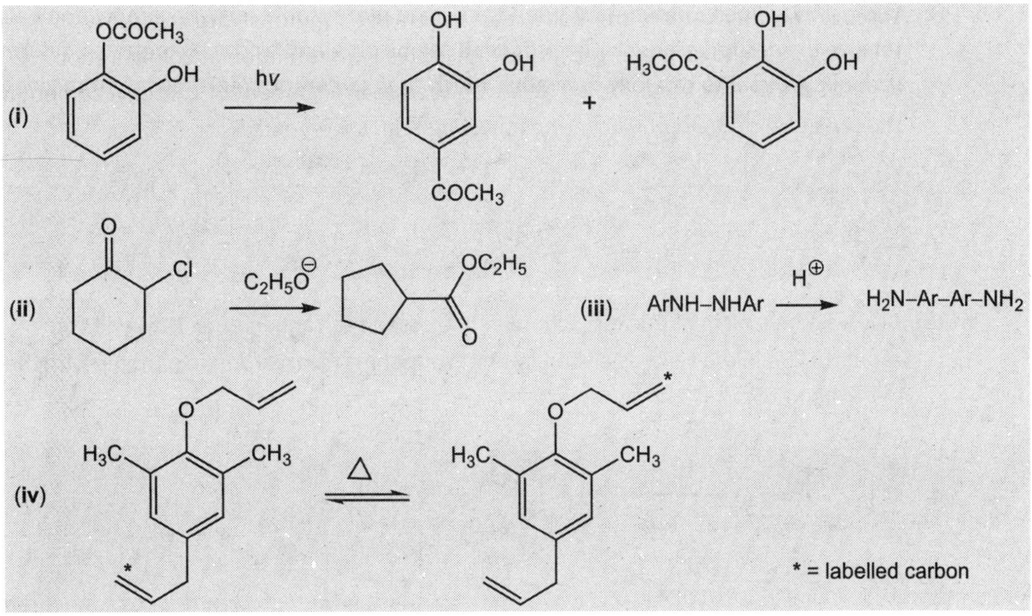

SOLUTIONS

1. Original compound, $(CH_3)_3CCH_2OH$; alkene, $(CH_3)_2C{=}CHCH_3$. The mechanism involves a methyl shift to generate a more stable carbocation followed by the loss of a proton.

2. 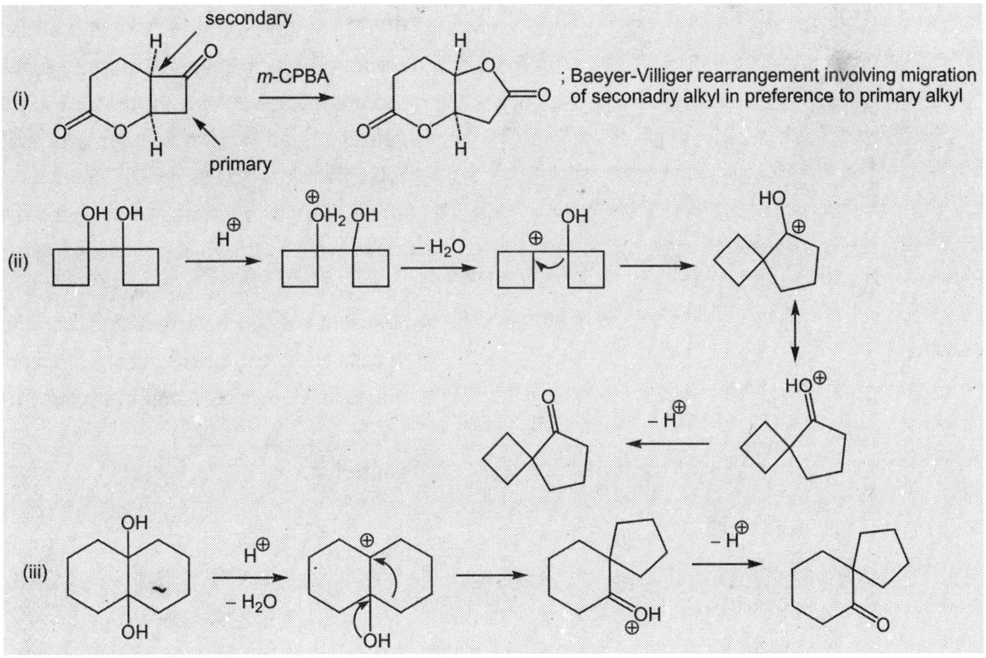

3. Acid-catalyzed dehydration of **A** would generate more stable tertiary carbocation which would not give the desired product. The product **B** can be obtained by converting less hindered primary alcoholic group into tosyloxy derivative which then undergoes rearrangement to give **B**.

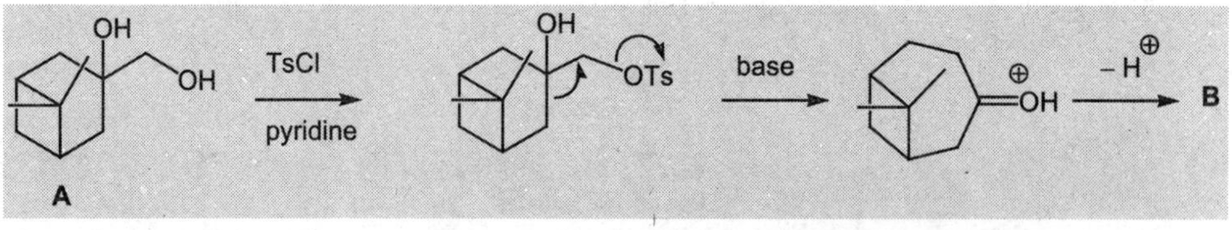

4. (i)

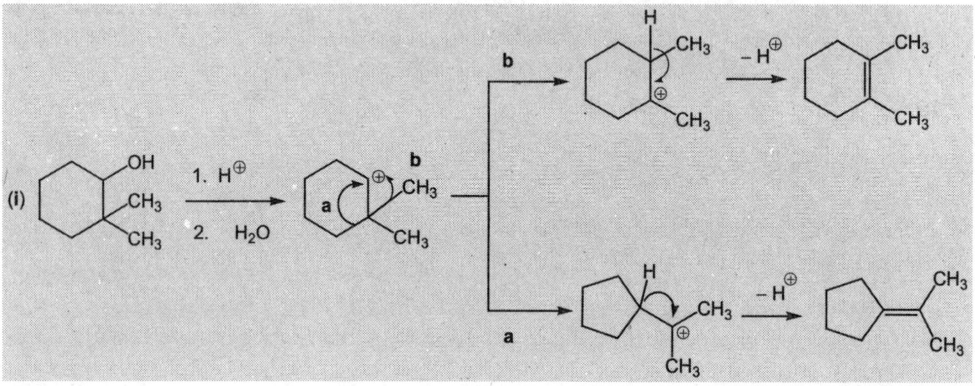

(ii) and (iii) are examples of pinacol-pinacolone rearrangements involving more stable carbocations formed by ethyl shift in (ii) and phenyl shift in (iii).

5.

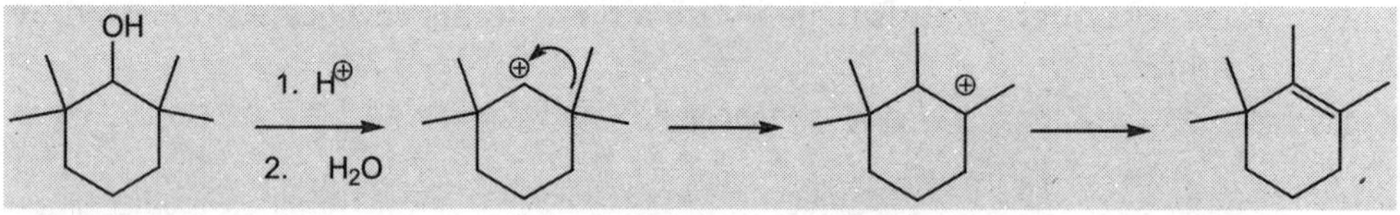

6.

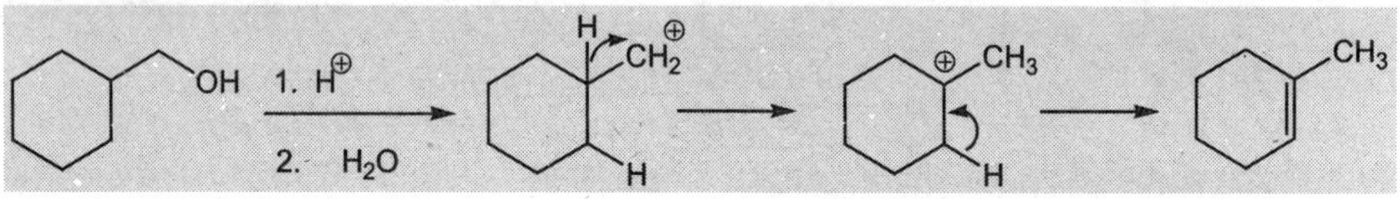

7.

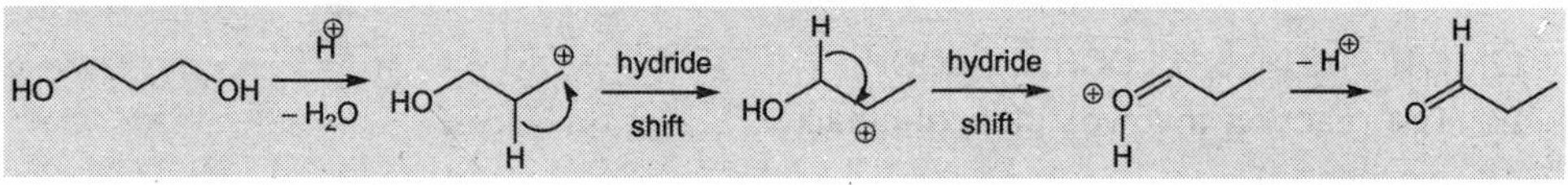

8.

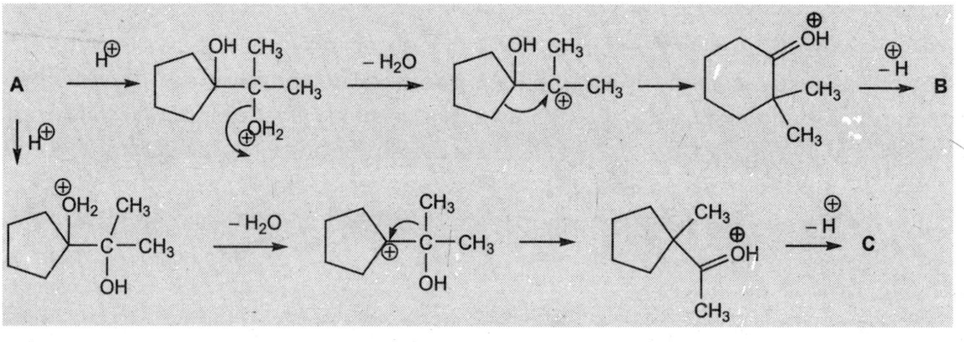

9. (i)

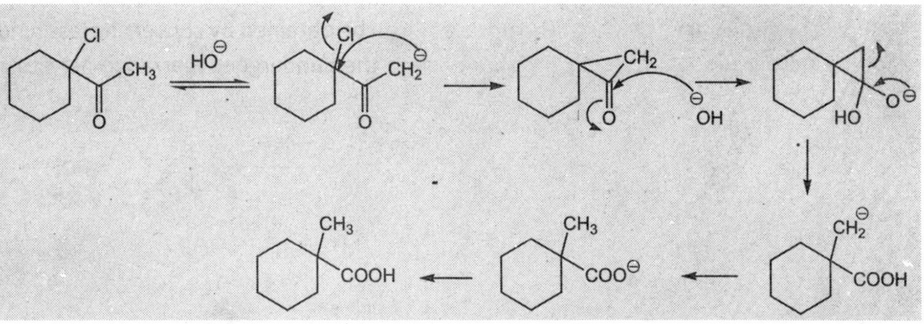

10.

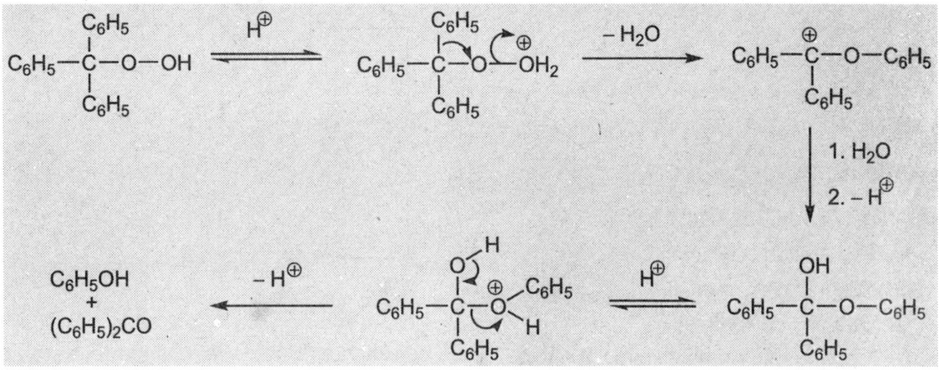

11. (i)

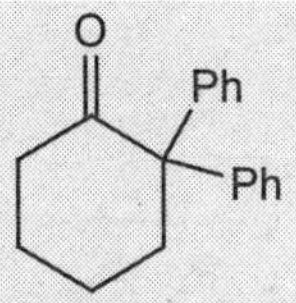

(Pinacol rearrangement) (ii) $NaNO_2$, HCl

(iii)

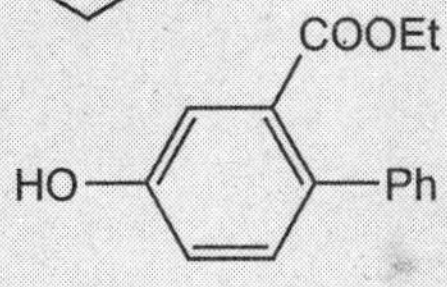

(Dienone-phenol rearrangement)

The COOEt migartes in preference to Ph because it destablizes the cation generated by migration of Ph group.

(iv) KOH (Favorskii rearrangement)

12. $C_6H_5SO_2OH$, C_6H_5COOH, CH_3COOH.
13. The reaction involves pinacol-pinacolone rearrangement. See **solutions 2(ii)** and **2(iii)** for mechanism.

Chapter 13

Free Radicals

Chapter Outline

13.1 SHORT-LIVED RADICALS
- A. Production of Radicals
- B. Detection of Radicals
- C. Stability of Radicals
- D. Configuration of Radicals
- E. Reactions of Radicals

13.2 SOME SYNTHETICALLY USEFUL REACTIONS
- A. Coupling of Alkynes
- B. Arylation of Aromatic Rings
- C. Sandmeyer Reaction
- D. Hunsdiecker Reaction

13.3 STABLE FREE RADICALS

13.4 BIRADICALS

In addition to heterolytic cleavage of a covalent bond which produces carbocations and carbanions, rupture of the covalent bond may proceed by another way, *homolytic cleavage*, in which both the atoms possess unpaired electrons due to the symmetrical rupture of the bond.

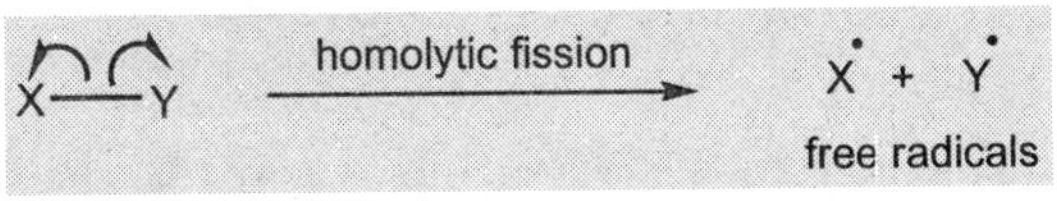

These fragments, which are neutral species, are called *free radicals*. We thus define a free radical as an atom or group of atoms with an unpaired electron. Free radical intermediates are often involved in the reactions carried out at high temperatures or under the influence of light. Even at room temperature many reactions proceed by a free radical mechanism, particularly in nonpolar solvents, and in the presence of substances, known to produce radicals, such as peroxides. Most common radicals are extremely reactive species and are difficult to isolate. Such radicals are referred to as *short-lived radicals* in contrast to *stable free radicals* which are not so reactive and exist in equilibrium with the normal compounds.

13.1 SHORT-LIVED RADICALS

A. Production of Radicals

Short-lived free radicals are often produced when a molecule is supplied with sufficient energy (thermal or photochemical) to cause homolysis of a covalent bond. In addition, oxidation –reduction reactions resulting in the gain or loss of a single electron can also be used for the generation of radicals.

I Thermal Reactions
II Photochemical Reactions
III Redox Reactions

I. Thermal Reactions

Historically, the first work to demonstrate the formation and existence of free radicals was that of Paneth and Hofeditz in 1929-31. Their procedure, often referred to as the *Paneth technique*, consists in passing a stream of nitrogen and the vapours of tetramethyllead, through a quartz tube (**Fig. 13.1**). The pressure in the quartz tube is held at about 2 mm by the action of a vacuum pump.

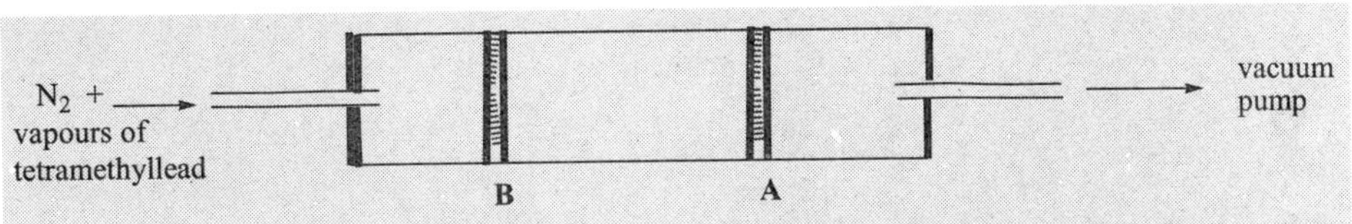

Fig. 13.1 Production of radicals (Paneth technique)

On heating the tube strongly at any point **A**, a lead mirror is deposited due to the decomposition of tetramethyllead. The gaseous product of the decomposition was found to be ethane. Subsequent heating of the tube at another point **B** on the upstream side resulted in the simultaneous disappearance and formation of mirrors at points **A** and **B**, respectively.

Tetramethyllead was detected this time in the effluent gas. This unique experiment established that tetramethyllead decomposes to form lead and methyl radicals.

$$(CH_3)_4Pb \longrightarrow 4\,\dot{C}H_3 + \underset{\text{mirror}}{Pb}$$

Methyl radicals, produced as a result of heating at point **B**, react with the mirror at point **A** to regenerate the volatile tetramethyllead.

$$4\dot{C}H_3 + Pb \longrightarrow (CH_3)_4Pb$$

In the absence of the lead mirror the free methyl radicals combine with each other to form ethane.

$$2\,\dot{C}H_3 \longrightarrow CH_3{-}CH_3$$

Further, the rate of the disappearance of the mirror at **A** was found to decrease with the increase in the distance between **A** and **B**, thereby suggesting that the methyl radicals react very rapidly to form ethane. Paneth estimated the half-life of methyl radicals to be about 0.006 sec. under the above experimental conditions.

Thermal production of the radicals is not confined to the decomposition of tetramethyllead; rather they are readily generated whenever a compound having a weak covalent bond is heated. The O—O bond of peroxide or a peracid, for example, has strength of only 36 kcal/mol or 151.2 kJ/mol and so it undergoes rupture even at room temperature.

$$C_6H_5{-}\overset{O}{\overset{\|}{C}}{-}O{-}O{-}\overset{O}{\overset{\|}{C}}{-}C_6H_5 \longrightarrow 2\,C_6H_5CO\dot{O} \longrightarrow 2\,\dot{C}_6H_5 + 2\,CO_2$$

Sometimes, the formation of radicals is facilitated by the simultaneous release of a very stable molecule such as nitrogen.

$$CH_3{-}N{=}N{-}CH_3 \xrightarrow{300^\circ C} 2\,\dot{C}H_3 + N_2$$

$$(CH_3)_3C{-}N{=}N{-}C(CH_3)_3 \xrightarrow{100^\circ C} (CH_3)_3\dot{C} + N_2$$

II. Photochemical Reactions

Absorption of visible or ultraviolet light provides a molecule with sufficient energy to break covalent bonds and thus photochemical dissociation to yield radicals may occur. Very frequently photochemical dissociation is preferred to thermolysis as the former results in more selective reactions. Irradiation of gaseous acetone, for example, yields ethane, biacetyl and carbon monoxide presumably *via* free radicals.

$$CH_3COCH_3 \xrightarrow{h\nu} \dot{C}H_3 + \dot{C}OCH_3$$

$$2\,\dot{C}H_3 \longrightarrow C_2H_6$$

$$CH_3\dot{C}O \longrightarrow \dot{C}H_3 + CO$$

$$2\,CH_3\dot{C}O \longrightarrow CH_3COCOCH_3$$

Many examples of the formation of radicals by irradiation are given in Chapter 14.

III. Redox Reactions

One-electron oxidation-reduction reactions are very often employed to produce radicals. One such reaction that has considerable synthetic importance is the Kolbe electrolysis of the salts of organic acids.

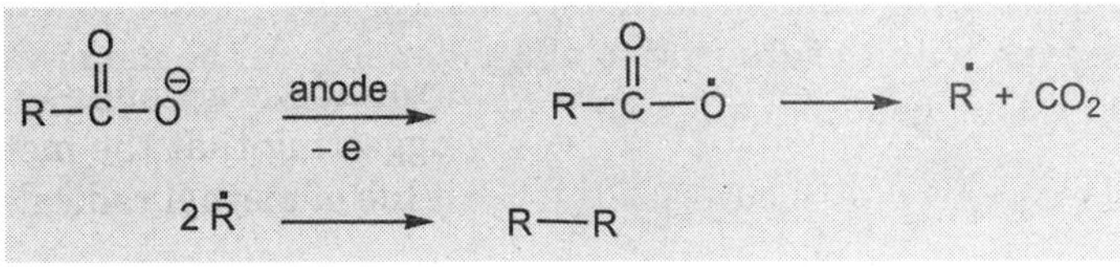

Another well-known example is that of Fenton's reagent produced by the reaction of hydrogen peroxide with ferrous ions.

$$Fe^{++} + H_2O_2 \longrightarrow Fe^{+++} + \overset{\ominus}{O}H + \dot{O}H$$

The generated hydroxyl radicals then oxidize organic substances.

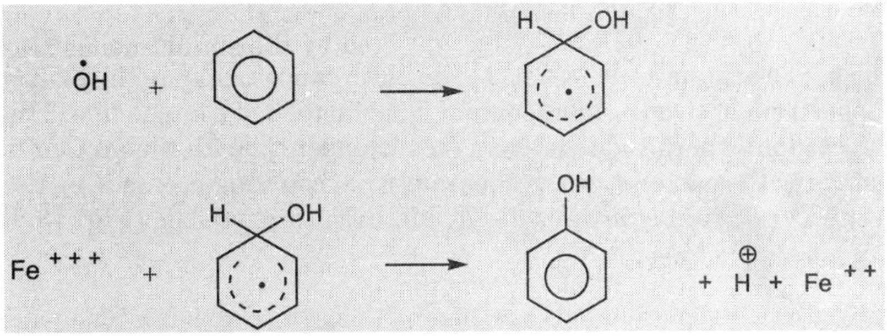

B. Detection of Radicals

In addition to the classical method based on the disappearance of lead mirror, there are now several other methods available for the detection of radicals. One of the relatively simpler methods to know whether a particular reaction proceeds *via* a radical intermediate or not is to introduce a small quantity of a *radical inhibitor*, such as hydroquinone. The kinetics of the reaction is then studied with or without inhibitor. If the reaction proceeds *via* radical intermediates, its rate should be seriously affected by the presence of inhibitors. An alternate indication of a radical mechanism is the dependence of the reaction rate on a small amount of *initiator* such as organic peroxides. There are many radical reactions that proceed only on the addition of such initiators to reaction mixture.

The direct evidence for the presence of an odd electron is usually obtained from its magnetic properties. An earlier method made use of the magnetic susceptibility of the substances that depended on the paramagnetic contribution of the odd electron. However, this method of detecting radicals has been superseded by a technique which directly measures the paramagnetism due to an unpaired electron. This method, known as *electron spin resonance* (ESR) or *electron paramagnetic resonance* (EPR) spectroscopy is applicable to the study of radicals even at concentrations as low as 10^{-8} M. The basic principle of this form of spectroscopy is that an unpaired electron, like a proton, has spin and this spin has an associated magnetic moment.

When a free radical is placed in a magnetic field and exposed to electromagnetic radiation, the magnetic moment generated by the spinning electron can be lined up *with* or *against* the external magnetic field. These two orientations define two energy states which differ in energy contents; the one aligned with the field obviously has less energy and is, therefore, more stable than the other. However, with the absorption of light energy in the microwave region, the spin state of the electron is changed from alignment with the field to against the field (**Fig. 13.2**).

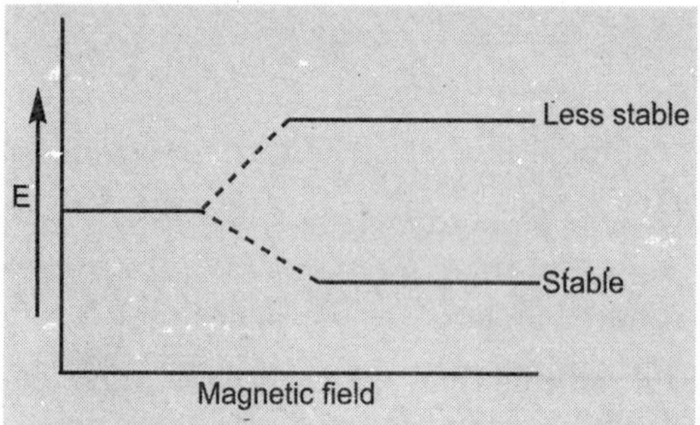

Fig. 13.2 Principle of ESR spectroscopy

Transition between stable and less stable levels produces a line in the ESR spectrum. Very often the ESR spectrum of a radical has several lines instead of a single line. This splitting of a line is due to the coupling of the odd electron, most frequently with the protons situated on the same atom that carry the odd electron. If the number of coupling protons is n then the number of lines in the ESR spectrum is always $n + 1$. Methyl radical ($\dot{C}H_3$) (**Fig. 13.3**) for instance, shows 3 + 1 = 4 resonance lines.

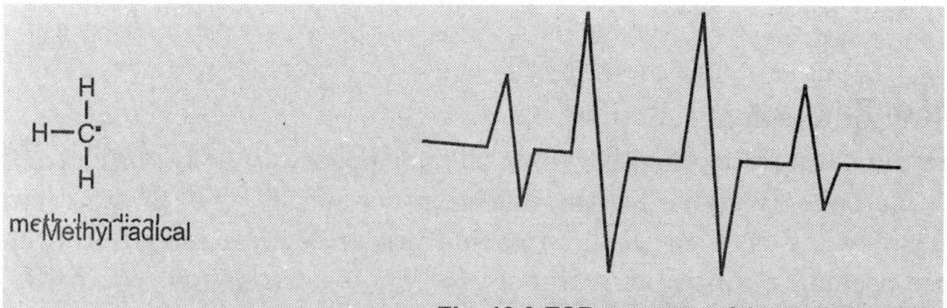

Fig. 13.3 ESR spectrum of the methyl radical

The ESR of cycloheptatrienyl radical ($\dot{C}_7H_7$) (Fig. 13.4) generated by X-irradiation of 1,3,5-cycloheptatriene shows eight equally spaced lines indicative of the coupling of the odd electron with seven equivalent protons. It thus confirms the assumption that the electron is uniformly distributed around the rings.

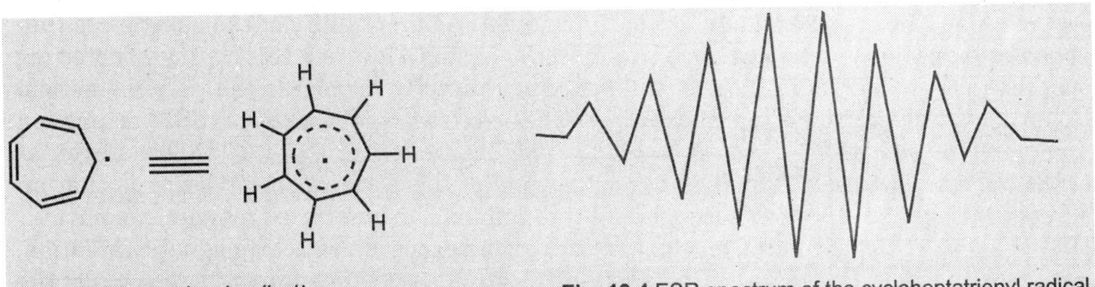

Cycloheptatrienyl radicall

Fig. 13.4 ESR spectrum of the cycloheptatrienyl radical

Thus the usefulness of ESR spectroscopy lies not only in its great sensitivity in detecting the free radicals in very low concentrations, but it also provides a wonderful method of knowing about their structures and the way the odd electron is spread over the molecule.

C. Stability of Radicals

Usually the radicals react rapidly either with themselves or with other substances. Under a given set of conditions, the rate at which radicals disappear depends on the inherent stability associated with them. A simple way to compare the stabilities of various radicals is to calculate the bond dissociation energies which may be defined as the energy required for homolytic cleavage of the covalent bond in the molecule to form two radicals.

In general, the smaller the amount of energy required for bond rupture, the more stable is the radical. Methods for determination of bond dissociation energies are to be found in physical chemistry texts. **Table 13.1** provides bond dissociation energies (D) of some compounds of interest.

However, some care is needed to derive the relative stabilities of different radicals from this **Table** precisely because of the different environment around the covalent bonds that are being broken. It is clear from the **Table 13.1** that the stability of the radicals increases in the order primary < secondary < tertiary and also with the number of the alkyl groups that are attached to the carbon atom carrying the unpaired electron. The reason for the stabilizing influence of alkyl groups is not far to seek if we consider that in going from methyl to *tert*-butyl, there is an increasing tendency for electron release by a *hyperconjugative* mechanism. The carbon atom, carrying an unpaired electron, has a half-filled orbital that gives the radical its characteristic instability; the *hyperconjugative* mechanism tends to fill this orbital and thus stabilizes it to some extent. Hyperconjugation in a *tert*-butyl radical may be represented as follows:

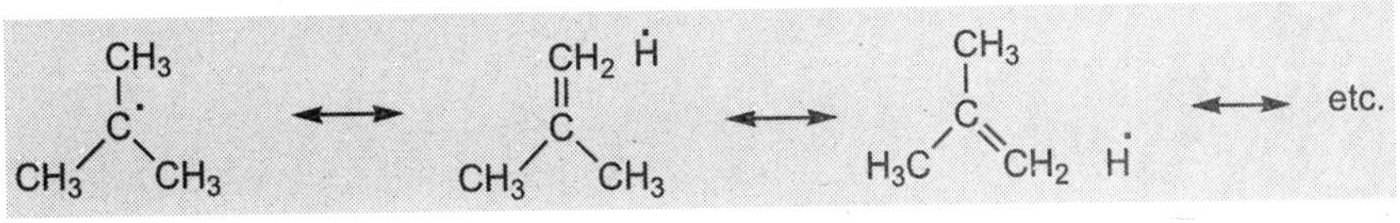

Table 13.1 Bond dissociation energies of some compounds

Bond	D kcal/mol (kJ/mol)	Bond	D kcal/mol (kJ/mol)
CH_3—H	102 (428.4)	H—H	104 (436.8)
C_2H_5—H	98 (411.6)	F—H	135 (567)
$CH_3CH_2CH_2$—H	95 (399)	Cl—H	103 (432.6)
$CH_3CH_2CH_2CH_2$—H	94 (394.8)	Cl—Cl	58 (243.6)
$(CH_3)_2CH$—H	89 (373.8)	Br—Br	46 (193.2)
$(CH_3)_3C$—H	86 (361.2)	F—F	37 (155.4)
HO—OH	52 (218.4)	C_6H_5—H	102 (428.4)
CH_3—Cl	81 (340.2)	$C_6H_5CH_2$—H	78 (327.6)
CH_3—Br	67 (281.4)	I—I	36 (151.2)
H_2N—H	102 (428.4)	CH_3—I	53 (222.6)
Br—H	87 (365.4)	CH_3—F	108 (453.6)
I—H	71 (298.2)	H_2N—NH_2	60 (252)
CH_3—NH_2	79 (331.8)	CH_3—OH	90 (378)
CH_3—CH_3	83 (348.6)	Br—H	87 (365.4)

We can thus say that the energy difference between *iso*butane and the corresoponding radical, *tert*-butyl radical is smaller as compared to that between propane and *iso*propyl radical. We can also presume that as the stability of the transition state parallels stability of radicals, the transition state leading to the formation of *tert*-butyl radical will be of lower energy. Hence we find that the more stable the radical, the more stable the transition state leading to its formation, and the faster the radical is formed. Let us consider the transition states leading to the formation of *tert*-butyl and *iso*propyl radicals in the bromination reaction.

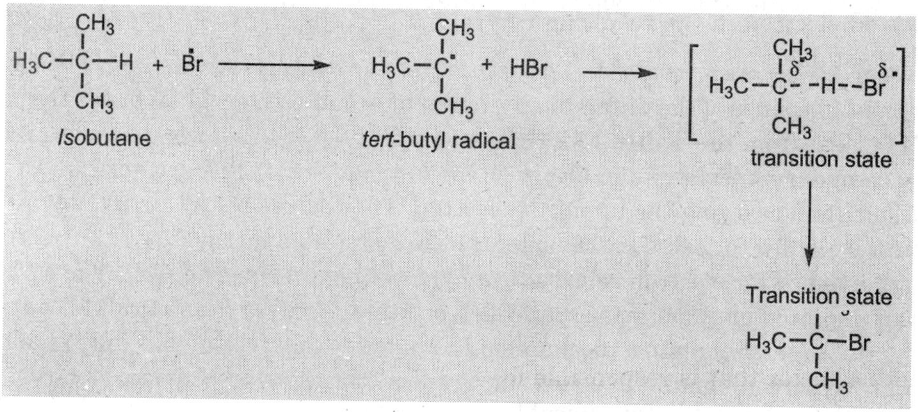

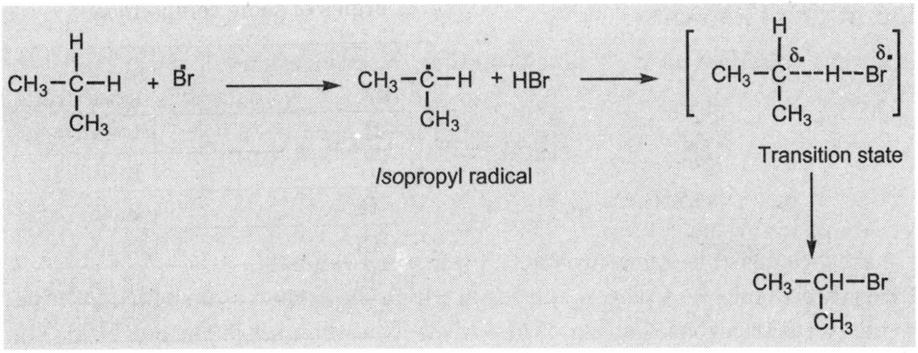

We know that in the transition states, hydrogen-bromine bond is partly formed and the carbon-hydrogen bond is partly broken. It is reasonable to believe that the energy difference between *iso*butane and incipient *tert*-butyl radical in the transition state to be *smaller* than between propane and the incipient *iso*propyl radical in similar situation.

Let us consider the bond dissociation energies for C–H bonds in four simple alkanes in **Table 13.2**.

Table 13.2 Bond dissociation energies for C–H bonds in simple alkanes.

Bond	**H—CH$_3$**	**H—CH$_2$CH$_3$**	**H—CH(CH$_3$)$_2$**	**H—C(CH$_3$)$_3$**
Bond dissociation energy kcal/ mol (kJ/mol)	102 (428.4)	98 (411.6)	89 (373.8)	86 (361.2)

It becomes apparent that the order of reactivity is, *iso*butane > propane > ethane > methane or in other words the order is tertiary C–H > secondary C–H > primary C–H.

The allylic and benzylic C-H bonds are weaker than saturated hydrocarbons because the unpaired electron in the resulting radicals is delocalized. Allyl and benzylic radicals are particularly stable.

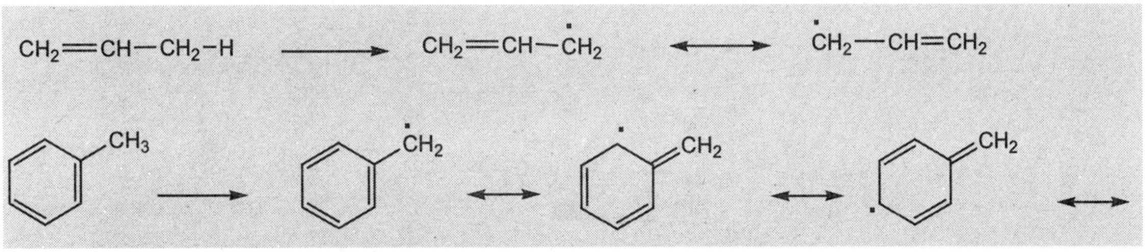

Another factor that is responsible for the increased stability of tertiary radicals may be steric. Formation of a *tert*-butyl radical by the removal of a hydrogen atom from *iso*butane is favoured as the steric repulsion between the methyl groups is relieved to a certain extent by an increase in bond angles from about 109° to about 120°. Clearly, this steric relief should be the greatest in the formation of a tertiary radical.

D. Configuration of Radicals

Free radicals are classified as 1°, 2° and 3° similar to carbocations

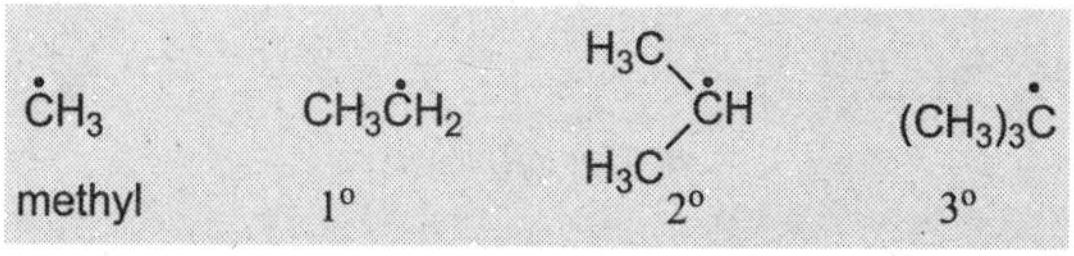

There are two possible structures for simple alkyl radicals.

Planar: A free radical may be a planar species in which the carbon atom bearing the odd electron is sp^2-hybridized (as in carbocation) and the odd electron remains in the p-orbital. The p-orbital containing this odd electron is always perpendicular to the plane of sp^2 carbon.

Pyramidal: Free radical can also acquire a shape resembling a shallow pyramid which is neither planar nor tetrahedral, and the orbital containing unpaired electron is some sort of a hybrid between a p and sp^3 orbital.

One of the most elegant experiments to determine the configuration of free radicals is due to Brown, Kharasch and Chao who found that free radical chlorination of (+)-1-chloro-2-methylbutane yields *racemic* 1,2-dichloro-2-methylbutane.

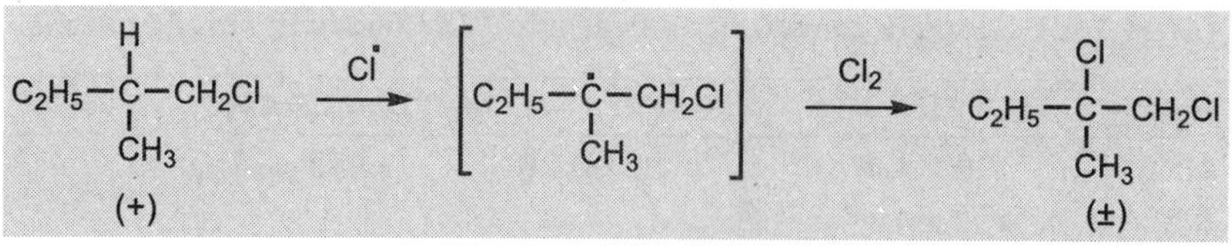

Two reasonable pictures of free radicals fit in with the above experimental data.

(*i*). A free radical is a planar species in which the carbon atom bearing the odd electron is sp^2 hybridized (as in a carbocation) and the odd electron remains in the p orbital. Existence of such a planar radical, in which attachment of chlorine atom to either face is equally likely, explains the equal formation of the two optical isomers of 1,2-dichloro-2-methylbutane leading to racemization.

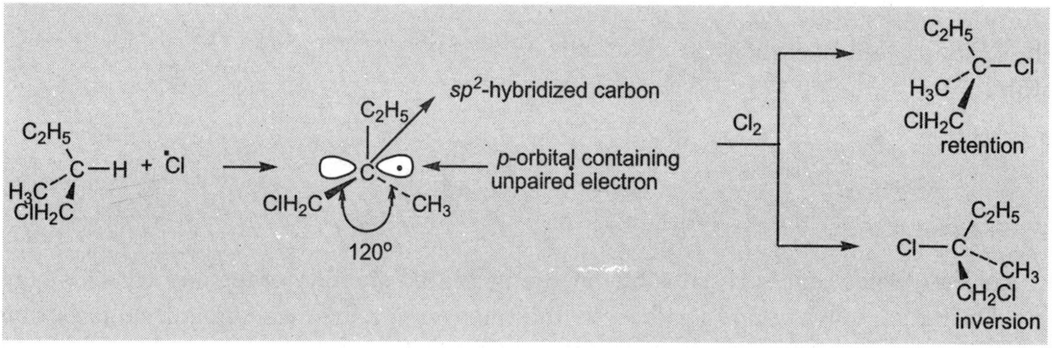

(*ii*). A free radical could also have a shape resembling a shallow pyramid which is neither planar nor tetrahedral and the orbital containing unpaired electron is some sort of hybrid between a *p* and an sp^3 orbital. If we assume that the inversion of such a pyramid is faster than the transfer reaction with chlorine molecule, we should expect a racemic product (**Fig. 13.5**).

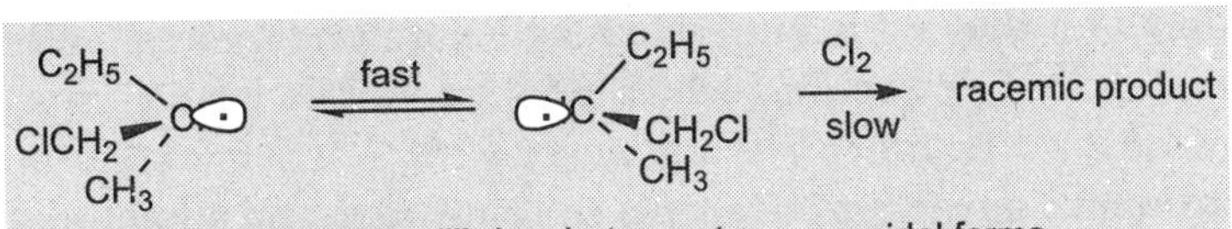

Fig. 13.5 Rapid equilibrium between two pyramidal forms

The question as to whether free radicals exist in a *planar configuration*, or in a rapid equilibrium between the two *pyramidal forms* is difficult to answer. Since the energy barrier between the planar conformation and shallow pyramid is rather small, ultraviolet, infrared and ESR studies have not been able to distinguish between these two configurations. However, available evidence indicates that in radicals stabilized by the unpaired electron, the three bonds of the carbon are coplanar.

Radicals at bridgehead systems have been investigated to explore the geometric requirements for their formation. We know that formation of carbocations at bridgehead positions is strongly resisted because the skeletal geometry prevents attainment of the preferred planarity and S_N1 solvolysis is greatly retarded. However, there is also rate-retardation for the formation of a norbornyl radical, but it is much less than that observed for the generation of carbocation. Apocamphyl peroxide (**1**), for instance, readily dissociates in carbon tetrachloride yielding the apocamphyl radical (**2**) and then the corresponding chloride.

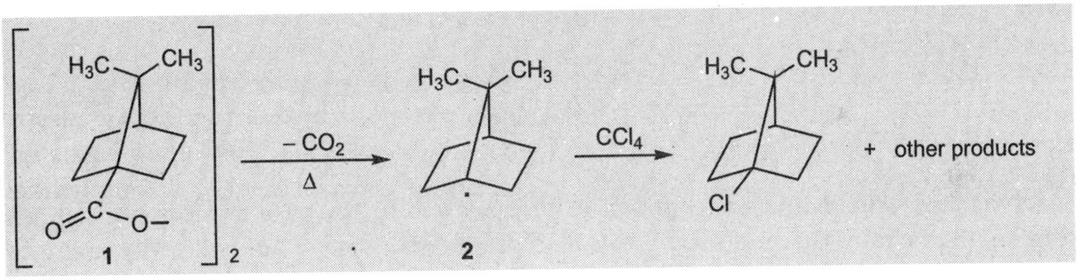

Rate of retardation for the formation of a radical is still smaller for less strained bicyclic systems and decarbonylation of such bridgehead aldehydes was found to proceed smoothly.

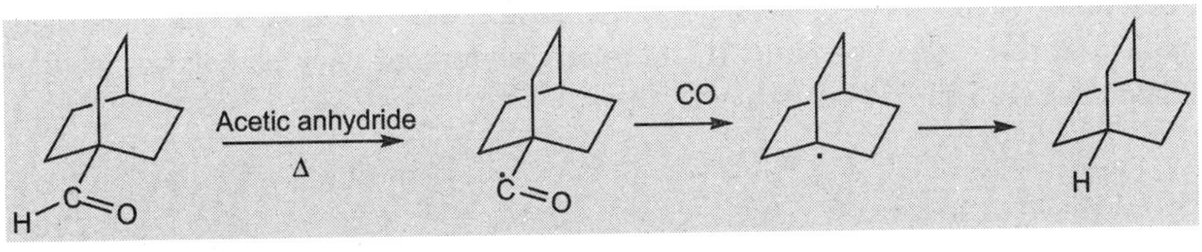

A study of ESR spectra of radicals suggests a pyramidal geometry at the bridgehead carbon atoms. As expected, the allyl radical is planar in order to maximize delocalization of the π-electrons. Planarity of this system has been confirmed by a study of ESR, IR and electron-diffraction measurements.

E. Reactions of Radicals

Unpaired electrons are desperate to be paired up again. This means that radicals usually have a very short lifetime and they do not survive long before undergoing a chemical reaction. It is the abstraction step that determines which product will be formed in a chain reaction. *A free radical always abstracts a univalent* and never a tetra or tercovalent atom, and seldom a divalent one. For example, a reaction between a chlorine atom and ethane gives an ethyl radical and not a hydrogen atom:

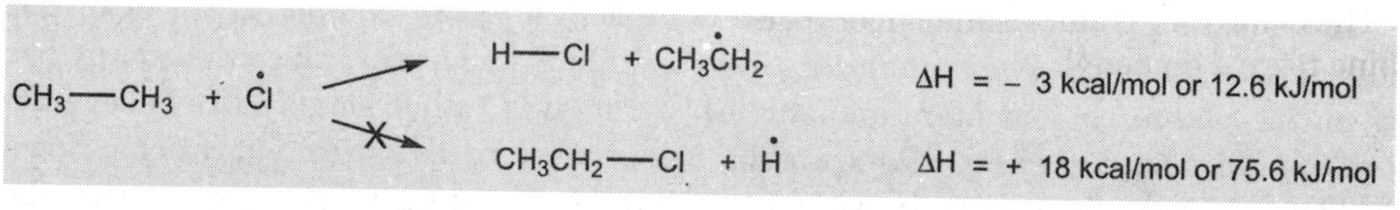

The main reason for this is steric. A univalent atom is much more exposed to the attack by the incoming radical than an atom with a higher valency. Another reason is that in many cases, abstraction of a univalent atom is energetically more favoured.

In cases of alkanes, the following is the decreasing order of the ease of abstraction of different kinds of hydrogens: tertiary H > secondary H > primary H > methane H. This is according to the relative stabilities of the free radicals formed after abstraction of hydrogen. In the case of alkenes, vinylic hydrogens are never abstracted and the abstraction of allylic hydrogens is greatly preferred to other positions of the molecule because of the resonance stabilization of the allylic radical:

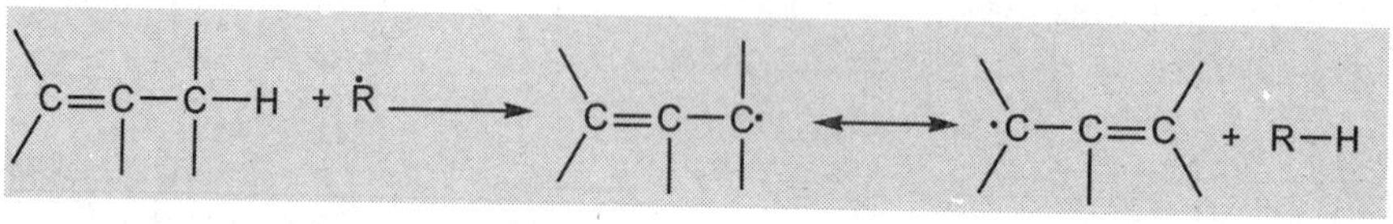

Allylic rearrangements are common in these cases. In the case of aromatic compounds having an alkyl chain, the preferred position of attack on a side chain is usually alpha to the ring. Aromatic hydrogens are seldom abstracted if there are aliphatic hydrogens present in the molecule.

Many radical reactions are initiated by peroxides. The decomposition of one such peroxide, benzoyl peroxide, has been investigated most thoroughly and it illustrates the complexity of radical reactions. A molecule of benzoyl peroxide may be looked upon as consisting of two dipoles joined at their negative ends. It, therefore, undergoes easy homolytic cleavage due to the repulsion between the two negative charges.

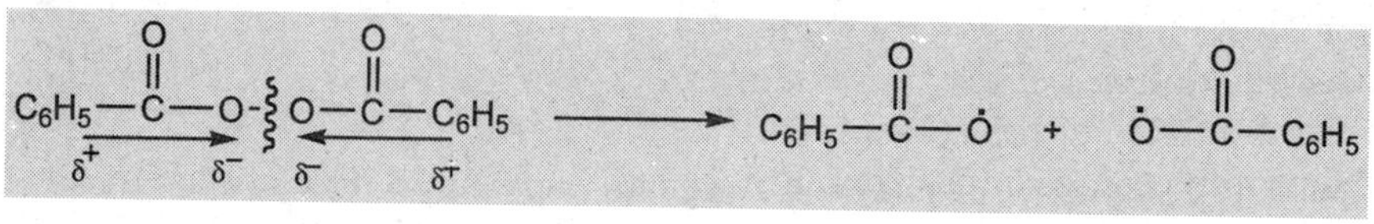

Accordingly, substitution of the benzene ring with electron-donating groups decreases stability of benzoyl peroxide while electron-withdrawing groups slow down the rate of decomposition. Resonance stabilization of the product benzoyloxy radicals helps further in lowering the bond-dissociation energy of the already weak O–O bond.

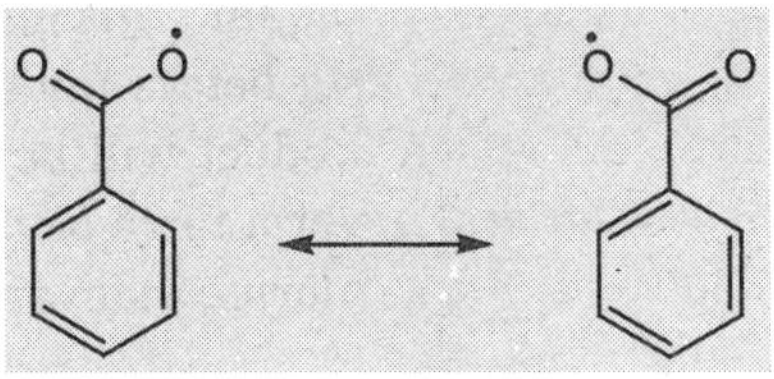

The benzoyloxy radicals may liberate carbon dioxide yielding phenyl radicals, which then combine to give biphenyl.

$$C_6H_5CO\dot{O} \longrightarrow \dot{C}_6H_5 + CO_2 \quad ; \quad 2\,\dot{C}_6H_5 \longrightarrow C_6H_5{-}C_6H_5$$

The intermediacy of benzoyloxy radicals in the decomposition of benzoyl peroxide is further established by the isolation of benzoic acid on carrying out the decomposition in wet carbon tetrachloride containing iodine.

$$C_6H_5CO\dot{O} + I_2 \longrightarrow [\,C_6H_5COOI\,] \xrightarrow{H_2O} C_6H_5COOH$$

In the presence of benzene, the decomposition of benzoyl peroxide results in the formation of substitution products in addition to carbon dioxide, biphenyl and benzoic acid.

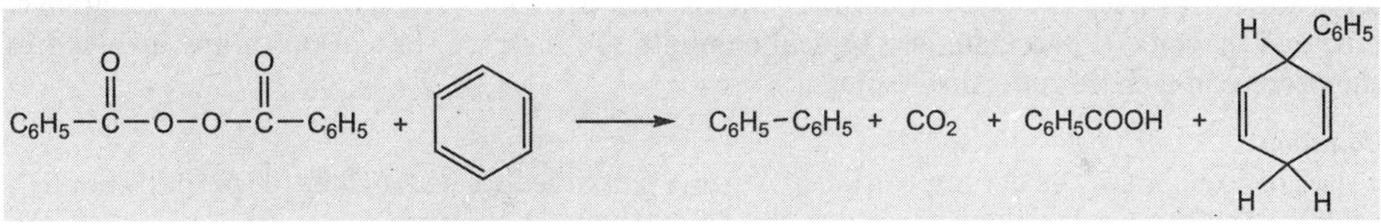

The decomposition of benzoyl peroxide is a chain process in which intermediate radicals, formed by the reaction of benzoyloxy or phenyl radicals with substrates, attack the peroxide molecule to induce further decomposition.

Another class of compounds that are widely used as radical initiators are azo compounds. Azomethane, for example, decomposes to yield ethane, presumably *via* methyl radicals

$$H_3C{-}N{=}N{-}CH_3 \longrightarrow 2\,\dot{C}H_3 + N_2 \quad ; \quad 2\,\dot{C}H_3 \longrightarrow CH_3{-}CH_3$$

Even though the C—N bond in azo compound is of normal strength 70 kcal/mol (or 294 kJ/mol) the drive for the dissociation is provided by the nitrogen molecule whose heat of formation is 225 kcal/mol (or 945 kJ/mol). Study of the commercially useful initiator, azoisobutyronitrile(AIBN) (**3**), has shown that unlike benzoyl peroxide, the rate of decomposition of azo compounds is independent of the nature of the solvents suggesting thereby that the generated radicals are unable to produce induced decomposition. Furthermore, these studies have revealed that after the initial decomposition, the products are held in a solvent cage from where either they slowly diffuse out as free radicals or the two radicals combine to form tetramethylsuccinonitrile (**4**). AIBN is frequently used as an initiator in radical reactions.

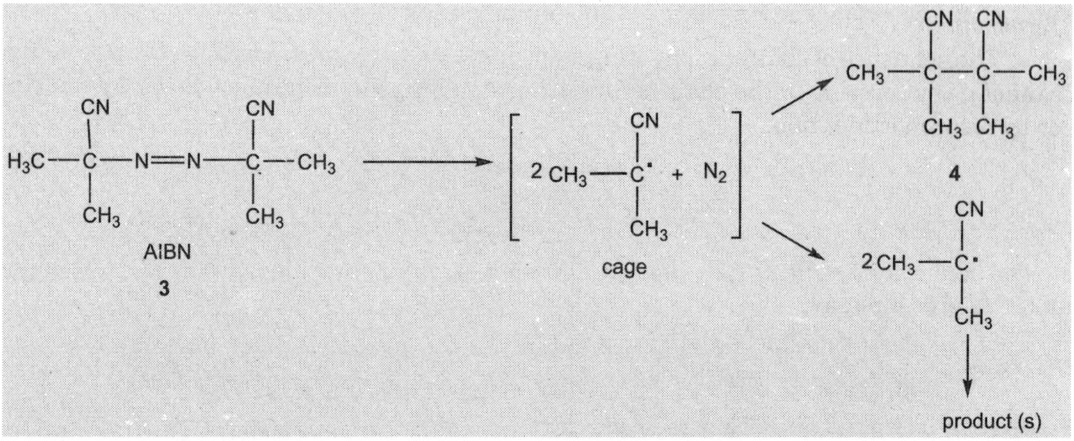

Free radicals undergo many reactions such as polymerization, halogenation, addition, autoxidation, substitution and rearrangements.

I. Polymerization

Many synthetic polymers such as plastics, man-made fibers and structural materials are made by radical processes. The polymerization of olefins, which proceeds by a chain mechanism, is one of the most extensively studied radical reactions. This reaction, which has great commercial importance, involves the treatment of the olefin with peroxide. The polymerization of styrene, initiated by benzoyl peroxide, is a typical example to illustrate the various steps involved in the process of a chain radical reaction.

Initiation

This consists of the decomposition of benzoyl peroxide into benzoyloxy (or perhaps phenyl) radicals and their subsequent attack on the double bond of styrene.

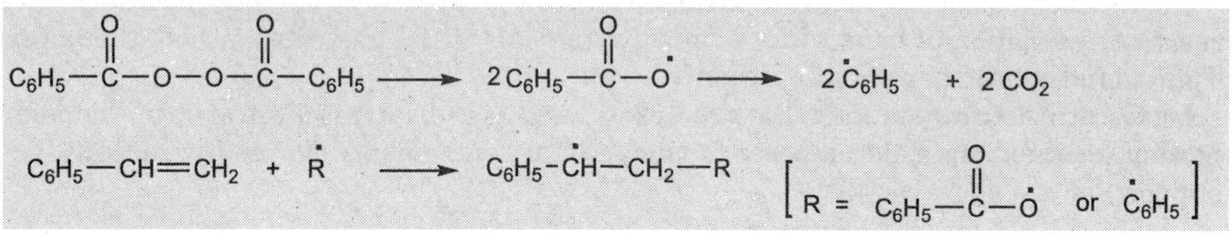

Propagation

In the chain-propagating step, which is extremely rapid, the above radical adds to the styrene double bond. This process is again feasible as a carbon-carbon single bond is being formed while a reactive π-bond is being opened.

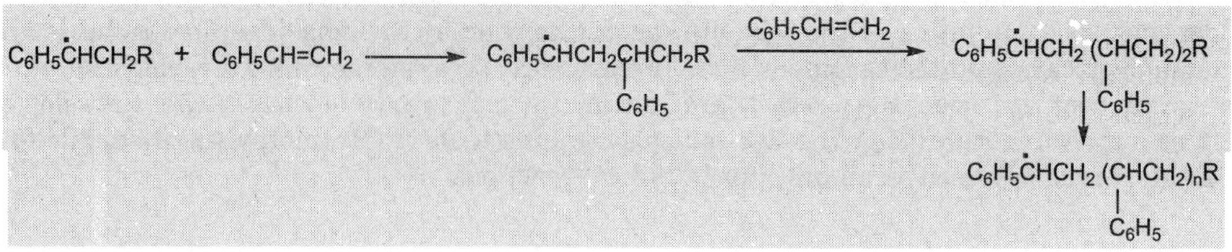

Termination

The sequence of chain-propagating steps must stop at some time, since the polymerization cannot go on forever. In the chain terminating step, two chains interact either by combination or by disproportionation.

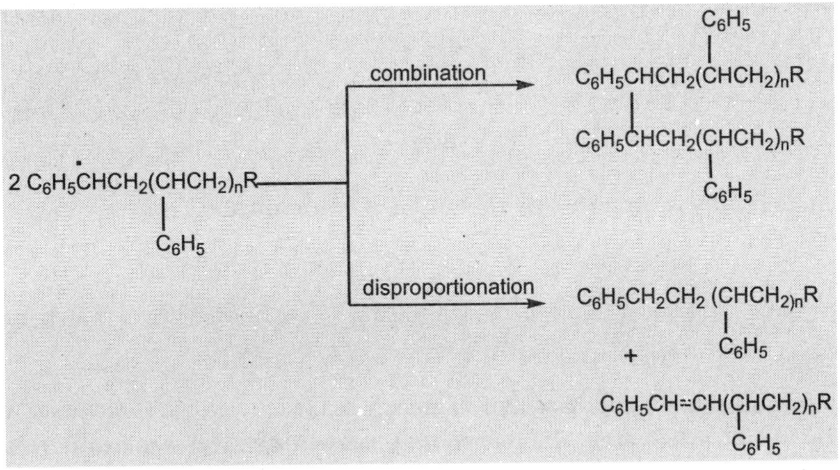

When we need to control the molecular weight of the polymer, *chain-terminators* (HX) are added to the reaction mixture. These compounds are capable of donating a hydrogen atom to the chain radical and thus terminating the chain.

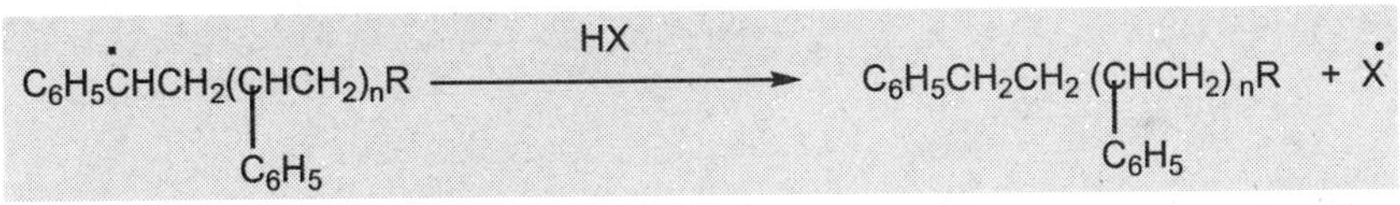

$\dot{X}$ is usually a radical of low reactivity which is not capable of further initiating any radical reaction. Usually polymers formed in radical processes have molecular weight of 10^5 to 10^7. The existence of free radicals in polymerization processes has been demonstrated by ESR spectroscopy. Moreover, when the reaction is initiated by *p*-bromobenzoyl peroxide, instead of benzoyl peroxide, the final product contains *p*-bromophenyl residues thus providing an additional support for the above sequence.

II. Halogenation

Halogenation of alkanes is again a chain-reaction consisting of initiation, propagation and termination. For the sake of clarity we shall consider halogenation by different halogens one by one.

(i) Chlorination: Chlorination of alkanes takes place either thermally or photochemically and it has been shown that under either of these conditions, the first step is the breaking of a chlorine molecule into two chlorine atoms. This process requires 58 kcal/mol (or 243.6 kJ/mol) of energy which is supplied by heating the reaction mixture up to 250 °C or irradiating it at 487.5 nm.

$$\text{Initiation:}\quad \underset{(58)}{\text{Cl—Cl}} \xrightarrow[\text{or } \Delta]{h\nu} \dot{\text{Cl}} + \dot{\text{Cl}} \qquad (\Delta H = +\ 58 \text{ kcal/mol or } 243.6 \text{ kJ/mol})$$

The chain-propagating step consists of the attack by the chlorine atom on the alkane molecule to abstract a hydrogen atom and thereby produces an alkyl radical. This radical, on collision with a chlorine molecule, abstracts a chlorine atom to form a molecule of alkyl chloride.

Propagation:

$$\dot{Cl} + \underset{(102)}{H-CH_3} \longrightarrow \underset{(103)}{H-Cl} + \dot{C}H_3 \quad (\Delta H = -1 \text{ kcal/mol or } 4.2 \text{ kJ/mol})$$

$$\underset{(58)}{Cl-Cl} + \dot{C}H_3 \longrightarrow \underset{(81)}{CH_3Cl} + \dot{Cl} \quad (\Delta H = -23 \text{ kcal/ mol or } 96.6 \text{ kJ/mol})$$

The chain is finally terminated by the union of two radicals.

Termination:

$$\dot{Cl} + \dot{Cl} \longrightarrow Cl_2$$
$$\dot{C}H_3 + \dot{C}H_3 \longrightarrow CH_3-CH_3$$
$$\dot{C}H_3 + \dot{Cl} \longrightarrow CH_3Cl$$

A consideration of the bond-dissociation energy makes it clear that although energy must be supplied for the *initiation* step, the two *chain propagating* steps are exothermic and hence proceed readily. As expected, the hydrogen atoms in an alkane are substituted in the general order, primary < secondary < tertiary. Thus, chlorination of propane yields *n*-propyl and *iso*propyl chlorides in the ratio of 1:1.

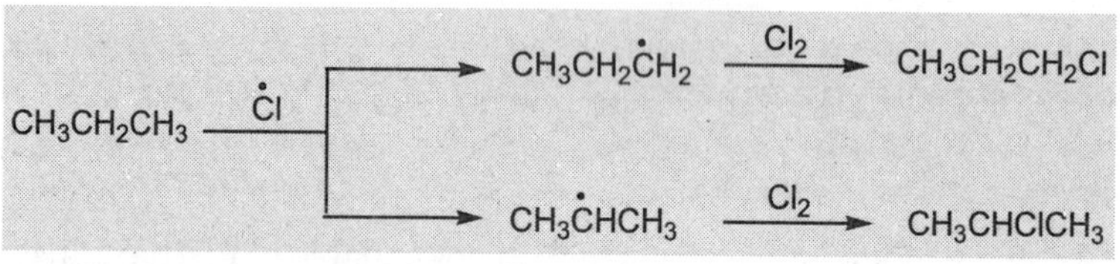

Since there are six primary hydrogens against only two secondary hydrogens in a molecule of propane, the probability factor favours the formation of a primary radical by the ratio of 3:1. The actual composition of the product mixture, however, shows that the ease of formation of a secondary radical is three times more than that of a primary radical.

It may be noted that due to the greater reactivity of the chlorine atom, abstraction of primary, secondary and tertiary hydrogens are all exothermic. As a consequence, there is *diminished selectivity* in the reactions involving chlorination. For example, chlorination of 2,3-dimethylbutane gives a mixture of two products and a mixture is again obtained when ethylbenzene is chlorinated. There is chlorination at methyl and methylene positions in spite of the fact that a resonance stabilized benzyl radical is far more stable than a primary radical.

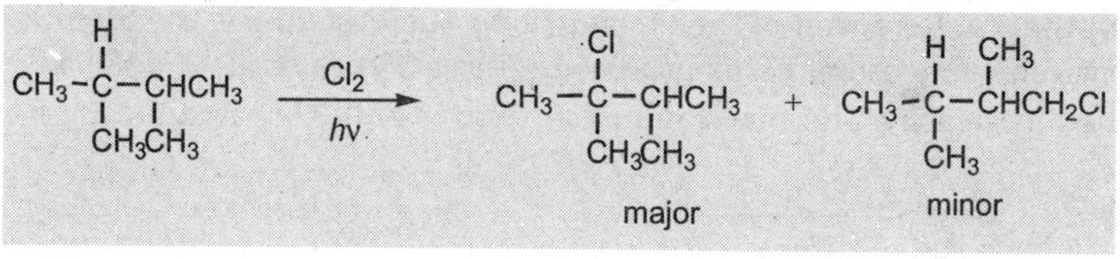

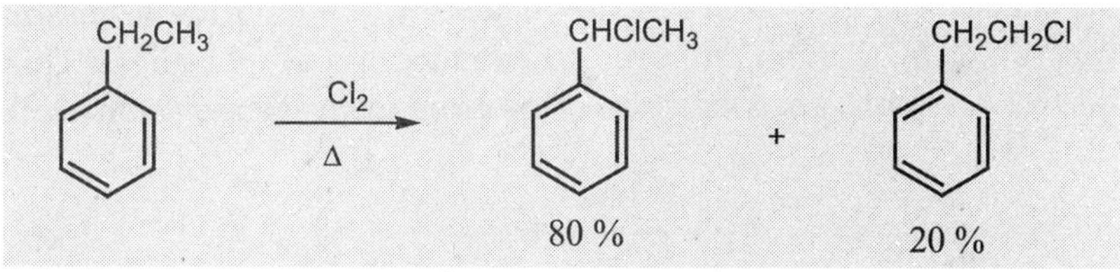

(ii) Bromination: Bromination of alkanes has close similarities to chlorination except for the fact that the chain reaction initiated by bromine atoms proceeds relatively slowly. In other words, more bromine molecules are to be dissociated to produce the same yield. This difference is mainly due to different bond-dissociation energies of the molecules involved in the reaction.

$$\underset{(46)}{Br{-}Br} \xrightarrow[\text{or } \Delta]{h\nu} 2\,\dot{Br} \qquad (\Delta H = +46 \text{ kcal/mol (or 193.2 kJ/mol)}$$

$$\dot{Br} + \underset{(102)}{H{-}CH_3} \longrightarrow H{-}Br + \dot{C}H_3 \qquad (\Delta H = +15 \text{ kcal/mol (or 63 kJ/mol)}$$

$$\dot{C}H_3 + \underset{(46)}{Br{-}Br} \longrightarrow CH_3{-}Br + \dot{Br} \qquad (\Delta H = -21 \text{ kcal/mol (or 88.2 kJ/mol)}$$

Although the last step of the reaction is exothermic (and, therefore, occurs readily) the attack of bromine atom on the alkane is endothermic unlike the analogous step in the chlorination. Thus it is somewhat difficult for a bromine atom to acquire the *activation energy* required to abstract a hydrogen atom in competition with the chain-terminating steps such as combining with another bromine atom or an alkyl radical. Overall bromination is, therefore, a slower process than chlorination. An important consequence of the *low reactivity* of bromine atom is its *high selectivity* for tertiary and secondary positions in comparison to a chlorine atom. This explains why gaseous bromination of *iso*butane yields mainly the tertiary bromides.

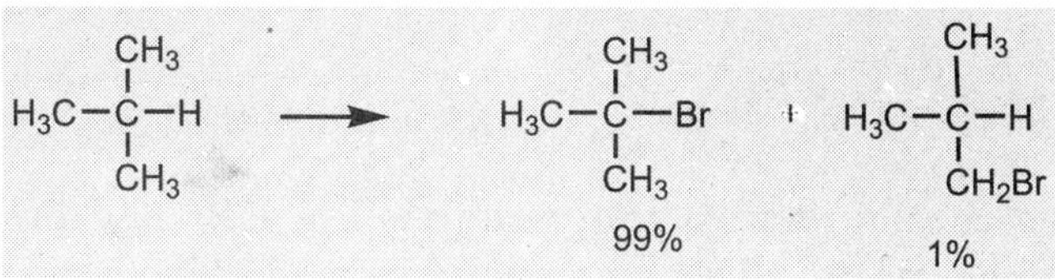

Bromination by N-Bromosuccinimide

N-Bromosuccinimide (NBS) deserves a special consideration since it is a very specific reagent for allylic and benzylic bromination. The reagent is prepared by treatment of succinamide in alkaline solution with bromine and usually reacts in the presence of a peroxide as initiator by a chain mechanism.

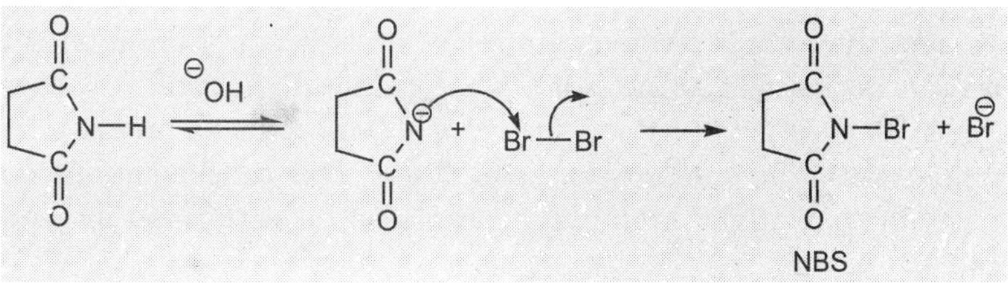

The following examples illustrate the versatility of this reagent.

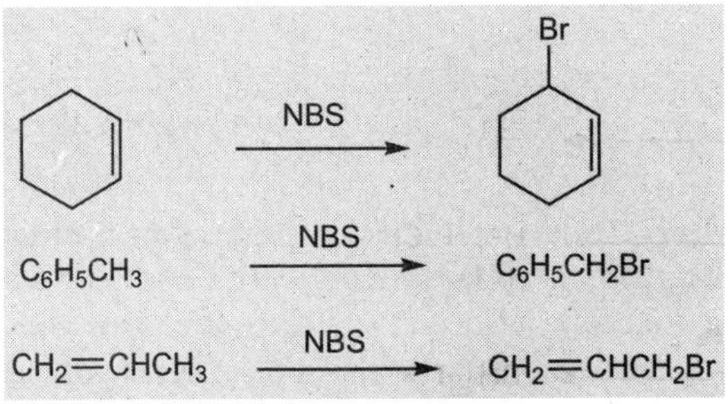

A mechanism was proposed in which the function of NBS is merely to act as a bromine reservoir maintaining a low concentration of molecular bromine by reacting with HBr that is initially formed in side reactions.

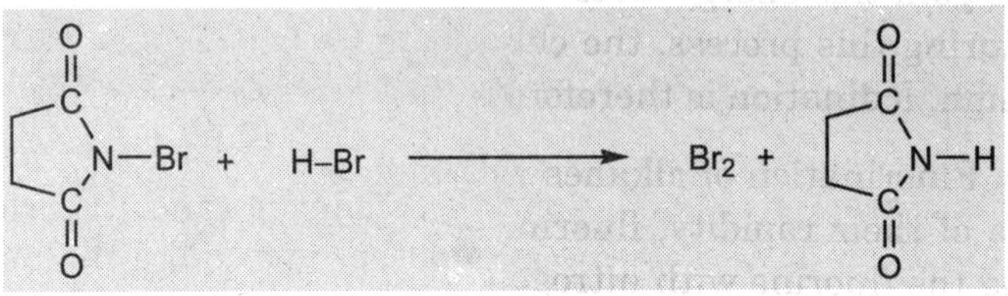

The bromine molecule dissociates into bromine atoms in presence of light or radical initiators. This explains why bromination by NBS is initiated by a peroxide or ultraviolet light. Hydrogen abstraction by bromine atom generates an allylic radical which then reacts with the molecular bromine to yield the product.

$$\dot{Br} + CH_2{=}CH{-}CH_3 \longrightarrow HBr + CH_2{=}CH{-}\dot{C}H_2$$

$$CH_2{=}CH{-}\dot{C}H_2 + Br_2 \longrightarrow CH_2{=}CH{-}CH_2Br + \dot{Br}$$

The chain then continues with the production of HBr and bromine atoms. A low concentration of bromine favours allylic bromination over addition to the double bond since the addition of bromine atoms to the double bond is a reversible process.

$$\dot{Br} + CH_2{=}CH{-}CH_3 \rightleftharpoons Br{-}CH_2\dot{C}HCH_3$$

$$Br_2 + Br{-}CH_2\dot{C}H{-}CH_3 \longrightarrow \dot{Br} + Br{-}CH_2{-}CHBr{-}CH_3$$

In other words, the radical formed as a result of addition of bromine will revert to the alkene if the concentration of bromine is low whereas an allylic radical is capable of waiting (since the hydrogen abstraction reaction is irreversible) till it gets a bromine molecule. Thus a low bromine concentration favours allylic bromination. This conclusion has received support from the observation that allylic bromination can also be performed by slow photolysis of molecular bromine.

(iii) Iodination: Since iodine atom is also very stable, kinetic chain in iodination is much shorter. The heats of reaction for the initiation and propagation steps in the iodination of methane are the following:

$$\underset{(36)}{\text{I–I}} \longrightarrow 2\dot{\text{I}} \qquad (\Delta H = +36 \text{ kcal/mol or } 151.2 \text{ kJ/mol})$$

$$\dot{\text{I}} + \underset{(102)}{\text{H–CH}_3} \longrightarrow \underset{(71)}{\text{H–I}} + \dot{\text{C}}\text{H}_3 \qquad (\Delta H = +31 \text{ kcal/mol or } 130.2 \text{ kJ/mol})$$

$$\underset{(36)}{\text{I–I}} + \dot{\text{C}}\text{H}_3 \longrightarrow \text{CH}_3\text{–I} + \dot{\text{I}} \qquad (\Delta H = -17 \text{ kcal/mol or } 71.4 \text{ kJ/mol})$$

It is evident from the above data that hydrogen abstraction by iodine atom is appreciably endothermic and so the energy of activation for this step must be higher than 31 kcal/mol (or 130.2 kJ/mol). To gain this energy an iodine atom must collide with a large number of alkane molecules and during this process, the chances of its getting combined with a second iodine atom are quite high. Iodination is therefore a slow process.

(iv) Fluorination: Fluorination of alkanes with fluorine is an extremely rapid and exothermic process. Because of their rapidity, fluorination reactions are usually carried out in the gas phase by diluting the fluorine with nitrogen or other inert gases. The heat of reaction for the propagation steps in the fluorination of methane explains the experimental observations.

$$\dot{\text{F}} + \underset{(102)}{\text{H–CH}_3} \longrightarrow \underset{(135)}{\text{H–F}} + \dot{\text{C}}\text{H}_3 \qquad \Delta H = -33 \text{ kcal/mol (or } 38.6 \text{ kJ/mol)}$$

$$\dot{\text{C}}\text{H}_3 + \underset{(37)}{\text{F–F}} \longrightarrow \underset{(108)}{\text{CH}_3\text{–F}} + \dot{\text{F}} \qquad \Delta H = -71 \text{ kcal/mol (or } 298.2 \text{ kJ/mol)}$$

Since fluorination proceeds in the dark even at –80° and in the absence of chemical chain-initiators, initial dissociation of a fluorine molecule (requiring 37 kcal/mol or 155.4 kJ/mol) of energy is not an essential requirement for the success of a reaction. Instead, chain initiation proceeds by the reaction of fluorine molecule with an alkane producing a fluorine atom along with an alkyl radical. Such a reaction requires only 5 kcal/mol or 21 kJ/mol of energy and so takes place even at low temperatures.

$$\underset{(37)}{\text{F–F}} + \underset{(102)}{\text{H–CH}_3} \longrightarrow \underset{(134)}{\text{H–F}} + \dot{\text{F}} + \dot{\text{C}}\text{H}_3 \qquad (\Delta H = +5 \text{ kcal/mol or } 21 \text{ kJ/mol})$$

(v) Polar Effects in Halogenation: We have already seen that in free radical reactions, there is always a tendency for the formation of the most stable radical and accordingly the order of reactivity in aliphatic hydrocarbons has been found to be tertiary > secondary > primary. However, there is now an evidence that in addition to radical stability, *polar factors* often play a role in deciding which hydrogen of the reactant is to be abstracted by the attacking radical.

The relative reactivity of various hydrogen atoms of *n*-butyl chloride in photochlorination are shown as follows:

$$\begin{array}{cccc} \delta & \gamma & \beta & \alpha \\ H_3C- & CH_2- & CH_2- & CH_2-Cl \\ 8 & 17 & 5.5 & 1 \end{array}$$

In spite of the fact that all the three carbon atoms (α, β and γ) are secondary in nature, there is a striking difference in the reactivity of hydrogen situated on them. If the radical stability is the only factor, then hydrogen of α-carbon should be the most reactive as chlorine atom is expected to stabilize the radical on this carbon. These experimental observations are explained in terms of the polar factor due to the electron-withdrawing inductive effect of chlorine. The attacking chlorine radical is *strongly electronegative* and so it prefers to approach hydrogens which are away from the electronegative chlorine substituent. Predictably α and β carbon atoms are chlorinated to a lesser extent than the γ-carbon atom. The lower reactivity of the methyl group than the adjacent methylene is obviously due to the formation of a less stable radical. In contrast, alkyl radicals are *nucleophilic* which react preferentially at sites where the electron density is low. Another example of the operation of a polar effect is the chlorination of the propionic acid. Chlorine atom attacks a primary carbon rather than a secondary carbon, even though the secondary radical would be conjugated with the carbonyl group.

$$\begin{array}{ll} CH_3- & CH_2-COOH \\ 30 & 1 \end{array}$$

Apparently, the polar effect due to the electronegative carboxyl group overcomes the radical-stabilization effect.

We can thus summarize that in compounds having electron-withdrawing substituents, such as Z-CH_2CH_3 (Z = COOH, COCl, COOR or CX_3), the β positions are attacked predominantly or exclusively in free radical halogenations. This is because electron-withdrawing groups deactivate adjacent alpha positions. Compounds like acetic acid and acetyl chloride are not attacked by the free radicals as halogen atoms are *electrophilic* radicals and look for the positions of high electron density. Hydrogens on the carbon atom next to the electron-withdrawing groups have low electron density and, therefore, the attack is avoided at alpha position. Radicals that are *not electrophilic* do not show this behaviour for example, the methyl radical attacks at the α position. Some radicals, for example, *tert*-butyl, benzyl, cyclopropyl and phenyl are nucleophilic and tend to abstract electron-poor hydrogens.

A study of the relative reactivity of substituted toluenes toward bromine atom supports the above conclusions.

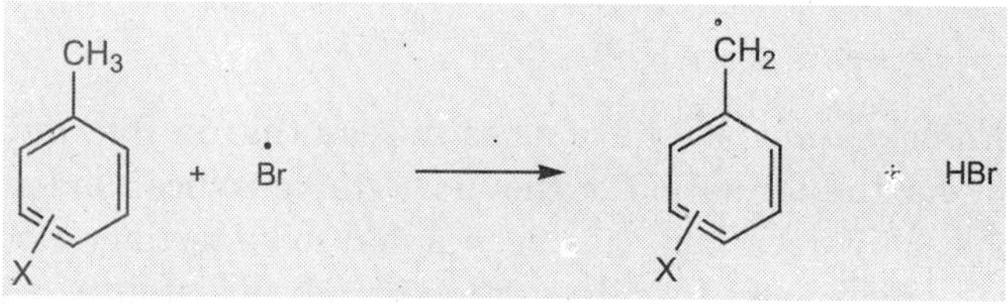

It was observed that an electron-donating substituent such as *p*-methoxy or *p*-methyl group favours the reaction while an electron-withdrawing substituent such as *p*-cyano group, deactivates the toluene. Radical stability alone cannot explain this difference as a radical will be stabilized by both types of substituents:

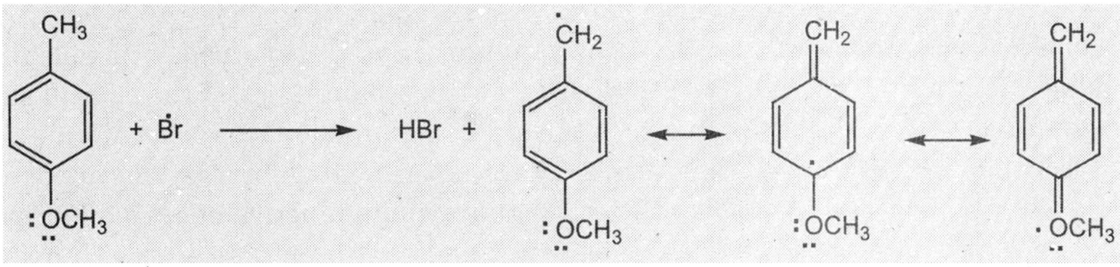

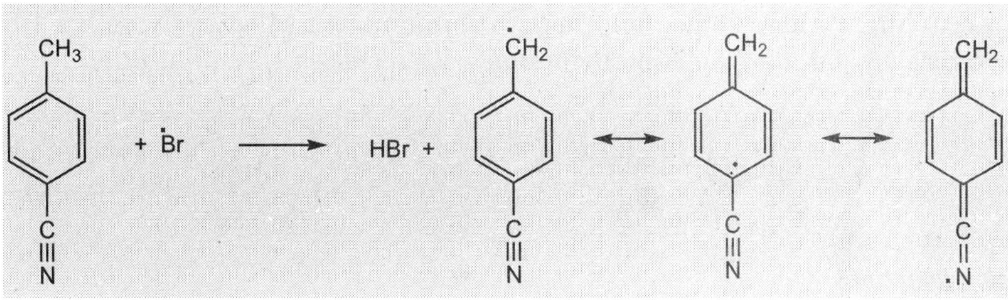

However, a consideration of the resonating structures of the transition state for hydrogen abstraction clearly explains the above observations. As the bromine atom approaches hydrogen, three electrons get distributed over two partial bonds and since bromine is more electronegative than carbon, one of the resonating structures will be ionic in which carbon bears a positive charge.

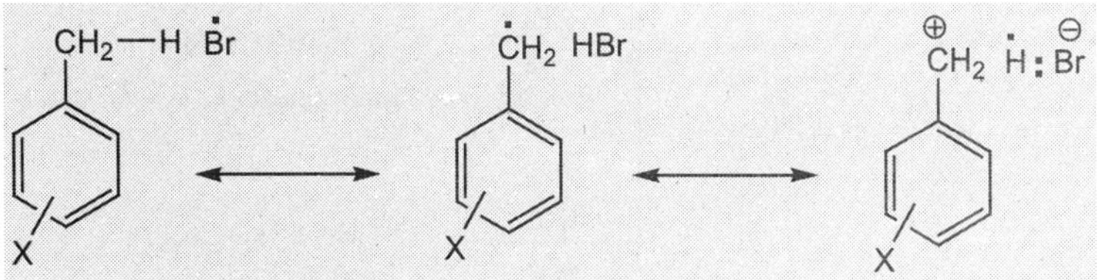

Since electron-donating substituents stabilize a positive charge, they are expected to enhance the reactivity of a substituted toluene whereas the converse is true for the electron-withdrawing substituents.

(vi) Neighbouring Group Assistance: In certain cases it has been shown that free radical reactions are accelerated by the presence of neighbouring groups. For example, bromination of alkyl bromides gives 84 to 94% substitution at the carbon adjacent to the bromine already present in the molecule. This is surprising because positions close to a –I group such as bromine should actually be deactivated by its electron-withdrawing effect. This unusual regioselectivity is explained by the following mechanism in which the abstraction of hydrogen is assisted by a neighbouring group and the reaction proceeds through a *bridged free radical.*

$$Br_2 \longrightarrow \dot{Br} + \dot{Br}$$

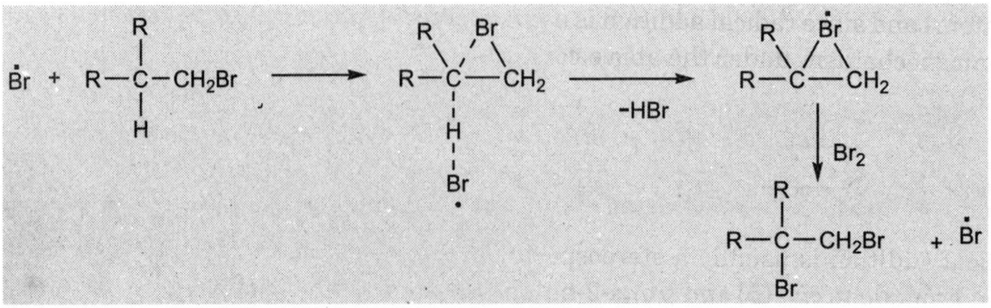

The above mechanism is supported by the fact that as expected, it proceeds with the retention of configuration. Thus photochemical bromination of optically active 1-bromo-2-methylbutane gives 1,2-dibromo-2-methylbutane with *retention of configuration*.

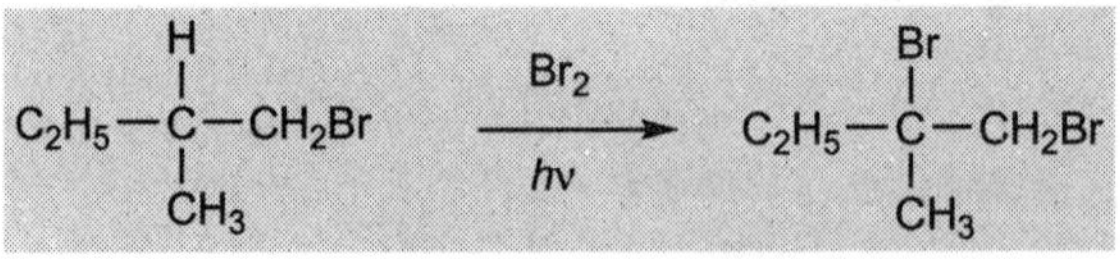

III. Addition Reactions

(i) Addition of Halogen Acids: Perhaps the most celebrated example of free radical addition is the 'abnormal' addition of the hydrogen bromide to olefins. This reaction was studied by Kharasch and Mayo who showed that it proceeds by either an ionic or a radical mechanism. In the dark and in the absence of radical initiators, the reaction proceeds in an ionic fashion while in the presence of light or a peroxide, a much faster radical addition occurs.

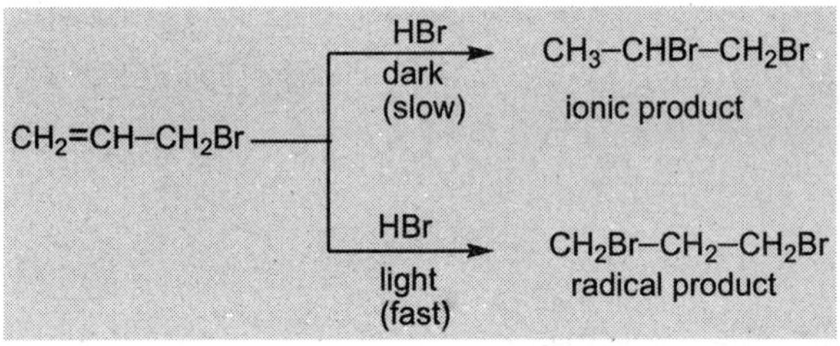

The difference in the nature of the products can be accounted for by a consideration of the stabilities of radicals and carbocations which follow the order primary < secondary < tertiary. Initial addition of a proton produces a relatively stable carbocation in the ionic reaction while a bromine atom adds first to generate the more stable secondary radical.

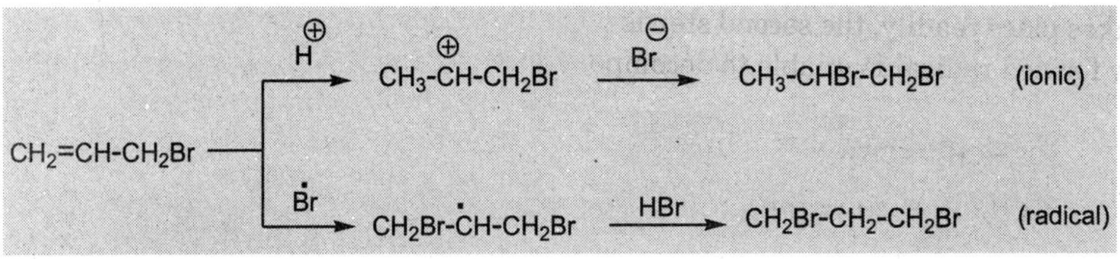

In the presence of oxygen, light or peroxide, hydrogen bromide dissociates to generate bromine atoms and since radical addition is a faster process, it overtakes the addition proceeding *via* the ionic mechanism under the above conditions.

$$HBr + O_2 \longrightarrow H\dot{O}_2 + \dot{B}r \quad ; \quad HBr \xrightarrow{h\nu} \dot{H} + \dot{B}r$$

$$ROOH \longrightarrow R\dot{O} + \dot{O}H \quad ; \quad R\dot{O} + HBr \longrightarrow ROH + \dot{B}r$$

Radical addition is usually a stereospecific process yielding a *trans* product. Addition of deuterium bromide to *cis-* (**5**) and *trans*-2-butene (**6**), for example, yields *threo* (**7**) and *erythro*-3-deutero-2-bromobutanes (**8**).

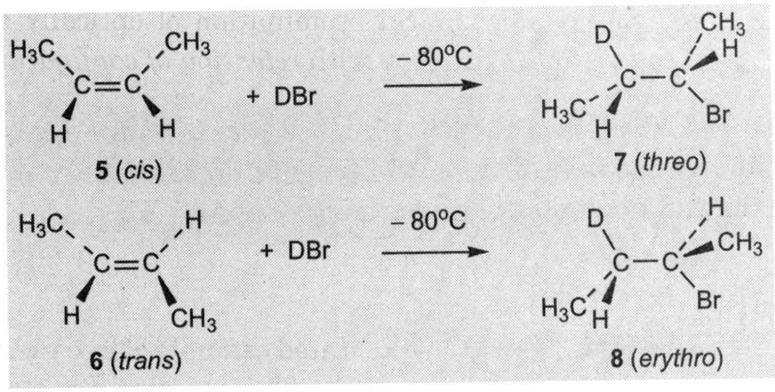

In contrast to the addition of HBr, radical addition of HI and HCl to olefins normally does not take place. Consideration of the energy changes which would be associated with each propagating step of the radical chain process makes this absolutely clear.

$$\dot{X} + CH_2{=}CH_2 \longrightarrow XCH_2\dot{C}H_2 \quad \begin{bmatrix} X = Cl; & \Delta H = -26 \text{ kcal/mol or } 109.2 \text{ kJ/mol} \\ X = Br; & \Delta H = -5 \text{ kcal/mol or } 21 \text{ kJ/mol} \\ X = I; & \Delta H = +7 \text{ kcal/mol or } 29.4 \text{ kJ/mol} \end{bmatrix}$$

$$XCH_2\dot{C}H_2 + HX \longrightarrow XCH_2CH_3 + \dot{X} \quad \begin{bmatrix} X = Cl; & \Delta H = +5 \text{ kcal/mol or } 21 \text{ kJ/mol} \\ X = Br; & \Delta H = -11 \text{ kcal/mol or } 46.4 \text{ kJ/mol} \\ X = I; & \Delta H = -27 \text{ kcal/mol or } 113.4 \text{ kJ/mol} \end{bmatrix}$$

For the addition of HCl, although the first step is exothermic and so the addition of chlorine atom takes place readily, the second step is endothermic by 5 kcal/mol (or 21 kJ/mol). Thus, the initially formed radical is unable to decompose HCl and polymerization begins to complete.

$$\dot{C}l + CH_2{=}CH_2 \longrightarrow ClCH_2\dot{C}H_2$$

$$ClCH_2\dot{C}H_2 + CH_2{=}CH_2 \longrightarrow ClCH_2CH_2CH_2\dot{C}H_2.$$

$$ClCH_2\dot{C}H_2CH_2CH_2 + n\ CH_2{=}CH_2 \longrightarrow ClCH_2(CH_2CH_2)_{n+1}\dot{C}H_2$$
$$ClCH_2(CH_2CH_2)_{n+1}\dot{C}H_2 + HCl \longrightarrow ClCH_2(CH_2CH_2)_{n+1}CH_3 + \dot{C}l$$
$$(n = 4 - 10)$$

These low polymers are known as *telemers* and are produced in the addition of hydrogen chloride to ethylene in the presence of benzoyl peroxide under pressure. For the addition of HBr, both the steps are exothermic and so the addition takes place readily. The homolytic addition of HI to ethylene is difficult as the first step of the chain-propagating step is endothermic by 7 kcal/ mol. In fact no HI has so far been added to an olefin in a free radical chain process.

In summary, we can say that it is only for HBr that both these steps are exothermic. The reaction with HCl is slow because of the endothermicity of the reaction between the carbon radical and HCl. HI fails to add because the reaction of iodine atoms with the alkene is endothermic.

(i) Addition of Thiols: The energies of the RS–H and R–Br bonds are rather close and so both the steps of propagation in the radical-chain addition of thiols to the olefins are expected to be exothermic. Indeed the addition takes place in the presence of light, air or peroxides.

$$R\dot{S} + CH_2{=}CHC_6H_5 \longrightarrow RS{-}CH_2\dot{C}HC_6H_5$$
$$RS{-}CH_2\dot{C}HC_6H_5 + RSH \longrightarrow RS{-}CH_2CH_2C_6H_5 + R\dot{S}$$

(ii) Addition of Halogens: Under the catalytic influence of light, it is often possible to bring about the addition of halogens to the carbon-carbon double bond by a free-radical chain reaction. To avoid addition through ionic mechanism, these reactions are very often studied either in the vapour phase or in a non-polar solvent like carbon tetrachloride. A consideration of the energies of the various steps involved in the chain-propagating steps of the addition of halogens to ethylene clearly demonstrates that the homolytic addition of chlorine and bromine would be exothermic and should, therefore, take place readily. Addition of iodine, on the other hand, is an endothermic process and hence difficult to achieve under normal experimental conditions.

$$X_2 \xrightarrow{h\nu} 2\dot{X}$$
$$\dot{X} + CH_2{=}CH_2 \longrightarrow XCH_2\dot{C}H_2 \qquad \left[\begin{array}{l} X = Cl;\ \Delta H = -26 \text{ kcal/mol or } 109.2 \text{ kJ/mol} \\ X = Br;\ \Delta H = -5 \text{ kcal/mol or } 21 \text{ kJ/mol} \\ X = I;\ \Delta H = +7 \text{ kcal/mol or } 29.4 \text{ kJ/mol} \end{array}\right]$$
$$XCH_2\dot{C}H_2 + X_2 \longrightarrow XCH_2CH_2X + \dot{X} \qquad \left[\begin{array}{l} X = Cl;\ \Delta H = -19 \text{ kcal/mol or } 79.8 \text{ kJ/mol} \\ X = Br;\ \Delta H = -17 \text{ kcal/mol or } 71.4 \text{ kJ/mol} \\ X = I;\ \Delta H = -13 \text{ kcal/mol or } 54.6 \text{ kJ/mol} \end{array}\right]$$

Presence of the oxygen inhibits the addition as it attacks the radical adduct and thereby terminates chain-propagation. Isolation of trichloroacetyl chloride and phosgene in the free-radical chlorination of the tetrachloroethylene in the presence of oxygen confirms the above conclusion.

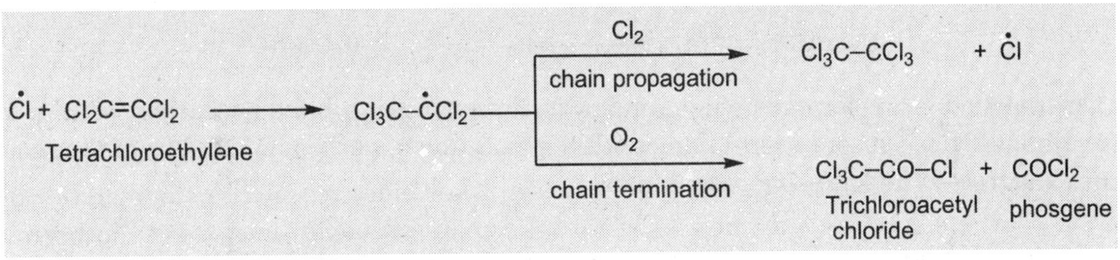

An important consequence of the addition of bromine to olefins, which is a reversible reaction, is the *cis-trans* isomerization of the olefins. It has been shown, for example, that addition of bromine to *cis*-dichloroethylene is accompanied by isomerization and the kinetic data suggest that a common intermediate is involved.

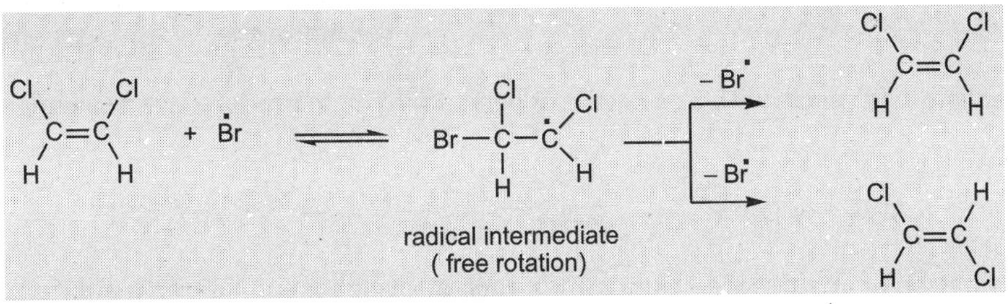

As the *trans* isomer is more stable, it is formed predominantly by the rotation about the carbon-carbon bond of the intermediate radical.

Because of the greater strength of the H–F bond (135 kcal/mol), attempted addition of fluorine to olefins often results in the formation of HF by abstraction of the hydrogen atom. However, halogenated olefins may undergo homolytic addition of fluorine by a mechanism which does not necessitate the prior formation of fluorine atoms.

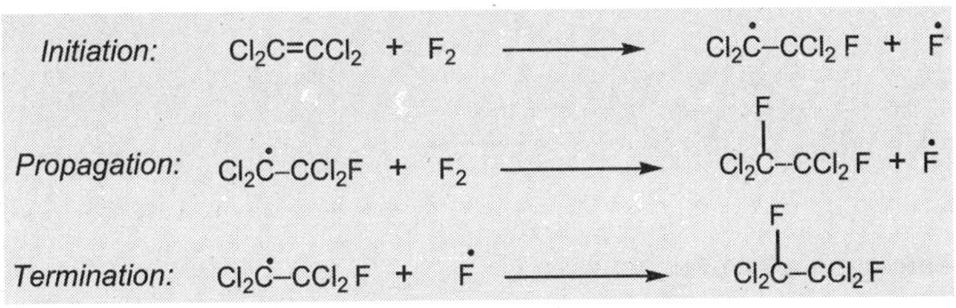

The driving force for the initiation of the above sequence is the formation of a strong C–F bond (102 kcal/mol or 428.4 kJ/mol) at the cost of the weak F–F bond (37 kcal/mol or 155 kJ/mol).

IV. Autoxidation

Slow oxidation of organic substance by atmospheric oxygen is termed as autoxidation. The oxygen molecule frequently behaves as a biradical and it is, therefore, not surprising that this process involves a radical mechanism. Perhaps the most thoroughly studied example of autoxidation is the conversion of benzaldehyde to benzoic acid.

$$C_6H_5CHO \xrightarrow{O_2} C_6H_5COOH$$

The reaction is accelerated by light and by initiators and inhibited by 'antioxidants' such as hydroquinone. These observations along with the isolation of perbenzoic acid from the reaction mixture suggested the following mechanism.

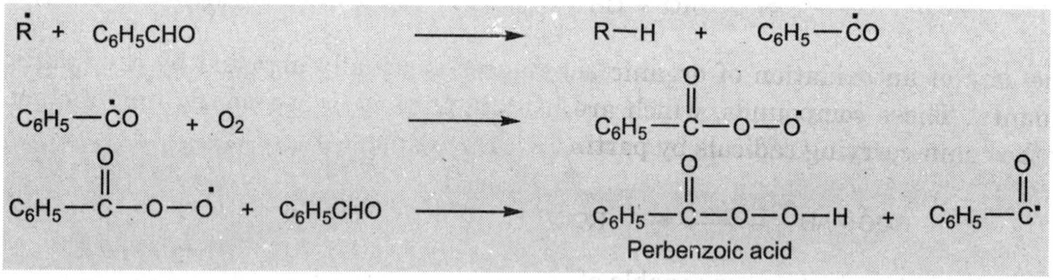

Perbenzoic acid reacts with a molecule of benzaldehyde to produce two molecules of benzoic acid.

$$C_6H_5{-}COOOH + C_6H_5CHO \longrightarrow 2\ C_6H_5COOH$$

Autoxidation of unsaturated compounds such as cyclohexene follows a similar course and results in the formation of hydroperoxides. In this case, resonance stabilization of an allylic radical greatly facilitates the removal of hydrogen.

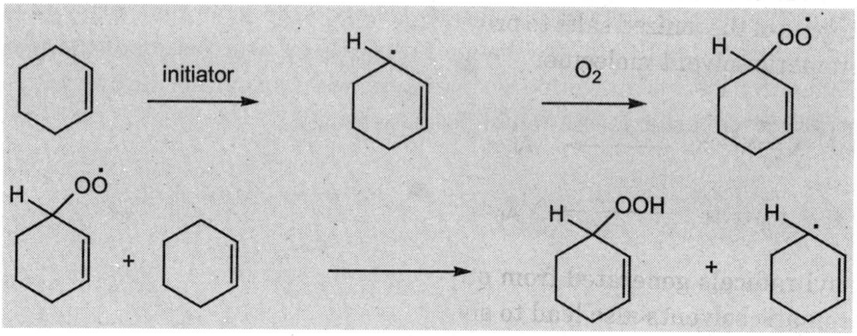

Tetralin behaves in a similar manner.

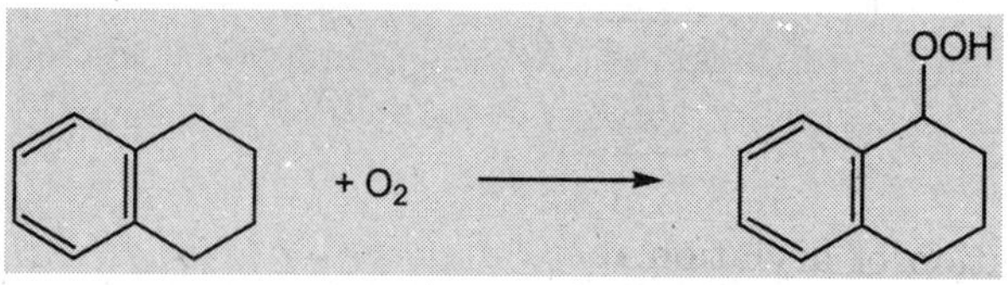

Ethers with α-hydrogen are autoxidized to *bis*-hydroperoxides.

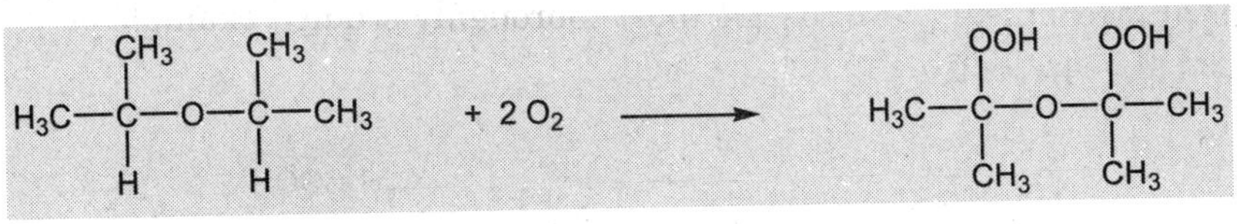

Very frequently hydroperoxides decompose to generate radicals which start new chain reactions.

$$ROOH \longrightarrow R\dot{O} + \dot{O}H \quad ; \quad R\dot{O} + RH \longrightarrow ROH + \dot{R}$$

The rate of autoxidation of organic substances is usually arrested by the addition of 'antioxidants'. These compounds, which are usually phenols or secondary aromatic amines, destroy the chain-carrying radicals by parting with a hydrogen atom.

$$RO\dot{O} + AH \longrightarrow ROOH + \dot{A} \qquad (AH = \text{antioxidant})$$

The radical ($\dot{A}$), which is not capable of initiating fresh chain, is resonance-stabilized and relatively unreactive. They are destroyed either by combination with a similar radical or with peroxy radical.

$$\dot{A} + \dot{A} \longrightarrow A\text{–}A \quad ; \quad RO\dot{O} + \dot{A} \longrightarrow ROOA$$

V. Aromatic Substitution

Hey observed in 1934 that the thermal decomposition of diazonium salts in aromatic solvents leads to the arylation of the aromatic rings. These results were interpreted at that time as involving *homolysis* of the ionized salts to provide aryl radicals which directly displace a hydrogen atom in the aromatic solvent molecule.

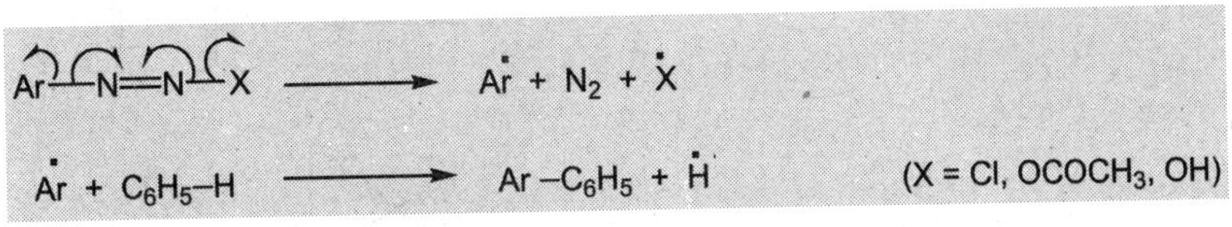

The phenyl radicals generated from other sources such as the decomposition of benzoyl peroxide in aromatic solvents also lead to similar results.

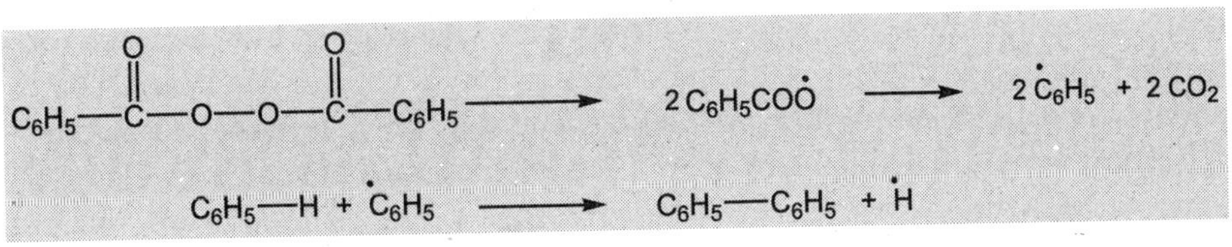

Further investigation of arylation reactions, however, resulted in the accumulation of evidence against the direct displacement mechanism. It was noted, for instance, that as the reaction involves the formation of carbon-carbon bond at the expense of a relatively stronger carbon-hydrogen bond, it should require very high activation energy which is actually not the case. Furthermore, the direct displacement mechanism does not explain products such as **9** and **10**.

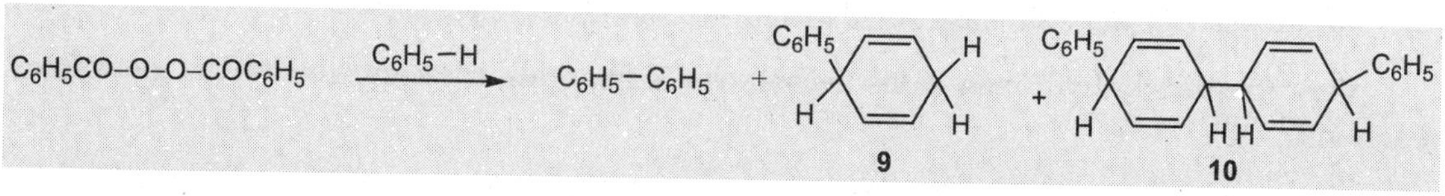

A two-step mechanism, which is not only exothermic but also explains the formation of **9** and **10**, was subsequently proposed.

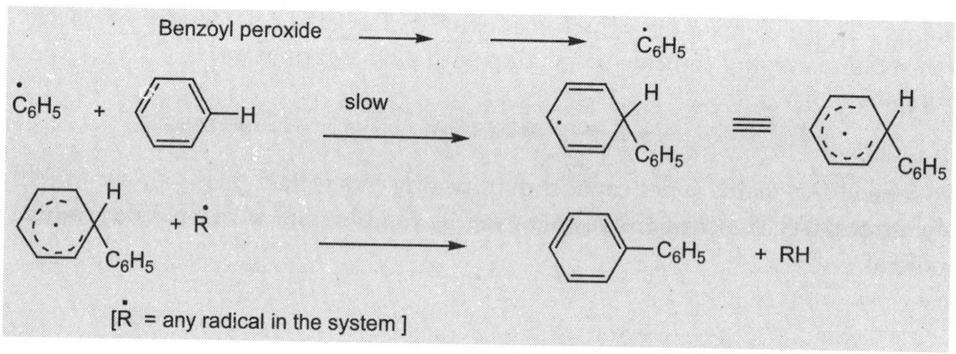

Products **9** and **10** arise due to disproportionation and dimerization of the intermediate radical, respectively.

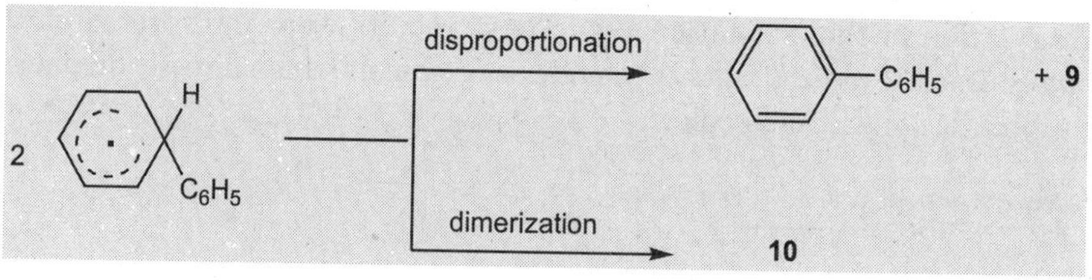

Relative rates of phenylation of different aromatic compounds have been determined by the competitive method in which a phenyl radical is generated in a solvent consisting of two substances (benzene and chlorobenzene) and the proportion of each compounds phenylated is then determined. Such studies have shown that nitrobenzene, chlorobenzene and toluene react faster than benzene in contrast to electrophilic aromatic substitution in which electron-donation or withdrawal by the substituents is an important factor. This is to be expected since radicals such as phenyl are neutral species and are not influenced by the polar properties of the substrate to any significant extent. Furthermore, it has been demonstrated that both electron-donating and electron-withdrawing substituents stabilize a free radical. The following generalizations have been made regarding the reactivity of free radicals with aromatic substrates.

1. All substituents increase reactivity at *ortho* and *para* positions as compared to that of benzene. There is no great difference between electron-donating and electron-withdrawing groups.

2. Reactivity at *meta* position is almost equal to that of the benzene. This fact, coupled with the preceding one, means that all substituents are activating and none are deactivating.
3. Reactivity at *ortho* position is greater than at *para* position.
4. Electron-withdrawing groups exert a greater and activating effect than electron-donating groups.
5. Substituents exert less influnce than in aromatic electrophilic or nucleophilic substitution reactions.
6. Although hydrogen is the leaving group in most of the free radical aromatic substitution reactions, *ipso* attack and *ipso* substitution (*e.g.* with Br, NO_2, or CH_3CO as the leaving group) have been observed in many cases.

VI. Rearrangement of Radicals

Although a primary radical may sometimes produce a tertiary radical by the migration of a substituent from the neighbouring carbon atom, in practice, radical rearrangements have been found to be distinctly less common than carbocation rearrangements. For example, *neopentyl cation* undergoes rearrangement to a more stable carbocation unlike *neopentyl radical* which does not rearrange.

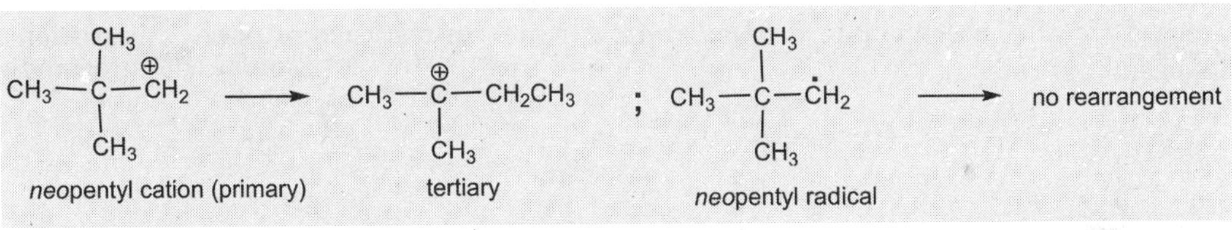

This diminished tendency for rearrangements is due to the fact that the difference in stability between a *primary* and a *tertiary* radical is not as much as between a primary and a tertiary carbocation. Most of the reported radical rearrangements involve the shift of an aryl group from one atom to the adjacent atom. One illustrative example is the rearrangement accompanying decarbonylation of β-phenyl*iso*valeraldehyde (**11**). The reaction, which is induced by *tert*-butyl peroxide, yields a product mixture consisting of almost equal amounts of *iso*butylbenzene (**12**) and *tert*-butylbenzene (**13**).

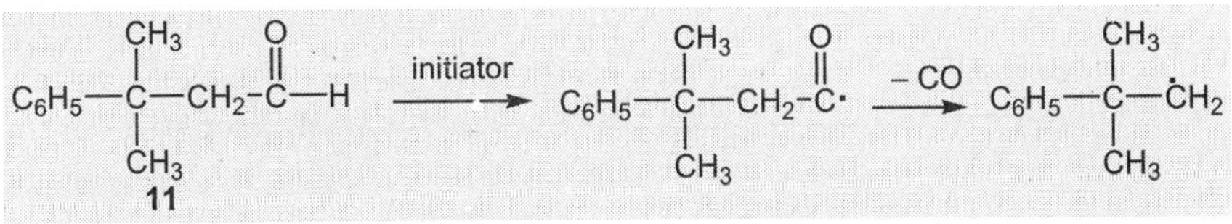

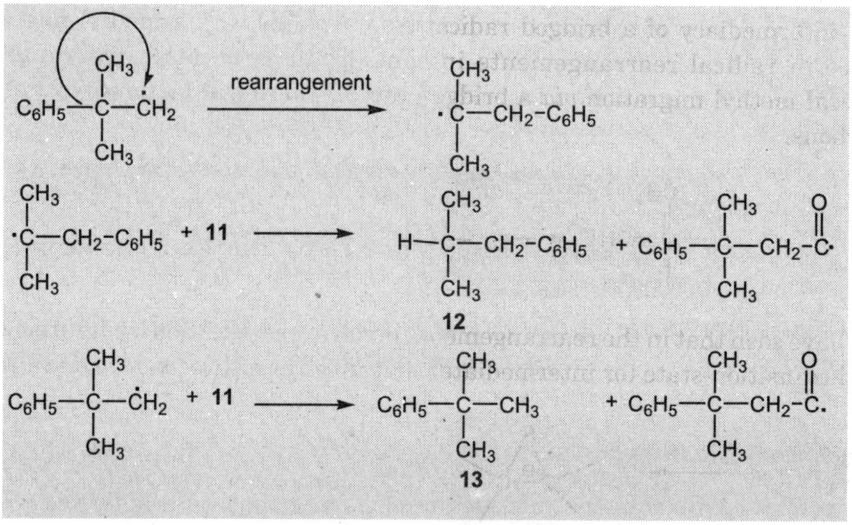

The yield of the rearranged product increases with the dilution of the reaction mixture. This appears quite consistent with the mechanism formulated above as decrease in the concentration of the hydrogen donor (**11**) would increase rearrangement at the expense of hydrogen abstraction by the radical. Further, the rate of the rearrangement is inhibited by the addition of an effective hydrogen donor such as thiophenol. The rearrangement is believed to proceed *via* the intermediate *bridged radical* which in the case of phenyl migration is a cyclohexadienyl radical.

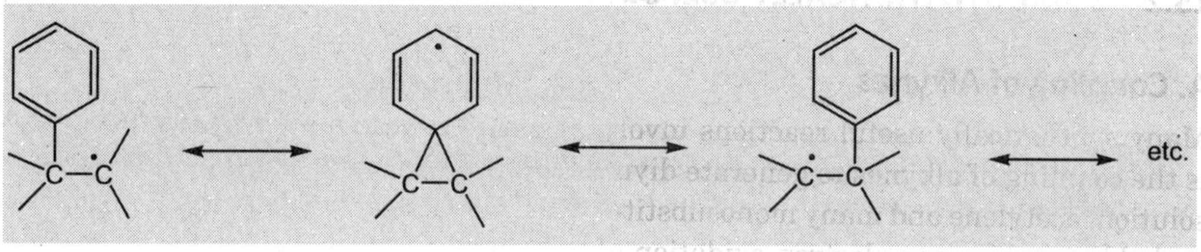

In many cases, vinyl and acyl substituents also migrate in reactions involving free radical rearrangements. Obviously, both these processes involve intermediacy of cyclopropyl intermediates.

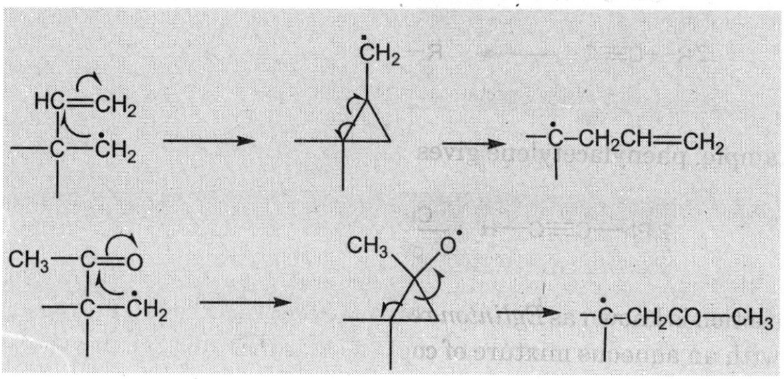

The intermediacy of a bridged radical also explains why a methyl migration does not take place in radical rearrangements in contrast to carbocation rearrangements. For a hypothetical methyl migration *via* a bridged radical there will be *three electrons* spread over three carbons.

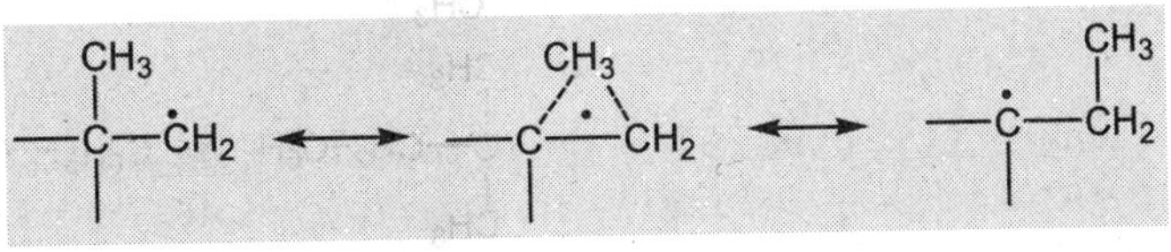

We have seen that in the rearrangements involving carbocations, migration occurs through a bridged transition-state (or intermediate) that involves a three-centre *two-electron* bond.

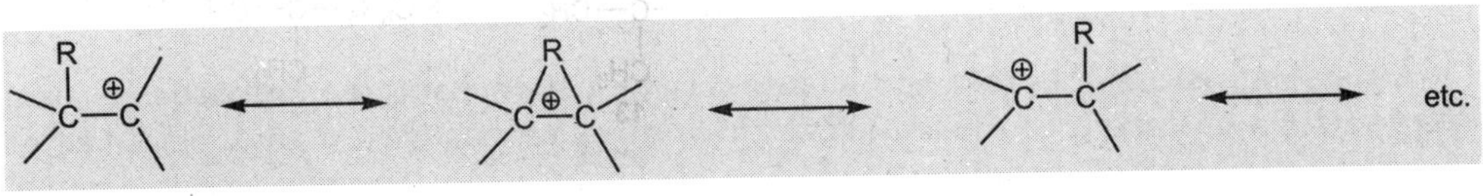

In a free radical, there is a third electron in the system that cannot occupy the same orbital and must go to an antibonding orbital. It is thus clear that the transition-state for migration is less favourable than for the corresponding carbocation. Molecular orbital calculations have also shown that such a system is very unstable in comparison to a two electron system of a carbocation rearrangement.

13.2 SOME SYNTHETICALLY USEFUL REACTIONS

A. Coupling of Alkynes

Many synthetically useful reactions involve intermediacy of free radicals. One such reaction is the coupling of alkynes to generate diynes. In the presence of copper (II) acetate in pyridine solution, acetylene and many monosubstituted alkynes give corresponding diynes. The reaction probably involves one-electron oxidation of the acetylide anion by copper (II) ion followed by dimerization of the resulting acetylide radicals:

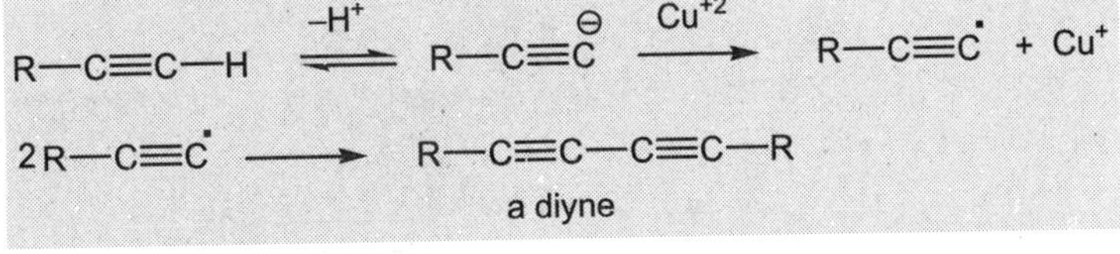

For example, phenylacetylene gives 1,4-diphenylbutadiyne in excellent yields.

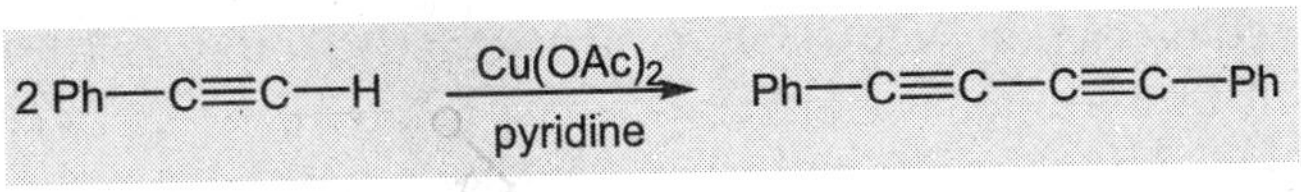

This reaction is known as *Eglinton reaction*. The reaction can also be carried out by treating the alkyne with an aqueous mixture of copper(I) chloride and ammonia or ammonium chloride in air. In the acidic condition, removal of the proton may be facilitated by complexing between the C≡C bond and copper (I) ion,

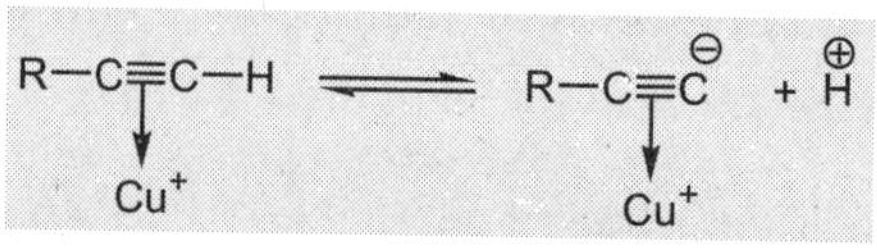

Some of the copper(I) ion is oxidized by air to copper(II) ion which in turn oxidizes the acetylide ion to the radical while returning to the copper(I) state. Yields are normally high, although for acetylene itself the yield is quite low, the main product being vinylacetylene. Atmospheric oxygen or some other oxidizing agent, such as permanganate or H_2O_2 is required in this method. This reaction is often reffered to as *Glaser reaction*.

Synthesis of Unsymmetrical Di- and Poly-ynes

A coupling reaction between the mixture of two monosubstituted acetylenes normally gives a mixture of three products. Although this procedure, has been used to obtain several unsymmetrical di-and poly-ynes in low yield, it is more satisfactory to employ a related coupling process which can give only one product. This procedure (*Cadiot-Chodkiewicz coupling*) consists of treating a monosubstituted acetylene with a 1-bromoalkyne in the presence of copper (I) ion: The copper(I) ion acts catalytically, probably *via* the complex which it forms with C≡C bonds.

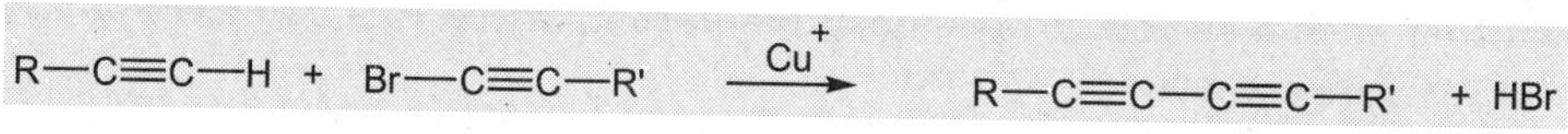

B. Arylation of Aromatic Rings

In this synthetically useful reaction, homolytic arylation of aromatic compounds is carried out by generating aryl radicals. In one of the procedures, aromatic diazoacetates are decomposed to produce aryl radicals which then react with the aromatic substrate present in the liquid form. The reaction sequence may be written as:

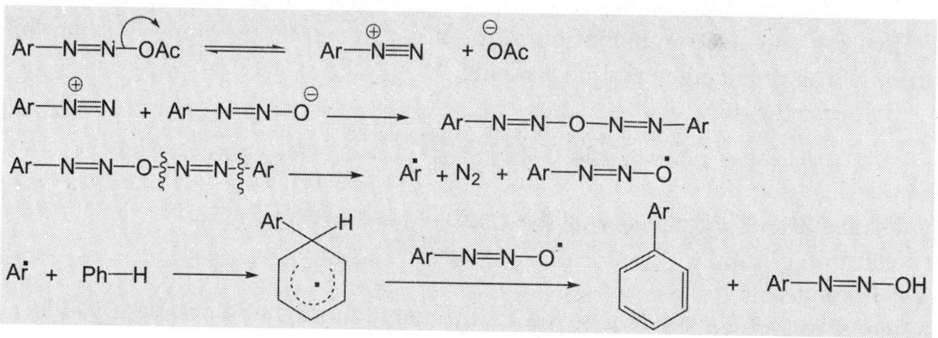

In Gomberg reaction (or Gomberg-Bachmann reaction), a mixture consisting of the solution of the diazonium salt and aromatic liquid, which is to be arylated, is treated with aqueous solution of NaOH resulting in the formation of diazohydroxide.

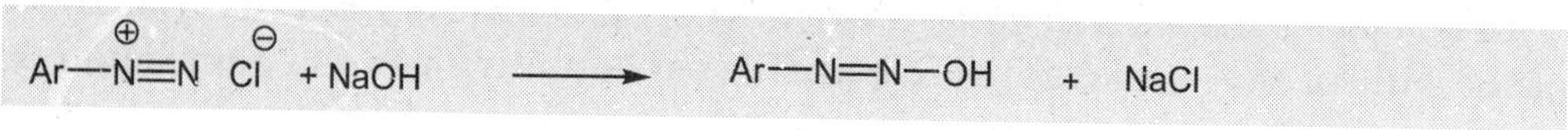

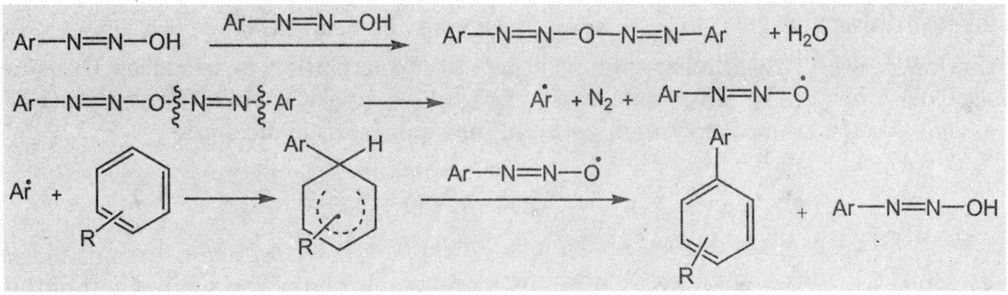

In case of an unsymmetrical aromatic compound, arylation in Gomberg conditions provides a mixture of all the possible products. For example,

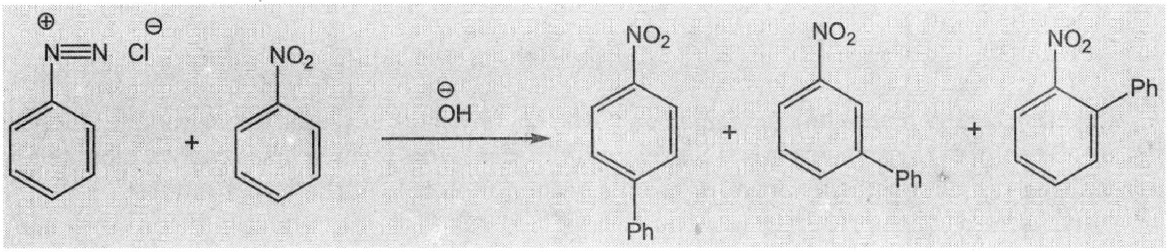

With this limitation, the Gomberg reaction is more valuable when benzene itself or a symmetrical *p*-disubstituted benzene is involved.

C. Sandmeyer Reaction

The arenediazonium halides, obtained by treating aromatic amines with nitrous acid, upon decomposition by warming with cuprous chloride or bromide in presence of excess of hydrogen acids yield aryl halides. It is the halogen attached to copper which goes to benzene ring. This is known as Sandmeyer reaction.

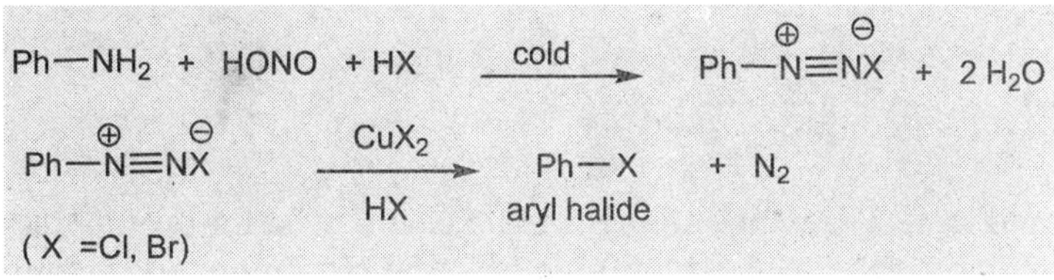

Aryl chlorides are usually prepared by adding a cold aqueous solution of the diazonium chloride to a solution of copper (I) chloride in HCl and then followed by heating to obtain aryl chlorides. Aryl bromides are prepared from diazonium hydrogen sulphates and copper (I) bromide in HBr. The mechanism involves one-electron reduction of the diazonium ion.

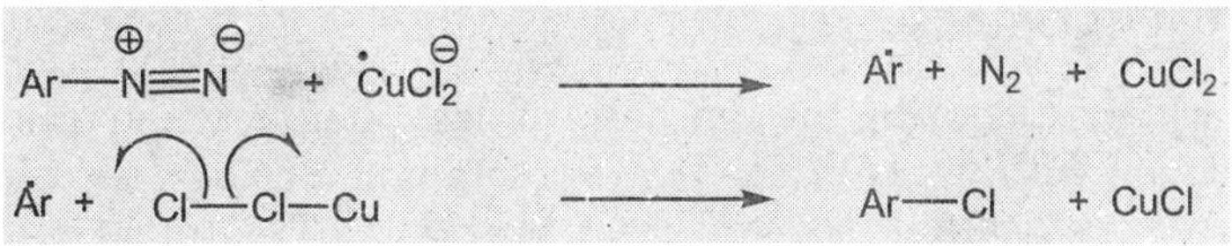

Hunsdiecker Reaction

A synthetically useful reaction involving bromo-decarboxylation is called as Hunsdiecker reaction. Treatment of the silver salt of a carboxylic acid with bromine in refluxing carbon tetrachloride gives a bromo compound with elimination of carbon dioxide:

$$RCO_2Ag + Br_2 \xrightarrow[\Delta]{CCl_4} RBr + AgBr + CO_2$$

The silver salt is prepared by treating an aqueous solution of the sodium salt of the acid with silver nitrate. Yields are variable but in many cases these are high. For instance, *neo*pentyl bromide is obtained in good yields from the corresponding silver salt.

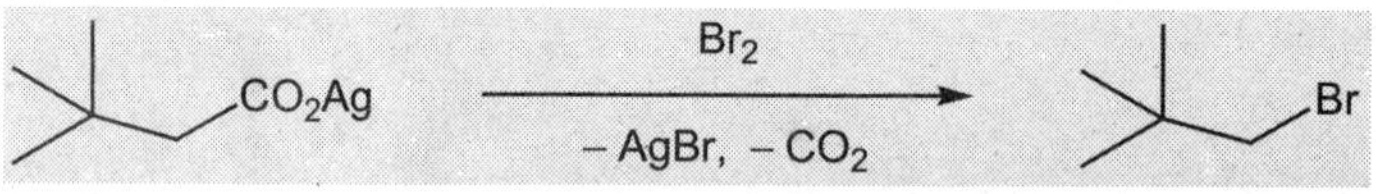

Reaction occurs by the initial formation of the acyl hypobromite and subsequent homolysis of the O–Br bond of the hypobromite. The carboxy radical thus formed loses carbon dioxide and the resulting radical abstracts bromine from a second molecule of the hypobromite:

Various steps of the reaction can thus be written as:

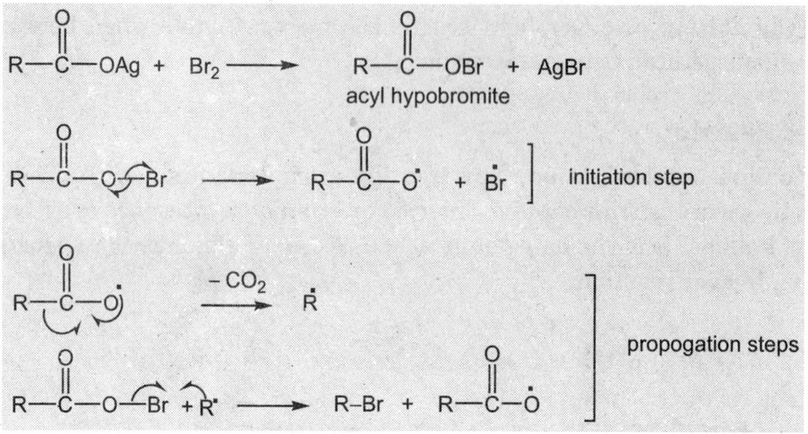

Mercury(II) and lead (IV) compounds have also been used for bromo-decarboxylation. For instance, bromocyclopropane has been obtained by the treatment of cyclopropane carboxylic acid with mercury (II) oxide and bromine.

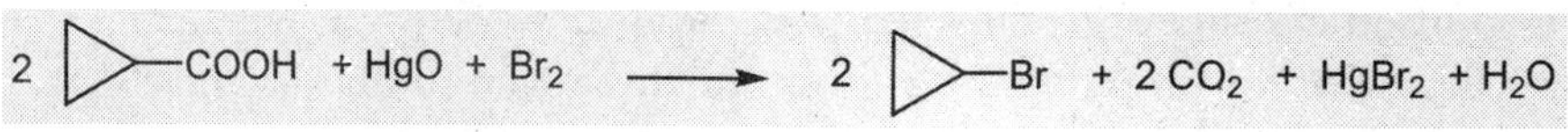

13.3 STABLE FREE RADICALS

In 1900 Gomberg, while attempting the synthesis of hexaphenylethane from triphenylmethyl chloride by the reaction with finely divided silver, obtained a *colourless solid* whose solution in benzene was *bright yellow* and showed great reactivity. Treatment of the solution with iodine, for instance, yielded triphenyliodomethane [$(C_6H_5)_3C$–I]. Similarly, it combined with nitric oxide and oxygen to yield triphenylnitrosomethane [$(C_6H_5)_3C(NO)$] and the peroxide $(C_6H_5)_3C$–O–O–$C(C_6H_5)_3$, respectively.

Although the molecular weight of the solid substance (dimeric product) was much nearer to hexaphenylethane than that of triphenylmethyl, the reactivity and colour of its solution led Gomberg to believe that the solid he obtained was triphenylmethyl. Subsequent work, however, showed that the solid dimer in benzene solution slowly dissociates reversibly into the highly reactive yellow triphenylmethyl radical.

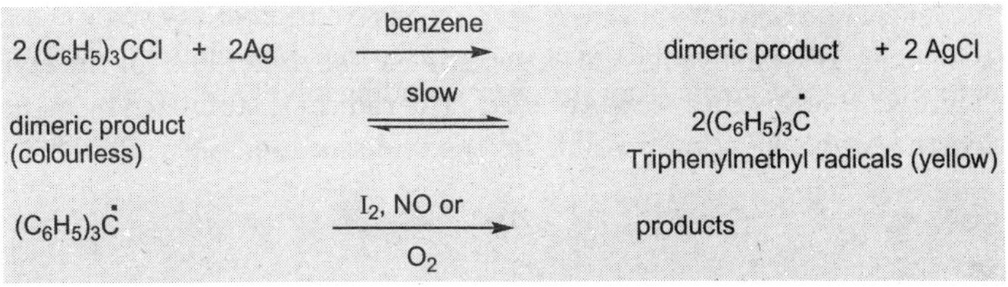

After being consumed in any of the above reactions, more triphenylmethyl radicals are formed by the dissociation of the dimer until finally the whole substance has reacted. A simple experiment to demonstrate equilibrium between them may be carried out by exposing a solution to the air for a short time when the colour of the solution is rapidly discharged due to the conversion of the radicals into peroxide. When air is excluded, radicals begin to accumulate and so the colour of the solution returns in a few minutes. The colour can be discharged again by admitting oxygen. The cycle can be repeated several times before entire quantity is destroyed. It was eventually shown that the structure of the dimeric product is a cyclohexadienes derivative (**18**).

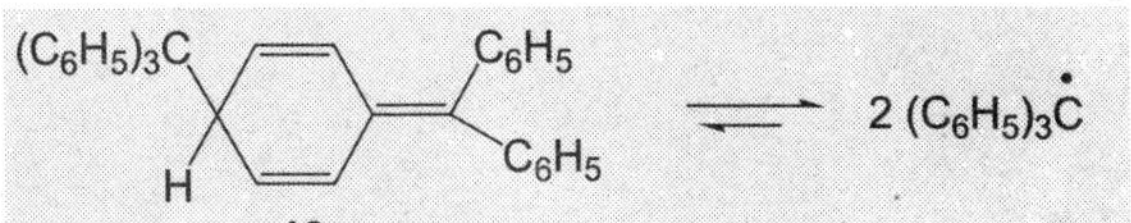

18

It may be noted that the dimerization of triphenylmethyl radicals gives the product which does not have the hexaphenylethane structure. Rather the aromatic stabilization energy of one ring has been sacrificed to avoid sterically congested dimer, $(C_6H_5)_3C$–$C(C_6H_5)_3$. It is interesting that for 70 years, the product of dimerization of triphenylmethyl radical was believed to be hexaphenylethane which indeed was found to be a cyclohexadiene dimer (**18**) by NMR studies.

original suggested structure of the dimer (1900)

correct structure of the dimer (1970) (**18**)

Mechanism for the dimerization of triphenylmethyl radical to give **18** may be postulated as:

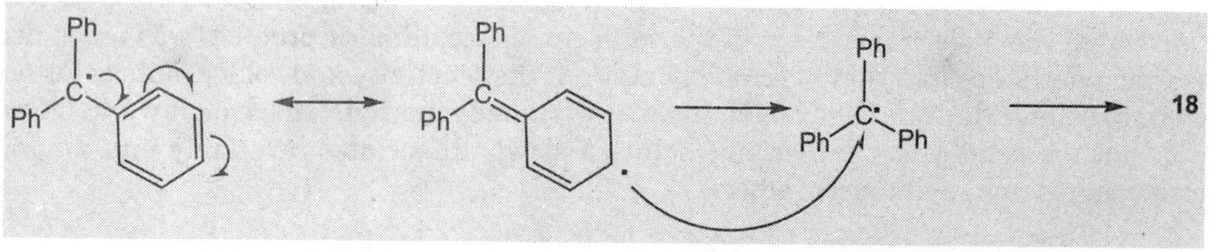

One of the factors responsible for the dissociation of the dimer and for the stabilization of the triphenylmethyl radical is the *resonance effect*. The radical is stabilized by many additional resonating structures which are not possible for the undissociated molecule (**18**).

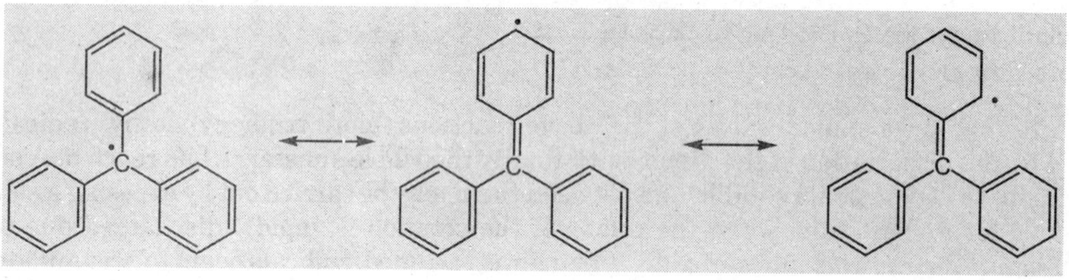

Resonance energy of cyclohexadiene derivative (**18**) is, therefore, less than that of the radicals formed from it and hence dissociation is favoured. Furthermore, the extent of dissociation increases in the substituted dimer as they permit additional resonating structures. Thus for a *p*-alkyl substituent there is hyperconjugative stabilization.

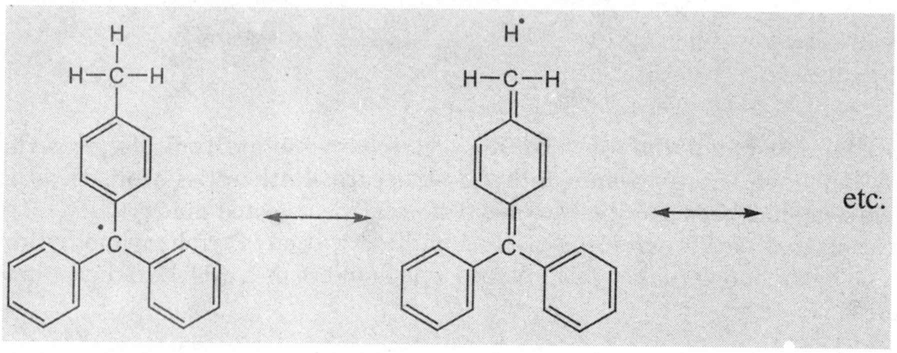

The degree of dissociation can also be increased by introducing the phenyl groups to polynuclear systems. Successive introduction of *p*-biphenylyl groups (*p*-$C_6H_5.C_6H_4$–) in place of phenyls, for example, increases the percentage of dissociation to 15, 80, and 100%. Evaporation of the benzene solution of the radical with three biphenyl groups offered the first crystalline triarylmethyl radical. Electron-withdrawing substituents such as nitro groups also increase radical stability and hence promote dissociation.

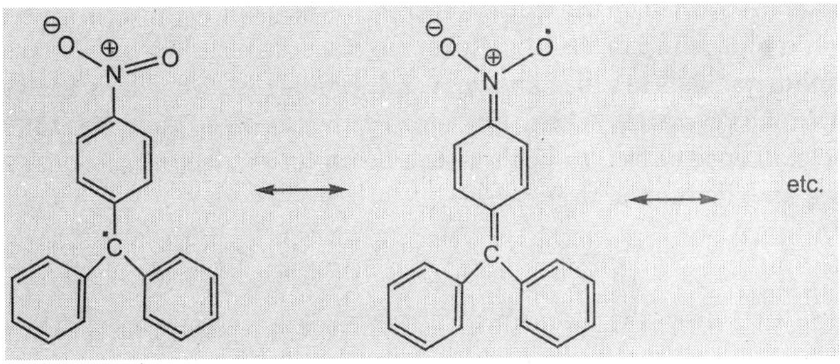

Steric effects are sometimes invoked to explain the stability of these radicals. It is argued, for example, that as the three phenyl groups in a triphenylmethyl radical occupy large volume, it is difficult for two such radicals to come closer for a bond formation. This assumption is further supported by the observation that *ortho*-substituted 'hexaphenylethanes' (**18**) are more highly dissociated than their *meta*-and *para*-substituted analogues. It has indeed been found that the three phenyl rings in the triphenylmethyl radical are not coplanar but are twisted by about 30°, like a *propeller* thereby reducing the extent of delocalization. The central carbon, having the radical, is thus sterically shielded by the twisted phenyl groups and is less available for chemical reactions.

However the steric effect does not explain the greater dissociation of *para*-substituted **18** in comparison to the unsubstituted molecule. Similarly, it is difficult to explain on this basis, the greater increase in dissociation which occurs when phenyl substituents are replaced by *p*-phenylyl substituents. It, therefore, looks almost certain that the major factor responsible for the ready dissociation of **18** is the *resonance-stabilization* of the resultant triphenylmethyl radicals.

Stable radicals with unpaired nitrogen are also known. Wieland in 1911 prepared tetraphenylhydrazine (**19**) which forms colourless solution at room temperature, but on heating at 100°C develops a green colour. Nitric oxide was found to react with tetraphenylhydrazine yielding diphenylnitrosamine (**20**) the reaction being first order with a rate independent of the nitric oxide concentration. The most obvious explanation is that **19** dissociates to diphenylamino radical in a rate-determining step which then reacts with nitric oxide to give **20**.

$$\underset{\mathbf{19}}{(C_6H_5)_2N{-}N(C_6H_5)_2} \xrightleftharpoons{\text{slow}} (C_6H_5)_2\dot{N} + \dot{N}(C_6H_5)_2$$

$$(C_6H_5)_2N^{\bullet} + \dot{N}O \xrightarrow{\text{fast}} \underset{\mathbf{20}}{(C_6H_5)_2NNO}$$

Probably the best known stable aroxyl radical is 2,4,6-tri-*tert*-butylphenoxyl radical.

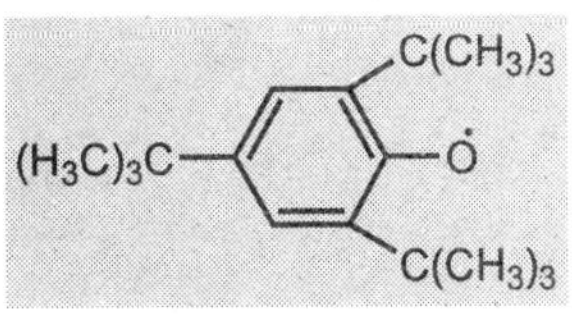

This substance is formed by the oxidation of the corresponding phenol and is a black, stable solid which gives a blue solution. Its structure has been confirmed by ESR studies.

Stable sulphur radicals are also known. A solution of *bis*-(2-benzothiazolyl) disulphide **21** dissociates in toluene reversibly when the temperature is raised to above 100°C. ESR studies have shown the existence of free radicals in the hot solution. Again, dissociation is promoted by the resonance stabilization of the radical.

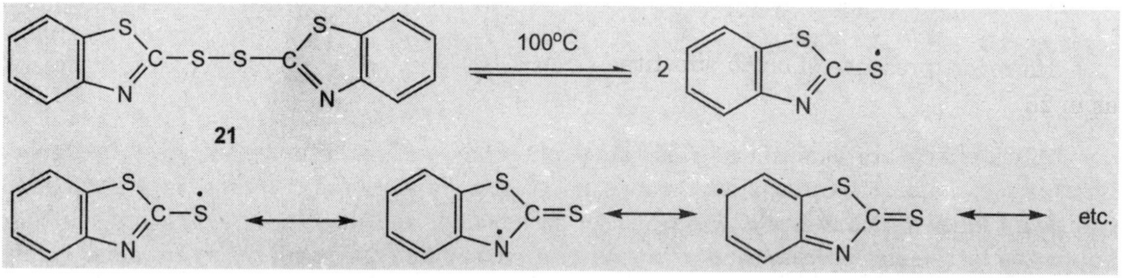

An interesting example of stable free radical is Galvinoxyl (**22**) which is quite stable and can be stored as a solid. The major factors for its unusual stability are the equivalence of two oxygens and delocalization of the electron over the two aromatic rings.

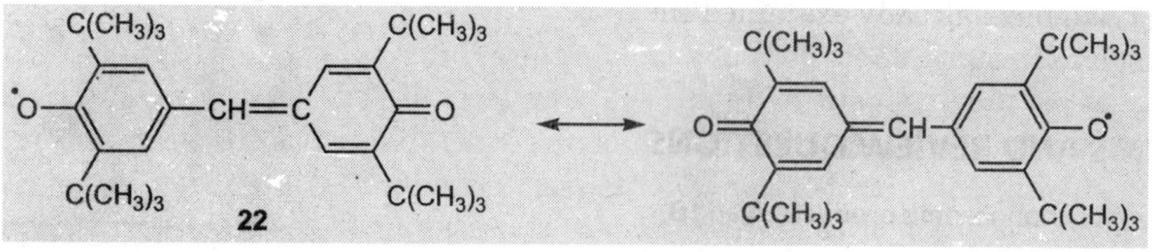

13.4 BIRADICALS

A molecule possessing two unpaired electrons is known as a biradical. The most familiar example is the oxygen molecule, O_2, which has two unpaired electrons with parallel spins.

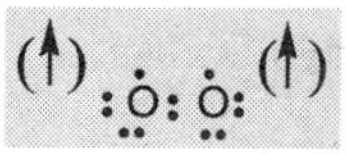

Biradicals such as $\dot{C}H_2CH_2CH_2CH_2\dot{C}H_2$ are prepared by treating polymethylene dihalides with sodium vapours. These types of biradicals display characteristic properties of free radicals.

A number of biradicals having sufficiently long lives are reported. One interesting example is the Chichibabin hydrocarbon (**24**) obtained by the action of zinc upon the corresponding dichloro compound **23**.

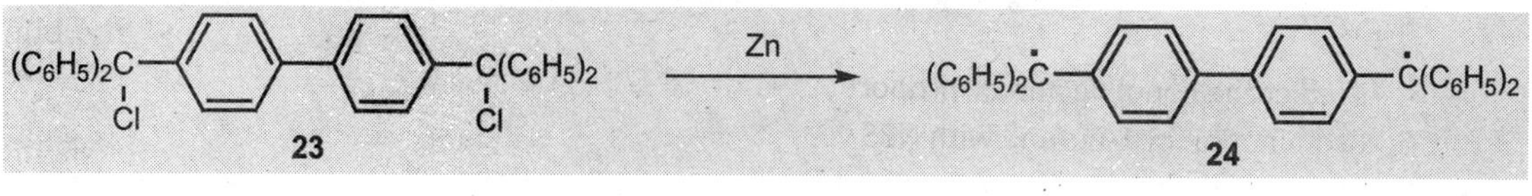

It has a violet-red colour that rapidly fades in the presence of oxygen. ESR studies have revealed that this biradical is in equilibrium with a larger amount of the quinonoid substance.

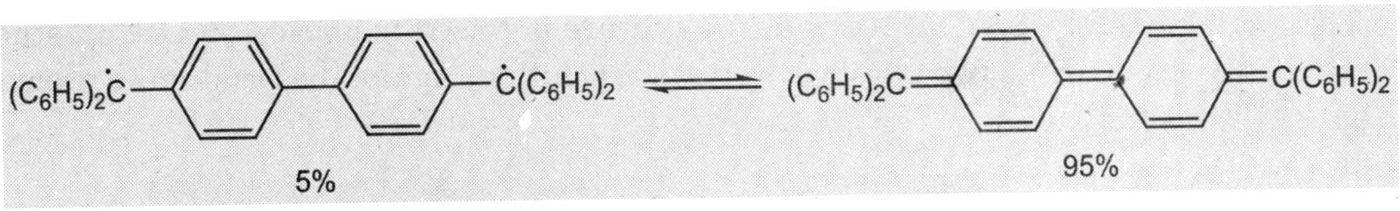

However, presence of *ortho* substituents prevents the formation of quinonoid structure as in **25**.

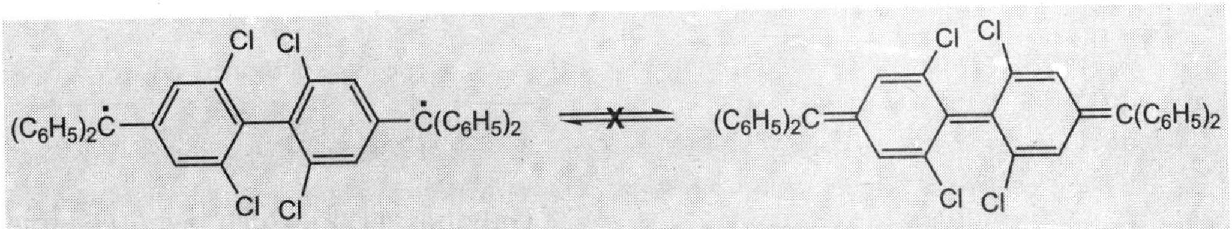

25

Derivatives of methylene, commonly referred to as *carbenes*, are also considered to be biradicals. We have already examined their chemistry in Chapter 8.
closing to give B.

PROBLEMS AND REVIEW QUESTIONS

1. Two monobromo products, **A** and **B**, are obtained by the bromination of *n*-pentane in the presence of light. On treatment with base, both these compounds undergo elimination reaction yielding an alkene, **C**. What are the structures of **A, B** and **C**?
2. Which of the two isomeric compounds, **A** and **B**, will readily dissociate to yield free radicals?

A **B**

3. How do you explain the existence of kinetic isotope effect in the free radical bromination of $C_6H_5CD_3$? Will this isotope effect increase (or decrease), if HBr is removed from the reaction mixture as soon as it is formed?
4. Consider radicals **A** and **B**:

$CH_3CH_2CH_2\dot{C}H_2$ $CH_3-\dot{C}H-CH_3$

A   **B**

Predict the coupling and disproportionation products of **A** and **B**.

5. Bromination of 1-butene with NBS yields two products, **A** and **B**.

$CH_3CHBrCH{=}CH_2$	$CH_3CH{=}CHCH_2Br$
A	**B**

Which will be the major product?

6. How do you account for the formation of three products on the treatment of labelled (* = ^{14}C) cyclohexene with NBS?

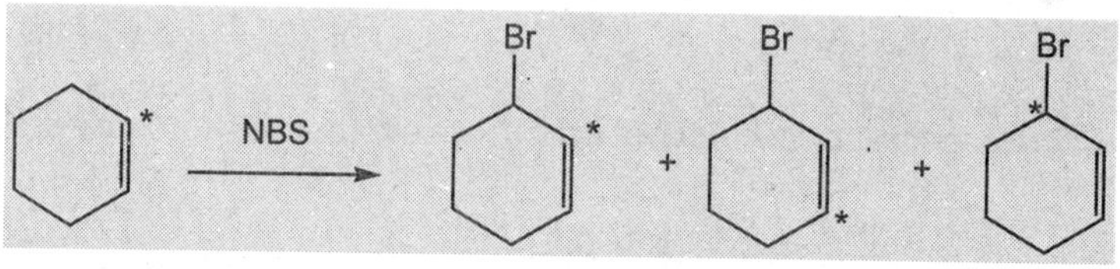

7. An alkane having molecular formula C_5H_{12} gives only one monochloro derivative. Write its structure.
8. Two isomeric hydrocarbons **A** and **B** have the molecular formula C_6H_{12} and do not show the presence of a double bond in their NMR and IR spectra. Both these compounds react with chlorine in sunlight to afford the monochloro derivatives having molecular formula $C_6H_{11}Cl$, whereas only one product is obtained in case of **A**, several compounds are isolated in the case of **B**. Write structures of **A** and **B** and explain the reactions.
9. Sketch a mechanistic scheme for the following transformation:

$$C_6H_5CH_3 + BrCCl_3 \xrightarrow{h\nu} C_6H_5CH_2Br + CHCl_3$$

10. Tell the number and write the structures of all the chloroalkanes that could be produced by the free radical chlorination of ethane.
11. Free radical monochlorination of optically active 1,2-dichloropropane yields two optically inactive and one optically active products. Write their structures.
12. Rank the following free radicals in order of decreasing stability (most stable first):

$C_6H_5\dot{C}HC_6H_5$	$C_6H_5\dot{C}HCH{=}CH_2$	$CH_3-\dot{C}H-CH_3$
A	**B**	**C**
$C_6H_5\dot{C}HCH_3$	$CH_3CH{=}CHCH_2\dot{C}H_2$	$CH_3CH_2-\dot{C}(CH_3)-CH_3$
D	**E**	**F**

13. Rank the following compounds in terms of decreasing selectivity in free-radical halogenation (most selective first):

N-Bromosuccinimide	N-Chlorosuccinimide	Cl_2	F_2
A	**B**	**C**	**D**

14. Predict the product(s) in the following transformations:

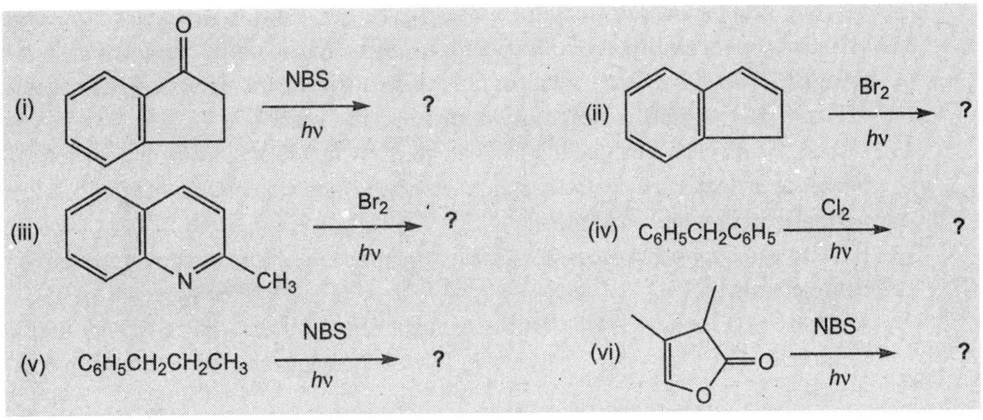

15. Explain formation of the major product in the given reaction along with the mechanism.

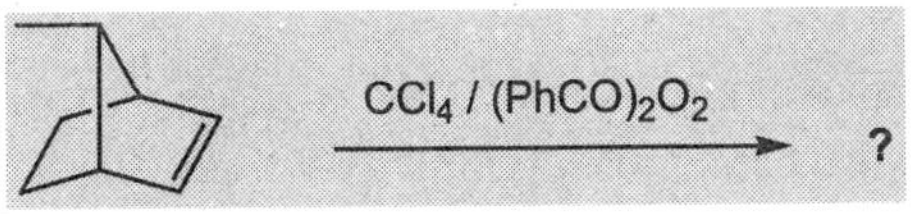

16. Write the structure of product **X** which is formed in the given reaction.

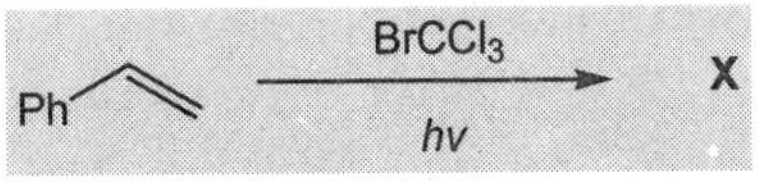

17. Identify the structures of **A** and **B** in the following reactions:

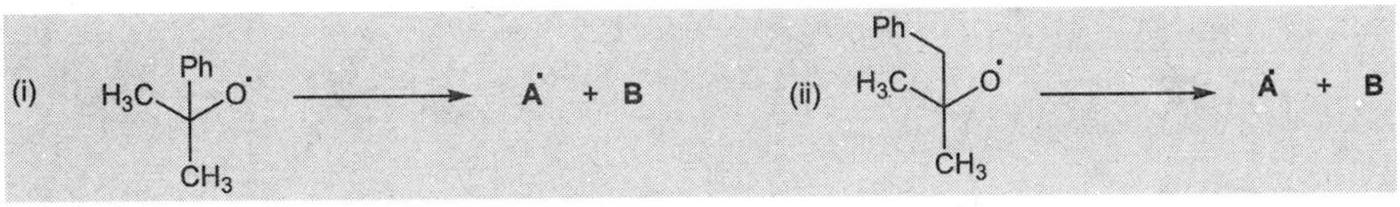

18. Vitamin E plays an important role in taming of harmful free radicals being generated in living tissues. Explain?
19. Which of the following easily generates free radicals?

Sugars	Peptides	Nucleosides	Toluene
A	**B**	**C**	**D**

20. Illustrate the role of peroxides in initiating free radical reactions. Can azo compounds be used as radical initiators? What is the driving force for the dissociation of both these classes of compounds?
21. Discuss the mechanism of polymerization of olefins. What is the role of chain-terminators in controlling the molecular weight of a polymer?
22. Discuss the radical mechanism of halogenation of alkanes with particular reference to the energy involved in the initiation, propagation and termination steps. What is the role of light (or heat) in the halogenation of alkanes?
23. It is observed that the polar factors play an important role in deciding which hydrogen of the

reactant is to be abstracted by the attacking radicals. How do you explain this observation?

24. Discuss the mechanism of addition of bromine to the olefins in dark and in the presence of light. How do you account for the observation that unlike HBr, radical addition of HI and HCl to olefins normally does not take place?
25. Discuss the mechanism of addition of halogens to the carbon-carbon double bond under catalytic influence of light. How do you explain the fact that addition of iodine is difficult to achieve under normal experimental conditions?
26. How do you define autoxidation? Discuss the mechanism of the conversion of benzaldehyde to benzoic acid.
27. Arylation of an aromatic ring can be brought about by aryl radicals generated from a variety of sources. Is it a one-step or two-step process? Discuss the formation of side products in this reaction.
28. How do you explain the observation that the radical rearrangement is distinctly less common than carbocation rearrangements? Explain the formation of *iso*butylbenzene and *t*-butylbenzene during the decarbonylation of β-phenylisovaleraldehyde. Why does the yield of the rearranged product increase with the dilution of the reaction mixture?
29. How do you explain the unusual stability of triphenylmethyl radicals? What is the driving force for the decomposition of hexaphenylethane? Discuss the role of steric and resonance effects in the stabilization of triphenylmethyl radicals.
30. Write the product(s) in the following reactions:

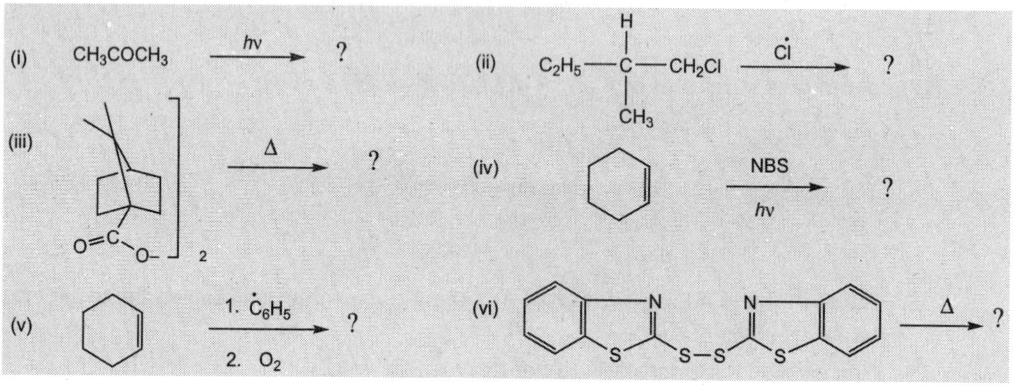

SOLUTIONS

1. **A**, $CH_3CHBrCH_2CH_2CH_3$; **B**, $CH_3CH_2CHBrCH_2CH_3$; **C**, $CH_3CH{=}CHCH_2CH_3$
2. Because of greater steric hindrance, **A** undergoes fission more readily than **B**.
3. As the C-H (C-D) bond is broken in the rate-determining step of the reaction, it will show a kinetic isotope effect. This isotope effect will be greater if the HBr is removed from the reaction mixture as the presence of HBr will result in the formation of $C_6H_5CH_3$ due to successive exchange reactions between the benzyl radical and HBr.
4. Coupling products: **A**, $CH_3(CH_2)_6CH_3$; **B**, $(CH_3)_2CH\text{-}CH(CH_3)_2$

Disproportionation: **A**, $CH_3CH_2CH_2CH_3$ and $CH_3CH_2CH{=}CH_2$
B, $CH_3CH_2CH_3$ and $CH_3CH{=}CH_2$

5. The intermediate free radical is a resonance hybrid. However, a secondary free radical is more stable and hence the major product is **A**.

$$\underset{\text{secondary}}{CH_3\dot{C}HCH{=}CH_2} \longleftrightarrow \underset{\text{primary}}{CH_3CH{=}CH\dot{C}H_2}$$

6. Hint: Write all the resonance structures of the radical generated by the removal of hydrogen atom from both the allylic positions. Addition of bromine radical then generates observed products.
7. $(CH_3)_4C$.
8. **A**, cyclohexane ; **B**, methylcyclopentane.
9.

$$BrCCl_3 \xrightarrow{h\nu} \dot{B}r + \dot{C}Cl_3 \;;\quad C_6H_5CH_3 + \dot{C}Cl_3 \longrightarrow \underset{\text{(resonance-stabilized)}}{C_6H_5\dot{C}H_2} + CHCl_3$$

$$C_6H_5\dot{C}H_2 + BrCCl_3 \longrightarrow C_6H_5CH_2Br + \dot{C}Cl_3$$

10. Nine compounds are possible (one monochloro, two dichloro, two trichloro, two tetrachloro, one pentachloro and one hexachloro derivatives).
11. $ClCH_2CHClCH_2Cl$, $ClCH_2CCl_2CH_3$ and $CH_3CHClCHCl_2$ (optically active).
12. **A, B, D, F, C, E.**
13. **A, B, C, D.**
14.

(i) 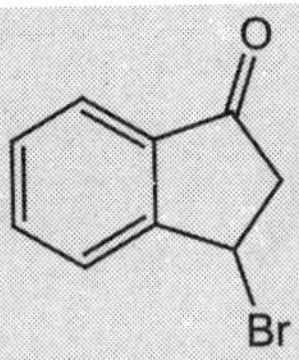

(ii) 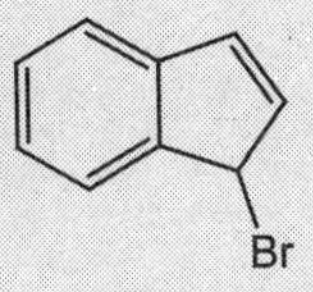

(iii) 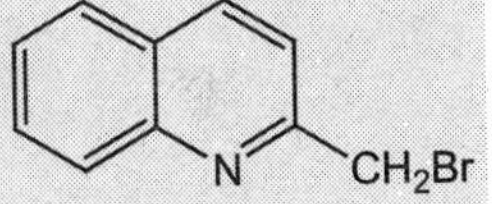

(iv) $C_6H_5CHClC_6H_5$

(v) $C_6H_5CHBrCH_2CH_3$

(vi) 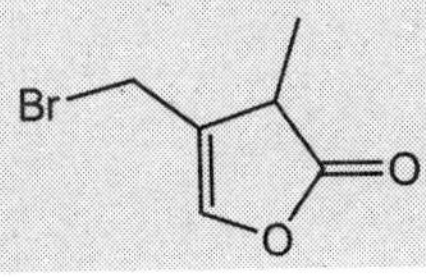

15. The phenyl radical generated on the decomposition of benzoyl peroxide reacts with CCl_4 to produce $^{\cdot}CCl_3$ radical which further reacts on double bond and furnishes trichloromethyl substituted norbornyl radical. As the *exo* face of the radical so generated is shielded by CCl_3 group, chlorine radical subsequently attacks from the *endo* face to give product **A**.

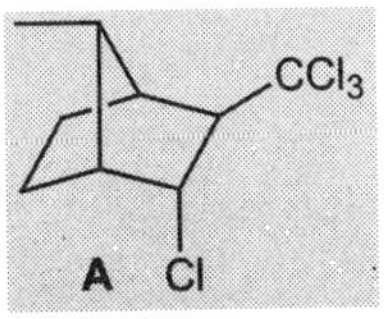

A

16. $BrCCl_3$ has a weak C–Br bond which on homolytic cleavage provides highly stabilized $^{\bullet}CCl_3$ radical. Addition of this radical gives a more stable benzyl radical which on subsequent reaction with another molecule of $BrCCl_3$ gives the product and $^{\cdot}CCl_3$ radical in the chain propagating step. Thus, the final product is $PhCH(Br)CH_2CCl_3$.

17. (*i*) **A** = CH_3, **B** = $PhCOCH_3$ [Cleavage of such radicals is dependent on the stability of generated radical after the cleavage.]

    (*ii*) **A** = $PhCH_2$, **B** = CH_3COCH_3; [Alkoxyl radicals, particularly *tert*-alkoxyl radicals, induce a β-cleavage reaction to generate the stable alkyl radicals and ketones].

18. Vitamin E is converted into a species which is highly stable and cannot react further to damage tissue.

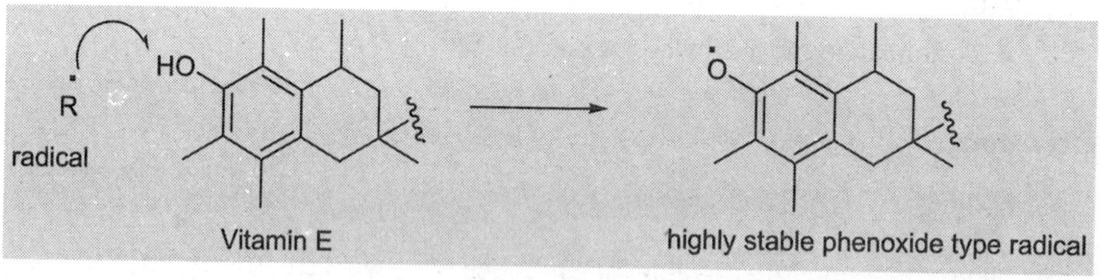

19. (D). Due to strong bond dissociation energy, carbonyl, ester, amino, and hydroxy groups are not affected by the radical reactions. That is why sugars, nucleosides, and peptides can be easily used in free radical reactions without the need of the protection of functional groups.

Chapter 14

Organic Photochemistry

Chapter Outline

14.1 PRINCIPLES OF PHOTOCHEMISTRY
- A. Photochemical Energy
- B. Electronic Excitation
- C. Excited States, Modes of Dissipation of Energy
- D. Energy Transfer
- E. Quantum Efficiency

14.2 PHOTOCHEMISTRY OF CARBONYL COMPOUNDS
- A. Photoreduction
- B. Norrish Type I Reaction
- C. Norrish Type II Reaction
- D. Paterno-Büchi Reaction
- E. Photochemistry of α,β-Unsaturated Ketones

14.3 PHOTOCHEMISTRY OF OLEFINS
- A. Cis-trans Isomerization
- B. Dimerization Reaction
- C. Photochemistry of Butadiene

14.4 PHOTOCHEMISTRY OF AROMATIC COMPOUNDS
- A. Isomerization
- B. Oxidative Coupling and Cycloaddition Reactions
- C. Substitution Reaction

14.5 PHOTOREARRANGEMENT OF CYCLOHEXADIENONES
- A. Photochemistry of α-Santonin

Photochemistry is the study of reactions that are brought about by the action of visible or ultraviolet light. One of the most common examples of photochemical reactions is the process of photosynthesis in the plants by means of sunlight.

$$6\,CO_2 + 6\,H_2O \xrightarrow[\text{chlorophyll}]{\text{sunlight}} C_6H_{12}O_6 + 6\,O_2$$

The carbohydrates so obtained form the basis of life on our planet. In spite of its tremendous potential, the photochemistry remained a field of only incidental and sporadic interest for organic chemists until the late 1950s, when a period of remarkable renaissance began. Very soon its importance was realized in many areas of organic chemistry, especially in the synthesis of certain strained compounds hitherto inaccessible. In addition, interesting mechanisms have come to light during the study of photochemical reactions of organic compounds. According to Chapman, *organic photochemistry is the science arising from application of photochemical methods to organic chemistry and organic chemical methods to photochemistry*. Organic photochemistry is now a fertile field for organic chemists.

14.1 PRINCIPLES OF PHOTOCHEMISTRY

A. Photochemical Energy

Generally speaking, a certain amount of energy, known as *activation energy*, must be supplied to the molecules in order to bring about a chemical transformation. Sometimes, the activation energy requirements are so low that molecules undergo spontaneous transformations even at room temperature. But in most cases, additional energy should be supplied to the molecules before they undergo a chemical change. Very commonly, energy is supplied by increasing the reaction temperature which results in an energy distribution in such a way that most molecules present in the system have about the same amount of energy. At a given temperature, molecules which have acquired sufficient energy react. A second method of introducing energy in a system involves the absorption of electromagnetic radiation in the visible or ultraviolet region. Unlike thermal conditions the *absorption of light excites an individual molecule to an excited electronic state without immediately affecting the surrounding molecules*. This selective excitation of individual molecules is a unique character of photochemical reactions. The chemistry of excited molecules differs from the chemistry of those in the ground state and so it is often possible to change the course of a reaction by activating the reactants by light rather than by heat.

The basic relationships of energy of radiation to frequency and wavelength are given by expressions:

$$\varepsilon = h\nu \qquad (14.1)$$

ε = energy per quantum

h = Planck's constant = 6.63×10^{-27} erg.sec.

ν = frequency of radiation in cps or Hz

$$\nu = c/\lambda \qquad (14.2)$$

λ = wavelength in cm

c = velocity of light (3×10^{10} cm sec^{-1})

From Eq. 14.2 we derive that the frequency and wavelength are inversely related and so radiations of shorter wavelength should contain more energy (Eq. 14.1). The energy absorbed per mole at a specific wavelength may be readily calculated from the equation,

$$E = N h\nu = N hc / \lambda \text{ where, } N \text{ is Avogadro's number} = 6.02 \times 10^{23}$$

$$E = \text{energy per mol}$$

$$E = \frac{6.02 \times 10^{23} (6.63 \times 10^{-27}) (3 \times 10^{10})}{4.184 \times 10^{7} \lambda \text{ (nm)}} \times 10^{7} \text{ cal/mol}$$

$$= \frac{2.86 \times 10^{4}}{\lambda \text{ (nm)}} \text{ kcal/mol}$$

Radiations in the infrared region of the spectrum correspond to 1-10 kcal/mol (or 4.2-42 kJ/mol) of energy which is capable of producing only vibrationally or rotationally excited molecules. On the other hand, light in the visible and ultraviolet region has sufficient energy to cover the range of chemical bond energies and is often able to induce chemical changes by exciting molecules to *higher electronic states*. The energy of visible light varies from 38 kcal/mol (or 159.6 kJ/mol) (750 nm) to 71 kcal/mol (or 298.2 kJ/mol) (400 nm), whereas ultraviolet light is still more effective as it provides energy upto143 kcal/mol (or 597.8 kJ/mol) (200 nm).

B. Electronic Excitation

We know that an orbital can accommodate a maximum of two electrons and that a combination of two atomic orbitals (to form a bond) produces two molecular orbitals. One of these is a low energy orbital and is referred to as a *bonding orbital* which accommodates both the electrons in the ground state of the molecule. The other orbital, called an *antibonding orbital*, has higher energy than either of the atomic orbitals and it remains vacant in the ground state (**Fig. 14.1**).

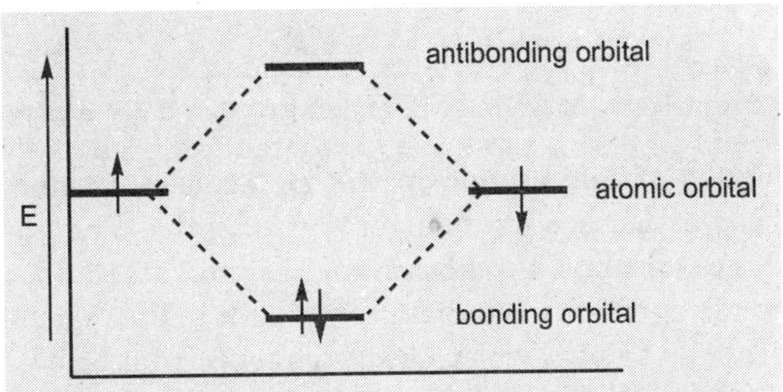

Fig. 14.1 Relative energies of atomic and molecular orbitals

The absorption of light energy by molecules in the visible and ultraviolet region involves *promotion of an electron from the bonding orbital to the corresponding antibonding orbital*. The antibonding orbitals corresponding to σ and π bonds are called σ^* (sigma starred) and π^* (pi starred), respectively. As the nonbonding electrons (n) are not involved in the bond formation, there are no antibonding orbitals corresponding to them. Excitation promotes a nonbonding electron either to a σ* or a π*.

We have the following types of electronic transitions involved in the visible and ultraviolet regions: σ, σ*; *n*, σ*; *n*, π*; and π, π*. The energies required for σ, σ* and *n*, σ* transitions are usually very high, requiring a wavelength of far less than 200 nm (~150 nm). Although photochemical excitation may be affected using such light, this is relatively uncommon because strong absorption by atmospheric oxygen below 200 nm creates complications and hence irradiation must be done in vacuum. Butadiene provides a typical example of a π, π* transition in which an electron is promoted from ψ_2 (bonding orbital) to ψ_3 (antibonding orbital) on excitation (Chapter 15).

By far the most interesting types of electronic excitation in organic molecules are the *n*, π* and the π, π* transitions. These transitions have lower energy requirements and occur at longer wavelengths, usually well within the regions of the commercial ultraviolet lamps. Both types may be illustrated with a saturated aldehyde or ketone which absorbs at two wavelengths, one around 285 nm and the other around 180 nm. The absorptions at 285 nm is due to the excitation of an electron from the nonbonding orbital (*n*) to the lowest unoccupied π antibonding orbital (π*) and *since this transition requires less energy than other electronic transitions (σ, σ*; π, π*), it occurs at the highest wavelength.* The other absorption around 180 nm is attributed to the promotion of an electron from the π bonding orbital to the π* antibonding orbital. Obviously, it requires more energy than the *n*, π* transition. **Figure 14.2** schematically displays *n*, π* and π, π* transition energies in such a system.

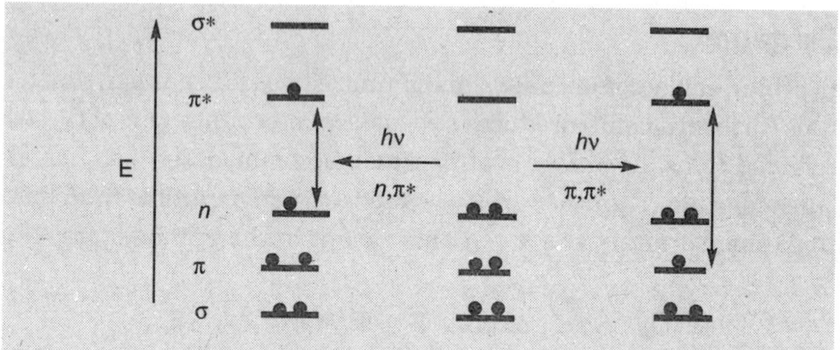

Fig. 14.2 Schematic molecular orbital diagram showing n, π* and π, π* transitions

With the introduction of conjugation in the molecule *n*, π* and π, π* transitions of the carbonyl group occur at increasingly longer wavelengths. The absorption spectrum of benzophenone, for instance, displays two absorption maxima at 345 (*n*, π*) and at 245 (π, π*) nm. For exciting a molecule, it must be irradiated with light in the region where it absorbs, and so it is helpful to determine the ultraviolet absorption spectrum before irradiation. Mercury arc lamps, commonly used for photochemical reactions, have three principal emission lines at 253.7, 313.0 and 366.0 nm. It is then clear that for π, π* excitation, one should use the 253.7 nm line while the 313.0 and 366.0 nm lines are needed for *n*, π* transitions. Filters are available which allow the selection of any of these lines. Irradiations with 313.0 and 366.0 nm lines are conveniently performed in pyrex vessels (pyrex glass is opaque to light of wavelength less than 300 nm) while an experiment using 253.7 nm light necessitates the use of a quartz vessel.

C. Excited States, Modes of Dissipation of Energy

There are even numbers of electrons in a typical organic molecule, and these are usually paired in the ground state. Absorption of light of the correct energy occurs without spin inversion and so the electron spins remain paired in the excited state. This state is commonly called *excited singlet state* or S_1. However, in many cases, electronic excitation is followed by spin inversion thus giving rise to a new excited state with two unpaired electrons. This state is called an *excited triplet state* or T_1. Direct excitation of the ground state (S_o) to a *triplet state is not possible, hence* T_1 *is invariably populated via* S_1. These processes can be visualized for the ethylene molecule, in which absorption of light of the correct wavelength promotes one of the π electrons to the π *antibonding* state, π* (**Fig. 14.3**).

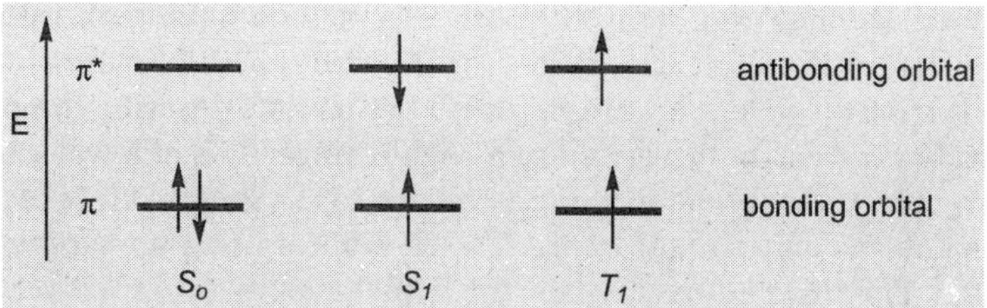

Fig. 14.3 Singlet and triplet states

The terms *singlet* and *triplet* arise from the fact that singlet states do not split in a magnetic field, while triplet states have three possible energy states in an applied magnetic field. A triplet state is more stable than the corresponding excited singlet state because the two electrons do not get very close together in space when their spins are *parallel and hence lesser inter-electronic repulsion is expected with unpaired than with the paired electrons.*

Let us now describe the major events that occur following light absorption by a molecule thus raising it to a higher electronic state, say from S_o to S_2 (S_2 represents the higher vibration level of the upper excited singlet state) in detail. The lifetime of upper excited singlet states is usually less than 10^{-11} second and they rapidly decay to the lowest level (S_1) by a process called *internal conversion.* Decay of S_1 to S_o may also occur but it is usually a slower process than decay from higher electronic singlet states (S_2 etc.) to S_1. Consequently, S_1 has the longest life (~10^{-8} s) of all the excited singlet states and is very important in photochemistry. S_1 may undergo one of the following four energy-degrading processes. (i) it may emit the energy as a photon and drop to S_o. This process is called *fluorescence* and generally occurs within 10^{-9} to 10^{-6} s. (ii) it may return to S_o by non-radiative processes in which excess of energy of the excited state is shuffled into vibration modes. (iii) S_1 may undergo chemical reactions, and finally (iv) its decay to a triplet state (T_1) may occur before any of the above processes takes place.

Although the conversion of S_1 to T_1 is energetically gainful, it is rather a slow process in accordance with the spectroscopic rule that *change in multiplicity is formally a forbidden process.* However, if the singlet state is *sufficiently long-lived*, as in the aromatic and carbonyl compounds, the S_1 to T_1 change (called intersystem crossing; ISC) occurs with a high degree of efficiency. Intersystem crossing is a very important phenomenon in photochemistry as T_1 exists for a much longer period of time (10^{-3} to 1s) than a singlet state and, therefore, photochemical reactions are favoured by the longer life of T_1 state. Intersystem crossing is more efficient in the molecules where the energy gap between S_1 and T_1 is rather small. A typical example is benzophenone. In this case, intersystem crossing has 100% efficiency because of the small energy gap (5 kcal/mol or 21 kJ/mol; **Fig. 14.4**).

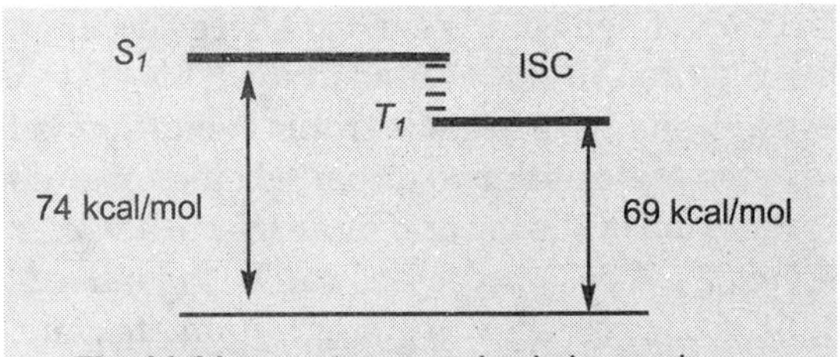

Fig. 14.4 Intersystem crossing in benzophenone

Many aromatic ketones and other aromatic compounds have high intersystem crossing efficiencies while due to the high energy gap in the two states of olefins (50-60 kcal/mol or 210-252 kJ/mol) the latter have a lower tendency. Let us now describe the probable paths for T_1 states. A T_1 state may not be converted to S_1 again since T_1 is of lower energy. However, it may either undergo a chemical reaction or may return to the ground state (S_o) by emission of a photon, a process known as *phosphorescence*. The lifetime of phosphorescence is usually in the range of 10^{-3} to 10s. The third alternative for the triplet state is to return to S_0 by radiationless decay (RD). All these consequences of the absorption of light are summarized in **Fig. 14.5** and also through a Jablonskii diagram (**Fig. 14.6**).

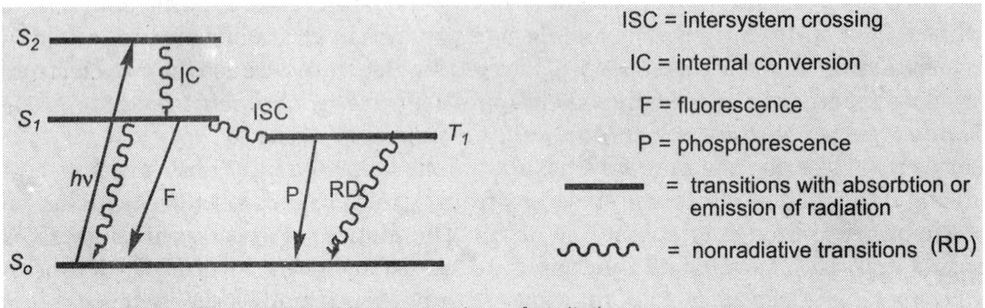

Fig. 14.5 Possible routes to loss of excitation energy

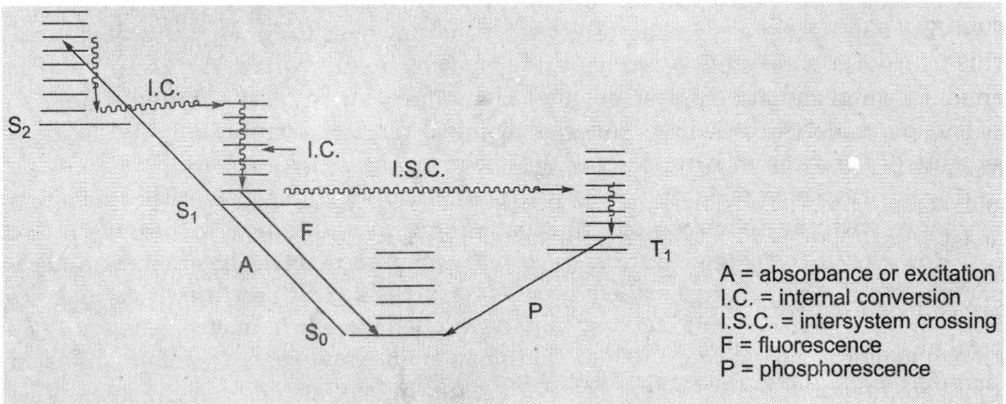

Fig. 14.6 Jablonskii diagram

In addition to the above-mentioned processes, excitation may be transferred to another molecule (**A**), which is thereby raised to an excited singlet (or triplet) state depending upon the nature of the excited molecule (singlet or triplet).

These processes are summarized as follows:

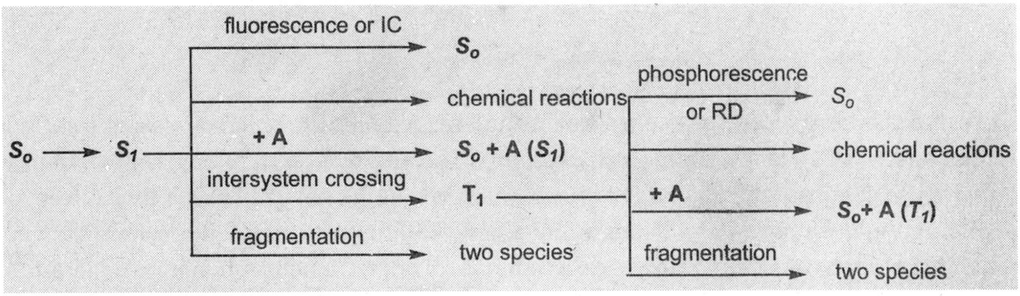

Since S_1 and T_1 states differ in energy contents and electronic distributions, they are expected to possess different chemical and photochemical properties. For example, the S_1 state of butadiene (**1**) undergoes valence isomerization (**2**,**3**) while the T_1 state, *attained by energy transfer, prefers to dimerize* (**4**,**5**,**6**).

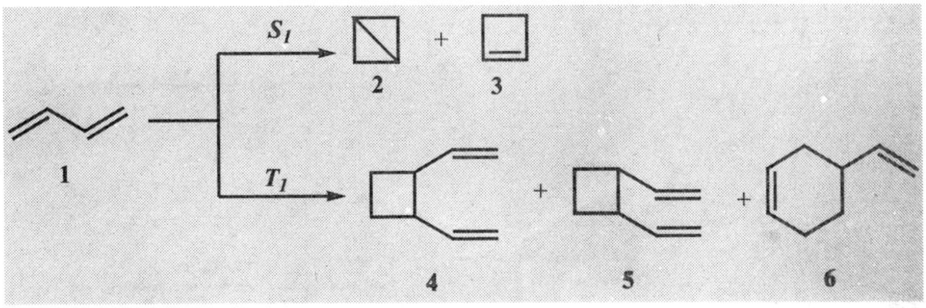

D. Energy Transfer

We have referred to energy transfer as a means of populating the T_1 state. Energy transfer refers specifically to the one-step radiationless transfer of excitation energy from an electronically excited molecule (donor) to the ground state of another molecule (acceptor). As a consequence of this transfer, the donor molecule returns to the ground state and the acceptor molecule gets excited. Depending upon the multiplicity (singlet or triplet state) of the donor molecule, there may be a *triplet-triplet* or a *singlet-singlet* energy transfer. However, the lifetime of S_1 is short compared to that of T_1 and hence there is greater probability of energy transfer through a triplet state. The basic requirement for energy transfer is that the donor molecule should have at least 5 kcal/mol (or 21 kJ/mol) more energy than the energy required to excite the acceptor molecule to its lowest excited state. Under these conditions, energy transfer does take place on every collision between the excited donor molecule and the ground state acceptor molecule. A typical mechanism for triplet energy transfer involves the excitation to S_1 of the donor molecule on irradiation, and the subsequent decay of S_1 to T_1. Collision of the T_1 donor with the ground state acceptor molecule produces a T_1 acceptor and ground state donor molecule.

$$D \xrightarrow{h\nu} {}^{1}D \qquad D = \text{donor}$$
$${}^{1}D \xrightarrow{ISC} {}^{3}D \qquad A = \text{acceptor}$$
$$A + {}^{3}D \xrightarrow{\text{collision}} D + {}^{3}A \qquad 1 = \text{singlet}$$
$${}^{3}A \longrightarrow \text{Products} \qquad 3 = \text{triplet}$$

Let us consider the specific example of butadiene, a molecule which possesses a sufficiently low triplet energy (E_T = 60 kcal/mol or 252 kJ/mol) to be an excellent triplet acceptor. In practice, the only way to generate the T_1 state of butadiene is through energy transfer by a triplet donor. Absorption of light by butadiene is negligible above 250 nm and so no photochemistry is to be expected under such conditions. However, when it is mixed with benzophenone and irradiated at 366 nm, which is a region where benzophenone absorbs appreciable amount of light, butadiene undergoes a change to yield dimers, **4, 5** and **6**.

Apparently, the light is being absorbed by benzophenone while the reaction is taking place with butadiene, or in other words benzophenone is acting as a *photosensitizer*. The mechanism of the energy transfer becomes clear when we view the energy relationships of the various excited states of the two substances (**Fig. 14.7**).

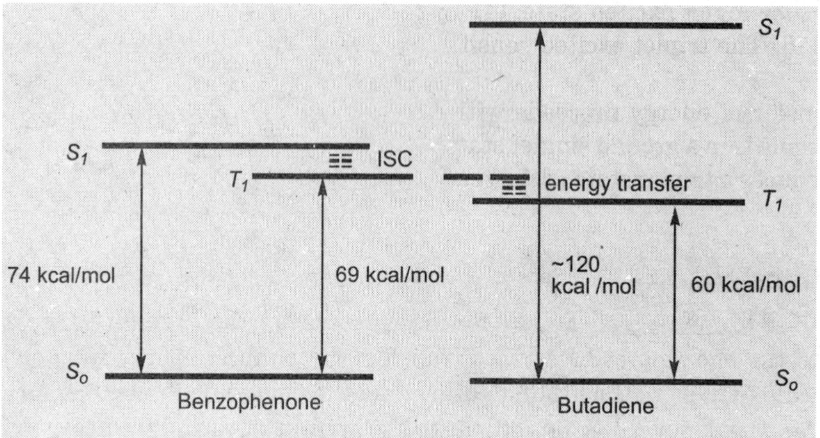

Fig. 14.7 Mechanism of the energy transfer

Benzophenone absorbs light at about 366 nm and is excited to its S_1 state and since the energy gap between S_1 and T_1 state is only 5 kcal/mol (or 21 kJ/mol), there is a very efficient intersystem crossing which populates the T_1 state. When the benzophenone triplet collides with butadiene, a triplet energy transfer occurs resulting in the formation of the triplet state of butadiene which then reacts to produce dimers.

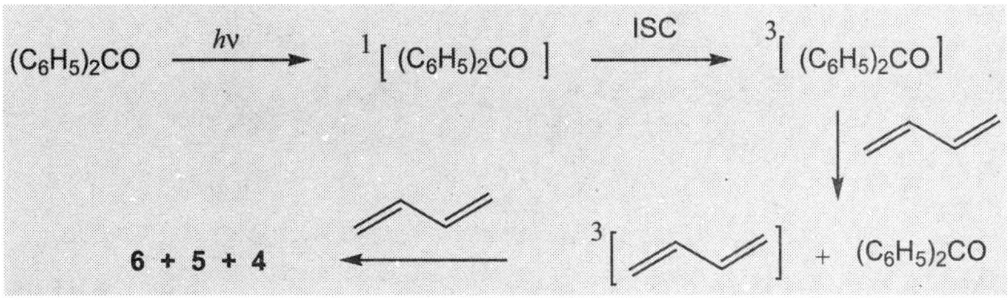

Triplet energy transfer is possible here only because the triplet state of butadiene is of lower energy than that of benzophenone. On the other hand, singlet excitation energy of benzophenone is less than that of butadiene and so singlet energy transfer from benzophenone to butadiene is not possible in the above reaction.

Photosensitization is an important alternative to the direct excitation of molecules and this method usually results in reactions occurring *via* a triplet excited state. If a reaction is to be carried out by photosensitization, a substance known as the sensitizer is present in the system. Each sensitizer has a characteristic energy, E_T, which it can transfer to a reactant. The sensitizer is chosen to meet the following criteria: (1) It must be excited by the irradiation to be used. (2) It must be present in sufficient concentration and absorb more strongly than the other reactants under the conditions of experiment so that it is the major light absorber. (3) Its intersystem crossing rate must be faster than energy transfer to the reactant or solvent from the singlet excited state. (4) The energy of its triplet state must be greater than the reactant. (5) The triplet excited sensitizer must be able to transfer energy to the desired reactant.

The transfer of energy proceeds with net conservation of spin. In the usual cases, the reactant molecule is in a ground singlet state, and its reaction with the triplet excited state of the sensitizer produces a triplet state of the reactant. The mechanism for triplet photosensitization is as follows:

Sens	⟶	^{1}Sens*	Sensitizer singlet formed
^{1}Sens*	⟶	^{3}Sens*	Intersystem crossing of sensitizer
^{3}Sens* + Reactant	⟶	Sens + ^{3}Reactant*	Energy transfer to the reactant molecule

Intramolecular Energy Transfer: It has been observed that certain substituted aromatic ketones and aldehydes are capable of transferring excitation energy from one part of the molecule to the other. An interesting example is that of the series of 4-(1-naphthylalkyl)benzophenones (**7**) which display two independent absorption bands for the benzophenone and the naphthalene components. Irradiation of these compounds with light of the wavelength absorbed by benzophenone showed that there is an efficient transfer of triplet excitation from the benzophenone moiety to the naphthalene moiety, a fact which is further supported by a study of emission spectrum of the system. Again, there is no possibility of a singlet-singlet energy transfer as S_1 of naphthalene is at a much higher energy level than that of benzophenone. In addition, when the compound is irradiated with light of wavelength absorbed by naphthalene, an efficient singlet-singlet energy transfer to benzophenone takes place.

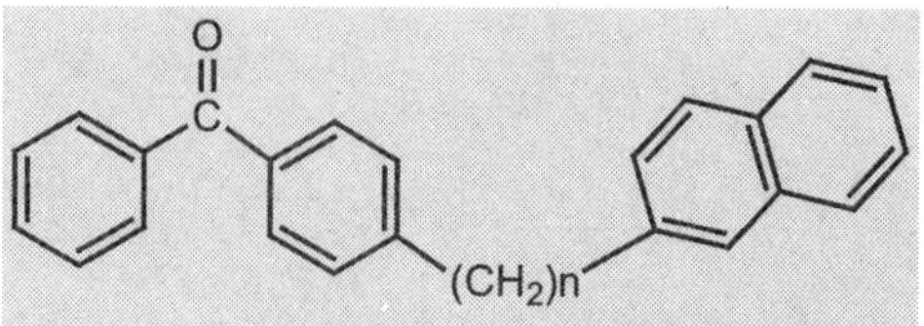

7

Another example of intramolecular energy transfer is the *cis-trans* isomerization of 4-hexen-2-one (**8**) on irradiation with the light absorbed by the carbonyl group.

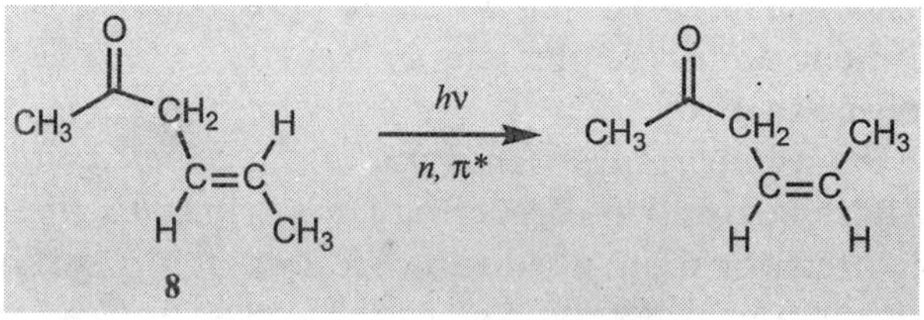

Apparently, the above photoreaction takes place through energy transfer from the carbonyl group to the carbon-carbon double bond.

Energy transfer plays a key role in many photochemical reactions where the molecules do not absorb light in a convenient wavelength range. Even if they do absorb light of high energy, intersystem crossing may be quite inefficient and S_1 may not tend to form much of T_1. In all such cases, T_1 is made available for reactions through the use of triplet photosensitizers such as benzophenone. Finally, triplet energy transfer has been used to form pure triplet species for comparison with the singlet species formed by direct irradiation.

E. Quantum Efficiency

The efficiency of a photochemical reaction is usually expressed in terms of quantum efficiency or quantum yield, Φ, which is defined as

$$\Phi = \frac{\text{Number of molecules reacting or formed in unit time}}{\text{Number of quanta absorbed in unit time}}$$

Quantum yields may be measured using a ferrioxalate actinometer which permits calculation of the amount of the light falling on the compound undergoing a photochemical reaction. If each quantum of light absorbed leads to the formation of a product molecule, then Φ will be 1.0, but in practice it is usually less than 1 because of the decay of some light energy by fluorescence, phosphorescence or radiationless processes. If the quantum yield is 0.01, then only 1 out of 100 molecules that are excited, undergoes photochemical reaction. In the case of chain reactions, quantum yields may have value as high as several thousands. Quantum yields of hydrogen chloride formation, for instance, may be 10^6-10^7 at room temperature. Here a single photoexcitation initiates a series of repeating reactions leading to the formation of many molecules of product per initiation step. Obviously over a million molecules of HCl are formed for every one quantum of light absorbed by chlorine. A chain mechanism explains this observation.

$$Cl_2 \xrightarrow{h\nu} Cl^{\bullet} + Cl^{\bullet}$$
$$Cl^{\bullet} + H_2 \longrightarrow HCl + H^{\bullet}$$
$$H^{\bullet} + Cl_2 \longrightarrow HCl + Cl^{\bullet}$$

The chain is eventually terminated by a combination of two hydrogen atoms, or a hydrogen and a chlorine atom, or two chlorine atoms.

14.2 PHOTOCHEMISTRY OF CARBONYL COMPOUNDS

Carbonyl compounds especially ketones, undergo many interesting and versatile reactions on irradiation and consequently have received more attention than any other class of organic compounds. Two absorption regions of interest in the ultraviolet spectra of ketones are around 285 nm (*n*, π*) and 180 nm (π, π*). Most commonly, the lowest energy transitions is *n*, π* in which a nonbonding oxygen electron is promoted to π* orbital. For many ketones, S_1 and T_1 are derived from *n*, π* excitation and so usually a vessel made of pyrex may be used for carrying out typical photoreactions.

The molecular orbital construction of *n*, π* transition may be visualized by the following description. The sigma skeleton of the ketone is formed by the sp^2 hybrid orbitals leaving an unused *p* electron on the carbon. The carbon atom of the carbonyl group is sp^2 hybridized. The three sp^2 orbitals form *three* σ bonds-one with the oxygen and two with the other substituents. The carbonyl oxygen is also sp^2 hybridized using one of its sp^2 orbitals to form a σ bond with the carbonyl carbon. The other two sp^2 orbitals contain a lone pair. The π bond between carbon and oxygen is formed by the overlap of the *p* orbital of the carbonyl oxygen with the remaining *p* orbital of the carbon. The sp^2 hybridization implies that the carbonyl group is planar, and the angle between the substituents is approximately 120°. Excitation promotes an electron of the lone pair to a vacant π* orbital (**Fig. 14.8**).

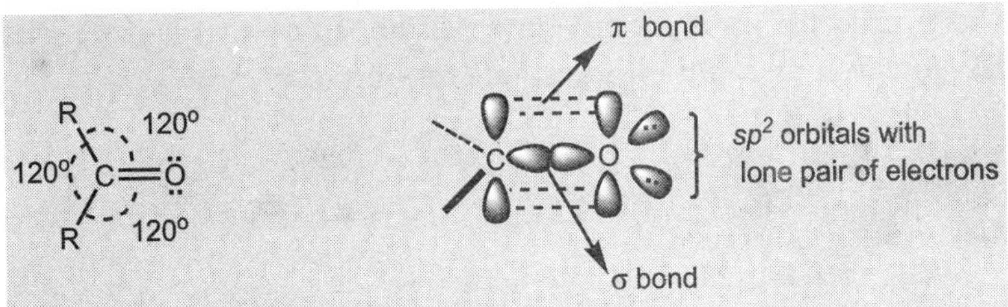

Fig. 14.8 Structure of the carbonyl group

The *n*, π* excited state has a singly occupied *n* orbital and it, therefore, undergoes reactions that place an electron in this orbital.

Although the *n*, π* transition occurs at a longer wavelength, it is always less intense than a π, π* transition. This may be explained in the following way. The electrons in the *n* orbital are situated perpendicular to the plane of the π bond (and hence to the plane of the π* orbital) and consequently the probability of the jump of an electron from the *n* to π* is quite low and in fact zero according to *symmetry selection rules*. However, vibrations of the atoms bring about a partial overlap between the perpendicular planes and so n, π^* transition does occur, but only to a limited extent. In contrast, π and π^* orbitals are situated in the same plane and so π, π^* transition occurs with a high probability on irradiating the molecule with light of suitable wavelength.

Due to the small energy gap between S_1 and T_1 states, ketones undergo intersystem crossing with a high degree of efficiency. As a result, most of the photochemistry of ketones arises from their triplet states. Very frequently, the structure of the ketones along with the medium of irradiation, dictate the nature of the products.

A. Photoreduction

The photoreduction of benzophenone (**9**) leading to benzpinacol (**10**) is one of the oldest and most thoroughly investigated photochemical reactions. The reaction is carried out by irradiating a solution of benzophenone in *iso*propyl alcohol with light of wavelength of 345 nm. *Iso*propyl alcohol does not absorb light at this wavelength but benzophenone undergoes an n, π^* transition.

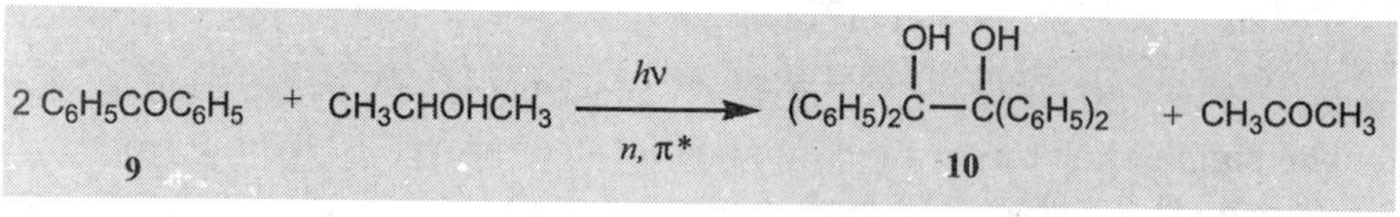

9 10

The reaction is very general, and it can be performed with a large number of aliphatic and aromatic ketones.

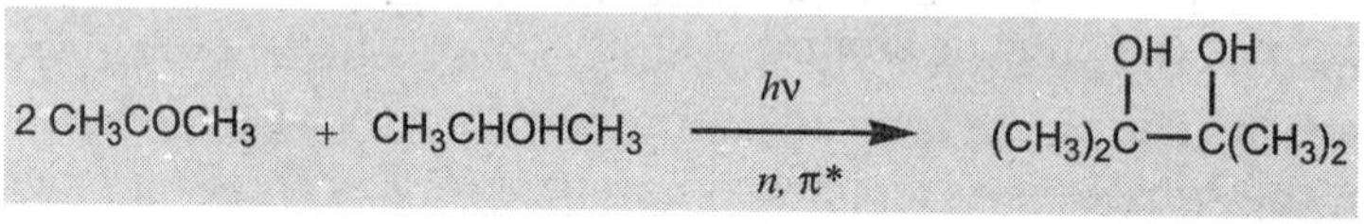

In addition to *iso*propyl alcohol, a variety of solvents such as toluene, cyclohexane and methanol have been used for carrying out the reaction. In case of benzophenone, the initial light absorption is followed by rapid intersystem crossing from S_1 to T_1.

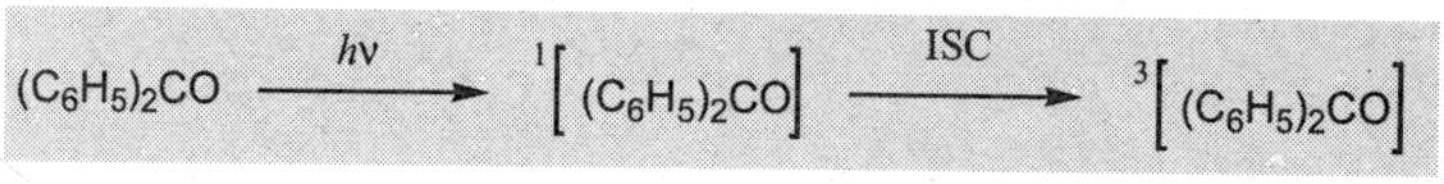

The carbonyl n, π^* triplet (**11**) has a structure which closely resembles the structure of alkoxy radicals (**12**).

11 12

Like alkoxy radicals, the n, π^* triplet is capable of abstracting the α-hydrogen of *iso*propyl alcohol thus giving rise to two radicals, **13** and **14**. Combination of two units of **13** leads to benzpinacol (**10**).

$$^{3}[(C_6H_5)_2CO] + CH_3CHOHCH_3 \longrightarrow \underset{\mathbf{13}}{(C_6H_5)_2\dot{C}OH} + \underset{\mathbf{14}}{(CH_3)_2\dot{C}OH}$$

$$2\,\underset{\mathbf{13}}{(C_6H_5)_2\dot{C}(OH)} \longrightarrow \underset{\mathbf{10}}{(C_6H_5)_2C(OH){-}C(OH)(C_6H_5)_2}$$

Since the quantum yield, Φ, for benzpinacol formation is about 1, the second unit of **13** does not form as a result of the absorption of a fresh quantum of light; rather it is generated by the reaction between benzophenone and **14**.

$$\underset{\mathbf{14}}{(CH_3)_2\dot{C}OH} + (C_6H_5)_2CO \longrightarrow \underset{\mathbf{13}}{(C_6H_5)_2\dot{C}OH} + CH_3COCH_3$$

Benzpinacol formation can also be achieved by irradiating benzophenone and benzhydrol in benzene solution. The mechanism of this reaction is similar to that of the above except that there is simultaneous formation of two radicals, which subsequently dimerize.

$$(C_6H_5)_2CO + (C_6H_5)_2CHOH \xrightarrow{h\nu} \mathbf{10}$$

$$^{3}[(C_6H_5)_2CO] + (C_6H_5)_2CHOH \longrightarrow 2\,\underset{\mathbf{13}}{(C_6H_5)_2\dot{C}OH} \longrightarrow \mathbf{10}$$

The entire process can thus be summarized as:

$$(C_6H_5)_2CO \xrightarrow{h\nu} {}^{*}[(C_6H_5)_2CO]$$

$$^{*}[(C_6H_5)_2CO] + CH_3CHOHCH_3 \longrightarrow \underset{\mathbf{13}}{(C_6H_5)_2\dot{C}OH} + \underset{\mathbf{14}}{(CH_3)_2\dot{C}OH}$$

$$\underset{\mathbf{14}}{(CH_3)_2\dot{C}OH} + (C_6H_5)_2CO \longrightarrow \underset{\mathbf{13}}{(C_6H_5)_2\dot{C}OH} + CH_3COCH_3$$

$$2\,\underset{\mathbf{13}}{(C_6H_5)_2\dot{C}OH} \longrightarrow \underset{\mathbf{10}}{(C_6H_5)_2C(OH){-}C(OH)(C_6H_5)_2}$$

There is a good deal of evidence for the participation of excited triplet states in benzpinacol formation. One evidence is the observation that the rate of benzpinacol formation is not appreciably decreased when the reaction is carried out with lesser amounts of benzhydrol in benzene solution. This result can be interpreted to mean that the excited state of benzophenone involved in the reaction is a sufficiently long-lived species capable of waiting for benzhydrol molecule. It should, therefore, be a T_1 rather than an S_1 species. Another evidence rests on *quenching studies*. Addition of small amount of naphthalene to the irradiation mixture reduces the quantum yield to a significant extent. Since naphthalene does not absorb light of the wavelength needed for an *n*, π* excitation in benzophenone, it is obviously an example of *energy transfer* in which excited benzophenone molecules return to ground state with the simultaneous excitation of naphthalene. Naphthalene thus acts as a *quencher*. Benzophenone must be in a triplet state (not a singlet) which only has a sufficiently long life to undergo collisions for the observed energy transfer with very low concentration of naphthalene.

The hydrogen-abstraction reaction is much more efficient in the *n*, π* state of ketones than in the π, π* state as the latter does not have an odd electron as in an oxygen *n* orbital in the *n*, π* state and, therefore, lacks the characteristics of the *n*, π* triplet. The most striking example is Michler's ketone, which does not undergo photoreduction on irradiation in *iso*propyl alcohol. Conjugation of the lone pair of electrons situated on nitrogen atom with the aromatic rings lowers the energy required for the π, π* transition in this molecule. The π, π* transition thus becomes the lowest triplet of the system and occurs more readily than the *n*, π* transition. The relative energies of *n*, π* and π, π* transitions of benzophenone and Michler's ketone are shown schematically in **Fig. 14.9**.

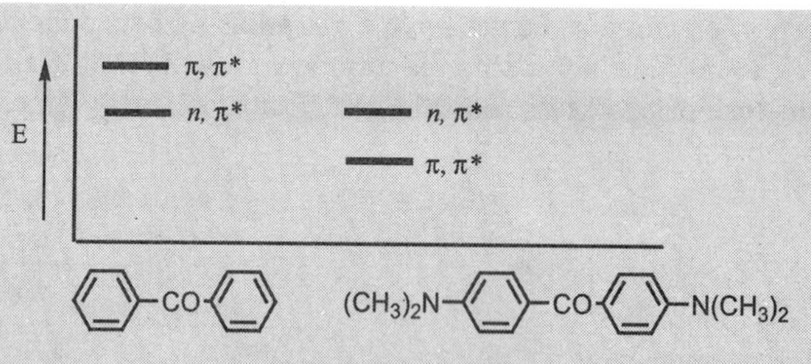

Fig. 14.9 The relative excitation energies of Michler's ketone and benzophenone

α-Keto esters such as **15** also undergo photoreduction to yield pinacols:

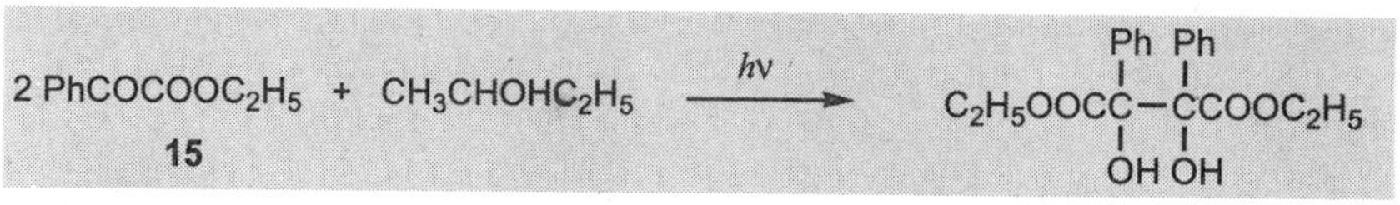

However, ethyl acetoacetate yields solvent adduct (**16**) as the major product instead of the expected dimeric pinacol, which is subsequently converted into a lactone (**17**).

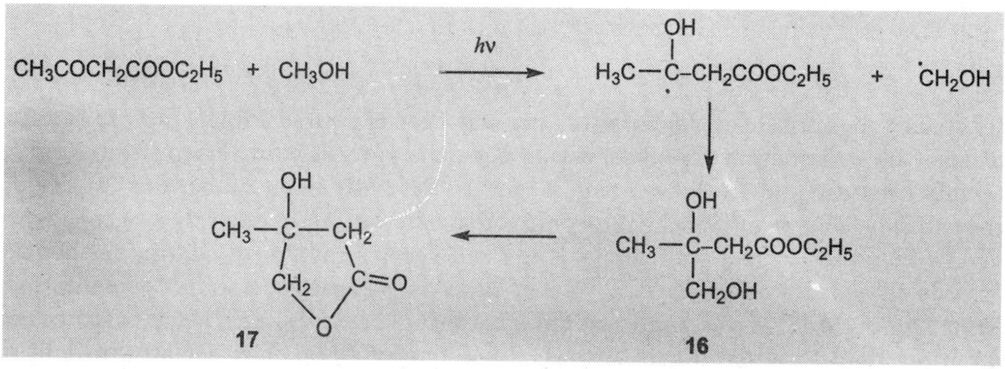

Because of intramolecular hydrogen atom transfer, the efficiency of photoreduction of benzophenone derivatives is greatly reduced in *ortho*-alkyl substituted benzophenones. There is hydrogen abstraction from the benzylic position generating an unstable enol which reverts to the original benzophenone derivative without photoreduction. The evidence for this process is provided by performing experiments with deuterated hydroxylic solvents in which deuterium has been found to be incorporated in the starting material.

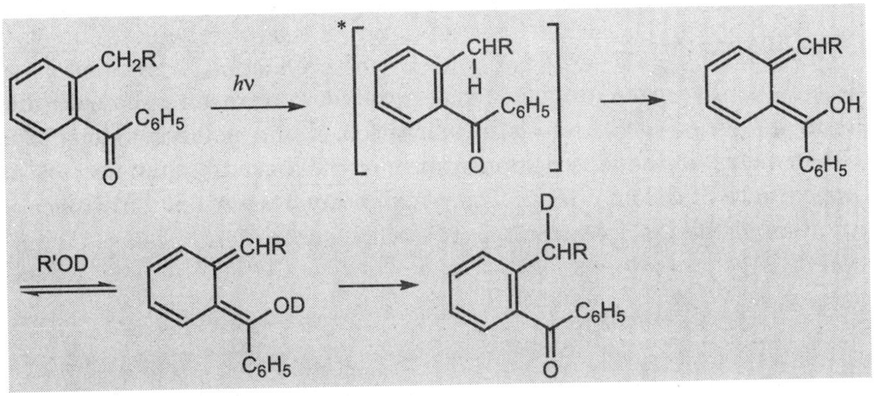

B. Norrish Type I Reaction

The bond dissociation energy of a carbon-carbon bond adjacent to a carbonyl is comparatively small and consequently photochemical excitation of ketones usually results in the homolytic fission of the α carbon-carbon bond. This process is known as α-cleavage or *Norrish type I reaction*. The most thoroughly studied example is acetone which has been photolyzed in the vapour phase as well as in the liquid phase. Absorption of light results in the formation of an *n*, π* excited state of acetone which undergoes a carbon-carbon cleavage yielding a methyl radical and an acetyl radical.

$$CH_3-\overset{O}{\overset{\|}{C}}-CH_3 \xrightarrow{h\nu} \left[CH_3-\overset{O}{\overset{\|}{C}}-CH_3\right]^* \longrightarrow \dot{C}H_3 + CH_3-\dot{C}O$$

At room temperature, two acetyl radicals combine to form biacetyl. At temperatures above 100°C, acetyl radicals decarbonylate with the ultimate production of ethane and carbon monoxide.

$$2\,CH_3-\dot{C}O \longrightarrow CH_3CO-COCH_3$$

$$CH_3-\dot{C}O \longrightarrow \dot{C}H_3 + CO \quad ; \quad 2\,\dot{C}H_3 \longrightarrow CH_3-CH_3$$

In the case of unsymmetrical ketones, splitting takes place in a way so as to generate the more stable of the two possible radicals.

$$CH_3COCH_2CH_3 \xrightarrow{h\nu} CH_3-\dot{C}O + CH_3\dot{C}H_2 \quad ; \quad CH_3COCH(CH_3)_2 \xrightarrow{h\nu} CH_3-\dot{C}O + (CH_3)_2\dot{C}H$$

Both the singlet and the triplet *n*, π* excited states are capable of undergoing α-cleavage. In the case of acetone, most of the excited singlet molecules undergo intersystem crossing to T_1 at room temperature, but with increase in temperature, there is less intersystem crossing and thus both the species participate in *Norrish type I reaction*. Similarly, at the room temperature nearly half of the molecules of diethylketone react in the S_1 state before the intersystem crossing to T_1 occurs.

C. Norrish Type II Reaction

Ketones possessing a γ-hydrogen atom (**18**) undergo a very interesting photochemical reaction which results in the formation of an olefin and the enol of a smaller ketone. Spectroscopic studies have demonstrated that the disappearance of the starting ketone occurs at the same rate as the appearance of the enol form. The reaction involves initial γ-hydrogen abstraction by the oxygen atom yielding a 1,4-biradical (**19**) which subsequently cleaves to an olefin (**20**) and an enol (**21**).

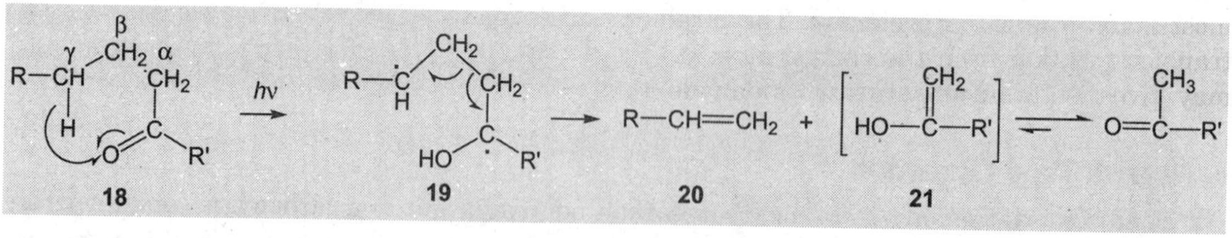

Both *n*, π* singlet and *n*, π* triplet participate in the intramolecular hydrogen abstraction reactions while ketones with a low-lying π, π* state do not undergo this reaction. γ-Hydrogen abstraction is sometimes followed by ring closure of the 1,4-biradical.

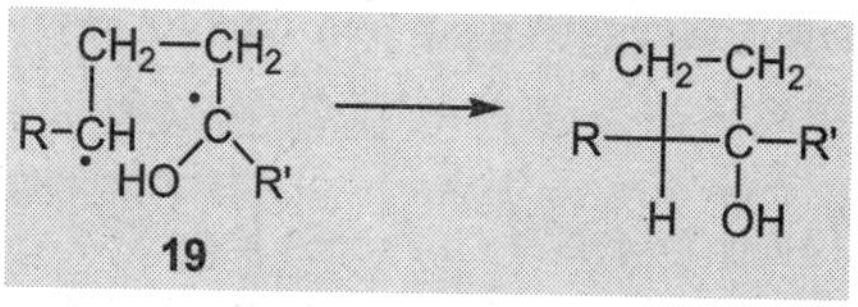

The formation of the 1,4-biradical is a reversible process and the equilibrium between the reactant and the products depends upon the structure of the starting ketone and the conditions of irradiation. When γ-hydrogens are replaced by deuterium, the resultant 1,4-biradical has an O–D bond and since more energy is required to break it than an O H bond, the reaction proceeds in the forward direction more smoothly.

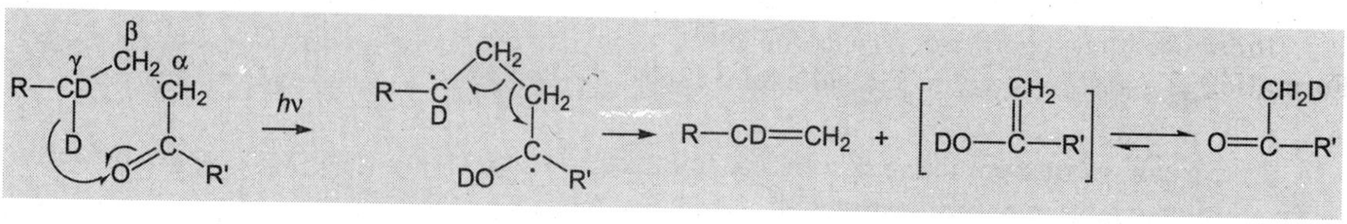

Norrish type II cleavage is often accompanied by a type I split; for instance, irradiation of 2-pentanone (**22**) yields the following products:

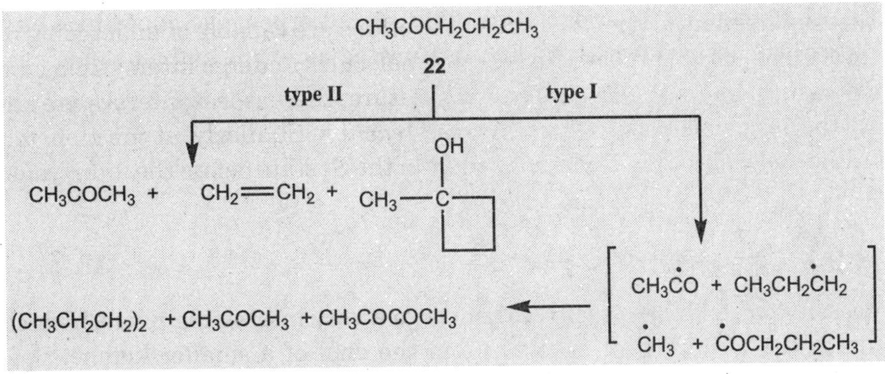

Photochemical Reactions of Cyclic Ketones

The photochemical reactions of saturated cyclic ketones resemble rather closely the photochemical reactions of open-chain analogues. Usually, the photochemistry begins with the n, π^* excitation of the carbonyl group followed by α-fission in which formation of the most stable radical is preferred. The biradical, thus formed, may then undergo a hydrogen transfer reaction from the carbon atom α to the carbonyl group yielding a ketene (**23**) or it may give rise to an unsaturated aldehyde (**24**) by hydrogen transfer to the carbonyl group.

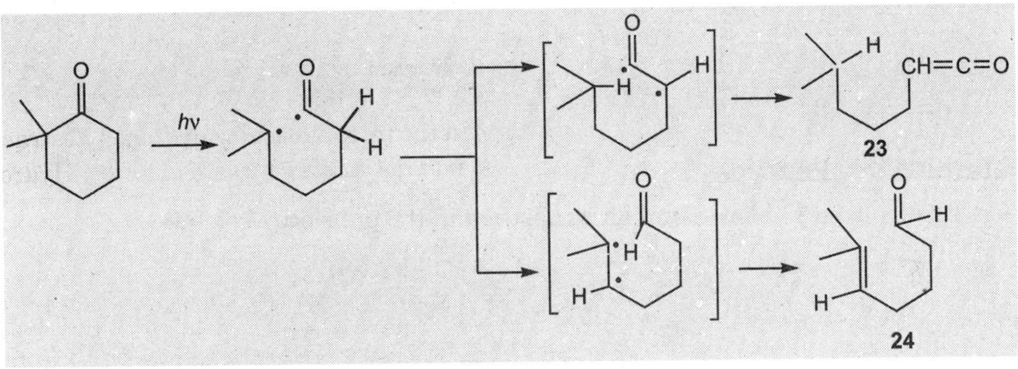

Alternatively, the biradical may prefer to decarbonylate.

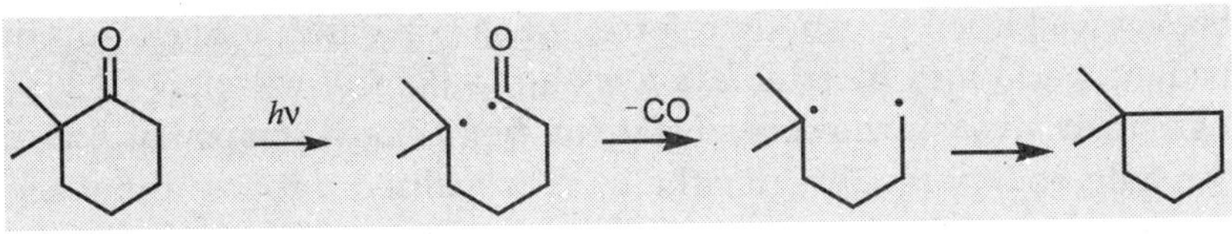

Indanols have been reported to be formed on the irradiation of aryl benzyl ketones. Formation of a stable benzylic radical explains this observation.

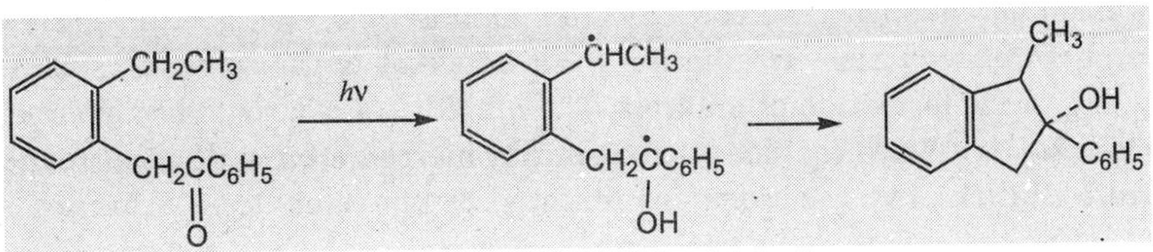

An interesting example is coronopiline (**25**; a sesquiterpene lactone) which on irradiation yields photoproducts derived from both an unsaturated aldehyde and a ketene.

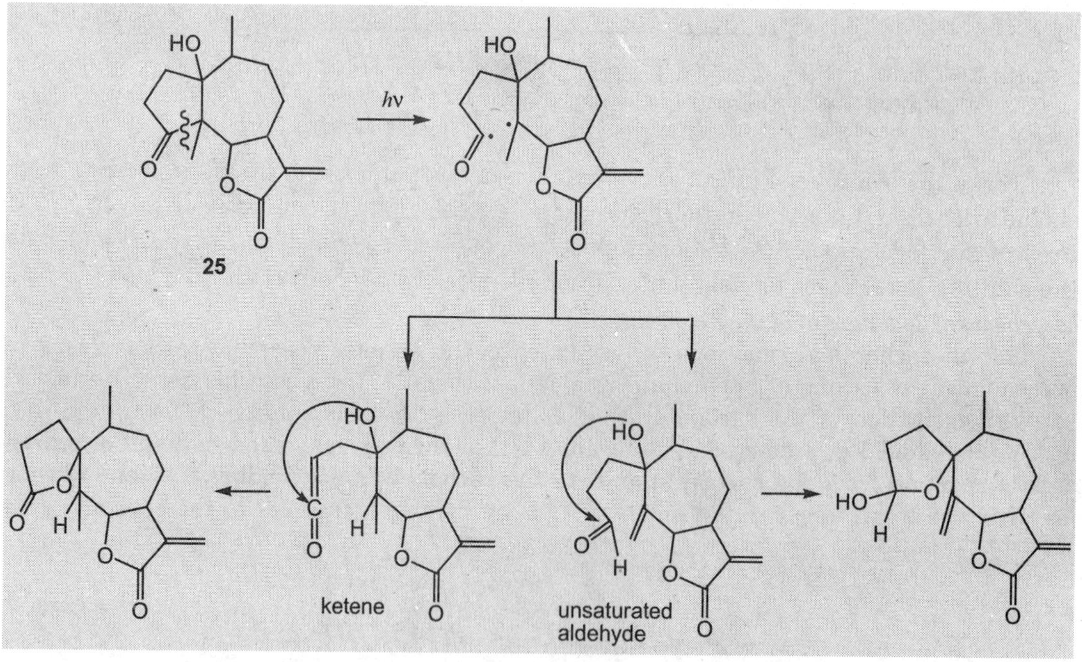

D. Paterno-Büchi Reaction

Carbonyl compounds yield oxetanes on irradiation in the presence of olefins.

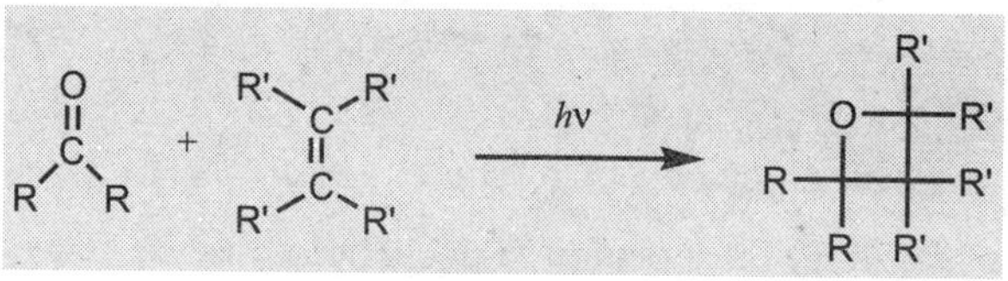

This photocycloaddition is usually referred to as the Paterno-Büchi reaction. Since the addition is usually performed by irradiation with the light of wavelength absorbed only by the carbonyl group, it involves the excited state of the carbonyl compound rather than that of the olefin. The light energy required for the n, π^* transition initiates the reaction in simple carbonyl compounds, while for some conjugated ketones, it is difficult to distinguish between n, π^* and π, π^* transitions. Available evidence favours participation by both the excited states in the latter case.

It has been established that carbonyl compounds undergoing Paterno-Büchi reaction can also be photoreduced upon irradiation in presence of protic solvents, such as *iso*propyl alcohol, leading to the formation of pinacols. The inability of a carbonyl compound to undergo photocycloaddition is always coupled with a similar inertness towards photoreduction. These facts illustrate that the two reactions are initiated by the same type of excited states. This excited state is a triplet species which has been shown in the case of photoreduction processes and on this basis the mechanism of the Paterno-Büchi reaction can be written as follows:

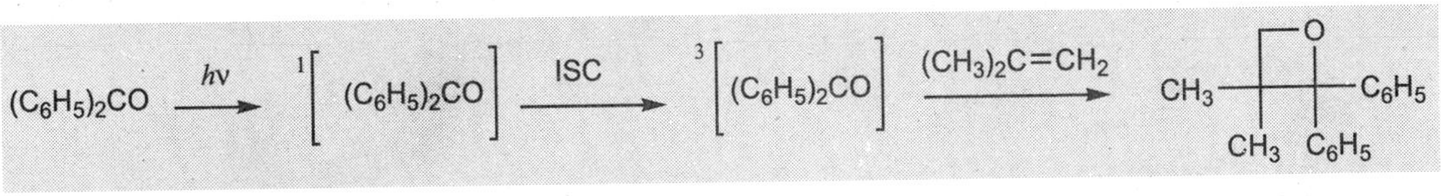

Since the olefin used in the above reaction is unsymmetrical, theoretically two products should arise from this cycloaddition. The other product is, indeed, obtained but only in 10% yield. Preferential formation of the major product indicates intermediacy of a biradical and logically the addition takes place to yield preferentially the *most stable biradical* intermediate (**26**).

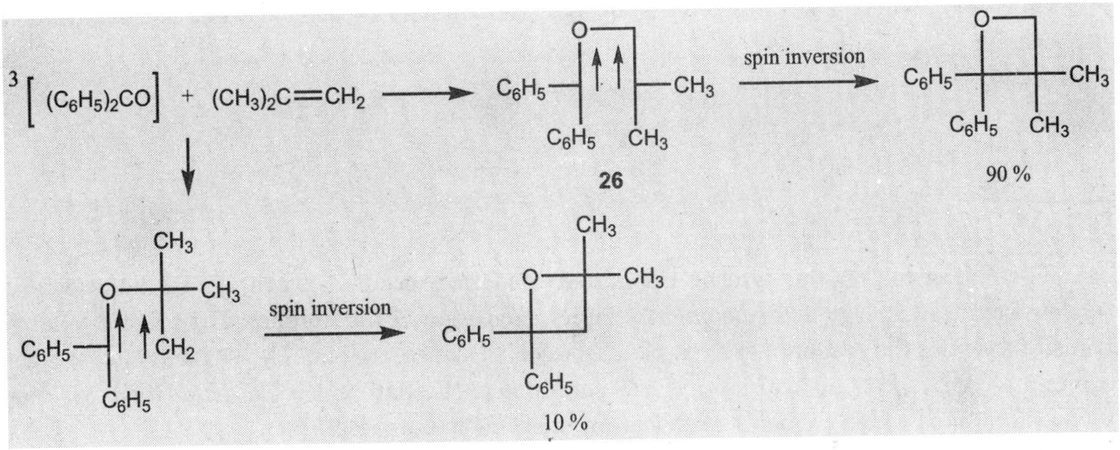

Moreover, irradiation of benzophenone with either *cis* or *trans*-2-butene yields the same mixture of isomeric oxetanes. Had it been a *concerted reaction* in which both the bonds were formed simultaneously, different products would have been formed in each case. The above result thus confirms intermediacy of a biradical in which the stereochemistry is lost.

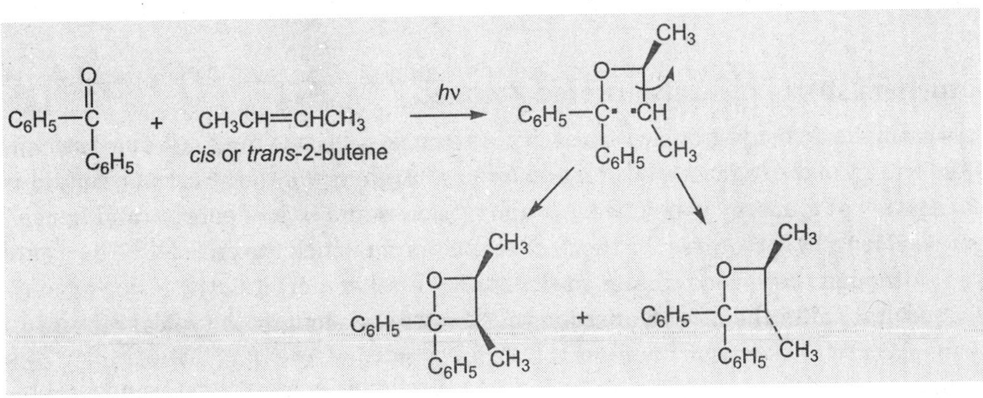

A serious drawback of the Paterno-Büchi reaction is the competing *energy transfer process* from the excited ketone to the olefin, in cases where the triplet energy of the former exceeds that of the latter. Irradiation of acetone with norbornene (**27**), for instance, yields dimers of norbornene rather than oxetanes.

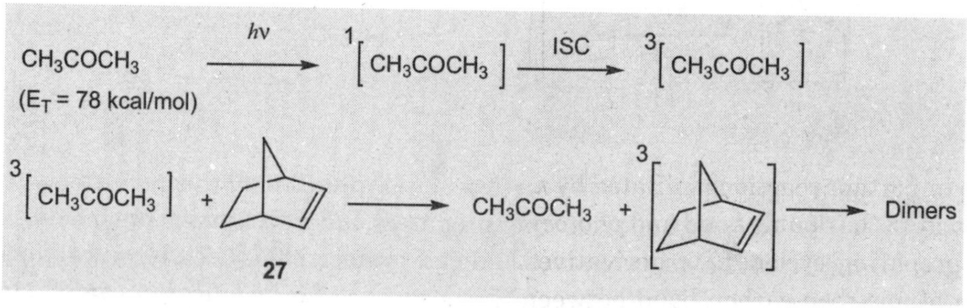

However, when a ketone having less triplet energy is used we get the oxetane of norbornene.

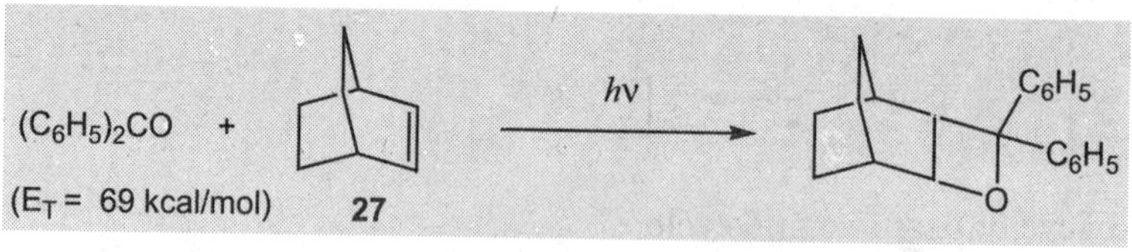

For similar reason, the Paterno-Büchi reaction fails to occur with conjugated dienes whose triplet excitation energy is *far below* the triplet excitation of the ketones. Here again, energy transfer becomes the predominant reaction between the ketone triplet ($E_T \sim 70$ kcal/mol) and the conjugated diene ($E_T \sim 60$ kcal/mol). Acetylenes also participate in the Paterno-Büchi reaction to yield products which probably arise *via* an unstable oxetene (**28**).

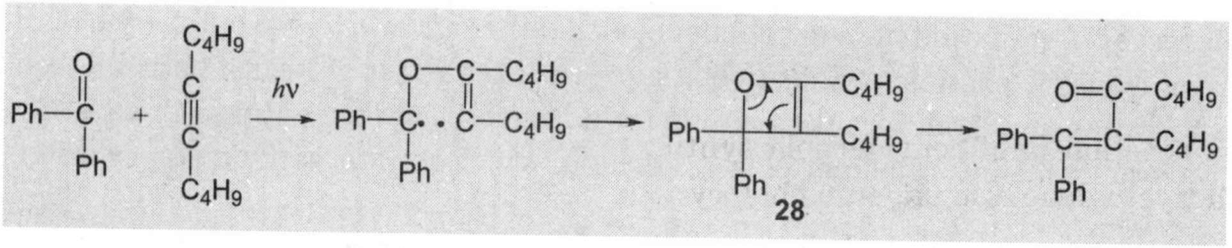

E. Photochemistry of α,β-Unsaturated Ketones

Conjugation of the carbonyl group with a carbon-carbon double bond in an α,β-unsaturated ketone *raises* the energy of the highest π orbital and *lowers* the energy of the π* orbital. In other words, there is now a lesser energy gap between π and π* states and so less energy (and hence light of a longer wavelength) is required for the π, π* transition in which the entire C=C-C=O system is involved. Although there is no change in the energy level of *n* orbital, the lowering of the level of the π* orbital shifts the *n*, π* transition to a longer wavelength. A typical α,β-unsaturated ketone displays two absorption maxima at 220 nm (π, π*) and at 310 nm (*n*, π*) *and irradiation may induce either of these transitions*. The excited state of such a ketone usually has more π, π* than *n*, π* triplet character even though there may be an initial *n*, π* excitation.

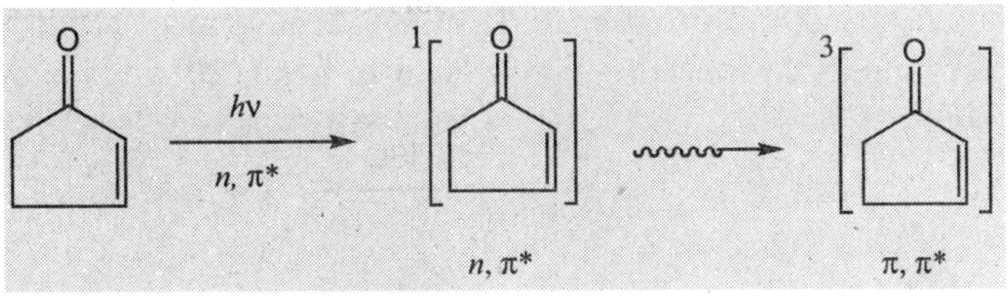

Two important reactions initiated by π, π* states of the α,β-unsaturated ketones are olefin *addition* across the double bond and *photocyclodimerization*. Both the reactions have long been used for preparing cyclobutane derivatives. Photocycloaddition reaction involves the initial formation of a carbon-carbon bond between the α carbon of α,β-unsaturated ketone and the olefin, followed by cyclization.

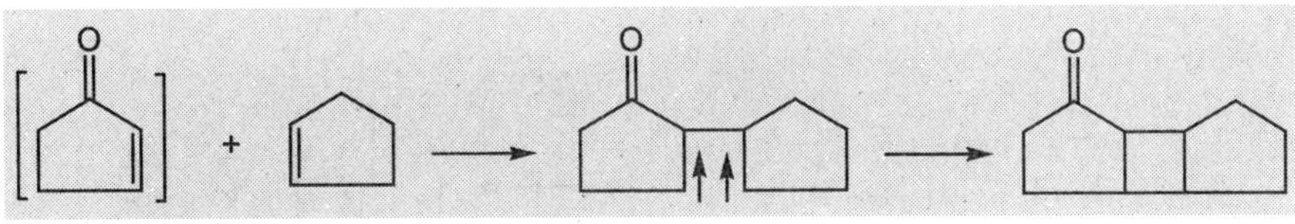

Irradiation of cyclohexenone and cyclopentene provides adducts with about 2 : 1 *cis* : *trans* ratio.

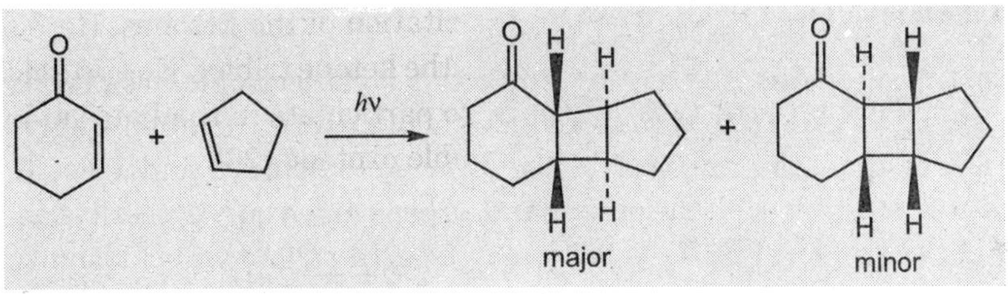

The reaction is of considerable synthetic utility since substituted olefins such as 1, 1-dimethoxyethylene also undergo photocycloaddition.

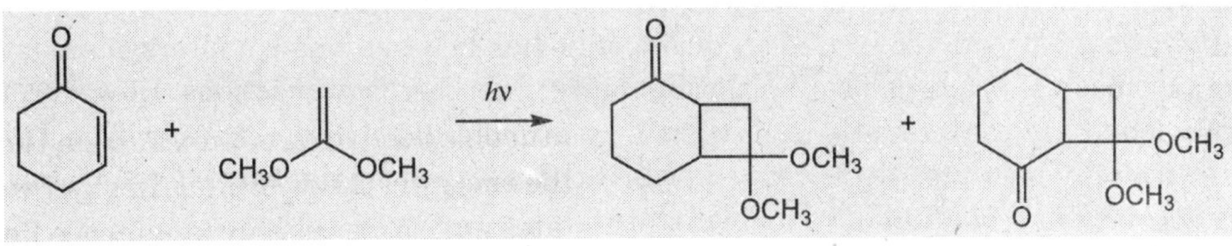

Acetylenes yield cyclobutenes in a similar fashion.

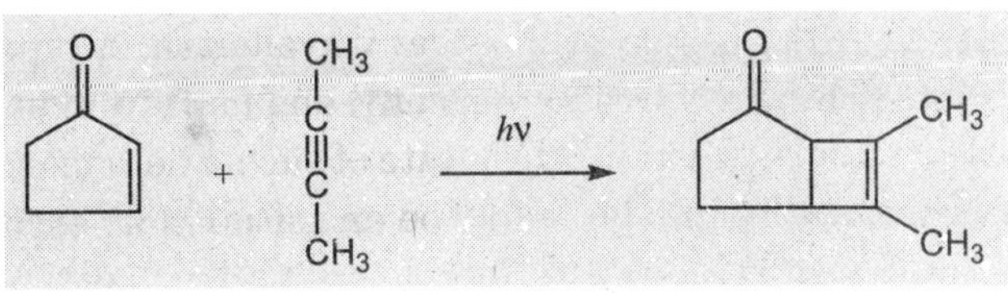

The *photodimerization* of α,β-unsaturated ketones occurs most readily in cyclic systems. Irradiation of cyclopentenone, for instance, yields head-to-head (**29**) and head-to-tail (**30**) dimers in almost equal amounts.

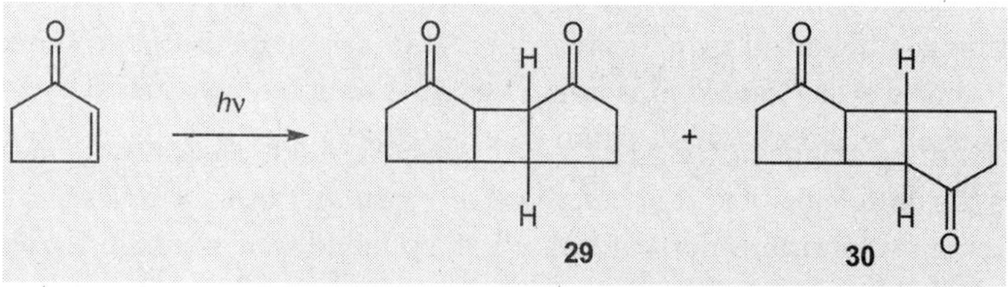

Head-to-head dimers are preferred in polar solvents, while non-polar media favour head-to-tail dimers. The dimerization of cyclohexenone derivatives occurs in a similar manner.

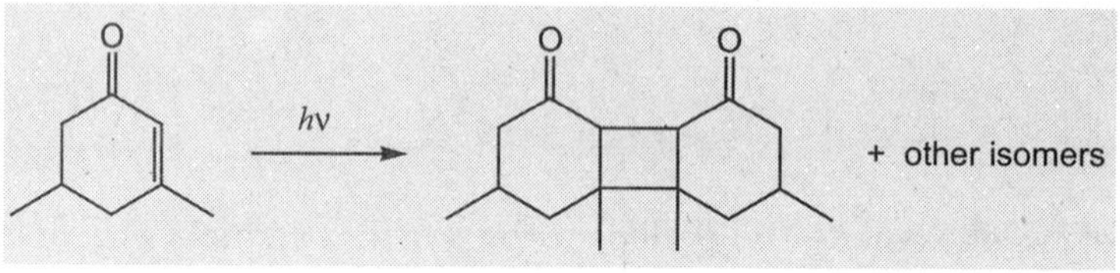

14.3 PHOTOCHEMISTRY OF OLEFINS

The photochemistry of olefins involves only two types of electronic excitations: σ, π* and π, π*. Promotion of an electron from σ to π* requires more energy (available only from the light of wavelength lower than 150 nm) and hence difficult to achieve under usual experimental conditions. The π, π* excitation, on the other hand, is experimentally accessible as it requires the absorption of light of about 180-210 nm for *nonconjugated olefins* and of about 220 nm or more for the conjugated olefins. It is precisely because of this reason that most of the photochemical reactions of olefins studied to date involve π, π* excitation. The initial excitation *(π, π*)* usually occurs with no change in multiplicity and so a singlet excited state is obtained. Unlike *n*, π* transition of ketones, this transition is *symmetry-allowed* and thus gives rise to a strong absorption band. The singlet excited state of olefins has less tendency to intersystem crossing and they themselves are capable of initiating many photochemical reactions. However, the T_1 states of olefins are conveniently generated by *intermolecular energy transfer* from a triplet donor to an olefin molecule. More often, the photochemistry of singlet excited state of an olefin differs considerably from that of its triplet state.

A. *Cis-Trans* Isomerization

Olefins usually undergo *cis-trans* isomerization upon irradiation with ultraviolet light. The transformation can be effected by direct irradiation of the olefins or by irradiation in the presence of a sensitizer, and it may either take place through a *singlet* or a *triplet* excited species. It is, however, known that isomerization in the triplet state possesses a *lower barrier to rotation* around the carbon-carbon bond. Triplets are indeed frequently involved since simple olefins absorb light at about 200 nm and experimentally it is more convenient to sensitize these reactions by using a triplet donor which usually absorbs at longer wavelengths.

Cis-trans isomerization of alkenes assumes particular significance as irradiation provides a simple way to convert a *trans* alkene to the less stable *cis* isomer. This is because of the higher extinction coefficient of *trans* isomer which absorbs substantially more light than the *cis* isomer. In other words, the *trans* isomer is excited more easily than the *cis,* and the *trans* to the *cis* conversion will occur faster than the converse process when the two isomers are in equal concentration.

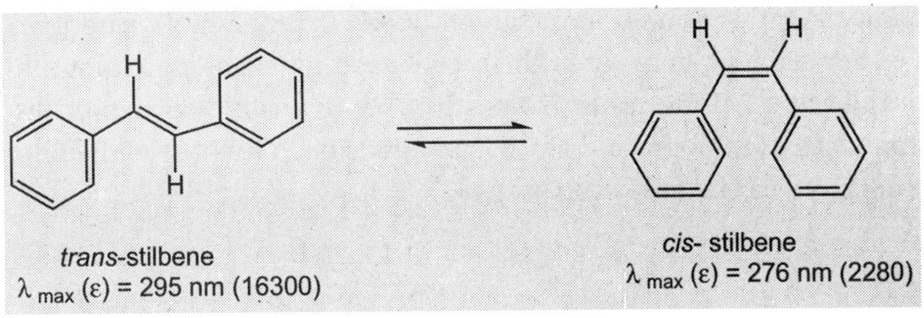

The *cis-trans* isomerization of alkene takes place *via* an excited state in which two sp^2 carbons are twisted by 90° with respect to one another for attaining the minimum energy geometry.

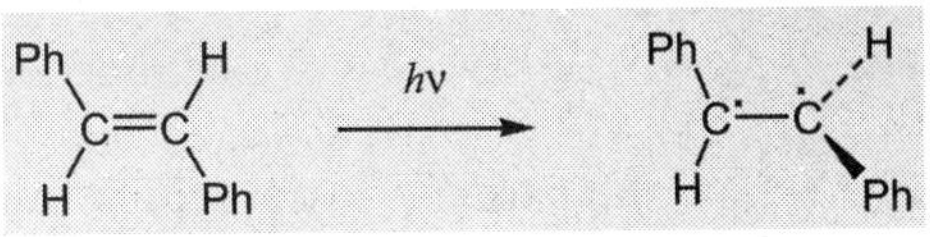

While returning to the ground state, both *cis* and *trans*-isomers are obtained by twisting in either direction. The return from the singlet excited state to the ground state involves re-pairing of the electrons while return from the triplet state requires intersysytem crossing before the re-pairing of the electrons.

The photoisomerization of the stilbenes is probably the most completely understood process of this type. Direct irradiation of solutions of either *cis*-or *trans*-stilbene (**31**) yields a constant mixture consisting of 93% *cis*-stilbene and 7% *trans*-stilbene.

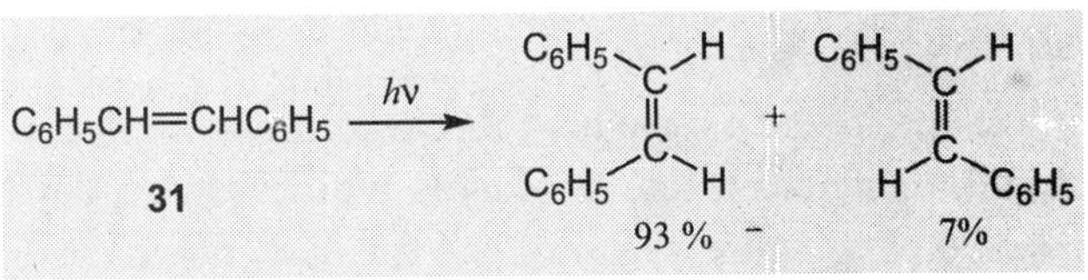

This condition is called a photostationary state. It is interesting that equilibrium favours the formation of thermodynamically less stable *cis* isomer. Initial absorption of light by either of these isomers is rapidly followed by *intersystem crossing* to the respective triplet states. Photoisomerization then occurs *via interconversion* or probably *via* a common triplet intermediate.

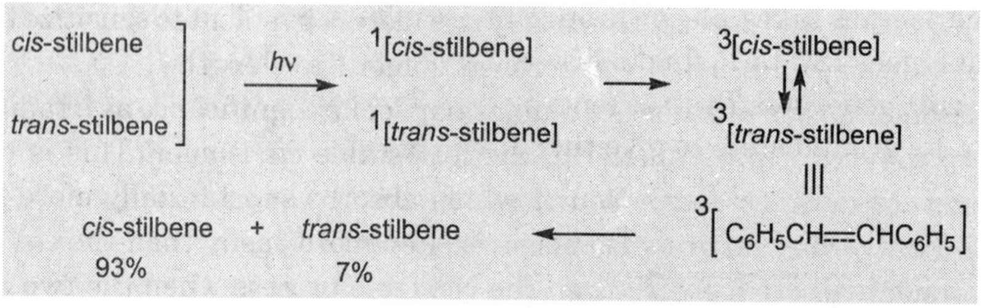

The photoisomerization can be sensitized by an excited triplet molecule having more triplet energy than *cis*- (50 kcal/mol) and *trans*-stilbene (57 kcal/mol). Energy transfer from T_1 of the donor to either stilbene generates the corresponding stilbene triplet state which either interconverts to the triplet of the other isomer or goes to a common triplet intermediate, and in either case final decay to the ground state isomer occurs. Thus the benzophenone sensitized process may be depicted in the following steps:

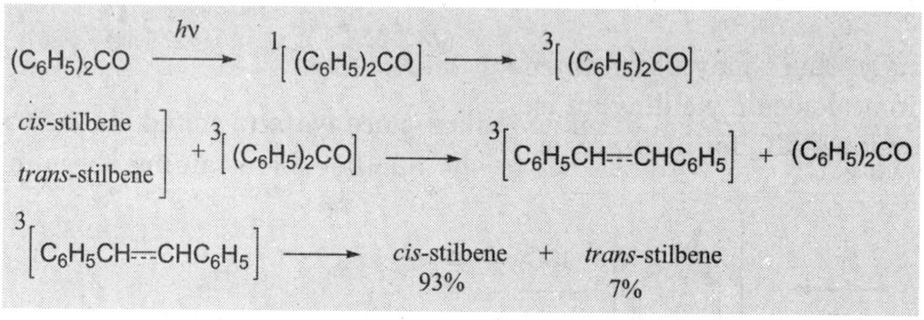

Sometimes photochemical *cis-trans* isomerization may take place in the presence of halogens. It appears that under these conditions there is a photochemical production of halogen atoms which add to the olefin yielding a radical. Elimination of the halogen atom from this radical yields a constant ratio of *cis* and *trans* isomers.

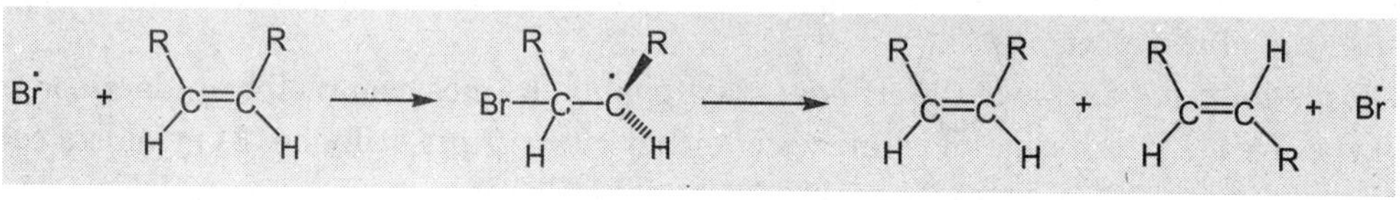

B. Dimerization Reaction

Like α,β-unsaturated ketones, olefins may also undergo dimerization reactions. The process involves the generation of an excited triplet molecule which subsequently reacts with ground state of an olefin molecule. A well-known example is the acetone-sensitized photodimerization of norbornene.

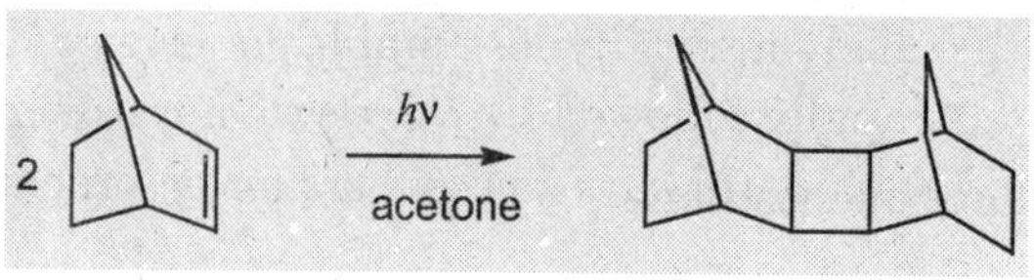

Photodimerization takes place through the triplet state. The sensitizer is excited to S_1 state which undergoes ISC to T_1 due to less energy gap. Interaction of T_1 of the sensitizer with an alkene molecule generates T_1 state (energy transfer) which reacts with the second molecule of the alkene in its ground state to give the dimer (**Fig. 14.10**).

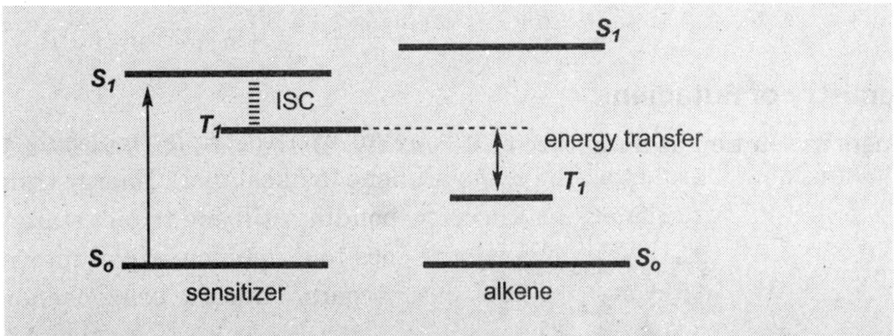

Fig. 14.10 Schematic diagram of photosensitized dimerization reaction of alkenes

Interestingly, there may be an *intramolecular reaction* between two properly situated double bonds in a molecule yielding an isomeric substance. The conversion of norbornadiene (**32**) to quadricyclane (**33**), for instance, occurs on irradiation.

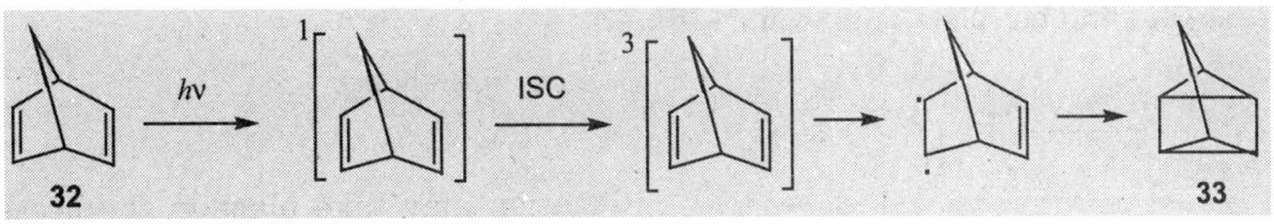

Irradiation of cyclopentadiene generates a highly strained compound bicyclo[2.1.0]pent-2-ene.

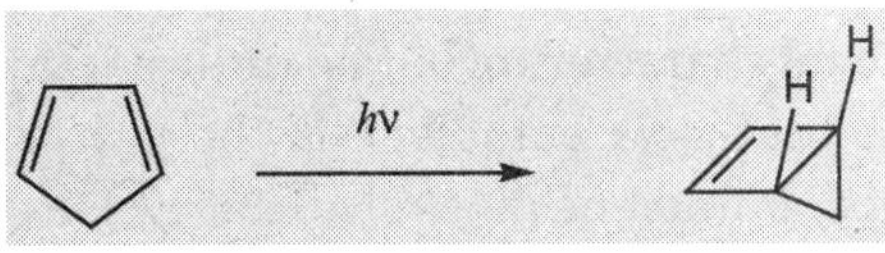

Cyclic olefins may undergo addition reactions, instead of dimerization on irradiation in methanol. The reaction of **34** with methanol is sensitized by xylene.

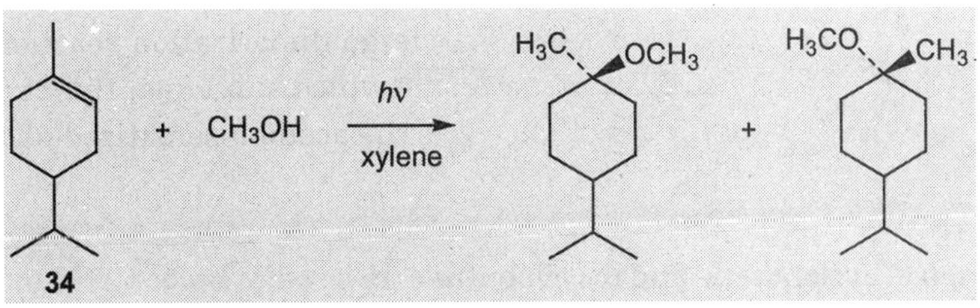

Photodimerization of cyclopentadiene using benzophenone as a triplet sensitizer gives a mixture of products.

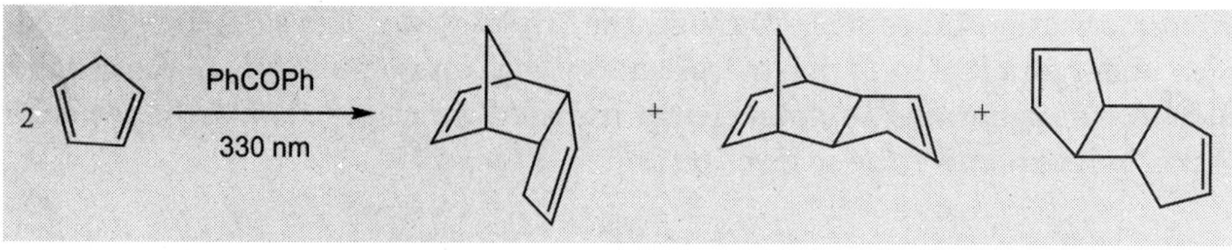

C. Photochemistry of Butadiene

Butadiene exists in solution as a mixture of *s-trans* (95%) and *s-cis* (5%) conformers.

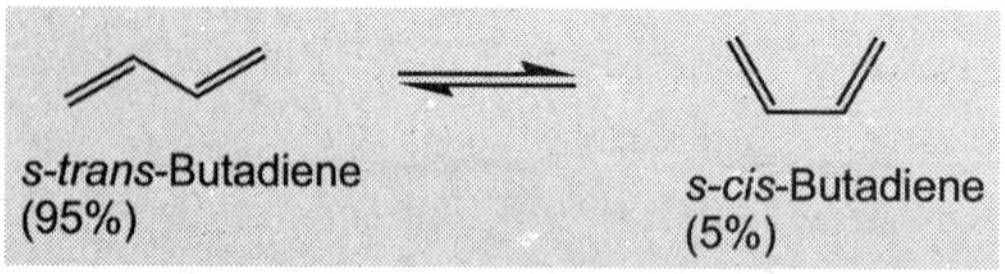

Irradiation of butadiene promotes an electron from ψ_2 to ψ_3 (π, π^* transition) which results in the increased bonding between C_2 and C_3 at the expense of C_1–C_2 and C_3–C_4. In other words, the lowest excited states of *s-trans* and *s-cis* butadiene should exhibit still larger barriers to rotation about the C_2–C_3 bond because of its double bond character. Thus conformational character of butadienes is retained in the excited states. Direct irradiation of butadienes yields cyclobutene (**35**) and bicyclo[1.1.0]butane (**36**).

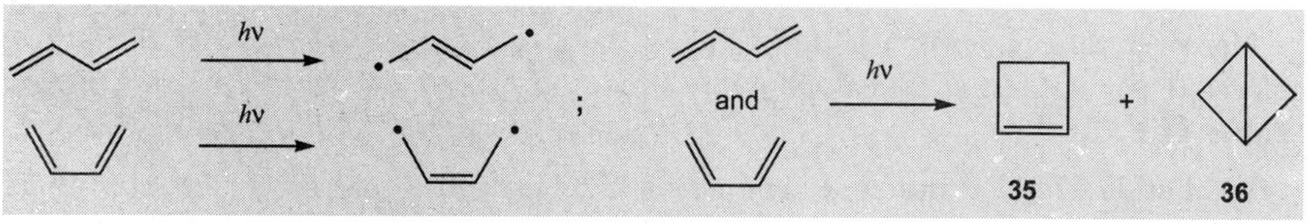

The formation of these products occurs directly from the S_1 state of the butadiene and assuming that the conformational characters of butadiene are retained in the S_1 state, it appears quite reasonable to speculate that *s-cis* butadiene is the precursor of cyclobutene while the excited state resembling *s-trans* butadiene gives rise to bicyclobutane.

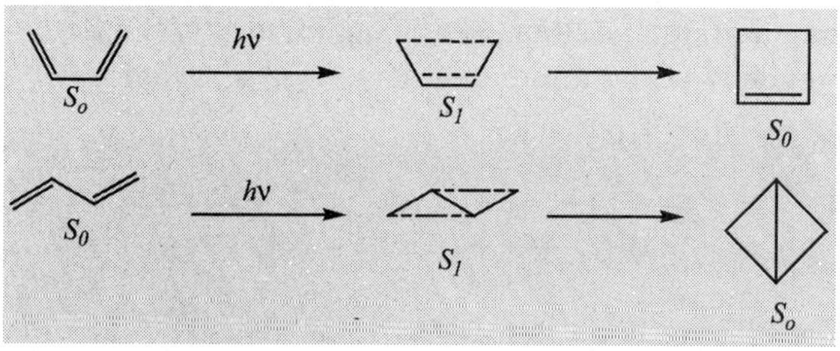

The formation of cyclobutene and bicyclobutane on irradiation of 1,3-butadienes is a fairly general photochemical reaction.

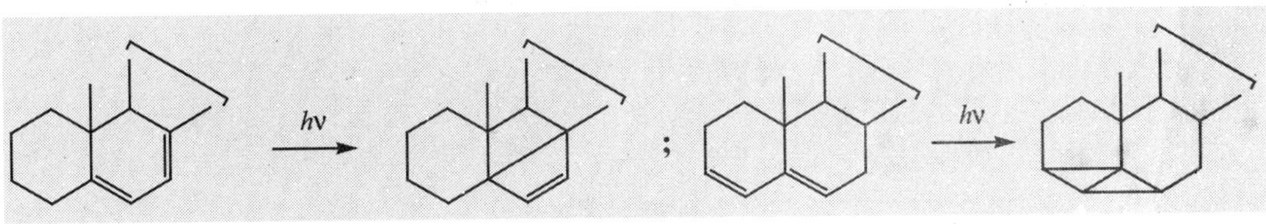

The singlet excitation energies of the two conformers of butadiene are not exactly known (roughly they are in the range of 110-120 kcal/mol), but it appears from the product composition (more cyclobutene than bicyclobutane) that the energy required for the singlet excitation of *S-cis* butadiene is lower than the required for *S-trans* isomer. The triplet excitation energies for the *S-cis* and *S-trans* isomers have been found to be 53 and 60 kcal/mol, respectively. The large energy gap between S_1 and T_1 accounts for the diminished intersystem crossing in butadienes.

The photochemistry of the T_1 states of butadienes has been investigated through energy transfer from triplet state sensitizers. A mixture of three dimers **37, 38** and **39** is obtained on sensitized irradiation of butadiene.

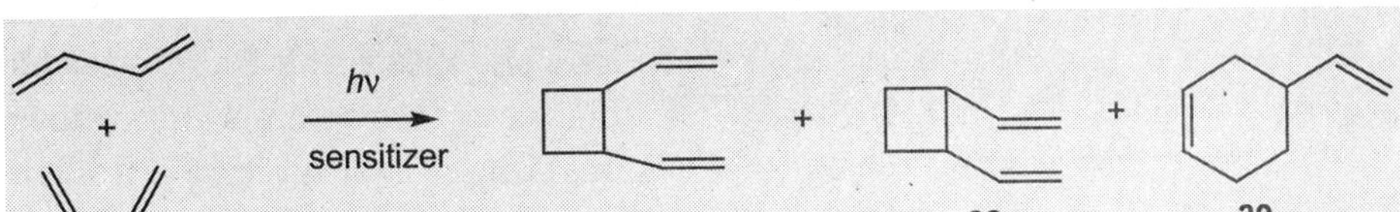

37 **38** **39**

Moreover, the composition of the dimer mixture varies with the triplet energy of sensitizers employed to populate the T_1 state of butadiene. Typical results from two sensitizers are given in **Table 14.1**.

Table 14.1 Composition of the products (**37, 38** and **39**) by using different sensitizers

Sensitizer	E_T	**37**	**38**	**39**
Benzophenone	69	80	16	4
Benzil	54	49	8	43

The T_1 states of butadienes are of lower energy than the T_1 of benzophenone, so triplet energy transfer takes place efficiently to both the butadiene conformers.

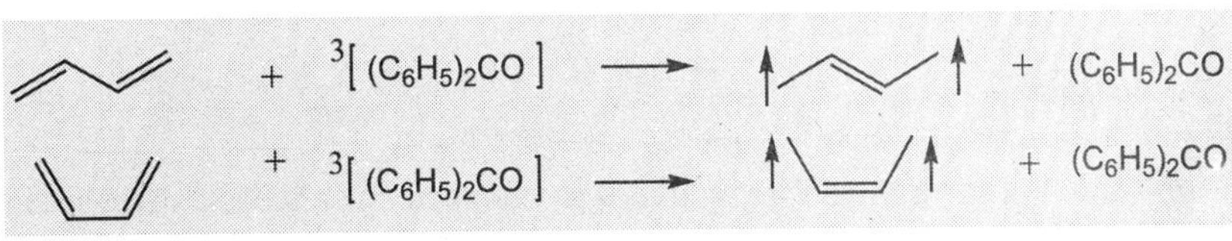

The *trans* triplet of butadiene is formed predominantly, because the diene is mostly *trans*. Subsequent reaction of the *trans* triplet with *S-trans* butadiene leads to an intermediate hexadienyl biradical in which both the allylic units have the *trans* configuration and so a cyclization leading to **37** and **38** is favoured.

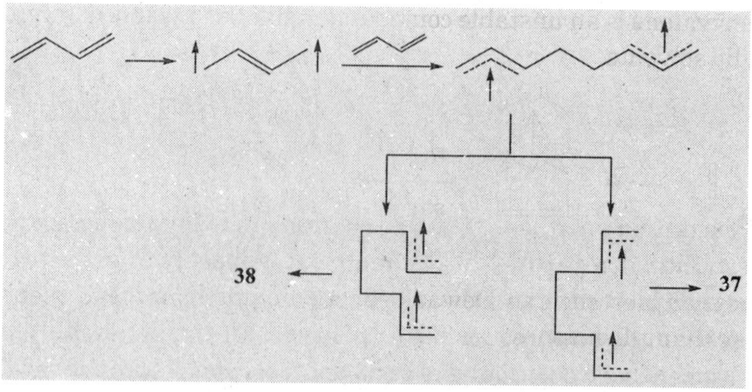

A similar reaction of the *cis* triplet, which is formed in small amounts, with a molecule of *S-trans* butadiene leads to a biradical in which one allylic unit has *cis* configuration, and in one conformation it is favourably disposed toward cyclization to **39**.

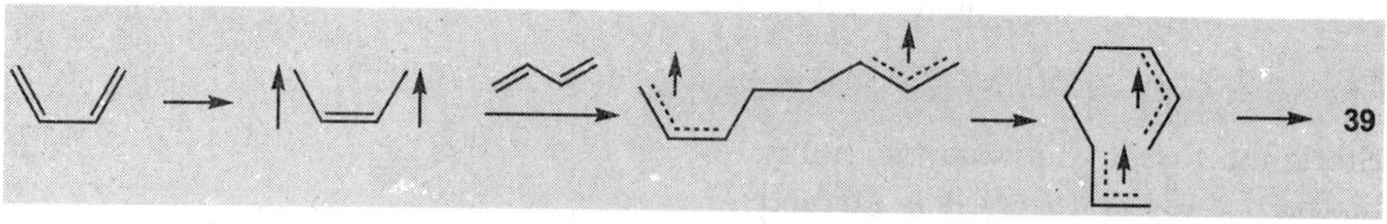

In contrast to benzophenone (E_T 69 kcal/mol), a triplet sensitizer with E_T below 60 kcal/mol but above 53 kcal/mol preferentially should form the *cis*-butadiene triplet by energy transfer to *S-cis* butadiene. This explains the predominance of **39** when benzyl is employed as a sensitizer. As the triplet energy of the photosensitizer is decreased below 53 kcal/mol, there will not be any energy transfer since excitation of the diene now requires more energy than the sensitizers can provide.

14.4 PHOTOCHEMISTRY OF AROMATIC COMPOUNDS

A. Isomerization

Irradiation of benzene gives a mixture consisting of fulvene and small amount of benzvalene.

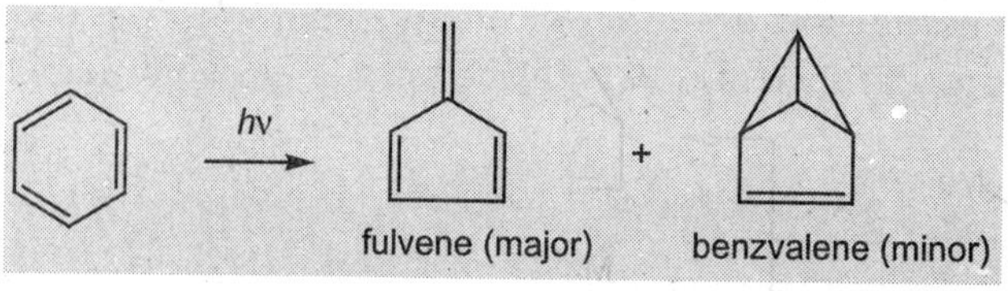

The reaction takes place by the formation of a resonance-stabilized diradical intermediate called perfulvene.

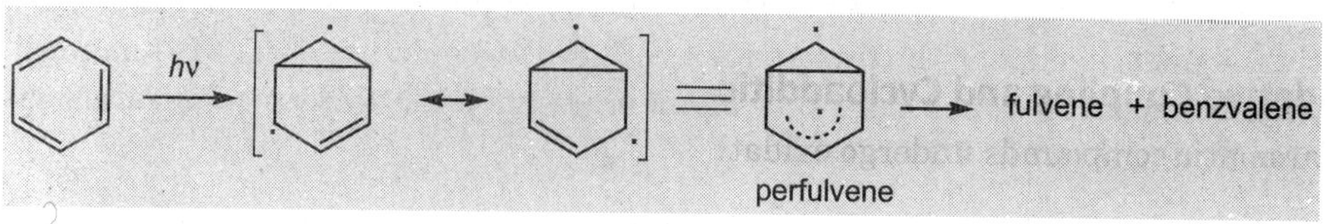

Although benzvalene is an unstable compound, it has been trapped by irradiating benzene in acidic hydroxylic solvents.

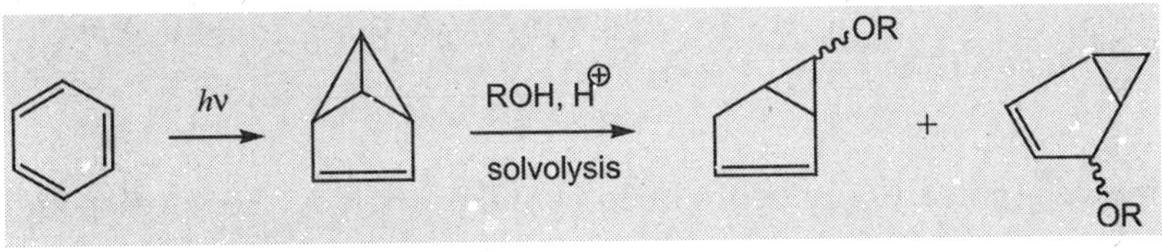

Interesting structures such as Dewar benzene or prismane have been obtained by the irradiation of tri-*tert*-butylbenzenes.

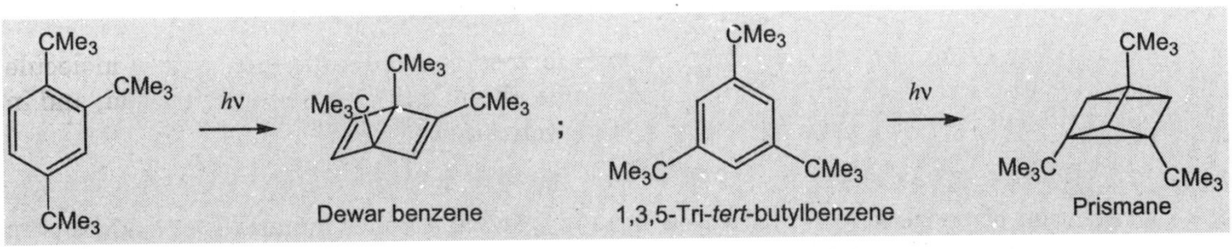

Steric interaction between two bulky *tert*-butyl groups are reduced on the formation of bicyclic compounds and thus it is attractive to speculate that the presence of these groups favours cyclization in tri-*tert*-butylbenzenes. It is presumed that such compounds are generated through diradical triplet state of benzene.

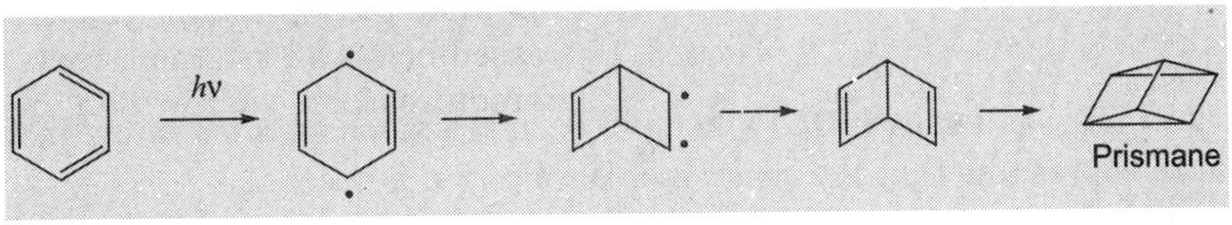

Photoisomerization of *o*-xylene provides *m*- and *p*-isomers. The reaction is believed to proceed *via* prismane-type intermediates.

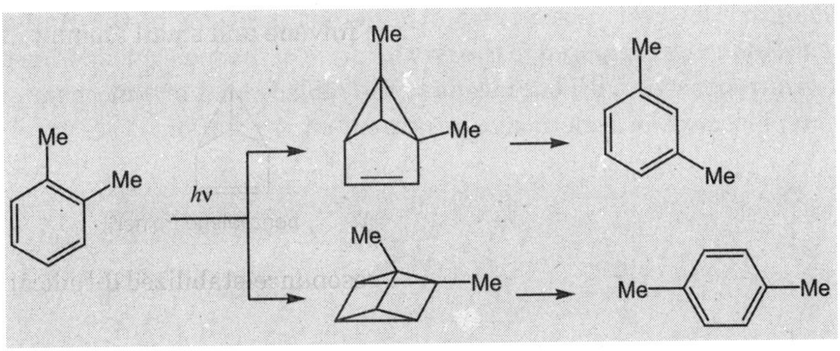

B. Oxidative Coupling and Cycloaddition Reactions

Many aromatic compounds undergo oxidative coupling in the presence of oxygen.

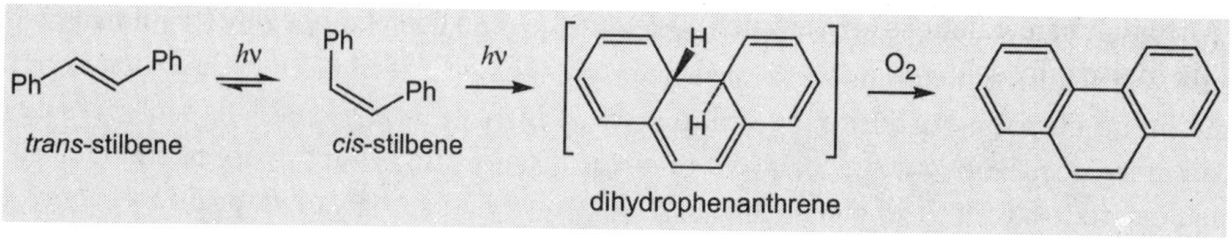

Irradiation of similarly constituted compounds provides the corresponding polycyclic structures, for example

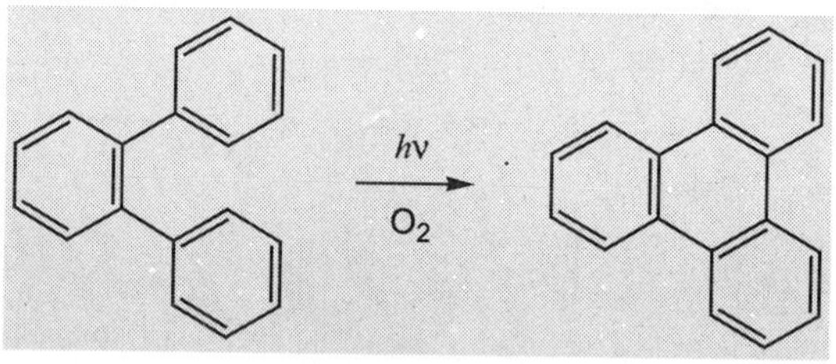

Irradiation of benzene (or substituted benzenes) with alkenes generates 1:1 adducts in which the alkene moiety bridges the *m*-position of the benzene ring.

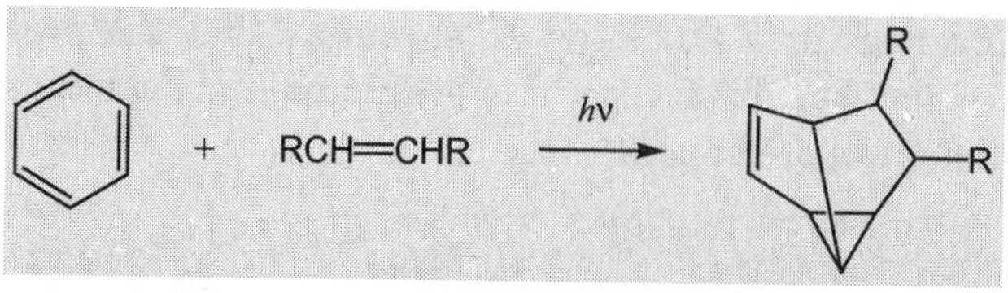

As the geometry of a substituted alkene in such reactions *remains intact* in the product, it is reasonable to presume that the reaction is a *concerted process*.

C. Substitution Reaction

Many aromatic compounds undergo substitution reactions on irradiation with the light of suitable wavelength. However, reactivities at different positions of the ring are somewhat reversed in *excited states* as compared to the *ground state* of the molecule. For instance, heating of 3,4-dimethoxynitrobenzene with base leads to the replacement of 4-methoxy substituent, in contrast to the replacement of 3-methoxy substituent on irradiation.

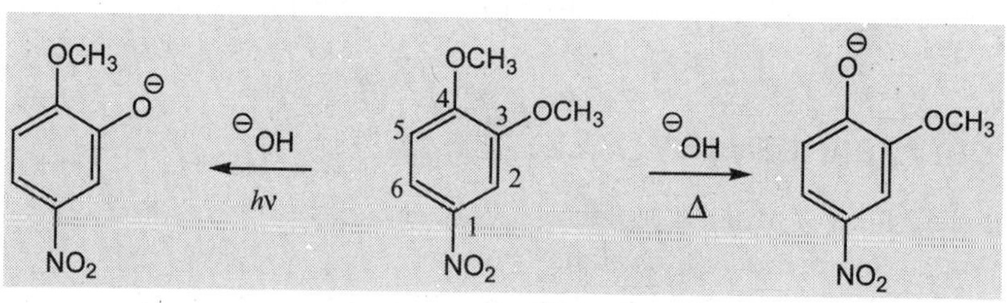

We have already seen that there is greater accumulation of negative charge at *o*- and *p*-positions in the presence of an electron-releasing substituent such as methoxy group on an

aromatic ring. The *m*-position has less negative charge. However, in the excited state of the molecule, the *o*- and *m*-positions have more negative charge as compared to the *p*-position. In the presence of an electron-withdrawing group such as nitro, *o*- and *p*-positions have less electron density (more positively charge) relative to *m*-position in the ground state of the molecule. In contrast, excited state of the molecule has less electron density at *o*- and *m*-positions as compared to the *p*-position.

Thermal and photochemical reactions of *p*-nitroanisole provides an illustrative example.

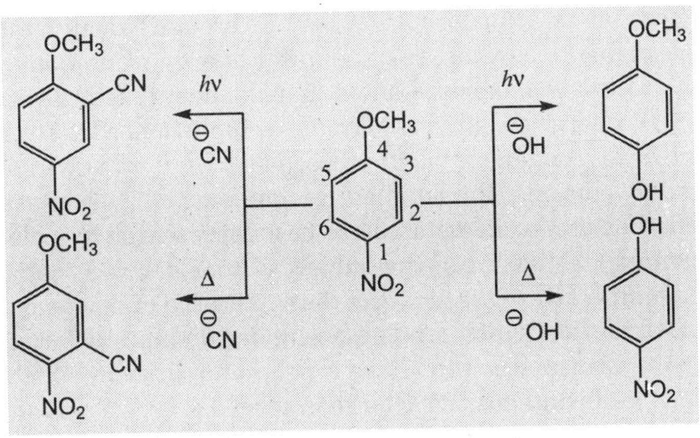

Similar change in the reactivity at *o*-, *m*- and *p*-positions has been observed in electrophilic deuteration of substituted benzenes using CH_3COOD.

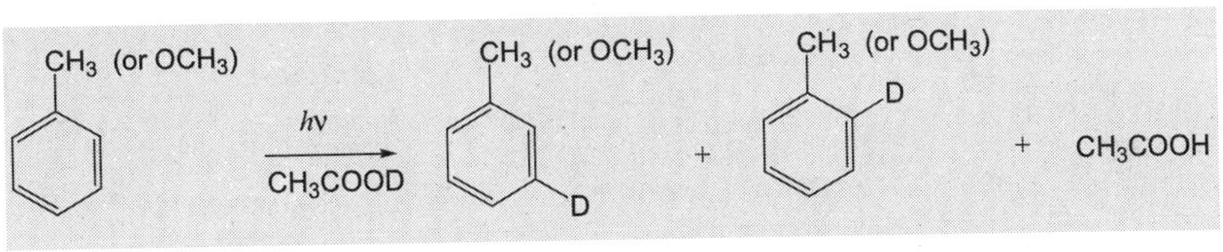

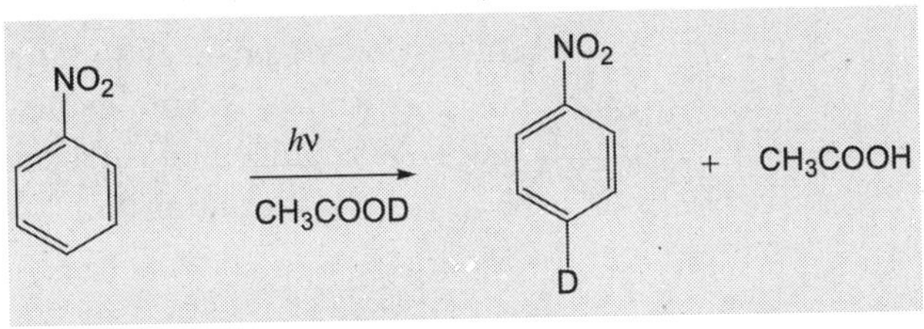

14.5 PHOTOREARRANGEMENT OF CYCLOHEXADIENONES

Photochemical reactions of organic molecules may sometimes prove to be complex processes. One such type of transformation, that has attracted attention for several decades, is the photochemical rearrangement of compounds containing the cyclohexadienone chromophore. 4, 4-Diphenylcyclohexadienone (**40**), for instance, is rapidly isomerized to a ketone (**41**) and two phenols (**42**) and (**43**) on irradiation.

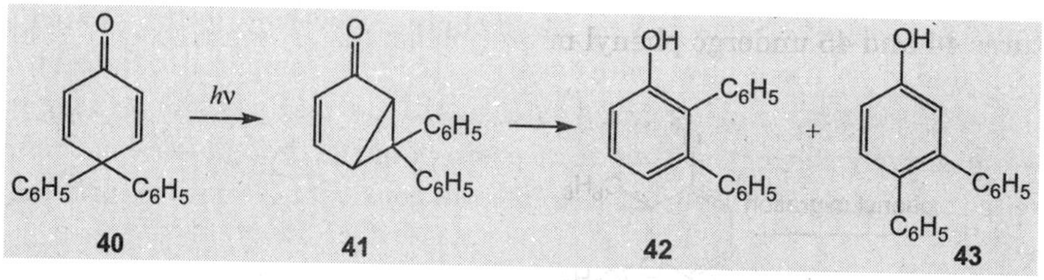

Cyclohexadienone displays absorption at about 300 nm (n, π^*) and at about 240 nm (π, π^*). Since the above photochemical transformation may be brought about by irradiation at either wavelength, it is reasonable to assume that the lower energy transition (n, π^*) is involved in the initial excitation process. It has been shown that the energy transfer takes place when **40** is irradiated with triplet energy sensitizers such as acetophenone. These observations suggest that the photochemical rearrangement of **40** proceeds *via* an n, π^* triplet state.

In spite of these conclusions, the detailed mechanisms of these rearrangements are not fully understood. Whether the excited states are best represented by polar structures or free radical types of intermediates has been the subject of considerable controversy. However, it appears logical *from the structure of the products* that excitation cf ketone moiety of the molecule involves electron redistribution which renders the molecule more polarizable with respect to the ground state. The intermediates thus have a polar character in which negative charge is localized on the oxygen atom and positive charge is spread over the carbon skeleton. Based on the polar state concept, as advanced by Chapman, we may write the mechanism of the above reaction in the following steps.

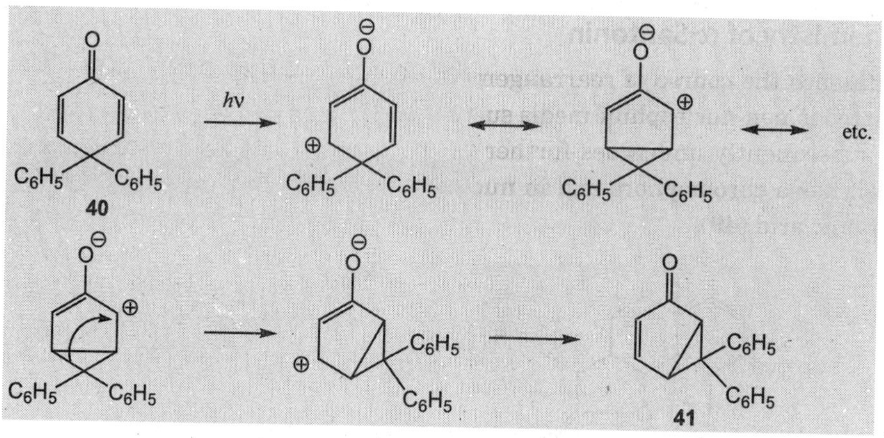

Since **41** still possesses a chromophore, it can undergo further n, π^* excitation to produce the following resonance hybrid.

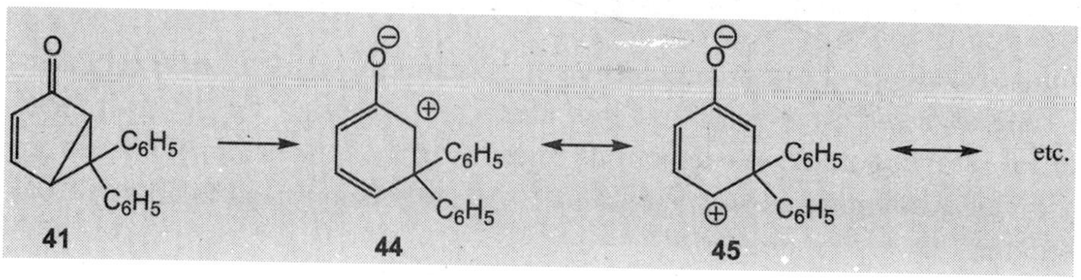

Structures **44** and **45** undergo phenyl migration leading, respectively, to **42** and **43**.

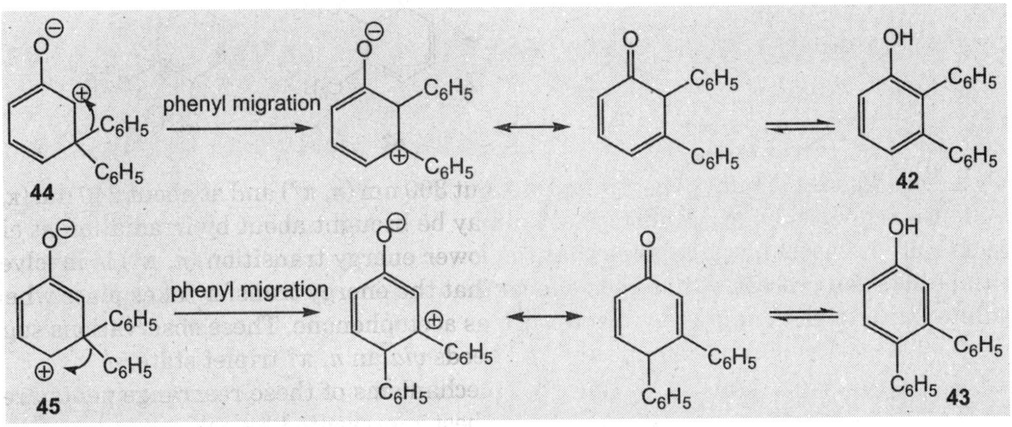

Zimmerman, on the other hand, has consistently advanced arguments in favour of participation by *free radical types of intermediates*. According to him, there occurs continuous electron redistribution after the absorption of light, culminating in the production of a series of excited states which finally deactivate to the ground state possessing a polar structure (like the one proposed by Chapman). It is this ground state that actually undergoes the rearrangement.

A. Photochemistry of α-Santonin

Solvents influence the course of rearrangement in dienones markedly. Thus α-santonin (**46**) when irradiated in non-nucleophilic media such as benzene is rapidly isomerized to lumisantonin (**47**) which subsequently undergoes further reaarangement to a dienone (**48**). However, this dienone itself has a chromophore and in nucleophilic media, such as water, it is transformed to photosantonic acid (**49**).

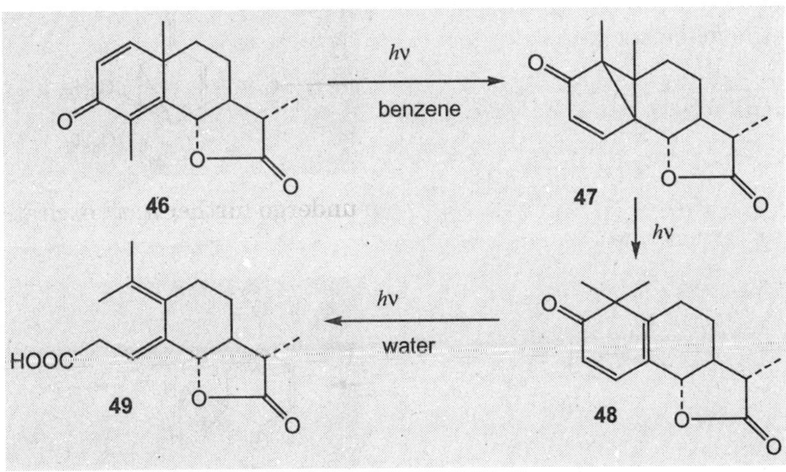

Based on Chapman's polar state concept, the mechanism of the above transformation may be written in the following steps.

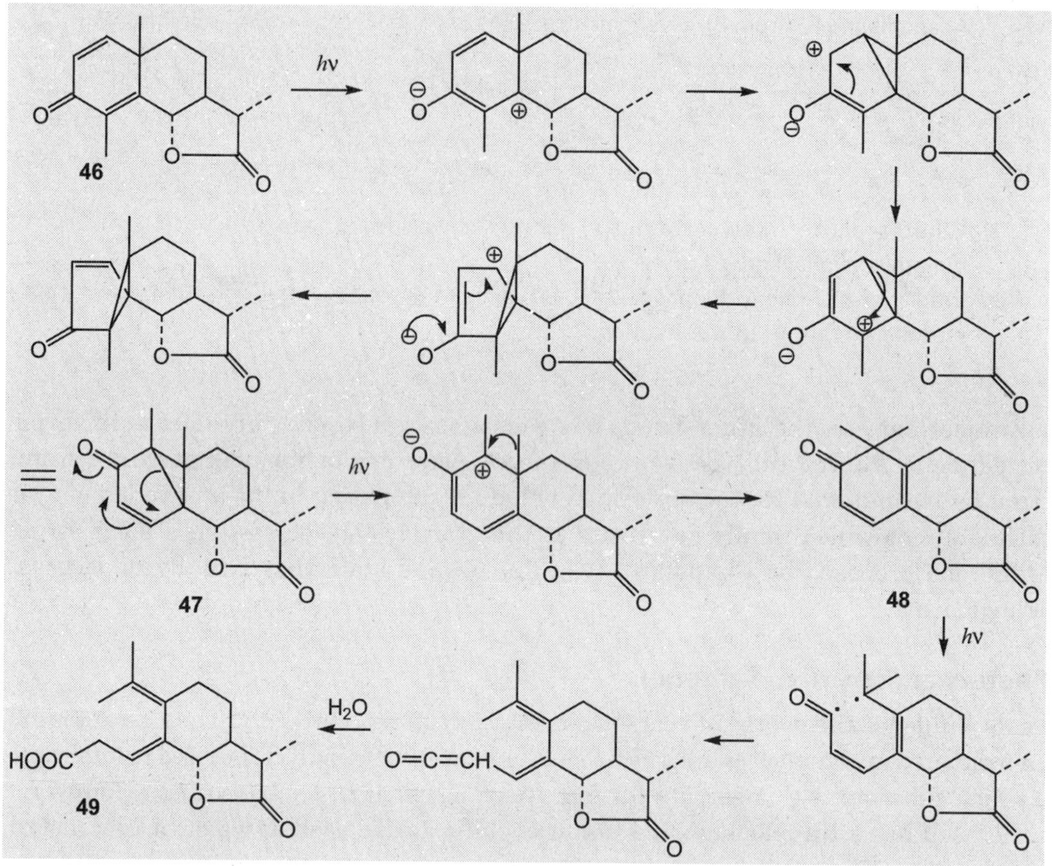

Irradiation of **46** in an aqueous acidic medium leads to the formation of isophotosantonic lactone (**50**) by the following path.

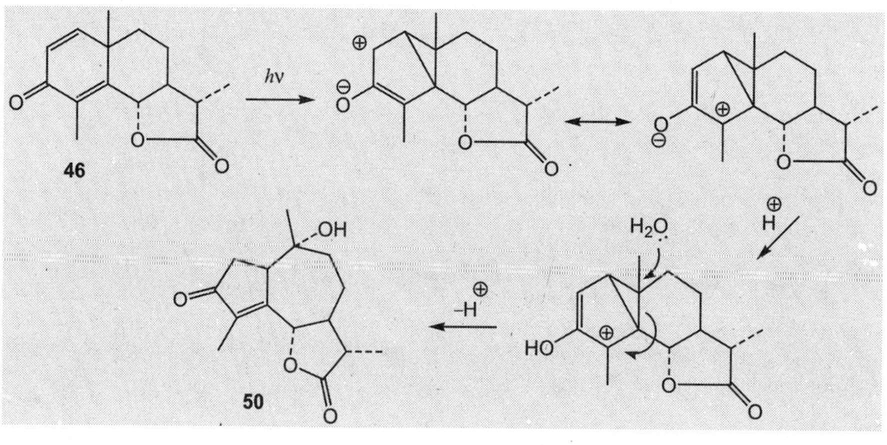

PROBLEMS AND REVIEW QUESTIONS

1. Compounds **A** and **B** undergo photodecarbonylation resulting in the formation of **I** and **II** (from **A**) and **III** and **IV** (from **B**). Give their structures and the mechanism of their formation.

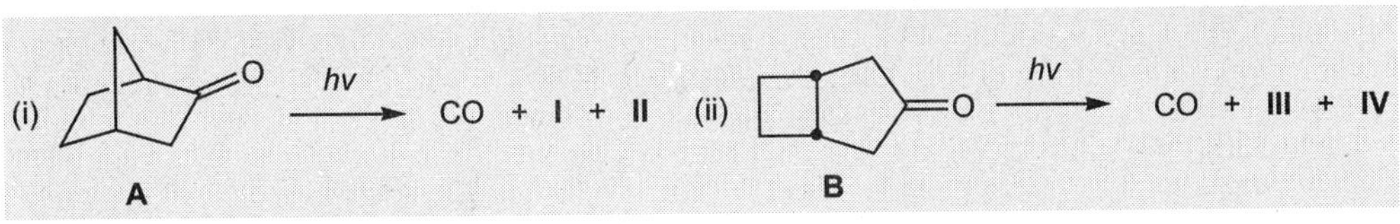

2. What are the likely product(s) from the vapour phase irradiation of *trans*-2,6-dimethylcyclohexanone?
3. Irradiation of 2-phenylthiophene-2-^{14}C (**A**) yields 3-phenylthiophene-3-^{14}C (**B**). Suggest a plausible mechanism for this transformation.

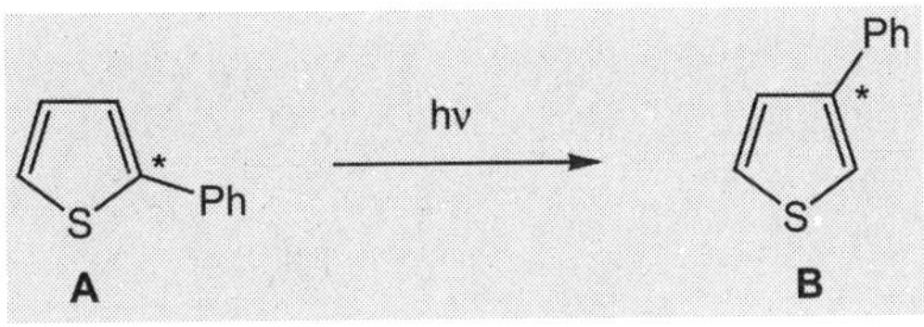

4. What are the products you expect to arise by the irradiation of

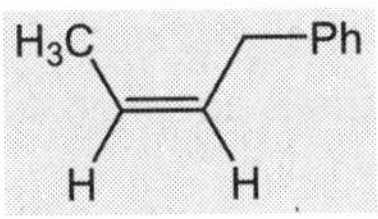

5. Write the structure of the intermediate **A** in the reaction.

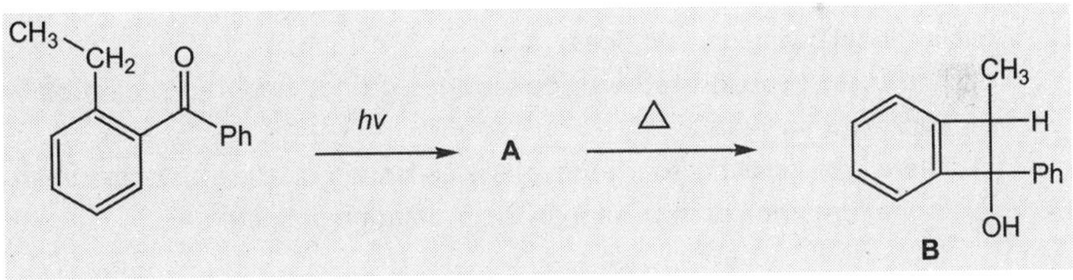

6. Predict the product in the given photochemical reactions.

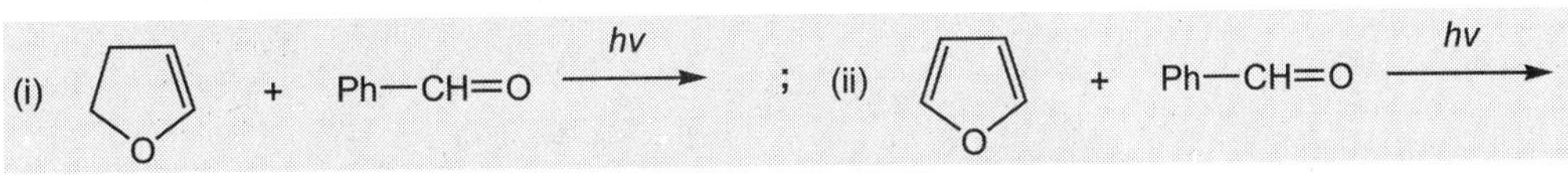

7. Predict the product(s) in the given reactions.

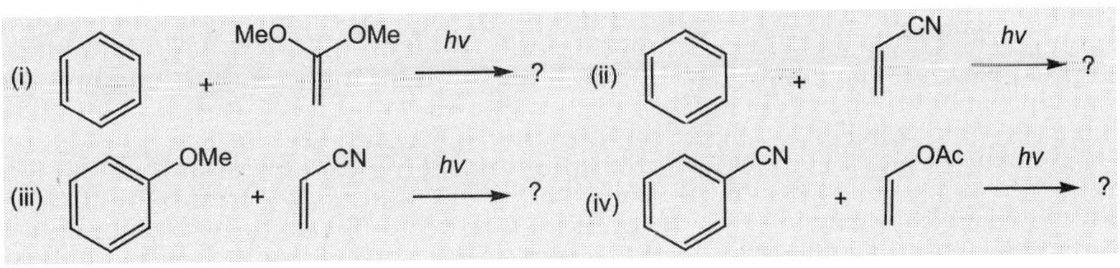

8. Indicate the products that could be formed on the irradiation of the following compounds:

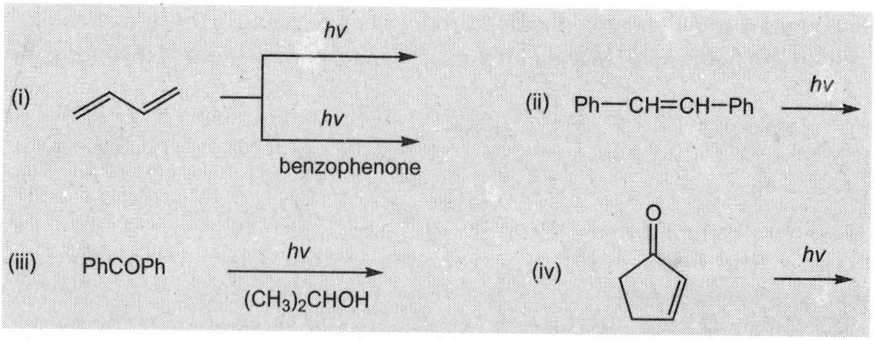

9. Draw the structure of possible products in the given reactions along with the mechanism.

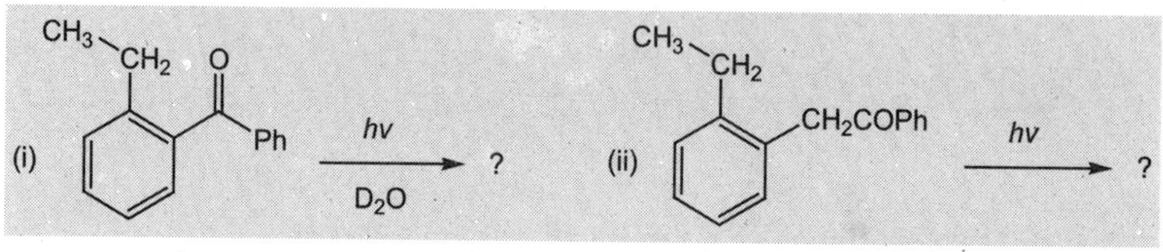

10. Irradiation of norbornene with benzophenone leads to the formation of the corresponding oxetane. However, dimer of norbornene is obtained on the irradiation of norbornene with acetone. How do you explain these results?
11. What are the product(s) obtained on the irradiation of benzophenone with (i) 2-methylpropene, and (ii) *cis*- and *trans*-2-butenes? Comment upon the mechanism of these reactions.
12. Explain the following observations:
    (i) *Ortho* alkyl substituted benzophenones do not undergo reduction when irradiated in hydroxylic solvents.
    (ii) There is preferential formation of the *cis* isomer on the irradiation of an alkene.
13. Explain the formation of indanol **B** on the irradiation of ketone **A**.

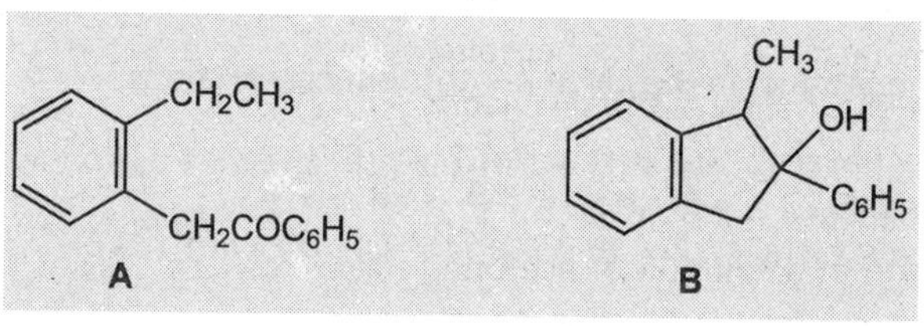

14. Outline the mechanism of formation of fulvene and benzvalene on irradiation of benzene.
15. Write various steps involved in the photochemical transformation of α-santonin (**A**) to lumisantonin (**B**).

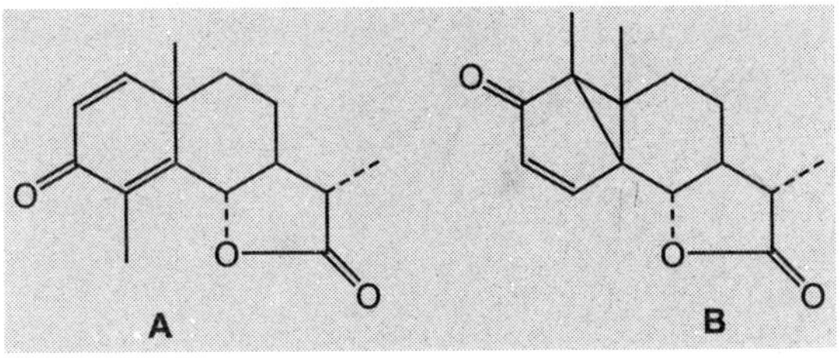

16. Draw the structure(s) of the products and comment on their formation.

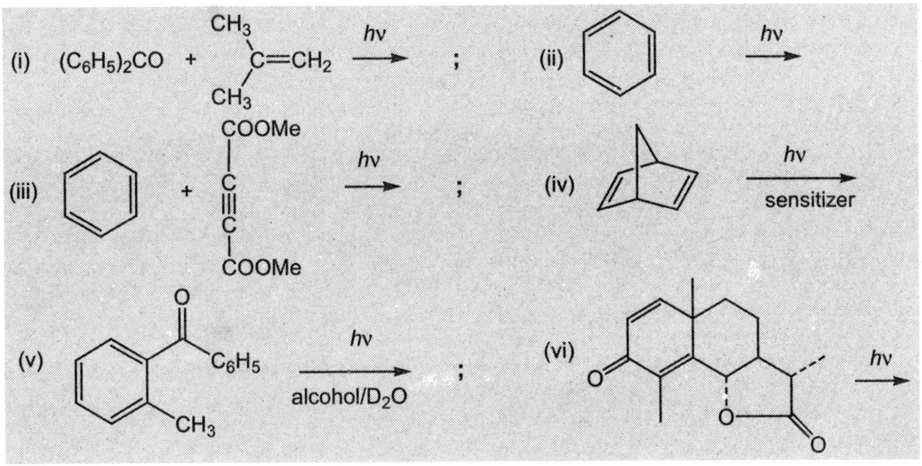

17. Discuss the mechanism of the photoreduction of benzophenone leading to the formation of benzpinacol.
18. Classify various types of electronic excitations giving their relative energy levels. It is said that *n*, π^* transition is a low energy but a forbidden process. Discuss.
19. Discuss the significance of (a) energy transfer and (b) intersystem crossing in organic photochemical transformations.
20. (i) What is meant by the quantum yield? What information can be drawn from this study?
    (ii) The quantum yield for benzpinacol formation is about 1. How can this information be used for elucidation of the mechanism of the reaction?
21. What is Paterno-Büchi reaction? Discuss its mechanism along with the stereochemical consequences.
22. Irradiation of a solution of acetone in *iso*propyl alcohol leads to the formation of pinacol. An attractive mechanism for this process involves the dimerization of initially formed acetone ketyl radicals with a quantum yield for the disappearance of acetone being 2. However, the quantum yield is only about 1. It has also been demonstrated that reaction of two ketyl radicals actually gives 1:1 mixture of pinacol and acetone. Suggest a mechanism consistent with the above data.

SOLUTIONS

1. (i)

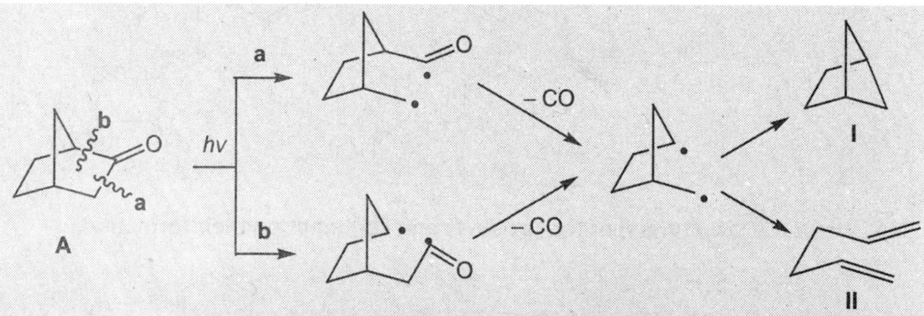

(ii)

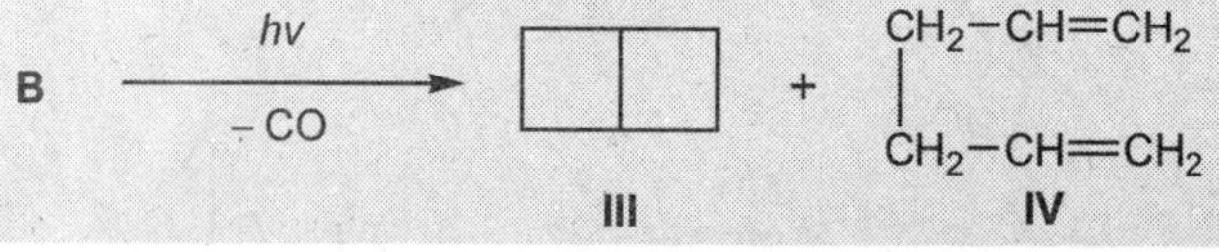

2.

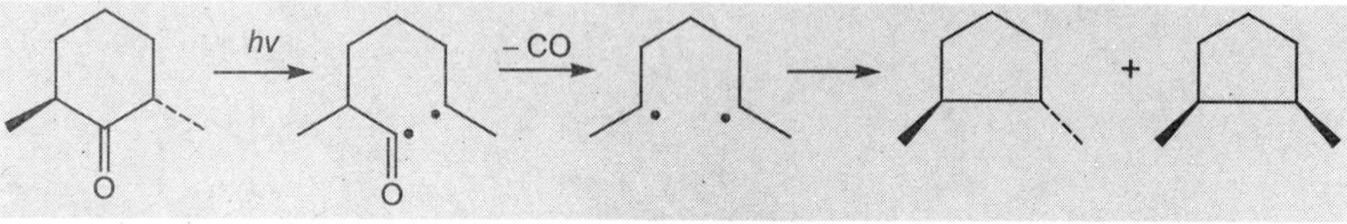

3.

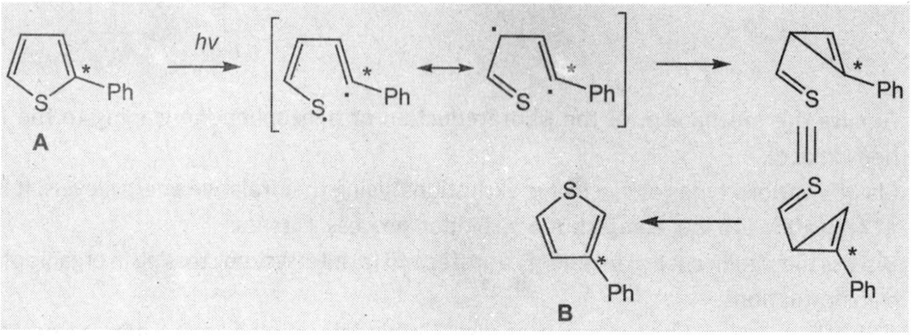

4.

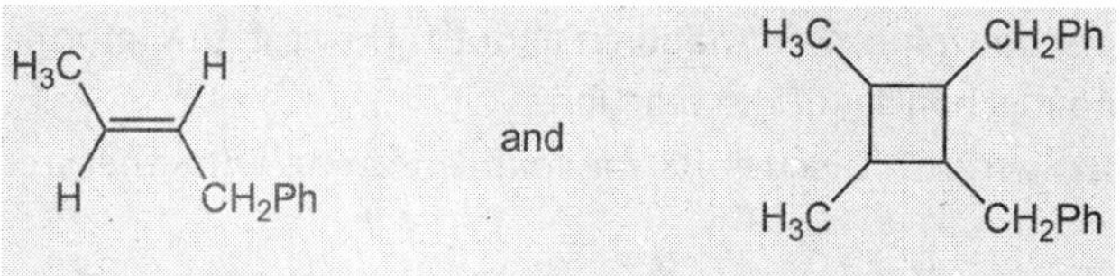

(isomerization) (dimerization)

5. The photoreduction efficiency of benzophenone derivatives is greatly reduced when an *ortho* alkyl substituent is present due to a new photoreaction in which *intramolecular* hydrogen atom transfer becomes the dominant process over *intermolecular* abstraction of hydrogen. As a result, hydrogen abstraction takes place from the benzylic position on the adjacent alkyl chain, giving an unstable enol (**A**) that can revert to the original benzophenone instead of pinacol derivative. The reactive dienol undergoes thermal cyclization to **B**.

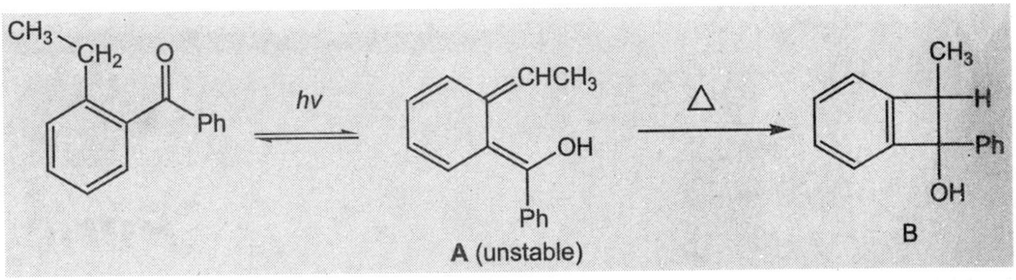
hv
Δ
A (unstable)
B

6. (i)

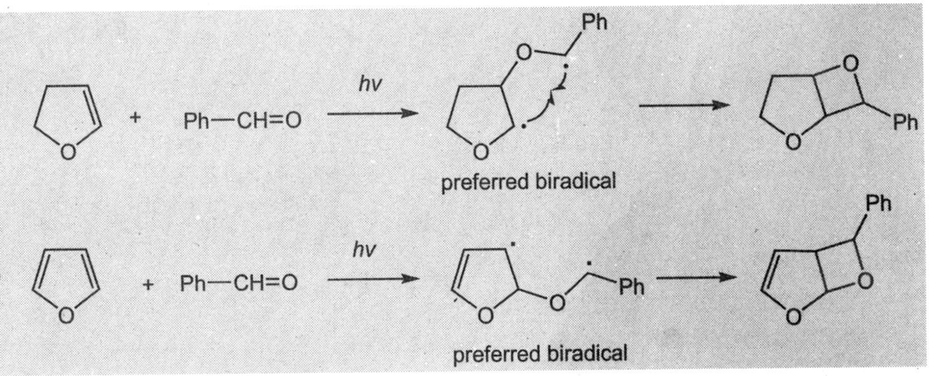
Ph—CH=O
hv
preferred biradical
(ii)
Ph—CH=O
hv
preferred biradical

7. (i)

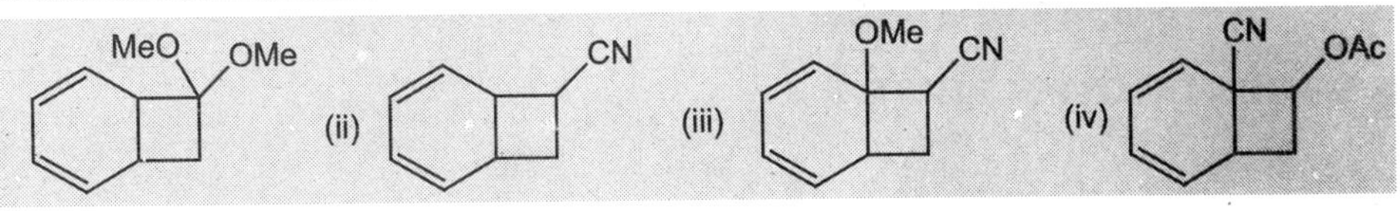
MeO
OMe
(ii)
CN
(iii)
OMe
CN
(iv)
CN
OAc

Chapter 15

Pericyclic Reactions

Chapter Outline

15.1 CONSERVATION OF MOLECULAR ORBITAL SYMMETRY
- A. Symmetry Properties of Molecular Orbitals

15.2 ELECTROCYCLIC REACTIONS
- A. Correlation Diagrams
- B. Frontier Molecular Orbital (FMO) Method

15.3 CYCLOADDITION REACTIONS
- A. Stereochemical Modes of Cycloaddition: Suprafacial and Antarafacial Processes
- B. Orbital Symmetry in Cycloaddition Reactions: Correlation Diagrams
- C. Frontier Molecular Orbital (FMO) Method
- D. 1,3-Dipolar Cycloaddition (1,3-DPCA) Reactions
- E. Cheletropic Reactions

15.4 SIGMATROPIC REARRANGEMENTS
- A. Suprafacial and Antarafacial Processes
- B. Analysis of Sigmatropic Rearrangements
- C. Cope and Claisen Rearrangements

15.5 APPLICATION OF PMO METHOD TO PERICYCLIC REACTIONS

There is a large body of chemical reactions of olefinic compounds which do not involve ionic or free radical intermediates and which are remarkably insensitive to the presence or absence of solvents and catalysts. Many of these reactions are characterized by the making and breaking of two or more bonds in a single concerted step through *cyclic transition-states*, wherein all first-order bonding changes occur. Because of their *cyclic transition-states*, they are known as *pericyclic reactions*. They are initiated by *heat* (thermolysis) and/or *light* (photolysis) and are highly stereospecific. The most remarkable observation about these reactions is that, very often, thermal and photochemical processes yield products with different stereochemistry. If *cis* isomer is obtained under thermolysis, *trans* isomer will be obtained under photolysis and *vice versa*. Pericyclic reactions include three different families of reactions namely, *electrocyclic reactions*, *cycloaddition* reactions and *sigmatropic rearrangements*.

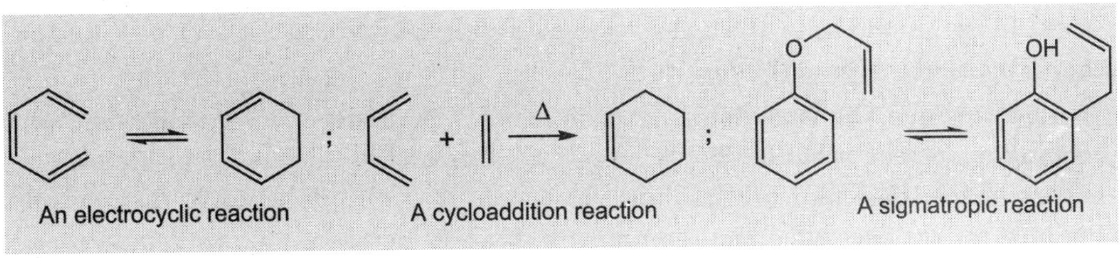

15.1 CONSERVATION OF MOLECULAR ORBITAL SYMMETRY

Many such reactions have been known for a long time, but it was in 1965 when Woodward and Hoffmann offered a reasonable explanation about the relationship between the structure and configuration of the reactant, the condition under which the reaction takes place and the stereochemistry of the products. This explanation is based on principle of *Conservation of Orbital Symmetry*. It states that orbital symmetry is conserved in *concerted reactions*.

Concerted reactions are those in which a reactant is converted into the product without the intervention of an intermediate and bond formation and bond breakage occur synchronously, though not necessarily symmetrically. In a cyclic concerted reaction, *i.e.* reaction involving a cyclic conjugated transition state, orbitals in the reactant can only transform into those orbitals in the product which have the same *symmetry properties* with respect to the elements of symmetry preserved in the reaction. Even if symmetry is slightly disturbed in a reactant by a trivial substituent or by asymmetry of the molecule, a concerted reaction may still be analyzed by mixing of interacting orbitals according to quantum mechanical principles and following them through the reaction. The energy of the transition state of a *symmetry-allowed process* will necessarily be lower than that of the alternative *symmetry-forbidden path* and even when this difference is small, a concerted reaction will take the path of least resistance, *i.e.* the *symmetry-allowed path*, if available.

A. Symmetry Properties of Molecular Orbitals

Let us consider molecular orbital (MO) theory with reference to the simplest π-molecule, *ethylene*. There are two *p* orbitals in the molecule (we ignore σ-bond skeleton in this treatment) and their linear combination gives us an approximation set of two molecular π orbitals having different energies than the *p* atomic orbitals. The energy of the one approximation (Ψ_1) is considerably lower than that of the other (Ψ_2). Therefore, the electrons in the ground state are described by the Ψ_1 orbital. Ψ_1 is bonding while Ψ_2 of higher energy is antibonding with one node* (**Fig. 15.1**). In the ground state, the highest occupied molecular orbital (HOMO) is Ψ_1 and lowest unoccupied molecular orbital (LUMO) is Ψ_2.

In 1,3-butadiene, the four atomic orbitals combine to produce four π molecular orbitals: two bonding molecular orbitals (Ψ_1, Ψ_2) and two antibonding molecular orbitals (Ψ_3, Ψ_4). The energy of the molecular orbitals increases from Ψ_1 to Ψ_4 (**Fig. 15.2**). The number of bonding interactions decreases and the number of nodes increases (Ψ_1 : three and zero; Ψ_2 : two and one; Ψ_3 : one and two; Ψ_4 : zero and three). In the ground state, the highest occupied molecular orbital (HOMO) is Ψ_2 and lowest unoccupied molecular orbital (LUMO) is Ψ_3. The molecule will be in excited state after absorption of light by promoting an electron from its ground state HOMO to LUMO. In excited state the HOMO is Ψ_3 and LUMO is Ψ_4. This means that the reactant is in ground state under thermal condition and in excited state under photochemical condition.

Similarly in 1,3,5-hexatriene, the six atomic orbitals combine to produce six π molecular orbitals: three bonding molecular orbitals (Ψ_1, Ψ_2, Ψ_3) and three antibonding molecular orbitals (Ψ_4, Ψ_5, Ψ_6). The energy of the molecular orbitals increases from Ψ_1 to Ψ_6 (**Fig. 15.3**). The number of bonding interactions decreases and the number of nodes increases (Ψ_1 : five and zero; Ψ_2 : four and one; Ψ_3 : three and two; Ψ_4 : two and three; Ψ_5 : one and four; Ψ_6 : zero and five). In the ground state, the highest occupied molecular orbital (HOMO) is Ψ_3 and lowest unoccupied molecular orbital (LUMO) is Ψ_4. The molecule will be in excited state after absorption of light by promoting an electron from its ground state HOMO to LUMO. In excited state the HOMO is Ψ_4 and LUMO is Ψ_5. This means that the reactant is in ground state under thermal condition and in excited state under photochemical condition.

---

* When two out of phase atomic orbitals interact, a node is created between the two nuclei. It is the region which has least probability of finding the electron(s). It is denoted by a dot (•).

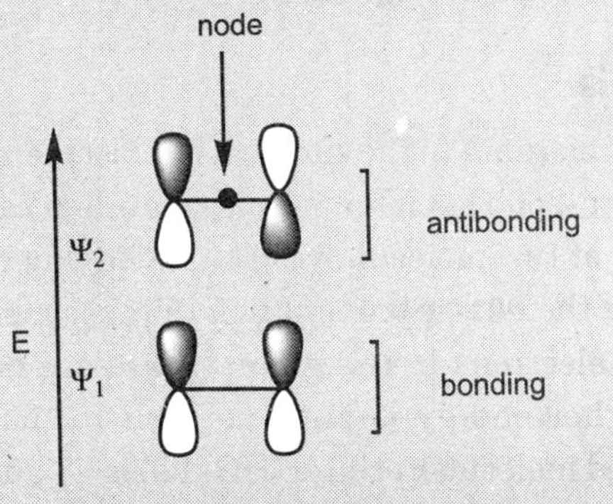

Fig.15.1 Molecular orbitals of ethylene

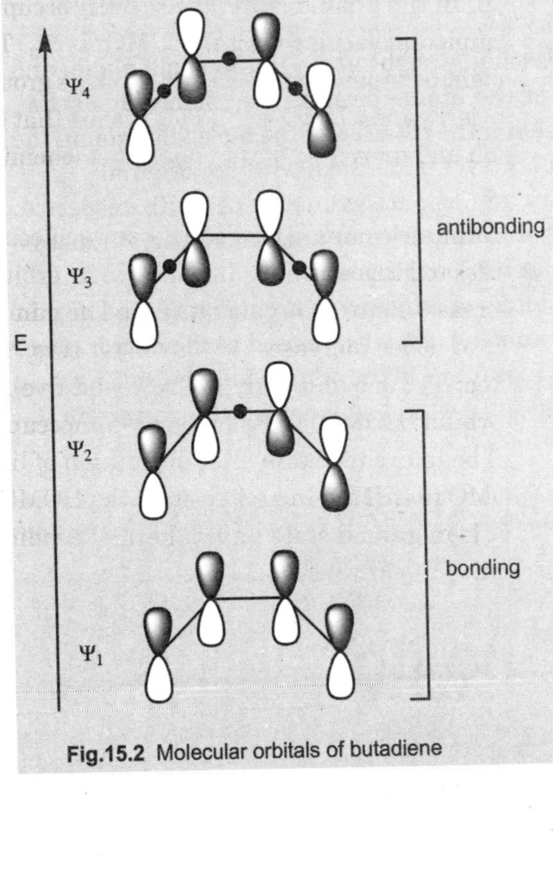

Fig.15.2 Molecular orbitals of butadiene

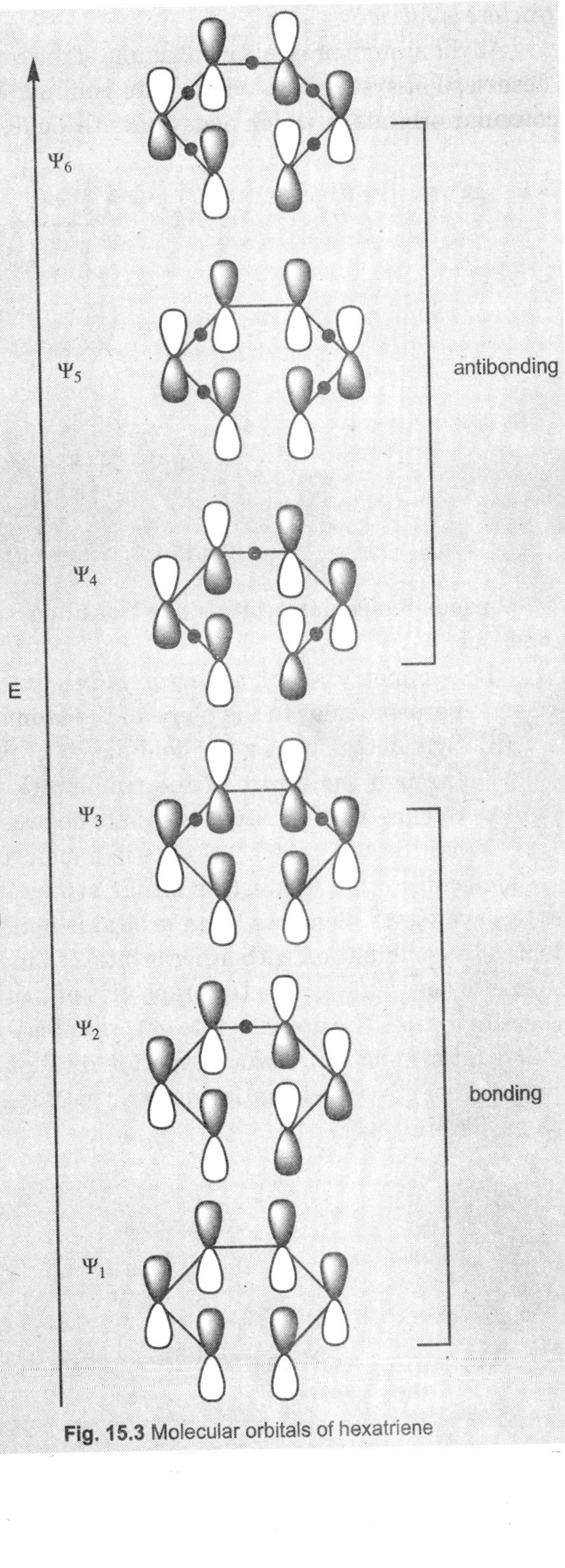

Fig. 15.3 Molecular orbitals of hexatriene

Structure of Allylic System

Allylic system presents an example of the conjugated system having odd number of carbons. These type of systems, in addition to bonding and antibonding orbitals, possess nonbonding molecular orbitals in which nodes pass through carbon atoms (**Fig. 15.4**).

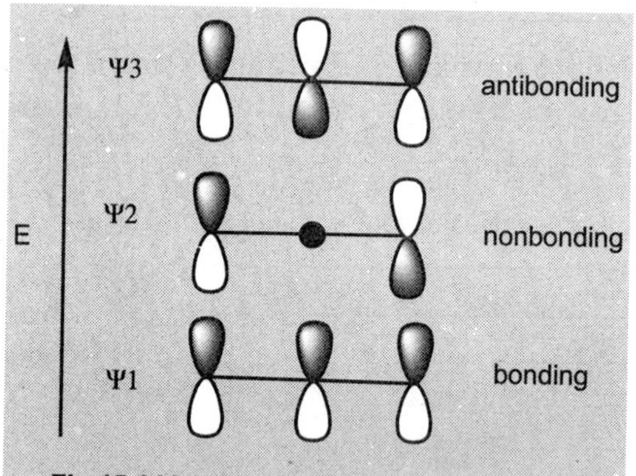

Fig.15.4 Molecular orbital of an allylic system

Various molecular orbitals are classified according to their two independent symmetry properties:

(i) Symmetry about a *mirror plane* (m) which bisects the molecular orbital and is perpendicular to the plane of the atoms.

(ii) Symmetry about a two-fold axis (C_2) passing at right angle in the same plane, and through the centre of the framework of the atoms forming the molecular orbital. Symmetry of this type is said to be present if the rotation of the molecule around the axis by 180° (360°/2) results in a molecular orbital identical with the original.

Molecular orbitals must be either symmetric (S) or antisymmetric (A) with respect to the two symmetry elements. If an orbital is neither symmetric nor antisymmetric, it must be adopted by combination with other orbitals to meet this requirement.

Let us examine symmetry properties of π orbitals of ethylene in ground state and also in the excited state. The ground state (π) orbital is symmetric (S) with respect to the mirror plane (m) and antisymmetric (A) with respect to rotation axis (C_2). On the other hand, the antibonding orbital (π^*) of ethylene is antisymmetric with respect to m and symmetric with respect to the C_2 axis (**Table 15.1**).

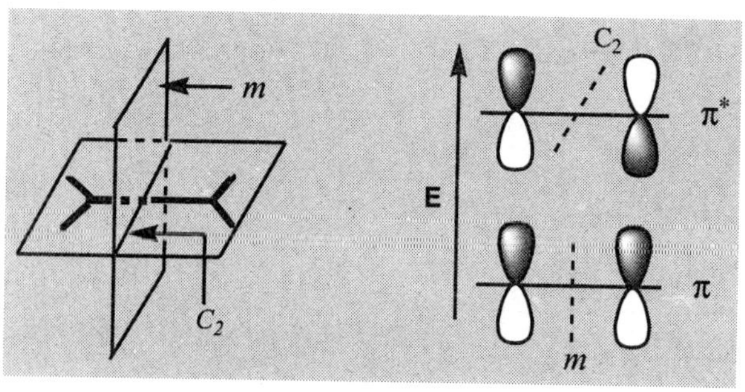

Table 15.1 Symmetry properties of π orbitals

Orbital (wave function)	m	C_2
(Ψ_2)	A	S
(Ψ_1)	S	A

A similar consideration leads to the following symmetry properties for the four molecular π orbitals of 1,3-butadiene and six molecular π orbitals of 1,3,5-hexatriene (**Tables 15.2** and **15.3**)

Table 15.2 Symmetry properties of four π molecular orbitals of 1,3-butadiene

Orbital (wave function)	m	C_2
Ψ_4	A	S
Ψ_3	S	A
Ψ_2	A	S
Ψ_1	S	A

Table 15.3 Symmetry properties of six π molecular orbitals of 1,3,5-hexatriene

Orbital (wave function)	m	C_2
Ψ_6	A	S
Ψ_5	S	A
Ψ_4	A	S
Ψ_3	S	A
Ψ_2	A	S
Ψ_1	S	A

The sigma orbital of a C–C covalent bond has mirror plane symmetry and since a rotation of 180° through its midpoint regenerates the same sigma orbital, it also has a C_2 symmetry. A σ* orbital is antisymmetric with respect to both m and C_2 (**Table 15.4**).

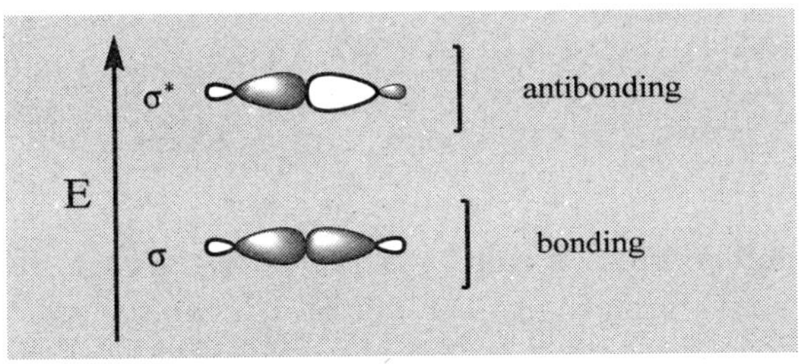

Table 15.4 Symmetry properties of sigma orbitals

Orbital (wave function)	m	C_2
σ*	A	A
σ	S	S

For a linear conjugated π system, the wave function Ψ_n will have $(n - 1)$ nodes. When $(n - 1)$ is zero or an even integer, Ψ_n will be symmetric with respect to m and antisymmetric with respect to C_2. When $(n - 1)$ is an odd integer, Ψ_n will have the symmetry exactly reversed (**Table 15.5**).

Table 15.5 Symmetry elements in the orbital Ψ_n of a linear conjugated π system

Nodes $(n - 1)$	m	C_2
0 or even integer	S	A
odd integer	A	S

15.2 ELECTROCYCLIC REACTIONS

An electrocyclic reaction is a process in which two π bonds are rearranged to a σ bond and a π bond leading to a cyclic product or *vice-versa*. Thus if the open-chain partner contains $k\pi$ electrons, the cyclic partner has $(k–2)$ π electrons and two electrons in a new σ bond.

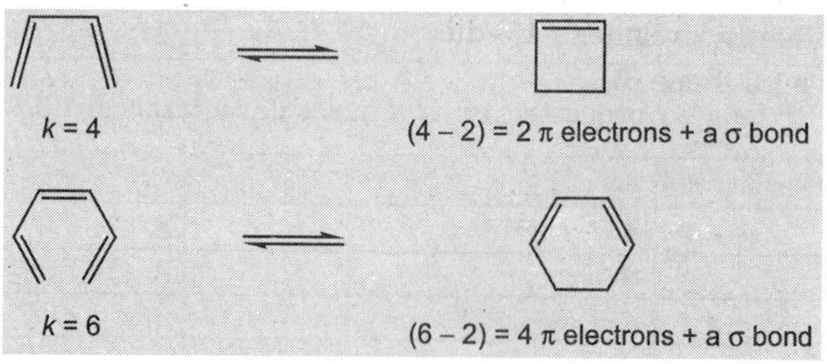

Let us first consider the simplest example in which a cyclobutene system (**1**) opens to a 1,3-butadiene system (**2** and **3**). The reaction can be performed thermally or photochemically and under either condition the rearrangement is completely stereospecific.

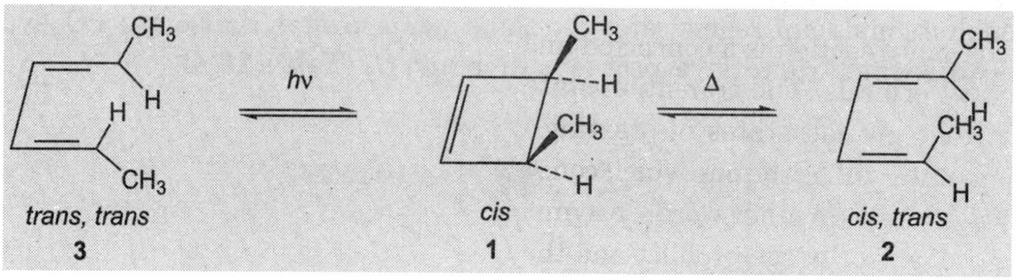

A σ bond of cyclobutene must break to yield the open-chain diene and obviously this bond may break in two ways. First, the two atomic orbital components of the σ bond may both rotate in the same direction, clockwise or anticlockwise (*conrotatory process*) as under thermal condition.

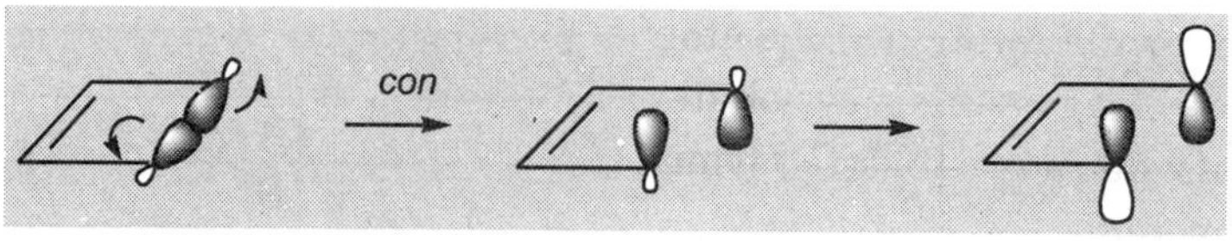

Second, the atomic orbitals may rotate in opposite directions, one clockwise and the other anticlockwise (*disrotatory process*) as under photochemical condition.

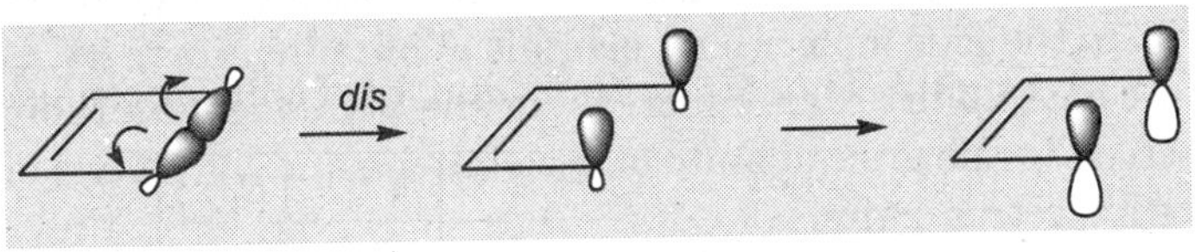

The stereochemical significance of these two modes of ring-opening (or ring-closing) becomes apparent when we consider substituted reactants. Thus depending upon these modes, the substituents may rotate in the same direction (*conrotatory*) or in opposite directions (*disrotatory*).

However, from the above analysis it is incorrect to infer that all electrocyclic reactions proceed in a conrotatory fashion under thermal condition and in a disrotatory mode on irradiation. For instance, irradiation of *trans*, *cis*, *trans*-1,6-dimethylhexa-1,3,5-triene (**4**) yields *trans*-5,6-dimethylcyclohexa-1,3-diene (**5**), while thermal ring-closure leads to *cis*-5,6-dimethylcyclohexa-1,3-diene (**6**).

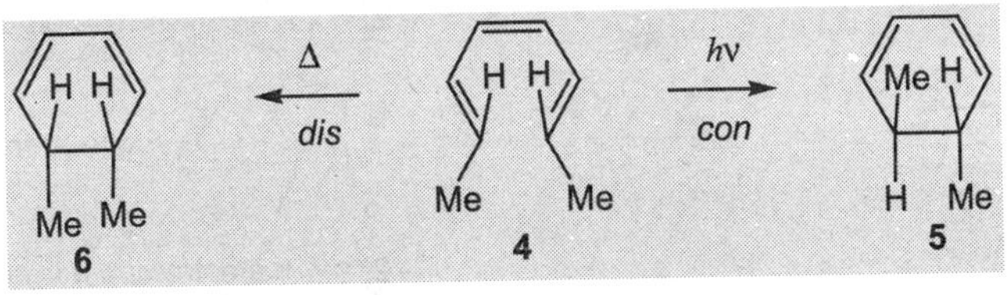

Electrocyclic reactions can be analyzed by correlation diagrams and Frontier Molecular Orbital (FMO) method.

A. Correlation Diagrams

An electrocyclic reaction is a concerted and cyclic process in which reactant orbitals transform into product orbitals. The transition-state of such reactions should be intermediate between the electronic ground states of starting material and product. Obviously, the most stable transition-state will be the one which conserves the *symmetry* of the *reactant* orbitals in passing to *product* orbitals. In other words, a symmetric (*S*) orbital in the reactant must transform into a *symmetric orbital* in the product and that an antisymmetric (*A*) orbital must transform into an *antisymmetric orbital*. If the symmetries of the reactant and product orbitals are not the same, the reaction will not take place in a concerted manner.

Let us exemplify the above principle by analyzing the cyclobutene-butadiene transformation. The ring-opening (or ring-closing) may be a *disrotatory process* in which the atom or group of atoms on the saturated carbons rotate in *opposite directions*, or, alternatively, it may proceed *via conrotation*, involving rotation of the atom or group of atoms in the *same direction*. In the case of the *disrotatory* ring-opening, cyclobutene preserves a plane of symmetry (m) throughout the transformation while a two-fold (C_2) symmetry axis is maintained at all times in the *conrotatory* mode of ring-opening.

We are now set to analyze the above transformation in terms of the fundamental rule of the conservation of orbital symmetry as proposed by Woodward and Hoffmann. The orbitals of cyclobutene involved directly are σ and π bonds and the related antibonding orbitals σ* and π*; these orbitals pass on to the four molecular π orbitals of butadiene *viz.*, ψ_1, ψ_2, ψ_3, ψ_4. All these orbitals are listed in **Fig. 15.5** in order of increasing energy along with their mirror plane and C_2 symmetry properties. In the ground state of cyclobutene and butadiene only σ, π and ψ_1, ψ_2 orbitals are filled with electron pairs.

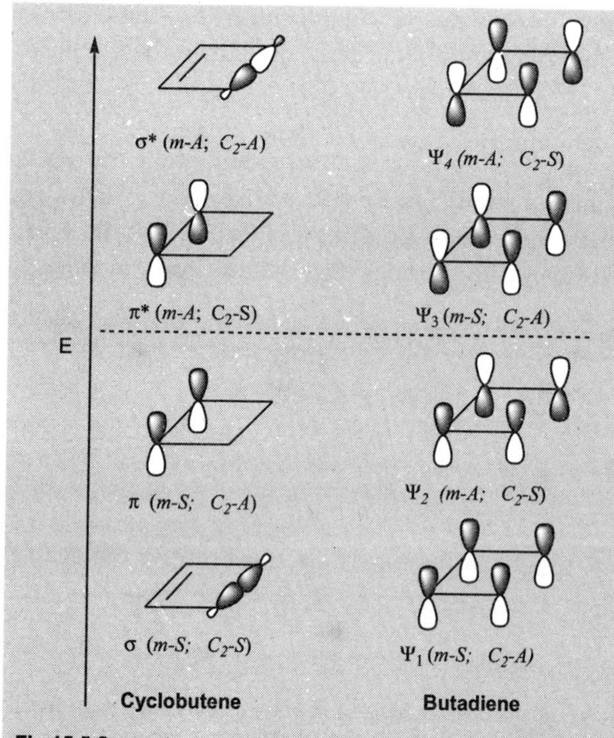

Fig.15.5 Symmetry properties of molecular orbitals of cyclobutene and butadiene

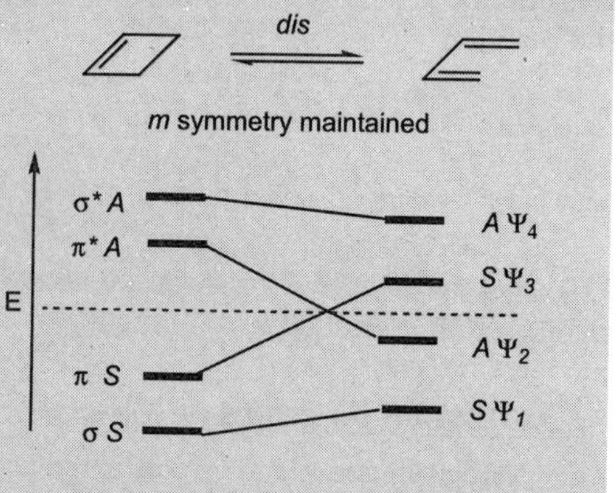

Fig.15.6 Correlation diagram for disrotatory interconversion of cyclobutene-butadiene system

It is easy to analyze an electrocyclic reaction by constructing a correlation diagram, which is simply a diagram showing the possible transformation of reactant orbitals to product orbitals. Let us first analyze a *disrotatory* opening of cyclobutene in which a mirror plane symmetry (*m*) is maintained.

In constructing this correlation diagram (**Fig. 15.6**) we have simply connected by lines the various orbitals of cyclobutene and butadiene keeping in view that there is correlation between orbitals of the same symmetry, having minimal energy differences. The following two conclusions can immediately be drawn by a close inspection.

A. We expect a thermal transformation to take place only when the ground state orbitals of the reactants correlate with the ground state orbitals of the products. Although in the above diagram the cyclobutene ground state orbital σ *correlates* with the butadiene ground state orbital Ψ_1, the π orbital of the former does not correlate with Ψ_2 of the latter. Instead, it correlates with Ψ_3 which is an excited state and an antibonding orbital. Thermal transformation of cyclobutene $\rightleftharpoons$ butadiene system by *disrotatory* process is thus *symmetry forbidden.*

$$\sigma^2\pi^2 \underset{dis}{\overset{\Delta}{\nrightarrow}} \Psi_1^2\Psi_3^2 \qquad\qquad \Psi_1^2\Psi_2^2 \underset{dis}{\overset{\Delta}{\nrightarrow}} \sigma^2\pi^{2*}$$

B. Irradiation of cyclobutene produces the first excited state in which an electron is promoted from π to π^* orbital and in this case σ, π and π^* orbitals of cyclobutene correlate with ψ_1, ψ_2 and ψ_3 orbitals of butadiene. In other words, the first excited state of cyclobutene correlates with the first excited state of butadiene and hence *disrotatory* ring-opening (ring-closing) is photochemically a *symmetry allowed process.*

$$\underset{\textit{ground state}}{\sigma^2\pi^2} \xrightarrow{h\nu} \underset{\textit{first excited state}}{\sigma^2\pi\pi^*} \rightleftharpoons \underset{\textit{first excited state}}{\Psi_1^2\Psi_2\Psi_3} \xleftarrow{h\nu} \underset{\textit{ground state}}{\Psi_1^2\Psi_2^2}$$

Working on similar lines, we can construct another correlation diagram (**Fig. 15.7**) for the conrotatory opening of cyclobutene in which a C_2 axis of symmetry is maintained throughout the reaction.

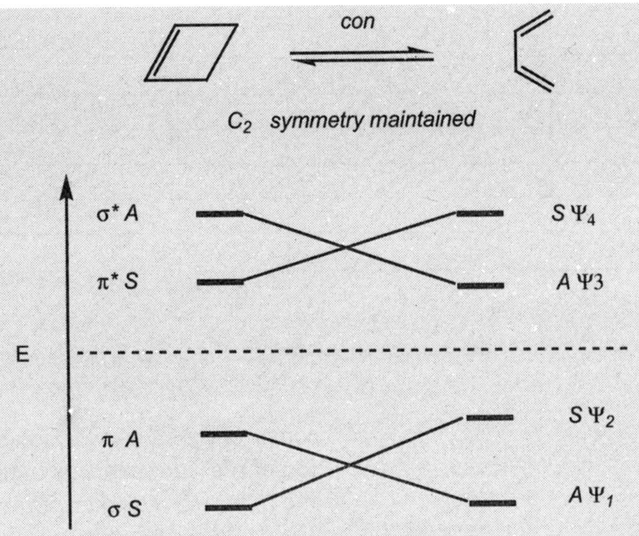

Fig.15.7 Correlation diagram for conrotatory interconversion of cyclobutene-butadiene system

Two conclusions may again be drawn from the above correlation diagram.

A. Since there is correlation between the ground state orbitals of cyclobutene and butadiene ($\sigma^2\pi^2 \rightleftharpoons \Psi_1^2\Psi_2^2$), a thermal conrotatory process in either direction is a *symmetry-allowed process.*

B. The first excited state of cyclobutene ($\sigma^2\,\pi\,\pi^*$) correlates with the upper excited state ($\psi_1^2\,\psi_2\,\psi_4$) of butadiene thus making it a high energy *symmetry-forbidden process.* Similarly, the first excited state of butadiene ($\psi_1^2\,\psi_2\,\psi_3$) correlates with a high energy upper excited state ($\sigma^2\,\pi\,\sigma^*$) of cyclobutene. In other words, a photochemical conrotatory process in either direction is *symmetry-forbidden.*

$$\sigma^2\pi^2 \xrightarrow{h\nu} \sigma^2\pi\pi^* \not\longrightarrow \Psi_1^2\Psi_2\Psi_4$$

$$\Psi_1^2\Psi_2^2 \xrightarrow{h\nu} \Psi_1^2\Psi_2\Psi_3 \not\longrightarrow \sigma^2\pi\sigma^*$$

Thus it becomes clear from the above considerations that thermal opening of the cyclobutene proceeds in a *conrotatory* process while photochemical interconversion involves a *disrotatory* mode. These generalizations are true for all systems containing $4q$ π electrons where $q = 0, 1, 2$, etc. However, for systems containing $(4q+2)$ π electrons theoretical predictions are entirely different and are in conformity with the actual observations.

A typical system of this type is the interconversion of cyclohexadiene and hexatriene. In this transformation six molecular orbitals (ψ_1 to ψ_6) of hexatriene and six molecular orbitals (four π and two σ) of cyclohexadiene are actually involved and, therefore, need to be considered. Symmetry properties of the six molecular orbitals of hexatriene have already been deduced earlier and working on similar lines, we have the following symmetry properties for the molecular orbitals of cyclohexadiene (**Table 15.6**).

Table 15.6 Symmetry properties of the molecular orbitals of cyclohexadiene

Orbital	***m***	C_2
σ^*	*A*	*A*
π_4	*A*	*S*
π_3	*S*	*A*
π_2	*A*	*S*
π_1	*S*	*A*
σ	*S*	*S*

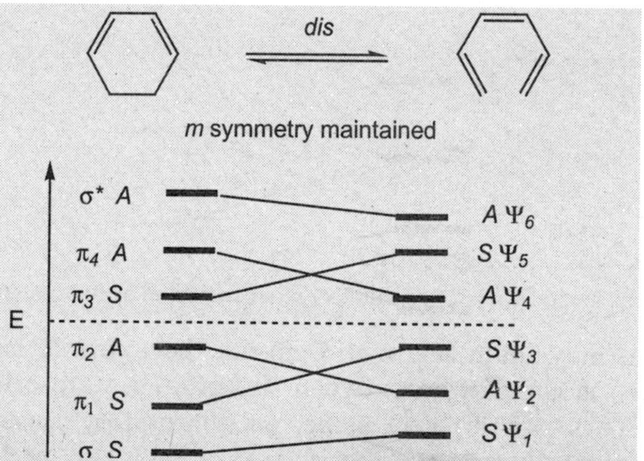

Fig. 15.7 Correlation diagram for disrotatory interconversion of cyclohexadiene-hexatriene system
cyclohexadiene-hexatriene system

The correlation diagram for the *disrotatory* pathway (**Fig. 15.7**) is constructed in the same way as in the case of cyclobutene-butadiene transformation.

The following conclusions may be drawn from these correlation diagrams.

A. In the disrotatory mode, ground state bonding orbitals of cyclohexadiene correlate with the ground state bonding orbitals of hexatriene and so it is a *thermally allowed* process.

$$\sigma^2 \pi_1^2 \pi_2^2 \xrightarrow{\Delta\ dis} \psi_1^2 \psi_2^2 \psi_3^2$$

B. But in the *conrotatory* mode (C_2 symmetry), ground state bonding orbitals of cyclohexadiene do not correlate with the ground state bonding orbitals of hexatriene (Fig. 15.8). Since the presence of two electrons in Ψ_4 is a very high energy process, a conrotatory mode is prohibited under thermal conditions.

$$\sigma^2 \pi_1^2 \pi_2^2 \xrightarrow[\times]{\Delta\ con} \psi_1^2 \psi_2^2 \psi_4^2$$

C. However, if we promote an electron to π_3 in cyclohexadiene (obviously by irradiation) then the orbitals of the reactant with C_2 symmetry correlate with the first excited state of the product.

$$\sigma^2 \pi_1^2 \pi_2^2 \xrightarrow{h\nu} \sigma^2 \pi_1^2 \pi_2 \pi_3 \rightleftharpoons \psi_1^2 \psi_2^2 \psi_3 \psi_4 \xleftarrow{h\nu} \psi_1^2 \psi_2^2 \psi_3^2$$

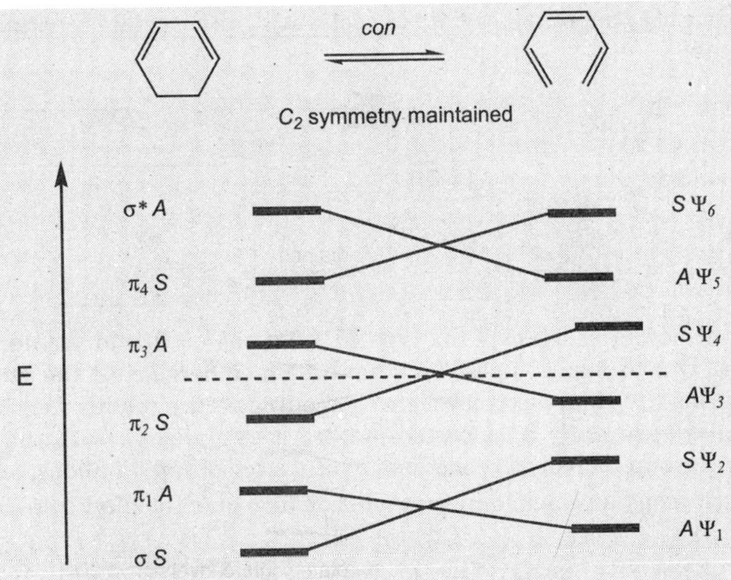

Fig.15.8 Correlation diagram for conrotatory interconversion of cyclohexadiene-hexatriene system

Therefore, photochemical interconversion is *allowed* in the conrotatory pathway. These generalizations are true for all the systems containing ($4q + 2$) π electrons, where q = 0,1,2, etc. Thus, Woodward-Hoffmann rules for electrocyclic reactions may be summed up as given in the **Table 15.7**.

Table 15.7 Selection rules for electrocyclic reactions

Number of π electrons	Thermal	Photochemical
$4q$	con	dis
$4q+2$	dis	con

Woodward and Hoffmann have further explained that under severe thermal conditions, symmetry-forbidden reactions may also take place but then they follow a non-concerted path and their energy of activation is 10-15 kcal/mol (or 42-63 kj/mol) higher than those for symmetry-allowed reactions. *It is because of this energy difference that most of the electrocyclic reactions follow Woodward-Hoffmann rules.*

Electrocyclic Reaction of Cyclopropyl Cations

The Woodward-Hoffmann orbital symmetry rules can also be applied to charged systems, the conversion of a cyclopropyl cation to an allylic cation being the simplest example because it involves only two π electrons. There is formation of the cyclopropyl cation which undergoes ring opening in a fast step.

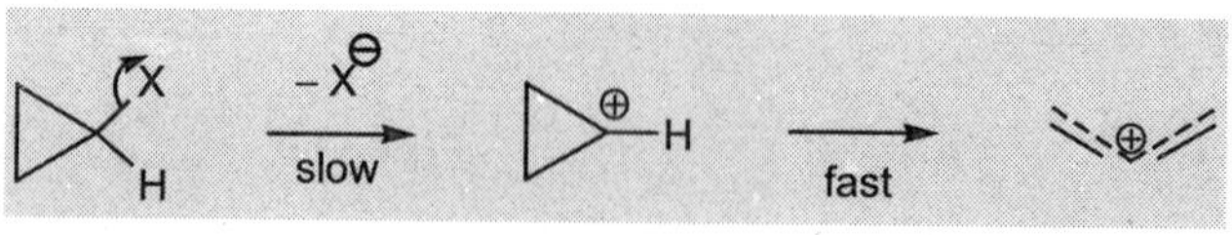

In the solvolysis of cyclopropyl tosylate in acetic acid, allyl acetate is obtained rather than cyclopropyl acetate.

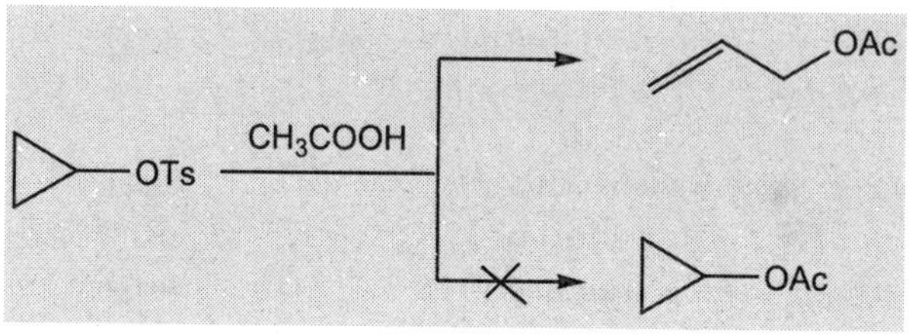

This is an electrocyclic reaction of the type $4q + 2$ type ($q = 0$) and should, therefore, be a *disrotatory* process. Due to strain imposed by the three-membered ring, the cyclopropyl cation is not a stable intermediate and electrocyclic ring opening occurs readily to give allyl acetate.

Ring opening is apparently a concerted process in conjunction with the rupture of the bond to the leaving group. Solvolysis reactions of other cyclopropyl halides, sulphonates and diazotization of cyclopropylamine in aqueous solution also give the allylic products.

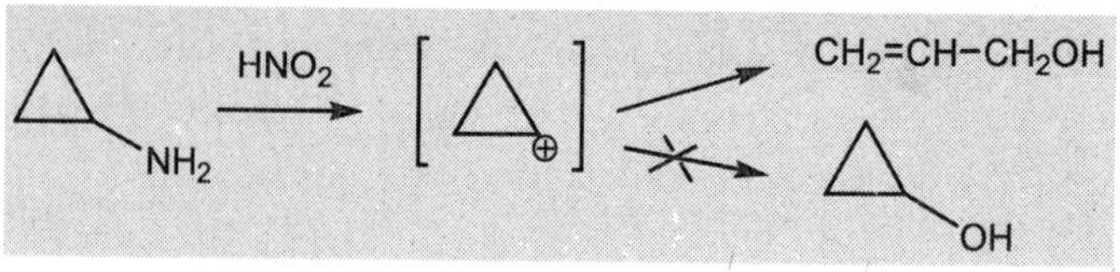

As with cyclobutenes, the allowed *disrotatory* ring opening can occur in two directions leading to the formation of two types of allylic cations which may have either a *u*-shape or *w*-shape.

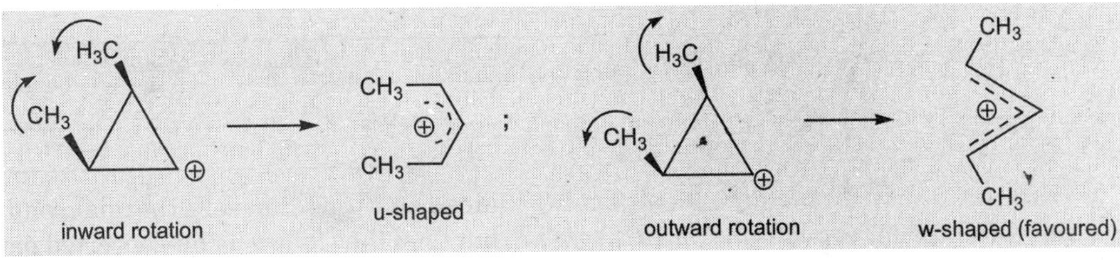

The *w*-shaped allylic cation is preferred over sterically less favourable *u*-shaped cation. Cyclopropyl tosylate **7b** has indeed been found to undergo solvolysis thousands of times faster than **7a**.

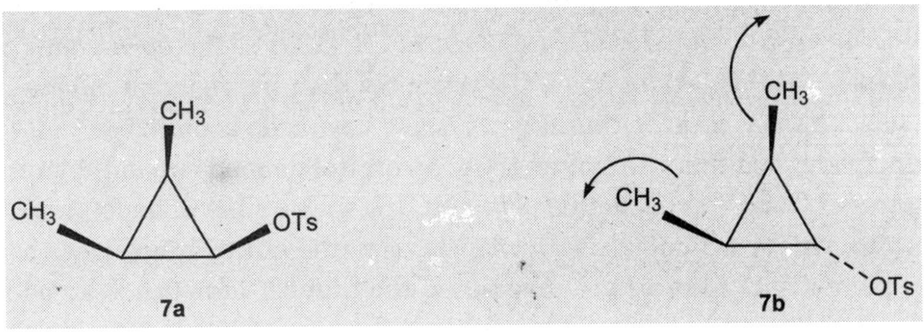

Formation of cyclopropyl cation is not the rate-determining step in such reactions. Had it been so, **7a** would have reacted much faster than **7b** as in the process of ionization, there would have been *relief in steric interaction* between the leaving group (tosylate) and the methyl substituents. It thus appears logical that it is a concerted process involving ionization and ring opening.

It has also been established by theoretical calculations that cyclopropyl cation is not a stable intermediate and readily undergoes electrocyclic ring opening as a concerted process following the departure of the leaving group. The departure of the leaving group is assisted by the electrons of the bond joining the two methyl groups provided they are located towards the back side. Such a situation also requires that the methyl substituents, which are located *anti* to the tosylate group must rotate *outward* to give *w*-shaped allylic ion.

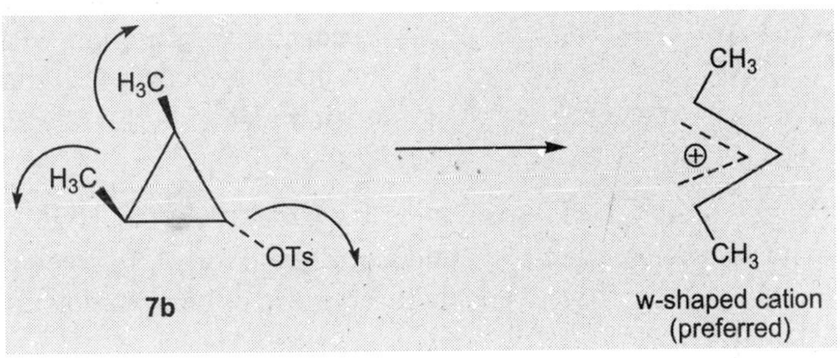

On the other hand, *inward* rotation in **7a** will bring methyl substituent closer and will not be preferred due to increased steric interaction.

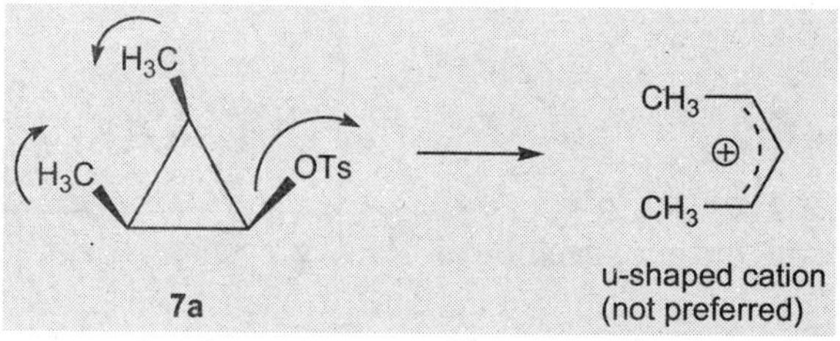

B. Frontier Molecular Orbital (FMO) Method

Although it is more fruitful to construct a correlation diagram for the detailed analysis of an electrocyclic reaction, there is, nevertheless, an alternative method which also enables us to reach similar conclusion. In this approach which is extremely simple, our only guide is the symmetry of *the highest occupied molecular orbital* (HOMO) of the *open-chain partner* in an electrocyclic reaction. If this orbital has a C_2 symmetry then the reaction follows a conrotatory path and if it has a mirror plane symmetry, a *disrotatory* mode is observed.

We have already seen that the symmetry of an orbital depends upon the number of nodes, which is equal to $n-1$ (Ψn = wave function of the MO). If the number of node(s) is zero or an even integer, the orbital will be symmetric with respect to m and antisymmetric with respect to C_2. However, the symmetry properties are reversed if the number of nodes is an odd integer. For example, in the ground state of butadiene which is the open-chain partner in the butadiene-cyclobutene interconversion, Ψ_2 is the highest occupied molecular orbital, and since it has one node and displays C_2 symmetry, thermal ring-closure is a conrotatory process. Irradiation of butadiene promotes an electron from Ψ_2 to Ψ_3 which then becomes the highest occupied molecular orbital, and since Ψ_3 has an m symmetry (two nodes), disrotation is required for photochemical ring-closure. Similar analyses may be applied to other electrocyclic reactions.

The explanation for this alternative approach is based on the fact that overlapping of wave functions of the same sign is essential for bond formation. For example, in the ground state of butadiene, HOMO is Ψ_2 and obviously cyclization is possible only through conrotation. Irradiation of butadiene promotes an electron to Ψ_3 which then becomes HOMO and bond formation is possible only through disrotation.

Similarly, in the hexatriene-cyclohexadiene transformation, the HOMO of the open-chain partner under thermal and photochemical conditions are Ψ_3 and Ψ_4, respectively. As may be expected the reaction proceeds by disrotation on heating and by conrotation under photochemical conditions.

We have noted that the *conrotatory* as well as the *disrotatory* ring-opening (or the reverse process) of systems like cyclobutene, cyclohexadiene or cyclooctatriene can take place in two directions each. The preferred direction will depend on the stereochemical factor involving the end groups in the transition state (**Fig. 15.10**).

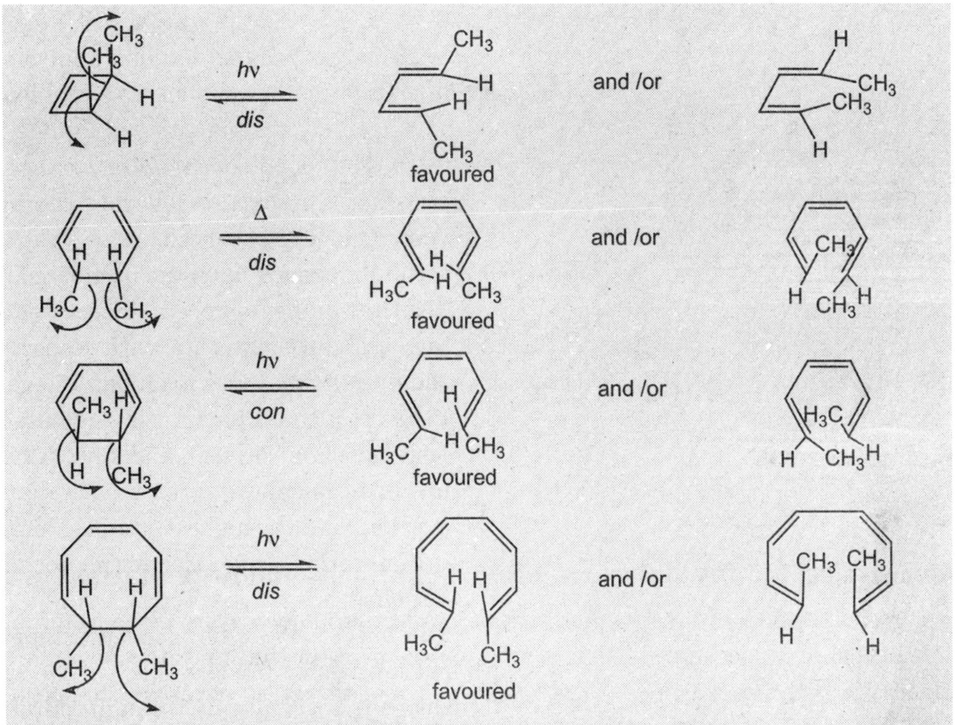

Fig.15.10 Conrotatory and disrotatory ring openings (only the favoured direction of opening of the rings are shown)

15.3 CYCLOADDITION REACTIONS

The most familiar example of a cycloaddition reaction is the Diels-Alder reaction which involves the formation of a cyclic compound from an alkene and a diene.

These reactions consist in the addition of a system of m π electrons to a system of n π electrons thereby forming a cyclic product. They offer a versatile route to the synthesis of cyclic compounds with a high degree of stereoselectivity under thermal and photochemical conditions. Depending upon the number of π electrons participating in the process, these reactions are termed $(m + n)$ or $(m + n + \ldots)$ cycloaddition reactions.

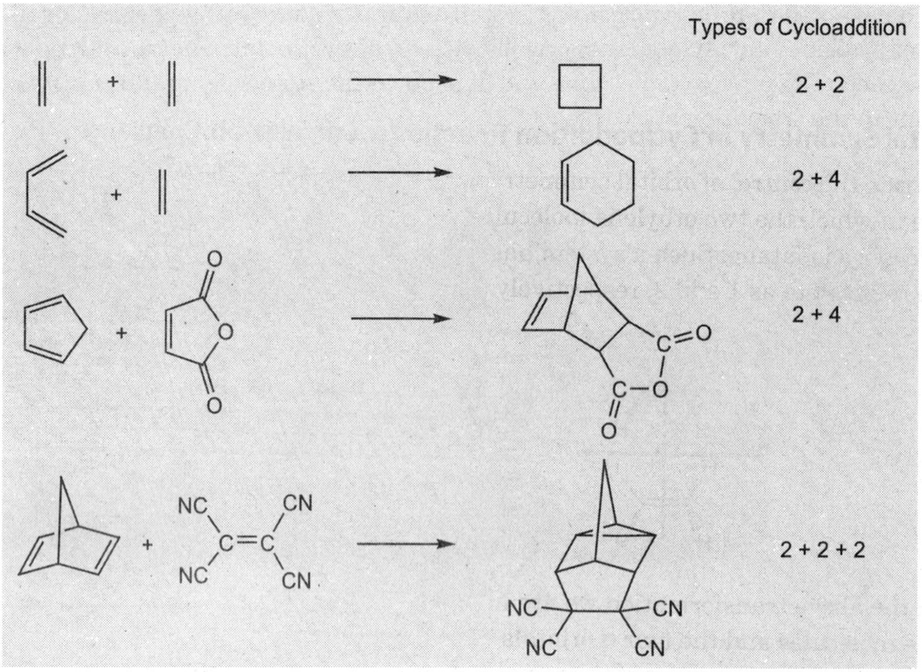

A. Stereochemical Modes of Cycloaddition: Suprafacial and Antarafacial Processes

Since in a typical cycloaddition reaction, there is addition of two systems containing double bonds, it is logical to expect the addition to occur on the same or the opposite side of the system. Furthermore, as both the π systems are undergoing addition, it is necessary to specify these modes of addition on each of them. These different modes have been termed *suprafacial* (on the same side) and *antarafacial* (on the opposite side).

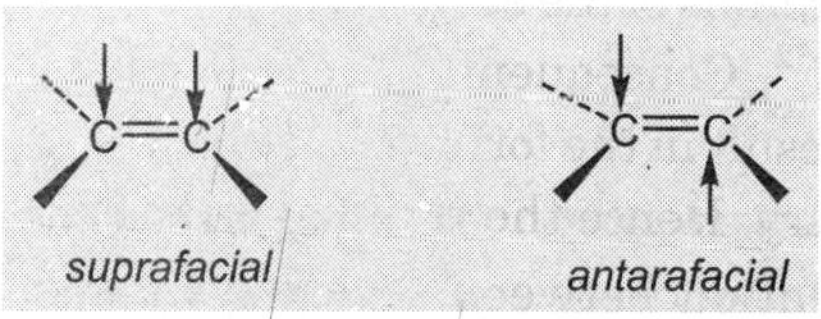

This specification is usually made by placing a suitable subscript (s or a) after the number referring to the pi component. The Diels-Alder reaction may be considered as a process involving $2s + 4s$ cycloaddition.

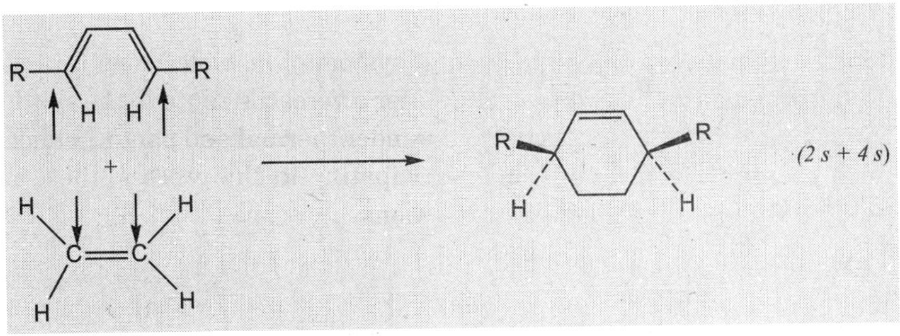

B. Orbital Symmetry in Cycloaddition Reactions: Correlation Diagrams

To illustrate the control of orbital symmetry on cycloaddition reactions, we choose the simplest example in which the two ethylene molecules approach each other vertically (2*s*+2*s*) to form a molecule of cyclobutane. Such a system has vertical and horizontal planes of symmetry which shall be referred to as 1 and 2, respectively.

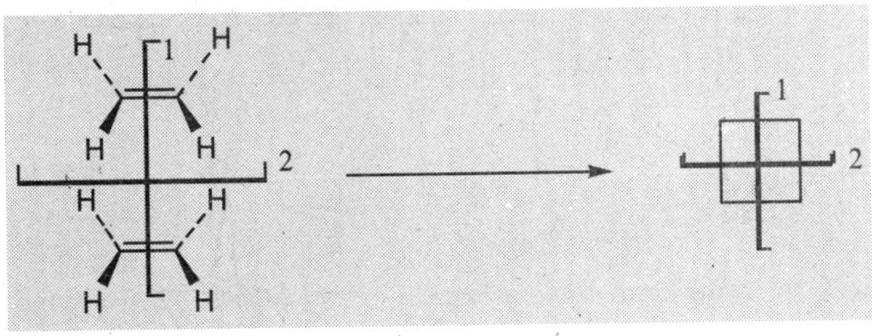

In the above transformation we are mainly concerned with the four π orbitals of the two ethylene molecules and the four σ orbitals of cyclobutane.

The symmetry properties of the remaining orbitals remain unchanged during the reaction and need not be considered further. The shapes and symmetry properties of these four π orbitals (π, π bonding; π*, π* antibonding) and four σ orbitals (σ, σ bonding; σ*, σ* antibonding) are listed in (**Fig. 15.11**). The symmetry classifications (*SS, SA, AS and AA*) are with respect firstly to plane of symmetry 1 and then to 2.

On the basis of the above information, a correlation diagram (**Fig. 15.12**) may be drawn in which the levels of like symmetry are connected by lines.

A close examination of the diagram leads us to the following two conclusions:

A. The ground state orbitals of the ethylene correlate with an excited state of cyclobutane, $\pi^2\,\pi^2 \rightleftharpoons \sigma_1^{\,2}\,\sigma_1^{\,*2}$. Consequently, the combination of two ground state ethylene molecules cannot result in the formation of ground state cyclobutane while conserving the orbital symmetry. Hence the thermal process is *symmetry-forbidden*.

B. As there is correlation between the first excited state of the ethylene system and cyclobutane, $\pi^2\,\pi\,\pi^* \rightleftharpoons \sigma_1^{\,2}\,\sigma_2\,\sigma_1^{\,*}$, the photochemical process is *symmetry-allowed*.

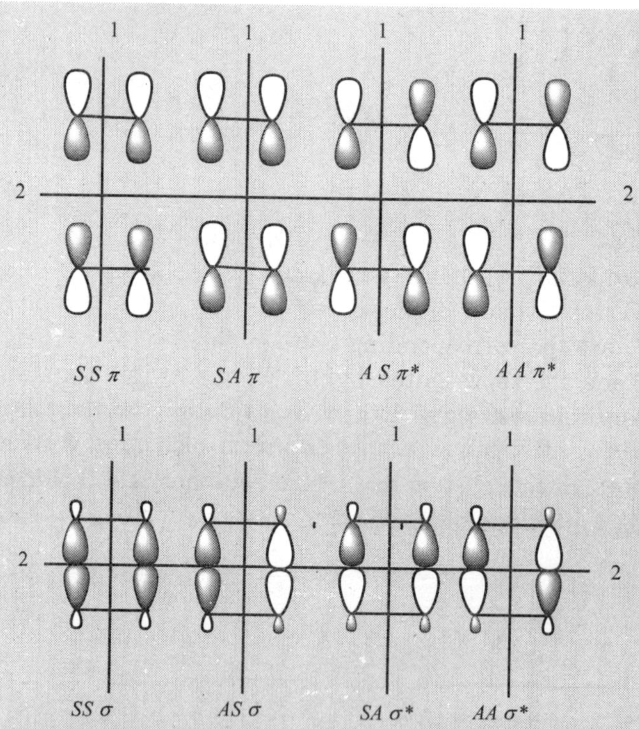

Fig. 15.11 Symmetry properties of interacting ethylene π orbitals and cyclobutane σ orbitals

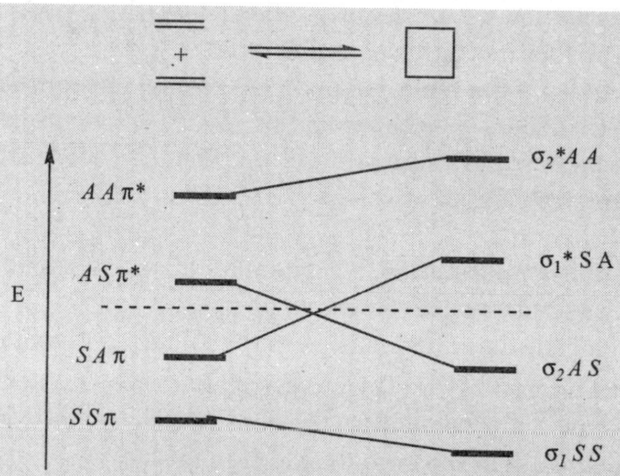

Fig.15.12 Correlation diagram for cycloaddition and cycloreversion of ethylene-cyclobutane system.

A similar correlation diagram may be constructed for the Diels-Alder reaction which is a 4*s* + 2*s* cycloaddition reaction. In this case there is only a single vertical plane of symmetry bisecting the carbon framework of two reactants and the product.

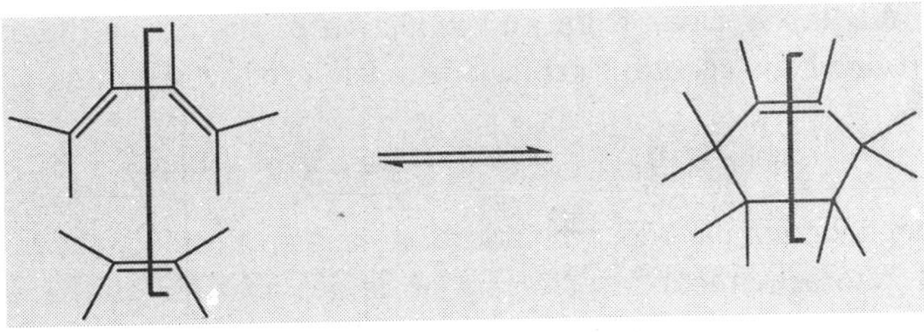

In this transformation, we have to consider six orbitals each of the reactants and the product. The ground state orbitals of the reactants are Ψ_1 Ψ_2 (of butadiene) and π (of ethylene) while Ψ_3, π^* and Ψ_4 are the corresponding antibonding orbitals. Similarly, the ground state orbitals of cyclohexene are represented by Ψ_1, Ψ_2 and π; the remaining three orbitals are antibonding. All these orbitals along with their symmetry properties are shown in the correlation diagram (**Fig. 15.13**).

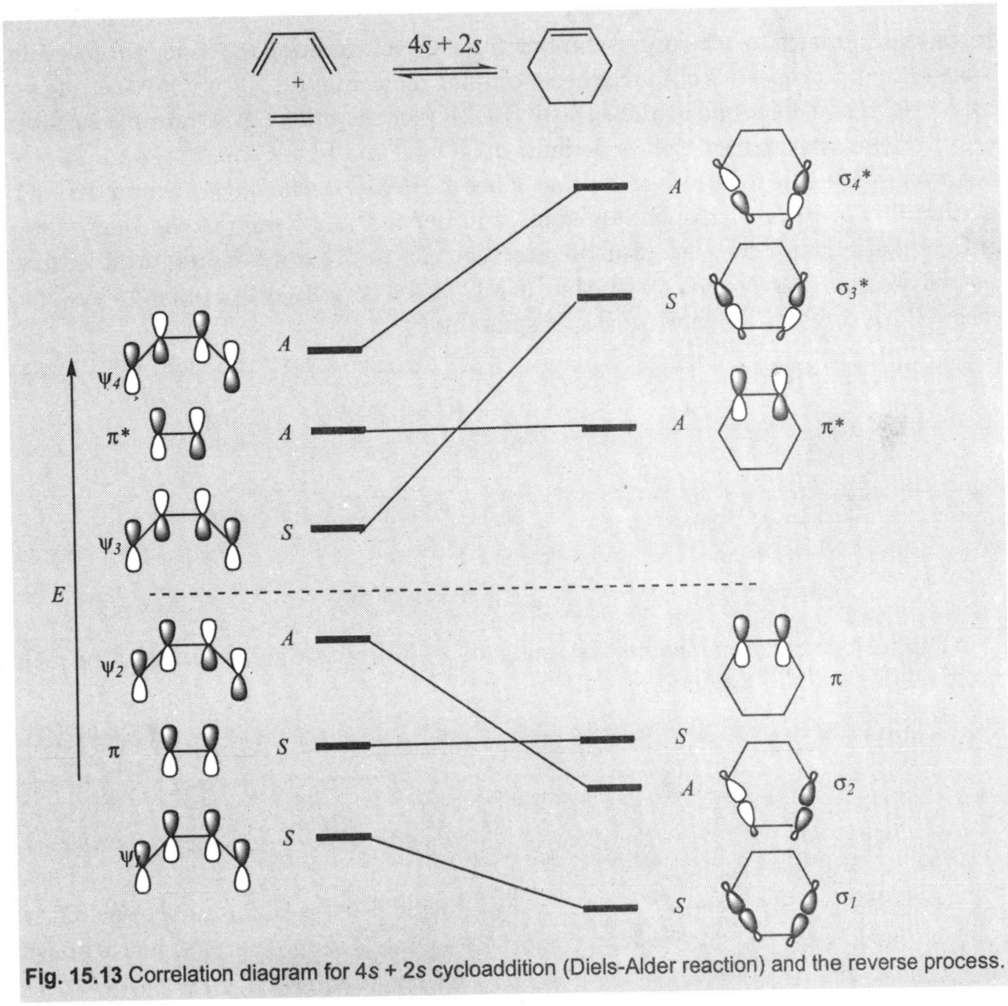

Fig. 15.13 Correlation diagram for 4*s* + 2*s* cycloaddition (Diels-Alder reaction) and the reverse process.

It becomes immediately clear from an inspection of the above diagram that there is a smooth transformation of the reactant orbitals into the product orbitals.

$$\psi_1^2 \pi^2 \psi_2^2 \longrightarrow \sigma_1^2 \sigma_2^2 \pi^2$$

The Diels-Alder reaction (4*s* + 2*s* cycloaddition) is, therefore, a *thermally allowed* process. On the other hand, photochemical transformation is not possible as the first excited state of the reactant does not correlate with the first excited state of the product. Rather it correlates with the upper excited state of the product.

$$\psi_1^2 \pi^2 \psi_2 \psi_3 \not\longrightarrow \sigma_1^2 \sigma_2^2 \pi \sigma_3^*$$

Hence there is a symmetry-imposed barrier to photochemical reactions of (4*s* + 2*s*) type.

C. Frontier Molecular Orbital (FMO) Method

An alternative approach to determine whether or not a cycloaddition reaction is allowed depends upon the symmetry properties of the highest occupied molecular orbital (HOMO) of one reactant and the lowest unoccupied molecular orbital (LUMO) of the other. A favourable interaction is possible only when the signs of the coefficients of HOMO and LUMO are the same. In the 2*s*+2*s* cycloaddition of ethylene to form cyclobutane, lobes of HOMO in one molecule and that of LUMO in the other do not have corresponding signs and hence the reaction is thermally forbidden. Irradiation of ethylene, however, promotes an electron to the antibonding orbital which now becomes HOMO and corresponds with the LUMO of the second unexcited ethylene molecule. As expected, the combination now proceeds smoothly.

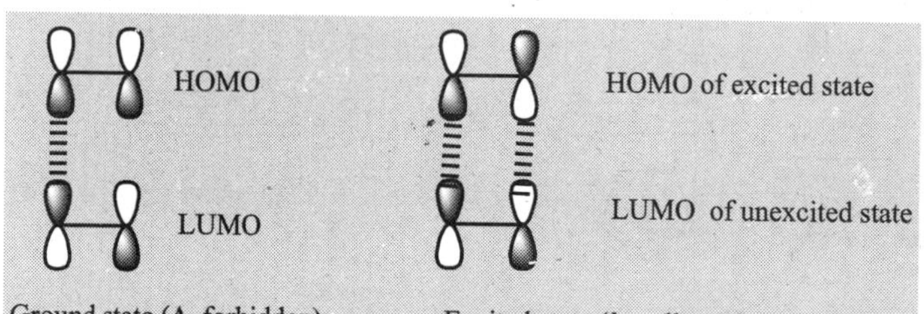

Ground state (Δ forbidden) Excited state ($h\nu$ allowed)

The Diels-Alder reaction may also be analyzed by a similar consideration of the molecular orbitals of ethylene and butadiene.

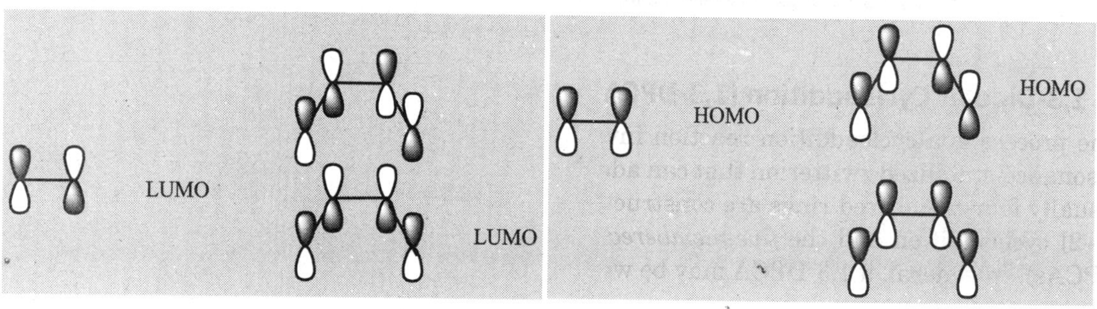

Since signs of the 1,4-lobes of butadiene HOMO match those in the LUMO of ethylene, the addition is thermally allowed. We reach a similar conclusion by considering the signs of butadiene LUMO and HOMO of ethylene.

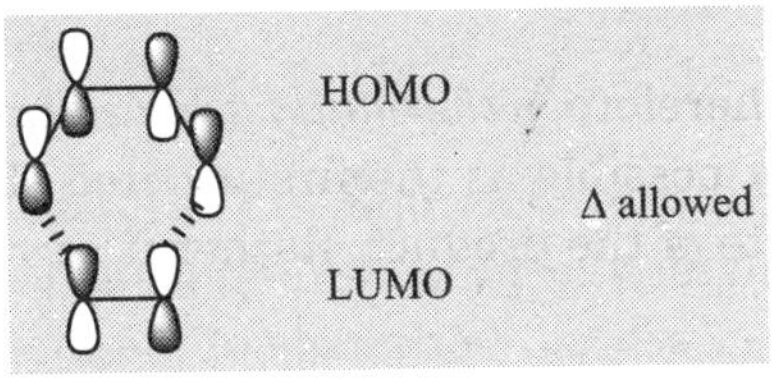

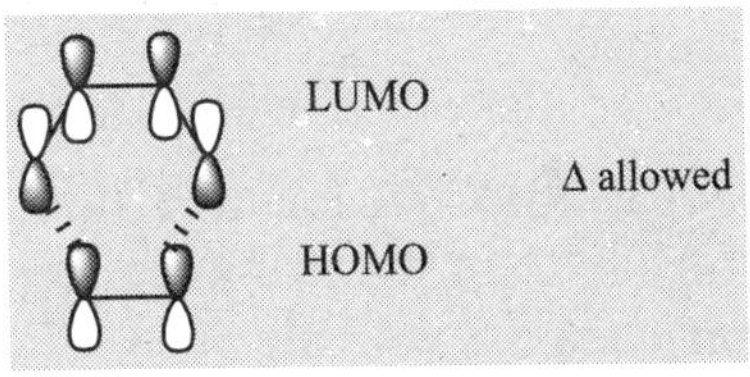

No such correspondence is obtained by the irradiation of the reactants and consequently the Diels-Alder reaction is a photochemically forbidden process.

An interesting example of the role of frontier orbitals in determining the product is the Diels-Alder reaction of cyclopentadiene forming dicyclopentadiene. Invariably, the *endo* dimer is formed rather than the *exo*. This is due to the favourable secondary forces as a result of interaction of frontier orbitals of diene and dienophile components which lower the energy of the transition state. In the following figure (**Fig. 15.14**) orbitals involved in actual bonding are connected by full lines and the secondary interactions shown by broken lines. An inspection of the transition states reveals that these interactions are present in the *endo* transition state while they are absent in the *exo*. Thus, the *endo* transition state for this reaction is stabilized *vis-à-vis* the *exo* and, therefore, the *endo* attack should be favoured. However, in some cases steric factors may be of a greater magnitude than this effect.

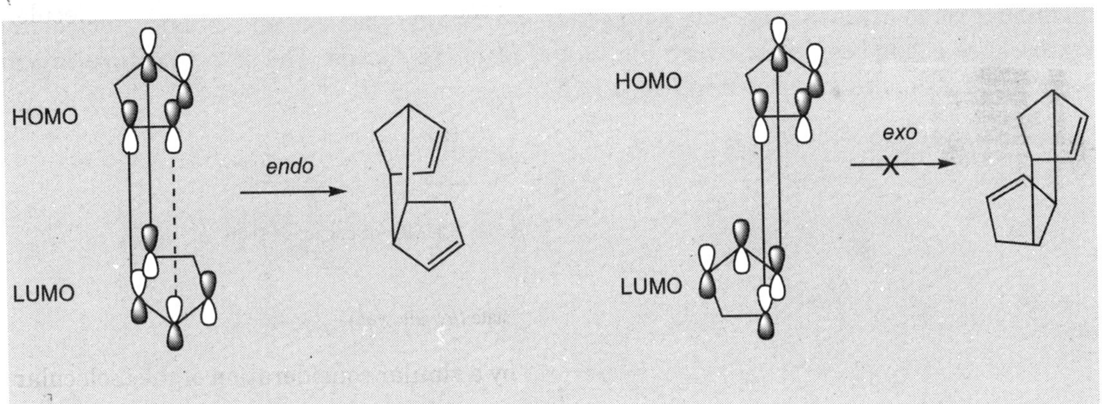

Fig. 15.14 Orbital interactions in Diels-Alder reaction of cyclopentadiene

D. 1,3-Dipolar Cycloaddition (1,3-DPCA) Reactions

The process is a cycloaddition reaction involving atoms with formal charges. 1,3-Dipole is a resonance-stabilized zwitterion that can add in a concerted fashion to an olefinic dipolarophile. Usually four-membered rings are constructed by [2 + 2] cycloaddition, six-membered rings by [4+2] cycloaddition and the *five-membered rings by 1,3-dipolar cycloaddition reactions* (1,3-DPCAs). In general, a 1,3-DPCA may be written as:

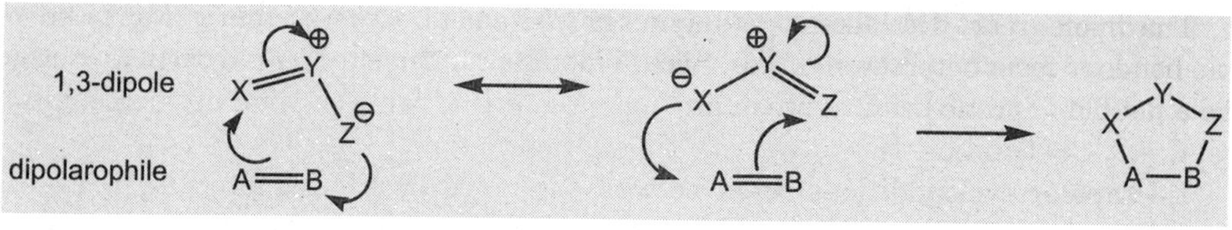

The reactant X-Y-Z is called the 1,3-dipole and A-B is the dipolarophile. The kind of dipoles that feature in the 1,3-DPCAs are isoelectronic with an allyl or propargyl anion as they have π-electron system consisting of two filled and one empty orbital, both ends of the dipole have nucleophilic as well as electrophilic properties. Each dipole has a charge-separated resonance structure with opposite charges in a 1,3 relationship.

Substituted alkenes and alkynes are typical dipolarophiles, however, other systems containing a π bond such as carbonyl, imine and nitroso also act as dipolarophiles.

Many dipoles are stable compounds such as diazomethane and ozone.

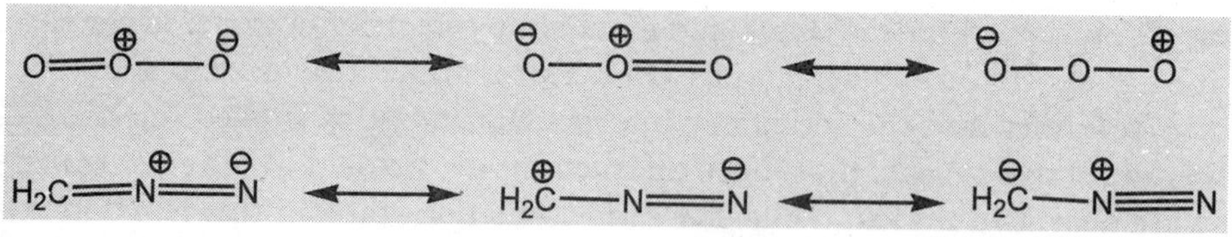

Others like the nitrile oxides, ylides and imines are reaction intermediates that have to be generated *in situ*. The following reactions illustrate some of the processes used for preparing 1,3-dipolar compounds *in situ*. It is not correct to conclude that one end is more nucleophilic just because it has been drawn with the formal negative charge. The ccre structure of some important 1,3-dipoles are shown in **Fig. 15.15**.

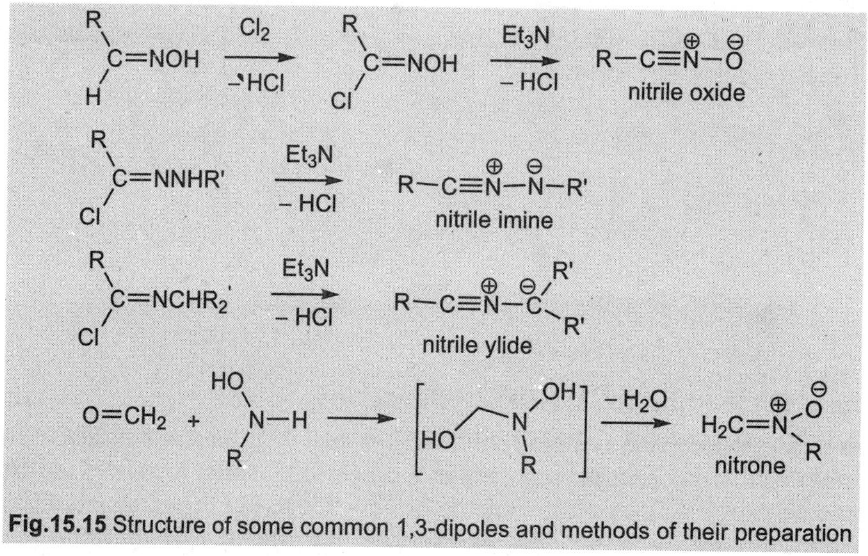

Fig.15.15 Structure of some common 1,3-dipoles and methods of their preparation

The dipoles react with alkenes or alkynes or with a heteroatom system containing double or triple bonds to form heterocyclic rings. Like dienophiles in Diels-Alder reaction, dipolarophiles have a double or triple bond between any pair A and B of same or different elements such as C=C, C—C, C=O and C—N.

1,3-Dipolar cycloadditions could be well explained by FMO approach. In 1,3-dipoles, four π-electrons are distributed over three atoms and they can be considered as the structural variant of the diene component in the Diels-Alder reaction. Moreover, the HOMO and LUMO of a 1,3-dipole are of similar symmetry to those of a diene with respect to the two-fold axis and to the mirror plane which bisects the molecule. The HOMO of the 1,3-dipole is Ψ_2 and LUMO of the dipolarophile is π*. Both components are antisymmetric with respect to mirror symmetry and, therefore, maximum overlap could be obtained for the reaction to take place in forward direction.

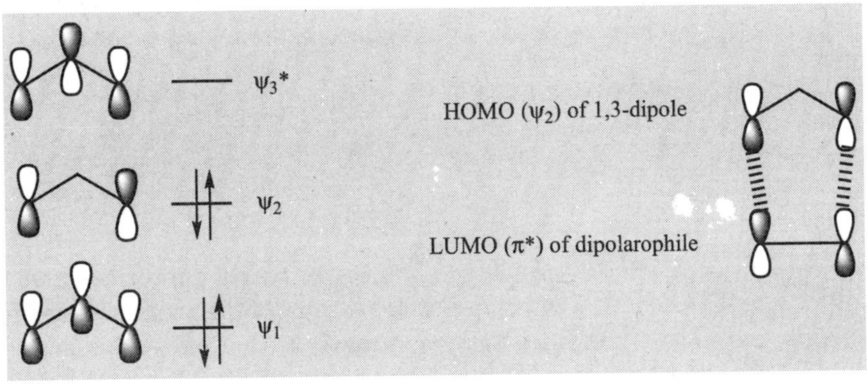

Since 1,3-DPCA reactions involve 4 electrons from the 1,3-dipole and two from the dipolarophile, it may be considered as [π2s + π4s] cycloaddition resembling Diels-Alder reaction. There is HOMO-LUMO interaction in which either reactant can be the electrophilic or nucleophilic component.

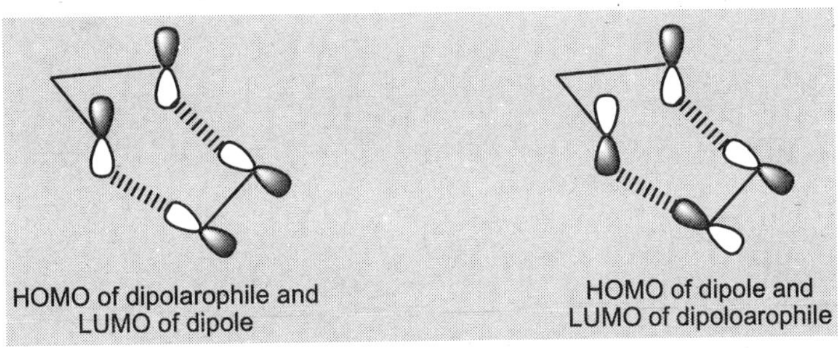

The 1,3-dipolar addition of diazomethane with ethylene can thus be depicted as

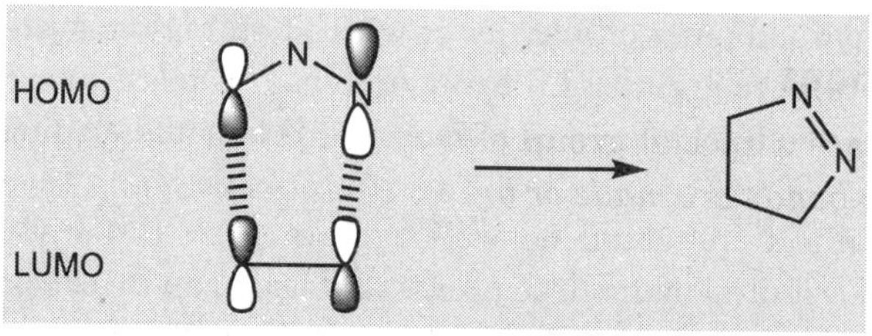

1,3-DPCAs are sometimes accompanied by proton transfer, *e.g.* diazopropane reacts with methyl acrylate to give the product which undergoes rapid transfer of a proton. As expected, the normal product is obtained when methyl methacrylate is used in place of methyl acrylate (**Fig. 15.16**).

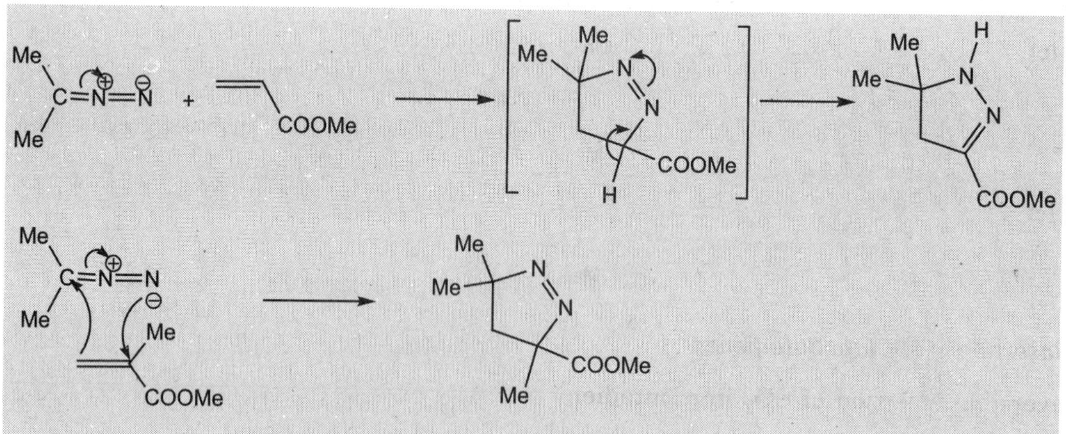

Fig. 15.16 Reaction of diazopropane with acrylate esters

It has been found that the addition of diazomethane to olefins is stereospecific. For instance, isomeric pyrazolines have been synthesized by choosing an appropriate olefin.

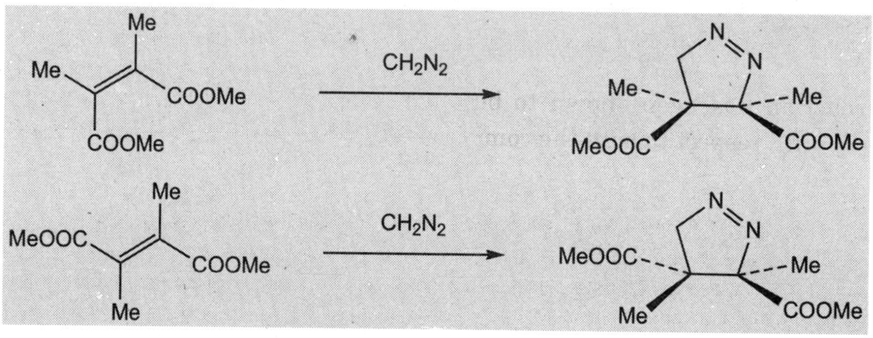

Structural variations in 1,3-dipole and dipolarophile make 1,3-DPCAs a very convenient method for synthesizing 5-membered heterocyclic compounds. Many such examples are given in Chapter 7.

E. Cheletropic Reactions

Cheletropic reactions are a special group of concerted cycloadditions or cycloreversions with the difference that *two bonds are made or broken on the same atom*. Some common cheletropic reactions are:

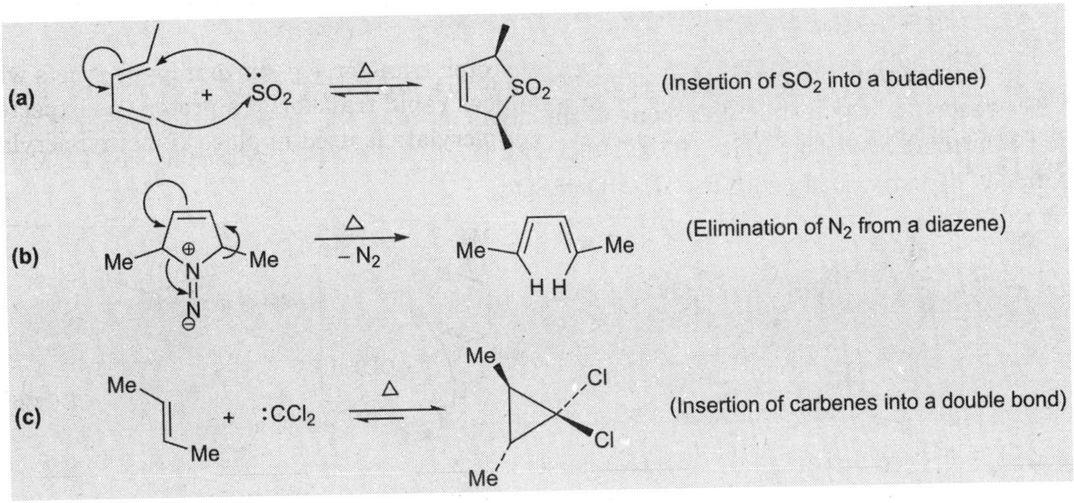

I. Insertion of SO_2 into Butadienes

Reversible insertion of SO_2 into butadiene presents an interesting example of cheletropic reactions.

The reaction has been shown to be stereospecific, thermally allowed and proceeds *suprafacially* with respect to both the components.

The reaction proceeds in such a manner that SO_2 lies in a plane bisecting *S-cis* conformation of the diene. On analyzing this reaction on the basis of orbital symmetry rules, we find that HOMO of SO_2 has a lone pair of electrons in the plane of the atoms, whereas the LUMO is a *p*-orbital located in a perpendicular fashion to HOMO.

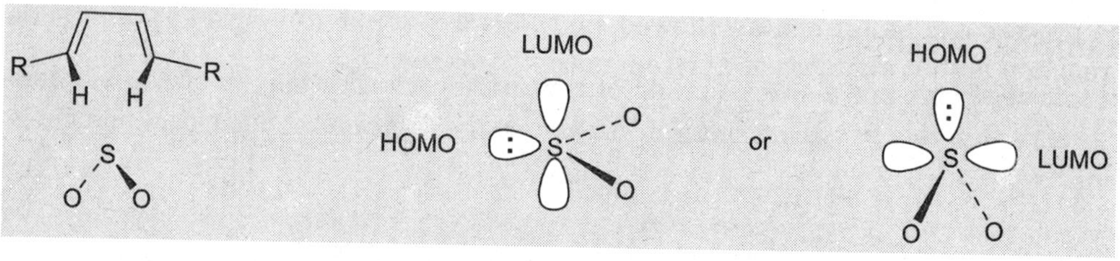

As the thermally-allowed reaction between a diene and SO_2 is *suprafacial* with respect to the two reactants, the terminal carbons of the diene move in the disrotatory manner in such a way that there is overlap of its HOMO (Ψ_2) with the LUMO of SO_2. Alternatively, LUMO of the diene (Ψ_3) can overlap with the HOMO of SO_2.

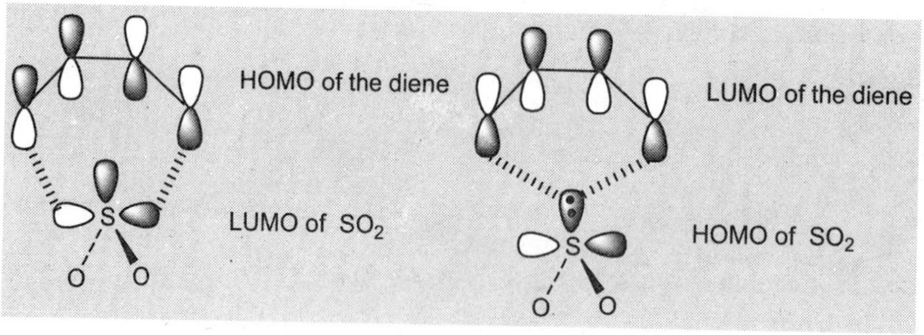

These reversible reactions are stereospecific, *trans*, *trans*-1,4-dimethyl butadiene gives the corresponding *cis* substituted 3-sulpholene dioxide *via* disrotation. As may be predicted, the reaction proceeds *via* conrotation under photochemical condition.

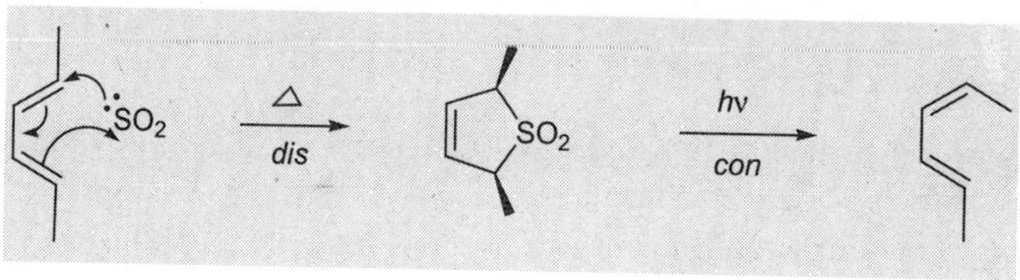

The reversible thermal reaction between butadiene and SO_2 is often used to carry butadiene to a reaction site requiring butadiene which is released at elevated temperatures. For example, in Diels-Alder reaction the diene is protected and the adduct is heated in the presence of dienophiles as a convenient source of butadiene.

II. Elimination of N_2 from Diazenes

Irreversible extrusion of nitrogen from diazenes is also a *suprafacial* process proceeding in a stereospecific manner.

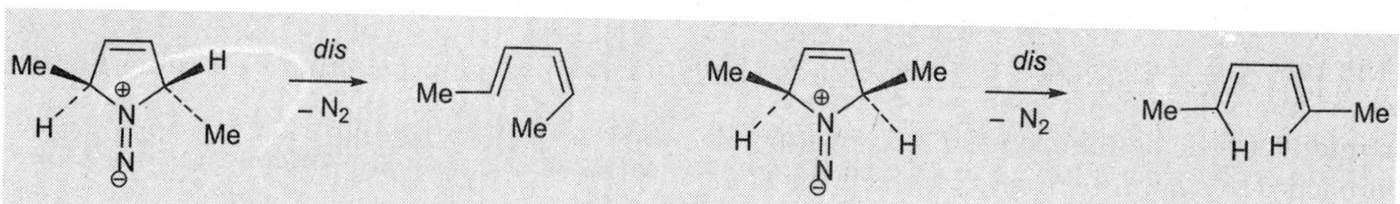

The process can be termed as allowed *suprafacial* retro-[4+2]-cycloaddition reaction. A similar reaction is the rapid loss of CO from norbornadienone.

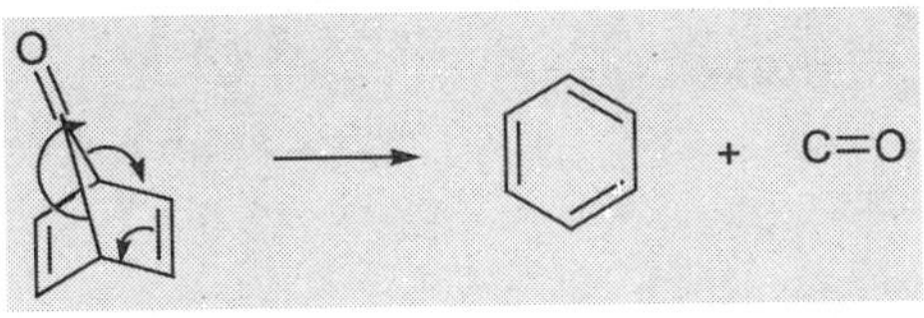

Both these processes are straightforward 6-electron cheletropic reactions and have been shown to be *suprafacial* on the conjugated system.

III. Insertion of Carbenes into a Double Bond

An 'anomalous' example of concerted cheletropic reaction is the insertion of a carbene into an alkene to give a cyclopropane.

The addition is stereospecific when singlet carbene is used. On using a triplet carbene this stereospecificity is lost.

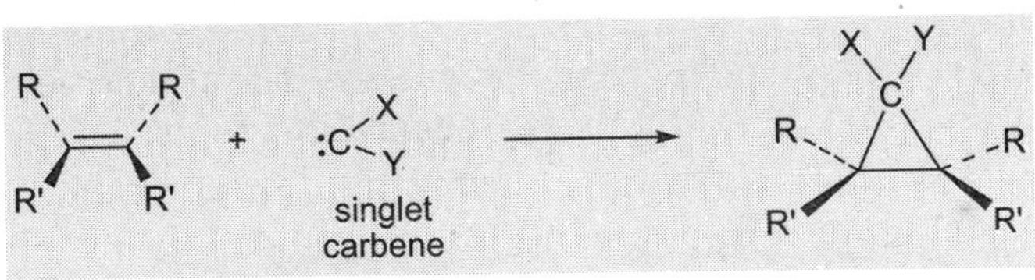

In the singlet state, carbon of the carbene is presumed to have approximate sp^2 hybridization. Two of the three sp^2 hybrid orbitals are utilized in two covalent bonds with the substituents. The remaining third orbital contains the unshared pair of electrons. The *p*-orbital remains vacant.

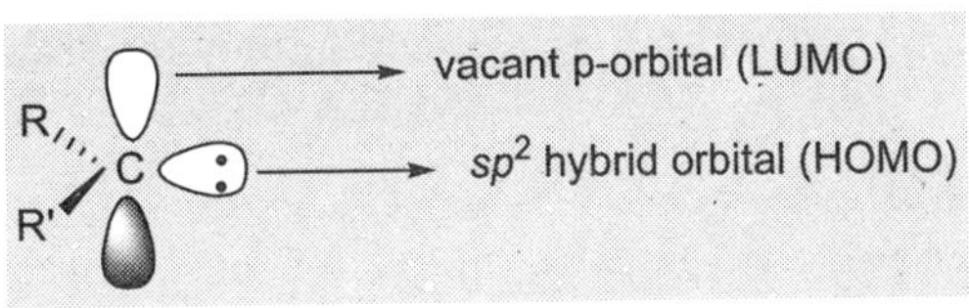

Calculations have shown that this reaction involves interaction of the HOMO of alkene with the LUMO of carbene. However, the stereospecific insertion of carbenes (or dihalocarbenes) involves only four electrons and is suprafacial on the alkene thus preserving stereospecificity of the reaction. Apparently the reaction is a forbidden process if the carbene along with its two sp^2 hybridized substituents approaches the π-molecular orbital of the alkene directly into the middle of the double bond in a *perpendicular fashion* (linear approach). However, carbene approaches in a *nonlinear* way in which the plane of the two substituents on carbene is *parallel* to that of the double bond of pi system.

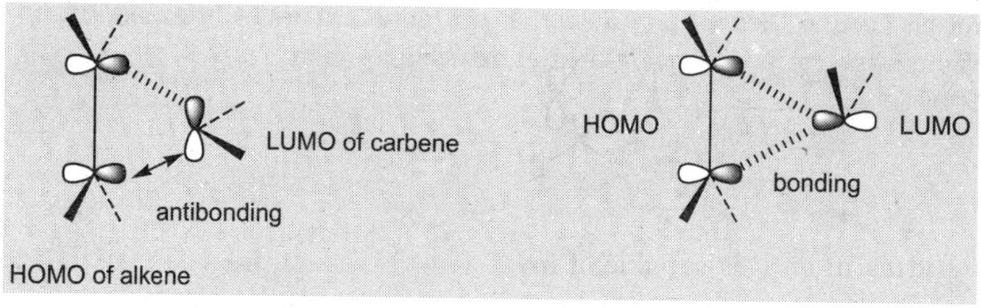

The reason for the loss of stereospecificity involving triplet carbenes is that initially formed diradical is sufficiently long-lived to undergo spin inversion.

15.4 SIGMATROPIC REARRANGEMENTS

Many thermal (or photochemical) rearrangements involve the shifting of a σ bond, flanked by one or more π electron systems, to a new position [*i*, *j*] within the molecule in an uncatalyzed intramolecular process. Since it is rearrangement of a σ bond, these reactions are called *sigmatropic rearrangements* of order [*i*,*j*]. These reactions are often classified with two numbers *i* and *j*, set in brackets [*i*, *j*] and the system is numbered by starting with the atoms forming the migrating σ bond. These numbers [*i* and *j*] indicate the new positions of the σ bond whose termini are *i*-1 and *j*-1 atoms removed from the original bonded loci.

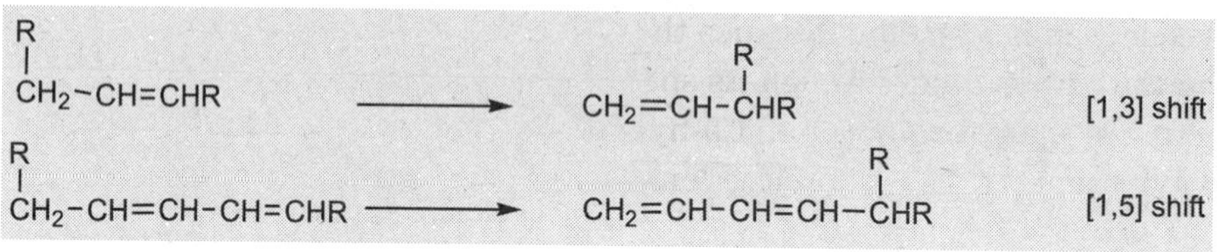

Very often the migrating σ bond is situated in between two conjugated π bond system as in Cope and Claisen rearrangements.

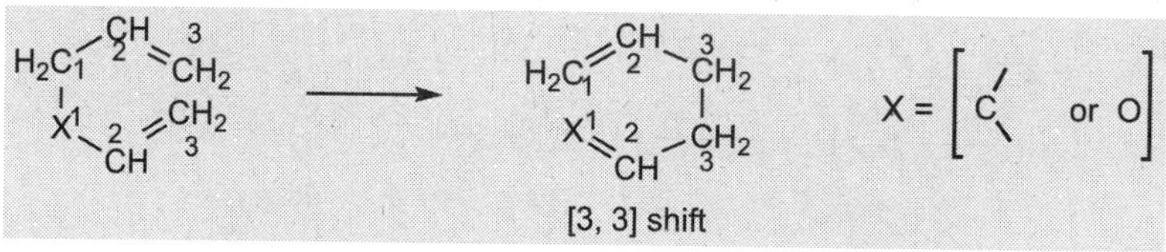

A. Suprafacial and Antarafacial Processes

Since a sigmatropic reaction involves the migration of a σ bond across the π electron system, there are two different stereochemical courses by which the process may occur. When the migrating σ bond moves across the same face of the conjugated system, it is called a *suprafacial* process whereas in *antarafacial* process the migrating σ bond is reformed on the opposite π electron face of the conjugated system. The following [1, 5] sigmatropic shift illustrates both these processes and their stereochemical consequences.

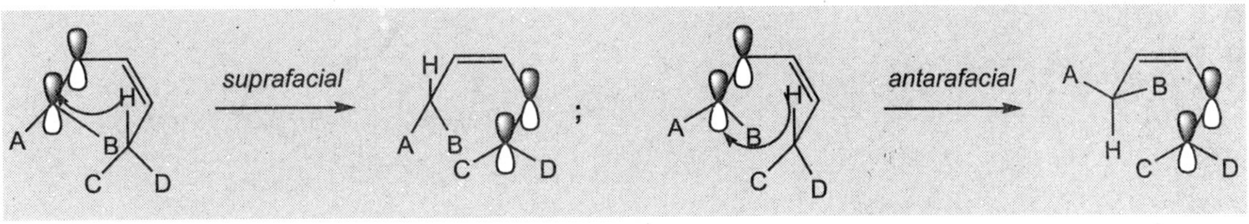

If migrating atom R is capable of inversion, there is a possibility of four stereochemical modes: suprafacial (retention, s_r or inversion, s_i) and antarafacial (retention, a_r or inversion, a_i).

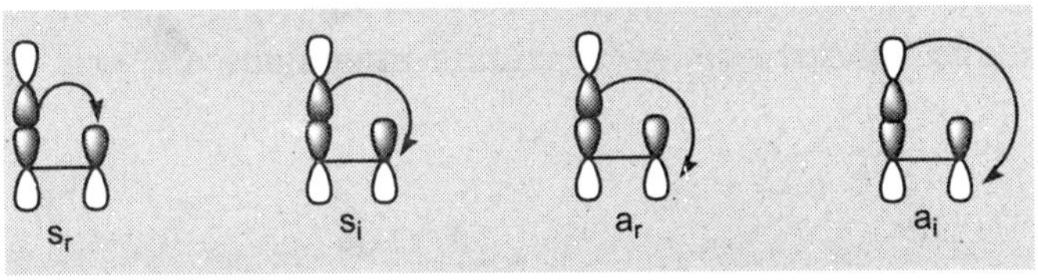

Due to steric reasons, *suprafacial* migrations are more common than *antarafacial shifts*. However, with the lengthening of the conjugated system, sometimes it is possible for a σ bond to migrate to the opposite π electron face.

It is to be noted that *antarafacial* [1,3] hydrogen shift is allowed only by the principles of orbital symmetry. The transition state is a highly deformed species and the reaction is forbidden because of the steric inhibition involved in such a process. *Suprafacial* [1,3] hydrogen shift under thermal condition is forbidden because there is no question of inversion at this atom which is bonded to the carbon atom through its spherically symmetrical 1s orbital.

There are many examples of 1,5-hydrogen migrations in molecules that incorporate a pentadienyl fragment. The example given below is showing the transformation to a more stable diene.

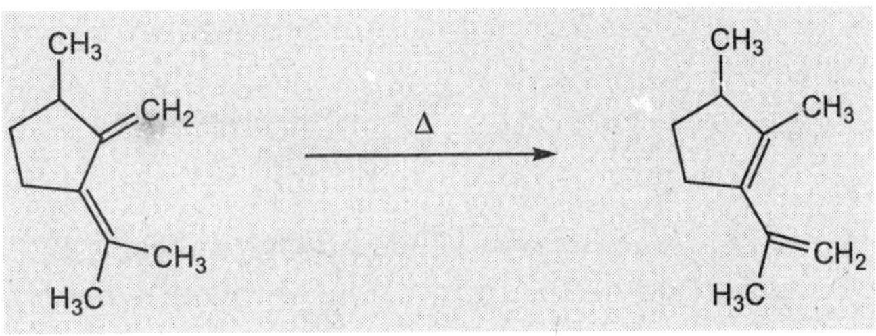

The equilibration among the methylcyclopentadienes and methylcycloheptadienes can be explained by a 1,5-hydrogen shift.

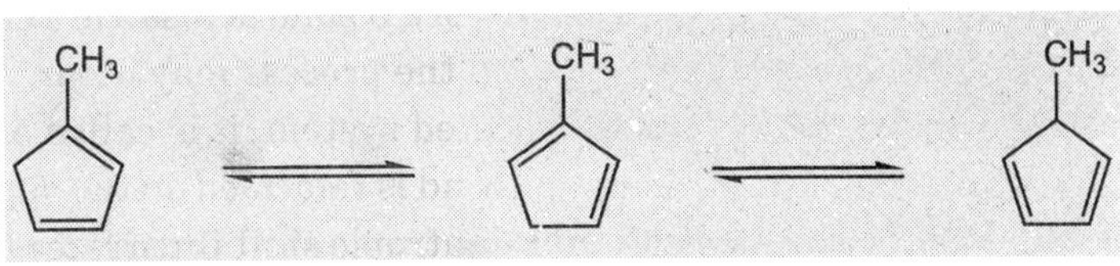

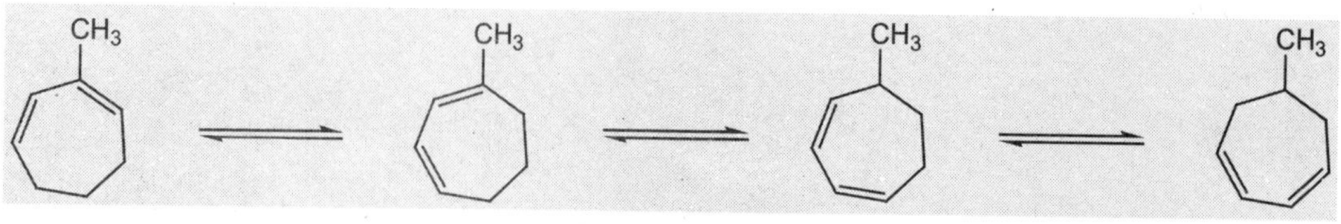

Like the thermal 1,3-hydrogen shift, a 1,7-hydrogen shift follows the same symmetry rules and is allowed when *antarafacial*. The geometrical restrictions on the *antarafacial* transition state of 1,7-hydrogen shift are not as severe as in the 1,3-case, because a π-system involving seven carbon atoms is more flexible than one involving only three carbon atoms. 1,7-Hydrogen shift can be illustrated in the *thermal* equilibrium between precalciferol and calciferol.

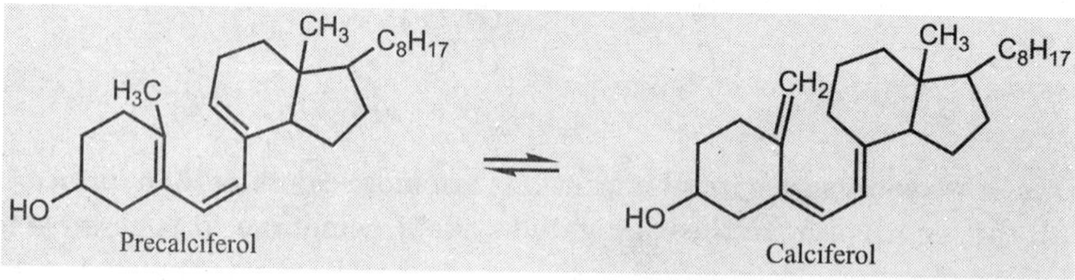

B. Analysis of Sigmatropic Rearrangements

One of the ways to analyze sigmatropic rearrangements is to assume that the migrating bond undergoes homolytic cleavage resulting in the formation of a pair of radicals. As bonding characters are to be maintained throughout the course of the rearrangements, the most important bonding interaction will be between the HOMOs of the two species produced by this cleavage. This is to be expected as it is these orbitals that contain the unpaired electrons. We shall illustrate this analysis by examining a suprafacial [1,5] *sigmatropic shift* of hydrogen in which the homolytic cleavage results in the production of a hydrogen atom and a pentadienyl radical.

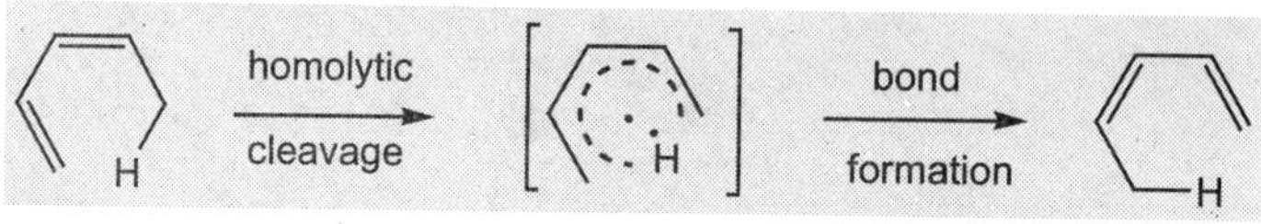

The ground state electronic configuration of pentadienyl radical is $\Psi_1^2\,\Psi_2^2\,\Psi_3$. Since HOMO ($\Psi_3$) of this radical has similar sign on the terminal lobes (plane symmetry), [1,5] *suprafacial* migration will be a *thermally allowed* process.

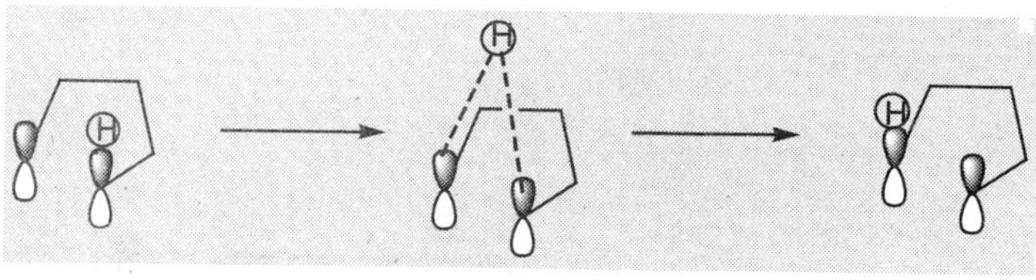

The first excited state of the pentadienyl radical has the configuration $\Psi_1^2\,\Psi_2^2\,\Psi_4$ (**Fig. 15.17**) and the symmetry characteristics of HOMO (Ψ_4) are thus reversed (C_2 symmetry).

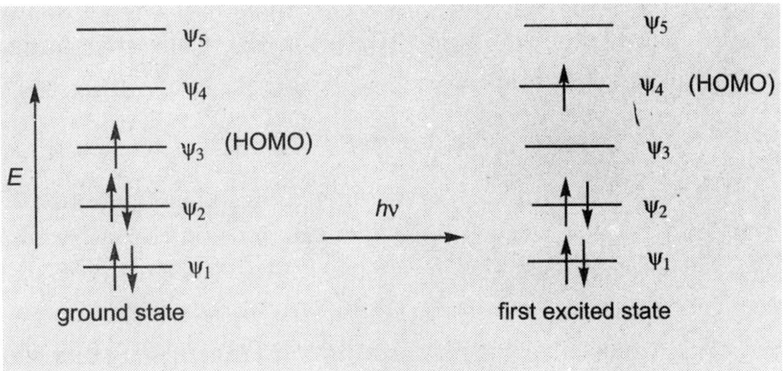

Fig. 15.17 Excitation of pentadienyl radical

[1,5] *Suprafacial* migration is no longer possible and the shift has to be an *antarafacial* process.

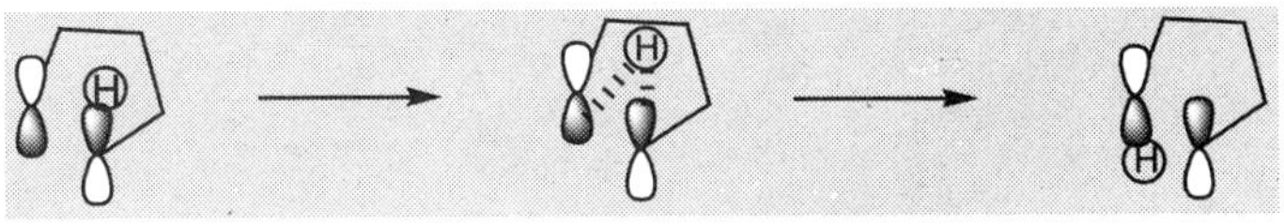

A similar analysis of such systems has led to the formulation of selection rules which state that if a sigmatropic reaction of the order $[i, j]$ (for hydrogen migration $i = 1$) has $i + j = 4q + 2$, then thermal reaction is *suprafacial* and photochemical reaction will be *antarafacial.* However, for those cases in which $i + j = 4q$, the rules are reversed and the thermal reactions are *antarafacial* while the photochemical reaction will be *suprafacial.* These rules are summarized in **Tables 15.8** and **15.9**.

Table 15.8 Selection rules if there is no inversion in R

$[i + j]$	Δ allowed, *hv* forbidden	*hv* allowed, Δ forbidden
$4q$	*antara*	*supra*
$4q + 2$	*supra*	*antara*

Table 15.9 Selection rules if R is capable of inversion

$[i + j]$	Δ allowed, *hv* forbidden	*hv* allowed, Δ forbidden
$4q$	*supra*-inversion (s_i) *antara*-retention (a_r)	*supra*-retention (s_r) *antara*-inversion (a_i)
$4q + 2$	*supra*-retention (s_r) *antara*-inversion (a_i)	*supra*-inversion(s_i) *antara*-retention (a_r)

C. Cope and Claisen Rearrangements

Cope and Claisen rearrangements are all sigmatropic changes of order $[i, j]$ where i and j are both 3 and predictions of the *stereochemical variations* in these reactions following the selection rules in **Table 15.8** (where $m + n = i + j$) conform to those made by other methods.

We have mentioned earlier that the Cope rearrangement is an intramolecular process proceeding through a cyclic transition state (Chapter 12). Cope rearrangement is the 3,3-sigmatropic rearrangement of 1,5-dienes.

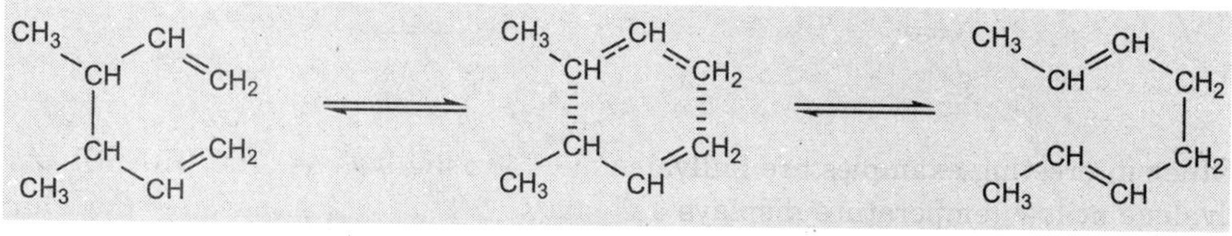

It is, therefore, an example of a [3, 3] sigmatropic shift in which a *chair-like transition state* is energetically preferred to a *boat-like structure*. Rearrangement of the *meso*-diene through such a transition state then would give the *cis-trans* isomer, while in the case of the rearrangement of the racemic mixture the *trans-trans* isomer is the major product.

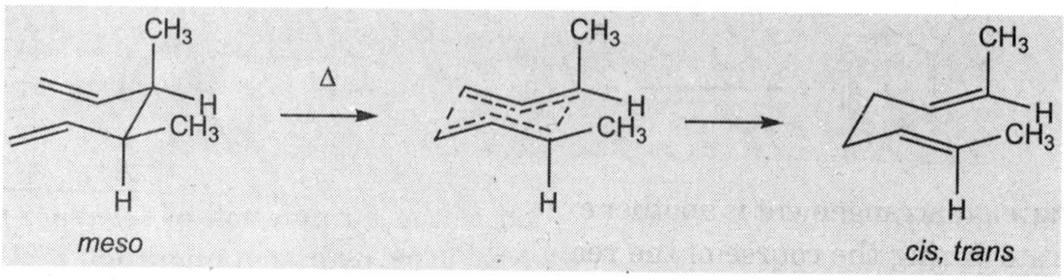

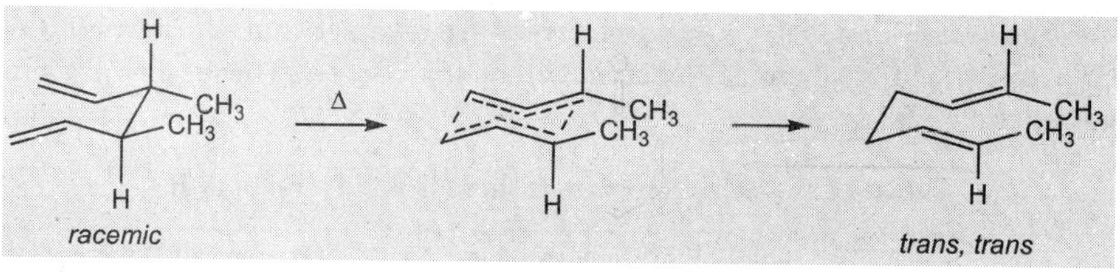

The equilibrium in the Cope reaction is controlled by the conjugation of the carbonyl and the cyano group in the product.

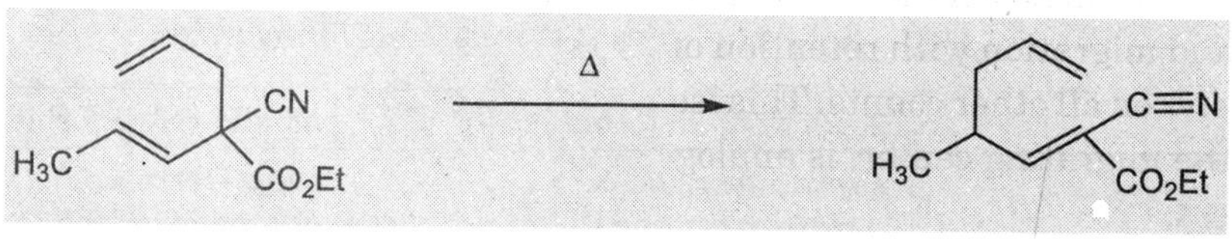

Sometimes in the reaction process, no overall change in structure occurs. It is a degenerate rearrangement and the product of rearrangement is identical to the starting material. Depending upon the rate of the reaction, the existence of the rearrangement can be detected by use of isotopic labels or by interpretation of the temperature dependence of NMR spectra. As in homotropilidene, the occurrence of dynamic equilibrium is evident from the NMR spectrum.

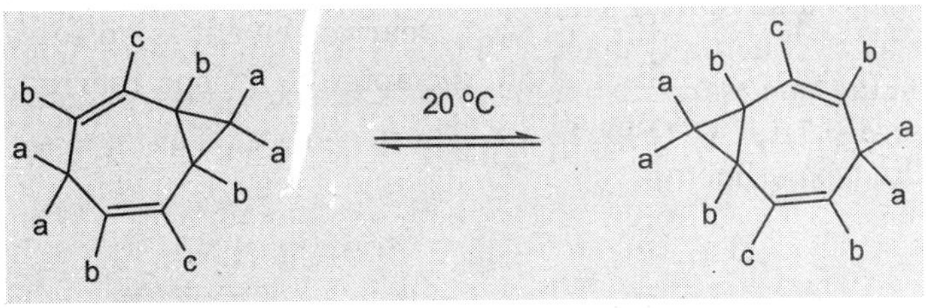

Other interesting examples are bullvalene and semibullvalene. The ^{1}H NMR spectrum of bullvalene at low temperature displays a signal at δ 4.22, which indicates the "fluxional" nature of the molecule. Due to the three fold axis of symmetry in the molecule, the degenerate rearrangement results in all the carbons having an identical averaged environment.

A variation in the Cope rearrangement is the oxy-Cope reaction. In the 1,5-diene system, a hydroxyl substituent at C-3 gives the enol product. The enol undergoes tautomerization.

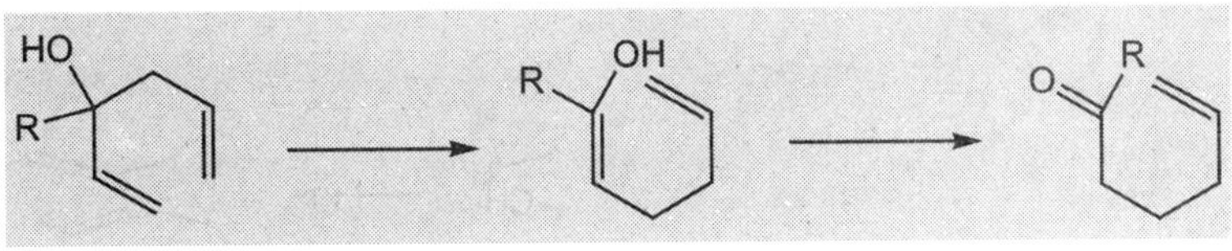

The Claisen rearrangement is another example of a thermal reaction in which the fragments do not separate during the course of the rearrangement. It is a [3,3] sigmatropic shift of allyl vinyl ether.

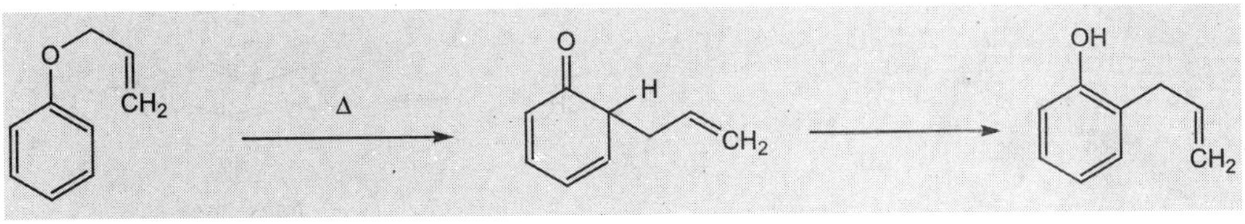

The following thermal isomerization (**8–9**), a [1,3] shift (**Fig. 15.18**), though apparently forbidden, proceeds with inversion at the carbon atom that moves. This may be regarded as a triumph of the orbital symmetry rules because the transition-state has severe non-bonded interactions and migration with retention of configuration at the migrating carbon atom would seem to be easier on all other counts. This isomerization, that is [1,3] *suprafacial* migration with inversion at the migrating center, is analogous to [1,3] *antarafacial* shift which is a *thermally allowed* process.

Perhaps the most dramatic example of orbital symmetry control of uncatalyzed unimolecular reactions is the thermal isomerization of 1-methoxy-bicyclo[3.2.0]hepta-3,6-diene-2-one (**10**) to 1-methoxy-bicyclo[3.2.0]hepta-1,4-diene-2-one system (**12**). The starting ketone, being a strained isomer of the stable aromatic tropolone system (**11**), would be expected to undergo facile isomerization to the latter merely by the disrotatory opening of the cyclobutene ring since conrotatory opening would lead to a *trans* double bond which is impossible for reasons related to the stability of tropolone. But thermal disrotatory opening of a cyclobutene ring is symmetry forbidden. Thus conservation of orbital symmetry forbids what appears to be a facile process. However, molecular orbital symmetry arguments suggest that the starting bicyclic ketone could undergo, as it actually does, a thermally allowed *antarafacial-antarafacial* [3,3] sigmatropic reaction (a Cope rearrangement) of the 1,5-hexadiene system made up of atoms in the following sequence: 6,7,1,5,4,3 (**Fig. 15.19**). The movement of atoms has been demonstrated by deuterium labelling experiments.

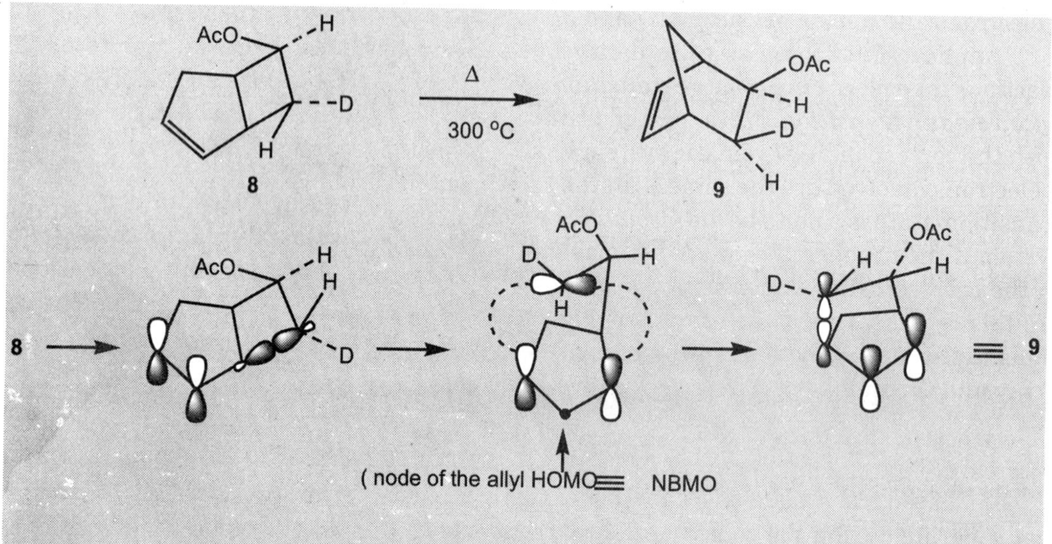

Fig. 15.18 Thermal [1,3] sigmatropic shift with inversion at the migrating centre

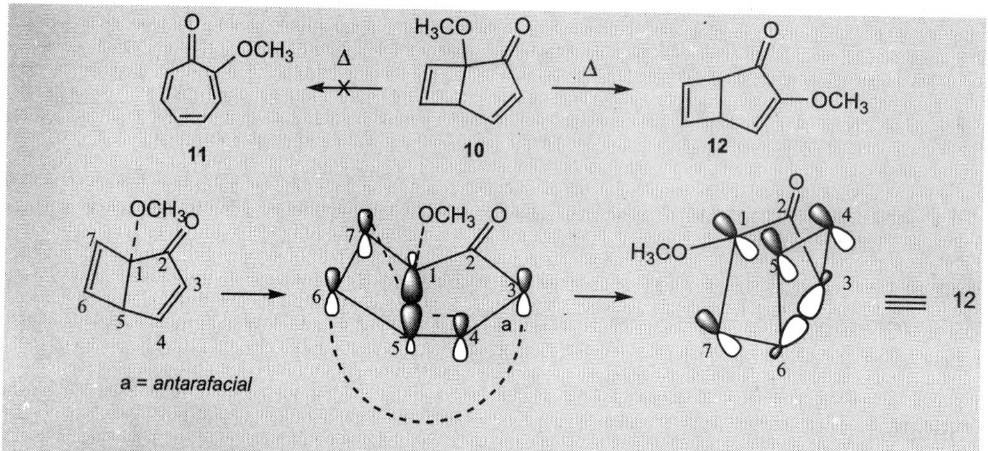

Fig.15.19 Thermal *antarafacial-antarafacial* [3,3] sigmatropic shift in 1,5-hexadiene system

15.5 APPLICATION OF PMO METHOD TO PERICYCLIC REACTIONS

There is yet another qualitative molecular orbital approach developed by M.J.S. Dewar which yields simple mnemonics to predict the same stereochemical variations in pericyclic reactions as do the other methods.

Huckel's rule of aromaticity states that a monocyclic planar conjugated system with $(4q + 2)$ π electrons is *aromatic* and consequently stable in the ground state. A system with $(4q)$ π electrons is unstable and is referred to as *antiaromatic*. However, further calculations show that these rules are reversed by the presence of a node in the array of atomic orbitals. In other words, systems with $(4q + 2)$ π electrons and a node are *antiaromatic* while systems with $(4q)$ π electrons and a node are aromatic and hence stable (in theory) in the ground state.

Application of these rules to pericyclic reactions led to the generalization that thermal reactions take place *via aromatic transition-states* whereas photochemical reactions proceed *via antiaromatic transition-states*. A cyclic transition-state is said to be aromatic or isoconjugate with the corresponding aromatic system if the number of the conjugated atoms and that of the π electrons involved are the same as in the corresponding aromatic system. Conversely, a cyclic transition-state is said to be *antiaromatic* or isoconjugate with the corresponding *antiaromatic* system if the number of conjugate atoms and that of the π electrons involved are the same as in the corresponding *antiaromatic* system. We have only to consider a cyclic array of atomic orbitals representing those orbitals which undergo change in the transition-state and assign signs to the wave functions in the best manner for overlap. Then the number of nodes in the array and the number of electrons involved are counted.

I. Electrocyclic Reactions

Let us consider the disrotatory and conrotatory processes for the system, *cis*-1,3,5-hexatriene $\rightleftharpoons$ cyclohexadiene.

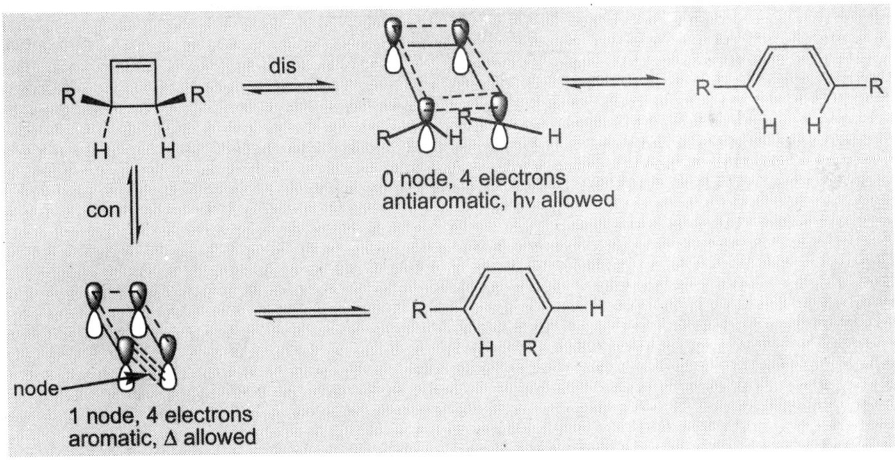

Similarly, for cyclobutene $\rightleftharpoons$ 1,3-butadiene system

Thus we reach the same conclusion as described earlier under other approaches. For convenience, the selection rules by this approach to electrocyclic reactions are tabulated in **Table 15.10**.

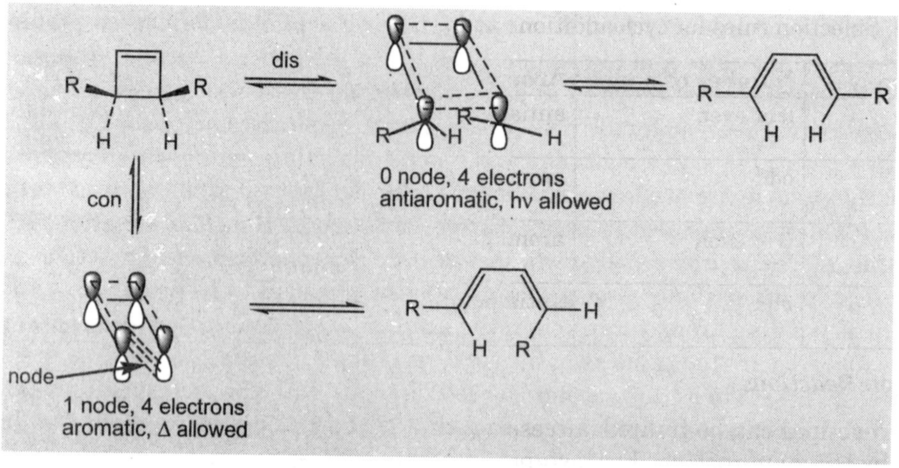

Table 15.10 Selection rules for electrocyclic reactions by PMO method q is an integer.

Array of electrons involved	Number of nodes	Aromaticity	Δ allowed	*hν* allowed
$4q$	0 or even	antiaromatic	–	disrotatory
$4q$	odd	aromatic	conrotatory	–
$4q + 2$	0 or even	aromatic	disrotatory	–
$4q + 2$	odd	antiaromatic	–	conrotatory

II. Cycloaddition

The transition-state for $2s + 2s$ cycloaddition has 4 electrons with zero node and is, therefore, antiaromatic and hence the reaction is *thermally forbidden*.

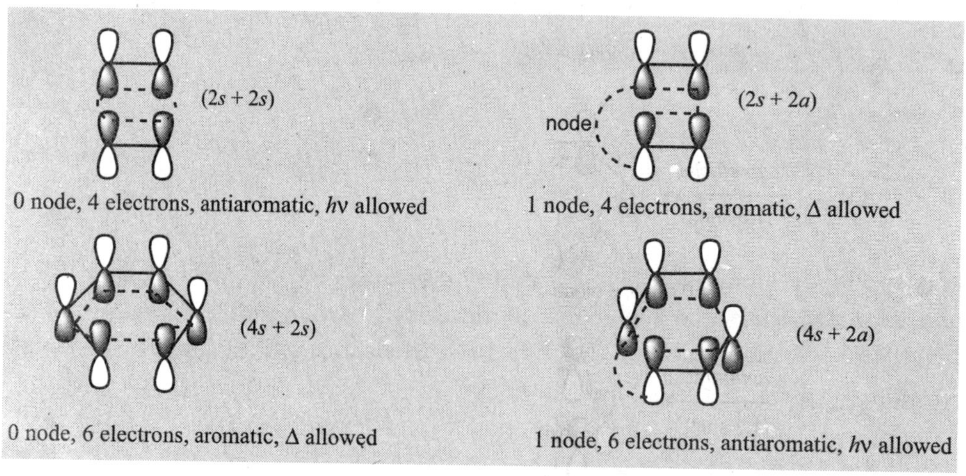

Thus, we come to the same conclusions as by other methods. The selection rules for cycloadditions are summarized in **Table 15.11**.

Table 15.11 Selection rules for cycloadditions and cycloreversions by PMO method *q* is an integar

***m + n* electrons**	**Number of nodes**	**Aromaticity**	**Δ allowed**	***hν* allowed**
4*q*	0 or even	antiaromatic	–	*supra-supra* *antara-antara*
4*q*	odd	aromatic	*supra-antara* *antara-supra*	–
4*q* + 2	0 or even	aromatic	*supra-supra* *antara-antara*	–
4*q* + 2	odd	antiaromatic	–	*supra-antara* *antara-supra*

III. Sigmatropic Reactions

Sigmatropic reactions can be treated successfully by PMO method and similar conclusions are arrived at as by other approaches. For instance, the [1, 3] *suprafacial* shift occurs *via* a transition state with 0 node and 4 electrons (antiaromatic) and thus is an *hν allowed* process, whereas the [1, 3] *antarafacial* shift can occur *via* a transition state with 1 node, 4 electrons (aromatic) and is thus a Δ *allowed* process.

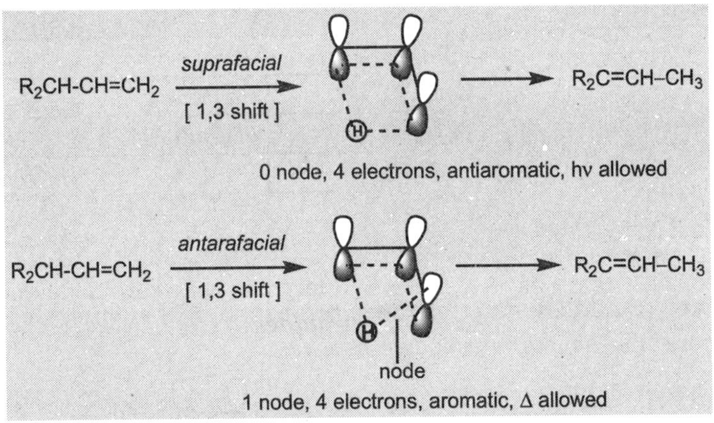

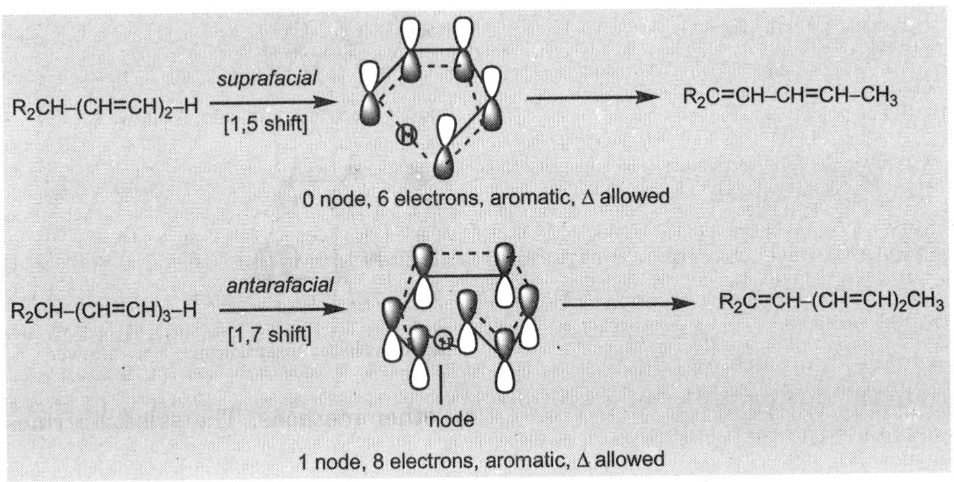

The selection rules for sigmatropic reactions of order $[i, j]$ by this method are summarized in **Table 15.12**.

Table 15.12 Selection rules for sigmatropic change by PMO method

Number of electrons involved $[i + j]$	Number of nodes	Aromaticity	Shift mode
$4q$	0 or even	antiaromatic	*supra* ($h\nu$)
$4q$	odd	aromatic	*antara* (Δ)
$4q + 2$	0 or even	aromatic	*supra* (Δ)
$4q + 2$	odd	antiaromatic	*antara* ($h\nu$)

Concerted Cope and Claisen rearrangements are all sigmatropic changes of order $[i, j]$ where i and j are both 3, and predictions of the stereochemical variations in these recations following the selection rules in **Table 15.9** (where $m + n = i + j$) conform to those made by other methods.

In the Cope rearrangement, for instance, there can be two stereochemical variations, *suprafacial-suprafacial*, (similarly *antarafacial-antarafacial*) and *suprafacial-antarafacial* as shown in the following transition states.

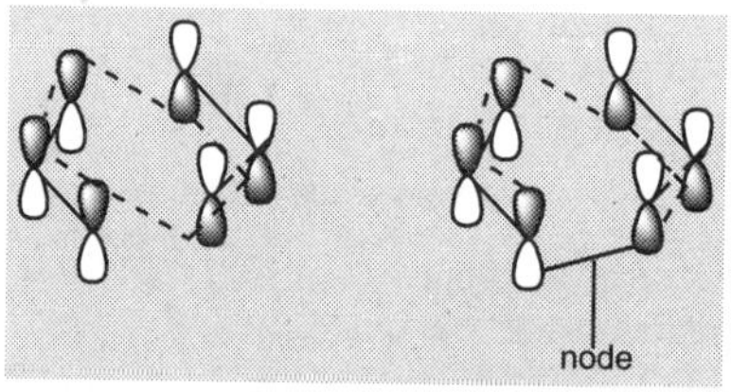

The chair shape transition-state for the *supra-supra* [3,3] sigmatropic shift is favoured over the boat shape transition state as shown below.

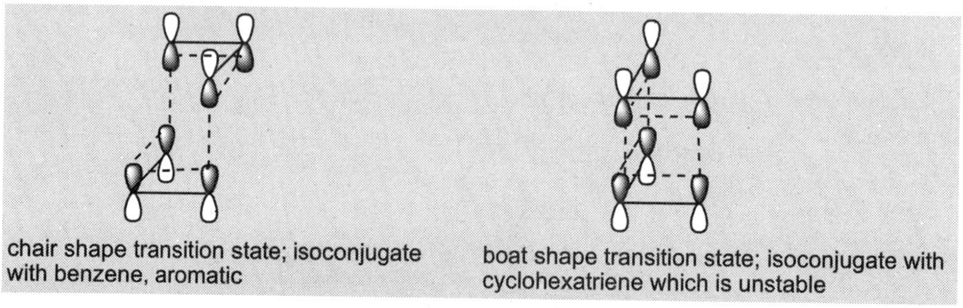

chair shape transition state; isoconjugate with benzene, aromatic

boat shape transition state; isoconjugate with cyclohexatriene which is unstable

The two examples may also be explained with the help of the PMO method. In the first example, the transition state has 4 electrons and 1 node, and is, therefore, an aromatic system. The [1,3] shift with inversion at the migrating carbon atom is thus thermally allowed. In the second example, there are six electrons and 2 nodes and, therefore, the transition-state is an aromatic system. The *antarafacial–antarafacial* sigmatropic change of order [3,3] of the 1,5-hexadiene is thus thermally allowed.

In conclusion, it must be said that while the prediction rules under different methods have dramatic success for thermolytic reactions (symmetry-allowed reactions on heating), photolytic reactions are more complex. Photolysis is initiated by light absorption first to a high-energy excited state which may also collapse *via* other, not necessarily concerted, pathways.

PROBLEMS AND REVIEW QUESTIONS

1. What is the product of concerted cycloaddition of ethylene and *cis*-2-butene, addition being *suprafacial* on both components? Is it symmetry allowed under thermal or photochemical conditions?
2. Draw the stereostructure of the products of the following cycloaddition reactions. Indicate the reaction condition under which these are symmetry allowed:
   (i) ethylene (*antara*) + *cis*-2-butene (*supra*)
   (ii) ethylene (*supra*) + *cis*-2-butene (*antara*)
3. Determine whether the following *supra-supra* cycloaddition reactions are symmetry allowed under thermal or photochemical conditions?
   (i) pentadienyl anion + ethylene → cycloheptenyl anion
   (ii) pentadienyl cation + ethylene → cycloheptenyl cation
4. What will be the stereochemical course of thermal cyclization of (i) pentadienyl anion, and (ii) pentadienyl cation?
5. Why concerted 1,3-sigmatropic shift of hydrogen is thermally forbidden? Explain.
6. The butadienes, **A** and **B**, are easily interconverted under thermal conditions. Suggest a suitable mechanism.

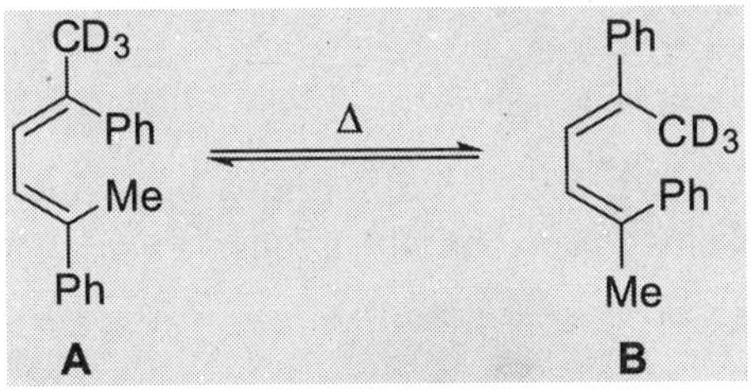

7. Explain how α-pyrone undergoes photolytic ring closure to **A**. Predict the stereostructure of the product.

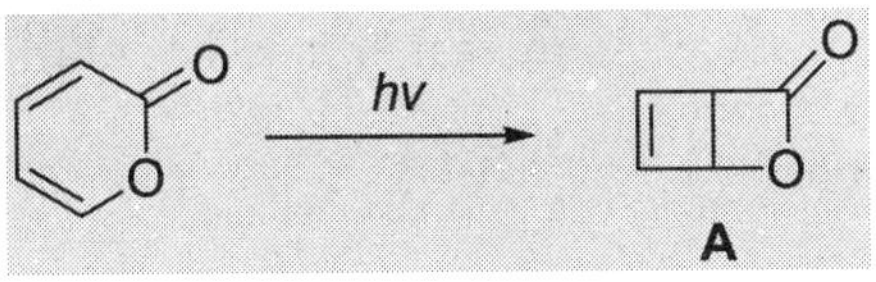

8. Explain the following observations:
   (i) The following isomer of toluene has been synthesized. Although much less stable than toluene, it has a surprisingly long life.

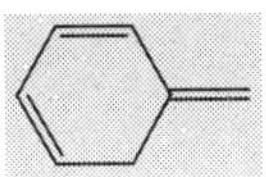

(ii) 3-Deuterioindene scrambling of the label to all three positions in the five-membered ring takes place on heating.

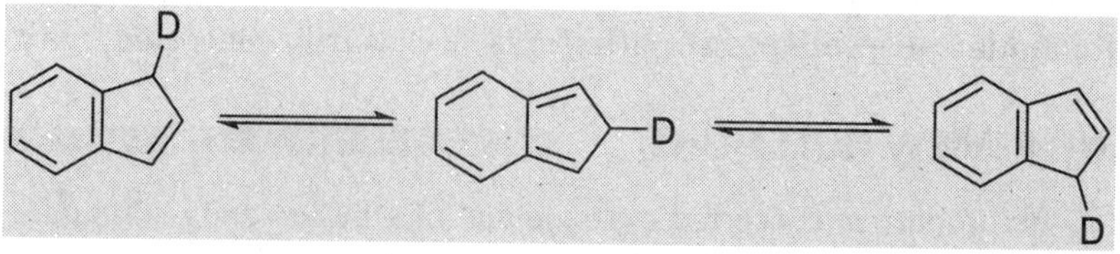

9. Write the products along with the stereochemistry in the following reactions:

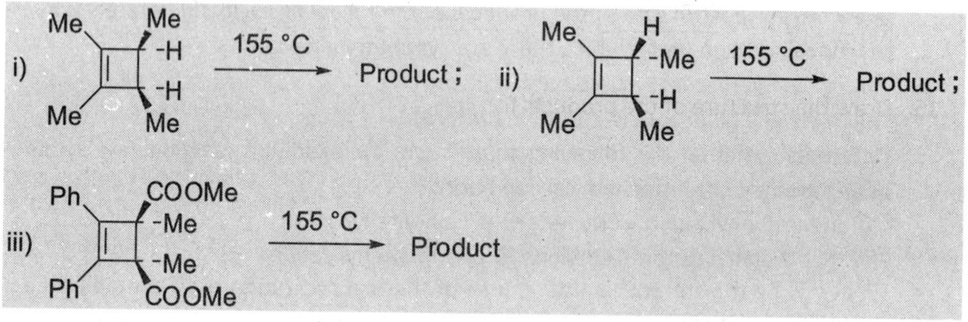

10. Complete the given reaction sequence.

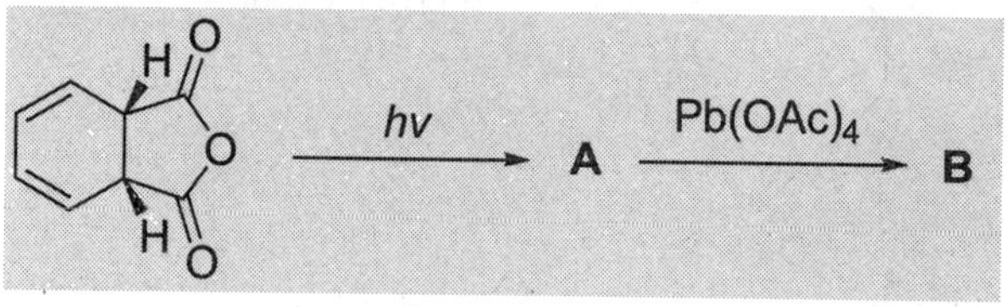

11. Write the structure of the product and explain its formation with a suitable mechanism.

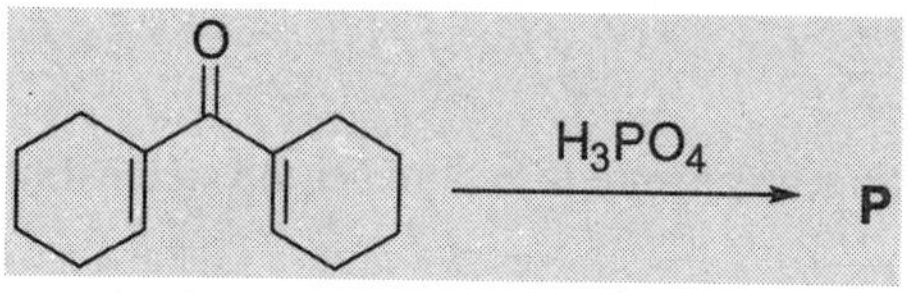

12. What will happen when propanoyl chloride is subjected to heat in presence of triethylamine followed by the subsequent treatment with ethylene?

13. Outline mechanism of the given transformation:

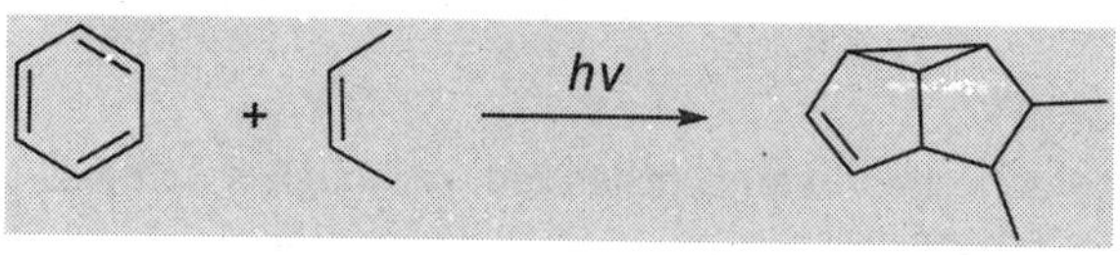

14. Predict the products **A** and **B** in the following reactions, and explain how these are formed.

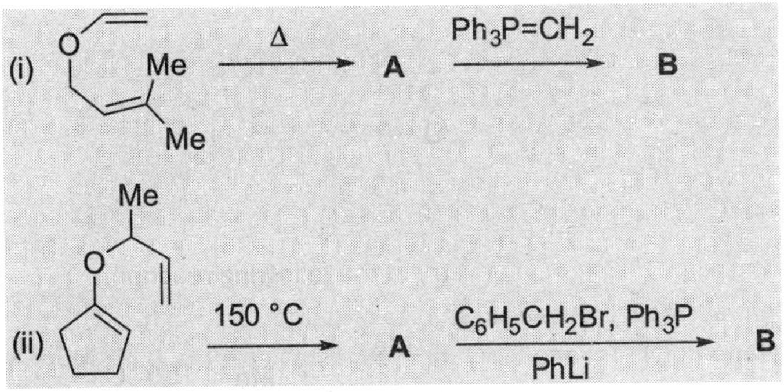

15. Draw the structure of the products formed in the following thermal electrocyclic reactions:

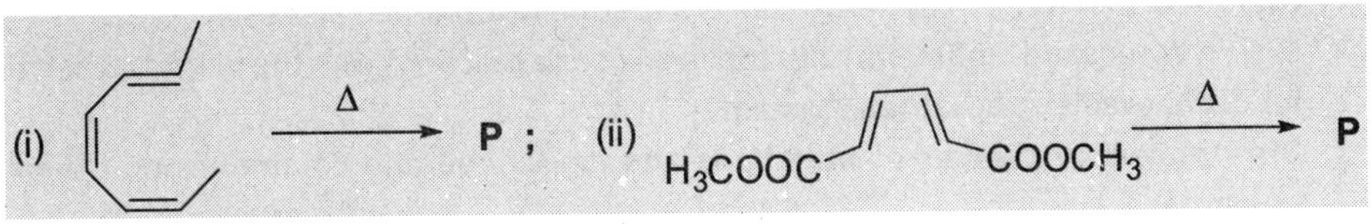

16. Complete the following reactions:

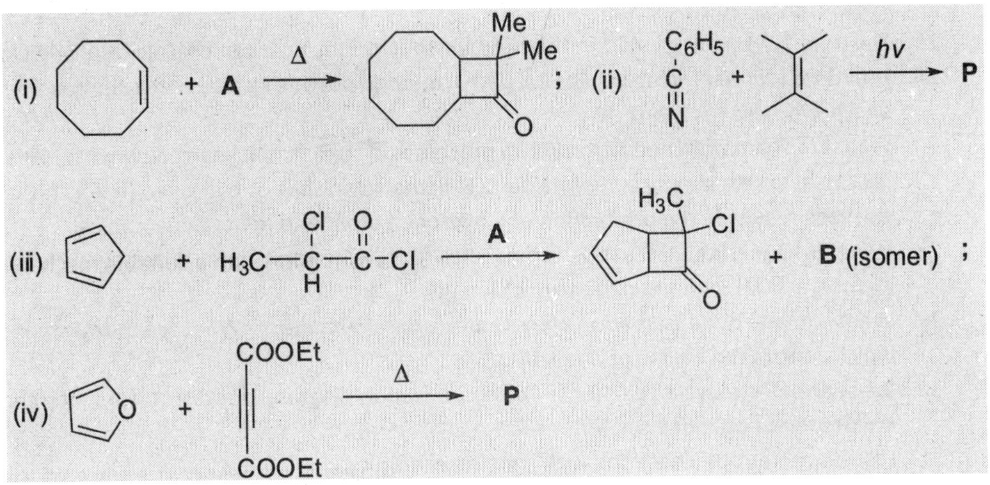

17. Explain how compound **A** can give **B** and **C** under thermal and photochemical conditions, respectively. Are they the same?

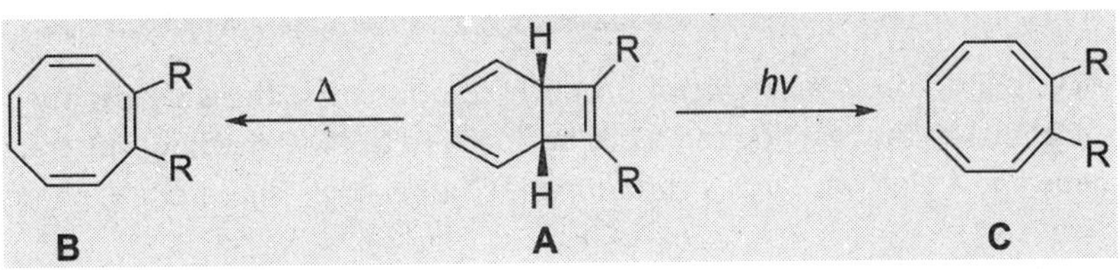

18. Cyclo-octadienyl anion (**A**) undergoes thermal ring-closure to give anion (**B**). Explain.

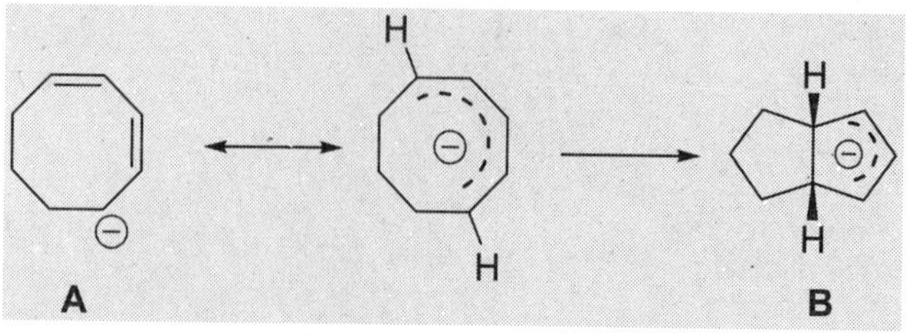

19. Depict Cope and Claisen rearrangements as examples of sigmatropic shift.
20. With the help of correlation diagram and FMO method, show that the Diels-Alder reaction is a thermally allowed process.
21. With the help of symmetry properties of the molecular orbitals of cyclohexadiene show why its *conrotatory* conversion to hexatriene is a thermally forbidden process?
22. Explain Woodward-Hoffmann rules for electrocyclic reactions with the help of correlation diagram for cyclohexadiene-hexatriene system.
23. Draw molecular orbitals of butadiene and hexatriene and comment upon their symmetry properties.
24. In the solvolysis of cyclopropyl tosylate in acetic acid, allyl acetate is obtained rather than cyclopropyl acetate. Explain.
25. Illustrate 1,7-hydrogen shift in the *thermal* equilibrium between precalciferol and calciferol.
26. How does Frontier Molecular Orbital (FMO) method provide an alternative approach for analyzing an electrocyclic reaction?
27. What are *suprafacial* and *antarafacial* processes in sigmatropic rearrangements? Illustrate these processes by examining a *suprafacial* 1,5-sigmatropic shift of hydrogen in which the homolytic cleavage results in the production of a hydrogen atom and pentadienyl radical.
28. Using PMO method, discuss selection rules for (i) electrocyclic reactions (ii) cycloadditions and cycloreversions, and (iii) sigmatropic changes.
29. What is meant by conservation of molecular orbital symmetry? What are the symmetry elements which control the course of a reaction?
30. What are electrocyclic reactions? Draw correlation diagrams for disrotatory interconversion of cyclobutene-butadiene system.
31. Discuss cheletropic reactions with suitable examples.
32. Draw the HOMO of ground state in the case of a conjugated diene and a triene, and compare the two states.

SOLUTIONS

1. Since the cycloaddition is *suprafacial* on both components, the sole product is *cis*-1,2-dimethylcyclobutane. The reaction can easily be analyzed by PMO method. The transition state has zero node and 4 electrons and is antiaromatic. The reaction is, therefore, photochemically allowed.

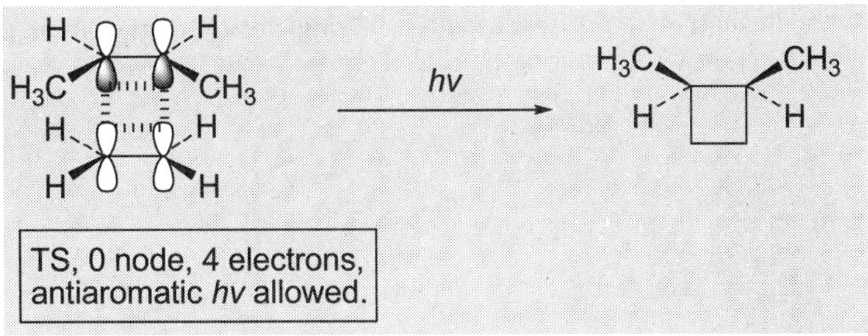

2. Both reactions are thermally allowed and give the isomerically different products, *cis* or *trans*-1,2-dimethylcyclobutane.

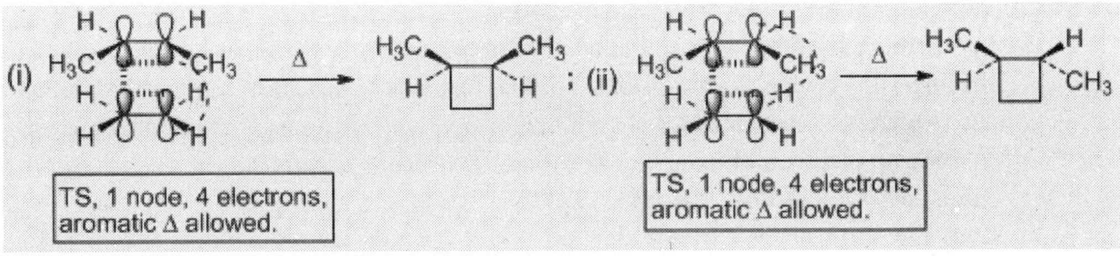

3. (*i*) It is $\pi^6s + \pi^2s$ process and is, therefore, allowed under photochemical condition (FMO). The PMO method also tells us that the TS for *supra-supra* cycloaddition in this case has 8 electrons and zero node and is, therefore, antiaromatic. Thus, it is photochemically allowed.

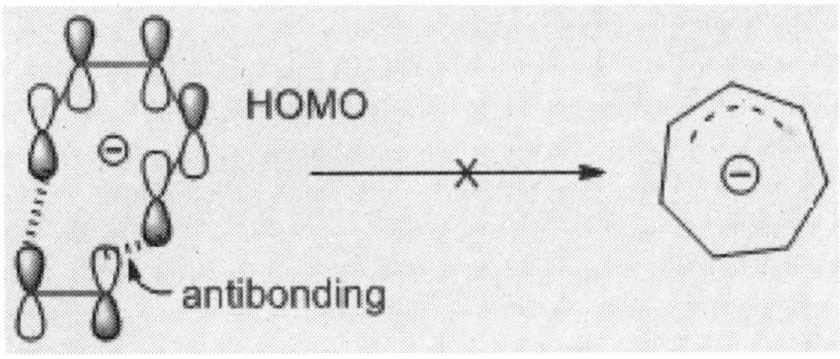

(*ii*) It is $\pi^4s + \pi^2s$ process and is, therefore, thermally allowed (FMO). The PMO method also tells us that the TS for *supra-supra* cycloaddition consists of 6 electrons and zero node and is, therefore, aromatic. The reaction is thus thermally allowed.

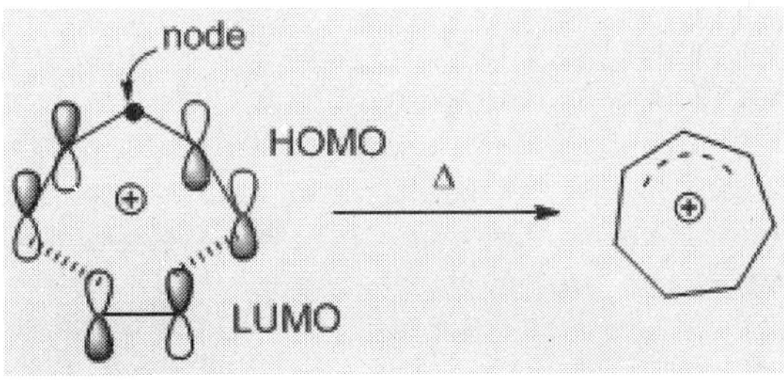

4. (*i*) PMO: The TS for *disrotatory* ring closure has 6 electrons with zero node and is aromatic. Therefore, the reaction is thermally allowed. In other words, under thermal condition, the cyclization will be *disrotatory*.

(*ii*) PMO: The TS for *conrotatory* process has 4 electrons with one node and is aromatic. Therefore, under thermal condition, the *conrotatory* ring closure will take place.

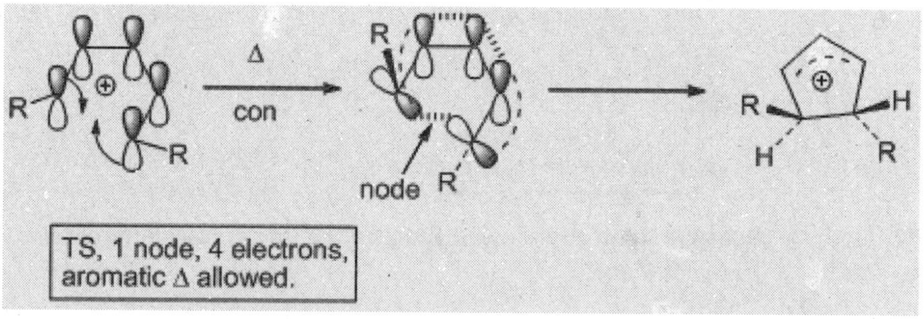

5. Concerted 1,3-sigimatropic shift is allowed only if it is *antarafacial* or *suprafacial* with inversion at the migration centre. *Antarafacial* 1, 3-shift is forbidden because of the steric inhibition involved in such a process. *Suprafacial* 1,3-shift of hydrogen under the thermal condition is forbidden because there is no question of inversion at this atom which is bonded to the carbon atom through its spherically symmetrical 1*s* orbital.
6. Both the steps are symmetry-allowed under thermal condition since the HOMO (ψ_2) of the open-chain partner will have C_2 symmetry. Therefore, both ring-closure and ring-opening will be *conrotatory* processes.

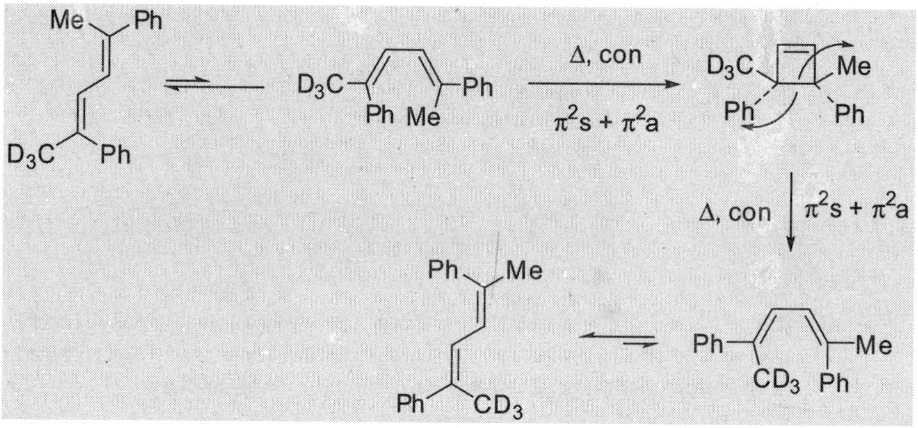

PMO: The transition states in both the steps of *conrotatory* ring-closure and ring- opening have 4 electrons with 1 node and are aromatic. Thus, both the steps are thermally allowed.

7. The compound has a 1,3-butadiene system and under photochemical condition, it undergoes *disrotatory* ring-closure as the HOMO of the excited state of butadiene is ψ_3 which has similar signs of the lobes at the extreme ends. Therefore, cyclization takes place in a *disrotatory* manner. We should expect a *cis* structure for the product.
8. (**i**) This is explained by the fact that thermal 1, 3-shift of any of the two methylene hydrogens is symmetry forbidden (see problem 5). If the isomer of toluene is to transform to toluene, it cannot do so through a facile symmetry-allowed concerted process. Hence, resistance to isomerization and long life.

   (ii) **Hint**: A series of [1,5] shifts result in the scrambling.

9. 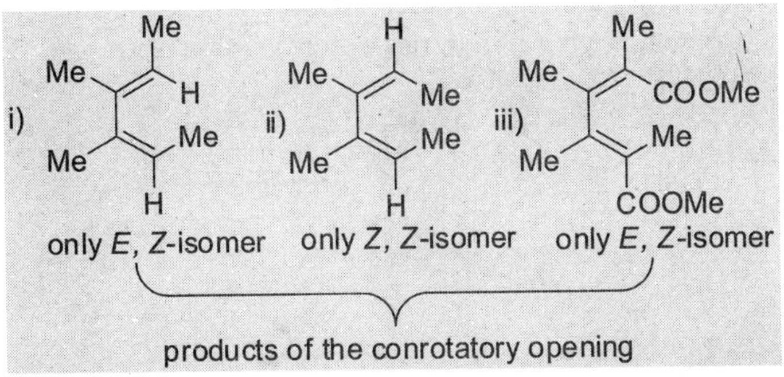
i)
ii)
iii)
Me
H
COOMe
only E, Z-isomer
only Z, Z-isomer
only E, Z-isomer
products of the conrotatory opening

10. 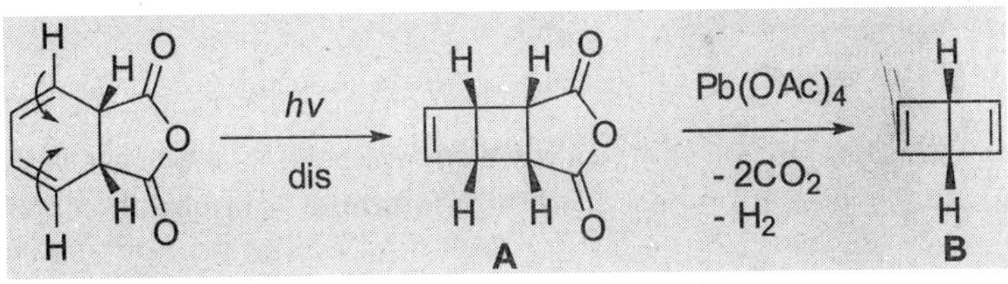
hv
dis
A
Pb(OAc)4
- 2CO2
- H2
B

11. 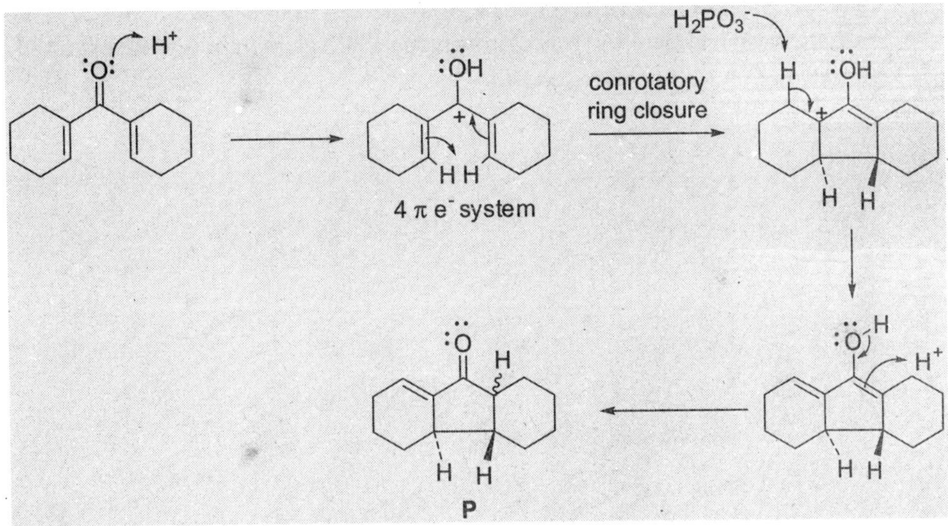
H+
4 π e- system
H2PO3-
conrotatory
ring closure
P

12. 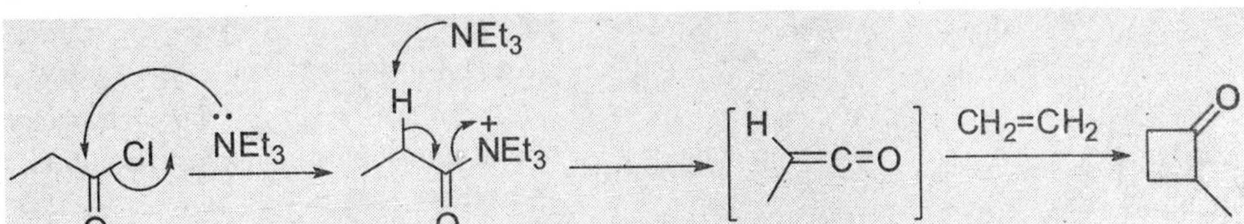
NEt3
Cl
C=O
CH2=CH2

13. 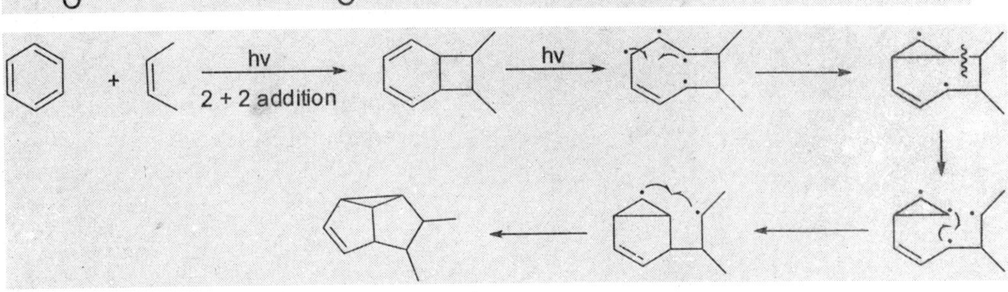
+
hv
2 + 2 addition
hv

14. (i) 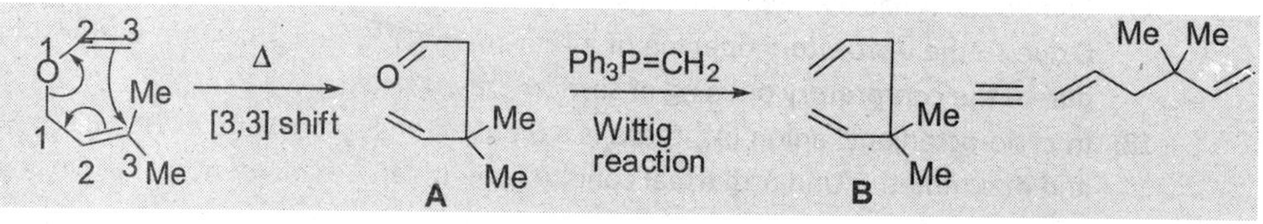

(ii) 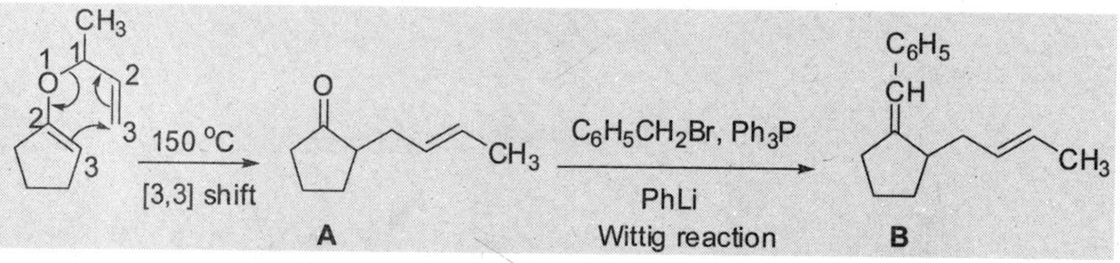

15. (i) Under thermal condition, HOMO for the given diene system is ψ_3 which has ***m*** symmetry. So the cyclization occurs in disrotatory mode which gives *trans*-3, 4-disubstitutedcyclohexadiene derivative.

(ii) In thermal condition, HOMO for the given diene system is ψ_2 which has C_2 symmetry. So the cyclization can only occurs in conrotatory mode which gives *trans*-3,4-disubstituted cyclobutene derivative.

(i) 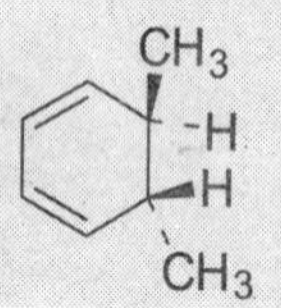

(ii) 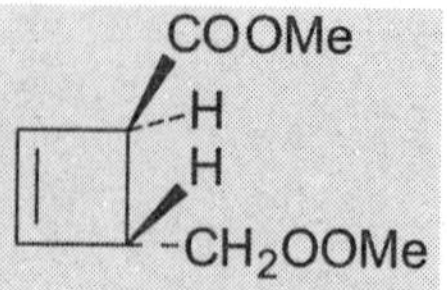

16. (i) 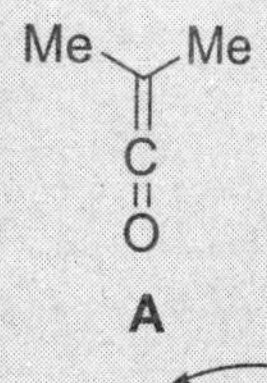

(ii) 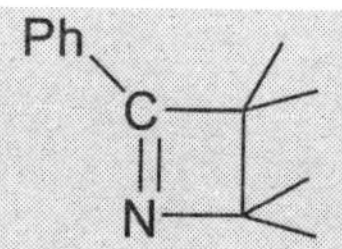

(iii) 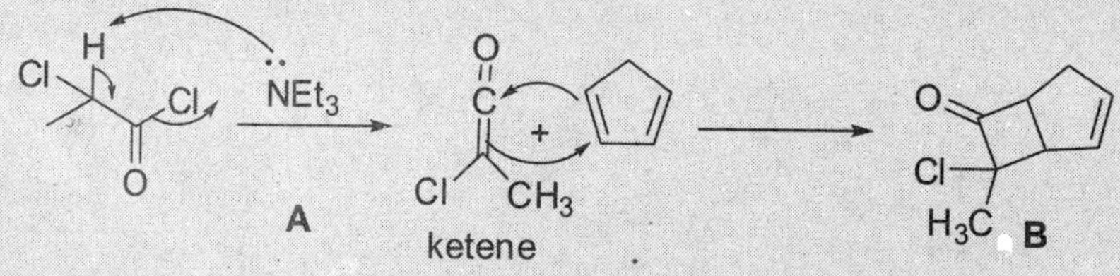

(iv) 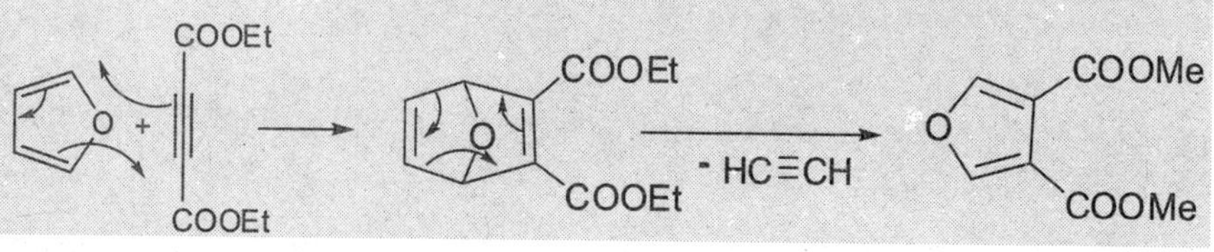

singlet-singlet, 613
triplet-triplet, 613
Enolates, 432
Enthalpy, 248
Entropy, 248
Epoxidation and hydroxylation, 342
Epoxides,
rearrangement, 527
ring-opening, 343
Equatorial bonds, 162
Equilibrium,
constant, 248
process, 248
Erythro form, 80,
Erythrose, 80
Esterification,
acid-catalyzed, 436
ortho-disubstituted benzoic acids, 537
reversibility, 436
Excitation,
butadiene, 613
carbonyl group, 617
Excited states, 611
E/Z isomers, 101

F

Favorskii rearrangement, 545
Felkin-Anh model in asymmetric synthesis, 134
Fenton's reagent, 567
Fluorescence, 611
FMO method,
cycloaddition reactions, 666
electrocyclic reactions, 660
sigmatropic reactions, 674
Free radicals, see radicals
Friedel-Crafts reaction, 468
acylation, 470
acylium ions in, 471
alkylation, 468
carbonium ions in, 469
Lewis acids in, 468
rearrangements in, 469
Fries rearrangement, 542
cross products in, 542
photo-Fries reaction, 544
Fullerenes, 32

G

Gatterman-Koch reaction, 489
General acid and base catalysis, 217
Geometrical isomers, 63, 100
definition, 100,
nomenclature, E/Z system
of alicyclic compounds, 103
Glaser reaction, 594
Gomberg reaction, 594
Glucose, configuration and conformation, 185, 189
Glycidic acid, decarboxylation,445
Glycidic esters,
formation, 422
rearrangement, 266
1,2-Glycols, rearrangement, 520
Grignard additions, 422

H

Halonium ion, 330, 336
Haloform reaction, 430
Halogenation,
aromatic electrophilic, 464
ketones, 429
radical, 577
Hammett equation, 257
modified, 485
rate constant (ρ), 259
substituent constant (σ), 258
Hammond,s postulate, 254
Heat of hydrogenation and relative stability of alkenes, 34
Helical dissymmetry or helicity, 151
Helicity due to molecular overcrowding, 152
Heterolytic fission, 200
Heteromorphic ligands/faces, 119
Hexatriene,
cyclization,
symmetry properties of, 651
molecular orbitals, 649
Hofmann rearrangement, 530
correlation with Beckmann rearrangement, 539
intramolecularity, 531
isolation of intermediates, 531
retention of configuration, 531
Hofmann rule, 375
Homolytic fission, 200
Homomorphic ligands/faces, 119
Homomorphus, 188
Homotopic faces, 120
Homotopic ligands, 120
Houben-Hoesch reaction, 490
Hunsdicecker reaction, 596
Huckel's rule, 24
HMO π-energy level diagram,
allyl system, 21

benzene, 25
cyclobutadiene, 25
cyclodecapentaene, 25
cyclooctatetraene, 25
cyclopropenyl cation, 25
HSAB principle, 218
Hybridization, 3
and bond properties, 5
and pKa values, 215
Hydride,
lithium aluminium, 403
sodium boro-, 408
Hydroboration reaction, 338
applications, 341
sterechemical aspects, 342
Hydrocarbons, alternant and nonalternant, 31
Hydrogenation of alkenes and alkynes, 347
Hydrogen bond, 47
and pka values, 215
effects on physical properties, 49
inter- and intramolecular, 48
Hydrolysis,
of acid chlorides, 443
of amides, 441
anhydrides, 443
of esters with acids, 436
of esters with base, 43
Hydroxylation, 342
Hyperconjugation, 21

I

Inclusion compounds, 48
Inductive effect, 7
and pKa values, 214
Intermediates, 219
isolation, 267
spectroscopic detection, 267
trapping, 268
Interconversions of projection formulae, 67
Internal conversion, 612
Intersystem crossing, 612
Intramolecular energy transfer, 615
Inversion of configuration, 94
Isotope effects, 271
primary, 271
secondary, 272
Isotopic labelling, 270

J

Jablonskii diagram, 612

K

Ketoacids, decarboxylation, 444
Ketoester, rearrangement, 12,
Keto-enol tautomerism, 33
Ketones,
addition of carbanions, 414
condensation with derivatives of ammonia, 402
halogenations of, 429
reactions with hydride donors, 403
reactions with water and alcohols, 399
Ketoximes,
Beckmann rearrangement of, 539
syn/anti (E/Z) isomers, 3
Kinetic control, 255,
Kinetic evidence, 275
Kinetics, 249
molecularity, 250
order of reaction, 249
Kolbe electrolysis, 567
Knoevenagel condensation, 417

L

LCAO treatment, 8
Lewis acids, 4,
Lewis bases, 4,
Linear free energy relationship (LFER), 5
Lithium aluminium hydride, 403
in reduction of carbonyl compounds, 397
Lossen rearrangement, 532

M

Mannich reaction, 427
Markownikoff's rule, 334
Meerwein-Ponndorf reduction, 410,
Meisinheimer salt, 502
Menthyl chlorides, in eliminations, 219
Mesitoic acid, esterification of, 437
Meso compounds, 81
Mesomerism, 14
Methyl radical,
esr spectrum ,568
structure, 3
Michael addition, 356
Migratory aptitude,523
Molecularity of the reaction, 5, 250
Molecular orbital,
antibonding, 8
bonding, 8
method, 8
nonbonding, 8

of butadiene,649
of hexatriene,649
π, 11
π*, 11
σ, 9
σ*, 9
symmetry properties,648
Molecular rearrangements, 518
Molecular representation of stereoisomers, 65
Fischer projection, 65
Flying wedge projection, 67
Newman projection, 66
Sawhorse projection, 66
Molozonide, 350
Mustard gas, hydrolysis of,290
Mutarotation, 401

N

Neighbouring group participation, 289
Nitrating mixture, 462
Nitration of benzene, 462
Nitrene intermediates, 238
classification and structure, 239
detection, 239
formation, 240
reactions 240
Nitronium ions, 463
Nonaromatic, 29, 30
Non-classical carbocation, 305
Norbornene, reaction with deuterated diborane, 340
Norrish type I reaction, 621
of acetone, 621
Norrish type II reaction, 622
formation of 1,4-biradical, 622
Nucleophiles, 205
Nucleophilic additions, 202
to acrylonitrile, 355
to carbonyl compounds,391
to α,β-unsaturated carbonyl compounds, 356
Nucleophilic substitution, aliphatic, 280
carbocation process, 281
direct displacement process, 283
epoxides, 291
factors affecting reactivity, 311
neighbouring group participation, 289
stereochemistry, 285
Nucleophilic substitution aromatic, 498
addititon-elimination mechanism, 500
benzyne mechanism, 502
bimolecular mechanism, 500
cine-substitution, 503
elimination-addition mechanism, 502
unimolecular mechanism, 499

O

Optical activity, 70
Optical isomers, 63
Optical purity, 100
Orbitals,
antibonding, 8
atomic, 9
bonding, 8
hybridization of, 3
molecular, 8
nonbonding, 8
Orbital symmetry,647
conservation of,
cycloaddition reactions, 663
electrocyclic reactions, 653
sigmatropic reactions, 680
Order of reaction, 249
Osmium tetraoxide, addition to alkenes, 345
Oxymercuration, 346
Ozonides, 350
Ozonolysis, 350
π bond, 2, 11
π*, 11
π-complex,330

P

Paneth technique, 565
Partial rate factor,484
Paterno-Buchi reaction, 624
and photoreduction, 618
energy transfer, 613
intermediacy of biradical, 622
Peracids,
in epoxidition, 342
Pericyclic reactions, 646
Perkin condensation, 417
pKa, 208
pKb, 208
PMO method, 680
application to Huckel's rule, 681
cycloaddition reactions, 682
electrocyclic reactions, 681
sigmatropic rearrangements, 683
treatment of aromaticity, 681
Pentaerythritol, formation of, 413
Phenonium ion, 299
Phosphorescence, 612

Photochemical,
energy, 608
reactions of cyclic ketones,
Photochemistry, 607
in *cis-trans* isomerizaion, 628
of butadienes, 632
of carbonyl compounds, 617
of olefins, 628
of α-santonin, 639
of α,β-unsaturated ketones, 626
principle of,608
Photocyclodimerization, 627
Photoisomerization of stilbenes, 629
Photorearrangement of cyclohexadienones, 627
Photoreduction,
and Paterno-Buchi reaction, 624
benzophenone,618
ethyl acetoacetate,620
α-keto ester,620
mechanism,621
Polar effects on radical reactions, 581
Polymerization of styrene, 576
Prelog's rule (1,4-asymmetric induction), 135
Principle of stereoselctivity, 130
Probabilty factor, 251
Product analysis, 265
Product development control, 404
Prostereoisomerism, definition classification, 119
Prostereoisomerism and prochirality, 122
Protium ion, 334
Prototropy, 33
acetylacetone, 36
azomethines, 38
amides, 37
concerted *vs* stepwise mechanism, 37, 38
nitroparaffins, 37
nitroso compounds, 37

Q

Quantam,
efficiency, 616
yield, 619
Quencher, 620
Quenching studies in photoreduction, 619
Quinuclidine,
base strength, 216
nucleophilicity, 216

R

Racemic mixture or racemic modification, 96
Racemization,
definition, 96
in S_N1 reactions,
of biphenyls, 146
of ketones, 235
Radicals, 238, 564
addition reactions, 584
apocamphyl, 573
benzoyloxy, 574
biradicals, 600
configuration of, 572
detection of, 567
dimerization of, 590
disproportionation of, 577
from photochemical reactions, 566
from redox reactions, 567
from thermal reaction, 565
in aromatic substitution, 589
in arylation of aromatic rings, 594
in autoxidation, 587
in coupling of alkynes, 593
in halogenations, 577
in polymerization, 576
inhibitors in, 567
initiators in, 568, 575
in neighboring group assistance, 583
production of, 565
rearrangements, 591
short-lived, 565
stable, 565, 596
stability, 569
triphenylmethyl, 596
Rate determining step, 250
Rate of reaction, and catalyst, 250
Reaction,
energy considerations, 248
intermediate and transition state, 250
molecularity, 250
order of, 249
rate determining step of, 250
reactivity and rates of, 250
transition state theory, 251
types of, 201
Reaction mechanisms,
definition of, 248
general principles of, 248
methods of determination of, 265
Reagents, types of, 204
Rearrangement
allylic, 39
Baeyer-Villiger, 534
Beckmann, 537

benzidine, 548
benzilic acid, 421
carboaction, 520
Claisen, 552
Cope, 555
Curtius, 532
Demjanov, 527
dienone-phenol, 525
Favorskii, 545
Fries, 542
Hofmann, 530
in bicyclic compounds, 528
in α,β-diol esters, 522
in epoxides, 527
in Friedel-Crafts alkylation, 469
in glycidic esters, 266
in 1,2-glycols, 520
in α-keto esters, 525
Lossen, 532
migratory aptitude in, 523
of free radical, 591
photo-Fries, 544
pinacol-pinacolone, 520
Schmidt, 532
sigmatropic, 674, 683
semipinacol, 523
Smiles, 511
Sommelet-Hauser, 508
Stevens, 510
transannular, 528
von-Richter, 507
Wagner-Meerwein, 520
Wolf, 529
Reduction
Birch, 350
Meerwein-Ponndorf reduction, 410
Reformatsky reaction, 423
Reimer-Tiemann reaction, 391
Resonance,
concept of, 14
effect on pKa values, 209
energy, 15
hybrid, 15
method, rules for, 16
Resonance integral, 19
Resolution,
definition, 97
methods, 97
of racemic modification of tartaric acid, 98
Retention of configuration, 94
Rho, 259
Robinson-Mannich ring annulations, 428
R/S system, 86

S

Sandmeyer reaction, 595
Sanger's reagent 502
α-Santonin, photochemistry of, 639
Saytzeff rule, 374
Schimidt rearrangement, 532
SET mechanism, 310
Sharpless enantioselective epoxidation, 140
Short-lived radicals, 565
σ, substituent constant, 258
Sigmatropic rearrangement,
analysis of, 676
definition, 674
FMO method, 674
PMO method, 683
Smiles rearrangement, 511
S_N1 reaction,
energy diagram for, 283
ion effects on, 284
kinetic evidence, 283
rearrangements, 288
solvent effects, 284
steric orientation, 287
Sommelet-Hauser rearrangement, 508
S_N2 reaction,
energy diagram for, 282
kinetic evidence, 281
solvent effects, 313
steric hindrance, 311
steric orientation, 285
S'_N1 mechanism, 40
S'_N2 mechanism, 40
S_{Ni} mechanism, 308
S'_{Ni} mechanism, 41
Solvated ion,
symmetrical, 287
unsymmetrical, 287
Solvent cage, 544
Solvolysis reactions, 284
Stable free radicals, 596
Steric acceleration,
Steric approach control, 404
Steric effects,
on pKa values, 216
Steric orientation, 285
Stereochemical evidence, 274
Stereochemistry,
of allenes, 148

of biphenyls, 142
of cyclohexanes, 162
of decalins, 175
of nitrogen compounds, 153
of phosphorus compounds, 157
of spiranes, 149
of sulphur compounds, 159
of monosachharides, 185
Stereogenic centre, 72
Stereoheterotopic ligands, 120
NMR spectral characteristics of, 125
Stereoisomerism, classification and definition, 63
Stereoselectivity, 127
enantioselectivity, 131
diastereoselectivity, 131
principle of, 130
Stereospecificity, 128
Stevens rearrangement
Stobbe condensation, 411
Substituent constant (σ), 258
Sulphonation,
Suprafacial, 662
Symmetry properties of molecular orbitals, 648

T

Taft equation, 263
Tautomerism, 33
anionotropy, 39
prototropy, 33
Telemers, 586
Tetramethyllead, for radical generation, 566
Thermodynamic control, 255
Transannular rearrangement, 528
Transition state, theory, 251
Triphenylmethyl radical, 596
Tschugaev reaction, 383

V

van der Waals interactions, 52
Verley-Meerwein-Ponndorf reduction, 410
Vilsmeir-Haack reaction, 488
Vinyl chloride, addition of HBr, 336
Vinyl system, stabilization through resonance, 20
von Richter rearrangement, 507

W

Wagner-Meerwein rearrangement, 520
Walden inversion, 274, 285
Wittig reaction, 424
reagent, 424
modified, 425
Wolff rearrangement, 529
Woodward-Hoffmann rules, 654

X

Xanthates, pyrolysis of, 383